Lecture Notes on Data Engineering and Communications Technologies

Volume 17

Series editor

Fatos Xhafa, Technical University of Catalonia, Barcelona, Spain
e-mail: fatos@cs.upc.edu

The aim of the book series is to present cutting edge engineering approaches to data technologies and communications. It publishes latest advances on the engineering task of building and deploying distributed, scalable and reliable data infrastructures and communication systems.
The series has a prominent applied focus on data technologies and communications with aim to promote the bridging from fundamental research on data science and networking to data engineering and communications that lead to industry products, business knowledge and standardisation.

More information about this series at http://www.springer.com/series/15362

Leonard Barolli · Fatos Xhafa
Nadeem Javaid · Evjola Spaho
Vladi Kolici
Editors

Advances in Internet, Data & Web Technologies

The 6th International Conference on Emerging Internet, Data & Web Technologies (EIDWT-2018)

Editors
Leonard Barolli
Department of Information and Communication Engineering
Fukuoka Institute of Technology
Fukuoka-shi
Japan

Fatos Xhafa
Technical University of Catalonia
Barcelona
Spain

Nadeem Javaid
Department of Computer Science
COMSATS Institute of Information Technology
Islamabad
Pakistan

Evjola Spaho
Polytechnic University of Tirana
Tirana
Albania

Vladi Kolici
Polytechnic University of Tirana
Tirana
Albania

ISSN 2367-4512 ISSN 2367-4520 (electronic)
Lecture Notes on Data Engineering and Communications Technologies
ISBN 978-3-319-75927-2 ISBN 978-3-319-75928-9 (eBook)
https://doi.org/10.1007/978-3-319-75928-9

Library of Congress Control Number: 2018934341

Printed on acid-free paper

This Springer imprint is published by the registered company Springer International Publishing AG part of Springer Nature
The registered company address is: Gewerbestrasse 11, 6330 Cham, Switzerland

Welcome Message of EIDWT-2018 International Conference Organizers

Welcome to the sixth International Conference on Emerging Internet, Data and Web Technologies (EIDWT-2018), which will be held from March 15 to March 17, 2018, at Polytechnic University of Tirana, Tirana, Albania.

EIDWT is dedicated to the dissemination of original contributions that are related to the theories, practices, and concepts of emerging internetworking and data technologies yet most importantly of their applicability in business and academia toward a collective intelligence approach.

In EIDWT-2018 will be discussed topics related to Information Networking, Data Centers, Data Grids, Clouds, Crowds, Mashups, Social Networks, Security Issues, and other Web 2.0 implementations toward a collaborative and collective intelligence approach leading to advancements of virtual organizations and their user communities. This is because, current and future Web and Web 2.0 implementations will store and continuously produce a vast amount of data, which if combined and analyzed through a collective intelligence manner will make a difference in the organizational settings and their user communities. Thus, the scope of EIDWT-2018 includes methods and practices which bring various emerging Internet and data technologies together to capture, integrate, analyze, mine, annotate, and visualize data in a meaningful and collaborative manner. Finally, EIDWT-2018 aims to provide a forum for original discussion and prompt future directions in the area. For EIDWT-2018 International Conference, we accepted for presentation 98 papers (about 28% acceptance ratio).

An international conference requires the support and help of many people. A lot of people have helped and worked hard for a successful EIDWT-2018 technical program and conference proceedings. First, we would like to thank all authors for submitting their papers. We are indebted to Program Area chairs, Program Committee members, and reviewers who carried out the most difficult work of carefully evaluating the submitted papers. We would like to give our special thanks to Prof. Makoto Takizawa, Hosei University, Japan, as Honorary Chair of EIDWT-2018, for his guidance and support. We would like to express our appreciation to our keynote speakers: Prof. Vincenzo Loia, University of Salerno, Italy, and Prof. Schahram Dustdar, Vienna University of Technology (TU Wien),

Austria, for accepting our invitation and delivering very interesting keynotes at the conference.

We would like as well to thank the Local Arrangements Chairs for making excellent local arrangements for the conference. We hope you will enjoy the conference and have a great time in Tirana, Albania.

EIDWT-2018 International Conference Organizers

EIDWT-2018 Steering Committee Co-chairs

Leonard Barolli	Fukuoka Institute of Technology (FIT), Japan
Fatos Xhafa	Technical University of Catalonia, Spain

EIDWT-2018 General Co-chairs

Vladi Kolici	Polytechnic University of Tirana, Albania
Tomoya Enokido	Rissho University, Japan
Santi Caballé	Open University of Catalonia, Spain

EIDWT-2018 Program Committee Co-chairs

Evjola Spaho	Polytechnic University of Tirana, Albania
Nadeem Javaid	COMSATS Institute of Information Technology, Pakistan
Flora Amato	University of Naples Federico II, Italy

EIDWT-2018 Organizing Committee

Honorary Chair

Makoto Takizawa	Hosei University, Japan

General Co-chairs

Vladi Kolici	Polytechnic University of Tirana, Albania
Tomoya Enokido	Rissho University, Japan
Santi Caballé	Open University of Catalonia, Spain

Program Co-chairs

Evjola Spaho	Polytechnic University of Tirana, Albania
Nadeem Javaid	COMSATS Institute of Information Technology, Pakistan
Flora Amato	University of Naples Federico II, Italy

International Advisory Committee

Janusz Kacprzyk	Polish Academy of Sciences, Poland
Vincenzo Loia	University of Salerno, Italy
Akio Koyama	Yamagata University, Japan
Arjan Durresi	IUPUI, USA
Hiroaki Nishino	Oita University, Japan

Publicity Co-chairs

Elinda Kajo	Polytechnic University of Tirana, Albania
Keita Matsuo	Fukuoka Institute of Technology, Japan
Admir Barolli	Alexander Moisiu University of Durres, Albania

Klodiana Goga	Istituto Superiore Mario Boella, Italy
Omar Hussain	University of New South Wales, Canberra, Australia

International Liaison Co-chairs

Algenti Lala	Polytechnic University of Tirana, Albania
Francesco Palmieri	University of Salerno, Italy
Elis Kulla	Okayama University of Science, Japan
Farookh Hussain	University of Technology Sydney, Australia

Local Organizing Co-chairs

Olimpjon Shurdi	Polytechnic University of Tirana, Albania
Bexhet Kamo	Polytechnic University of Tirana, Albania
Ilir Shinko	Polytechnic University of Tirana, Albania
Enida Sheme	Polytechnic University of Tirana, Albania
Renalda Kushe	Polytechnic University of Tirana, Albania

Web Administrators

Shinji Sakamoto	Fukuoka Institute of Technology (FIT), Japan
Donald Elmazi	Fukuoka Institute of Technology (FIT), Japan
Yi Liu	Fukuoka Institute of Technology (FIT), Japan
Miralda Cuka	Fukuoka Institute of Technology (FIT), Japan

Finance Chair

Makoto Ikeda	Fukuoka Institute of Technology (FIT), Japan

Steering Committee Co-chairs

Leonard Barolli	Fukuoka Institute of Technology (FIT), Japan
Fatos Xhafa	Technical University of Catalonia, Spain

Track Area Chairs

1. Databases, Knowledge Discovery, Semantics, and Mining

Track Chairs

Giuseppe Fenza	University of Salerno, Italy
Jugappong Natwichai	Chiang Mai University, Thailand

PC Members

Pruet Boonma	Chiang Mai University, Thailand
Xue Li	The University of Queensland, Australia
Bowonsak Srisungsittisunti	University of Phayao, Thailand
Xingzhi Sun	IBM Research, Australia
Pornthep Rojanavasu	University of Phayao, Thailand
Alex Pongpech	National Institute of Development Administration, Thailand
Panachit Kittipanya-ngam	Electronic Government Agency, Thailand
Francesco Orciuoli	University of Salerno, Italy
Mariacristina Gal	University of Salerno, Italy
Néstor Álvarez-Díaz	University of La Laguna, Spain
Moisés Lodeiro-Santiago	University of La Laguna, Spain
Francisco Javier Cabrerizo	Distance Learning University of Spain (UNED), Spain

2. Ontologies, Metadata Representation, and Digital Libraries

Track Chairs

Titela Vilceanu	University of Craiova, Romania
Salvatore Ventiqincue	University of Campania Luigi Vanvitelli, Italy

PC Members

Beniamino Di Martino	Università della Campania “Luigi Vanvitelli”, Italy
Alba Amato	National Research Council (CNR) - Institute for High-Performance Computing and Networking (ICAR), Italy
Tatiana A. Gavrilova	Saint Petersburg University, Russia
Pavel Smrž	Brno University of Technology, Czech Republic
Teodor Florin Fortis	West University of Timisoara, Romania

3. P2P, Grid, and Cloud Computing

Track Chairs

Hiroshi Shigeno	Keio University, Japan
Klodiana Goga	ISMB, Italy

PC Members

Harold Castro	Universidad de Los Andes, Bogotá, Colombia
Olivier Terzo	ISMB, Turin, Italy
Philip Moore	Lanzhou University, China
Pietro Ruiu	ISMB, Turin, Italy
Alberto Scionti	ISMB, Turin, Italy
Giuseppe Caragnano	ISMB, Turin, Italy
Giovanni Masala	Plymouth University, UK
Fumiaki Sato	Toho University, Japan
Tomoya Kawakami	NAIST, Japan
Gen Kitagata	Tohoku University, Japan
Akimitsu Kanzaki	Shimane University, Japan
Tomoki Yoshihisa	Osaka University, Japan
Mauro Marcelo Mattos	FURB Universidade Regional de Blumenau, Brazil

4. Parallel and Distributed Systems

Track Chairs

Giovanni Morana	C3DNA, USA
Naohiro Hayashibara	Kyoto Sangyo University, Japan

PC Members

Lucian Prodan	Polytechnic University Timisoara, Romania
Md. Abdur Razzaque	University of Dhaka, Bangladesh
Ji Zhang	The University of Southern Queensland, Australia
Ragib Hasan	The University of Alabama at Birmingham, USA
Antonella Di Stefano	University of Catania, Italy
Rao Mikkilineni	C3dna, USA
Douglas D. J. de Macedo	Federal University of Santa Catarina, Florianópolis, Brasil
Ilias Savvas	TEI of Larissa, Greece
Andrea Araldo	Massachusetts Institute of Technology, USA

5. Networked Data Centers, IT Virtualization Technologies, and Clouds

Track Chairs

Gabriele Mencagli	University of Pisa, Italy
Mennan Selimi	University of Cambridge (Computer Laboratory), UK

PC Members

Felix Freitag	Universitat Politecnica de Catalunya (UPC), Spain
Amin M. Khan	Pentaho, Hitachi Data Systems, Japan
Albin Ahmeti	TU Wien, Austria
Besim Bilalli	University Politecnica de Catalunya (UPC), Spain
Leila Sharifi	Urmia University, Iran
Christoph Hochreiner	TU Wien (Distributed System Group), Austria
Ruben Mayer	University of Stuttgart, Germany
Stefano Forti	University of Pisa, Italy
Dalvan Griebler	Pontifícia Universidade Católica do Rio Grande do Sul, Brazil
Massimo Torquati	University of Pisa, Italy

6. Network Protocols, Modeling, Optimization and Performance Evaluation

Track Chairs

Bhed Bista	Iwate Prefectural University, Japan
Fabrizio Messina	University of Catania, Italy

PC Members

Éric Renault	Institut Mines-Télécom, Télécom SudParis, France
Douglas Macedo	Federal University of Santa Catarina (UFSC), Brazil
Lidia Fotia	Università Mediterranea di Reggio Calabria (DIIES), Italy
Ilias Savvas	Department of Computer Science and Engineering, TEI of Thessaly, Greece
Matthias Steinbauer	Onlinegroup.at creative online systems, Austria
Jiahong Wang	Iwate Prefectural University, Japan
Shigetomo Kimura	University of Tsukuba, Japan
Chotipat Pornavalai	King Mongkut's Institute of Technology Ladkrabang, Thailand
Danda B. Rawat	Howard University, USA
Gongjun Yan	University of Southern Indiana, USA
Sachin Shetty	Old Dominion University, USA

7. Data Security, Trust and Reputation

Track Chairs

Mirang Park	Kanagawa Institute of Technology, Japan
Francesco Palmieri	University of Salerno, Italy

PC Members

Shigetomo Kimura	University of Tsukuba, Japan
Toshihiro Yamauchi	Okayama University, Japan
Hiroaki Yamamoto	Shinshu University, Japan
Naonobu Okazaki	University of Miyazaki, Japan
Arcangelo Castiglione	University of Salerno, Italy
Raffaele Pizzolante	University of Salerno, Italy
Ugo Fiore	Federico II University of Napoli, Italy
Sergio Ricciardi	Barcelonatech, Spain

8. Web Science, Learning and Business Intelligence

Track Chairs

Ana Azevedo	ISCAP, Porto, Portugal
Filipe Portela	University of Minho, Portugal

PC Members

Alvaro Figueira	University of Porto, Portugal
Marina Ribaudo	University of Genoa, Italy
Sotirios Kontogiannis	University of Ioannina, Greece
Dumitru Burdescu	University of Craiova, Romania
Feng Xia	Dalian University of Technology, China
Hugo Peixoto	University of Minho, Portugal
Inna Skarga-Bandurova	East Ukrainian National University, Ukraine
Jorge Bernardino	Polytechnic Institute of Coimbra, Portugal
Julio Duarte	University of Minho, Portugal
Teresa Guarda	State University of Santa Elena Peninsula, Ecuador

9. Data Analytics for Learning and Virtual Organisations

Track Chairs

Nobuo Funabiki	Okayama University, Japan
Marcello Trovati	Edge Hill University, UK

PC Members

Shinji Sugawara	Chiba Institute of Technology, Japan
Kazuyoshi Kojima	Saitama University, Japan
Tomoya Kawakami	Nara Institute of Science and Technology, Japan
Makoto Fujimura	Nagasaki University, Japan
Kiyoshi Ueda	Nihon University, Japan
Yoshinobu Tamura	Tokyo City University, Japan
Mohsen Farid	University of Derby, Japan
Richard Conniss	University of Derby, UK
Georgios Kontonatsios	Edge Hill University, UK
Jeffrey Ray	Edge Hill University, UK

10. Data Management and Information Retrieval

Track Chairs

Farookh Hussain	University of Technology Sydney, Australia
Carmen de Maio	University of Salerno, Italy

PC Members

Francesco Orciuoli	University of Salerno, Italy
Stefania Tomasiello	University of Salerno, Italy
Neha Warikoo	Academia Sinica, Taiwan
Stefania Boffa	Università dell'Insubria, Italy
Chang Yung-Chun	Taipei Medical University, Italy
Salem Alkhalaf	Qassim University, Saudi Arabia
Osama Alfarraj	King Saud University, Saudi Arabia
Thamer AlHussain	Saudi Electronic University, Saudi Arabia
Mukesh Prasad	University of Technology Sydney, Australia

11. Massive Processing and Machine Learning on Large Data Sets

Track Chairs

Paolo Trunfio	University of Calabria, Italy
Douglas Macedo	Federal University of Santa Catarina (UFSC), Brazil

PC Members

Yue Zhao	National Institutes of Health, USA
Pelle Jakovits	University of Tartu, Estonia
Gustavo Medeiros de Araujo	Universidade Federal de Santa Catarina, Brazil
Diego Kreutz	Universidade Federal do Pampa, Brazil
Fabrizio Marozzo	University of Calabria, Italy
Marwan Hassani	TU Eindhoven, The Netherlands
Sofian Maabout	Bordeaux University, France
Anirban Mondal	Shiv Nadar University, India

12. Data Modeling, Visualization and Representation Tools

Track Chairs

Omar Hussain	UNSW Canberra, Australia
Matthias Steinbauer	Johannes Kepler University Linz, Austria

PC Members

Walayat Hussain	University of Technology Sydney, Australia
Naeem Janjua	Edith Cowan University, Australia
Jamshaid Ashraf	Data Processing Services, Kuwait
Farookh Hussain	University of Technology, Australia
Zia Rehman	COMSATS Institute of Information Technology (CIIT), Pakistan
Sazia Parvin	Deakin University, Australia
Morteza Saberi	UNSW, Canberra, Australia
Saqib Ali	Sultan Qaboos University, Australia

13. IoT and Fog Computing

Track Chairs

Benoît Parrein	LS2N, University of Nantes, France
Luiz Fernando Bittencourt	UNICAMP—Universidade Estadual de Campinas, Brazil

PC Members

Paolo Bellavista	University of Bologna, Italy
Victor Kardeby	RISE Acreo, Sweden

Dimitri Pertin	INRIA, Nantes, France
Stefano Secci	LIP6, University Paris 6, France
Suayb Arslan	MEF University, Istanbul, Turkey
Rafael Tolosana-Calasanz	University of Zaragoza, Spain
Congduc Pham	University of Pau, France
Vlado Stankovski	University of Ljubljana, Slovenia
Dana Petcu	West University of Timisoara, Romania

14. Mobile and Wireless Networks

Track Chairs

Bexhet Kamo	Polytechnic University of Tirana, Albania
Elis Kulla	Okayama University of Science, Japan

PC Members

Kengo Katayama	Okayama University of Science, Japan
Isaac Woungang	Ryerson University, Canada
Bhed Bista	Iwate Prefectural University, Japan
Akira Uejima	Okayama University of Science, Japan
Tetsuya Oda	Okayama University of Science, Japan
Algenti Lala	Polytechnic University of Tirana, Albania
Olimpjon Shurdi	Polytechnic University of Tirana, Albania
Elson Agastra	Polytechnic University of Tirana, Albania
Alban Rakipi	Polytechnic University of Tirana, Albania
Shkelzen Cakaj	Polytechnic University of Tirana, Albania

15. Big Data Management and Scalable Storage Technologies

Track Chairs

Shadi Ibrahim	Inria Rennes, France
Pruet Boonma	Chiang Mai University, Thailand

PC Members

Sivadon Chaisiri	University of Waikato, New Zealand
Long Cheng	Eindhoven University of Technology, The Netherlands
Houssem Chihoub	Grenoble Institute of Technology, France
Chonho Lee	Osaka University, Japan
Catalin Leordeanu	University Politehnica of Bucharest, Romania

Dusit Niyato	NTU, Singapore
Anis Yazidi	Oslo and Akershus University College of Applied Sciences, Norway
Amelie Chi Zhou	Shenzhen University, China
Yunbo Li	IMT Atlantique, France
Matthieu Dorier	Argonne National Laboratory, USA

16. Energy-Aware and Green Computing Systems

Track Chairs

Evangelos Pournaras	ETH Zurich, Switzerland
Tomoya Enokido	Rissho University, Japan

PC Members

Omar Khadeer Hussain	University of New South Wales, Australia
Akio Koyama	Yamagata University, Japan
Eric Pardede	La Trobe University, Australia
Minoru Uehara	Toyo University, Japan
Motoi Yamagiwa	University of Yamanashi, Japan
Florin Pop	University Politehnica of Bucharest, Romania
Akshay Uttama Nambi S. N.	Microsoft Research India
Stefan Bosse	University of Bremen, Germany
Venkatesha Prasad	Delft University of Technology, The Netherlands
Per Ola Kristensson	University of Cambridge, UK

17. Multimedia Networking and Medical Applications

Track Chairs

Hiroaki Nishino	Oita University, Japan
Sajal Mukhopadhyay	National Institute of Technology, Durgapur, India

PC Members

Makoto Fujimura	Nagasaki University, Japan
Nobukazu Iguchi	Kindai University, Japan
Makoto Nakashima	Oita University, Japan
Yoshihiro Okada	Kyushu University, Japan
Toshiya Takami	Oita University, Japan
Kenzi Watanabe	Hiroshima University, Japan

Subhrabrata Choudhury	National Institute of Technology, Durgapur, India
Debashis Nandi	National Institute of Technology, Durgapur, India
Jaydeep Howlader	National Institute of Technology, Durgapur, India
Animesh Dutta	National Institute of Technology, Durgapur, India
Mansaf Alam	Jamia Millia Islamia, New Delhi, India
P. Sakthivel	Anna University, Chennai, India
Dipankar Das	Jadavpur University, Kolkata, India

18. Applied Cryptography and Cloud Security

Track Chairs

Xu An Wang	Engineering University of CAPF, China
Mingwu Zhang	Hubei University of Technology, China

PC Members

Guangquan Xu	Tianjin University, China
Yang Lei	Engineering University of CAPF, China
Yuechuan Wei	Engineering University of CAPF, China
Minghu Wu	Hubei University of Technology, China
Zhiqiang Gao	Engineering University of CAPF, China
Xuan Guo	Officer's University of CAPF, China
Xiaoou Song	Engineering University of CAPF, China
Yongqiang Li	Xi'an High Technology Research Institute, China
Ling Chen	Yangtze University, China
Leyou Zhang	Xidian University, China
Zhenhua Chen	Xi'an University of Technology, China

EIDWT-2018 Reviewers

Ali Khan Zahoor	Di Martino Beniamino	Ikeda Makoto
Barolli Admir	Dobre Ciprian	Ishida Tomoyuki
Barolli Leonard	Enokido Tomoya	Kikuchi Hiroaki
Bessis Nik	Ficco Massimo	Kolici Vladi
Bista Bhed	Fiore Ugo	Koyama Akio
Caballé Santi	Fun Li Kin	Kulla Elis
Castiglione Aniello	Gotoh Yusuke	Lee Kyungroul
Chellappan Sriram	Hachaj Tomasz	Loia Vincenzo
Chen Hsing-Chung	Hussain Farookh	Matsuo Keita
Chen Xiaofeng	Hussain Omar	Koyama Akio
Cui Baojiang	Javaid Nadeem	Kryvinska Natalia

Nishino Hiroaki
Oda Tetsuya
Ogiela Lidia
Ogiela Marek
Palmieri Francesco
Paruchuri Vamsi Krishna
Pop Florin
Rahayu Wenny
Rawat Danda
Shibata Yoshitaka
Sato Fumiaki
Spaho Evjola
Sugita Kaoru
Takizawa Makoto
Taniar David
Terzo Olivier
Uchida Noriki
Uehara Minoru
Venticinque Salvatore
Waluyo Agustinus Borgy
Wang Xu An
Woungang Isaac
Xhafa Fatos
Yim Kangbin
Zhang Mingwu

EIDWT-2018 Keynote Talks

Smart Cities and Safe Cities by Situational Awareness and Computational Intelligence

Vincenzo Loia, University of Salerno, Salerno, Italy

Abstract. Situation Awareness is usually defined in terms of what information is important for a particular job or goal. Most of the problems with Situation Awareness occur at the level "Perception" and "Comprehension" because of missing information, information overload, information perceived in a wrong way (e.g., noise) or also information not pertinent with respect to the specific goal. Thus, the current situation must be identified, in general, in uncertainty conditions and within complex and critical environments. In this case, it is needed an effective hybridization of the human component with the technological (automatic) component to succeed in tasks related to Situation Awareness. Situation Awareness-oriented systems have to organize information around goals and provide a proper level of abstraction of meaningful information. To answer these issues, we propose a Cognitive Architecture, for defining Situation Awareness-oriented systems, that is defined by starting from the well-known Endsley's Model and integrating a set of Computational Intelligence techniques (e.g., Fuzzy Cognitive Maps and Formal Concept Analysis) to support the three main processes of the model (perception, comprehension, and projection). One of these techniques is Granular Computing that makes information observable at different levels of granularity and approximation to allow humans to focus on specific details, overall picture, or on any other level with respect to their specific goals, constraints, roles, characteristics, and so on. Furthermore, the proposed Cognitive Architecture considers some enabling technologies like multi-agents systems and semantic modeling to provide a solution to face the complexity and heterogeneity of the monitored environment and the capability to represent, in a machine-understandable way, procedural, factual and other kind of knowledge and all the memory facilities that could be required.

Cyber-Human Partnerships—Toward a Resilient Ecosystem in Smart Cities

Schahram Dustdar, Vienna University of Technology (TU Wien), Vienna, Austria

Abstract. In this talk, I will explore one of the most relevant challenges for a decade to come: How to integrate the Internet of Things (IoT) with software, people, and processes, considering modern Cloud Computing and the IoT with Big Data. I will present a fresh look at this problem and examine how to integrate people, software services, and things with their data, into one novel resilient ecosystem, which can be modeled, programmed, and deployed on a large scale in an elastic way. This novel paradigm has major consequences on how we view, build, design, and deploy ultra-large-scale distributed systems and establishes a novel foundation for an "architecture of value" driven Smart City.

Contents

Implementation of a New Function for Preventing Short Reconnection in a WLAN Triage System

Kosuke Ozera[1(✉)], Takaaki Inaba[1], Kevin Bylykbashi[2], Shinji Sakamoto[1], and Leonard Barolli[1]

[1] Department of Information and Communication Engineering, Fukuoka Institute of Technology (FIT), 3-30-1 Wajiro-Higashi, Higashi-Ku, Fukuoka 811–0295, Japan
kosuke.o.fit@gmail.com, g.takaaki.inaba@gmail.com, shinji.sakamoto@ieee.org, barolli@fit.ac.jp

[2] Faculty of Information Technologies, Polytechnic University of Tirana, Bul. "Dëshmorët e Kombit", "Mother Theresa" Square, Nr. 4, Tirana, Albania
kevini_95@hotmail.com

Abstract. The IEEE 802.11e standard for Wireless Local Area Networks (WLANs) is an important extension of the IEEE 802.11 standard focusing on QoS that works with any PHY implementation. The IEEE 802.11e standard introduces EDCF and HCCA. Both these schemes are useful for QoS provisioning to support delay-sensitive voice and video applications. EDCF uses the contention window to differentiate high priority and low priority services. However, it does not consider the priority of users. In order to deal with this problem, in our previous work, we proposed a Fuzzy-based Admission Control System (FACS), which is used in a WLAN triage testbed. In this paper, we present a new function for preventing short reconnection in a WLAN Triage system. These experimental results show that in previous system, all clients reconnect to AP between 12 and 24 s. In the proposed system, when UP is 100, the clients reconnect to AP after 22 s. However, when UP is 0, the clients are reconnected to AP after 166 s. We found that if UP is higher, the reconnected time is shorter compared with the case when UP is low.

1 Introduction

With the development of wireless technology and Internet, there is an increasing need towards portable and mobile computers such as smart phones [25, 28]. The wireless networks need to provide communications between mobile terminals. The Wireless Local Area Networks (WLANs) provide high bandwidth access for users in a limited geographical area. With the popularization of mobile devices, many device communicate together over WLANs [29]. WLANs have a lot of restriction on communication resources comparing wired LANs. Therefore, it is more difficult to guarantee the Quality of Service (QoS).

L. Barolli et al. (Eds.): EIDWT 2018, LNDECT 17, pp. 1–17, 2018.
https://doi.org/10.1007/978-3-319-75928-9_1

In our previous work, the admission decision is done by a Fuzzy-based Admission Control System (FACS) [7–9], which is used in a WLAN Triage testbed. In this paper, we implement a new function for preventing short reconnection in a WLAN Triage system.

The paper is organized as follows. In Sect. 2, we give introduce of CSMA/CA. In Sect. 3, we discuss the application of Fuzzy Logic (FL) for control. In Sect. 4, we present the implemented testbed. In Sect. 5, we show the experimental results. Finally, in Sect. 6, we conclude the paper.

2 CSMA/CA

WLAN is standardized by IEEE 802.11 which uses CSMA/CA as shown in Fig. 1. The nodes check whether other devices are communicating or not before starting communication [5]. If other nodes are communicating with the AP, the node waits for a period of time, called Distributed Inter Frame Space (DIFS). After that, it waits an additionally random time called back-off time. After the back-off time, if other devices are not communicating, the node starts to send data [26]. AP which received the data waits a constant time, called Short Inter Frame Space (SIFS) and sends ACK to the node which sent the data. Because ACK frame should be sent soon, it is shorter than DIFS.

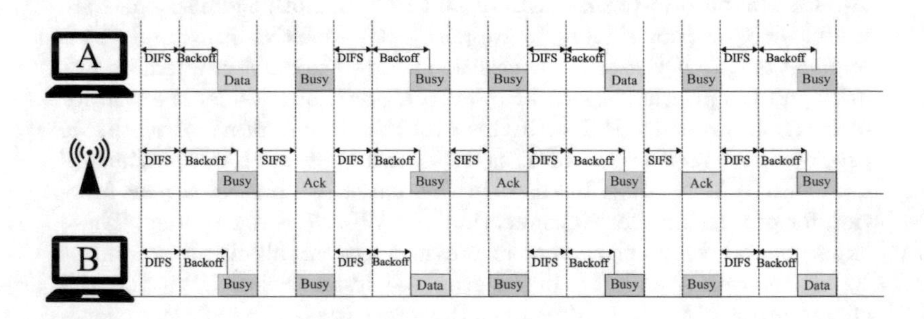

Fig. 1. CSMA/CA.

The IEEE 802.11e standard is an important extension of the IEEE 802.11 standard focusing on QoS [15] that works with any PHY implementation. Wireless nodes equipped with IEEE 802.11e features are now known as QoS stations (QSTAs) and they are associated with a QoS access point (QAP) to form a QoS basic service set (QBSS). The main feature of the IEEE 802.11e standard is that it improves the MAC layer for QoS provisioning by providing support for: segregation of data packets based on priority requirements; negotiation of QoS parameters through a central coordinator or AP; and admission control.

The IEEE 802.11e standard introduces EDCF. This scheme is useful for QoS provisioning to support delay-sensitive voice and video applications [2].

In the DCF configuration, a contention window is set after a frame is transmitted. This is done to avoid any collisions. The window defines the contention time of various stations who contend with each other for access to channel. However, each of the stations cannot size the channel immediately, rather the MAC protocol uses a randomly chosen time period for each station after that channel has undergone transmission [30].

The EDCF uses this contention window to differentiate between high priority and low priority services [24]. The central coordinator assigns a contention window of shorter length to the stations with higher priority that helps them to transmit before the lower priority ones [23]. To differentiate further, Inter Frame Spacing (IFS) can be varied according to different traffic categories. Instead of using a DIFS as for the DCF traffic, a new inter-frame spacing called Arbitration Inter Frame Spacing (AIFS) is used. The AIFS used for traffic has a duration of a few time slots longer than the DIFS duration. Therefore, a traffic category having smaller AIFS gets higher priority.

IEEE 802.11e provides EDCA and Hybrid coordination function controlled channel access (HCCA) as a priority control method [2,16,32]. Mainly, EDCA is used because of easy implementation and compatibility of CSMA/CA.

3 Application of Fuzzy Logic for Control

The ability of fuzzy sets and possibility theory to model gradual properties or soft constraints whose satisfaction is matter of degree, as well as information pervaded with imprecision and uncertainty, makes them useful in a great variety of applications [4,12–14,20,27].

The most popular area of application is Fuzzy Control (FC), since the appearance, especially in Japan, of industrial applications in domestic appliances, process control, and automotive systems, among many other fields [1,3,6,11,17].

3.1 FC

In the FC systems, expert knowledge is encoded in the form of fuzzy rules, which describe recommended actions for different classes of situations represented by fuzzy sets.

In fact, any kind of control law can be modeled by the FC methodology, provided that this law is expressible in terms of "if ... then ..." rules, just like in the case of expert systems. However, FL diverges from the standard expert system approach by providing an interpolation mechanism from several rules. In the contents of complex processes, it may turn out to be more practical to get knowledge from an expert operator than to calculate an optimal control, due to modeling costs or because a model is out of reach.

3.2 FC Rules

FC describes the algorithm for process control as a fuzzy relation between information about the conditions of the process to be controlled, x and y, and the output for the process z. The control algorithm is given in "if ... then ..." expression, such as:

If x is small and y is big, then z is medium;
If x is big and y is medium, then$z$ is big.

These rules are called *FC rules.* The "if" clause of the rules is called the antecedent and the "then" clause is called consequent. In general, variables x and y are called the input and z the output. The "small" and "big" are fuzzy values for x and y, and they are expressed by fuzzy sets.

Fuzzy controllers are constructed of groups of these FC rules, and when an actual input is given, the output is calculated by means of fuzzy inference.

3.3 Control Knowledge Base

There are two main tasks in designing the control knowledge base. First, a set of linguistic variables must be selected which describe the values of the main control parameters of the process. Both the input and output parameters must be linguistically defined in this stage using proper term sets. The selection of the level of granularity of a term set for an input variable or an output variable plays an important role in the smoothness of control. Second, a control knowledge base must be developed which uses the above linguistic description of the input and output parameters. Four methods [18,22,31,33] have been suggested for doing this:

- expert's experience and knowledge;
- modelling the operator's control action;
- modelling a process;
- self organization.

Among the above methods, the first one is the most widely used. In the modeling of the human expert operator's knowledge, fuzzy rules of the form "If Error is *small* and Change-in-error is *small* then the Force is *small*" have been used in several studies [10,19]. This method is effective when expert human operators can express the heuristics or the knowledge that they use in controlling a process in terms of rules of the above form.

3.4 Defuzzification Methods

The defuzzification operation produces a non-FC action that best represent the membership function of an inferred FC action. Several defuzzification methods have been suggested in literature. Among them, four methods which have been applied most often are:

- Tsukamoto's Defuzzification Method;
- The Center of Area (COA) Method;
- The Mean of Maximum (MOM) Method;
- Defuzzification when Output of Rules are Function of Their Inputs.

4 Implemented Testbed

In the implemented tesbed, the admission control is done by FACS which is based on FL.

4.1 FACS

Fuzzy Logic Controller (FLC) is the main part of FACS and its components are shown in Fig. 2. The Fuzzy Rule Base (FRB) and the membership functions are shown in Table 1 and Fig. 3, respectively. In FACS, we consider 3 input parameters: User Priority (UP), Received Signal Strength Indication (RSSI) and User Connected Time (UCT). The output is Connection Priority (CP). The term sets of *UP*, *RSSI* and *UCT* are defined respectively as:

$$UP = \begin{pmatrix} Low\ priority \\ Middle\ priority \\ High\ priority \end{pmatrix} = \begin{pmatrix} Lp \\ Mp \\ Hp \end{pmatrix};$$

$$RSSI = \begin{pmatrix} Very\ Low \\ Low \\ Middle \\ High \\ Very\ High \end{pmatrix} = \begin{pmatrix} VL \\ L \\ M \\ H \\ VH \end{pmatrix};$$

$$UCT = \begin{pmatrix} Very\ Very\ Short \\ Very\ Short \\ Short \\ Middle \\ Long \\ Very\ Long \\ Very\ Very\ Long \end{pmatrix} = \begin{pmatrix} VVSH \\ VSH \\ SH \\ MI \\ LO \\ VLO \\ VVLO \end{pmatrix}.$$

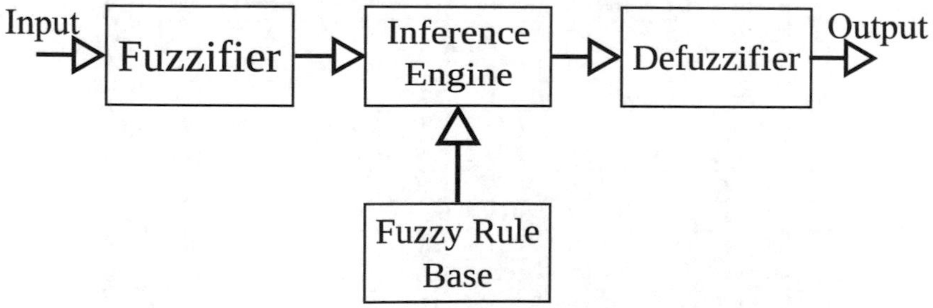

Fig. 2. Fuzzy logic controller.

The term set for the output CP is defined as:

$$CP = \begin{pmatrix} Level\ 1 \\ Level\ 2 \\ Level\ 3 \\ Level\ 4 \\ Level\ 5 \\ Level\ 6 \\ Level\ 7 \\ Level\ 8 \\ Level\ 9 \end{pmatrix} = \begin{pmatrix} L1 \\ L2 \\ L3 \\ L4 \\ L5 \\ L6 \\ L7 \\ L8 \\ L9 \end{pmatrix}.$$

Table 1. Fuzzy rule base

Rule	UP	RSSI	UCT	CP	Rule	UP	RSSI	UCT	CP	Rule	UP	RSSI	UCT	CP
1	LP	VL	VVSH	L4	36	MP	VL	VVSH	L5	71	HP	VL	VVSH	L5
2	LP	VL	VSH	L3	37	MP	VL	VSH	L4	72	HP	VL	VSH	L4
3	LP	VL	SH	L2	38	MP	VL	SH	L3	73	HP	VL	SH	L3
4	LP	VL	MI	L1	39	MP	VL	MI	L3	74	HP	VL	MI	L3
5	LP	VL	LO	L1	40	MP	VL	LO	L2	75	HP	VL	LO	L2
6	LP	VL	VLO	L1	41	MP	VL	VLO	L1	76	HP	VL	VLO	L4
7	LP	VL	VVLO	L1	42	MP	VL	VVLO	L1	77	HP	VL	VVLO	L4
8	LP	L	VVSH	L4	43	MP	L	VVSH	L6	78	HP	L	VVSH	L8
9	LP	L	VSH	L3	44	MP	L	VSH	L5	79	HP	L	VSH	L7
10	LP	L	SH	L2	45	MP	L	SH	L4	80	HP	L	SH	L7
11	LP	L	MI	L2	46	MP	L	MI	L3	81	HP	L	MI	L6
12	LP	L	LO	L1	47	MP	L	LO	L3	82	HP	L	LO	L6
13	LP	L	VLO	L1	48	MP	L	VLO	L2	83	HP	L	VLO	L5
14	LP	L	VVLO	L1	49	MP	L	VVLO	L1	84	HP	L	VVLO	L4
15	LP	M	VVSH	L5	50	MP	M	VVSH	L7	85	HP	M	VVSH	L8
16	LP	M	VSH	L4	51	MP	M	VSH	L6	86	HP	M	VSH	L8
17	LP	M	SH	L3	52	MP	M	SH	L5	87	HP	M	SH	L7
18	LP	M	MI	L3	53	MP	M	MI	L4	88	HP	M	MI	L7
19	LP	M	LO	L2	54	MP	M	LO	L4	89	HP	M	LO	L7
20	LP	M	VLO	L1	55	MP	M	VLO	L3	90	HP	M	VLO	L6
21	LP	M	VVLO	L1	56	MP	M	VVLO	L2	91	HP	M	VVLO	L5
22	LP	H	VVSH	L7	57	MP	H	VVSH	L8	92	HP	H	VVSH	L9
23	LP	H	VSH	L6	58	MP	H	VSH	L7	93	HP	H	VSH	L9
24	LP	H	SH	L5	59	MP	H	SH	L6	94	HP	H	SH	L8
25	LP	H	MI	L4	60	MP	H	MI	L6	95	HP	H	MI	L8
26	LP	H	LO	L4	61	MP	H	LO	L5	96	HP	H	LO	L8
27	LP	H	VLO	L3	62	MP	H	VLO	L4	97	HP	H	VLO	L7
28	LP	H	VVLO	L2	63	MP	H	VVLO	L4	98	HP	H	VVLO	L7
29	LP	VH	VVSH	L8	64	MP	VH	VVSH	L9	99	HP	VH	VVSH	L9
30	LP	VH	VSH	L7	65	MP	VH	VSH	L8	100	HP	VH	VSH	L9
31	LP	VH	SH	L7	66	MP	VH	SH	L8	101	HP	VH	SH	L9
32	LP	VH	MI	L6	67	MP	VH	MI	L7	102	HP	VH	MI	L9
33	LP	VH	LO	L6	68	MP	VH	LO	L7	103	HP	VH	LO	L9
34	LP	VH	VLO	L5	69	MP	VH	VLO	L6	104	HP	VH	VLO	L8
35	LP	VH	VVLO	L4	70	MP	VH	VVLO	L6	105	HP	VH	VVLO	L8

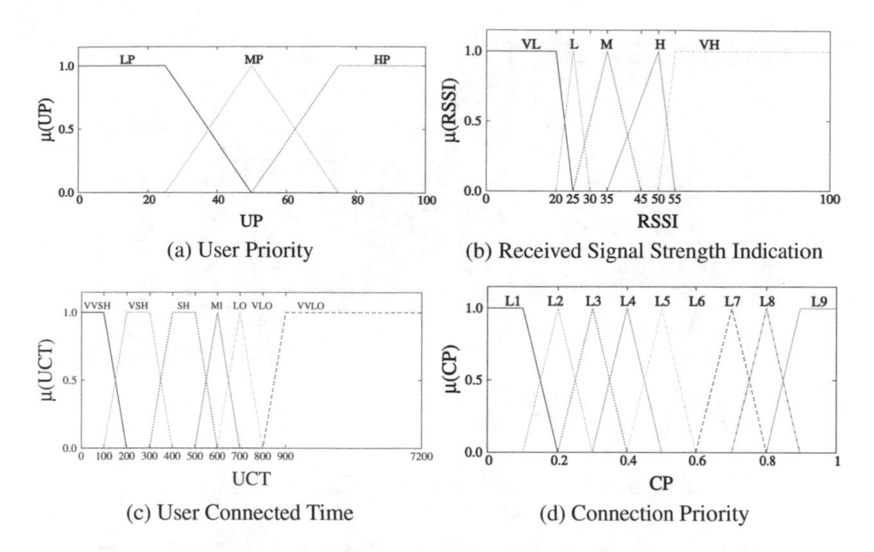

(a) User Priority (b) Received Signal Strength Indication

(c) User Connected Time (d) Connection Priority

Fig. 3. Membership functions of FACS.

4.2 Testbed Structure

The structure of the implemented testbed is shown in Fig. 4. In our previous system, the disconnected clients are reconnected to AP immediately. In order to improve the FACS performance, we propose CCST (Cut Client Saving Table) to prevent the clients to reconnect in a short interval.

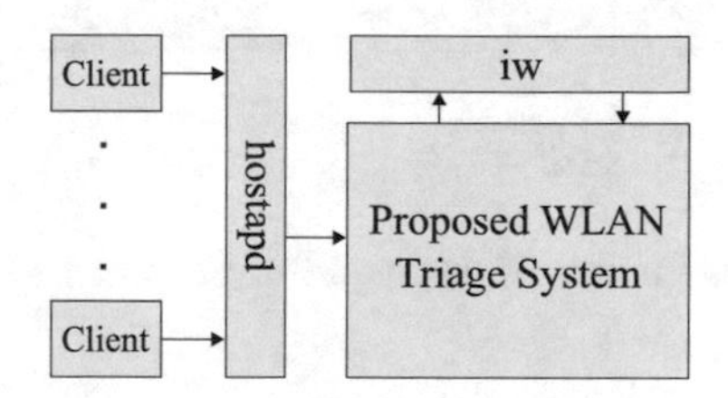

Fig. 4. Testbed structure.

In the CCST, we set up the Refresh Time (RT) as shown in Eq. (1).

$$RT = Basic\,Refresh\,Time\,(BRT) * Prioritized\,Refresh\,Time\,(PRT) \quad (1)$$

If the UP of a client is high, the reconnection time is short, while UP is low, the reconnection time is long.

We show the process of "Accept" in previous system and proposed system, in Figs. 5 and 6, respectively. At the beginning when a user sends a connection request to AP, the UCT for this user is zero.

In the Classification, the user request is classified connect or disconnect. Moreover, in the proposed system, the system checks if the user is registered in CCST.

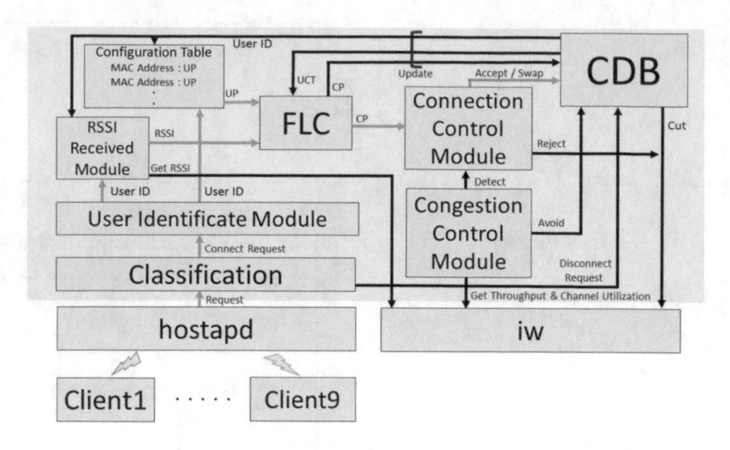

Fig. 5. Process of "Accept" in previous system.

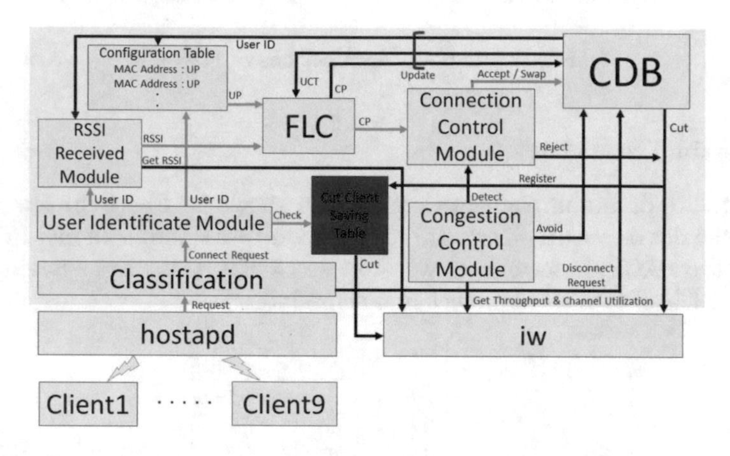

Fig. 6. Process of "Accept" in proposed system.

The User Identification Module (UIM) distinguishes MAC address. Then, the input parameters UCT, RSSI and UP are used by FLC to get CP. The user will be accepted if congestion does not occur in AP. When the congestion occurs, the Congestion Control Module (CCM) is activated to avoid the congestion.

In the testbed, the implemented system uses FLC to make the admission decision. Our testbed uses "iw" command to get the information of clients and make disconnection. We show the communication flow for the proposed system from Figs. 6, 7, 8 and 9.

Figure 6 shows "Accept" process. The clients requests are classified connect or disconnect in "Classification" module. The system checks it the client is registered in CCST. If the client registered to CCST, the client cannot connect to AP. When the clients are accepted by AP, the client information is stored in Connection DataBase (CDB).

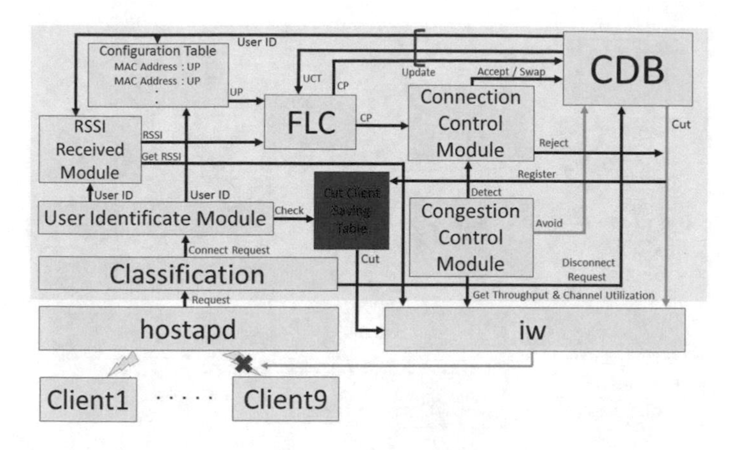

Fig. 7. Process of "Swap" in proposed system.

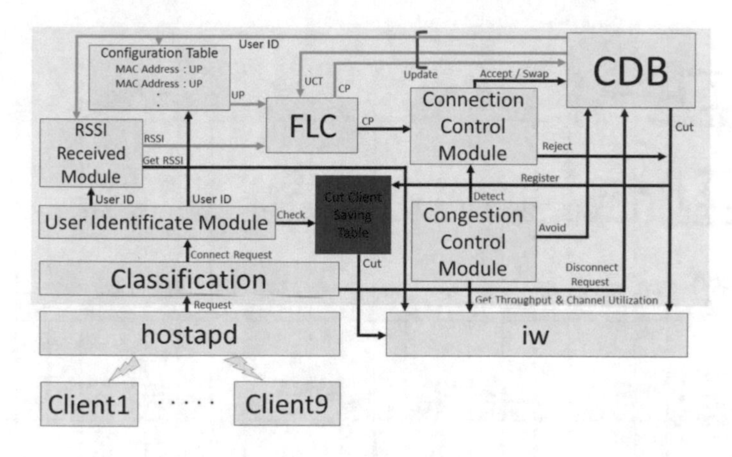

Fig. 8. Process of "Update" in proposed system.

We show the "Swap" process in Fig. 7. After detecting the congestion, the CCM enters the Monitoring phase and notifies the Connection Control Module (CONCM). When CONCM receives information on congestion detection, it swaps or rejects connection requests, thus preventing the increase of the number of clients connected to the AP. Thus, the clients are swapped as shown in Fig. 7.

In general, the CDB needs to updates the information. Figure 8 shows the "Update" process. Figure 9 shows the process of "Avoid". In the Avoid phase, a client will be disconnected from AP when the clients' CP is the lowest in the CDB. Moreover, the client is registered to CCST.

4.3 CCM

In wireless LAN, the radio wave and the distance between a client and AP effect the communication speed of a client. In our experiment, we try to decrease

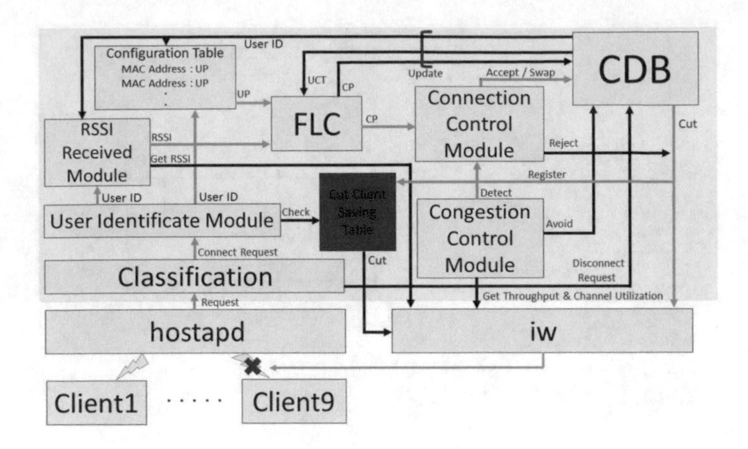

Fig. 9. Process of "Avoid" in proposed system.

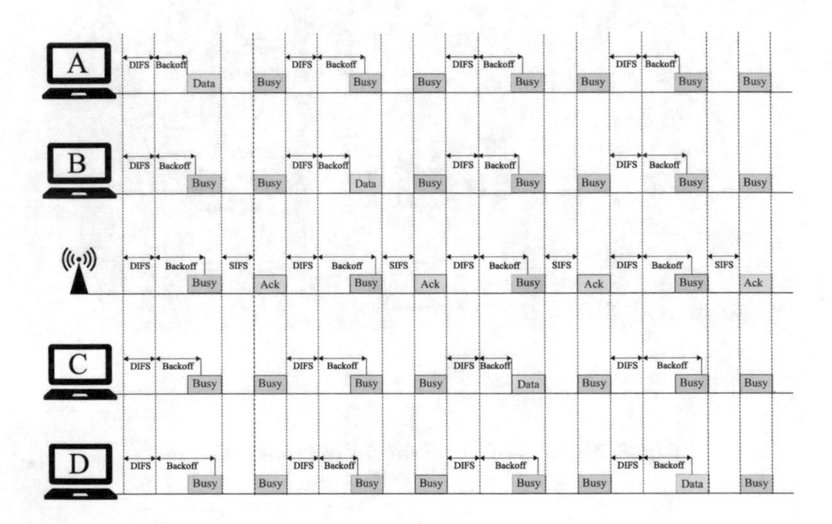

Fig. 10. NCD is 4

the communication speed to allow devices to connect to the AP. We show the behavior when a client sends data and the number of connected devices are changed. When the Number of Connected Devices (NCD) is 2, each client can send date twice (see Fig. 1). But, when NCD is 4, each client can send date only once (see Fig. 10). In general, some of clients are connected to the AP, but may be they do not send the data. So, it is better to consider the real number of the clients that are sending the data.

In [21], we decided the congestion definition and the congestion detection by NCD. This was a simple method adopted for the current wireless LAN. However, by this method the AP bandwidth is not used efficiently. Therefore, in this paper, we consider a new congestion condition.

In order to decide the congestion condition, we considered the "Congestion Throughput (CT)", "Channel Utilization (CU)" and "Number of Minimum Communicating Device (NMCD)". By using these parameters, we know the situation of clients that are communicating with AP and the congestion situation of the AP.

The CT is the average throughput of communicating devices per unit time as shown in Fig. 11. In CCM, the sending date size of each client is measured at fixed intervals. We use the average throughput for making decision on the congestion situation, because the clients connected to the AP do not have the same throughput.

The CU shows the percentage of AP bandwidth usage for unit of time. The CU and NMCD parameters are used for congestion mis-detection. We show the relation of these three parameters in Fig. 12.

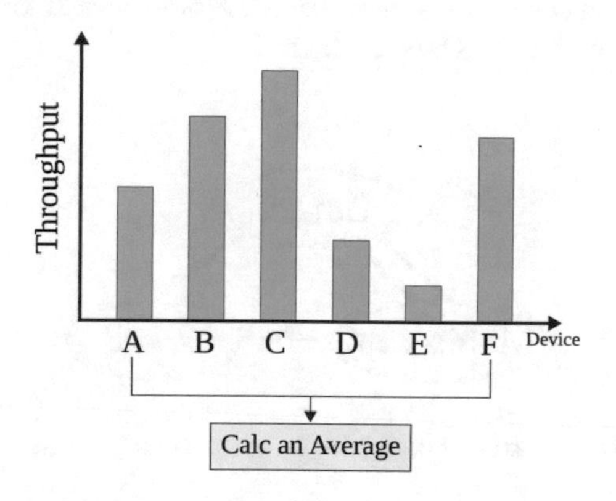

Fig. 11. Calculation method of congestion throughput.

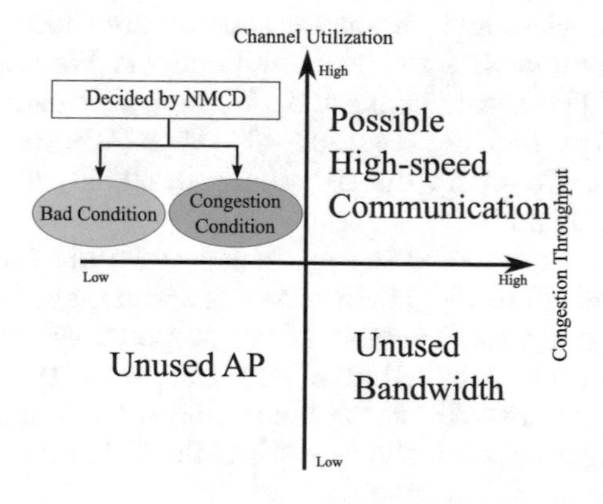

Fig. 12. Relation of parameters of congestion.

4.4 Congestion Process Phases

In Fig. 13, we show the relation among three phases. The three phases process is as follows:

1. In Detect phase, the CCM checks congestion state every Detect interval. If congestion is detected, the connection control will be enabled and move to Monitoring phase.
2. In Monitoring phase, the CCM checks the congestion state every Monitoring interval. If the congestion state will be over during Monitoring Time, it returns to Detect phase.
3. If congestion state is not over until the Monitoring phase has finished, it moves to Avoid phase.
4. In Avoid phase, the clients are disconnected every Avoid interval. The congestion state is checked for every disconnection and if the congestion state will be over, it returns to Detect phase.

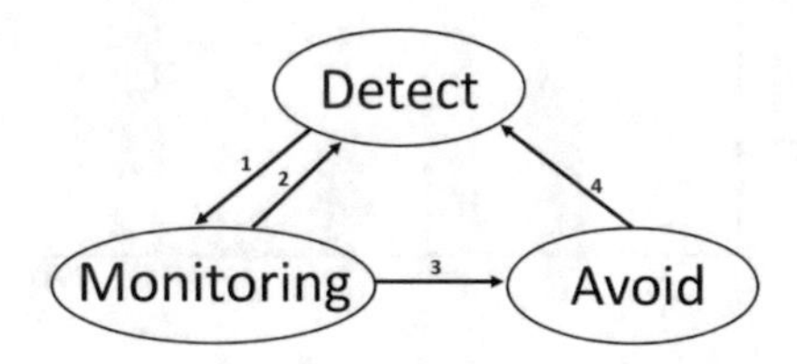

Fig. 13. Relation among three congestion process phases.

5 Experimental Results

The experimental devices and parameter settings are shown in Tables 2 and 3, respectively. We use 9 devices as clients and one AP. We conducted the experiments for 90 min. The clients connect to AP and watch streaming video from YouTube with 720 px and bitrate 1500–4000 kbps. We set up the congestion throughput 1 Mbps. We set up the UP values: 0, 30, 50, 80 and 100, and PRT values: 100, 70, 50, 20 and 0.

The experimental results for the previous system are shown in Fig. 14 and for the proposed system in Fig. 15. In previous system, all clients can reconnect to AP between 12 and 24 s. However, in the proposed system, if the client UP is 100, the client reconnect to AP after 22 s. But, when the client UP is 0, the client reconnect to AP after 166 s. We found that if UP is higher, the reconnect time is shorter compared when the case when the UP is low.

Table 2. Experimental devices.

Purpose	Device	CPU	NIC	Memory	OS
AP	Raspberry Pi 2	ARM Cortex-A7	Buffalo WLI-UC-GNM	1GB	Raspbian 8.0-kernel 3.16.7
Client	Laptop	Intel Pentium 1.2GHz	Buffalo WLI-UC-GNM	750MB	Cent OS 6.8-kernel 2.6.32-642
	Laptop	Intel Pentium 1.3GHz	Buffalo WLI-UC-GNM	1GB	Cent OS 6.8-kernel 2.6.32-643
	Laptop	Intel Pentium 1.5GHz	Buffalo WLI-UC-GNM	1.5GB	Cent OS 6.8-kernel 2.6.32-644
	Laptop	Intel Core i5-3230M 2.60GHz*2	Intel Centrino Wireless-N+ 6150	8GB	Ubuntu 16.04 LTS-kernel 4.4
	Laptop	Intel Core i5-4200U 1.60GHz*4	Buffalo WLI-UC-GNM	4GB	Ubuntu 14.04 LTS-kernel 3.13
	Laptop	Intel Core i5 2.80GHz*2	AirMac Extreme	16GB	Mac OS X 10.11.4
	Desktop	Intel Core i5-6500 3.20GHz*4	Buffalo WLI-UC-GNM	10GB	Ubuntu 16.04 LTS-kernel 4.4
	Desktop	Intel Core i5-3571 3.40GHz*4	Buffalo WLI-UC-GNM	16GB	Ubuntu 14.04 LTS-kernel 3.13
	Desktop	Intel Core i5-661 3.30GHz*4	Buffalo WLI-UC-GNM	4GB	Ubuntu 16.04 LTS-kernel 4.4

Table 3. Parameter settings.

Parameters	Values
Number of Clients	9
CDB Update Interval	5 [sec]
Detect Interval	2 [sec]
Monitoring Interval	2 [sec]
Avoid Interval	8 [sec]
Monitoring Time	6 [sec]
UP	0, 30, 50, 80, 100
PRT	100, 70, 50, 20, 0 [sec]
BRT	10 [sec]
Congestion ThresholdCongestion Throughput	1 [Mbps]
Congestion ThresholdChannel Utilization	80 [%]
Congestion ThresholdNumber of Minimum Clients	3

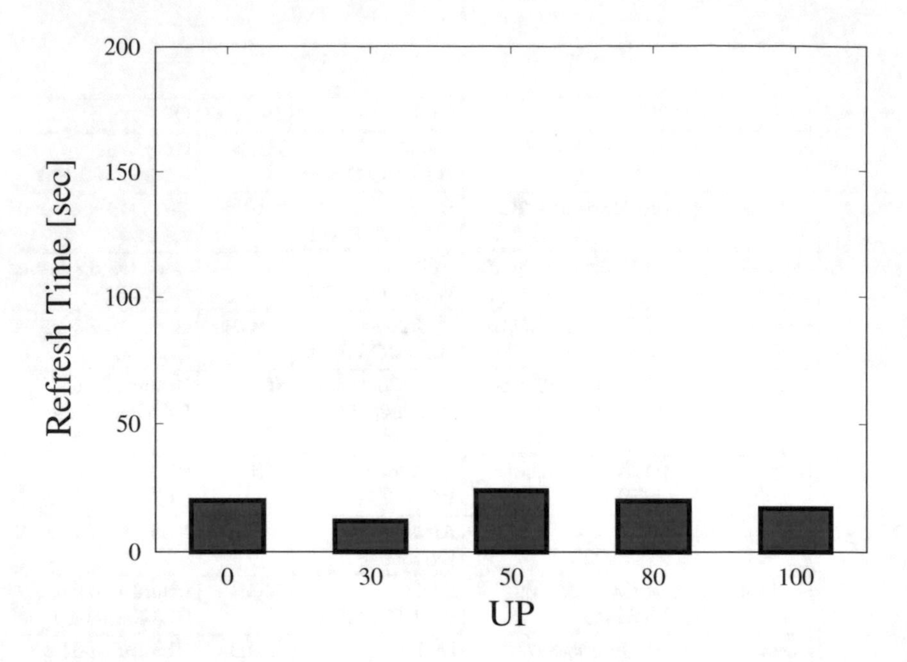

Fig. 14. Experimental result of previous system.

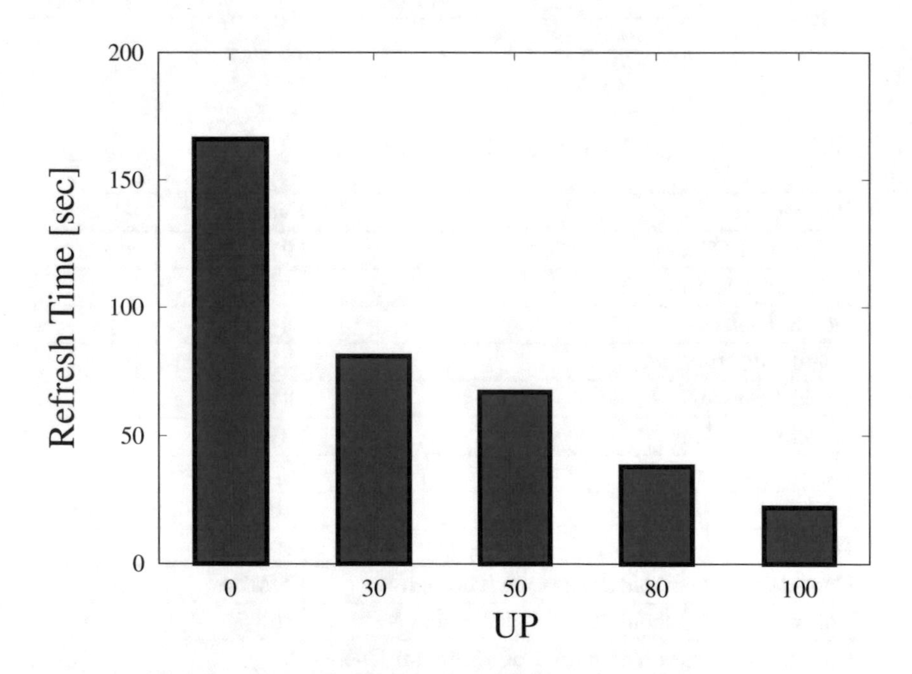

Fig. 15. Experimental result of proposed system.

6 Conclusions

In this paper, we presented the implementation of a WLAN Triage testbed. We implemented of a new function for preventing short reconnection in a WLAN Triage system.

These experimental results show that in previous system, all clients reconnect to AP between 12 and 24 s. In the proposed system, when UP is 100, the clients reconnect to AP after 22 s. However, when UP is 0, the clients are reconnected to AP after 166 s. We found that if UP is higher, the reconnect time is shorter compared with the case when UP is low.

In our future work, we would like to carry out extensive experiments. Moreover, we would like to compare the performance of implemented testbed with other systems.

References

1. Cao, Q., Fujita, S.: Load-balancing schemes for a hierarchical peer-to-peer file search system. Int. J. Grid Util. Comput. **2**(2), 164–171 (2011)
2. Choi, S., Del Prado, J., Mangold, S., et al.: IEEE 802.11e Contention-based Channel Access (EDCF) performance evaluation. In: International Conference on Communications (ICC-2003), vol. 2, pp. 1151–1156 (2003)
3. Elmazi, D., Inaba, T., Sakamoto, S., Oda, T., Ikeda, M., Barolli, L.: Selection of secure actors in wireless sensor and actor networks using fuzzy logic. In: The 10th International Conference on Broadband and Wireless Computing, Communication and Applications (BWCCA-2015), pp. 125–131 (2015)
4. Elmazi, D., Sakamoto, S., Oda, T., Kulla, E., Spaho, E., Barolli, L.: Two fuzzy-based systems for selection of actor nodes inwireless sensor and actor networks: a comparison study considering security parameter effect. Mob. Netw. Appl. **21**(1), 53–64 (2016)
5. Gao, D., Cai, J., Ngan, K.N.: Admission control in IEEE802.11e wireless LANs. IEEE Netw. **19**(4), 6–13 (2005)
6. Inaba, T., Elmazi, D., Sakamoto, S., Oda, T., Ikeda, M., Barolli, L.: A secure-aware call admission control scheme for wireless cellular networks using fuzzy logic and its performance evaluation. J. Mob. Multimed. **11**(3&4), 213–222 (2015)
7. Inaba, T., Sakamoto, S., Oda, T., Ikeda, M., Barolli, L.: A QoS-aware admission control system for WLAN using fuzzy logic. In: The 30th International Conference on Advanced Information Networking and Applications Workshops (WAINA-2016), pp. 499–505 (2016)
8. Inaba, T., Sakamoto, S., Oda, T., Ikeda, M., Barolli, L.: A testbed for admission control in WLAN: a fuzzy approach and its performance evaluation. In: The 11th International Conference on Broadband and Wireless Computing, Communication and Applications (BWCCA-2016), pp. 559–571 (2016)
9. Javanmardi, S., Shojafar, M., Shariatmadari, S., Ahrabi, S.S.: Fr trust: a fuzzy reputation-based model for trust management in semantic P2P grids. Int. J. Grid Util. Comput. **6**(1), 57–66 (2014)
10. Klir, G.J., Folger, T.A.: Fuzzy Sets, Uncertainty, and Information. Prentice Hall, Englewood Cliffs (1988)

11. Kolici, V., Inaba, T., Lala, A., Mino, G., Sakamoto, S., Barolli, L.: A fuzzy-based CAC scheme for cellular networks considering security. In: 2014 17th International Conference on Network-Based Information Systems, pp. 368–373. IEEE (2014)
12. Kulla, E., Mino, G., Sakamoto, S., Ikeda, M., Caballé, S., Barolli, L.: FBMIS: a fuzzy-based multi-interface system for cellular and ad hoc networks. In: The 28th IEEE International Conference on Advanced Information Networking and Applications (AINA-2014), pp. 180–185 (2014)
13. Liu, Y., Sakamoto, S., Matsuo, K., Ikeda, M., Barolli, L., Xhafa, F.: Improving reliability of JXTA-overlay P2P platform: a comparison study for two fuzzy-based systems. J. High Speed Netw. **21**(1), 27–42 (2015)
14. Liu, Y., Sakamoto, S., Matsuo, K., Ikeda, M., Barolli, L., Xhafa, F.: A comparison study for two fuzzy-based systems: improving reliability and security of jxta-overlay p2p platform. Soft Comput. **20**(7), 2677–2687 (2016)
15. Mangold, S., Choi, S., Hiertz, G.R., Klein, O., Walke, B.: Analysis of IEEE 802.11e for QoS support in wireless LANs. IEEE Wirel. Commun. **10**(6), 40–50 (2003)
16. Mangold, S., Choi, S., May, P., Klein, O., Hiertz, G., Stibor, L.: IEEE802.11e wireless LAN for quality of service. Proc. Eur. Wirel. **2**, 32–39 (2002)
17. Matsuo, K., Elmazi, D., Liu, Y., Sakamoto, S., Mino, G., Barolli, L.: FACS-MP: a fuzzy admission control system with many priorities for wireless cellular networks and its performance evaluation. J. High Speed Netw. **21**(1), 1–14 (2015)
18. McNeill, F.M., Thro, E.: Fuzzy Logic: A Practical Approach. Academic Press, London (1994)
19. Munakata, T., Jani, Y.: Fuzzy systems: an overview. Commun. ACM **37**(3), 68–76 (1994)
20. Ozera, K., Inaba, T., Elmazi, D., Sakamoto, S., Oda, T., Barolli, L.: A fuzzy approach for secure clustering in manets: effects of distance parameter on system performance. In: The 31st IEEE International Conference on Advanced Information Networking and Applications Workshops (WAINA-2017), pp. 251–258. Tamkang University, Taipei, Taiwan, 27–29 March 2017 (2017)
21. Ozera, K., Inaba, T., Sakamoto, S., Barolli, L.: Implementation of a wlan triage testbed using fuzzy logic: Evaluation for different number of clients
22. Procyk, T.J., Mamdani, E.H.: A linguistic self-organizing process controller. Automatica **15**(1), 15–30 (1979)
23. Qashi, R., Bogdan, M., Hänssgen, K.: Evaluating the QoS of WLANs for the IEEE802.11 EDCF in real-time applications. In: International Conference on Communications and Information Technology (ICCIT-2011), pp. 32–35 (2011)
24. Romdhani, L., Ni, Q., Turletti, T.: Adaptive EDCF: enhanced service differentiation for IEEE802.11 wireless ad-hoc networks. Wirel. Commun. Netw. (WCNC-2003) 2, 1373–1378 (2003)
25. Sakamoto, S., Oda, T., Ikeda, M., Barolli, L., Xhafa, F.: Implementation and evaluation of a simulation system based on particle swarm optimisation for node placement problem in wireless mesh networks. Int. J. Commun. Netw. Distrib. Syst. **17**(1), 1–13 (2016)
26. Song, N.O., Kwak, B.J., Song, J., Miller, L.E.: Enhancement of IEEE 802.11 distributed coordination function with exponential increase exponential decrease backoff algorithm. In: The 57th IEEE Semiannual Vehicular Technology Conference, vol. 4, pp. 2775–2778 (2003)
27. Spaho, E., Sakamoto, S., Barolli, L., Xhafa, F., Ikeda, M.: Trustworthiness in P2P: performance behaviour of two fuzzy-based systems for JXTA-overlay platform. Soft Comput. **18**(9), 1783–1793 (2014)

28. Uchida, K., Takematsu, M., Lee, J.H., Honda, J.: A particle swarm optimisation algorithm to generate inhomogeneous triangular cells for allocating base stations in urban and suburban areas. Int. J. Space-Based Situated Comput. **3**(4), 207–214 (2013)
29. Wu, H., Peng, Y., Long, K., Cheng, S., Ma, J.: Performance of reliable transport protocol over IEEE 802.11 wireless LAN: analysis and enhancement. In: The 21st Annual Joint Conference of the IEEE Computer and Communications Societies, vol. 2, pp. 599–607 (2002)
30. Yang, X., Vaidya, N.H.: Priority scheduling in wireless ad hoc networks. In: Proceedings of the 3rd ACM International Symposium on Mobile Ad Hoc Networking & Computing, pp. 71–79 (2002)
31. Zadeh, L.A., Kacprzyk, J.: Fuzzy Logic for the Management of Uncertainty. John Wiley & Sons Inc., New York (1992)
32. Zhu, J., Fapojuwo, A.O.: A new call admission control method for providing desired throughput and delay performance in IEEE 802.11e Wireless LANs. IEEE Trans. Wirel. Commun. **6**(2), 701–709 (2007)
33. Zimmermann, H.J.: Fuzzy Set Theory and Its Applications. Springer Science & Business Media, Netherlands (1991)

A Web-Based English Listening System for Learning Different Pronunciations in Various Countries

Kohei Kamimura(✉) and Kosuke Takano

Department of Information and Computer Sciences, Faculty of Information Technology, Kanagawa Institute of Technology, Atsugi, Japan
s1421172@cco.kanagawa-it.ac.jp,
takano@ic.kanagawa-it.ac.jp

Abstract. The rapid progress of globalization has been increasing our opportunity to have English communication with various people living in different countries. In this study, we focus on ‘local-pronunciation English’, which is English that has characteristics in pronunciation and accent depending on countries and areas, and propose an English listening system for learning different local-pronunciation English in various countries and areas. The feature of our system is that our system can extract countries and areas where English speech is relatively easy to understand to a learner by the sound characteristics, based on the listening learning history of local-pronunciation English. This function allows each learner to learn English listening step by step according to the learner’s listening capability. In this study, we evaluate the feasibility of our system using a Web-based prototype.

1 Introduction

The rapid progress of globalization has been increasing our opportunity to have English communication with various people living in different countries. Ministry of Education, Culture, Sports, Science and Technology in Japan has also promoted English educational reforms for developing global human resources so far.

Meanwhile, social, cultural, educational, and economic exchanges with Asian countries are getting more important in our country. Our university is also forming partnerships with educational institution in Asian countries such China, Thailand, Indonesia, Vietnam, and Malaysia, and encouraging exchanges among students. Thus, our country and Asian countries have close relationships, and English spoken in Asian countries are becoming popular as well as English in Europe and United States.

Although English voice in Europe and United States are often used for English listening learning materials, it is pointed out that there are different pronunciations and accents in English for different countries and areas [2]. We call English that has characteristics in pronunciation and accent depending on countries and areas ‘local-pronunciation English’.

In this study, we focus on such ‘local-pronunciation English’, since we assume that the easiness in speech hearing of English depends on countries and areas where English

L. Barolli et al. (Eds.): EIDWT 2018, LNDECT 17, pp. 18–28, 2018.
https://doi.org/10.1007/978-3-319-75928-9_2

are spoken, to some degree. For example, some learners will find easiness to listen to English spoken in Asian countries than in Europe and United States. Here, we define the easiness in speech hearing of English as the degree in understating of meaning and expression by a listener, which is varied by the difference of sound characteristics such as pronunciation and accent. However, it is difficult to determine or predict the degree of easiness in listening to a given English, since it depends on the individual.

In this study, we propose an English listening system for learning different pronunciations and accents in various countries and areas. The feature of our system is that our system can extract countries and areas where English speech is relatively easy to understand to a learner by the sound characteristics, based on the listening learning history of pronunciations and accents in various countries and areas. This function allows each learner to learn English listening step by step according to the learner's listening capability.

In this study, we evaluate the feasibility of our system using a Web-based prototype.

2 Motivation

Social and cultural exchanges with Asian countries such as China, Thailand, Malaysia, and Indonesia have been active in Japan, and the importance of developing English communication skills with these countries is increasing more and more.

Thorough the authors' own experiences of exchanges with international students, we have often felt that English spoken by students from South-East Asian country is easier to understand than by students from Europe and United States. Figure 1 shows

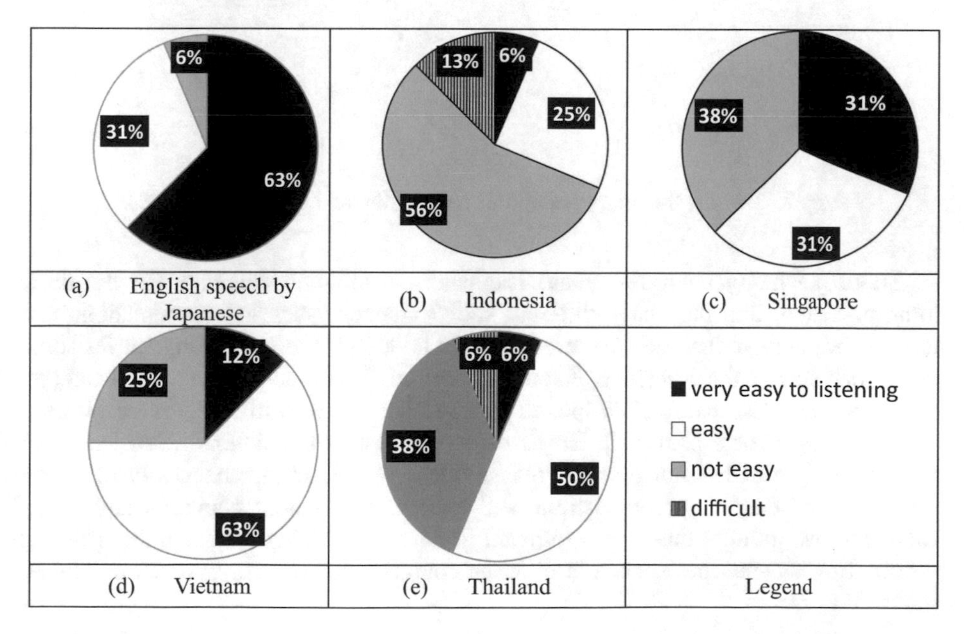

Fig. 1. Result of preliminary examination for easiness in speech hearing of English

the result of preliminary examination to sixteen Japanese students in our university for easiness in speech hearing of English in five countries: Japan, Indonesia, Singapore, Vietnam, and Thailand. In this examination, we collected English speech data that were recorded by ourselves. In addition, there are different speakers for each country, and each speaker speech the same English sentences for the recording. From the result as shown in Fig. 1, we can confirm that these Japanese subjects has a tendency that they could relatively catch up English spoken by people from Japan and Singapore, but found difficulty understanding English spoken by people from Indonesia.

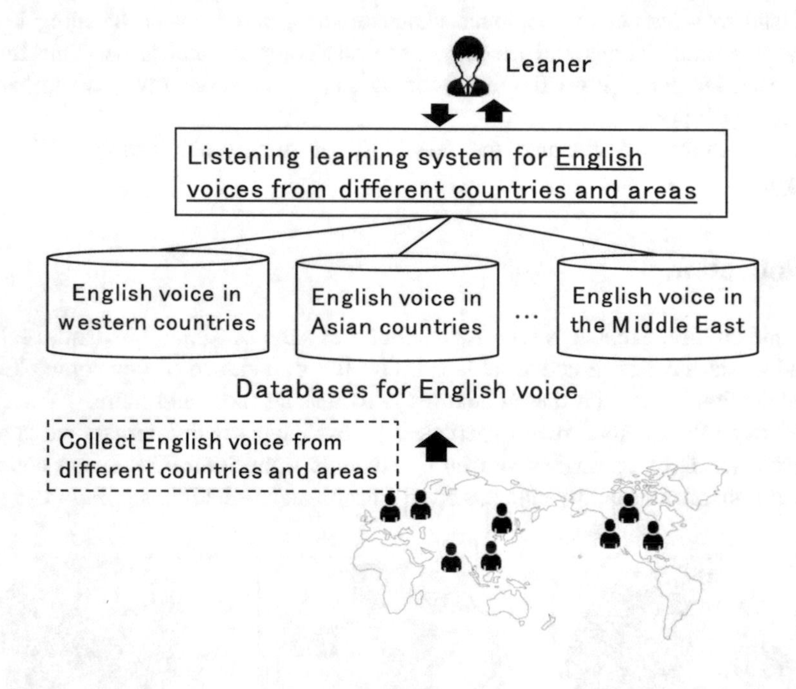

Fig. 2. English listening system for learning different countries and areas

There are many countries where English is a common language for the daily communication, and they have different sound characteristics in pronunciations and accents. We believe that the degree of easiness in speech hearing of English for some people depends on the difference of sound characteristics in each country. For example, there is a possibility that English spoken by a speaker A is easy to catch up for a learner B, but not easy for a learner C. However, to the best of the authors' knowledge, an e-Learning system has not existed so far, where we can comprehensively learn the differences of English pronunciations and accents in different countries and areas. Therefore, we thought that it is significant to develop learning materials for listening various English speeches spoken in different countries, not only in Europe and United States (Fig. 2).

Meanwhile, if there are English easy to catch up to some learners and English not easy based on the sound characteristics to learners, the e-Learning system can personalize the order of listening questions according to the learners' listening abilities. Therefore, we focus on realizing a learning platform where learners can learn English listening learning step by step according to their listening abilities (Fig. 3), by extracting countries and areas where English speech is relatively easy to understand to the individual learners.

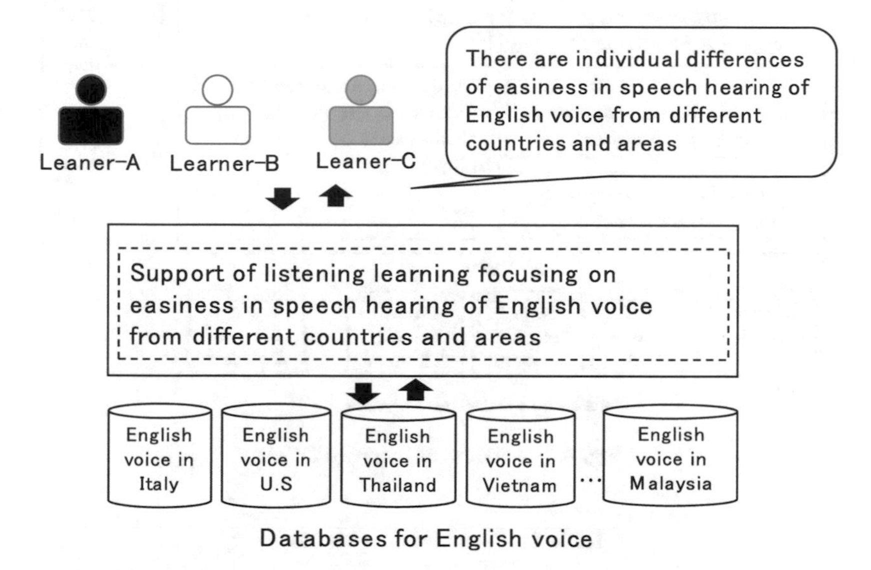

Fig. 3. English learning platform where listening questions are given according to easiness in speech hearing of English

3 Proposed System

We propose an e-Learning system that supports English listening learning by providing various English speech in different countries including Asian countries, not only Europe and United States (Fig. 4). The feature of proposed system is to provide a function of extracting countries and areas where English speech is relatively easy to understand to a learner by the sound characteristics, based on the learner's correct answer rates for listening questions of English speeches that are attached region labels in advance.

3.1 English Speech Database

English speech databases, which stores English speech data collected from people in different countries such as Europe, United States, and the Middle East, are constructed for realizing a listening learning environment of English speech spoken in various countries and areas.

Each English speech data v_x is classified into regions. Here, the classification ID is referred to as 'region label' c_y. Table 1 shows an example of region labels.

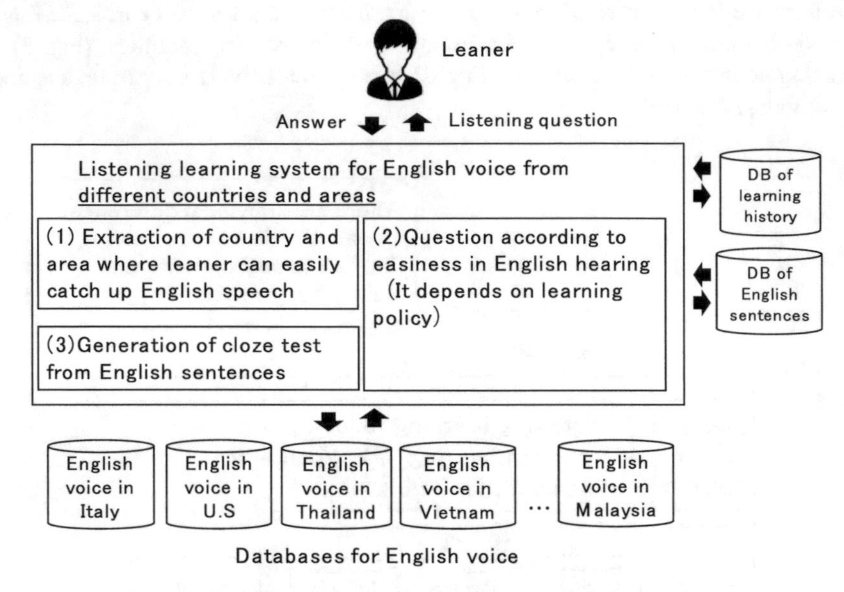

Fig. 4. Overview of proposed system

Table 1. Example of region labels

Region label	Region name
c_1	Thailand
c_2	Malaysia
c_3	Indonesia
c_4	Japan
c_5	Vietnam

3.2 A Function of Extracting Regions Based on Easiness in English Speech Hearing

This function extracts countries and areas where English is easy to catch up to a learner by the sound characteristics, based on correct answer rates for each region from a listening learning history. Here, the correct answer rates for each region can be calculated based on the number of correct answer for each region using 'region labels' assigned to each English speech data. This function is executed in the following steps.

Step-1: A set of questions P, which consists of n-listening-questions p_x, is prepared for a learner u. Here, each listening question p_x consists of a set of English speech data v_x, English sentence data s_x, and region label c_x: p_x (v_x, s_x, c_x).

Step-2: A close test is generated using English sentence data s_x. Suppose the number of blanks to fill in and the number of correct answers be b_x and $a_{u,x}$, respectively, the correct answer rate $r_u(s_x, b_x)$ of the learner u is calculated as follows:

$$r_u(s_x, b_x) = \frac{a_{u,x}}{b_x} \tag{1}$$

Step-3: After the learner u answers n-listening-questions, the correct answer rate $r_u(c_x)$ for English speech data associated with region label c_x is calculated. Suppose the number of English speech data associated with region label c_x in P be m_x, $r_u(c_x)$ is calculated as follows:

$$r_u(c_x) = \sum\nolimits_{i \in x} r_u(s_i, b_i)/m_x \tag{2}$$

Step-4: After calculating $r_u(c_x)$ for all regions, regions with high correct answer rate $r_u(c_x)$ is extracted, of which English is easy for the leaner u to catch up.

Suppose that correct answer rates for a cloze test by a learner u be ones as shown in Table 2. In this case, the correct answer rates for each region are calculated as below. This result shows that for the learner u, the correct answer rates for region c_3 is the highest, and the region c_3 will be extracted as a region where the learner u can hear the most easily.

$$r_u(c_1) = \frac{60 + 70 + 50}{3} = 60\%$$
$$r_u(c_2) = \frac{70 + 90}{2} = 80\%$$
$$r_u(c_3) = \frac{80 + 90}{2} = 85\%$$

Table 2. Example of correction answer rates for cloze test (learner u)

Question	English speech data	Region label	Correction answer rates
p_1	v_1	c_1	60%
p_2	v_2	c_2	70%
p_3	v_3	c_3	80%
p_4	v_4	c_2	90%
p_5	v_5	c_1	70%
p_6	v_6	c_3	90%
p_7	v_7	c_1	50%

4 Prototype System

We implemented our prototype system as Web-based application that can be used from Web browser on a PC, smartphone, and tablet. Figure 5 shows a system architecture of the prototype system. The prototype system consists of four main functions: (1) a function of generating cloze tests, (2) a function of calculating correct answer rates, (3) a function of calculating correct answer rates for regions, and (4) a function of balancing number of questions for each region. In addition, three databases, English speech database, English sentence, and learning history database, are included in the prototype system.

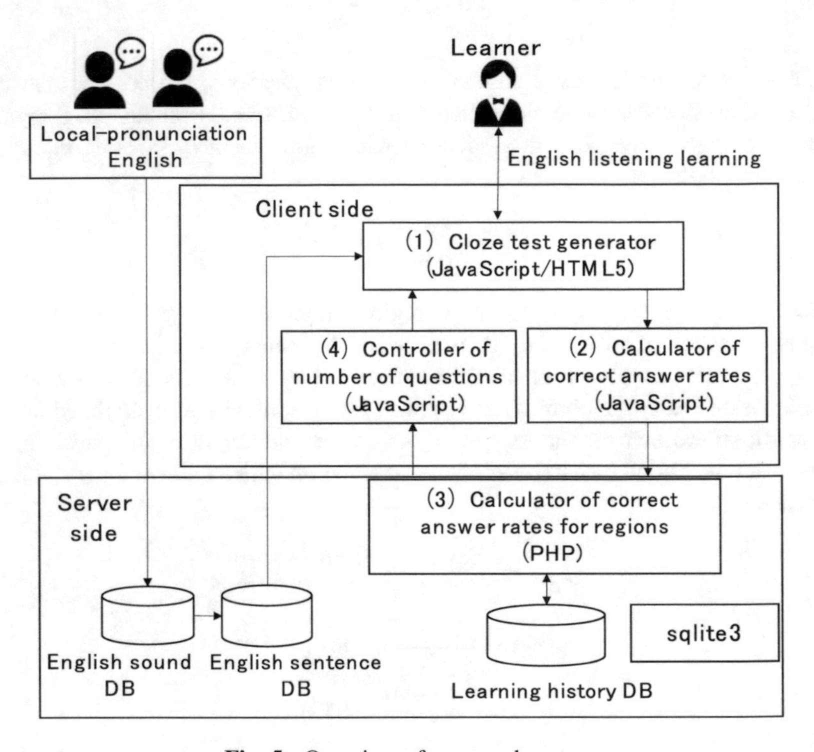

Fig. 5. Overview of proposed system

Table 3. Example of correction answer rates for cloze test (learner *u*)

ID	Title	Time (sec)	Number of words
1	Hong Kong kids will be opened	40	100
2	Summer outside festival	50	91
3	Welcome to a factory	45	93
4	Air conditioner was broken	38	88
5	Database of a cup	34	106

Table 3 show the example of listening learning contents. We assign 30 to 40-s English sounds to English sentences including 50 to 100 words. In addition, currently, the prototype system provides English sounds from three countries: Japan, Thailand, and Indonesian for each English sentence.

4.1 A Function of Generating Cloze Tests

This function generates cloze tests by randomly selecting blanks to fill in from English sentences stored in English sentence database. Figure 6 shows an example of screen shot of cloze test.

Output

Ladies , welcome to the Sanford Theater.

curtain will go up on the musical production Kong Boys in minutes.

All the seats of this reserved.

check your ticket stub and sit in your seat.

Our staff are distributing the theater programs in aisles.

food and during the performance, so please finish refreshments in the lobby.

two-act musical lasts two and a half hours, and we will have a 15-minute .

Again, the curtain go up in five .

Please proceed to your seats.

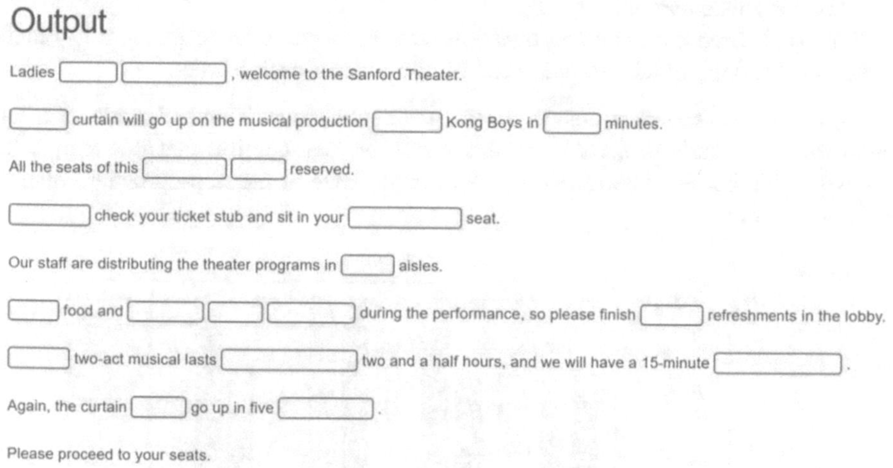

Fig. 6. Example of screen shot of cloze test

4.2 A Function of Calculating Correct Answer Rates

A learner answers a question by clicking blank to fill in, when the learner could not hear a word in the blank. This function calculates the correct answer rate by the method based on the number of blanks and the number of blanks clicked, as described in Step-2 of Sect. 3.2.

4.3 A Function of Calculating Correct Answer Rates for Regions

The correct answer rate by region is calculated based on the correct answer rates in cloze tests and region labels associated with English speech data. The detail of this function is described in Sect. 3.2.

4.4 A Function of Balancing Number of Questions for Each Region

This function calculates the number of listening questions for each region based on the correct answer rates for each region. This function extracts countries and areas where a learner can feel easiness to catch up English speech, and provides a learning environment according to the learner's listening ability. The detail of this function is described in Sect. 3.2.

5 Experiment

We conducted experiments for confirming the feasibility of our proposed system by comparing two patterns of the order of questions as follows:

(Pattern 1) Random: Listening questions are randomly set without consideration of a learner's listening ability.
(Pattern 2) Step-up: Listening questions are set in the order of easiness in English speech hearing, which are measured in advance for each learner.

Figure 7 shows correct answer rates for three regions: Thailand, Indonesia, and Japan for one female subject. From this result, we can confirm that this learner can achieve higher correct answer rate in the average value of the step-up pattern than the random pattern.

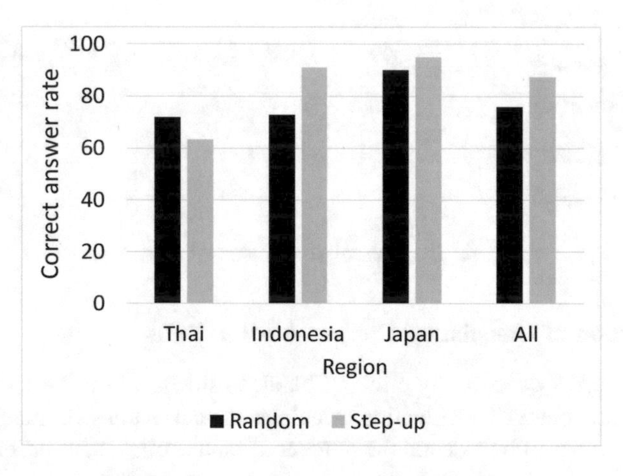

Fig. 7. Correct answer rate for each region

Figures 8, 9 and 10 shows learning progresses for English Speeches in Indonesia, Thailand, and Japan, respectively. From these results, it seems that the learner can progress listening learning while keeping high correct answer rates in the step-up pattern, and on the contrary, correct answer rates are decreasing for English speech in two countries: Thailand and Japan, in the case of using the random pattern.

From these results, it is deemed that the learner can continue the listening learning without much stress, since she can start from listening questions that is easy to catch up for her. Thus, our system can motivate a learner to continue English listening learning, and this is the one of advantages derived from the usage of our proposed system.

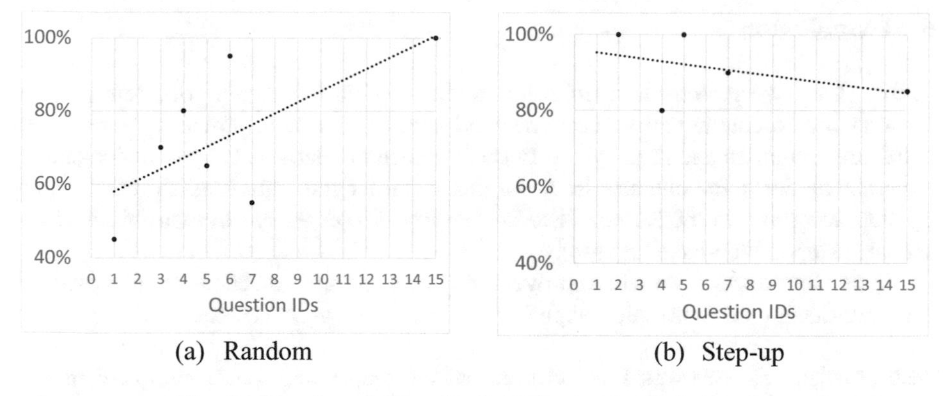

(a) Random (b) Step-up

Fig. 8. Learning progress for English speeches in Indonesia

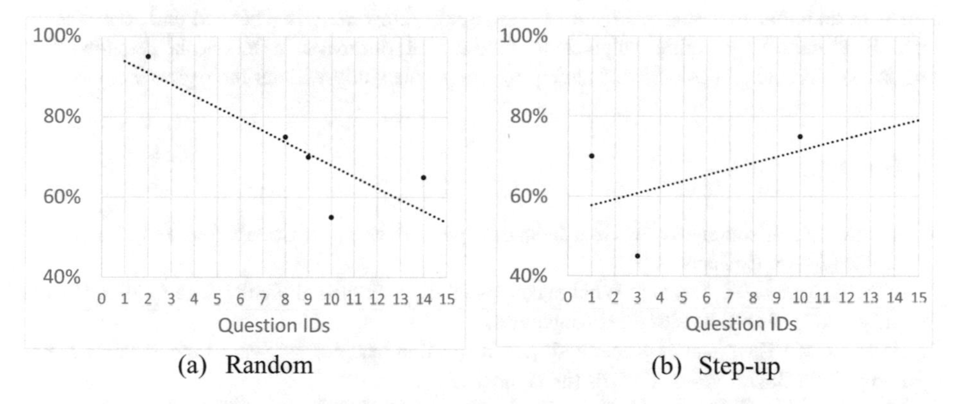

(a) Random (b) Step-up

Fig. 9. Learning progress for English speeches in Thailand

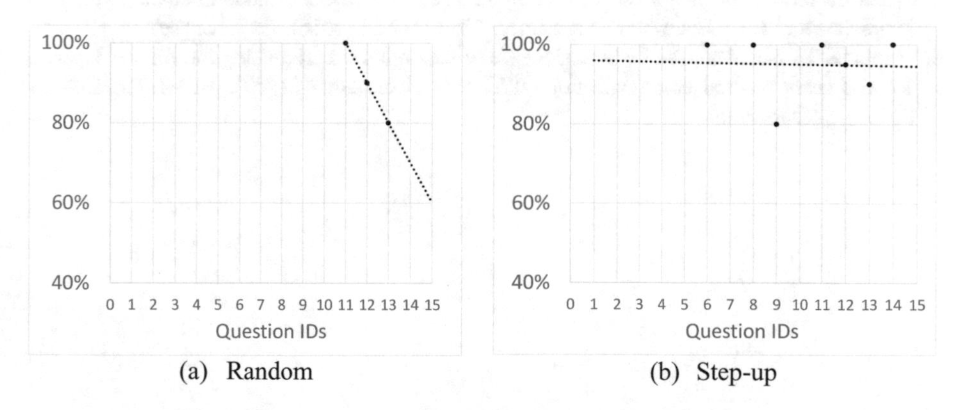

(a) Random (b) Step-up

Fig. 10. Learning progress for English speeches in Japan

6 Conclusion

In this study, we presented an English listening system for learning different pronunciations and accents in various countries and areas. Our system provides a function of extracting countries and areas where English speech is relatively easy to understand to a learner by the sound characteristics, so that a learner can learn English learning step by step according to the learners listening ability. We evaluated the feasibility of our system using a Web-based prototype.

In our future work, we will improve our system for more practical use and evaluate it by conducting demonstration experiments in actual English classes.

Acknowledgments. We would like to thank all the people who cooperated to record their speech in English for our experiment. Thanks to Ms. Rie Okuda, Ms. Satomi Maki, and other members from International Affair Office, Prof. Tatsuya Inaba, and Lecture Mr. Matsuo from Department of Information and Computer Sciences, we could collect English speech data from Japan. In addition, four students from Chulalongkorn University in Thailand and four students from Electronic Engineering Polytechnic Institute of Surabaya in Indonesia provided their English speech data. We could not have progressed this study without their great help.

References

1. Nozawa, K.: A sound of vowel that Japanese speakers interpret. Collab. Res. Proj. Lang. Cult. 61–72 (2016). (in Japanese)
2. O'ki, T., Maeda, H., Oka, H.: What makes listening to English difficult? J. Fac. Educ. Hakuho Univ. **10**(2), 511–530 (2016). (in Japanese)
3. Ikegami, M.: The effect of pauses and speech speed in teaching English listening skills. J. Res. Lang. Cult. **35**(2), 33–54 (2016). (in Japanese)
4. Mizusawa, Y.: Utilization of movie for improving English listening skill. Proc. Lingua **26**, 97–111 (2016). (in Japanese)
5. Kawano, S., Hayashi, Y., Mouri, F.: An investigation into effects of a shadowing of song lyrics on foreign language listening comprehension by Japanese undergraduate learners of English. Proc. Prac. Educ. Res. Saga Univ. **33**, 111–118 (2016). (in Japanese)
6. Yamada, T., Mizutani, A., Ichimura, S.: Creating English voice intelligible for the Japanese for the cross cultural communication. IPSJ SIG Notes **2007**(91(2007-EIP-037)), 159–164 (2007). (in Japanese)

FOG Computing and Low Latency Context-Aware Health Monitoring in Smart Interconnected Environments

Philip Moore[1(✉)] and Hai Van Pham[2]

[1] School of Information Science and Engineering, Lanzhou University, Lanzhou 730030, Gansu Sheng, China
ptmbcu@gmail.com

[2] School of Information Technology and Communication, Hanoi University of Science and Technology, Hanoi, Vietnam
haipv@soict.hust.edu.vn

Abstract. Treatment and management of the increasing complexity in medical conditions experienced by an ageing demographic requires increased use of medical resources and patient management. Effective management may be achieved using autonomic health monitoring systems, a topic much discussed in the literature, however such monitoring has generally been limited to the 'smart-home' environment. In this paper we consider extending patient monitoring from only the 'smart-home' to a wider 'smart-environment' which conflates 'smart-homes' with the 'smart-city' based on the FOG computing paradigm. FOG is a term created by Cisco systems and is also known as edge computing or 'fogging'. We introduce a model which incorporates FOG and cloud-based computing for a low latency healthcare monitoring system which enables comprehensive 'real-time' monitoring with situational awareness and data analytic solutions. Two illustrative scenarios are presented predicated on the monitoring of patients with dementia, however, the posited approach will generalise to other medical conditions where monitoring is required.

1 Background

Healthcare providers globally must address many challenges in the treatment and management of the increasingly complex medical conditions experienced by an ageing demographic. A particularly challenging medical condition is Alzheimer's disease and dementias [21]; we defer an introduction to dementia and it's management to Sect. 6. The treatment and management of patients with dementia will ideally provide for effective patient management while maximising the quality-of-life (QoL) for both patients and carers alike. Realising these twin aims requires constant and consistent monitoring on a 24/7 basis where a patient's dynamically changing state (or *context*) is measured with feedback appropriate to the current context as it relates to the current medical context. Moreover, patient monitoring must maintain continuous situational awareness relating to a patient's location.

L. Barolli et al. (Eds.): EIDWT 2018, LNDECT 17, pp. 29–40, 2018.
https://doi.org/10.1007/978-3-319-75928-9_3

Such monitoring may be achieved using autonomic health monitoring systems which has been a topic much discussed in the literature [24,26,28]. However, monitoring has generally been limited to the 'smart-home' environment using a relatively limited range of health indicators and metrics. Additionally, the use of health monitoring in the external environment has gained traction [albeit again using a limited range of health indicators] using spacial and temporal data implemented with mobile devices.

To address the requirements implicit in the realisation of constant monitoring with the maintenance of situational awareness, we propose extending patient monitoring from only the 'smart-home' to a wider 'smart-environment' which conflates 'smart-homes' with the 'smart-city' based on the FOG computing paradigm. We introduce a model which incorporates FOG and cloud-based computing for a low latency healthcare monitoring system which enables comprehensive 'real-time' monitoring with situational awareness and big-data analytics.

Two illustrative scenarios predicated on the monitoring of patients with dementia are presented to illustrate the proposed approach in 'real-world' situations. While the focus of this paper lies in monitoring patients with dementia, the posited approach will generalise to other medical conditions [such as mental disorders] where monitoring is required for the benefit of all stakeholders in healthcare systems.

The remainder of this paper is structured as follows: in Sect. 2 we provide an brief overview of 'smart cities', 'smart homes', and our proposed 'smart environment'. Latency is considered in Sect. 3 with the proposed FOG health monitoring model described in Sect. 4. Networks and communications are central to FOG computing, we briefly consider these components in Sect. 5. In Sect. 6, using illustrative scenarios, we demonstrate the proposed FOG health monitoring model in action. Additionally we introduce a 'real-world' research project, the ultimate goal of which is the creation of 'smart-environments'. The paper concludes with a discussion and concluding observations with projected future directions for research in Sect. 7.

2 Smart Environments

We have considered the demographic challenges faced by healthcare providers. In this section we consider the concept of a 'smart home' and 'smart city' followed by an introduction to our proposed conflation of these concepts into a combined 'smart environment' designed to enable consistent situational awareness and patient monitoring.

The concept of a 'smart-city' is difficult to define as there are many interpretations of the term with no common generally agreed definition [3] as definitions are generally related to the domain of interest (the functional requirements); moreover, in the current digitised intelligent connectivity domains function on the basis of 'real-time' response. What is generally agreed is the nature of the constituent parts that combine to create a 'smart-city', these being: (a) an urban environment, (b) the use and integration of information and communication technologies (ICT), and (c) the use of sensor-based technologies which include the

Internet-of-things (IoT) in complex networks [24,26] designed to capture data for processing into information useful to systems and individual users [7]. Moreover, a 'smart-city' provides a basis upon which managers may interact in 'real-time' with interconnected systems and infrastructures to monitor, manage and control a heterogeneous range of city functions ranging from traffic management to environmental monitoring.

Deakin and Wear [11] define the smart city as "one that utilises ICT to meet the demands of the citizens of the city" and "uses ICT to positively impact the local community"; we may also observe that the 'smart-city' will have a positive impact on all stakeholders living in and managing an urban environment. We may therefore conclude that, while a commonly agreed definition remains elusive, the overall goal of improving the environment to the benefit of the population is generally agreed. As this goal relates to the topic this paper addresses (health monitoring with situational awareness), the overall aim of addressing the aims of all stakeholders in healthcare provision concurs with the general aim of 'smart-cities'.

Turning to the 'smart-home', as for 'smart-cities' there is no commonly agreed definition of the 'smart-home' concept, this is due to the differing pre-determined aims and objectives for a 'smart-home'. For example, a 'smart-home' may be designed to implement: (a) security, (b) environmental control, and/or (c) health monitoring systems which include activity tracking and health related physiological metrics. The focus of this paper relates to the capture and processing of physiological, spacial, and temporal data to measure changes in a monitored patient's dynamically changing context [19] as discussed in [18].

In this paper we propose a FOG Health Monitoring Model which is essentially a conflation of the 'smart-home' and 'smart-city', a concept designed to provide a 'smart-environment'. Such an environment may provide an effective basis upon which consistent 24/7 patient monitoring with situational awareness may be realised as discussed in Sect. 6.

3 Latency

The preceding section has considered smart environments, in the following section we introduce the FOG computing model. However, prior to introducing the model we must consider latency, an essential feature of 'real-time' autonomous health monitoring systems. While latency is not generally an issue for 'smart-homes' [where monitoring is achieved using localised body-area networks] as shown in Fig. 1, the traditional cloud-based paradigm experiences issues where data is transmitted over networked systems for processing in the cloud [2,4,9,14,17]. In this paper, our proposed FOG computing model extends the conventional FOG paradigm by introducing a conflation of the 'smart-home' and the 'smart-city' concepts to realise a 'smart-environment' which enables the monitoring of individuals and the maintenance of situational awareness while located in their 'smart-home' and also while the individuals are in the external environment as demonstrated in Sect. 6.

Latency is domain specific, in human computer interaction (HCI) and machine-to machine interaction (M2M), latency may be viewed in terms of usability, i.e., the time interval between the request and the result. Applied to health monitoring systems, latency relates to the time interval between an action (i.e., a sensor signal) and the receipt of the data by a system for processing and the return of a result (the time delay between the capture of patient data and the return of the dynamic prognosis).

For HCI latency may be merely an irritation, however in autonomous health monitoring systems [a M2M interaction] which are 'real-time' systems the speed of data transmission and processing is a critical feature and the latency experienced in cloud-based systems is not acceptable. The proposed FOG model as shown in Fig. 1 addresses the latency problem by the use of local processing which employs a body-area-network (BAN) [8] with bulk loading of data into the cloud for later analysis as discussed in [20].

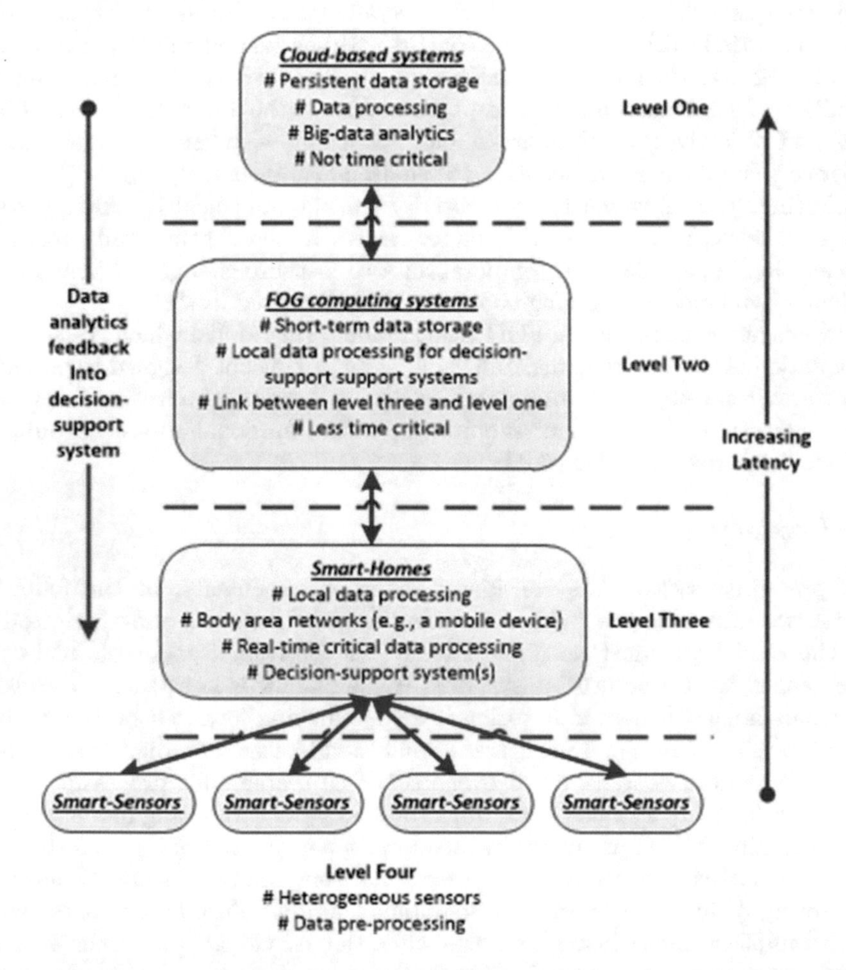

Fig. 1. A conceptual model showing the architectural relationships between heterogeneous 'smart-sensors', the 'smart-home' (using body-area networks), FOG computing systems, and cloud-based systems. The model shows the deployment of a homogeneous system comprising data capture, computing, storage, and networking components

4 The FOG Health Monitoring Model

In Fig. 1 it can be seen that the proposed architecture uses four levels which are designed to address: (i) the latency issue, (ii) data processing in 'real-time' systems, (iii) 'big-data' analytic solutions, and (iv) in-memory and persistent data storage. The differing requirements are reflected in the FOG model with the data derived from 'real-time' (in-memory data store) processing fed into cloud-based analytic solutions (persistent data store).

As shown in Fig. 1, FOG nodes in level three provides localisation which enables both low latency and context-awareness while cloud-based systems in level one provide global centralisation [4]. The results derived from the analytic solutions are, following collaborative verification and checking in the form of revisions to parameters and system rules, fed back into a 'real-time' decision-support-system (DSS) via a continuous collaborative development feedback loop [22]. A DSS implemented in a 'smart-home' leverages low latency and localisation to enable 'real-time' data processing with the results of 'big-data' analytics being fed back into the patient monitoring and prognosis processing.

Level one: represents cloud-based systems which are designed to provide persistent data storage and processing for both traditional relational database management systems (RDMS) and unstructured data which is generally termed *NoSQL* [20]. Cloud-based systems offer many potential benefits which include the capability to leverage data processing facilities and maximise resource utilisation while minimising operational expenditure (OPEX). However, as shown in Fig. 1, cloud-based systems may experience significant latency issues which are not acceptable in 'real-time' health monitoring systems. Data analytics are a positive feature of cloud-based systems where data processing is not time-critical. Moreover, the results of data analytic solutions may produce important feedback into 'real-time' DSS in the form of revised diagnostic scales [22] and/or rules in context-processing and context-matching [19].

Level two: is in effect an intermediate layer which is more locally based, Bonomi et al. refer to the "FOG, simply because the fog is a cloud close to the ground". In the patient monitoring scenario, it forms an intermediate layer between the 'smart-home' and the cloud, as such it may be one stage removed from the cloud and may monitor multiple patients based on the results derived from their 'smart-homes'. Such a facility provides reductions in latency and the capability to monitor patients' current context (their dynamic medical state), maintain a situational awareness, and implement a reactive response dependent on the reported context from level three.

Level three: represents the 'smart-home' environment. In this layer, local low latency data processing is achieved using a BAN which captures data from heterogeneous sensors in 'real-time' and processes the data in a DSS as discussed in [18,19]. In this layer, the data storage utilises both 'in-memory' and persistent (albeit short-term) data storage. The results from the DSS are sent to a FOG computing systems (level two) for implementation and interventions as required by the patient's dynamically changing context. As discussed in [18], the 'in-memory' (real-time) data storage is limited to the current context data [19] with the data stored locally and at pre-determined intervals bulk loaded to the cloud (level one) for analytic processing and feedback. Such intervals may take advantage of periods when data flows in networks are light thus reducing latency.

We have considered levels one to three which are the essential functions implemented in a 'real-time' low-latency data processing system with big-data analytics implemented in level one. ***Level four*** is in effect an extension of level three and implements the heterogeneous sensors which capture health related metrics data. The sensors are envisaged as intelligent networking components which (where required) may pre-process the captured data and feed the results into the 'smart-home' DSS (level three) to monitor a patient's dynamically changing context.

5 Communications and Networking

Networking and communications represent an extremely homogeneous range of technologies and a comprehensive discussion is beyond the scope of this paper. There is a large body of published research addressing network design and implementation, for example see: [4,5,8,13,15,16,23,25,27,31].

In summary, considering the topic addressed in this paper, as shown in Fig. 1 there are a number of data flows within the FOG computing model which are reflected in the acceptable latency levels; as can be seen the level of latency increases as we move from level one to level four. Moreover, the model requires the use of a range of networking models: (i) wide area networks (WAN), (ii) local; area networks (LAN), and BAN [31].

In considering data speeds, as they relate to the systemic requirements [4], data transfer in levels three and four may be measured in milliseconds. Data transfer from level three to level two will range from milliseconds to sub-seconds. Data transfer from level three to level four will range from seconds to minutes, this is not time critical and as noted in Sect. 4 data transfer may be scheduled for times when network traffic is low thus reducing congestion in the network. Considering our model (see Fig. 1), in the 'smart-home' we propose the utilisation of a BAN which communicates with the 'smart-sensors' in level four to capture the raw and pre-processed data for use in a context-aware DSS. The results derived from the DSS in level three are passed to level two over a LAN. Finally, the data flows into the cloud-based system (level four) will utilise a WAN. The networking systems may be hard wired of implemented in software defined networks.

Clearly, for a health monitoring DSS application reliability is critical. Therefore, while WiFi networking offers the lowest latency, we must build in redundancy by providing a *failover* capability which will provide for a handover from WiFi to mobile networking using 3G and 4G (and 5G when available). It may in future be the case that 5G surpasses WiFi but currently this is a matter of informed conjecture, in such a case the use of 5G networks may be the default technology with WiFi being the failover networking solution.

6 Illustrative Scenarios

We have introduced the proposed FOG health monitoring model which is designed to enable low latency health monitoring in smart environments while enabling 'real-time' data processing and data' analytic solutions. The focus of this paper lies in the monitoring and management of dementia. Two prototypical scenarios are considered which address the monitoring for patients' with dementia in: (i) a domestic setting, and (ii) a residential nursing home setting. These settings are relevant as dementias may be classified under a patient's prognosis which degrades over time with increasing problems frequently related short-term memory loss and increased agitation and aggression [10,21] which is measured in terms of the behavioural and psychological symptoms of dementia (BPSD) [12,21,30]. The illustrative scenarios show how the progression of dementia may be managed while maximising the QoL for patients and carers alike.

BPSD represents an important consideration in the prognosis and thus the projected treatment and care pathway [21]. Figure 2 models the progression of dementia from a BPSD perspective, for a detailed discussion on the topic see [6, 12,21,30]. However, in summary while the demarcation between differing levels is essentially fuzzy and patient specific, patients categorised in levels 1 to 4 may be accommodated in a domestic setting (ideally their own home) while for levels 5 to 7 residential nursing home accommodation is a requirement for the safety of patients' and their carers [21,30].

Scenario One is applicable to BPSD levels 1–4 (see Fig. 2) and relates to the treatment and management of patients in a domestic setting. This setting is desirable as retention of a familiar setting aids QoL for patients and carers with significant benefits for all stakeholders in the provision of healthcare services. Patient monitoring in such a setting is designed to test for change in a patient's current dynamic context. Such checking relates to daily tasks which include taking prescribed medicines and eating regularly. Additionally, such monitoring will include monitoring: (i) a patient's weight, (ii) excessive movement, (iii) lack of movement, and (iv) issues such as falls or illness using spatial, temporal, and physical metrics. The results of the monitoring are reported to the FOG node in level two, the uploading of data to the cloud for analysis remains as discussed in Sect. 4.

Scenario Two is applicable to BPSD levels 5–7 (see Fig. 2) and relates to the treatment and management of patients in a residential nursing home setting. This setting is required as, once dementia develops to levels 5–7, living in a domestic setting becomes so challenging (for both patients and carers) that a residential nursing home becomes the only realistic option given the extreme mood shifts and elevated BPSD. Moreover, while monitoring within the residential environment becomes less of a priority, it remains a valuable component in managing the patient with dementia while providing an element of freedom to move around commensurate with effective management of the patient. Within the confines of the nursing home patient monitoring may be achieved using a similar approach and metrics to that used in a domestic setting. However, the

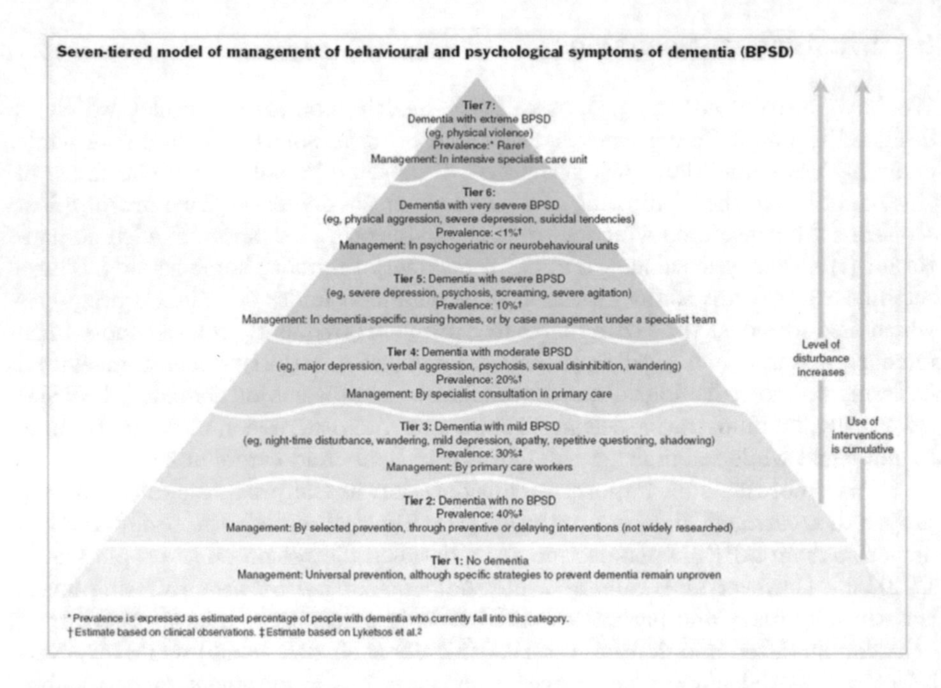

Fig. 2. Seven tiered model of management of behavioural and psychological symptoms of dementia (BPSD). Seven tiers (levels) are shown ranging from the least severe (tier one) to the most severe (critical) (tier seven). The model shows the development of the condition and may be used in the patient prognosis and the identification of the appropriate patient management regime. (source: [6,30])

location of the FOG node (located in level three for a domestic setting) may be more localised to the nursing home, the uploading of data to the cloud for analysis remains as discussed in Sect. 4.

In considering the two scenarios, in an enclosed 'smart-home' environment monitoring may be achieved using a range of sensors with 'real-time' data processing [28]. However, research has shown that physical exercise as a preventive or disease-modifying treatment for dementias and brain ageing [1] is an important contributor in minimising the progression of dementia. Therefore, if we can facilitate patient access to an external environment we may provide controlled levels of freedom exercise in a 'smart-environment' which in scenario one may be a 'smart-city' and in scenario two may be controlled grounds around a nursing home. We argue that this promotes therapeutic benefits. Of course, in facilitating such freedom of movement, the greater freedom enabled in scenario one and the limited freedom in scenario two requires consistent situational awareness. This may be achieved using our proposed 'smart-environment' as discussed in Sect. 2.

A related benefit in the proposed approach lies in the implementation of choice architectures and nudging using mobile information technologies [29]. Thomas et al. consider this topic and note that, as modelled in Fig. 1, multiple servers, sensors, devices and connections, are likely to make it a highly complex system in full operation. Moreover "good" decisions could be considered emergent properties in a 'sea of chaos', potentially forming strange attractors around which associated norms develop". In the context of a patient with dementia, the adoption of choice architectures and nudging technologies may realise positive cognitive benefits for patients.

The ***Mekong Delta Environmental Project***[1]: is a world-Bank funded project which represents ongoing international research into the management of urban environments. The project is designed to monitor and provide scientific evidence which will provide an important contribution to assist regional authorities (in Vietnam) in restructuring agriculture, improving the quality of water resources and infrastructure systems, and the development of appropriate livelihood models.

A data centre, using heterogeneous sensor-derived data and human indicators (environmental factors are important in health related problems), forms a component part in the Mekong Delta Integrated Climate Resilience and Sustainable Livelihoods Project. This project aims to: (i) enhance the regions capacity to adapt to climate change by improving planning and promoting resilient rural livelihoods, and (ii) construct a climate smart infrastructure in the region. This is a nascent project which illustrates the concept of the 'smart-environment' in action where heterogeneous data sources are leveraged to provide multi-factorial benefits (which include healthcare provision) for all stakeholders in an urban environment.

7 Discussion and Concluding Observations

There is a demographic challenge represented by the increasingly complex medical conditions experienced by an ageing population and a particularly significant challenge lies in the treatment and management of patients with Alzheimer's disease and dementias. In this paper we argue that the effective treatment, management, and monitoring of patients with dementia provides improved QoL for both patients and carers alike. However, such monitoring demands constant and consistent monitoring on a 24/7 basis where a patient's dynamically changing *state* (or *context*) is measured with feedback appropriate to the current context (as it relates to the current medical state). Moreover, patient monitoring must maintain continuous situational awareness relating to a patient's location.

To address the demands of constant monitoring with the maintenance of situational awareness, we have proposed extending patient monitoring in a 'smart-home' to a wider 'smart-environment' which conflates 'smart-homes' with the 'smart-city' based on the FOG computing paradigm with cloud-computing. We have proposed a model which incorporates FOG and cloud-based computing for a low latency healthcare monitoring system which provides an effective basis upon which comprehensive 'real-time' monitoring with situational awareness and big-data analytics may be realised.

[1] Mekong Eye: https://www.mekongeye.com/2016/10/11/wb-helps-build-environment-data-centre-in-mekong-delta/.

To demonstrate the implementation of the proposed model, we have presented two illustrative scenarios predicated on the monitoring of patients with dementia in 'real-world' situations. While the focus of this paper lies in monitoring patients with dementia, the posited approach will generalise to other medical conditions [such as mental disorders] where monitoring is required for the benefit of all stakeholders in healthcare systems. Moreover, the proposed model will generalise to other domains of interest where low-latency monitoring is required in 'real-time' system where rapid response with the ability to implement data analytic solutions form a systemic requirement. As an example of the capability to generalise to a broad range of domains and systems, we have introduced the Mekong Delta Environmental Project which illustrates the application of FOG and cloud-based computing solutions as envisaged in our proposed latency controlled FOG computing model.

Clearly, as discussed in [22] where context-aware evidence-based data driven development of diagnostic scales for depression is considered as it relates to collaborative development is addressed, the development of our model into a working system requires a collaborative approach where domain specialists collaborate with computer scientists. However, a central component is the realisation of context-processing (CP) and context-matching (CM) to identify changes in a patients dynamically changing context. The fuzzy rule-base approach [19] has been shown to provide a basis for CP and CM with constraint satisfaction compliance, the proposed future directions for research will address extending the current approach to addressing the needs of the proposed FOG computing paradigm while realising the recognised benefits of cloud-computing.

Effective implementation of our proposed model offers the potential to realise low-latency 'real-time' monitoring systems is a broad and diverse range of domains and systems to the benefit of all stakeholders.

References

1. Ahlskog, J.E., Geda, Y.E., Graff-Radford, N.R., Petersen, R.C.: Physical exercise as a preventive or disease-modifying treatment of dementia and brain aging. Mayo Clinic Proc. **86**(9), 876–884 (2011). https://doi.org/10.4065/mcp.2011.0252. http://www.sciencedirect.com/science/article/pii/S0025619611652191
2. Ahmed, A., Ahmed, E.: A survey on mobile edge computing. In: 2016 10th International Conference on Intelligent Systems and Control (ISCO), pp. 1–8. IEEE (2016). https://doi.org/10.1109/ISCO.2016.7727082
3. Al-Doghman, F., Chaczko, Z., Ajayan, A.R., Klempous, R.: A review on Fog computing technology. In: 2016 IEEE International Conference on Systems, Man, and Cybernetics (SMC), pp. 001525–001530. IEEE, IEEE Computer Society (2016). https://doi.org/10.1109/SMC.2016.7844455
4. Bonomi, F., Milito, R., Zhu, J., Addepalli, S.: Fog computing and its role in the internet of things. In: Proceedings of the First Edition of the MCC Workshop on Mobile Cloud Computing, MCC 2012, pp. 13–16. ACM, New York (2012). https://doi.org/10.1145/2342509.2342513

5. Braham, R., Douma, F., Nahali, A.: Medical body area networks: mobility and channel modeling. In: 2016 7th International Conference on Sciences of Electronics, Technologies of Information and Telecommunications (SETIT), pp. 1–6. IEEE (2016)
6. Brodaty, H., Draper, B.M., Low, L.F.: Behavioural and psychological symptoms of dementia: a seven-tiered model of service delivery. Med. J. Aust. **178**(5), 231–235 (2003)
7. Checkland, P., Holwell, S.: Data, Capta, Information and Knowledge. Elsevier, London (2006)
8. Chen, M., Gonzalez, S., Vasilakos, A., Cao, H., Leung, V.C.M.: Body area networks: a survey. Mob. Netw. Appl. **16**(2), 171–193 (2011). https://doi.org/10.1007/s11036-010-0260-8
9. Cisco: Fog computing and the internet of things: Extend the cloud to where the things are (2017). http://www.cisco.com/c/dam/en_us/solutions/trends/iot/docs/computing-overview.pdf
10. Cohen-Mansfield, J., Billig, N.: Agitated behaviors in the elderly: I. a conceptual review. J. Am. Geriatr. Soc. **34**(10), 711–721 (1986)
11. Deakin, M., Waer, H.A.: From intelligent to smart cities. Intell. Buildings Int. **3**(3), 140–152 (2011). https://doi.org/10.1080/17508975.2011.586671
12. Finkel, S., Burns, A.: Behavioral and psychological symptoms of dementia (BPSD): a clinical and research update. Int. Psychogeriatr. **12**(S1), 9–12 (2000). https://doi.org/10.1017/S1041610200006694
13. Hassan, M.M., Albakr, H., Al-Dossari, H., Mohamed, A.: Resource provisioning for cloud-assisted body area network in a smart home environment. IEEE Access **5**, 13213–13224 (2017)
14. Hu, Y.C., Patel, M., Sabella, D., Sprecher, N., Young, V.: Mobile edge computing–a key technology towards 5g. ETSI White Pap. **11**(11), 1–16 (2015)
15. Intanagonwiwat, C., Estrin, D., Govindan, R., Heidemann, J.: Impact of network density on data aggregation in wireless sensor networks. In: Proceedings 22nd International Conference on Distributed Computing Systems, pp. 457–458. IEEE, Vienna (2002). https://doi.org/10.1109/ICDCS.2002.1022289
16. Kim, U.H., Kong, E., Choi, H.H., Lee, J.R.: Analysis of aggregation delay for multisource sensor data with on-off traffic pattern in wireless body area networks. Sensors **16**(10), 1622 (2016). https://doi.org/10.3390/s16101622. http://www.mdpi.com/1424-8220/16/10/1622
17. Luan, T.H., Gao, L., Li, Z., Xiang, Y., Wei, G., Sun, L.: Fog computing: Focusing on mobile users at the edge. arXiv preprint arXiv:1502.01815 (2015)
18. Moore, P., Liu, H.: Modelling uncertainty in health care systems. In: International Conference on Brain Informatics and Health, Part of the Lecture Notes in Computer Science book series (LNCS), The 8th International Conference on Brain Informatics and Health (BIH 2015), pp. 410–419. Springer, Cham (2015). https://doi.org/10.1007/978-3-319-23344-4_40
19. Moore, P., Pham, H.V.: Personalization and rule strategies in human-centric data intensive intelligent context-aware systems. Knowl. Eng. Rev. **30**(2), 140–156 (2015). https://doi.org/10.1017/S0269888914000265
20. Moore, P., Qassem, T., Xhafa, F.: NoSQL and electronic patient records: Opportunities and challenges. In: Proceedings of the 9th International Conference on P2P, Parallel, Grid, Cloud, and Internet Computing (3PGCIC 2014), pp. 300–307. IEEE, Guangzhou (2014). https://doi.org/10.1109/3PGCIC.2014.81

21. Moore, P., Thomas, A., Tadros, G., Xhafa, F., Barolli, L.: Detection of the onset of agitation in patients with dementia: real-time monitoring and the application of big-data solutions. Int. J. Space-Based Situated Comput. **3**(3), 136–154 (2013). https://doi.org/10.1504/IJSSC.2013.056405
22. Moore, P.T., Pham, H.V.: On context-aware evidence-based data driven development of diagnostic scales for depression. In: Barolli, L., Terzo, O. (eds.) The 11th International Conference on Complex, Intelligent, and Software Intensive Systems (CISIS-2017): The 7th International Workshop on Intelligent Computing in Large-Scale Systems (ICLS-2017), no. 611 in Advances in Intelligent Systems and Computing, pp. 929–942. Springer, Torino (2017). https://doi.org/10.1007/978-3-319-61566-0
23. Naik, M.R.K., Samundiswary, P.: Wireless body area network security issues–survey. In: 2016 International Conference on Control, Instrumentation, Communication and Computational Technologies (ICCICCT), pp. 190–194. IEEE (2016)
24. Patsakis, C., Venanzio, R., Bellavista, P., Solanas, A., Bouroche, M.: Personalized medical services using smart cities' infrastructures. In: 2014 IEEE International Symposium on Medical Measurements and Applications (MeMeA), pp. 1–5 (2014). https://doi.org/10.1109/MeMeA.2014.6860145
25. Peng, Y., Peng, L.: A cooperative transmission strategy for body-area networks in healthcare systems. IEEE Access **4**, 9155–9162 (2016)
26. Solanas, A., Patsakis, C., Conti, M., Vlachos, I.S., Ramos, V., Falcone, F., Postolache, O., Pérez-Martínez, P.A., Di Pietro, R., Perrea, D.N., et al.: Smart health: a context-aware health paradigm within smart cities. IEEE Commun. Mag. **52**(8), 74–81 (2014)
27. Stojmenovic, I., Wen, S.: The Fog computing paradigm: scenarios and security issues. In: 2014 Federated Conference on Computer Science and Information Systems, pp. 1–8 (2014). https://doi.org/10.15439/2014F503
28. Thomas, A., Moore, P., Evans, C., Shah, H., Sharma, M., Mount, S., Xhafa, F., Pham, H., Barolli, L., Patel, A., et al.: Smart care spaces: pervasive sensing technologies for at-home care. Int. J. Ad Hoc Ubiquit. Comput. **16**(4), 268–282 (2014). https://doi.org/10.1504/IJAHUC.2014.064862
29. Thomas, A.M., Parkinson, J., Moore, P., Goodman, A., Xhafa, F., Barolli, L.: Nudging through technology: choice architectures and the mobile information revolution. In: 2013 Eighth International Conference on P2P, Parallel, Grid, Cloud and Internet Computing (3PGCIC), pp. 255–261. IEEE, University of Technology of Compiegne, Compiegne (2013). https://doi.org/10.1109/3PGCIC.2013.44
30. Vickland, V., Brodaty, H.: Visualisation of clinical and non-clinical characteristics of patients with behavioural and psychological symptoms of dementia. In: 2008 Fifth International Conference BioMedical Visualization: Information Visualization in Medical and Biomedical Informatics, pp. 23–28 (2008). https://doi.org/10.1109/MediVis.2008.20
31. Wu, T., Wu, F., Redouté, J.M., Yuce, M.R.: An autonomous wireless body area network implementation towards IoT connected healthcare applications. IEEE Access **5**, 11413–11422 (2017)

Application of Fuzzy Logic for Improving Human Sleeping Conditions in an Ambient Intelligence Testbed

Kevin Bylykbashi[1(✉)], Ryoichiro Obukata[2], Yi Liu[2], Evjola Spaho[1], Leonard Barolli[3], and Makoto Takizawa[4]

[1] Faculty of Information Technology, Polytechnic University of Tirana, Mother Teresa Square, No. 4, Tirana, Albania
kevin.bylykbashi@fti.edu.al, evjolaspaho@hotmail.com

[2] Graduate School of Engineering, Fukuoka Institute of Technology (FIT), 3-30-1 Wajiro-Higashi, Higashi-Ku, Fukuoka 811-0295, Japan
obukenkyuu@gmail.com, ryuui1010@gmail.com

[3] Department of Information and Communication Engineering, Fukuoka Institute of Technology (FIT), 3-30-1 Wajiro-Higashi, Higashi-Ku, Fukuoka 811-0295, Japan
barolli@fit.ac.jp

[4] Department of Advanced Sciences, Faculty of Science and Engineering, Hosei University, Kajino-Machi, Koganei-Shi, Tokyo 184-8584, Japan
makoto.takizawa@computer.org

Abstract. Ambient Intelligence (AmI) deals with a new world of ubiquitous computing devices, where physical environments interact intelligently and unobtrusively with people. AmI environments can be diverse, such as homes, offices, meeting rooms, schools, hospitals, control centers, vehicles, tourist attractions, stores, sports facilities, and music devices. In our previous work, we presented the implementation and evaluation of actor node for AmI testbed. In this paper, we introduce the implementation of the AmI testbed. We present the simulation results of the proposed Fuzzy-based Sleeping Condition System (FSCS) considering four parameters: room lighting, humidity, temperature and noise. The simulation results show that different parameters have different effects on human sleeping condition.

1 Introduction

Ambient Intelligence (AmI) is the vision that technology will become invisible, embedded in our natural surroundings, present whenever we need it, enabled by simple and effortless interactions, attuned to all our senses, adaptive to users and context and autonomously acting [1]. High quality information and content must be available to any user, anywhere, at any time, and on any device.

In order that AmI becomes a reality, it should completely envelope humans, without constraining them. Distributed embedded systems for AmI are going to change the way we design embedded systems, in general, as well as the way

L. Barolli et al. (Eds.): EIDWT 2018, LNDECT 17, pp. 41–50, 2018.
https://doi.org/10.1007/978-3-319-75928-9_4

we think about such systems. But, more importantly, they will have a great impact on the way we live. Applications ranging from safe driving systems, smart buildings and home security, smart fabrics or e-textiles, to manufacturing systems, and rescue and recovery operations in hostile environments, are poised to become part of society and human lives.

There are a lot of works done for AmI. In [2], the authors present a simulation environment that offers a library of Networked Control Systems (NCS) blocks. Thus, the constraints can be considered and integrated in the design process. They describe a real process, an inverted pendulum, which is automated based on Mica nodes. These nodes were designed especially for AmI purposes. This real NCS serves as a challenging benchmark for proving the AmI suitability of the controllers.

In [3], the authors present the development of an adaptive embedded agent, based on a hybrid PCA-NFS scheme, able to perform true real-time control of AmI environments in the long term. The proposed architecture is a single-chip HW/SW architecture. It consists of a soft processor core (SW partition), a set of NFS cores (HW partition), the HW/SW interface, and input/output (I/O) peripherals. An application example based on data obtained in an ubiquitous computing environment has been successfully implemented using an FPGA of Xilinx's Virtex 5 family [4].

In [5], the authors describe a framework to Context Acquisition Services and Reasoning Algorithms (CASanDRA) to be directly consumed by any type of application needing to handle context information. CASanDRA decouples the acquisition and inference tasks from the application development by offering a set of interfaces for information retrieval. The framework design is based on a data fusion-oriented architecture. CASanDRA has been designed to be easily scalable. It simplifies the integration of both new sensor access interfaces and fusion algorithms deployment, as it also aims at serving as a testbed for research.

In this work, we present an AmI testbed. We investigate the effects of the bed temperature, room lighting, humidity and noise on human sleeping condition by using the fuzzy based model.

The structure of the paper is as follows. In Sect. 2, we present a short description of AmI. In Sect. 3, we introduce Fuzzy Logic used for control. In Sect. 4, we present the proposed fuzzy-based system. In Sect. 5, we discuss the simulation results. Finally, conclusions and future work are given in Sect. 6.

2 Ambient Intelligence (AmI)

In the future, small devices will monitor the health status in a continuous manner, diagnose any possible health conditions, have conversation with people to persuade them to change the lifestyle for maintaining better health, and communicates with the doctor, if needed [6]. The device might even be embedded into the regular clothing fibers in the form of very tiny sensors and it might communicate with other devices including the variety of sensors embedded into the home to monitor the lifestyle. For example, people might be alarmed about

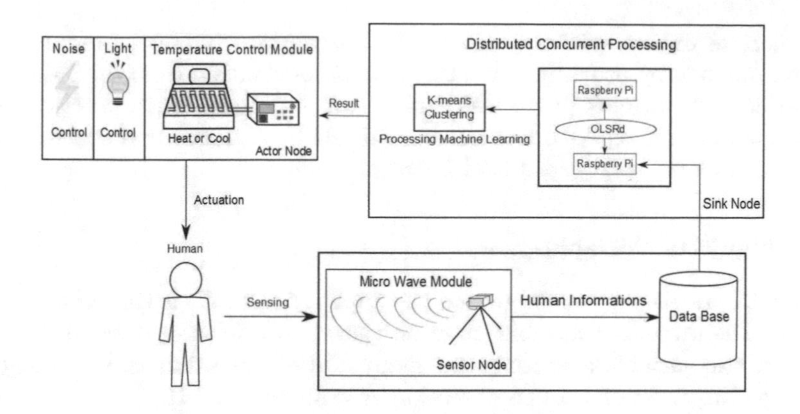

Fig. 1. Structure of AmI testbed.

the lack of a healthy diet based on the items present in the fridge and based on what they are eating outside regularly.

The AmI paradigm represents the future vision of intelligent computing where environments support the people inhabiting them [7–9]. In this new computing paradigm, the conventional input and output media no longer exist, rather the sensors and processors will be integrated into everyday objects, working together in harmony in order to support the inhabitants [10]. By relying on various artificial intelligence techniques, AmI promises the successful interpretation of the wealth of contextual information obtained from such embedded sensors, and will adapt the environment to the user needs in a transparent and anticipatory manner.

3 Application of Fuzzy Logic for Control

The ability of fuzzy sets and possibility theory to model gradual properties or soft constraints whose satisfaction is matter of degree, as well as information pervaded with imprecision and uncertainty, makes them useful in a great variety of applications.

The most popular area of application is Fuzzy Control (FC), since the appearance, especially in Japan, of industrial applications in domestic appliances, process control, and automotive systems, among many other fields.

3.1 FC

In the FC systems, expert knowledge is encoded in the form of fuzzy rules, which describe recommended actions for different classes of situations represented by fuzzy sets.

In fact, any kind of control law can be modeled by the FC methodology, provided that this law is expressible in terms of "if ... then ..." rules, just like in the case of expert systems. However, FL diverges from the standard expert system approach by providing an interpolation mechanism from several rules. In the contents of complex processes, it may turn out to be more practical to get knowledge from an expert operator than to calculate an optimal control, due to modeling costs or because a model is out of reach.

3.2 Linguistic Variables

A concept that plays a central role in the application of FL is that of a linguistic variable. The linguistic variables may be viewed as a form of data compression. One linguistic variable may represent many numerical variables. It is suggestive to refer to this form of data compression as granulation [11].

The same effect can be achieved by conventional quantization, but in the case of quantization, the values are intervals, whereas in the case of granulation the values are overlapping fuzzy sets. The advantages of granulation over quantization are as follows:

- it is more general;
- it mimics the way in which humans interpret linguistic values;
- the transition from one linguistic value to a contiguous linguistic value is gradual rather than abrupt, resulting in continuity and robustness.

3.3 FC Rules

FC describes the algorithm for process control as a fuzzy relation between information about the conditions of the process to be controlled, x and y, and the output for the process z. The control algorithm is given in "if ... then ..." expression, such as:

If x is small and y is big, then z is medium;
If x is big and y is medium, then z is big.

These rules are called *FC rules*. The "if" clause of the rules is called the antecedent and the "then" clause is called consequent. In general, variables x and y are called the input and z the output. The "small" and "big" are fuzzy values for x and y, and they are expressed by fuzzy sets.

Fuzzy controllers are constructed of groups of these FC rules, and when an actual input is given, the output is calculated by means of fuzzy inference.

3.4 Control Knowledge Base

There are two main tasks in designing the control knowledge base. First, a set of linguistic variables must be selected which describe the values of the main control parameters of the process. Both the input and output parameters must be linguistically defined in this stage using proper term sets. The selection of the level of granularity of a term set for an input variable or an output variable plays an important role in the smoothness of control. Second, a control knowledge base must be developed which uses the above linguistic description of the input and output parameters. Four methods [12–15] have been suggested for doing this:

- expert's experience and knowledge;
- modelling the operator's control action;
- modelling a process;
- self organization.

Among the above methods, the first one is the most widely used. In the modeling of the human expert operator's knowledge, fuzzy rules of the form "If Error is small and Change-in-error is small then the Force is small" have been used in several studies [16,17]. This method is effective when expert human operators can express the heuristics or the knowledge that they use in controlling a process in terms of rules of the above form.

3.5 Defuzzification Methods

The defuzzification operation produces a non-FC action that best represent the membership function of an inferred FC action. Several defuzzification methods have been suggested in literature. Among them, four methods which have been applied most often are:

- Tsukamoto's Defuzzification Method;
- The Center of Area (COA) Method;
- The Mean of Maximum (MOM) Method;
- Defuzzification when Output of Rules are Function of Their Inputs.

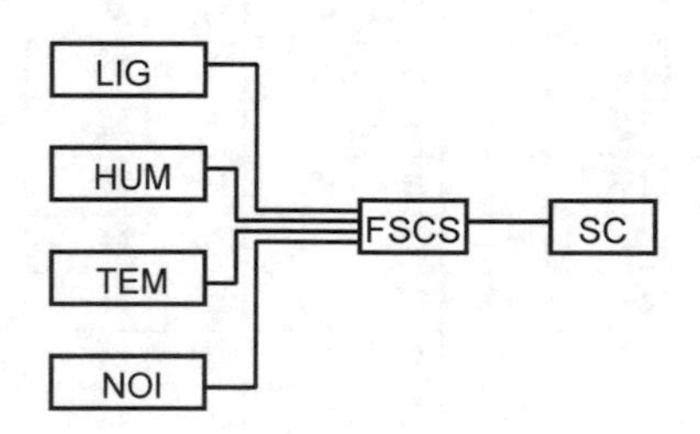

Fig. 2. Proposed FSCS structure.

4 Proposed Fuzzy-Based System

The structure of the implemented testbed is shown in Fig. 1. We implement the fuzzy-based system in the actor node of the testbed. In this work, we consider four input parameters: Light (LIG), Humidity (HUM), Temperature (TEM),

Table 1. FRB.

Rule	LIG	HUM	TEM	NOI	SC	Rule	LIG	HUM	TEM	NOI	SC
1	Da	L	Lo	Q	N	41	Mi	M	Me	N	FG
2	Da	L	Lo	N	FG	42	Mi	M	Me	VN	N
3	Da	L	Lo	VN	N	43	Mi	M	Hi	Q	N
4	Da	L	Me	Q	B	44	Mi	M	Hi	N	FG
5	Da	L	Me	N	N	45	Mi	M	Hi	VN	B
6	Da	L	Me	VN	B	46	Mi	H	Lo	Q	B
7	Da	L	Hi	Q	B	47	Mi	H	Lo	N	N
8	Da	L	Hi	N	N	48	Mi	H	Lo	VN	B
9	Da	L	Hi	VN	VB	49	Mi	H	Me	Q	VB
10	Da	M	Lo	Q	G	50	Mi	H	Me	N	B
11	Da	M	Lo	N	VG	51	Mi	H	Me	VN	VB
12	Da	M	Lo	VN	G	52	Mi	H	Hi	Q	EB
13	Da	M	Me	Q	FG	53	Mi	H	Hi	N	VB
14	Da	M	Me	N	G	54	Mi	H	Hi	VN	EB
15	Da	M	Me	VN	FG	55	Br	L	Lo	Q	B
16	Da	M	Hi	Q	FG	56	Br	L	Lo	N	N
17	Da	M	Hi	N	G	57	Br	L	Lo	VN	B
18	Da	M	Hi	VN	N	58	Br	L	Me	Q	VB
19	Da	H	Lo	Q	N	59	Br	L	Me	N	B
20	Da	H	Lo	N	FG	60	Br	L	Me	VN	VB
21	Da	H	Lo	VN	N	61	Br	L	Hi	Q	EB
22	Da	H	Me	Q	B	62	Br	L	Hi	N	VB
23	Da	H	Me	N	N	63	Br	L	Hi	VN	EB
24	Da	H	Me	VN	B	64	Br	M	Lo	Q	FG
25	Da	H	Hi	Q	VB	65	Br	M	Lo	N	G
26	Da	H	Hi	N	B	66	Br	M	Lo	VN	FG
27	Da	H	Hi	VN	VB	67	Br	M	Me	Q	N
28	Mi	L	Lo	Q	B	68	Br	M	Me	N	FG
29	Mi	L	Lo	N	N	69	Br	M	Me	VN	N
30	Mi	L	Lo	VN	B	70	Br	M	Hi	Q	B
31	Mi	L	Me	Q	VB	71	Br	M	Hi	N	N
32	Mi	L	Me	N	B	72	Br	M	Hi	VN	B
33	Mi	L	Me	VN	VB	73	Br	H	Lo	Q	VB
34	Mi	L	Hi	Q	VB	74	Br	H	Lo	N	B
35	Mi	L	Hi	N	B	75	Br	H	Lo	VN	VB
36	Mi	L	Hi	VN	EB	76	Br	H	Me	Q	EB
37	Mi	M	Lo	Q	FG	77	Br	H	Me	N	VB
38	Mi	M	Lo	N	G	78	Br	H	Me	VN	EB
39	Mi	M	Lo	VN	FG	79	Br	H	Hi	Q	EB
40	Mi	M	Me	Q	N	80	Br	H	Hi	N	VB
						81	Br	H	Hi	VN	EB

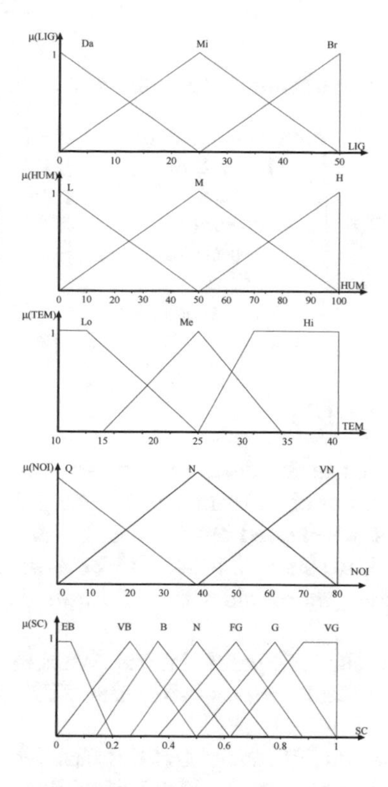

Fig. 3. Membership functions.

Noise (NOI) to decide the Sleeping Condition (SC). The structure of this system called Fuzzy-based Sleeping Condition System (FSCS) is shown in Fig. 2. These four parameters are fuzzified using fuzzy system, and based on the decision of fuzzy system a sleeping condition is calculated. The membership functions for our system are shown in Fig. 3. In Table 1, we show the Fuzzy Rule Base (FRB) of our proposed system, which consists of 81 rules.

The input parameters for FSCS are: LIG, HUM, TEM and NOI, and the output linguistic parameter is SC. The term sets of *LIG*, *HUM*, *TEM* and *NOI* are defined respectively as:

$$\begin{aligned} LIG &= \{Dark,\ Middle,\ Bright\} \\ &= \{Da,\ Mi,\ Br\}; \\ HUM &= \{Low,\ Middle,\ High\} \\ &= \{L,\ M,\ H\}; \\ TEM &= \{Low,\ Medium,\ High\} \\ &= \{Lo,\ Me,\ Hi\}; \\ NOI &= \{Quiet,\ Noisy,\ Very\ Noisy\} \\ &= \{Q,\ N,\ VN\}. \end{aligned}$$

and the term set for the output SC is defined as:

$$SC = \begin{pmatrix} Extremely\ Bad \\ Very\ Bad \\ Bad \\ Normal \\ Few\ Good \\ Good \\ Very\ Good \end{pmatrix} = \begin{pmatrix} EB \\ VB \\ B \\ N \\ FG \\ G \\ VG \end{pmatrix}$$

5 Simulation Results

In this section, we present the simulation results for our proposed system. In our system, we decided the number of term sets by carrying out many simulations. These simulation results are carried out in MATLAB.

From Figs. 4, 5 and 6, we show the relation between LIG, HUM, TEM, NOI and SC. In this simulation, we consider the LIG and HUM as constant parameters.

In Fig. 4, we consider the LIG value 0 units. We change the HUM value from 10 to 90 units. From the result, we see when the HUM is 50 units, the sleeping condition is the best.

In Fig. 5, we increase the LIG value to 25 units. We see that when the LTG is 0 the sleeping condition is better. In Fig. 6, the LTG values is 50 units. When the LIG increases, the SC is decreased.

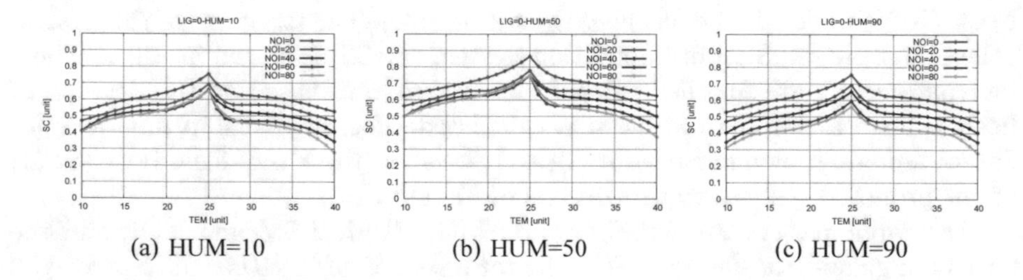

(a) HUM=10 (b) HUM=50 (c) HUM=90

Fig. 4. Relation between SC and TEM for different HUM when LIG = 0.

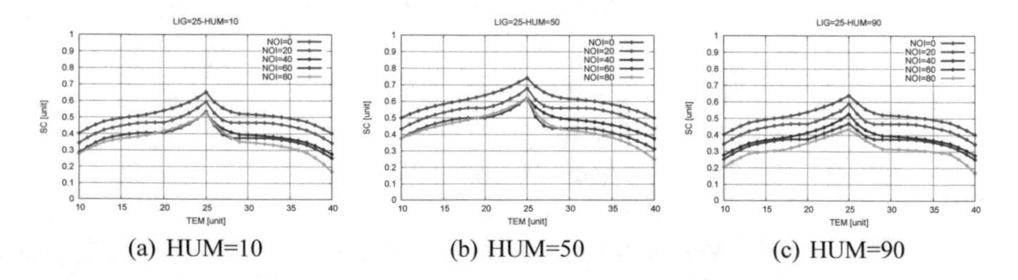

(a) HUM=10 (b) HUM=50 (c) HUM=90

Fig. 5. Relation between SC and TEM for different HUM when LIG = 25.

From Figs. 4, 5 and 6, when the TEM is 25 units, the sleeping condition is very good. We change the NOI values from 0 to 80 units. When the NOI is 0, the sleeping condition is the best.

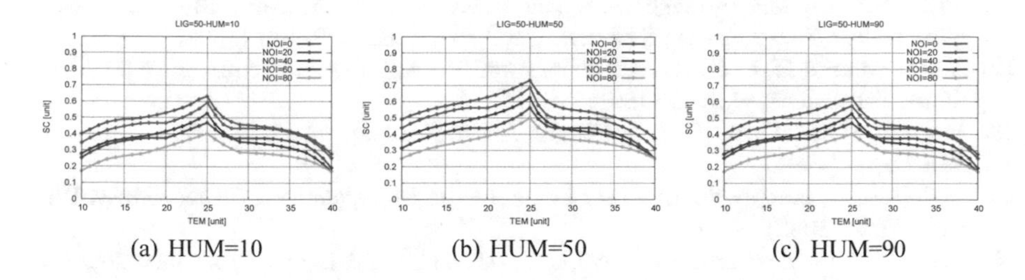

(a) HUM=10 (b) HUM=50 (c) HUM=90

Fig. 6. Relation between SC and TEM for different HUM when LIG = 50.

6 Conclusions

In this paper, we proposed a fuzzy-based system to decide the SC. We took into consideration four parameters: LIG, HUM, TEM and NOI. We evaluated the performance of proposed system by computer simulations. From the simulations results, we conclude that the room lighting, humidity, temperature and noise have different effects to human during sleeping.

In the future, we would like to make extensive simulations to evaluate the proposed systems and compare the performance with other systems.

References

1. Lindwer, M., Marculescu, D., Basten, T., Zimmermann, R., Marculescu, R., Jung, S., Cantatore, E.: Ambient intelligence visions and achievements: linking abstract ideas to real-world concepts. In: Design, Automation and Test in Europe Conference and Exhibition, pp. 10–15 (2003)
2. Gabel, O., Litz, L., Reif, M.: NCS testbed for ambient intelligence. In: IEEE International Conference on Systems, Man and Cybernetics, vol. 1, pp. 115–120 (2005)
3. del Campo, I., Martinez, M.V., Echanobe, J., Basterretxea, K.: A hardware/software embedded agent for realtime control of ambient-intelligence environments. In: IEEE International Conference on Fuzzy Systems (FUZZ-IEEE), pp. 1–8 (2012)
4. Virtex 5 Family Overview. Xilinx Inc., San Jose, CA (2009)
5. Bernardos, A.M., Tarrio, P., Casar, J.R.: CASanDRA: a framework to provide context acquisition services and reasoning algorithms for ambient intelligence applications. In: International Conference on Parallel and Distributed Computing, Applications and Technologies, pp. 372–377 (2009)
6. Acampora, G., Cook, D., Rashidi, P., Vasilakos, A.V.: A survey on ambient intelligence in health care. Proc. IEEE **101**(12), 2470–2494 (2013)
7. Aarts, E., Wichert, R.: Ambient intelligence. In: Bullinger, H.J. (ed.) Technology Guide. Springer, Heidelberg (2009). https://doi.org/10.1007/978-3-540-88546-7_47

8. Aarts, E., de Ruyter, B.: New research perspectives on ambient intelligence. J. Ambient Intell. Smart Environ. **1**(1), 5–14 (2009)
9. Vasilakos, A., Pedrycz, W.: Ambient Intelligence, Wireless Networking, and Ubiquitous Computing. Norwood, Artech House Inc, MA, USA (2006)
10. Sadri, F.: Ambient intelligence: a survey. ACM Comput. Surv. **43**(4), 66 (2011)
11. Kandel, A.: Fuzzy Expert Systems. CRC Press, Boca Raton (1992)
12. Zimmermann, H.J.: Fuzzy Set Theory and its Applications, 2nd edn. Kluwer Academic Publishers, Boston (1991)
13. McNeill, F.M., Thro, E.: Fuzzy Logic: A Practical Approach. Academic Press Inc., San Diego (1994)
14. Zadeh, L.A., Kacprzyk, J.: Fuzzy Logic for the Management of Uncertainty. Wiley, New York (1992)
15. Procyk, T.J., Mamdani, E.H.: A linguistic self-organizing process controller. Automatica **15**(1), 15–30 (1979)
16. Klir, G.J., Folger, T.A.: Fuzzy Sets, Uncertainty, and Information. Prentice Hall, Englewood Cliffs (1988)
17. Munakata, T., Jani, Y.: Fuzzy Systems: an overview. Commun. ACM **37**(3), 69–76 (1994)

Performance Evaluation of WMN-PSOSA Considering Four Different Replacement Methods

Shinji Sakamoto[1(✉)], Kosuke Ozera[1], Admir Barolli[2], Leonard Barolli[3], Vladi Kolici[4], and Makoto Takizawa[5]

[1] Graduate School of Engineering, Fukuoka Institute of Technology, 3-30-1 Wajiro-Higashi, Higashi-Ku, Fukuoka 811-0295, Japan
shinji.sakamoto@ieee.org, kosuke.o.fit@gmail.com

[2] Department of Information Technology, Aleksander Moisiu University of Durres, L.1, Rruga e Currilave, Durres, Albania
admir.barolli@gmail.com

[3] Department of Information and Communication Engineering, Fukuoka Institute of Technology, 3-30-1 Wajiro-Higashi, Higashi-Ku, Fukuoka 811-0295, Japan
barolli@fit.ac.jp

[4] Faculty of Information Technology, Polytechnic University of Tirana, Mother Teresa Square, No. 4, Tirana, Albania
vkolici@fti.edu.al

[5] Department of Advanced Sciences, Faculty of Science and Engineering, Hosei University, Kajino-Machi, Koganei-Shi, Tokyo 184-8584, Japan
makoto.takizawa@computer.org

Abstract. Wireless Mesh Networks (WMNs) have many advantages such as low cost and increased high speed wireless Internet connectivity, therefore WMNs are becoming an important networking infrastructure. In our previous work, we implemented a Particle Swarm Optimization (PSO) based simulation system for node placement in WMNs, called WMN-PSO. Also, we implemented a simulation system based on Simulated Annealing (SA) for solving node placement problem in WMNs, called WMN-SA. In this paper, we implement a hybrid simulation system based on PSO and SA, called WMN-PSOSA. We evaluate the performance of WMN-PSOSA by conducting computer simulations considering four different replacement methods. The simulation results show that LDIWM have better performance than CM, RIWM and LDVM replacement methods.

1 Introduction

The wireless networks and devises are becoming increasingly popular and they provide users access to information and communication anytime and anywhere [11,13–15,29–31]. Wireless Mesh Networks (WMNs) are gaining a lot of attention because of their low cost nature that makes them attractive for

L. Barolli et al. (Eds.): EIDWT 2018, LNDECT 17, pp. 51–64, 2018.
https://doi.org/10.1007/978-3-319-75928-9_5

providing wireless Internet connectivity. A WMN is dynamically self-organized and self-configured, with the nodes in the network automatically establishing and maintaining mesh connectivity among them-selves (creating, in effect, an ad hoc network). This feature brings many advantages to WMNs such as low up-front cost, easy network maintenance, robustness and reliable service coverage [1]. Moreover, such infrastructure can be used to deploy community networks, metropolitan area networks, municipal and corporative networks, and to support applications for urban areas, medical, transport and surveillance systems.

Mesh node placement in WMN can be seen as a family of problems, which are shown (through graph theoretic approaches or placement problems, e.g. [8,18]) to be computationally hard to solve for most of the formulations [35]. In fact, the node placement problem considered here is even more challenging due to two additional characteristics:

(a) Locations of mesh router nodes are not pre-determined, in other wards, any available position in the considered area can be used for deploying the mesh routers.
(b) Routers are assumed to have their own radio coverage area.

Here, we consider the version of the mesh router nodes placement problem in which we are given a grid area where to deploy a number of mesh router nodes and a number of mesh client nodes of fixed positions (of an arbitrary distribution) in the grid area. The objective is to find a location assignment for the mesh routers to the cells of the grid area that maximizes the network connectivity and client coverage. Node placement problems are known to be computationally hard to solve [16,17,36]. In some previous works, intelligent algorithms have been recently investigated [2–4,6,7,9,10,19,21,24–26,37].

In our previous work, we implemented a Particle Swarm Optimization (PSO) based simulation system, called WMN-PSO [27]. Also, we implemented a simulation system based on Simulated Annealing (SA) for solving node placement problem in WMNs, called WMN-SA [22,23].

In this paper, we implement a hybrid simulation system based on PSO and SA. We call this system WMN-PSOSA. We evaluate the performance of hybrid WMN-PSOSA system considering four different replacement methods.

The rest of the paper is organized as follows. The mesh router nodes placement problem is defined in Sect. 2. We present our designed and implemented

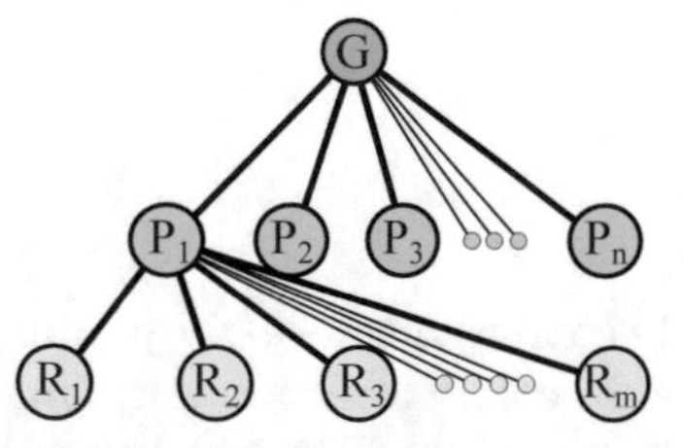

Fig. 1. Relationship among global solution, particle-patterns and mesh routers.

hybrid simulation system in Sect. 3. The simulation results are given in Sect. 4. Finally, we give conclusions and future work in Sect. 5.

2 Node Placement Problem in WMNs

For this problem, we have a grid area arranged in cells we want to find where to distribute a number of mesh router nodes and a number of mesh client nodes of fixed positions (of an arbitrary distribution) in the considered area. The objective is to find a location assignment for the mesh routers to the area that maximizes the network connectivity and client coverage. Network connectivity is measured by Size of Giant Component (SGC) of the resulting WMN graph, while the user coverage is simply the number of mesh client nodes that fall within the radio coverage of at least one mesh router node and is measured by Number of Covered Mesh Clients (NCMC).

An instance of the problem consists as follows.

- N mesh router nodes, each having its own radio coverage, defining thus a vector of routers.
- An area $W \times H$ where to distribute N mesh routers. Positions of mesh routers are not pre-determined and are to be computed.
- M client mesh nodes located in arbitrary points of the considered area, defining a matrix of clients.

It should be noted that network connectivity and user coverage are among most important metrics in WMNs and directly affect the network performance.

In this work, we have considered a bi-objective optimization in which we first maximize the network connectivity of the WMN (through the maximization of the SGC) and then, the maximization of the NCMC.

In fact, we can formalize an instance of the problem by constructing an adjacency matrix of the WMN graph, whose nodes are router nodes and client nodes and whose edges are links between nodes in the mesh network. Each mesh node in the graph is a triple $\boldsymbol{v} = <x, y, r>$ representing the 2D location point and r is the radius of the transmission range. There is an arc between two nodes $\boldsymbol{u}$ and $\boldsymbol{v}$, if $\boldsymbol{v}$ is within the transmission circular area of $\boldsymbol{u}$.

3 Proposed and Implemented Simulation System

3.1 PSO Algorithm

In PSO a number of simple entities (the particles) are placed in the search space of some problem or function and each evaluates the objective function at its current location. The objective function is often minimized and the exploration of the search space is not through evolution [20]. However, following a widespread practice of borrowing from the evolutionary computation field, in this work, we consider the bi-objective function and fitness function interchangeably. Each particle then determines its movement through the search space by combining some aspect of the history of its own current and best (best-fitness) locations with those of one or more members of the swarm, with some random perturbations. The next iteration takes place after all particles have been moved. Eventually the swarm as a whole, like a flock of birds collectively foraging for food, is likely to move close to an optimum of the fitness function.

Each individual in the particle swarm is composed of three $\mathcal{D}$-dimensional vectors, where $\mathcal{D}$ is the dimensionality of the search space. These are the current position $\mathbf{x}_i$, the previous best position $\mathbf{p}_i$ and the velocity $\mathbf{v}_i$.

The particle swarm is more than just a collection of particles. A particle by itself has almost no power to solve any problem; progress occurs only when the particles interact. Problem solving is a population-wide phenomenon, emerging from the individual behaviors of the particles through their interactions. In any case, populations are organized according to some sort of communication structure or topology, often thought of as a social network. The topology typically consists of bidirectional edges connecting pairs of particles, so that if j is in i's neighborhood, i is also in j's. Each particle communicates with some other particles and is affected by the best point found by any member of its topological neighborhood. This is just the vector $\mathbf{p}_i$ for that best neighbor, which we will denote with $\mathbf{p}_g$. The potential kinds of population "social networks" are hugely varied, but in practice certain types have been used more frequently.

In the PSO process, the velocity of each particle is iteratively adjusted so that the particle stochastically oscillates around $\mathbf{p}_i$ and $\mathbf{p}_g$ locations.

3.2 Simulated Annealing

3.2.1 Description of Simulated Annealing

SA algorithm [12] is a generalization of the metropolis heuristic. Indeed, SA consists of a sequence of executions of metropolis with a progressive decrement of the temperature starting from a rather high temperature, where almost any move is accepted, to a low temperature, where the search resembles Hill Climbing. In fact, it can be seen as a hill-climber with an internal mechanism to escape local optima. In SA, the solution s' is accepted as the new current solution if $\delta \leq 0$ holds, where $\delta = f(s') - f(s)$. To allow escaping from a local optimum, the movements that increase the energy function are accepted with a decreasing probability $\exp(-\delta/T)$ if $\delta > 0$, where T is a parameter called the "temperature". The decreasing values of T are controlled by a *cooling schedule*, which specifies the temperature values at each stage of the algorithm, what represents an important decision for its application (a typical option is to use a proportional method, like $T_k = \alpha \cdot T_{k-1}$). SA usually gives better results in practice, but uses to be very slow. The most striking difficulty in applying SA is to choose and tune its parameters such as initial and final temperature, decrements of the temperature (cooling schedule), equilibrium and detection.

In our system, cooling schedule (α) will be calculated as

$$\alpha = \left(\frac{SA\ Ending\ temperature}{SA\ Starting\ temperature} \right)^{1.0/Total\ iterations}.$$

3.2.2 Acceptability Criteria

The acceptability criteria for newly generated solution is based on the definition of a threshold value (accepting threshold) as follows. We consider a succession t_k such that $t_k > t_{k+1}$, $t_k > 0$ and t_k tends to 0 as k tends to infinity. Then, for any two solutions s_i and s_j, if $fitness(s_j) - fitness(s_i) < t_k$, then accept solution s_j.

For the SA, t_k values are taken as accepting threshold but the criterion for acceptance is probabilistic:

- If $fitness(s_j) - fitness(s_i) \leq 0$ then s_j is accepted.
- If $fitness(s_j) - fitness(s_i) > 0$ then s_j is accepted with probability $\exp[(fitness(s_j) - fitness(s_i))/t_k]$ (at iteration k the algorithm generates a random number $R \in (0, 1)$ and s_j is accepted if $R < \exp[(fitness(s_j) - fitness(s_i))/t_k]$).

In this case, each neighbour of a solution has a positive probability of replacing the current solution. The t_k values are chosen in way that solutions with large increase in the cost of the solutions are less likely to be accepted (but there is still a positive probability of accepting them).

3.3 WMN-PSOSA Hybrid Simulation System

In this subsection, we present the initialization, particle-pattern and fitness function.

Initialization
Our proposed system starts by generating an initial solution randomly, by *ad hoc* methods [37]. We decide the velocity of particles by a random process considering the area size. For instance, when the area size is $W \times H$, the velocity is decided randomly from $-\sqrt{W^2 + H^2}$ to $\sqrt{W^2 + H^2}$.

Particle-pattern
A particle is a mesh router. A fitness value of a particle-pattern is computed by combination of mesh routers and mesh clients positions. In other words, each particle-pattern is a solution as shown in Fig. 1. Therefore, the number of particle-patterns is a number of solutions.

Fitness function
One of most important thing in PSO algorithm is to decide the determination of an appropriate objective function and its encoding. In our case, each particle-pattern has an own fitness value and compares other particle-pattern's fitness value in order to share information of global solution. The fitness function follows a hierarchical approach in which the main objective is to maximize the SGC in WMN. Thus, the fitness function of this scenario is defined as

$$\text{Fitness} = 0.7 \times \text{SGC}(\boldsymbol{x}_{ij}, \boldsymbol{y}_{ij}) + 0.3 \times \text{NCMC}(\boldsymbol{x}_{ij}, \boldsymbol{y}_{ij}).$$

3.3.1 WMN-PSOSA Web GUI Tool and Pseudo Code

The Web application follows a standard Client-Server architecture and is implemented using LAMP (Linux + Apache + MySQL + PHP) technology (see Fig. 2). Remote users (clients) submit their requests by completing first the parameter setting. The parameter values to be provided by the user are classified into three groups, as follows.

- Parameters related to the problem instance: These include parameter values that determine a problem instance to be solved and consist of number of router nodes, number of mesh client nodes, client mesh distribution, radio coverage interval and size of the deployment area.
- Parameters of the resolution method: Each method has its own parameters.
- Execution parameters: These parameters are used for stopping condition of the resolution methods and include number of iterations and number of independent runs. The former is provided as a total number of iterations and depending on the method is also divided per phase (e.g., number of iterations in a exploration). The later is used to run the same configuration for the same problem instance and parameter configuration a certain number of times.

We show the WMN-PSOSA Web GUI tool in Fig. 3. The pseudo code of our implemented system is shown in Algorithm 1.

3.3.2 WMN Mesh Routers Replacement Methods

A mesh router has x, y positions and velocity. Mesh routers are moved based on velocities. There are many moving methods in PSO field, such as:

Constriction Method (CM)
: CM is a method which PSO parameters are set to a week stable region ($\omega = 0.729$, $C_1 = C2 = 1.4955$) based on analysis of PSO by Clerc et al. [5,33].

Random Inertia Weight Method (RIWM)
: In RIWM, the ω parameter is changing randomly from 0.5 to 1.0. The C_1 and C_2 are kept 2.0. The ω can be estimated by the week stable region. The average of ω is 0.75 [33].

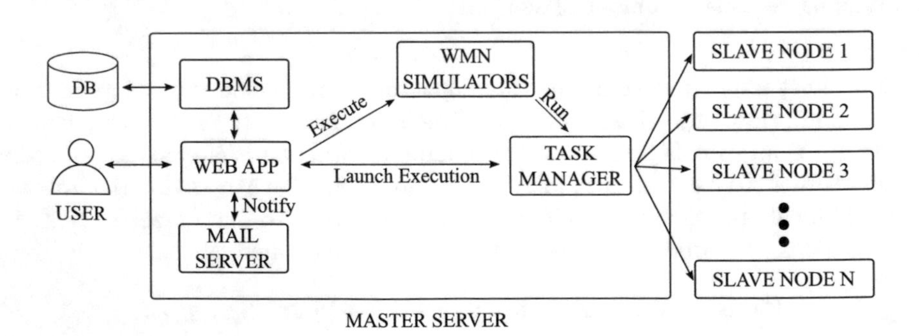

Fig. 2. System structure for web interface.

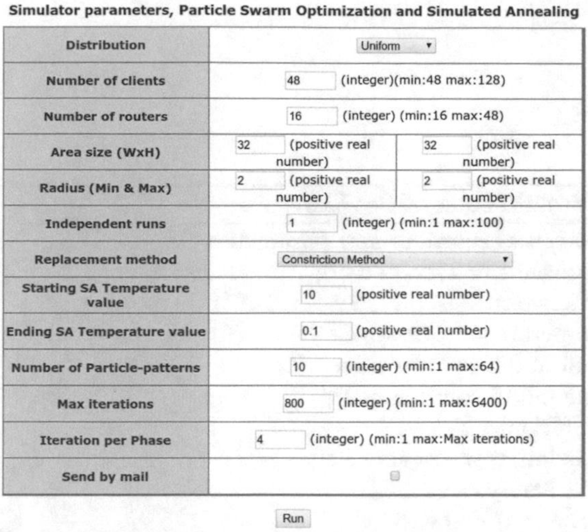

Simulator parameters, Particle Swarm Optimization and Simulated Annealing

Parameter	Value	
Distribution	Uniform	
Number of clients	48 (integer)(min:48 max:128)	
Number of routers	16 (integer) (min:16 max:48)	
Area size (WxH)	32 (positive real number)	32 (positive real number)
Radius (Min & Max)	2 (positive real number)	2 (positive real number)
Independent runs	1 (integer) (min:1 max:100)	
Replacement method	Constriction Method	
Starting SA Temperature value	10 (positive real number)	
Ending SA Temperature value	0.1 (positive real number)	
Number of Particle-patterns	10 (integer) (min:1 max:64)	
Max iterations	800 (integer) (min:1 max:6400)	
Iteration per Phase	4 (integer) (min:1 max:Max iterations)	
Send by mail		

Run

Fig. 3. WMN-PSOSA web GUI tool.

Linearly Decreasing Inertia Weight Method (LDIWM)

In LDIWM, C_1 and C_2 are set to 2.0, constantly. On the other hand, the ω parameter is changed linearly from unstable region ($\omega = 0.9$) to stable region ($\omega = 0.4$) with increasing of iterations of computations [33,34].

Linearly Decreasing Vmax Method (LDVM)

In LDVM, PSO parameters are set to unstable region ($\omega = 0.9$, $C_1 = C_2 = 2.0$). A value of V_{max} which is maximum velocity of particles is considered. With increasing of iteration of computations, the V_{max} is kept decreasing linearly [32].

Rational Decrement of Vmax Method (RDVM)

In RDVM, PSO parameters are set to unstable region ($\omega = 0.9$, $C_1 = C_2 = 2.0$). The V_{max} is kept decreasing with the increasing of iterations as

$$V_{max}(x) = \sqrt{W^2 + H^2} \times \frac{T - x}{x}.$$

Where, W and H are the width and the height of the considered area, respectively. Also, T and x are the total number of iterations and a current number of iteration, respectively [28].

Algorithm 1. Pseudo code of PSOSA.

```
/* Generate the initial solutions and parameters */
Computation maxtime:= T_max, t := 0;
Number of particle-patterns:= m, 2 ≤ m ∈ N^1;
Starting SA temperature:= Temp;
Decreasing speed of SA temperature:= T_d;
Particle-patterns initial solution:= P_i^0;
Global initial solution:= G^0;
Particle-patterns initial position:= x_ij^0;
Particles initial velocity:= v_ij^0;
PSO parameter:= ω, 0 < ω ∈ R^1;
PSO parameter:= C_1, 0 < C_1 ∈ R^1;
PSO parameter:= C_2, 0 < C_2 ∈ R^1;
/* Start PSO-SA */
Evaluate(G^0, P^0);
while t < T_max do
  /* Update velocities and positions */
  v_ij^{t+1} = ω · v_ij^t
        +C_1 · rand() · (best(P_ij^t) − x_ij^t)
        +C_2 · rand() · (best(G^t) − x_ij^t);
  x_ij^{t+1} = x_ij^t + v_ij^{t+1};
  /* if fitness value is increased, a new solution will be accepted. */
  if Evaluate(G^(t+1), P^(t+1)) >= Evaluate(G^(t), P^(t)) then
    Update_Solutions(G^t, P^t);
    Evaluate(G^(t+1), P^(t+1));
  else
    /* a new solution will be accepted, if condition is true. */
    if Random() > e^((Evaluate(G^(t+1),P^(t+1)) − Evaluate(G^(t),P^(t))) / Temp) then
      /* "Reupdate_Solutions" makes particle back to previous position */
      Reupdate_Solutions(G^{t+1}, P^{t+1});
    end if
  end if
  Temp = Temp × t_d;
  t = t + 1;
end while
Update_Solutions(G^t, P^t);
return Best found pattern of particles as solution;
```

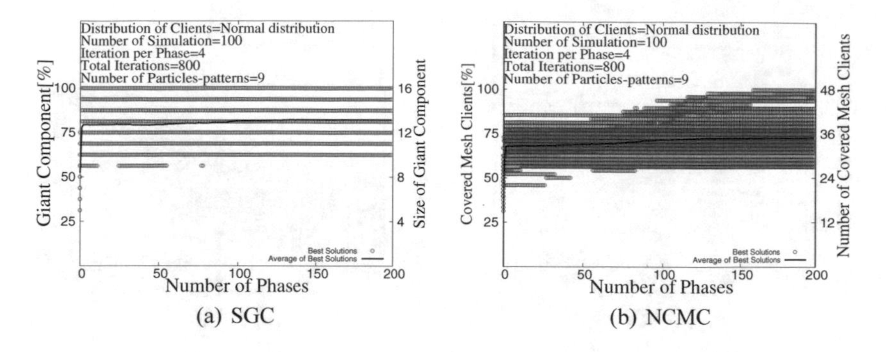

Fig. 4. Simulation results of WMN-PSOSA considering CM.

4 Simulation Results

In this section, we show simulation results using WMN-PSOSA system. In this work, we consider normal distribution of mesh clients. The number of mesh routers is considered 16 and the number of mesh clients 48, respectively. The total number of iterations is considered 800 and the iterations per phase is

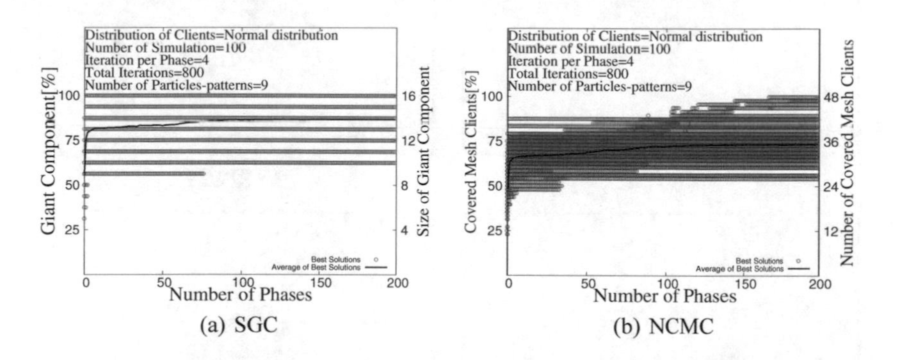

Fig. 5. Simulation results of WMN-PSOSA considering RIWM.

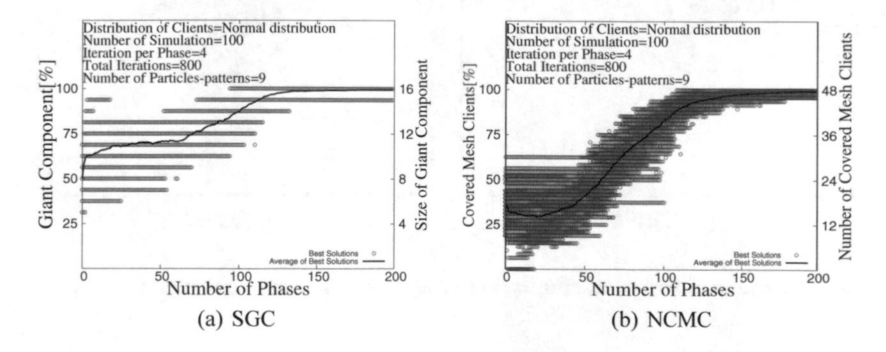

Fig. 6. Simulation results of WMN-PSOSA considering LDIWM.

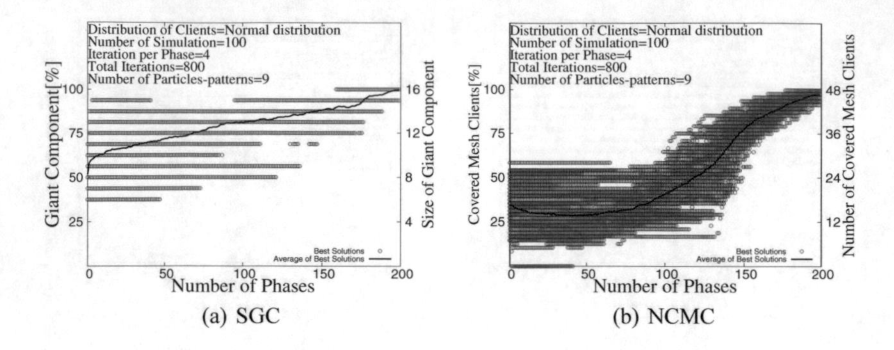

(a) SGC (b) NCMC

Fig. 7. Simulation results of WMN-PSOSA considering LDVM.

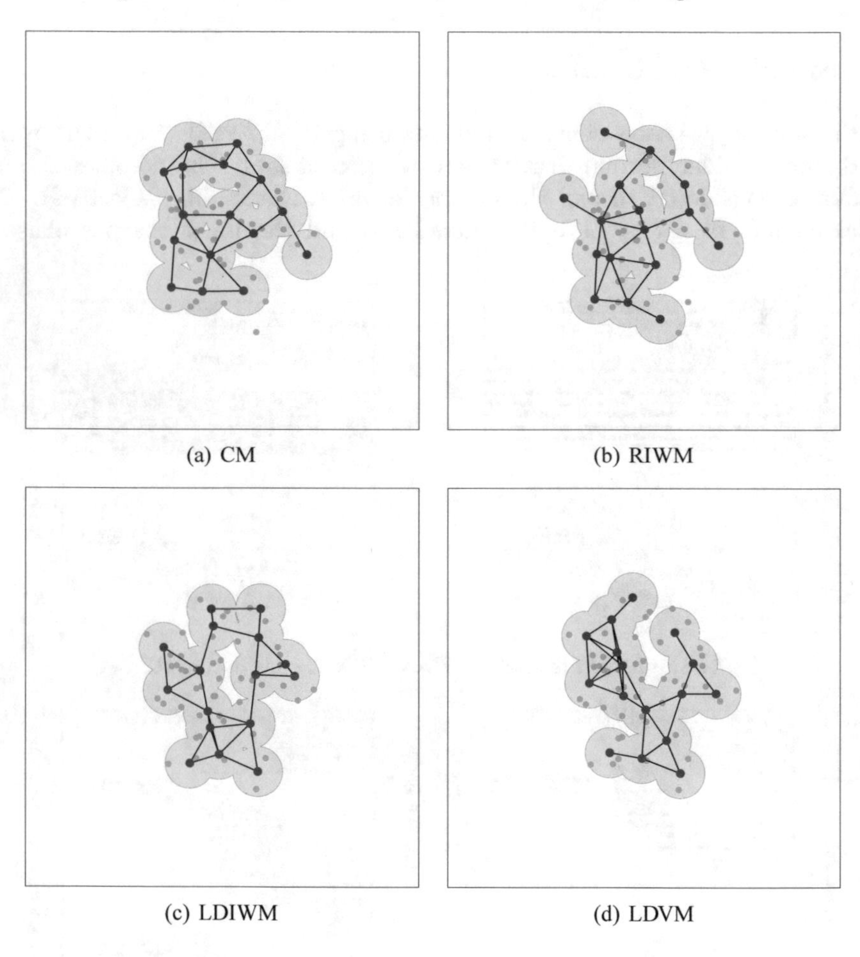

(a) CM (b) RIWM

(c) LDIWM (d) LDVM

Fig. 8. Visualized image of simulation results for different replacement methods.

Table 1. WMN-PSOSA parameters.

Parameters	Values
Clients distribution	Normal distribution
Area size	32.0 × 32.0
Number of mesh routers	16
Number of mesh clients	48
Total iterations	800
Iteration per phase	4
Number of particle-patterns	9
Radius of a mesh router	2.0
SA starting temperature value	10.0
SA ending temperature value	0.1
Temperature decreasing speed (α)	0.99426
Replacement method	CM, RIWM, LDIWM, LDVM

considered 4. We consider the number of particle-patterns 9. We conducted simulations 100 times, in order to avoid the effect of randomness and create a general view of results. We show the parameter setting for WMN-PSOSA in Table 1.

We show the simulation results from Figs. 4, 5, 6, 7 and 8. In Figs. 4 and 5, we show results when the replacement method is CM and RIWM. We see that the two methods are similar, also solutions converge before 100 phases and NCMC does not reach maximum (100%). We show the results of LDIWM and LDVM in Figs. 6 and 7. That results show that the SGC increases gradually and all solution reaches maximum value. In particular, LDIWM converges faster among them and the average of best solutions is the highest for both SGC and NCMC at the end of simulations (at 200 phase). We show the visualized results for WMN-PSOSA in Fig. 8. We see that all mesh nodes are connected when we use LDIWM or LDVM replacement methods. However, the LDIWM has better performance than other replacement methods.

5 Conclusions

In this work, we evaluated the performance of a hybrid simulation system based on PSO and SA (called WMN-PSOSA) considering different replacement methods. Simulation results show that the LDIWM converges faster and has better performance than CM, RIWM and LDVM replacement methods.

In our future work, we would like to evaluate the performance of the proposed system for different parameters and scenarios.

Acknowledgement. This work is supported by a Grant-in-Aid for Scientific Research from Japanese Society for the Promotion of Science (JSPS KAKENHI Grant Number 15J12086). The authors would like to thank JSPS for the financial support.

References

1. Akyildiz, I.F., Wang, X., Wang, W.: Wireless mesh networks: a survey. Comput. Netw. **47**(4), 445–487 (2005)
2. Amaldi, E., Capone, A., Cesana, M., Filippini, I., Malucelli, F.: Optimization models and methods for planning wireless mesh networks. Comput. Netw. **52**(11), 2159–2171 (2008)
3. Barolli, A., Spaho, E., Barolli, L., Xhafa, F., Takizawa, M.: QoS routing in ad-hoc networks using GA and multi-objective optimization. Mob. Inf. Syst. **7**(3), 169–188 (2011)
4. Behnamian, J., Ghomi, S.F.: Development of a PSO-SA hybrid metaheuristic for a new comprehensive regression model to time-series forecasting. Expert Syst. Appl. **37**(2), 974–984 (2010)
5. Clerc, M., Kennedy, J.: The particle swarm-explosion, stability, and convergence in a multidimensional complex space. IEEE Trans. Evol. Comput. **6**(1), 58–73 (2002)
6. Cunha, M.C., Sousa, J.: Water distribution network design optimization: simulated annealing approach. J. Water Resour. Plann. Manag. **125**(4), 215–221 (1999)
7. Del Valle, Y., Venayagamoorthy, G.K., Mohagheghi, S., Hernandez, J.C., Harley, R.G.: Particle swarm optimization: basic concepts, variants and applications in power systems. IEEE Trans. Evol. Comput. **12**(2), 171–195 (2008)
8. Franklin, A.A., Murthy, C.S.R.: Node placement algorithm for deployment of two-tier wireless mesh networks. In: Proceedings of Global Telecommunications Conference, pp. 4823–4827 (2007)
9. Ge, H., Du, W., Qian, F.: A hybrid algorithm based on particle swarm optimization and simulated annealing for job shop scheduling. In: Third International Conference on Natural Computation (ICNC 2007), vol. 3, pp. 715–719 (2007)
10. Girgis, M.R., Mahmoud, T.M., Abdullatif, B.A., Rabie, A.M.: Solving the wireless mesh network design problem using genetic algorithm and simulated annealing optimization methods. Int. J. Comput. Appl. **96**(11), 1–10 (2014)
11. Goto, K., Sasaki, Y., Hara, T., Nishio, S.: Data gathering using mobile agents for reducing traffic in dense mobile wireless sensor networks. Mob. Inf. Syst. **9**(4), 295–314 (2013)
12. Hwang, C.R.: Simulated annealing: theory and applications. Acta Applicandae Mathematicae **12**(1), 108–111 (1988)
13. Inaba, T., Elmazi, D., Sakamoto, S., Oda, T., Ikeda, M., Barolli, L.: A secure-aware call admission control scheme for wireless cellular networks using fuzzy logic and its performance evaluation. J. Mob. Multimedia **11**(3–4), 213–222 (2015)
14. Inaba, T., Obukata, R., Sakamoto, S., Oda, T., Ikeda, M., Barolli, L.: Performance evaluation of a QoS-aware fuzzy-based CAC for LAN access. Int. J. Space Based Situated Comput. **6**(4), 228–238 (2016)
15. Inaba, T., Sakamoto, S., Oda, T., Ikeda, M., Barolli, L.: A testbed for admission control in WLAN: a fuzzy approach and its performance evaluation. In: International Conference on Broadband and Wireless Computing, Communication and Applications, pp. 559–571. Springer (2016)
16. Lim, A., Rodrigues, B., Wang, F., Xu, Z.: k-center problems with minimum coverage. In: Computing and Combinatorics, pp. 349–359 (2004)
17. Maolin, T., et al.: Gateways placement in backbone wireless mesh networks. Int. J. Commun. Netw. Syst. Sci. **2**(1), 44 (2009)
18. Muthaiah, S.N., Rosenberg, C.P.: Single gateway placement in wireless mesh networks. In: Proceedings of 8th International IEEE Symposium on Computer Networks, pp. 4754–4759 (2008)

19. Naka, S., Genji, T., Yura, T., Fukuyama, Y.: A hybrid particle swarm optimization for distribution state estimation. IEEE Trans. Power Syst. **18**(1), 60–68 (2003)
20. Poli, R., Kennedy, J., Blackwell, T.: Particle swarm optimization. Swarm Intell. **1**(1), 33–57 (2007)
21. Sakamoto, S., Kulla, E., Oda, T., Ikeda, M., Barolli, L., Xhafa, F.: A comparison study of simulated annealing and genetic algorithm for node placement problem in wireless mesh networks. J. Mob. Multimedia **9**(1–2), 101–110 (2013)
22. Sakamoto, S., Kulla, E., Oda, T., Ikeda, M., Barolli, L., Xhafa, F.: A comparison study of hill climbing, simulated annealing and genetic algorithm for node placement problem in WMNs. J. High Speed Netw. **20**(1), 55–66 (2014)
23. Sakamoto, S., Kulla, E., Oda, T., Ikeda, M., Barolli, L., Xhafa, F.: A simulation system for WMN based on SA: performance evaluation for different instances and starting temperature values. Int. J. Space Based Situated Comput. **4**(3–4), 209–216 (2014)
24. Sakamoto, S., Kulla, E., Oda, T., Ikeda, M., Barolli, L., Xhafa, F.: Performance evaluation considering iterations per phase and SA temperature in WMN-SA system. Mob. Inf. Syst. **10**(3), 321–330 (2014)
25. Sakamoto, S., Lala, A., Oda, T., Kolici, V., Barolli, L., Xhafa, F.: Application of WMN-SA simulation system for node placement in wireless mesh networks: a case study for a realistic scenario. Int. J. Mob. Comput. Multimedia Commun. (IJMCMC) **6**(2), 13–21 (2014)
26. Sakamoto, S., Oda, T., Ikeda, M., Barolli, L., Xhafa, F.: An integrated simulation system considering WMN-PSO simulation system and network simulator 3. In: International Conference on Broadband and Wireless Computing, Communication and Applications, pp. 187–198. Springer (2016)
27. Sakamoto, S., Oda, T., Ikeda, M., Barolli, L., Xhafa, F.: Implementation and evaluation of a simulation system based on particle swarm optimisation for node placement problem in wireless mesh networks. Int. J. Commun. Netw. Distrib. Syst. **17**(1), 1–13 (2016)
28. Sakamoto, S., Oda, T., Ikeda, M., Barolli, L., Xhafa, F.: Implementation of a new replacement method in WMN-PSO simulation system and its performance evaluation. In: The 30th IEEE International Conference on Advanced Information Networking and Applications (AINA 2016), pp. 206–211 (2016). https://doi.org/10.1109/AINA.2016.42
29. Sakamoto, S., Obukata, R., Oda, T., Barolli, L., Ikeda, M., Barolli, A.: Performance analysis of two wireless mesh network architectures by WMNSA and WMN-TS simulation systems. J. High Speed Netw. **23**(4), 311–322 (2017)
30. Sakamoto, S., Ozera, K., Barolli, A., Ikeda, M., Barolli, L., Takizawa, M.: Implementation of an intelligent hybrid simulation systems for WMNs based on particle swarm optimization and simulated annealing: performance evaluation for different replacement methods. Soft Comput. (2017). http://doi.org/10.1007/s00500-017-2948-1. Accessed 11 Dec 2017
31. Sakamoto, S., Ozera, K., Ikeda, M., Barolli, L.: Implementation of intelligent hybrid systems for node placement problem in WMNs considering particle swarm optimization, hill climbing and simulated annealing. Mob. Netw. Appl. (2017). http://doi.org/10.1007/s11036-017-0897-7. Accessed 06 Sep 2017
32. Schutte, J.F., Groenwold, A.A.: A study of global optimization using particle swarms. J. Global Optim. **31**(1), 93–108 (2005)
33. Shi, Y.: Particle swarm optimization. IEEE Connections **2**(1), 8–13 (2004)
34. Shi, Y., Eberhart, R.C.: Parameter selection in particle swarm optimization. In: Evolutionary Programming VII, pp. 591–600 (1998)

35. Vanhatupa, T., Hannikainen, M., Hamalainen, T.: Genetic algorithm to optimize node placement and configuration for WLAN planning. In: Proceedings of 4th IEEE International Symposium on Wireless Communication Systems, pp. 612–616 (2007)
36. Wang, J., Xie, B., Cai, K., Agrawal, D.P.: Efficient mesh router placement in wireless mesh networks. In: Proceedings of IEEE International Conference on Mobile Adhoc and Sensor Systems (MASS 2007), pp. 1–9 (2007)
37. Xhafa, F., Sanchez, C., Barolli, L.: Ad hoc and neighborhood search methods for placement of mesh routers in wireless mesh networks. In: Proceedings of 29th IEEE International Conference on Distributed Computing Systems Workshops (ICDCS 2009), pp. 400–405 (2009)

Improving Team Collaboration in MobilePeerDroid Mobile System: A Fuzzy-Based Approach Considering Four Input Parameters

Yi Liu[1(✉)], Kosuke Ozera[1], Keita Matsuo[2], Makoto Ikeda[2], Leonard Barolli[2], and Vladi Kolici[3]

[1] Graduate School of Engineering, Fukuoka Institute of Technology (FIT), 3-30-1 Wajiro-Higashi, Higashi-Ku, Fukuoka 811-0295, Japan
ryuui1010@gmail.com, kosuke.o.fit@gmail.com

[2] Department of Information and Communication Engineering, Fukuoka Institute of Technology (FIT), 3-30-1 Wajiro-Higashi, Higashi-Ku, Fukuoka 811-0295, Japan
kt-matsuo@fit.ac.jp, makoto.ikd@acm.org, barolli@fit.ac.jp

[3] Faculty of Information Technology, Polytechnic University of Tirana, Mother Theresa Square, No. 4, Tirana, Albania
vkolici@fti.edu.al

Abstract. In this work, we present a distributed event-based awareness approach for P2P groupware systems. Unlike centralized approaches, several issues arise and need to be addressed for awareness in P2P groupware systems, due to their large-scale, dynamic and heterogenous nature. In our approach, the awareness of collaboration will be achieved by using primitive operations and services that are integrated into the P2P middleware. We propose an abstract model for achieving these requirements and we discuss how this model can support awareness of collaboration in mobile teams. We present a fuzzy-based model, in which every member has the task accomplishment according to four parameters: state of workflow, number of exchanged messages, available resources and sustained communication time. This model will be implemented in MobilePeerDroid system to give more realistic view of the collaborative activity and better decisions for the groupwork, while encouraging peers to increase their reliability in order to support awareness of collaboration in MobilePeerDroid Mobile System.

1 Introduction

Peer to Peer technologies has been among most disruptive technologies after Internet. Indeed, the emergence of the P2P technologies changed drastically the concepts, paradigms and protocols of sharing and communication in large scale distributed systems. As pointed out since early 2000 years [1–5], the nature of the sharing and the direct communication among peers in the system, being these

L. Barolli et al. (Eds.): EIDWT 2018, LNDECT 17, pp. 65–78, 2018.
https://doi.org/10.1007/978-3-319-75928-9_6

machines or people, makes possible to overcome the limitations of the flat communications through email, newsgroups and other forum-based communication forms.

The usefulness of P2P technologies on one hand has been shown for the development of stand alone applications. On the other hand, P2P technologies, paradigms and protocols have penetrated other large scale distributed systems such as Mobile Ad hoc Networks (MANETs), Groupware systems, Mobile Systems to achieve efficient sharing, communication, coordination, replication, awareness and synchronization. In fact, for every new form of Internet-based distributed systems, we are seeing how P2P concepts and paradigms again play an important role to enhance the efficiency and effectiveness of such systems or to enhance information sharing and online collaborative activities of groups of people. We briefly introduce below some common application scenarios that can benefit from P2P communications.

Awareness is a key feature of groupware systems. In its simplest terms, awareness can be defined as the system's ability to notify the members of a group of changes occurring in the group's workspace. Awareness systems for online collaborative work have been proposed since in early stages of Web technology. Such proposals started by approaching workspace awareness, aiming to inform users about changes occurring in the shared workspace. More recently, research has focussed on using new paradigms, such as P2P systems, to achieve fully decentralized, ubiquitous groupware systems and awareness in such systems. In P2P groupware systems group processes may be more efficient because peers can be aware of the status of other peers in the group, and can interact directly and share resources with peers in order to provide additional scaffolding or social support. Moreover, P2P systems are pervasive and ubiquitous in nature, thus enabling contextualized awareness.

Fuzzy Logic (FL) is the logic underlying modes of reasoning which are approximate rather then exact. The importance of FL derives from the fact that most modes of human reasoning and especially common sense reasoning are approximate in nature [6]. FL uses linguistic variables to describe the control parameters. By using relatively simple linguistic expressions it is possible to describe and grasp very complex problems. A very important property of the linguistic variables is the capability of describing imprecise parameters.

The concept of a fuzzy set deals with the representation of classes whose boundaries are not determined. It uses a characteristic function, taking values usually in the interval [0, 1]. The fuzzy sets are used for representing linguistic labels. This can be viewed as expressing an uncertainty about the clear-cut meaning of the label. But important point is that the valuation set is supposed to be common to the various linguistic labels that are involved in the given problem.

The fuzzy set theory uses the membership function to encode a preference among the possible interpretations of the corresponding label. A fuzzy set can be defined by examplification, ranking elements according to their typicality with respect to the concept underlying the fuzzy set [7].

In this paper, we propose a fuzzy-based system for task accomplishment in MobilePeerDroid system considering four parameters: State of Workflow (SW), Number of Exchanged Messages (NEM), Available Resources (AR) and Sustained Communication Time (SCT) to decide the Prediction of Task Accomplishment (PTA). We evaluated the proposed system by simulations. The simulation results show that with increasing of SW, NEM, AR and SCT, the PTA is increasing.

The structure of this paper is as follows. In Sect. 2, we introduce the scenarios of collaborative teamwork. In Sect. 3, we introduce the group activity awareness model. In Sect. 4, we introduce FL used for control. In Sect. 5, we present the proposed fuzzy-based system. In Sect. 6, we discuss the simulation results. Finally, conclusions and future work are given in Sect. 7.

2 Scenarios of Collaborative Teamwork

In this section, we describe and analyse some main scenarios of collaborative teamwork for which P2P technologies can support efficient system design.

2.1 Collaborative Teamwork and Virtual Campuses

Collaborative work through virtual teams is a significant way of collaborating in modern businesses and online learning systems [8]. Collaboration in virtual teams requires efficient sharing of information (both data sharing among the group members as well as sharing of group processes) and efficient communication among members of the team. Additionally, coordination and interaction are crucial for accomplishing common tasks through a shared workspace environment. P2P systems can enable fully decentralized collaborative systems by efficiently supporting different forms of collaboration [9]. One such form is using P2P networks, with super-peer structure as show in Fig. 1.

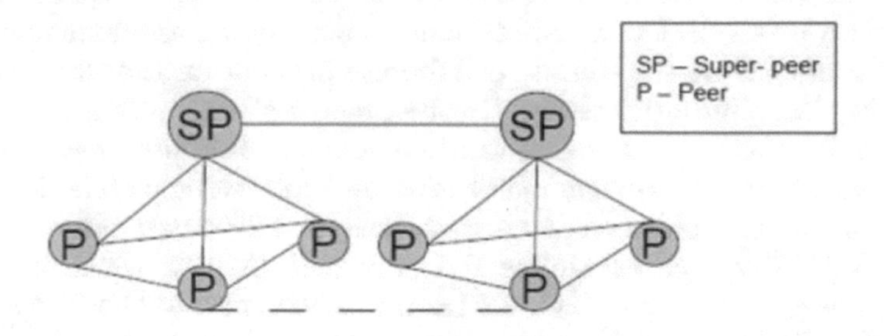

Fig. 1. Super-peer P2P group network.

During the last two decades, online learning has become very popular and there is a widespread of virtual campuses or combinations of face-to-face with semi-open teaching and learning. Virtual campuses are now looking at ways to effectively support learners, especially for online courses implemented as PBL-Project Based Learning or SBL Scenario Based Learning there is an increasing need to develop mobile applications that support these online groupwork learning paradigms [10]. In such setting, P2P technologies offer interesting solutions for (a) decentralizing the virtual campuses, which tend to grow and get further centralized with the increase of number of students enrolled, new degrees, and increase in academic activity; (b) in taking advantage of resources of students and developing volunteerbased computing systems as part of virtual campuses and (c) alleviating the communication burden for efficient collaborative teamwork. The use of P2P libraries such as JXTA have been investigated to design P2P middleware for P2P eLearning applications. Also, the use of P2P technologies in such setting is used for P2P video synchronization in a collaborative virtual environment [11]. Recently, virtual campuses are also introducing social networking among their students to enhance the learning activities through social support and scaffolding. Again the P2P solutions are sought in this context [12] in combination with social networking features to enhance especially the interaction among learners sharing similar objectives and interest or accomplishing a common project.

2.2 Mobile Ad hoc Networks (MANETs)

Mobile ad-hoc networks are among most interesting infrastructureless network of mobile devices connected by wireless having self-configuring properties [13]. The lack of fixed infrastructure and of a centralized administration makes the building and operation in MANETS challenging. P2P networks and mobile ad hoc networks (MANETs) follow the same idea of creating a network without a central entity. All nodes (peers) must collaborate together to make possible the proper functioning of the network by forwarding information on behalf of others in the network [14]. P2P and MANETs share many key characteristics such as self-organization and decentralization due to the common nature of their distributed components. Both MANETs and P2P networks follow a P2P paradigm characterized by the lack of a central node or peer acting as a managing server, all participants having therefore to collaborate in order for the whole system to work. A key issue in both networks is the process of discovering the requested data or route efficiently in a decentralized manner. Recently, new P2P applications which uses wireless communication and integrates mobile devices such as PDA and mobile phones is emerging. Several P2P-based protocols can be used for MANETs such as Mobile P2P Protocol (MPP), which is based on Dynamic Source Routing (DSR), JXTA prtotocols, and MANET Anonymous Peer-to-peer Communication Protocol (MAPCP), which serves as an efficient anonymous communication protocol for P2P applications over MANET.

3 Group Activity Awareness Model

The awareness model considered here focuses on supporting group activities so to accomplish a common group project, although it can also be used in a broader scope of teamwork. The main building blocks of our model (see also [15,16] in the context of web-based groupware) are described below.

Activity awareness: Activity awareness refers to awareness information about the project-related activities of group members. Project-based work is one of the most common methods of group working. Activity awareness aims to provide information about progress on the accomplishment of tasks by both individuals and the group as a whole. It comprises knowing about actions taken by members of the group according to the project schedule, and synchronization of activities with the project schedule. Activity awareness should therefore enable members to know about recent and past actions on the project's work by the group. As part of activity awareness, we also consider information on group artifacts such as documents and actions upon them (uploads, downloads, modifications, reading). Activity awareness is one of most important, and most complex, types of awareness. As well as the direct link to monitoring a group's progress on the work relating to a project, it also supports group communication and coordination processes.

Process awareness: In project-based work, a project typically requires the enactment of a workflow. In such a case, the objective of the awareness is to track the state of the workflow and to inform users accordingly. We term this process awareness. The workflow is defined through a set of tasks and precedence relationships relating to their order of completion. Process awareness targets the information flow of the project, providing individuals and the group with a partial view (what they are each doing individually) and a complete view (what they are doing as a group), thus enabling the identification of past, current and next states of the workflow in order to move the collaboration process forward.

Communication awareness: Another type of awareness considered in this work is that of communication awareness. We consider awareness information relating to message exchange, and synchronous and asynchronous discussion forums. The first is intended to support awareness of peer-to-peer communication (when some peer wants to establish a direct communication with another peer); the second is aimed at supporting awareness about chat room creation and lifetime (so that other peers can be aware of, and possibly eventually join, the chat room); the third refers to awareness of new messages posted at the discussion forum, replies, etc.

Availability awareness: Availability awareness is useful for provide individuals and the group with information on members' and resources' availability. The former is necessary for establishing synchronous collaboration either in peer-to-peer mode or (sub)group mode. The later is useful for supporting members' tasks requiring available resources (e.g. a machine for running a software program). Groupware applications usually monitor availability of group members by simply looking at group workspaces. However, availability awareness encompasses not only knowing who is in the workspace at any given moment but also who is available when, via members' profiles (which include also personal calendars) and information explicitly provided by members. In the case of resources, awareness is achieved via the schedules of resources. Thus, both explicit and implicit forms of gathering availability awareness information should be supported.

4 Application of Fuzzy Logic for Control

The ability of fuzzy sets and possibility theory to model gradual properties or soft constraints whose satisfaction is matter of degree, as well as information pervaded with imprecision and uncertainty, makes them useful in a great variety of applications.

The most popular area of application is Fuzzy Control (FC), since the appearance, especially in Japan, of industrial applications in domestic appliances, process control, and automotive systems, among many other fields.

4.1 FC

In the FC systems, expert knowledge is encoded in the form of fuzzy rules, which describe recommended actions for different classes of situations represented by fuzzy sets.

In fact, any kind of control law can be modeled by the FC methodology, provided that this law is expressible in terms of "if ... then ..." rules, just like in the case of expert systems. However, FL diverges from the standard expert system approach by providing an interpolation mechanism from several rules. In the contents of complex processes, it may turn out to be more practical to get knowledge from an expert operator than to calculate an optimal control, due to modeling costs or because a model is out of reach.

4.2 Linguistic Variables

A concept that plays a central role in the application of FL is that of a linguistic variable. The linguistic variables may be viewed as a form of data compression. One linguistic variable may represent many numerical variables. It is suggestive to refer to this form of data compression as granulation [17].

The same effect can be achieved by conventional quantization, but in the case of quantization, the values are intervals, whereas in the case of granulation the values are overlapping fuzzy sets. The advantages of granulation over quantization are as follows:

- it is more general;
- it mimics the way in which humans interpret linguistic values;
- the transition from one linguistic value to a contiguous linguistic value is gradual rather than abrupt, resulting in continuity and robustness.

4.3 FC Rules

FC describes the algorithm for process control as a fuzzy relation between information about the conditions of the process to be controlled, x and y, and the output for the process z. The control algorithm is given in "if ... then ..." expression, such as:

If x is small and y is big, then z is medium;
If x is big and y is medium, then z is big.

These rules are called *FC rules.* The "if" clause of the rules is called the antecedent and the "then" clause is called consequent. In general, variables x and y are called the input and z the output. The "small" and "big" are fuzzy values for x and y, and they are expressed by fuzzy sets.

Fuzzy controllers are constructed of groups of these FC rules, and when an actual input is given, the output is calculated by means of fuzzy inference.

4.4 Control Knowledge Base

There are two main tasks in designing the control knowledge base. First, a set of linguistic variables must be selected which describe the values of the main control parameters of the process. Both the input and output parameters must be linguistically defined in this stage using proper term sets. The selection of the level of granularity of a term set for an input variable or an output variable plays an important role in the smoothness of control. Second, a control knowledge base must be developed which uses the above linguistic description of the input and output parameters. Four methods [18–21] have been suggested for doing this:

- expert's experience and knowledge;
- modelling the operator's control action;
- modelling a process;
- self organization.

Among the above methods, the first one is the most widely used. In the modeling of the human expert operator's knowledge, fuzzy rules of the form "If Error is small and Change-in-error is small then the Force is small" have been used in several studies [22, 23]. This method is effective when expert human operators can express the heuristics or the knowledge that they use in controlling a process in terms of rules of the above form.

4.5 Defuzzification Methods

The defuzzification operation produces a non-FC action that best represent the membership function of an inferred FC action. Several defuzzification methods have been suggested in literature. Among them, four methods which have been applied most often are:

- Tsukamoto's Defuzzification Method;
- The Center of Area (COA) Method;
- The Mean of Maximum (MOM) Method;
- Defuzzification when Output of Rules are Function of Their Inputs.

5 Proposed Fuzzy-Based System for Peer Task Accomplishment

The P2P group-based model considered is that of a superpeer model. In this model, the P2P network is fragmented into several disjoint peergroups (see Fig. 2). The peers of each peergroup are connected to a single superpeer. There is frequent local communication between peers in a peergroup, and less frequent global communication between superpeers.

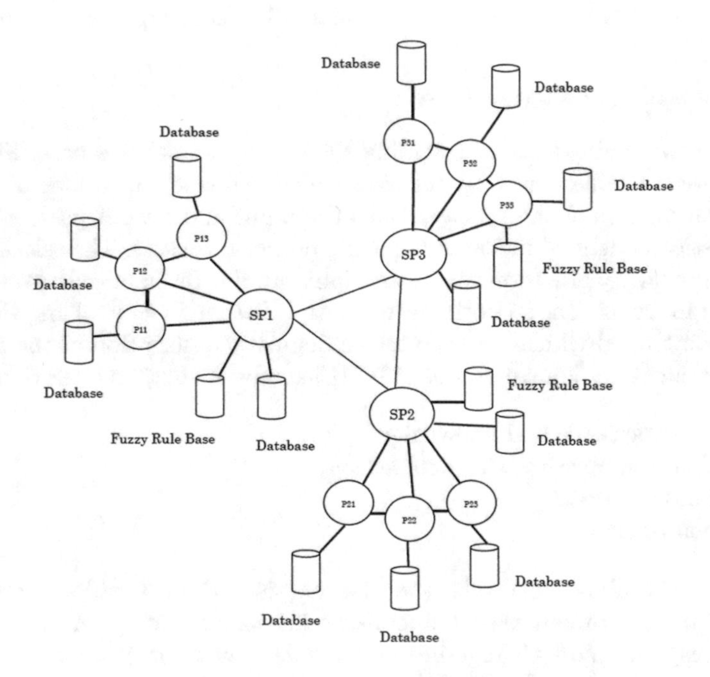

Fig. 2. P2P group-based model.

To complete a certain task in P2P mobile collaborative team work, peers often have to interact with unknown peers. Thus, it is important that group members must select reliable peers to interact.

In this work, we consider four parameters: State of Workflow (SW), Number of Exchanged Messages (NEM), Available Resources (AR), Sustained Communication Time (SCT) to decide the Prediction of Task Accomplishment (PTA). The structure of this system called Fuzzy-based Task Accomplishment System (FTAS) is shown in Fig. 3. These four parameters are fuzzified using fuzzy system and based on the decision of fuzzy system a task accomplishment is calculated. The membership functions for our system are shown in Fig. 4. In Table 1, we show the Fuzzy Rule Base (FRB) of our proposed system, which consists of 108 rules.

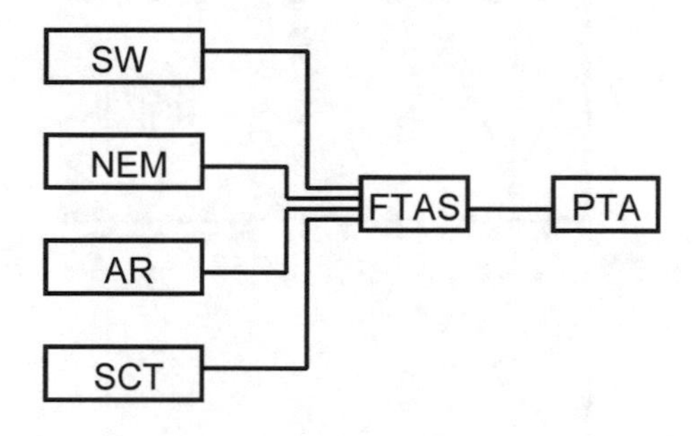

Fig. 3. Proposed of structure.

The input parameters for FTAS are: SW, NEM, AR and SCT. The output linguistic parameter is PTA. The term sets of *SW*, *NEM*, *AR* and *SCT* are defined respectively as:

$$\begin{aligned}
SW &= \{Slow\ Progress,\ Normal,\ Fast\ Progress\} \\
&= \{SP,\ NL,\ FP\}; \\
NEM &= \{Very\ Few,\ Few,\ Middle,\ Many\} \\
&= \{Vf,\ Fe,\ Mi,\ Ma\}; \\
AR &= \{Few,\ Average,\ Many\} \\
&= \{F,\ A,\ M\}; \\
SCT &= \{Short,\ Middle,\ Long\} \\
&= \{S,\ M,\ L\}
\end{aligned}$$

and the term set for the output *PTA* is defined as:

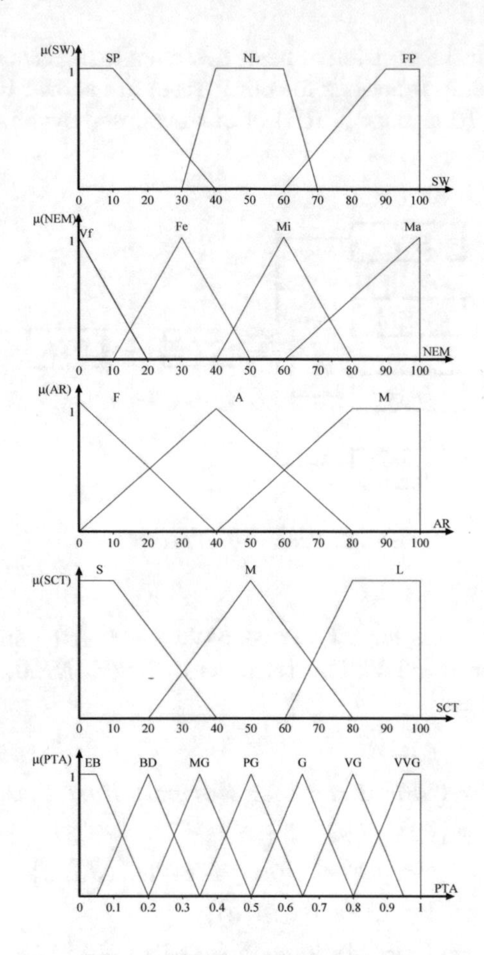

Fig. 4. Membership functions.

$$PTA = \begin{pmatrix} Extremely\ Bad \\ Bad \\ Minimally\ Good \\ Partially\ Good \\ Good \\ Very\ Good \\ Very\ Very\ Good \end{pmatrix} = \begin{pmatrix} EB \\ BD \\ MG \\ PG \\ G \\ VG \\ VVG \end{pmatrix}$$

6 Simulation Results

In this section, we present the simulation results for our proposed system. In our system, we decided the number of term sets by carrying out many simulations.

From Figs. 5, 6, and 7, we show the relation between SW, NEM, AR, SCT and PTA. In this simulation, we consider the SW and SCT as constant parameters.

In Fig. 5, we consider the SCT value 10 units. We change the SW value from 10 to 90 units. When the SW increases, the PTA is increased. Also, when the NEM and AR increase, the PTA is increased.

Table 1. FRB.

Rule	SW	NEM	AR	SCT	PTA	Rule	SW	NEM	AR	SCT	PTA	Rule	SW	NEM	AR	SCT	PTA
1	Sl	Vf	F	S	EB	37	Mi	Vf	F	S	EB	73	Fa	Vf	F	S	EB
2	Sl	Vf	F	M	EB	38	Mi	Vf	F	M	EB	74	Fa	Vf	F	M	BD
3	Sl	Vf	F	L	EB	39	Mi	Vf	F	L	BD	75	Fa	Vf	F	L	MG
4	Sl	Vf	A	S	EB	40	Mi	Vf	A	S	EB	76	Fa	Vf	A	S	BD
5	Sl	Vf	A	M	EB	41	Mi	Vf	A	M	BD	77	Fa	Vf	A	M	PG
6	Sl	Vf	A	L	BD	42	Mi	Vf	A	L	PG	78	Fa	Vf	A	L	G
7	Sl	Vf	B	S	EB	43	Mi	Vf	B	S	MG	79	Fa	Vf	B	S	PG
8	Sl	Vf	B	M	MG	44	Mi	Vf	B	M	PG	80	Fa	Vf	B	M	G
9	Sl	Vf	B	L	PG	45	Mi	Vf	B	L	G	81	Fa	Vf	B	L	VG
10	Sl	Fe	F	S	EB	46	Mi	Fe	F	S	EB	82	Fa	Fe	F	S	BD
11	Sl	Fe	F	M	EB	47	Mi	Fe	F	M	BD	83	Fa	Fe	F	M	MG
12	Sl	Fe	F	L	BD	48	Mi	Fe	F	L	MG	84	Fa	Fe	F	L	PG
13	Sl	Fe	A	S	EB	49	Mi	Fe	A	S	BD	85	Fa	Fe	A	S	MG
14	Sl	Fe	A	M	BD	50	Mi	Fe	A	M	MG	86	Fa	Fe	A	M	G
15	Sl	Fe	A	L	MG	51	Mi	Fe	A	L	G	87	Fa	Fe	A	L	VG
16	Sl	Fe	B	S	BD	52	Mi	Fe	B	S	PG	88	Fa	Fe	B	S	G
17	Sl	Fe	B	M	PG	53	Mi	Fe	B	M	G	89	Fa	Fe	B	M	VG
18	Sl	Fe	B	L	G	54	Mi	Fe	B	L	VG	90	Fa	Fe	B	L	VVG
19	Sl	Mi	F	S	EB	55	Mi	Mi	F	S	EB	91	Fa	Mi	F	S	MG
20	Sl	Mi	F	M	EB	56	Mi	Mi	F	M	MG	92	Fa	Mi	F	M	PG
21	Sl	Mi	F	L	MG	57	Mi	Mi	F	L	PG	93	Fa	Mi	F	L	G
22	Sl	Mi	A	S	BD	58	Mi	Mi	A	S	MG	94	Fa	Mi	A	S	PG
23	Sl	Mi	A	M	MG	59	Mi	Mi	A	M	PG	95	Fa	Mi	A	M	VG
24	Sl	Mi	A	L	PG	60	Mi	Mi	A	L	VG	96	Fa	Mi	A	L	VG
25	Sl	Mi	B	S	MG	61	Mi	Mi	B	S	G	97	Fa	Mi	B	S	VG
26	Sl	Mi	B	M	G	62	Mi	Mi	B	M	VG	98	Fa	Mi	B	M	VVG
27	Sl	Mi	B	L	VG	63	Mi	Mi	B	L	VVG	99	Fa	Mi	B	L	VVG
28	Sl	Ma	F	S	EB	64	Mi	Ma	F	S	BD	100	Fa	Ma	F	S	PG
29	Sl	Ma	F	M	BD	65	Mi	Ma	F	M	PG	101	Fa	Ma	F	M	G
30	Sl	Ma	F	L	PG	66	Mi	Ma	F	L	G	102	Fa	Ma	F	L	VG
31	Sl	Ma	A	S	MG	67	Mi	Ma	A	S	PG	103	Fa	Ma	A	S	G
32	Sl	Ma	A	M	PG	68	Mi	Ma	A	M	G	104	Fa	Ma	A	M	VG
33	Sl	Ma	A	L	G	69	Mi	Ma	A	L	VG	105	Fa	Ma	A	L	VVG
34	Sl	Ma	B	S	PG	70	Mi	Ma	B	S	VG	106	Fa	Ma	B	S	VG
35	Sl	Ma	B	M	VG	71	Mi	Ma	B	M	VG	107	Fa	Ma	B	M	VVG
36	Sl	Ma	B	L	VG	72	Mi	Ma	B	L	VVG	108	Fa	Ma	B	L	VVG

In Figs. 6 and 7, we increase the SCT values to 50 and 90 units, respectively. We see that, when the SCT increases, the PTA is increased.

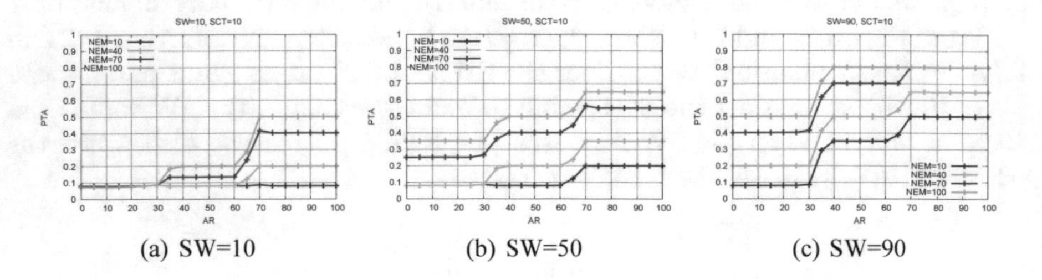

(a) SW=10 (b) SW=50 (c) SW=90

Fig. 5. Relation of PTA with AR and NEM for different SW when SCT = 10.

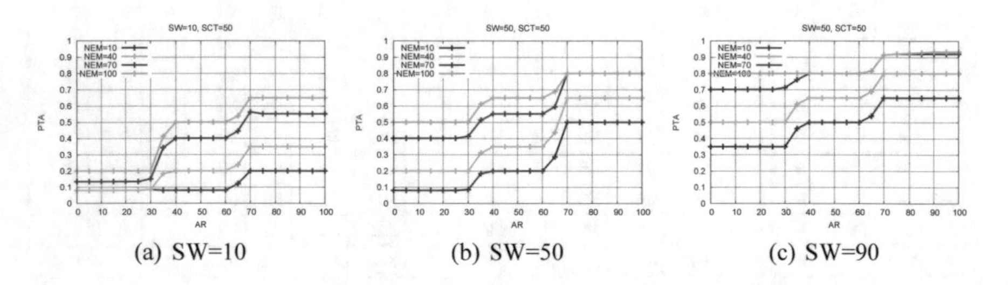

(a) SW=10 (b) SW=50 (c) SW=90

Fig. 6. Relation of PTA with AR and NEM for different SW when SCT = 50.

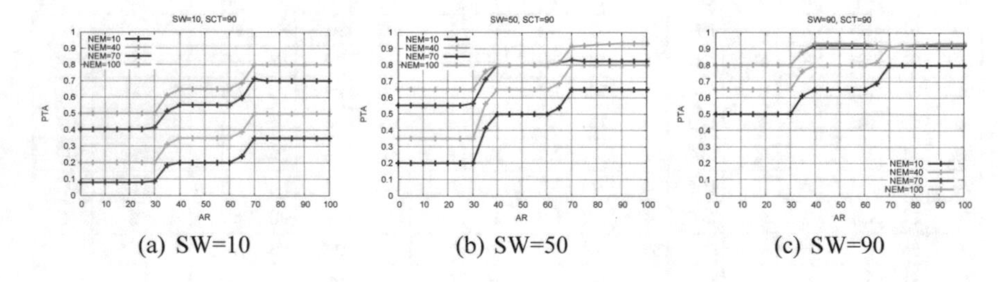

(a) SW=10 (b) SW=50 (c) SW=90

Fig. 7. Relation of PTA with AR and NEM for different SW when SCT = 90.

7 Conclusions and Future Work

In this paper, we proposed a fuzzy-based system to decide the PTA. We took into consideration four parameters: SW, NEM, AR and SCT. We evaluated the performance of proposed system by computer simulations. From the simulations results, we found that when SW, NEM and AR values are increased, the prediction of task accomplishment value is increased.

In the future, we would like to make extensive simulations to evaluate the proposed systems and compare the performance with other systems.

References

1. Oram, A. (ed.): Peer-to-Peer: Harnessing the Power of Disruptive Technologies. O'Reilly and Associates, Sebastopol (2001)
2. Sula, A., Spaho, E., Matsuo, K., Barolli, L., Xhafa, F., Miho, R.: A new system for supporting children with autism spectrum disorder based on IoT and P2P technology. Int. J. Space-Based Situated Comput. **4**(1), 55–64 (2014)
3. Di Stefano, A., Morana, G., Zito, D.: QoS-aware services composition in P2PGrid environments. Int. J. Grid Util. Comput. **2**(2), 139–147 (2011)
4. Sawamura, S., Barolli, A., Aikebaier, A., Takizawa, M., Enokido, T.: Design and evaluation of algorithms for obtaining objective trustworthiness on acquaintances in P2P overlay networks. Int. J. Grid Util. Comput. **2**(3), 196–203 (2011)
5. Higashino, M., Hayakawa, T., Takahashi, K., Kawamura, T., Sugahara, K.: Management of streaming multimedia content using mobile agent technology on pure P2P-based distributed e-learning system. Int. J. Grid Util. Comput. **5**(3), 198–204 (2014)
6. Inaba, T., Obukata, R., Sakamoto, S., Oda, T., Ikeda, M., Barolli, L.: Performance evaluation of a QoS-aware fuzzy-based CAC for LAN access. Int. J. Space-Based Situated Comput. **6**(4), 228–238 (2016)
7. Terano, T., Asai, K., Sugeno, M.: Fuzzy Systems Theory and Its Applications. Academic Press, Inc., Harcourt Brace Jovanovich, Publishers, SanDiego (1992)
8. Mori, T., Nakashima, M., Ito, T.: SpACCE: a sophisticated ad hoc cloud computing environment built by server migration to facilitate distributed collaboration. Int. J. Space-Based Situated Comput. **2**(4), 230–239 (2012)
9. Xhafa, F., Poulovassilis, A.: requirements for distributed event-based awareness in P2P groupware systems. In: Proceedings of AINA 2010, pp. 220–225 (2010)
10. Xhafa, F., Barolli, L., Caballé, S., Fernandez, R.: Supporting scenario-based online learning with P2P group-based systems. In: Proceedings of NBiS 2010, pp. 173–180 (2010)
11. Gupta, S., Kaiser, G.: P2P video synchronization in a collaborative virtual environment. In: Proceedings of the 4th International Conference on Advances in Web-Based Learning (ICWL 2005), pp. 86–98 (2005)
12. Martnez-Alemn, A.M., Wartman, K.L.: Online Social Networking on Campus Understanding What Matters in Student Culture. Taylor and Francis, Routledge (2008)
13. Puzar, M., Plagemann, T.: Data sharing in mobile ad-hoc networks - a study of replication and performance in the MIDAS data space. Int. J. Space-Based Situated Comput. **1**(2–3), 137–150 (2011)
14. Spaho, E., Kulla, E., Xhafa, F., Barolli, L.: P2P solutions to efficient mobile peer collaboration in MANETs. In: Proceedings of 3PGCIC 2012, pp. 379–383, November 2012
15. Gutwin, C., Greenberg, S., Roseman, M.: Workspace awareness in real time distributed groupware: framework, widgets, and evaluation. In: BCS HCI 1996, pp. 281–298 (1996)

16. You, Y., Pekkola, S.: Meeting others - supporting situation awareness on the WWW. Decis. Support Syst. **32**(1), 71–82 (2001)
17. Kandel, A.: Fuzzy Expert Systems. CRC Press, Boca Raton (1992)
18. Zimmermann, H.J.: Fuzzy Set Theory and Its Applications. Kluwer Academic Publishers, Boston (1991). Second Revised Edition
19. McNeill, F.M., Thro, E.: Fuzzy Logic: A Practical Approach. Academic Press Inc., San Diego (1994)
20. Zadeh, L.A., Kacprzyk, J.: Fuzzy Logic For The Management of Uncertainty. Wiley, New York (1992)
21. Procyk, T.J., Mamdani, E.H.: A linguistic self-organizing process controller. Automatica **15**(1), 15–30 (1979)
22. Klir, G.J., Folger, T.A.: Fuzzy Sets, Uncertainty, and Information. Prentice Hall, Englewood Cliffs (1988)
23. Munakata, T., Jani, Y.: Fuzzy systems: an overview. Commun. ACM **37**(3), 69–76 (1994)

Design and Implementation of a Hybrid Intelligent System Based on Particle Swarm Optimization and Distributed Genetic Algorithm

Admir Barolli[1](✉), Shinji Sakamoto[2], Kosuke Ozera[2], Leonard Barolli[3], Elis Kulla[4], and Makoto Takizawa[5]

[1] Department of Information Technology, Aleksander Moisiu University of Durres, L.1, Rruga e Currilave, Durres, Albania
admir.barolli@gmail.com

[2] Graduate School of Engineering, Fukuoka Institute of Technology, 3-30-1 Wajiro-Higashi, Higashi-Ku, Fukuoka 811-0295, Japan
shinji.sakamoto@ieee.org, kosuke.o.fit@gmail.com

[3] Department of Information and Communication Engineering, Fukuoka Institute of Technology, 3-30-1 Wajiro-Higashi, Higashi-Ku, Fukuoka 811-0295, Japan
barolli@fit.ac.jp

[4] Department of Information and Computer Engineering, Okayama University of Science, 1-1 Ridai-cho, Kita-Ku, Okayama 700-0005, Japan
kulla@ice.ous.ac.jp

[5] Department of Advanced Sciences, Faculty of Science and Engineering, Hosei University, Kajino-Machi, Koganei-Shi, Tokyo 184-8584, Japan
makoto.takizawa@computer.org

Abstract. Wireless Mesh Networks (WMNs) have many advantages such as low cost and increased high speed wireless Internet connectivity, therefore WMNs are becoming an important networking infrastructure. In our previous work, we implemented a Particle Swarm Optimization (PSO) based simulation system, called WMN-PSO, and a simulation system based on Genetic Algorithm (GA), called WMN-GA, for solving node placement problem in WMNs. In this paper, we implement a hybrid simulation system based on PSO and distributed GA (DGA), called WMN-PSODGA. We evaluate WMN-PSODGA by computer simulations. The simulation results show that the WMN-PSODGA has good performance when the number of GA islands is 64.

1 Introduction

The wireless networks and devises are becoming increasingly popular and they provide users access to information and communication anytime and anywhere [11–14,22,28–30]. Wireless Mesh Networks (WMNs) are gaining a lot of attention because of their low cost nature that makes them attractive for

L. Barolli et al. (Eds.): EIDWT 2018, LNDECT 17, pp. 79–93, 2018.
https://doi.org/10.1007/978-3-319-75928-9_7

providing wireless Internet connectivity. A WMN is dynamically self-organized and self-configured, with the nodes in the network automatically establishing and maintaining mesh connectivity among them-selves (creating, in effect, an ad hoc network). This feature brings many advantages to WMNs such as low up-front cost, easy network maintenance, robustness and reliable service coverage [1]. Moreover, such infrastructure can be used to deploy community networks, metropolitan area networks, municipal and corporative networks, and to support applications for urban areas, medical, transport and surveillance systems.

Mesh node placement in WMN can be seen as a family of problems, which are shown (through graph theoretic approaches or placement problems, e.g. [8,17]) to be computationally hard to solve for most of the formulations [34]. In fact, the node placement problem considered here is even more challenging due to two additional characteristics:

(a) Locations of mesh router nodes are not pre-determined, in other wards, any available position in the considered area can be used for deploying the mesh routers.
(b) Routers are assumed to have their own radio coverage area.

Here, we consider the version of the mesh router nodes placement problem in which we are given a grid area where to deploy a number of mesh router nodes and a number of mesh client nodes of fixed positions (of an arbitrary distribution) in the grid area. The objective is to find a location assignment for the mesh routers to the cells of the grid area that maximizes the network connectivity and client coverage. Node placement problems are known to be computationally hard to solve [15,16,35]. In some previous works, intelligent algorithms have been recently investigated [2–4,6,7,9,10,18,20,23–25,36].

In our previous work, we implemented a Particle Swarm Optimization (PSO) based simulation system, called WMN-PSO [26]. Also, we implemented another simulation system based on Genetic Algorithm (GA), called WMN-GA [21], for solving node placement problem in WMNs.

In this paper, we design and implement a hybrid simulation system based on PSO and distributed GA (DGA). We call this system WMN-PSODGA. We evaluate the implemented WMN-PSODGA system by computer simulations.

The rest of the paper is organized as follows. The mesh router nodes placement problem is defined in Sect. 2. We present our designed and implemented hybrid simulation system in Sect. 3. The simulation results are given in Sect. 4. Finally, we give conclusions and future work in Sect. 5.

2 Node Placement Problem in WMNs

For this problem, we have a grid area arranged in cells we want to find where to distribute a number of mesh router nodes and a number of mesh client nodes of fixed positions (of an arbitrary distribution) in the considered area. The objective is to find a location assignment for the mesh routers to the area that maximizes the network connectivity and client coverage. Network connectivity is measured by Size of Giant Component (SGC) of the resulting WMN graph, while the user coverage is simply the number of mesh client nodes that fall within the radio coverage of at least one mesh router node and is measured by Number of Covered Mesh Clients (NCMC).

An instance of the problem consists as follows.

- N mesh router nodes, each having its own radio coverage, defining thus a vector of routers.
- An area $W \times H$ where to distribute N mesh routers. Positions of mesh routers are not pre-determined and are to be computed.
- M client mesh nodes located in arbitrary points of the considered area, defining a matrix of clients.

It should be noted that network connectivity and user coverage are among most important metrics in WMNs and directly affect the network performance.

In this work, we have considered a bi-objective optimization in which we first maximize the network connectivity of the WMN (through the maximization of the SGC) and then, the maximization of the NCMC.

In fact, we can formalize an instance of the problem by constructing an adjacency matrix of the WMN graph, whose nodes are router nodes and client nodes and whose edges are links between nodes in the mesh network. Each mesh node in the graph is a triple $\boldsymbol{v} = <x, y, r>$ representing the 2D location point and r is the radius of the transmission range. There is an arc between two nodes $\boldsymbol{u}$ and $\boldsymbol{v}$, if $\boldsymbol{v}$ is within the transmission circular area of $\boldsymbol{u}$.

3 Proposed and Implemented Simulation System

3.1 Particle Swarm Optimization

In PSO a number of simple entities (the particles) are placed in the search space of some problem or function and each evaluates the objective function at its current location. The objective function is often minimized and the exploration of the search space is not through evolution [19]. However, following a widespread practice of borrowing from the evolutionary computation field, in this work, we consider the bi-objective function and fitness function interchangeably. Each particle then determines its movement through the search space by combining some aspect of the history of its own current and best (best-fitness) locations with those of one or more members of the swarm, with some random perturbations. The next iteration takes place after all particles have been moved. Eventually the swarm as a whole, like a flock of birds collectively foraging for food, is likely to move close to an optimum of the fitness function.

Each individual in the particle swarm is composed of three $\mathcal{D}$-dimensional vectors, where $\mathcal{D}$ is the dimensionality of the search space. These are the current position $\mathbf{x}_i$, the previous best position $\mathbf{p}_i$ and the velocity $\mathbf{v}_i$.

The particle swarm is more than just a collection of particles. A particle by itself has almost no power to solve any problem; progress occurs only when the

Algorithm 1. Pseudo code of PSO.

/* Initialize all parameters for PSO */
Computation maxtime:= Tp_{max}, $t := 0$;
Number of particle-patterns:= m, $2 \leq m \in \boldsymbol{N}^1$;
Particle-patterns initial solution:= $\boldsymbol{P}_i^0$;
Particle-patterns initial position:= $\boldsymbol{x}_{ij}^0$;
Particles initial velocity:= $\boldsymbol{v}_{ij}^0$;
PSO parameter:= ω, $0 < \omega \in \boldsymbol{R}^1$;
PSO parameter:= C_1, $0 < C_1 \in \boldsymbol{R}^1$;
PSO parameter:= C_2, $0 < C_2 \in \boldsymbol{R}^1$;
/* Start PSO */
Evaluate($\boldsymbol{G}^0, \boldsymbol{P}^0$);
while $t < Tp_{max}$ **do**
  /* Update velocities and positions */
  $\boldsymbol{v}_{ij}^{t+1} = \omega \cdot \boldsymbol{v}_{ij}^t$
    $+C_1 \cdot \text{rand}() \cdot (best(P_{ij}^t) - x_{ij}^t)$
    $+C_2 \cdot \text{rand}() \cdot (best(G^t) - x_{ij}^t)$;
  $\boldsymbol{x}_{ij}^{t+1} = \boldsymbol{x}_{ij}^t + \boldsymbol{v}_{ij}^{t+1}$;
  /* if fitness value is increased, a new solution will be accepted. */
  Update_Solutions($\boldsymbol{G}^t, \boldsymbol{P}^t$);
  $t = t + 1$;
end while
Update_Solutions($\boldsymbol{G}^t, \boldsymbol{P}^t$);
return Best found pattern of particles as solution;

particles interact. Problem solving is a population-wide phenomenon, emerging from the individual behaviors of the particles through their interactions. In any case, populations are organized according to some sort of communication structure or topology, often thought of as a social network. The topology typically consists of bidirectional edges connecting pairs of particles, so that if j is in i's neighborhood, i is also in j's. Each particle communicates with some other particles and is affected by the best point found by any member of its topological neighborhood. This is just the vector $\mathbf{p}_i$ for that best neighbor, which we will denote with $\mathbf{p}_g$. The potential kinds of population "social networks" are hugely varied, but in practice certain types have been used more frequently. We show the pseudo code of PSO in Algorithm 1.

In the PSO process, the velocity of each particle is iteratively adjusted so that the particle stochastically oscillates around $\mathbf{p}_i$ and $\mathbf{p}_g$ locations.

3.2 Distributed Genetic Algorithm

Distributed Genetic Algorithm (DGA) has been focused from various fields of science. DGA has shown their usefulness for the resolution of many computationally hard combinatorial optimization problems. We show the pseudo code of DGA in Algorithm 2.

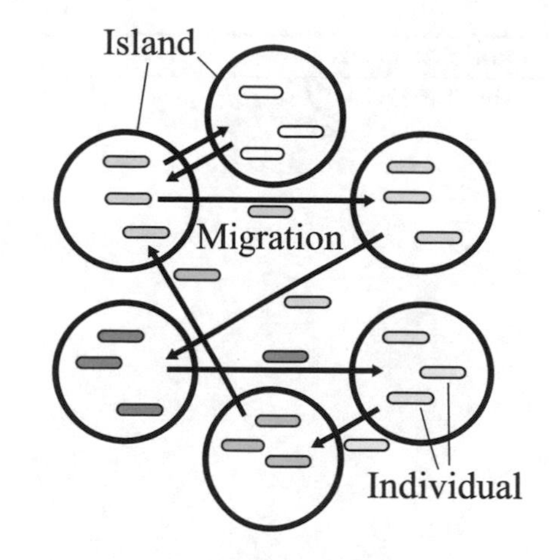

Fig. 1. Model of migration in DGA.

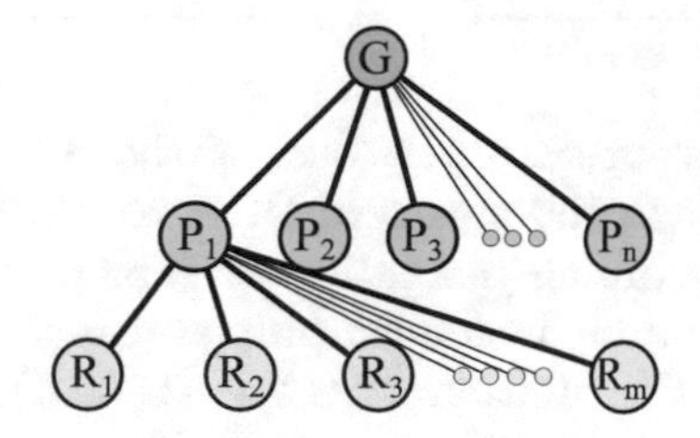

Fig. 2. Relationship among global solution, particle-patterns and mesh routers in PSO part.

Population of individuals: Unlike local search techniques that construct a path in the solution space jumping from one solution to another one through local perturbations, DGA use a population of individuals giving thus the search a larger scope and chances to find better solutions. This feature is also known as "exploration" process in difference to "exploitation" process of local search methods.

Fitness: The determination of an appropriate fitness function, together with the chromosome encoding are crucial to the performance of DGA. Ideally we would construct objective functions with "certain regularities", i.e. objective functions that verify that for any two individuals which are close in the search space, their respective values in the objective functions are similar.

Selection: The selection of individuals to be crossed is another important aspect in DGA as it impacts on the convergence of the algorithm. Several selection schemes have been proposed in the literature for selection operators trying to cope with premature convergence of DGA. There are many selection methods in GA. In our system, we implement 2 selection methods: Random method and Roulette wheel method.

Algorithm 2. Pseudo code of DSA.

/* Initialize all parameters for DGA */
Computation maxtime:= Tg_{max}, $t := 0$;
Number of islands:= n, $1 \leq n \in \boldsymbol{N}^1$;
initial solution:= $\boldsymbol{P}_i^0$;
/* Start DGA */
Evaluate($\boldsymbol{G}^0, \boldsymbol{P}^0$);
while $t < Tg_{max}$ **do**
  for all islands **do**
    Selection();
    Crossover();
    Mutation();
  end for
  $t = t + 1$;
end while
Update_Solutions($\boldsymbol{G}^t, \boldsymbol{P}^t$);
return Best found pattern of particles as solution;

Crossover operators: Use of crossover operators is one of the most important characteristics. Crossover operator is the means of DGA to transmit best genetic features of parents to offsprings during generations of the evolution process. Many methods for crossover operators have been proposed such as Blend Crossover (BLX-α), Unimodal Normal Distribution Crossover (UNDX), Simplex Crossover (SPX).

Mutation operators: These operators intend to improve the individuals of a population by small local perturbations. They aim to provide a component of randomness in the neighborhood of the individuals of the population. In our system, we implemented two mutation methods: uniformly random mutation and boundary mutation.

Escaping from local optima: GA itself has the ability to avoid falling prematurely into local optima and can eventually escape from them during the search process. DGA has one more mechanism to escape from local optima by considering some islands. Each island computes GA for optimizing and they migrate its gene to provide the ability to avoid from local optima (see Fig. 1).

Convergence: The convergence of the algorithm is the mechanism of DGA to reach to good solutions. A premature convergence of the algorithm would cause that all individuals of the population be similar in their genetic features and thus the search would result ineffective and the algorithm getting stuck into local optima. Maintaining the diversity of the population is therefore very important to this family of evolutionary algorithms.

3.3 WMN-PSODGA Hybrid Simulation System

In this subsection, we present the initialization, particle-pattern, fitness function and replacement methods.

Initialization
Our proposed system starts by generating an initial solution randomly, by *ad hoc* methods [36]. We decide the velocity of particles by a random process considering the area size. For instance, when the area size is $W \times H$, the velocity is decided randomly from $-\sqrt{W^2 + H^2}$ to $\sqrt{W^2 + H^2}$.

Particle-pattern
A particle is a mesh router. A fitness value of a particle-pattern is computed by combination of mesh routers and mesh clients positions. In other words, each particle-pattern is a solution as shown is Fig. 2.

Gene coding
A gene describes a WMN. Each individual has its own combination of mesh nodes. In other words, each individual has a fitness value. Therefore, the combination of mesh nodes is a solution.

Fitness function
One of most important thing in PSO algorithm is to decide the determination of an appropriate objective function and its encoding. In our case, each particle-pattern has an own fitness value and compares it with other particle-pattern's fitness value in order to share information of global solution. The fitness function follows a hierarchical approach in which the main objective is to maximize the SGC in WMN. Thus, the fitness function of this scenario is defined as

$$\text{Fitness} = 0.7 \times \text{SGC}(\boldsymbol{x}_{ij}, \boldsymbol{y}_{ij}) + 0.3 \times \text{NCMC}(\boldsymbol{x}_{ij}, \boldsymbol{y}_{ij}).$$

Routers replacement method for PSO part.
A mesh router has x, y positions and velocity. Mesh routers are moved based on velocities. There are many moving methods in PSO field, such as:

Constriction Method (CM)
CM is a method which PSO parameters are set to a week stable region ($\omega = 0.729$, $C_1 = C2 = 1.4955$) based on analysis of PSO by Clerc et al. [5,32].

Random Inertia Weight Method (RIWM)
In RIWM, the ω parameter is changing ramdomly from 0.5 to 1.0. The C_1 and C_2 are kept 2.0. The ω can be estimated by the week stable region. The average of ω is 0.75 [32].

Linearly Decreasing Inertia Weight Method (LDIWM)
In LDIWM, C_1 and C_2 are set to 2.0, constantly. On the other hand, the ω parameter is changed linearly from unstable region ($\omega = 0.9$) to stable region ($\omega = 0.4$) with increasing of iterations of computations [32,33].

Linearly Decreasing Vmax Method (LDVM)
In LDVM, PSO parameters are set to unstable region ($\omega = 0.9$, $C_1 = C_2 = 2.0$). A value of V_{max} which is maximum velocity of particles is considered. With increasing of iteration of computations, the V_{max} is kept decreasing linearly [31].

Rational Decrement of Vmax Method (RDVM)
In RDVM, PSO parameters are set to unstable region ($\omega = 0.9$, $C_1 = C_2 = 2.0$). The V_{max} is kept decreasing with the increasing of iterations as

$$V_{max}(x) = \sqrt{W^2 + H^2} \times \frac{T - x}{x}.$$

Where, W and H are the width and the height of the considered area, respectively. Also, T and x are the total number of iterations and a current number of iteration, respectively [27].

3.4 WMN-PSODGA Web GUI Tool and Pseudo Code

Algorithm 3. Pseudo code of WMN-PSODSA system.

Computation maxtime:= T_{max}, $t := 0$;
Initial solutions: $\boldsymbol{P}$.
Initial global solutions: $\boldsymbol{G}$.
/* Start PSODGA */
while $t < T_{max}$ **do**
  Subprocess(PSO);
  Subprocess(DGA);
  WaitSubprocesses();
  Evaluate($\boldsymbol{G}^t, \boldsymbol{P}^t$)
  /* Migration() swaps solutions (see Fig. 3). */
  Migration();
  $t = t + 1$;
end while
Update_Solutions($\boldsymbol{G}^t, \boldsymbol{P}^t$);
return Best found pattern of particles as solution;

The Web application follows a standard Client-Server architecture and is implemented using LAMP (Linux + Apache + MySQL + PHP) technology (see Fig. 4). Remote users (clients) submit their requests by completing first the parameter setting. The parameter values to be provided by the user are classified into three groups, as follows.

- Parameters related to the problem instance: These include parameter values that determine a problem instance to be solved and consist of number of router nodes, number of mesh client nodes, client mesh distribution, radio coverage interval and size of the deployment area.
- Parameters of the resolution method: Each method has its own parameters.

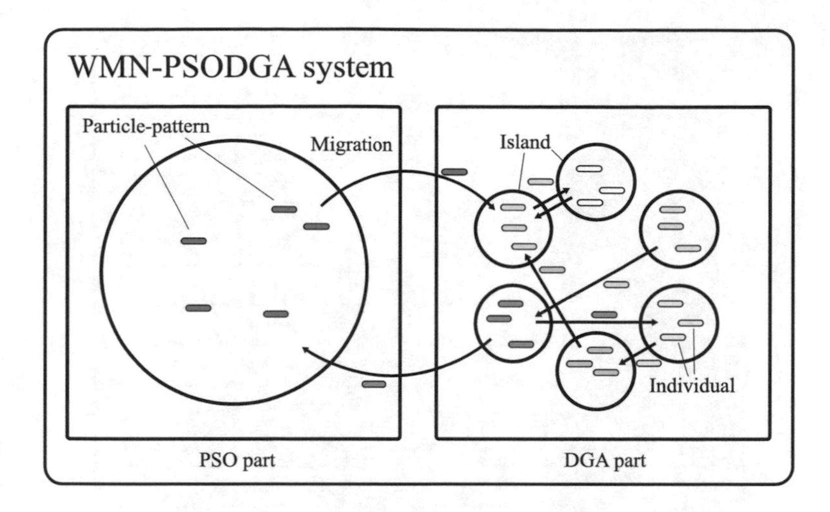

Fig. 3. Model of WMN-PSODGA migration.

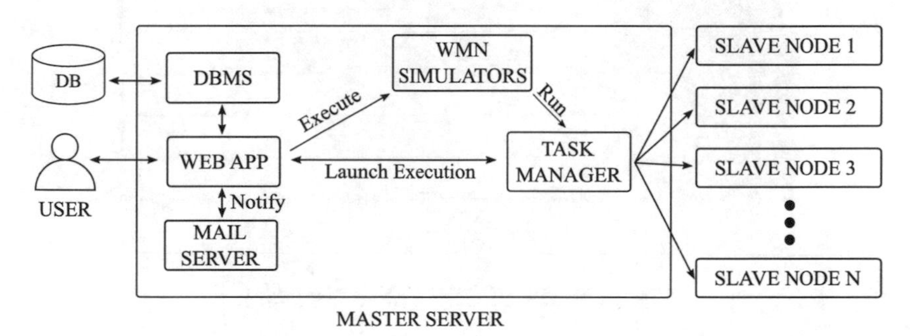

Fig. 4. System structure for web interface.

- Execution parameters: These parameters are used for stopping condition of the resolution methods and include number of iterations and number of independent runs. The former is provided as a total number of iterations and depending on the method is also divided per phase (e.g., number of iterations in a exploration). The later is used to run the same configuration for the same problem instance and parameter configuration a certain number of times.

We show the WMN-PSODGA Web GUI tool in Fig. 5. The pseudo code of our implemented system is shown in Algorithm 3.

4 Simulation Results

In this section, we show simulation results using WMN-PSODGA system. In this work, we consider normal distribution of mesh clients. The number of mesh

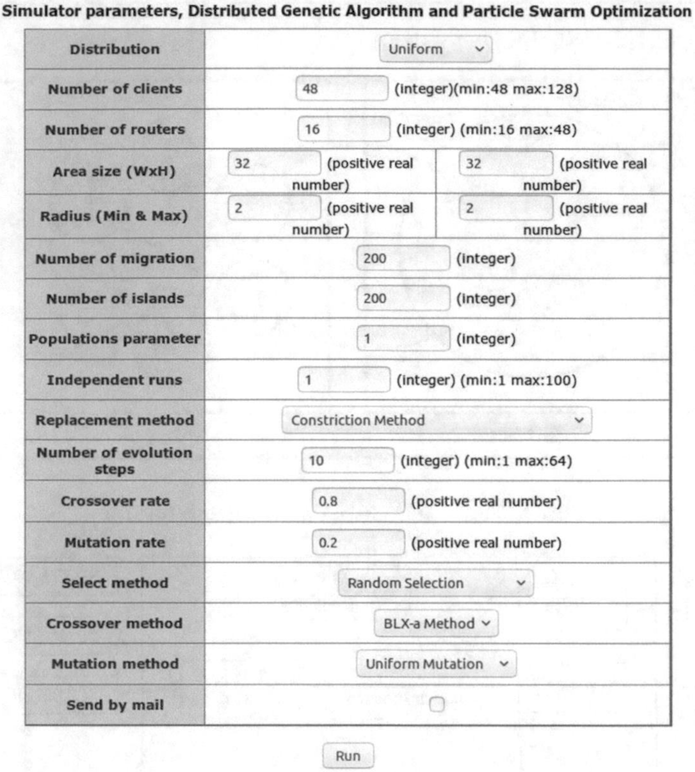

Simulator parameters, Distributed Genetic Algorithm and Particle Swarm Optimization

Distribution	Uniform	
Number of clients	48 (integer)(min:48 max:128)	
Number of routers	16 (integer) (min:16 max:48)	
Area size (WxH)	32 (positive real number)	32 (positive real number)
Radius (Min & Max)	2 (positive real number)	2 (positive real number)
Number of migration	200 (integer)	
Number of islands	200 (integer)	
Populations parameter	1 (integer)	
Independent runs	1 (integer) (min:1 max:100)	
Replacement method	Constriction Method	
Number of evolution steps	10 (integer) (min:1 max:64)	
Crossover rate	0.8 (positive real number)	
Mutation rate	0.2 (positive real number)	
Select method	Random Selection	
Crossover method	BLX-a Method	
Mutation method	Uniform Mutation	
Send by mail		

Run

Fig. 5. WMN-PSODSA Web GUI tool.

routers is considered 16 and the number of mesh clients 48. We conducted simulations 100 times, in order to avoid the effect of randomness and create a general view of results. We show the parameter setting for WMN-PSODGA in Table 1.

We show the simulation results in Figs. 6 and 7. In Fig. 6(a) and (b), we see that two mesh routers are not connected with other mesh routers when the number of GA islands is 16 and 32. We increase the number of GA islands to 48, 64, and 80. When the number of GA islands is increased, the fitness value is increased as shown in Fig. 7(a). However, when the number of GA islands is increased, the calculation time also increased linearly as shown in Fig. 7(b). Therefore, for these scenarios, the best performance is for 64 islands.

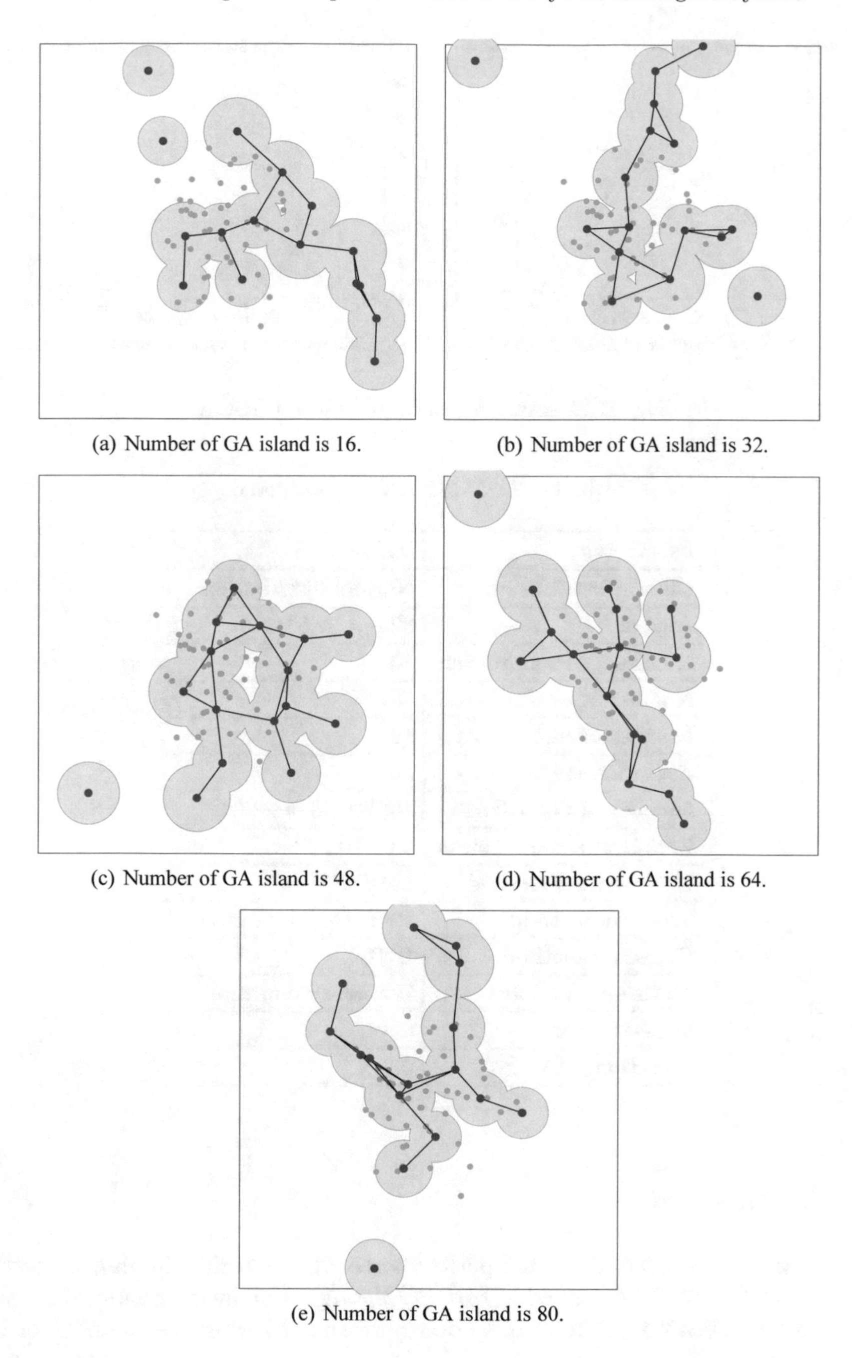

(a) Number of GA island is 16.

(b) Number of GA island is 32.

(c) Number of GA island is 48.

(d) Number of GA island is 64.

(e) Number of GA island is 80.

Fig. 6. Visualized simulation results of WMN-PSOGA for different number of GA islands.

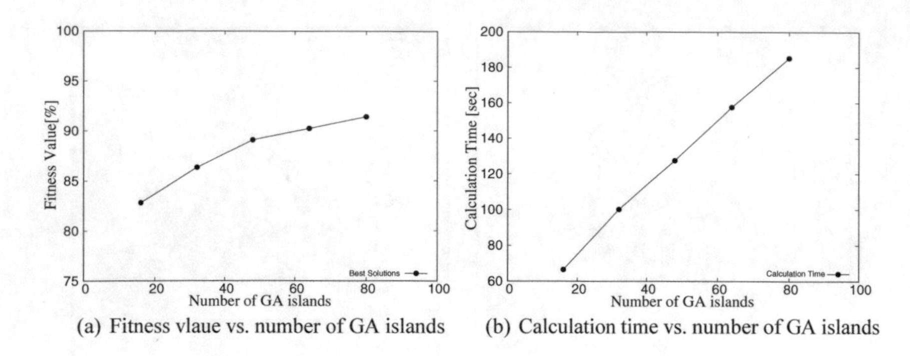

(a) Fitness vlaue vs. number of GA islands (b) Calculation time vs. number of GA islands

Fig. 7. Simulation results of WMN-PSOGA.

Table 1. WMN-PSODGA parameters.

Parameters	Values
Clients distribution	Normal distribution
Area size	32.0 × 32.0
Number of mesh routers	16
Number of mesh clients	48
Number of migrations	10
Evolution steps	40
Number of GA islands	16, 32, 48, 64, 80
Radius of a mesh router	2.0–3.0
Replacement method	LDVM
Selection method	Roulette wheel method
Crossover method	UNDX
Mutation method	Boundary mutation
Crossover rate	0.8
Mutation rate	0.2

5 Conclusions

In this work, we evaluated the performance of a hybrid simulation system based on PSO and DGA (called WMN-PSODGA). Simulation results show that the proposed WMN-PSODGA has good performance when the number of GA islands is 64.

In our future work, we would like to evaluate the performance of the proposed system for different parameters and patterns.

References

1. Akyildiz, I.F., Wang, X., Wang, W.: Wireless mesh networks: a survey. Comput. Netw. **47**(4), 445–487 (2005)
2. Amaldi, E., Capone, A., Cesana, M., Filippini, I., Malucelli, F.: Optimization models and methods for planning wireless mesh networks. Comput. Netw. **52**(11), 2159–2171 (2008)
3. Barolli, A., Spaho, E., Barolli, L., Xhafa, F., Takizawa, M.: QoS routing in ad-hoc networks using GA and multi-objective optimization. Mob. Inf. Syst. **7**(3), 169–188 (2011)
4. Behnamian, J., Ghomi, S.F.: Development of a PSO-SA hybrid metaheuristic for a new comprehensive regression model to time-series forecasting. Expert Syst. Appl. **37**(2), 974–984 (2010)
5. Clerc, M., Kennedy, J.: The particle swarm-explosion, stability, and convergence in a multidimensional complex space. IEEE Trans. Evol. Comput. **6**(1), 58–73 (2002)
6. Cunha, M.C., Sousa, J.: Water distribution network design optimization: simulated annealing approach. J. Water Resour. Plan. Manag. **125**(4), 215–221 (1999)
7. Del Valle, Y., Venayagamoorthy, G.K., Mohagheghi, S., Hernandez, J.C., Harley, R.G.: Particle swarm optimization: basic concepts, variants and applications in power systems. IEEE Trans. Evol. Comput. **12**(2), 171–195 (2008)
8. Franklin, A.A., Murthy, C.S.R.: Node placement algorithm for deployment of two-tier wireless mesh networks. In: Proceedings of Global Telecommunications Conference, pp. 4823–4827 (2007)
9. Ge, H., Du, W., Qian, F.: A hybrid algorithm based on particle swarm optimization and simulated annealing for job shop scheduling. In: Third International Conference on Natural Computation (ICNC 2007), vol. 3, pp. 715–719 (2007)
10. Girgis, M.R., Mahmoud, T.M., Abdullatif, B.A., Rabie, A.M.: Solving the wireless mesh network design problem using genetic algorithm and simulated annealing optimization methods. Int. J. Comput. Appl. **96**(11), 1–10 (2014)
11. Goto, K., Sasaki, Y., Hara, T., Nishio, S.: Data gathering using mobile agents for reducing traffic in dense mobile wireless sensor networks. Mob. Inf. Syst. **9**(4), 295–314 (2013)
12. Inaba, T., Elmazi, D., Sakamoto, S., Oda, T., Ikeda, M., Barolli, L.: A secure-aware call admission control scheme for wireless cellular networks using fuzzy logic and its performance evaluation. J. Mob. Multimedia **11**(3&4), 213–222 (2015)
13. Inaba, T., Obukata, R., Sakamoto, S., Oda, T., Ikeda, M., Barolli, L.: Performance evaluation of a QoS-aware fuzzy-based CAC for LAN access. Int. J. Space Based Situat. Comput. **6**(4), 228–238 (2016)
14. Inaba, T., Sakamoto, S., Oda, T., Ikeda, M., Barolli, L.: A testbed for admission control in WLAN: a fuzzy approach and its performance evaluation. In: International Conference on Broadband and Wireless Computing, Communication and Applications, pp. 559–571. Springer (2016)
15. Lim, A., Rodrigues, B., Wang, F., Xu, Z.: k-Center problems with minimum coverage. In: Computing and Combinatorics, pp. 349–359 (2004)
16. Maolin, T., et al.: Gateways placement in backbone wireless mesh networks. Int. J. Commun. Netw. Syst. Sci. **2**(1), 44 (2009)
17. Muthaiah, S.N., Rosenberg, C.P.: Single gateway placement in wireless mesh networks. In: Proceedings of 8th International IEEE Symposium on Computer Networks, pp. 4754–4759 (2008)

18. Naka, S., Genji, T., Yura, T., Fukuyama, Y.: A hybrid particle swarm optimization for distribution state estimation. IEEE Trans. Power Syst. **18**(1), 60–68 (2003)
19. Poli, R., Kennedy, J., Blackwell, T.: Particle swarm optimization. Swarm Intell. **1**(1), 33–57 (2007)
20. Sakamoto, S., Kulla, E., Oda, T., Ikeda, M., Barolli, L., Xhafa, F.: A comparison study of simulated annealing and genetic algorithm for node placement problem in wireless mesh networks. J. Mob. Multimedia **9**(1–2), 101–110 (2013)
21. Sakamoto, S., Kulla, E., Oda, T., Ikeda, M., Barolli, L., Xhafa, F.: A comparison study of hill climbing, simulated annealing and genetic algorithm for node placement problem in WMNs. J. High Speed Netw. **20**(1), 55–66 (2014)
22. Sakamoto, S., Kulla, E., Oda, T., Ikeda, M., Barolli, L., Xhafa, F.: A simulation system for WMN based on SA: performance evaluation for different instances and starting temperature values. Int. J. Space Based Situat. Comput. **4**(3–4), 209–216 (2014)
23. Sakamoto, S., Kulla, E., Oda, T., Ikeda, M., Barolli, L., Xhafa, F.: Performance evaluation considering iterations per phase and SA temperature in WMN-SA system. Mob. Inf. Syst. **10**(3), 321–330 (2014)
24. Sakamoto, S., Lala, A., Oda, T., Kolici, V., Barolli, L., Xhafa, F.: Application of WMN-SA simulation system for node placement in wireless mesh networks: a case study for a realistic scenario. Int. J. Mob. Comput. Multimedia Commun. (IJMCMC) **6**(2), 13–21 (2014)
25. Sakamoto, S., Oda, T., Ikeda, M., Barolli, L., Xhafa, F.: An integrated simulation system considering wmn-pso simulation system and network simulator 3. In: International Conference on Broadband and Wireless Computing, Communication and Applications, pp. 187–198. Springer (2016)
26. Sakamoto, S., Oda, T., Ikeda, M., Barolli, L., Xhafa, F.: Implementation and evaluation of a simulation system based on particle swarm optimisation for node placement problem in wireless mesh networks. Int. J. Commun. Netw. Distrib. Syst. **17**(1), 1–13 (2016)
27. Sakamoto, S., Oda, T., Ikeda, M., Barolli, L., Xhafa, F.: Implementation of a new replacement method in WMN-PSO simulation system and its performance evaluation. In: The 30th IEEE International Conference on Advanced Information Networking and Applications (AINA 2016), pp. 206–211 (2016). https://doi.org/10.1109/AINA.2016.42
28. Sakamoto, S., Obukata, R., Oda, T., Barolli, L., Ikeda, M., Barolli, A.: Performance analysis of two wireless mesh network architectures by WMN-SA and WMN-TS simulation systems. J. High Speed Netw. **23**(4), 311–322 (2017)
29. Sakamoto, S., Ozera, K., Barolli, A., Ikeda, M., Barolli, L., Takizawa, M.: Implementation of an intelligent hybrid simulation systems for WMNs based on particle swarm optimization and simulated annealing: performance evaluation for different replacement methods. Soft Comput. (2017). http://doi.org/10.1007/s00500-017-2948-1. Accessed 11 Dec 2017
30. Sakamoto, S., Ozera, K., Ikeda, M., Barolli, L.: Implementation of intelligent hybrid systems for node placement problem in WMNs considering particle swarm optimization, hill climbing and simulated annealing. Mob. Netw. Appl. (2017). http://doi.org/10.1007/s11036-017-0897-7. Accessed 06 Sep 2017
31. Schutte, J.F., Groenwold, A.A.: A study of global optimization using particle swarms. J. Glob. Optim. **31**(1), 93–108 (2005)
32. Shi, Y.: Particle swarm optimization. IEEE Connect. **2**(1), 8–13 (2004)
33. Shi, Y., Eberhart, R.C.: Parameter selection in particle swarm optimization. In: Evolutionary Programming VII, pp. 591–600 (1998)

34. Vanhatupa, T., Hannikainen, M., Hamalainen, T.: Genetic algorithm to optimize node placement and configuration for WLAN planning. In: Proceedings of 4th IEEE International Symposium on Wireless Communication Systems, pp. 612–616 (2007)
35. Wang, J., Xie, B., Cai, K., Agrawal, D.P.: Efficient mesh router placement in wireless mesh networks. In: Proceedings of IEEE Internatonal Conference on Mobile Adhoc and Sensor Systems (MASS 2007), pp. 1–9 (2007)
36. Xhafa, F., Sanchez, C., Barolli, L.: Ad hoc and neighborhood search methods for placement of mesh routers in wireless mesh networks. In: Proceedings of 29th IEEE International Conference on Distributed Computing Systems Workshops (ICDCS 2009), pp. 400–405 (2009)

A Fuzzy-Based System for Selection of IoT Devices in Opportunistic Networks Considering IoT Device Storage, Waiting Time and Security Parameters

Miralda Cuka[1(✉)], Donald Elmazi[1], Kevin Bylykbashi[2], Evjola Spaho[2], Makoto Ikeda[3], and Leonard Barolli[3]

[1] Graduate School of Engineering, Fukuoka Institute of Technology (FIT), 3-30-1 Wajiro-Higashi, Higashi-Ku, Fukuoka 811-0295, Japan
mcuka91@gmail.com, donald.elmazi@gmail.com

[2] Department of Electronics and Telecommunication, Polytechnic University of Tirana, Bul. Deshmoret e Kombit, Mother Theresa Square, Nr. 4, Tirana, Albania
evjolaspaho@hotmail.com

[3] Department of Information and Communication Engineering, Fukuoka Institute of Technology (FIT), 3-30-1 Wajiro-Higashi, Higashi-Ku, Fukuoka 811-0295, Japan
makoto.ikd@acm.org, barolli@fit.ac.jp

Abstract. The opportunistic networks are a subclass of delay-tolerant networks where communication opportunities (contacts) are intermittent and there is no need to establish an end-to-end link between the communication nodes. The Internet of Things (IoT) present the notion of large networks of connected devices, sharing data about their environments and creating a diverse ecosystem of sensors, actuators, and computing nodes. IoT networks are a departure from traditional enterprise networks in terms of their scale and consist of heterogeneous collections of resource constrained nodes that closely interact with their environment. There are different issues for these networks. One of them is the selection of IoT devices in order to carry out a task in opportunistic networks. In this work, we implement a Fuzzy-Based System for IoT device selection in opportunistic networks. For our system, we use three input parameters: IoT Device Storage (IDST), IoT Device Waiting Time (IDWT) and IoT Device Security (IDSC). The output parameter is IoT Device Selection Decision (IDSD). The simulation results show that the proposed system makes a proper selection decision of IoT-devices in opportunistic networks.

1 Introduction

The Internet is dramatically evolving and creating various connectivity methodologies. The Internet of Things (IoT) is one of those methodologies which transforms current Internet communication to Machine-to-Machine (M2M) basis. Hence, IoT can seamlessly connect the real world and cyberspace via physical objects that embed with various types of intelligent sensors. A large number

L. Barolli et al. (Eds.): EIDWT 2018, LNDECT 17, pp. 94–105, 2018.
https://doi.org/10.1007/978-3-319-75928-9_8

of Internet-connected machines will generate and exchange an enormous amount of data that make daily life more convenient, help to make a tough decision and provide beneficial services. The IoT probably becomes one of the most popular networking concepts that has the potential to bring out many benefits [1,2].

Opportunistic Networks are the variants of Delay Tolerant Networks (DTNs). It is a class of networks that has emerged as an active research subject in the recent times. Owing to the transient and un-connected nature of the nodes, routing becomes a challenging task in these networks. Sparse connectivity, no infrastructure and limited resources further complicate the situation. These networks can be useful for routing in places where there are few base stations and connected routes for long distances [3,4]. Hence, the challenges for routing in opportunistic networks are very different from the traditional wireless networks and their utility and potential for scalability makes them a huge success. However, this leads to security as a matter of concern for the protection of the personal data collected by such IoT systems since it is necessary to provide full awareness and control of the automatic data flow to the end user. Security issue is emphasized by the lack of standards specifically designed for devices with limited resources and heterogeneous technologies. In addition, these devices, due to many vulnerabilities, represent a fertile ground for existing cyber threats.

In an opportunistic network, when nodes move away or turn off their power to conserve energy, links may be disrupted or shut down periodically. These events result in intermittent connectivity. When there is no path existing between the source and the destination, the network partition occurs. Therefore, nodes need to communicate with each other via opportunistic contacts through store-carry-forward operation. Since these types of networks require the IoT devices to store some information, storage is an important parameter in evaluation of their performance. However, the storage capacity of the device is limited which makes storage a requirement to be considered.

The Fuzzy Logic (FL) is unique approach that is able to simultaneously handle numerical data and linguistic knowledge. It is a nonlinear mapping of an input data (feature) vector into a scalar output. Fuzzy set theory and FL establish the specifics of the nonlinear mapping.

In this paper, we propose and implement a simulation system for selection of IoT devices in opportunistic networks. The system is based on fuzzy logic and considers three parameters for IoT device selection. We show the simulation results for different values of parameters.

The remainder of the paper is organized as follows. In the Sect. 2, we present a brief introduction of IoT. In Sect. 3, we describe the basics of opportunistic networks including research challenges and architecture. In Sect. 4, we introduce the proposed system model and its implementation. Simulation results are shown in Sect. 5. Finally, conclusions and future work are given in Sect. 6.

2 IoT

2.1 IoT Architecture

The typical IoT architecture can be divided into five layers as shown in Fig. 1. Each layer is briefly described below.

Perception Layer: The perception layer is similar to physical layer in OSI model which consists of the different types of sensor devices and environmental elements. This layer generally deals with identification and collection of specific information by each type of sensor devices. The gathered information can be location, wind speed, vibration, pH level, humidity, amount of dust in the air and so on. The gathered information is transsmited through Network layer toward central information processing system.

Network Layer: The Network layer plays an important role in securely transferring and keeping the sensitive information confidential from sensor devices to the central information processing system through 3G, 4G, UMTS, WiFi, WiMAX, RFID, Infrared and Satellite dependent on the type of sensors devices. Thus, this layer is mainly responsible for transferring the information from Perception layer to upper layer.

Middleware Layer: The devices in the IoT system may generate various type of services when they are connected and communicate with others. Middleware layer has two essential functions, including service management and store the lower layer information into the database. Moreover, this layer has capability to retrieve, process, compute information, and then automatically decide based on the computational results.

Application Layer: Application layer is responsible for inclusive applications management based on the processed information in the Middleware layer. The IoT applications can be smart postal, smart health, smart car, smart glasses, smart home, smart independent living, smart transportation, etc.

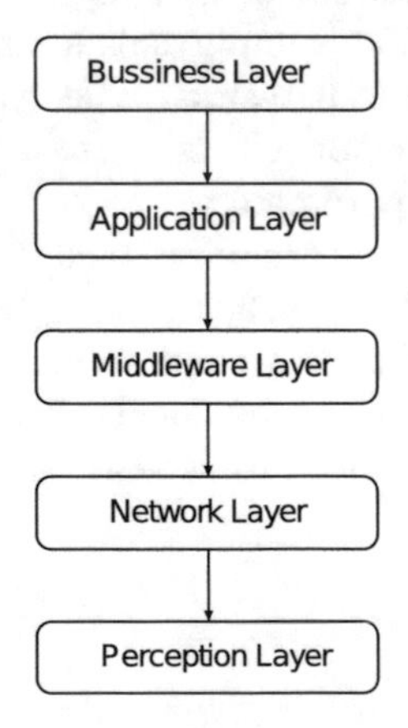

Fig. 1. IoT architecture layers

Business Layer: This layer functions cover the whole IoT applications and services management. It can create practically graphs, business models, flow chart and executive report based on the amount of accurate data received from lower layer and effective data analysis process. Based on the good analysis results, it will help the functional managers or executives to make more accurate decisions about the business strategies and roadmaps.

2.2 IoT Protocols

In following we will briefly describe about the most frequently used protocols for Machine-to-Machine (M2M) communication.

The Message Queue Telemetry Transport (MQTT) is a Client Server publishes or subscribes messaging transport protocol. It is light weight, open, simple and designed so as to be easy to implement. The protocol runs over TCP/IP or over other network protocols that provide ordered, lossless, bi-directional connections. The MQTT features include the usage of the publish/subscribe message pattern which provides one-to-many message distribution, a messaging transport that is agnostic to the content of the payload. Furthermore, the MQTT protocol has not only minimized transport overhead and protocol exchange to reduce network traffic but also has an extraordinary mechanism to notify interested parties when an abnormal disconnection occurs as well.

The Constraint Application Protocol (CoAP) is a specialized web transfer protocol for use with constrained nodes and constrained networks. The nodes often have 8-bit microcontroller with small amounts of ROM and RAM, while constrained network often have high packet error rate and typical throughput is 10 kbps. This protocol designed for M2M application such as smart city and building automation. The CoAP provides a request and response interaction model between application end points, support build-in discovery services and resources, and includes key concepts of the Web such as URIs and Internet media types. CoAP is designed to friendly interface with HTTP for integration with the Web while meeting specialized requirements such as multicast support, very low overhead and simplicity for constrained environments.

3 Opportunistic Networks

3.1 Opportunistic Networks Challenges

In an opportunistic network, when nodes move away or turn off their power to conserve energy, links may be disrupted or shut down periodically. These events result in intermittent connectivity. When there is no path existing between the source and the destination, the network partition occurs. Therefore, nodes need to communicate with each other via opportunistic contacts through store-carry-forward operation. In this section, we consider two specific challenges in an opportunistic network: the contact opportunity and the node storage.

- *Contact Opportunity:* Due to the node mobility or the dynamics of wireless channel, a node can make contact with other nodes at an unpredicted time. Since contacts between nodes are hardly predictable, they must be exploited opportunistically for exchanging messages between some nodes that can move between remote fragments of the network. The routing methods for opportunistic networks can be classified based on characteristics of participants movement patterns. The patterns are classified according to two independent properties: their inherent structure and their adaptiveness to the demand in the network. Other approaches proposed message ferries to provide communication service for nodes in the deployment areas. In addition, the contact capacity needs to be considered [5,6].
- *Node Storage:* As described above, to avoid dropping packets, the intermediate nodes are required to have enough storage to store all messages for an unpredictable period of time until next contact occurs. In other words, the required storage space increases as a function of the number of messages in the network. Therefore, the routing and replication strategies must take the storage constraint into consideration [7].

3.2 Opportunistic Networks Architectures

In an opportunistic network, a network is typically separated into several network partitions called regions. Traditional applications are not suitable for this kind of environment because they normally assume that the end-to-end connection must exist from the source to the destination.

The opportunistic network enables the devices in different regions to interconnect by operating message in a store-carry-forward fashion. The intermediate nodes implement the store-carry-forward message switching mechanism by overlaying a new protocol layer, called the bundle layer, on top of heterogeneous region-specific lower layers.

In an opportunistic network, each node is an entity with a bundle layer which can act as a host, a router or a gateway. When the node acts as a router, the bundle layer can store, carry and forward the entire bundles (or bundle fragments) between the nodes in the same region. On the other hand, the bundle

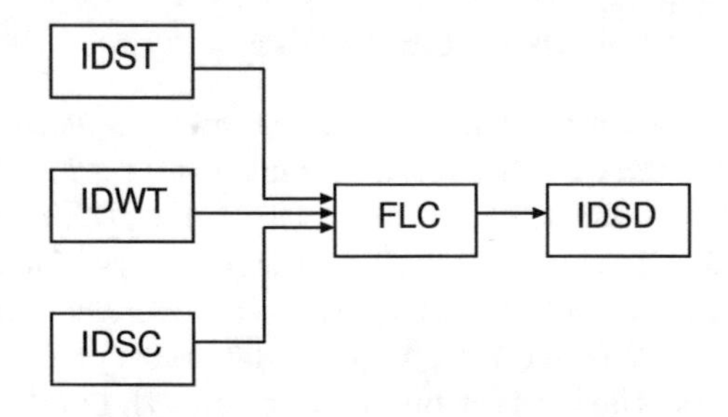

Fig. 2. Proposed system model.

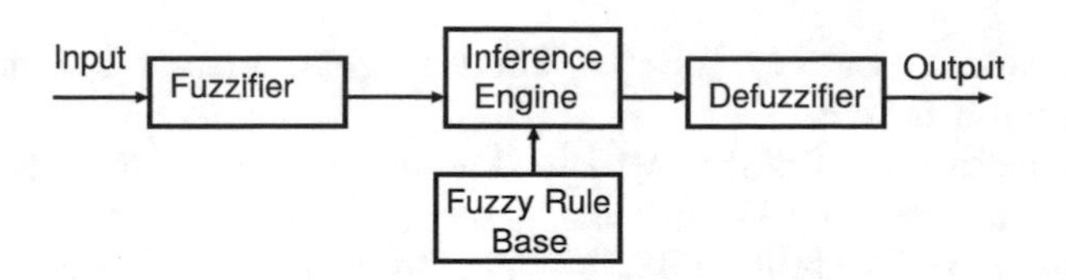

Fig. 3. FLC structure.

layer of gateway is used to transfer messages across different regions. A gateway can forward bundles between two or more regions and may optionally be a host, so it must have persistent storage and support custody transfers.

4 Proposed System

4.1 System Parameters

Based on opportunistic networks characteristics and challenges, we consider the following parameters for implementation of our proposed system.

IoT Device Storage (IDST): In delay tolerant networks data is carried by the IoT device until a communication opportunity is available. Considering that different IoT devices have different storage capabilities, the selection decision should consider the storage capacity.

IoT Device Waiting Time for sending data (IDWT): Considering network congestion, some IoT devices wait longer and some wait less for sending data. The IoT devices that have been waiting longer have a high possibility to be selected.

IoT Device Security (IDSC): IoT devices have many vulnerabilities. So, security measures against attacks or illegal requests should be considered. An Iot device which is more secure will be selected for a required task.

IoT Device Selection Decision (IDSD): The proposed system considers the following levels for IoT device selection:

- Very Low Selection Possibility (VLSP) - The IoT device will have very low probability to be selected.

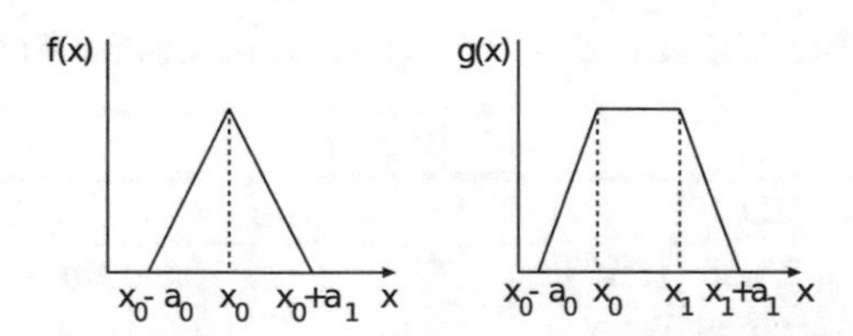

Fig. 4. Triangular and trapezoidal membership functions.

- Low Selection Possibility (LSP) - There might be other IoT devices which can do the job better.
- Middle Selection Possibility (MSP) - The IoT device is ready to be assigned a task, but is not the "chosen" one.
- High Selection Possibility (HSP) - The IoT device takes responsibility of completing the task.
- Very High Selection Possibility (VHSP) - The IoT device has almost all the required information and potential to be selected and then allocated in an appropriate position to carry out a job.

4.2 System Implementation

Fuzzy sets and fuzzy logic have been developed to manage vagueness and uncertainty in a reasoning process of an intelligent system such as a knowledge based system, an expert system or a logic control system [8–21]. In this work, we use fuzzy logic to implement the proposed system.

The structure of the proposed system is shown in Fig. 2. It consists of one Fuzzy Logic Controller (FLC), which is the main part of our system and its basic elements are shown in Fig. 3. They are the fuzzifier, inference engine, Fuzzy Rule Base (FRB) and defuzzifier.

As shown in Fig. 4, we use triangular and trapezoidal membership functions for FLC, because they are suitable for real-time operation [22]. The x_0 in $f(x)$ is the center of triangular function, $x_0(x_1)$ in $g(x)$ is the left (right) edge of trapezoidal function, and $a_0(a_1)$ is the left (right) width of the triangular or trapezoidal function. We explain in details the design of FLC in following.

4.3 Description of FLC

We use three input parameters for FLC: IoT Device Storage (IDST), IoT Device Waiting Time (IDWT) and IoT Device Security (IDSC).

The term sets for each input linguistic parameter are defined respectively as shown in Table 1.

$$T(IDST) = \{Small(Sm), Medium(Me), High(Hi)\}$$
$$T(IDWT) = \{Short(Sho), Medium(Mi), Long(Lg)\}$$
$$T(IDSC) = \{Weak(We), Moderate(Mo), Strong(St)\}$$

Table 1. Parameters and their term sets for FLC.

Parameters	Term sets
IoT Device Storage (IDST)	Small (Sm), Medium (Me), High (Hi)
IoT Device Waiting Time (IDWT)	Short (Sho), Medium (Mi), Long (Lg)
IoT Device Security (IDSC)	Weak (We), Moderate (Mo), Strong (St)
IoT Device Selection Decision (IDSD)	VLSP, LSP, MSP, HSP, VHSP

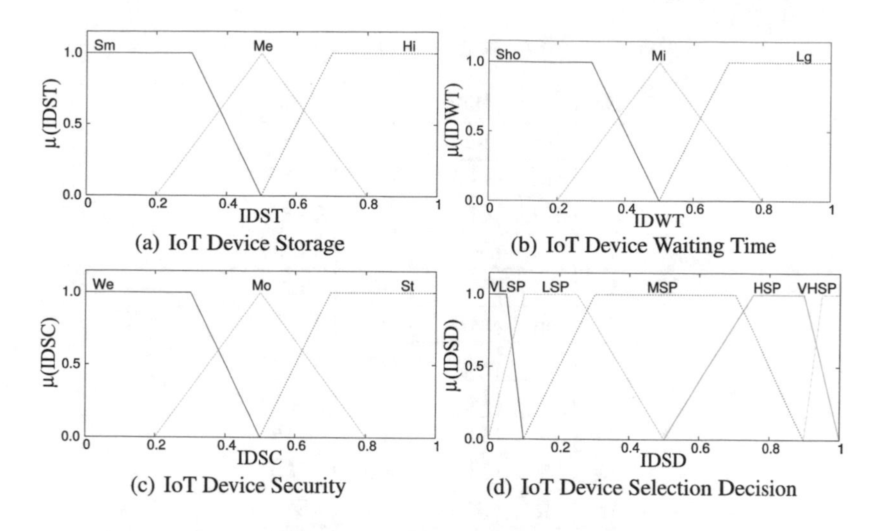

Fig. 5. Fuzzy membership functions.

The membership functions for input parameters of FLC are defined as:

$$
\begin{aligned}
\mu_{Sm}(IDST) &= g(IDST; Sm_0, Sm_1, Sm_{w0}, Sm_{w1}) \\
\mu_{Me}(IDST) &= f(IDST; Me_0, Me_{w0}, Me_{w1}) \\
\mu_{Hi}(IDST) &= g(IDST; Hi_0, Hi_1, Hi_{w0}, Hi_{w1}) \\
\mu_{Sho}(IDWT) &= g(IDWT; Sho_0, Sho_1, Sho_{w0}, Sho_{w1}) \\
\mu_{Mi}(IDWT) &= f(IDWT; Mi_0, Mi_{w0}, Mi_{w1}) \\
\mu_{Lg}(IDWT) &= g(IDWT; Lg_0, Lg_1, Lg_{w0}, Lg_{w1}) \\
\mu_{We}(IDSC) &= g(IDSC; We_0, We_1, We_{w0}, We_{w1}) \\
\mu_{Mo}(IDSC) &= f(IDSC; Mo_0, Mo_{w0}, Mo_{w1}) \\
\mu_{St}(IDSC) &= g(IDSC; St_0, St_1, St_{w0}, St_{w1}).
\end{aligned} \tag{1}
$$

The small letters *w0* and *w1* mean left width and right width, respectively.

The output linguistic parameter is the IoT device Selection Decision (IDSD). We define the term set of IDSD as:

$$
\begin{aligned}
\{&Very\ Low\ Selection\ Possibility\ (VLSP), \\
&Low\ Selection\ Possibility\ (LSP), \\
&Middle\ Selection\ Possibility\ (MSP), \\
&High\ Selection\ Possibility\ (HSP), \\
&Very\ High\ Selection\ Possibility\ (VHSP)\}.
\end{aligned}
$$

The membership functions for the output parameter *IDSD* are defined as:

$$
\begin{aligned}
\mu_{VLSP}(IDSD) &= g(IDSD; VLSP_0, VLSP_1, VLSP_{w0}, VLSP_{w1}) \\
\mu_{LSP}(IDSD) &= g(IDSD; LSP_0, LSP_1, LSP_{w0}, LSP_{w1}) \\
\mu_{MSP}(IDSD) &= g(IDSD; MSP_0, MSP_1, MSP_{w0}, MSP_{w1}) \\
\mu_{HSP}(IDSD) &= g(IDSD; HSP_0, HSP_1, HSP_{w0}, HSP_{w1}) \\
\mu_{VHSP}(IDSD) &= g(IDSD; VHSP_0, VHSP_1, VHSP_{w0}, VHSP_{w1}).
\end{aligned}
$$

Table 2. FRB of proposed fuzzy-based system.

No.	IDST	IDWT	IDSC	IDSD
1	Sm	Sho	We	VLSP
2	Sm	Sho	Mo	VLSP
3	Sm	Sho	St	LSP
4	Sm	Mi	We	VLSP
5	Sm	Mi	Mo	LSP
6	Sm	Mi	St	MSP
7	Sm	Lg	We	VLSP
8	Sm	Lg	Mo	MSP
9	Sm	Lg	St	HSP
10	Me	Sho	We	VLSP
11	Me	Sho	Mo	LSP
12	Me	Sho	St	MSP
13	Me	Mi	We	LSP
14	Me	Mi	Mo	MSP
15	Me	Mi	St	HSP
16	Me	Lg	We	MSP
17	Me	Lg	Mo	HSP
18	Me	Lg	St	VHSP
19	Hi	Sho	We	LSP
20	Hi	Sho	Mo	HSP
21	Hi	Sho	St	VHSP
22	Hi	Mi	We	MSP
23	Hi	Mi	Mo	VHSP
24	Hi	Mi	St	VHSP
25	Hi	Lg	We	HSP
26	Hi	Lg	Mo	VHSP
27	Hi	Lg	St	VHSP

The membership functions are shown in Fig. 5 and the Fuzzy Rule Base (FRB) for our system are shown in Table 2. The FRB forms a fuzzy set of dimensions $|T(IDST)| \times |T(IDWT)| \times |T(IDSC)|$, where $|T(x)|$ is the number of terms on $T(x)$. We have three input parameters, so our system has 27 rules. The control rules have the form: IF "conditions" THEN "control action".

5 Simulation Results

We present the simulation results in Fig. 6. In this figure, we show the relation between the probability of an IoT device to be selected (IDSD) to carry out a task, versus IDSC, IDST and IDWT. We consider IDST as a constant and change the values of IDWT and IDSC. We see that when the security is increased, the possibility of the present IoT device to be selected for carrying out a job is increased. By increasing the IDWT value, the IDSD is also increased. This means that the IoT device with a longer waiting time will be selected. In Fig. 6(a), when IDWT is 0.1 and IDSC is 0.6, the IDSD is 0.2. For IDWT 0.5, the IDSD is 0.4 and for IDWT 0.9, IDSD is 0.5, thus the IDSD is increased about 20% and 30%, for IDWT 0.5 and IDWT 0.9, respectively.

In Fig. 6(b) and (c), we increase the IDST value to 0.5 and 0.9, respectively. We see that with the increase of the IDST parameter, the possibility of an IoT device to be selected is increased. In Fig. 6(a), for IDSC 0.7 and IDWT 0.9, IDSD increases sharply. For IDSC = 0.7 and IDWT = 0.9, comparing Fig. 6(b) and (a) and Fig. 6(c) and (a), the IDSD is increased 20% and 38%, respectively.

6 Conclusions and Future Work

In this paper, we proposed and implemented a fuzzy-based IoT device selection system for opportunistic networks, which is used to select an IoT device for a required task.

We evaluated the proposed system by computer simulations. The simulation results show that the highest the security, the greater is the possibility of IoT

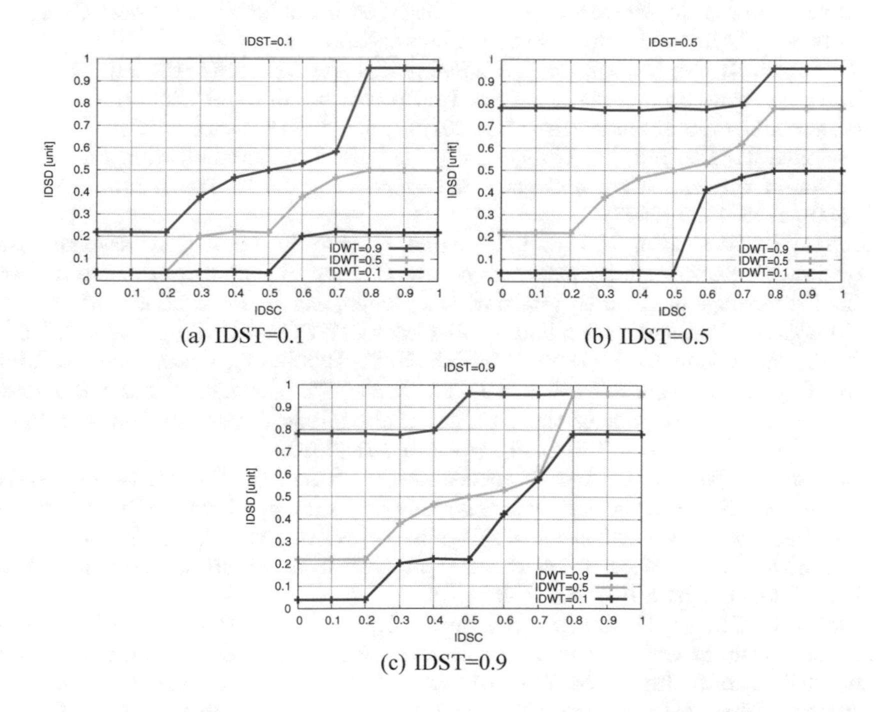

(a) IDST=0.1

(b) IDST=0.5

(c) IDST=0.9

Fig. 6. Results for different values of *IDST*.

device to be selected for carrying out a job. We can see that by increasing IDST and IDWT, the IDSD is also increased.

In the future work, we will also consider other parameters for IoT device selection in opportunistic networks and make extensive simulations to evaluate the proposed system.

References

1. Kraijak, S., Tuwanut, P.: A survey on internet of things architecture, protocols, possible applications, security, privacy, real-world implementation and future trends. In: Proceedings of the IEEE 16th International Conference on Communication Technology (ICCT), pp. 26–31. IEEE (2015)
2. Arridha, R., Sukaridhoto, S., Pramadihanto, D., Funabiki, N.: Classification extension based on IoT-big data analytic for smart environment monitoring and analytic in real-time system. Int. J. Space-Based Situated Comput. **7**(2), 82–93 (2017)
3. Dhurandher, S.K., Sharma, D.K., Woungang, I., Bhati, S.: HBPR: history based prediction for routing in infrastructure-less opportunistic networks. In: Proceedings of the IEEE 27th International Conference on Advanced Information Networking and Applications (AINA), pp. 931–936. IEEE (2013)
4. Spaho, E., Mino, G., Barolli, L., Xhafa, F.: Goodput and PDR analysis of AODV, OLSR and DYMO protocols for vehicular networks using CAVENET. Int. J. Grid Util. Comput. **2**(2), 130–138 (2011)
5. Akbas, M., Turgut, D.: APAWSAN: actor positioning for aerial wireless sensor and actor networks. In: Proceedings of the IEEE 36th Conference on Local Computer Networks (LCN-2011), pp. 563–570, October 2011
6. Akbas, M., Brust, M., Turgut, D.: Local positioning for environmental monitoring in wireless sensor and actor networks. In: Proceedings of the IEEE 35th Conference on Local Computer Networks (LCN-2010), pp. 806–813, October 2010
7. Melodia, T., Pompili, D., Gungor, V., Akyildiz, I.: Communication and coordination in wireless sensor and actor networks. Proc. IEEE Trans. Mob. Comput. **6**(10), 1126–1129 (2007)
8. Inaba, T., Sakamoto, S., Kolici, V., Mino, G., Barolli, L.: A CAC scheme based on fuzzy logic for cellular networks considering security and priority parameters. In: Proceedings of the 9-th International Conference on Broadband and Wireless Computing, Communication and Applications (BWCCA-2014), pp. 340–346 (2014)
9. Spaho, E., Sakamoto, S., Barolli, L., Xhafa, F., Barolli, V., Iwashige, J.: A fuzzy-based system for peer reliability in JXTA-overlay P2P considering number of interactions. In: Proceedings of the 16th International Conference on Network-Based Information Systems (NBiS-2013), pp. 156–161 (2013)
10. Matsuo, K., Elmazi, D., Liu, Y., Sakamoto, S., Mino, G., Barolli, L.: FACS-MP: a fuzzy admission control system with many priorities for wireless cellular networks and its performance evaluation. J. High Speed Netw. **21**(1), 1–14 (2015)
11. Grabisch, M.: The application of fuzzy integrals in multicriteria decision making. Eur. J. Oper. Res. **89**(3), 445–456 (1996)
12. Inaba, T., Elmazi, D., Liu, Y., Sakamoto, S., Barolli, L., Uchida, K.: Integrating wireless cellular and ad-hoc networks using fuzzy logic considering node mobility and security. In: Proceedings of the 29th IEEE International Conference on Advanced Information Networking and Applications Workshops (WAINA-2015), pp. 54–60 (2015)

13. Kulla, E., Mino, G., Sakamoto, S., Ikeda, M., Caballé, S., Barolli, L.: FBMIS: a fuzzy-based multi-interface system for cellular and ad hoc networks. In: Proceedings of the International Conference on Advanced Information Networking and Applications (AINA-2014), pp. 180–185 (2014)
14. Elmazi, D., Kulla, E., Oda, T., Spaho, E., Sakamoto, S., Barolli, L.: A comparison study of two fuzzy-based systems for selection of actor node in wireless sensor actor networks. J. Ambient Intell. Humaniz. Comput. **6**(5), 635–645 (2015)
15. Zadeh, L.: Fuzzy logic, neural networks, and soft computing. Commun. ACM **37**, 77–84 (1994)
16. Spaho, E., Sakamoto, S., Barolli, L., Xhafa, F., Ikeda, M.: Trustworthiness in P2P: performance behaviour of two fuzzy-based systems for JXTA-overlay platform. Soft. Comput. **18**(9), 1783–1793 (2014)
17. Inaba, T., Sakamoto, S., Kulla, E., Caballe, S., Ikeda, M., Barolli, L.: An integrated system for wireless cellular and ad-hoc networks using fuzzy logic. In: Proceedings of the International Conference on Intelligent Networking and Collaborative Systems (INCoS-2014), pp. 157–162 (2014)
18. Matsuo, K., Elmazi, D., Liu, Y., Sakamoto, S., Barolli, L.: A multi-modal simulation system for wireless sensor networks: a comparison study considering stationary and mobile sink and event. J. Ambient Intell. Humaniz. Comput. **6**(4), 519–529 (2015)
19. Kolici, V., Inaba, T., Lala, A., Mino, G., Sakamoto, S., Barolli, L.: A fuzzy-based CAC scheme for cellular networks considering security. In: Proceedings of the International Conference on Network-Based Information Systems (NBiS-2014), pp. 368–373 (2014)
20. Liu, Y., Sakamoto, S., Matsuo, K., Ikeda, M., Barolli, L., Xhafa, F.: A comparison study for two fuzzy-based systems: improving reliability and security of JXTA-overlay P2P platform. Soft. Comput. **20**(7), 2677–2687 (2015)
21. Matsuo, K., Elmazi, D., Liu, Y., Sakamoto, S., Mino, G., Barolli, L.: FACS-MP: a fuzzy admission control system with many priorities for wireless cellular networks and its perforemance evaluation. J. High Speed Netw. **21**(1), 1–14 (2015)
22. Mendel, J.M.: Fuzzy logic systems for engineering: a tutorial. Proc. IEEE **83**(3), 345–377 (1995)

Selection of Actor Nodes in Wireless Sensor and Actor Networks Considering Failure of Assigned Task as New Parameter

Donald Elmazi[1(✉)], Miralda Cuka[1], Kevin Bylykbashi[2], Evjola Spaho[2], Makoto Ikeda[3], and Leonard Barolli[3]

[1] Graduate School of Engineering, Fukuoka Institute of Technology (FIT), 3-30-1 Wajiro-Higashi, Higashi-Ku, Fukuoka 811-0295, Japan
donald.elmazi@gmail.com, mcuka91@gmail.com

[2] Faculty of Information Technology, Polytechnic University of Tirana, Mother Teresa Square, No. 4, Tirana, Albania
kevini_95@hotmail.com, evjolaspaho@hotmail.com

[3] Department of Information and Communication Engineering, Fukuoka Institute of Technology (FIT), 3-30-1 Wajiro-Higashi, Higashi-Ku, Fukuoka 811-0295, Japan
makoto.ikd@acm.org, barolli@fit.ac.jp

Abstract. Wireless Sensor and Actor Network (WSAN) is formed by the collaboration of micro-sensor and actor nodes. Whenever there is any special event i.e., fire, earthquake, flood or enemy attack in the network, sensor nodes have responsibility to sense it and send information towards an actor node. The actor node is responsible to take prompt decision and react accordingly. In order to provide effective sensing and acting, a distributed local coordination mechanism is necessary among sensors and actors. In this work, we consider the actor node selection problem and propose a fuzzy-based system that based on data provided by sensors and actors selects an appropriate actor node. We use 4 input parameters: Job Type (JT), Distance to Event (DE), Remaining Energy (RE) and different from our previous work we consider the Failure of Assigned Task (FAT) parameter. The output parameter is Actor Selection Decision (ASD). Based on these parameters, the simulation results show that the proposed system makes a proper selection of actor nodes.

1 Introduction

Recent technological advances have lead to the emergence of distributed Wireless Sensor and Actor Networks (WSANs) wich are capable of observing the physical world, processing the data, making decisions based on the observations and performing appropriate actions [1].

Wireless Sensor Networks (WSNs) can be defined as a collection of wireless self-configuring programmable multi-hop tiny devices, which can bind to each

L. Barolli et al. (Eds.): EIDWT 2018, LNDECT 17, pp. 106–118, 2018.
https://doi.org/10.1007/978-3-319-75928-9_9

other in an arbitrary manner, without the aid of any centralized administration, thereby dynamically sending the sensed data to the intended recipient about the monitored phenomenon. WSNs are comprised of multiple sensors which are connected to each other in order to perform collaborative or cooperative functions. These nodes are typically connected as a multi-hop mesh network [2,3].

In WSAN, the devices deployed in the environment are sensors able to sense environmental data, actors able to react by affecting the environment or have both functions integrated. Actor nodes are equipped with two radio transmitters, a low data rate transmitter to communicate with the sensor and a high data rate interface for actor-actor communication. For example, in the case of a fire, sensors relay the exact origin and intensity of the fire to actors so that they can extinguish it before spreading in the whole building or in a more complex scenario, to save people who may be trapped by fire [4].

Unlike WSNs, where the sensor nodes tend to communicate all the sensed data to the sink by sensor-sensor communication, in WSANs, two new communication types may take place. They are called sensor-actor and actor-actor communications. Sensed data is sent to the actors in the network through sensor-actor communication. After the actors analyse the data, they communicate with each other in order to assign and complete tasks. To provide effective operation of WSAN, it is very important that sensors and actors coordinate in what are called sensor-actor and actor-actor coordination. Coordination is not only important during task conduction, but also during network's self-improvement operations, i.e. connectivity restoration [5,6], reliable service [7], Quality of Service (QoS) [8,9] and so on.

Sensor-Actor (SA) coordination defines the way sensors communicate with actors, which actor is accessed by each sensor and which route should data packets follow to reach it. Among other challenges, when designing SA coordination, care must be taken in considering energy minimization because sensors, which have limited energy supplies, are the most active nodes in this process. On the other hand, Actor-Actor (AA) coordination helps actors to choose which actor will lead performing the task (actor selection), how many actors should perform and how they will perform. Actor selection is not a trivial task, because it needs to be solved in real time, considering different factors. It becomes more complicated when the actors are moving, due to dynamic topology of the network.

In this paper, different from our previous work [10], we propose and implement a simulation system which considers also the Failure of Assigned Task (FAT) parameter. The system is based on fuzzy logic and considers four input parameters for actor selection. We show the simulation results for different values of parameters.

The remainder of the paper is organized as follows. In Sect. 2, we describe the basics of WSANs including research challenges and architecture. In Sect. 3, we describe the system model and its implementation. Simulation results are shown in Sect. 4. Finally, conclusions and future work are given in Sect. 5.

2 WSAN

2.1 WSAN Challenges

Some of the key challenges in WSAN are related to the presence of actors and their functionalities.

- *Deployment and Positioning*: At the moment of node deployment, algorithms must consider to optimize the number of sensors and actors and their initial positions based on applications [11,12].
- *Architecture*: When important data has to be transmitted (an event occurred), sensors may transmit their data back to the sink, which will control the actors' tasks from distance or transmit their data to actors, which can perform actions independently from the sink node [13].
- *Real-Time*: There are a lot of applications that have strict real-time requirements. In order to fulfill them, real-time limitations must be clearly defined for each application and system [14].
- *Coordination*: In order to provide effective sensing and acting, a distributed local coordination mechanism is necessary among sensors and actors [13].
- *Power Management*: WSAN protocols should be designed with minimized energy consumption for both sensors and actors [15].
- *Mobility*: Protocols developed for WSANs should support the mobility of nodes [6,16], where dynamic topology changes, unstable routes and network isolations are present.
- *Scalability*: Smart Cities are emerging fast and WSAN, as a key technology will continue to grow together with cities. In order to keep the functionality of WSAN applicable, scalability should be considered when designing WSAN protocols and algorithms [12,16].

2.2 WSAN Architecture

A WSAN is shown in Fig. 1. The main functionality of WSANs is to make actors perform appropriate actions in the environment, based on the data sensed

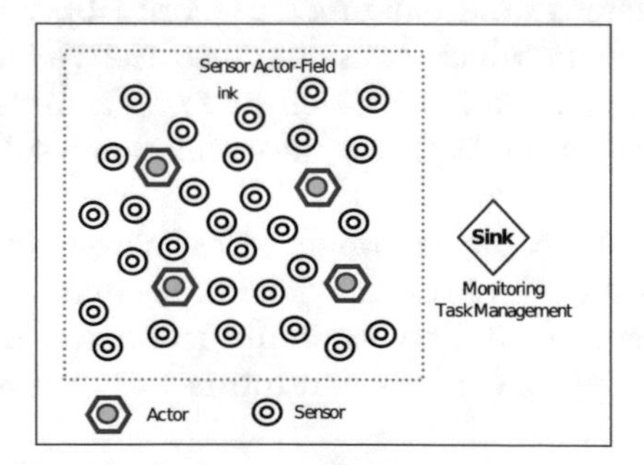

Fig. 1. Wireless Sensor Actor Network (WSAN).

from sensors and actors. When important data has to be transmitted (an event occurred), sensors may transmit their data back to the sink, which will control the actors' tasks from distance, or transmit their data to actors, which can perform actions independently from the sink node. Here, the former scheme is called Semi-Automated Architecture and the latter one Fully-Automated Architecture (see Fig. 2). Obviously, both architectures can be used in different applications. In the Fully-Automated Architecture are needed new sophisticated algorithms in order to provide appropriate coordination between nodes of WSAN. On the other hand, it has advantages, such as *low latency, low energy consumption, long network lifetime* [1], *higher local position accuracy, higher reliability* and so on.

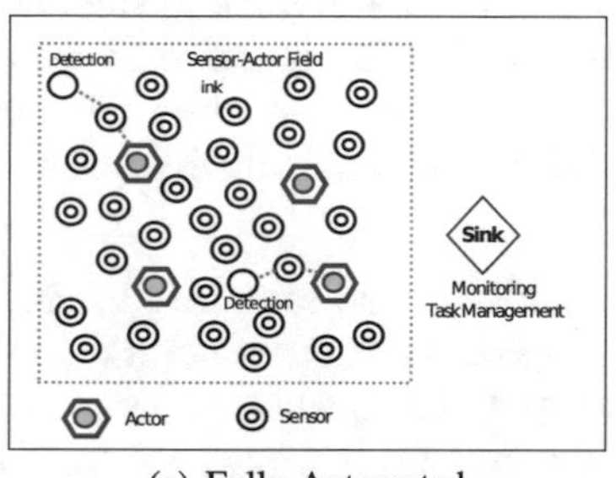

(a) Fully-Automated

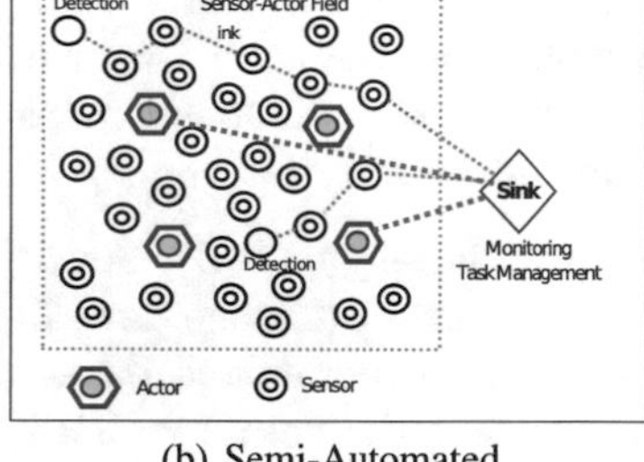

(b) Semi-Automated

Fig. 2. WSAN architectures.

3 Proposed System Model

3.1 Problem Description

After data has been sensed from sensors, they are collected to the sink for semi-automated architecture or spread to the actors for fully-automated architecture. Then a task is assigned to actors. In general, one or more actors take responsibility and perform appropriate actions. Different actors may be chosen for acting, depending on their characteristics and conditions. For example, if an intervention is required in a building, a flying robot can go there faster and easier. While, if a kid is inside a room in fire, it is better to send a small robot. The issue here is which of the actors will be selected to respond to critical data collected from the field (actor selection).

If WSAN uses semi-automated architecture, the sinks are used to collect data and control the actors. They may be supplied with detailed information about actors characteristics (size, ability etc.). If fully-automated architecture is being used, the collected data are processed only by actors, so they first have to decide whether they have the proper ability and right conditions to perform. Soon after that, actors coordinate with each-other, to decide more complicated procedures like acting multiple actors, or choosing the most appropriate one from several candidates. In this work, we propose a fuzzy-based system in order to select an appropriate actor node for a required task.

3.2 System Parameters

Based on WSAN characteristics and challenges, we consider the following parameters for implementation of our proposed system.

Job Type (JT): A sensed event may be triggered by various causes, such as when water level passed a certain height of the dam. Similarly, for solving a problem, actors need to perform actions of different types. Actions may be classified regarding time duration, complexity, working force required etc., and then assign a priority to them, which will guide actors to make their decisions. The hardest the task, the more likely an actor is to be selected.

Distance to Event (DE): The number of actors in a WSAN is smaller than the number of sensors. Thus, when an actor is called for action near an event, the distance from the actor to the event is important because when the distance is longer, the actor will spend more energy. Thus, an actor which is close to an event, should be selected.

Remaining Energy (RE): As actors are active in the monitored field, they perform tasks and exchange data in different ways. Thus some actors may have a lot of remained power and other may have very little, when an event occurs. It is better that the actors which have more power are selected to carry out a task.

Failure of Assigned Task (FAT): Due to node failure and bad communication between sensor and actors, an actor may fail to carry out an assigned task. If an actor fails many times, it is better not to select for the next task.

Actor Selection Decision (ASD): Our system is able to decide the willingness of an actor to be assigned a certain task at a certain time. The actors respond in five different levels, which can be interpreted as:

- Very Low Selection Possibility (VLSP) - It is not worth assigning the task to this actor.
- Low Selection Possibility (LSP) - There might be other actors which can do the job better.
- Middle Selection Possibility (MSP) - The Actor is ready to be assigned a task, but is not the "chosen" one.
- High Selection Possibility (HSP) - The actor takes responsibility of completing the task.
- Very High Selection Possibility (VHSP) - Actor has almost all required information and potential and takes full responsibility.

3.3 System Implementation

Fuzzy sets and fuzzy logic have been developed to manage vagueness and uncertainty in a reasoning process of an intelligent system such as a knowledge based system, an expert system or a logic control system [17–30]. In this work, we use fuzzy logic to implement. We consider three levels of RP for actor selection.

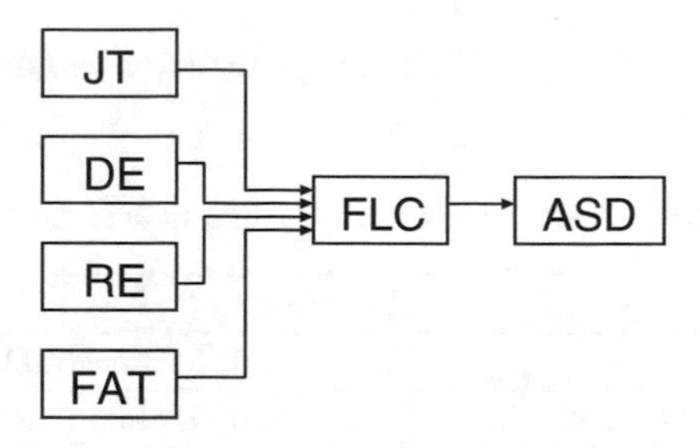

Fig. 3. Proposed system.

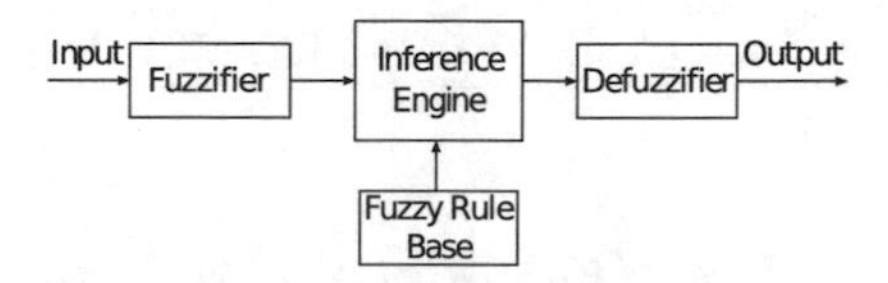

Fig. 4. FLC structure.

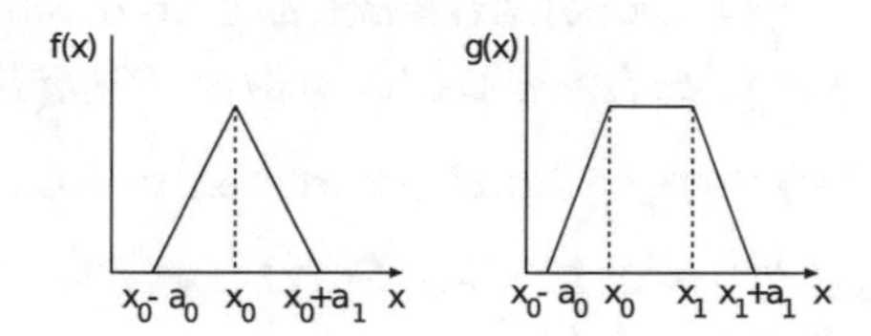

Fig. 5. Triangular and trapezoidal membership functions.

The structure of the proposed system is shown in Fig. 3. It consists of one Fuzzy Logic Controller (FLC), which is the main part of our system and its basic elements are shown in Fig. 4. They are the fuzzifier, inference engine, Fuzzy Rule Base (FRB) and defuzzifier.

As shown in Fig. 5, we use triangular and trapezoidal membership functions for FLC, because they are suitable for real-time operation [31]. The x_0 in $f(x)$ is the center of triangular function, $x_0(x_1)$ in $g(x)$ is the left (right) edge. We consider three levels of RP for actor selection.

The x_0 in $f(x)$ is the center of triangular function, $x_0(x_1)$ in $g(x)$ is the left (right) edge of trapezoidal function, and $a_0(a_1)$ is the left (right) width of the triangular or trapezoidal function. We explain in details the design of FLC in following.

3.4 Description of FLC

We use four input parameters for FLC:

- Job Type (JT);
- Distance to Event (DE);
- Remaining Energy (RE);
- Failure of Assigned Task (FAT).

The term sets for each input linguistic parameter are defined respectively as shown in Table 1.

Table 1. Parameters and their term sets for FLC.

Parameters	Term sets
Job Type (JT)	Easy (Ea), Moderate (Mo), Hard (Hd)
Distance to Event (DE)	Near (Ne), Middle (Md), Far (Fa)
Remaining Energy (RE)	Low (Lo), Medium (Mdm), High (Hi)
Failure of Assigned Task (FAT)	Low (Lw), Medium (Me), High (Hg)
Actor Selection Decision (ASD)	VLSP, LSP, MSP, HSP, VHSP

$$\begin{aligned}
T(JT) &= \{Easy(Ea), Moderate(Mo), Hard(Hd)\} \\
T(DE) &= \{Near(Ne), Middle(Md), Far(Fa)\} \\
T(RE) &= \{Low(Lo), Medium(Mdm), High(Hi)\} \\
T(FAT) &= \{Low(Lw), Medium(Me), High(Hg)\}
\end{aligned}$$

The membership functions for input parameters of FLC are defined as:

$$\begin{aligned}
\mu_{Ea}(JT) &= g(JT; Ea_0, Ea_1, Ea_{w0}, Ea_{w1}) \\
\mu_{Me}(JT) &= f(JT; Me_0, Me_{w0}, Me_{w1}) \\
\mu_{Hi}(JT) &= g(JT; Hd_0, Hd_1, Hd_{w0}, Hd_{w1}) \\
\mu_{Ne}(DE) &= g(DE; Ne_0, Ne_1, Ne_{w0}, Ne_{w1}) \\
\mu_{Md}(DE) &= f(DE; Md_0, Md_{w0}, Md_{w1}) \\
\mu_{Fa}(DE) &= g(DE; Fa_0, Fa_1, Fa_{w0}, Fa_{w1}) \\
\mu_{Lo}(RE) &= g(RE; Lo_0, Lo_1, Lo_{w0}, Lo_{w1}) \\
\mu_{Mdm}(RE) &= f(RE; Mdm_0, Mdm_{w0}, Mdm_{w1}) \\
\mu_{Hi}(RE) &= g(RE; Hi_0, Hi_1, Hi_{w0}, Hi_{w1}). \\
\mu_{Lw}(FAT) &= g(FAT; Lw_0, Lw_1, Lw_{w0}, Lw_{w1}) \\
\mu_{Me}(FAT) &= f(FAT; Me_0, Me_{w0}, Me_{w1}) \\
\mu_{Hg}(FAT) &= g(FAT; Hg_0, Hg_1, Hg_{w0}, Hg_{w1})
\end{aligned}$$

The small letters *w0* and *w1* mean left width and right width, respectively.

The output linguistic parameter is the Actor Selection Decision (ASD). We define the term set of ASD as:

$$\begin{aligned}
\{&Very\ Low\ Selection\ Possibility\ (VLSP), \\
&Low\ Selection\ Possibility\ (LSP), \\
&Middle\ Selection\ Possibility\ (MSP), \\
&High\ Selection\ Possibility\ (HSP), \\
&Very\ High\ Selection\ Possibility\ (VHSP)\}.
\end{aligned}$$

Table 2. FRB of proposed fuzzy-based system.

No.	JT	DE	RE	FAT	ASD	No.	JT	DE	RE	FAT	ASD
1	Ea	Ne	Lo	Lw	MSP	41	Mo	Md	Mdm	Md	MSP
2	Ea	Ne	Lo	Me	VHSP	42	Mo	Md	Mdm	Hg	HSP
3	Ea	Ne	Lo	Hg	VHSP	43	Mo	Md	Hi	Lw	MSP
4	Ea	Ne	Mdm	Lw	MSP	44	Mo	Md	Hi	Md	MSP
5	Ea	Ne	Mdm	Me	VHSP	45	Mo	Md	Hi	Hg	VHSP
6	Ea	Ne	Mdm	Hg	VHSP	46	Mo	Fa	Lo	Lw	VLSP
7	Ea	Ne	Hi	Lw	VHSP	47	Mo	Fa	Lo	Md	VLSP
8	Ea	Ne	Hi	Me	VHSP	48	Mo	Fa	Lo	Hg	MSP
9	Ea	Ne	Hi	Hg	VHSP	49	Mo	Fa	Mdm	Lw	LSP
10	Ea	Md	Lo	Lw	VLSP	50	Mo	Fa	Mdm	Md	LSP
11	Ea	Md	Lo	Me	HSP	51	Mo	Fa	Mdm	Hg	MSP
12	Ea	Md	Lo	Hg	VHSP	52	Mo	Fa	Hi	Lw	LSP
13	Ea	Md	Mdm	Lw	MSP	53	Mo	Fa	Hi	Md	MSP
14	Ea	Md	Mdm	Me	HSP	54	Mo	Fa	Hi	Hg	VHSP
15	Ea	Md	Mdm	Hg	VHSP	55	Hd	Ne	Lo	Lw	VLSP
16	Ea	Md	Hi	Lw	VHSP	56	Hd	Ne	Lo	Md	LSP
17	Ea	Md	Hi	Me	VHSP	57	Hd	Ne	Lo	Hg	MSP
18	Ea	Md	Hi	Hg	VHSP	58	Hd	Ne	Mdm	Lw	LSP
19	Ea	Fa	Lo	Lw	HSP	59	Hd	Ne	Mdm	Md	MSP
20	Ea	Fa	Lo	Md	MSP	60	Hd	Ne	Mdm	Hg	MSP
21	Ea	Fa	Lo	IIg	IISP	61	IId	Ne	Hi	Lw	MSP
22	Ea	Fa	Mdm	Lw	LSP	62	Hd	Ne	Hi	Md	HSP
23	Ea	Fa	Mdm	Md	MSP	63	Hd	Ne	Hi	Hg	VHSP
24	Ea	Fa	Mdm	Hg	VHSP	64	Hd	Md	Lo	Lw	VLSP
25	Ea	Fa	Hi	Lw	HSP	65	Hd	Md	Lo	Md	VLSP
26	Ea	Fa	Hi	Md	VHSP	66	Hd	Md	Lo	Hg	LSP
27	Ea	Fa	Hi	Hg	VHSP	67	Hd	Md	Mdm	Lw	VLSP
28	Mo	Ne	Lo	Lw	HSP	68	Hd	Md	Mdm	Md	VLSP
29	Mo	Ne	Lo	Md	VHSP	69	Hd	Md	Mdm	Hg	MSP
30	Mo	Ne	Lo	Hg	VHSP	70	Hd	Md	Hi	Lw	LSP
31	Mo	Ne	Mdm	Lw	VLSP	71	Hd	Md	Hi	Md	MSP
32	Mo	Ne	Mdm	Md	MSP	72	Hd	Md	Hi	Hg	HSP
33	Mo	Ne	Mdm	Hg	VHSP	73	Hd	Fa	Lo	Lw	VLSP
34	Mo	Ne	Hi	Lw	MSP	74	Hd	Fa	Lo	Md	VLSP
35	Mo	Ne	Hi	Md	VHSP	75	Hd	Fa	Lo	Hg	VLSP
36	Mo	Ne	Hi	Hg	VHSP	76	Hd	Fa	Mdm	Lw	LSP
37	Mo	Md	Lo	Lw	VLSP	77	Hd	Fa	Mdm	Md	VLSP
38	Mo	Md	Lo	Md	LSP	78	Hd	Fa	Mdm	Hg	MSP
39	Mo	Md	Lo	Hg	HSP	79	Hd	Fa	Hi	Lw	VLSP
40	Mo	Md	Mdm	Lw	LSP	80	Hd	Fa	Hi	Md	LSP
						81	Hd	Fa	Hi	Hg	MSP

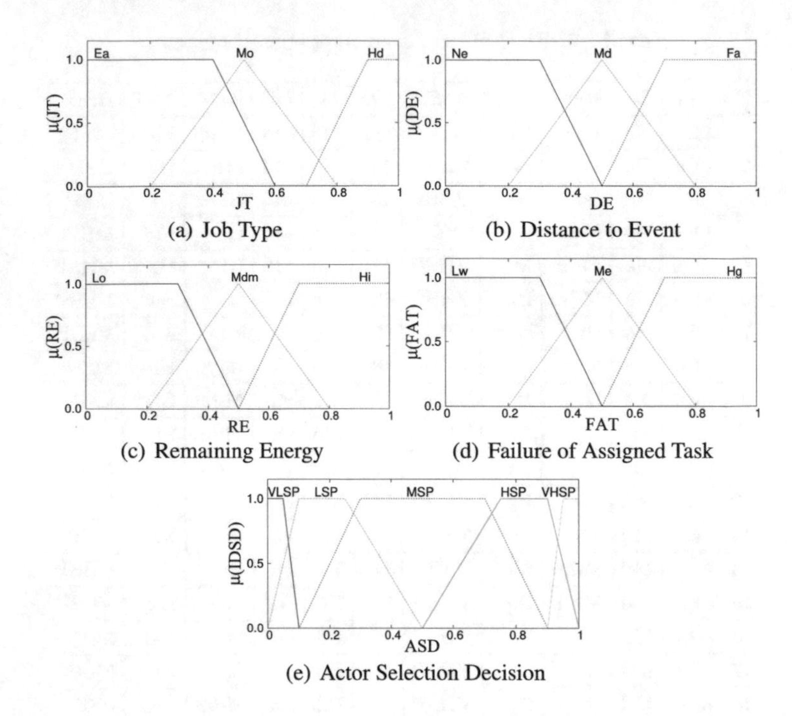

Fig. 6. Fuzzy membership functions.

The membership functions for the output parameter ASD are defined as:

$$\mu_{VLSP}(ASD) = g(ASD; VLSP_0, VLSP_1, VLSP_{w0}, VLSP_{w1})$$
$$\mu_{LSP}(ASD) = g(ASD; LSP_0, LSP_1, LSP_{w0}, LSP_{w1})$$
$$\mu_{MSP}(ASD) = g(ASD; MSP_0, MSP_1, MSP_{w0}, MSP_{w1})$$
$$\mu_{HSP}(ASD) = g(ASD; HSP_0, HSP_1, HSP_{w0}, HSP_{w1})$$
$$\mu_{VHSP}(ASD) = g(ASD; VHSP_0, VHSP_1, VHSP_{w0}, VHSP_{w1}).$$

The membership functions are shown in Fig. 6 and the Fuzzy Rule Base (FRB) is shown in Table 2. The FRB forms a fuzzy set of dimensions $|T(JT)| \times |T(DE)| \times |T(RE)| \times |T(FAT)|$, where $|T(x)|$ is the number of terms on $T(x)$. The FRB has 81 rules. The control rules have the form: IF "conditions" THEN "control action".

4 Simulation Results

We present the simulation results in Figs. 7, 8 and 9. From simulation results, we found that as JT becomes difficult, the ASD becomes higher, because actors are programmed for different jobs. In Fig. 7 are shown the simulation results for DE = 0.1, we can see that the performance decreases from 0.6 to 1 unit as the FAT parameter increases. But, when RE is increased, the ASD is increased.

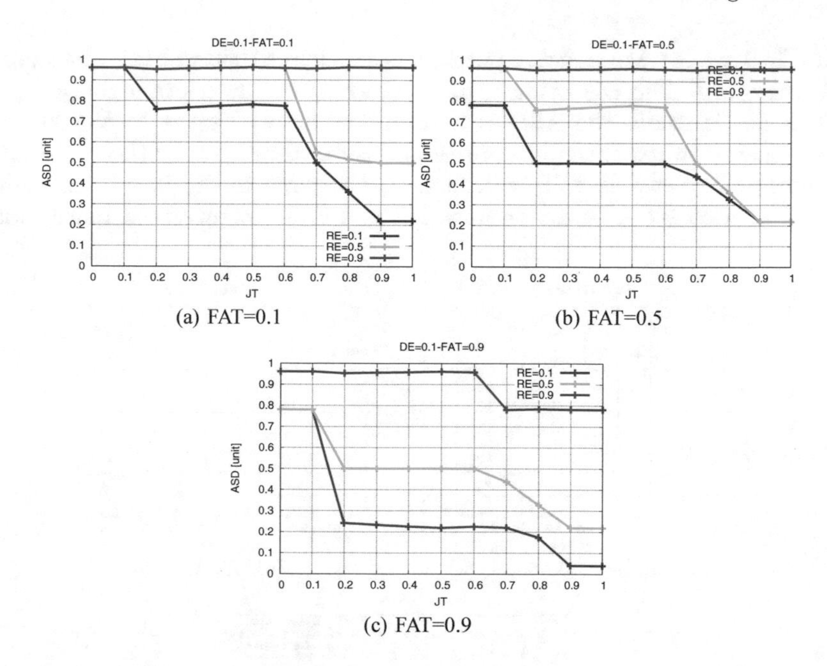

(a) FAT=0.1

(b) FAT=0.5

(c) FAT=0.9

Fig. 7. Results for $DE = 0.1$.

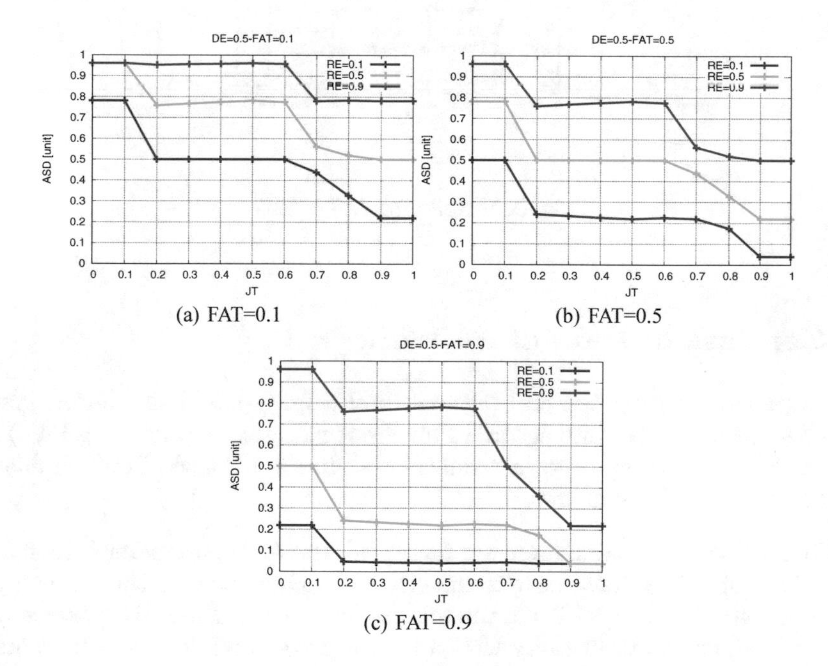

(a) FAT=0.1

(b) FAT=0.5

(c) FAT=0.9

Fig. 8. Results for $DE = 0.5$.

In Fig. 8, we can see that the performance is lower than in Fig. 7, because the DE is increased. The DE defines the distance of the actor from the job place, so when DE is small, the ASD is higher. The actors closest to the job place use less energy to reach the job position. Also in these graphs the performance continues to decrease as FAT increases. Furthermore in Fig. 9, we can see that the performance is the lowest because DE and FAT are increased much more.

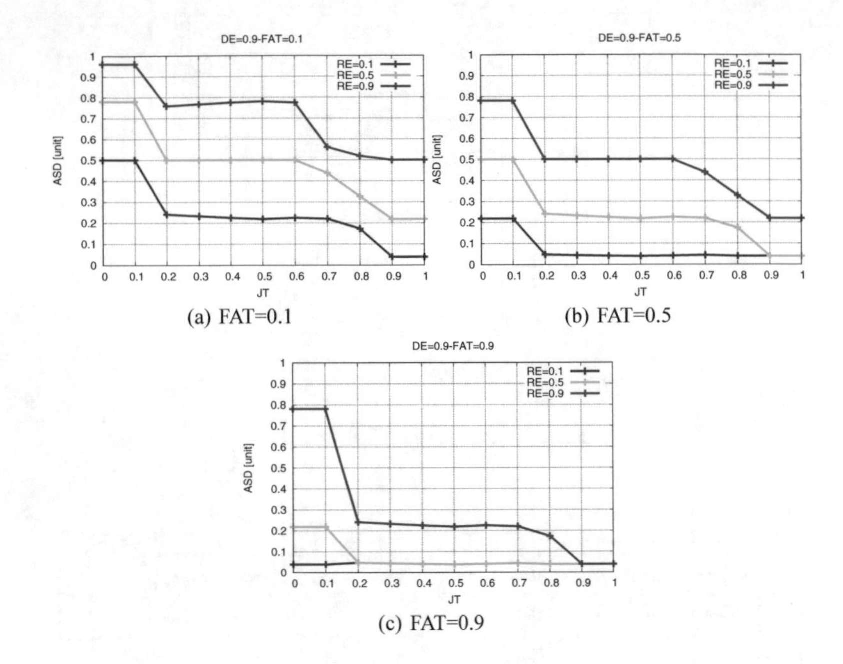

(a) FAT=0.1

(b) FAT=0.5

(c) FAT=0.9

Fig. 9. Results for $DE = 0.9$.

5 Conclusions and Future Work

In this paper, we proposed and implemented a fuzzy-based simulation system for WSAN, which takes into account four input parameters (including FAT) and decides the actor selection for a required task in the network. From simulation results, we conclude as follows.

- When JT and RE parameters are increased, the ASD parameter is increased, so the probability that the system selects an actor node for the job is high.
- When the DE and FAT parameter are increased, the ASD parameter is decreased, so the probability that an actor node is selected for the required task is low.

In the future work, we will consider also other parameters for actor selection and make extensive simulations to evaluate the proposed system.

References

1. Akyildiz, I.F., Kasimoglu, I.H.: Wireless sensor and actor networks: research challenges. Ad Hoc Netw. J. **2**(4), 351–367 (2004)
2. Akyildiz, I., Su, W., Sankarasubramaniam, Y., Cayirci, E.: Wireless sensor networks: a survey. Comput. Netw. **38**(4), 393–422 (2002)
3. Boyinbode, O., Le, H., Takizawa, M.: A survey on clustering algorithms for wireless sensor networks. Int. J. Space-Based Situated Comput. **1**(2/3), 130–136 (2011)
4. Bahrepour, M., Meratnia, N., Poel, M., Taghikhaki, Z., Havinga, P.J.: Use of wireless sensor networks for distributed event detection in disaster managment applications. Int. J. Space-Based Situated Comput. **2**(1), 58–69 (2012)
5. Haider, N., Imran, M., Saad, N., Zakariya, M.: Performance analysis of reactive connectivity restoration algorithms for wireless sensor and actor networks. In: IEEE Malaysia International Conference on Communications (MICC-2013), November 2013, pp. 490–495 (2013)
6. Abbasi, A., Younis, M., Akkaya, K.: Movement-assisted connectivity restoration in wireless sensor and actor networks. IEEE Trans. Parallel Distrib. Syst. **20**(9), 1366–1379 (2009)
7. Li, X., Liang, X., Lu, R., He, S., Chen, J., Shen, X.: Toward reliable actor services in wireless sensor and actor networks. In: 2011 IEEE 8th International Conference on Mobile Adhoc and Sensor Systems (MASS), October 2011, pp. 351–360 (2011)
8. Akkaya, K., Younis, M.: Cola: a coverage and latency aware actor placement for wireless sensor and actor networks. In: IEEE 64th Conference on Vehicular Technology (VTC-2006) Fall, September 2006, pp. 1–5 (2006)
9. Kakarla, J., Majhi, B.: A new optimal delay and energy efficient coordination algorithm for WSAN. In: 2013 IEEE International Conference on Advanced Networks and Telecommuncations Systems (ANTS), December 2013, pp. 1–6 (2013)
10. Elmazi, D., Cuka, M., Oda, T., Ikeda, M., Barolli, L.: Effect of node density on actor selection in wsans: a comparison study for two fuzzy-based systems. In: 2017 IEEE 31st International Conference on Advanced Information Networking and Applications (AINA), pp. 865–871. IEEE (2017)
11. Akbas, M., Turgut, D.: APAWSAN: actor positioning for aerial wireless sensor and actor networks. In: 2011 IEEE 36th Conference on Local Computer Networks (LCN), October 2011, pp. 563–570 (2011)
12. Akbas, M., Brust, M., Turgut, D.: Local positioning for environmental monitoring in wireless sensor and actor networks. In: 2010 IEEE 35th Conference on Local Computer Networks (LCN), October 2010, pp. 806–813 (2010)
13. Melodia, T., Pompili, D., Gungor, V., AkyildizZX, I.: Communication and coordination in wireless sensor and actor networks. IEEE Trans. Mob. Comput. **6**(10), 1126–1129 (2007)
14. Gungor, V., Akan, O., Akyildiz, I.: A real-time and reliable transport (RT2) protocol for wireless sensor and actor networks. IEEE/ACM Trans. Networking **16**(2), 359–370 (2008)
15. Selvaradjou, K., Handigol, N., Franklin, A., Murthy, C.: Energy-efficient directional routing between partitioned actors in wireless sensor and actor networks. IET Commun. **4**(1), 102–115 (2010)
16. Nakayama, H., Fadlullah, Z., Ansari, N., Kato, N.: A novel scheme for wsan sink mobility based on clustering and set packing techniques. IEEE Trans. Autom. Control **56**(10), 2381–2389 (2011)

17. Inaba, T., Sakamoto, S., Kolici, V., Mino, G., Barolli, L.: A CAC scheme based on fuzzy logic for cellular networks considering security and priority parameters. In: The 9-th International Conference on Broadband and Wireless Computing, Communication and Applications (BWCCA-2014), pp. 340–346 (2014)
18. Spaho, E., Sakamoto, S., Barolli, L., Xhafa, F., Barolli, V., Iwashige, J.: A fuzzy-based system for peer reliability in JXTA-overlay P2P considering number of interactions. In: The 16th International Conference on Network-Based Information Systems (NBiS-2013), pp. 156–161 (2013)
19. Matsuo, K., Elmazi, D., Liu, Y., Sakamoto, S., Mino, G., Barolli, L.: FACS-MP: a fuzzy admission control system with many priorities for wireless cellular networks and its performance evaluation. J. High Speed Netw. **21**(1), 1–14 (2015)
20. Grabisch, M.: The application of fuzzy integrals in multicriteria decision making. Eur. J. Oper. Res. **89**(3), 445–456 (1996)
21. Inaba, T., Elmazi, D., Liu, Y., Sakamoto, S., Barolli, L., Uchida, K.: Integrating wireless cellular and ad-hoc networks using fuzzy logic considering node mobility and security. In: The 29th IEEE International Conference on Advanced Information Networking and Applications Workshops (WAINA-2015), pp. 54–60 (2015)
22. Kulla, E., Mino, G., Sakamoto, S., Ikeda, M., Caballé, S., Barolli, L.: FBMIS: a fuzzy-based multi-interface system for cellular and ad hoc networks. In: International Conference on Advanced Information Networking and Applications (AINA-2014), pp. 180–185 (2014)
23. Elmazi, D., Kulla, E., Oda, T., Spaho, E., Sakamoto, S., Barolli, L.: A comparison study of two fuzzy-based systems for selection of actor node in wireless sensor actor networks. J. Ambient Intell. Humanized Comput. **6**(5), 635–645 (2015)
24. Zadeh, L.: Fuzzy logic, neural networks, and soft computing. Commun. ACM **37**(3), 77–84 (1994)
25. Spaho, E., Sakamoto, S., Barolli, L., Xhafa, F., Ikeda, M.: Trustworthiness in P2P: performance behaviour of two fuzzy-based systems for JXTA-overlay platform. Soft. Comput. **18**(9), 1783–1793 (2014)
26. Inaba, T., Sakamoto, S., Kulla, E., Caballe, S., Ikeda, M., Barolli, L.: An integrated system for wireless cellular and ad-hoc networks using fuzzy logic. In: International Conference on Intelligent Networking and Collaborative Systems (INCoS-2014), pp. 157–162 (2014)
27. Matsuo, K., Elmazi, D., Liu, Y., Sakamoto, S., Barolli, L.: A multi-modal simulation system for wireless sensor networks: a comparison study considering stationary and mobile sink and event. J. Ambient Intell. Humanized Comput. **6**(4), 519–529 (2015)
28. Kolici, V., Inaba, T., Lala, A., Mino, G., Sakamoto, S., Barolli, L.: A fuzzy-based CAC scheme for cellular networks considering security. In: International Conference on Network-Based Information Systems (NBiS-2014), pp. 368–373 (2014)
29. Liu, Y., Sakamoto, S., Matsuo, K., Ikeda, M., Barolli, L., Xhafa, F.: A comparison study for two fuzzy-based systems: improving reliability and security of JXTA-overlay P2P platform. Soft Comput. **20**(7), 2677–2687 (2015)
30. Matsuo, K., Elmazi, D., Liu, Y., Sakamoto, S., Mino, G., Barolli, L.: FACS-MP: a fuzzy admission control system with many priorities for wireless cellular networks and its perforemance evaluation. J. High Speed Netw. **21**(1), 1–14 (2015)
31. Mendel, J.M.: Fuzzy logic systems for engineering: a tutorial. Proc. IEEE **83**(3), 345–377 (1995)

Malicious Information Flow in P2PPS Systems

Shigenari Nakamura[1(✉)], Lidia Ogiela[2], Tomoya Enokido[3], and Makoto Takizawa[1]

[1] Hosei University, Tokyo, Japan
nakamura.shigenari@gmail.com, makoto.takizawa@computer.org
[2] AGH University of Science and Technology, Krakow, Poland
logiela@agh.edu.pl
[3] Rissho University, Tokyo, Japan
eno@ris.ac.jp

Abstract. We consider the peer-to-peer (P2P) type of topic-based publish/subscribe (P2PPS) model where each process (peer) can publish and subscribe event messages with no centralized coordinator. An event message e_1 may carry information of another event message e_2 causally preceding e_1, which are denoted by topics. If a peer receiving e_1 is not allowed to subscribe the topics, illegal information flow to the peer occur. In our previous studies, the subscription-based synchronization (SBS), topic-based synchronization (TBS), and flexible synchronization for hidden topics (FS-H) protocols are proposed to prevent illegal information flow. In this paper, we newly consider malicious information flow among peers where a source peer does not give related topics or gives unrelated topics to event messages. If a source peer p_i publishes an event message e without including some related topic or with including some unrelated topic, the peer p_i is malicious. In this paper, we newly define types of malicious information flow among peers based on a set of topics which every peer can publish and subscribe.

1 Introduction

A distributed system is composed of peer processes (peers) which are cooperating with one another by exchanging messages in networks. Through the cooperation of peers, information in objects flow to other objects. Even if a peer is not allowed to read data of an object, the peer can read the data in another object [3]. Here, illegal information flow occur. Types of synchronization protocols [6–9,13] are proposed based on the role-based access control (RBAC) model [5] to prevent illegal information flow among objects. On the other hand, context-based systems like publish/subscribe (PS) systems [1,2,4,19] are getting more important in various applications.

We consider a peer-to-peer (P2P) model [20] of topic-based PS system [18] (P2PPS model) [16,17] where each peer can play both publisher and subscriber roles with no centralized coordinator. The topic-based access control (TBAC)

L. Barolli et al. (Eds.): EIDWT 2018, LNDECT 17, pp. 119–129, 2018.
https://doi.org/10.1007/978-3-319-75928-9_10

model is proposed as an access control model in PS systems [10]. Here, only a peer granted publish and subscribe rights is allowed to publish and subscribe topics, respectively. The publication $p_i.P$ and subscription $p_i.S$ of a peer p_i are subsets of topics which the peer p_i is allowed to publish and subscribe, respectively. An event message e published by a peer p_i is delivered to a target peer p_j if the subscription $p_j.S$ includes at least one common topic with the publication $e.P$. Here, topics in the subscription $p_i.S$ but not in the subscription $p_j.S$ are *hidden* in the event message e. Topics in the publication $e.P$ but not in the subscription $p_j.S$ are *forgotten* in the event message e [12].

In our previous studies, the subscription-based synchronization (SBS) [10], topic-based synchronization (TBS) [12], and flexible synchronization for hidden topics (FS-H) [14] protocols are proposed based on the TBAC model to prevent illegal information flow. Here, event messages which cause illegal information flow are referred to as *illegal*. In the SBS protocol, it is checked whether or not an event message is illegal in terms of publication and subscription (PS) rights. Here, every illegal event message is banned, but even legal event message may be banned. In the TBS protocol, it is checked whether or not an event message is illegal in terms of only topics which each peer really manipulates. Hence, all and only illegal event messages are banned. In the protocols, it is difficult, maybe impossible for each peer to know about every topic, especially in a scalable P2PPS system. A learning mechanism is introduced so that a peer obtains new topics through communicating with other peers, in our previous studies [15]. In the FS-H protocol [14], even if an event message carries hidden topics which are strongly related with some subscription topics, the peer accepts the event message and adds the hidden topics to its subscription.

In this paper, we newly consider malicious information flow among peers in addition to illegal information flow. Each event message is characterized in terms of topics. If a source peer p_i publishes an event message e without giving some related topic or with some unrelated topic, a destination peer p_j may misunderstand the meaning of the event message e. If a source peer p_i publishes an event message e without including every topic which denotes information of the event message e, the peer p_i is considered to be *malicious*. In this paper, we newly define types of malicious event messages based on a set of topics which every peer can publish and subscribe.

In Sect. 2, we discuss the information flow relation among peers in the TBAC model. In Sect. 3, we discuss the protocols to prevent illegal information flow. In Sect. 4, we newly define malicious information flow among peers.

2 Information Flow in the TBAC Model

2.1 TBAC Model

In this paper, we consider a peer-to-peer (P2P) model [20] of a publish/subscribe (PS) system [1,2,4,19] (P2PPS model) [16,17]. Let P be a set of peer processes (peers) $p_1, \ldots, p_{pn}$ ($pn \geq 1$). Here, each peer p_i can play both publisher and subscriber roles and a pair of event messages published by different peers are independently delivered to every common target peer. In this paper, we consider a topic-based PS system [18]. Let T be a set $\{t_1, \ldots, t_{tn}\}$ $(tn \geq 1)$ of all topics in a system. A peer p_i publishes an event message e with publication $e.P$ $(\subseteq T)$. A peer p_i specifies the subscription $p_i.S$ $(\subseteq T)$. An event message e is delivered to a peer p_i if $e.P \cap p_i.S \neq \phi$. Here, p_i is a *target* peer of the event message e.

In the *topic-based access control* (TBAC) model [10], an access right is specified in a pair $\langle t, op \rangle$ of a topic t $(\in T)$ and an operation *op* which is a publish (*pb*) or subscribe (*sb*). A peer p_i is allowed to publish an event message e with publication $e.P$ $(\subseteq T)$ only if the peer p_i is granted a publication right $\langle t, pb \rangle$ for every topic t in the publication $e.P$. The subscription $p_i.S$ $(\subseteq T)$ of a peer p_i is a subset of topics which the peer p_i is allowed to subscribe. If a peer p_i is a target peer of an event message e from a peer p_j, topics of the event message e in the subscription $p_j.S$ but not in the subscription $p_i.S$ are *hidden* topics $e.H$. Here, a target peer p_i may receive an event message including the hidden topics which the peer p_i is not allowed to subscribe.

2.2 Information Flow Relations

First, the information flow relation $p_i \rightarrow p_j$ means an event message published by a peer p_i is allowed to be delivered to a peer p_j. The information flow relation $(p_i \rightarrow p_j)$ is defined as follows [10,11]:

Definition 1. A peer p_i *precedes* a peer p_j with respect to information flow $(p_i \rightarrow p_j)$ iff (if and only if) $p_i.P \cap p_j.S \neq \phi$.

If a peer p_i precedes a peer p_j $(p_i \rightarrow p_j)$, an event message published by the peer p_i can be delivered to the peer p_j. Otherwise, no information from the peer p_i flow into the peer p_j.

Definition 2. [10,11]

1. A peer p_i *legally precedes* a peer p_j $(p_i \Rightarrow p_j)$ iff one of the following conditions holds:
   a. $p_i.S \neq \phi$, $p_i \rightarrow p_j$, and $p_i.S \subseteq p_j.S$.
   b. For some peer p_k, $p_i \Rightarrow p_k$ and $p_k \Rightarrow p_j$.
2. A pair of peers p_i and p_j are *legally equivalent* with each other $(p_i \Leftrightarrow p_j)$ iff $p_i \Rightarrow p_j$ and $p_j \Rightarrow p_i$.
3. A peer p_i *illegally precedes* a peer p_j $(p_i \mapsto p_j)$ iff $p_i \rightarrow p_j$ but $p_i \not\Rightarrow p_j$.

The information flow relation $\rightarrow$ is not transitive. The legal information flow relation $\Rightarrow$ is transitive but not symmetric.

Suppose a peer p_i publishes an event message e_2 with publication $e_2.P$ $(\subseteq p_i.P)$ after another event message e_1 published by a peer p_j is delivered to the peer p_i. Here, the event message e_2 might carry information in the event message e_1. The information is characterized by topics in the subscription $p_i.S$. Topics in the subscription $p_i.S$ but not in the subscription $p_k.S$ of a destination peer p_k are *hidden* topics which the event message e_2 carries to the destination peer.

A target peer of the event message e_2 does not know about the hidden topics. Let $e.H$ be a set of hidden topics of an event message e published by a peer p_i with respect to the target peer p_j [10–12]. Hidden topics of an event message e published by a peer p_i might be related with the topics which the peer p_i may so far obtain but are not included in the subscription $p_j.S$, i.e. $\{t \mid t \in p_i.S \wedge t \notin p_j.S\}$. Here, even if a target peer p_j receives an event message e, the peer p_j does not recognize the event message e might be related with the hidden topics. Next, topics in the publication $e.P$ but not in the subscription $p_j.S$, i.e. $\{t \mid t \in e.P \wedge t \notin p_j.S\}$, are *forgotten* topics $e.F$ [10–12] of the event message e with respect to the target peer p_j. A target peer p_j recognizes an event message e to be only related with topics in $e.P \cap p_j.S$ but forgets that the event message e is related with the forgotten topics in $e.F$.

If an event message e is delivered to a target peer p_i, the event message e is related with not only forgotten topics $e.F$ but also hidden topics $e.H$. *Implicit* topics of a peer p_i are hidden or forgotten topics of event messages which the peer p_i receives. Let $p_i.I$ indicate a set of implicit topics of a peer p_i [10–12]. $p_i.I$ is manipulated by a peer p_i as follows:
[A peer p_i]

1. Initially $p_i.I = \phi$;
2. [Receipt] Each time a peer p_i receives an event message e from a peer p_j, $e.F = e.P - p_i.S$; $e.H = p_j.S - p_i.S$; $p_i.I = p_i.I \cup e.H \cup (e.F - e.H)$;
3. [Publication] The peer p_i publishes an event message e with publication $e.P$.

Thus, implicit topics are accumulated in the variable $p_i.T$ of the peer p_i each time the peer p_i receives an event message.

Example 1. Suppose there are three topics t_1, t_2, and t_3, $T = \{t_1, t_2, t_3\}$ in a system. We also suppose a peer p_i is granted access rights $\langle t_2, pb\rangle$, $\langle t_3, pb\rangle$, $\langle t_1, sb\rangle$, and $\langle t_2, sb\rangle$, another peer p_j is granted access rights $\langle t_1, pb\rangle$, $\langle t_1, sb\rangle$, and $\langle t_2, sb\rangle$, and the other peer p_k is granted access rights $\langle t_3, pb\rangle$, $\langle t_1, sb\rangle$, and $\langle t_3, sb\rangle$, i.e. $p_i.P$ $(= \{t_2, t_3\})$, $p_i.S$ $(= \{t_1, t_2\})$, $p_j.P$ $(= \{t_1\})$, $p_j.S$ $(= \{t_1, t_2\})$, $p_k.P$ $(= \{t_3\})$, and $p_k.S$ $(= \{t_1, t_3\})$. First, the peer p_i publishes an event message e_1 with publication $e_1.P = \{t_2, t_3\}$ $(\subseteq p_i.P)$. Here, the peer p_i precedes the peer p_j $(p_i \rightarrow p_j)$ since $p_i.P$ $(= \{t_2, t_3\}) \cap p_j.S$ $(= \{t_1, t_2\}) \neq \phi$. In this case, the topic t_3 is a forgotten topic of the event message e since $\{t_3\} \in e.P$ but $\{t_3\} \notin p_j.S$. $p_i \Rightarrow p_j$ since $p_i.S \neq \phi$, $p_i \rightarrow p_j$, and $p_i.S \subseteq p_j.S$. Hence, the event message e_1 is delivered to the peer p_j.

Next, suppose a peer p_j publishes an event message e_2 with publication $e_2.P = \{t_1\}$ $(\subseteq p_j.P)$. Here, $p_j \rightarrow p_k$ since $p_j.P$ $(= \{t_1\}) \cap p_j.S$ $(= \{t_1, t_3\}) \neq \phi$. However, the peer p_j illegally precedes the peer p_k $(p_j \mapsto p_k)$ since $p_j.T$ $(= \{t_1, t_2\}) \not\subseteq p_k.S$ $(= \{t_1, t_3\})$. This means, an event message on the topic t_2 which the peer p_k is not allowed to subscribe can be delivered to the peer p_k. In this case, a topic t_2 is hidden topic of the event message e since $\{t_2\} \in p_j.S$ but $\{t_2\} \notin p_k.S$. Here, information illegally flows to the peer p_k from the peer p_j.

3 Synchronization Protocols for Hidden Topics

In our previous studies, subscription-based synchronization (SBS) [10], topic-based synchronization (TBS) [12], and flexible synchronization for hidden topics (FS-H) [14] protocols are proposed to check whether or not illegal information flow to occur when an event message is delivered to a target peer. Here, each peer p_i has a topic set $p_i.T$. The set $p_i.T$ is composed of topics which the peer so far obtains. Each time the peer p_i receives an event message, the topics which the event message carries are added to the set $p_i.T$.

In the SBS protocol, it is checked whether or not an event message may cause illegal information flow in terms of access rights of each peer. If a peer p_i publishes an event message e and $p_i \rightarrow p_j$, a peer p_j manipulates the topic set $p_j.T$ on receipt of the event message e as follows:

[Subscription-based Synchronization (SBS) Protocol]

1. If $p_i.T \subseteq p_j.S$, the event message e is delivered to the peer p_j and $p_j.T = p_i.S \cup p_j.T$;
2. Otherwise, the event message e is banned at the peer p_j.

In the TBS protocol, it is checked whether or not an event message may cause illegal information flow in terms of only topics which each peer really manipulates. Suppose a peer p_i publishes an event message e and there is a peer p_j such that $p_i \rightarrow p_j$.

[Topic-based Synchronization (TBS) Protocol]

1. If $p_i.T \subseteq p_j.S$, the event message e is delivered to a peer p_j and $p_j.T = p_i.T \cup p_j.T$;
2. Otherwise, the event message e is banned at the peer p_j.

To more reduce the number of event messages banned, the flexible synchronization for hidden topics (FS-H) protocol is proposed [14]. In the FS-H protocol, on receipt of an event message e, a peer p_i accepts the event message e only if some hidden topic of the event message e is strongly related with topics in which the peer p_i is interested. In addition, each peer p_i learns other topics which p_i should subscribe from event messages.

The relevance concept of topics [15] shows how much a pair of topics are related. Here, topics are totally ordered as $t_1, \ldots, t_{tn}$ as follows:

- For three topics t_i, t_j, and t_k, if $|i - j| < |i - k|$, the topic t_j is more related with the topic t_i than the topic t_k.

Then, we define the relevance RbT_{kl} ($0 \leq RbT_{kl} \leq 1$) between a kth topic t_k and an lth topic t_l as follows:

$$RbT_{kl} = e^{-\{(l-k)^2/(2\times10^2)\}}. \quad (1)$$

We assume there is a main topic $p_i.mt$ in the subscription $p_i.S$ of each peer p_i. The main topic t_k ($= p_i.mt$) of a peer p_i means that the peer p_i subscribes topics related with the topic t_k.

[Flexible Synchronization for Hidden Topics (FS-H) Protocol]. A peer p_i publishes an event message e and $p_i \rightarrow p_j$ where t_k is the main topic of p_j.

1. If $p_i.T \subseteq p_j.S$, the event message e is delivered to the peer p_j and $p_j.T = p_i.S \cup p_j.T$;
2. If $p_i.T \not\subseteq p_j.S$ and $BWT(e,\ k) \leq AWT(e,\ p_j,\ k)$, the event message e is delivered to a peer p_j. Each topic t_l ($\{t_l \mid t_l \in e.H\}$) is added to $p_j.S$ with probability RbT_{kl}. $p_j.T = p_i.S \cup p_j.T$.
3. Otherwise, the event message e is banned at the peer p_j.

The summation of acceptance weights $AWT(e,\ p_j,\ k)$ and ban weights $BWT(e,\ k)$ are given as follows:

$$AWT(e,\ p_j,\ k) = \sum_{t_m \in p_i.T \cap p_j.S} RbT_{km}. \tag{2}$$

$$BWT(e,\ k) = \sum_{t_l \in e.H} (1 - RbT_{kl}). \tag{3}$$

4 Malicious Information Flow

In our previous studies, the SBS [10], TBS [12], and FS-H [14] protocols are proposed to prevent illegal information flow from occurring. If an implicit topic t exists in a publication from a source peer p_i, some information on the topic t which a target peer p_j is not allowed to subscribe can be delivered to the target peer p_j. Here, illegal information flow from the source peer p_i to the target peer p_j occur. In this paper, we newly consider a malicious source peer which does not give related topics or gives unrelated topics to an event message which the peer publishes. Here, *malicious information* flow occur from the source peer to the destination pees in the P2PPS model.

Suppose a peer p_i publishes an event message e_2 with publication $e_2.P$ to the peer p_j after another event message e_1 is delivered to the peer p_i as shown in Fig. 1. Here, the event message e_1 causally precedes the event message e_2. We assume some topic t related with the event message e_1 are not included in the publication $e_2.P$. Here, the peer p_j recognizes the event message e_2 is only related with the topics in the publication $e_2.P$. However, the event message e_2 might carry information on the topic t included in the event message e_1. As discussed in the SBS, TBS, and FS-H protocols, the peer p_i obtains new publication topics of the event message e_1 to the set $p_i.T$ on receipt of the event message e_1. That is, some information which the event message e_2 carries is not related with topics in the publication $e_2.P$. Here, the peer p_i may publish an event message e_2 with malice since the peer p_i does not give every topic to the event message e_2.

Suppose a peer p_j receives an event message e published by a peer p_i. As presented in the preceding sections, each peer p_i accumulate topics newly obtained in the variable $p_i.T$ as the peer p_i receives an event message. On receipt of an event message e, the peer p_i accepts the event message e if $e.P \cap p_i.T = \phi$. There are cases on conditions of the topic sets $p_i.T$, $p_i.P$, and $e.P$ as shown in Table 1.

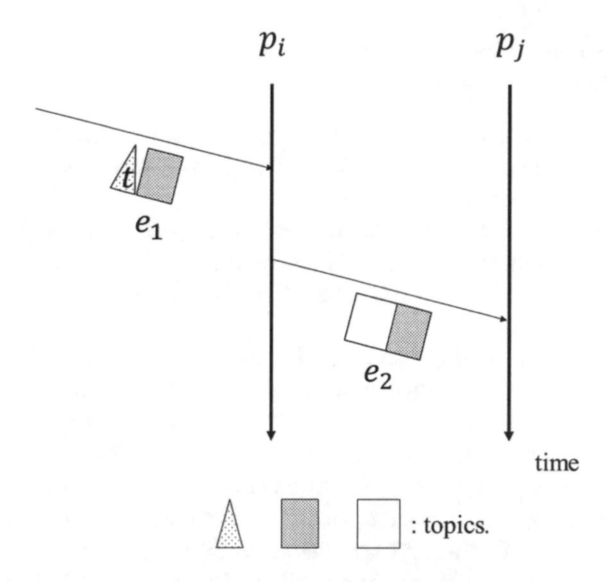

Fig. 1. p_i publishes e_2 to p_j after delivery of e_1.

Table 1. Conditions to decide the feature of each information flow.

Case #	$p_i.T \subseteq p_i.P$	$p_i.T \subseteq e.P$	$p_i.T \subseteq p_j.S$
1	○	○	○
2	○	○	×
3	○	×	○
4	○	×	×
5	×	×	○
6	×	×	×

○ : satisfy, × : not satisfy.

1. The first condition "$p_i.T \subseteq p_i.P$" means that the peer p_i can publish an event message with every topic which the peer p_i subscribes. If this condition does not hold, the source peer p_i may include some topic t in the event message e, where the topic t is not related with the event message e.
2. The second condition "$p_i.T \subseteq e.P$" shows that an event message e which the peer p_i publishes includes every topic which the peer p_i subscribes. If this condition is not satisfied, the source peer p_i does not give related topics or gives unrelated topics to event message e. Here, malicious information flow from the peer p_i to the peer p_j may occur.
3. The third condition "$p_i.T \subseteq p_j.S$" denotes that the peer p_j can accept an event message with every topic which the peer p_i subscribes. If this condition does not hold, the event message e may carry information on topics which the target peer p_j is not allowed to subscribe. Here, illegal information flow

from the peer p_i to the peer p_j may occur as discussed in the SBS [10], TBS [12], and FS-H [14] protocols.

In the case 1, a pair of the conditions "$p_i.T \subseteq e.P$" and "$p_i.T \subseteq p_j.S$" hold in addition to the condition "$p_i.T \subseteq p_i.P$". Here, information flow is not only legal but also unmalicious.

In the cases 2, 4, and 6, the third condition "$p_i.T \subseteq p_j.S$" is not satisfied. This means, the event message e may carry information on some topics which the peer p_j can not subscribe. On the other hand, no illegal information flow occur in cases 1, 3, and 5.

In the cases 3, 4, 5, and 6, the second condition "$p_i.T \subseteq e.P$" does not hold. This means, the event message e may carry information which the peer p_i does not subscribe. The peer p_i may add topics to the event message e, which are not related with the event message e. Here, a malicious information flow from the peer p_i to the peer p_j may occur. On the other hand, no malicious information flow from the peer p_i to the peer p_j occur in the cases 1 and 2.

In the cases 5 and 6, the first condition "$p_i.T \subseteq p_i.P$" is not satisfied. Here, there are topics which the peer p_i is not allowed to publish. This means, the peer p_i cannot give topics which are not included in the publication $p_i.P$ of the peer p_i to the publication $e.P$ of the event message e. Hence, the second condition "$p_i.T \subseteq e.P$" always dons not hold.

In the cases 4 and 6, neither the second condition "$p_i.T \subseteq e.P$" nor the third condition "$p_i.T \subseteq p_j.S$" are satisfied. Here, information flow is not only illegal but also malicious.

Figure 2 shows relations among illegal and malicious event messages. Each area shows a subset of possible event messages. The numbers in Fig. 2 show the cases in Table 1.

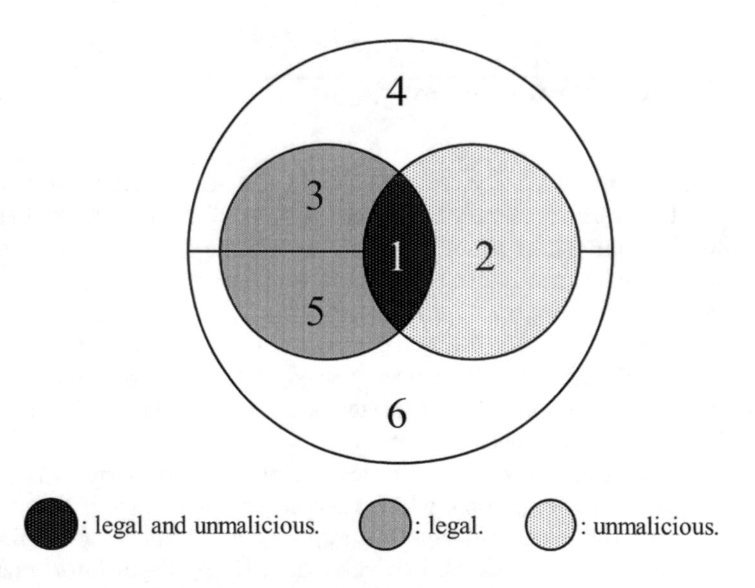

Fig. 2. Feature of each case of an information flow.

Table 2. Summary of effect brought about by banning information flows in the cases.

Case # banned	Effect	Protocol
2, 3, 4, 5, 6	Illegal or malicious flows are prevented	–
3, 4, 5, 6	Malicious flows are prevented	–
2, 4, 6	Illegal flows are prevented	SBS, TBS
4, 6	Illegal and malicious flows are prevented	–

– : not exist.

Table 2 summarizes what kinds of information flow, illegal or malicious flow, are prevented by banning event messages in each of the cases. If types of event messages are banned in the cases 2, 3, 4, 5, or 6, illegal or malicious information flow is prevented. On the other hand, if event messages are banned in the cases 4 or 6, illegal and malicious information flow is prevented. If event messages are banned in the cases 3, 4, 5, or 6, only malicious information flow is prevented. If event messages are banned in the cases 2, 4, or 6, only illegal information flow is prevented like the SBS and TBS protocols.

Example 2. We consider Example 1 again. Suppose the peer p_i already obtains information on a pair of topics t_1 and t_2, i.e. $p_i.T = \{t_1, t_2\}$. First, the peer p_i publishes the event message e_1 with publication $e_1.P = \{t_2, t_3\}$. The peer p_j receives the event message e_1 since $e_1.P$ $(= \{t_2, t_3\}) \cap p_j.S$ $(\{t_1, t_2\}) \neq \phi$. Here, the information flow from the peer p_i to the p_j occur in the case 5 since $p_i.T$ $(\{t_1, t_2\}) \not\subseteq p_i.P$ $(\{t_2, t_3\})$, $p_i.T$ $(\{t_1, t_2\}) \not\subseteq e_1.P$ $(\{t_2, t_3\})$, and $p_i.T$ $(\{t_1, t_2\}) \subseteq p_j.S$ $(\{t_1, t_2\})$. Hence, the information flow is not illegal but malicious since $p_i.T$ $(\{t_1, t_2\}) \subseteq p_j.S$ $(\{t_1, t_2\})$ and $p_i.T$ $(\{t_1, t_2\}) \not\subseteq e_1.P$ $(\{t_2, t_3\})$. If the information flow is not banned, a pair of topics t_1 and t_2 are added to the topic set $p_j.T$ of the peer p_j, i.e. $p_j.T = \{t_1, t_2\}$.

Next, the peer p_j publishes the event message e_2 with publication $e_2 = \{t_1\}$. The event message e_2 is delivered to the peer p_k since $e_2.P$ $(= \{t_1\}) \cap p_k.S$ $(\{t_1, t_3\}) \neq \phi$. Here, types of information flow from the peer p_j to the peer p_k occur in the case 6 since $p_j.T$ $(\{t_1, t_2\}) \not\subseteq p_j.P$ $(\{t_1\})$, $p_j.T$ $(\{t_1, t_2\}) \not\subseteq e_2.P$ $(\{t_1\})$, and $p_j.T$ $(\{t_1, t_2\}) \not\subseteq p_k.S$ $(\{t_1, t_3\})$. Hence, the information flow is illegal and malicious since $p_j.T$ $(\{t_1, t_2\}) \not\subseteq p_k.S$ $(\{t_1, t_3\})$ and $p_j.T$ $(\{t_1, t_2\}) \not\subseteq e_2.P$ $(\{t_1\})$.

In order to implement the protocols to prevent malicious information flow, each destination peer p_j has to get topic sets $p_j.T$ of each source peer p_i. A source peer p_i includes $p_i.T$ in an event message e on publishing the event message e, $e.T$. In order to reduce the length of each event message e, $e.T$ is realized in a bit map where the each bit shows a topic in a system.

5 Concluding Remarks

In our previous studies, we defined illegal information flow based on the TBAC model. In the SBS [10], TBS [12], and FS-H [14] protocols, an event message whose hidden topics may cause illegal information flow is banned by each target peer. In this paper, we newly defined malicious information flow based on sets of topics which are assigned to each peer. If the peer p_i publishes an event message e to the peer p_j without adding topics of information of the event message e into the publication $e.P$, the peer p_j receives information which the peer p_j does not expect to get. In this case, the peer p_i may publish the event message e with malice. Here, a malicious information flow from the peer p_i to the peer p_j occurs. Information flow is classified into six cases. Some types of information flow is not only illegal but also malicious.

We are designing classes of protocols to prevent malicious information flow as well as illegal information flow.

Acknowledgements. This work was supported by Japan Society for the Promotion of Scienc (JSPS) KAKENHI 15H0295 and Grant-in-Aid for JSPS Research Fellow grant 17J00106.

References

1. Google alert. http://www.google.com/alerts
2. Blanco, R., Alencar, P.: Event models in distributed event based systems. In: Principles and Applications of Distributed Event-Based Systems, pp. 19–42 (2010)
3. Denning, D.E.R.: Cryptography and Data Security. Addison Wesley, Reading (1982)
4. Eugster, P.T., Felber, P.A., Guerraoui, R., Kermarrec, A.M.: The many faces of publish/subscribe. ACM Comput. Surv. **35**(2), 114–131 (2003)
5. Ferraiolo, D.F., Kuhn, D.R., Chandramouli, R.: Role-Based Access Controls, 2nd edn. Artech, Boston (2007)
6. Nakamura, S., Duolikun, D., Enokido, T., Takizawa, M.: A flexible read-write abortion protocol to prevent illegal information flow among objects. J. Mob. Multimed. **11**(3&4), 263–280 (2015)
7. Nakamura, S., Duolikun, D., Enokido, T., Takizawa, M.: A write abortion-based protocol in role-based access control systems. Int. J. Adapt. Innov. Syst. **2**(2), 142–160 (2015)
8. Nakamura, S., Duolikun, D., Enokido, T., Takizawa, M.: A read-write abortion (RWA) protocol to prevent illegal information flow in role-based access control systems. Int. J. Space Based Situat. Comput. **6**(1), 43–53 (2016)
9. Nakamura, S., Duolikun, D., Takizawa, M.: Read-abortion (RA) based synchronization protocols to prevent illegal information flow. J. Comput. Syst. Sci. **81**(8), 1441–1451 (2015)
10. Nakamura, S., Enokido, T., Takizawa, M.: Information flow control models in peer-to-peer publish/subscribe systems. In: Proceedings of the 10th International Conference on Complex, Intelligent, and Software Intensive Systems (CISIS-2016), pp. 167–174 (2016)
11. Nakamura, S., Enokido, T., Takizawa, M.: Subscription initialization (SI) protocol to prevent illegal information flow in peer-to-peer publish/subscribe systems. In: Proceedings of the 19th International Conference on Network-Based Information Systems (NBiS-2016), pp. 42–49 (2016)

12. Nakamura, S., Enokido, T., Takizawa, M.: Topic-based synchronization (THS) protocols to prevent illegal information flow in peer-to-peerpublish/subscribe systems. In: Proceedings of the 11th International Conference on Broadband and Wireless Computing, Communication and Applications (BWCCA-2016), pp. 57–68 (2016)
13. Nakamura, S., Enokido, T., Takizawa, M.: Sensitivity-based synchronization protocol to prevent illegal information flow among objects. Int. J. Web Grid Serv. (IJWGS) **13**(3), 315–333 (2017)
14. Nakamura, S., Ogiela, L., Enokido, T., Takizawa, M.: A flexible synchronization protocol for hidden topics to prevent illegal information flow in P2PPS systems. In: Proceedings of the 12th International Conference on Broad-Band Wireless Computing, Communication and Applications (BWCCA-2017), pp. 138–148 (2017)
15. Nakamura, S., Ogiela, L., Enokido, T., Takizawa, M.: Flexible synchronization protocol to prevent illegal information flow in peer-to-peer publish/subscribe systems. In: Proceedings of the 11th International Conference on Complex, Intelligent, and Software Intensive Systems (CISIS-2017), pp. 82–93 (2017)
16. Nakayama, H., Duolikun, D., Enokido, T., Takizawa, M.: Selective delivery of event messages in peer-to-peer topic-based publish/subscribe systems. In: Proceedings of the 18th International Conference on Network-Based Information Systems (NBiS-2015), pp. 379–386 (2015)
17. Nakayama, H., Duolikun, D., Enokido, T., Takizawa, M.: Reduction of unnecessarily ordered event messages in peer-to-peer model of topic-based publish/subscribe systems. In: Proceedings of IEEE the 30th International Conference on Advanced Information Networking and Applications (AINA-2016), pp. 1160–1167 (2016)
18. Setty, V., van Steen, M., Vitenberg, R., Voulgaris, S.: PolderCast: fast, robust, and scalable architecture for P2P topic-based pub/sub. In: Proccedings of ACM/IFIP/USENIX 13th International Conference on Middleware (Middleware 2012), pp. 271–291 (2012)
19. Tarkoma, S.: Publish/Subscribe System: Design and Principles, 1st edn. Wiley, Hoboken (2012)
20. Waluyo, A.B., Taniar, D., Rahayu, W., Aikebaier, A., Takizawa, M., Srinivasan, B.: Trustworthy-based efficient data broadcast model for P2P interaction in resource-constrained wireless environments. J. Comput. Syst. Sci. (JCSS) **78**(6), 1716–1736 (2012)

Eco Migration Algorithms of Processes with Virtual Machines in a Server Cluster

Ryo Watanabe[1(✉)], Dilawaer Duolikun[1], Cuiqin Qin[1], Tomoya Enokido[2], and Makoto Takizawa[1]

[1] Hosei University, Tokyo, Japan
ryo.watanabe.4h@stu.hosei.ac.jp, dilewerdolkun@gmail.com, qcuiqin@gmail.com, makoto.takizawa@computer.org
[2] Rissho University, Tokyo, Japan
eno@ris.ac.jp

Abstract. It is critical to reduce the electric energy consumption of servers in a cluster. In this paper, we discuss a migration approach to reducing the electric energy consumption where a virtual machine with application processes migrates to a more energy-efficient server. We newly propose an ISEAM2 algorithm to select a pair of a virtual machine on a host server to perform a process issued by an application and a guest server to which the virtual machine migrates so that the electric energy consumption of the host and guest servers can be minimized. In the evaluation, we show the total electric energy consumption and active time of the servers and the average execution time of processes can be reduced in the ISEAM2 algorithm.

1 Introduction

We have to reduce electric energy consumption of servers in clusters like cloud computing systems [1] in order to realize eco society [3]. Energy-efficient hardware devices like CPUs [2] are developed in the hardware-oriented approach. In our macro-level approach [9,10,16], we aim at reducing the total electric energy consumed by a server to perform application processes. Here, if an application process is issued to a cluster, one energy-efficient server is selected to perform the application process in types of algorithms [8,10,12]. In the migration approach [5–7], an application process on a host server migrates to an energy-efficient guest server which is expected to consume smaller electric energy than the host server after the application process is started on the host server. Cloud computing systems support applications with virtual computation service by using virtual machines [4]. A virtual machine can migrate from a host server to another guest server like the live migration [4]. The SEAM (Simple Energy-aware Migration) [18,19] and ISEAM (Improved SEAM) [17] algorithms are proposed to migrate virtual machines to energy-efficient guest servers. Here, a server which is expected to mostly consume electric energy is selected and just a smallest virtual machine where the fastest number of processes are performed is selected to migrate. In this paper, we make clear the mathematical formula to calculate

L. Barolli et al. (Eds.): EIDWT 2018, LNDECT 17, pp. 130–141, 2018.
https://doi.org/10.1007/978-3-319-75928-9_11

the electric energy consumption of a host server and a guest sever to migrate a virtual machine. Then, we propose an ISEAM2 algorithm to select a virtual machine on a host server and a guest server so as to minimize the total electric energy to be consumed by the host and guest servers.

We evaluate the ISEAM2 algorithm compared with other non-migration algorithms like SGEA [15] and the migration algorithm ISEAM. We show the electric energy consumption of servers can be reduced in the ISEAM2 algorithm.

In Sect. 2, we present a system model. In Sect. 3, we propose the ISEAM2 algorithm. In Sect. 4, we evaluate the ISEAM2 algorithm.

2 System Model

A cluster S is composed of servers $s_1, \ldots, s_m$ $(m \geq 1)$ and supports applications on clients with virtual service on computation resources like CPUs and storages by using virtual machines $vm_1, \ldots, vm_v$ $(v \geq 1)$ [4]. If a client issues an application process p_i to the cluster S, one virtual machine vm_h is selected to perform the process p_i. Here, the process p_i is *resident* on the virtual machine vm_h. A virtual machine is *active* if and only if (iff) at least one process is performed, otherwise *idle*. An active server is one where there is at least one active resident virtual machine. A server which hosts at least one virtual machine is *engaged*, otherwise *free*. nv_h is a number of resident processes on the virtual machine vm_h at time τ. A virtual machine vm_h on a host server s_t is *resident* on the host server s_t. A virtual machine vm_h is *smaller* than a virtual machine vm_k $(vm_h < vm_k)$ iff $nv_h < nv_k$.

In this paper, a *process* means a computation type of application process which uses CPU resource on a server. A server s_t is composed of np_t (≥ 1) homogeneous CPUs. Each CPU is composed of cc_t (≥ 1) homogeneous cores. There are nc_t $(= np_t \cdot cc_t)$ cores in the server s_t. Each core supports the same number ct_t (≤ 2) of threads. The total number nt_t of homogeneous threads on a server s_t is $np_t \cdot cc_t \cdot ct_t$. An *active* thread is a thread where at least one process is performed. Let $CP_t(\tau)$ be a set of processes performed on a server s_t at time τ.

The electric power consumption $NE_t(n)$ of a server s_t to perform $n(= |CP_t(\tau)|)$ processes is $minE_t + nap_t(n) \cdot bE_t + nac_t(n) \cdot cE_t + nat_t(n) \cdot tE_t$ where $nap_t(n) = n$ if $n \leq np_t$, else np_t, $nac_t(n) = n$ if $n \leq nc_t$, else nc_t, and $nat_t(n) = n$ if $n \leq nt_t$, else nt_t in the MLPCM model [12,13].

Each process p_i is at a time performed on a thread of a host server s_t. It takes T_{ti} time units [tu] to perform a process p_i on a thread of a server s_t. If only a process p_i is performed on a server s_t without any other process, the execution time T_{ti} of the process p_i is shortest, $minT_{ti}$. In the cluster S, $minT_i$ shows a shortest one of $minT_{1i}, \ldots,$ and $minT_{mi}$. If $minT_{fi} = minT_i$, a thread of a server s_f is *fastest* in the cluster S. A server s_f with a fastest thread is *fastest*.

We assume one virtual computation step [vs] is performed on a thread of a fastest server s_f for one time unit [tu], i.e. the thread computation rate

CRT_f = 1[vs/tu]. The total number VC_i of virtual computation steps of a process p_i is defined to be $minT_i$ [tu] $\cdot$ CRT_f [vs/tu] = $minT_i$ [vs]. Thus, $minT_i$ shows the total amount of computation of a process p_i. The maximum computation rate $maxCR_{ti}$ of a process p_i on a server s_t is $VC_i/minT_{ti} = minT_i/minT_{ti}$ [vs/tu] ($\leq$1).

The computation rate $CR_t(\tau)$ of a server s_t of time τ is $at_t(\tau) \cdot CRT_t$. The maximum computation rate $maxCR_t$ [vs/tu] ($\leq$1) of a server s_t is $nt_t \cdot CRT_t$. As presented here, $at_t(\tau) = n$ if $n \leq nt_t$ and $at_t(\tau) = nt_t$ if $n > nt_t$ for $n = |CP_t(\tau)|$. The computation rate $NCR_t(n)$ [vs/tu] of a server s_t to concurrently perform n processes is given in the MLCM model [12, 13] as follows:

$$NCR_t(n) = \begin{cases} n \cdot CRT_t & \text{if } n \leq nt_t. \\ maxCR_t \ (= nt_t \cdot CRT_t) & \text{if } n > nt_t. \end{cases} \quad (1)$$

Here, $CR_t(\tau) = NCR_t(|CP_t(\tau)|)$. The server computation rate $NCR_t(n)$ is fairly allocated to each process p_i of n current processes. That is, the process computation rate $NPR_{ti}(n)$ of each process p_i performed with $(n-1)$ processes on a server s_t is $NCR_t(n)/n$.

If a process p_i starts at time st and ends at time et, $\sum_{\tau=st}^{et} NPR_{ti}(|CP_t(\tau)|)$ = VC_i [vs]. The computation laxity $lc_{ti}(\tau)$ [vs] is the number of virtual computation steps [vs] to be performed in the process p_i on a server s_t after time τ. At time τ a process p_i starts, $lc_{ti}(\tau) = VC_i$. Then, the computation laxity $lc_{ti}(\tau)$ is decremented by the computation rate $NPR_{ti}(|CP_t(\tau)|)$, i.e. $lc_{ti}(\tau + 1) = lc_{ti}(\tau) - NPR_{ti}(|CP_t(\tau)|)$. If $lc_{ti}(\tau + 1) \leq 0$, the process p_i terminates at time τ.

3 Energy-Efficient Migration of Virtual Machines

3.1 Estimation Model

A client issues a request process p_i to a set VM of virtual machines $vm_1, \ldots,$ vm_v ($v \geq 1$) in a cluster S of servers $s_1, \cdots s_m$. It is not easy to estimate the termination time ET_i of each current process p_i on every virtual machine vm_h of a server s_t by using the computation laxity $lc_{ti}(\tau)$ and computation rate $NPR_{ti}(n)$ of each process p_i where $n = |CP_t(\tau)|$ [6, 7, 12–14]. Hence, we assume VC_i of each process p_i to be a constant, $VC = 1$. We assume every current process finishes the half VC/2 of the total computation.

Suppose n processes are currently performed on a server s_t and k new processes are issued to the server s_t. Here, the total numbers of virtual computation steps to finish the n current processes and the k new processes are n / 2 and k. respectively. The computation rate of the server s_t is $NCR_t(n + k)$ since $(n + k)$ processes are concurrently performed. The expected termination time $SET_t(n, k)$ and electric energy consumption $SEE_t(n, k)$ of a server s_t to perform n current processes and k new processes are given as follows:

$$SET_t(n, k) = (n/2 + k)/NCR_t(n + k). \quad (2)$$

$$SEE_t(n, k) = SET_t(n,k) \cdot NE_t(n+k) = (n/2+k) \cdot NE_t(n+k)/NCR_t(n+k). \quad (3)$$

Variables n_t and nv_h denote the numbers $|CP_t(\tau)|$ and $|VCP_h(\tau)|$ of processes on each server s_t and on each virtual machine vm_h, respectively, at current time τ. Here, $nv_h \leq n_t$. Suppose a virtual machine vm_h resides on a host server s_t. First, suppose no virtual machine on the server s_t migrates to the server s_u. A pair of the servers s_t and s_u consume electric energy $EE_t = SEE_t(n_t, 0)$ by time $ET_t = SET_t(n_t, 0)$ and $EE_u = SEE_u(n_u, 0)$ by time $ET_u = SET_u(n_u, 0)$ to perform every current process. The servers s_t and s_u totally consume the electric energy EE_{tu}:

$$EE_{tu} = \begin{cases} EE_t + EE_u + (ET_t - ET_u) \cdot minE_u \text{ if } ET_t \geq ET_u. \\ EE_t + EE_u + (ET_u - ET_t) \cdot minE_t \text{ if } ET_t < ET_u. \end{cases} \quad (4)$$

If every process on the server s_u terminates by the time ET_u before ET_t, i.e. $ET_t \geq ET_u$, the server s_u consumes the minimum electric energy $minE_u$ from the time ET_u to the time ET_t. Hence, a pair of the servers s_t and s_u consume the electric energy $EE_t + EE_u + (ET_t - ET_u) \cdot minE_u$. If $ET_t < ET_u$, the server s_t consumes the minimum electric energy $minE_t$ from the time ET_t to the time ET_u. Hence, a pair of the servers s_t and s_u consume the electric energy $EE_t + EE_u + (ET_u - ET_t) \cdot minE_t$.

Next, suppose a virtual machine vm_h resident on the server s_t migrates to the server s_u. Here, $(n_t - nv_h)$ processes are performed on the server s_t while $(n_u + nv_h)$ processes are performed on the server s_u. Hence, the servers s_t and s_u consume the electric energy $NE_t = SEE_t(n_t - nv_h, 0)$ by time $NT_t = SET_t(n_t - nv_h, 0)$ and the electric energy $NE_u = SEE_u(n_u + nv_h, 0)$ by time $NT_u = SET_u(n_u + nv_h, 0)$, respectively, to perform every process on the servers s_t and s_u. Hence, the servers s_t and s_u totally consume the electric energy NE_{tu} as follows:

$$NE_{tu} = \begin{cases} NE_t + NE_u + (NT_t - NT_u) \cdot minE_u \text{ if } NT_t \geq NT_u. \\ NE_t + NE_u + (NT_u - NT_t) \cdot minE_t \text{ if } NT_t < NT_u. \end{cases} \quad (5)$$

If $EE_{tu} > NE_{tu}$, the virtual machine vm_h can migrate from the host server s_t to the guest server s_u since the total electric energy consumption of the servers s_t and s_u can be reduced by migrating the virtual machine vm_h.

In order to make simple the computation to calculate the expected termination time $SET_t(n_t, 0)$ of a server s_t, we assume the computation rate $NCR_t(n_t)$ of the server s_t to be the maximum server computation rate $maxCR_t$. In addition, we assume a server s_t consumes the maximum electric power $CE_t(n_t) = maxE_t$. That is, we take the SPC model [11]. Hence, the expected termination time $ET_t = n_t/maxCR_t$, the expected electric energy consumption $EE_t = ET_t \cdot maxE_t = n_t \cdot maxE_t/maxCR_t$, $ET_u = n_u/maxCR_u$, and $EE_u = ET_u \cdot maxE_u = n_u \cdot maxE_u/maxCR_u$ where no virtual machine migrates. Next, the expected termination time $NT_t = (n_t - nv_h)/maxCR_t$ and $NT_u = (n_u + nv_h)/maxCR_u$ and the expected electric energy consumption

$NE_t = (n_t - nv_h) \cdot maxE_t/maxCR_t$ and $NE_u = (n_u + nv_h) \cdot maxE_u/maxCR_u$ are obtained where a virtual machine vm_h migrates from the server s_t to s_u.

The total electric energy consumption $TE_{tu}(n_t, n_u)$ of the servers s_t and s_u is given as follows, where no virtual machine migrates from the server s_t to s_u:

$$TE_{tu}(n_t, n_u) = \begin{cases} n_t \cdot maxE_t/maxCR_t + n_u \cdot maxE_u/maxCR_u \\ +(n_t/maxCR_t - n_u/maxCR_u) \cdot minE_u & \text{if } n_t/maxCR_t \geq n_u/maxCR_u. \\ n_t \cdot maxE_t/maxCR_t + n_u \cdot maxE_u/maxCR_u \\ +(n_u/maxCR_u - n_t/maxCR_t) \cdot minE_t & \text{if } n_t/maxCR_t < n_u/maxCR_u. \end{cases} \quad (6)$$

The electric energy consumption $EC_{tu}(n_t, n_u, nv_h)$ of the servers s_t and s_u is given as follows, where a virtual machine vm_h migrates from the server s_t to s_u:

$$EC_{tu}(n_t, n_u, nv_h) = \begin{cases} A_{1tu} \cdot nv_h + C_1(n_t, n_u) \text{ if } NT_t \geq NT_u. \\ A_{2tu} \cdot nv_h + C_2(n_t, n_u) \text{ if } NT_t < NT_u. \end{cases} \quad (7)$$

$$A_{1tu} = (maxE_u - minE_u)/maxCR_u - (maxE_t + minE_u)/maxCR_t. \quad (8)$$
$$A_{2tu} = (maxE_u + minE_t)/maxCR_u - (maxE_t - minE_t)/maxCR_t. \quad (9)$$
$$C_1(n_t, n_u) = n_u \cdot (maxE_u - minE_u)/maxCR_u + n_t \cdot (maxE_t + minE_u)/maxCR_t. \quad (10)$$
$$C_2(n_t, n_u) = n_u \cdot (maxE_u + minE_t)/maxCR_u + n_t \cdot (maxE_t - minE_t)/maxCR_t. \quad (11)$$
$$NT_t = (n_t - nv_h)/maxCR_t. \quad (12)$$
$$NT_u = (n_u + nv_h)/maxCR_u. \quad (13)$$

In formula (7), $EC_{tu}(n_t, n_u, 0) = TE_{tu}(n_t, n_u)$ for $nv_h = 0$, i.e. no virtual machine migrates from a server s_t to the server s_u. Hence, $TE_{tu}(n_t, n_u)$ is given as follows:

$$TE_{tu}(n_t, n_u) = \begin{cases} C_1(n_t, n_u) \text{ if } NT_t \geq NT_u. \\ C_2(n_t, n_u) \text{ if } NT_t < NT_u. \end{cases} \quad (14)$$

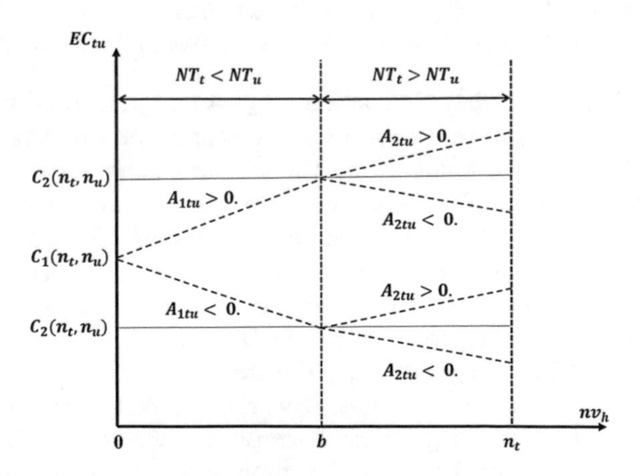

Fig. 1. Energy consumption EC_{tu}.

Let b be $(n_t \cdot maxCR_u - n_u \cdot maxCR_t)/(maxCR_t + maxCR_u)$. If $nv_h = b$, $NT_t = NT_u$. If $nv_h > b$, $NT_t > NT_u$ and if $nv_h < b$, $NT_t < NT_u$.

Figure 1 shows the total electric energy consumption $EC_{tu}(n_t, n_u, nv_h)$ for $nv_h(\leq n_t)$. If $A_{1tu} \geq 0$ and $A_{2tu} \geq 0$, the total electric energy consumption $EC_{tu}(n_t, n_u, nv_h)$ linearly increases as the number nv_h on a virtual machine vm_h increases. If $A_{1tu} < 0$ and $A_{2tu} < 0$, $EC_{tu}(n_t, n_u, nv_h)$ linearly decreases as the number nv_h increases.

3.2 VM Selection (VMS) Algorithm

First, suppose a client issues a process p_i to a cluster S of servers $s_1, \ldots, s_m$ ($m \geq 1$) with a set VM of virtual machines $vm_1, \ldots, vm_v$ ($v \geq 1$) at time τ. One virtual machine vm_h on a server s_t is selected to perform and the process p_i.
[**VM selection**] A client issues a process p_i to a cluster S.

1. **Select** an engaged host server s_t for the process p_i where at least one virtual machine resides, which satisfies a selection conditions in the cluster S.
2. **Select** a smallest virtual machine vm_h on the selected host server s_t.
3. **Perform** the process p_i on the virtual machine vm_h of the host server s_t.

First, we have to discuss how to select a host server s_t in a cluster S. In the random (RD) algorithm, an engaged server s_t is randomly selected for each process p_i. In the round-robin (RR) algorithm, a server s_t is selected after s_{t-1} is selected. As discussed in the SGEA algorithm [15], not only a host server s_t of a new process p_i but also the other servers consume electric energy even if the servers are idle. Here, let NT_t and NE_t be the expected termination time $SET_t(n_t, 1)$ and expected electric energy $SEE_t(n_t, 1)$ of a host server s_t, respectively, where a new process is assumed to be performed. Let NT_u and NE_u be $SET_u(n_t, 0)$ and $SEE_u(n_t, 0)$ of a server s_u, respectively, where only current processes are performed on another server s_u. We introduce a following function:

$$FE_t(E, TT, TU) = \begin{cases} E + minE_t \cdot (TT - TU) & \text{if } TT \geq TU. \\ E \cdot TU/TT & \text{if } TT < TU. \end{cases} \quad (15)$$

The expected electric energy consumption $NSEE_{tu}$ of another server s_u ($\neq s_t$) for a host server s_t of a process p_i is $FE_u(NE_u, NT_t, NT_u)$ which the server s_u consumes by the time NT_t. The expected total electric energy consumption $SGEE_t$ of the servers $s_1, \ldots, s_m$ for a host server s_t is $NE_t + \sum_{u=1,\ldots,m(\neq t)} NSEE_{tu}$.

If a process p_i is issued to a virtual machine vm_h on a server s_t, the expected termination time $ET_t = SET_t(n_t, 1)$ to finish both the process p_i and n_t current processes is given as $(n_t/2 + 1)/NCR_t(n_t + 1)$. The server s_t is considered to consume the electric power $NE_t(n_t + 1)$[W] to perform the n_t current processes and the process p_i. Here, the expected electric energy $EE_t = SEE_t(n_t, 1)$ to be consumed by a server s_t is $(n_t/2 + 1) \cdot NE_t(n_t + 1)/NCR_t(n_t + 1)$ [W tu]. The processes on the server s_t are expected to terminate by time $ET_t (= SET_t(n_t, 1))$.

For each server $s_u(\neq s_t)$, the expected electric energy $NSEE_u$ is calculated by formula (15). A server s_t whose expected total electric energy $SGEE_t$ is minimum is selected as a host server of the process p_i.
[Server selection algorithms]

1. Random (RD): Randomly select a host server s_t in a cluster S.
2. Round-robin (RR): select a server s_t after s_{t-1} is selected.
3. SLEA: select a server s_t where NE_t is minimum.
4. SGEA: select a server s_t where $SGEE_t$ is minimum.

Then, a smallest virtual machine vm_h has to be selected in the selected server s_t and the process p_i is performed on vm_h.

3.3 VM Migration (VMM) Algorithm

Each engaged server s_t is periodically checked. For each server $s_u(\neq s_t)$, a virtual machine vm_{tu} resident on the server s_t is selected to migrate to the server s_u, where the expected electric energy to be consumed by the host server and the guest server s_u is minimum as follows:
[VMM algorithm]
For each server $s_u(\neq s_t)$, a virtual machine vm_{tu} resident on the server s_t is selected as follows:

1. **if** $b(= (n_t \cdot maxCR_u - n_u \cdot maxCR_t)/(maxCR_t + maxCR_u)) \leq 0$, i.e. $n_t/maxCR_t \leq n_u/maxCR_u$,
   if $A_{1tu} < 0$, **select** a largest virtual machine vm_{tu} on the server s_t;

2. **if** $0 < b < n_t$,
   if $A_{1tu} < 0$ and $A_{2tu} < 0$, **select** a largest virtual machine vm_{tu} on the server s_t;
   if $A_{1tu} < 0$ and $A_{2tu} \geq 0$, **select** a virtual machine vm_{tu} where $|nv_{tu} - b|$ is minimum;
   if $A_{1tu} \geq 0$ and $A_{2tu} < 0$,
   select a smallest virtual machine vm_{tu} where $A_{2tu} \cdot nv_u < C_2(n_t, n_u)$;

Here, a resident virtual machine vm_{tu} on a server s_t is selected for each server s_u. Then, a guest server s_u and a virtual machine are selected where $EC_{tu}(n_t, n_u, nv_{tu})$ is minimum. If selected, the virtual machine vm_{tu} migrates from the host server s_t to the guest server s_u. If vm_{tu} is not found, no virtual machine migrates from the server s_t.

4 Evaluation

We consider a cluster S of four real servers s_1, s_2, s_3, and s_4 $(m = 4)$, respectively, in our laboratory and eight virtual machines vm_1, ..., vm_8 $(v = 8)$. Initially, each server s_t hosts two virtual machines. The servers s_1 and s_2 are equipped with two and one Intel Xeon E5-2667 v2 CPU, respectively. The servers s_3

and s_4 are equipped with one Intel Xeon E5-2620 and Intel Corei7-6700K CPU, respectively. The performance parameters like thread computation rate CRT_t and electric energy parameters like $minE_t$ of each server s_t are shown in Table 1. The threads of the servers s_1 and s_2 are the fastest, $CRT_1 = CRT_2 = 1$ [vs/tu]. In the simulation, the electric energy consumption EE_t and active time AT_t of each server s_t are obtained. The active time AT_t is time when the server s_t is active. The total electric energy consumption TEE of servers is $EE_1 + \cdots + EE_4$ and the total active time TAT is $AT_1 + \cdots + AT_4$.

There are n (>0) processes $p_1, \ldots, p_n$. In our previous studies, the starting time $stime_i$ of each process p_i is randomly taken from time 0 to $xtime$ −1. Here, $xtime$ is 1,000 time units [tu]. In this paper, $3n/4$ processes are randomly taken from time 0 to $xtime$ -1. Then, $stime_i$ of each process p_i of the other $n/4$ processes is randomly taken from $xtime/4 - 10$ to $xtime/4 + 10$. In fact, one time unit [tu] shows 100 [msec] [14]. The minimum execution time $minT_i$ is randomly taken from 5 to 20 [tu]. This means, the VC_i of computation steps of each process p_i is 5.0 to 20.0 [vs]. The parameters of each process p_i are shown in Table 2. Each process p_i starts at time $stime_i$ and terminates at time $etime_i$. The termination time $etime_i$ is obtained in the simulation. The execution time T_i of a process p_i is $etime_i - stime_i + 1$ [tu]. The average execution time AET is $(T_1 + \cdots + T_n)/n$.

We consider the random (RD), round robin (RR), SGEA [15], ISEAM [17], and ISEAM2 algorithms. In the RD and RR algorithms, virtual machines do not migrate and stay on host servers. In the SGEA algorithm, a server s_t whose expected total electric energy consumption is minimum is first selected. Then, a the process is performed on the smallest resident virtual machine vm_h of the server s_t. Virtual machines do not migrate. In the ISEAM algorithm, a smallest virtual machine migrates to a more energy efficient server in addition to the SGEA algorithm. In the ISEAM2 algorithm, a virtual machine and a guest server are found for each host server so that the total electric energy to be consumed by the servers is minimum.

Figure 2 shows the total electric energy consumption TEE [W · tu] of the four servers for number n of processes. The total electric energy consumption TEE of the ISEAM2 algorithm is smaller than the non-migration RD and RR algorithms. For example, TEE is about 20% smaller than the RD and RR algorithms. The ISEAM2 algorithm supports the same TEE as the ISEAM algorithm. The TEE of the ISEAM2 algorithm is smaller than the SGEA algorithm for $n \leq$ 1,500 but lager for $n >$ 2,000.

Figure 3 shows the TAT [tu] of the four servers. The total active time TAT of the ISEAM2 algorithm is the same as the ISEAM algorithm. For example, the total active time TAT of the ISEAM2 algorithm is about 60% smaller than the RD and RR algorithms. However, the TAT of ISEAM2 algorithm is larger than the SGEA algorithm. TAT of the SGEA algorithm is the shortest in the algorithms.

Figure 4 shows the average execution time AET [tu] of n processes. The AET of the ISEAM2 algorithm is shorter than the RD and RR algorithms but longer than SGEA algorithm.

In the ISEAM2 algorithm, processes are issued to a smallest resident virtual machine on a server. This means, processes are uniformly allocated to virtual machines and the number of processes on each virtual machine is almost the same on each server. In the VMM algorithm, the total electric energy consumption $EC_{tu}(n_t, n_u, nv_h)$ is almost the same for any virtual machine nv_h on a host server s_t. We have to allocate processes to virtual machines so that different number of processes are performed on resident virtual machines.

Table 1. Parameters of servers.

Parameters	s_1	s_2	s_3	s_4
np_t	2	1	1	1
nc_t	8	8	6	4
nt_t	32	16	12	8
CRT_t [vs/tu]	1.0	1.0	0.5	0.7
$maxCR_t$ [vs/tu]	32	16	6	5.6
$minE_t$ [W]	126.1	126.1	87.2	41.3
$maxE_t$ [W]	301.1	207.3	131.2	89.5
bE_t [W]	30	30	16	15
cE_t [W]	5.6	5.6	3.6	4.7
tE_t [W]	0.8	0.8	0.9	1.1

Table 2. Parameters of processes.

Parameters	Values
n	Number of processes $p_1, \ldots, p_n$.
$minT_i$ [tu]	Minimum computation time of a process p_i.
VC_i [vs]	5.0 – 20.0 ($VC_i = minT_i$).
$stime_i$ [tu]	Starting time of p_i ($0 \leq st_i < xtime - 1$).
$xtime$ [tu]	Simulation time (= 1,000 (= 100 [sec])).

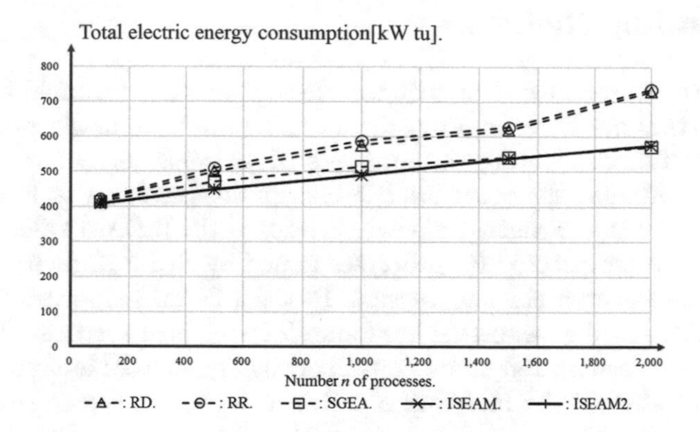

Fig. 2. Total electric energy consumption TEE ($m = 4$, $v = 8$).

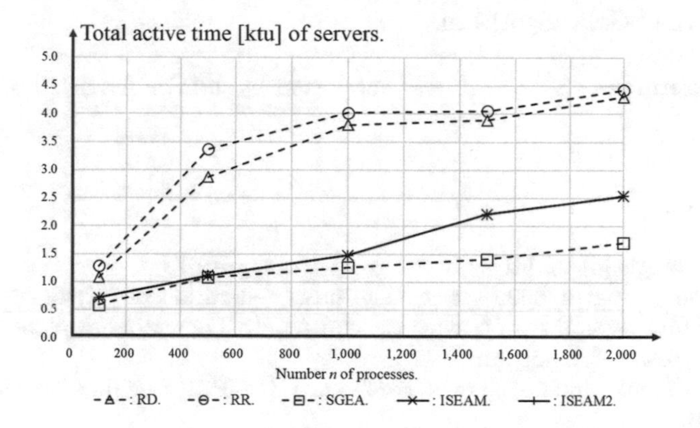

Fig. 3. Total active time TAT ($m = 4$, $v = 8$).

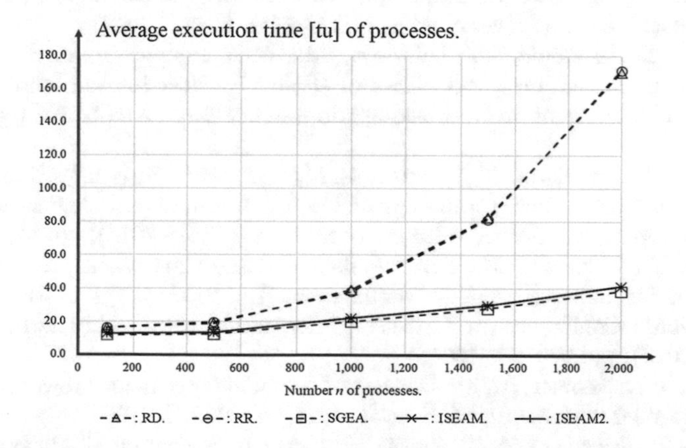

Fig. 4. Average execution time of processes AET ($m = 4$, $v = 8$).

5 Concluding Remarks

It is critical to discuss how to reduce the electric energy consumed by servers by migrating virtual machines in a cluster. In this paper, we newly made clear the mathematical formula among the numbers of processes on servers and virtual machines to minimize the total electric energy consumption of host and guest servers. Based on the formula, we newly proposed the ISEAM2 algorithm where virtual machines are selected for processes issued by clients and virtual machines migrate to more energy efficient servers. In the ISEAM2 algorithm, a pair of a virtual machine and a guest server are found for each host server so that the total electric energy consumption to be consumed by the servers is minimum. In the evaluation, we showed the ISEAM2 algorithm implies the same resident as the ISEAM algorithm. We also showed the active time of servers and the average execution time of processes in the ISEAM2 algorithm can be more reduced. However, the total electric energy consumption of the ISEAM2 algorithm is larger than the SGEA algorithm.

Acknowledgements. This work was supported by JSPS KAKENHI grant number 15H0295.

References

1. Google, Google green. http://www.google.com/green/
2. Intel xeon processor 5600 series: The next generation of intelligent server processors white paper. http://www.intel.com/content/www/us/en/processors/xeon/xeon-processor-e5-family.html
3. United nations climate change conference (COP21). https://en.wikipedia.org/wiki/2015
4. A virtualization infrastructure for the linux kernel (kernel-based virtual machine). https://en.wikipedia.org/wiki/Kernel-based_Virtual_Machine
5. Doulikun, D., Enokido, T., Takizawa, M.: An energy-aware algorithm to migrate virtual machines in a server cluster. IJSSC **7**(1), 32–42 (2017)
6. Duolikun, D., Enokido, T., Takizawa, M.: Asynchronous migration of process replica in a cluster. In: Proceedings of IEEE the 29th International Conference on Advanced Information Networking and Applications (AINA-2015), pp. 271–278 (2015)
7. Duolikun, D., Enokido, T., Takizawa, M.: Asynchronous migration of process replica in a cluster. In: Proceedings of the 9th International Conference on Complex, Intelligent, and Software Intensive Systems (CISIS-2015), pp. 118–125 (2015)
8. Enokido, T., Takizawa, M.: Energy-efficient delay time-based process allocation algorithm for heterogeneous server clusters. In: Proceedings of IEEE the 29th International Conference on Advanced Information Networking and Applications (AINA-2015), pp. 279–286 (2015)
9. Enokido, T., Aikebaier, A., Takizawa, M.: A model for reducing power consumption in peer-to-peer systems. IEEE Syst. J. **4**(2), 221–229 (2010)
10. Enokido, T., Aikebaier, A., Takizawa, M.: Process allocation algorithms for saving power consumption in peer-to-peer systems. IEEE Trans. Industr. Electron. **58**(6), 2097–2105 (2011)

11. Enokido, T., Aikebaier, A., Takizawa, M.: An extended simple power consumption model for selecting a server to perform computation type processes in digital ecosystems. IEEE Trans. Industr. Electron. **10**(2), 1627–1636 (2014)
12. Kataoka, H., Duolikun, D., Enokido, T., Takizawa, M.: Evaluation of energy-aware server selection algorithm. In: Proceedings of the 9th International Conference on Complex, Intelligent, and Software Intensive Systems (CISIS-2015), pp. 318–325 (2015)
13. Kataoka, H., Duolikun, D., Enokido, T., Takizawa, M.: Multi-level computation and power consumption models. In: Proceedings of the 18th International Conference on Network-Based Information Systems (NBiS -2015), pp. 40–47 (2015)
14. Kataoka, H., Duolikun, D., Enokido, T., Takizawa, M.: Power consumption and computation models of a server with a multi-core CPU and experiments. In: Proceedings of IEEE the 29th International Conference on Advanced Information Networking and Applications Workshops (WAINA-2015), pp. 217–223 (2015)
15. Kataoka, H., Duolikun, D., Enokido, T., Takizawa, M.: Simple energy-aware algorithms for selecting a server in a scalable cluster. In: Proceedings of IEEE the 31st International Conference on Advanced Information Networking and Applications Workshops (WAINA-2017), pp. 146–153 (2017)
16. Kataoka, H., Nakamura, S., Doulikun, D., Enokido, T., Takizawa, M.: Multi-level power consumption model and energy-aware server selection algorithm. IJGUC **8**(3), 201–210 (2017)
17. Watanabe, R., Duolikun, D., Cuiqin, Q., Enokido, T., Takizawa, M.: A simple energy-aware virtual machine migration algorithm in a server cluster. In: Proceedings of the 19th International Conference on Network-Based Information Systems (NBiS-2017), pp. 55–65 (2017)
18. Watanabe, R., Duolikun, D., Enokido, T., Takizawa, M.: An energy-efficient migration algorithm of virtual machines in server clusters. In: Proceedings of the 11th International Conference on Complex, Intelligent, and Software Intensive Systems (CISIS-2017), pp. 94–105 (2017)
19. Watanabe, R., Duolikun, D., Enokido, T., Takizawa, M.: A simply energy-efficient migration algorithm of processes with virtual machines in server clusters. J. Wirel. Mobile Netw. Ubiquitous Comput. Dependable Appl. (JoWUA) **8**(2), 1–18 (2017)

Collision Avoidance for Omnidirectional Automated Transportation Robots Considering Entropy Approach

Keita Matsuo(✉) and Leonard Barolli

Department of Information and Communication Engineering, Fukuoka Institute of Technology (FIT), 3-30-1 Wajiro-Higashi, Higashi-Ku, Fukuoka 811-0295, Japan
{kt-matsuo,barolli}@fit.ac.jp

Abstract. In recent years, the labor shortage accompanying with the declining birthrate and aging of population is becoming a big social problem. Therefore, it is important to consider other labor resources. To deal with this problem, there are many research work that consider application of robots. The robots can play an active role in factories or plants. It can be used also in hotels, offices or medical institutions. In this paper, we present the implementation of an omnidirectional automated transportation robot. We describe a method for collision avoidance considering entropy approach for ensuring safety of robots.

1 Introduction

Robots are being steadily introduced into modern everyday life and are expected to play a key role in the near future. Typically, the robots are deployed in situations where it is too dangerous, expensive, tedious, and complex for humans to operate.

Recently, the labor shortage accompanying with the declining of birthrate and aging of population is becoming a big social problem. Also, the older age population is increased. According to WHO (World Health Organization) by 2025, the population over 60 years is predicted to reach 23% in North America, 17% in East Asia, 12% in Latin America and 10% in South Asia. Thus, it is an urgent problem to raise productivity among limited labor resources. In order to deal with this problem, the Automated Transportation Robots (ATR) can be a good approach.

In this paper, we introduce the implementation of proposed omnidirectional ATR and present a collision avoidance method considering entropy approach for ensuring the safety of ATRs.

The rest of the paper is organized as follows. In Sect. 2, we introduce the related work. In Sect. 3, we present the proposed system for omnidirectional ATR.

Finally, conclusions and future work are given in Sect. 4.

L. Barolli et al. (Eds.): EIDWT 2018, LNDECT 17, pp. 142–151, 2018.
https://doi.org/10.1007/978-3-319-75928-9_12

2 Related Work

The conventional robots used in factories or plants move on the decided line. These robots are called Automated Guided Vehicle (AGV). In the case when line of AGV should be changed, it is needed to reset the magnet tape. This will increase the load work in the factory when will be frequently changes [4].

Recently, there are some research work that consider the movement of ATRs by using sensor and image processing [8,9,11]. For example, an ATR can be used to help a person carrying the baggages when he/she is going in a hotel or is moving in the airport. Also, the ATRs can be used to keep a determined distance with humans in order to have a safe communication. In addition, there are some studies that consider the navigation between ATRs to find the location of robots [2,3,10].

In our previous work, we presented the implementation and evaluation of an omnidirectional wheelchair [5–7]. We have shown some cases for using omnidirectional wheelchair for moving in narrow spaces of a room, playing tennis or badminton.

There are also some other works related with guide-path design, ATR scheduling, ATR positioning, battery management, ATR routing and deadlock resolution [1,13]. In [12], the authors proposed a control method for cooperation of AGVs in order to avoid collision.

3 Proposed System

3.1 Omnidirectional Automated Transportation Robot

In this section, we will describe the implementation of omnidirectional ATR. In the future is expected the ATRs will play a more important role in factories, plants or everyday life. Therefore, it is important to consider safely move of ATRs. For instance, when ATR is carrying the baggage of a person going to the hotel or is moving in the airport, it should avoid the collision with people. Also, in the case when there are many ATRs at same location, it is needed a method to determine the route of an ATR in order to avoid collision and deadlock between ATRs.

In the case when there are dynamic obstacles, the rate of collision between ATRs will increase because it is difficult to predict whether the obstacle is dynamic or not. The ATR needs to take an appropriate action for each obstacle.

In our work, by addressing the problems of previous research, we will implement an ATR that can move in all directions (Omnidirectional ATR). We show the moving of a omnidirectional ATR in Fig. 1.

In Fig. 2 is shown a real implementation of omnidirectional ATR (Length: 580 mm and Width: 580 mm). The proposed omnidirectional ATR has 3 omniwheels.

We show the implementation of control circuit for omnidirectional ATR consisting of motor driver, control board and CPU in Fig. 3. In addition, the brushless motor (the output power of 100 W) is used for driving. The brushless motor

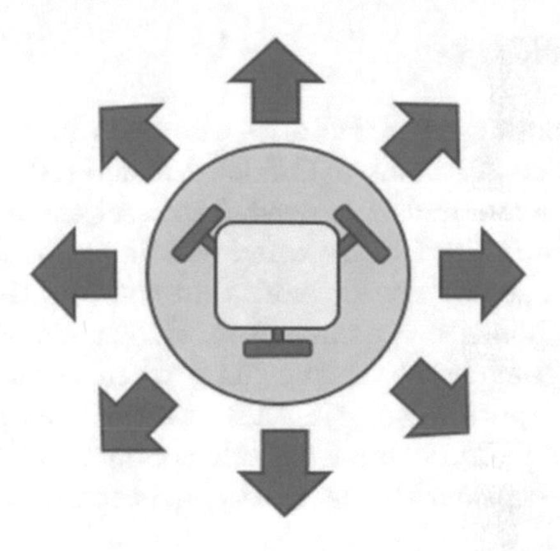

Fig. 1. Moving of omnidirectional ATR.

Fig. 2. Real implementation of omnidirectional ATR.

can reduce consumption energy for the omnidirectional ATR. Also, it can detect the rotation number for the 3 motors.

3.2 Kinematics

For the control of the omnidirectional ATR are needed: omniwheel speed, movement speed and direction to decide the control value.

Let us consider the movement of the omnidirectional ATR in 2 dimensional space. In Fig. 4, we show the model of omniwheel. In this figure, there are 3 onmiwheels which are placed 120° with each other. The omniwheels are moving in clockwise direction. We consider the speed for each omniwheel M1, M2 and M3, respectively.

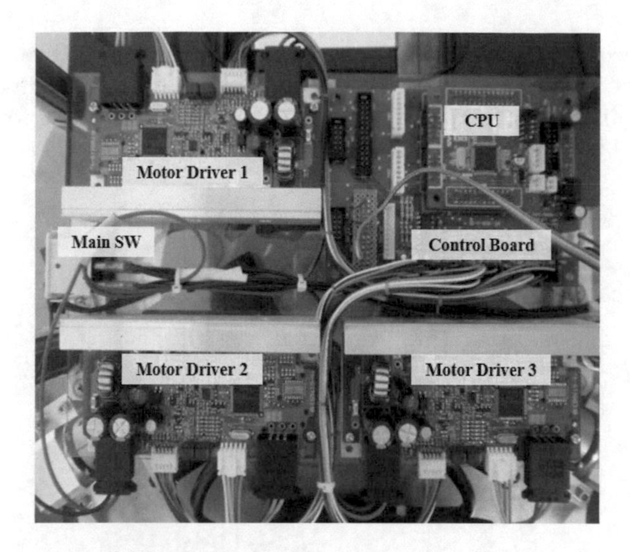

Fig. 3. Real implementation of control circuit for omnidirectional ATR.

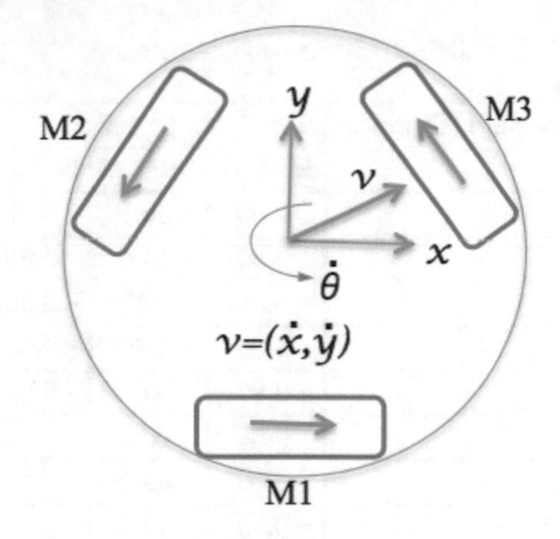

Fig. 4. Model of omniwheel.

As shown in Fig. 4, the axis of the omnidirectional ATR are x and y, the speed is $v = (\dot{x}, \dot{y})$ and the rotating speed is $\dot{\theta}$.

In this case, the moving speed of the omnidirectional ATR can be expressed by Eq. (1).

$$V = (\dot{x}, \dot{y}, \dot{\theta}) \quad (1)$$

Based on the Eq. (1), the speed of each omniwheel can be decided. By considering the control value of the motor speed ratio of each omniwheel as linear and synthesising the vector speed of 3 omniwheels, we can get Eq. (2) by using Reverse Kinematics, where (d) is the distance between the center and the omniwheels. Then, from the rotating speed of each omniwheel based on Forward

Kinematics, we get the omnidirectional ATR moving speed. If we calculate the inverse matrix of Eq. (2), we get Eq. (3). Thus, when the omnidirectional ATR moves in all directions (omnidirectional movement), the speed for each motor (theoretically) is calculated as shown in Table 1.

$$\begin{vmatrix} M_1 \\ M_2 \\ M_3 \end{vmatrix} = \begin{vmatrix} 1 & 0 & d \\ -\frac{1}{2} & -\frac{\sqrt{3}}{2} & d \\ -\frac{1}{2} & \frac{\sqrt{3}}{2} & d \end{vmatrix} \begin{vmatrix} \dot{x} \\ \dot{y} \\ \dot{\theta} \end{vmatrix} \tag{2}$$

$$\begin{vmatrix} \dot{x} \\ \dot{y} \\ \dot{\theta} \end{vmatrix} = \begin{vmatrix} \frac{2}{3} & -\frac{1}{3} & -\frac{1}{3} \\ 0 & -\frac{1}{\sqrt{3}} & \frac{1}{\sqrt{3}} \\ \frac{1}{3d} & \frac{1}{3d} & \frac{1}{3d} \end{vmatrix} \begin{vmatrix} M_1 \\ M_2 \\ M_3 \end{vmatrix} \tag{3}$$

Table 1. Motor speed ratio.

Direction (Degrees)	Motor speed ratio		
	Motor1	Motor2	Motor3
0	0.00	−0.87	0.87
30	0.50	−1.00	0.50
60	0.87	−0.87	0.00
90	1.00	−0.50	−0.50
120	0.87	0.00	−0.87
150	0.50	0.50	−1.00
180	0.00	0.87	−0.87
210	−0.50	−1.00	−0.50
240	−0.87	0.87	0.00
270	−1.00	0.50	0.50
300	−0.87	0.00	0.87
330	−0.50	−0.50	1.00
360	0.00	−0.87	0.87

3.3 Control System of the Proposed Omnidirectional Automated Transportation Robot

For the control of the proposed omnidirectional ATR, we considered R8C38 CPU board from Renesas Electronics Corporation. This CPU board has a small size and high speed processing time. The core of the CPU has a maximum frequency of 20 MHz. It is equipped with a flash memory, which is easy to rewrite. The R8C38 board has the following features:

- 8bit multi functions timer: 2,
- 16bit output competition timer: 5,
- Real time clock timer: 1,
- UART/clock synchronization type serial interface: 3 channels,
- 10bit A/D converter: 20 channels,
- 8bit D/A converter: 2 circuits,
- Voltage detected circuit,
- Number of output and input port: 75,
- External interrupt input: 9.

The proposed omnidirectional ATR can be controlled remotely using WiFi communication system embedded on it as shown in Fig. 5. Using this communication system, we can get the number of rotation for each motor and the direction of the omnidirectional ATR. In particular, real speed of each motor is very important because when there is difference between the value of control signal and real speed, it causes incorrect moving of the omnidirectional ATR. So, we have to modify the control signal based on this value.

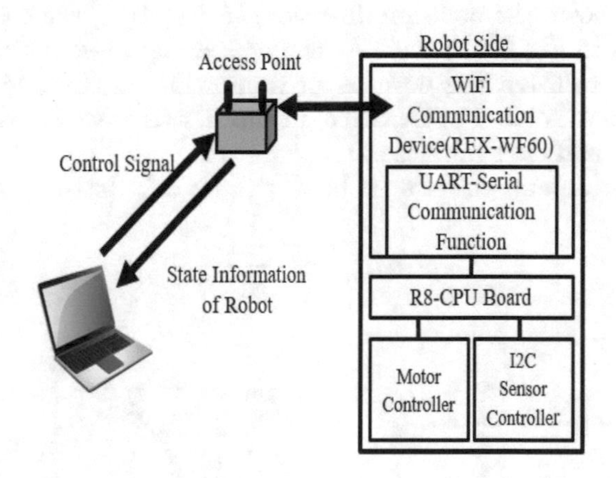

Fig. 5. Control system for omnidirectional ATR.

3.4 Collision Avoidance Method Considering Position Entropy

We present in following the collision avoidance method considering position entropy for omnidirectional ATR. As shown in Fig. 6, first is calculated the probability of collision. The ATR is considered in the center of the figure and the figure is divided in several blocks. Then, the probability of collision is calculated by multiplying the values of Y axis and X axis. The arrow in the left side of

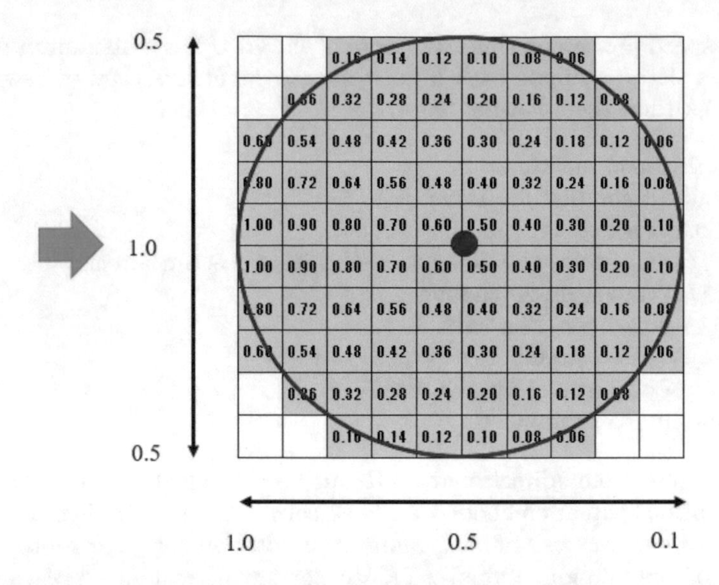

Fig. 6. X axis and Y axis of scale for collision.

Fig. 6 shows the direction of the obstacle, which is moving from the left to the center approaching the omnidirectional ATR.

In Fig. 7 is shown the collision direction. In Fig. 7(a), the collision direction angle is 45° and in Fig. 7(b) is 90°. As can be seen by these figures, by changing the direction of collision the position of value "1.0" is changed. The maximal value for each axis is 1.0. For instance, in the X axis if we set the left end side to 1.0, the right end side value is 0.0.

The information amount of each block can be calculated by Eq. (4).

$$I(a_j) = \log_2 \frac{1}{P(a_j)} \tag{4}$$

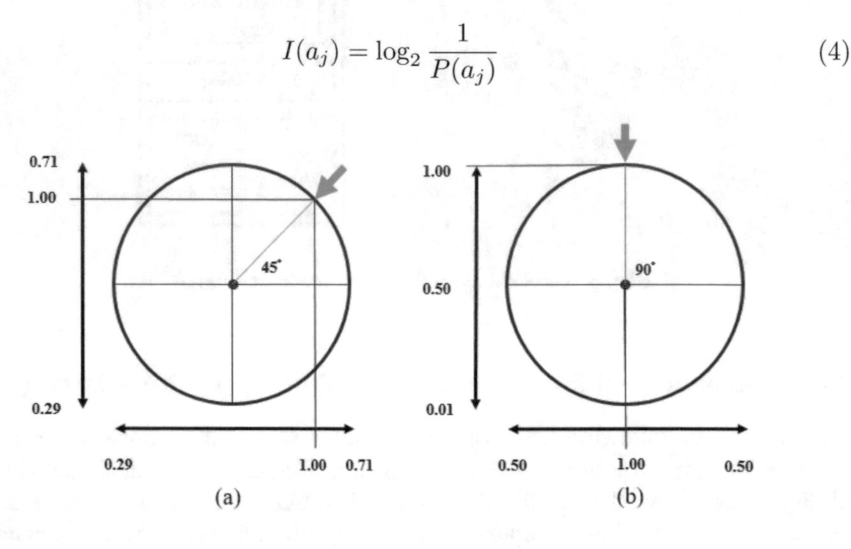

Fig. 7. Collision direction.

In Fig. 8 is shown the collision avoidance direction of a omnidirectional ATR using the proposed method. In oder to calculate the collision avoidance, four blocks are grouped into one group. The A is one of the event and J is the number of blocks, where A is $\{a_1, a_2, a_3, ..., a_J\}$. The value of entropy $H(A)$ (position entropy) can be calculated by Eq. (5). The event can be one of position of ATR (from 1 to 7) as shown in Fig. 8. The position entropy values for each ATR position are shown in Table 2. When the value of position entropy is higher, the probability to avoid collision is higher. For instance, the direction number 5 of ATR has the highest position entropy value. This shows that position number 5 of ATR is more safely than other positions.

$$H(A) = \sum_{j=1}^{J} P(a_j) \log_2 \frac{1}{P(a_j)} \tag{5}$$

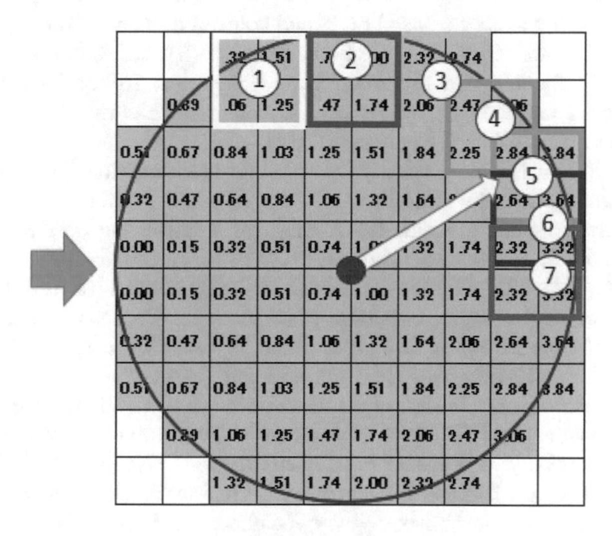

Fig. 8. Direction of collision avoidance.

Table 2. Value of position entropy.

No	Position entropy value [bit]
1	1.27
2	1.71
3	2.36
4	2.59
5	3.07
6	2.80
7	2.66

4 Conclusions and Future Work

In this paper, we presented the implementation of an omnidirectional ATR, which can be used in factories, plants, hotels or airports. These kind of robots can improve the labor resources. First, we introduced the related work and have shown their problems and issues. After that we presented the implementation issues and kinematics of proposed omnidirectional ATR. Finally, we described in details a collision avoidance method by considering entropy approach for ensuring safety of robots.

In the future, we will carry out experiments to evaluate the performance of the implemented omnidirectional ATR.

References

1. Baruwa, O.T., Piera, M.A., Guasch, A.: Deadlock-free scheduling method for flexible manufacturing systems based on timed colored petri nets and anytime heuristic search. IEEE Trans. Syst. Man Cybern. Syst. **45**(5), 831–846 (2015)
2. Chessa, S., Gallicchio, C., Guzman, R., Micheli, A.: Robot localization by echo state networks using RSS. In: Recent Advances of Neural Network Models and Applications, pp. 147–154. Springer (2014)
3. Cucchiara, R., Perini, E., Pistoni, G.: Efficient stereo vision for obstacle detection and AGV navigation. In: ICIAP, vol. 7, pp. 291–296 (2007)
4. Maki, S., Shirane, K., Masaki, R.: A map for logistics support mobile robot and its application. J. Robot. Soc. Jpn. **33**(10), 732–737 (2015)
5. Matsuo, K., Barolli, L.: Design and implementation of an omnidirectional wheelchair: control system and its applications. In: Proceedings of the IEEE 9th International Conference on Broadband and Wireless Computing, Communication and Applications (BWCCA-2014), pp. 532–535 (2014)
6. Matsuo, K., Barolli, L.: Design of an omnidirectional wheelchair for playing tennis. In: Proceedings of the IEEE 10th International Conference on Complex, Intelligent, and Software Intensive Systems (CISIS-2016), pp. 377–381 (2016)
7. Matsuo, K., Liu, Y., Elmazi, D., Barolli, L., Uchida, K.: Implementation and evaluation of a small size omnidirectional wheelchair. In: Proceedings of the IEEE 29th International Conference on Advanced Information Networking and Applications Workshops (WAINA-2015), pp. 49–53 (2015)
8. Prasad, G.V., Bindu, C.S.: Robot localization and object detection with fish-eye vision system and sensors. i-manager's J. Patt. Recogn. **2**(3), 16–23 (2015)
9. Rahmani, B., Putra, A.E., Harjoko, A., Priyambodo, T.K.: Review of vision-based robot navigation method. IAES Int. J. Robot. Autom. **4**(4), 254–261 (2015)
10. Ronzoni, D., Olmi, R., Secchi, C., Fantuzzi, C.: AGV global localization using indistinguishable artificial landmarks. In: IEEE International Conference on Robotics and Automation (ICRA), pp. 287–292 (2011)
11. Sibley, G.T., Rahimi, M.H., Sukhatme, G.S.: Robomote: a tiny mobile robot platform for large-scale ad-hoc sensor networks. In: IEEE International Conference on Robotics and Automation (ICRA), vol. 2, pp. 1143–1148 (2002)

12. Watanabe, M., Furukawa, M., Kakazu, Y.: Acquisition of communication protocol for autonomous multi-AGVs driving. In: Proceedings of the IEEE 2nd International Conference on Intelligent Processing and Manufacturing of Materials, (IPMM 1999), vol. 2, pp. 1115–1121 (1999)
13. Zhou, Y., Hu, H., Liu, Y., Ding, Z.: Collision and deadlock avoidance in multirobot systems: a distributed approach. IEEE Trans. Syst. Man Cybern. Syst. **47**, 1712–1726 (2017)

Performance Evaluation of an Active Learning System Using Smartphone: A Case Study for High Level Class

Noriyasu Yamamoto(✉) and Noriki Uchida(✉)

Department of Information and Communication Engineering, Fukuoka Institute of Technology, 3-30-1 Wajiro-Higashi, Higashi-Ku, Fukuoka 811-0295, Japan
{nori,n-uchida}@fit.ac.jp

Abstract. In our previous work, it was presented an interactive learning process in order to increase the students learning motivation and the self-learning time. We proposed an Active Learning System (ALS) for student's self-learning. For low and middle level class, we showed that the students could keep high concentration by using traditional ALS. However, for high level class, many student could not keep their concentration, because of they have a lot of question that they did not understand. In this paper, to solve this problem, we propose a method of group discussion to deal with students' questions for high level class. We also present the performance evaluation of ALS for high level class. The evaluation results show that when the lecture use proposed ALS for high level class, the average dropping out was half compared with the conventional ALS. Also, the method of group discussion increases the students' concentration for high level class.

1 Introduction

There is a total of 777 universities today in Japan. The "middle-level" universities (which are ranked at around the middle in terms of academic level) form the largest group in Japan and their undergraduates students are the main workforces. Because the group represents the "middle-level", academic capacities of students often vary significantly. For this reason, the teaching speed should be controlled when deciding the understanding level. When a lesson offers advanced contents or is fast-paced, good students find it satisfying, while middle level or low level students fail to catch up. If a lesson is designed too simple or slow, the middle level and low level students find it satisfying, while good students get frustrated. To solve this problem, there are many e-learning systems [1–9]. Also, today's classes increasingly utilize information terminals, such as notebook PCs and projectors. These information tools have enabled lecturers to offer more information to students. But, their effectiveness has no positive impact on students who lack concentration and motivation. The students just look at the projector screen and get the information materials. Their satisfaction level for the lecture is increased, but their scores do not improve.

Usage of the desk-top computers and notebook PCs for lecture may be inconvenient and they occupy a lot of space. Therefore, it will be better that students use small and

L. Barolli et al. (Eds.): EIDWT 2018, LNDECT 17, pp. 152–160, 2018.
https://doi.org/10.1007/978-3-319-75928-9_13

lightweight terminals like Personal Digital Assistant (PDA) devices. Also, because of wireless networks are spread over university campuses, it is easy to connect mobile terminals to the Internet and now the students can use mobile terminals in many lecture rooms, without needing a large-scale network facility. Our idea has been to use various information terminals and tools to boost students' concentration and motivation.

We considered a method of acquiring/utilizing the study record using smartphone in order to improve the students learning motivation [10]. During the lecture the students use the smartphone for learning. The results showed that the proposed study record system has a good effect for improving students' motivation for learning.

For the professors of the university, it is difficult to offer all necessary information to the students. In addition, they cannot provide the information to satisfy all students because the quantity of knowledge of each student attending a lecture is different. Therefore, for the lectures of a higher level than intermediate level, the students should study the learning materials by themselves.

In our previous work, it was presented an interactive learning process in order to increase the students learning motivation and the self-learning time [11]. However, the progress speed of a lecture was not good. To solve this problem, we proposed an Active Learning System (ALS) for student's self-learning [12, 13]. Also, to improve student's self-learning procedure, we proposed some functions for the ALS [14]. We performed experimental evaluation of our ALS and showed that the proposed ALS can improve student self-learning and concentration [15]. Although, the self-learning time and the examination score were increased, the number of student that passed examination didn't increase significantly. To solve this problem, we proposed the group discussion procedure to perform discussion efficiently [16].

The learning system proposed so far only gives a feedback of students' learning record to lecturer. We propose a mechanism to enhance the learning effects by using a record of students' learnings in the whole process of learning system [17].

The previous interfaces for interactive learning had limited adequacy in maintaining participants' concentration when their skill levels vary. Thus, we proposed the interface of the mobile devices (Smartphone/Pad) on our ALS to improve the learning concentration [18]. We presented the performance evaluation of ALS when using the improved human interface on the mobile devices [19]. When the lecture uses ALS, the average dropping out was half compared with the conventional lecture. The evaluation results show that the interactive lecture by using ALS increases the students' concentration. In this paper, we show the flow of the group discussion and the performance evaluation of ALS for high level class.

The paper structure is as follows. In Sect. 2, we introduce the related work on learning systems and present ALS with Interactive Lecture. In Sect. 3, we indicate the flow of the group discussion for high level class. Then, in Sect. 4, we show the performance evaluation of ALS, which was improved by the group discussion. Finally, in Sect. 5, we give some conclusions and future work.

2 Active Learning System with Interactive Lecture

Our goal is to develop of Active Learning System (ALS). In Fig. 1, we show ALC (Active Learning Cycle) of the proposed ALS. The system facilitates a learning cycle consisting of a lecture, reviews at home and group discussions. The student and the lecturer confirm the movement by setting each cycle and information by the smartphone. At the beginning of ALC, the lecturer performs the interactive lecture by confirming the understanding degree of the student using their smartphone in real time. Prior to the lecture, the lecturer prepares "study points". "Small examination" refers to a mini quiz prepared for each study point. A mini quiz consists of simple multiple-choice questions. "Understanding level" is set by the result of the "Small examination". "Lecture speed" suggests whether students find the lecture progress too fast or too slow. By "Understanding level" and "Lecture speed", students' understanding can be judged. These two functions are used as feedbacks through the application on students' smartphone (students' application). During the lecture, these data are recorded in the database to be reflected in the application for the lecturer.

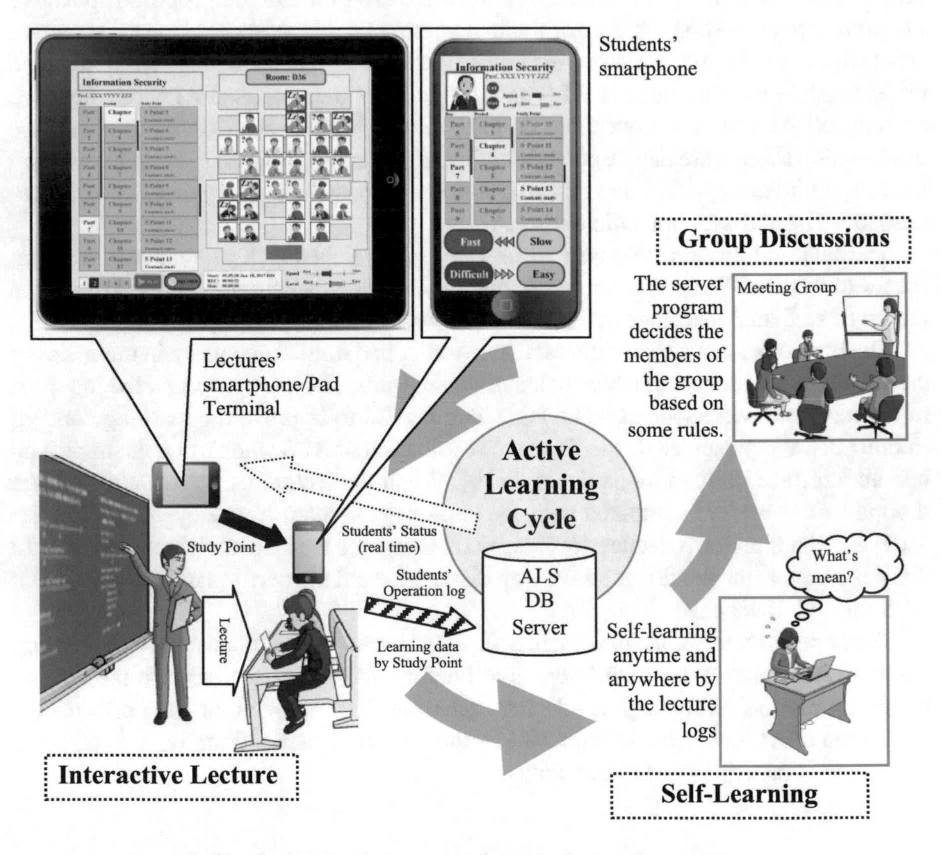

Fig. 1. ALC using a smartphone in the previous work.

The application for the lecturer on the left side of the screen shows the learning points and the control panel. When the lecturer operates with the study points, the class moves on to the next step. The right side of the screen shows real-time student responses.

The application for the student on the top of the screen shows lecture information. The center of the screen shows information of study points currently in progress. These elements are controlled by the lecturer through the control panel operation. The bottom of the screen features control buttons for students. When students find the lecture speed too fast, they press "Fast" button. When they feel it too slow, they press "Slow". Likewise, they choose "Difficult" or "Easy" button depending on how they feel. Students can press these buttons anytime. Based on the record accumulated through button operations, the system creates logs and updates the database on real-time. Furthermore, the log automatically updates the study points in the database to be reflected on the lecturer's screen.

Therefore, a lecturer can transmit the knowledge to the student effectively. After the lecture is finished, the student can read the lecture log by their smartphone. The student can review the lecture content using the lecture log anywhere and anytime.

Then, the student discusses the lecture content in small groups. In the group discussion, the students discuss what they did not understand in the lecture. Then, they submit the result of the group discussion to a lecturer as reports.

At the beginning of the next lecture, the student groups and the lecturer carry out open discussions based on the submitted reports and solve the problems that students may have. After the open discussion, the lecturer performs the next interactive lecture.

In this ALC, by adding the group learning and the open discussion, the understanding is increased and the lecturer can keep a fixed progress speed of the lecture. For low and middle level class, we showed that the students could keep high concentration on the traditional ALS [18, 19]. For high level class, however, many student could not keep their concentration, because they have a lot of question that they did not understand. To solve this problem, we propose a method for group discussion to deal with the students' questions for high level class.

3 Flow of Group Discussion for High Level Class

In this section, we explain the proposed flow of the group discussion after students' self-learning for high level class. In the traditional system, the students formed automatically the random member group, and they had short meeting and discussion. For high level class, the students who didn't understand the lecture, should discuss with each other. To solve this problem, we improve the traditional group discussion flow on ALS.

In Fig. 2, we show the flow of group discussion for high level class. It looks as a three step discussion, because in there is one week until the next lecture. Firstly, the formation of the group are randomly performed by the ALS server and the students are informed of the group member by their smartphone application. In the group discussion, the students discuss what they did not understand in the lecture. Then, they submit the result of the group discussion to the lecturer as short reports.

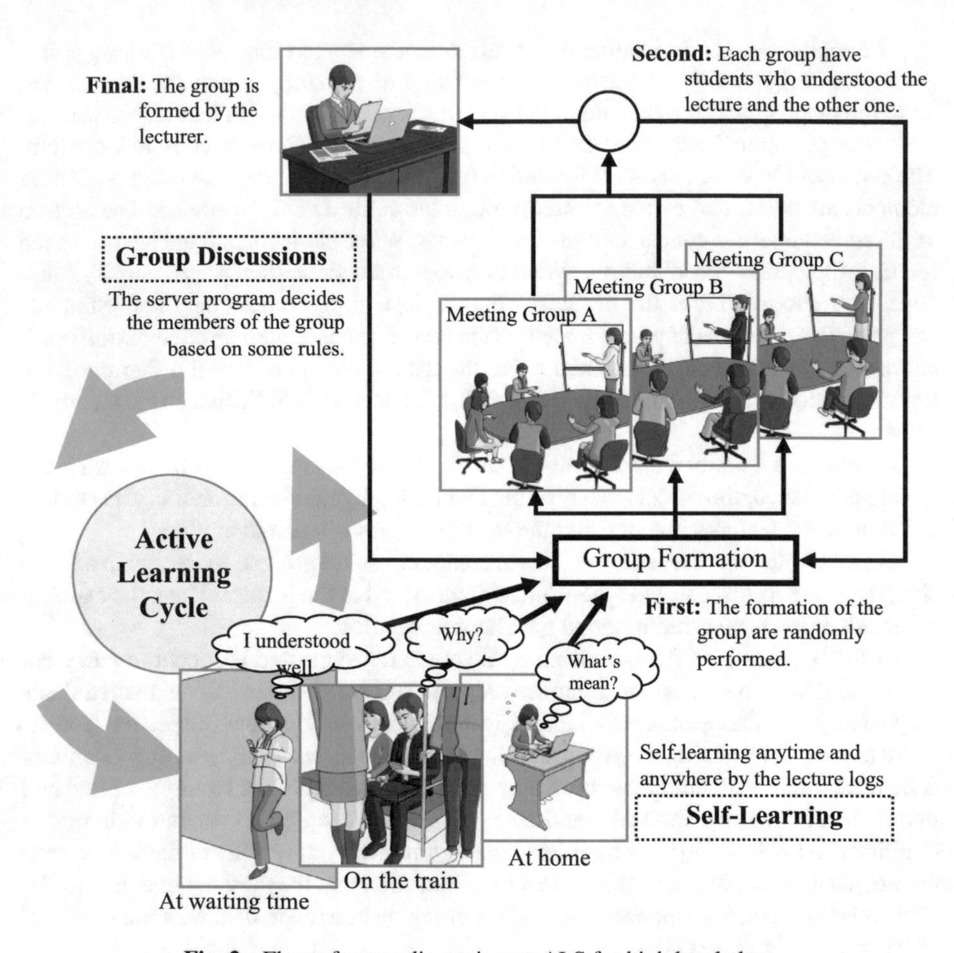

Fig. 2. Flow of group discussion on ALS for high level class.

Secondly, the group is automatically formed with the students who understood the lecture and the students who did not understand the lecture by the ALS server. Finally, the group is formed by the lecturer with the students' discussion reports. Most of students can solve their problem in the lecture using the group discussion and they can keep their learning motivation.

4 Performance Evaluation of ALS for High Level Class

We performed experimental evaluation using the improved group discussion method for ALS on high level class. In the evaluation experiment, we used two programing lectures: low level class and high level class. The number of students of these lectures were 22 and 18, for low level class and high level class, respectively. One of the lectures was performed for 90 min. We performed the lecture 14 times.

We investigated the pass rate of the conventional lectures and we found that it was 65% and 31%, for low level class and high level class, respectively. About 20% and 50% students could not keep their concentration and dropped out of the conventional lectures (not using ALS).

We compared the students' concentration as their self-learning time (when including their group discussion time) when using conventional ALS and new ALS with the improved group discussion method. Figure 3 shows the time that students spend on self-learning for low level class when using conventional method and ALS. The time spent on self-learning constantly increased. This is because the lecture contents get more difficult as the course progresses. By using ALS on low level class, it seems that students can maintain their motivation, because they use more time on self-learning.

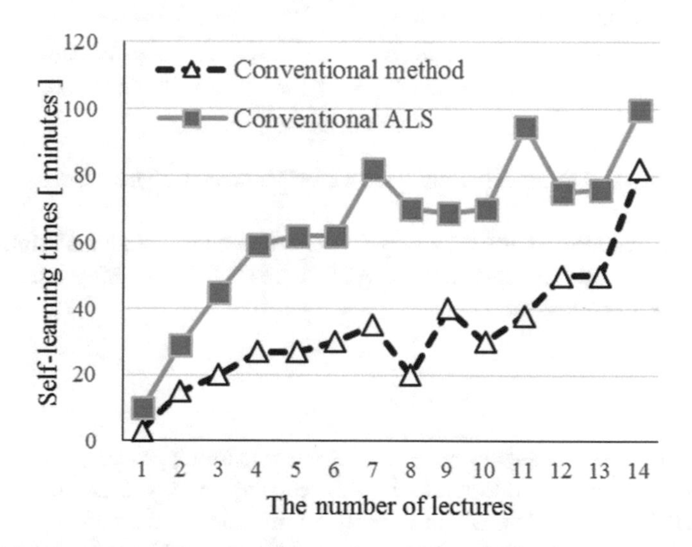

Fig. 3. The time that students spend on self-learning in low level class.

Figure 4 shows the time that students spend on self-learning for high level class when using conventional ALS and the proposed new ALS with improved group discussion. The time spent on self-learning it is almost constant. This is because all lecture contents are difficult on high level class. When using the conventional ALS, the self-learning decreases, because many student dropped out on high level lecture and they could not keep their concentration. By using the new ALS, the students can maintain their concentration and the time they spent on self-learning is increased (including group discussions). The evaluation results show that when the ALS was not used about 50% students dropped out the high level lecture and when using conventional ALS about 40% students dropped out the high level lecture. However, when using the new ALS with the improved group discussion about 20% students drop out the high level lecture.

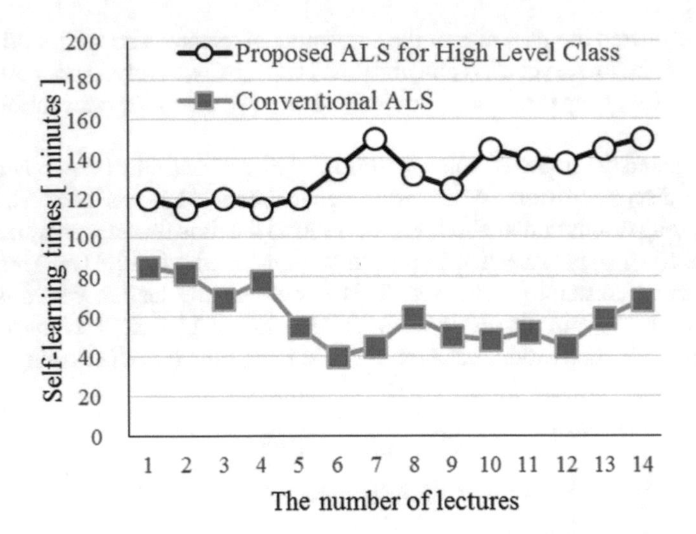

Fig. 4. The time students spend on self-learning in high level class.

Thus, ALS with the improved group discussion can keep students' high concentration, and these results indicate that the ALS increased the learning efficiently of the students on high level class.

5 Conclusions

In our previous work, it was presented an interactive learning process in order to increase the students learning motivation and the self-learning time. We proposed an ALS for student's self-learning. Also, we introduced the interface of the mobile devices on our ALS to improve the learning concentration. For low and middle level class, we showed that the students could keep high concentration by using traditional ALS. However, for high level class, many student could not keep the concentration, because they have a lot of question that they did not understand.

In this paper, to solve this problem, we proposed a method for group discussion to deal the with students' questions for high level class. Then, we presented the performance evaluation using new ALS for high level class. The evaluation results show that when using the new ALS, the average dropping out for the lecture was half compared with the conventional ALS. Also, the method of group discussion increased the students' concentration for high level class.

As a future work, there are plans to develop an automatically group formation algorithm in order to increase students' concentration.

References

1. Underwood, J., Szabo, A.: Academic offences and e-learning: individual propensities in cheating. Br. J. Edu. Technol. **34**(4), 467–477 (2003)
2. Harashima, H.: Creating a blended learning environment using Moodle. In: The Proceedings of the 20th Annual Conference of Japan Society of Educational Technology, pp. 241–242, September 2004
3. Brandl, K.: Are you ready to "Moodle"? Lang. Learn. Technol. **9**(2), 16–23 (2005)
4. Dagger, D., Connor, A., Lawless, S., Walsh, E., Wade, V.P.: Service-oriented e-learning platforms: from monolithic systems to flexible services. IEEE Internet Comput. **11**(3), 28–35 (2007)
5. Patcharee, B., Achmad, B., Achmad, H.T., Okawa, K., Murai, J.: Collaborating remote computer laboratory and distance learning approach for hands-on IT education. J. Inf. Process. **22**(1), 67–74 (2013)
6. Emi, K., Okuda, S., Kawachi, Y.: Building of an e-learning system with interactive whiteboard and with smartphones and/or tablets through electronic textbooks. Information Processing Society of Japan (IPSJ), IPSJ SIG Notes 2013-CE-118(3), 1–4, 2013-02-01 (2013)
7. Yamaguchi, S., Ohnichi, Y., Nichino, K.: An efficient high resolution video distribution system for the lecture using blackboard description. Technical report of IEICE, vol. 112(190), pp. 115–119 (2013)
8. Hirayama, Y., Hirayama, S.: An analysis of the two-factor model of learning motivation in university students. Bulletin of Tokyo Kasei University, 1, Cultural and Social Science, vol. 41, pp. 101–105 (2001)
9. Ichihara, M., Arai, K.: Moderator effects of meta-cognition: a test in math of a motivational model. Jpn. J. Educ. Psychol. **54**(2), 199–210 (2006)
10. Yamamoto, N., Wakahara, T.: An interactive learning system using smartphone for improving students learning motivation. In: Park, J., Barolli, L., Xhafa, F., Jeong, H.Y. (eds.) Information Technology Convergence. Lecture Notes in Electrical Engineering, vol. 253, pp. 305–310. Springer, Dordrecht (2013)
11. Yamamoto, N.: An interactive learning system using smartphone: improving students' learning motivation and self-learning. In: Proceeding of the 9th International Conference on Broadband and Wireless Computing, Communication and Applications (BWCCA-2014), pp. 428–431, November 2014
12. Yamamoto, N.: An active learning system using smartphone for improving students learning concentration. In: Proceeding of International Conference on Advanced Information Networking and Applications Workshops (WAINA-2015), pp. 199–203, March 2015
13. Yamamoto, N.: An interactive e-learning system for improving students motivation and self-learning by using smartphones. J. Mob. Multimedia (JMM) **11**(1&2), 67–75 (2015)
14. Yamamoto, N.: New functions for an active learning system to improve students self-learning and concentration. In: Proceeding of the 18th International Conference on Network-Based Information Systems (NBIS-2015), pp. 573–576, September 2015
15. Yamamoto, N.: Performance evaluation of an active learning system to improve students self-learning and concentration. In: Proceeding of the 10th International Conference on Broadband and Wireless Computing, Communication and Applications (BWCCA-2015), pp. 497–500, November 2015
16. Yamamoto, N.: Improvement of group discussion system for active learning using smartphone. In: Proceeding of the 10th International Conference on Innovative Mobile and Internet Services in Ubiquitous Computing (IMIS-2016), pp. 143–148, July 2016

17. Yamamoto, N.: Improvement of study logging system for active learning using smartphone. In: Proceedings of the 11th International Conference on P2P, Parallel, Grid, Cloud and Internet Computing (3PGCIC–2016), pp. 845–851, November 2016
18. Yamamoto, N., Uchida, N.: Improvement of the interface of smartphone for an active learning with high learning concentration. In: Proceeding of the 31st International Conference on Advanced Information Networking and Applications Workshops (AINA-2017), pp. 531–534, March 2017
19. Yamamoto, N., Uchida, N.: Performance evaluation of a learning logger system for active learning using smartphone. In: Proceeding of the 20th International Conference on Network-Based Information Systems (NBiS-2017), pp. 443–452, August 2017

Performance Evaluation of an Enhanced Message Suppression Controller Considering Delayed Ack Using Different Road Traffic Conditions

Daichi Koga[1], Yu Yoshino[1], Shogo Nakasaki[2], Makoto Ikeda[2(✉)], and Leonard Barolli[2]

[1] Graduate School of Engineering, Fukuoka Institute of Technology (FIT), 3-30-1 Wajiro-higashi, Higashi-ku, Fukuoka 811-0295, Japan
daichi.kg@outlook.com, mgm17107@bene.fit.ac.jp

[2] Department of Information and Communication Engineering, Fukuoka Institute of Technology, 3-30-1 Wajiro-higashi, Higashi-ku, Fukuoka 811-0295, Japan
tshogonakasakit@gmail.com, makoto.ikd@acm.org, barolli@fit.ac.jp

Abstract. In recent years, inter-vehicle communication has attracted attention because can be applicable not only to alternative networks but also to various communication systems. In our previous work, we proposed a method to reduce the duplicated bundle messages in Vehicular Ad-hoc Networks (VANETs). In this paper, we evaluate the performance of our proposed message suppression controller with delayed ack using different road traffic conditions. From the simulation results, we found that our proposed method with delayed ack has good performance for these traffic conditions.

Keywords: Inter-vehicle communication · VANET
Message suppression · Delayed ack

1 Introduction

In recent years, a number of disasters have been occurred around the world. The technologies of disaster management system are improved, however the communication system does not work well in disaster area due to the traffic concentration, device failure, and so on. A key for creating a valuable disaster rescue plan is to prepare alternative communication systems. In disaster area, mobile devices are often disconnected in the network due to network traffic congestion and access point failure. Inter-vehicle communication has attracted attention as an alternative network in disaster situations. In this case, Delay/Disruption/Disconnection Tolerant Networking (DTN) are used as one of a key alternative option to provide the network services [23].

L. Barolli et al. (Eds.): EIDWT 2018, LNDECT 17, pp. 161–170, 2018.
https://doi.org/10.1007/978-3-319-75928-9_14

The DTN aims to provide seamless communications with a wide range of network, which have not good performance characteristics [6]. DTN has the potential to interconnect vehicles in regions that current networking protocol cannot reach the destination. For inter-vehicle communications, a number of communication types are comprised such as Vehicle-to-Vehicle (V2V), Vehicle-to-Infrastructure (V2I), Vehicle-to-Pedestrian (V2P) and Vehicle-to-X (V2X) communications [3,8,9,11,17]. IEEE 802.11p supports these communications in outdoor environments. It defines enhancements to 802.11 required to support Intelligent Transport System (ITS) applications. The technology operates at 5.9 GHz in various propagation environments to high-speed moving vehicles.

We have proposed message suppression algorithms to improve the performance of conventional DTN protocols. In [12], we proposed a Message Suppression Controller (MSC) for V2V and V2I communications. We considered some parameters to control the message suppression dynamically. However, a fixed parameter still is used to calculate the duration of message suppression. To solve this problem, we proposed an Enhanced Message Suppression Controller (EMSC) [13] for Vehicular-DTN (V-DTN). The EMSC is an expanded version of MSC [12] and can be used for various network conditions. But, many control packets were delivered in the network. To solve this problem, we proposed a delayed ack for improving the EMSC [15,16]. The proposed method can increase the efficiency with less reply of MS-ACKs, however the paper lacked the performance evaluation.

In this paper, we evaluate the performance of our proposed EMSC with delayed ack for different road traffic conditions.

The structure of the paper is as follows. In Sect. 2, we summarize an overview of DTN protocol. Section 3 provides a detailed description of proposed EMSC with delayed ack approach. In Sect. 4 describes a system design and application setting. In Sect. 5, we show the simulation results. Finally, conclusions and future work are given in Sect. 6.

2 DTN

DTN are occasionally connected networks, characterized by the absence of a continuous path between the source and destination [1,10]. The data can be transmitted by storing them at nodes and forwarding them later when there is a working link. This technique is called message switching. Eventually the data will be relayed to the destination. The inspiration for DTNs came from an unlikely source: efforts to send packets in space. Space networks must deal with intermittent communication and very long delays [22]. In [10], the author observed the possibility to apply these ideas for other applications.

The main assumption in the Internet that DTNs seek to relax is that an End-to-End (E2E) path between a source and a destination exists for the entire duration of a communication session. When this is not the case, the normal Internet protocols fail. The DTN architecture is based on message switching. It is also intended to tolerate links with low reliability and large delays. The architecture is specified in RFC 4838 [7].

Bundle protocol has been designed as an implementation of the DTN architecture. A bundle is a basic data unit of the DTN bundle protocol. Each bundle comprises a sequence of two or more blocks of protocol data, which serve for various purposes. In poor conditions, bundle protocol works on the application layer of some number of constituent Internet, forming a store-and-forward overlay network to provide its services. The bundle protocol is specified in RFC 5050 [20]. It is responsible for accepting messages from the application and sending them as one or more bundles via store-carry-forward operations to the destination DTN node. The bundle protocol provides a transport service for many different applications.

A number of DTN protocols have been proposed, such as Epidemic [18,24], Spray and Wait (SpW) [21], MaxProp [5] and so on.

Epidemic is flooding-based DTN routing protocol [18,24]. Vehicles continuously replicate and send messages to newly discovered nodes that do not already possess a copy of the message. In the most simple case, Epidemic routing is flooding, but more elaborated techniques can be used to limit the number of message transfers.

SpW achieves resource efficiency by setting a strict upper bound on the number of copies per message allowed in the network [21]. The protocol is composed of two phases: the spray phase and the wait phase. Spray phase is terminated when the number of replications reaches the upper limit. When a relay node receives the replica, it enters the wait phase, where the relay simply holds that particular message until the destination is encountered directly. It is possible to suppress the communication cost compared with Epidemic.

MaxProp addresses the storage issue by prioritizing both the schedule of packets transmitted to other vehicles and the schedule of packets to be dropped [5]. When contacts with an adjacent vehicle, MaxProp transmits packets with high priority, which are sorted according to their hop count. When the storage is full, MaxProp dropped the packets with low priority, which are sorted according to their delivery likelihood.

We focused on the high delivery rate characteristics of conventional Epidemic protocol. Thus, we use the delayed ack approach with Epidemic to suppress the communication cost for different road traffic conditions.

3 Delayed Ack for EMSC

3.1 Overview

The EMSC provides message suppression including communication cost, for V2V and V2I communications. The module of EMSC works together with Epidemic, SpW and MaxProp [14,25]. We present the flowchart of EMSC algorithms for two states of vehicles in Fig. 1(a) and (b). EMSC is embedded in Vehicles (EMSCV).

3.2 Control Messages

When EMSCV receives HELLO message from other vehicles, the EMSCV stores the bundle ID and source node ID in the memory (see Fig. 1(a)). In addition,

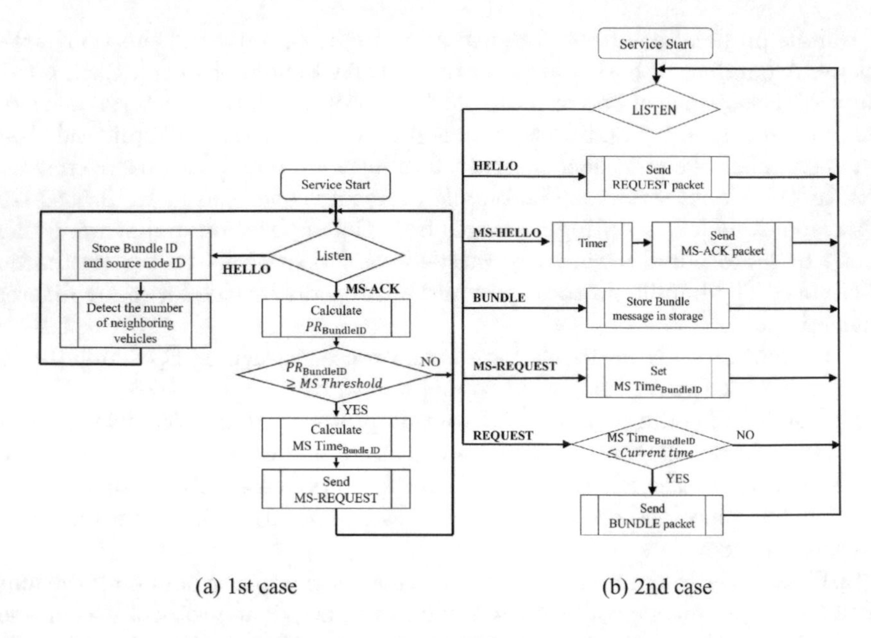

Fig. 1. Flowcharts of EMSC with delayed ack.

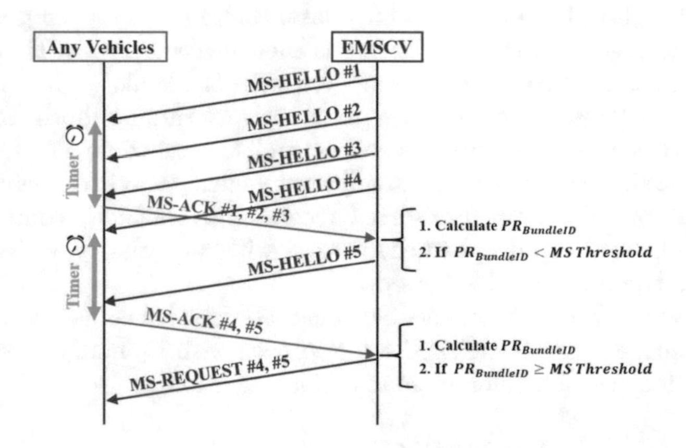

Fig. 2. Sequence chart of enhanced message suppression controller with delayed ack.

EMSCV detects the number of neighboring vehicles N to calculate the duration of message suppression.

EMSCV sends the MS-HELLO packet to vehicles in intervals of 1 s. In our previous work, when vehicle receives the MS-HELLO from EMSCV, the vehicle sends a MS-ACK to the EMSCV. In this case, vehicle sent many MS-ACKs, thus increasing communication cost. Based of this study, we have proposed a delayed ack approach for EMSC to decrease the communication cost. The delayed ack has a timer as shown in Figs. 1 and 2. The method increases efficiency by sending fewer reply packets than the conventional EMSC. When the EMSCV receives the MS-ACK from a vehicle, the EMSCV calculates Possession Rate of the Bundle ID ($PR_{BundleID}$). The formula of $PR_{BundleID}$ is:

$$PR_{BundleID} = \frac{\text{Number of same detected bundle IDs}(N_{BundleID})}{N_{new}}, \quad (1)$$

where N_{new} indicates the number of newly discovered vehicles from the EMSCV, which is calculates from the number of received HELLO packets. If $PR_{BundleID} \geq$ *MS Threshold*, EMSCV sends MS-REQUEST to the vehicle. The *MS Threshould* (MST) is threshold value. In this paper, we set the MST 1.

3.3 Message Suppression Time

Before EMSCV sends MS-REQUEST, the EMSCV calculates the MS Time based on $N_{BundleID}$, N_{new} and N values as shown in Eqs. (2) and (3).

$$\text{MS Time}_{Bundle\ ID} = \text{Current time} + (N_{BundleID} \times N_{new} \times R_{new}) \quad (2)$$

$$R_{new} = \frac{N_{new}}{N} \quad (3)$$

MS $\text{Time}_{BundleID}$ will be used in MS-REQUEST to suppress the bundle message of its bundle ID. After that, the vehicle does not send the bundle message of its bundle ID to other vehicles until the MS $\text{Time}_{BundleID}$ will be expired. As an exception, when vehicle moves to near end-point, the vehicle sends the bundle message to end-point even if MS $\text{Time}_{BundleID}$ is not expired. R_{new} indicates the ratio of the newly discovered vehicles at each EMSCV.

The stored N and N_{new} will be reset every second. In order to reduce the storage usage in each vehicle, we use five functions as shown in Fig. 1(b). When a vehicle receives HELLO packet from other vehicles, the other vehicles send REQUEST packet to the vehicle. Then, if Current time is greater than MS $\text{Time}_{BundleID}$, the vehicle sends the bundle message of its bundle ID to other vehicles. When vehicle receives bundle message, the vehicle stores the bundle message in the memory. In this way, the amount of traffic can be reduced.

4 System Design

4.1 Simulation Scenario

In this work, we consider an urban area (see Fig. 3) with different road traffic conditions. We consider two types of road traffic conditions such as Okinawa and Fukuoka, which are prefectures in Japan. A list of study cases is shown in Table 1. The road traffic conditions are the same as in [4]. The simulations are conducted using Scenargie network simulator considering 802.11p standard [19]. In each scenario, from 10% to 30% of vehicles are embedded in EMSC with delayed ack method. We consider the buildings, where some EMSCVs are present which are located on the road for forwarding the bundle messages considering the

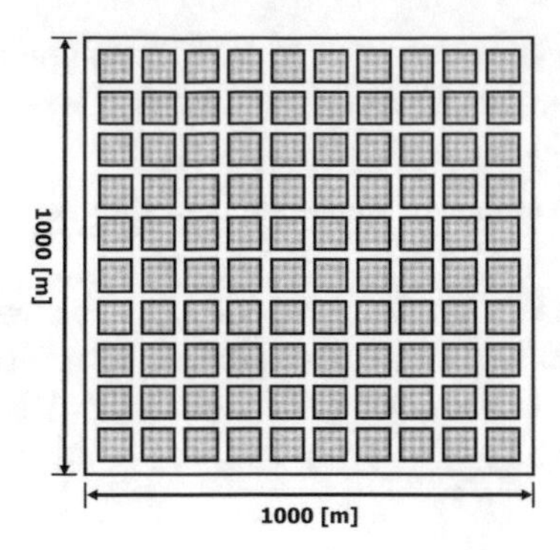

Fig. 3. Road model.

Table 1. Study cases.

Case ID	Prefecture	Vehicle density [vehicle/m^2]	Number of vehicles [vehicles]
V-DTN-Okinawa	Okinawa	0.001	330
V-DTN-Fukuoka	Fukuoka	0.0015	495

EMSC with delayed ack method to all vehicles. We used ITU-R P.1411 propagation model [2] for outdoor environment because it considers the interference of buildings.

Both start-point and end-point are static. The other vehicles and EMSCVs move according to Geographic Information System (GIS)-based random waypoint mobility model. In this model, the vehicles move roads defined by map data for real cities and limits their mobility according to vehicular congestion and simplified traffic control mechanisms.

4.2 Application Setting

The start-point sends a bundle message to end-point for different road traffic conditions. The distance between start-point to end-point is 1.414 [km]. Every vehicle has an on-board unit for broadcasting the bundle messages and to display the advertisement information. Simulation parameters are shown in Table 2.

We evaluate the network performance considering received bundles, sent MS-ACK and E2E delay.

5 Simulation Results

We discuss the simulation results for V-DTN-Okinawa and V-DTN-Fukuoka scenarios. We assumed that the road vehicle density of V-DTN-Fukuoka is higher than V-DTN-Okinawa. In Fig. 4 are shown the results of number of received bundles for different scenarios. For both scenarios and each timer, the results of

Table 2. Simulation parameters.

Parameter	Value
Simulation time	600 s
Area dimensions	1000 m × 1000 m
Grid size	10 × 10
Building: height	10 m
Mobility model	GIS-based random waypoint
Minimum speed ($V_{\min}$)	40 km/h
Maximum speed ($V_{\max}$)	60 km/h
Routing algorithm	EMSC with delayed ack
EMSC: timer of delayed ack	1.0, 1.5, 2.0 s
EMSC: MS threshold	1
EMSC: refresh interval	1 s
Application	Bundle message
Bundle: start and end time	10–590 s
Bundle: message sent interval	10 s
Bundle: message size	500 bytes
PHY model	IEEE 802.11p
Frequency	5.9 GHz
Propagation model	ITU-R P.1411
Antenna model	Omni-directional
Antenna height	1.5 m

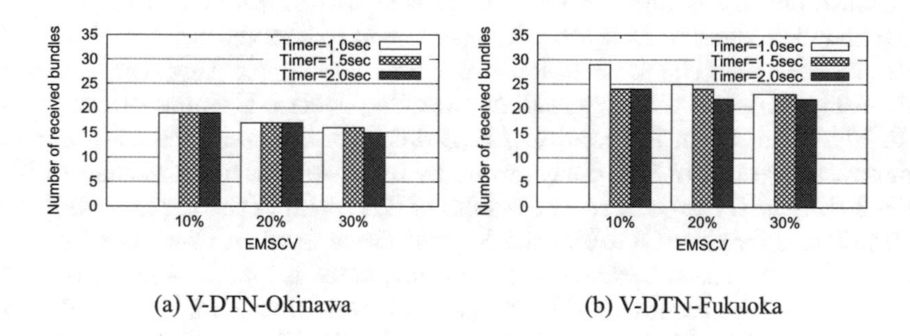

(a) V-DTN-Okinawa (b) V-DTN-Fukuoka

Fig. 4. Median received bundles for different cases.

received bundles decrease with increase of the EMSCVs. When 30% EMSCVs are active, MS-REQUEST is increased, thus the performance of received bundles is smaller than other cases. In case of V-DTN-Okinawa, the results of received bundles are almost same for each EMSCV. On the other hand, the difference was increased in case of V-DTN-Fukuoka. V-DTN-Fukuoka has delivered a lot

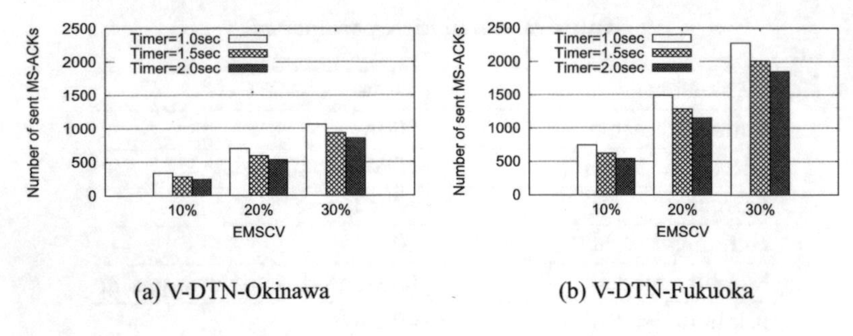

(a) V-DTN-Okinawa (b) V-DTN-Fukuoka

Fig. 5. Median sent MS-ACKs.

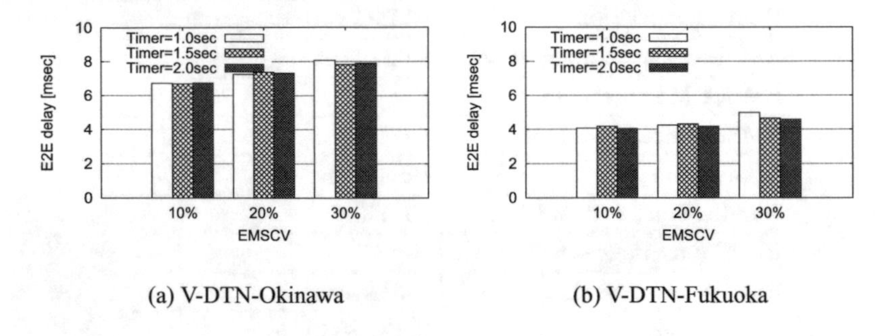

(a) V-DTN-Okinawa (b) V-DTN-Fukuoka

Fig. 6. Median E2E delay.

of bundle messages to the end-point compared with V-DTN-Okinawa. This was influenced by a large number of intermediate vehicles.

In Fig. 5 are shown the results of sent MS-ACKs for different scenarios. When the vehicle density is high, the results of sent MS-ACKs become 2 to 3 times higher than low density. Delayed ack method can reduce the number of MS-ACK transmissions when the suppression timer becomes longer. However, due to the relationship with the vehicle speed, suppressing control becomes difficult.

In Fig. 6 are shown the results of E2E delay for different scenarios. For both scenarios, the results of E2E delay gradually increase with increase the EMSCVs. When 30% EMSCV are active, the results of E2E delay is the highest when timer is 1.0 s. Then, for EMSCV 20%, the highest delay is when timer is 1.5 s. When EMSCV is 10%, the difference of the performance is small. When the vehicle density is low, the results of E2E delay is higher. In case of V-DTN-Fukuoka, there are many MS-ACKs, but our proposed method has good performance.

6 Conclusions

In this paper, we evaluated the performance of our proposed EMSC with delayed ack for different road traffic conditions. The simulations were conducted in urban environment considering vehicle densities of Okinawa and Fukuoka prefectures. From simulation results, we found that our proposed EMSC with delayed ack has good performance for these scenarios.

In the future, we would like to use the imported map data from open street map to evaluate the performance of our proposed method compared with other approaches.

References

1. Delay- and disruption-tolerant networks (DTNs) tutorial. NASA/JPL's Interplanetary Internet (IPN) Project (2012). http://www.warthman.com/images/DTN_Tutorial_v2.0.pdf
2. Rec. ITU-R P.1411-7: Propagation data and prediction methods for the planning of short-range outdoor radiocommunication systems and radio local area networks in the frequency range 300 MHz to 100 GHz. ITU (2013)
3. Araniti, G., Campolo, C., Condoluci, M., Iera, A., Molinaro, A.: LTE for vehicular networking: a survey. IEEE Commun. Mag. **21**(5), 148–157 (2013)
4. Automobile Inspection and Registration Information Association: Number of registered automobile in Japan, August 2017. http://www.airia.or.jp/publish/statistics/mycar.html
5. Burgess, J., Gallagher, B., Jensen, D., Levine, B.N.: MaxProp: routing for vehicle-based disruption-tolerant networks. In: Proceedings of the 25th IEEE International Conference on Computer Communications (IEEE INFOCOM 2006), pp. 1–11, April 2006
6. Burleigh, S., Hooke, A., Torgerson, L., Fall, K., Cerf, V., Durst, B., Scott, K., Weiss, H.: Delay-tolerant networking: an approach to interplanetary internet. IEEE Commun. Mag. **41**(6), 128–136 (2003)
7. Cerf, V., Burleigh, S., Hooke, A., Torgerson, L., Durst, R., Scott, K., Fall, K., Weiss, H.: Delay-tolerant networking architecture. IETF RFC 4838 (Informational), April 2007
8. Cheng, X., Yao, Q., Wen, M., Wang, C.X., Song, L.Y., Jiao, B.L.: Wideband channel modeling and intercarrier interference cancellation for vehicle-to-vehicle communication systems. IEEE J. Sel. Areas Commun. **31**(9), 434–448 (2013)
9. Dias, J.A.F.F., Rodrigues, J.J.P.C., Xia, F., Mavromoustakis, C.X.: A cooperative watchdog system to detect misbehavior nodes in vehicular delay-tolerant networks. IEEE Trans. Ind. Electron. **62**(12), 7929–7937 (2015)
10. Fall, K.: A delay-tolerant network architecture for challenged Internets. In: Proceedings of the International Conference on Applications, Technologies, Architectures, and Protocols for Computer Communications, SIGCOMM 2003, pp. 27–34 (2003)
11. Grassi, G., Pesavento, D., Pau, G., Vuyyuru, R., Wakikawa, R., Zhang, L.: VANET via named data networking. In: Proceedings of the IEEE Conference on Computer Communications Workshops (INFOCOM WKSHPS 2014), pp. 410–415, April 2014
12. Honda, T., Ikeda, M., Ishikawa, S., Barolli, L.: A message suppression controller for vehicular delay tolerant networking. In: Proceedings of the 29th IEEE International Conference on Advanced Information Networking and Applications (IEEE AINA 2015), pp. 754–760, March 2015
13. Ikeda, M., Ishikawa, S., Barolli, L.: An enhanced message suppression controller for vehicular-delay tolerant networks. In: Proceedings of the 30th IEEE International Conference on Advanced Information Networking and Applications (IEEE AINA 2016), pp. 573–579, March 2016

14. Ikeda, M., Ishikawa, S., Honda, T., Barolli, L.: Performance evaluation of message suppression method for dtn routing protocols. In: Proceedings of the 9th International Conference on Complex, Intelligent, and Software Intensive Systems (CISIS 2015), pp. 225–230, July 2015
15. Koga, D., Ikeda, M., Barolli, L.: An improved message suppression controller considering delayed acknowledgment for VANETs. In: Proceedings of the 19th International Conference on Network-Based Information Systems (NBiS 2016), pp. 420–424, September 2016
16. Koga, D., Ikeda, M., Barolli, L.: Performance evaluation of delayed ACK method for message suppression in VANETs. In: Proceedings of the 31st IEEE International Conference on Advanced Information Networking and Applications (IEEE AINA 2017), pp. 743–748, March 2017
17. Ohn-Bar, E., Trivedi, M.M.: Learning to detect vehicles by clustering appearance patterns. IEEE Trans. Intell. Transp. Syst. **16**(5), 2511–2521 (2015)
18. Ramanathan, R., Hansen, R., Basu, P., Hain, R.R., Krishnan, R.: Prioritized epidemic routing for opportunistic networks. In: Proceedings of the 1st International MobiSys Workshop on Mobile Opportunistic Networking (MobiOpp 2007), pp. 62–66 (2007)
19. Scenargie: Space-time engineering, LLC. http://www.spacetime-eng.com/
20. Scott, K., Burleigh, S.: Bundle protocol specification. IETF RFC 5050 (Experimental), November 2007
21. Spyropoulos, T., Psounis, K., Raghavendra, C.S.: Spray and wait: an efficient routing scheme for intermittently connected mobile networks. In: Proceedings of the ACM SIGCOMM Workshop on Delay-Tolerant Networking 2005 (WDTN 2005), pp. 252–259 (2005)
22. Tanenbaum, A.S., Wetherall, D.J.: Computer Networks, 5th edn. Pearson Education Inc., Prentice Hall (2011)
23. Uchida, N., Ishida, T., Shibata, Y.: Delay tolerant networks-based vehicle-to-vehicle wireless networks for road surveillance systems in local areas. Int. J. Space Based Situat. Comput. **6**(1), 12–20 (2016)
24. Vahdat, A., Becker, D.: Epidemic routing for partially-connected ad hoc networks. Technical report, Duke University (2000)
25. Yoshino, Y., Koga, D., Nakasaki, S., Ikeda, M., Barolli, L.: A message suppression method considering priority for inter-vehicle communications. In: Proceedings of the 12th International Conference on Broad-Band Wireless Computing, Communication and Applications (BWCCA 2017), pp. 754–763, November 2017

Improved Energy-Efficient Quorum Selection Algorithm by Omitting Meaningless Methods

Tomoya Enokido[1(✉)], Dilawaer Duolikun[2], and Makoto Takizawa[2]

[1] Faculty of Business Administration, Rissho University, Tokyo, Japan
eno@ris.ac.jp

[2] Department of Advanced Sciences, Faculty of Science and Engineering, Hosei University, Tokyo, Japan
dilewerdolkun@gmail.com, makoto.takizawa@computer.org

Abstract. Distributed applications are composed of multiple objects and each object is replicated in order to increase reliability, availability, and performance. On the other hand, the larger amount of electric energy is consumed in a system since multiple replicas of each object are manipulated on multiple servers. In our previous studies, the energy efficient quorum selection (EEQS) algorithm is proposed to construct a quorum for each method in the quorum based locking protocol so that the total electric energy of servers to perform methods can be reduced. In this paper, the improved energy efficient quorum selection (IEEQS) algorithm is proposed to furthermore reduce the total electric energy of servers by omitting meaningless methods. Evaluation results show the total electric energy of servers, the average execution time of each transaction, and the number of aborted transactions can be reduced in the IEEQS algorithm than the EEQS algorithm.

Keywords: Energy-aware information systems
Quorum-based locking protocol · Object-based systems
IEEQS algorithm · Data management

1 Introduction

In object-based systems [1,6], each object is a unit of computation resource like a file and is an encapsulation of data and methods to manipulate the data in the object. In order to provide reliable application services [2,3], each object is replicated on multiple servers. A transaction is an atomic sequence of methods [4] to manipulate objects. Conflicting methods issued by multiple transactions have to be serialized [5] to keep the replicas of each object mutually consistent. In the two-phase locking (2PL) protocol [4], one of the replicas of an object for a *read* method and all the replicas for a *write* method are locked before manipulating the object to keep the replicas mutually consistent. Since all the replicas have to be locked for every write method, the 2PL protocol is not efficient

L. Barolli et al. (Eds.): EIDWT 2018, LNDECT 17, pp. 171–182, 2018.
https://doi.org/10.1007/978-3-319-75928-9_15

in write-dominated application. In the quorum-based protocol [6,7], subsets of replicas locked for *read* and *write* methods are referred to as *read* and *write quorums*, respectively. The quorum numbers nQ^r and nQ^w for read and write methods have to be "$nQ^r + nQ^w > N$" where N is the total number of replicas. In the quorum-based protocol, the more number of write methods are issued, the fewer number of write quorums can be taken. As a result, the overhead to perform write methods can be reduced. However, the total amount of electric energy consumed in a system is larger than non-replication systems since each method issued to an object is performed on multiple replicas. In our previous studies, the *energy efficient quorum selection* ($EEQS$) algorithm [9] is proposed to construct a quorum for each method issued by a transaction so that the total electric energy of servers to perform the method is the minimum. Here, the total electric energy of a server cluster can be reduced in the EEQS algorithm than the traditional quorum-based locking protocol.

In this paper, we first define meaningless methods which are not required to be performed on each replica of an object based on the precedent relation and semantics of methods. Next, the *improved energy efficient quorum selection* ($IEEQS$) algorithm is proposed to furthermore reduce the total electric energy of a server cluster to perform methods by omitting meaningless methods on each replica. We evaluate the IEEQS algorithm compared with the EEQS algorithm. The evaluation results show the total electric energy of a server cluster, the average execution time of each transaction, and the number of aborted transactions in the IEEQS algorithm can be more reduced than the EEQS algorithm.

In Sect. 2, we discuss the data access model and power consumption model of a server. In Sect. 3, we discuss the IEEQS algorithm. In Sect. 4, we evaluate the IEEQS algorithm compared with the EEQS algorithm.

2 System Model

2.1 Objects and Transactions

A system is composed of multiple servers s_1, ..., s_n ($n \geq 1$) interconnected in reliable networks. Let S be a cluster of servers s_1, ..., s_n ($n \geq 1$). Let O be a set of objects o_1, ..., o_m ($m \geq 1$) [1]. Each object o_h is a unit of computation resource like a file and is an encapsulation of data and methods to manipulate the data in the object o_h. In this paper, methods are classified into *read* (r) and *write* (w) methods. Write methods are furthermore classified into *full* write (w^f) and *partial* write (w^p) methods, i.e. $w \in \{w^f, w^p\}$. In a full write method, a whole data in an object is fully written. In a partial write method, a part of data in an object is written. Suppose a file object F supports *modify*, *insert*, *delete*, and *read* methods. Here, a modify method is a full write method. Insert and delete methods are partial write methods. Let $op(o_h)$ be a state obtained by performing a method op ($\in \{r, w\}$) on an object o_h. A pair of methods op_1 and op_2 on an object o_h are *compatible* if and only if (iff) $op_1 \circ op_2(o_h) = op_2 \circ op_1(o_h)$. Otherwise, a method op_1 *conflicts* with another method op_2. In this paper, conflicting relations among methods are as shown in Table 1.

Each object o_h is replicated on multiple servers to make the system more reliable and available. Let $R(o_h)$ be a set of replicas o_h^1, ..., o_h^l $(1 \leq l \leq n)$ [2] of an object o_h. Let $nR(o_h)$ be the total number of replicas of an object o_h, i.e. $nR(o_h) = |R(o_h)|$. Replicas of each object o_h are distributed on multiple servers in a server cluster S. Let S_h be a subset of servers which hold a replica of an object o_h in a server cluster S $(S_h \subseteq S)$.

Table 1. Conflicting relation among methods.

		read (r)	write (w)	
			full (w^f)	partial (w^p)
read (r)		Compatible	Conflict	Conflict
write (w)	full (w^f)	Conflict	Conflict	Conflict
	partial (w^p)	Conflict	Conflict	Conflict

2.2 Quorum-Based Locking Protocol

A *transaction* is an atomic sequence of methods [4]. A transaction T_i issues r and w methods to manipulate replicas of objects. Multiple conflicting transactions are required to be *serializable* [4,5] to keep replicas of each object mutually consistent. Let $\mathbf{T}$ be a set of $\{T_1, ..., T_k\}$ $(k \geq 1)$ of transactions. Let H be a schedule of the transactions in $\mathbf{T}$. A transaction T_i *precedes* another transaction T_j $(T_i \rightarrow_H T_j)$ in a schedule H iff (if and only if) a method op_i from the transaction T_i is performed before a method op_j from the transaction T_j and op_i conflicts with op_j. A schedule H is serializable iff the precedent relation $\rightarrow_H$ is acyclic [4].

In this paper, multiple conflicting transactions are serialized based on the *quorum-based locking* protocol [6,7]. Let $\mu(op)$ be a *lock mode* of a method op $(\in \{r, w\})$. In this paper, a lock mode $\mu(w)$ is adapted to a full and partial write methods. A lock mode $\mu(r)$ is adapted to a read method. If op_1 is compatible with op_2 on an object o_h, the lock mode $\mu(op_1)$ is compatible with $\mu(op_2)$. Otherwise, a lock mode $\mu(op_1)$ conflicts with another lock mode $\mu(op_2)$. Let Q_h^{op} $(op \in \{r, w\})$ be a subset of replicas of an object o_h to be locked by a method op, named a *quorum* of the method op $(Q_h^{op} \subseteq R(o_h))$. Let nQ_h^{op} be the *quorum number* of a method op on a object o_h, i.e. $nQ_h^{op} = |Q_h^{op}|$. The quorums have to satisfy the following constraints: (1) $Q_h^r \subseteq R(o_h)$, $Q_h^w \subseteq R(o_h)$, and $Q_h^r \cup Q_h^w = R(o_h)$. (2) $nQ_h^r + nQ_h^w > nR(o_h)$, i.e. $Q_h^r \cap Q_h^w \neq \phi$. (3) $nQ_h^w > nR(o_h)/2$.

In the quorum-based locking protocol, a transaction T_i locks replicas of an object o_h by the following procedure [6]:

1. A quorum Q_h^{op} for a method op is constructed by selecting nQ_h^{op} replicas in a set $R(o_h)$ of replicas.
2. If every replica in a quorum Q_h^{op} can be locked by a lock mode $\mu(op)$, the replicas in the quorum Q_h^{op} are manipulated by the method op.

3. When the transaction T_i commits or aborts, the locks on the replicas in the quorum Q_h^{op} are released.

Each replica o_h^q has a *version number* v_h^q. Suppose a transaction T_i reads an object o_h. The transaction T_i selects nQ_h^r replicas in the set $R(o_h)$, i.e. *read* (r) quorum Q_h^r. If every replica in the r-quorum Q_h^r can be locked by a lock mode $\mu(r)$, the transaction T_i reads data in a replica o_h^q whose version number v_h^q is the maximum in the r-quorum Q_h^r. Every r-quorum surely includes at least one newest replica since $nQ_h^r + nQ_h^w > nR(o_h)$. Next, suppose a transaction T_i writes data in an object o_h. The transaction T_i selects nQ_h^w replicas in the set $R(o_h)$, i.e. *write* (w) quorum Q_h^w. If every replica in the w-quorum Q_h^w can be locked by a lock mode $\mu(w)$, the transaction T_i writes data in a replica o_h^q whose version number v_h^q is maximum in the w-quorum Q_h^w and the version number v_h^q of the replica o_h^q is incremented by one. The updated data and version number v_h^q of the replica o_h^q are sent to every other replica in the w-quorum Q_h^w. Then, data and version number of each replica in the w-quorum Q_h^w are replaced with the newest values.

2.3 Data Access Model

Methods which are being performed and already terminate are *current* and *previous* at time τ, respectively. Let $RP_t(\tau)$ and $WP_t(\tau)$ be sets of current *read* (r) and *write* (w) methods on a server s_t at time τ, respectively. Let $P_t(\tau)$ be a set of current r and w methods on a server s_t at time τ, i.e. $P_t(\tau) = RP_t(\tau) \cup WP_t(\tau)$. Let $r_{ti}(o_h^q)$ and $w_{ti}(o_h^q)$ be methods issued by a transaction T_i to read and write data in a replica o_h^q on a server s_t, respectively. By each method $r_{ti}(o_h^q)$ in a set $RP_t(\tau)$, data is read in a replica o_h^q at rate $RR_{ti}(\tau)$ [B/sec] at time τ. By each method $w_{ti}(o_h^q)$ in a set $WP_t(\tau)$, data is written in a replica o_h^q at rate $WR_{ti}(\tau)$ [B/sec] at time τ. Let $maxRR_t$ and $maxWR_t$ be the maximum read and write rates [B/sec] of r and w methods on a server s_t, respectively. The read rate $RR_{ti}(\tau)$ ($\leq maxRR_t$) and write rate $WR_{ti}(\tau)$ ($\leq maxWR_t$) are $fr_t(\tau) \cdot maxRR_t$ and $fw_t(\tau) \cdot maxWR_t$, respectively. Here, $fr_t(\tau)$ and $fw_t(\tau)$ are degradation ratios. $0 \leq fr_t(\tau) \leq 1$ and $0 \leq fw_t(\tau) \leq 1$. The degradation ratios $fr_t(\tau)$ and $fw_t(\tau)$ are $1/(|RP_t(\tau)| + rw_t \cdot |WP_t(\tau)|)$ and $1/(wr_t \cdot |RP_t(\tau)| + |WP_t(\tau)|)$, respectively. $0 \leq rw_t \leq 1$ and $0 \leq wr_t \leq 1$.

The *read laxity* $lr_{ti}(\tau)$ [B] and *write laxity* $lw_{ti}(\tau)$ [B] of methods $r_{ti}(o_h^q)$ and $w_{ti}(o_h^q)$ show how much amount of data are read and written in a replica o_h^q by the methods $r_{ti}(o_h^q)$ and $w_{ti}(o_h^q)$ at time τ, respectively. Suppose that methods $r_{ti}(o_h^q)$ and $w_{ti}(o_h^q)$ start on a server s_t at time st_{ti}, respectively. At time st_{ti}, the read laxity $lr_{ti}(\tau) = rb_h^q$ [B] where rb_h^q is the size of data in a replica o_h^q. The write laxity $lw_{ti}(\tau) = wb_h^q$ [B] where wb_h^q is the size of data to be written in a replica o_h^q. The read laxity $lr_{ti}(\tau)$ and write laxity $lw_{ti}(\tau)$ at time τ are rb_h^q - $\Sigma_{\tau=st_{ti}}^{\tau} RR_{ti}(\tau)$ and wb_h^q - $\Sigma_{\tau=st_{ti}}^{\tau} WR_{ti}(\tau)$, respectively.

2.4 Power Consumption Model of a Server

Let $E_t(\tau)$ be the electric power [W] of a server s_t at time τ. $maxE_t$ and $minE_t$ show the maximum and minimum electric power [W] of the server s_t, respectively. The *power consumption model for a storage server* (PCS model) [8] to perform storage and computation processes are proposed. In this paper, we assume only r and w methods are performed on a server s_t. According to the PCS model, the electric power $E_t(\tau)$ [W] of a server s_t to perform multiple r and w methods at time τ is given as follows:

$$E_t(\tau) = \begin{cases} WE_t & \text{if } |WP_t(\tau)| \geq 1 \text{ and } |RP_t(\tau)| = 0. \\ WRE_t(\alpha) & \text{if } |WP_t(\tau)| \geq 1 \text{ and } |RP_t(\tau)| \geq 1. \\ RE_t & \text{if } |WP_t(\tau)| = 0 \text{ and } |RP_t(\tau)| \geq 1. \\ minE_t & \text{if } |WP_t(\tau)| = |RP_t(\tau)| = 0. \end{cases} \quad (1)$$

A server s_t consumes the minimum electric power $minE_t$ [W] if no method is performed on the server s_t, i.e. the electric power in the idle state of the server s_t. The server s_t consumes the electric power RE_t [W] if at least one r method is performed on the server s_t. The server s_t consumes the electric power WE_t [W] if at least one w method is performed on the server s_t. The server s_t consumes the electric power $WRE_t(\alpha)$ [W] $= \alpha \cdot RE_t + (1 - \alpha) \cdot WE_t$ [W] where $\alpha = |RP_t(\tau)|/(|RP_t(\tau)| + |WP_t(\tau)|)$ if both at least one r method and at least one w method are concurrently performed. Here, $minE_t \leq RE_t \leq WRE_t(\alpha) \leq WE_t \leq maxE_t$.

The total electric energy $TE_t(\tau_1, \tau_2)$ [J] of a server s_t from time τ_1 to τ_2 is $\Sigma_{\tau=\tau1}^{\tau_2} E_t(\tau)$. The processing power $PE_t(\tau)$ [W] of a server s_t at time τ is $E_t(\tau) - minE_t$. The total processing electric energy $TPE_t(\tau_1, \tau_2)$ of a server s_t from time τ_1 to τ_2 is given as $TPE_t(\tau_1, \tau_2) = \Sigma_{\tau=\tau1}^{\tau_2} PE_t(\tau)$. The total processing electric energy laxity $tpecl_t(\tau)$ shows how much electric energy a server s_t has to consume to perform every current r and w methods on the server s_t at time τ. The total processing energy consumption laxity $tpecl_t(\tau)$ of a server s_t at time τ is obtained by the following $\boldsymbol{TPECL_t}$ procedure:

$\boldsymbol{TPECL_t(\tau)}$ {
  if $RP_t(\tau) = \phi$ and $WP_t(\tau) = \phi$, **return**(0);
  $laxity = E_t(\tau) - minE_t$; /* $PE_t(\tau)$ of a server s_t at time τ */
  for each r-method $r_{ti}(o_h^q)$ in $RP_t(\tau)$, {
    $lr_{ti}(\tau + 1) = lr_{ti}(\tau) - RR_{ti}$;
    if $lr_{ti}(\tau + 1) = 0$, $RP_t(\tau + 1) = RP_t(\tau) - \{r_{ti}(o_h^q)\}$;
  } /* for end */
  for each w-method $w_{ti}(o_h^q)$ in $WP_t(\tau)$, {
    $lw_{ti}(\tau + 1) = lw_{ti}(\tau) - WR_{ti}$;
    if $lw_{ti}(\tau + 1) = 0$, $WP_t(\tau + 1) = WP_t(\tau) - \{w_{it}(o_h^q)\}$;
  } /* for end */
  return($laxity + \boldsymbol{TPECL_t}(\tau + 1)$);
}

In the $\boldsymbol{TPECL_t}$ procedure, each time τ data is read in a replica o_h^q by a method $r_{ti}(o_h^q)$, the read laxity $lr_{ti}(\tau)$ of the method $r_{ti}(o_h^q)$ is decremented by read rate RR_{ti}. Similarly, the write laxity $lw_{ti}(\tau)$ of a method $w_{ti}(o_h^q)$ is decremented by write rate WR_{ti} each time τ data is written in a replica o_h^q by the method $w_{ti}(o_h^q)$. If the read laxity $lr_{ti}(\tau + 1)$ and write laxity $lw_{ti}(\tau + 1)$ get 0, every data is read and written in the replica o_h^q by the methods $r_{ti}(o_h^q)$ and $w_{ti}(o_h^q)$, respectively, and the methods terminate at time τ.

3 Improved EEQS (IEEQS) Algorithm

3.1 Quorum Selection

In the *improved energy-efficient quorum selection* (*IEEQS*) algorithm, replicas to be members of a quorum of each method are selected so that the total electric energy of a server cluster S to perform the method is the minimum. Suppose a transaction T_i issues a method op ($op = \{r, w\}$) to manipulate an object o_h at time τ. Each transaction T_i selects a subset S_h^{op} ($\subseteq S_h$) of nQ_h^{op} servers whose total processing electric energy laxity is the minimum for each method op by following quorum selection ($\boldsymbol{QS}$) procedure [9]:

QS(op, o_h, τ) { /* $op \in \{r, w\}$ */
  $S_h^{op} = \phi$;
  while ($nQ_h^{op} > 0$) {
    for each server s_t in S_h, {
      if $op = r$, $RP_t(\tau) = RP_t(\tau) \cup \{op\}$;
      else $WP_t(\tau) = WP_t(\tau) \cup \{op\}$; /* $op = w$ */
      $TPE_t(\tau) = \boldsymbol{TPECL_t}(\tau)$;
    } /* for end */
    $server$ = a server s_t where $TPE_t(\tau)$ is the minimum;
    $S_h^{op} = S_h^{op} \cup \{server\}$; $S_h = S_h - \{server\}$; $nQ_h^{op} = nQ_h^{op} - 1$;
  } /* while end */
  return(S_h^{op});
}

3.2 Meaningless Methods

A method op_1 *precedes* op_2 in a schedule H ($op_1 \rightarrow_H op_2$) iff (1) the methods op_1 and op_2 are issued by the same transaction T_i and op_1 is issued before op_2, (2) the method op_1 issued by a transaction T_i conflicts with the method op_2 issued by a transaction T_j and $T_i \rightarrow_H T_j$, or (3) $op_1 \rightarrow_H op_3 \rightarrow_H op_2$ for some method op_3. Let H_h be a *local schedule* of methods which are performed on an object o_h in a schedule H.

[Definition]. A method op_1 *locally precedes* another method op_2 in a local schedule H_h ($op_1 \rightarrow_{H_h} op_2$) iff $op_1 \rightarrow_H op_2$.

A partial write method op_1 locally precedes another full write method op_2 in a local schedule H_h ($op_1 \rightarrow_{H_h} op_2$) on the object o_h. Here, the partial write method op_1 is not required to be performed on the object o_h if the full write method op_2 is surely performed on the object o_h just after the method op_1, i.e. the method op_2 can *absorb* the method op_1.

[Definition]. A full write method op_1 *absorbs* another partial or full write method op_2 in a local subschedule H_h on an object o_h if $op_2 \rightarrow_{H_h} op_1$, and there is no read method op' such that $op_2 \rightarrow_{H_h} op' \rightarrow_{H_h} op_1$, or op_1 absorbs op'' and op'' absorbs op_2 for some method op''.

[Definition]. A method op is *meaningless* iff the method op is absorbed by another method op' in the local subschedule H_h of an object o_h.

3.3 Omitting Meaningless Methods

Suppose three replicas o_1^1, o_1^2, and o_1^3 of an object o_1 are stored in three servers s_1, s_2, and s_3, respectively, i.e. $S_1 = \{s_1, s_2, s_3\}$. The version numbers v_1^1, v_1^2, and v_1^3 of replicas o_1^1, o_1^2, and o_1^3 are 2, 1, and 2, respectively, as shown in Fig. 1. The quorum numbers nQ_1^w and nQ_1^r for the object o_1 are two, respectively. Let $T_i.Q_h^{op}$ be a quorum to perform a method op issued by a transaction T_i. Let $T_i.S_h^{op}$ be a subset of servers which hold replicas in a quorum $T_i.Q_h^{op}$.

Suppose a pair of replicas o_1^1 and o_1^2 are locked by a transaction T_1 with lock mode $\mu(w)$ and a partial write methods $w_1^p(o_1)$ is issued to a w-quorum $T_1.Q_1^{w_1^p(o_1)} = \{o_1^1, o_1^2\}$ as shown in Fig. 1. The partial write method $w_{11}^p(o_1^1)$ is performed on the replica o_1^1 since the version number v_1^1 of the replica o_1^1 is the maximum in the w-quorum $T_1.Q_1^{w_1^p(o_1)}$, i.e. v_1^1 $(= 2) > v_1^2$ $(= 1)$. Then, the version number v_1^1 is incremented by one, i.e. $v_1^1 = 3$. The updated data and version number v_1^1 $(= 3)$ are sent to the replica o_1^2. In the traditional quorum-based locking protocol, data and version number of the replica o_1^2 are replaced with the newest values as soon as the replica o_1^2 receives the updated data and version number, i.e. the partial write method $w_{21}^p(o_1^2)$ is performed on the replica o_1^2. In the IEEQS algorithm, the version number of the replica o_1^2 is replaced with the newest value but data of the replica o_1^2 is not replaced with the newest values until the next method is performed on the replica o_1^2. This means that the partial write method $w_{21}^p(o_1^2)$ to replace data of a replica o_1^2 is delayed until the next method is performed on the replica o_1^2. Suppose a pair of replicas o_1^2 and o_1^3 are locked by a transaction T_2 with lock mode $\mu(w)$ and a full write methods $w_2^f(o_1)$ is issued to a w-quorum $T_2.Q_1^{w_2^p(o_1)} = \{o_1^2, o_1^3\}$ after the transaction T_1 commits. The full write method $w_{22}^f(o_1^2)$ issued by the transaction T_2 is performed on the replica o_1^2 since the version number v_1^2 is the maximum in the w-quorum $T_2.Q_1^{w_2^p(o_1)}$, i.e. v_1^2 $(= 3) > v_1^3$ $(= 2)$. Here, the partial write method $w_{21}^p(o_1^2)$ issued by the transaction T_1 is meaningless since the full write method $w_{22}^f(o_1^2)$ issued by the transaction T_2 absorbs the partial write method $w_{21}^p(o_1^2)$ on the replica o_1^2. Hence, the full write method $w_{22}^f(o_1^2)$ is performed on the replica o_1^2 without performing the partial write method $w_{21}^p(o_1^2)$ and the version number of the replica o_1^2 is incremented by one, i.e. $v_1^2 = 4$. That is, the meaningless method $w_{21}^p(o_1^2)$ is omitted on the replica o_1^2.

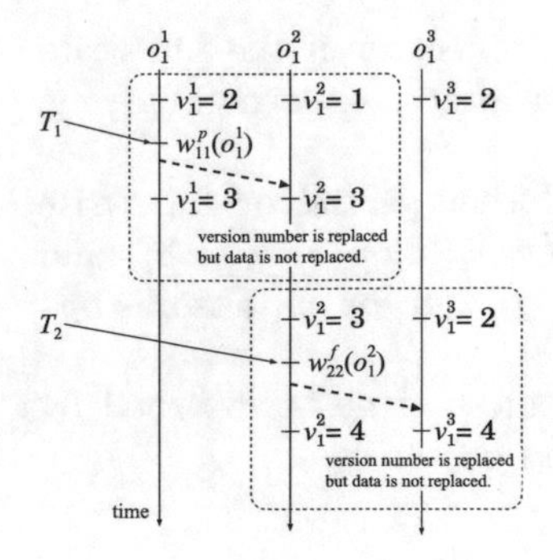

Fig. 1. Example of meaningless methods

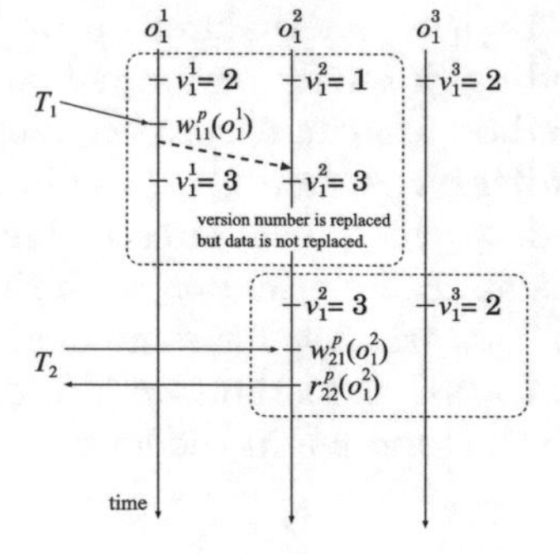

Fig. 2. Execution of read methods.

Suppose a pair of replicas o_1^2 and o_1^3 are locked by a transaction T_2 with lock mode $\mu(r)$ and a read method $r_1(o_1)$ is issued to a r-quorum $T_2.Q_1^{r_1(o_1)} = \{o_1^2, o_1^3\}$ after the transaction T_1 commits as shown in Fig. 2. The transaction T_2 reads data in the replica o_1^2 since the version number v_1^2 is the maximum in the r-quorum $T_2.Q_1^{r_1(o_1)}$, i.e. v_1^2 $(= 3) > v_1^3$ $(= 2)$. Here, the partial write method $w_{21}^p(o_1^2)$ issued by a transaction T_1 has to be performed before the read method $r_{22}(o_1^2)$ is performed since the read method $r_{22}(o_1^2)$ has to read data written by the partial write method $w_{21}^p(o_1^2)$.

Let $o_h^q.DW$ be a write method $w_{ti}(o_h^q)$ issued by a transaction T_i to replace data of a replica o_h^q in a server s_t with the updated data d_h, which is waiting for the next method *op* to be performed on the replica o_h^q. Suppose a transaction T_i issues a method *op* to a quorum Q_h^{op} for manipulating an object o_h. The method $op(o_h)$ is performed on a replica o_h^q whose version number is the maximum in the quorum Q_h^{op}. In the IEEQS algorithm, the method *op* is performed on the replica o_h^q whose version number is the maximum in the quorum Q_h^{op} by the following **IEEQS_Perform** procedure:

```
IEEQS_Perform(op(o_h^q)) {
  if op(o_h^q) = r, {
    if o_h^q.DW = φ, perform(op(o_h^q));
    else { /* o_h^q.DW ≠ φ */
      perform(o_h^q.DW); perform(op(o_h^q)); o_h^q.DW = φ;
    }
  }
  else { /* op = w */
    v_h^q = v_h^q + 1;
    if o_h^q.DW ≠ φ and o_h^q.DW is not meaningless, {
      perform(o_h^q.DW); perform(op(o_h^q)); o_h^q.DW = φ;
    }
    else perform(op(o_h^q)); /* o_h^q.DW = φ or o_h^q.DW method is omitted */
    v_h^q and updated data d_h are sent to every replica o_h^q' in a quorum Q_h^op;
  }
}
```

Each time a replica $o_h^{q'}$ in a write quorum Q_h^w receives the newest version number v_h^q and updated data d_h from another replica o_h^q as a result obtained by performing a write method $w_{ti}(o_h^q)$, the replica $o_h^{q'}$ manipulate the version number v_h^q and updated data d_h by the following **IEEQS_Replace** procedure:

IEEQS_Replace(v_h^q, d_h, $w_{ti}(o_h^q)$) {
  $v_h^{q'} = v_h^q$;
  if $o_h^{q'}.DW = \phi$, $o_h^{q'}.DW = w_{ti}(o_h^q)$;
  else { /* $o_h^{q'}.DW \neq \phi$ */
    if $w_{ti}(o_h^q)$ absorbs $o_h^{q'}.DW$, $o_h^{q'}.DW = w_{ti}(o_h^q)$;
    else {
      perform($o_h^{q'}.DW$); $o_h^{q'}.DW = w_{ti}(o_h^q)$;
    }
  }
}

4 Evaluation

4.1 Environment

We evaluate the IEEQS algorithm in terms of the average execution time of each transaction, the average number of aborted transactions, and the total electric energy of a server cluster S compared with the EEQS algorithm. A homogeneous server cluster S which is composed of ten homogeneous servers $s_1, \ldots, s_{10}$ ($n = 10$) is considered. In the server cluster S, every server s_t ($t = 1, \ldots, 10$) follows the same data access model and power consumption model as shown in Table 2. Parameters of each server s_t are given based on the experimentations [8]. There are fifty objects $o_1, \ldots, o_{50}$ in a system. The size of data in each object o_h is randomly selected between 50 and 100 [MB]. Each object o_h supports *read* (r), *full write* (w^f), and *partial write* (w^p) methods. The total number of replicas for every object is five, i.e. $nR(o_h) = 5$. Replicas of each object are randomly distributed on five servers in the server cluster S. The quorum numbers nQ_h^w and nQ_h^r on every object o_h are three, respectively, i.e. $nQ_h^w = nQ_h^r = 3$.

Table 2. Homogeneous cluster S

Server s_t	$maxRR_t$	$maxWR_t$	rw_t	wr_t	$minE_t$	WE_t	RE_t
s_t	80 [MB/sec]	45 [MB/sec]	0.5	0.5	39 [W]	53 [W]	43 [W]

The number m of transactions are issues to manipulate objects. Each transaction issues three methods randomly selected from one-hundred fifty methods on the fifty objects. The total amount of data of an object o_h is fully written by each full write (w^f) method. On the other hand, a half size of data of an object o_h is written and read by each partial write (w^p) and read (r) methods. The starting time of each transaction T_i is randomly selected in a unit of one second between 1 and 360 [sec].

4.2 Average Execution Time of Each Transaction

Let ET_i be the execution time [sec] of a transaction T_i where the transaction T_i commits. Suppose a transaction T_i starts at time st_i and commits at time et_i. The execution time ET_i of the transaction T_i is $et_i - st_i$ [sec]. The execution time ET_i for each transaction T_i is measured ten times for each total number m of transactions ($0 \leq m \leq 1{,}000$). Let ET_i^{tm} be the execution time ET_i obtained in tm-th simulation. The average execution time AET [sec] of each transaction for each total number m of transactions is $\sum_{tm=1}^{10}\sum_{i=1}^{m} ET_i^{tm}/(m \cdot 10)$.

Figure 3 shows the average execution time AET [sec] of the m transactions in the IEEQS and EEQS algorithms. In the IEEQS and EEQS algorithms, the average execution time AET increases as the total number m of transactions increases since more number of transactions are concurrently performed. For $0 < m \leq 1{,}000$, the average execution time AET can be more reduced in the IEEQS algorithm than the EEQS algorithm. In the IEEQS algorithm, each transaction can commit without waiting for performing meaningless methods. Hence, the average execution time of each transaction can be more reduced in the IEEQS algorithm than the EEQS algorithm.

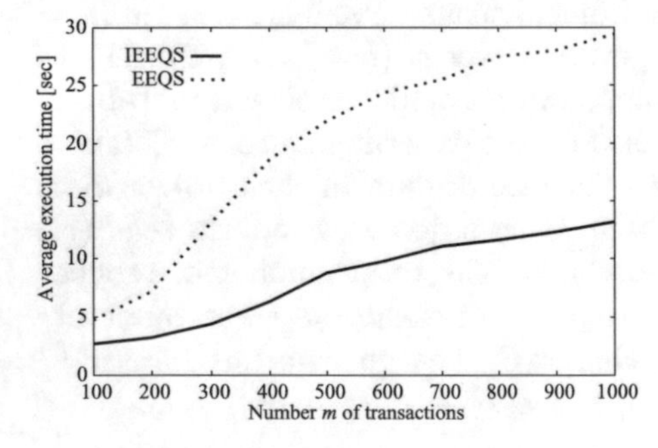

Fig. 3. Average execution time AET [sec] of each transaction.

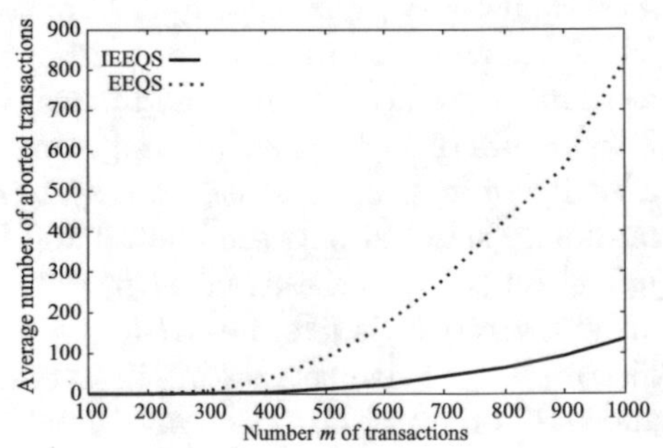

Fig. 4. Average number of aborts for each transaction

4.3 Average Number of Aborted Transaction Instances

If a transaction T_i could not lock every replica in an r-quorum Q_h^r or w-quorum Q_h^w, the transaction T_i aborts. Then, the transaction T_i is restarted after δ time units. The time units δ [sec] is randomly selected between twenty and thirty seconds in this evaluation. Every transaction T_i is restarted until the transaction T_i commits. Each execution of a transaction is referred to as transaction *instance*. We measure how many number of transaction instances are aborted until each transaction commits. Let AT_i be the number of aborted instances of a transaction T_i. The number AT_i of aborted instances for each transaction T_i is measured ten times for each total number m of transactions ($0 \leq m \leq 1{,}000$). Let AT_i^{tm} be the number AT_i of aborted transaction instances obtained in tmth simulation. The average number AAT of aborted instances of each transaction for each total number m of transactions is $\sum_{tm=1}^{10}\sum_{i=1}^{m} AT_i^{tm}/(m \cdot 10)$.

Figure 4 shows the average number AAT of aborted transaction instances to perform the total number m of transactions in the IEEQS and EEQS algorithms. The more number of transactions are concurrently performed, the more number of transactions cannot lock replicas. Hence, the number of aborted transactions instance increases in the IEEQS and EEQS algorithms as the total number m of transactions increases. For $0 < m \leq 1{,}000$, the average number AAT of aborted instances of each transaction can be more reduced in the IEEQS algorithm than the random algorithm. The average execution time of each transaction can be reduced in the IEEQS algorithm than the EEQS algorithm. As a result, the number of aborted transactions can be more reduced in the IEEQS algorithm than the EEQS algorithm since the number of transaction to be concurrently performed can be reduced.

4.4 Average Total Energy Consumption of a Server Cluster

Let TEC_{tm} be the total electric energy [J] to perform the number m of transactions ($0 \leq m \leq 1{,}000$) in the server cluster S obtained in the tm-th simulation. The total electric energy TEC_{tm} is measured ten times for each number m of transactions. Then, the average total electric energy $ATEC$ [J] of the server cluster S is calculated as $\sum_{tm=1}^{10} TEC_{tm}/10$ for each number m of transactions.

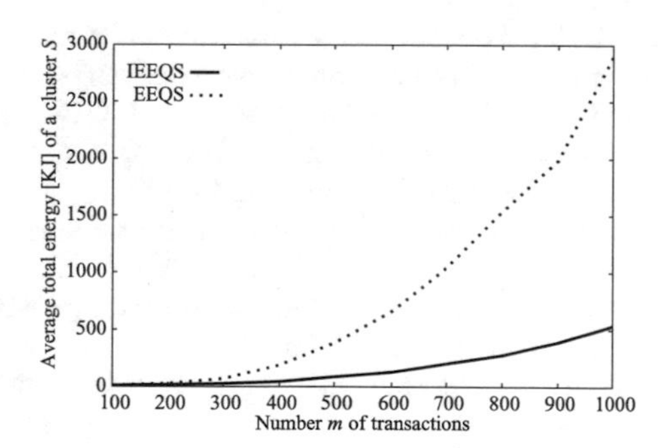

Fig. 5. Average total energy consumption (ATEC) [KJ].

Figure 5 shows the average total electric energy $ATEC$ of the server cluster S to perform the number m of transactions in the IEEQS and EEQS algorithms. For $0 \leq m \leq 1{,}000$, the average total electric energy $ATEC$ of the server cluster S can be more reduced in the IEEQS algorithm than the EEQS algorithm.

In the IEEQS algorithm, meaningless methods are omitted on each replica. In addition, the average execution time and the number of aborted instances of each transaction can be more reduced in the IEEQS algorithm than the EEQS algorithm. As a result, the average total electric energy $ATEC$ of the server cluster S can be more reduced in the IEEQS algorithm than the EEQS algorithm.

Following the evaluation, the total electric energy of a server cluster, the average execution time of each transaction, and the number of aborted transactions in the IEEQS algorithm can be reduced than the EEQS algorithm, respectively. Hence, the IEEQS algorithm is more useful than the EEQS algorithm.

5 Concluding Remarks

In this paper, we newly proposed the Improved EEQS (IEEQS) algorithm to reduce the total electric energy of a server cluster in the quorum-based locking protocol by omitting meaningless methods. We evaluated the IEEQS algorithm compared with the EEQS algorithm. The evaluation results show the total electric energy of a server cluster, the average execution time of each transaction, and the number of aborted transactions can be more reduced in the IEEQS algorithm than the EEQS algorithm. Following the evaluation, the IEEQS algorithm is more useful than the EEQS algorithm.

References

1. Object Management Group Inc.: Common object request broker architecture (CORBA) specification, version 3.3, part 1 - interfaces (2012). http://www.omg.org/spec/CORBA/3.3/Interfaces/PDF
2. Schneider, F.B.: Replication management using the state-machine approach. In: Distributed Systems, 2nd edn. ACM Press (1993)
3. Sawamura, S., Barolli, A., Aikebaier, A., Enokido, T., Takizawa, M.: Design and evaluation of algorithms for obtaining objective trustworthiness on acquaintances in P2P overlay networks. Int. J. Grid Utility Comput. (IJGUC) **2**(3), 196–203 (2011)
4. Bernstein, P.A., Hadzilacos, V., Goodman, N.: Concurrency Control and Recovery in Database Systems. Addison-Wesley, Reading (1987)
5. Gray, J.N.: Notes on database operating systems. In: Operating Systems, vol. 60, pp. 393–481 (1978)
6. Tanaka, K., Hasegawa, K., Takizawa, M.: Quorum-based replication in object-based systems. J. Inf. Sci. Eng. **16**(3), 317–331 (2000)
7. Garcia-Molina, H., Barbara, D.: How to assign votes in a distributed system. J. ACM **32**(4), 814–860 (1985)
8. Sawada, A., Kataoka, H., Duolikun, D., Enokido, T., Takizawa, M.: Energy-aware clusters of servers for storage and computation applications. In: Proceedings of the 30th IEEE International Conference on Advanced Information Networking and Applications (AINA-2016), pp. 400–407 (2016)
9. Enokido, T., Duolikun, D., Takizawa, M.: Energy-efficient Quorum selection algorithm for distributed object-based systems. In: Proceedings of the 11th International Conference on Conference on Complex, Intelligent, and Software Intensive Systems, (CISIS-2017), pp. 32–42 (2017)

Improving Data Loss Prevention Using Classification

Brunela Karamani(✉)

Faculty of Information Technology, Polytechnic University of Tirana, Tirana, Albania
bkaramani@fti.edu.al

Abstract. The financial institutions provide the resources to protect their sensitive data and information by trying to prevent unauthorized leakage. They approve policies and realize technical restrictions to block the loss and revelation of sensitive data and information by external attackers as well as careless insiders. One example of Data Loss Prevention (DLP) restrictions consists of endpoint protection solutions to block data transmissions to USB storage devices. Nevertheless, financial institutions approve exceptions to these policies, based on the business need for the specific user, in order to be able to fulfill their job-related tasks. But from these exceptions derive the following questions: How an approval for an exception can create impact over the risk of data leakage for the financial institution? What is the particular risk for according an individual user a confident exception? This paper introduces a new concept to risk depending on exception management, which will provide the financial institution to assign exceptions derived from on basic DLP. Initially, the paper presents an approach for evaluating and classification users based on their access to sensitive data and information, and afterward, a standard of rights is decided for assigning exceptions to derive from the classification of users, which allows specific approvers to prepare knowledgeable decisions concerning exception requests.

Keywords: Security · DLP · Information · Risk · Classification

1 Introduction

This DLP implements most significantly to blocking the flow of sensitive data and information outside a secure zone of the financial institution. It as well-known as data leak or simply leak protection, describes systems and technologies designed to detect potential data breaches, or attempts to move data outside an organization's secure storage and systems, and beyond its control [1]. The financial institution is progressively getting dynamic on how they manage their business and clients data on and off premises. With increasing cyber threat area, data must be protected at all times from unapproved operation, modification, and depository. Protecting the data, especially sensitive data is getting challenging given the sophistication of cyber-attacks. DLP objective is to minimize the data loss and business impact at all times due to a data breach event that could potentially become an incident [2]. The data which needs to be protected according to departments are:

L. Barolli et al. (Eds.): EIDWT 2018, LNDECT 17, pp. 183–189, 2018.
https://doi.org/10.1007/978-3-319-75928-9_16

Legal: contracts, intellectual property, memos, communications, internal investigation, corporate governance, internal legal presentations etc.
Business: roadmap, business plans, forecasts, competitive data, client pricing and volumes, client sales quotations etc.
IT: network diagrams, configuration files (networks, systems, application, and database), wireless access keys, software source code etc.
Finance: Pre-earnings release, financial statements, payroll and equity data etc.
HR: all employee data, recruiting lists, organization reporting structure etc.

DLP systems support three different types of protection:

In-use protection is implemented when sensitive data and information are in use by applications and mainly depends on various types of user authentication to establish the identity for those requesting access to the data [3].

In-motion protection is implemented when sensitive data and information is in transportation on a network, and mainly depends on adequately powerful encryption mechanisms and technologies to reduce the risk of spying and to considerably lower the prospect of a successful decryption attack [4]. At-rest protection is implemented to data and information as it resides on some type of persistent storage medium. This usually involves access controls to limit access to programs and users with a legitimate need to know and strong encryption to protect against theft or attack against the physical media where such data is stored [1].

The exceptions which have been chosen to demonstrate our method are: authorized and encrypted USB mobile storage devices, unrestricted USB connectivity, web upload and administrative privileges. Blocking web upload to websites at least reduces the risk of users transferring whole documents or files [5]. Financial institutions may not only want to block websites with inappropriate and undesired content, but also websites that provide the ability to share potentially sensitive information, such as social media sites, blogs, forums, and public file storage sites [6]. Many businesses today require storage and transportation of data via mobile storage devices. In such cases, organizations might allow a user to request and use authorized and encrypted USB mobile storage devices provided by the organization. Although the information is encrypted and therefore protected while transferred on the authorized mobile storage device, the data is out of control of the organization as soon as it is copied from the device to a third party system [7].

To improve the accuracy of DLP solution the financial institution can use other security tools such as firewalls and anti-virus products [2]. Below are shown in Table 1 the enterprises for DLP solution, which are leaders and visionaries in the market with their advantages and disadvantages [8].

Table 1. The enterprises for DLP solution in the market.

Vendors	Descriptions	Advantages(A)/Disadvantages(D)
Symantec	Has product components for: DLP Enforce Platform DLP IT Analytics Cloud Storage and Cloud Prevent for MSOffice 365 DLP for Endpoint, for Mobile, for Network, for Storage	A: offers the most comprehensive sensitive data detection techniques in the market Supports a hybrid deployment model for several of its DLP products
		D: monitoring and discovery of sensitive data in cloud applications requires both DLP endpoint discovery and the required Symantec CASB Connectors
Digital guardian	Covers DLP, advanced threat protection, and endpoint detection and response (EDR) in a single agent form factor installed on desktops, laptops and servers	A: has integrations with broader security products, including threat intelligence, network sandbox, user and entity behavior analytics, cloud data protection, and security information and event management
		D: lacks a common policy across the endpoint and network products and concern with the integration speed of the CGN acquisition Structured data fingerprinting is not supported on the endpoint agent
Intel security	Is a suitable choice for organizations that have considerable resources invested in McAfee ePO and want a unified vendor that can provide DLP, device control and encryption capabilities	A: supports decryption and re-encryption of web traffic for on-the-fly content inspection Has a basic level of data classification in the DLP and DLP endpoint rules are location-aware
		D: supports a native API-based integration with Box for cloud data discovery Support for email, web and cloud are still lacking. Linux is not supported
CoSoSys	Is famous for its endpoint DLP product and purposes encryption products, mobile device management for iOS and Android, and specific DLP products that cover Linux, Mac OSX, printers, terminal servers and other VDI thin clients	A: has a wide variety of endpoint DLP platform and has supports for all versions of Windows, Mac OS X and Linux The management platform is very intuitive and easy to navigate
		D: does not have strong market recognition and redaction support is limited to text only Lacks integration with third-party encryption Is limited to agent-based DLP discovery only
InfoWatch	Is a suitable choice for clients that need strong network DLP capabilities, strong network	A: cover a broad range of languages and supports inspection of mobile text messages, Skype voice calls and communications via mobile messengers
		D: Native, API-based cloud support is absent and lacks of discovery of sensitive data at rest in the cloud and support for Mac OS X

2 The Proposal Method

The proposal method presented in this paper for achieving a risk depend on exception management, which will provide the financial institution to assign exceptions derived from on basic data leakage risk, is a two-phase concept as introduced in Fig. 1. The idea is to establish transparency by representing a concept that evaluates the risk of data leakage on an individual user basis.

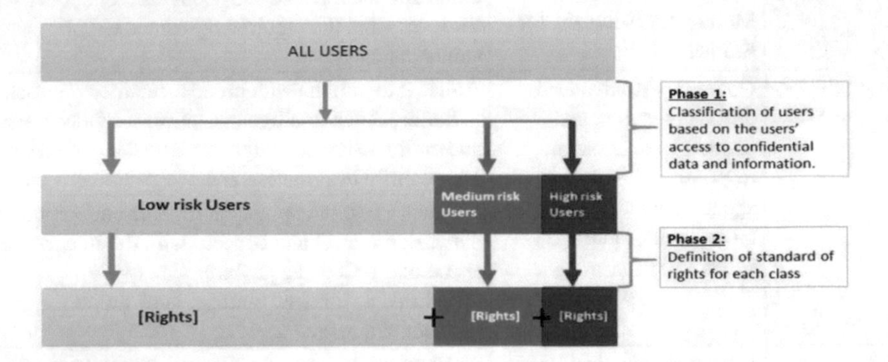

Fig. 1. Rights based on proposal method

At the beginning, all users might be classified in three different classes (high, medium, low) based on their degree of data leakage risk, where the classification is referring to the users' access to confidential data and information.

After that, a standard of rights is decided which derive the exceptions agreeable for each of the classes of classified users. Rights for the low-risk class as well implement to high-risk class, which means a user in the high-risk class is also objected to rights for the high and low-risk classes.

Based on the classification of users and the definition of standard of rights for each class, managers are thus authorized to allow or refuse DLP exclusions derived on the data leakage risk of the user. The high-risk class shows that there is a higher risk for the financial institution regarding data leakage if that user has a mobile storage device privilege. The motive could be that this user. At the beginning, all users might be classified in three different classes (high, medium, low) based on their degree of data leakage risk, where the classification is referring to the users' access to confidential data and information.

After that, a standard of rights is decided which derive the exceptions agreeable for each of the classes of classified users. Rights for the low-risk class as well implement to high-risk class, which means a user in the high-risk class is also objected to rights for the high and low-risk classes.

Based on the classification of users and the definition of standard of rights for each class, managers are thus authorized to allow or refuse DLP exclusions derived on the data leakage risk of the user. The high-risk class shows that there is a higher risk for the financial institution regarding data leakage if that user has a mobile storage device privilege. The motive could be that this user has the right to specific sensitive legal information that cannot be flowed to competitors.

3 The Classification Criteria

The risk of data leakage is evaluated for each user by assessing the user's access to confidential data and information. The classification of users is significant information and aims attention at on confidential information and who has access to it. The financial institution has to determine their own data classification strategy and adopted this proposal method respectively. The financial institution may select to describe more or less risk classes when achieving this proposal method to perform their specifications and objectives.

At the beginning, we have classified the users into three risk classes, as in Fig. 2. First criteria considered for classification is confidential information which has two types: highly sensitive data and information and standard sensitive data and information, where the users with highly sensitive data and information are in high-risk class.

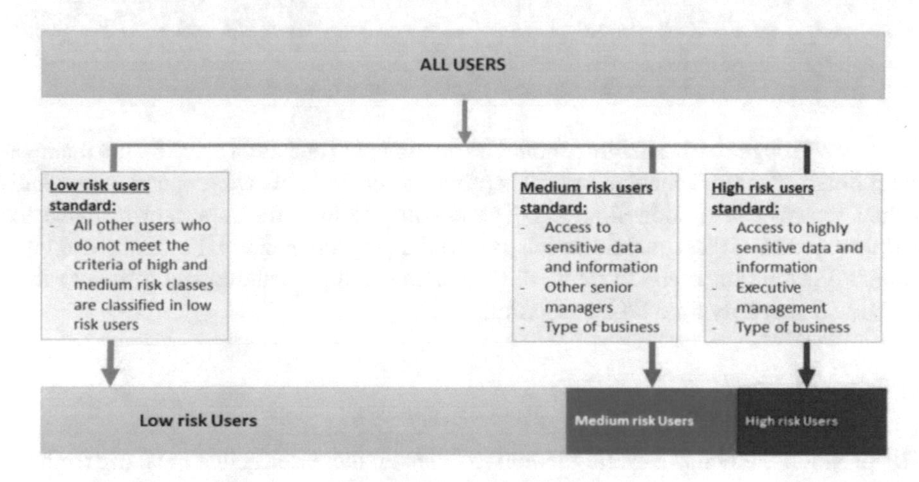

Fig. 2. Classification criteria into three risk classes

Second criteria added for classification is user's position defined by two types: Executive management for the users which are in high-risk class and other senior managers for the users which are in medium risk class. Third criteria used for classification is the type of business, where again, financial institution have to determine their business risks.

4 Implementation of the Specific Rights

Every user is granted to have any privilege if there is a valid business requirement. If a user is examined to be in high-risk users as shown in Fig. 3 description of standards of rights for three risk classes, specific types of privileges are forbidden. The specific right varies based on their individual needs for each financial institution, as well as the particular privileges in scope. In this figure as shown, users in the medium risk class are denied administrative privileges, web upload, and unrestricted USB access. In addition to that, users in the high-risk class are denied the use of encrypted USB sticks and all rights for medium and low classes.

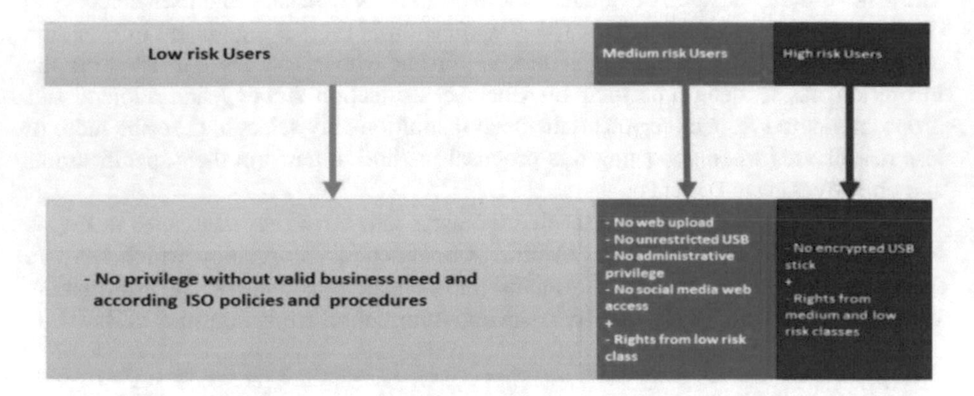

Fig. 3. The specific rights for three risk classes

Any privileges to these rights should be documented and authorized by the management board. The implementation of this proposal method idea will depend on the individual financial institution situation. The documentations are important to make the framework and are essential on which to base the proposal method. For that reason, the financial institution needs to improve all policies and procedures and to assess their individual necessity for a DLP resolution.

5 Conclusion

The proposal method is a useful method to financial increase decision making from the institutional need to a risk knowledgeable agreement. Even though the idea defined in this paper is essential, it supplies as framework for the financial institution to construct their own personalized solution. The principal concept stays the same; users are classified according to their authorizations to confidential data and information. In addition the financial institution decides specific rights for each class of users to restrict their competence to leak the data they have admission to.

References

1. Tomsitpro. http://www.tomsitpro.com/articles/threat_management-utm-it_security-it_certification-infosec,2-473.html. Accessed 15 Oct 2017
2. Radwan, T., Yousef, S.: Data leakage/loss prevention systems (DLP). NNGT J. Int. J. Inf. Syst. (2014)
3. Shabtai, A., Elovici, Y., Rokach, L.: A Survey of Data Leakage Detection and Prevention Solutions. Springer, New York (2012)

4. Gugelmann, D., Studerus, P., Lenders, V., Ager, B.: Can Content-Based Data Loss Prevention Solutions Prevent Data Leakage in Web Traffic? IEEE Security Privacy (2015). ISSN 1540-7993
5. Chitchyan, D.R.: Detecting and Preventing Data Exfiltration (2014). www.cpni.gov.uk/documents/publications
6. Tischer, M., Durumeric, Z., Foster, S., Duan, S., Mori, A., Bursztein, E., Bailey, M.: Users really do plug in USB drives they find. In: Proceedings of the 37th IEEE Symposium on Security and Privacy (S&P 2016), San Jose, California, USA, May 2016
7. Silowash, G.J., Lewellen, T.B.: Insider Threat Control: Using Universal Serial Bus (USB) Device Auditing to Detect Possible Data Exfiltration by Malicious Insiders (2013)
8. Reed, B., Kish, D.: Magic Quadrant for Enterprise DLP. Gartner, Inc. (2017)

Integrated Model of the Wavelet Neural Network Based on the Most Similar Interpolation Algorithm and Pearson Coefficient

Hong Zhao[1(✉)] and Yi Wang[2(✉)]

[1] Nankai University, Weijin Road, Tianjin, China
zhaoh@nankai.edu.cn
[2] Nankai University, Tongyan Road, Tianjin, China
wynk@mail.nankai.edu.cn

Abstract. Environmental monitoring departments in China adopt multiple air quality prediction models, each behaving differently depending on the scenario. Integrated methods are needed to obtain an integrated model with higher accuracy and adaptability. In relation to this, the most similar interpolation algorithm is often used to deal with missing data. We assigned different weights to six existing models based on the most similar value and Pearson coefficient. Then, we used the bootstrap algorithm for data augmentation. Next, we used multiple air quality prediction models for the proposed integrated model called the SIM-PB-Wavelet model. The experiment results show that the mean square root error and the Theil inequality coefficient of the model are lower than those of the six other models. Specifically, the RMSE is reduced by 36% and the TIC is reduced by 33.3% compared with the best results of the six models, thus indicating the ability of the integrated model to improve prediction accuracy.

1 Introduction

Numerical weather prediction (NWP) uses mathematical models of the atmosphere and oceans to predict the weather based on current weather conditions [1]. Two factors affect the accuracy of numerical predictions: the density and quality of observations used as forecast inputs and the deficiencies of current numerical models. Accurate pollution source parameters, emission intensity, spatial distribution, and dynamic information are needed to source emission inventory in the NWP model. However, some meteorological processes are too small-scale or complex to be explicitly included in the NWP models. Therefore, current source emission inventory has difficulty in objectively describing the spatial and temporal characteristics of urban pollution sources in China [2]. Hence, statistical prediction models based on historical data are also widely studied. In China, environmental monitoring departments adopt multiple air quality prediction models, but the performance of each model varies [3]. Therefore, research on multimode integration is necessary so that an integrated model with high accuracy can be obtained.

L. Barolli et al. (Eds.): EIDWT 2018, LNDECT 17, pp. 190–201, 2018.
https://doi.org/10.1007/978-3-319-75928-9_17

Holstein and Carlaxel [4] first proposed the idea of multimode integration, but relatively few studies were conducted on multimode integration. Krishnamurti et al. [5] proposed an integrated model arising from a study of the statistical properties. They used multiple regression to determine coefficients from multimode prediction. They also used coefficients in the multimode integration technique. The integrated model outperformed all forecasting models for multiple seasons, medium-range weather, and hurricane forecasts. In addition, the integrated model proved to have better predictions than those based solely on ensemble averaging. Monache and Stull [6] used an unweighted average method for multimode integration of four models and proved that the prediction effect is better than those of each model. Wang et al. [7] developed an air quality multimodal prediction system and proved that the effect of the integrated model through the weighted average is better than that of arithmetic mean method. Huang et al. [8] used non-equal weight linear regression method to study the multimode integration of wrf-chem, CAMx, and CMAQ and proved that its prediction accuracy is higher than those of three models. Chen et al. [9] proposed that the average weight, neural network, and multiple regression may improve the accuracy of air quality prediction model in the future. Zhang et al. [10] used Back Propagation (BP) neural network to establish the nonlinear mapping relationship between the prediction and the measured value and showed improved accuracy. Bai et al. [11] proposed the w-bpnn model established by the wavelet transform and BP neural network and also reported improved accuracy. Wavelet analysis has good localization characteristics in both time and frequency domains [12]. In this study, we adopted a wavelet neural network to study the integrated model for multiple air quality prediction models.

Missing data are common in air quality prediction model, and this problem makes multimode integration difficult. We proposed the most similar interpolation algorithm to deal with the missing data. We assigned the weights of six models based on Pearson coefficient and the most similar value and found that they have various contributions to the measured value. We also used the bootstrap algorithm for data augmentation. Finally, to avoid overfitting, we adopted the strategy of early stopping [13] and learning rate gradient descent [14] when training.

2 Related Work

In this section, we mainly introduce the wavelet neural network and the bootstrap algorithm, which are used for the integrated model.

2.1 Wavelet Neural Network

Wavelet analysis [15] and neural network are combined in two ways. The first one is the loose type of wavelet neural network. The signal is pretreated with wavelet analysis and then sent to the neural network (Fig. 1). The second one is the compact wavelet neural network, which uses the wavelet function to replace the activation function of neurons in the hidden layer, fully inheriting the characteristics of the time frequency localization and self-learning function of the neural network (Fig. 2). In this study, we used a compact wavelet neural network to study the multimode integration.

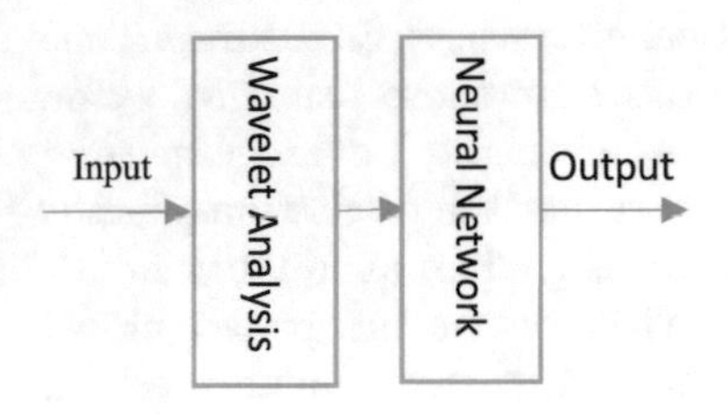

Fig. 1. Loose type of wavelet neural network

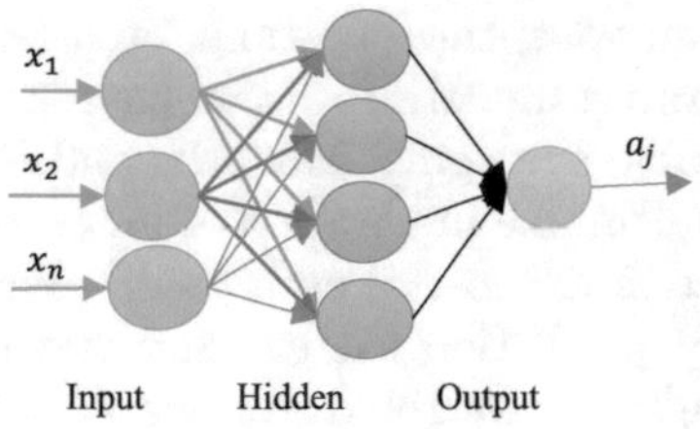

Fig. 2. Compact type of wavelet neural network

2.2 Bootstrap Algorithm

Bootstrap [16] is a resampling algorithm based on its own resources and is more suitable for a small sample compared with others. Table 1 shows the basic steps.

Table 1. Steps for the bootstrap method

ID	Description
1	Use the resampling technique to extract a certain number of samples from the original sample, thus allowing repeated sampling
2	Count the statistics T according to the sample
3	Repeat the above N times and obtain statistic T

3 SIM-PB-Wavelet Model

Multiple air quality prediction models are integrated by wavelet neural networks. We develop the most similar interpolation algorithm to deal with missing data and assign the weight for each model based on Pearson coefficients and the most similar value. We also use the bootstrap algorithm for data augmentation. This integrated model is called the SIM-PB-Wavelet model.

3.1 Most Similar Interpolation Algorithm

The similarity between two objects is the numerical measure of these two objects [17]. In general, the similarity is often valued between 0 and 1. The methods of numerical similarity include the mismatch ratio, Manhattan distance, and cosine similarity algorithm.

The result in (1) is the mismatch rate, where x is the prediction vector of each model, and y is the measured vector and this is calculated as

$$d_1(x, y) = \sum \frac{(x_i - y_i)}{x_i}. \tag{1}$$

The smaller the result of (1), the more similar the objects are. The result in (2) is the distance of Manhattan and this is calculated as

$$d_2(x, y) = |x_1 - y_1| + \cdots + |x_i - y_i|. \tag{2}$$

The smaller the result of (2), the more similar the objects are. The result in (3) is the cosine similarity algorithm and this is calculated as

$$d_3(x, y) = \frac{(x \times y)}{(||x|| \times ||y||)}. \tag{3}$$

The larger the result in (3), the more similar the objects are. We propose an algorithm to deal with the missing value and this is calculated as

$$simest = \frac{1}{d_1(x, y)} + \frac{1}{d_2(x, y)} + d_3(x, y) \overset{Max(simest)}{\longrightarrow} Avg(season, model) \tag{4}$$

The simest value is the similarity between the predicted and measured values of each model. The greater the result in (4), the closer to the measured value it is. The average of the most similar prediction model is used as the missing value for different seasons.

3.2 Weight of Each Model

The Pearson coefficient [18] is used to show the correlation between two variables. Equation (5) is the Pearson correlation coefficient, and it is calculated as

$$r = \frac{\sum (x - \bar{x})(y - \bar{y})}{\sqrt{\sum (x - \bar{x})^2 \sum (y - \bar{y})^2}} \tag{5}$$

The Pearson coefficient ranges 0.8–1.0, 0.6–0.8, 0.4–0.6, 0.2–0.4, and 0.0–0.2 indicate extremely strong, strong, moderate, weak and very weak or no correlation, respectively.

In Eq. (6), the Pearson coefficient and the most similar value proposed in this study are used to calculate the weight for each model, and this is calculated as

$$w_i = \frac{r_i}{\sum_{i=1}^{n} r_i} + \frac{simest(i)}{\sum_{i=1}^{n} simest(i)} \tag{6}$$

Here, a higher value of r and a larger simest value indicate greater weight.

3.3 Training of the Wavelet Neural Network

We adopt a three-layer wavelet neural network structure with input, implicit, and output layers. The number of hidden layer is calculated as

$$N = \sqrt[2]{m + n} + a \tag{7}$$

where m represents the number of input nodes, n indicates the number of output nodes, and a is a constant between 1 and 10 [19].

$$RMSE = \sqrt{\frac{1}{n}\sum_{i=1}^{n}(y_i - \widehat{y_i})^2} \tag{8}$$

Equation (9) is used to calculate the TIC value, and this is calculated as

$$TIC = \frac{\sqrt{\frac{1}{n}\sum_{i=1}^{n}(y_i - \widehat{y_i})^2}}{\sqrt{\frac{1}{n}\sum_{i=1}^{n}(\widehat{y_i})^2} + \sqrt{\frac{1}{n}\sum_{i=1}^{n}(y_i)^2}} \tag{9}$$

The TIC value is between 0 and 1; the smaller the value, the closer it is to the measured value.

4 Modeling and Methods

In this section, we introduce the modeling methods employed. Three models are established, namely, SIM-Wavelet, SIM-P-Wavelet, and SIM-PB-Wavelet models, to verify the accuracy of the SIM-PB-Wavelet model.

4.1 SIM-Wavelet Model

In this study, the missing data of different seasons for the different model are calculated by Eqs. (1) and (4). The interpolated predictions of the six models are used as inputs to the SIM-Wavelet model. Table 2 shows the steps of the SIM-Wavelet model.

Table 2. Steps of the SIM-Wavelet model

Steps	Description
1	Input: Data D = $\{X_1 X_2 \cdots X_p\}$
2	According to (1) and (4), we use the most similar interpolation algorithm to deal with the missing data, and a new data set Ω is obtained
3	Data Ω are used as the input to the wavelet neural network
4	Train the wavelet neural network until the RMSE is smaller than the limited value
5	Output: prediction

4.2 SIM-P-Wavelet Model

The missing data of different seasons for the different models are calculated by (1) and (4). Equation (6), which is based on Pearson coefficient and the proposed most similar value, are used to calculate the weight for each model. We have established the integrated model called SIM-P-Wavelet.

The prediction data of the six models are interpolated and weighted as inputs to the SIM-P-Wavelet model. Table 3 shows the steps of the SIM-P-Wavelet model.

Table 3. Steps of the SIM-P-Wavelet model

Steps	Description
1	Input: Data $D = \{X_1 X_2 \cdots X_p\}$
2	According to (1) and (4), we use the most similar interpolation algorithm to deal with the missing data, and a new data set Ω is obtained
3	According to (5), we calculate the Pearson coefficient
4	According to (6), we calculate the weight based on the Pearson coefficient and the most similar value for each model, and the new data ϕ are obtained
5	Data ϕ are used as the input to the wavelet neural network
6	Train the wavelet neural network until the RMSE is smaller than the limited value
7	Output: prediction

Table 4. Steps of the SIM-PB-Wavelet model

Steps	Description
1	Input: Data $D = \{X_1 X_2 \cdots X_p\}$
2	According to (1) and (4), we use the most similar interpolation algorithm to deal with the missing data, and a new data set Ω is obtained
3	According to (5), we calculate the Pearson coefficient
4	According to (6), we calculate the weight based on the Pearson coefficient and the most similar value for each model, and the new data ϕ are obtained
5	Use bootstrap algorithm for data augmentation, and the new data set φ is obtained
6	Data φ are used as the input to the wavelet neural network
7	Train the wavelet neural network until the RMSE is smaller than the limited value
8	Output: prediction

4.3 SIM-PB-Wavelet Model

Equation (6) is used to calculate the weight for each model. We use bootstrap algorithm for data augmentation to reduce the error of the wavelet neural network. We have established the SIM-PB-Wavelet model. Table 4 shows the steps of the SIM-PB-Wavelet model.

5 Experiments

We performed the modeling in Matlab7.0. In this section, we introduce the data used and the experiment results. Three models are established, namely, SIM-Wavelet, SIM-P-Wavelet, and SIM-PB-Wavelet models, to verify the accuracy of the SIM-PB-Wavelet model. The same parameters are set: the learning efficiency is 0.001 and the largest number of training times is 235. The number of input nodes is 6 and that of the output nodes is 1. When a is 4, according to Eq. (7), the hidden layer node is 6.

5.1 Data

In this study, six air quality prediction models are used. The data include the daily air quality index (AQI) of six air quality prediction models and the measured value (AQI) over the same period. Multiple air quality prediction models have different adaptabilities in different scenarios. Missing data are common in prediction models. Table 5 calculates the missing rate between August 20, 2015, and December 20, 2016.

Table 5. Missing rate of multiple air quality models

Name	Model1	Model2	Model3	Model4	Model5	Model6
Missing rate	11.2%	46.4%	38.2%	18.4%	2.2%	54.2%

5.2 Results and Discussions

Here, we introduce the effectiveness of the proposed model after dealing with the missing data. We also present the results of the experiments of prediction.

5.2.1 Dealing with the Missing Data

We calculate the similarities among the six prediction models and the measured value by using Eqs. (1) and (3), and the simest value is calculated by using Eq. (4). Table 6 shows the result.

The results show that the missing data of seasons 1, 2, and 4 are calculated by the fifth prediction model and those of season 3 are calculated by the first

Table 6. Similarities among the six prediction models and the measured value

Season	Model	Cosine similarity	Mismatch rate	Distance of Manhattan	Simest	(Season, model)
1	1	0.89	21.60	2424.00	0.93	
1	2	0.59	17.22	1406.00	0.65	
1	3	0.78	20.22	1886.00	0.83	
1	4	0.70	23.14	2038.00	0.75	
1	**5**	**0.94**	**43.45**	**3499.00**	**0.97**	**(1,5)**
1	6	0.82	30.02	2486.00	0.85	
2	1	0.83	35.40	3423.00	0.86	
2	2	0.67	15.56	1953.00	0.74	
2	3	0.78	39.82	2016.00	0.80	
2	4	0.91	24.37	2676.00	0.95	
2	**5**	**0.94**	**30.38**	**2547.00**	**0.98**	**(2,5)**
2	6	0.79	20.43	1693.00	0.84	
3	**1**	**0.94**	**45.46**	**5728.00**	**0.97**	**(3,1)**
3	2	0.61	46.88	4222.00	0.64	
3	3	0.58	19.29	2288.00	0.63	
3	4	0.85	48.52	4928.00	0.87	
3	5	0.92	48.25	5536.00	0.94	
3	6	0.34	7.16	690.00	0.48	
4	1	0.76	28.01	4474.00	0.80	
4	2	0.83	39.58	4050.00	0.86	
4	3	0.82	27.85	3724.00	0.86	
4	4	0.86	515.31	5372.00	0.86	
4	**5**	**0.92**	**31.64**	**5069.00**	**0.96**	**(4,5)**
4	6	0.65	21.27	2917.00	0.69	

prediction model according to Eq. (4). The result of interpolation is analyzed as follows.

Figure 3 shows the multivariate linear regression analysis [20] between the interpolated results and the measured values. The six prediction models show approximately normal distribution and are positively correlated with the measured value; hence, the most similar interpolation algorithm is effective.

According to Eq. (5), we calculate the Pearson coefficient between the interpolation result and the measured value (Table 7). According to Eq. (6), we calculate the weight based on the Pearson coefficient and the most similar value, and new data ϕ are obtained.

Table 7. Pearson coefficients between the interpolation results and measured values

ID	Model1	Model2	Model3	Model4	Model5	Model6	Season
1	0.45	0.45	0.47	0.46	0.55	0.29	1
2	0.53	0.52	0.57	0.48	0.68	0.43	2
3	0.45	0.48	0.49	0.48	0.63	0.29	3
4	0.56	0.59	0.63	0.54	0.72	0.46	4

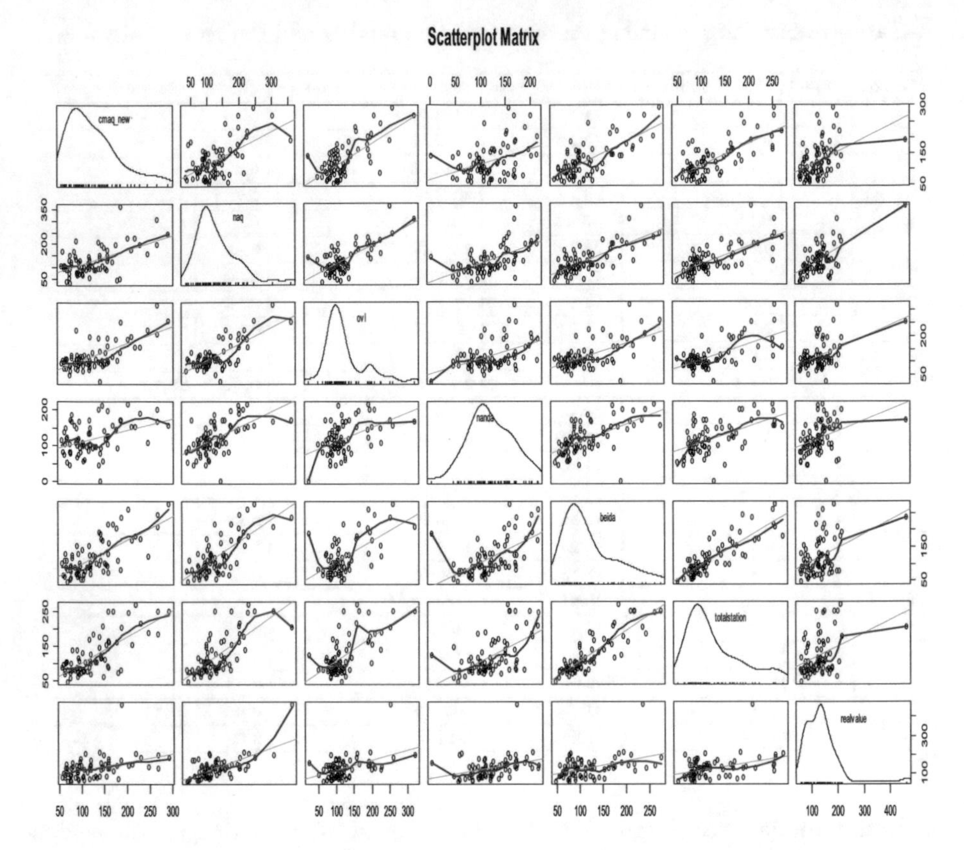

Fig. 3. Multivariate linear regression between interpolation result and measured value

5.2.2 Prediction Experiments

We use bootstrap algorithm for data augmentation, and new data are obtained. The new data are used as inputs to the wavelet neural network. We train the wavelet neural network until the RMSE is smaller than the limited value, thus avoiding the problem of overfitting.

During training, the data of a city from August 20, 2015 to November 31, 2016 are used as training data. Data from December 120, 2016 are used for prediction data. Figure 4 shows the prediction results of the SIM-PB-Wavelet model. Table 8 shows the prediction results of the six models, SIM-Wavelet, SIM-P-Wavelet, and SIM-PB-Wavelet models. Among these models, the prediction results of the SIM-PB-Wavelet model are the most accurate.

Among these models, the predictions of the SIM-PB-Wavelet model are the most accurate. Moreover, the RMSE is decreased by 36%, and the TIC is decreased by 33.3% compared with the best results of the six prediction models. Thus, among the models, the SIM-PB-Wavelet model obviously improved the accuracy of air quality prediction.

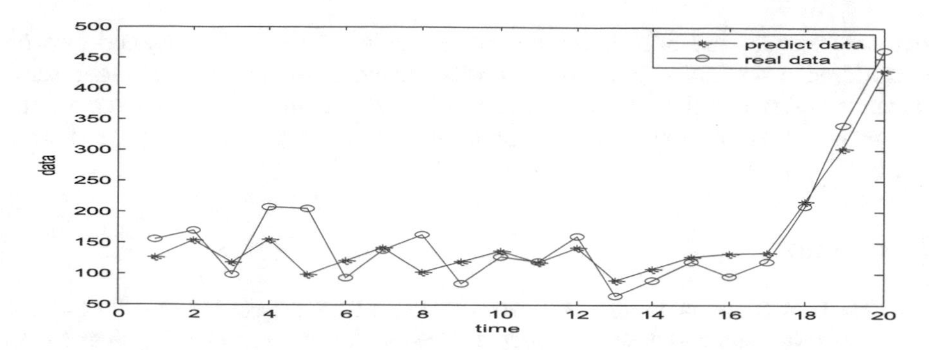

Fig. 4. Prediction results of the SIM-PB-Wavelet model

Table 8. Predictions of multiple models

Model name	RMSE	TIC
Model1	**57.30**	**0.15**
Model2	85.71	0.23
Model3	63.64	0.17
Model4	60.05	0.17
Model5	66.33	0.17
Model6	117.63	0.43
SIM-Wavelet model	85.49	0.27
SIM-P-Wavelet model	68.73	0.21
SIM-PB-Wavelet model	**36.65**	**0.10**

6 Conclusions

Environmental monitoring departments in China generally adopt multiple air quality prediction models, but the performance and adaptability of each model vary. In this study, we established the SIM-PB-Wavelet model to study a multi-modal integration.

First, in this study, the most similar interpolation algorithm has improved the effectiveness of interpolation. Second, we calculate the weight based on the Pearson coefficients and the most similar values, thus improving the prediction precision of the wavelet neural network. Third, the SIM-PB-Wavelet model is established with the combination of the most similar interpolation algorithm mentioned, the algorithm of assigning weight, and bootstrap algorithm, resulting in the decrease of the RMSE and the TIC coefficient by 36% and 33.3%, respectively, compared with the best results of the six other models. Hence, the SIM-PB-Wavelet model is a simple, practical, and predictive model for the integration of air quality prediction models.

This study, however, has some deficiencies in that we can verify the performance and stability of the SIM-PB-Wavelet model in a broader field weather, climate, oceans, and hurricanes. Other interpolation algorithms should also be considered in the future. The SIM-PB-Wavelet model is a prediction model based on the wavelet neural network. Future studies can also focus on multiple methods of machine learning towards a multimodal integration of air quality prediction models.

References

1. Garcia, A.R., Mar-Morales, B.E., Ruiz-Surez, L.G.: Numerical air quality forecast modeling system: Mexico case study. In: Recent Advances in Fluid Dynamics with Environmental Applications. Springer (2016)
2. Im, U., Bianconi, R., Solazzo, E., et al.: Evaluation of operational on-line-coupled regional air quality models over Europe and North America in the context of AQMEII phase 2. Part I: Ozone. Atmos. Environ. **115**(2), 404–420 (2015)
3. Eckhardt, S., Quennehen, B., Olivi, D.J.L., et al.: Current model capabilities for simulating black carbon and sulfate concentrations in the Arctic atmosphere: a multi-model evaluation using a comprehensive measurement data set. Atmos. Chem. Phys. Discuss. **15**(7), 10425–10477 (2015)
4. Holstein, S.V., Carlaxel, S.: An experiment in probabilistic weather forecasting. J. Appl. Meteorol. **10**(4), 635–645 (1971)
5. Krishnamurti, T.N., Kishtawal, C.M., Larow, T.E., et al.: Improved weather and seasonal climate forecasts from multimodel superensemble. Science **285**(5433), 1548 (1999)
6. Monache, L.D., Stull, R.B.: An ensemble air-quality forecast over western Europe during an ozone episode. Atmos. Environ. **37**(25), 3469–3474 (2003)
7. Wang, Z.F., Wu, Q.Z., Gbaguidi, A., et al.: Ensemble air quality multi-model forecast system for Beijing (EMS-Beijing): model description and preliminary application. J. Nanjing Univ. Inf. Sci. Technol. **1**, 19–26 (2009)
8. Huang, S., Zhang, M., Xie, B.G., et al.: Multi-model blending method and application on reducing the uncertainty of air quality models. In: EGU General Assembly Conference. EGU General Assembly Conference Abstracts (2017)
9. Chen, H., Wang, Z., Qizhong, W.U., et al.: Application of air quality multi-model forecast system in Guangzhou: model description and evaluation of PM10 forecast performance. Clim. Environ. Res. **18**(4), 427–435 (2013)
10. Zhang, W., Wang, Z., Junling, A.N., et al.: Update the ensemble air quality modeling system with BP model during Beijing Olympics. Clim. Environ. Res. **15**(5), 595–601 (2010)
11. Bai, Y., Li, Y., Wang, X., et al.: Air pollutants concentrations forecasting using back propagation neural network based on wavelet decomposition with meteorological conditions. Atmos. Pollut. Res. **7**(3), 557–566 (2016)
12. Chen, Y., Yang, B., Dong, J.: Time-series prediction using a local linear wavelet neural network. Neurocomputing **69**(4–6), 449–465 (2006)
13. Gwo-Ching, L.: Hybrid improved differential evolution and wavelet neural network with load forecasting problem of air conditioning. Int. J. Electr. Power Energy Syst. **61**(1), 673–682 (2014)
14. Wang, X., Zhang, Y., Zhao, S.: Air quality forecasting based on dynamic granular wavelet neural network. Comput. Eng. Appl. **49**(6), 221–224 (2013)
15. Zhang, Q.G., Benveniste, A.: A wavelet networks. IEEE Trans. Neural Netw. **3**(6), 889–898 (1992)

16. Goltsev, A.V., Dorogovtsev, S.N., Mendes, J.F.: k-core (bootstrap) percolation on complex networks: critical phenomena and nonlocal effects. Phys. Rev. E Stat. Nonlinear Soft Matter Phys. **73**(5 Pt 2), 056101 (2006)
17. Yong, K.L., Lee, S.J., Park, J.: Tag-based object similarity computation using term space dimension reduction. In: International ACM SIGIR Conference on Research and Development in Information Retrieval, pp. 790–791. ACM (2009)
18. Benesty, J., Chen, J., Huang, Y., et al.: Pearson correlation coefficient. In: Noise Reduction in Speech Processing, pp. 1–4. Springer, Heidelberg (2009)
19. Chen, Z., Feng, T.J., Chen, G.: A kind of BP algorithm-learning wavelet neural network. J. Ocean Univ. Qingdao **1**, 122–128 (2001)
20. Ul-Saufie, A., Yahya, A., Ramli, N.: Improving multiple linear regression model using principal component analysis for predicting PM10 concentration in Seberang Prai, Pulau Pinang. Int. J. Environ. Sci. **2**, 403 (2011)

Stochastic Power Management in Microgrid with Efficient Energy Storage

Itrat Fatima[1], Nadeem Javaid[1(✉)], Abdul Wahid[1], Zunaira Nadeem[3], Muqqadas Naz[1], and Zahoor Ali Khan[2]

[1] Department of Computer Science, COMSATS Institute of Information Technology, Islamabad 44000, Pakistan
nadeemjavaidqau@gmail.com

[2] Internetworking Program, Faculty of Engineering, Dalhousie University, Halifax, NS B3J 4R2, Canada
zahoor.khan@dal.ca

[3] National University of Science and Technology, Islamabad 44000, Pakistan
http://www.njavaid.com

Abstract. In order to mitigate the extra cost and to reduce the energy consumption, distributive power system are widely accepted in recent years. The reason of adaptation of distributive power system is the scalability of power supply and demand which helps in reliable power supply and optimizes the annual expenditures. Moreover, the integration of power distributive systems with renewable energy sources enabled the optimal utilization of photovoltaic arrays for effective and cost efficient power supply. To exploit the integration of distributive power and renewable sources, we solve the power dispatch problem with heuristic optimization techniques. We have performed scheduling for supply side management. For this purpose, we have formulate our problem using chance constrained optimization and transformed the problem into mixed integer linear programming. Finally, simulation results demonstrate that the proposed scheduling method for microgrid performs efficiently and effectively.

Keywords: Smart grid · Microgrid · Renewable energy sources
Supply side management · Chance constrained optimization
Mixed integer linear programming

1 Introduction

With the increase in worlds population, the demand of electricity has also been increased. Due to technological advancements, this issue is not the big problem. To fulfill electricity demand, renewable energy resources (RESs) and non-RESs are most widely and frequently used. In the research area, lots of research has been done by combining the microgrid (MG) with smart grid. The expert energy management system (EEMS) with integrated MG for optimal usage of

L. Barolli et al. (Eds.): EIDWT 2018, LNDECT 17, pp. 202–213, 2018.
https://doi.org/10.1007/978-3-319-75928-9_18

distributed energy sources (DERs) and wind turbine (WT) has been proposed in [1]. Bee colony optimization (BCO) has been proposed first time in 2001 and its modification is done in [2]. To use non-RESs and RESs, the size and location is needed to be defined, like how many micro turbines (MTs) are used to fulfill demand. Location and size of non-RESs are analysed in [3] using teaching learning based optimization (TLBO) technique.

The authors in [4], proposed an incentive based program. The user has to curtail his/her load demand according to bids, to take incentive or pay penalty cost otherwise. However, when power is excessive, the storage devices are charged. If the power is still excessive after storage, then power is sold to macrogrid.

Due to rapid increase of complexity in unit commitments and the economic dispatch problems, number of algorithms are used to solve these problems. These algorithm includes programming and heuristic techniques. Furthermore, heuristic techniques include ACO, BPSO and GA and programming techniques include linear programming (LP), integer LP (ILP) and mixed ILP (MILP). In this paper, for economic dispatch we have used chance constrained optimization (CCO) for only problem formulation.

In this paper, the generators are scheduled to fulfill the load demand by minimizing the generation cost. To solve the scheduling problem on SSM, CCO method is used for problem formulation by transforming the problem into MILP. Further, to solve this problem we have used branch and bound method of MILP. Remaining paper is organized as; in Sect. 2 literature review has been discussed, Sect. 3 illustrates the proposed system model, Sect. 4 presents problem formulation, Sect. 5 contains simulation results and discussion, Sect. 6 is based on conclusion.

2 Literature Review

MG plays an essential role to make our society smart. It is an electricity provision system which includes DERs and load. DERs contains storage system, controllable loads, RESs and the non-RESs. It works in controlled and coordinated environment.

In [5], the author discussed two case studies of different countries. The objectives are cost minimization, PAR and load reduction. The results are shown for both DSM and SSM. For SSM, RESs are used to fulfill the load demand. The results are compared between case studies of Netherlands and Burundi. The objective function defined in the paper could be modified.

A control strategy for MG using grid connected distributed generator and RESs is used to eliminate the disturbance in harmonics [6]. The results are based on two case studies for comparison between proposed strategy and active power filters (APFs) in which cost, complexity and size has been reduced. Also, they have features of fast dynamic response, design simplicity, analysis stability and fast transient response. Point of common coupling (PCC) is a current system, used to improve power quality and reduce disturbance in harmonics.

The paper is based on multiple environment dispatch problem (MEDP) [7]. This problem is solved using ACO by comparing the results between two case studies with RESs and without RESs. Also, the presented method is compared with gradient method. The objective is to minimize the overall operational cost of RESs and emission reduction. Results for without RESs, MEDP-ACO techniques saved 4.50% over gradient method and with RESs saved 6.20%.

Travelling salesman problem (TSP) is used as a benchmark for the proposed system in [8]. The authors used interval type-2 fuzzy system with ACO and applied on mobile robot optimization (MRO) and TSP without changing their parameters. To solve the optimization problem, self adaptive GA is proposed with fuzzy decisions [9]. Objectives of the paper are voltage offset, transmission loss, construction, purchase and environmental cost minimization. The results are presented for better progress to improve the efficiency and speed of algorithm. Which is improved effectively and objectives are highly appreciated in real time control.

Research on fuzzy system with MG and optimization techniques is evolving frequently. The authors proposed an adaptive neuro fuzzy interference system (ANFIS) [10]. The technique is trained with the inputs like, instant energy of available sources and required load demand. They aimed to use PV, WT, ultra capacitor, batteries and FC for supply side management (SSM) to fulfill the demanded load in their research. From the results, MG is fully used with fuzzy systems and got better results applied on different case studies. However, the power flow is managed and proficient.

Economic dispatch problem of MG system is solved using enhanced BCO (EBCO). An EBCO is based on time of use (TOU) price tariff and constraints defined in [11]. It is an economically optimized power dispatch to satisfy the load demand. Two scenarios are considered, where energy is managed in both islanded and grid-connected MG scenarios. To check the performance and effectiveness of the proposed algorithm, tests are applied on low voltage distribution systems. From the results, the proposed algorithm is robust, feasible and more effective in comparison with other developed algorithms.

For energy storage, batteries and other storage devices are used. For SSM, batteries are used to store energy for later use to fulfill the load demand. However, batteries are used to consume energy when more energy is required for DSM. The geographic information system (GIS) tool is used to store the information of PV for mini grid system from geographical area. Performance of PV is analysed from the stored information in [12]. The objective of the paper is to reduce cost by increasing reliability. The authors use Li-on batteries instead of lead-acid batteries for energy storage from PV arrays. The objectives are achieved by choosing reliability threshold to save electricity cost. The drawback of the proposed technique is that, it is not suitable in some conditions due to large differences in countries.

In [13], the authors work in both grid connected and islanded modes of MG. In this paper, the authors proposed hybrid cuckoo optimization algorithm (COA) with linear programming (LP). PCC is used to find the optimal solution of fault current limiter (FCL). Optimize the protection coordination of directional over-current relays (DORs). COA is proposed to optimize the behaviour pattern of cuckoo. The results are compared with GA and PSO to check the efficiency and effectiveness of proposed hybrid COA-LP. BPSO and ILP is proposed in [14]. The paper has the objective of cost minimization and thermal comfort maximization for interval number analysis to handle uncertain hot water demand and ambient temperature. But waiting time is not calculated in the paper. The author proposed multi-objective mixed integer non-linear programming (MO-MINLP) [15]. The technique is proposed for comprehensive scheduling of appliances with objective to minimize the trade-off between energy saving and user comfort.

From the above literature, it is analysed that the optimization techniques play an important role for practical solution. RESs are used in real world environment to fulfill the overall load demand of the buildings, homes and commercial areas as well. In this paper, RESs are used with non-RESs and the resultant generated power is used to fulfill the load demand. Excessive power is sold to macrogrid after fulfillling the load demand and battery storage.

3 Proposed System Model

A MG includes multiple loads, DERs and the storage system to fulfill the demand of an area, the area may be residential, commercial and industrial area. Mainly, a MG works in two different mode, islanded mode and grid connected mode. In grid connected mode, MG works by establishing the connection with macrogrid. However, MG work without involving the macrogrid in islanded mode and work independently.

For SSM, we have used five generators which include RESs and non-RESs with storage batteries. The MG used in this paper is in grid connected mode, this means that the macrogrid is used for power exchange. However, MG is used to fulfill the load demand of residential area, which is based on fifteen homes. The diagrammatic representation of SSM is shown in Fig. 1.

4 Problem Formulation

Another objective of this paper is to minimize the generation cost for economical power supply. Cost is minimized by solving the scheduling problem of DERs. For this purpose, we have used CCO for problem formulation. However, B&B method is used to solve the problem by transforming it into MILP.

4.1 Photo Voltaic

PV produces electricity with the help of solar radiations during day time when sunlight falls on the surface of PV panels. The cells on panel produce electricity and charge the batteries. Users can fulfill their demand through these batteries. The PV power completely depends on solar irradiance. However, output power of PV is calculated by using Eq. 1 [16]:

$$P_{PV}(t) = \frac{G(t)}{1000} \times \rho_r \times \eta_{pv} \tag{1}$$

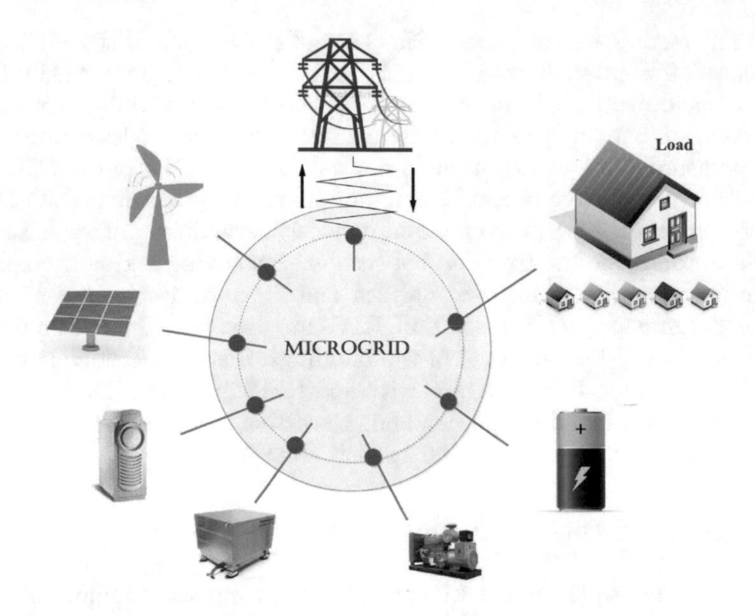

Fig. 1. Proposed system model

The constraint defined for the PV output power is shown in Eq. 2, the output power lies between the maximum and minimum limit of the power obtained from Eq. 1.

$$P_{PV}^{min} \leq P_{PV}(t) \leq P_{PV}^{max} \tag{2}$$

4.2 Wind Turbine

WT produces electricity by moving its rotors with wind blow. As rotor moves, electricity is produced and stored in batteries. The WT is a device which produces electricity by converting kinetic energy into electrical energy. To calculate the wind power, Eq. 3 is used [16]:

$$P_{WT}(t) = \psi_r \times \frac{v(t) - v_{in}}{v_r - v_{in}} \tag{3}$$

$$P_{WT}^{min} \leq P_{WT}(t) \leq P_{WT}^{max} \tag{4}$$

4.3 Micro Turbine

MT is used for power generation and produce both heat and electricity on a comparatively small scale [17]. These generators have advantages of light-weight, higher efficiency and minimum emission. Due to small size, they have minimum capital, maintenance and operation cost. The output power is determined according to the constraints defined in Eq. 5.

$$0 \leq P_{MT}(t) \leq P_{MT}^{max} \tag{5}$$

4.4 Diesel Engine

DEs are the electrical generators used for homes, industrial and commercial area with different sizes and specifications [18]. These power sources are also used for the back-up option. Small size DEs are sufficient for home use. In this paper, we have used small size DE homes and generate within a specified limits as defined in Eq. 6.

$$P_{DE}^{min} \leq P_{DE}(t) \leq P_{DE}^{max} \tag{6}$$

4.5 Fuel Cells

FCs are useful for both traditional and remote locations. They are also used for backup power sources. FCs are used to generate power with minimum cost and methane emissions. In California, a 2.8 MW FC plant used is called to be the biggest type of FCs [19]. The output power of FC is obtained according to the constraint defined in Eq. 7.

$$P_{FC}^{min} \leq P_{FC}(t) \leq P_{FC}^{max} \tag{7}$$

4.6 Energy Storage System

The operating time of WT and PV are fully depend on the climatic conditions. To store the excessive power generated from DERs, we have used batteries for efficient energy storage. Excessive energy also is used to charge the batteries. The energy storage level should be lie between the maximum and minimum capacity of the battery. To compute charged and discharged power of batteries, we use the Eqs. 8, 9, 10 in [20].

$$P_B(t) = P_B(t-1) + P_{ch}(t)\eta_{ch} - \frac{P_{disch}}{\eta_{disch}}, \forall t \in 24 \tag{8}$$

$$P_{ch}^{max} \leq 0.1 E_B^{max}(1 - \zeta_B(t)) \tag{9}$$

$$P_{disch}^{max} \leq 0.1 E_B^{max} \zeta_B(t) \tag{10}$$

Maximum charging and discharging power is determined using Eqs. 11 and 12. Whereas, charging and discharging power for a day is determined using Eqs. 13 and 14.

$$P_{ch}^{max} \leq (P_B^{max} - P_B(t-1))\eta_{ch} \tag{11}$$

$$P_{disch}^{max} \leq \frac{(P_B(t-1) - P_B^{min})}{\eta_{disch}} \tag{12}$$

$$P_{ch}^{min} \leq P_{ch}(t) \leq P_{ch}^{max} \tag{13}$$

$$P_{disch}^{min} \leq P_{disch}(t) \leq P_{disch}^{max} \tag{14}$$

4.7 Macrogrid Exchange

Power exchanged with macrogrid is determined as given in Eq. 15. Macrogrid is used to exchange power when more power is required or there exist surplus power.

$$P_G(t) = P_{Buy}(t) - P_{Sell}(t) \tag{15}$$

Power exchange limit to buy or sell power to or from the macrogrid is determined within the defined limits as given in Eqs. 16 and 17.

$$0 \leq P_{Buy}(t) \leq P_{Buy}^{max} \tag{16}$$

$$0 \leq P_{Sell}(t) \leq P_{Sell}^{max} \tag{17}$$

4.8 Chance Constrained Optimization

CCO is used to solve the problems under various uncertainties [21]. The confidence level is high by restricting the feasible region while finding a solution.

4.9 Branch and Bound Method

B&B method depends on the effective evaluation of upper and lower bounds of a search space branches or variables [22]. The concept of B&B is like divide and conquer concept. Where, branch is like dividing and bound is like conquering. So, to minimize the total generation cost, the objective function is presented in Eq. 18. Generation cost is minimized by using CCO method for formulation and transformed into MILP to solve this problem.

$$minimize \sum_{x=1}^{T} \sum_{y=1}^{ngen} TC_{xy} \tag{18}$$

subject to

$$\sum_{y=1}^{ngen} P_y(t) + P_G(t) + P_B = Demand(t) \tag{19}$$

$$TC_{xy} \geq 0 \tag{20}$$

(2), (4), (5), (6), (7), (9), (10), (11), (12), (13), (14), (16), (17)

5 Simulation Results and Discussion

On SSM, the problem is to schedule the output power of DERs. These are schedule to minimize the generation cost. The scheduling problem is formulated using CCO and transformed into MILP and the problem is solved using B&B method. DERs are scheduled for a day based on hourly timeslots.

Table 1 represents the operating cost of each generator. The power of PV is calculated using Eq. 1 and power of WT using Eq. 3. The power profiles of FC, DG and MT are taken from [1,16,23] and the operating cost is taken as fixed from [24].

For scheduling of DERs, we have used CCO for formulation and transformed into MILP using branch and bound method. The generators are scheduled to minimize the generation cost. We have considered two cases for the scheduling of DERs to check different cases for minimizing cost. In this paper, the power generated from DERs used before scheduling are shown in Figs. 2 and 3 and cost in Fig. 4.

Table 1. Operating Cost of Each Generator

Generators	Operating cost ($/h)
Photovoltaic arrays	0.05
Wind turbine	0.007
Micro turbine	0.08
Fuel cell	0.2
Diesel engine	0.022

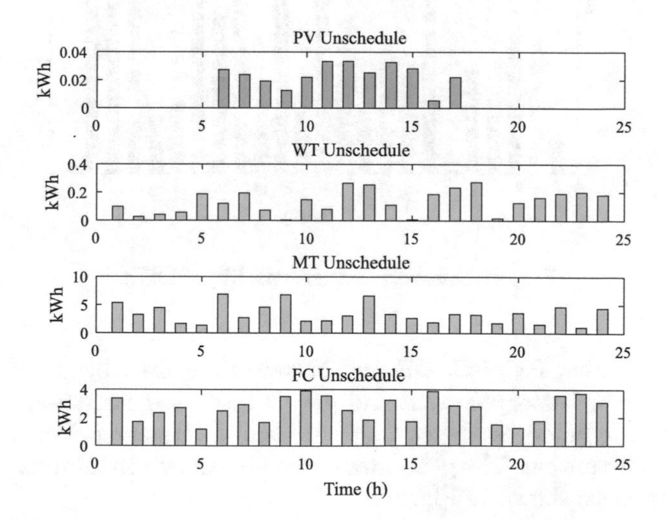

Fig. 2. Unscheduled power profile of PV, WT, MT and FC

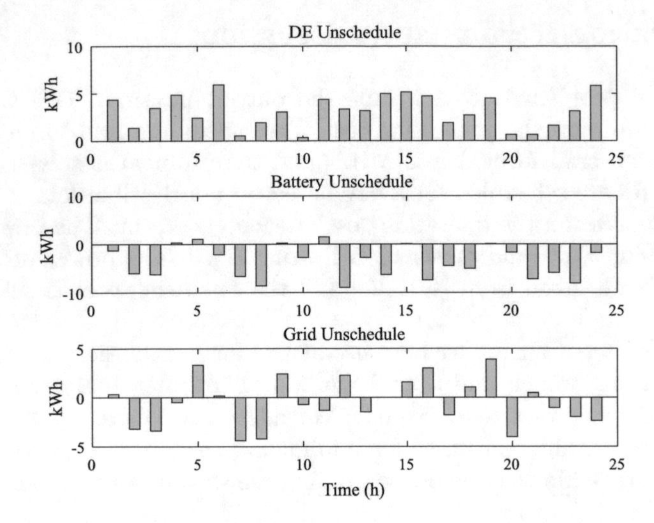

Fig. 3. Unscheduled power profile of DE, battery and grid

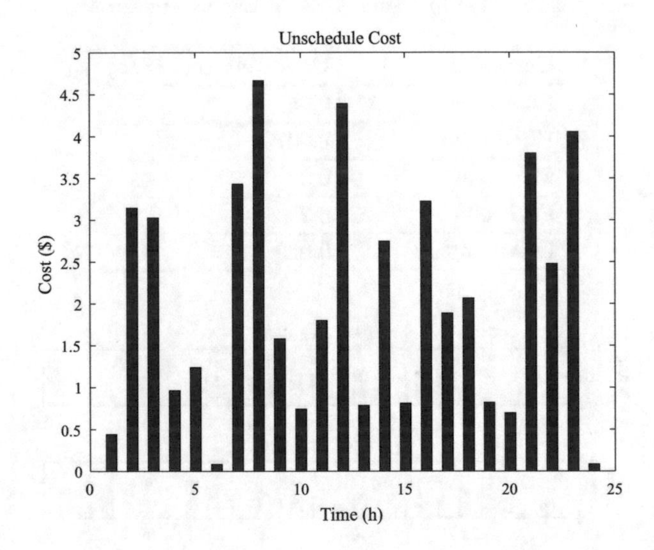

Fig. 4. Unscheduled cost profile of DERs

After scheduling, PV, FC, DE and batteries are used by prioritizing these power generators, battery system and grid is also used for power exchange to fulfill the load demand as shown in Figs. 5 and 6. However, WT and MT are saved for backup option. The generators used are enough to minimize the overall operational cost as shown in Fig. 7.

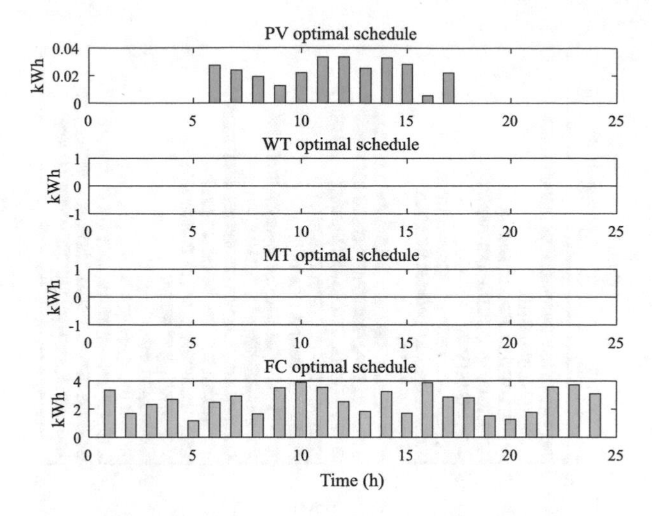

Fig. 5. Schedule power profile of PV, WT, MT and FC

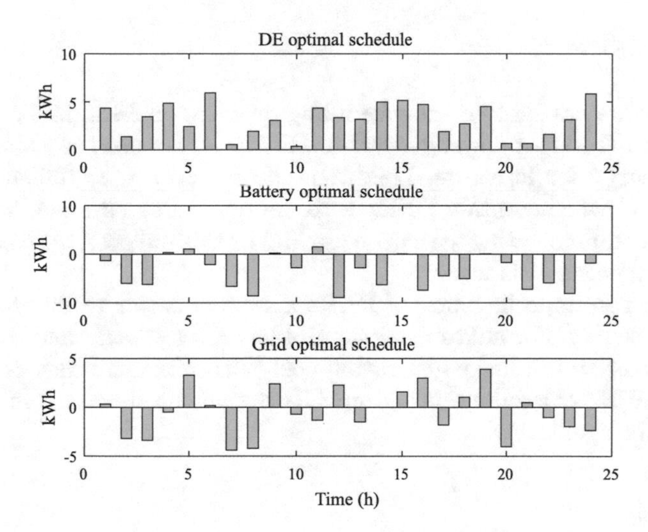

Fig. 6. Schedule power profile of DE, battery and grid

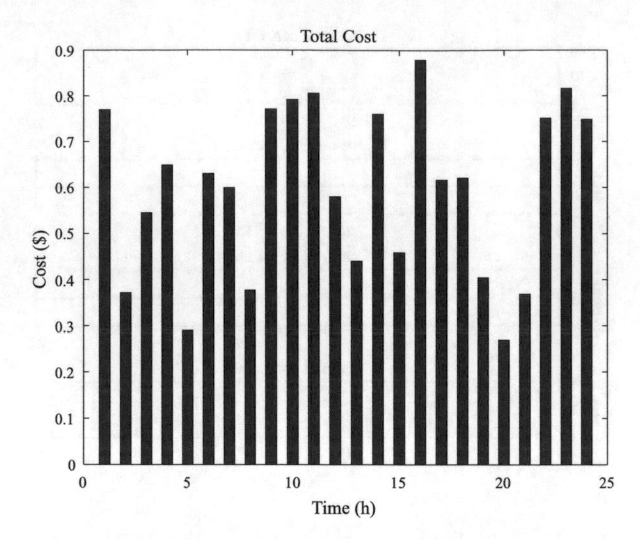

Fig. 7. Scheduled cost profile of DERs

6 Conclusion

In this paper, we have solved the scheduling problem on SSM. MG contains both RESs and non-RESs for power generation. Batteries are used as a storage system and store energy for later use. Load demand of users are fulfilled using grid-connected MG which contain DERs with energy storage system. The power is balanced by communicating with utility grid. For this purpose, we have scheduled the DERs for power balance.

From the results, scheduling of DERs also give better results to satisfy the objective efficiently. Formulation of the problem after transformation into MILP performs better to minimize generation cost with efficient usage of power generators. In DERs scheduling, using only RES gives better result in comparison with the other results.

References

1. Motevasel, M., Seifi, A.R.: Expert energy management of a micro-grid considering wind energy uncertainty. Energy Convers. Manage. **83**, 58–72 (2014)
2. Davidović, T.: Bee colony optimization Part I: the algorithm overview. Yugoslav J. Oper. Res. **25**(1) (2016)
3. Mohanty, B., Tripathy, S.: A teaching learning based optimization technique for optimal location and size of DG in distribution network. J. Electr. Syst. Inf. Technol. **3**(1), 33–44 (2016)
4. Hasanpor Divshali, P., Choi, B.J.: Electrical market management considering power system constraints in smart distribution grids. Energies **9**(6), 405 (2016)
5. Varela Souto, A.: Optimization and Energy Management of a Microgrid Based on Frequency Communications (2016)

6. Naderipour, A., Mohd Zin, A.A., Habibuddin, M.H., Moradi, M., Miveh, M., Afrouzi, H.N.: A new compensation control strategy for grid-connected wind turbine and fuel cell inverters in a microgrid. Int. J. Power Electron. Drive Syst. (IJPEDS) **8**(1), 272–278 (2017)
7. Trivedi, I.N., Thesiya, D.K., Esmat, A., Jangir, P.: A multiple environment dispatch problem solution using ant colony optimization for micro-grids. In: International Conference on Power and Advanced Control Engineering (ICPACE), 2015, pp. 109–115. IEEE (2015)
8. Olivas, F., Valdez, F., Castillo, O., Gonzalez, C.I., Martinez, G., Melin, P.: Ant colony optimization with dynamic parameter adaptation based on interval type-2 fuzzy logic systems. Appl. Soft Comput. **53**, 74–87 (2017)
9. Li, X., Wang, Y., Wang, Z., Shu, X., Zhang, Y.: The open electrical and electronic engineering journal. Open Electr. Electron. Eng. J. **10**, 46–57 (2016)
10. Thirumalaisamy, B., Jegannathan, K.: A novel energy management scheme using ANFIS for independent microgrid. Int. J. Renew. Energy Res. (IJRER) **6**(3), 735–746 (2016)
11. Lin, W.-M., Tu, C.-S., Tsai, M.-T.: Energy management strategy for MG by using enhanced bee colony optimization. Energies **9**(1), 5 (2015)
12. Huld, T., Moner-Girona, M., Kriston, A.: Geospatial analysis of photovoltaic mini-grid system performance. Energies **10**(2), 218 (2017)
13. Dehghanpour, E., Karegar, H., Kheirollahi, R., Soleymani, T.: Optimal coordination of directional overcurrent relays in microgrids by using cuckoo-linear optimization algorithm and fault current limiter. IEEE Trans. Smart Grid (2016)
14. Wang, J., Li, Y., Zhou, Y.: Interval number optimization for household load scheduling with uncertainty. Energy Buildings **130**, 613–624 (2016)
15. Moon, S., Lee, J.-W.: Multi-residential demand response scheduling with multi-class appliances in smart grid. IEEE Trans. Smart Grid (2016)
16. Dong, W., Li, Y., Xiang, J.: Optimal sizing of a stand-alone hybrid power system based on battery/hydrogen with an improved ant colony optimization. Energies **9**(10), 785 (2016)
17. https://www.wbdg.org/resources/microturbines. Accessed 21 Oct 2017
18. http://www.dieselserviceandsupply.com/why_use_diesel.aspx. Accessed 21 Oct 2017
19. World's Largest Carbon Neutral Fuel Cell Power Plant, 16 October 2012. Accessed 21 Oct 2017
20. Zachar, M., Daoutidis, P.: Microgrid/Macrogrid energy exchange: a novel market structure and stochastic scheduling. IEEE Trans. Smart Grid **8**(1), 178–189 (2017)
21. https://optimization.mccormick.northwestern.edu/index.php/Chance-constraint_method. Accessed 24 Oct 2017
22. https://en.wikipedia.org/wiki/Branch_and_bound. Accessed 24 Oct 2017
23. Roy, K., Mandal, K.K., Mandal, A.C.: Modeling and managing of MG connected system using improved artificial bee colony algorithm. Int. J. Electr. Power Energy Syst. **75**, 50–58 (2016)
24. Cheng, Y.-S., Chuang, M.-T., Liu, Y.-H., Wang, S.-C., Yang, Z.-Z.: A particle swarm optimization based power dispatch algorithm with roulette wheel redistribution mechanism for equality constraint. Renew. Energy **88**, 58–72 (2016)

A Metaheuristic Scheduling of Home Energy Management System

Anila Yasmeen[1], Nadeem Javaid[1(✉)], Itrat Fatima[1], Zunaira Nadeem[2], Asif Khan[1], and Zahoor Ali Khan[3]

[1] COMSATS Institute of Information Technology, Islamabad 44000, Pakistan
nadeemjavaidqau@gmail.com
[2] National University of Sicence and Technology, Islamabad 44000, Pakistan
[3] CIS, Higher Colleges of Technology, Fujairah Campus, 4114, Fujairah, UAE
http://www.njavaid.com

Abstract. Smart grid (SG) provides a prodigious opportunity to turn traditional energy infrastructure into a new era of reliability, sustainability and robustness. The outcome of new infrastructure contributes to technology improvements, environmental health, grid stability, energy saving programs and optimal economy as well. One of the most significant aspects of SG is home energy management system (HEMS). It encourages utilities to participate in demand side management programs to enhance efficiency of power generation system and residential consumers to execute demand response programs in reducing electricity cost. This paper presents HEMS on consumer side and formulates an optimization problem to reduce energy consumption, electricity payment, peak load demand, and maximize user comfort. For efficient scheduling of household appliances, we classify appliances into two types on the basis of their energy consumption pattern. In this paper, a meta-heuristic firefly algorithm is deployed to solve our optimization problem under real time pricing environment. Simulation results signify the proposed system in reducing electricity cost and alleviating peak to average ratio.

Keywords: Smart grid · Firefly algorithm
Renewable energy sources · Real time pricing signal
Demand side management · Demand response

1 Introduction

In the current epoch, electricity demand is increasing rapidly. At the same time, electricity generation from different burning fuels is increasing global warming effect; it causes tremendous change in the environment. To handle these problems, consumption of electricity should be minimized, and it can be controlled by optimal consumption of energy. Traditional power grids can not handle effectively power demand challenges effectively. So, to meet these challenges, a new infrastructure is required. In this regard, SG efficiently meet these challenges.

L. Barolli et al. (Eds.): EIDWT 2018, LNDECT 17, pp. 214–224, 2018.
https://doi.org/10.1007/978-3-319-75928-9_19

SG is the advance form of traditional grid. SG integrates advanced metering infrastructure, control systems, distributed renewable energy generation, bi-directional flow of electricity and communication technologies [1]. The two-way information exchange creates an automated energy delivery network.

Demand side management (DSM) is an important strategy of SG, it maintains balance between load demand and energy supply. To implement effective demand response (DR) strategies, utilities instigate the consumers through incentives, to maintain their load pattern efficiently. Smart meters communicate with control systems and collect information about electricity usage of households. The major objectives of SG are minimization of electricity cost, curtailment of peak to average ratio (PAR) and maximization of user comfort. In order to achieve above mentioned objectives, various DSM techniques have been proposed in literature. Authors in [2], focused on daily electricity payment reduction at consumer end by using integer linear programming (ILP). In [3], authors used particle swarm optimization (PSO) to reduce PAR and electricity cost of residential consumer in the presence of renewable energy sources (RESs). Authors in [4], used teaching learning based optimization (TLBO) and differential evolution (DE) techniques to minimize electricity bill and maximize user comfort. These techniques face difficulties to handle multiobjectives and high dimensional problems. Swarm based heuristic techniques for global optimization are mostly used due to low complexity, fast convergence and processing time. We have used metaheuristic firefly algorithm in our HEMS to achieve our objectives. The control parameters of our system are power rating of household appliances, their operational time interval (OTI), length of operational time (LOT) and price signals [5].

The rest of the paper is organized as follows: Sect. 2 presents literature review. Section 3 describes system models and problem formulation. Simulation results and discussion are given in Sect. 4. This paper is concluded in Sect. 5.

2 Literature Review

In smart grid, DSM plays an important role in energy management. It supports SG functionalities in many areas such as electricity market, distributed energy sources, utility grid stability and management of whole infrastructure. In [6,7] authors proposed a new system architecture as HEMS. The HEMS is integrated with RES. Control algorithm (CA) is used to control the temperature of thermal appliances [6] and the home energy management (HEM) algorithm is proposed to curtail electricity cost and peak load [7]. The controller used TOU pricing signal to calculate electricity bill and reduce peak load demand. Simulation results show that proposed algorithm significantly reduces electricity cost. However, installation and maintenance cost of photovoltaic system and battery storage system is neglected.

A dynamic pricing strategy is used to curtail the peak load demand of residential users and electricity bill [8,9]. The proposed system used a game theoretical approach to converge the dynamic response to efficient Nash equilibrium solutions. A game theoretic approach is used for the interaction and trading among users with the excess power. The proposed algorithm reduced energy cost of the users. It also encourages users to utilize surplus energy generated from RESs or sell to nearby users rather than sell back to utility. The authors in [10], evaluate the performance of home energy management controller on the basis of heuristic algorithms; genetic algorithm (GA), binary particle swarm optimization (BPSO) and ant colony optimization (ACO). The objectives of this work are minimization of energy consumption, user comfort maximization, electricity bill and PAR reduction. A combined model of time of use (TOU) and inclined block rate (IBR) is used to calculate electricity bill. Simulation results demonstrate that GA based energy management controller (EMC) performs better than BPSO and ACO based EMC. However, user comfort is compromised in order to reduce electricity cost.

The authors in [11], proposed an optimized HEMS. The proposed system also incorporated RES and energy storage system (ESS). Multiple knapsack problem is used to mathematically formulate the problem and it is solved by using heuristic algorithms: GA, BPSO, WDO, BFO and hybrid GAPSO (HGPO). In paper [12], a mixed integer linear programming (MILP) approach is used to schedule appliances of a smart home. For optimal scheduling of energy consumption, RTP is used. The proposed technique in [11,12] achieved the objectives of electricity cost reduction and PAR curtailment. User comfort maximization is ignored in this work. In [2,13], Day-ahead pricing signal is used for forecasting the electricity payments. The authors applied an optimization power scheduling technique in the response of load demand from residential area. A heuristic evolutionary algorithm (EA) is deployed for solving minimization problem [13]. Appliances are categorized according to power scheduling problem. The proposed strategy reduced the peak load demand of smart grid.

From literature [2,6–13], it can be sum up that all DSM strategies have electricity cost and user comfort objectives. We use FA to solve load management problem because of its convergence and optimal results. The FA gives significant results in order to achieve electricity cost and PAR reduction; and user comfort maximization.

3 Problem Formulation

Our objectives are to reduce electricity cost and PAR while maintaining user comfort. To minimize total cost, optimization algorithm reschedules the residential load and shifts the load from on-peak hours. This also helps in reducing PAR. Load consumption is scheduled by employing metaheuristic technique FA. Objective function is as follow:

$$min \sum_{t=1}^{T} (\sum_{\wp=1}^{n} (P_t^{\wp} \times EP_t)) \quad (1)$$

Subjected to:

$$P(t) = P^e(t) + P^f(t) \tag{1a}$$

$$P_t \leq \lambda_{max} \tag{1b}$$

$$E^{sch}_{cost} < E^{unsch}_{cost} \tag{1c}$$

$$t_s < t < t_e \tag{1d}$$

Total energy consumption after optimization should be equal to the energy consumption without optimization. We have defined an upper limit of load in each time slot as in Eq. (1b). During scheduling process, we have shifted some load from on-peak hour to off-peak hour, so that electricity monetary cost should be less than or equal to the cost of unscheduled load as given in Eq. (1c). These constraints helps to achieve our objective function efficiently. Some performance parameters are elaborated as following to achieve the defined objective function.

3.1 Energy Consumption

Each HEMS has two kinds of appliances as mentioned above, i.e. E and F. Essential appliances contain set of appliances E = {x1, x2, x3, x4, ...} and the set of flexible appliances is F = {y1, y2, y3, y4, ...}. The scheduling time period contains set of 120 time slots. Each time slot is of 12 min T = {t1, t2, t3, t4, ..., t120}. The total energy consumption for a day is given by Eq. (2):

$$P^{total} = \sum_{t=1}^{T}(\sum_{e=1}^{E} P^e_t + \sum_{f=1}^{F} P^f_t) \tag{2}$$

We generalize the formula for total energy consumption of all appliances $\wp$ as given by Eq. (3):

$$P^{total} = \sum_{t=1}^{T}(\sum_{\wp=1}^{n} P^{\wp}_t)) \tag{3}$$

3.2 Electricity Cost and Price Signal

The objective of this paper is to minimize electricity monetary cost by controlling energy consumption. The RTP signal is used to calculate the cost of energy consumption in 24 h. We multiply power rating of appliance $\wp$ and electricity price EP to calculate electricity cost at time slot t:

$$E_{cost} = \sum_{t=1}^{T}(\sum_{\wp=1}^{n}(P^{\wp} \times EP_t) \tag{4}$$

3.3 PAR

In our proposed system model, PAR is defined as the ratio between peak to average load in a given time t. Balancing the PAR, is beneficial for both utility and consumer. It improves the stability of the grid. PAR is calculated by equation given below in Eq. (5):

$$PAR = \frac{max(P_t^{\wp})}{\frac{1}{T}(\sum_{t=1}^{T}(P_t^{\wp}))} \tag{5}$$

3.4 User Comfort

We calculate user comfort as the delay time. It is the difference between user defined start time of an appliance and its scheduled time. It is important that consumer should set some parameters for residential appliances. Let $s_{\wp}$ and $e_{\wp}$ be the start and end time of time slots, $lot_{\wp}$ represents LOT and variable $t_{\wp}$ can be defined as operational start time (OST) of an appliance $\wp$. We can define delay time $\omega_{\wp}$ as:

$$\omega = \frac{t_{\wp} - s_{\wp}}{e_{\wp} - lot_{\wp} - s_{\wp}} \tag{6}$$

4 System Architecture

In this section, we elucidate our proposed system model. We also describe energy consumption calculation, pricing signal and cost calculation procedure.

4.1 System Model

Smart grid is an electrical grid which includes energy management, advanced sensing technologies, two way communication, control methods and energy efficient resources. Our main focus is on DSM in residential area. In this regard, we propose a HEMS. Our HEMS of a smart home associated with energy management controller (EMC), different smart appliances and a smart meter. Smart appliances are connected (wireless) with EMC and exchange information about their status and power ratings. The consumer uses in-home display to enter information about when smart appliances to be scheduled, e.g., their operational time interval and parameters related to user's preference. The EMC interacts with smart meter to exchange information related to electricity demand and price signal. Moreover, smart meter interacts with utility for receiving RTP signals. EMC controls smart appliances and schedules them according to the requirements of the consumer and information from main utility. The proposed system model is shown in Fig. 1. In this system models, appliances are classified into two categories: Essential and flexible appliances. Essential appliances have fixed operational time which include refrigerator, washing machine, cloth dryer and oven. Flexible appliances can be switched on in any time during a day contains rice cooker, dishwasher, electric kettle and humidifier. These appliances are scheduled for one day. We use RTP signal to calculate electricity cost. Start $s_\wp$ and end time $e_\wp$ of appliances, power rating $\rho_\wp$ and LOT $lot_\wp$ are taken from [5]. To schedule these appliances, a time interval of 12 min is used, because some appliances like electric kettle and dish washer use not more than 12 min a day. So, half hour or one hour operational time of such appliances causes wasting of time slots.

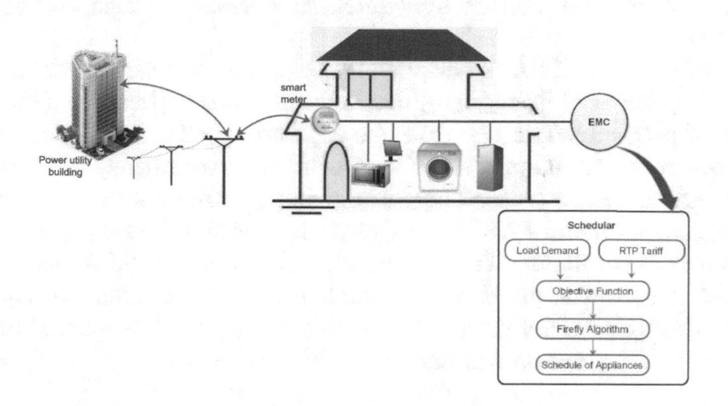

Fig. 1. System model

4.2 Optimization Technique

The firefly algorithm (FA) is an optimization technique that is inspired by the social behaviour of fireflies in finding their mates. It is a novel technique based on swarm intelligence methods developed by Yang in 2008 [14] and inspired by the flashing behaviour of flash lights of fireflies. It is proposed for optimization problems with continuous variables. The FA is formulated based on two key issues: the variation in light intensity I and firefly attractiveness ξ which vary with distance υ under a fixed light absorption coefficient η. Firefly algorithm steps are performed to find the best schedule for 24 h. A physical formula shows that light intensity I is inversely proportional to the square of distance υ^2 under a specific light absorption coefficient η, given as in Eq. (7):

$$I_s \alpha 1/\upsilon^2 \tag{7}$$

This formula depicts that if distance υ increases from the light source, light intensity of a firefly becomes weak. Equation (8) shows the variation in light intensity I with distance υ.

$$I = I_o exp(-\eta \upsilon^2) \tag{8}$$

Some assumptions are made about the flashing behaviour of fireflies to formulate FA, given as follow:

- All fireflies are of same sex as per assumption.
- Their attractiveness is directly proportional to the light intensities of fireflies.
- The light intensity of a firefly is associated with the fitness function.

The movement of fireflies towards nearby firefly is determined by Eq. (9). Fireflies move to the firefly which have more light intensity I.

$$p_a = p_a + \xi exp(-\eta v_{ab}^2)(p_a - p_b) + \mu\varsigma \tag{9}$$

At the end of each iteration, all fireflies are ranked according to their light intensities. Firefly with highest light intensity is considered as best solution of the problem.

The performance of FA is based on some parameters which are: population size (nPop), number of fireflies (n), maximum iterations (MaxIt), light intensity (I), value of attractiveness (ξ), distance (v) and specific absorption coefficient (η). The control parameters of algorithm include: power rating, LOT, start time, end time and electricity price. The performance parameters are: electricity cost, energy consumption and PAR. This algorithm runs 100 iterations, at each iteration, it evaluates the fitness of the firefly. The constraints of the objective function are applied to solve the problem. In peak hours, maximum load is scheduled to the hours where electricity price is low or the load demand is within limit. Eventually, energy consumption demand is acquired and electricity cost is calculated for a day.

5 Simulation Results and Discussion

In the underlying section, performance of our proposed HEMS is validated through simulation results in MATLAB. For simulations, we consider nine different residential appliances in a single home [5]. For scheduling, we use FA for scheduling of smart appliances in a single home. The performance measuring parameters considered in our work are: energy consumption, electricity cost and PAR. We use RTP signal as shown in Fig. 2. This figure illustrates that 71–90 time slots are on-peak hours, 61–70 and 90–100 time slots are shoulder-peak hours and off-peak hours are 1–60 and 100–120 time slots. Our simulations show the results of energy consumption, electricity cost, PAR and user comfort. Our main objectives are to minimize electricity cost and PAR by anticipating load at consumer end.

Simulation results are divided in three sections on the basis of energy consumption, electricity cost and PAR.

5.1 Energy Consumption

This section illustrates the energy consumption of unscheduled and scheduled load demand per day in a single home. Figure 3 presents unscheduled and scheduled energy consumption for 24 h. It is worth mentioning here that the unscheduled load pattern shows high user activities during on-peak hours leading to high electricity cost and PAR. While user activities are low during off-peak hours, electricity cost is minimal in off-peak hours. In case of our optimization technique FA, maximum energy consumption during on-peak hours is shifted to off-peak hours and peaks are also avoided. The maximum energy consumption according

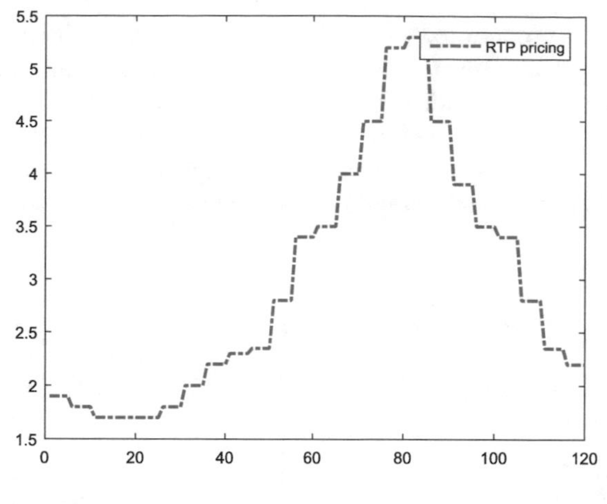

Fig. 2. RTP

to this figure is 16 kWh in unscheduled case, it is shifted to off-peak hours by scheduling load with FA. This shifting of load elucidates the reduction of cost by managing the load consumption in an efficient manner. It is concluded that our scheme FA has reduced peak load leading to stable load pattern. Overall power consumption by scheduling is optimal and relatively low.

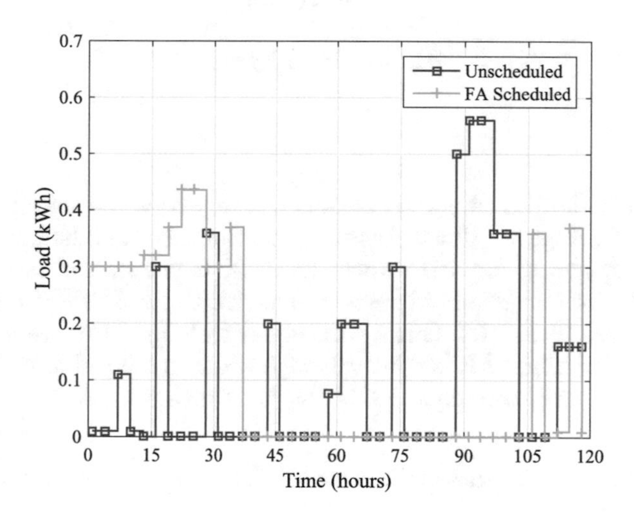

Fig. 3. Energy consumption

5.2 Electricity Cost

From simulations, we elucidate how firefly effectively reduces electricity monetary cost by controlling energy consumption. During peak hours 15 pm–18 pm, high energy consumption leads to high electricity cost. Figure 4 illustrates the electricity cost of unscheduled and scheduled appliances. The main focus here is to curtail the cost and shift demand of energy requirements in off-peak hours using HEMS with our optimization technique. In this figure, unscheduled cost is 52 cents while scheduled cost is reduced to 30 cents. It also shows that FA minimizes almost 50% cost of unscheduled energy consumption. It is noted that FA shows best results in reducing cost because percentage cost reduction is relatively high.

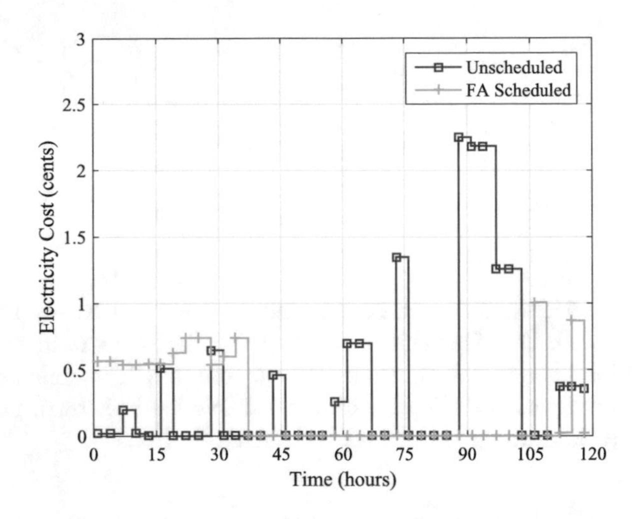

Fig. 4. Electricity cost

5.3 PAR

PAR is useful for both utility and consumer, it improves the efficiency and capability of the grid. Figure 5 illustrates the performance of unscheduled and scheduled energy consumption with respect to PAR. It shows the unscheduled and scheduled PAR of our proposed system. The PAR in unscheduled is 0.038 and in scheduled case, it is 0.02. Our optimization technique FA presents the PAR decrement of 47%. The PAR is relatively high when cost is reduced. Result shows that our optimization technique FA is able to reduce the PAR.

5.4 Performance Trade-off Between Electricity Cost and User Comfort

Generally, we calculate user comfort in terms of delay or waiting time. We define delay time as the difference between user defined start time of an appliance and its scheduled start time. There is inverse relationship between electricity monetary cost and delay time. Hence, there is a trade-off between cost and delay time. In unscheduled case, consumer turn on appliances when required; regardless of the cost and utility grid sustainability. So, there is no delay in unscheduled case.

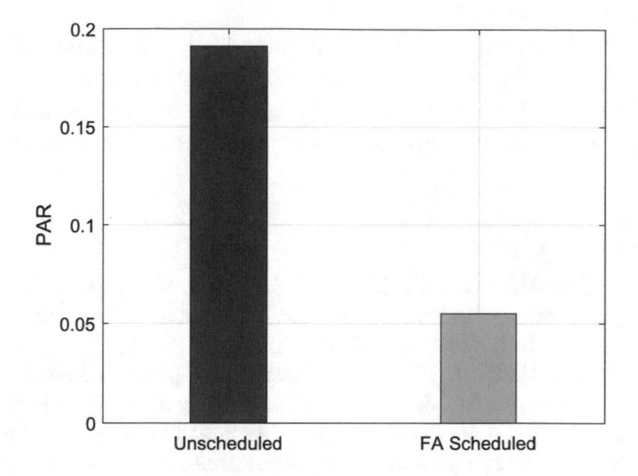

Fig. 5. PAR

Figure 6 demonstrates the relationship between electricity cost and average delay. It shows the trade-off between these two performance parameters. More specifically, the value of waiting time increases when value of electricity cost decreases, and vice versa. However, if the consumer prefers user comfort, he must have to compromise on electricity cost.

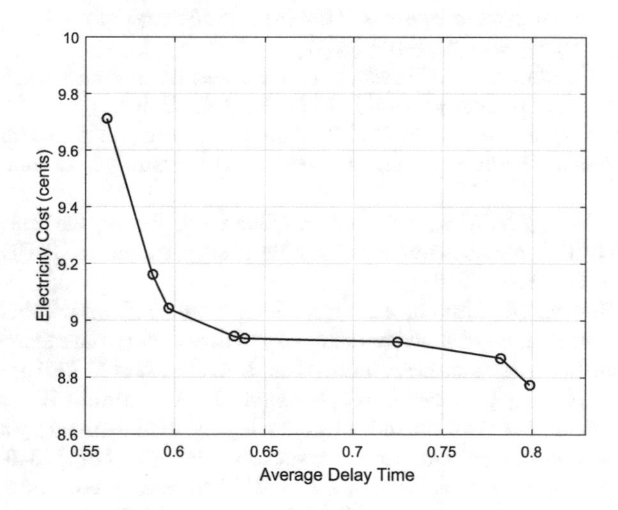

Fig. 6. Relationship between electricity cost and average delay time

6 Conclusion

In this paper, we have proposed a HEMS to schedule household appliances. Firefly algorithm is deployed to schedule household appliances. HEMS scheduled with FA reduces cost and PAR. Simulation results show that our proposed system with the integration of RES has achieved our defined objectives and shows better results. In future, we will extend our work for whole residential sector and incorporate power trading among multiples homes under large share of RES for optimal cost minimization and consumer to be prosumer.

References

1. Tushar, W., Chai, B., Yuen, C., Smith, D.B., Wood, K.L., Yang, Z., Vincent Poor, H.: Three-party energy management with distributed energy resources in smart grid. IEEE Trans. Ind. Electron. **62**(4), 2487–2498 (2015)
2. Ma, K., Yao, T., Yang, J., Guan, X.: Residential power scheduling for demand response in smart grid. Int. J. Electr. Power Energy Syst. **78**, 320–325 (2016)
3. Hakimi, S.M., Moghaddas-Tafreshi, S.M.: Optimal planning of a smart microgrid including demand response and intermittent renewable energy resources. IEEE Trans. Smart Grid **5**(6), 2889–2900 (2014)
4. Garcia, J.A.M., Mena, A.J.G.: Optimal distributed generation location and size using a modified teaching learning based optimization algorithm. Int. J. Electr. Power Energy Syst. **50**, 65–75 (2013)
5. Zhao, Z., Lee, W.C., Shin, Y., Song, K.-B.: An optimal power scheduling method for demand response in home energy management system. IEEE Trans. Smart Grid **4**(3), 1391–1400 (2013)
6. Shakeri, M., Shayestegan, M., Abunima, H., Reza, S.M.S., Akhtaruzzaman, M., Alamoud, A.R.M., Sopian, K., Amin, N.: An intelligent system architecture in home energy management systems (HEMS) for efficient demand response in smart grid. Energy Build. **138**, 154–164 (2017)
7. Rajalingam, S., Malathi, V.: HEM algorithm based smart controller for home power management system. Energy Build. **131**, 184–192 (2016)
8. Barbato, A., Capone, A., Chen, L., Martignon, F., Paris, S.: A distributed demand-side management framework for the smart grid. Comput. Commun. **57**, 13–24 (2015)
9. Samadi, P., Wong, V.W.S., Schober, R.: Load scheduling and power trading in systems with high penetration of renewable energy resources. IEEE Trans. Smart Grid **7**(4), 1802–1812 (2016)
10. Rahim, S., Javaid, N., Ahmad, A., Khan, S.A., Khan, Z.A., Alrajeh, N., Qasim, U.: Exploiting heuristic algorithms to efficiently utilize energy management controllers with renewable energy sources. Energy Build. **129**, 452–470 (2016)
11. Ahmad, A., Khan, A., Javaid, N., Hussain, H.M., Abdul, W., Almogren, A., Alamri, A., Niaz, I.A.: An optimized home energy management system with integrated renewable energy and storage resources. Energies **10**(4), 549 (2017)
12. Zhang, D., Shah, N., Papageorgiou, L.G.: Efficient energy consumption and operation management in a smart building with microgrid. Energy Convers. Manage. **74**, 209–222 (2013)
13. Logenthiran, T., Srinivasan, D., Shun, T.Z.: Demand side management in smart grid using heuristic optimization. IEEE Trans. Smart Grid **3**(3), 1244–1252 (2012)
14. Yang, X.-S.: Firefly algorithms for multimodal optimization. In: International Symposium on Stochastic Algorithms, pp. 169–178. Springer, Heidelberg (2009)

Void Hole and Collision Avoidance in Geographic and Opportunistic Routing in Underwater Wireless Sensor Networks

Aasma Khan[1], Nadeem Javaid[1(✉)], Ghazanfar Latif[1], Obaida Abdul Karim[2], Faisal Hayat[3], and Zahoor Ali Khan[4]

[1] COMSATS Institute of Information Technology, Islamabad 44000, Pakistan
nadeemjavaidqau@gmail.com
[2] International Islamic University, Islamabad 44000, Pakistan
[3] The University of Lahore, Islamabad 44000, Pakistan
[4] CIS, Higher Colleges of Technology, Fujairah Campus, Fujairah 4114, UAE
http://www.njavaid.com

Abstract. Underwater Wireless Sensor Networks (UWSNs) facilitate an extensive variety of aquatic applications such as military defense, monitoring aquatic environment, disaster prevention, etc. However UWSNs routing protocols face many challenges due to adverse underwater environment such as high propagation and transmission delays, high deployment cost, nodes movement, energy constraints, expensive manufacture, etc. Due to random deployment of nodes void holes may occur that results in the failure of forwarding data packet. In this research work we propose two schemes, Geographic and Opportunistic Routing using Backward Transmission (GEBTR) and Geographic and Opportunistic Routing using Collision Avoidance (GECAR) for UWSNs. In aforesaid scheme fall back recovery mechanism is used to find an alternative route to deliver the data when void occurs. In later, fall along with nomination of forwarder node which has minimum number of neighbor nodes is selected. Simulation results show that our techniques outperform compared with baseline solution in terms of packet delivery ratio by 5% in GEBTR and 45% in GECAR, fraction of void nodes by 8% and 11% in GECAR and energy consumption by 8% in GEBTR and 10% in GECAR.

Keywords: Underwater Wireless Sensor Networks
Geographic and opportunistic routing
Void hole · Backward Transmission · Collision

1 Introduction

Almost 70% of earth is covered with water. Underwater atmospheres are very essential for human beings because of their effect on primary global production of CO_2 involvement. Due to the reduction of terrestrial means, researchers are giving more attention to the consideration of underwater resources. UWSNs have

L. Barolli et al. (Eds.): EIDWT 2018, LNDECT 17, pp. 225–236, 2018.
https://doi.org/10.1007/978-3-319-75928-9_20

gained a lot of attention from researchers because of their potential to monitor underwater environment. In underwater atmosphere protocols premeditated for Terrestrial Wireless Sensor Networks (TWSNs) cannot work perfectly. TWSNs and UWSNs have differences in many aspects such as use of acoustic links instead of radio links. UWSNs topology is more dynamic than TWSNs because nodes move freely with water currents and change their position frequently. Localization of nodes is difficult as compared to TWSNs. Deployment of nodes is comparatively sparse, moreover energy is limited. After deployment it is difficult to recharge the batteries. In addition, UWSNs face great challenges like low bandwidth, high propagation delay, high communication cost. The acoustic signal transmission delay for underwater is five times less than radio signals [3]. However, UWSNs have extensive variety of uses such as military defense, monitoring aquatic environment, disaster prevention, oil/gas extraction, offshore exploration, commercial and scientific purposes, etc. [1,2].

There are many protocols of UWSNs however, due to the above stated limitations and tests, an efficient protocol is required that is energy efficient, with low packet drop, by having a mechanism for void hole avoidance and less collision as well.

In this perspective, geographic routing appears the best capable for scheming the routing protocols for UWSNs. Geographic routing is simple and scalable. Geographic routing protocol is not required to create or maintain the complete routing table from source to destination [3].

Communication void regions occur when a forwarder node does not find the next forwarder closer to sink in its transmission range [4]. Each time a packet gets jammed in a node that is in a void region the routing protocol should direct the stucked packet by using some retrieval technique otherwise the packet should be thrown away.

Motivated by above consideration, we present geo-opportunistic routing with the mechanism of one hop backward transmission (GEBTR). In proposed scheme, when a void region occurs instead of depth adjustment as in Geographic and opportunistic routing with Depth Adjustment (GEDAR) routing protocol [2], one hop backward transmission is done, as depth adjustment of each node increases the energy consumption and our second scheme (GECAR) selects that node having minimum neighboring nodes to avoid collision. In this way communication over void regions, packet deliver ratio and energy consumption is improved. Simulation results show that proposed schemes beat the GEDAR scheme.

The contributions of proposed work related to UWSNs are (i) using anycast geo-opportunistic based routing practice (ii) a routing scheme based on neighbor node selection taking in to account two metrics that selects the next node on the basis of depth difference of two hops and if void region occurs than it uses one hop backward transmission (GEBTR) (iii) selection of that node with route to sink with minimum nodes to avoid collision (GECAR) (iv) simulations are done to validate the efficiency of our schemes.

The rest of this paper is organized as follows: In Sect. 2, we review the related works about routing protocols in UWSNs. Section 3 defines the problem statement. Section 4 defines the system model of our proposed schemes. In Sect. 5 feasible region of energy minimization using linear programming is calculated. Finally, we present the performance evaluation of our schemes using simulation results in Sect. 6, followed by our conclusion in Sect. 7.

2 Related Work

In this section the related work on routing protocols in UWSNs is presented with their features, advantages and disadvantages.

In UWSNs, routing protocols are categorized as depth based, location based and location free routing protocols. Depth Based Routing (DBR) [5] is a depth based routing protocol that handles dynamic networks efficiently. It requires local depth information to forward packets greedily towards the sink. In this protocol, each eligible node forwards the packet on the basis of priority. Priority is defined as the forwarder node's depth is less than current forwarder node and also that it has not sent the same packet previously. DBR provides improved network lifetime and packet delivery ratio. Moreover, it has taken advantage of multi sink architecture without using extra cost, however, due to greedy approach it causes void holes, increased energy consumption and end to end delay due to holding time. It is inefficient for sparse network because DBR only has a greedy mode.

Improved Adaptive Mobility of Courier nodes in Threshold-optimized DBR (iAMCTD) [6] is a location free routing protocol specially designed for time critical applications. In this paper, to handle flooding, latency and path loss. It provides improved network lifetime, minimized E2ED due to efficient movement of courier nodes. However, void hole problem still exists and overhead due to control packets exchange.

Balanced Load Distribution (BLOAD) [7] is a two dimensional network protocol. BLOAD tackles the problem of energy holes. In this scheme, data is divided in to three portions. These portions of data are then forwarded using direct and multihop transmission to the destination. To achieve balanced energy consumption using balanced distribution of the total data traffic, each node uses both direct transmission and hop by hop transmission that improves network stability and lifetime. The limitation of BLOAD is increased consumption of energy and excessive load near the sink.

Delay sensitive schemes [1] provide an advancement of localization free routing protocols of DBR, EEDBR and AMCTD as improved delay sensitive versions. In this paper, authors make these routing protocols adaptable to time critical applications. This paper achieves minimum E2ED transmission loss and minimum consumption of energy. However duplication of packets occurs, energy consumption increases and void hole problem exists.

In Adaptive Relay Chain Routing (ARCR) [8], the authors introduced to use mobile sensor nodes to reduce the problem of energy hole and keeping sink fixed in its particular location. To achieve this, clusters are used in the network and for collection of data mobile nodes are used and then the data is forwarded towards the sink. This routing mechanism achieves energy efficiency, maximum network lifetime and load balancing. However, the network disconnects when the sensor nodes are unsuitable to forward the data.

Energy efficient Channel Aware Routing Protocol (E-CARP) [9] is improved version of CARP it is distributed cross layer reactive routing protocol. It is important for data collecting and transmission in UWSNs. It provides improved network lifetime and reduced energy consumption by avoiding control packets however, reduced throughput and high path loss due to mobility of nodes.

Hydraulic-pressure-based anyCast routing (HydroCast) [3] is a pressure based routing protocol its objective is to design an effective routing algorithm for consistent broadcasting to any of the sink and to solve the problem of void hole. Thus, it directs the data upward to lower depths. However the gauge used has to perfectly guess the depth of that node. The depth information can be used for geographic anycast routing. It has improved packet delivery ratio and its limitations are its low performance and increased energy consumption.

Weighting Depth and Forwarding Area Division-DBR (WDFAD-DBR) [10] not only ponder depth of present node but also next forwarding hop depth that help to reduce the presence of void holes. However forwarding using two hops do not remove the void hole problem.

Adaptive Hop-by-Hop Vector-Based forwarding routing protocol (AHH-VBF) [11] uses a scheme that changes radius of pipeline adaptively for amending forwarding area for packets that help in reducing duplicate packets. This scheme results in improved packet delivery ratio, less consumption of energy and E2ED. However it still does not eliminate the void hole problem when the network is sparse.

Opportunistic Routing based on Residual energy (ORR) [12] is an opportunistic routing protocol for asynchronous duty-cycled wireless sensor networks. Their objective is to calculate the finest forwarders that are based on forwarding score calculation and considering residual energy while selecting forwarder sets. This protocol addresses the problems of load balancing, losing coverage and connectivity with residual energy for selecting forwarder node. However, duty cycling causes additional delay due to waiting for the next hop node to wake up. This protocol is not low as required. Number of forwarders can wake up simultaneously that result in duplication of packets.

In Energy Efficient and Load Balanced distributed Routing (ELBAR) [13] sensor nodes collaborate to find the expected polygon of a particular hole and then the sensor nodes inform one another about this particular polygon depending on hole view angle and hole covering parallelogram data is forwarded along escape route that surrounds hole. Authors objective is to route data in the presence of these holes. This routing scheme increases the lifetime of the network and also minimizes the energy consumption. However, this scheme may lead to longer routing path, end to end delay and more energy consumption.

3 Problem Statement

The avoidance of void holes in a demand to minimize the energy consumption is one of most researched topics in UWSNs. In this regard, various routing mechanisms have been proposed [2]. However, the proposed strategies are not as efficient as required. In existing scheme the neighbor nearby to sink is selected as the next forwarder node. In case of void node recovery, void node is moved to new depths to continue the forwarding data greedily. However, movement of nodes to new depth causes excessive energy consumption, extra cost and holding time at each hop to discover routes in void regions. Moreover, communication overhead increases due to continuous message beaconing. In order to tackle the aforementioned problems we have proposed schemes for void hole avoidance and collision minimization, to minimize the packet drop ratio and consumption of energy. We have proposed two schemes: GEBTR checks neighbor information of nodes up to two hops and if void region occurs it uses backward transmission to deliver data packets towards its successor nodes. In GECAR, we use the same feature of GEBTR however; a node with minimum number neighbor nodes is selected as a forwarder to avoid collision.

4 System Model

We consider UWSNs architecture as follows. In the proposed multi sink architecture, nodes are randomly deployed at the ocean bottom and sink nodes at the surface of ocean as shown in Fig. 1. This system model consists two types of nodes relay nodes and anchored nodes. The nodes that are anchored are static at the bottom of ocean whereas relay nodes are arbitrarily placed at different locations that are attached to pumpup buoys or pulley based apparatus. Relay nodes use acoustic signals to communicate with each other. They transmit the data from bottom nodes to the sink at surface. Anchored nodes get the essential data and forward it to sink nodes using the relay nodes. In sink nodes both acoustic modems and radio modems are used for communication. After receiving data sink forwards it to satellite using the radio link, satellite then forward it to the monitoring center. It is assumed that packets received at a single sink are considered to be received at all the sinks successfully.

4.1 Proposed Schemes

In this section we describe our both proposed routing protocols for monitoring underwater sensors. GEBTR and GECAR finds the set of next hop forwarders using the greedy opportunistic forwarding mechanism and fall back mechanism is used to find an alternative route to deliver the data in case of void region.

The main steps of our protocol are as follows, first if a node is in fall back recovery mechanism new data packets will be queued and the greedy forwarding mechanism is rescheduled to resend these data packets. If node is not in the

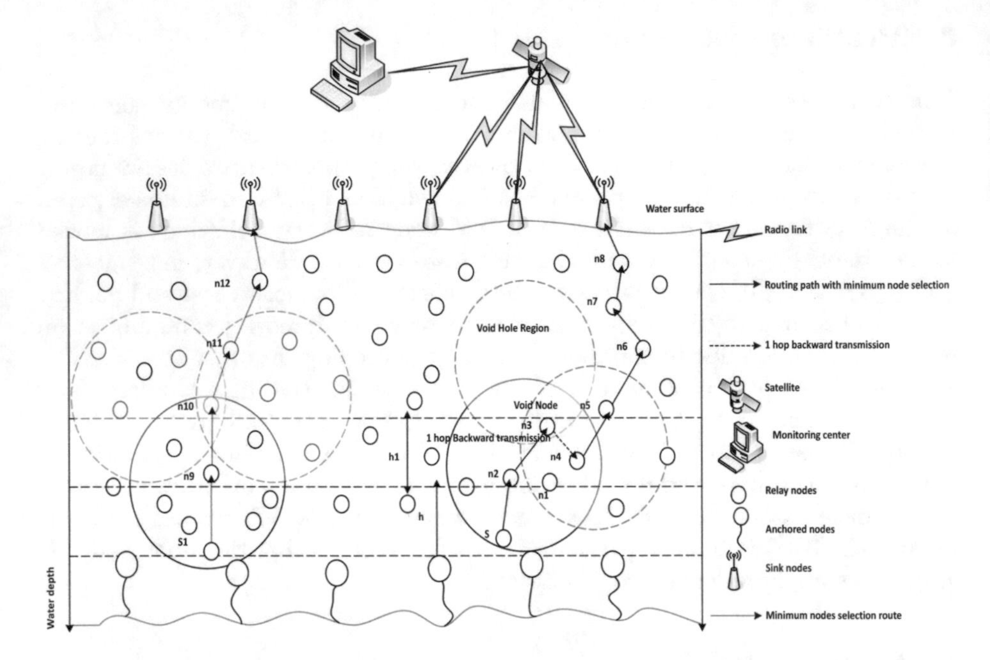

Fig. 1. System model of proposed schemes illustrating void hole and collision avoidance

communication void region then it will continue to forward the data greedily towards the sink. Otherwise, it will switch to fall back recovery mechanism.

In proposed scheme GEBTR as shown on the right side of Fig. 1 the node 'S' looks up to two hop neighbor information. In case of void region GEBTR uses backward transmission at node 'n3' instead of depth adjustment and forward it to node 'n4' that continues to forward packet to the sink node. To further improve the packet delivery ratio of network, GECAR selects those nodes which have minimum number of neighbor nodes to avoid the collision as shown on the left side of Fig. 1 at node S1.

5 Energy Minimization

High energy consumption definitely influences the performance of the network. For the minimization of energy consumption, proposition of linear programming based mathematical technique is used to achieve best possible results. This approach begins with an objective function followed by linear constraints. The objective function must satisfy the constraints in order to achieve best results. For achieving best results, we develop objective function to reduce the energy consumption per node as follows:

$$Min\Sigma_{r=1}^{r_{max}} Energy_{tax}(r) \quad \forall r \in r_{max} \tag{1}$$

where $Energy_{tax}$ is the energy tax calculated for all the nodes.

$$Energy_{tax}(r) = \Sigma_{r=1}^{r_{max}}(Energy_{consumed}(r)) \quad \forall r \in r_{max} \tag{2}$$

GEBTR and GECAR use Eq. 2 to calculate the energy consumption per node. Where $Energy_{consumed}$ includes both reception and transmission energies of the nodes in the network i.e.

$$Energy_{consumed}(r) = E_{tx} + E_{rcv} \quad \forall r \in r_{max}. \tag{3}$$

Where,

$$E_{tx} = P_{tx}\left(\frac{Data_{size}}{Data_{rate}}\right) \tag{4}$$

E_{tx} is the energy consumed during the transmission of data, P_{tx} is the transmission power.

$$E_{rcv} = P_{rcv}\left(\frac{Data_{size}}{Data_{rate}}\right) \tag{5}$$

E_{rcv} is the energy consumed in receiving the data and P_{rcv} is the receiving power. The objective function is to minimize the energy consumption per node. Constraints of the objective function are given as follows:

Constraints

$$C_1\colon E_{tx}, E_{rcv} \leq E_i \tag{6}$$

$$C_2\colon E_F \leq E_F^{min} \tag{7}$$

$$C_3\colon TX_n \leq TX_{max} \tag{8}$$

Equation 6 ensures that the energy required for transmission and reception should be less than the initial energy E_i of the node i.e. energy provided to all the nodes. Equation 7 shows the constraint for the selection of the forwarder node, which have minimum energy consumption. Where E_F is the forwarder node with minimum energy consumption. Equation 8 ensures that in order to receive a good quality signal, the data should be transmitted within its transmission range. Where TX_n is the transmission range of the node and TX_{max} is the maximum transmission range of the node.

Graphical Analysis: To provide clear visualization of the proposed problem, graphical analysis is presented to compute all possible values within the feasible region. Assuming $Data_{size}$ = 100 byte, $Data_{rate}$ = 50 kbps, $P_{tx} = \{0.5, 1..., 2\}$ and $P_{rcv} = \{0.025, 0.05, ..., 0.1\}$, feasible solution for energy minimization is computed as follows from aforementioned constraints in (Eqs. 6–8):

$$1 \leq E_{tx} \leq 4 \tag{9}$$

$$0.05 \leq E_{rcv} \leq 0.2 \tag{10}$$

$$0.105 \leq E_{tx} + E_{rcv} \leq 0.42 \tag{11}$$

Feasible region is plotted in Fig. 2 via points extracted from (Eqs. 9–11) and the points on the boundary of this feasible region are:

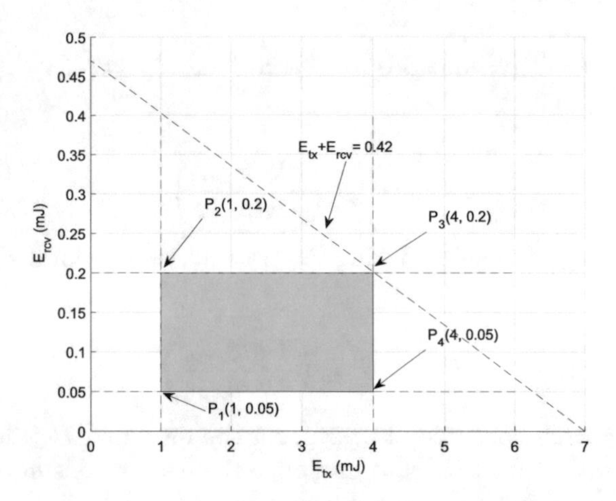

Fig. 2. Feasible region: energy minimization

$$P_1\,(1, 0.05) = 1.05 \text{ mJ}$$

$$P_2\,(1, 0.2) = 1.2 \text{ mJ}$$

$$P_3\,(4, 0.2) = 4.2 \text{ mJ}$$

$$P_4\,(4, 0.05) = 4.05 \text{ mJ}$$

An optimal solution is validated from these feasible points. Hence selecting any value from these points results in minimization of energy consumption per node.

6 Performance Evaluation

To evaluate the performance of our proposed system, we conduct our simulations in MATLAB. For performance evaluation, proposed schemes are compared with GEDAR protocol. In simulations, number of nodes varies from 150–450 and number of sinks are 45. They are randomly deployed in a 3-D region of 1500 m × 1500 m × 1500 m volume. In all experiments sensor nodes have a transmission range of 250 m and data rate of 50 kbps. Payload of the data packet is set to be 150 bytes. The values of the energy consumption are $P_t = 2$ W, $P_r = 0.1$ W and

$P_i = 10$ mW for transmission, reception and idle energies respectively. Therefore mentioned parameters are taken from [2]. Simulation parameters are given in Table 1.

Following metrics are considered to assess the performance of proposed scheme:

1. Fraction of void Nodes: Fraction of local maximum nodes is the proportion of void nodes occurred during communication.
2. Packet Delivery Ratio (PDR): Packet delivery ratio is the fraction of data successfully received by sink node to the sum of data sent by source nodes.
3. Energy Tax: Energy tax is the regular consumption of energy during the transmission of packet from source node to sink node. Unit used for measuring energy tax is joule.

Table 1. Simulation parameters UWSNs

Parameter	Value
Nodes	150–450
Sinks	45
Area length (m)	1500
Area width (m)	1500
Area depth (m)	1500
Initial energy (J)	70
Transmission range (m)	250
Transmission power (W)	2
Reception power (W)	0.1
Idle power (W)	0.01
Frequency (kHz)	10
Packet Size (bytes)	100
Data rate (kbps)	50

Figure 3 shows the fraction of void nodes of existing and proposed schemes. The plot shows that the fraction of local maximum nodes of GEBTR, GECAR and GEDAR increases when the network is sparse and decreases with increase in node density in the network. However, both proposed schemes GEBTR and GECAR achieve better results as compared to GEDAR, both in sparse and dense nodes deployment. This happens because backward transmissions reduce the fraction of void nodes in the sparse network.

Figure 4 shows that PDR of GEDAR, GEBTR and GECAR increases with the increase in the network density. The probability of void holes is high in sparse network due to which PDR is low. Figure 4 shows that GEBTR and GECAR

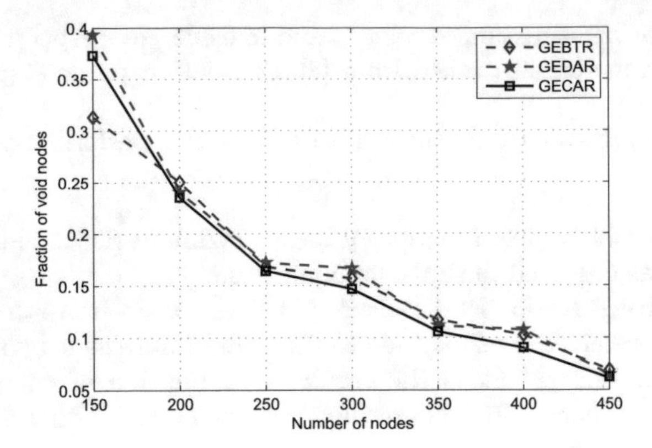

Fig. 3. Fraction of void nodes of existing scheme and proposed schemes

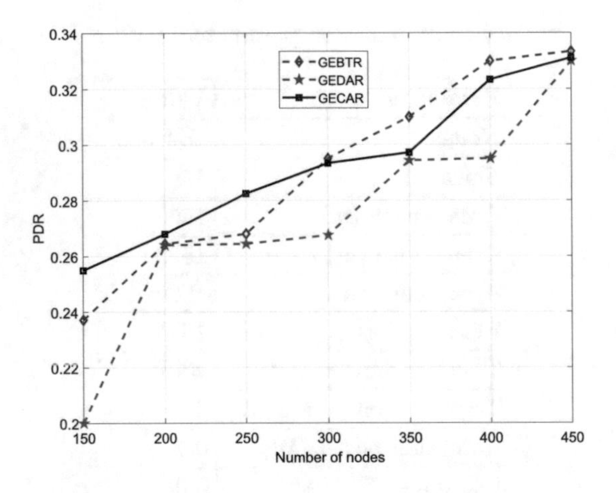

Fig. 4. PDR of existing scheme and proposed schemes

outperform GEDAR in term of PDR for both sparse and dense network. During the depth adjustment, in GEDAR sensor nodes exchange beacon messages that causes communication overhead. This communication overhead results in decreased PDR. GECAR scheme is better than GEBTR and GEDAR in sparse network because it looks up to two hop neighbor information. It avoids collision by selecting nodes with minimum neighbors along with backward transmission that further improves the PDR. Furthermore, GEDAR has no mechanism for collision avoidance.

Figure 5 shows that the energy tax decreases with the increase in network density for GECAR and GEDAR. GEDAR shows high energy consumption in sparse network compared to other two schemes, as the mechanism of depth adjustment and continuous beaconing of nodes increase its energy consumption. GECAR consumes more energy than GEBTR because it considers a node with minimum neighbors to avoid collision. Both GEBTR and GECAR perform better than GEDAR in terms of energy tax in sparse networks but as the density of network increases both proposed schemes behave same as GEDAR. This happens because average displacement per node decreases.

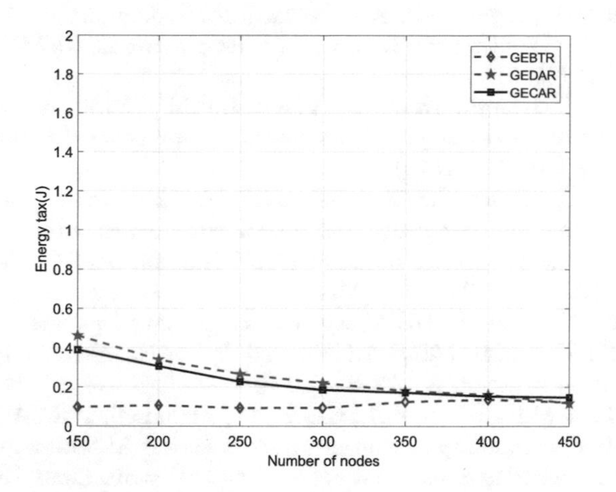

Fig. 5. Energy tax of existing and proposed schemes

7 Conclusion

In this paper we have proposed and assessed the GEBTR and GECAR routing protocols to enhance the routing mechanism in UWSNs. Both these protocols are simple geographic routing protocols that uses the location information of the nodes and uses greedy forwarding mechanism to forward data packets towards the sinks. Furthermore, both schemes provide mechanism to get better communication over void regions. In the proposed schemes, selecting nodes with minimum neighboring nodes to avoid collision, GECAR out performs GEDAR and GEBTR with respect to PDR. Furthermore, GEBTR shows improvement in reducing energy consumption of nodes, increased PDR and reduced fraction of void nodes. Our simulation results verify the effectiveness of proposed schemes in term of PDR, energy tax and fraction of void nodes. As a future work, we plan to apply these schemes to the design of geographic and opportunistic routing protocol for UWSNs, considering end to end delay and life time of the network.

References

1. Javaid, N., Jafri, M.R., Ahmed, S., Jamil, M., Khan, Z.A., Qasim, U.: Delay-sensitive routing schemes for underwater acoustic sensor networks. Int. J. Distrib. Sens. Networks **11**(3), 532–676 (2015)
2. Coutinho, R.W.L., Boukerche, A., Vieira, L.F.M., Loureiro, A.A.F.: Geographic and opportunistic routing for underwater sensor networks. IEEE Trans. Comput. **65**(2), 548–561 (2016)
3. Noh, Y., Lee, U., Lee, S., Wang, P., Vieira, L.F.M., Cui, J., Gerla, M., Kim, K.: HydroCast: pressure routing for underwater sensor networks. IEEE Trans. Veh. Technol. **65**(1), 333–347 (2016)
4. Coutinho, R.W.L., Boukerche, A., Vieira, L.F.M., Loureiro, A.A.F.: GEDAR: geographic and opportunistic routing protocol with depth adjustment for mobile underwater sensor networks. In: 2014 IEEE International Conference on Communications (ICC), pp. 251–256 (2014)
5. Yan, H., Shi, Z.J., Cui, J.: DBR: depth-based routing for underwater sensor network. In: NETWORKING 2008 Ad Hoc and Sensor Networks, Wireless Networks, Next Generation Internet, pp. 72–86 (2008)
6. Javaid, N., Jafri, M.R., Khan, Z.A., Qasim, U., Alghamdi, T.A., Ali, M.: iAMCTD: improved adaptive mobility of courier nodes in threshold-optimized DBR protocol for underwater wireless sensor networks. Int. J. Distrib. Sens. Networks **10**(11), 213-012 (2014)
7. Azam, I., Javaid, N., Ahmad, A., Abdul, W., Almogren, A., Alamri, A.: Balanced load distribution with energy hole avoidance in underwater WSNs. IEEE Access **5**, 15206–15221 (2017)
8. Kong, L., Ma, K., Qiao, B., Guo, X.: Adaptive relay chain routing with load balancing and high energy efficiency. IEEE Sens. J. **16**, 1–10 (2016)
9. Zhou, Z., Yao, B., Xing, R., Shu, L., Bu, S.: E-CARP: an energy efficient routing protocol for UWSNs in the internet of underwater things. IEEE Sens. J. **16**(11), 4072–4082 (2016)
10. Yu, H., Yao, N., Wang, T., Li, G., Gao, Z., Tan, G.: Ad Hoc networks WDFAD-DBR: weighting depth and forwarding area division DBR routing protocol for UASNs. Ad Hoc Netw. **37**, 256–282 (2016)
11. Yu, H., Yao, N., Liu, J.: Ad Hoc networks an adaptive routing protocol in underwater sparse acoustic sensor networks. Ad Hoc Netw. **34**, 121–143 (2015)
12. So, J., Byun, H.: Load-balanced opportunistic routing for duty-cycled wireless sensor networks. IEEE Trans. Mob. Comput. **16**(7), 1940–1955 (2017)
13. Nguyen, K., Le, P., Huy, Q., Van Do, T.: An energy efficient and load balanced distributed routing scheme for wireless sensor networks with holes. J. Syst. Softw. **123**, 92–105 (2017)

Optimized Energy Management Strategy for Home and Office

Saman Zahoor[1], Nadeem Javaid[1(✉)], Anila Yasmeen[1], Isra Shafi[2], Asif Khan[1], and Zahoor Ali Khan[3]

[1] Department of Computer Science, COMSATS Institute of Information Technology, Islamabad 44000, Pakistan
nadeemjavaidqau@gmail.com
[2] Abasyn University, Islamabad Campus, Islamabad 44000, Pakistan
[3] CIS, Higher Colleges of Technology, Fujairah Campus, Fujairah 4114, UAE
http://www.njavaid.com

Abstract. In smart grid, Demand Side Management (DSM) plays a vital role in dealing with consumer's demand and making communication efficient. DSM not only reduces electricity cost but also increases the stability of the grid. In this regard, we introduce an energy management system model for a home and office, then propose efficient scheduling techniques for power usage in both. This system schedule the appliances on the basis of four different optimization techniques to achieve objectives that are electricity cost minimization, reduction in Peak to Average Ratio and energy consumption management. Moreover, we use Real Time Pricing because it is highly flexible and provides an understanding to consumer about price signal variations. Simulation results show that the proposed model for energy management work efficiently to achieve the objectives and provide cost-effective solution to increase the stability of smart grid.

Keywords: Energy management system · Real time pricing
Smart grid · Demand response · Appliances · Optimization techniques

1 Introduction

Traditional grid is not enough to meet the requirements of power grid such as reliability, robustness, security and stability [1]. Therefore, a new infrastructure is introduced that deals smartly with challenges and reduces the pressure of global environment is known as smart grid. It is the way of advance communication between utility and consumer. To meet the increasing demand of consumers the concept of zero energy buildings, Micro Grids (MGs) with Energy Storage Systems (ESS) is growing. MGs are low voltage energy systems which consist of different distributed energy resources and manageable loads, that can be operated in grid connected mode or islanded mode [2]. MG distributed energy resources are categorized into Renewable Energy Sources (RES) and Non RES (NRES).

L. Barolli et al. (Eds.): EIDWT 2018, LNDECT 17, pp. 237–246, 2018.
https://doi.org/10.1007/978-3-319-75928-9_21

However, for demand management a strategy known as DSM is used. In DSM, Demand Response (DR) refers to an opportunity for consumers to play a vital role in the main utility operations by decreasing or shifting their electricity usage patterns in response to market price signals or incentives. It has changed the previous concepts, in which end users do not have options to participate in the operations of energy systems.

Initial research on DR programs was limited, then it is considered in research as in [3] and many other studies. In this era, most of researchers considered only residential area and few of them considered optimization of commercial and industrial area. Scheduling of loads can achieve demand and supply energy balance by changing the load patterns. It increases energy efficiency and reduces burden of grid. Hence, it is a real consequence to continue the optimal scheduling of appliances for smart (home, office or industry). As mentioned above, researchers identified the significances of good scheduling. Many policies and strategies are available for optimal scheduling of household appliances and some of them for office appliances. Moreover, reduction in PAR is achieved by consumer in reaction to incentives defined by utility. When utility react in DR (Time of Use (TOU), Real Time Pricing (RTP)), it defines incentives for consumers. The problem is that the consumer may not act in this situation due to lack of knowledge or laziness or understanding. So, in response to incentives provided by utility there is a need of highly intelligent automatic systems to take energy aware decisions. Such systems need to modify energy consumption patterns of appliances. To tackle this issue there is a need of optimization and scheduling techniques.

In state of the art work, already existing Energy Management Systems (EMSs) as in [4–6], uses different optimization techniques for efficient usage of energy. The control parameters of our work are same as in literature. The selected control parameters are Length of Operational Time (LOT), Power Rating (PR) of appliances, Operational Time Interval (OTI) (start time and end time interval) and price signal of each interval. In addition to existing work, we select two areas, i.e., residential area and commercial area. In both areas, we focus on single office and single home scheduling. To optimize the system we use bio-inspired Genetic Algorithm (GA) and swarm inspired Artificial Bee Colony (ABC) optimization algorithm. By using the above techniques our major contributions are as follows: energy management of residential and commercial area, reduction in PAR and electricity cost minimization.

The rest of this paper is organized as follows: Sect. 2 presented the state of the art work. Section 3 proposes the energy management system model. Then problem formulation is discussed in Sect. 4. In Sect. 5 simulation results are discussed in detail. Finally, conclusions are drawn in Sect. 6.

2 State of the Art Work

A lot of research has been conducted in the area of EMS using different optimization techniques. In this regard, few papers are discussed below.

In paper [6], a heuristic methodology is used called as Signaled Particle Swarm Optimization (SPSO) can be applied to a large number of end users. In this work, a comparison is performed between SPSO, PSO, mixed integer non linear programming and evolutionary PSO revealed the superiority of the new scheme in terms of better robustness, cost, time execution, absolute error and faster convergence. The authors in [7], defined a mechanism known as optimization power scheduling scheme where price of electricity is predefined to implement the demand side response in residential unit. As a result, there is a trade off between the cost and comfort using the defined strategy. In study [8], authors presented the PSO methodology that can be applied to many energy resources like, distributed generators, DR, etc. for minimization of cost. The method used the Gaussian mutation of strategic parameters and according to the context self parameterization of minimum and maximum velocities. An efficient DSM system for residential energy management in order to reduce peak formations, electricity cost and maximize the user comfort is presented in [9]. For optimization authors used the Ant Colony Optimization (ACO) and TOU with Inclined Block Rate (IBR) price tariff.

Authors in [10], a bio-inspired optimization algorithm is proposed to solve the energy optimization problem on demand side. The optimization algorithm none-dominated sorting (NSGA-II) is the variation of GA. The objective of this paper is cost minimization and PAR reduction. Authors in [11], presented an expert energy management system to optimize the operations of MG for reduction of operating and maintenance cost. The Modified Bacterial Foraging Optimization (MBFO) is used for scheduling of distributed generators. This work also checked the distributed energy resources and ESS impact on economy and environment. DSM provides many benefits to smart grid, especially at distributed network. In paper [12], a strategy for load shifting is proposed, which is formulated as a minimization problem. The problem is solved by heuristic based evolutionary algorithm. The main objective of the given study is to reduce the peak load demand and operational cost of the smart grid, which is achieved by using the above mentioned algorithm. The authors of [13], elaborate a fully DSM system to reduce the peak demand of residential consumers. Dynamic pricing approach is used. This system based on game theory where appliances are act as a player. In domestic sector, an Artificial Neural Network/GA (ANN-GA) approach for appliance scheduling is presented in [14]. Where objective of the authors are to reduce the peak load demand and maximize the use of renewable resources (PV and WT) instead of utility.

In [15], an optimal scheduling model of home appliances for energy management in terms of DR is considered. The objective of this is electricity cost reduction and user comfort maximization by using TOU pricing scheme. This study is conducted for one day and the working DR application of one house hold. There is a trade off between the cost and comfort. DR program is used for shifting the peak load into off peak load. In [16], a method is defined for residential area DR using hybrid Lighting Search Algorithm (LSA)-based ANN to find the optimum status of appliances (ON/OFF). The results of hybrid LSA-ANN is compared with hybrid of PSO-ANN. The results of hybrid LSA-ANN is better as compared to hybrid PSO-ANN.

3 System Model

In smart grid, an EMS is used for efficient utilization of energy. Mainly EMS contains energy management controllers, smart meters, smart appliances and advance communication systems. In this paper, we work on DSM. It is also known as the DR or Demand Side Response (DSR). By using DR, we do energy management efficiently. Here, we consider different areas to make them smart. These areas are Commercial Areas (CAs), Residential Areas (RAs) and Industrial Areas (IA). Here, for energy management we consider the RA and CA which is shown in Fig. 1. Both office and home are managed by EMS. In Fig. 1, we represent a office, utility, a MG, sub-utility of CA and RA and a single home. A smart office/home is comprised of smart appliances, Energy Management Smart Controller (EMSC), smart meter and HAN. Each office/home has its own EMSC and appliances. The components of office works same as home except appliances and usage patterns. Sub-utility acts as a intermediary between main utility, MG and smart meters for efficient utilization of energy. The EMSC, smart meter and sub-utility are capable of exchanging their information with each other by using WAN. A smart meter is usually installed outside of each building between the sub utility and EMSC. The smart meter reads and processes the energy consumption data to be transfered to the sub-utility and concurrently send the DR signal to the EMSC for more analysis. In this paper, we follow the appliance classification from paper [3] while for offices we select appliances from [17]. There are two types of appliances: Manually Operated Appliances (MOAs) and Automatically Operated Appliances (AOAs). The MOAs can operate only if the consumer is using it manually. MOAs can be switched ON or OFF only in the presence of consumer, due to which, we can only schedule the AOAs. Here, in a smart home nine type of AOAs, while in a office eight AOAs are used. These AOAs cannot operate manually, it will be operated on its own. AOAs are only capable of interacting with controller they do not interact with each other. The EMSC act as a central window for consumers to control all smart AOAs in a smart home/office. All the operations of smart appliances needed to be scheduled at the beginning of the day by EMSC. For home we perform scheduling for full day, while for office we perform scheduling for day time only according to its usage pattern. The set of appliances either it is office/home appliances is denoted by A. Let a denotes a single appliance, so, $a \in A = \{a_1, a_2, ..., a_n\}$. We assume, a scheduling vector of energy consumption E_a^s, that is

$$E_a^s = \{E_a^1, E_a^2, ...E_a^{120}\} \tag{1}$$

Where $\{E_a^1, E_a^2, ...E_a^{120}\}$ represents energy consumption of an appliance for 120 time slots. Each slot has time resolution of twelve minutes. The total energy consumption for a day E_T for all the appliances is calculated as:

$$E_T = \sum_{s=1}^{S} \sum_{i=1}^{A} E_i^s \tag{2}$$

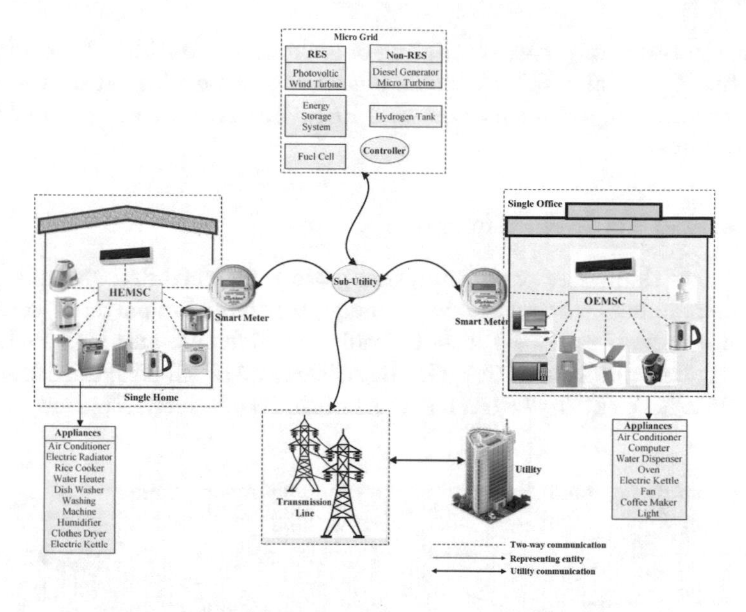

Fig. 1. Residential area system model

4 Problem Formulation

In this paper, our main objectives are electricity cost minimization and PAR reduction. The electricity cost is minimized by optimizing the patterns of energy consumption of consumers and maintain balance between demand and supply. We minimize cost by scheduling the peak loads into off-peak loads. The objective function to minimize electricity cost for single home/office is written as:

$$min \sum_{s=1}^{S} \sum_{a=1}^{A} \rho_a^s \times \chi_a^s \times \zeta^s \tag{3}$$

Equation 3 represents our objective function. In given equation, the power rating ρ of an appliance in slots s is multiplied with the electricity price ζ and appliance status χ at time slot s. The main concern of Eq. 3 is to minimize the obtained value. A denotes the set of smart home/office appliances. PAR is calculated as:

$$PAR = \frac{max(E_T)}{mean(E_T)} \tag{4}$$

PAR of a home/office is calculated as: maximum value of energy consumption vector divided by mean of energy consumption vector.

5 Simulation Results

Extensive simulations have been performed in MATLAB considering the predefined objectives. Here, we evaluate the performance of different optimization techniques in terms of power scheduling of appliances, considered techniques are ABC, ABC-R, GA and GA-R. These algorithms are evaluated on the basis of power consumption, electricity cost and PAR. Simulations are conducted for a home and office.

5.1 Scenario 1: Single Home

Figure 2 shows daily energy consumption pattern of scheduled and unscheduled load for a single home. In unscheduled energy consumption patterns, user activities are high in 1–27 and 73–106 slots leading to high cost and high PAR. Maximum energy consumption of GA, GA-R, ABC and ABC-R scheduled load is limited to 4.2 kWh, 3.9 kWh, 2.9 kWh and 2.1 kWh, respectively. Figure 3, shows the

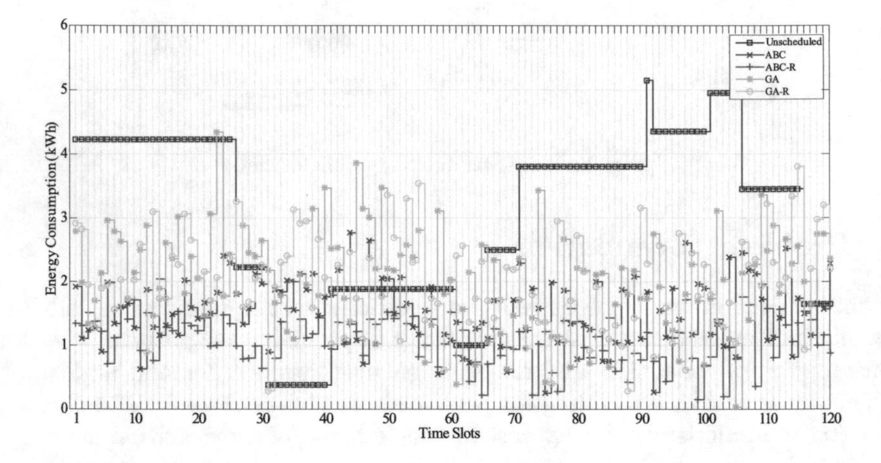

Fig. 2. Energy consumption of single home

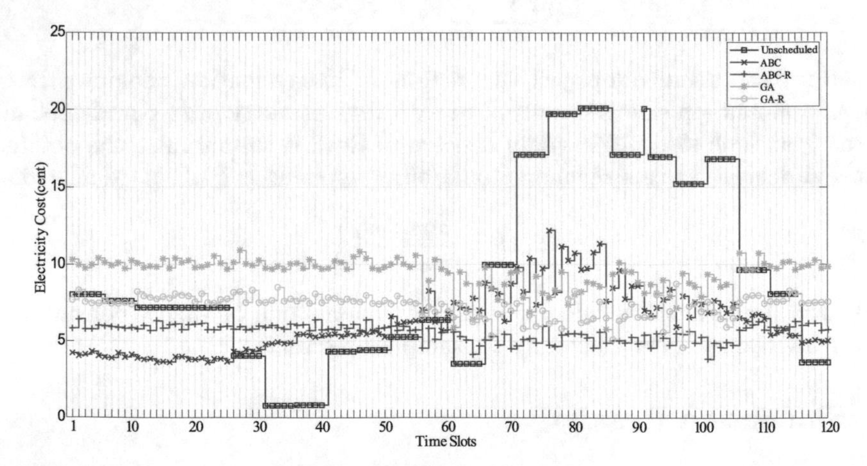

Fig. 3. Electricity cost of single home

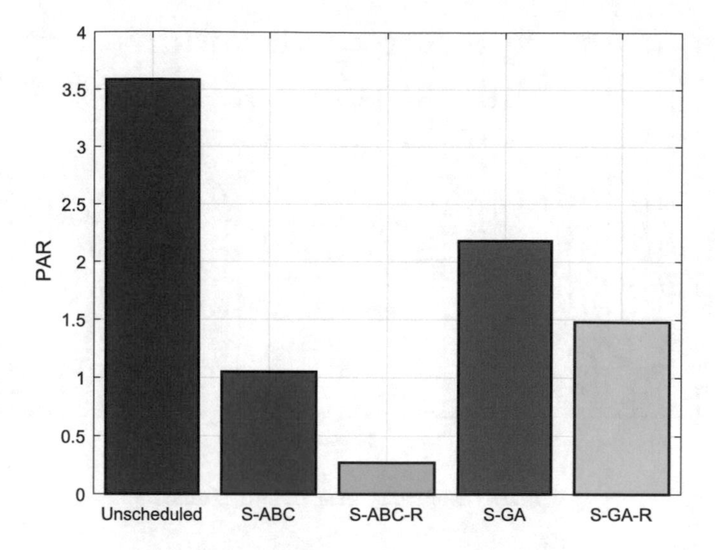

Fig. 4. PAR of single home

scheduled and unscheduled cost for one day. In unscheduled case the highest electricity cost is 20 cents and in scheduled case maximum cost of ABC, ABC-R, GA and GA-R is 11.5 cents, 8 cents, 12.3 cents and 7 cents, respectively. It is concluded that the cost of scheduled with respect to unscheduled is reduced. By shifting of load from on peak to off peak slots, PAR is also decreased as shown in Fig. 4.

5.2 Scenario 2: Single Office

Office consumers fulfill their requirements by efficient scheduling, they cannot shift their load from morning to evening or day to night. However, they can achieve their objectives by using different optimization techniques. Figure 5, shows daily

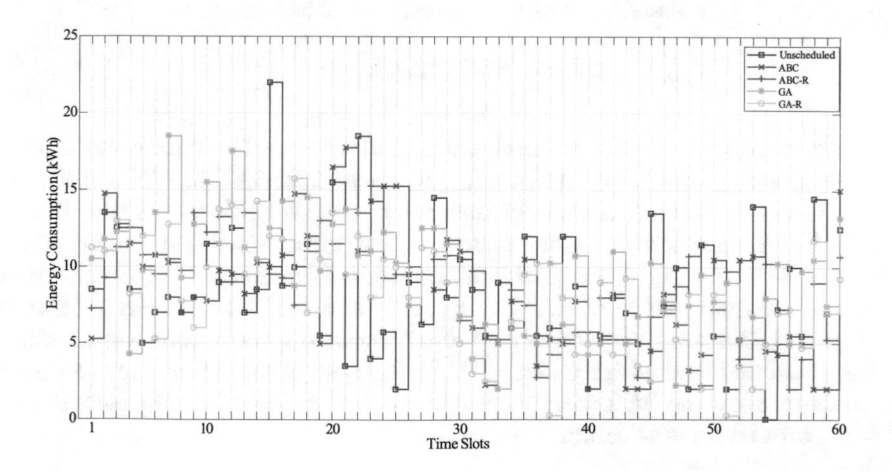

Fig. 5. Energy consumption of single office

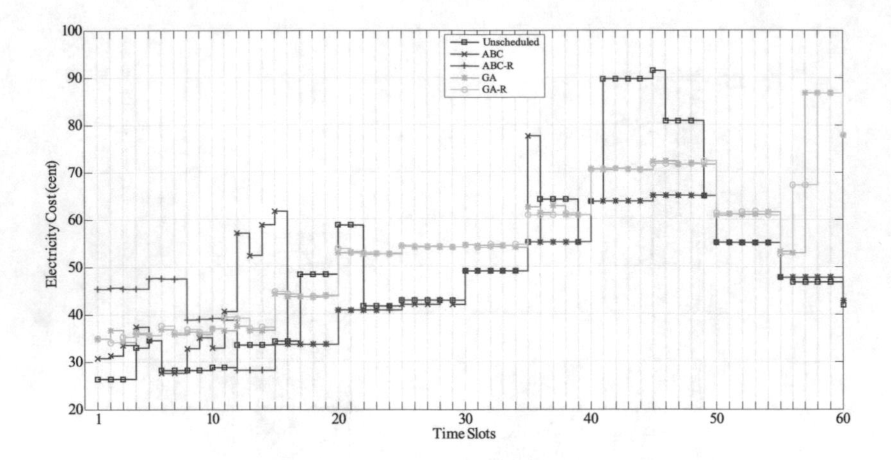

Fig. 6. Electricity cost of single office

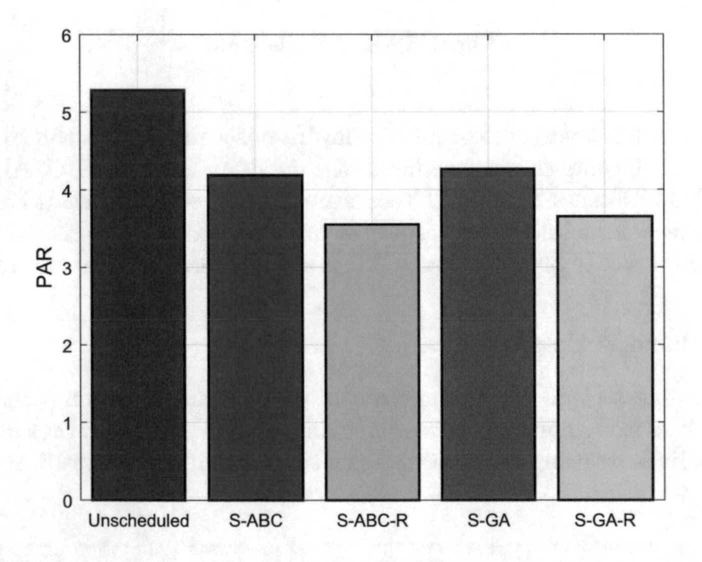

Fig. 7. PAR of single office

energy consumption pattern of unscheduled and scheduled load for a single office. Maximum energy consumption of unscheduled load is 22.5 kWh. While in the case of GA, GA-R, ABC and ABC-R the maximum value is 19 kWh, 16 kWh, 18 kWh and 15.2 kWh, respectively. Comparison of cost for unscheduled and scheduled energy is shown in Fig. 6. In unscheduled case, maximum cost is 91 cents. However, the maximum value of GA, GA-R, ABC and ABC-R is 88 cents, 88 cents, 79 cents and 62.5 cents, respectively. Figure 7, shows the performance of scheduled and unscheduled load with respect to PAR. Here, we observe that each technique is capable of reducing PAR with respect to unscheduled load but the performance of ABC with roulette selection is better.

6 Conclusion

In this work, we first introduce the EMS model for office and home and then presented approaches for power scheduling of both with respect to consumer preferences and RTP signal. The worthwhile parameters obtained by applying our proposed approaches are the minimization of electricity cost, management of energy consumption and PAR reduction for consumers. Furthermore, the benefit rewarded to utility companies is the PAR reduction, which increases the stability of whole utility system. By using RTP, our approach achieve all performance parameters not only for consumers but also for utility. From simulation results, it can be concluded that for scheduling our proposed approach is efficient in terms of cost saving which is the preference of consumer. In future, we can check the efficiency of any other optimization technique by applying on the same scenario in smart grid.

References

1. Gungor, V.C., Sahin, D., Kocak, T., Ergut, S., Buccella, C., Cecati, C., Hancke, G.P.: Smart grid technologies: communication technologies and standards. IEEE Trans. Ind. Inf. **7**(4), 529–539 (2011)
2. Shi, W., Xie, X., Chu, C.-C., Gadh, R.: A distributed optimal energy management strategy for microgrids. In: 2014 IEEE International Conference on Smart Grid Communications (SmartGridComm), pp. 200–205. IEEE (2014)
3. Zhao, Z., Lee, W.C., Shin, Y., Song, K.-B.: An optimal power scheduling method for demand response in home energy management system. IEEE Trans. Smart Grid **4**(3), 1391–1400 (2013)
4. Ju, L., Li, H., Zhao, J., Chen, K., Tan, Q., Tan, Z.: Multi-objective stochastic scheduling optimization model for connecting a virtual power plant to wind-photovoltaic-electric vehicles considering uncertainties and demand response. Energy Convers. Manage. **128**, 160–177 (2016)
5. Mary, G.A., Rajarajeswari, R.: Smart grid cost optimization using genetic algorithm. Int. J. Res. Eng. Technol. **3**(07), 282–287 (2014)
6. Soares, J., Silva, M., Sousa, T., Vale, Z., Morais, H.: Distributed energy resource short-term scheduling using Signaled Particle Swarm Optimization. Energy **42**(1), 466–476 (2012)
7. Ma, K., Yao, T., Yang, J., Guan, X.: Residential power scheduling for demand response in smart grid. Int. J. Electr. Power Energy Syst. **78**, 320–325 (2016)
8. Faria, P., Soares, J., Vale, Z., Morais, H., Sousa, T.: Modified particle swarm optimization applied to integrated demand response and DG resources scheduling. IEEE Trans. Smart Grid **4**(1), 606–616 (2013)
9. Rahim, S., Iqbal, Z., Shaheen, N., Khan, Z.A., Qasim, U., Khan, S.A., Javaid, N.: Ant colony optimization based energy management controller for smart grid. In: 2016 IEEE 30th International Conference on Advanced Information Networking and Applications (AINA), pp. 1154–1159. IEEE (2016)
10. Mehrshad, M., Tafti, A.D., Effatnejad, R.: Demand-side management in the smart grid based on energy consumption scheduling by NSGA-II. Int. J. Eng. Pract. Res. **2**, 197–200 (2013)

11. Motevasel, M., Seifi, A.R.: Expert energy management of a micro-grid considering wind energy uncertainty. Energy Convers. Manage. **83**, 58–72 (2014)
12. Logenthiran, T., Srinivasan, D., Shun, T.Z.: Demand side management in smart grid using heuristic optimization. IEEE Trans. Smart Grid **3**(3), 1244–1252 (2012)
13. Barbato, A., Capone, A., Chen, L., Martignon, F., Paris, S.: A distributed demand-side management framework for the smart grid. Comput. Commun. **57**, 13–24 (2015)
14. Yuce, B., Rezgui, Y., Mourshed, M.: ANN-GA smart appliance scheduling for optimised energy management in the domestic sector. Energy Build. **111**, 311–325 (2016)
15. Lu, X., Zhou, K., Chan, F.T.S., Yang, S.: Optimal scheduling of household appliances for smart home energy management considering demand response. Nat. Hazards **88**, 1–15 (2017)
16. Ahmed, M.S., Mohamed, A., Homod, R.Z., Shareef, H.: Hybrid LSA-ANN based home energy management scheduling controller for residential demand response strategy. Energies **9**(9), 716 (2016)
17. Awais, M., Javaid, N., Shaheen, N., Iqbal, Z., Rehman, G., Muhammad, K., Ahmad, I.: An efficient genetic algorithm based demand side management scheme for smart grid. In: 2015 18th International Conference on Network-Based Information Systems (NBiS), pp. 351–356. IEEE (2015)

Energy Balanced Load Distribution Through Energy Gradation in UWSNs

Ghazanfar Latif[1], Nadeem Javaid[1(✉)], Aasma Khan[1], Faisal Hayat[2], Umar Rasheed[2], and Zahoor Ali Khan[3]

[1] COMSATS Institute of Information Technology, Islamabad 44000, Pakistan
nadeemjavaidqau@gmail.com
[2] University of Lahore, Islamabad 44000, Pakistan
[3] Computer Information Science, Higher Colleges of Technology, Fujairah, UAE
http://www.njavaid.com

Abstract. Underwater wireless sensor networks (UWSNs) find applications in various aspect of life like the tsunami and earthquake monitoring, pollution monitoring, ocean surveillance for defense strategies, seismic monitoring, equipment monitoring etc. The sensor node consumes more energy and load distribution suffer from imbalance at long distance. In this paper, we present an energy balanced load distribution through energy gradation (EBLOAD-EG) technique to minimize the energy consumption in direct transmission. The proposed scheme aims to balance the load distribution among different coronas of network field. In this scheme, the numbers of sensor nodes are uniformly deployed in a circular network field and the sink is located at the center of network field. In EBLOAD-EG, the accumulated data is partitioned into data fractions like small, medium and large. Simulation results show that our scheme outperforms the existing scheme in terms of energy efficiency, balanced load distribution, stability period and network lifetime.

1 Introduction

Recently, researchers have been more interested in the field of UWSNs. In UWSNs, there are different types of application like tsunami and earthquake monitoring, pollution monitoring, ocean surveillance for defense strategies, seismic monitoring, equipment monitoring etc. Various UWSNs devices communicate with themselves in these application areas. The routing protocols designed for land-based communication cannot work properly in UWSNs due to the peculiarity of that environment. Efficient routing of UWSNs in the presence of long propagation delay, high attenuation error where radio signal does not propagate well in water is almost impossible. Acoustic signals are used for acoustic communication.

1. The speed of acoustic signal is 1500 m/s while the speed of radio signal is 300,000,000 m/s.
2. High bit error rate (BER) is due to the multi-path fading, path loss, and transmission loss.

L. Barolli et al. (Eds.): EIDWT 2018, LNDECT 17, pp. 247–257, 2018.
https://doi.org/10.1007/978-3-319-75928-9_22

There are two types of data transmissions used in UWSNs. These are direct transmission and multi-hop transmissions. In Bload scheme [3], nodes can balance their energy expenditure by sending some fractions of data through dtx i.e. directly send data to sink and remaining fractions of data sent by multi-hop transmission. Data is divided into three fractions like small, medium and large. These fractions are then transmitted in hop-by-hop and direct transmission to the sink which is located at the water surface. All sensor nodes of the network act as one hop away neighbor of the sink due to direct transmission range (dtx). Thus, sensor nodes closest to the sink have less of relaying data which avoids the energy hole problem. The proposed routing protocol is used to minimize the energy consumption and balances the load of each sensor node in the network. Our aim is to balance the energy consumption in direct transmission using energy gradation technique. Our proposed scheme also intends to improve the network lifetime, energy efficiency and also avoid the energy hole problem near the sink. The rest of the paper is organized as follows. Related work is discussed in Sect. 2. The problem statement of our system is given in Sect. 3. System model and proposed scheme EBLOAD-EG is discussed in Sect. 4. Simulation and Results are presented in Sect. 5 and Conclusion is in Sect. 6 and finally, references are listed.

2 Related Work

In this section, we explain some related work on routing protocols with respect to feature, achievement, and limitation.

In [4], the authors proposed a spherical hole repair technique (SHORT) for overcoming the coverage and energy hole problem in UWSNs. Their scheme is made up of three phases like knowledge sharing phase (KSP), network operation phase (NOP) and hole repair phase (HRP). This protocol achieved better performance of energy consumption and network lifetime. This is a trade-off of the greater end-to-end delay.

Haitao et al., proposed a weighting depth and forwarding area division depth based routing protocol (WDFAD-DBR). The selection of forwarder node is based on the two-hop neighbor. This protocol achieved improvement in the reliability of communication, increased the packet delivery ratio and decreased the end-to-end delay due to the increase of node density in the network. However, the manufacturer deployment and cost is high [5].

Zidi et al., proposed a balanced routing (BR) protocol for underwater communication. The goal is to minimize the energy hole problem and maximize the network lifetime. This protocol achieves higher reliability. However, the load distribution is an imbalance [6].

Zonglin et al., proposed a relative distance based forwarding protocol (RDBF). In this protocol, a suitable forwarder node exists, which work with a fitness function. The forwarder nodes are those nodes having the minimum. This paper achieved the better packet delivery ratio, an end-to-end delay, and energy efficiency. However, due to the minimum hop count, the load imbalance is present [7] (Table 1).

Table 1. Comparison of different routing protocols

Techniques	Features	Achievements	Limitations
SHORT [4]	Coverage hole and energy hole repair, hop-by-hop transmission	Higher throughput, hole recovered area, coverage area	Greater end-to-end delay, higher energy consumption
WDFAD-DBR [5]	Weighting depth and forwarding area division, minimized energy consumption	Void hole reduces, minimized energy consumption	Manufacturer deployment and cost is high
BR [6]	Two hop transmission using 2r. To minimize the energy sink hole issue	Prevents energy sink hole problem at the one-hop neighbor. Increases network lifetime	Due to corona 2 act as one hop neighbor the total load is imbalanced
RDBF [7]	Relative distance based forwarding	Energy efficiency, minimize end to end delay	Due to small hop count the load is imbalance, and distance is small from sink
ARCR [8]	Clustering, relay node and mobile sink	Network lifetime, energy balances, load balances	Higher end-to-end delay
GEDAR [9]	Geographic and opportunistic routing, forwarder node based on Greedy forwarding scheme	Void node decreases, better network performance, minimizes collision	Higher energy consumption due to the lower number of node density
HMR-LEACH [10]	Hierarchical multi-path routing	Balances energy consumption, network lifetime	Higher computational complexity
DBR [11]	Hop-by-hop transmission based on depth	Minimizes energy consumption, high throughput	Higher end to end delay
EBH [12]	Hybrid transmission for balanced energy consumption	Higher network lifetime, energy efficiency	Inefficient for dense network and non-linear network
BTM [13]	One hop and multi-hop transmission based on energy level (EL)	Network lifetime, energy efficiency	Energy consumption due to long distance, loop formation
EEBET [14]	To resolve problem in BTM. Energy balances at long distance	Increases network lifetime, higher throughput and energy efficiency	Due to increasing the network radius to minimize the network performance
L2-ABF [15]	Location free routing, angle based flooding	Higher packet delivery ratio, energy efficiency	Higher end-to-end delay due to greater number of layer
AHH-VBF [16]	End-to-end delay, network lifetime	Reliability, minimized energy consumption, throughput	Due to increasing pipeline radius the void hole occur

In [8], the authors proposed an adaptive relay chain routing (ARCR) protocol. Major objectives were to avoid the energy hole problem and also prevent the localization problem. Their paper achieved the load balancing with the motion of mobile relay nodes, energy efficiency and network lifetime. However, the relay node can not transmit the data packet until the network is in connection mode.

In [9], the authors proposed a geographic and opportunistic routing for UWSNs (GEDAR). This paper describes the selection of forwarder node on the basis of greedy forwarding and opportunistic routing for data forwarding. This paper achieved avoidance of the void hole using depth adjustment technique. The limitation of this protocol is that the energy consumption is higher due to the lower node density.

In [10], the authors proposed a hierarchical multipath routing LEACH (HMR-LEACH) routing protocol. This paper considered the multi-hop routing scheme except for the direct transmission. This protocol selects the transmission path based on energy and transmission distance. The data packet sends to the sink and assigns the weight corresponding transmission path. Each transmission path based on its probability achieves improved the network lifetime and balances energy consumption.

In [11], the authors proposed a depth based routing (DBR) protocol, it is also known as the localization free protocol. In a dense network, DBR selects the forwarder node with the greedy forwarding approach. This protocol achieves high data delivery ratio for the dense network. However, in the sparse network, the packet delivery ratio is relatively low.

In [12], the authors proposed a balanced hybrid (EBH) and differential initial battery (DIB) which describe the balanced energy consumption per node in the sparse network. This paper achieved improvement in the network lifetime of UWSNs. However, both techniques like EBH and DIB are inefficient for the dense environment. It is also inefficient for a non-linear network.

Jiabao et al., proposed a balanced transmission mechanism (BTM) which is called the hybrid mechanism and two-dimensional network model. This protocol achieves the balance load distribution among sensors through partitioning the energy of sensor nodes into energy levels. When a node consumes higher energy in multi-hop then sensor node communicates directly to the sink. The limitation of this protocol is higher energy consumption and also the energy consumption is no balanced among sensor nodes [13].

Javaid et al., proposed a routing protocol called enhanced efficient and balancing energy technique (EEBET) to overcome the issues in BTM. The efficient and balancing energy consumption technique (EBET) achieves the improvement of energy efficiency and also calculate the appropriate energy level to increase the network lifetime. The energy hole problem is not resolved by this routing protocol [14].

In [15], the authors proposed a routing protocol like layer by layer angle based flooding (L2-ABF) and its deployment in the form of layers and calculate the depth of the sensor. This protocol is a localization free protocol, where all sensor nodes transmit data to sink through calculated the flooding angle. This paper achieved lower energy consumption and higher packet delivery ratio.

In [16], the authors proposed an adaptive hop-by-hop vector based forwarding routing protocol (AHH-VBF) based on the hop-by-hop vector based forwarding routing protocol (HH-VBF). This paper achieved the energy efficiency and reliability in a sparse network. However, in a dense network, the number of duplicate packets are effectively reduced because the pipeline radius adaptively changes hop-by-hop to restrict the forwarding range of packet. The limitation of this protocol is in increasing the pipeline radius which causes the void hole problem in the network. The manufacturing and deployment cost is maximum (Table 2).

Table 2. Performance parameters achieved

Performance parameters	BLOAD	EBLOAD-EG	Percentage achievement in EBLOAD-EG
Packet load distribution (%)	10	7	30
Energy consumption (%)	0.01	0.003	70
Number of dead nodes (%)	45	35	22.22
Number of alive nodes (%)	5	15	66.66

3 Problem Statement

The higher energy consumption and energy hole problem degrade the performance of the network in terms of load distribution, network lifetime etc. This problem is resolved by mixed routing transmission among various coronas. In this paper, we are mitigating the energy consumption, energy hole problem and manages the balanced load distribution. The sensor nodes are transmitting data via directly and multi-hop transmissions. Directly transmission is an important way for sending data to sink if transmission range is larger to reach the sink then there will require much energy consumption for data transmission. In BLOAD scheme [3], there are three transmission ranges for data forwarding among coronas; one hop, two hops, and direct transmission. The reason behind these scenarios, the total load, and energy consumption is balanced but in direct transmission, much energy is consumed due to long distance. In this process, the total load is minimized by the distribution of total data among transmission ranges. However, the energy consumption in direct transmission, which is a long distance from sink increases because the sensor nodes of direct transmission become away from sink using transmission range dtx. This leads to maximum energy consumption and load imbalance is inevitable long distance. So, balanced load distribution is necessary to solve balanced energy consumption and to also avoid energy hole problem.

4 Proposed Scheme

In the proposed scheme, the number of sensor nodes is uniformly deployed in a circular network area around the number of coronas with equal width r. The detail of network model and proposed scheme is as given below:

4.1 Network Model

In UWSNs, mostly the sensor nodes are sparsely deployed. Our proposed system, consider a two-dimensional model where the number of sensor nodes is anchored at the bottom of water can move horizontally up to certain limits. Our system considers a circular monitoring area of radius R and a sink located at the center of the circular area. The network field is partitioned into the number of coronas of equal width r as shown in Fig. 1. Each corona contains the same number of sensor nodes have some distance from sink. Each corona node is responsible for transmitting its data using three transmission ranges (r, 2r, dtx) based on energy gradation. The distribution of transmission is necessary for balancing and minimizing the energy consumption. The data transmission among sensor nodes based on energy gradation is explained as given below.

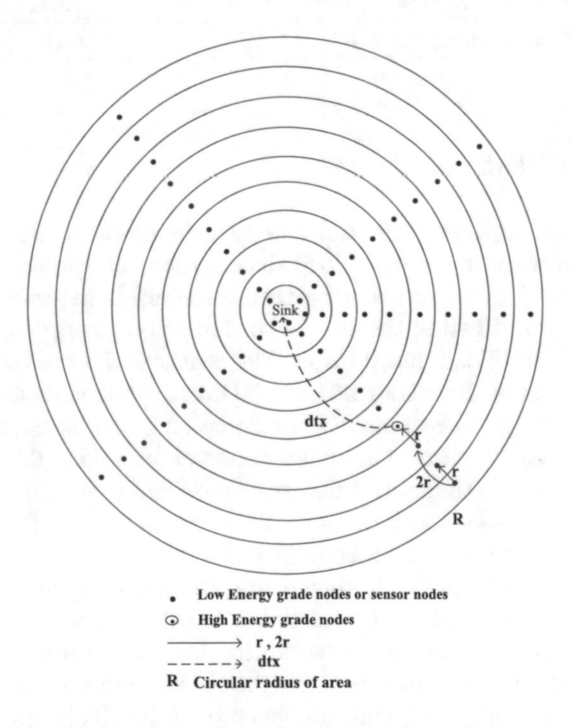

Fig. 1. Network model of proposed scheme

4.2 Energy Gradation

We propose an energy gradation scheme to achieve the balanced energy consumption among sensor nodes. The data prevents are transmitted via direct and multi-hop using transmission power P. To utilize the energy optimally, the battery power is divided into segments m. The mechanism is defined that node deployed near to the sink consume one unit of energy and in multi-hoping, the dissipation depends on a number of hops. However, determining the suitable energy gradation of a node is necessitating to balanced energy consumption among nodes. Now, we evaluate the appropriate grades m among nodes. If the grade number of the first node is greater than the other grade number of the node then the first node is directly transmitted to sink. However, if energy grade is small then they transmit to hop by hop using transmission ranges (r, 2r) to sink. By this process, the nodes dissipate energy evenly, so network lifetime, stability period and energy efficiency can be increased.

5 Simulation Results

In this section, we evaluate the performance of our protocol and compare the results with the existing protocol in terms of energy balancing of the network. The numbers of sensor nodes are uniformly deployed in a circular network area with equal width $r = 100$ m. We assume that the sink is located at the center of the network field. The initial energy of sensor node is set to be E$o = 1$ J, $Etx = 0.0005$ J and $Erx = 0.00005$ J. The number of coronas is set to be $C = 10$ and circular monitor area is 1000 m respectively. The control parameters used in this paper is given below:

In performance evaluation, we assume that the following metrics are given below [3]:

5.1 Performance Metrics

1. **Packet Load Distribution:** The division of total data at each sensor node using three transmission ranges (r, 2r, dtx).
2. **First Node Dead Time (FNDT):** The time starts from network establishment to the time when the first node dead is known as FNDT. It is also called stability period.
3. **All Nodes Dead Time (ANDT):** The time starts from FNDT to all the sensor nodes are dead which is known as ANDT.
4. **Network Lifetime:** The time starts from initializes the network up to ANDT or network is completely dead. It is also known as network lifetime.

We evaluate the following results Fig. 2 depicts the packet load distribution per corona. According to Fig. 2, our proposed protocol performs better than BLOAD. The reason is that we are dividing the initial energy among nodes by energy gradation scheme and compares the energy of nodes for data transmission from the source node to sink. Our proposed protocol not only minimized the load but also balanced the packet load among all the coronas. Our scheme increased 50% in Corona 9 and 30% in corona 10 respectively. Overall, our proposed protocol achieved 55% better performance than BLOAD scheme.

In the proposed scheme, the balanced load distribution and also to improve the energy efficiency of each corona which is better than the existing scheme. However, the balanced load distribution and energy consumption on each corona are as shown in Figs. 2 and 3. This is because the total data directly transmits to sink using energy gradation scheme (Table 3).

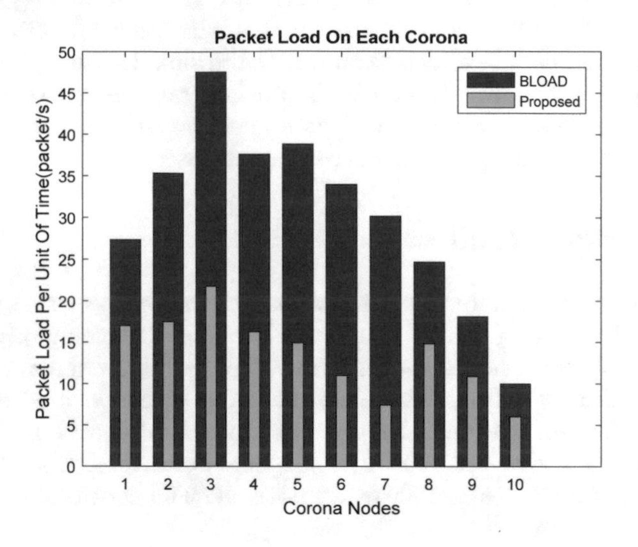

Fig. 2. Load distribution per corona

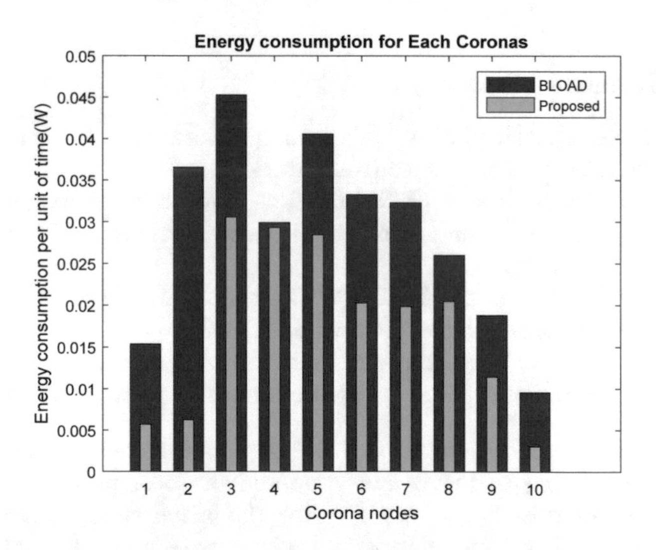

Fig. 3. Energy consumption of different scheme

Table 3. Simulation parameters

Control parameters	Values	Units
Initial energy (E*o*)	1	Joules
Transmission energy (Etx)	0.0005	Joules
Reception energy (Erx)	0.00005	Joules
Circular area (A)	1000	Meters
Radius of circular area (r)	100	Meters

Figure 3 depicts the energy consumption per corona. The performance of our proposed protocol is better than the BLOAD because data transmission of the sensor node is based upon energy gradation. As energy consumption does not depend on packet load and energy consumption is minimized due to energy gradation in data transmissions. However, our protocol consumes only 0.005 energy in the first corona and 0.006 energy in the second corona so achieved 67% in the first corona and 83.33% in the second corona and so on. Also, our protocol achieved 37% in nine corona and 70% in the last corona respectively. Our proposed protocol achieved 45% better performance than BLOAD scheme.

Figure 4 depicts the number of dead nodes per unit time. The proposed protocol performs better than BLOAD. The reason is that we are dividing the initial energy among nodes by energy gradation scheme and compares the energy of nodes for data transmission from the source node to sink. By the process of energy gradation, less number of nodes are dead than BLOAD. In our proposed protocol, simulation results showed that at 100 time (s) all the nodes of BLOAD

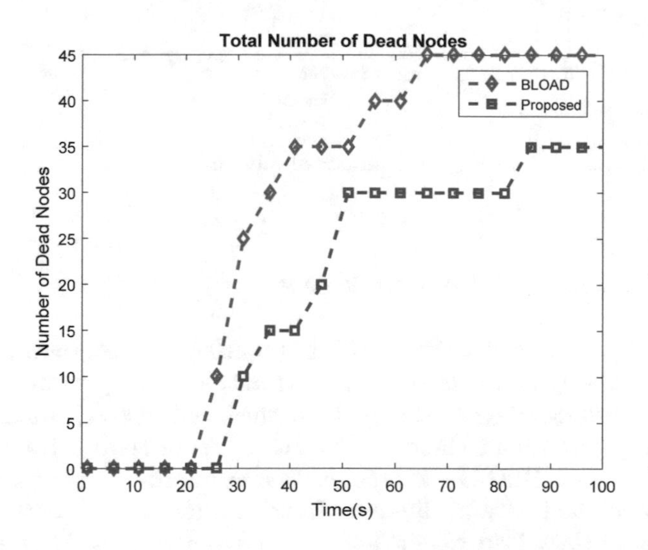

Fig. 4. Number of dead nodes

are dead while in the proposed protocol the number of dead nodes is 35 and at last 10 nodes are alive. Our proposed protocol achieved 32% better performance than BLOAD scheme.

Figure 5 depicts the number of alive nodes per unit time. The proposed protocol performs better than BLOAD. The reason is that we are dividing the initial energy among nodes by energy gradation scheme and compares the energy of nodes for data transmission from the source node to sink. By the process of energy, gradation number of nodes are alive than BLOAD. In our proposed protocol, simulation results showed that at 100-time number of nodes of BLOAD are alive while in the proposed protocol the number of alive nodes is 15. Our proposed protocol achieved 35% better performance than BLOAD scheme. Using these simulation results, it shows that the number of dead nodes decreases and number of alive nodes increase.

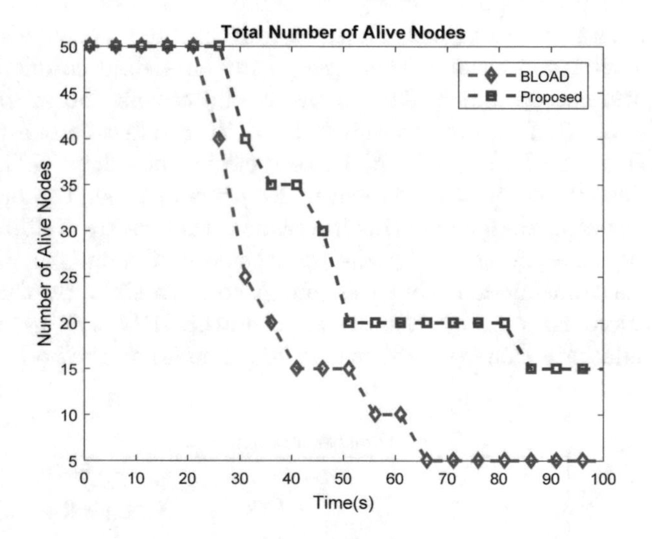

Fig. 5. Number of alive nodes

6 Conclusion and Future Work

In this paper, we proposed an EBLOAD-EG routing protocol towards improving the energy efficiency and load balance distribution among coronas in UWSNs. If the node consumes higher energy then they will directly transmit to sink through energy gradation technique. Simulation results show better performance as compared to the BLOAD technique. It also improves the stability period, energy efficiency and network lifetime. In the future, we will work on different combinations of data load weight and transmission ranges. Moreover, we will plan to solve these major challenges in the UWSNs.

References

1. Akyildiz, I.F., Wang, P., Lin, S.-C.: SoftWater: software-defined networking for next-generation underwater communication systems. Ad Hoc Netw. **46**, 1–11 (2016)
2. Ayaz, M., Abdullah, A.: Underwater wireless sensor networks: routing issues and future challenges. In: Proceedings of the 7th International Conference on Advances in Mobile Computing and Multimedia, pp. 370–375. ACM (2009)
3. Azam, I., Javaid, N., Ahmad, A., Wadood, A., Almogren, A., Alamri, A.: Balanced load distribution with energy hole avoidance in underwater WSNs. IEEE Access **5**, 15206–15221 (2017)
4. Latif, K., Javaid, N., Ahmad, A., Khan, Z.A., Alrajeh, N., Khan, M.I.: On energy hole and coverage hole avoidance in underwater wireless sensor networks. IEEE Sensors J. **16**(11), 4431–4442 (2016)
5. Yu, H., Yao, N., Wang, T., Li, G., Gao, Z., Tan, G.: WDFAD-DBR: weighting depth and forwarding area division DBR routing protocol for UASNs. Ad Hoc Netw. **37**, 256–282 (2016)
6. Zidi, C., Bouabdallah, F., Boutaba, R.: Routing design avoiding energy holes in underwater acoustic sensor networks. Wirel. Commun. Mobile Comput. **16**(14), 2035–2051 (2016)
7. Li, Z., Yao, N., Gao, Q.: Relative distance based forwarding protocol for underwater wireless networks. Int. J. Distrib. Sensor Netw. **10**(2), 173089 (2014)
8. Kong, L., Ma, K., Qiao, B., Guo, X.: Adaptive relay chain routing with load balancing and high energy efficiency. IEEE Sensors J. **16**(14), 5826–5836 (2016)
9. Coutinho, R.W.L., Boukerche, A., Vieira, L.F.M., Loureiro, A.A.F.: Geographic and opportunistic routing for underwater sensor networks. IEEE Trans. Comput. **65**(2), 548–561 (2016)
10. Liu, G., Wei, C.: A new multi-path routing protocol based on cluster for underwater acoustic sensor networks. In: 2011 International Conference on Multimedia Technology (ICMT), pp. 91–94. IEEE (2011)
11. Yan, H., Shi, Z., Cui, J.-H.: DBR: depth-based routing for underwater sensor networks. In: Networking 2008 Ad Hoc and Sensor Networks, Wireless Networks, Next Generation Internet, pp. 72–86 (2008)
12. Luo, H., Guo, Z., Kaishun, W., Hong, F., Feng, Y.: Energy balanced strategies for maximizing the lifetime of sparsely deployed underwater acoustic sensor networks. Sensors **9**(9), 6626–6651 (2009)
13. Cao, J., Dou, J., Dong, S.: Balance transmission mechanism in underwater acoustic sensor networks. Int. J. Distrib. Sensor Netw. **11**, 429340 (2015)
14. Javaid, N., Shah, M., Ahmad, A., Imran, M., Khan, M.I., Vasilakos, A.V.: An enhanced energy balanced data transmission protocol for underwater acoustic sensor networks. Sensors **16**(4), 487 (2016)
15. Ali, T., Jung, L.T., Faye, I.: End-to-end delay and energy efficient routing protocol for underwater wireless sensor networks. Wirel. Pers. Commun. **79**(1), 339–361 (2014)
16. Yu, H., Yao, N., Liu, J.: An adaptive routing protocol in underwater sparse acoustic sensor networks. Ad Hoc Netw. **34**, 121–143 (2015)

Routing Protocol with Minimized Load Distribution for UASNs

Faisal Hayat[1], Nadeem Javaid[2(✉)], Mehreen Shah[1], Umar Rasheed[1], Aasma Khan[2], and Zahoor Ali Khan[3]

[1] The University of Lahore, Islamabad 44000, Pakistan
[2] COMSATS Institute of Information Technology, Islamabad 44000, Pakistan
nadeemjavaidqau@gmail.com
[3] CIS, Higher Colleges of Technology, Fujairah Campus, Fujairah 4114, UAE
http://www.njavaid.com

Abstract. Underwater Acoustic Sensor Networks (UASNs) have gained interest of many researches due to its challenges like long propagation delay, high bit error rate, limited battery power and bandwidth. Node mobility and the uneven load distribution of sensor nodes results in creation of void holes in UASNs. Avoiding void holes benefits better coverage over an area, less energy consumption and high throughput. Therefore, in our proposed scheme, the sleep awake scheduling of corona nodes is done in order to minimize the data traffic load as well as balance the energy consumption in each corona. After network initialization, nodes in the even numbered coronas are set to sleep mode whereas nodes in odd coronas are in active mode. When the nodes in a corona deplete certain amount of energy, the nodes in sleep mode are switched to active operation mode. Thus, by scheduling data traffic load on each corona node is minimized also the energy of corona nodes is balanced. Simulation results verify the effectiveness of our proposed scheme in terms of traffic load distribution and energy consumption in sparse network.

Keywords: Underwater wireless sensor networks · Load balancing
Energy consumption

1 Introduction

Underwater environment are extremely important for human life due to their roles on the primary global production, carbon dioxide and earth's climate regulation. The underwater environment, especially deep ocean is not suitable for human beings due to its huge pressure and low visibility. Therefore, Underwater Acoustic Sensor Networks (UASNs) are mostly used in coastline surveillance, ocean disaster prevention, pollution monitoring, marine aquatic environment monitoring and resource exploration etc.

UASNs consist of variable number of sensor nodes which report their generated data to the sink periodically. Acoustic signals are usually preferred for

L. Barolli et al. (Eds.): EIDWT 2018, LNDECT 17, pp. 258–269, 2018.
https://doi.org/10.1007/978-3-319-75928-9_23

underwater communication, as high frequency radio waves are strongly absorbed in water and optical waves experiences heavy scattering and are only used for short range applications. The speed of acoustic signals is 1500 m/s whereas radio signals travels at a speed of 3×10^8 m/s. Therefore, large and variable delays are observed in UASNs. The underwater acoustic communication also faces multi-path fading, path loss, transmission loss and high Bit Error Rate (BER). In addition, as sensor nodes are expensive due to their heavy size and they require the aid of ships for deployment in underwater, mostly they are deployed sparsely within UASNs.

Sensor nodes normally adopt multi-hop mode to transmit their sensed data towards the sink node, however, data can also be transmitted directly to the sink node. The multi-hop mode puts high data transmission load on the nodes nearer to the sink, whereas, the farthest nodes deplete energy earlier while using direct transmission mode. This ultimately results in unbalanced energy consumption and shorter network lifespan. However, a combination of multi-hop and direct transmission modes, termed as the hybrid transmission mode can also be used for balancing the energy consumption among the network nodes. Thus, void hole problem occurs when a sensor node do not get any forwarder node in its vicinity to forward its data.

We propose a novel data load balancing technique; Routing Protocol with Minimized Load Distribution (RP-MLD) for UASNs. In RP-MLD, we define sleep awake schedule for each corona nodes to minimize the data transmission load on specific nodes. Thus, energy consumption is balance within the whole network.

This rest of the paper is organized as follows. Section 2 presents the state-of-the-art work. Section 3 presents the problem statement whereas Sect. 4 provides the network model of our proposed system. In Sect. 5, we compare our scheme with the Balanced Routing scheme [7] and performance evaluation is presented. Finally overall work is concluded in Sect. 6.

2 Related Work

In the past few decades, many researchers worked for balancing the energy consumption and balanced load distribution in UASNs. Unbalanced energy consumption which results in energy holes is one of the key issues in UASNs. Unbalanced energy consumption which results in energy holes is one of the key issue in UASNs. In this section, we discuss some of the latest work on balancing energy consumption and the limitations are identified.

In Balanced Transmission Mechanism (BTM) [1], a technique composed of both direct and multi-hop transmission have been proposed. BTM considers a two-dimensional network field in which energy of sensor nodes energy is divided into Energy Levels (EL). When the energy levels of successor and predecessor node varies, the successor node starts transmitting data directly to the sink. However, when network radius increases, BTM consumes high energy when it transmits the packet at longer distance.

In Enhanced Efficient and Balanced Energy Consumption Technique (EEBET) [2], works on the limitations of BTM. EEBET avoids direct data transmissions over long distances to save excessive energy consumption. To enhance the lifetime of the network, optimum energy levels are calculated. However, problem occur due to energy hole at 1-hop and 2-hop is not covered in this scheme.

In Weighting Depth Forwarding Area Division (WDFAD-DBR) [3], the forwarder nodes are selected on the basis of two hops depth difference, where the current depth and the depth of expected hop is considered by which void holes and the energy consumption caused by duplicate packets and neighbor request are reduced. However, considering two hops does not eliminate void hole. So to avoid void holes and balance the energy consumption new routing protocols need to be proposed.

In Adaptive Hop-by-Hop VBR (AHH-VBR) [4], during packet transmission pipeline radius is adaptively adjusted at each hop according to the three-dimensional space position of the neighbor nodes. Thus reduced end-to-end delay, energy consumption and improved data delivery ratio. However, increasing pipeline's radius does not overcome the void hole problem and energy is imbalanced due to the depletion of the nodes energy and network performance is reduced.

In Geographic and Opportunistic Depth Adjustment based Routing (GEDAR) [5], a set of neighbors is selected according to the distance of sensor nodes from the sink. The node having highest advancement forwards data to the next hop node. If a void hole occurs, depth adjustment is done by moving the node away from the sink. Thus, it achieves high packet delivery ratio and decreases the end-to-end delay. However, node depth adjustment results in high and imbalance energy consumption and again void hole occurrence due to depth adjustment.

In Avoiding Void Holes and Collision with Reliable Interface-Aware Routing (Re-INTAR) [6], the forwarder nodes are selected on the basis of two hops depth difference, where the current depth as well as the depth of expected hop is considered. If there exist a void hole, one hop backward transmission at source node is done. Thus improved packet delivery ratio and void holes are minimized. However, when void hole occurs, one hop backward transmission at source node consumes more energy to transmit the data to the sink as well as again void occurrence is not considered in this protocol.

In Homogeneous Balanced Routing (Homo-BR) [7] scheme, traffic load on corona 1 is reduced by increasing the transmission range of a sensor node, i.e., r, 2r. However, the data traffic load at corona 2 increases as it becomes 1-hop neighbor to sink. when corona 2 dies, load on corona 1 increases which results in depletion of its energy faster which results in sink hole problem.

In Balanced Load Distribution (BLOAD) [8], nodes transmission range is adjusted as to balance the load on the next hop neighbor nodes. This protocol uses three transmission ranges for transmitting the data such that r transmission range for the next hop neighbor 2r transmission range for the two hop neighbor and thirdly uses dtx for sending the data directly to the sink. When the farthest corona nodes uses transmission range dtx it consumes more energy to directly send its generated packet to the sink. Thus energy consumption is not balanced and farthest node dies quicker than the nodes nearer to the sink. Thus network lifetime degrades.

In Energy Efficient DBR (EEDBR) [9], a forwarder node is selected whose residual energy is higher and depth difference to the sink is minimum. However, the protocol does not consider the sink hole problem. As load on the node closer to the sink increases it depletes its energy quicker which results in sink hole. Thus network performance is degraded.

In Enhanced Energy Efficient DBR (EEEDBR) [10], selection of forwarder node is depends upon the nodes residual energy, minimum depth to the sink. In EEEDBR, network life is achieved by deploying idle nodes in the medium depth region. When a node in the medium depth region die an idle node takes its place and starts receiving and transmitting the data towards the sink. However, nodes that are in the lower depth region consumes more energy than the middle depth region nodes and higher depth region nodes, which is not considered in this technique.

Table 1 enlists some of the existing work along with their limitations, features and the parameters achieved.

3 Problem Statement

The sensor nodes that are direct neighbors of sink are used for forwarding the data received from far away nodes. As a result they consume more energy when they relay data. The 1-hop neighbor nodes deplete their energy earlier as compared to the other nodes in the network. So the death of such 1-hop neighbors introduces energy sink hole problem in the network. However, by adjusting the transmission range of sensor nodes, load on each corona can be balanced and the lifetime of the network can be enhanced.

In Homo-BR [7] scheme, a node transmits a portion of its data at a transmission range r while the remaining portion of the data is transmitted at $2r$. By this process, the load on corona 1 is minimized by making corona 2 one hop neighbor to sink, so the corona 1 should be relieved of sending all the data to the sink. However, as corona 2 nodes become 1-hop neighbor to sink, the data transmission load on corona 2 increases and after nodes in corona 2 dies corona 1 again become one hop neighbor to sink. Thus, the corona 1 starts relaying traffic of the other coronas and its energy starts to deplete faster. Hence, when corona 2 dies, the sink hole problem occurs and the whole network collapse, where the other corona nodes still have a lot of residual energy. Thus, an enhanced and energy efficient routing protocol is needed by which sink hole problem can be avoided and network lifetime is maximized.

4 Proposed Network Model

We considered a two dimensional circular network field with radius R where the nodes are bottom anchored. In order to keep the sensors depth same a floating

Table 1. Comparison of the state-of-the-art work

Techniques	Features	Parameters achieved	Limitation
BTM [1]	Divides energy into energy levels, Changes transmission modes	Reduced End-to-end delay, Throughput maximized	High energy consumption using direct transmission, creation of loops
EEBET [2]	Avoids direct transmission at longer distance, optimum energy levels	Throughput maximize, Network lifetime enhanced	Poor performance when network increases
WDFAD-DBR [3]	Two hop forwarder selection, Forwarding area changed adaptively	Avoided void holes, Duplicate packets suppressed	Void hole occurrence at two hop
AHH-VBF [4]	Change pipeline direction, Transmission range adaptively changed	Throughput increased, Reduced energy consumption	Inappropriate forwarder selection
GEDAR [5]	Geographic and Opportunistic Routing, Communication void region problem	Avoid routing holes, Network performance	high and imbalance energy consumption, Again void hole occurrence due to depth adjustment
RE-INTAR [6]	Underwater wireless sensor networks, Energy consumption, Void hole, Interference	Packet delivery ratio, void holes are minimized	Energy consumption due to backward transmission, Again void hole occurrence due to backward transmission
Homo-BR [7]	Load balance, Performance analysis, Energy conservation	Overcome energy hole problem, Network lifetime is improved, Balanced load on each Corona	Imbalance load on coronas, Corona 2 dies quickly due to direct transmission
BLOAD [8]	Energy hole, Balance load distribution, Energy balancing, Stability	Network lifetime, Energy Balancing	High and imbalance energy consumption
EEDBR [9]	Depth Utilization, Considers residual energy	Network lifetime, Reduces energy consumption, Reduced delay	Energy imbalanced, Reduced delivery ratio, Packet droped
EEEDBR [10]	Depth based routing, Control packet, Idle nodes, Network lifetime	Improved Network lifetime, Reduced end-to-end delay	Only applicable for small scale networks

buoy is attached with them. Sink is static and deployed at the center of the network field. The field is divided into disconnected coaxial sets named coronas of which have same width r as shown in Fig. 1. Number of nodes in each corona are same as well as the distance of each node from sink is same in a specific corona. Each sensor node can transmit its sensed data using two transmission ranges, r for its 1-hop neighbor and $2r$ for its 2-hop neighbor. Whereas, only the first corona transmits its data using transmission range r.

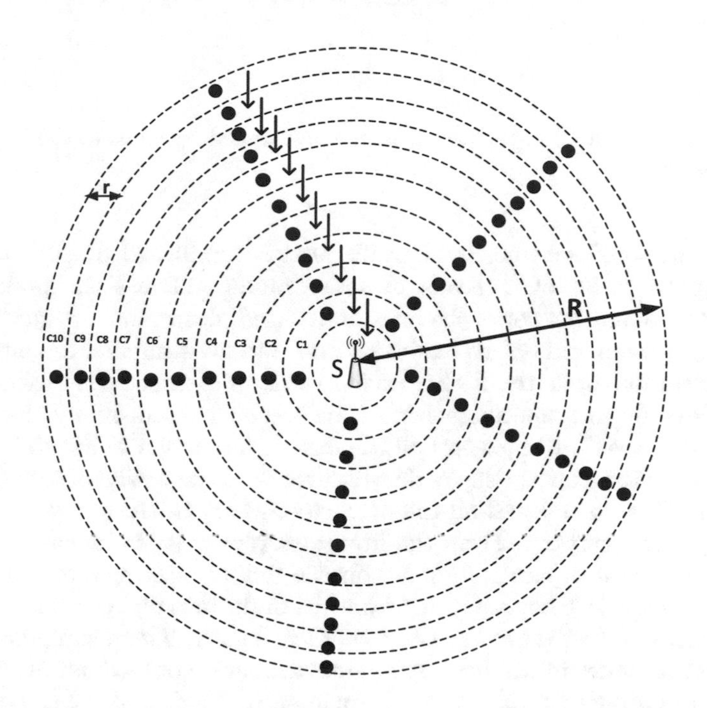

Fig. 1. Proposed network model

4.1 RP-MLD

Sensor nodes are deployed uniformly in the network, where the sink is placed in the center of the network field. Uniform means, that number of nodes in each corona are same. Total number of coronas are 10, where the distance of all nodes in a corona to its next hop neighbor corona nodes is equal. Each sensor node knows its own location as well as of the its neighbor nodes location. This network follows following assumptions:

- Sensor nodes always have data to send.
- Periodically, each node send its generated data to the sink. Sensor nodes report to the sink periodically.
- Sink is placed at the surface of the water.

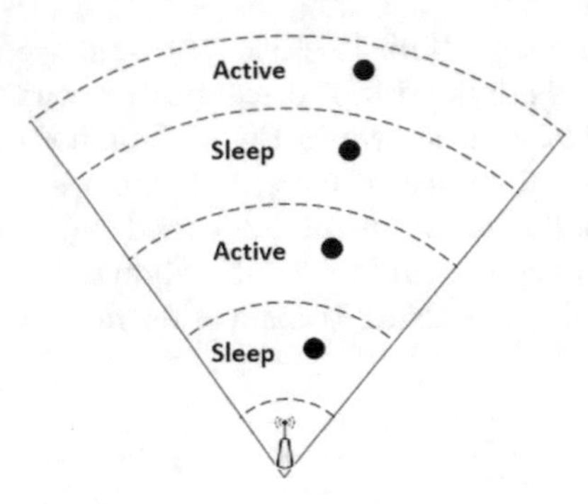

Fig. 2. Status of every corona when it is in active and sleep mode at the initialization of the network

In the proposed scheme, total traffic load is minimized on each corona by scheduling the nodes of a corona in active mode and in sleep mode. During initialization phase network schedules active and sleep mode. Figure 2 depicts scheduling of sleep and active modes on odd and even number of coronas. The distance from one node to another node as well as distance from each nodes to the sink is same in a specific corona. The reason for scheduling the corona 1 into sleep mode is that in general all the data is forwarded to the sink using the nodes closer to the sink as relay node which results in depletion of energy quickly. Also corona 2, corona 4 and all the even coronas are in the active mode at the beginning of the network. From the literature review it is observed that nodes closer to sink get comparatively more burden due to relaying data. In proposed scheme, corona 1 is 1-hop neighbor from the sink. Initially, corona 1 nodes are kept in sleep mode so that they can save their energy. There is uniform and fix deployment of nodes in the field, and nodes are aware of location of each other.

The even coronas i.e., 2, 4, 6, 8, 10 are in active mode and are responsible for transmitting the data towards the sink while the odd coronas i.e., 1, 3, 5, 7, 9 are in the sleep mode. The benefit of scheduling the odd coronas into sleep mode is that in this scheme each sensor node can transmit its own data as well as the data received to its 2-hop neighbor easily by adjusting its transmission range. So in order to save energy and enhance the network lifetime the odd coronas are scheduled into sleep mode.

In route phase, all corona nodes except corona 1 selects those nodes that are 2-hop away from them by adjusting its transmission range $2r$ because the 1-hop neighbor of each corona are in the sleep mode. Whereas, data is directly transmitted to the sink by corona 1 using transmission range r. In data transmission phase, all coronas send their sensed data by transmission range $2r$ to the node selected in route phase except corona 1 because it directly transmit the data by transmission range r. The scheduling is done by switching the active coronas into sleep mode and their adjacent sleeping coronas into active when the active nodes depletes a certain level of energy.

In RP-MLD, scheduling is done in such a way that when the network initializes, the active coronas start the process of receiving and transmitting the data. As soon as energy of nodes fall below a defined threshold i.e. (0.5 J), odd numbered corona nodes are switched from active mode to sleep mode and their adjacent lower coronas (even) number of corona nodes are switched from sleep mode to active mode.

As the time evolves, energy of odd numbered corona nodes fall below 0.5 J, they switch their status form active mode into sleep mode. The corona nodes that were in sleep mode (even corona nodes) again switch their status from sleep mode to active mode. When (even) corona nodes completely drain their energy, their upper adjacent (odd) corona nodes switch their status form sleep mode to active mode unless these nodes die. Thus, scheduling minimizes the relaying of data from other corona nodes and energy sink hole problem is minimized. Simulation results show that RP-MLD outperforms the Homo-BR scheme and distribution of load as well as energy consumption is minimized as compared with Homo-BR scheme.

5 Simulation Results

In this section, simulation results are performed in comparison with Homo-BR scheme in term of transmission ranges r and 2r. Circular network field is partitioned into disconnected coaxial coronas of fixed width r. Sensor nodes periodically reports their generated data to the sink. Each sensor node reports its generated data to the sink over multiple hops. By crossing corona of same wedge, each sensor node forwards its generated packet from source to the sink. Our proposed scheme focuses on balancing energy consumption and data traffic load at each corona by scheduling the node status in a corona by switching them to sleep and active phase.

There are a total number of 50 nodes, which are uniformly deployed inside a circular coaxial field of over a total radius of 1000 m. There are a total number of 10 coronas are and their width is fixed by 100 m. Transmission ranges r, $2r$ are studied on both data packet and energy consumption for each corona. The nodes in the odd coronas send their data packets to their neighbor corona using transmission range $2r$, Only the 1st corona transmits data using transmission range r.

The simulation parameters taken in this paper are shown in Table 2.

Figure 3 shows the load distribution per corona. Figure represents that the proposed scheme performs better when compared with the Homo-BR scheme. In order to reduce load on each corona nodes are scheduled. When energy of active corona falls below a certain limit the active corona switch its status from active to sleep mode and the corona that was in the sleep mode turn its status to active mode. Our proposed scheme minimized the packet load at corona 1 by 73% and on corona 2 by 46%. Our scheme improved 18% in corona 9 and 22% in corona 10. Our scheme achieved a total of 40% better than Homo-BR scheme.

Table 2. Simulation parameters

Simulation parameters	Value
Network radius	1000 m
Total number of coronas	10
Number of sensor nodes	50
Initial energy of nodes	1 J
1-hop distance	100 m
2-hop distance	200 m

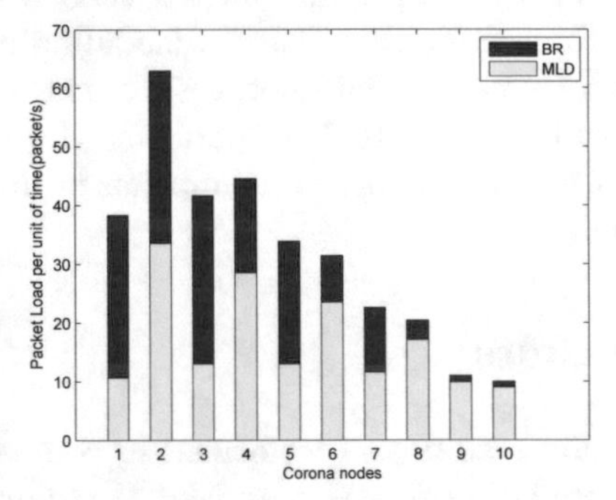

Fig. 3. Packet load per corona

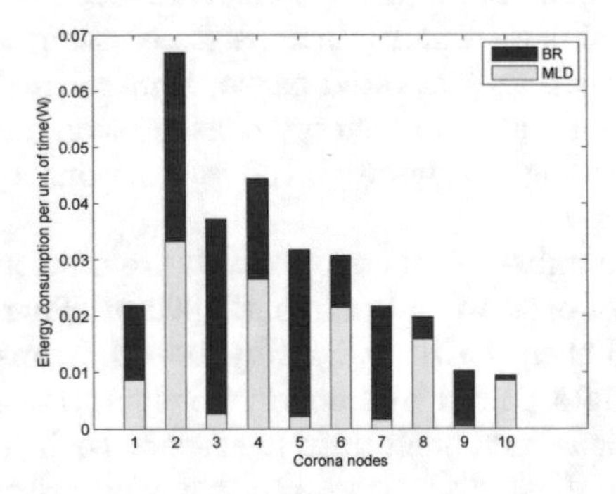

Fig. 4. Energy consumption per corona

Figure 4 shows the energy consumption per corona. Our technique shows better results than the homo-BR. As energy consumption is defined as the total energy consumed by each corona in transmission and reception of data, energy consumption is not dependent upon packet load. Our scheme consume only 0.009% energy in the first corona and 0.033% energy in the second corona thus obtained 59% in the first corona and 51.4 in the second corona. Also our scheme achieved 90% and 5% in the last two coronas respectively. Overall our scheme achieved 56% better result than Homo-BR scheme.

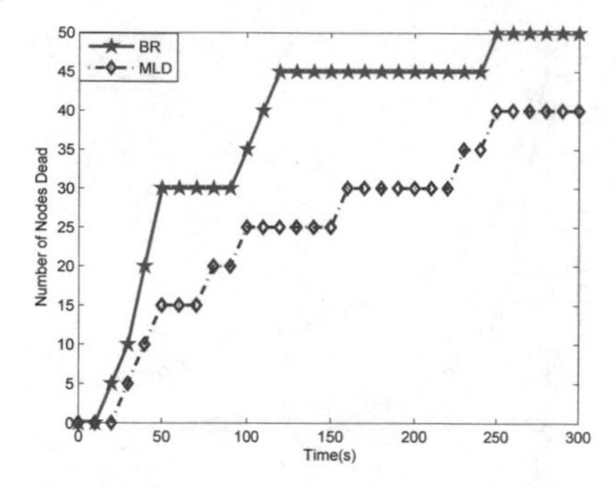

Fig. 5. Number of dead nodes

Proposed scheme minimizes load on corona that are close to sink as well as those corona which are far from sink. Figure 5 shows better results than Homo-BR in term of number of dead nodes. By the process of scheduling the corona nodes less number of nodes are dead than Homo-BR scheme. Simulation results showed that at 250 time (s) all the nodes of Homo-BR are dead whereas, in RP-MLD the total number of dead nodes are 40 and still 10 nodes are alive. Also our scheme achieved an average of 33% better results than Homo-BR scheme.

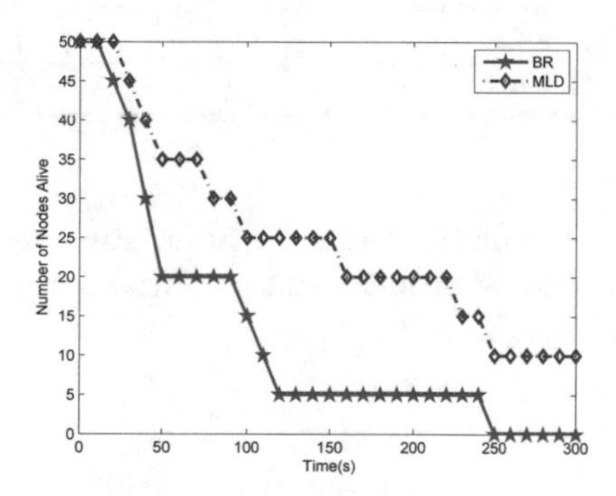

Fig. 6. Number of nodes alive

Figure 6 shows the number of nodes alive per unit time. As it is clearly shown that Modified-BR shows better result than the Homo-BR as more number of nodes are alive in Modified-BR. Scheduling of corona befits the network and thus, more number of nodes are alive in our proposed scheme. Simulation results showed that at time 250 (s), no nodes of Homo-BR are alive whereas in RP-MLD there are still 10 number of nodes alive at time 300 s. Our scheme achieved an average of 72% better than Homo-BR scheme.

Figure 7 shows residual energy per unit time. This figure shows that Modified-BR scheme has greater residual energy than the Homo-BR scheme. The reason is that the nodes are switched into active sleep mode, by which nodes energy

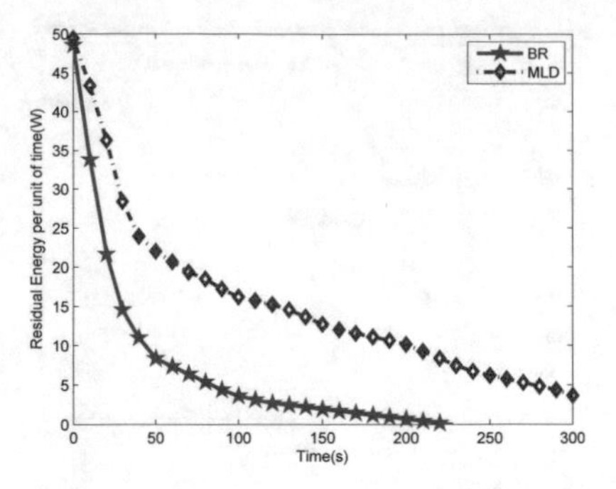

Fig. 7. Residual energy per time

Table 3. Parameters achieved

Performance parameters	Average Homo-BR	Average Modified-BR	Percentage achievement by Modified-BR over Homo-BR
NODN (%)	43	29	33
NONA (%)	8	20	72
Residual energy (%)	3	12	86
Packet load (%)	32	17	40
Energy consumption (%)	0.3	0.1	56

is saved and network lifetime is enhanced. Simulation result showed that our scheme outperforms Homo-Br scheme and achieved a total of 55.49% residual energy as provided in Table 3.

6 Conclusion

We proposed an energy balancing data transmission technique in which the data transmission load on each sensor node is managed in order to avoid the energy holes in UASNs. In our scheme, the nodes sleep awake schedule of nodes is defined. When a node consumes a certain amount of energy its operation mode is switched from active to sleep mode and its adjacent node's status is switched from sleep mode into active mode. The simulation results showed that RP-MLD scheme improves the performance of the network in terms of energy conservation and traffic load distribution.

References

1. Cao, J., Dou, J., Dong, S.: Balance transmission mechanism in underwater acoustic sensor networks. Int. J. Distrib. Sens. Netw. **11**(3), 429340 (2015)
2. Javaid, N., Shah, M., Ahmad, A., Imran, M., Khan, M.I., Vasilakos, A.V.: An enhanced energy balanced data transmission protocol for underwater acoustic sensor networks. Sensors **16**(4), 487 (2016)
3. Yu, H., Yao, N., Wang, T., Li, G., Gao, Z., Tan, G.: WDFAD-DBR: weighting depth and forwarding area division DBR routing protocol for UASNs. Ad Hoc Netw. **37**, 256–282 (2016)
4. Yu, H., Yao, N., Liu, J.: An adaptive routing protocol in underwater sparse acoustic sensor networks. Ad Hoc Netw. **34**, 121–143 (2015)
5. Coutinho, R.W., Boukerche, A., Vieira, L.F., Loureiro, A.A.: Geographic and opportunistic routing for underwater sensor networks. IEEE Trans. Comput. **65**(2), 548–561 (2016)
6. Majid, A., Azam, I., Khan, T., Khan, Z.A., Qasim, U., Javaid, N.: A reliable and interference-aware routing protocol for underwater wireless sensor networks. In: 2016 10th International Conference on Complex, Intelligent, and Software Intensive Systems (CISIS), pp. 246–255. IEEE, July 2016
7. Zidi, C., Bouabdallah, F., Boutaba, R.: Routing design avoiding energy holes in underwater acoustic sensor networks. Wirel. Commun. Mob. Comput. **16**(14), 2035–2051 (2016)
8. Azam, I., Javaid, N., Ahmad, A., Wadood, A., Almogren, A., Alamri, A.: Balanced Load Distribution with Energy Hole Avoidance in Underwater WSNs. IEEE Access (2017)
9. Wahid, A., Lee, S., Jeong, H.J., Kim, D.: EEDBR: energy-efficient depth-based routing protocol for underwater wireless sensor networks. In: Advanced Computer Science and Information Technology, pp. 223–234 (2011)
10. Khizar, M., Wahid, A., Pervaiz, K., Sajid, M., Qasim, U., Khan, Z.A., Javaid, N.: Enhanced energy efficient depth based routing protocol for underwater WSNs. In: 2016 10th International Conference on Innovative Mobile and Internet Services in Ubiquitous Computing (IMIS), pp. 70–77. IEEE, July 2016

Transmission Range Adjustment for Void Hole Avoidance in UWSNs

Mehreen Shah[1], Nadeem Javaid[2](✉), Umar Rasheed[1], Faisal Hayat[1], Ghazanfar Latif[2], and Zahoor Ali Khan[3]

[1] The University of Lahore, Islamabad 44000, Pakistan
[2] COMSATS Institute of Information Technology, Islamabad 44000, Pakistan
nadeemjavaidqau@gmail.com
[3] CIS, Higher Colleges of Technology, Fujairah Campus, Abu Dhabi 4114, UAE
http://www.njavaid.com

Abstract. Underwater Wireless Sensor Networks (UWSNs) have captured interest of many researchers with the desire to control the large portion of the world overspread by water. Energy efficiency is one of the major concerns in UWSNs due to the limited energy of the underwater sensor nodes. In order to enhance the network lifetime, efficient and reliable protocols must be presented while considering the underwater acoustic communication challenges like low bandwidth, longer propagation delays and limited battery life of sensor nodes. In this paper, we present Modified Geographic and Opportunistic Depth Adjustment based Routing (MGEDAR) protocol to minimize the energy hole problem in UWSNs. Our protocol works by adaptively adjusting the transmission range of sensor nodes in case of void holes. Each node selects its forwarder on the basis of a cost function. Simulation results showed that our proposed scheme improves network performance in terms of maximum throughput, minimum energy consumption and reduced void holes.

1 Introduction

Almost seventy percent of the earth is covered with water. The harsh underwater environment especially deep ocean regions is not suitable for human life due to high pressure. Therefore, Underwater Wireless Sensor Networks (UWSNs) are needed to be deployed in order to explore the marine resources. The UWSNs consists of variable number of underwater sensor and sink nodes to preform monitoring tasks over a given area. UWSNs gather data from the underwater environment and periodically reports that data to the sink. The UWSNs came up with many applications like disaster prevention, mineral extractions, pollution monitoring etc.

Acoustic communication is preferred in UWSNs as the radio waves get absorbed in aqueous environment. On the other hand, the use of optical waves is restricted in UWSNs as the optical waves gets scattered due to reflection and refraction. The inherent challenges of UWSNs like high bit error rate, limited bandwidth, temporary connection loss and long end-to-end delay affects

L. Barolli et al. (Eds.): EIDWT 2018, LNDECT 17, pp. 270–280, 2018.
https://doi.org/10.1007/978-3-319-75928-9_24

the underwater communication [1]. Another major challenge for UWSNs is the limited battery power of sensor nodes which is difficult to recharge or replace in harsh aquatic environment. Sensor nodes communicate with each other to transmit their data towards the sink. This routing process consumes abundance of energy. In multi-hop communication, due to high data load on nodes nearer to the sink, batteries of such nodes deplete earlier. On the other hand, nodes farthest from the sink die early while using direct data transmission technique. Many routing protocols are proposed to achieve energy efficiency and balance the energy consumption in UWSNs. In [2], the authors used clustering and multi-hopping technique to prolong the lifetime of the network. The network is divided into sparse and dense regions where clustering and multi-hopping is used in dense regions and two mobile sinks are used to gather data in sparse regions. However, this scheme prolongs the network lifetime at the cost of low throughput.

In this paper, we propose Modified GEographic and opportunistic Depth Adjustment based Routing (MGEDAR) with cost function based topology to recover the void regions. MGEDAR utilizes the location information of neighbor nodes and the sink. The forwarder set is calculated from the set of neighbor nodes in order to continue forwarding the packet towards the destination. To avoid unnecessary transmissions, low priority nodes suppress their transmissions whenever they detect the same packet sent by a high priority node. The most important aspect of MGEDAR is the void node recovery mechanism. Instead of the traditional message based void node recovery procedure, we propose a forwarder selection procedure using cost function. Moreover, for void hole recovery, we introduce the transmission range adjustment. The simulation results show that MGEDAR is able to reduce the amount of void nodes, improves the packet delivery ratio and decreases the end-to-end delay for the critical scenarios of low and high densities and diverse network traffic load.

The rest of the paper is organized as follow. Section 2 provides related work while in Sect. 3 problem statement is illustrated. In Sect. 4, proposed network model is presented and discussed. Section 5 presents the performance evaluation of our proposed work via simulations, whereas, the conclusion of this work is given in Sect. 6.

2 Related Work

In the past few years, many researchers have proposed routing protocols for energy balancing and avoiding void holes to enhance the network lifetime in UWSNs. The void hole creation problem occurs due to one of the following reasons: High data load on specific nodes and random deployment of sensor nodes. However, routing protocols plays a key role to overcome the void holes due to unbalanced energy consumption problem in UWSNs. In this section, we have discussed some existing routing protocols in UWSNs.

In Balanced Routing (BR) protocol [3], two-dimensional network model is proposed. The BR balances the data load of all the sensor nodes, avoids energy holes and minimizes the energy consumption. The total data traffic load is minimized at corona 1 by distributing data traffic load among all coronas of the

Table 1. Comparison of state-of-the-art of work

Techniques	Features	Achievements	Limitations
BR [3]	Load balancing, Energy conservation	Overcome energy hole problem, Network lifetime is improved	Corona 2 dies quickly due to direct transmissions
MEES [4]	Balanced energy consumption	Increases throughput, Prolong network lifetime	High energy consumption in direct transmissions
BTM [5]	Balanced energy consumption, Energy efficient	Network lifetime, Balanced energy consumption	High energy consumption while transmitting at long distance, Formation of transmission loops
EEBET [6]	Balance energy consumption, Avoid direct transmissions over long distance, Minimizes energy consumption	Improves network lifetime, Minimize number of hops	Energy holes problem still exist
IAEEDBR [7]	Interference aware, Energy efficiency	Improved network lifetime, Improved data delivery ratio	Unbalanced energy consumption, Void holes occur
WDFAD-DBR [8]	Adaptively changes the forwarding area	Void hole avoidance, Suppression of duplicate packets	Unnecessary energy consumption due to retransmissions
VBF [9]	Geographic and position based routing	Achieved robustness, Energy efficiency, High data delivery ratio	Impractical for sparse networks
HH-VBF [10]	Adaptive hop-by-hop location based routing	Improves robustness and packet delivery ratio in sparse networks	Poor performance in dense networks
AHH-VBF [11]	Dynamically adjusts forwarding region	Reduced end-to-end delay and energy consumption	void hole problem
GEDAR [12]	Depth adjustment	Avoids routing holes	High and imbalanced energy consumption, Highly dynamic topology
DFR [13]	Controlled flooding to increase reliability	High reliability with less communication overhead, Improves packet delivery ratio	High energy consumption, Void Holes

network using transmission range (r, 2r). The disadvantage of Homo-BR is high energy consumption at corona 2 because of receiving data from the upper coronas and sending the received as well as its own data to the sink over 2r transmission range. Thus, corona 2 consumes high energy as compered to other coronas.

In Mobile Energy Efficient Square (MEES) routing [4], 2-D network model is proposed. In this technique, authors use clustering technique to balance the energy consumption in the network. Network filed is divided into two regions sparse and dense regions on the basis of node density where clustering is done in dense network regions and mobile sinks are deployed in sparse network regions. The achievement of this technique is the increased throughput, prolonged network lifetime and balanced load over sensor nodes. The disadvantage of MEES is high energy consumption due to direct transmissions (Table 1).

In Balance Transmission Mechanism (BTM) [5] in UWSNs, a two dimensional network model is proposed. In BTM, the initial energy of each sensor node is divided into energy levels to balance the load among all sensor nodes in the network. The sensor nodes communicate with the sink using both multi-hop and direct transmission modes. When a node consumes one energy level earlier than its predecessor node, it starts transmitting its data via direct transmission mode. The main objective of BTM is balancing the load on sensor nodes to minimize the energy consumption. However, BTM consumes more energy in direct transmissions over long distance and node die quickly in large scale UWSNs.

Enhanced Efficient and Balance Energy consumption Transmission (EEBET) protocol [6] is proposed to reduce the limitations of BTM. The EEBET efficiently balances the energy consumption and avoids direct transmissions over long distances. The major reasons behind improved results in EEBET are: Minimizing the number of hops towards sink and overcomes the data load on nodes nearer to the sink.

In [7], an Interference Aware Energy Efficient Depth Based Routing (IAR-RDBE) protocol has been proposed. The IAEEDBR works for energy efficiency and prolongs the network lifetime. Each sensor node select a node as a forwarder whose number of neighbor nodes are minimum. Thus, congestion and interference is minimized, ultimately energy efficiency achieved which prolongs network lifetime. However, IAEEDBR provides no mechanism to balance the energy consumption and the hot spot nodes dies earlier. Thus, Energy unbalancing and void holes problem occurs.

In Weighting Depth and Forwarding Area Division Depth Based Routing (WDFAD-DBR) [8], authors proposed two techniques. Firstly, the weighting depth and secondly, forwarding area division DBR routing protocol. Node selection is based on the weighting sum depth difference between two hops. In this scheme, void holes are decreased. Secondly, the mechanism of forwarding area division and expected neighbor prediction is presented. The technique reduced energy consumption duplicate packets.

In Vector-Based Forwarding (VBF) [9], the forwarding range of the sending packet is limited inside a virtual pipeline and is based on distance between sender and the receiver node. The advantage of this process is, it can save energy by limiting the direction of node and the range of the forwarding node. The limitation of this scheme is constant forwarding direction of node and the constant radius of virtual pipeline.

To improve the performance of VBF, a Hop by Hop VBF (HH-VBF) protocol [10] is proposed. The HH-VBF changes the direction of the forwarding pipeline hop by hop. In this way, every forwarding node can make routing decision according to the current local topology information, so the better forwarding area can be found. Although the performance of the network can be improved by dynamically changing the direction of flooding pipeline. As compared to VBF, HH-VBF has low performance in uneven distributed networks as the radius of the pipeline keeps constant during entire lifetime of the network.

Adaptive Hop-by-Hop routing (AHH-VBF) protocol [11], adjusts the radius of virtual pipeline adaptively at every hop based on the three-dimensional positions of the neighbor nodes. The advantage of the scheme is it overcomes the energy consumption and the transmission power of each node is dynamically adjusted. Moreover, the holding time of every transmission packet is calculated based on the distance between each forwarder node and destination location. The advantage of this scheme is that it reduces end-to-end delay and improve the throughput. The disadvantage in AHH-VBF is when the radius of virtual pipeline is increased, it still does not overcome the void hole problem in the network.

In Geographic and Opportunistic Depth Adjustment based Routing (GEDAR) protocol [12], a set of neighbors are selected on the basis of depth difference of sender and the receiver node and the distance of neighbor node from the sink. Only one node is responsible for forwarding data which has the minimum distance to the sink. If a void hole occurs, depth adjustment is performed by moving the node vertically towards its predecessor node. The scheme achieves packet delivery ratio and reduces the end-to-end delay. The limitation of this protocol is the node depth adjustment results in high and imbalanced energy consumption and again void hole occurs due to depth adjustment.

Directional Flooding based Routing (DFR) [13], is proposed to overcome the flooding in UWSNs. In DFR, limited number of nodes are selected to transmit data packet from source to destination. The flooding region is decided based on the link quality of neighbor nodes and the angle selected between sender and receiver. DFR reduces amount of energy consumption through controlled flooding. The shortcoming of DFR is that it uses fixed transmission power level which results in energy wastage.

3 Problem Statement

GEDAR [12], presents a mechanism to overcome void holes in UWSNs using depth adjustment technology. The scheme greedily selects the forwarder node and forward its sensed data towards the sink through this node. If there occurs a void hole, GEDAR performs depth adjustment where the position of the node is changed vertically towards its predecessor node. If a sensor node finds a node after depth adjustment, it transmits its data towards the sink.

However, a large amount of energy is needed when a sensor node performs depth adjustment. While adjusting the depth, the node may still find no neighbor node in its vicinity. Also, due to dynamic topology, transmission of control packets increases which ultimately leads to higher energy consumption. Thus, the void hole recovery mechanism is not efficient. When a sensor node could not find any neighbor node after depth adjustment, the same depth adjustment procedure is repeatedly performed. As a consequence, the node completely depletes its energy and dies. Thus, an efficient and reliable as well as balanced energy consumption scheme is needed in order to recover the void holes and enhance the network lifetime. We propose a routing protocol to overcome void holes and prolong network lifetime. In our MGEDAR protocol, transmission range adjustment is introduced incase of void regions.

4 Proposed Network Model

In our proposed technique, we consider three dimensional network architecture (3-D) where each sensor node knows its location. The sensor nodes are randomly deployed in the network field to sense data of monitoring field and then transmit it towards the sink. Also, all the sensor nodes are homogeneous. There are 45 sinks deployed at the surface of the water which are responsible to gather data from whole network field and transmit it to the onshore data centre. We have assumed that all sensor nodes always have data to transmit. Each sensor node periodically transmit its sensed data. A data packet received at any sink is considered as successfully delivered. Each sensor node use acoustic signals to share its information with other nodes while the sink nodes use acoustic signal to

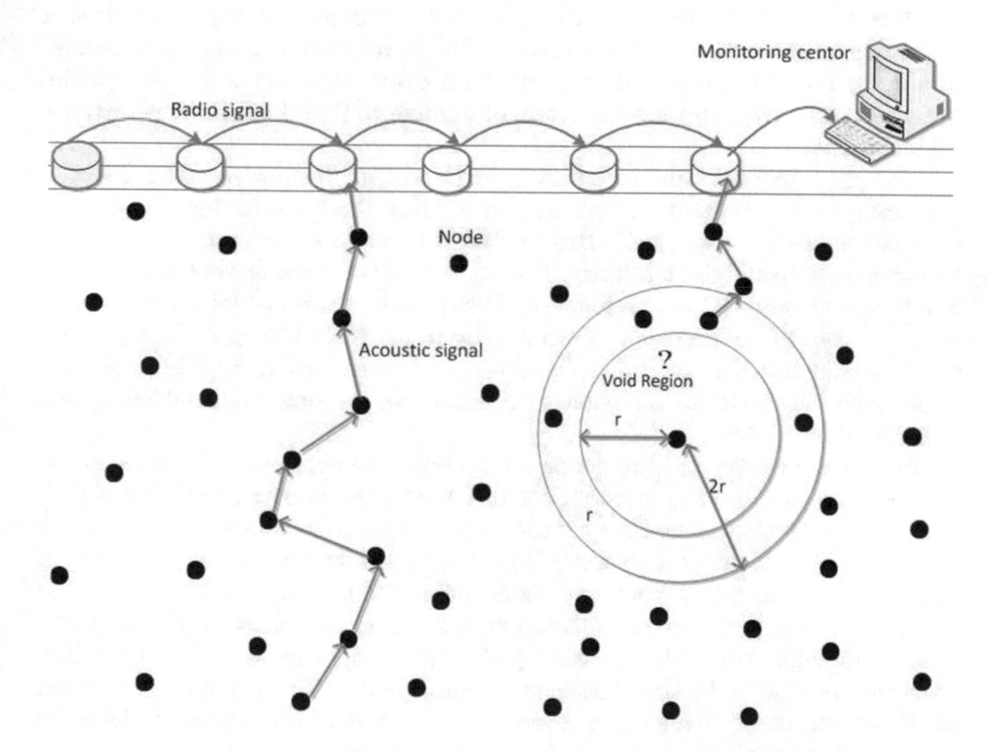

Fig. 1. Proposed network model

receive data from the sensor nodes and transmit to monitoring center using radio signals. Beacon packets are generated in order to get the position of neighbors and the sink (Fig. 1).

4.1 MGEDAR

In our proposed protocol, each node selects its neighbor node by calculating its distance with the neighbor node as well as the distance of neighbor node with the sink. The procedure helps a node to select appropriate neighbor whose advancement towards sink is higher. Thus, the energy consumption can be minimized by restricting those nodes to become neighbors, whose inclusion leads to increase in the number of hops with the sink. All the nodes whose distance with the sink is minimum than the distance of sender with the sink are considered as neighbor nodes. However, we also check that the depth difference of sender and the neighbor node should be above a defined threshold value.

When the neighbor search process completes, we need to select set of forwarder nodes to find an eligible forwarder node for the data transmission phase. During data transmission phase, we select appropriate forwarder nodes. From the set of neighbor nodes, we select a node whose number of neighbors are maximum among all other neighbor nodes of a sender node. Also, we check the distance parameter as well. A node which leads to minimum number of hops towards sink is the winner node for data transmission. This helps to reduce energy consumption as well as the end-to-end delay in the network. The list of neighbor nodes are sorted according to the parameters of number of neighbors and the distance with the sink.

The data packets are transmitted to the set of forwarder nodes selected according to the mechanism defined above. Only the node having highest priority is supposed to transmit data packet. The node at the top of the list in forwarder set has highest priority. If a node fails to transmit that data packet due to any reason, the second highest priority node transmits data. This is done with the help of holding time as defined by [7]. A node having highest priority will have shortest holding time so it transmits packet earlier. All other neighbor nodes suppress their transmission by overhearing the data transmitted by the highest priority node.

In case of void regions, we define a mechanism to overcome void holes in the network and improve the performance of UWSNs. As opposed to GEDAR [12], which only adjusts the depth of sensor nodes in case of void holes, we adjust the transmission range of sensor nodes if no eligible forwarder found or the region surrounding a sender is void. If a node fails to detect any neighbor node in its vicinity, it adjusts its transmission range and again checks for the neighbor node. This helps reduce the dynamicity of network topology as well as the energy consumption due to depth adjustment of sensor nodes. By changing the position of the sensor node, many other nodes gets affected which ultimately leads to high energy consumption and shorter network lifetime (Table 2).

Table 2. Parameters Achieved

Parameter	Value
Area	1500 m × 1500 m × 1500 m
Transmission energy	2 W
Reception energy	0.1 W
Transmission range	250 m

5 Simulation

In this section, we analyze the performance of MGEDAR and compare its results with the GEDAR protocol. We have compared the results of our proposed work against GEDAR protocol on the basis of following performance parameters: Fraction of void nodes, energy consumption and the packet delivery ratio. We have randomly deployed sensor nodes in a field having dimensions 1500 m × 1500 m × 1500 m. The transmission range of sensor nodes is 250 m with a data rate of 50 kbps. The values of transmission, reception and idle state energy consumption are taken as 2 W, 0.1 W and 10 mW respectively. Our objective is to investigate how the greedy forwarding strategy behaves as the network density ranges from low to high.

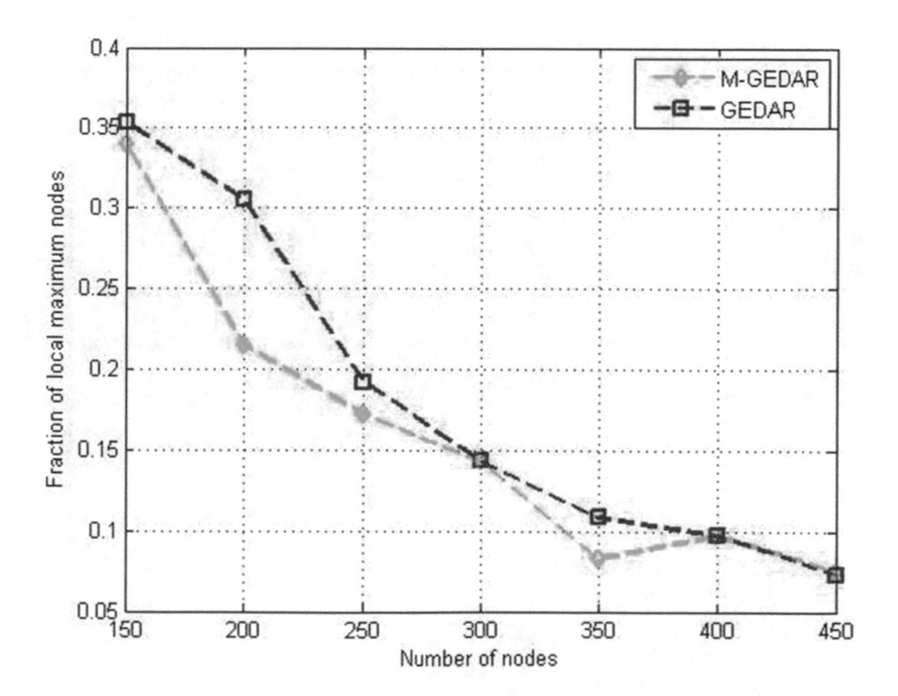

Fig. 2. Fraction of void holes with varying nodes densities

Fig. 2 shows the fraction of void nodes. As shown in the figure, when the node density increases, the total number of void nodes decreases in both protocols (GEDAR and MGEDAR). However, the MGEDAR protocol achieves better results as compared to GEDAR in terms of minimum number of void holes. The reason behind this result is the transmission range adjustment of the sensor nodes in case of void holes. Instead of changing the vertical position of sensor nodes, we search for neighbors by adjusting the transmission range of the sensor nodes. The GEDAR employs highly dynamic topology due to depth adjustment in case of void hole detection. This negatively impacts the network performance, thus, chances of void holes increases. The proposed cost function based topology control mechanism reduces 48 percent void nodes when compared to GEDAR.

Fig. 3 shows the results concerning the packet delivery ratio of GEDAR and MGEDAR. As shown in the figure, packet delivery ratio increases when the network density increases. The MGEDAR improves packet delivery ratio due to its void hole recovery procedure. The more the number of void holes are recovered, minimum is the packet drop ratio ultimately network throughput increases. However, the packet delivery ratio of MGEDAR affects due to selection of a forwarder node whose neighbors are maximum. It introduces congestion and the collisions in the network which increases the packet drop ratio. On the other hand, the GEDAR incorporates depth adjustments which leads to high dynamicity, thus, packet drop increases. However, the overall behaviour of both the schemes show that the packet delivery ratio increases increasing node density.

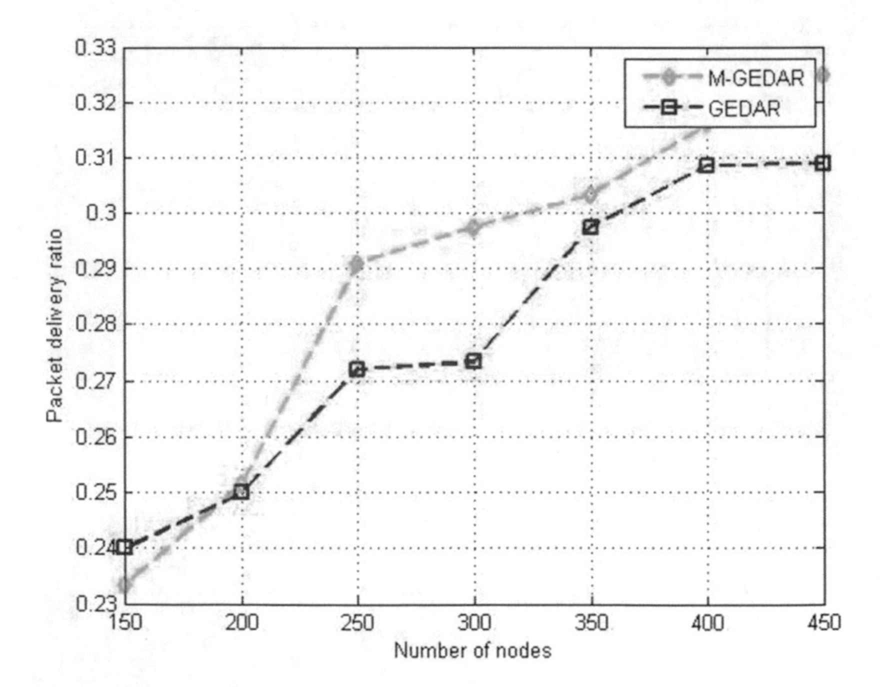

Fig. 3. Packet delivery ratio with varying number of nodes

In Fig. 4 we can see the average amount of energy consumed by a sensor node. The figure clearly shows that the energy consumption of MGEDAR is minimum as compared to the GEDAR. GEDAR performs depth adjustment in case of void holes which consumes excessive energy. Thus, the energy consumption in GEDAR is high as compared to the MGEDAR. Although the MGEDAR performs transmission range adjustment which is also a costly process in terms of energy consumption. However, due to depth adjustment chances of void holes increases as the whole topology of the network changes. Such a behaviour of GEDAR affects the energy consumption of the network. We can also observe from the Fig. 4 that the energy consumption of MGEDAR slightly varies by increasing the node density. This proves that the MGEDAR also plays a key role to balance the energy consumption in the network.

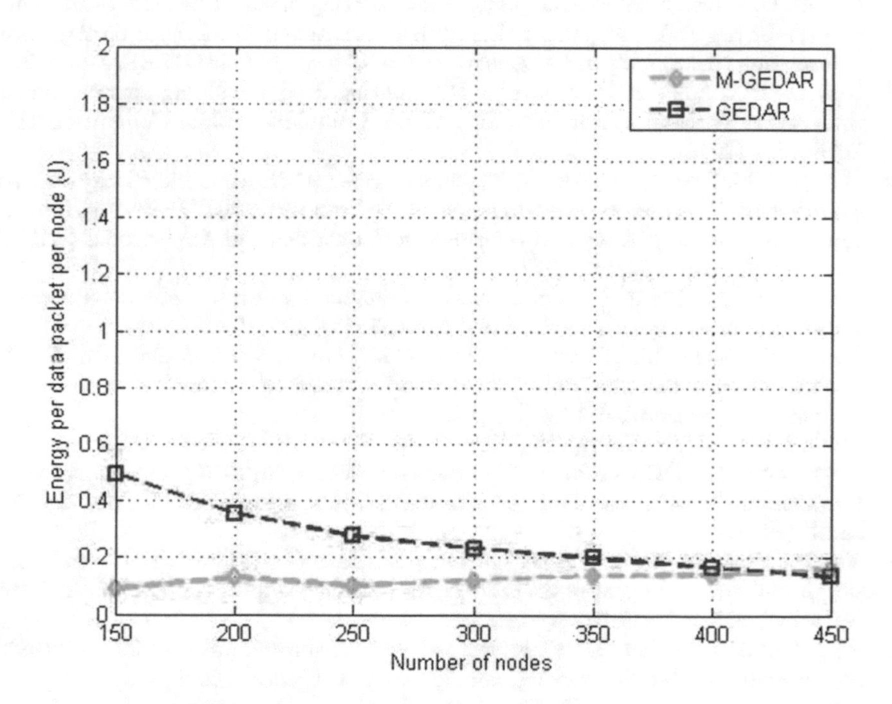

Fig. 4. Energy consumption by varying number of nodes in the network

6 Conclusion

In this paper, we have proposed MGEDAR routing protocol to minimize void holes in UWSNs. MGEDAR is a geographic and opportunistic routing protocol that utilizes the position information of the nodes. The transmission range adjustment mechanism defined by the MGEDAR controls the number of void holes in the network. This ultimately improves the network performance in terms of minimum energy consumption, maximum data delivery ratio and most importantly minimum number of void holes in the network. We have analyzed the performance of our proposed work through simulations. The simulation results proves that MGEDAR improves network performance in terms of said parameters.

References

1. Bouabdallah, F., Zidi, C., Boutaba, R.: Joint routing and energy management in underwater acoustic sensor networks. IEEE Trans. Netw. Serv. Manage. **14**(2), 456–471 (2017)
2. Azam, I., Majid, A., Ahmad, I., Shakeel, U., Maqsood, H., Khan, Z. A., Javaid, N.: SEEC: sparsity-aware energy efficient clustering protocol for underwater wireless sensor networks. In: 2016 IEEE 30th International Conference on Advanced Information Networking and Applications (AINA), pp. 352–361. IEEE, March 2016
3. zidi, C., Bouabdallah, F., Boutaba, R.: Routing design avoiding energy holes in underwater acoustic sensor networks. Wirel. Commun. Mobile Comput. **16**(14), 2035–2051 (2016)
4. Walayat, A., Javaid, N., Akbar, M., Khan, Z.A.: MEES: mobile energy efficient square routing for underwater wireless sensor networks. In: 2017 IEEE 31st International Conference on Advanced Information Networking and Applications (AINA), pp. 292–297. IEEE, March 2017
5. Cao, J., Dou, J., Dong, S.: Balance transmission mechanism in underwater acoustic sensor networks. Int. J. Distrib. Sens. Netw. **11**(3), 429340 (2015)
6. Javaid, N., Shah, M., Ahmad, A., Imran, M., Khan, M.I., Vasilakos, A.V.: An enhanced energy balanced data transmission protocol for underwater acoustic sensor networks. Sensors **16**(4), 487 (2016)
7. Shah, M., Javaid, N., Imran, M., Guizani, M., Khan, Z. A., Qasim, U.: Interference aware inverse EEDBR protocol for underwater WSNs. In: 2015 International Wireless Communications and Mobile Computing Conference (IWCMC), pp. 739–744. IEEE (2015)
8. Yu, H., Yao, N., Wang, T., Li, G., Gao, Z., Tan, G.: WDFAD-DBR: weighting depth and forwarding area division DBR routing protocol for UASNs. Ad Hoc Netw. **37**, 256–282 (2016)
9. Xie, P., Cui, J.-H., Lao, L.: VBF: vector-based forwarding protocol for underwater sensor networks. In: Networking, vol. 3976, pp. 1216–1221 (2006)
10. Nicolaou, N., See, A., Xie, P., Cui, J. H., Maggiorini, D.: Improving the robustness of location-based routing for underwater sensor networks. In: Oceans 2007, Europe, pp. 1–6. IEEE (2007)
11. Yu, H., Yao, N., Liu, J.: An adaptive routing protocol in underwater sparse acoustic sensor networks. Ad Hoc Netw. **34**, 121–143 (2015)
12. Coutinho, R.W., Boukerche, A., Vieira, L.F., Loureiro, A.A.: GEDAR: geographic and opportunistic routing protocol with depth adjustment for mobile underwater sensor networks. In: 2014 IEEE International Conference on Communications (ICC), pp. 251–256. IEEE (2014)
13. Hwang, D., Kim, D.: DFR: directional flooding-based routing protocol for underwater sensor networks. In: Oceans 2008, pp. 1–7. IEEE, September 2008

Exploiting Meta-heuristic Technique for Optimal Operation of Microgrid

Saman Zahoor[1], Nadeem Javaid[1(✉)], Ayesha Zafar[1], Anila Yasmeen[1], Asad-ur-rehman[2], and Zahoor Ali Khan[3]

[1] Department of Computer Science, COMSATS Institute of Information Technology, Islamabad 44000, Pakistan
nadeemjavaidqau@gmail.com
[2] National University of Science and Technology, Islamabad 44000, Pakistan
[3] CIS, Higher Colleges of Technology, Fujairah Campus, Fujairah 4114, UAE
http://www.njavaid.com

Abstract. A power system with different types of micro-sources are very popular in recent years. The aim of the paper is to make the environment green by reducing green house gases and meet the load demand in an efficient way. However, we propose a grid-connected microgrid system which meets the load demand in an efficient manner to achieve our objectives. The objective of this work is to find the optimal set points of controllable micro-sources in terms of cost minimization. The grid-connected microgrid also helps to exchange power with utility during different intervals of a day to meet the load demand. The significance and performance of the proposed strategy is proved through performing simulations in MATLAB. However, the overall cost of MG is less, while in schedulable microsources the cost of FC is less as compared to MT and DE.

Keywords: Energy management system · Micro grid · Microsources
Optimization technique · Load demand

1 Introduction

Microgrids (MG) are the integrated systems which is composed of multiple loads, different types of microsources and Energy Storage Systems (ESS). MG is basically low-voltage level distribution network, which is either work in grid connected mode or islanded mode. Microsources include variety of sources; such as Renewable Energy Sources (RES) and non-RES. In RES, Photo-Voltaic (PV) and Wind Turbines (WT) are considered. In non-RES, Micro Turbine (MT), Fuel Cell (FC) and Diesel Engines (DE) are considered. The vital operations of a MG require an efficient energy management system which handle the flow of power in MG by adjusting the power sold/bought to/from the utility. However, the nature of RES is both time varying and hard to predict. Whereas, non-RES and ESS can be linked to any point in the network and have predictive output.

L. Barolli et al. (Eds.): EIDWT 2018, LNDECT 17, pp. 281–291, 2018.
https://doi.org/10.1007/978-3-319-75928-9_25

Generally, if non-RES are properly planned and controlled, it improves power quality and decreases dependency on utility. However, if non-RES with RES and ESS are used, it also reduces dependency on utility, environmental concerns and overall cost of system. Usually a control system architecture of MG is composed of three control levels; microsources controller, a MG central controller and a distribution management system controller. The microsources controllers receive demand from MG central controller and generate output according to defined constraints. The MG controller is responsible for optimization of its operations and maximization of its usage instead of utility. One of the main scheduling problem for grid connected MG is the energy and operation management at minimum cost.

The energy and operation management of MG at low cost is categorized into two approaches, deterministic and probabilistic. In deterministic approach, it is assumed that the output power of RES as well as load demand and price signals are equal to their assumed value. The problem becomes probabilistic, when some of variables are uncertain. As for the probabilistic nature of solar irradiance and wind speed, the output power of RES units are random variables as well. Furthermore, due to these types of unexpected disturbances, prediction errors or load demand variations are occurred.

Currently, to solve deterministic and probabilistic energy and operation management of microgrids, various methods have been proposed as in [1,2]. In this article, for efficient and reliable optimization of energy and operation management of a MG, artificial bee colony algorithm is used. The performance of the proposed algorithm is tested on a typical grid connected microgrid by considering PV, WT, MT, FC, DE and ESS.

The rest of paper is organized as; in Sect. 2 related work is explained, detailed discription of system model is given in Sect. 3, problem formulation of this work is presented in Sect. 4, simulation results are discussed in Sect. 5 and at the end, conclusion is drawn.

2 Related Work

To make our society smart MG plays a vital role. Numerous strategies for energy management have been presented in the previous studies depending upon the generation and utilization of electric power. Authors in [3], proposed a strategy for energy exchange. While a chance constrained optimization is used to minimize the expected operational cost and energy exchange commitments. An optimal power dispatch formulation is presented in [4], where a stochastic weight tradeoff particle swarm optimization based backward-forward sweep is used, to fined the online optimal schedules of microsources. The residential feeder, in terms of reduction in emission and fuel cost and voltage deviation is investigated to check the effectiveness of proposed strategy.

RES integration is very help full in terms of fuel cost and emission minimization. Home energy management system for optimization with integration of RES is proposed in [5–8]. A approach for generation scheduling consist of traditional generators, RES, electric vehicle and storage battery is presented in [5].

Whereas, hybrid differential evaluation and harmony search algorithm is used for scheduling. EV also consumes energy, when it works as a storage system. In [6], authors optimized system by using binary Particle Swarm Optimization (PSO), wind driven optimization, bacterial foraging optimization, Genetic Algorithm (GA) and hybrid GA with PSO. Load curtailment and cost minimization is the objectives of the discussed study. While in [7], system performance evaluated by using GA, binary PSO and Ant Colony Optimization (ACO). The objectives of this work are; electricity bill and PAR reduction, user comfort maximization and minimization of energy consumption. Authors in [8], used GA, teaching learning based optimization, Enhanced Differential Evaluation (EDE) and hybrid EDE with TLBO for scheduling. By integrating RES, minimize electricity bill, PAR, discomfort and emission are the objectives of this work. The RESs are based on uncertainties thats why we use RES, non-RES and ESS in our MG to meet the load demand in an efficient manner.

An expert energy management system to schedule the operation of MG for minimization of operating and maintenance cost is presented in [9]. The objective of both [10,11] is to minimize the overall operational cost. In [10], a control strategy to discard the disturbance in harmonics for MG in grid connected mode is proposed. The results are compared between active power flow and proposed strategy in which size, complexity and cost has been reduced. While [11] based on multiple environment dispatch problem. By using ACO results are compared between two scenarios with and without RES. Authors in [12] proposed a multi-objective energy management for MG. An improved PSO is proposed for energy management in this work. Results show that MG performs better, when electric vehicle work with coordinated mode. However, no method is proposed for load fluctuation management which effect the operating cost.

3 System Model

The microgrid considered in this work includes WT, PV, FC, MT, DG, battery, a bi-directional connection to the utility and the scheduled load demand of a home as shown in Fig. 1. The scheduled load demand is calculated by applying ABC algorithm which helps to find the best schedule of each appliance of resident. Basically, MG works in either grid connected mode or islanded mode. In paper [13], MG works in islanded mode and load demand is assumed. Here, it is working in grid connected mode due to stochasticity in load demand and renewable uncertainty. It is difficult to design a powerful self-sufficient MG system without proper oversizing of generation and energy storage system, so the power exchange with the utility is vital. The MG is operated to reduce the cost of meeting the load demand through DGs, dispatch of storage systems and utility. This work focuses on the scheduling problem to meet the scheduled load demand in an efficient way and excessive power is sold back to utility instead of wastage. ABC optimization technique is used to find the optimal set points of generating units in MG.

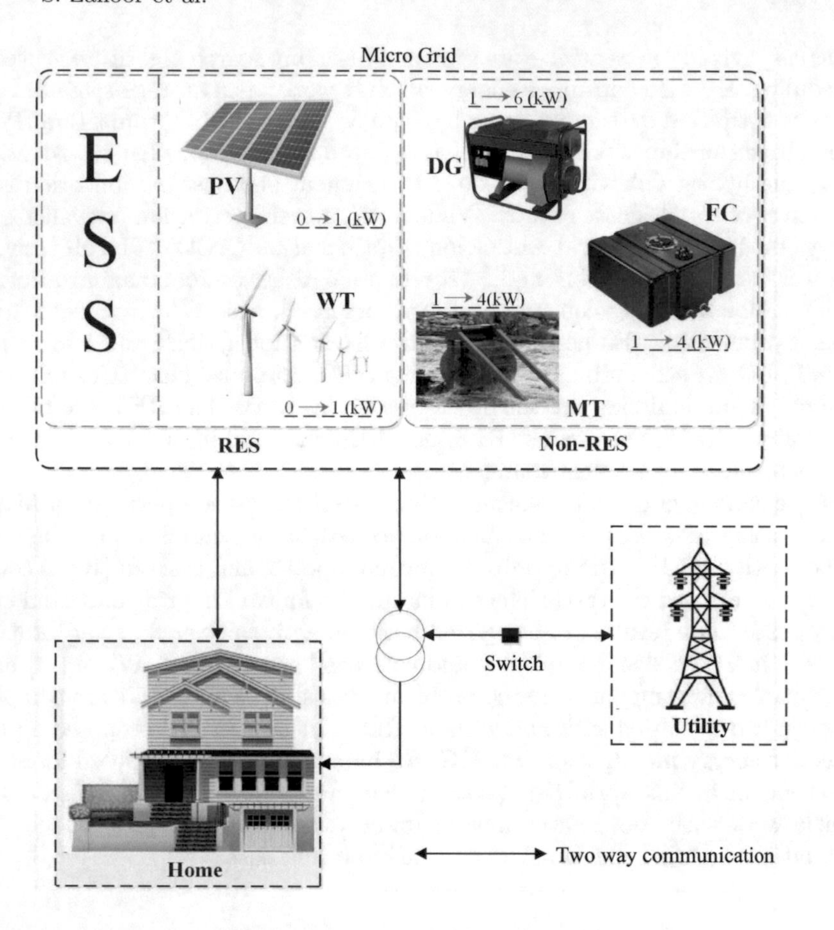

Fig. 1. System model of micro grid

3.1 Modeling the System Components of MG

For efficient energy management of MG, mathematical modeling of WT, PV, MT, DG and ESS are essential.

Wind Turbine. For a WT generation, the wind speed plays a vital role. It deduces energy from wind and transforms it into mechanical power. The extracted power from the WT depends on the wind speed. When the speed of wind exceeds the cut-in limit, the WT generator will start working. If the speed of wind exceeds the rated capacity of wind, it generates persistent output power and if the speed of wind exceeds the cut-out limit the wind generator stops working to protect the generator [14]. The output power of WT at time t is calculated as:

$$P_{WT} = \begin{cases} 0, & v \leq v_{in} > v_{out} \\ P_r \times \dfrac{v(t) - v_{in}}{v_r - v_{in}}, & v_{in} < v(t) \leq v_r \\ P_r, & v_r < v(t) \leq v_{out} \end{cases} \quad (1)$$

Where P_{WT} is the output power of WT, $v(t)$ is the speed of wind in (m/s) at time t, P_r is the rated output power of generator in (kW), v_{out} is cut-out, v_{in} is cut-in and v_r is rated speed of Wind (m/s), respectively.

Photo-Voltaic. A PV is a way of generating electricity from renewable source. PV system transforms light into energy by using semi-conductor material. Each PV panel output power at time t could be obtained from solar radiation which can be calculated as in [15]

$$P_{PV} = A_{PV} \times P_r \times \eta_{PV} \tag{2}$$

Where P_{PV} is the output power of PV, A_{PV} is the area of PV array (w/m^2) and P_r is the rated power of the PV.

Microturbine. MT is a controllable micro-source which generates a constant output power. Its power generation depends on the fuel cost and this cost is related to its working efficiency as shown in (3):

$$P_{MT} = (\chi^t_{MT} \times P^t_{MT} \times C_{ng})/\eta_{MT} \times LHV \tag{3}$$

Where C_{ng} represents the price of natural gas; P^t_{MT} is the output power of MT; χ^t_{MT} represents the status of MT in time t; LHV is the low-hot value of gas, its value is 9.7 kWh/m^3 and η_{MT} is the working efficiency.

Diesel Engine. DE is a distributed generator and its output power is variable according to minimum and maximum limits. We can compute the cost of P_{DE} by [16]:

$$C_{DE} = \alpha + \beta(P^t_{DE}) + c(P^t_{DE})^2 \tag{4}$$

where α, β and c are the constant co-efficients and P^t_{DE} is the output power of DE in time t.

Fuel Cell. Due to dependency of PV and WT on natural resources, its generation is intermittent. While FC overcomes this issue. It converts chemical energy into electrical energy. FC works with high efficiency and less emission of pollutants. The FC generation depends on current and voltage as given in equation:

$$P_{FC} = I_{FC} \times V_{FC} \tag{5}$$

$$P_{FC} = E - V_{ohmic} - V_{conc} - V_{act} \tag{6}$$

Where E is the reversible voltage; V_{ohmic} represents the ohmic voltage; V_{conc} represents the concentration voltage and V_{act} is the active voltage. Using Ohm's law

$$I_{FC} = V_{FC}/R_{FC} \tag{7}$$

Where R_{FC} is the resistance, which is $6.105 \times 10^{-12}\Omega$.

ESS. ESS plays a vital role in energy management of MG. It improves the efficiency and performance of system. So, to optimize a MG, a proper model for ESS (battery) should be developed. The essential physical properties of the ESS are its voltage and the State-of-Charge (SOC). When the energy generated from renewable microsources is excessive then it is stored in battery, if the generation is insufficient then it is discharged. However, the ESS SOC is expressed as:

$$SOC = E_{SB}(t)/E_{SB,cap} \times 100 \tag{8}$$

Where E_{SB} represents the energy stored in battery at time t and $E_{SB,cap}$ represents the energy when the storage is fully charged.

The parameters of controllable microsources are (Table 1):

Table 1. Specifications of microsources

Type	P_{min} (kW)	P_{max} (kW)	$C_m^{SU/SD}$ ($)	K_m ($/kW)
MT	1	4	0.48	0.005
FC	1	4	0.32	0.004
DE	1	6	0.24	0.022

4 Problem Formulation

The motive of the proposed energy management strategy is to find optimal set points of controllable microsources with respect to the economical criteria. A generalized formulation to minimize the cost of MG is given as follow:

$$min(TC) = C_{SB} + \sum_{t=1}^{T}(C_S^t + C_M^t + C_F^t + C_P^t) \tag{9}$$

Where t is the time interval; C_{SB}, C_S^t, C_M^t, C_F^t, C_P^t, respectively, represent the SB cost, startup/shutdown cost, operation and maintenance cost, fuel cost and pollutant emission conversion cost.

5 Simulation Results

In this section, the system under study is a typical low voltage grid-connected MG. To schedule the operations of MG according to load demand of each hour, which is shown in Fig. 1, ABC algorithm is used. The working flow of MG is shown in Fig. 2, which shows that if load demand of a home is equal to RES generation, then our load demand is satisfied. However, when RES generation is excessive than demand, energy is stored in battery, if power is still available then it is sold out to microgrid. If RES generation is insufficient to meet the load demand then battery is discharged, if power is still insufficient then we optimize

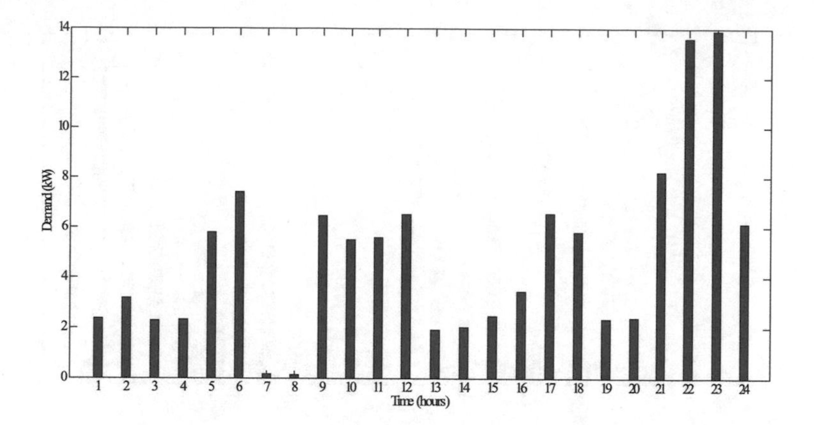

Fig. 2. Load demand

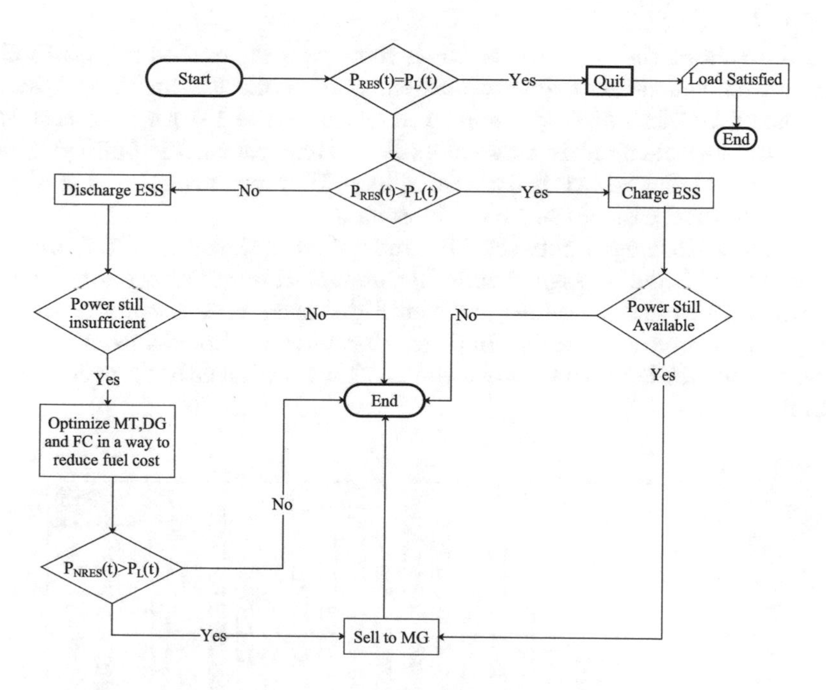

Fig. 3. Flow chart of grid-connected mode

the controllable sources in such a way to reduce cost and carbon dioxide emission. According to this description the obtained results are discussed below.

The output power of PV, WT and load demand is depicted in Fig. 3. While the output power of MT, FC and DE is shown in Fig. 4.

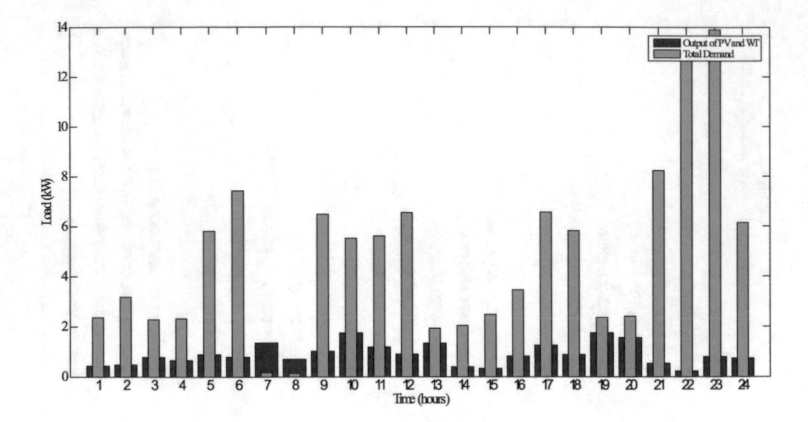

Fig. 4. Output power of PV, WT and net load

Figure 5 shows that the cost of DE is more as compared to MT and FC due to its greater generation capacity and emission, while FC has least cost. The total cost of MG for one day is shown in Fig. 6, where 1–10th hour cost is low because the load demand in these slots is less and demand is fulfilled by RES and in remaining hours cost is little more high. However, overall cost is less. The value for the total cost of the MG is in dollars.

The power exchanged between MG and utility is shown in Fig. 7, where, we observe that with the passage of time the amount of purchasing power from grid is reduced. The negative values in given Fig. 8 show that the power is sold to utility, while the positive values indicate that power is bought from the utility. Hence, the simulation results prove that MG work efficiently according to load demand.

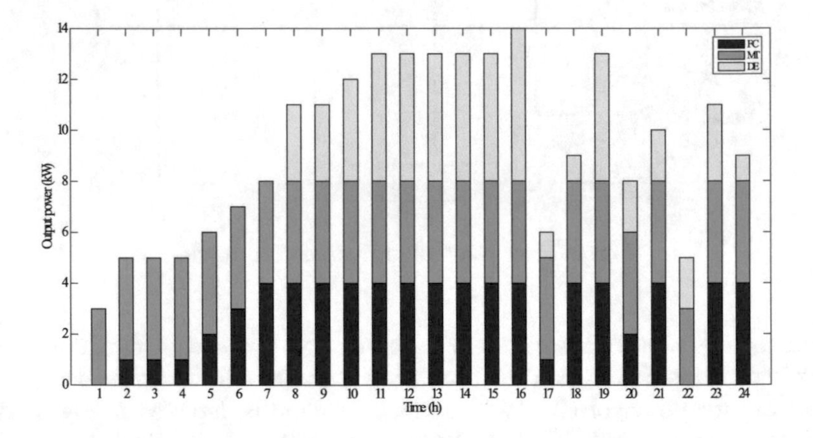

Fig. 5. Output power of controllable microsources

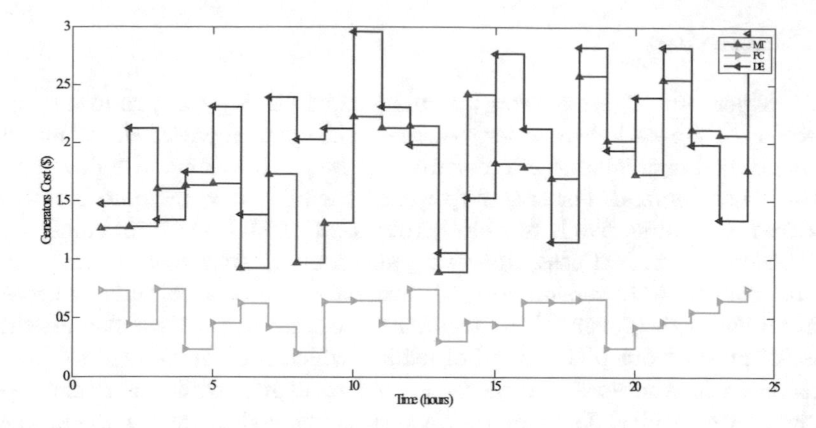

Fig. 6. Cost of each controllable microsource

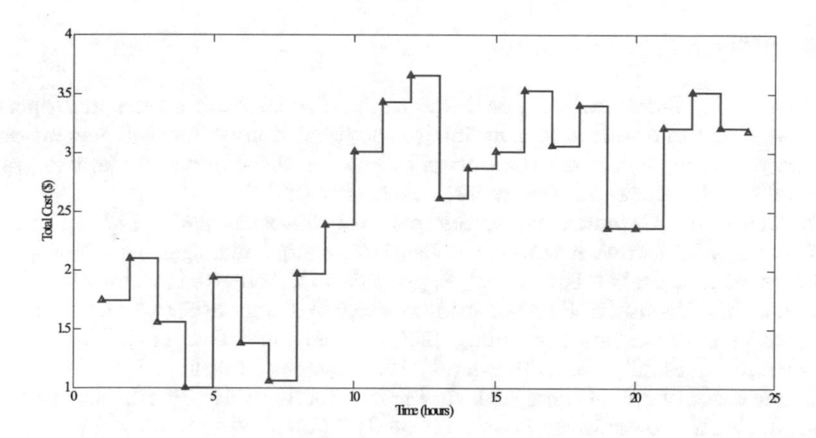

Fig. 7. Total cost of MG

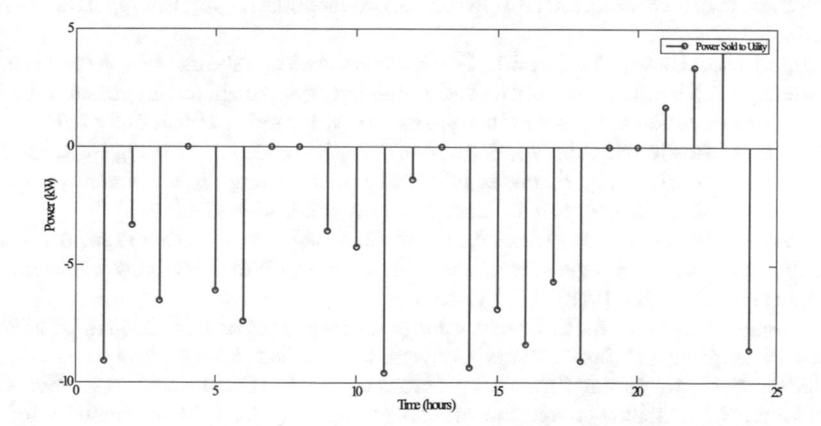

Fig. 8. Power exchange with utility

6 Conclusion

In this paper, an effective energy management strategy is proposed for grid connected MG operations in order to achieve the optimal power allocation for the aforementioned controllable microsources. It helps to balance the power supply and consumer demand. For this purpose, a scheduled load demand of a home is considered. Our objective is to reduce total cost of MG which includes; startup and shutdown cost, fuel cost, operating and maintenance cost, storage battery cost and pollutant conversion cost. To achieve our objective, microsources of a MG are scheduled by using ABC algorithm. Moreover, we have also maximized the use of power from MG instead of utility, which leads us to achieve low cost. In this scenario, the excessive power is sold to utility, while insufficient power is bought from utility. The simulation results show that we have achieved our objective significantly and in most of the hours, MG power is sold to the utility.

References

1. Niknam, T., Golestaneh, F., Malekpour, A.: Probabilistic energy and operation management of a microgrid containing wind/photovoltaic/fuel cell generation and energy storage devices based on point estimate method and self-adaptive gravitational search algorithm. Energy **43**(1), 427–437 (2012)
2. Tomoiaga, B., Chindris, M.D., Sumper, A., Marzband, M.: The optimization of microgrids operation through a heuristic energy management algorithm. In: Advanced Engineering Forum, vol. 8, pp. 185–194. Trans Tech Publications (2013)
3. Zachar, M., Daoutidis, P.: Microgrid/Macrogrid energy exchange: a novel market structure and stochastic scheduling. IEEE Trans. Smart Grid **8**(1), 178–189 (2017)
4. Mohan, V., Singh, J.G., Ongsakul, W., Reshma Suresh, M.P.: Performance enhancement of online energy scheduling in a radial utility distribution microgrid. Int. J. Electr. Power Energy Syst. **79**, 98–107 (2016)
5. Zhang, J., Yihong, W., Guo, Y., Wang, B., Wang, H., Liu, H.: A hybrid harmony search algorithm with differential evolution for day-ahead scheduling problem of a microgrid with consideration of power flow constraints. Appl. Energy **183**, 791–804 (2016)
6. Ahmad, A., Khan, A., Javaid, N., Hussain, H.M., Abdul, W., Almogren, A., Alamri, A., Niaz, I.A.: An optimized home energy management system with integrated renewable energy and storage resources. Energies **10**(4), 549 (2017)
7. Rahim, S., Javaid, N., Ahmad, A., Khan, S.A., Khan, Z.A., Alrajeh, N., Qasim, U.: Exploiting heuristic algorithms to efficiently utilize energy management controllers with renewable energy sources. Energy Build. **129**, 452–470 (2016)
8. Javaid, N., Hussain, S.M., Ullah, I., Noor, M.A., Abdul, W., Almogren, A., Alamri, A.: Demand side management in nearly zero energy buildings using heuristic optimizations. Energies **10**(8), 1131 (2017)
9. Motevasel, M., Seifi, A.R.: Expert energy management of a micro-grid considering wind energy uncertainty. Energy Convers. Manag. **83**, 58–72 (2014)
10. Naderipour, A., Mohn Zin, A.A., Habibuddin, M.H., Moradi, M., Miveh, M., Afrouzi, H.N.: A new compensation control strategy for grid-connected wind turbine and fuel cell inverters in a microgrid. Int. J. Power Electron. Drive Syst. (IJPEDS) **8**(1), 272–278 (2017)

11. Trivedi, I.N., Thesiya, D.K., Esmat, A., Jangir, P.: A multiple environment dispatch problem solution using ant colony optimization for micro-grids. In: 2015 International Conference on Power and Advanced Control Engineering (ICPACE), pp. 109–115. IEEE (2015)
12. Liu, H., Ji, Y., Zhuang, H., Hongbin, W.: Multi-objective dynamic economic dispatch of microgrid systems including vehicle-to-grid. Energies **8**(5), 4476–4495 (2015)
13. Roy, K., Mandal, K.K., Mandal, A.C.: Modeling and managing of micro grid connected system using improved artificial bee colony algorithm. Int. J. Electr. Power Energy Syst. **75**, 50–58 (2016)
14. Garcia, R.S., Weisser, D.: A wind-diesel system with hydrogen storage: joint optimisation of design and dispatch. Renew. Energy **31**(14), 2296–2320 (2006)
15. Maleki, A., Pourfayaz, F.: Optimal sizing of autonomous hybrid photovoltaic/wind/battery power system with LPSP technology by using evolutionary algorithms. Sol. Energy **115**, 471–483 (2015)
16. Shi, W., Xie, X., Chu, C.-C., Gadh, R.: Distributed optimal energy management in microgrids. IEEE Trans. Smart Grid **6**(3), 1137–1146 (2015)

Appliances Scheduling Using State-of-the-Art Algorithms for Residential Demand Response

Rasool Bukhsh[1,4], Zafar Iqbal[2], Nadeem Javaid[1(✉)], Usman Ahmed[4], Asif Khan[1], and Zahoor Ali Khan[3]

[1] COMSATS Institute of Information Technology, Islamabad 44000, Pakistan
nadeemjavaidqau@gmail.com
[2] PMAS Agriculture University, Rawalpindi 46000, Pakistan
[3] CIS, Higher Colleges of Technology, Fujairah Campus, Abu Dhabi 4114, UAE
[4] NFC Institute of Engineering and Fertlizer Research, Jaranwal Road, Faisalabad, Pakistan
http://www.njavaid.com

Abstract. Smart Grid (SG) plays vital role to utilize electric power with high optimization through Demand Side Management (DSM). Demand Response (DR) is a key program of DSM which assist SG for optimization. Smart Home (SH) is equipped with smart appliances and communicate bidirectional with SG using Smart Meter (SM). Usually, appliances considered as working for specific time-slot and scheduler schedule them according to tariff. If actual run and power consumption of appliances are observed closely, appliances may run in phases, major tasks, sub-tasks and run continuously. In the paper, these phases have been considered to schedule the appliances using three optimization algorithms. In one way, appliances were scheduled to reduce the cost considering continuous run for given time slot according to their power load given by company's manual. In other way, actual running of appliances with major and sub-tasks were paternalized and observed the actual consumption of load by the appliances to evaluate true cost. Simulation showed, Binary Particle Swarm Optimization (BPSO) scheduled more optimizing scheduling compared to Fire Fly Algorithm (FA) and Bacterial Frogging Algorithm (BFA). A hybrid technique of FA and GA have also been proposed. Simulation results showed that the technique performed better than GA and FA.

Keywords: Day Ahead Pricing (DAP) · Smart Grid (SG)
Demand Side Management (DSM) · Smart House (SH)
Bacteria Forging Algorithm (BFA) · Firefly Algorithm (FA)
Binary Particle Swarm Optimization (BPSO)

1 Introduction

Globally, power generation systems and plants are expending almost every year because of increasing of electricity demand. Power generation and consumption

L. Barolli et al. (Eds.): EIDWT 2018, LNDECT 17, pp. 292–302, 2018.
https://doi.org/10.1007/978-3-319-75928-9_26

must be balance because there is lack of proper power storage system. A huge amount of power is consumed by households. Cost of house demands can be reduced by shifting electric appliances from peak-cost timings to off-peak timings and this is achieved by Demand Response (DR) program. Demand Response is a change of pattern of electricity usage by consumer according to the fluctuation of electricity prices over time [1]. Both, consumer and utility can benefit from DR program. Consumer can reduce cost by shifting electricity usages [2,3] and on utility side by reducing peak-to-average ration. DR improves the reliability of grid and utilization of utility while protect from outages of the grid [4]. Demand Side Management (DSM) in-corporates with DR for quicker and intelligent response.

Smart grids provide two-way communication opportunity. Utility receives demand from consumer by DR system and consumer alerts with pricing by utility. Pricing pattern fluctuates with demand. This dynamic pricing scheme reflects more accurate relationship between supply and demand compared to flat-pricing schemes. Some basic dynamic pricing schemes are (1) Time of Use (ToU) rates, (2) Critical Peak Pricing (CPP), (3) Extreme Day Pricing, (4) Extreme Day CPP (ED-CPP), (5) Day Ahead Pricing (DAP), (6) Real Time Pricing (RTP). Most used pricing schemes are ToU [5,6] and DAP [7]. However, Demand Response with RTP have proved to be the efficient for market of electricity [8].

New trends of technologies being offered in new appliances by many companies. For example, big companies like LG, Panasonic, General Electronics, Bosch, Whirlpool, Electrolux and Samsung have started to produce smart appliances. These appliances will support DR automatically by networking consumers with the utility. These appliances shall collect information of RTP from smart meters or internet using built-in wireless equipment (e.g. WiFi, Zigbee). Smart appliance may able to schedule automatically according to RTP and also operated by user remotely using smart phones or other gadgets over the internet *. These huge companies are facing huge challenges in cost and compatibility for interoperability among variant brands issues. To understand such challenges, consider the failing of Internet DIOS refrigerator of LG due to high price (10,000) in 2008 [9].

Home energy management system have been proposed to reduce the cost in [22]. Multi-agent system was proposed using BPSO to schedule each appliance. Time of Use was taken as pricing scheme. Priority techniques and integration of electrical supply system were enabled for the proposed solution. The better result came up using priority with user comfort and BPSO. In this paper, DR of Home Energy Management (HEM) have been proposed by using FA, BFA and BPSO to reduce cost by shifting the appliances from peak-time slots to off-peak time slots. Appliances actual power consumption time have been extracted to calculate actual cost. This power curtailment have been considered to reduce more cost for the scheduled appliances. According to a survey 18 % of total energy is consumed by residential buildings and it is increasing annually. In America, 42% of total residential energy is consumed by household appliances [10]. The prime focus of researchers is to optimized these household appliances.

Automated optimizers incorporate with dynamic pricing patterns and schedule the appliances intelligently. It is reported in [10] that these systems prefer artificial intelligence and advance technology of communication.

In next section some related work have been discussed in Sect. 3 problem statement have been elaborated. In Sect. 4 results of simulation have been discussed in detail. Conclusion is summed up in Sect. 5.

2 Related Work

Researchers have been proposing, modeling and implementing various techniques for Smart Grid (SM) analysis and devising many algorithms for analysis and optimization of power consumption and cost for domestic buildings, non-domestic buildings and industries. Optimal scheduler would facilitate the utility and consumer/customer by balancing the usage of electricity to reduce the cost of a whole day. These algorithms considered different parameters like, appliances, user demand, pricing signals and many others.

Optimal scheduling for residential smart appliances using RTP is proposed in [11]. They targeted energy storage, cost minimization and unconventional use of thermal power. Proposed strategy reduced 22.6% cost at peak pricing and 11.7% for normal pricing scheme. It was not on optimization scheme. In [12] architecture was devised to store energy and minimize cost by buying electricity during off peak hours to store in energy storage banks. Optimal scheduling was attained by linear programming.

A model was proposed in [13] which determined the peak demand. Four scenarios were taken in account with limited number of appliances. Pricing tariff was RTP with all scenarios with varying demand of power. Quasi random process expressed the number of appliances while for peak demand calculation a recursive formula was amended. It was observed that, when there were infinite number of appliances it produced serious peak demand. Multiple users with priority load algorithm in an Energy Management System (EMS) was proposed in [14]. Proposed algorithm could reduce the power and cost using ToU scheme.

An algorithm based on mixed integer linear programing was proposed in [15,16] to schedule the appliances. A scheduler discussed in [15] for appliances within home area network to reduce cost and power peak. This scheduler simulated with actual tariffs defined in Czech Republic. Cost minimization for targeted task [16] by making allowance using job switching with multi-level preferences depending demand of lower cost. A scheduler design using Markov decision process based on reinforcement learning argued in [17]. An amendment of Q-learning algorithm was made in designing. This algorithm is free from predefining of function for consumer dissatisfaction when job is rescheduled.

A HEM described in [18] based on game theory of double cooperative which reduced the cost and deliberated utilities were the target. A model of two staged optimization using vehicular scheduler for a home was proposed in [19]. It used Binary Particle Swarm Optimization (BPSO). Power generation cost and stored power cost were calculated at the beginning. Power generation could vary due to wind and photovoltaic and it could reschedule which directly affects the cost determined at next stage. Researcher overcame the load by peak shaving and modification of load curve while cooperating photovoltaic and wind power.

3 Problem Statement

SG provides opportunity of bidirectional communication which supports the consumer to adjust his demand to minimize the cost as well as make demand request to utility. Utility also fulfills the customer request. Actually, SG maintains, improves maintainability and plans for operation of distribution of electricity [21]. So, a system for SG is called Supply Side Management (SSM). Demand Side Management (DSM) facilitates the customer/consumer of electric power to reduce the cost or cut down the power or both to avoid heavy bills and prevent waste of energy. Customers are usually educated to change their behavior toward energy usage with variant tariff of electricity.

With the advancement of technology new appliances are being introduced. These appliances are smarter than existing or previous ones. These are designed to prevent high electric bills and least power consumption is assured for DSM. For the purpose, many optimization techniques have been introduced. Usually, circle of appliances usage complete in a day and may repeat on same time for rest of days unless there are special changes made by consumer. Time of the day is divided into equal 24 time-slots. Usually, appliances complete their operation in one time-slot. The cost of appliance usage for one full day is calculated as in Eq. (1),

$$Cost = \sum_{h=1}^{24} \left(E_r^h \times P_r^{App} \right) \tag{1}$$

where E_r^h is electricity rate in specific hour/time-slot and P_r^{App} is the power rating of specific appliance. The price scheme for the proposed techniques is Day Ahead RTP (DA-RTP) and data was collected from [web]. Three appliances were used for the proposed research. These appliances have different pattern of run. For example, dishwasher completes its operation in given time-slot though it runs for first 5 min then go idle for 10 min and run for 16 min. Later, it goes idle for 23 min and run for 6 min for heated dry. This appliance consumed actual power for only 27 min in one hour. Similarly, cloth-dryer have 43 min of actual run in one hour. Refrigerator runs for full day but for 16 h it runs defrost cycle and consumes 50% less power because of running of air blower and heater during the cycle. Next 8 h it runs for ice-making cycle and consumes maximum power. Figures 1, 2 and 3 show running pattern of dishwasher, cloth-dryer and refrigerator.

In view of above duty cycles of appliances cost of actual run is calculated using Eq. (2).

$$Cost_{(h)} = E_r^h / 60 \times P_r^{App} \tag{2}$$

P_r^{App} of given appliance should in (kWm) instead of (kWh). In this proposed problem, appliances complete their task in given hour for total cost in a full day is given in Eq. (3).

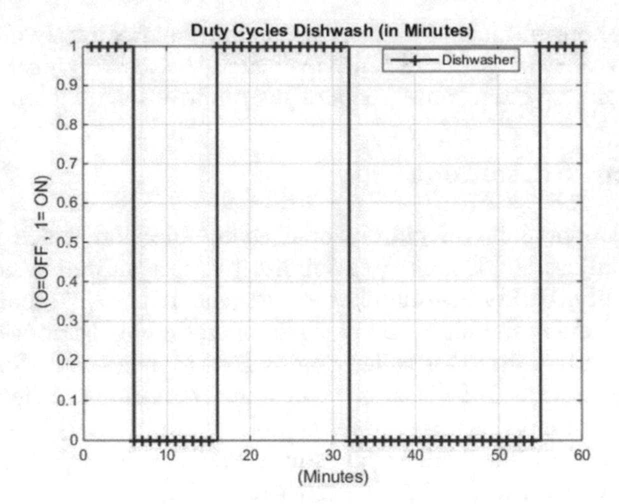

Fig. 1. Dish washer run

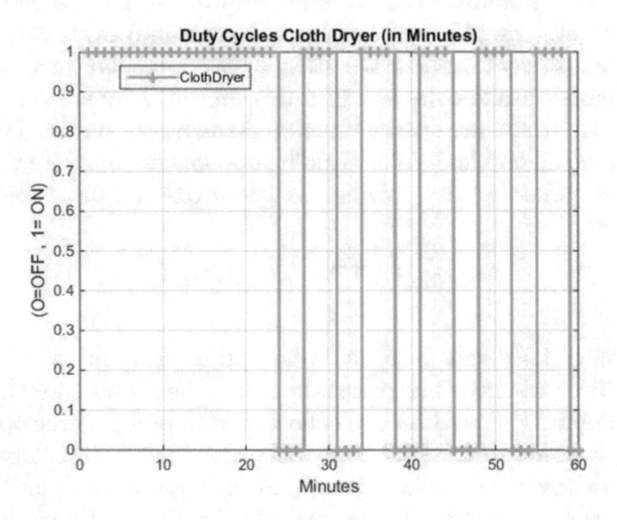

Fig. 2. Cloth-Dryer run

$$Cost_{(total)} = \sum_{h=1}^{24} \left(E_r^h / 60 \times P_r^{App} \right) \tag{3}$$

Total load of appliances are calculated with Eq. (4)

$$Load_{(total)} = \sum_{App}^{3} \left(\sum_{h=1}^{24} App_s^h \times Pr^{App} \right) \tag{4}$$

App_s^h is the status of particular appliance in specific hour. Where status is, $s = [0, 1]$ and Pr^{App} is the power rating of particular appliance. The total load

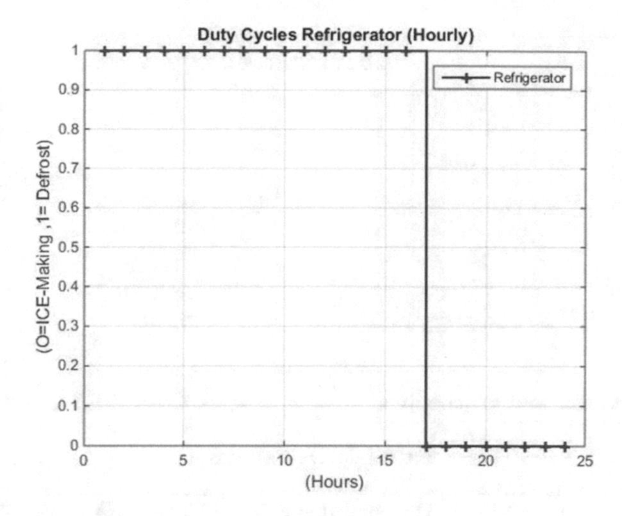

Fig. 3. Refrigerator cycles: defrostration and ice-making.

is curtailed when considering actual running time, load and discarding idle time. There is less power consumption while sub-parts of an appliance turned off for specific time period during the operation. Hence, this aspect of operations for appliances have been taken into account in the proposed research.

4 Simulation Results and Discussion

In proposed problem, three appliances e.g. dishwasher, cloth-dryer and refrigerator were taken in account. All of these were scheduled for cost optimization using three algorithms e.g. BFA, FA and BPSO. All of these showed decrease in cost compared to unscheduled cost.

In Fig. 4 BFA shifted appliances to reduce overall cost for full day. These appliances were taken as operating for complete hour as continuous operations (single tasked operation). In similar pattern is observed in FA as well as in BPSO. In early hours BFA have dominant graph line. These are the hours when tariff is low. After 15th hour BFA graph line is submissive to unscheduled graph line. These are the hours when tariff is high. Similar graph lines are observed in Fig. 5 for FA. FA almost, overlapping or submissive during 13th hour to 15th hour and from 22nd to 24th hour. This is almost, opposite to BFA graph line with unscheduled graph line. But, BFA have lesser total cost from FA. As shown in Table 1. In Fig. 6 BPSO shows unique pattern. BPSO have shifted more appliances before 15th hour where tariff is lesser. When most of the appliances operate in less tariff time-slots there is ultimately more reduced cost. This is why, BPSO is one of the most efficient compared to BFA and FA as shown in Table 1.

The operations of appliances may discontinue for few time and continue for other length of time. There are appliances which operate with variant lost by turning on and off sub-components with different time lengths. In proposed

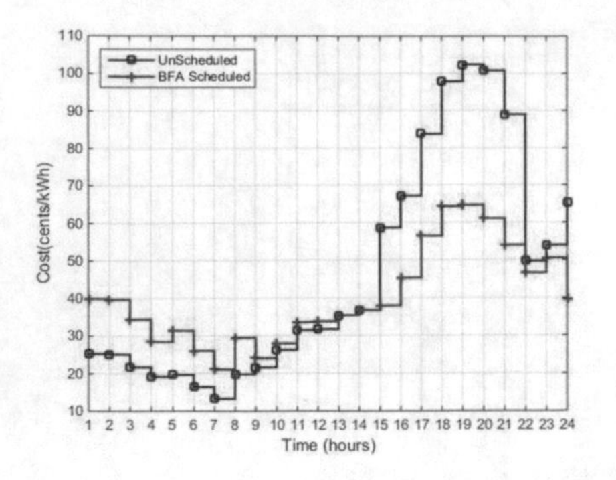

Fig. 4. Cost of BFA and unscheduled

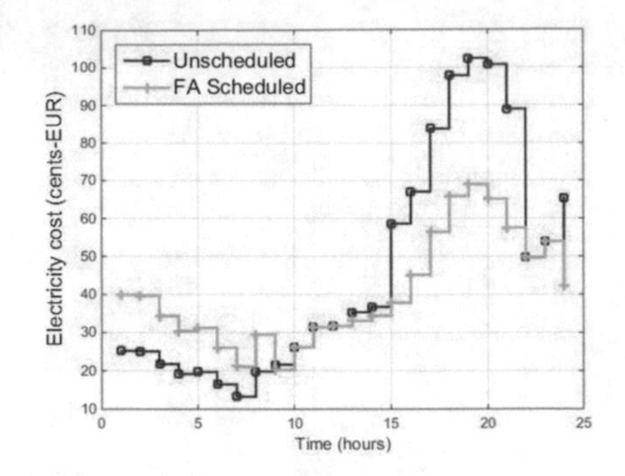

Fig. 5. Cost of FA and unscheduled

Table 1. Overall costs

Task	Unschedule	BFA	BPSO	FA
Continuous	1111	960.48	916.20	972.09
Duty-Cycle	816.56	716.33	687.93	734.83

problem, dishwasher and cloth-dryer have idle states and running states. While, refrigerator's load varies by putting the compressor in off state for 16 h during defrost. The lost also reduced to half compared to full run with compressor during ice-making cycle. In proposed research, dishwasher runs for first 5 min and it runs for 16 min after pause for 10 min. At the end runs for last 5 min after wait for 23 min. This is how it completes it task in one time-slot. While Cloth-dryer's heater runs for 23 min in the beginning and remains idle for 3 min and again run for 4 min. This on and idle state continue for one complete hour (time-slot). These running behavior of appliances consume less power as compared to continuous (single tasked operation) running of appliances.

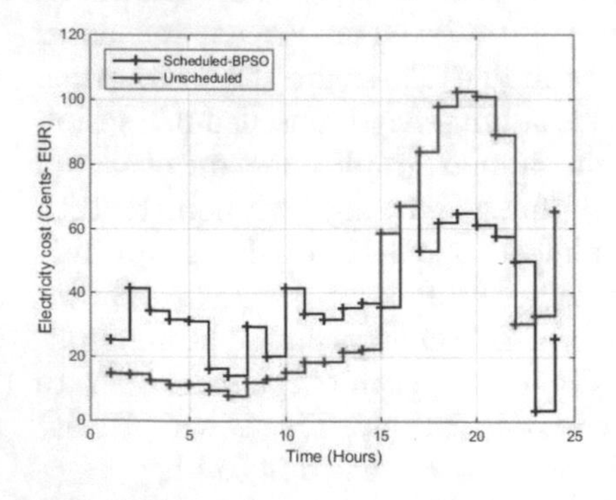

Fig. 6. Cost of BPSO and unscheduled

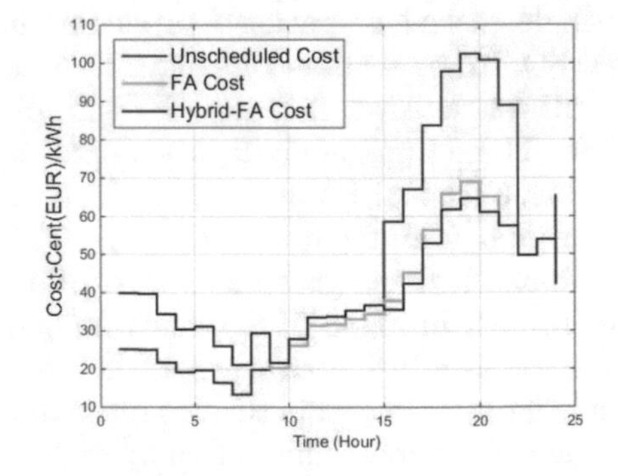

Fig. 7. Cost of Hybrid-FA, FA and unscheduled

In view of above scenario all of the techniques showed lesser power consumption and ultimately lesser cost. Table 1 shows costs for all the techniques used. BPSO with duty-cycle running of appliances showed 687.93 Cents compared to 816.56, 716.33, 734.83 Cents of unscheduled, BFA and FA respectively. An other technique have also been proposed and implemented, that is hybrid of FA. The population of FA made to be updated with Genetic Algorithm (GA). This hybrid approach have compelled the appliances to be scheduled in lesser tariff time-slots. Regular FA optimized the appliances according to their run in given time-slots by ignoring the values of tariff. Hybrid FA improves the regular FA by shifting the appliances run time from high tariff time-slots to lesser tariff time-slots. Table 2 summarizes the results for FA and hybrid FA.

Table 2. Overall total costs with Hybrid-FA

Unscheduled	FA	GA	Hybrid-FA
1111	972.09	1047.69	962.682

FA remained efficient by reducing 12.51% from unscheduled cost while hybrid-FA reduced 13.35% cost. Hybrid-FA reduced its cost for 0.97% compared to regular FA. A bigger percentage of difference may occurred if there were more the appliances used. This proposed approach is good for multiple appliances which schedules to reduce the total cost.

In Fig. 7 it is observed that appliances were shifted from high tariff hours (e.g. 15th hour to 21st hour) to low tariff hours (e.g. 9th hour to before 15th hour). This has definitely reduced over all cost of hybrid-FA. There were only three appliances in proposed research. It would have more percentage of cost reduction if there were more appliances which would shift to lower tariff to make significant difference of cost indeed.

4.1 Feasible Region

'Feasible region' is the region which is identified by set of points which meet certain constraints by satisfying the required result using objective function. In proposed research feasible regions of cost functions are identified for three appliances using BFA, FA and BPSO techniques. The pricing scheme DA-RTP' is used. Two level of objective functions for cost reduction have been used. In first level, appliances were scheduled/optimized to reduce the cost while running as major tasked or single tasked. In next level, duty cycles of appliances have been considered.

In Fig. 8 P_1, P_2,P_3 and P_4 identify the region for reduced cost using BPSO for all the three appliances. The highest cost stood 102.41 which is less than unscheduled cost of P_5 (106.722). But, it is higher cost when duty cycles of appliances were took into account using same technique e.g. P_6 where highest cost with highest load is 12.21. Which is the highest cost when duty cycle took into account for cost reduction and power curtailment.

In Fig. 9 the feasible region, P_1, P_2, P_3, P_4 identify the cost function when duty cycles of appliances were took into account and scheduling them using BFA technique. The highest cost is 57.8527 for scheduled appliances with duty-cycles. It is lesser than 104.1348 Cents which is highest cost when appliances are scheduled with continuous (single tasked) operations at point P_5. This scheduled highest cost is lesser than unscheduled cost at point P_7 with 106.722 Cents. Making is highest cost for compared to rest of two methods.

The highest cost of appliances, when unscheduled, is 106.722 at P_7 in Fig. 9. It reduced to 102.41 at P_5 when appliances were scheduled using Fire Fly (FA) algorithm. Duty cycles of appliances further squeeze the costs of appliances. The P_1, P_2, P_3, P_4 identify the region where duty-cycled scheduled appliances have optimized load and cost. The point P_3 has the highest cost '56.8944' when scheduled with duty-cycles.

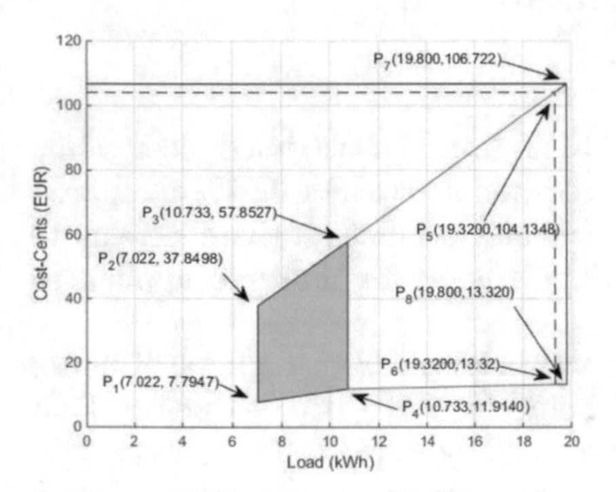

Fig. 8. Feasible region for BFA

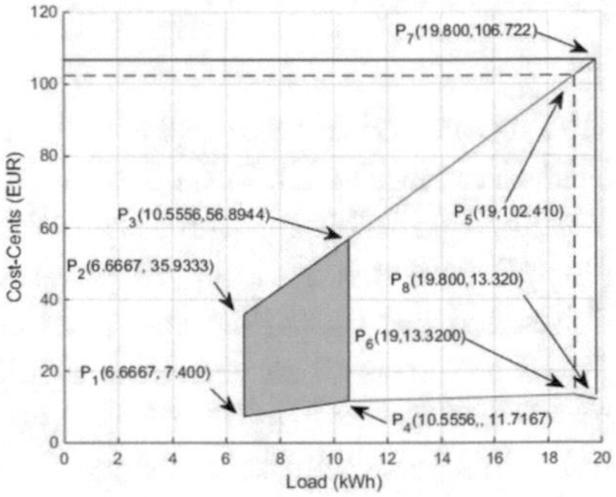

Fig. 9. Feasible region for FA

In Fig. 10 costs of BFA, FA and BPSO are lesser from unscheduled costs and higher than scheduled appliances with duty-cycle costs. SO, cost of appliances with duty-cycle scheduling are lesser because the actual loads of appliances were taken into account of observation during particular time-slots. As the appliances shift the status of their sub-components or their own status for limited time in given time-slot which curtailed the actual big load.

A righteous research may help the investor to take right decision. For that correct should be extracted. An erroneous research may discourage to take righteous decision rather betray the investor. Smart appliances switch on and off their sub-components or themselves using live sensors or data analysis, according to the situation. In proposed research, actual duty-cycles of appliances were taken. This strategy curtailed the full load of appliance for specific time length. It, ultimately, has reduced the cost. So, in first step, scheduling of appliances with single-tasked operation reduced cost. In second step, load curtailed in scheduled appliances due to consideration of duty-cycle which, further reduced the cost as

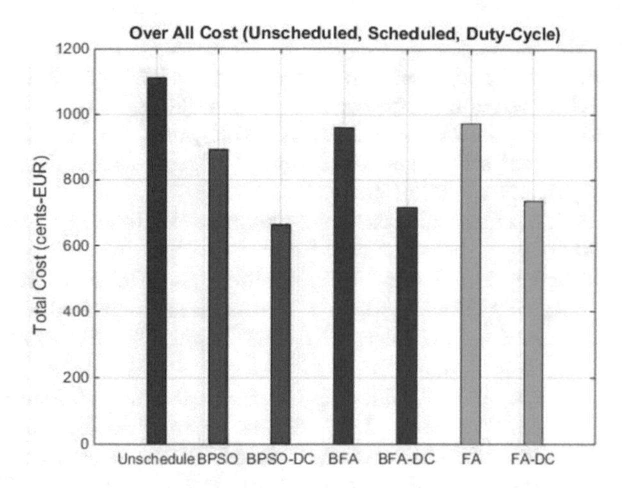

Fig. 10. Total costs of all

shown in Fig. 10. Appliances scheduled with BPSO have the least cost in single-tasked and sub-tasked. In future research, parameter like actual operation load should be considered to observe more accurate outcomes.

5 Conclusion

Appliances are not necessary consume power at their fullest all the time. Specially, intelligent and semi-intelligent appliances stop and shift the controls among sub-components, effecting the overall power consumption. In proposed scenario, refrigerator shifted control between fan and compressor while heater of cloth dryer stopped and run periodically even within the time-slot. Hence, true power consumption is lesser. The total cost with duty-cycle in all technique stood lesser. However, BPSO scheduled the appliances with most optimizing compared to FA, BFA and Unscheduled. The proposed hybrid of FA and GA performed 0.8% more efficient from FA and 0.96% from GA.

References

1. Albadi, M.H., El-Saadany, E.F.: A summary of demand response in electricity markets. Elect. Power Syst. Res. **78**(11), 1989–1996 (2008)
2. Jazayeri, P., Schellenberg, A., Rosehart, W.D., Doudna, J., Widergren, S., Lawrence, D., Mickey, J., Jones, S.: A survey of load control programs for price and system stability. IEEE Trans. Power Syst. **20**(3), 1504–1509 (2005)
3. Kirschen, D.S.: Demand-side view of electricity markets. IEEE Trans. Power Syst. **18**(2), 520–527 (2003)
4. Braithwait, S., Eakin, K.: The role of demand response in electric power market design. R. Christensen Asssociates Inc., Edison Electric Institute (2002)

5. Aalami, H., Yousefi, G.R., Moghadam, M.P.: Demand response model considering EDRP and TOU programs. In: Proceedings of IEEE/PES Transmission and Distribution Conference and Exposition, pp. 1–6 (2008)
6. Asano, H., Sagai, S., Imamura, E., Ito, K., Yokoyama, R.: Impacts of time-of-use rates on the optimal sizing and operation of cogenerationsystems. IEEE Trans. Power Syst. **7**(4), 1444–1450 (1992)
7. Bloustein, E.: Assessment of Customer Response to Real Time Pricing. Rutgers University (2005)
8. Zurn, H.H., Tenfen, D., Rolim, J.G., Richter, A., Hauer, I.: Electrical energy demand efficiency efforts in Brazil, past, lessons learned, present and future: a critical review. Renew. Sustain. Energy Rev. **67**(Suppl. C), 1081–1086 (2017). ISSN: 1364-0321
9. Yi, P., Dong, X., Iwayemi, A., Zhou, C., Li, S.: Real-time opportunistic scheduling for residential demand response. IEEE Trans. Smart Grid **4**(1), 227–234 (2013)
10. Energy Information Administration, United States Department of Energy, Washington, https://www.eia.gov/todayinenergy/detail.cfm?id=12251. Accessed 10 Oct 2017
11. Shirazi, E., Jadid, S.: Optimal residential appliance scheduling under dynamic pricing scheme via HEMDAS. Energy Build. **93**, 40–49 (2015)
12. Adika, C.O., Wang, L.: Smart charging and appliance scheduling approaches to DSM. Int. J. Electr. Power Energy Syst. **57**, 232–240 (2014)
13. Vardakas, J.S., Zorba, N., Verikoukis, C.V.: Power demand control scenarios for SG appliances. Appl. Energy **162**, 83–98 (2016)
14. Abushnaf, J., Rassau, A., Grnisiewicz, W.: Impact on electricity use of introducing time of use pricing to a multiuser home energy management system. Int. Trans. Electr. Energy Syst. (2015)
15. Bradac, Z., Kaczmarczyk, V., Fiedler, P.: Optimal scheduling of domestic appliances via MILP. Energies **8**(1), 217–232 (2014)
16. Jovanovic, R., Bousselham, A., Bayram, S.I.: Residential Demand Response Scheduling with Consideration of Consumer Preferences. Appl. Sci. **6**(1), 16 (2016)
17. Wen, Z., O'Neil, D., Maei, H.: Optimal demand response using device-based reinforcement learning. IEEE Trans. Smart Grid **6**(5), 2312–2324 (2015)
18. Gao, B., Liu, X., Zhang, W., Tang, Y.: Autonomous household energy management based on a double cooperative game approach in the smart grid. Energies **8**(7), 7326–7343 (2015)
19. Rabiee, A., Sadeghi, M., Aghaeic, J., Heidari, A.: Optimal operation of microgrids through simultaneous scheduling of electrical vehicles and responsive loads considering wind and PV units uncertainties. Renew. Sustain. Energy Rev. **57**, 721–739 (2016)
20. Rasheed, M.B., Javaid, N., Ahmad, A., Khan, Z.A., Qasim, U., Al-rajeh, N.: An efficient power scheduling scheme for residential load management in smart homes. Applied Sciences **5**(4), 1134–1163 (2015)
21. Chiu, W.-Y., Sun, H., Poor, H.V.: Energy imbalance management using a robust pricing scheme. IEEE Trans. Smart Grid **4**(2), 896–904 (2013)
22. Shah, S., Khalid, R., Zafar, A., Hussain, S.M., Rahim, H., Javaid, N.: An optimized priority enabled energy management system for smart homes. In: 2017 IEEE 31st International Conference on Advanced Information Networking and Applications (2017)

Optimal Energy Management in Microgrids Using Meta-heuristic Technique

Anila Yasmeen[1], Nadeem Javaid[1(✉)], Saman Zahoor[1], Hina Iftikhar[1], Sundas Shafiq[1], and Zahoor Ali Khan[2]

[1] COMSATS Institute of Information Technology, Islamabad 44000, Pakistan
nadeemjavaidqau@gmail.com
[2] CIS, Higher Colleges of Technology, Fujairah Campus, Fujairah 4114, UAE
https://www.njavaid.com

Abstract. The energy crisis and greenhouse gas emission are increasing around the world. In order to overcome these problems, distributed energy resources are integrated which introduce the concept of microgrid (MG). The microgrid exchanges power with utility to meet load demand with the help of common coupling point. An energy management strategy is proposed in this work, which helps to minimize the operating cost of MG while considering all constraints of the system. For this purpose, a firefly algorithm is used to schedule generators of MG to fulfill the consumer demand considering the desired objectives. The proposed scheme employs FA to minimize the operating cost of a MG. In both grid-connected and islanded modes of MG, proposed scheme is applied for scheduling of distributed generators. The Significance of the proposed strategy is verified through simulations and results.

Keywords: Energy management strategy · Microgrid
Firefly algorithm

1 Introduction

An electric system which consists of various loads and renewable generation units, operated with power grid in parallel makes a microgrid (MG). It raises system reliability with distributed generation and efficacy with lowered transmission time. A power supply system that incorporates renewable or non-renewable distributed power generation and energy storage system (ESS) to facilitate load demands [1]. It also helps policy makers and stockholders to increase choices for both consumers and sellers to fulfill their electricity demands and monetary cost reduction. Furthermore, the growing tendency towards high penetration of renewable energy sources forms the environment friendly and cost effective over traditional generation of power supply.

A MG usually operates in two modes: grid-connected and islanded mode. In grid-connected mode, MG operates with macrogrid in parallel through a point

L. Barolli et al. (Eds.): EIDWT 2018, LNDECT 17, pp. 303–314, 2018.
https://doi.org/10.1007/978-3-319-75928-9_27

of common coupling and it can purchase and sell the electricity. Moreover, for efficient energy operation of MG, energy management system is required that maintains all power generations in MG to enhance the performance of power system. The MG which satisfy the energy demands of various buildings (residential, commercial or industrial), can improve the performance of power systems, if energy management system (EMS) gathers data from these buildings to work collectively with efficacy, ultimately enhance the system reliability.

MGs are suitable for distributed renewable generation units, however, stochastic and intermittent nature of renewable power should be reduced with dispatchable power units such as fuel cell (FC), microturbine (MT) and ESS [2]. Due to stochasticity of renewables, a hybrid power system consists of various kinds of renewable power generation units is suitable and reliable than a single power system when power supply is provided [3]. Commonly, hybrid power system is built when power grid is far away such as in islands, mountains and so on. A hybrid power system with suitable size makes each generation unit to operate at its full capacity. Furthermore, it shows significance owing to reliable power supply and reduced system cost.

In this paper, the power generators are scheduled to fulfill the load demand by minimizing the generation cost. To solve the scheduling problem on supply side management (SSM), a meta-heuristic firefly algorithm (FA) is used. In this scenario, we utilize maximum energy from MG and then remaining from utility in case if required to consumer. This action leads us to mitigate environmental damage and power losses and low electricity cost. Exploiting above mentioned techniques, our major contributions are: scheduling of DERs of MG and operating cost minimization. The remaining paper is organized as; in Sect. 2 literature review is discussed, Sect. 3 illustrates the problem statement, Sect. 4 presents proposed system model, Sect. 5 contains simulation results and discussion, Sect. 6 concluded the paper.

2 Literature Review

MG scheduling is used to dispatch renewable resources and controllable loads. It also helps in coordination between charging and discharging of storage systems. In [4], authors reduced the troublesome impact of MG on the macrogrid in market structure using constrained energy exchange. This approach is useful to reduce the uncertainty and MG residual load depending on two tunable parameters, the schedule adaptability and schedule elasticity. The authors in [5] used an optimal sizing for power system to optimize the size of stand-alone hybrid renewable resources/battery/hydrogen for reliable economic supply. Maximizing the system reliability and minimizing annual cost of system is a major objective of this paper. To solve the problem of HS-BH an improved ant colony optimization (ACO) is used. Simulation result proves that minimizing system cost and maximizing reliability is paradoxical.

Authors used hybrid algorithm chaotic binary particle swarm optimization (CBPSO) to reduce economic cost [6]. BPSO is combined with chaos optimization algorithm to improve global search capability of BPSO. Fuzzy logic based mathematical theory is proposed to optimize model using some constraints to reduce economic cost and network loss of MG. Stochastic net resources are used to handle the negative load of uncertainty due to uncontrollable microsources. Stop and start strategy is applied to cooperate with optimal storage devices. Stochastic weight tradeoff (SWT-PSO) and modified backward-forward sweep BFS based optimal power flow (OPF) is used to enhance the performance of online energy scheduling of MGs to achieve in terms of emission cost, fuel cost and time set in DERs. Online OPF is found better than online economic dispatch (ED) in combination of grid trade and load curtailment for economic cost [7]. In [8], authors proposed an optimal centralized scheduling method to reduce electricity cost and electricity consumption. RESs and electrical vehicle (EV) are used and with the usage of these resources, MG becomes more stable, reliable and cost effective. Mixed integer linear programming (MILP) is used to solve problem optimally.

MG along with energy management system is used to attain the optimization goals. Multiple objectives have been achieved based on different constraints and decision criteria [9]. Two techniques genetic algorithm (GA) and modified MILP have been proposed under the network and unit constraints to schedule the unit commitment and economic dispatch of the MG. The modified GA is flexible, diversified and with better convergence behaviour provides the optimizing solution for MG along with MILP for handling the nonlinear network topology constraints. Furthermore, it is tested under different practical policies and applicable constraints.

MG works as an islanded mode or a grid-connected mode. The islanded mode can be explained simply when MG works as local energy provider without any help from MG. It helps to reduce energy expenses and pollutant emission by efficient use of DERs. The authors in [10] studied MILP approach for the optimal scheduling of a smart home's energy consumption. Moreover, for further reduction in energy consumption cost real-time pricing signal is adopted which gives more realistic results for peak reduction. Simulations are done on 30 and 90 homes with their own MG (a local energy provider) to get the possible cost saving in energy consumption. The principal goal of this study is to understand the weather conditions over Saharan area of south Algeria (Adrar) for analysing the performance of grid-connected photovoltaic (PV) system [11]. The region has low humidity rate, the high temperature in summers and strong potential for solar insulation. The experimental results show that the system efficiency and performance ratio caused by the high temperature is 41.1 °C in the daylight, and the continuous changing of the solar irradiance is due to variation in clouds behaviour and storm effects which may affect the stability and sustainability of PV system.

The scheduling for home appliances is done under real-time pricing with the integration of renewable energy resources [12]. Different users are assigned different priorities depending on their energy demand. The purpose is to propose a scheme which maximizes user comfort while reducing energy consumption cost. There exists a tradeoff between cost and waiting time. The priority enables early deadline first scheduling algorithm is proposed based on priority constraints. The simulation results validate the efficiency of the proposed technique in terms of user comfort maximization and cost reduction. In this paper, GA based energy management controller performs more efficiently than BPSO based energy management controller and ACO based energy management controller in terms of electricity bill reduction, peak to average ratio minimization and user comfort level maximization.

The authors in [13] designed the home energy controller on the basis of heuristic optimization techniques such as GA, BPSO and ACO. In addition to this, the problem is formulated via multiple knapsack problem. Simulation results show that GA based energy management controller performs more efficiently than BPSO and ACO based energy management controller in terms of cost saving while considering peak to average ratio minimization and user comfort level maximization.

An optimal resource management strategy is proposed in [14]. For this purpose, multi-agent system under consideration of finite time distributed optimization is applied. Without a central controller proposed framework in efficient manner. For simulation, IEEE 14-bus and 162-bus system and a real standalone MG depict that proposed framework is ideal for actual MG.

In [15], optimization techniques are proposed GA, teaching learning-based optimization (TLBO), enhanced differential evolution (EDE) algorithm and the proposed enhanced differential teaching-learning algorithm (EDTLA) to manage energy and user comfort, along with the integration of RESs into the MG. The desired objectives of this paper are cost minimization, integration of RESs, minimizing discomfort and minimizing the peak to average ratio (PAR) and carbon emission. The simulation results validate the efficiency of proposed techniques.

A demand response based energy management strategy is proposed for industrial sector using state task network (STN) and MILP in [16]. The pricing scheme used in this paper is day-ahead and proposed strategy demonstrates the scheduling of distributed generation units and schedulable tasks. Furthermore, implementation of proposed strategy insures the reliability of the energy system and saving the operational cost.

3 Problem Statement

Several distributed energy resources (DERs), ESS when integrated in a limited area, makes a MG to meet variable load demands. MG operates in two modes: Islanded/standalone mode and grid-connected mode. In islanded/standalone mode, main grid is not involved when MG operates whereas in grid-connected mode, MG operations are dependent on main grid. Pavan et al. [17] presented hybrid power system for green buildings. Sizing and management strategies of renewable energy based MG, are developed. With the support of hybrid optimization model for electric renewables (HOMER), technical feasibility and optimal economy are achieved by integrating clean energies to reduce carbon dioxide emissions. Additionally, Aduouane et al. [18] developed an interesting standalone

PV system integrated with LPG generator and storage batteries. After brief review of hybrid power systems, we observed another interesting adaptive power smoothing controller based on fuzzy logic is proposed by Nahidul et al. [19]. The proposed system efficiently manages the battery storage state of charge (SOC) by considering DC-DC buck converter for control scheme and frequency resynchronization technique to adopt switching between DC and AC buses for islanded/grid-connected MG systems. Authors in research articles [17–19] integrated RESs where they efficiently manage reliable supply of power, however, operating cost minimization is ignored and excess energy generated from RESs is dissipated as dump losses.

4 Proposed System Model

We have proposed a hybrid energy generation system (HEGS) in both modes: grid-connected mode and islanded mode. The HEGS is integrated with a supermarket. It constitutes different DERs, i.e., PV panel, wind turbines (WT), FC and an ESS. An energy management strategy for a HEGS integrated with a supermarket is shown in Fig. 1. The main purpose of the proposed strategy is to optimally schedule the power generation of different DERs in order to provide reliable power during uncertain conditions, keeping the operating cost under consideration. The operating cost involves the maintenance cost of DERs, fuel cost and the cost of electricity bought from the utility. The distributed generators of HEGS have been scheduled using the heuristic technique which helps to turn on the generators according to the load demand considering the economical criteria. A secondary frequency control for the converter between the AC and DC buses is considered for islanding and grid-connected operations. The excessive power is stored in batteries, if there is still surplus power, it is used to produce hydrogen through electrolyzer. The produced hydrogen is used as fuel for FC. A switch will enable islanded or grid-connected mode according to the load demand. In grid-connected mode, if there is surplus power, then it can be sold back to the grid. The system objective function is to minimize the operating cost with respect to the system security constraints ensuring reliable power supply. The overall objective is to minimize the power supply utility grid and meeting the overall load by maximizing renewable energy utilization. For System energy management, mathematical modeling of PV, WT, FC and ESS is essential.

4.1 PV System

A PV is a way of generating electricity from renewable source. PV system transforms light into energy by using semi-conductor material. The each PV panel output power at time t could be obtained from solar radiation which can be calculated as in [20]:

$$P_{PV} = A_{PV} \times P_r \times \eta_{PV} \tag{1}$$

where P_{PV} is the output power of PV, A_{PV} is the area of PV array (w/m^2) and P_r is the rated power of the PV.

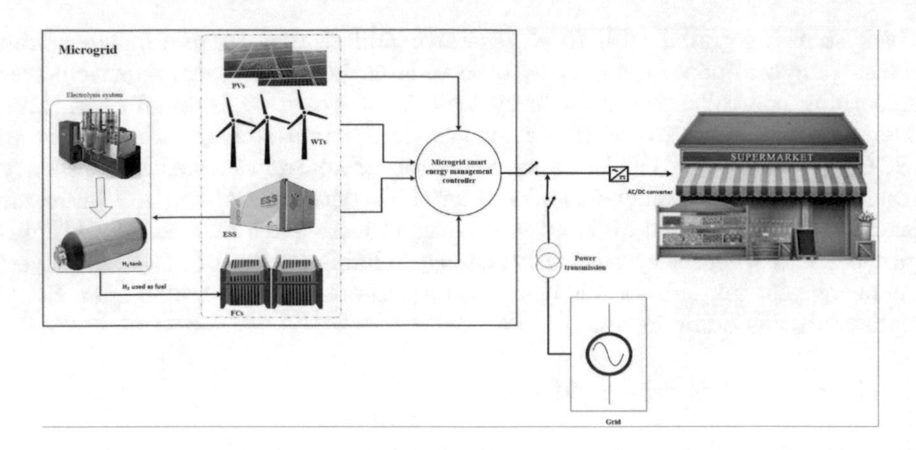

Fig. 1. Proposed system model

4.2 WT

For a WT generation, the wind speed plays a vital role. It deduces energy from wind and transform it into mechanical power. The extracted power from the WT depends on the wind speed. When the speed of wind exceeds the cut-in limit, the WT generator will start working. If the speed of wind exceeds the rated capacity of wind, it generates persistent output power and if the speed of wind exceeds the cut-out limit the wind generator stops working to protect the generator [21]. The output power of WT at time t, is calculated as:

$$P_{WT} = \begin{cases} 0, & v \leq v_{in} > v_{out} \\ P_r \times \dfrac{v(t) - v_{in}}{v_r - v_{in}}, & v_{in} < v(t) \leq v_r \\ P_r, & v_r < v(t) \leq v_{out} \end{cases} \tag{2}$$

Where P_{WT} is the output power of WT, $v(t)$ is the speed of wind in (m/s) at time t, P_r is the rated output power of generator in (kW), v_{out} is cut-out, v_{in} is cut-in and v_r is rated speed of Wind (m/s), respectively.

4.3 FC

Due to dependency of PV and WT on natural resources, its generation is intermittent. While FC overcome this issue. It converts chemical energy into electrical energy. FC works with high efficiency and less emission of pollutants. The FC generation depends on current and voltage as given in equation:

$$P_{FC} = I_{FC} \times V_{FC} \tag{3}$$

$$P_{FC} = E - V_{ohmic} - V_{conc} - V_{act} \tag{4}$$

where E is the reversible voltage; V_{ohmic} represents the ohmic voltage; V_{conc} represents the concentration voltage and V_{act} is the active voltage. Using ohm's law

$$I_{FC} = V_{FC}/R_{FC} \tag{5}$$

where R_{FC} is the resistance, which is $6.105 \times 10^{-12}\Omega$.

4.4 ESS

ESS plays a vital role in energy management of MG. It improves the efficiency and performance of system. So, to optimized a MG, a proper model for ESS (battery) should be developed. The essential physical properties of the ESS is its voltage and the SOC. When the energy generated from renewable microsources is excessive then it is stored in battery, if the generation is insufficient then it is discharged. However, the ESS SOC is expressed as:

$$SOC = E_{SB}(t)/E_{SB,cap} \times 100 \tag{6}$$

where E_{SB} represents the energy stored in battery at time t and $E_{SB,cap}$ represents the energy when the storage is fully charged.

5 Simulation Results

For the application of proposed strategy, we have considered a supermarket of 13,058 m^2 area and a MG. The MG considered in this work consists of a set of DERs which includes: a PV panel with maximum power rating of 1160 kW, two WTs, two FCs and an ESS. The power generation of different DERs is given in Fig. 2. In this work, we have evaluated the performance of proposed strategy in two modes: grid-connected mode, islanded mode. It is important to mention that, in this study, we have considered a period of one day which is further divided into the time intervals of one hour. According to the load demand as given in Fig. 3, the MG scheduler schedules the DERs using firefly optimization technique which is used to get the optimized states of DERs. The pricing signal is given in Fig. 4 which the bids for operating the MG generation units. The scheduling procedure is done in such a way to minimize total operating cost of DERs while creating a balance between demand and supply. In grid-connected mode the MG scheduler acquires the ON/OFF states of DG sources to minimize the total operating cost. While scheduling in grid-connected mode, RESs are given more priority to turn on because of their environment friendly nature and less bid costs. If consumer's demand at a specific time interval is less than the power supplied by MG, then ESS will be charged by the surplus power considering its constraints. Figure 5 demonstrates that if surplus power is more than the per hour charge rate of ESS, the remaining power will be sold back to the grid. Whereas, when power supplied by MG is less than the consumer demand, the required load demand will be fulfilled through the energy stored in ESS. However, if the required power is greater than the discharge rate of ESS, then remaining

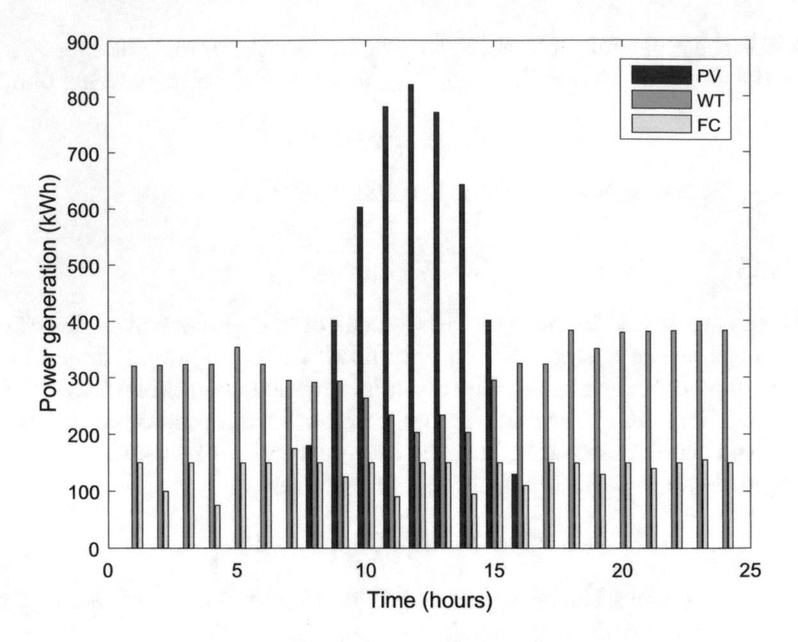

Fig. 2. Power generation

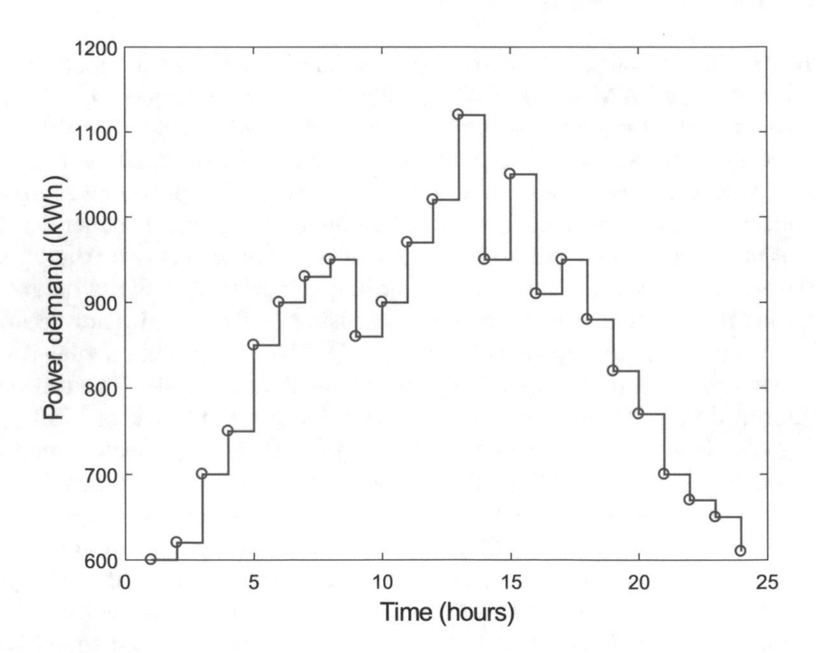

Fig. 3. Power demand

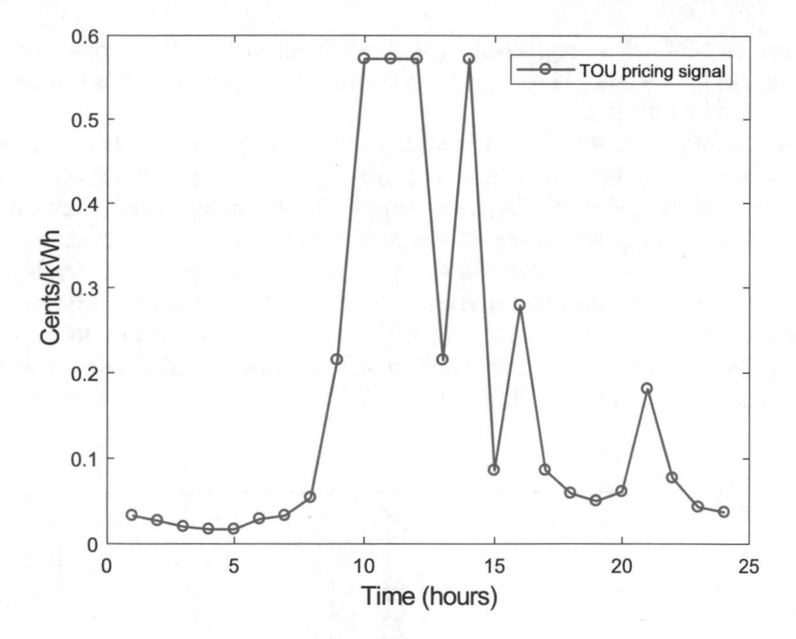

Fig. 4. Price signal

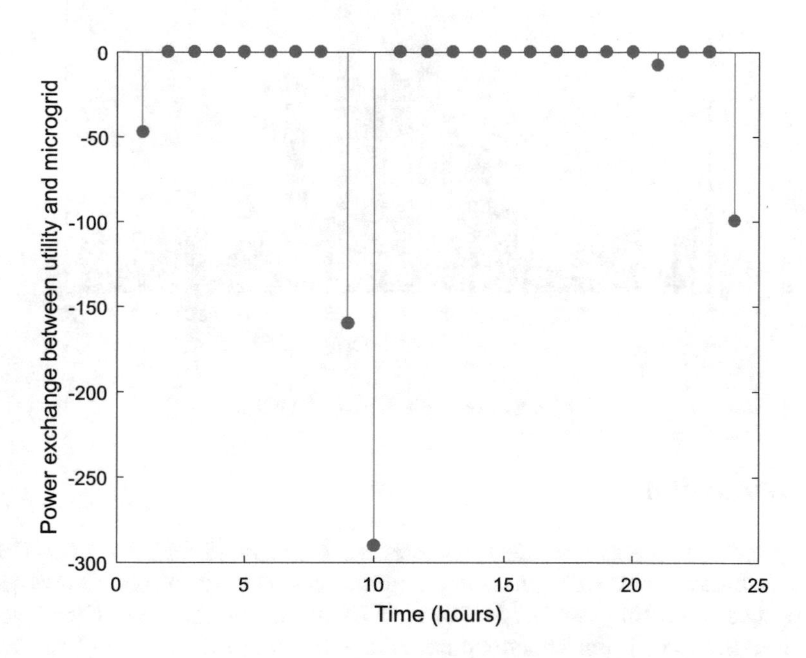

Fig. 5. Power exchange

load will be fulfilled through buying energy from utility. Here, positive values demonstrate that energy is bought from utility and negative values indicate that energy is sold to utility.

When a MG operates in the islanded mode, in separation from the utility grid, while focusing on the economical criteria the MG scheduler also aims at providing a reliable and stable power supply to the supermarket. In this mode MG generates energy according to the load demand of supermarket. As utility grid is not involved in this operational mode so a reliable power source is needed to assure the balance between demand and supply. Due to the intermittent nature of RES, the scheduler gives more priority to FC in order to provide a reliable source of power supply. After scheduling, the overall operational cost of DERs is minimized as given in Fig. 6.

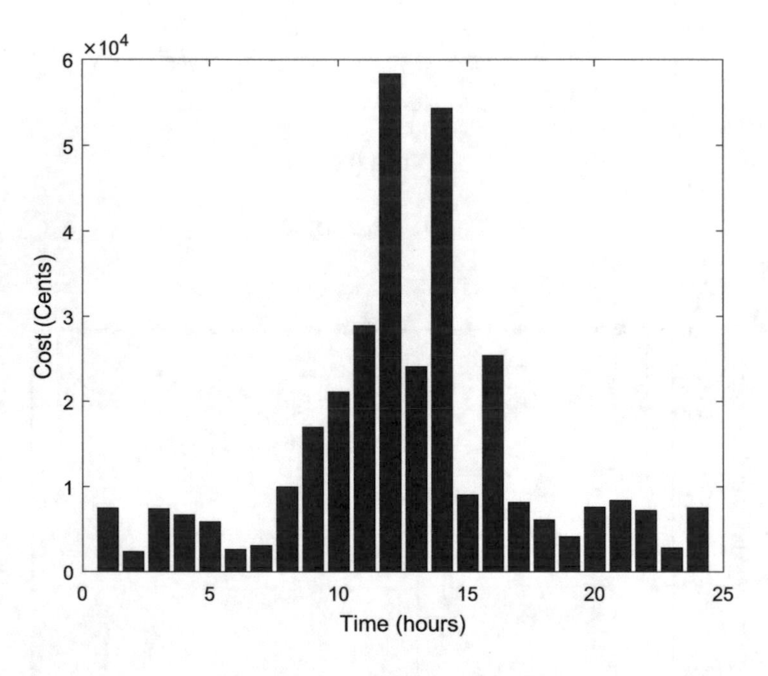

Fig. 6. Scheduled cost of DERs

6 Conclusion

In this paper, an energy management strategy is proposed for microgrid. It helps to make a balance between consumers demand and energy supply. Our objective is to reduce operating cost of DERs and to minimize the operational cost of DERs in a microgrid. For this purpose, a supermarket load is considered and to fulfill the load demand, DERs of microgrid are scheduled using firefly algorithm. Furthermore, we have calculated the operating cost of DERs by fulfilling load demand in two microgrid modes: grid-connected mode and islanded mode. The results conclude that the operating cost is significantly reduced as compared to the cost of fulfilling consumer demand through utility.

References

1. Wang, T., ONeill, D., Kamath, H.: Dynamic control and optimization of distributed energy resources in a microgrid. IEEE Trans. Smart Grid **6**(6), 2884–2894 (2015)
2. Soshinskaya, M., Crijns-Graus, W.H.J., Guerrero, J.M., Vasquez, J.C.: Microgrids: experiences, barriers and success factors. Renew. Sustain. Energy Rev. **40**, 659–672 (2014)
3. Zhang, Y., Gatsis, N., Giannakis, G.B.: Robust energy management for microgrids with high-penetration renewables. IEEE Trans. Sustain. Energy **4**(4), 944–953 (2013)
4. Zachar, M., Daoutidis, P.: Microgrid/Macrogrid energy exchange: a novel market structure and stochastic scheduling. IEEE Trans. Smart Grid **8**(1), 178–189 (2017)
5. Dong, W., Li, Y., Xiang, J.: Optimal sizing of a stand-alone hybrid power system based on battery/hydrogen with an improved ant colony optimization. Energies **9**(10), 785 (2016)
6. Li, P., Duo, X., Zhou, Z., Lee, W.-J., Zhao, B.: Stochastic optimal operation of microgrid based on chaotic binary particle swarm optimization. IEEE Trans. Smart Grid **7**(1), 66–73 (2016)
7. Mohan, V., Singh, J.G., Ongsakul, W., Suresh, M.P.R.: Performance enhancement of online energy scheduling in a radial utility distribution microgrid. Int. J. Electr. Power Energy Syst. **79**, 98–107 (2016)
8. Tushar, M.H.K., Assi, C., Maier, M., Uddin, M.F.: Smart microgrids: optimal joint scheduling for electric vehicles and home appliances. IEEE Trans. Smart Grid **5**(1), 239–250 (2014)
9. Mohsen, N., Braun, M., Tenbohlen, S.: Optimization of unit commitment and economic dispatch in microgrids based on genetic algorithm and mixed integer linear programming. In: Applied Energy (2017)
10. Zhang, D., Shah, N., Papageorgiou, L.G.: Efficient energy consumption and operation management in a smart building with microgrid. Energy Convers. Manage. **74**, 209–222 (2013)
11. Dabou, R., Bouchafaa, F., Arab, A.H., Bouraiou, A., Draou, M.D., Necaibia, A., Mostefaoui, M.: Monitoring and performance analysis of grid connected photovoltaic under different climatic conditions in south Algeria. Energy Convers. Manage. **130**, 200–206 (2016)
12. Rasheed, M.B., Javaid, N., Ahmad, A., Awais, M., Khan, Z.A., Qasim, U., Alrajeh, N.: Priority and delay constrained demand side management in real-time price environment with renewable energy source. Int. J. Energy Res. **40**(14), 2002–2021 (2016)
13. Javaid, N., Hussain, S.M., Ullah, I., Noor, M.A., Abdul, W., Almogren, A., Alamri, A.: Demand side management in nearly zero energy buildings using heuristic optimizations. Energies **10**(8), 1131 (2017)
14. Zhao, T., Ding, Z.: Distributed finite-time optimal resource management for microgrids based on multi-agent framework. In: IEEE Transactions on Industrial Electronics (2017)
15. Rahim, S., Javaid, N., Ahmad, A., Khan, S.A., Khan, Z.A., Alrajeh, N., Qasim, U.: Exploiting heuristic algorithms to efficiently utilize energy management controllers with renewable energy sources. Energy Build. **129**, 452–470 (2016)
16. Ding, Y.M., Hong, S.H., Li, X.H.: A demand response energy management scheme for industrial facilities in smart grid. IEEE Trans. Industr. Inf. **10**(4), 2257–2269 (2014)

17. Kumar, Y.V.P., Bhimasingu, R.: Renewable energy based microgrid system sizing and energy management for green buildings. J. Mod. Power Syst. Clean Energy. **3**(1), 1–13 (2015)
18. Adouane, M., Haddadi, M., Touafek, K., AitCheikh, S.: Monitoring and smart management for hybrid plants (photovoltaic-generator) in Ghardaia. J. Renew. Sustain. Energy **6**(2), 023112 (2014)
19. Ambia, M.N., Al-Durra, A.: Adaptive power smoothing control in grid-connected and islanding modes of hybrid micro-grid energy management. J. Renew. Sustain. Energy **7**(3), 033104 (2015)
20. Maleki, A., Pourfayaz, F.: Optimal sizing of autonomous hybrid photovoltaic/wind/battery power system with LPSP technology by using evolutionary algorithms. Sol. Energy **115**, 471–483 (2015)
21. Garcia, R.S., Weisser, D.: A wind-diesel system with hydrogen storage: joint optimisation of design and dispatch. Renew. Energy **31**(14), 2296–2320 (2006)

Usage Optimization of Mobile Devices Resources in Mobile Web

Nebojsha Ilijoski[1(✉)] and Vladimir Trajkovik[2(✉)]

[1] Vox Teneo Macedonia, Street Tome Arsovski 14, 1000 Skopje, Macedonia
n.ilijoski@voxteneo.com.mk

[2] Faculty of Computer Science and Engineering, Street Rugjer Boshkovikj 16, P.O. Box 393, 1000 Skopje, Macedonia
vladimir.trajkovik@finki.ukim.mk

Abstract. The continuous development of mobile devices and the huge number of mobile application users is implicating that somewhere in near future many mobile applications will be focusing on maximizing the usage of possibilities offered by the mobile devices. In that manner, effective usage of the data from the sensors embedded in the mobile devices is crucial. Today this data has more meaning for the mobile devices behavior than for the mobile applications. In this paper we are proposing a model based on the existing frameworks or content management systems to implement a service as a backend for lightweight mobile application. This does not mean that we might abandon the development of web applications as we know them, but in contrary, they could help in development of the mobile applications as decoupled (headless) web services focused on the data processing from the mobile devices sensors.

1 Introduction

Today we are witnessing the unstoppable progress of the mobile technology. Each new generation of mobile devices brings some improvements and introduces new features. What is more interesting the improvements are not just software or performances related, but also they are introducing new sensors and functionalities to the end users. Since the term mobile device refers to wide range of devices we'll be focusing on a subset of those, smartphones and tablets in particular. Today these devices are combining a range of different functions such as media player, camera, and GPS with advanced computing [1], accelerometer, gyroscope, NFC, proximity sensor, ambient light sensor, fingerprint scanner, Bluetooth, flashlight and others. With this on mind it seems like is impossible to keep up with the various new and updated old functionalities and the APIs for them in order to develop an application that optimize usage of the data available from them and give other purpose to the smartphones or the tablets as mobile devices besides their main usage as communication or gaming device.

On first look this looks very interesting, but if we get deeper into the world of mobile development we'll see a diversity of technologies and approaches regarding the mobile development. That means every step, every decision that a developer makes can be the right one or can be just a waste of time. There are many questions that are defining our decision. Are we working on an application dedicated to a platform or

L. Barolli et al. (Eds.): EIDWT 2018, LNDECT 17, pp. 315–325, 2018.
https://doi.org/10.1007/978-3-319-75928-9_28

we are looking for multi-platform solution? Are we going to use some tools like PhoneGap and jQueryMobile? Are we going to create another version for tablets? Should we focus only on particular functionality of the smartphones? Because of the scope of these questions sometimes it seems hard to find the correct answer.

For that purpose in our paper we are proposing an architectural model that combines existing technologies for mobile and web development into a single technology stack, minimizing the performance requirements of the mobile device (smartphone or tablet) and focusing on the processing power of the service. We make the decision of the type of mobile application and the technology for the service implementation based on analysis of results of other researchers. Also we introduce prototype application implementing the proposed model consisted of a hybrid mobile application based on PhoneGap (Cordova) and DrupalGap SDK, and service based on decoupled Drupal 7. And based on the prototype we discuss and generalize results in order to provide a decision advice.

The structure of this paper is following. In the Sect. 2 we study the related work that focuses on parts of our model, not the model in general. In Sect. 3 we are focusing on the service, making a comparison between the monolithic approach (because our model is based on this) and approach with microservices (as improvement), and also make an overview on mobile application development possibilities. Section 4 is dedicated to the prototype application as a base for the Sect. 5 where we give our analysis and discuss the achieved results from the observation. And finally in Sect. 6 we draw a conclusion.

2 Related Work

Most of the researches that are used as base for this paper are focusing on separate parts of the model proposed here, and which will be discussed in the appropriate sections. For the purpose to generalize the study achieved from different papers we grouped them in following groups:

- web services for mobile applications [2, 3, 12, 13]
- decoupling content management and mobile cloud computing [6, 7]
- mobile devices sensor-based context awareness [4, 5]
- and performance analysis of web services and cross-platform development for mobile applications [8, 9, 11]

The main motivation in the papers [2, 3] related with the web services for mobile applications is the idea of separating the user interface from the service, and overview of microservice versus monolith architecture for web services in [12, 13]. Exact same motivation provoked authors in [6, 7] to analyze the possibilities of decoupling the monolith content management systems and use them as a web services.

On the other hand, research papers [4, 5] have been motivated from expansion of sensors usage in the mobile devices and possibilities they offer. As we mentioned the focus in this paper is on smartphones and tablets and operations with the data from the provided sensors and optimizing their usage for every day tasks.

Performance analysis of web services and cross-platform mobile applications from the papers [8, 9, 11], helps us to make our technologies choices for the prototype application, and to complete the analysis.

3 Web Services for Mobile Applications

The model we are proposing here in its purest format is implementation of the basic service oriented architecture concept, but extended to mobile cloud computing (Fig. 1). Main parties in this model are the client (mobile user), the service provider and the infrastructure between them.

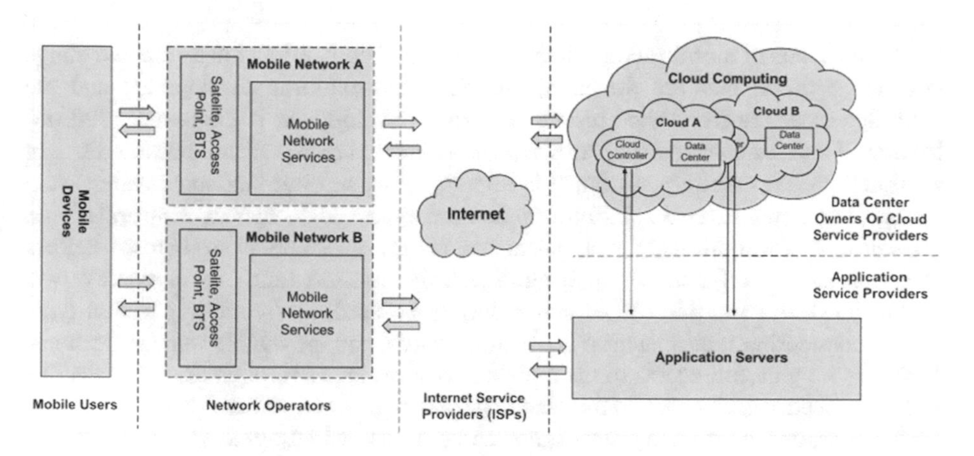

Fig. 1. General architecture of mobile cloud computing [7].

Our main focus is to propose optimization to the endpoints in this architectural representation, which means that we are looking at the mobile devices (applications for mobile devices in particular) on one side, and the web services on the other side. The Mobile Cloud Computing Forum defines "Mobile Cloud Computing at its simplest, refers to an infrastructure where both the data storage and the data processing happen outside of the mobile device. Mobile cloud applications move the computing power and data storage away from mobile phones and into the cloud, bringing applications and mobile computing to not just smartphone users but a much broader range of mobile subscribers". [7] Our case is focused on smartphones and tablets and in order to optimize development of mobile applications and the web services for them here we are proposing a set of existing technologies.

3.1 Mobile Applications

Three types of mobile application, their structure and how internal calls are made is presented on Fig. 2.

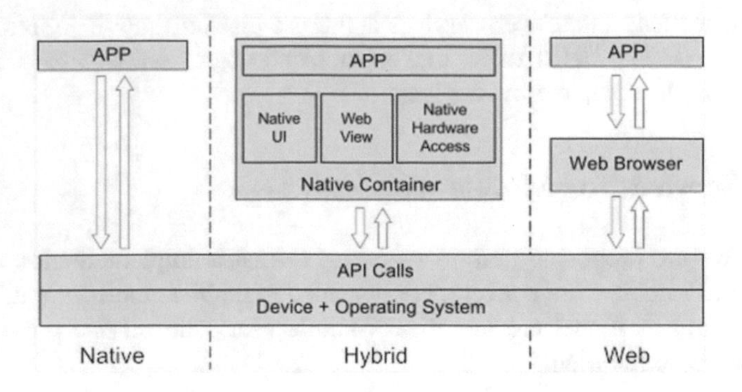

Fig. 2. Internal calls in native, hybrid and web mobile applications

All the types of mobile applications have pros and cons [9], but there are two things that define them, that are the possibility for cross-platform development and the unlimited access to the device resources. Native development is the best for full utilization of the device resources, but stays far behind all cross-platform because its tight relation with the mobile platform. It is important to point out that some authors recognize four types of cross-platform application types: *web*, *hybrid*, *interpreted* and *generated*. As most optimal development strategy for our model is used cross-platform development that offer native application package as end result (which exclude web applications). For the purpose of our prototype we used the hybrid application type, mainly considering it as a balance between the native and the mobile web applications. That way we gain full access to the mobile device resources through the API calls like native applications, and stick to the "develop once deploy everywhere" rule like the web applications, and their main advantage is ability to use widely used web development technologies (especially HTML5 and JavaScript). For comparison interpreted applications generate native code automatically to implement the user interface which is native for the platform but application logic is implemented independently using technologies like Java, Ruby, XML, which make them dependent on the software environment, and generated applications are native platform-specific applications but coded once in some software development environment which allow generating that native application from one source but require solid knowledge of that environment [11].

Another thing that require some consideration is what technologies and development platforms are we going to use for the hybrid application implementation. There are several platforms for that purpose among which the most used solutions are PhoneGap and Titanium[1]. Because of PhoneGap expanded community we decided to use it as development platform for the prototype application. In general what we ended up with is the case from Fig. 3. What the community offers is constant communication, knowledge and ready libraries exchange. What we receive from the community as

[1] According "The Drum" portal, http://www.thedrum.com/profile/news/247333/approaching-mobile-development-2016-part-2-5-popular-cross-platform-frameworks (accessed 06.11.2017), in 2016 PhoneGap was the most used platform for cross-platform development.

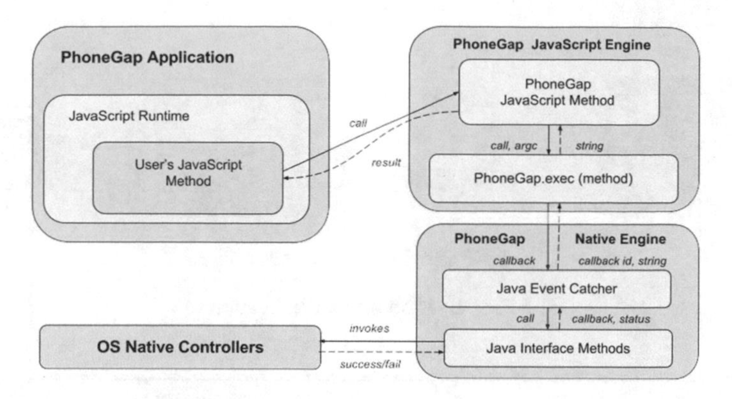

Fig. 3. PhoneGap's method call flow path [9]

ready to use semi products are plugins that are implementing native API calls for different platforms (Android, iOS, BlackBerry or Windows) into JavaScript callbacks.

3.2 Web Services

According to some authors web services were initially intended for dynamic business-to-business (B2B) interactions with services deployed on behalf of other enterprises or business entities. However with the advancement of web service technology the complexity of possible tasks and the availability of service at any time anywhere already move forward and will continue to increase [2].

There are several possibilities when it comes to the architectural choices for the web service like SOAP or REST, but we made our choice for REST web service architecture considering that this is more robust and the coasts are not a problem for implementation over the SOAP for which some authors are considering that offers more flexibility at lower integration costs [2].

When it comes to the options we have for the software architectural pattern for the web service, there have two possibilities to proceed with, *monolithic* and *microservices* approach. On Figs. 4 and 5 we can see the general structure of both monolithic and microservices structure. The microservices architecture is an approach to developing an application as a set of small independent services. Each of the services is running in its own independent process, can communicate with some lightweight mechanisms (usually based on HTTP) and can be deployed independently. The centralized management of these services is also a completely separate service and can be written in different programming languages, use own data models etc. On the other hand, as a completely opposite solution, monolithic architecture internally can have several services or components, but it is deployed as single application.

So the question is which approach is better? One of the benefits of using microservices is the ability to publish a large application as a set of small applications (microservices) that can be developed, deployed, scaled, operated and monitored independently. Microservices allow companies to manage large codebase applications using a more practical methodology where incremental improvements are executed by

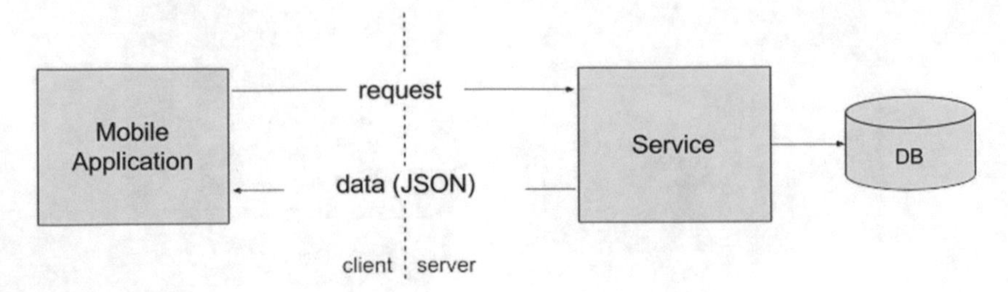

Fig. 4. Monolithic architecture

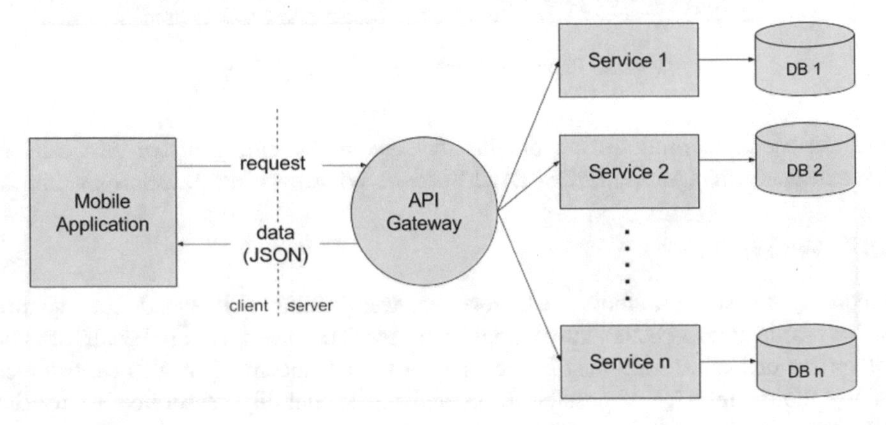

Fig. 5. Microservices architecture

small teams on independent codebases and deployments. The agility, cost reduction and granular scalability, brings some challenges of distributed systems and team development management practices that must be considered. But for applications with a small number of users (hundreds or thousands of users), the monolithic approach may be a more practical and faster way to start, so taking this in consideration and also the fact that in many practical cases microservice architectures started as monolithic applications and due to scaling problems at infrastructure and team management, they were incrementally modified to implement microservices using we propose and proceed with the monolithic architecture for the web service and will leave microservices approach for future work [13].

4 Prototype Application (Case Study)

As a result of the related work study regarding mobile application and web services for mobile applications we developed a prototype application that implements everything discussed in Sect. 3 and we use it as a base for the performance analysis and discussion in Sect. 5.

The application is an Employee Time Tracking solution that uses a mobile device (smartphone or tabled) as hardware solution. The requirements for the mobile device are minimalistic since the data is only proceeded to the web service where proper operations are made. From the sensors requirements it require NFC (read/write) support, camera, GPS and Internet access (3G, 4G or Wi-Fi regardless). A few screens from the mobile application are presented on Fig. 6.

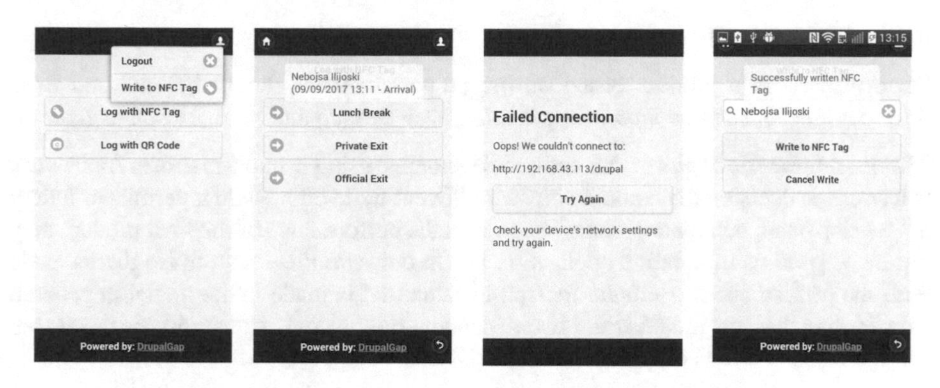

Fig. 6. Prototype application screens (hybrid mobile application)

The base for the monolithic service is a common CMS (Content Management System) solution, Drupal 7 in our case, which is used as decoupled (headless) CMS. For the decoupling we used three basic third party modules for Drupal 7, Services, View Datasources and DrupalGap. That way we gained all basic CRUD operations. It was configured to work as a REST service with authentication in a session. The data retrieved from this service is in JSON format and requests can be JSON, x-www-form-urlencoded, XML or form-data.

The mobile application for this solution is developed as a hybrid application using PhoneGap (the Cordova distribution in particular), on top of which we use DrupalGap SDK (solution developed for Drupal, that goes along with the module mentioned which provide basic service resources that can be called directly from the application developed with this SDK) with jQuery Mobile support, which is responsible for proper communication with the service.

What this application does is wait for NFC read event or request to scan QR code, retrieve the data which represents a user id of a user account stored on the web service, check if such account exists and create log entry. Once the entry is created, response with the new record id is retrieved and new update request with the GPS data is sent. From the this data nearest location (from the locations records stored on the service) is found and referenced in the log entry record.

Also it offers simple interface for administration from the mobile application which allow writing id to NFC Tag, and also NFC Tag UID to the user record.

What we really decoupled from the service application is the front end, and we keep the backend (administration) pages for configuration and data management.

5 Analysis and Discussion

With detailed observation of this case study we can make an overview of the optimization that was brought with it, and also we can take closer look at its advantages and disadvantages. For that purpose we made analysis and case study discussion on three levels: *performances, privacy and security* and *development methodology*.

5.1 Performances

Performances of a product based on the proposed model should be analyzed from different aspects such as *mobile application, web service* and *network connection*.

Mobile Application. For some authors developing using a multi-platform framework is a strategic decision that should consider different trade offs: while it permits to follow a "develop once, deploy anywhere" approach, the performance of the final product may not be as good as in a native application [9]. In our case the compromise that is made with the performances is crucial for optimization that is made in the model in general. Considering that our model is not that depending on the device performance as on web service resources the compensation is justified.

Web Service. This is the part where our processing power is focused. That means we should expect maximum performances from the web service. Performances of the web service are closely related with the server configuration, number of clients (especially because of the monolithic architecture) etc. That means it is important to keep the host infrastructure optimized.

When it comes to the performance analysis web service architecture is also important. RESTful web services are "Resources" that are identified by unique URIs. These resources are accessed and manipulated using a set of uniform methods (GET, POST, PUT and DELETE) where each resource has one or more representations (XML, JSON, Text, User-defined, etc.) that are transferred between the client and the service. RESTful Web services are simple because they are based on the existing well-known W3C/IETF standards (HTTP, XML, URI and MIME type) as well as the necessary infrastructure. Requests and responses for RESTful web services are typically HTTP messages that are far less in size compared to SOAP messages. Since in REST architecture a resource can be directly identified by its URI, therefore extensive SOAP parsing can be avoided that is required for invoking a service [15].

Network Connectivity. Network is also important factor for the general performances evaluation of the model. Most of all because it is required for proper data processing (request and response). Above we mentioned that RESTful messages are smaller in size than SOAP which is important because of the inconsistent network connection of the mobile device. Since the mobile device can move quickly from one to another network covered area or even to an area without a network coverage it is important to implement a backup plan, for example offline data storage, and process it after the network connection is back. But on the other hand, before proceeding with such solution it is important to consider if that data is useful after some period of time.

5.2 Privacy and Security

When it comes to the privacy and security for this model, that also needs to be observed from several aspects such as *data security, network security, mobile application security* and *service security*.

Data Security. Most of the mobile platforms today support hardware encryption which gives base level of protection to data on a mobile device when device is locked. Some devices also support an encrypted data storage where sensitive data can be stored in an encrypted state. But we'll have to consider a few things, first data is not secure if the user of the device does not enable locking of the device and second some of these hardware encryption approaches have been compromised.

Network Security. Network security is no different in a hybrid mobile application from a normal web application. In general network security can be violated because of insufficient security in the transport layer or storing not encrypted personal data.

Application Security. Since the hybrid application is based on HTML and JavaScript here also apply same conditions like for the rest of the web applications. Because of that biggest security risks in this part are XSS (Cross Site Scripting) and injection.

Service Security. Considering that the service is a separate application here the same security conditions are in force. This means that most of all application security is depending on the host security. Additionally, the fact that the service is build from decoupled CMS requires security risks that come with it to be identified and also the risks from the usage of the contributed libraries. In order to avoid such risks software should be always kept up to date, especially in case of security fixes.

5.3 Development Methodology and Efforts

The development of applications consisted of a client application represented by a hybrid mobile application and a web service based on a decoupled CMS (monolithic architecture) requires solid knowledge of the two areas with focus on the selected technologies. Because of the monolith architecture limitation in best case companies or software vendors could organize two teams, one for the client application and another for the service, and in worst case one (same that works on the CMS with solid HTML and JavaScript knowledge).

On the client side the advantage is that hybrid mobile applications in the most positive scenario would require development only once and produce application ready for all platforms considering that selected platform has big community with solid support. In the most worst scenario additional native development would be required for the missing APIs which would require platform oriented developers. But in both cases advantage is that this model require minimalistic development on the client side.

On the server side the idea of using CMS would lower requirements for development considering that all recent CMS have integrated RESTful services in the core or offer that as third party library. Because of that the focus would be on developing resources (functionalities) for services on using the CMS APIs.

6 Conclusion

The main goal of this paper is to propose a general model of a distributed architecture which main focus would be to optimize data usage from the smart mobile devices sensors and other peripherals. The idea was to propose a case study as base for observation and discussion. The general conclusion from the analysis is that this model has certain limitations, mostly because of the hybrid mobile application type and monolithic web service architecture. Also the results point out that the biggest improvement can be done within the web service by replacing it with microservices architecture. Microservices architecture was mentioned, described and compared with the monolithic in this paper but was not used as main choice. Because of that it will be considered for future work and will be used for comparative performance analysis with here proposed model based on monolithic architecture web service.

At the end we have to point out that the benefits and possibilities for this model in every day use are undoubtedly with great value for both the developers and the clients (end users). Since today number of users that own smart mobile devices is huge, this study would help to consider the need to optimization of the mobile devices sensors data usage in simple yet effective way. By taking advantage of the possibilities offered, companies and software vendors could significantly lower the development costs and introduce the existing mobile devices into new application areas (not just as personal devices for social interaction) that so far were not considered (like processing environment data retrieved from the available sensors or external sensors that communicate with the mobile device over any of the existing data exchange possibilities like Bluetooth, Wi-Fi, USB, Infrared etc.).

References

1. Heitkötter, H., Hanschke, S., Majchrzak, T.A.: Evaluating cross-platform development approaches for mobile applications. In: International Conference on Web Information Systems and Technologies, pp. 120–138. Springer, Berlin, Heidelberg, 18 April 2012
2. Dospinescu, O., Perca, M.: Web services in mobile applications. Inf. Econ. **17**(2), 17–26 (2013)
3. Champion, M., Ferris, C., Newcomer, E., Orchard, D.: Web services architecture, W3C Working Draft (2012). https://www.w3.org/TR/ws-arch/
4. Gellersen, H.W., Schmidt, A., Beigl, M.: Multi-sensor context-awareness in mobile devices and smart artifacts. Mob. Netw. Appl. **7**(5), 341–351 (2002)
5. Korpipää, P., Mäntyjärvi, J.: An ontology for mobile device sensor-based context awareness. In: Modeling and Using Context, pp. 451–458. Springer, Berlin, Heidelberg (2003)
6. Grünwald, S., Bergius, H.: Decoupling content management. In: Developer Track, WWW 2012 Conference, Lyon (2012)
7. Dinh, H.T., et al.: A survey of mobile cloud computing: architecture, applications, and approaches. Wireless Commun. Mob. Comput. **13**(18), 1587–1611 (2013)
8. Hamad, H., Saad, M., Abed, R.: Performance evaluation of RESTful web services for mobile devices. Int. Arab J. e-Technol. **1**(3), 72–78 (2010)
9. Corral, L., Sillitti, A., Succi, G.: Mobile multiplatform development: an experiment for performance analysis. Procedia Comput. Sci. **10**, 736–743 (2012)

10. Thu, E.E., Aung, T.N.: Developing mobile application framework by using RESTFul web service with JSON parser. In: International Conference on Genetic and Evolutionary Computing, pp. 177–184. Springer, Cham, 26 August 2015
11. Xanthopoulos, S., Xinogalos, S.: A comparative analysis of cross-platform development approaches for mobile applications. In: Proceedings of the 6th Balkan Conference in Informatics, pp. 213–220. ACM, 19 September 2013
12. Namiot, D., Sneps-Sneppe, M.: On micro-services architecture. Int. J. Open Inf. Technol. **2**(9), 7–24 (2014)
13. Villamizar, M., Garcés, O., Castro, H., Verano, M., Salamanca, L., Casallas, R., Gil, S.: Evaluating the monolithic and the microservice architecture pattern to deploy web applications in the cloud. In: 2015 10th Computing Colombian Conference (10CCC), pp. 583–590. IEEE, 21 September 2015
14. Dragoni, N., et al.: Microservices: yesterday, today, and tomorrow. arXiv preprint arXiv: 1606.04036 (2016)
15. Mohamed, K., Wijesekera, D.: Performance analysis of web services on mobile devices. Procedia Comput. Sci. **31**(10), 51–744 (2012)

The Performance Comparison for Low and Medium Earth Orbiting Satellite Search and Rescue Services

Bexhet Kamo(✉), Joana Jorgji(✉), Shkelzen Cakaj(✉), Vladi Kolici(✉), and Algenti Lala(✉)

Polytechnic University of Tirana, Tirana, Albania
{bkamo,jjorgji,shcakaj,vkolici,alala}@fti.edu.al

Abstract. LEOSAR (Low Earth Orbit Search and Rescue) is an international satellite system which operates continuously, detecting transmissions from emergency beacons carried by ships, aircrafts and individuals, providing location information related to worldwide distress events. LEOSAR based on Low Earth Orbits it is still limited on instantaneous alert and coverage. To improve the performance, this system is migrating towards MEOSAR (Medium Earth Orbit Search and Rescue), restructuring the capability to Medium Earth Orbit satellites. By real time 24-h satellite tracking and the simulation of the Almanac YUMA file using Trimble's Planning Software, a methodology of comparison is provided. For the MEOSAR system advantages in global coverage and instantaneous alert are evidenced. From the obtained results it is shown the limitation of the LEOSAR capability, but also the efficiency improvement of the search and rescue operations provided by MEOSAR.

1 Introduction

The COSPAS-SARSAT satellite constellation includes satellites in Low Earth Orbit (known as LEOSAR) and satellites in Geostationary Earth Orbit (GEOSAR). The system is originally sponsored by Canada, France, the former Soviet Union and the USA, and it was declared as fully operational by 1985 [1]. The LEOSAR system uses satellite-based payloads to detect and determine location of worldwide distress signals emitted by emergency beacons carried by ships, aircrafts or land users, based on Doppler Effect. Because of the lack of Doppler Effect in GEOs, the GEOSAR supports LEOSAR only on signal detection, without contribution on further location determination (calculation). The basic LEOSAR concept is illustrated in Fig. 1 [1].

The system is composed of:

1. Distress radio-beacons (ELT for aviation use, EPIRB for maritime use and PLB for personal use) which transmit the distress signal when manually or automatically activated;
2. Instruments on board satellites which detect and retransmit the distress signal;
3. Ground receiving stations, referred to as Local User Terminals (LUTs) which receive the satellite downlink signal and generate distress alerts;

L. Barolli et al. (Eds.): EIDWT 2018, LNDECT 17, pp. 326–338, 2018.
https://doi.org/10.1007/978-3-319-75928-9_29

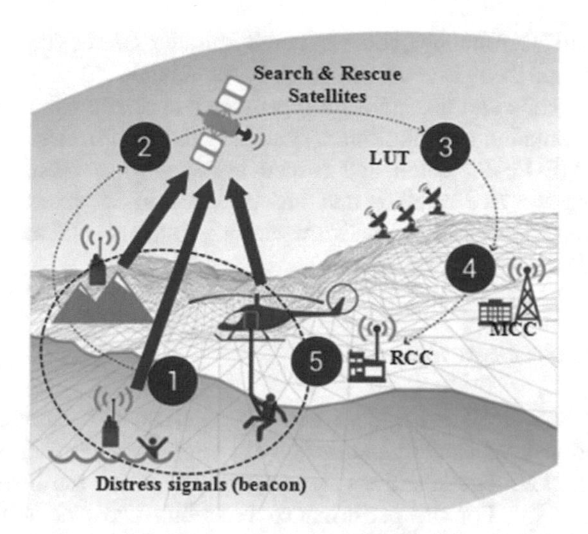

Fig. 1. The SARSAT LEO-based (LEOSAR) system concept [1].

4. Mission Control Centers (MCCs) which receive routed alerts from the LUTs and forward them to Rescue Coordination Centers (RCCs) or other MCCs after validation;

The LEO based system distress location determination relies on Doppler events, by measuring successive bursts generated by the beacon at the distress location. The beacon's position can be determined using the frequency of arrival (FOA) and employing the Doppler shift equations. Complete, yet non continuous coverage of the Earth by LEOSAR is achieved using emergency beacons that emit 406 MHz signals at distress location [2]. The location determination, accuracy and the time required to alert rescue authorities depends on satellite's visibility and availability between the ground segment and the satellites [3]. Because polar orbiting LEO satellites can only view a portion of the Earth at a given time, the system is unable to generate distress alerts unless the beacon is located under the satellite footprint. However, the *store and forward* mode of operation provides global coverage, with significant delay on location determination and time to access to respective location [4].

Considering the limitations of LEOSAR on coverage and instantaneous alert, in 2000, the United States, the European Commission (EC) and the Russian Federation began consultations within the COSPAS-SARSAT organization regarding the feasibility of installing new SAR (search and rescue) instruments on their respective GNSS satellites, thus, incorporating a 406 MHz repeaters on board of the navigation satellites (GPS, GLONASS and GALILEO), which operate at Medium Earth Orbits. Hence this system is known as Medium Earth Orbit Search and Rescue system, or MEOSAR. The U.S. MEOSAR system is known as SAR/GPS, the European system is SAR/Galileo and the Russian system is referred to as SAR/GLONASS. Recently, this migration process by placing search-and-rescue repeaters to MEO satellites is undergoing, and it is expected to be completed before the end of this decade, thus having the

MEOSAR system as fully operational. For continuity of services, the LEO based system will continue to be operational as well, for a while [5].

Firstly, the satellite and ground station geometry is briefly considered serving as a mathematical simulation's background. Further, the performance analysis of the LEOSAR and MEOSAR search and rescue systems with regard to visibility and availability using the live satellite tracking information is presented. The paper is concluded by the evidence of the performance improvements of the MEOSAR compared with LEOSAR system limitations.

2 Satellite Coverage Geometry

Communication via satellite begins when the satellite is positioned in its orbital position. Ground stations can communicate with satellites only when the satellite is in their visibility region. The satellite's coverage area on the Earth depends on orbit altitude and elevation. For the performance (visibility and availability) comparison purposes, between LEOSAR and MEOSAR, the satellite coverage geometry is further elaborated [6]. The basic geometry between a satellite and ground station is depicted in Fig. 2. The triangle vertices indicate the satellite (SAT), the ground station (P) and the Earth's center. The line crossing point P represents the horizon plane. The subsatellite point is indicated by T. Two sides of this triangle are usually known (the distance from the ground station to the Earth's center, $R_e = 6378 \times 10^3$ m and the distance from the satellite to Earth's center-orbital radius). There are four variables in this triangle: ε_0 - is elevation angle, α_0 - is nadir angle, β_0 - is central angle and d is slant range. As soon as two quantities are known, the others can be found with the following equations [7]:

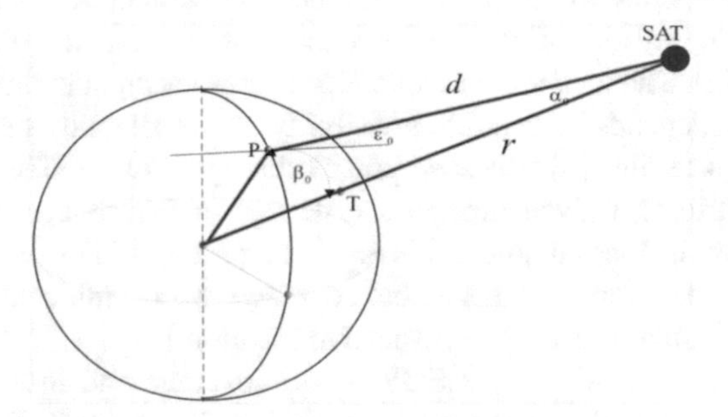

Fig. 2. Ground station and satellite geometry.

$$\varepsilon_0 + \alpha_0 + \beta_0 = 90 \tag{1}$$

$$d \cos \varepsilon_0 = r \sin \beta_0 \tag{2}$$

$$d \sin \alpha_0 = R_e \sin \beta_0. \tag{3}$$

The most crucial parameter is the slant range d (the distance from the ground station to the satellite). This parameter will be used during the link budget calculation and it is expressed through the elevation angle ε_0. Applying cosines law for above triangle at Fig. 2 yields out (4):

$$r^2 = R_e^2 + d^2 - 2R_e d\ \cos(90 + \varepsilon_0). \tag{4}$$

Solving (4) by d, applying (1), (2), (3) and substituting $r = H + R_e$ finally yields out the slant range as a function of the elevation angle ε_0 [7].

$$d(\varepsilon_0) = R_e \left[\sqrt{\left(\frac{H + R_e}{R_e} \right)^2 - \cos^2 \varepsilon_0} - \sin \varepsilon_0 \right] \tag{5}$$

H is the satellite's altitude above the Earth's surface. Transforming Fig. 2 to the coverage view, Fig. 3 derives.

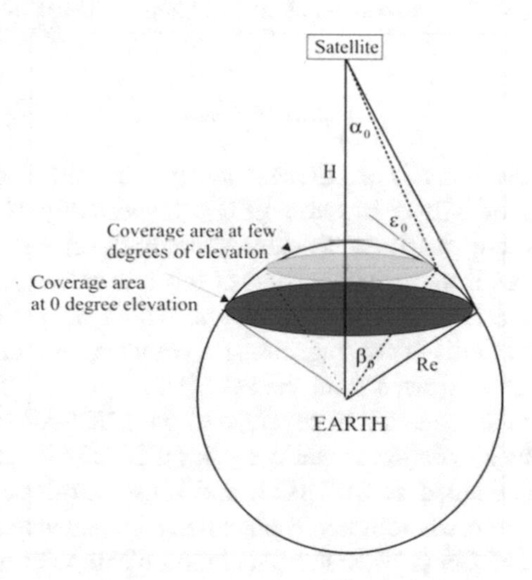

Fig. 3. Coverage geometry.

In Fig. 3, there are two triangles. The larger one represents the case of full coverage under elevation of 0°. The distance d under the elevation of 0° represents the radius of the largest circle of the coverage area. These coverage circles for LEO and MEO make the so-called coverage belts. The wideness of the coverage belt is twice the largest radius, given as follows (7) and presented in Fig. 4 [8].

$$d_{(\varepsilon_0=0)} = d_{\max} = R_e \left[\sqrt{\left(\frac{H + R_e}{R_e} \right)^2 - 1} \right] \tag{6}$$

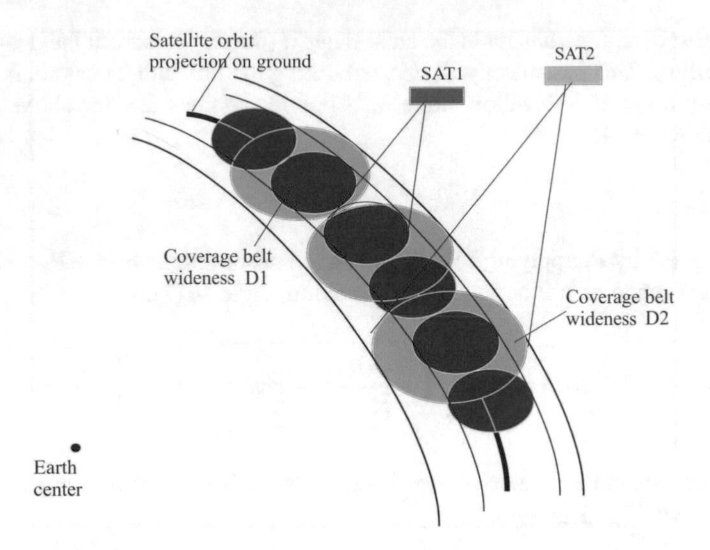

Fig. 4. The coverage belt of LEO and MEO satellites.

$$D_{BELT} = 2d_{max} \quad (7)$$

Considering that the altitudes of LEOs are lower than MEOs, it results an obviously larger coverage area of MEOs towards LEOs, respectively of MEOSAR towards LEOSAR. The largest coverage area is achieved under the elevation of 0º. However, in order to avoid obstacles from natural barriers at too low elevation, usually for the link budget calculations the minimal elevation angle, which ranges on (2–10)º, is determined. Generally, the satellite's coverage area (in percentage) is defined as a fraction of the Earth's area, which is viewed from the satellite.

In order to clarify the limitation in coverage of the LEOSAR system, an example is further elaborated. The simulation scenario applies a LEOSAR satellite, a hypothetical ground station (GS – defined as LUTKOS) and four radio-beacons labeled as BC1, BC2, BC3 and BC4 from where it could potentially be generated a distress signal, as presented in Fig. 5. The GS (LUTKOS) can communicate with the LEOSAR satellite since it is physically under the satellite's coverage area (satellite footprint). However, from Fig. 5, it is obvious that only (BC1 and BC4) can communicate with the satellite (being under the coverage area) and alert the distress towards the ground station (LUTKOS). The other two beacons (BC2 and BC3) have no communication with the satellite, since they are out of the coverage area, consequently cannot provide *immediate alert* to the satellite and further routing the distress signal to the ground station. Thus, the LEOSAR, under the simulated example is capable of providing immediate alert for beacons BC1 and BC4, yet not for BC2 and BC3. This is a drawback of LEOSAR, which is being improved or completely avoided by MEOSAR.

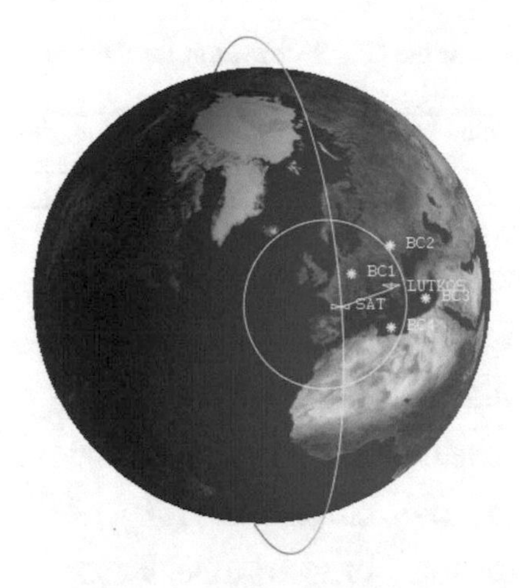

Fig. 5. The LEO coverage area (simulation scenario).

3 Simulated Performance Analysis

The performance analysis of the currently operational LEOSAR system [9] and the future planned MEOSAR system is viewed in terms of *visibility* and *availability*. Then, specifically in both cases, the coverage is issued under the circumstances considered for the simulation.

For the comparison purposes, live satellites tracking information of the LEOSAR and GPS constellations (U.S. MEOSAR) for over a period of 24 h is used. The tracking is made possible using satflare.com [10] and the ephemeris data of the Almanac YUMA file [11] on July 2, 2017. The LEOSAR system satellite constellation of six LEO satellites and the GPS satellites are observed from the hypothetical geographical location (coordinates of N 40.90028°, E 19.91667° and altitude of 100 m above the sea level, Tirana city, Albania) from where eventually the emergency beacon signal will be generated. Further processing of the Almanac YUMA file is run using Trimble's Planning Software.

3.1 Visibility and Availability

The live satellite tracking of the LEOSAR system, from the hypothetical location has given the results as in Table 1. The last column reports whether the satellite is visible or not (color represented) and which instrument could be required.

The obtained information highlights major drawbacks for the existing LEOSAR system. Results show that the visibility of the beacon at the hypothetical location, from the above listed LEOSAR satellites, is relatively low. The total communication duration with the LEOSAR system is no more than 113 min per 24 h, giving an availability of about 7.8% of the considered time interval and consequently causing delays on location determination, respectively on time to access at location of the distress event. For Doppler processing at least four events or bursts are required [12]. If this criterion is violated, in case of too short visibility with the beacon or the blockage from the

Table 1. Satellite passes of the LEOSAR system for the chosen location, during 24 h.

Satellite	Rise	Culminate	Set	Best time	Sat. Elev. at best	Duration	Visible
NOAA 15	-	-	-	-	-	-	-
NOAA 16	22:42:33	22:50:24	22:58:15	22:49:45	58^0	16 min	Binoculars
NOAA 18	20:51:09	20:56:48	21:02:26	20:54:49	7.7^0	11 min	Twilight
METOP A	20:29:26	20:36:27	20:43:28	20:35:59	25.5^0	14 min	Twilight
	22:08:28	22:16:19	22:23:49	22:15:19	37.7^0	15 min	Binoculars
	23:54:58	23:57:40	00:00:21	23:55:31	0.5^0	6 min	Low elev.
NOAA 19	02:42:37	02:48:33	02:50:29	02:50:19	9.2^0	8 min	Low elev.
	04:22:04	04:29:57	04:37:50	04:30:12	83.5^0	15 min	Naked eye
METOP B	21:22:33	21:30:14	21:37:55	21:29:51	70^0	15 min	Naked eye
	23:04:34	23:10:50	23:17:07	23:09:29	12.6^0	13 min	Binoculars
TOTAL						**113 min**	

terrain obstruction causing the masking under the low elevation [12], then the inaccuracy in location determination and the error margin increases. This critical case is evidenced for the satellite passes of METOP A having communication duration of 6 min, presented in Fig. 6 (above). These passes usually will not provide sufficient data for Doppler processing, considering that the communication duration is too short or obscured because of the low elevation (color represented in Table 1), having impact on accuracy on location determination. In Fig. 2 (below) is given the satellite pass of NOAA 16. While the maximal communication duration is 16 min, it can be seen that the satellite is obscured at its acquisition, consequently shortening the visibility and availability.

Using live tracking the satellite's visibility is recorded from the hypothetical beacon location for both LEOSAR and MEOSAR (GPS), as given in Fig. 7. Figure 7 shows that the communication duration of only one GPS (MEO) satellite varies from one to few hours, compared to a total of 113 min of communication duration of all the LEOSAR system. The considered interval between 20:00 to 00:00 was preferred in order to emphasize the difference in real-time of both systems when the LEOSAR system is at most available (7 out of 9 satellite passes viewed). Figure 8 further emphasizes the differences in availability. The basic principle of satellite based navigation requires having at least four GPS satellites in line-of-sight to the beacon (MEOSAR case) [13, 14]. The results in Fig. 8 show that the beacon is continuously seen from at least five GPS satellites thus meeting the criteria. As a result, location

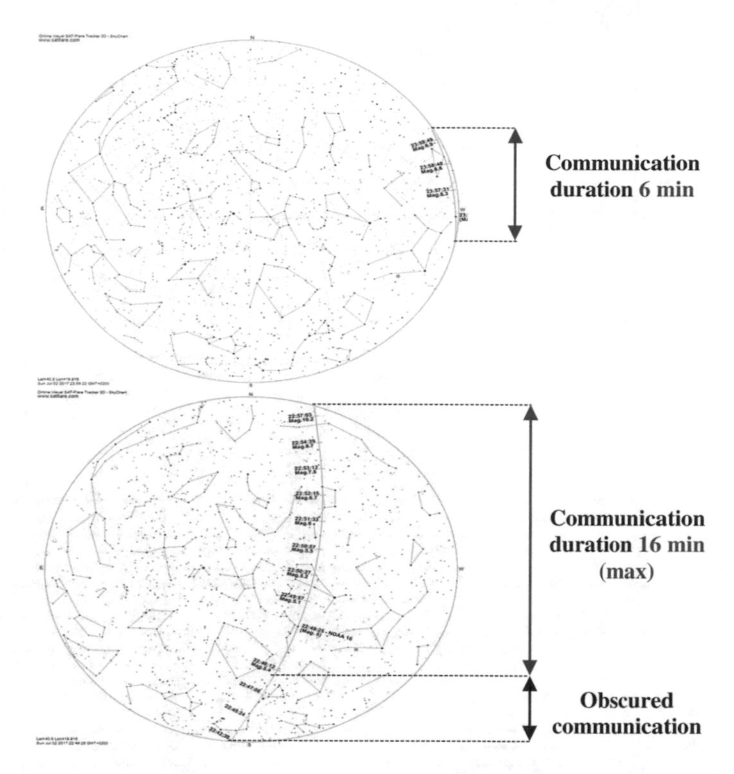

Fig. 6. Satellite passes and communication duration of METOP A (above) and NOAA 16 (below).

determination by MEOSAR is possible at any times. In contrast, although a single LEOSAR satellite using four Doppler events is often enough for positioning, it can be seen that generally no satellite passes appear during the considered 24-h interval. From Figs. 7 and 8 the advantage in visibility and availability of MEOSAR compared with LEOSAR can be concluded.

3.2 Coverage

Because of higher altitudes of the MEOSAR satellites, they can detect signals from beacons under a much larger area. With enough MEOLUTs and GPS (MEO) satellites in line-of-sight, instantaneous coverage will be possible for any point at Earth, what was not the case with LEOSAR. The analysis of the live tracking data is further presented.

Figure 9 shows the coverage of the USA 96 GPS (MEOSAR) satellite and METOP B (LEOSAR) satellite. It is clear that when both satellites are approximately above the same sub-satellite point, the USA 96 satellite footprint is much larger than that of METOP B. Thus, more distress events can be detected under the covered area. This coupled with the fact that usually there are more than four GPS satellites in line-of-sight, enables for immediate and too much quicker location estimation from

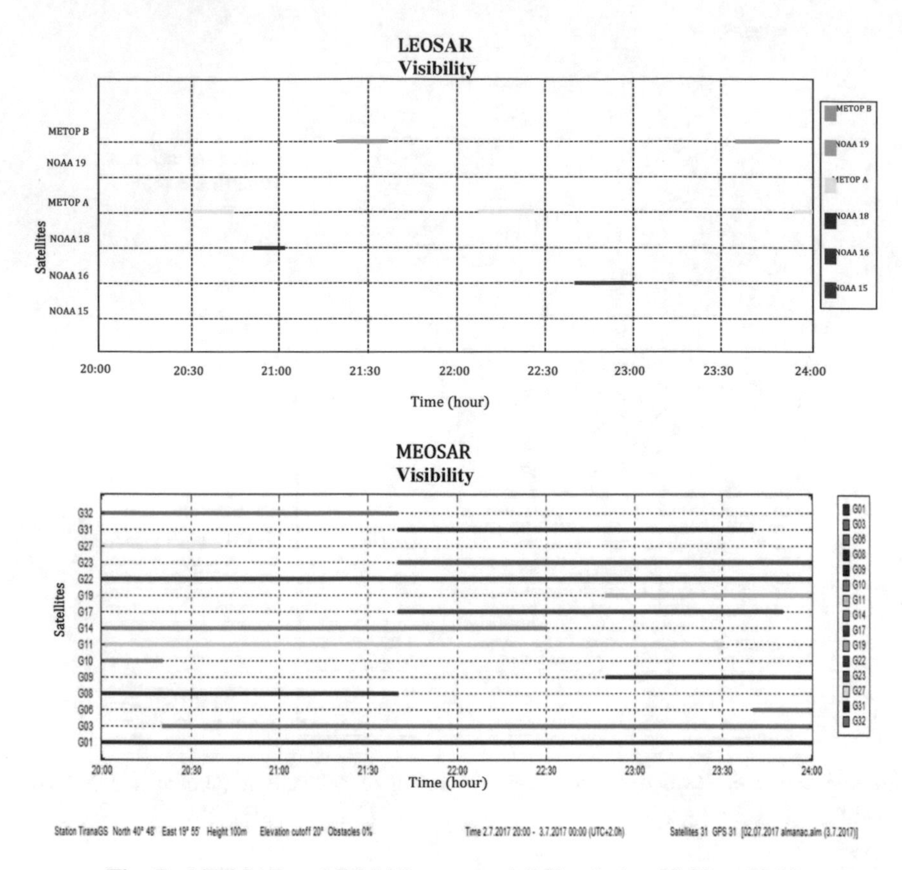

Fig. 7. MEOSAR vs LEOSAR system visibility during 20:00 to 00:00.

MEOSAR than the LEOSAR constellation. An even more critical issue with respect to the footprint of a LEOSAR satellite is evidenced in the case of NOAA 15. None of the two subsequent passes of the LEOSAR satellite appears visible to the supposed beacon location (Table 1), because of the satellite's small coverage. These passes do not establish communication and are considered as missed passes. In contrast, continuous coverage from MEOSAR enables detection and locating to be achieved much faster, which is often crucial for the rescue services.

3.3 Experimental MEOSAR versus LEO/GEOSAR Performance Comparison

On 4 May 2016 at 21:26 (UTC), a 406 MHz COSPAS-SARSAT EPIRB (Emergency Position-Indicating Radio Beacon) was activated at sea approximately 700NM west of the Galapagos Islands [15]. The distress situation was beneficial for saving lives and testing of the yet non fully-operational MEOSAR. The event is illustrated in Fig. 10. The results given in Table 2 [15] represent the chronological behavior of each system's satellite constellation to serve as a performance comparison.

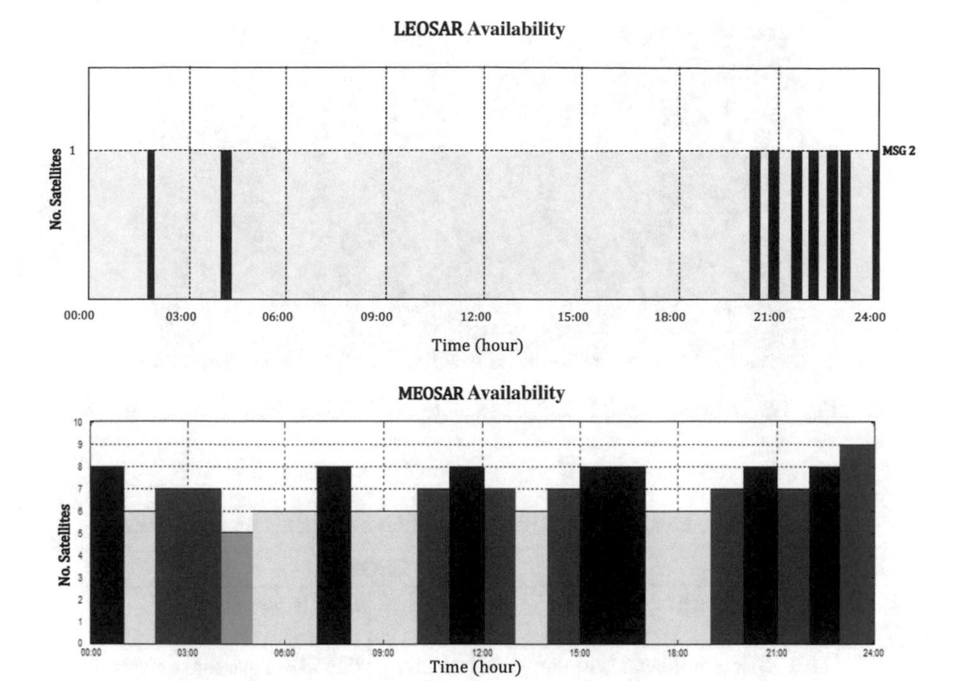

Fig. 8. Number of MEOSAR vs LEOSAR system satellites during 24 h seen from beacon.

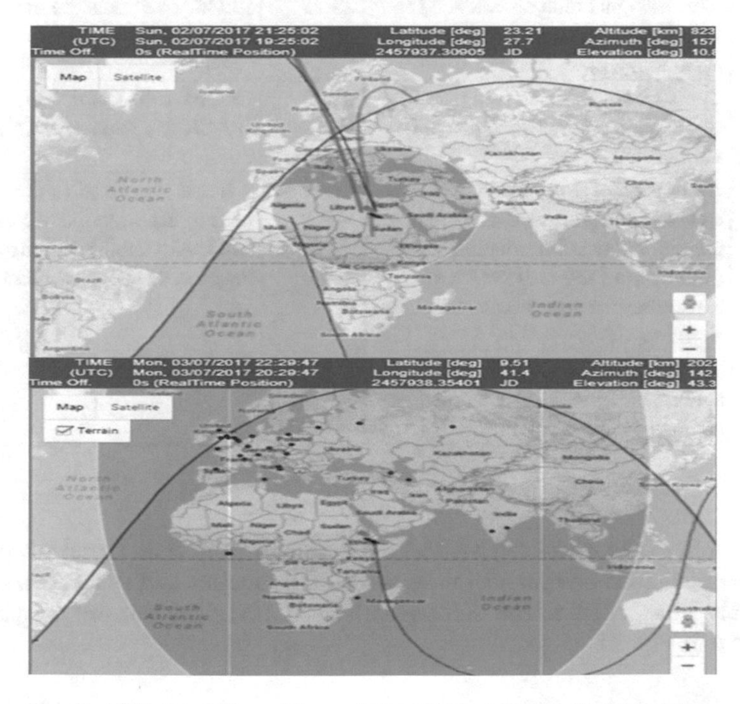

Fig. 9. USA 96 GPS satellite and METOP B SAR satellite footprints.

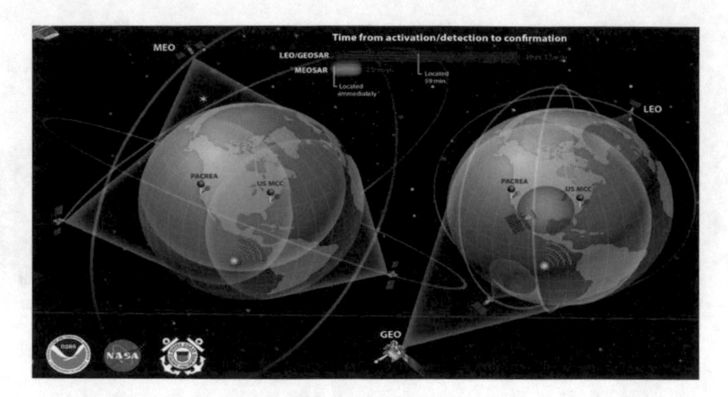

Fig. 10. MEOSAR and LEOSAR satellite constellation (real world case).

Table 2. Compare and contrast on each satellite's constellation performance.

MEOSAR		LEO/GEOSAR	
21:26 UTC	EPIRB was activated	21:26 UTC	EPIRB was activated
21:26 UTC	MEOSAR satellites detect the activated beacon with a Time Difference of Arrival (TDOA) position	21:26 UTC	GEOSAR satellite detects the activated beacon. No location is given
21:26 UTC	Beacon signal data downloaded to USMCC (Mission Control Center) at Suitland, Maryland	22:25 UTC	LEOSAR detects the activated beacon
21:26 UTC	Coast Guard District 11 (PACREA, Alameda, California) notified	22:25 UTC	Coast Guard District 11 (PACREA, Alameda, California) notified
21:32 UTC	An updated TDOA position was received and provided even more precise location information (diminishing the Expected Error) narrowing the search area significantly	23:43 UTC	2 h and 17 min after the initial beacon activation, a second LEOSAR satellite confirms location of activated beacon
21:46 UTC	20 min after the initial beacon activation, MEOSAR satellites confirm the location		

As a result of the calculated SARSAT coordinates on distress signal reception, the fishing vessel Cuidad de Portoviejo was successfully located and all 21 crew members were safely transported to the Galapagos Islands [15]. The timeline is graphically represented in Fig. 11.

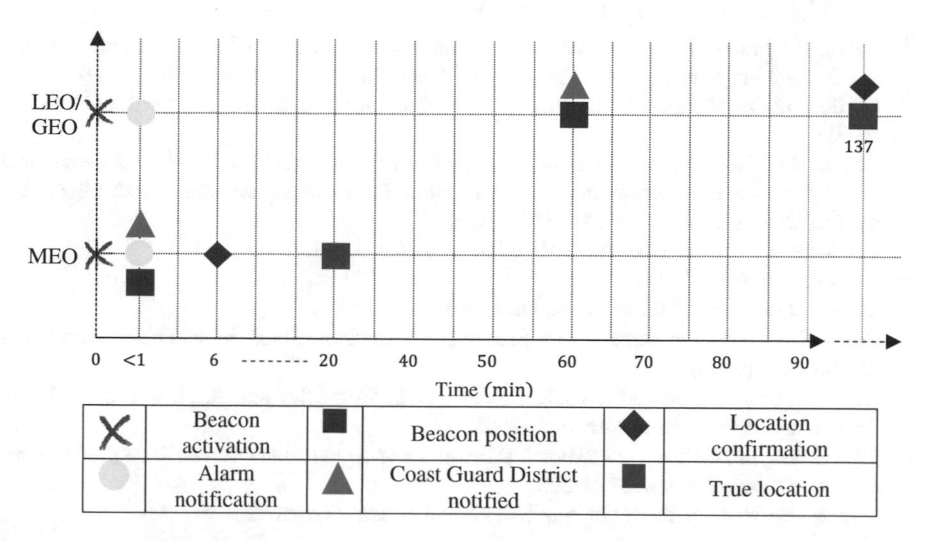

Fig. 11. From detection to location determination timeline (LEO/GEO vs MEO-based systems).

4 Conclusions

The comparison analysis, considering visibility and availability from the hypothetical location assumed as a potential location of distress signal generation, confirmed advantages in performance of MEOSAR towards LEOSAR. For such comparison purposes real time satellite tracking software is applied.

The implementation of MEOSAR provides larger coverage and instantaneous alert, compared to LEOSAR. Thus, the MEOSAR location estimation is much faster than with LEOSAR. A quicker distress alert directly contributes to more effective rescue services where timing is usually critical.

References

1. http://www.cospas-sarsat.org/
2. COSPAS – SARSAT 406 MHz Frequency Management Plan, C/T T.012, Issue 1 – Revision 5, Probability of Successful Doppler Processing and LEOSAR System Capacity (2008)
3. Taylor, W., Vigneault, O.M.: A neural network application to search and rescue satellite aided tracking (SARSAT). In: Proceedings of the Symposium/Workshop on Applications of Experts Systems in DND, pp. 189–201. Royal Military College of Canada (1992)
4. Cakaj, S., Fitzmaurice, M., Reich, J., Foster, E.: Simulation of local user terminal implementation for Low Earth Orbiting (LEO) search and rescue satellites. In: The Second International Conference on Advances in Satellite and Space Communications SPACOMM 2010, pp. 140–145. IARIA, Athens (2010)
5. Generic MEOSAR presentation, COSPAS-SARSAT secretariat, Montreal, Canada (2016)

6. Cakaj, S.: Practical horizon plane and communication duration for Low Earth Orbiting (LEO) satellite ground stations. WSEAS J. Trans. Commun. **4**(8), 373–383 (2009)
7. Gordon, G.D., Morgan, W.L.: Principles of Communication Satellites. Wiley, Hoboken (1993)
8. Cakaj, S.: The coverage belt for low earth orbiting satellites. In: 39th International Convention on Information and Communication Technology, Electronics and Microelectronics, Opatia, Croatia, pp. 554–557 (2016)
9. Information Bulletin, Issue 26 (2016). https://cospas-sarsat.int/en/
10. www.satflare.com (2017)
11. https://celestrak.com/GPS/almanac/Yuma/2017/ (2017)
12. Cakaj, S.: Local User Terminals for Search and Rescue Satellites. VDM Publishing House, Saarbrucken (2010)
13. Kaplan, D.E., Hegarty, J.C.: Understanding GPS: Principles and Applications, 2nd edn. Artech House Publishers, Norwood (2006)
14. Dye, S., Baylin, F.: The GPS Manual: Principles and Applications. Baylin/Gale Productions, Boulder (1997). ISBN 0917893298
15. MEOSAR, Medium Earth Orbiting Search and Rescue. Poster (2016)

An Integrated System Considering WLAN and DTN for Improving Network Performance: Evaluation for Different Scenarios and Parameters

Evjola Spaho[1(✉)], Kevin Bylykbashi[2], Leonard Barolli[3], and Makoto Takizawa[4]

[1] Department of Electronics and Telecommunication, Faculty of Information Technology, Polytechnic University of Tirana, Mother Teresa Square, No. 4, Tirana, Albania
evjolaspaho@hotmail.com

[2] Faculty of Information Technology, Polytechnic University of Tirana, Mother Teresa Square, No. 4, Tirana, Albania
kevin.bylykbashi@fti.edu.al

[3] Department of Information and Communication Engineering, Fukuoka Institute of Technology (FIT), 3-30-1 Wajiro-Higashi, Higashi-Ku, Fukuoka 811-0295, Japan
barolli@fit.ac.jp

[4] Department of Advanced Sciences, Hosei University, 3-7-2, Kajino-cho, Koganei-shi, Tokyo 184-8584, Japan
makoto.takizawa@computer.org

Abstract. In this paper, we integrate a Wireless Local Area Network (WLAN) with a Delay Tolerant Network (DTN) to improve the network performance when the network is congested or communication link problems occur. We evaluate the performance under different scenarios for three routing protocols. Simulations are conducted with the Opportunistic Network Environment (ONE) simulator. The simulation results shows that with the increase of the simulation time DTN hosts are in movement for longer time and the probability of the DTN hosts to meet and exchange the data is increased. The usage of DTN improves the performance of the network by sending information with high delivery probability.

1 Introduction

Advances in network access technologies and the widespread use of small portable devices such as smartphones, tablets and laptops, in the past few years has led in an unprecedented growth in the number of wireless users and applications.

Wireless Local Area Networks (WLANs) technology have emerged as a promising networking technology to extend network connectivity outside private

L. Barolli et al. (Eds.): EIDWT 2018, LNDECT 17, pp. 339–348, 2018.
https://doi.org/10.1007/978-3-319-75928-9_30

networks. Wireless technologies offer a very effective and inexpensive solution to bring wireless Internet at public venues. WLANs have been widely deployed in recent years at different areas such as libraries, cafeterias, classrooms, and residential houses, due to their inherently low implementation cost [1].

When users connect to one of the WLAN Access Points (APs) the following situations can occur: they can stay connected in the transmission range of one AP, can receive the signal of two APs, or stay isolated. Users connected with these APs can upload and download big amount of information creating network congestions. There are also scenarios where communication link problems occur. In order to solve this issue, Delay Tolerant Networks (DTNs) can be used. DTNs can be deployed in scenarios where infrastructure access is not available or to reduce the traffic load from congested infrastructure networks.

DTNs enable communication where connectivity issues like intermittent connectivity, high latency, asymmetric data rate, and no end-to-end connectivity exists. In order to handle disconnections and long delays, DTNs use store-carry-and-forward approach.

In this paper, we evaluate the performance of an integrated approach considering DTNs and WLANs technologies when there are communication link problems or WLAN is overloaded. We compare the performance of three different routing protocols: epidemic, spray and wait and prophet. For the simulations, we use the Opportunistic Network Environment (ONE) [2] simulator.

ONE is a simulation environment capable of generating node movement using different movement models. ONE offers various DTN routing algorithms for routing messages between nodes. Its graphical user interface visualize both mobility and message passing in real time. ONE can import mobility data from real-world traces or other mobility generators. It can also produce a variety of reports from node movement to message passing and general statistics.

Performance evaluation results, based on simulation, show that DTN is a good solution to improve the network performance by reducing the traffic load when WLAN is congested.

The remainder of the paper is organized as follows. Section 2 introduces DTN and routing protocols. The simulation system design is presented in Sect. 3. In Sect. 4 are shown the simulation results. Finally, the conclusions and future work are presented in Sect. 5.

2 DTNs and Routing Protocols

2.1 DTN Overview

DTNs are occasionally connected networks, characterized by the absence of a continuous path between the source and destination [3,4]. The data can be transmitted by storing them at nodes and forwarding them later when a link is established. This technique is called message switching. Eventually the data will be relayed to the destination. DTN is the "challenged computer network" approach that is originally designed from the Interplanetary Internet, and the data transmission is based upon the store-carry-and-forward protocol for the sake of carrying data packets under a poor network environment such as space [3]. Different copies of the same bundle can be routed independently to increase security and robustness, thus improving the delivery probability and reducing the delivery delay. However, such approach increases the contention for network resources (e.g., bandwidth and storage), potentially leading to poor overall network performance.

In [5], authors have studied this model and found that it can provide substantial capacity at little cost, and that the use of a DTN model often doubles that capacity compared with a traditional end-to-end model. The main assumption in the Internet that DTNs seek to relax is that an end-to-end path between a source and a destination exists for the entire duration of a communication session. When this is not the case, the normal Internet protocols fail. DTNs get around the lack of end-to-end connectivity with an architecture that is based on message switching. It is also intended to tolerate links with low reliability and large delays. The architecture is specified in RFC 4838 [6].

Bundle protocol has been designed as an implementation of the DTN architecture. A bundle is a basic data unit of the DTN bundle protocol. Each bundle comprises a sequence of two or more blocks of protocol data, which serve for various purposes. In poor conditions, bundle protocol works on the application layer of some number of constituent Internet, forming a store-and-forward overlay network to provide its services. The bundle protocol is specified in RFC 5050. It is responsible for accepting messages from the application and sending them as one or more bundles via store-carry-and-forward operations to the destination DTN node. The bundle protocol runs above the TCP/IP level.

2.2 Routing Protocols

In order to handle disconnections and long delays in sparse opportunistic network scenarios, DTN uses store-carry-and-forward approach. A network node stores a bundle and waits for a future opportunistic connection. When the connection is established, the bundle is forwarded to an intermediate node, according to a hop-by-hop forwarding/routing scheme. This process is repeated and the bundle will be relayed hop-by-hop until reaching the destination node. In [7–18,20,21] authors deal with routing in DTNs.

In this work, we will use three widely applicable DTN routing protocols Epidemic [14], Spray and Wait [19] and Prophet [21].

Epidemic routing protocol: Epidemic [14] is a protocol that is basically a flooding mechanism. Each message spreads like a disease in a population without priority and without limit. When two nodes encounter each other they exchange a list of message IDs and compare those IDs to decide which message is not already in storage in the other node. The next phase is a check of available buffer storage space, with the message being forwarded if the other node has space in its buffer storage. The main goals of this protocol are: maximize the delivery ratio, minimize the latency and minimize the total resources consumed in message delivery. It is especially useful when there is lack of information regarding network topology and nodes mobility patterns.

Spray and Wait routing protocol: Spray and Wait [19], is a routing protocol that attempts to gain the delivery ratio benefits of replication-based routing as well as the low resource utilization benefits of forwarding-based routing. The Spray and Wait protocol is composed of two phases: the spray phase and the wait phase. When a new message is created in the system, a number L is attached to that message indicating the maximum allowable copies of the message in the network. During the spray phase, the source of the message is responsible for "spraying", or delivery, one copy to L distinct "relays". When a relay receives the copy, it enters the wait phase, where the relay simply holds that particular message until the destination is encountered directly.

Prophet routing protocol: Prophet (Probabilistic Routing Protocol using History of Encounters and Transitivity) [21] is a variant of the epidemic routing protocol for intermittently connected networks that operates by pruning the epidemic distribution tree to minimize resource usage while still attempting to achieve the best case routing capabilities of epidemic routing. It uses a probabilistic metric: delivery predictability, that attempts to estimate, based on node encounter history, which node has the higher probability of successful delivery of a message to the final destination. When two nodes are in communication range, a new message copy is transferred only if the other node has a better probability of delivering it to the destination.

3 Simulation System Design

Simulations are carried out using the ONE simulator. When a network load or communication link problem is detected in WLAN, the DTN part will start to work. APs of WLAN are also DTN nodes. Mobile hosts moving near APs will receive the messages need to be delivered from APs, communicate to intermediate hosts to do the custodians of messages and deliver the data to destination host.

In Tables 1 and 2 are shown the simulation parameters and their values for APs and DTN hosts, respectively. Different interface types with different transmission speed and range are taken into consideration. We created three different simulation scenarios considering the simulation time 12 h, 24 h and 48 h. The simulation area is 4500 m × 3400 m. The coordinates where are placed the APs are shown in Table 3. DTN hosts move according to Random Waypoint mobility model with speed from 0.8 m/s to 1.8 m/s to model moving pedestrians equipped with a device. Hosts in the simulation area choose the random waypoint, then they move through this waypoint with a predefined speed. After the hosts arrive in the destination waypoint, they stop for a randomly chosen amount time and then they continue in the same way.

The event generator is responsible for generating bundles with sizes uniformly distributed in the ranges [50 kB, 200 kB]. A bundle is created every 15–25 s and data bundles ttl is 300 min. A screen-shot of the simulation environment is shown in Fig. 1.

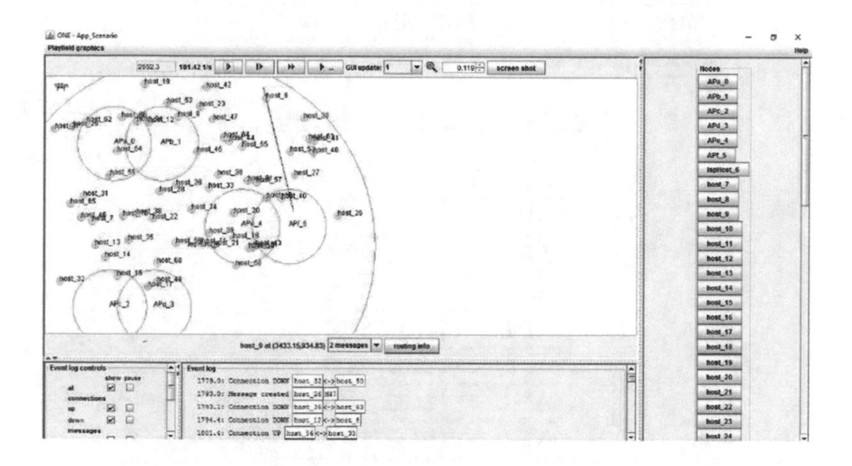

Fig. 1. A screen-shot of the simulation environment.

We evaluate the performance of the system for 3 different routing protocols: epidemic, spray and wait and prophet for different simulation times.

We use the following metrics to measure the performance of different routing protocols: delivery probability and average latency.

- **Delivery probability** is the ratio of number of delivered messages to that of created messages.
- **Average latency** is the average time elapsed from the creation of the messages at source to their successful delivery to the destination.
- **Average number of hops** is the average number of hops counts between the source and the destination nodes of bundles.

Table 1. Simulation parameters and their values for APs.

Parameters	Values
Number of APs	7
Mobility model	Stationary
Buffer size	500 MB
Network interface	WiFi 802.11g/n

Table 2. Simulation parameters and their values for DTN hosts.

Parameters	Values
Number of DTN hosts	100
Mobility model	Random Waypoint
Buffer size	100 MB
Network interface	WiFi 802.11g/n, Bluetooth v4

Table 3. Coordinates of APs.

AP name	Coordinates
AP1	1030, 990
AP2	1730, 990
AP3	970, 3400
AP4	1630, 3400
AP5	2930, 2200
AP6	3630, 2200
AP7	2130, 2500

4 Simulation Results

In this section, we present the simulation results. In Tables 4, 5 and 6 are shown the simulation results when the simulation time 12 h, 24 h and 48 h, respectively. The number of generated messages is the same because the simulation time is the same for all protocols. From the results, we can notice that epidemic protocol has the highest delivery probability, lowest average latency, but higher average hop count for all scenarios compared with other protocols.

Table 4. Simulation results for simulation time 12 h.

Parameter	Epidemic routing	Spray and wait routing	Prophet routing
Created messages	2209	2209	2209
Delivered messages	1604	1138	1351
Delivery probability	72.61%	51.52%	61.16%
Average delay	1.6 h	2.13 h	2.68 h
Average hop count	4.13	2.67	3

In Fig. 2 are shown the simulation results of delivery probability vs. simulation time for all considered routing protocols. With the increase of simulation time, delivery probability is increased for all protocols. DTN hosts are in movement for longer time and the probability of the DTN hosts to meet and exchange the data is higher.

Table 5. Simulation results for simulation time 24 h.

Parameter	Epidemic routing	Spray and wait routing	Prophet routing
Created messages	4427	4427	4427
Delivered messages	3531	2501	3092
Delivery probability	79.76%	56.49%	69.84%
Average delay	4.13 h	2.73 h	3 h
Average hop count	1.71	2.26	2.77

Table 6. Simulation results for simulation time 48 h.

Parameter	Epidemic routing	Spray and wait routing	Prophet routing
Created messages	8865	8865	8865
Delivered messages	7446	5275	6661
Delivery probability	83.99%	56.95%	75.14%
Average delay	4.13 h	2.36 h	3 h
Average hop count	1.81	2.26	2.83

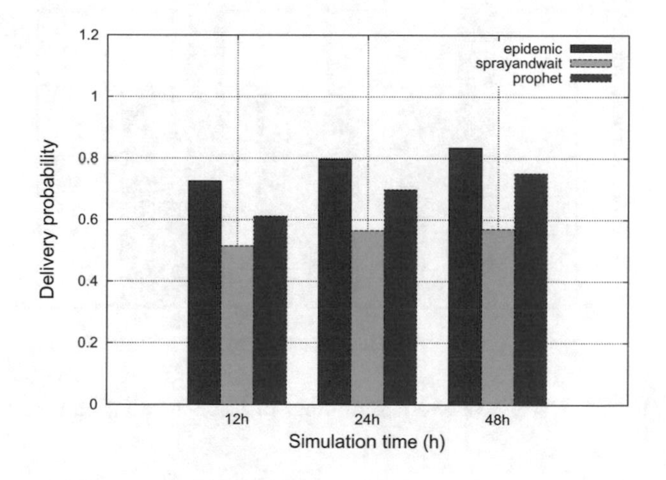

Fig. 2. Results of delivery probability for all scenarios.

The simulation results of average latency vs. simulation time for all considered routing protocols are shown in Fig. 3. Considering the time of the simulation, it can been seen that epidemic protocol presents better results compared with two other protocols.

In Fig. 4 are shown the simulation results of average number of hops vs. simulation time for three considered routing protocols. Spray and wait uses the smallest number of hops for communication compared with other protocols.

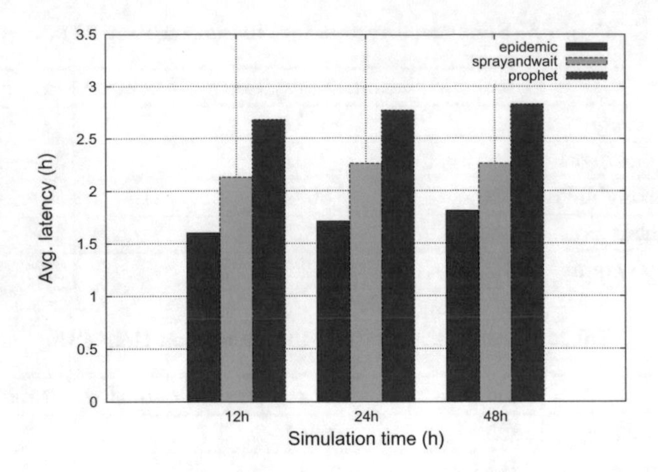

Fig. 3. Results of average latency for all scenarios.

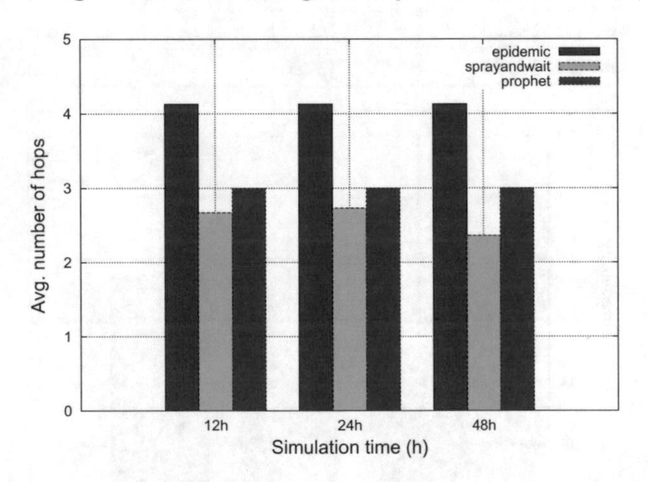

Fig. 4. Results of average number of hops for all scenarios.

5 Conclusions

In this work, we used DTN to support a WLAN in case of high network load or communication link problem. We evaluated the performance of this integrated WLAN-DTN architecture using stationary APs and DTN hosts moving according to Random Waypoint mobility model for three routing protocols (epidemic, spray and wait and prophet). The performance study showed the following results.

- DTN improves the performance of WLAN in cases of high network load or communication link problems.
- DTN makes possible delivering of the data that can not be sent using WLAN with a high delivery probability.

- Communication is realized using a small number of hops.
- The average latency is increased with the increase of the simulation time and the hybrid WLAN-DTN infrastructure is suitable only for applications that tolerates the delay in data transmission.
- Epidemic protocol achieves better performance in terms of delivery probability and delay compared with two other protocols.

In this work, we considered Random Waypoint mobility model for DTN hosts. In the future, we would like to consider other mobility models and make extensive simulations to evaluate the performance of WLAN-DTN for different routing protocols considering different scenarios and parameters.

References

1. Inaba, T., Obukata, R., Sakamoto, S., Oda, T., Ikeda, M., Barolli, L.: Performance evaluation of QoS-aware fuzzy-based CAC for LAN access. Int. J. Space-Based Situated Comput. **6**(4), 228–238 (2016)
2. Keranen, A., Ott, J., Karkkainen, T.: The ONE simulator for DTN protocol evaluation. In: Proceedings of the 2nd International Conference on Simulation Tools and Techniques (SIMUTools-2009) (2009). http://www.netlab.tkk.fi/tutkimus/dtn/theone/pub/the_one_simutools.pdf
3. Fall, K.: A delay-tolerant network architecture for challenged Internets. In: Proceedings of the International Conference on Applications, Technologies, Architectures, and Protocols for Computer Communications, ser. SIGCOMM 2003, p. 2734 (2003)
4. Delay- and disruption-tolerant networks (DTNs) tutorial, NASA/JPLs Interplanetary Internet (IPN) Project (2012). http://www.warthman.com/images/DTN_Tutorial_v2.0.pdf
5. Laoutaris, N., Smaragdakis, G., Rodriguez, P., Sundaram, R.: Delay tolerant bulk data transfers on the Internet. In: Proceedings of the 11-th International Joint Conference on Measurement and Modeling of Computer Systems (SIGMETRICS09), p. 22923 (2009)
6. Cerf, V., Burleigh, S., Hooke, A., Torgerson, L., Durst, R., Scott, K., Fall, K., Weiss, H.: Delay-tolerant networking architecture. IETF RFC 4838 (Informational), April 2007
7. Massri, K., Vernata, A., Vitaletti, A.: Routing protocols for delay tolerant networks: a quantitative evaluation. In: Proceedings of ACM Workshop PM2HW2N2012, pp. 107–114 (2012)
8. Ishikawa, S., Honda, T., Ikeda, M., Barolli, L.: Performance analysis of vehicular DTN routing under urban environment. In: Proceedings of CISIS-2014, July 2014
9. Demmer, M., Fall, K.: DTLSR: delay tolerant routing for developing regions. In: Proceedings of the 2007 ACM Workshop on Networked Systems for Developing Regions, 6 p. (2007)
10. Ilham, A.A., Niswar, M., Agussalim: Evaluated and optimized of routing model on Delay Tolerant Network (DTN) for data transmission to remote area. In: Proceedings of FORTEI, Indonesia University Jakarta, pp. 24–28 (2012)
11. Jain, S., Fall, K., Patra, R.: Routing in a delay tolerant network. In: Proceedings of ACM SIGCOMM 2004 Conference on Applications, Technologies, Architectures, and Protocols for Computer Communication, Portland, Oregon, USA, 30 August–3 September 2004, pp. 145–158 (2004)

12. Zhang, Z.: Routing in intermittently connected mobile ad hoc networks and delay. Commun. Surv. Tutor. **8**(1), 24–37 (2006)
13. Soares, V.N.G.J., Rodrigues, J.J.P.C., Farahmand, F.: GeoSpray: a geographic routing protocol for vehicular delay-tolerant networks. Inf. Fusion **15**(1), 102–113 (2014)
14. Vahdat, A., Becker, D.: Epidemic routing for partially connected ad hoc networks. Technical report CS-200006, Duke University, April 2000
15. Barros, M.: Impact of delay-tolerant network support in wireless local area networks. Master thesis (2014)
16. Uchida, N., Ishida, T., Shibata, Y.: Delay tolerant networks-based vehicle-to-vehicle wireless networks for road surveillance systems in local areas. Int. J. Space-Based Situated Comput. **6**(1), 12–20 (2016)
17. Shinko, I., Fouquet, Y., Nace, D.: Elastic routing for survivable networks. Int. J. Grid Util. Comput. **6**(2), 121–129 (2015)
18. Cui, B., Yan, X.: A review of data management and protocols for vehicular networks. Int. J. Web Grid Serv. **13**(2), 186–206 (2017)
19. Spyropoulos, T., Psounis, K., Raghavendra, C.S.: Spray and wait: an efficient routing scheme for intermittently connected mobile networks. In: Proceedings of ACM SIGCOMM 2005 Workshop on Delay Tolerant Networking and Related Networks (WDTN-05), Philadelphia, PA, USA, pp. 252–259 (2005)
20. Burgess, J., Gallagher, B., Jensen, D., Levine, B.N.: MaxProp: routing for vehicle-based disruption-tolerant networks. In: Proceedings of the IEEE Infocom, April 2006
21. Lindgren, A., Doria, A., Davies, E., Grasic, S.: Probabilistic routing protocol for intermittently connected networks, draft-irtf-dtnrg-prophet-09. http://tools.ietf.org/html/draft-irtf-dtnrg-prophet-09

An Efficient Algorithm to Energy Savings for Application to the Wireless Multimedia Sensor Networks

Astrit Hulaj(✉) and Adrian Shehu

Department of Electronic and Telecommunications,
Faculty of Information Technology, Polytechnic University of Tirana,
Tirana, Albania
astrit.hulaj@fti.edu.al, adshehu@tcn.al

Abstract. The lifetime of the Wireless Multimedia Sensor Network (WMSN), located along the green borderline, depends directly on their battery. Conversion of the captured image by sensor nodes in the black and white image directly affects in energy saving. However, in this case, there will be a loss of details of the objects in the image, respectively the loss of the structure of objects in the image. This will create problems in identifying various criminal groups during the illegal crossing of the state border from different animals which can be very active across the borderline. Therefore, in this paper, we will present an efficient algorithm, which will restore corrupted image pixels from various noises and will retain the image structure captured by WMSN, without the application of an algorithm for detection of the image edges.

1 Introduction

Nowadays, cross-border security in many countries around the world has become quite problematic and is becoming a global problem. Among the most problematic border sections is the security of green borderline. In particular, the security of those border areas with harsh terrain and the terrain covered with high and dense forests. As result of monitoring difficulties by security authorities, such areas can be easily used by criminal groups for illegal border crossing. Illegal border crossings from criminal groups can be for different purposes, such as illegal migration, terrorism, smuggling of arms, drugs and other valuables, guerrilla (destabilization of the neighboring country), etc. [1].

The state borders of one country have unique specifications from those of another country. So a system applied in one country may not be suitable for another country. Therefore, each country, depending on economic conditions and geopolitical factors, creates state strategies and policies for the protection of state borders [1]. The security of national borders is a fundamental element in the country's sovereignty itself.

Even today many countries, are still surveillance the green borderline through conventional systems [1], namely through police or military patrols. These patrols realize borderline control physically in designated time intervals. This kind of monitoring, cannot provide continuous control of the border, except that is also very troublesome. Border security, according to this method of monitoring depends directly on

L. Barolli et al. (Eds.): EIDWT 2018, LNDECT 17, pp. 349–358, 2018.
https://doi.org/10.1007/978-3-319-75928-9_31

the human factor. However, for cross-border security, more and more technologies are being applied. One of technologies that is under consideration to be applied along the green borderline is WMSN technology. However, the application of the WMSN technology for cross-border security is characterized by many problems of different nature. The main problem is the power supply of the sensor nodes. The operation of a sensor node is directly dependent on the battery. Therefore, for this reason, many researchers are now focused on finding alternatives that will affect of energy saving of the sensor node.

The objective of this paper is the proposal of an efficient algorithm, which will simultaneously eliminate the noises from images, which have degraded the image, such as Periodic Noise, Salt & Pepper Noise (SPN), Speckle Noise (SPKN), Gaussian noise, Poisson Noise, and FFT2 Phase, as well as extracting full structure of image converted to black and white image. This will directly affect the energy savings spent by the multimedia sensor nodes, during the image processing process.

2 Related Works

The application of technology to improve security across the state borderline is a key factor. In other words, state security now cannot be imagined without the application of the technology. In particular, a very important role has WMSN application along the green borderline. However, the application of this technology along the green borderline is characterized by challenges of different natures. Among the main challenges, as we have mentioned, is the electrical power supply of the sensor node. The lifetime of the sensor node is directly dependent on the battery. Therefore, continuously the most efficient methods are required that will affect the energy savings spent by the sensor node. Regarding the application of technology along the green borderline today some research and concrete proposals can be found.

In [2], a hybrid architecture of wireless network sensors for application along the state borderline has been presented. The architecture presented is called BorderSense. This system architecture of BorderSense has three layers. The unattended ground sensors and the underground sensors constitute the lower layer of the architecture, which provide higher granularity for monitoring. At the second layer, surveillance towers improve the accuracy of the system through visual information. Finally, mobile ground robots and unmanned aerial vehicles constitute the higher layer that provides additional coverage and flexibility. This architecture is appropriate to apply to those border areas, which are not characterized by harsh terrain and covered with dense forests. Applying such a system for cases when border areas are characterized by harsh terrain and covered with high forests is at an almost unreachable cost and many of the equipment involved in this system are unnecessary.

In [3], authors have studied the Turkish border security system via simulation to identify possible ways of increasing border control and security along the land borders.

In [4] authors a FemtoNode and an adaptable middleware platform for military surveillance have presented. In order to illustrate the use of the proposed platform infrastructure, i.e. the customizable FemtoNode and the adaptable middleware, an area surveillance application is studied by the authors. In this application, low-end sensors nodes are scattered on the ground along a borderline. In case of an unauthorized vehicle crossing of the borderline limit, the sensors issue an alarm which will trigger the use of Unmanned Aerial Vehicles (UAVs) equipped with more sophisticated sensors, such as radars or visible light cameras, in order to perform the recognition of the vehicle. However, this proposed platform by these, authors cannot be applied to cases when border zones are characterized by harsh terrain and the terrain covered with high and dense forests.

In [5], authors have proposed an Energy-efficient routing algorithm, for the cases when WSN are applied along the borderline. Routing algorithm will be extended to sleep mode and therefore a longer network lifetime can be achieved. Then the Degree of Aggregation algorithm DOA means the minimum number of reports about an event that a leader of a group waits to receive from its group members, before reporting the ship location to the base station.

In [1], a WMSN architecture is proposed for application across the green borderline, respectively in those areas where the application of surveillance towers and existing systems is almost impossible. The proposed architecture in [1] is shown in Fig. 1.

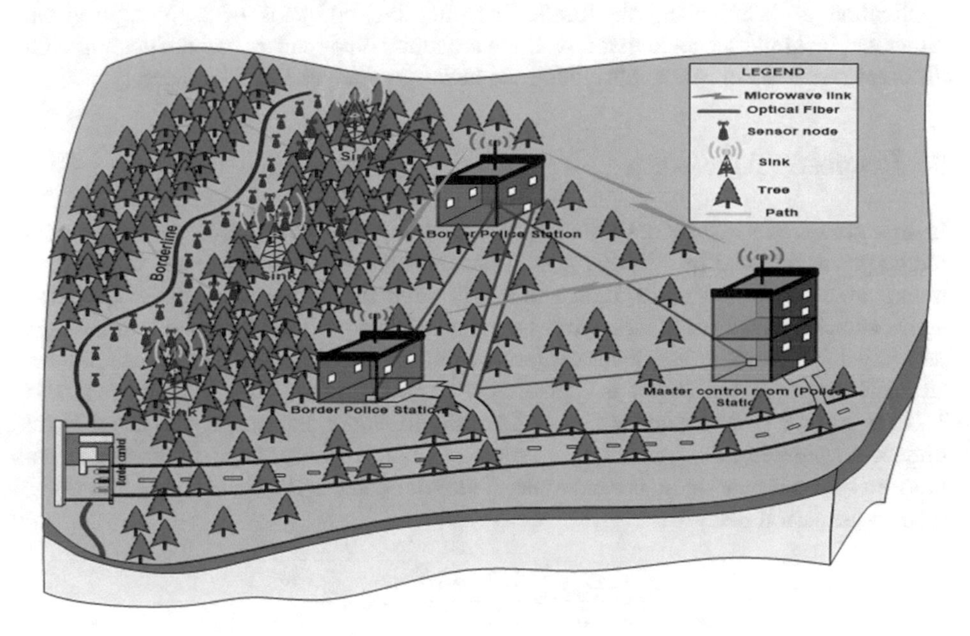

Fig. 1. The proposed network architecture for WMSN.

From Fig. 1, we can see that the network architecture is divided into three parts. The first part includes the green borderline or end network (Sensor nodes and Sink). In a WMSN, the sink is usually more robust than the sensor nodes, and it can be used as a gateway [9]. The second part includes the end buildings of the police stations, which are nearest along the borderline. The third part is the central monitoring room. This architecture can have an impact on improving the efficiency of police border patrols in detecting and preventing illegal crossings along the borderline. In other words, the WMSN technology proposed for borderline surveillance reduces the problem of sensor technology regarding false alarms.

In [6], an algorithm for energy savings of the sensor nodes is proposed. The logic of the algorithm's work is based on that, initially, the image captured by the WMSN is converted to a black and white image. After converting the image to white and black for the identification of objects in the image, the detection of pixels that corresponding to the edges of objects within the image is applied. Thus, this directly affects the energy savings spent by the sensor nodes during the process of processing and transmitting information to the state borderline management centre.

Also in [7], an algorithm for energy savings of the sensor nodes is proposed. This algorithm efficiently removes all types of noise that have corrupted the image captured by WMSN. In this case, there is no need to use any noise removal filters, but a single algorithm that is applied to enable efficient restoration of corrupted image pixels. This will directly affect savings of the energy spent by the sensor nodes because a single noise filtering algorithm will be applied [8].

In this section, we analyzed some of the papers which are dealing with the application of WSN along the borderline, and the problems of their application. However, it should be noted that now there are many other papers that are dealing with different problems of WSN, but which we have not analyzed in this paper.

3 Proposed Algorithm

In this sector, we will model an algorithm that will efficiently restore the corrupted image pixels captured by WMSN. Also, this algorithm will simultaneously convert the image to a black and white image and will save the image structure without the application of image edges detection. Image conversion in the black and white image will affect the image size reduction captured by multimedia sensors. For example, for an RGB image with of 300 $\times$ 600 pixels size is 540 kB, for a Greyscale image size is 180 kB and for a binary image size is 22.5 kB. Reducing the image size has a direct impact on reducing the transmission delay and the power required for image transmission by the sensor node. For example, if sensor node has Bandwidth B = 512 kbps, then transmission delay is:

$$T_t = \frac{L_{rgb}}{B} = \frac{4320 * 10^3}{512 * 10^3} = 8.44\,\mathrm{s} \tag{1}$$

$$T_t = \frac{L_{Gray}}{B} = \frac{1440 * 10^3}{512 * 10^3} = 2.8\,\mathrm{s} \tag{2}$$

$$T_t = \frac{L_{Binary}}{B} = \frac{180 * 10^3}{128 * 10^3} = 0.35\,\mathrm{s} \tag{3}$$

The working principle of this algorithm is based on the fact that, initially, the image is converted into a grayscale image of grace. Then it converts the matrix B to the intensity image I. The returned matrix I contains values in the range 0.0 (black) to 1.0 (full intensity or white). In other words, values are defined of amin and amax to the minimum and maximum values in B.

Then, different types of noise are added to the image, ranging from periodic noise to frequency spectrum. When noises in the black and white image are added, the algorithm for removing noise from the image and reconstructing the image structure is applied. For remove noises of the image, proposed algorithm using a 3 × 3 mask, which the pixel after the pixel in the image is applied. Then pixels within this mask, depending on the value are ordered by the pixel of lesser value or equal to the pixel of higher value. After ordering, the true value of the pixel is replaced with the value the fifth pixel of the filter. To better understand the logic of the algorithm, in Fig. 2 and the Eq. (4) is presented the principle of work of the proposed algorithm. While in Fig. 3, is presented the flowchart of the proposed algorithm.

Noisy pixels

Replaced value

0	127	129	118	123	130	132
118	255 127	130	103	113	124	128
116	117	**130**	108	107	108	113
115	117	118	108	104	102	104
120	120	108	102	108	102	95
126	125	110	104	97	95	97
133	131	130	129	114	109	104

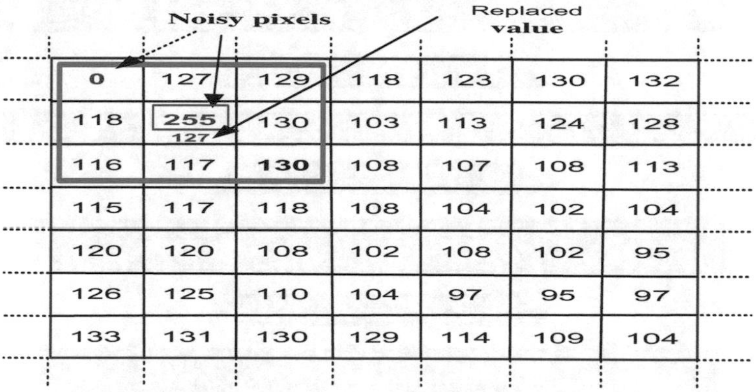

Fig. 2. Pixels matrix of the image.

$$\text{Neighbourhood values} = \begin{bmatrix} a_{11} & a_{12} & a_{13} \\ a_{21} & a_{22} & a_{23} \\ a_{31} & a_{32} & a_{33} \end{bmatrix} = a_{11} \leq a_{31} \leq a_{32} \leq a_{21} \leq a_{12} \leq a_{13} \leq a_{23} \leq a_{33} \leq a_{22} \tag{4}$$

$$= 0 \leq 0 \leq 116 \leq 117 \leq 118 \leq 127 \leq 129 \leq 130 \leq 130 \leq 255$$

$$\text{Replaced value} = 127$$

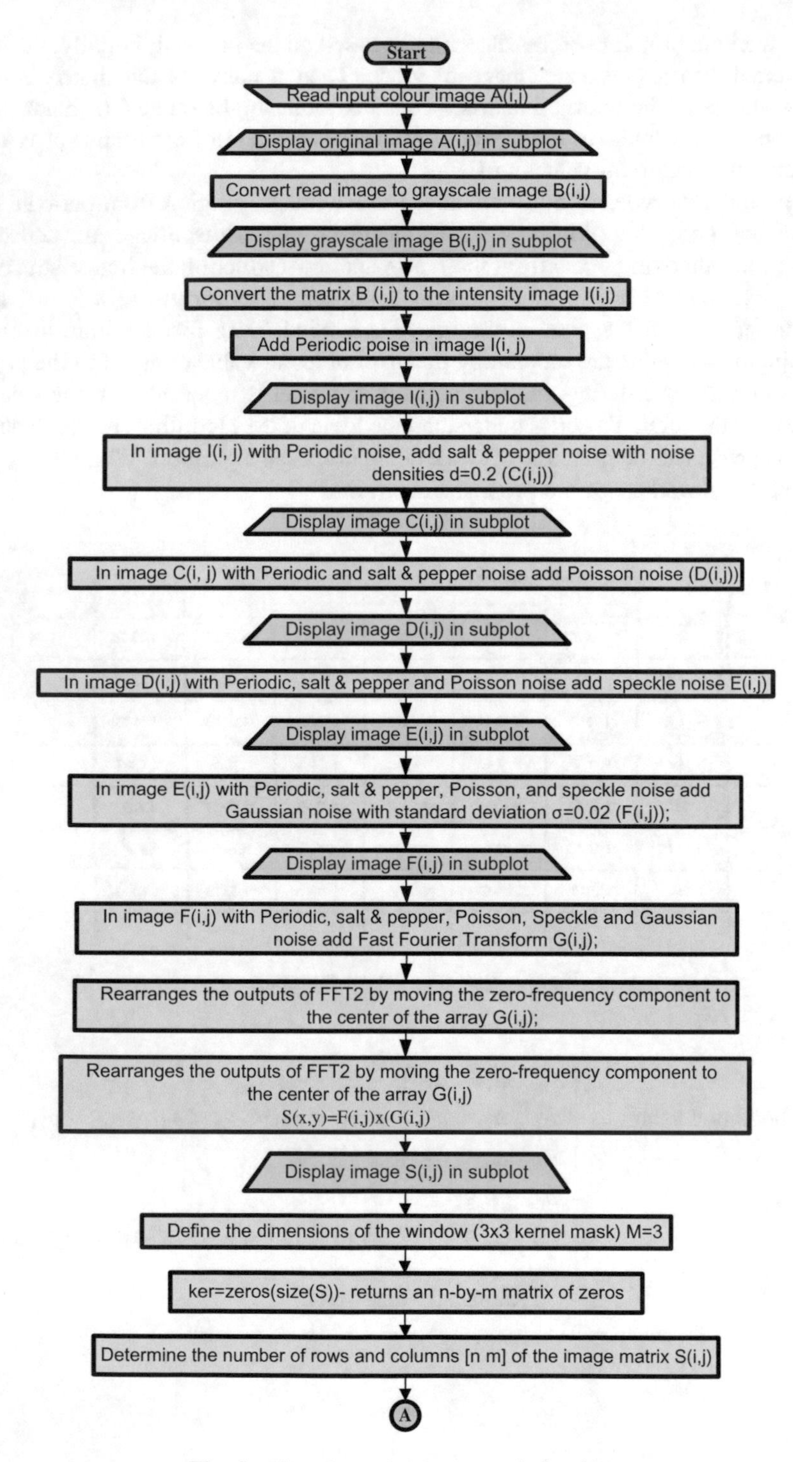

Fig. 3. Flowchart of the proposed algorithm.

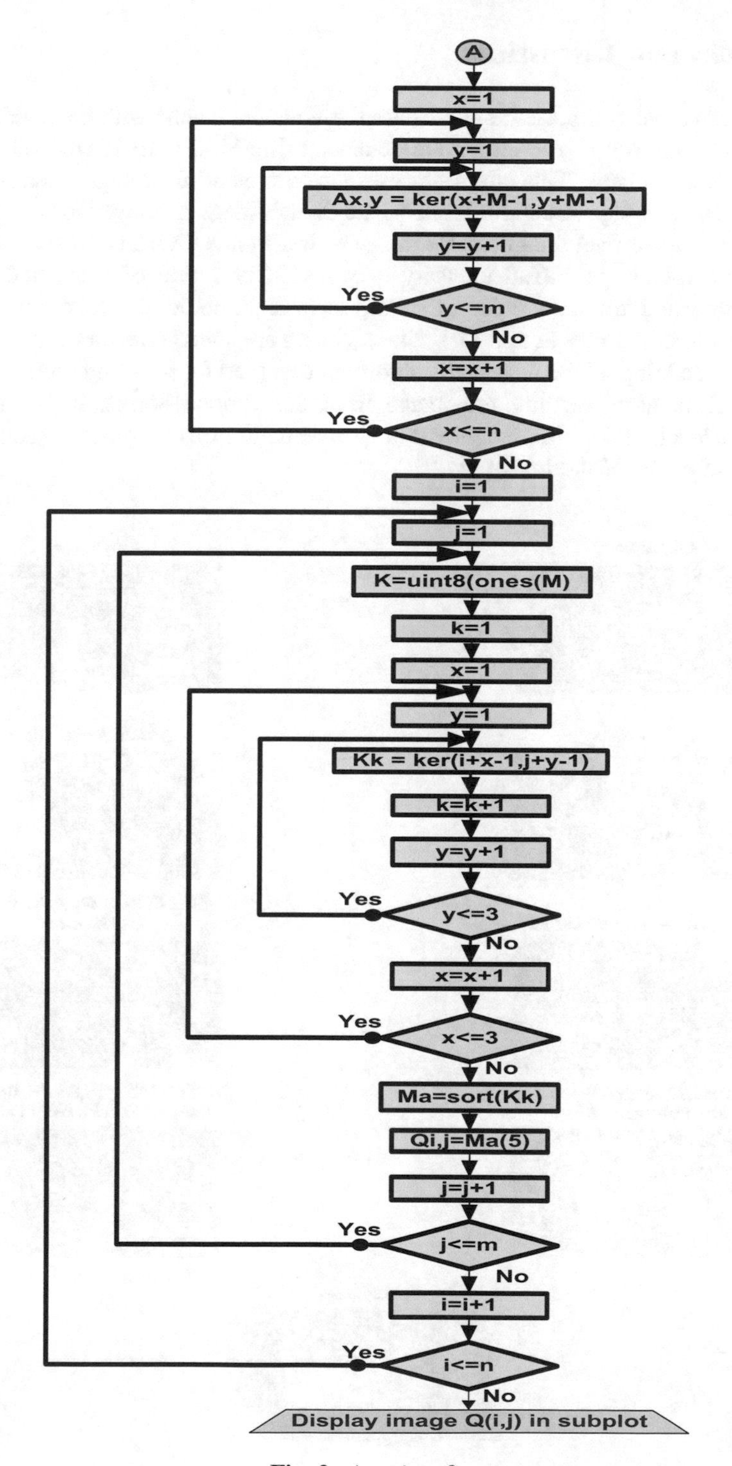

Fig. 3. (*continued*)

4 Results and Discussions

In this section, we will present and discuss the achieved results with the application of the proposed algorithm. The results are obtained using MATLAB 2016a and an image with 300 × 600 pixels. This algorithm uses a new method for image filtering and the preservation the image structure. Initially, the digital image is converted to a grayscale image. Then is returned the grayscale image matrix in a matrix that contains values in the range 0 (black) to 1 (full intensity or white). Corruption of black and white is realized by using different types of noise, such as Periodic, Salt&Pepper, Poisson, Speckle, Gaussian noises as and FFT Phase. Noises are added one after the other in the image. After adding all the noises, the algorithm designed for restoring corrupt pixels is applied. Then, after clearing the image from the various noises, in the output is obtained a black and white image with a clear structure of the image objects without detection of edges of the image.

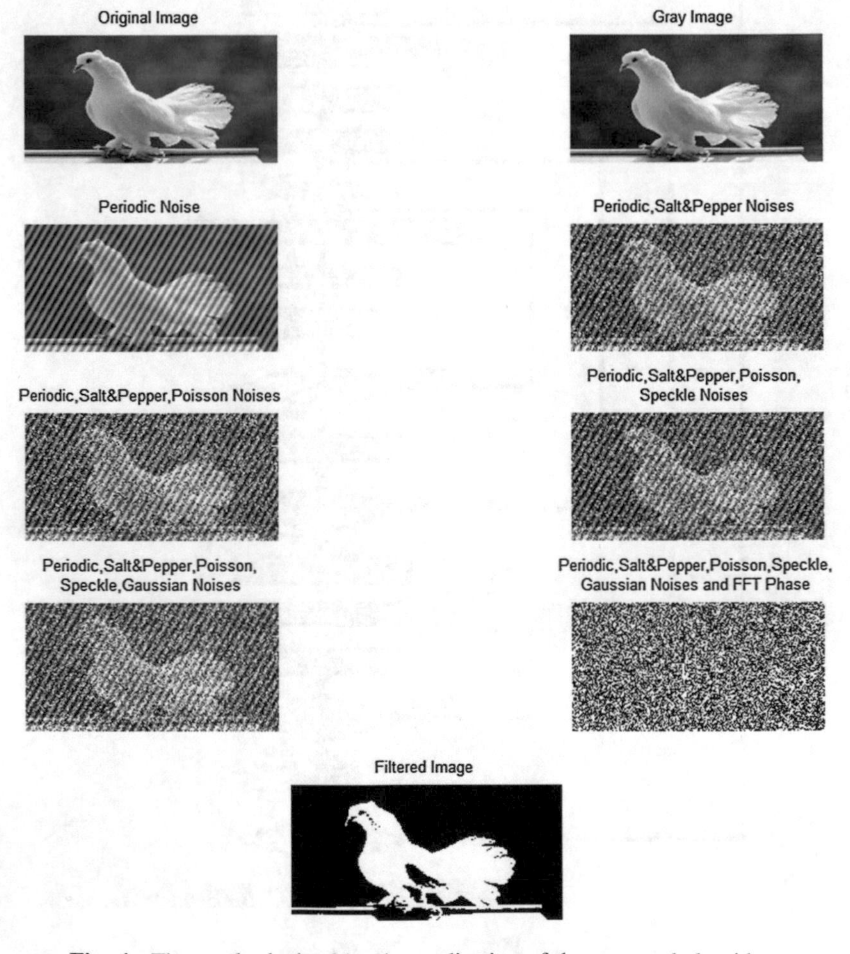

Fig. 4. The result obtained by the application of the proposed algorithm.

The results obtained by the applying this algorithm are shown in Fig. 4. Based on the results presented in Fig. 4, we can see that the proposed algorithm successfully performs the restoration of corrupted image pixels from the various noises. In particular, it should be noted that this algorithm successfully eliminates the FFT noise added to the image. Therefore, in this case, the application of a single algorithm is needed for removing various noises and preserving the image structure captured by WMSN.

Maintaining the structure of objects in a black and white image enables security authorities to easily identify and distinguish people from different animals that can be present along the green borderline. Also, the application of a single algorithm for this purpose, have a direct impact on energy saving.

In other words, instead of applying special filters for the removal of specific noises, a single algorithm is applied which will successfully enable the removal of different noises. This will cause the sensor node to spend less electricity during the processing and transmission of images to the monitoring centre. At the same time, it will affect the storage capacity of the transmitter link. In other words, this will increase the lifetime of the sensor node itself.

5 Conclusion

The lifetime of the sensor nodes located along the green borderline directly depends on their battery. Therefore, the battery charges should be as rational as possible.

In this paper, we proposed an algorithm which provides energy saving by sensor nodes, during the process of processing and transmission of the image captured by WMSN. The proposed algorithm for the purpose of energy saving, converts the initially captured image by the sensor node it into a black and white image. In order to preserve the structure of the objects in the black and white image, the image pixels are classified in two extreme values, respectively that correspond to the minimum and the maximum values of the image pixels. Then the corrupted image pixels are restored by the different noises, enabling the output to take a white and black image with a full structure as in Fig. 4.

The main conclusion is that the proposed algorithm preserves the structure of objects in the black and white image very well without the application of any additional algorithms for the detection of the edges. Therefore, based on the results presented, we can conclude that the proposed algorithm is very suitable for application in sensor nodes and other devices, in those cases where the rational use of electricity is needed.

In the future, we will make practical testing of the proposed algorithm using a Raspberry Pi Camera 3 Module, so that we can see the results of the practical application of this algorithm.

References

1. Hulaj, A., Shehu, A., Bajrami, X.: Application of wireless multimedia sensor networks for green borderline surveillance. In: 27th Annals of DAAAM & Proceedings (2016)
2. Sun, Z., Wang, P., Vuran, M.C., Al-Rodhaan, M.A., Al-Dhelaan, A.M., Akyildiz, I.F.: BorderSense: border patrol through advanced wireless sensor networks. Ad Hoc Netw. **9**(3), 468–477 (2011)

3. Çelik, G., Sabuncuoğlu, İ.: Simulation modelling and analysis of a border security system. Eur. J. Oper. Res. **180**(3), 1394–1410 (2007)
4. Freitas, E.P., Allgayer, R.S., Wehrmeister, M.A., Pereira, C.E., Larsson, T.: Supporting platform for heterogeneous sensor network operation based on unmanned vehicles systems and wireless sensor nodes. In: Intelligent Vehicles Symposium, pp. 786–791. IEEE (2009)
5. Razaque, A., Elleithy, K.M.: Energy-efficient boarder node medium access control protocol for wireless sensor networks. Sensors **14**(3), 5074–5117 (2014)
6. Shehu, A., Hulaj, A., Bajrami, X.: An algorithm for edge detection of the image for application in WSN. In: International Conference on Applied Physics, System Science and Computers, pp. 207–213. Springer, Cham, September 2017
7. Shehu, A., Hulaj, A., Bajrami, X.: The application of a single algorithm for filtering different noise in the image. In: European Conference on Electrical Engineering and Computer Science, Bern, Switzerland 2017. IEEE (in press)
8. Hulaj, A., Shehu, A., Bajrami, X.: Removal of various noises from digital images with the application of a single algorithm. Int. J. Civil Eng. Technol. (IJCIET) **8**, 804–816 (2017)
9. da Rocha Henriques, F., Lovisolo, L., Rubinstein, M.G.: DECA: distributed energy conservation algorithm for process reconstruction with bounded relative error in wireless sensor networks. EURASIP J. Wireless Commun. Networking **2016**(1), 1–18 (2016)

Endowing IoT Devices with Intelligent Services

Aneta Poniszewska-Maranda[1], Daniel Kaczmarek[1], Natalia Kryvinska[2](✉), and Fatos Xhafa[3]

[1] Lodz University of Technology, Lodz, Poland
aneta.poniszewska-maranda@p.lodz.pl, dkdaniel@vp.pl
[2] Comenius University in Bratislava, Bratislava, Slovakia
Natalia.Kryvinska@fm.uniba.sk
[3] Department of Computer Science, Technical University of Catalonia, Barcelona, Spain
fatos@cs.upc.edu

Abstract. The future of the Internet is to be found in Internet of Things (IoT) where every device communicates with others by making simple but intelligent decisions. To leverage the power of IoT, smart devices and objects need to be endowed with intelligent capabilities. Thus, the purpose of this paper is to investigate a selected Artificial Intelligence (AI) techniques/methods for use in the Internet of Things (IoT) concept. The main assumption is the use of a mobile device, typically mobile phone, as an intelligent object inside of IoT. To investigate the above issue, an IT system based on the concept of Internet of Things was built, and certain AI methods were implemented into this system. The paper covers also the issues of software engineering for building IT systems based on the IoT concept, using mobile devices and AI methods.

1 Introduction

The number of devices used in systems based on the IoT concept is exponentially growing. It is now becoming commonplace that a person is surrounded by many devices that can assist him and make some decisions either automatically or through an interaction process. To be able to achieve the best out of the IoT devices capabilities, they should be smart, therefore the field of Artificial Intelligence (AI) is crucial to achieve smart IoT. However, the nature and limitations of IoT devices require the right choice of AI methods in order to increase the intelligence and the accuracy of the decision, while ensuring intuitive and easy-to-use human interfaces and communications [1,6,9,15,16].

To make the discussion concrete, we consider here the use of artificial intelligence in IoT systems based on the mobile devices as smart objects. It should be noted that there is a great variety of smart devices, among which mobile devices are among most important ones. According to the IoT concept, mobile devices, e.g. mobile phones, can be treated as intelligent objects. Indeed, they

L. Barolli et al. (Eds.): EIDWT 2018, LNDECT 17, pp. 359–370, 2018.
https://doi.org/10.1007/978-3-319-75928-9_32

have built-in sensors, a place for local data storage, and communication technology elements, among others. Figure 1 shows the location of mobile devices in the context of the IoT architecture. It can be said that regardless of the architecture chosen, mobile devices fully support elements associated with the lowest layers of perception. Furthermore, without the need to add supplementary physical devices to the architecture, and only with internal software, components such as pre-processing and local storage can communicate with other intelligent devices in the system, and transmit data to the gateway (to deliver to relevant applications, services or servers). Thanks to this view of mobile devices, they can be easily integrated into information systems based on IoT [2,3,8,14].

In fact, mobile devices are hybrid devices in the IoT context since they can also play the role of a computer. In addition to using them to only collect sensor data they have a variety of sensors (up to a dozen) different applications that may be able to manage other IoT objects can be developed. This allows them to be aggregators of network traffic from multiple sensors. These devices have significant processing capabilities and can also act as gateways to platforms offering various services [18–20].

The reminder of the paper is structured as follows. In the Sect. 2 we present the main requirements on supporting of IoT systems with artificial intelligence. Some main use cases are considered as well. In Sect. 3, we propose a system to assess the suitability of selected artificial intelligence methods for use in the IoT paradigm. An experimental evaluation of the proposed system is presented as well. We end the paper in Sect. 4 with some conclusions and outlook for future work.

2 Smart-IoT Service System Built on AI Platform

Systems built on IoT concept use not only simple sensors transmitting information to systems that primarily operate on the basis of statistics and simple mathematical calculations. Such systems are increasingly being developed for making

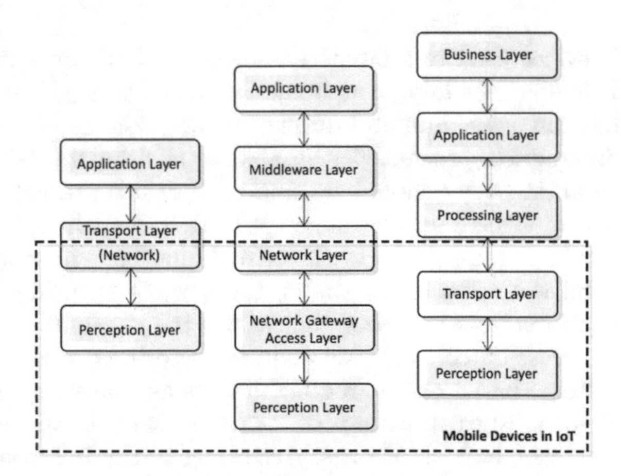

Fig. 1. The location of mobile devices in the context of IoT architectures.

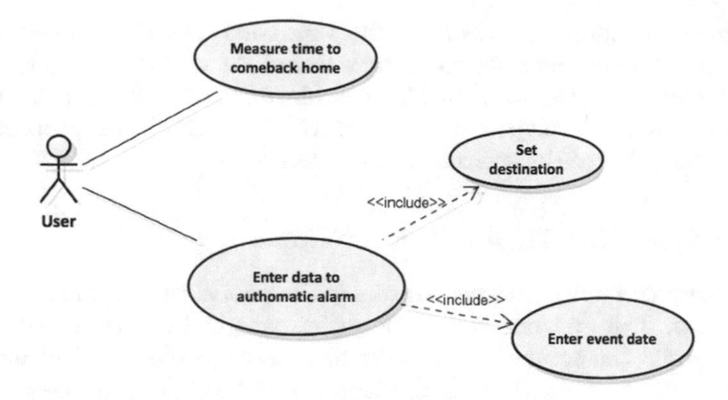

Fig. 2. General use case diagram of smart-IoT system.

decisions on different aspects and contexts. It is easy to imagine a system that turns on heating only on the basis of ambient temperature. However, it could regulate the temperature also in terms of the number of people in the room, the habits of a user, specific rooms or daytime. Therefore, endowing intelligence to these systems is an important issue, yet rather complex. In the aforementioned example of ambient temperature and user habits makes us aware that the IoT system should somehow learn about these users habits and adjust to them. Such actions are not to be achieved by simple statistics or simple equations. Here we need more refined and complex tools such ones based on artificial intelligence [4,5,7,8,10,13].

The idea of using artificial intelligence in the Internet of Things is a further step - namely, a question of the independence of machines in the context of their supervision. The use of AI methods can have advantages on decision making either autonomously by IoT system or in communication with humans [11,12, 17]. Thus, we deal here with a description of the functional and non-functional requirements and the technologies used for the smart-IoT system.

The location of mobile devices in the context of IoT architectures is graphically shown in Fig. 1.

The use case diagram in Fig. 2 illustrates the general functional requirements of the smart-IoT system.

Building a system has led to the creation of two main functionalities. The first is to measure the time to return home. This functionality is responsible for displaying the home-return time accurately to the minute in real-time, after having entered the location data of the place of residence. The second feature is the automatic setting of an alarm clock on the phone for a specific event. The alarm starts at a predetermined time and the proposed departure time is given so that the user can manage the event (for example, a job or meeting with a client). To determine the alarm and exit times, the system needs input data: the location of the place where the event occurs and the start time of the event.

The main functions (services) of the smart-IoT described above are used to calculate relevant information from selected AI methods (artificial neural networks). Consequently, an additional console application for learning neuronal networks has been developed with the method of error back propagation. A detailed description of the services is presented next.

2.1 Use Case: The Time to Return Home

The home-return service uses two sensors in the phone: the GPS module and the accelerometer. The first one collects location data in the form of latitude and longitude, while the second one is used to determine the speed of movement. This service needs input data as the length and latitude of a users residence. Figure 3 shows the use case diagram for the smart object in the context of the service described. According to the description of the system architecture, such objects must additionally prepare data for the central server.

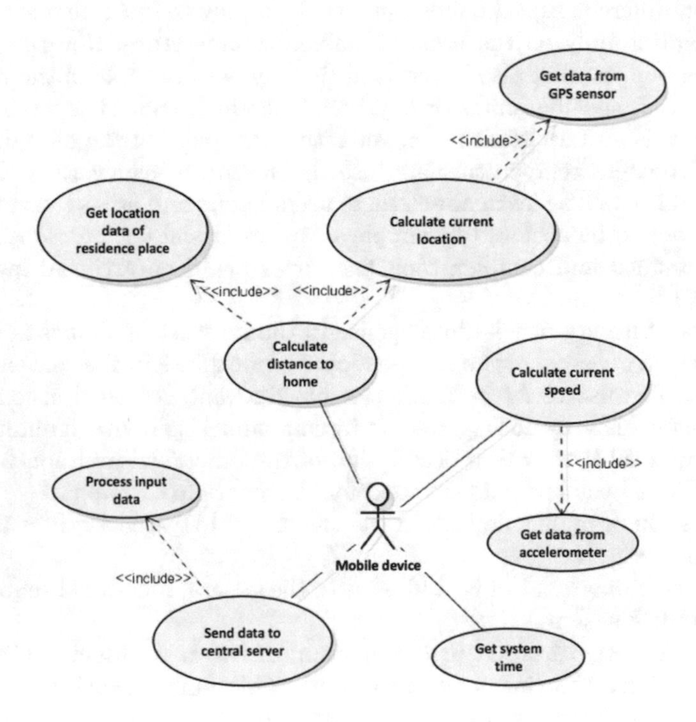

Fig. 3. Mobile device use-case diagram for a time-back-home service.

The data transmitted from the mobile device to the central server includes the following information:

- length and latitude of place of residence,
- the latitude and longitude of the current location,

- current speed of movement,
- system time

The device sends this information to the server intermittently (as specified a number of seconds). With this information, the central server is able to count the distance of residence from the current position, the direction of travel and the offset distance from the last measurement. The archival data from the previous readings is used for the last two items.

This allows the following input to the neural network to be prepared:

1. Distance of current position from the place of residence - numerical value in meters,
2. Current time - a numerical value specifying the number of minutes per day, counted from midnight,
3. Speed of movement - in km/h unit,
4. Shift from the last measurement - measured in meters,
5. Direction of movement - north, south, east, west,
6. Direction to the place of residence - as above.

Based on the above 6 input parameters, the artificial neural network corresponds to a specific numerical value that represents the estimated return time of the user to the home. This value, like the parameter, the current time, is given as the minute number counted from midnight. The server interprets the result accordingly and sends it to the mobile device where a specific response is displayed. In addition, this response is also displayed on the server itself. The process of this service is illustrated in the sequence diagram in Fig. 4.

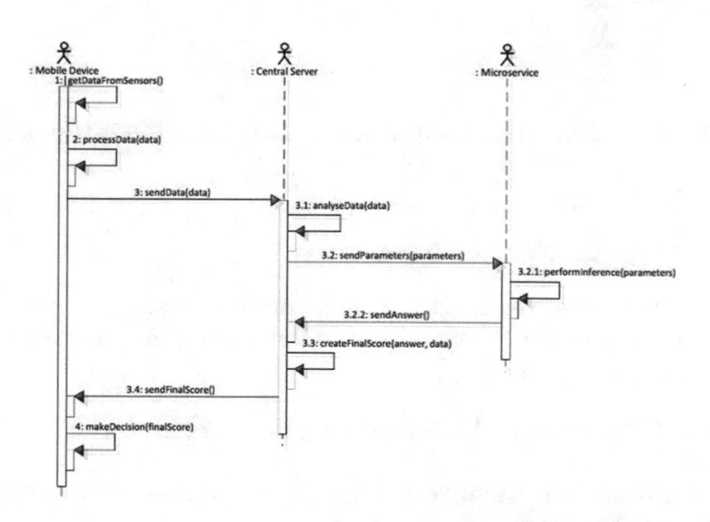

Fig. 4. Sequence diagram for home return time service.

2.2 Use Case: Automatic Alarm Setting

The auto-event alert service is designed to notify you when you should leave your home to reach a specific destination at your desired time. Data about the destination and the time of the planned arrival are needed, based on which, the mobile device will alert the user at the soonest time of the departure.

An additional element here is the so-called make sure you leave the house. A mobile device using the fact of connecting to a home Wi-Fi network (if any) and additionally location data resides periodically, using an alarm, to leave until you leave home. When the device has moved away from home, it is assumed that the user has left and goes to the destination. In this case, the phone informs about the so-called time indicator of arrival (messages: ahead of time, in standard, late). Figure 5 shows a use case diagram for an intelligent object in this service context.

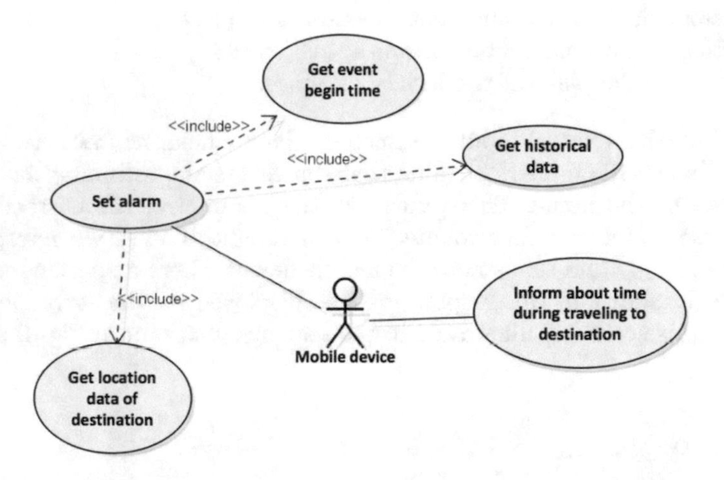

Fig. 5. Mobile device use case diagram for automatic alarm setting service.

2.3 The Artificial Neural Network

To set the alarm time, the service is based on historical data that is used to train an artificial neural network. The neural network thus constructed has the following input parameters:

- Distance of the target position from the place of residence - numerical value in meters,
- Time of arrival - the numerical value for the number of minutes per day, counted from the north,
- Average time of arrival at similar distance in minutes - data from previous journeys that represent the correct value,

- Distance difference in meters - the difference between the first parameter, i.e. the distance between the target position and the distance found in the archives.

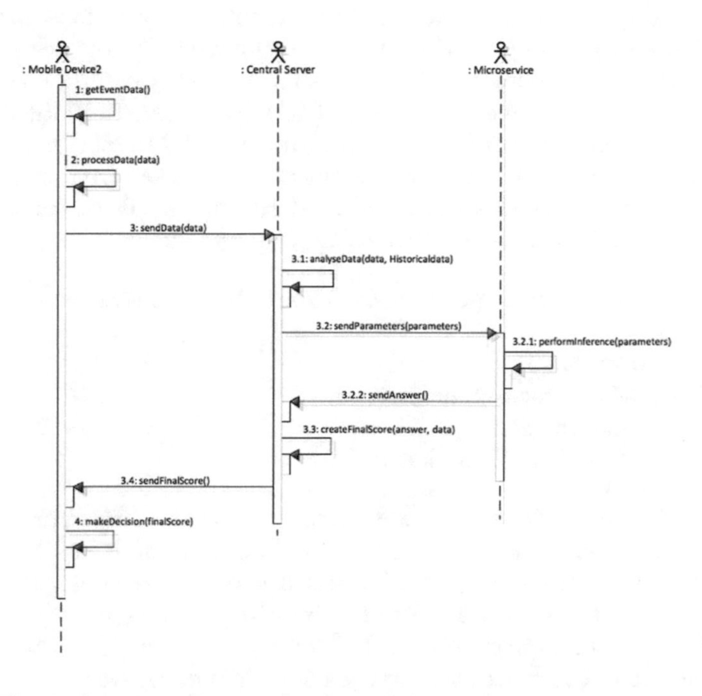

Fig. 6. Sequence diagram for the automatic alarm setting service.

Based on the above input parameters, the artificial neural network calculates a specific result and presents it as a time (minutes from midnight), representing the departure time. Once the server responds to the response, the appropriate time is sent to the mobile device that will start alerting the user at a certain time that the time of departure is approaching. Figure 6 shows the operation of this service illustrated in the sequence diagram. It is worth mentioning that the elements related to the transmission of the arrival time index use the sequence of action described in Fig. 4. For proper operation of the smart-IoT system it should meet the following standard requirements:

- Minimum Android 4.4 phone with GPS, accelerometer, and optional magnetometer - serves as an IoT device (we used a mobile device with 2 GB of RAM and a 4-core 2.2 GHz processor).
- A central server and micro-services installed.
- A wireless Wi-Fi router in a location that was previously described as a home.

3 Evaluation of IoT Service of Artificial Neural Networks with Back Propagation

In order to judge on the suitability of selected artificial intelligence methods in the smart-IoT system, namely, artificial neural networks with back propagation, an empirical study through various tests is conducted. For the tests a route of 4 Km was selected. To obtain training data for a neural network, measurements were made at different times of the day. There have been 10 routes (under the same configuration, some of them on foot and some by public transportation). From each run of one route 100 measurements were taken from different parts of it. As a result, the training data contained 1000 items. These consisted of the following (also previously described) six input parameters:

- Distance of your current position from your place of residence,
- Current time,
- Speed of movement,
- Shift from the last measurement,
- Direction of movement,
- Direction to the place of residence.

For the first measurement, it was assumed that the offset parameter from the last measurement was 0. For each element containing 6 input parameters, the correct result was known, i.e. the return time to the home. Having all of the above elements, two neural networks, multi-layer perceptrons, reverse-error propagation algorithms were studied. In both cases, a sigmoidal activation function was selected. The description of the network is as follows:

- A neural network with one hidden layer, called SN1 - a network containing 6 input neurons, 3 hidden neurons and 1 output neuron.
- Neural network with two hidden layers, called SN2 - a network containing 6 input neurons, 3 neurons in the first hidden layer, 2 neurons in the second hidden layer and 1 output neuron.

In addition, an SN3 neural network, which contains 4 input neurons, 3 neurons in the hidden layer and 1 neuron output, was taught for the automatic setting of the alarm. The above network contains the following input parameters (the same set of training data that was tailored to this problem was used):

- Distance of the target position from the place of residence,
- Time to reach the destination,
- Average time to reach a similar distance in minutes,
- Distance difference in meters.

Thus, using the neural network learning program, the SN1, SN2 and SN3 networks described above were used to carry out test routes. The first two were used in the context of the home-return time service, while the SN3 network

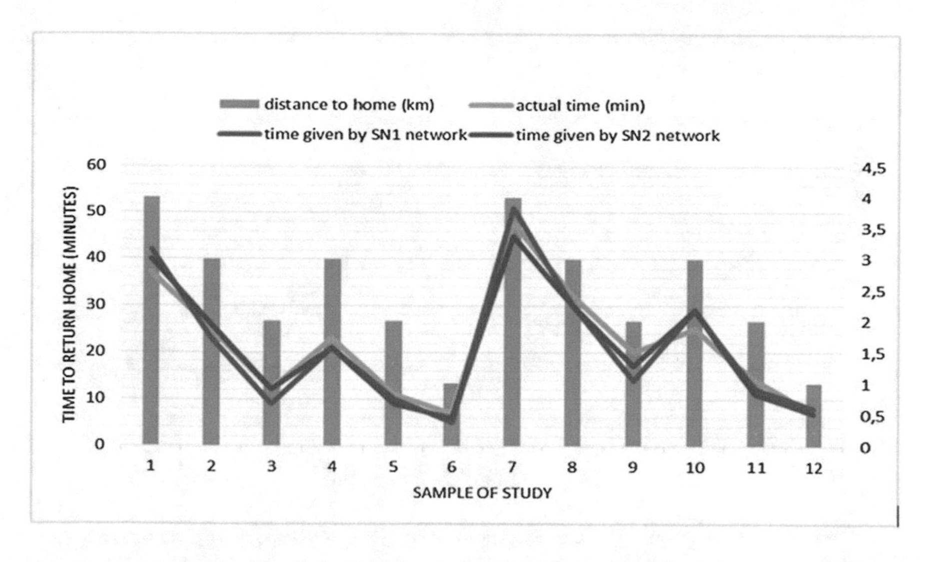

Fig. 7. Comparison of neural networks with actual data in IoT solutions.

was automatically set to alarms. The results of these measurements are given in Figs. 7, 8, 9 and 10.

Figures 7, 8 and 9 contain the results of the experiments conducted in the context of the time-to-return service. As can be seen, both neural networks achieved similar results and were similar to those of the actual ones. Also, in both cases, there is a tendency between morning time and afternoon hours. It turns out that morning measurements are usually more accurate than those from

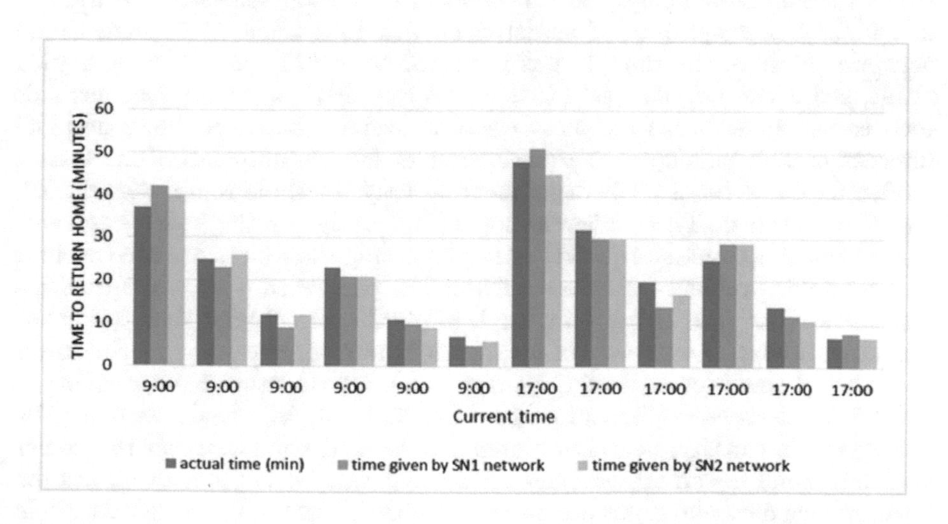

Fig. 8. Comparison of neural networks with actual data in IoT solutions for current time.

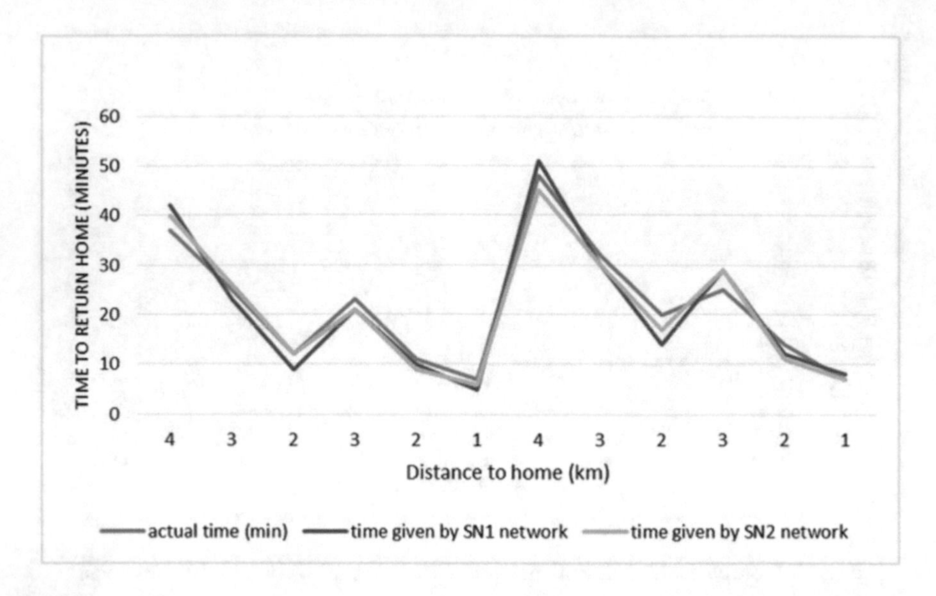

Fig. 9. Comparison of neural networks with actual data in IoT solutions for given distance to home.

a later hour. This may be due to the fact of traffic jams occurring in the city, which makes it harder to predict the return time.

Comparing SN1 results directly with SN2, we can conclude that in most cases, SN2 gave a slightly better result. In addition, some anomalies can be found, where the difference between the computed result and the real one is 6 min. Despite some shortcomings, both networks had a good tendency. Figures 7, 8 and 9 show a comparison of actual time with that achieved through neural networks. Most of the time, it was measured to within 3 min. This is a good result, and it can be safely said that in this case artificial intelligence methods such as neural networks are most useful in systems based on the concept of Internet of Things. Figure 10 presents results for the automatic time setting service for the event. In this experiment, exactly the same route was selected, but it was reversed. Three different hours (time of the meeting) were examined, which show the target arrival time. After analyzing the results, we can conclude that with this problem the neural network has performed a bit worse than the first service, but it is still good enough to be able to consider these results as generally satisfactory. As we can see, at noon time of arrival was much shorter than that planned (even 15 min ahead of time). On the other hand, arriving at 16:00 in most cases resulted in delays. This fact may be caused, as above, by traffic jams in the city; neural networks have learned some inference that is not always immune to certain anomalies. This may mean that the network has too little training data and input parameters. Probably increasing the dataset would improve the inference.

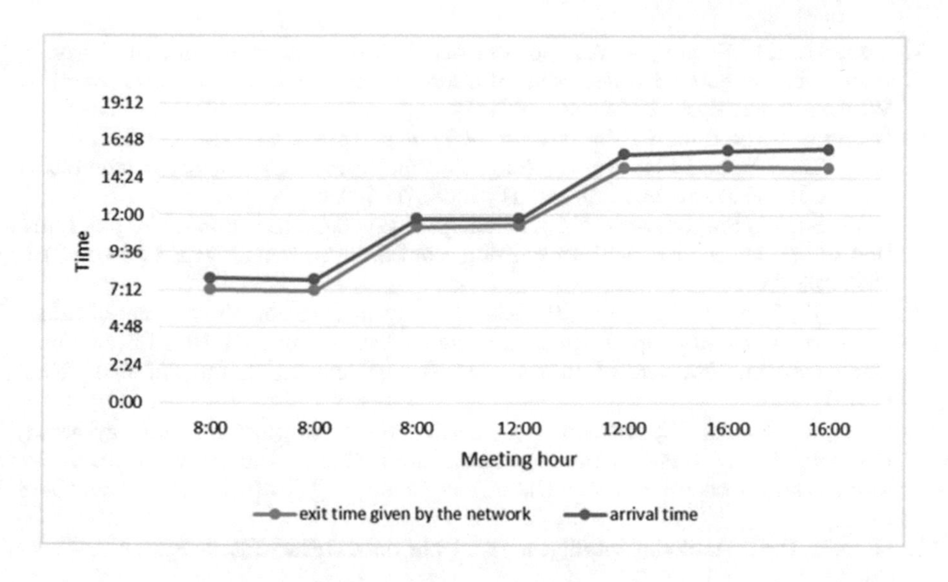

Fig. 10. Automatic alarm setup for SN3 neural network.

4 Conclusions and Future Works

In this paper we have investigated the complexity and usefulness of endowing IoT systems with artificial intelligence (AI). To that end, an IoT system, called smart-IoT, was built and several experiments were conducted to validate it with real data. The core AI methods are artificial neural networks with the reverse error propagation algorithm, although our approach can be an equally valid for other AI methods, to be used in the context of reasoning.

In our future works we plan to implement and evaluate more AI-based methods for intelligent services in IoT devices.

References

1. Anand, M., Susan, C.: Artificial intelligence meets internet of things. Int. J. Sci. Eng. Comput. Technol. **5**, 149 (2015)
2. Arsnio, A., Serra, H., Francisco, R., Nabais, F., Andrade, J., Serrano, E.: Internet of intelligent things: bringing artificial intelligence into things and communication networks. In: Inter-cooperative Collective Intelligence: Techniques and Applications, p. 137. Springer (2014)
3. Bessis, N., Xhafa, F., Varvarigou, D., Hill, R., Li, M.: Internet of Things and Inter-Cooperative Computational Technologies for Collective Intelligence. Springer, Heidelberg (2013)
4. Chen, F., Deng, P., Wan, J., Zhang, D., Vasilakos, A.V., Rong, X.: Data mining for the internet of things: literature review and challenges. Int. J. Distrib. Sens. Netw. **11**, 431047 (2015)

5. Chiuchisan, I., Geman, O.: An approach of a decision support and home monitoring system for patients with neurological disorders using internet of things concepts. WSEAS Trans. Syst. **13**, 460469 (2014)
6. DaCosta, F.: Rethinking the Internet of Things. Apress, Berkeley (2013)
7. Fan, Y.J., Yin, Y.H., Da Xu, L., Zeng, Y., Wu, F.: IoT-based smart rehabilitation system. IEEE Trans. Ind. Inform. **10**, 15681577 (2014)
8. Kelly, S.D.T., Suryadevara, N.K., Mukhopadhyay, S.C.: Towards the implementation of IoT for environmental condition monitoring in homes. IEEE Sens. J. **13**, 38463853 (2013)
9. Khan, R., Khan, S.U., Zaheer, R., Khan, S.: Future internet: the internet of things architecture, possible applications and key challenges. In: 2012 10th International Conference on Frontiers of Information Technology (FIT), pp. 257–260. IEEE (2012)
10. Miori, V., Russo, D.: Anticipating health hazards through an ontology-based, IoT domotic environment. In: 2012 Sixth International Conference on Innovative Mobile and Internet Services in Ubiquitous Computing (IMIS), pp. 745–750. IEEE (2012)
11. OLeary, D.E.: Artificial intelligence and big data. IEEE Intell. Syst. **28**, 96–99 (2013)
12. Poniszewska-Maranda, A., Kaczmarek, D.: Selected methods of artificial intelligence for Internet of Things conception. In: 2015 Federated Conference on Computer Science and Information Systems (FedCSIS), pp. 1343–1348. IEEE (2015)
13. Skouby, K.E., Lynggaard, P.: Smart home and smart city solutions enabled by 5G, IoT, AAI and CoT services. In: 2014 International Conference on Contemporary Computing and Informatics (IC3I), pp. 874–878. IEEE (2014)
14. Uckelmann, D., Harrison, M., Michahelles, F.: Architecting the Internet of Things, 1st edn. Springer Publishing Company Incorporated, Heidelberg (2011)
15. Valera, A.J.J., Zamora, M.A., Skarmeta, A.F.: An architecture based on internet of things to support mobility and security in medical environments. In: 2010 7th IEEE Consumer Communications and Networking Conference (CCNC), p. 15. IEEE (2010)
16. Vermesan, O., Friess, P.: Internet of Things: Converging Technologies for Smart Environments and Integrated Ecosystems. River Publishers, Aalborg (2013)
17. Wilks, Y.: An artificial intelligence approach to machine translation. Stanford University, Department of Computer Science (1972)
18. Wu, M., Lu, T.-J., Ling, F.-Y., Sun, J., Du, H.-Y.: Research on the architecture of Internet of Things. In: 2010 3rd International Conference on Advanced Computer Theory and Engineering (ICACTE). pp. V5-484–V5-487 (2010)
19. Yun, M., Yuxin, B.: Research on the architecture and key technology of Internet of Things (IoT) applied on smart grid. In: 2010 International Conference on Advances in Energy Engineering (ICAEE), pp. 69–72. IEEE (2010)
20. Zhang, X.M., Zhang, N.: An open, secure and flexible platform based on internet of things and cloud computing for ambient aiding living and telemedicine. In: 2011 IEEE International Conference on Computer and Management (CAMAN), p. 14 (2011)

VM Deployment Methods for DaaS Model in Clouds

Klodiana Goga[1](✉), Fatos Xhafa[2], and Olivier Terzo[1]

[1] Istituto Superiore Mario Boella, Turin, Italy
goga@ismb.it, terzo@ismb.it
[2] Universitat Politècnica de Catalunya, Barcelona, Spain
fatos@cs.upc.edu

Abstract. Big Data has become an enabling technology for many of the today's innovations. Given the exponential rate at which the data is produced there is a clear necessity for scalable solutions to control the overwhelming flow of new streams of information and extract information out of DaaS Clouds. In this paper we review and analyze some VM deployment methods and their suitability for Data as a Service (DaaS) model in Clouds. Then we approach some novel aspects of VM deployment, including VM migration.

1 Introduction

Data as a Service (DaaS) is an alternative service model in Cloud computing where data are made available to users as a service through network [16]. With the exponential increase of the data volumes [27], DaaS is gaining importance in many areas of scientific communities and also in the industrial context. DaaS provides new ways to control the overwhelming flow of new streams of information and eliminates much of the administrative work surrounding data. Almost every modern business has embraced data as a decision-making tool, but few companies have the in-house manpower and resources to fully leverage the power of the data they collect. Big data processing software like Hadoop [15] allow organizations to analyze data using commodity hardware and open source software but the costs of launching a big data initiative can still be remarkable if considering also the ongoing investment of time and resources needed to store and manage large data sets. By using DaaS platforms the organizations are able to outsource a variety of big data functions to the Cloud and pay only for the compute power they need. This solution eliminates many of the costs associated with a Hadoop [15] deployment and allows organizations to focus on gaining big data insights to drive business growth. And, when it comes to keeping data secure, DaaS providers vary greatly in strategy and collaboration with security experts by offering catered data streams tailored to client needs. In the market are available different solutions to support Data as a Service offered by the big players like Amazon, Google and Microsoft. This service eliminates the implementation and operational problems caused by big data. Proprietary systems management

L. Barolli et al. (Eds.): EIDWT 2018, LNDECT 17, pp. 371–382, 2018.
https://doi.org/10.1007/978-3-319-75928-9_33

of these products wraps proven configuration of standard open source technologies to provide a robust, scalable, secure and reliable service, giving customers a simple, developer-friendly environment. This can be deployed in public cloud, virtual private cloud, enterprise private cloud and dedicated clusters depending on business use case and data sensitivities [23,25].

Amazon Web Services [1] - AWS offers a comprehensive set of services to handle every step of the analytics process chain including data warehousing, business intelligence, batch processing, stream processing, machine learning, and data workflow orchestration. The AWS data analytics products include **Amazon Athena**, which is a serverless query service and allows to analyze data in Amazon S3, using standard SQL. **Amazon Elastic Search Service** is useful for the deployment, operating and scaling Elastic Search on AWS. **Amazon Kinesis** offers the possibility to work with streaming data on AWS. **Amazon QuickSight** is a very fast and easy-to-use Business Analytics Solution. **Amazon Redshift** is petabyte-scale data warehouse solution that allows users to analyze data using their existing business intelligence tools. **AWS Glue** is an ETL tool used to Prepare and load data to data stores. **AWS Data Pipeline** is a data workflow orchestration tool which allows to reliably process and move data between different AWS compute and storage services, as well as on-premise data sources, at specified intervals. The managed Hadoop service is called **Amazon Elastic MapReduce** runs on Amazons S3 storage infrastructure. It provides a managed Hadoop framework to process vast amounts of data quickly and cost-effectively. It runs on open source frameworks such as Apache Spark, HBase, Presto, and Flink. Amazon EMR is a managed cluster platform that simplifies running big data frameworks, such as Apache Hadoop and Apache Spark, on AWS to process and analyze vast amounts of data. By using these frameworks and related open-source projects, such as Apache Hive and Apache Pig, users can process data for analytics purposes and business intelligence workloads. Additionally, users can use Amazon EMR to transform and move large amounts of data into and out of other AWS data stores and databases, such as Amazon Simple Storage Service (Amazon S3) and Amazon DynamoDB.

Google [11] - Googles managed Big Data service has enjoyed fast growth. Google has been able to leverage its presence and reputation for Cloud innovation to offer a package industry has found attractive. It runs Hadoop and Spark on Googles Cloud Platform and integrates with the BigTable storage and BigQuery analytics frameworks. Google offers a set of services which can be combined in order to create a DaaS platform according to the end users needs. **Google BigQuery** [14] is Google's serverless, highly scalable, low cost enterprise data warehouse designed to increase productivity in data analysis. **Google Cloud Datalab** [13] is built on Jupyter (formerly IPython), which boasts a thriving ecosystem of modules and a robust knowledge base. Cloud Datalab enables analysis of the data on Google BigQuery, Cloud Machine Learning Engine, Google Compute Engine, and Google Cloud Storage using Python, SQL, and JavaScript (for BigQuery user-defined functions) and **Google Cloud Dataproc** [12] is a fast, easy-to-use, fully-managed cloud service for running Apache Spark and

Apache Hadoop clusters in a simpler, more cost-efficient way. Cloud Dataproc also easily integrates with other Google Cloud Platform (GCP) services, giving users a complete platform for data processing and analytics.

Microsoft Azure [2] - Microsofts strong presence in the business software market makes it play a big role in the DaaS offer. It has built on its Azure cloud framework by increasingly adding functionality and compatibility with open source technology such as Spark and Storm. Azure includes a set of services which can be used to create custom DaaS platforms. **Azure Search** [3] provides text search and a subset of OData's structured filters using REST or SDK APIs. **Cosmos DB** [4] is a NoSQL database service that implements a subset of the SQL SELECT statement on JSON documents. **Redis Cache** [5] is a managed implementation of Redis. **StorSimple** [6] manages storage tasks between on-premises devices and cloud storage. **SQL Database** [7], formerly known as SQL Azure Database, works to create, scale and extend applications into the cloud using Microsoft SQL Server technology. It also integrates with Active Directory and Microsoft System Center and Hadoop. **SQL Data Warehouse** [8] is a data warehousing service designed to handle computational and data intensive queries on datasets exceeding 1TB. **Azure Data Lake** is a scalable data storage and analytic service for big-data analytics workloads that require developers to run massively parallel queries. **Azure HDInsight** [9] is a big data relevant service, that deploys Hortonworks Hadoop on Microsoft Azure, and supports the creation of Hadoop clusters using Linux with Ubuntu. **Azure Stream Analytics** [10] is a serverless scalable event processing engine that enables users to develop and run real-time analytics on multiple streams of data from sources such as devices, sensors, web sites, social media, and other applications.

The rest of the paper is structured as follows. In Sect. 2 we describe some of the related work. In Sect. 3 some VM deployment models are analysed and some novel aspects are discussed. Finally, In Sect. 4 we give conclusions and indications for future work.

2 Related Work

Data as a Service Platforms are becoming a trend both in science and in industry. The implementation of DaaS platforms and Big Data applications in Cloud environments has been subject of research in recent years. In Terzo *et al.* [16] has been introduced a DaaS approach for intelligent sharing and processing of large data collections with the aim of abstracting the data location and to fully decouple the data from its processing. Assuncao *et al.* [28] discuss approaches and environments for carrying out analytics on Clouds for Big Data applications focusing on data management and supporting architectures; model development and scoring; visualization and user interaction; and business models. Demchenko *et al.* [22,24] have presented two bio-informatics use cases on Cloud using a Cloud application deployment and management automation platform. Zhao *et al.* [26] models ad formulates the profit optimization resource scheduling problem and proposes an optimization scheduling algorithm that maximizes profits for AaaS (Analytics as a Service) platforms and guarantees SLAs for query requests.

Toosi *et al.* [17] propose a new resource provisioning algorithm to support the deadline requirements of data-intensive applications in hybrid cloud environment. Bhimani *et al.* [21] investigates the performance of different Apache Spark applications using both Virtual Machines (VM) and Docker containers, comparing these different virtualization technologies for a big data enterprise cloud environment.

Finally, Dai *et al.* [19] has presented an algorithm based on dynamic programming to obtain the optimal deployment solution of VMs in cloud infrastructure for big data applications.

3 Virtual Machine Deployment

In Fig. 1 is represented a generic architecture of a DaaS system as described by Terzo *et al.* in [16].

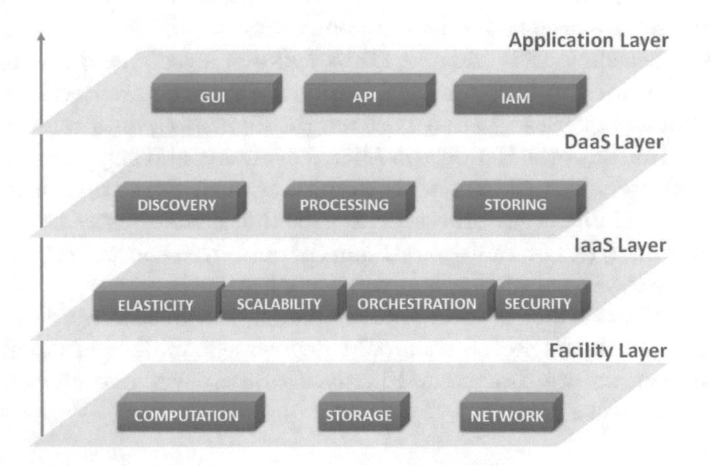

Fig. 1. DaaS generic architectural layers

The **Facility Layer** is composed by the physical systems necessary to process and store data. Are part of this layer the hardware components, storage, network and any other physical devices. The components of this layer can be owned, managed and operated by the end user, the service provider or a combination of them. It may exist on or off premises.

The **IaaS Layer** controls the computing resources by leveraging the potential of the virtualization technologies which allow the abstraction and pooling of the physical resources. This layer controls the resource allocation and the balancing between the running applications requirements and the effective capacity of the infrastructure. This layer offers also the necessary tools to increase scalability in order to face peak-loads. Also the security aspects should be considered in this layer which allows the integration of various security technologies.

The **DaaS Layer** includes all tools and technologies necessary to manage data with the purpose of discover(find and collect data from heterogeneous sources) process (data oriented distributed computing software or in-memory computing) and store the data.

The **Application Layer** allows users direct access through specific APIs in order to use the cloud computing infrastructure and the data provided and perform analysis through GUIs. In this layer are also included Identity and Access Management (IAM) functionalities which control the secure access to the DaaS service.

DaaS solutions have to face different challenges such as storage, sharing, analyzing and visualization. Another constraint regards the imbalance between the CPU and the I/O performance. The CPU performance is doubling each 18 months and the disk drives is also increasing at the same rate but the disks' rotational speed has slightly improved [17,18].

When considering the VM deployment in Cloud we have to evaluate three aspects as described by Dai *et al.* in [19], namely, cost, performance and availability searching to reach the best optimization solution for DaaS platforms achieving the highest availability and performance with lower costs.

In this paper we consider MapReduce [29] and its implementations like Hadoop [15,30] and Spark [31] as well as they offer a productive high level programming interface for large scale data processing and analytics. MapReduce is a framework which allows a cluster of computers to process a large set of data in parallel. It includes two main steps "map" which distribute pieces of huge data sets to single nodes in the cluster for processing and "reduce" which combines the intermediate data coming form "map" and creates the final output. The biggest disadvantage of Hadoop [21] consist in the fact that it is characterized by large amount of disk I/O operations for storing intermediate data during the map and reduce phase. In Spark this problem has been resolved by storing into memory all intermediate data (*Resilient Distributed Datasets (RDDs)*). The use of RDDs make Spark perform 100x better than Hadoop in multi-pass analytics. Spark (Fig. 2) is mainly implemented by setting up a cluster composed by a master and different worker nodes on hypervisor-based virtualization (e.g. Xen, KVM etc.). Spark uses a master/worker architecture. There is a driver that talks to a single coordinator called master that manages workers in which executors run. The driver and the executors run in their own Java processes. Users can run them all on the same (horizontal cluster) or separate machines (vertical cluster) or in a mixed machine configuration. Physical machines are called hosts or nodes. Apache Spark has a well-defined and layered architecture where all the spark components and layers are loosely coupled and integrated with various extensions and libraries. Apache Spark Architecture is based on two main abstractions: Resilient Distributed Datasets (RDD) and Directed Acyclic Graph (DAG). Container-based virtualization technologies such as Docker [32] are becoming an important option to speed-up big data applications due to their lightweight operation and better scaling if compared to Virtual Machines (VM). Containers and virtual machines have similar resource isolation and allocation benefits but different architectural approaches and resource management. Containers are more efficient and portable compared to bare metal and Virtual Machines [33].

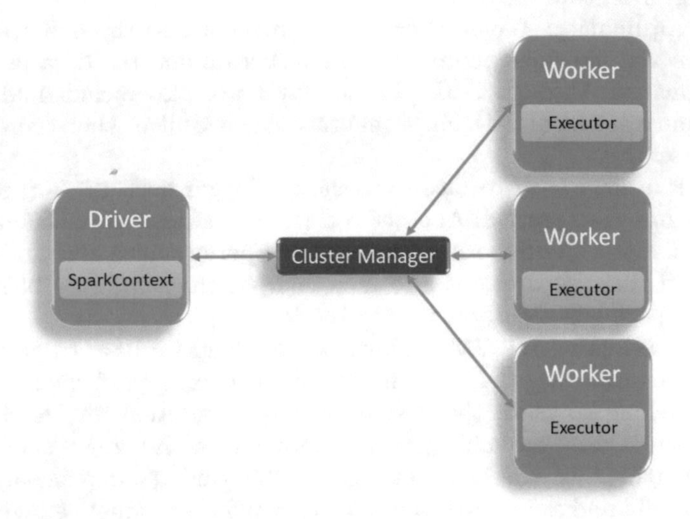

Fig. 2. Spark architecture

Containers are lightweight when compared to VMs as each container does not have to operate separate Operating Systems (OS). Containers perform shared resource management which enable more flexible sharing of resources which may increase the overall resource utilization while VMs perform distributed resource management which guarantees stability and security because each VM runs with its specific assigned set of resources.

As depicted in Fig. 3 Hypervisor is a virtual platform software operated as a middleware between the VM and OS that allows the implementation of multiple guest in a single physical server. Docker instead offers application virtualization by using a containerized environment. Application installation can only be performed inside Docker containers. In order to maintain lightweight characteristics, it is advisable to keep the installation stack within the container as small as possible. In both VM and Containers can be divided in three layers: Applications, OS & Drivers and Storage. Each container works in its own workspace in terms of file system and database. The application layer has multiple instances of different applications and workloads operating in multiple virtual machines or containers. The bigger difference between VM and Containers is that containers does not need to have an OS inside them which makes them lighters and lowers the overhead of managing device drivers in each instance. Containers can be launched faster with better performance. They also have less isolation as well as share the host's kernel. Both Hypervisor and Docker engine act as virtualization controller which decides the instantiations, terminations and check

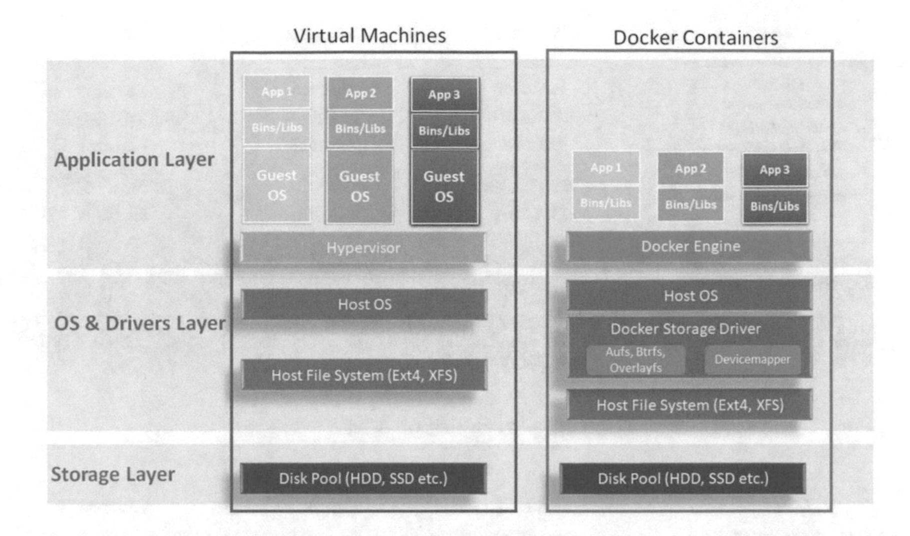

Fig. 3. System architecture

low level information. The Virtual Machine Hypervisor has distributed resource management among different virtual machines, where each VM is assigned a maximum limit odf resources it can sue. The Docker containers uses group to assign, allocate and manage resources like CPU, Memory, etc.

Containers uses shared resource management (e.g. memory, page cache) among different active and inactive containers. VM hypervisor distributes VM images, each including guest OS and individual library and application stack. This type of distribution results in better security and independence but larger overhead of start up and application performance. Docker containers are lightweight without guest OS and are preferred to have one or least possible number of applications in each container. The Docker daemon can only run one storage driver, and all containers created by that daemon instance use the same storage driver. Storage drivers operate with the copy-on-write technique [34], which provides advantages for read intensive applications and further reduces I/O overhead.

In both implementations, one VM or Docker container is used as the master node, and all others are worker nodes. Also, each node has its own IP address and port. For the VM setup, we have a number of virtual machines running on a physical server via VM Hypervisor. Each VM has its own guest OS as well as its own separate Spark data processing workspace to manage executor files and database as shown in Fig. 4. The Spark implementation on Docker comprises of multiple simultaneously operating containers. As shown in Fig. 5, we do not need to maintain a guest OS in each container.

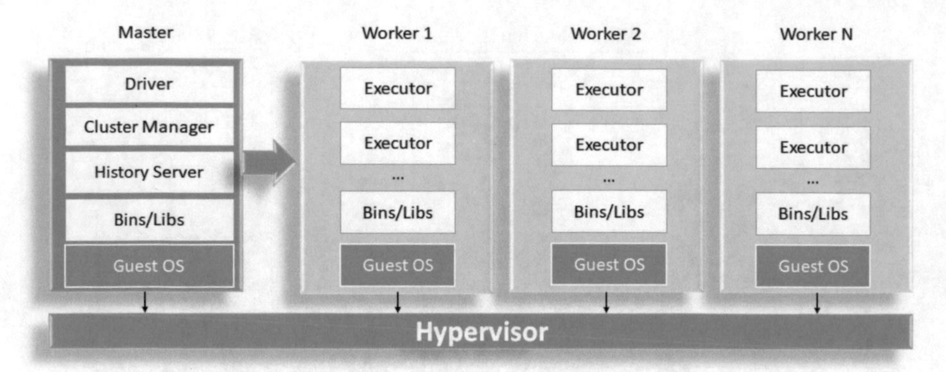

Fig. 4. Spark on VM

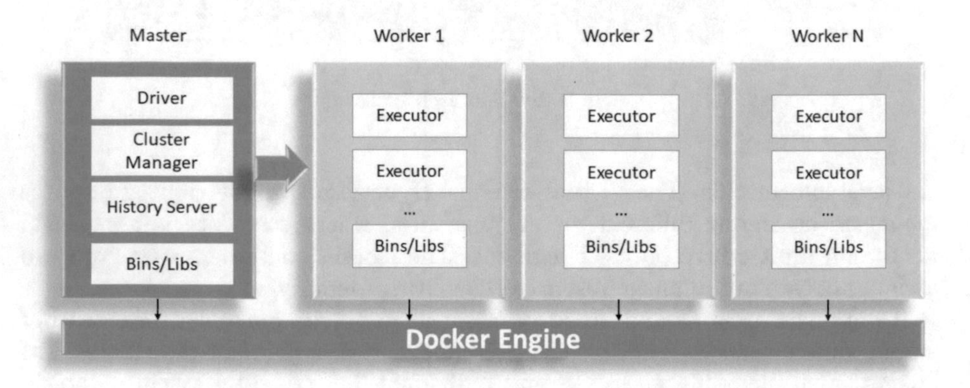

Fig. 5. Spark on Docker

3.1 VM Migration

Almost all the modern virtualization environments offers VM live migration feature, including Xen Server, VMware ESX Server, KVM, Microsoft Hyper-V, Oracle VM VirtualBox, and OpenVZ [35]. One of the most prominent features of the virtualization is the *Live Migration*, which allows for the transfer of a running VM from one physical machine to another, with little downtime of the services hosted by the VM. It transfers the current working state and memory of a VM across the network while it is still running. Live migration [20] has the advantage of transferring a VM across machines without disconnecting the clients from the services. Another approach for VM migration is the *Cold or Static VM Migration* in which the VM to be migrated is first shut down and a configuration file is sent from the source machine to the destination machine. The same VM can be started on the target machine by using the configuration file. This is a much faster and easier way to migrate a VM with negligible increase in the network traffic; however static VM migration incurs much higher downtime compared to live migration. Because of the obvious benefit of uninterrupted service and much less VM download time, live migration has been used as the most common VM migration technique in the production data centers. The Fig. 6 shows what exactly should be live-migrated, for both VMs and containers: The processing required to perform migration of a container is much more complicated, unlike VMs that can be seen as a black box with all the memory allocated inside this very box, containers are represented by a number of processes with memory distributed among them. In addition, a container requires more kernel objects to be stored and more actions to retrieve the list of processes and objects to be migrated.

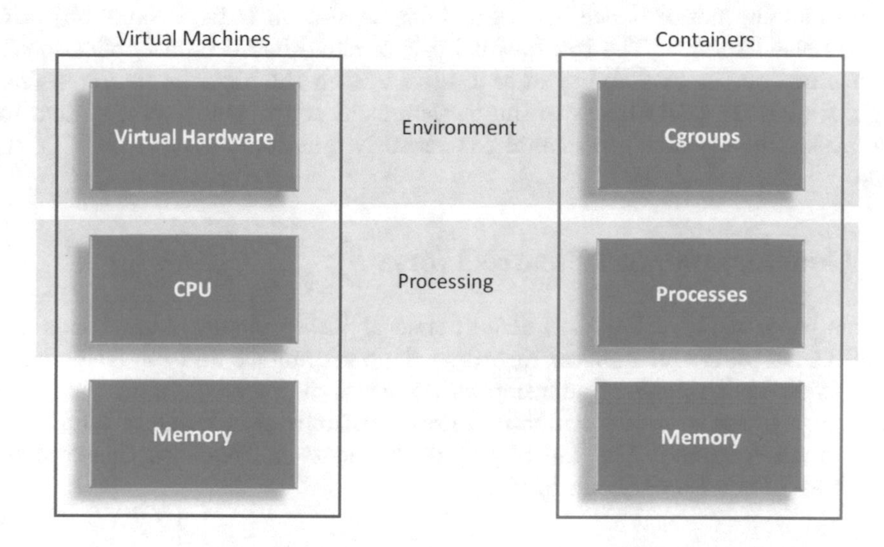

Fig. 6. Live migration

3.2 Performance and Availability

As well as for the other Cloud services, also for the DaaS Platforms the performance and availability requirements are stipulates in the Services Level Agreements (SLAs). In the SLA, the response time is usually used to evaluate the performance. The performance is directly determined by the amount and every individual performance of used VM. Virtual machines have to emulate hardware, while containerized applications run directly on the server that hosts them. That means containers should be faster than virtual machines, because they have less overhead. Containers do out-perform virtual machines. In particular, containers offer:

- Faster startup time - A containerized application usually starts in a couple of seconds. Virtual machines could take a couple of minutes.
- Better resource distribution - Containers use up only as many system resources as they need at a given time. Virtual machines usually require some

resources to be permanently allocated before the virtual machine starts. For this reason, virtual machines tie up resources on the host, even if they are not actually using them. Containers allow host resources to be distributed in an optimal way.
- Direct hardware access - Applications running inside virtual machines generally cannot access hardware like graphics cards on the host in order to speed processing. Containerized applications can.
- Less redundancy - With virtual machines, users have to install an entire guest operating system, which duplicates a lot of the components already running on your host server. Containers don't require this.

Besides the performance, the availability is also an indispensable objective for any cloud service. The total availability of cloud-based data applications is determined by the availability of each used VM an the distance between VMs. High availability (HA) refers to the measurement of the ability of a system (or application, component) to remain continuously functional with 99.999% of the time.

4 Conclusions and Future Works

In this paper as been analysed some potential VM deployments methods in a DaaS Cloud platform. Aspects regarding the performance and migration capability in different visualization frameworks are also discussed. As a future work, a performance test to better evaluate the right virtualization framework for DaaS Platforms is planned. The use of VM deployments in DaaS will increase the usefulness of the DaaS Cloud model.

References

1. Amazon Web Services - Big Data on AWS. https://aws.amazon.com/big-data/
2. Microsoft Azure - Big data and Analytics. https://azure.microsoft.com/en-us/solutions/big-data/
3. Azure Search. https://docs.microsoft.com/en-us/azure/search/search-what-is-azure-search
4. Azure Cosmos DB. https://azure.microsoft.com/en-us/services/cosmos-db/?v=17.45b
5. Azure Redis Cache. https://azure.microsoft.com/en-us/services/cache/
6. Azure StorSimple. https://azure.microsoft.com/en-us/services/storsimple/
7. Azure SQL Database. https://azure.microsoft.com/en-us/services/sql-database/
8. Azure SQL Data Warehouse. https://azure.microsoft.com/en-us/services/sql-data-warehouse/
9. Azure Data Lake. https://azure.microsoft.com/en-us/solutions/data-lake/
10. Azure Stream Analytics. https://azure.microsoft.com/en-us/services/stream-analytics/
11. Goolge Cloud Plartform - Big Data Solutions. https://cloud.google.com/solutions/big-data/?hl=en

12. Googel Cloud Dataproc. https://cloud.google.com/dataproc/?hl=en
13. Google Cloud Datalab. https://cloud.google.com/datalab/?hl=en
14. Google Bigquery. https://cloud.google.com/bigquery/?hl=en
15. Apache Hadoop. http://hadoop.apache.org/
16. Terzo, O., Ruiu, P., Bucci, E., Xhafa, F.: Data as a service (DaaS) for sharing and processing of large data collections in the cloud. In: Seventh International Conference on Complex, Intelligent, and Software Intensive Systems, July 2013, pp. 475–480 (2013). https://doi.org/10.1109/CISIS.2013.87, http://ieeexplore.ieee.org/document/6603936/
17. Toosi, A.N., Sinnott, R.O., Buyya, R.: Resource provisioning for data-intensive applications with deadline constraints on hybrid clouds using Aneka. Future Gener. Comput. Syst. **79**(Part 2), 765–775 (2018). https://doi.org/10.1016/j.future.2017.05.042, http://www.sciencedirect.com/science/article/pii/S0167739X17301863, ISSN 0167–739X
18. Philip Chen, C.L., Zhang, Ch.-Y.: Data-intensive applications, challenges, techniques and technologies: a survey on big data. Inf. Sci. **275**(Supplement C), 314–347 (2014). https://doi.org/10.1016/j.ins.2014.01.015, http://www.sciencedirect.com/science/article/pii/S0020025514000346, ISSN 0020–0255
19. Dai, W., Qiu, L., Wu, A., Qiu, M.: Cloud infrastructure resource allocation for big data applications. IEEE Trans. Big Data **PP**(99), 1 (2016). https://doi.org/10.1109/TBDATA.2016.2597149, http://ieeexplore.ieee.org/stamp/stamp.jsp?tp=&arnumber=7530891&isnumber=7153538
20. Fei, Zh., Xiaoming, F., Yahyapour, R.: CBase: a new paradigm for fast virtual machine migration across data centers. In: Proceedings of the 17th IEEE/ACM International Symposium on Cluster, Cloud and Grid Computing, CCGrid 2017, Madrid, Spain, pp. 284–293 (2017). https://doi.org/10.1109/CCGRID.2017.26, ISBN 978-1-5090-6610-0
21. Bhimani, J., Yang, Z., Leeser, M., Mi, N.: Accelerating big data applications using lightweight virtualization framework on enterprise cloud. In: 2017 IEEE High Performance Extreme Computing Conference (HPEC), Waltham, MA, pp. 1–7 (2017). https://doi.org/10.1109/HPEC.2017.8091086, http://ieeexplore.ieee.org/stamp/stamp.jsp?tp=&arnumber=8091086&isnumber=8091018
22. Demchenko, Y., Turkmen, F., de Laat, C., Blanchet, C., Loomis, C.: Cloud based big data infrastructure: architectural components and automated provisioning. In: 2016 International Conference on High Performance Computing & Simulation (HPCS), Innsbruck, pp. 628–636 (2016). https://doi.org/10.1109/HPCSim.2016.7568394, http://ieeexplore.ieee.org/stamp/stamp.jsp?tp=&arnumber=7568394&isnumber=7568299
23. Prodan, R., et al.: Use cases towards a decentralized repository for transparent and efficient virtual machine operations. In: 25th Euromicro International Conference on Parallel, Distributed and Network-Based Processing (PDP), St. Petersburg, pp. 478–485 (2017). https://doi.org/10.1109/PDP.2017.47, http://ieeexplore.ieee.org/stamp/stamp.jsp?tp=&arnumber=7912691&isnumber=7912607
24. Demchenko, Y., et al.: CYCLONE: a platform for data intensive scientific applications in heterogeneous multi-cloud/multi-provider environment. In: 2016 IEEE International Conference on Cloud Engineering Workshop (IC2EW), Berlin, pp. 154–159 (2016). https://doi.org/10.1109/IC2EW.2016.46, http://ieeexplore.ieee.org/stamp/stamp.jsp?tp=&arnumber=7527833&isnumber=7527789
25. Singh, S., Chana, I.: A survey on resource scheduling in cloud computing: issues and challenges. J. Grid Comput. **14**(2), 217–264 (2016). https://doi.org/10.1007/s10723-015-9359-2, ISSN 1572–9184

26. Zhao, Y., Calheiros, R.N., Bailey, J., Sinnott, R.: SLA-based profit optimization for resource management of big data analytics-as-a-service platforms in cloud computing environments. In: 2016 IEEE International Conference on Big Data (Big Data), Washington, DC, pp. 432–441 (2016). https://doi.org/10.1109/BigData.2016.7840634, http://ieeexplore.ieee.org/stamp/stamp.jsp?tp=&arnumber=7840634&isnumber=7840573
27. Abaker, I., Hashem, T., Yaqoob, I., Badrul Anuar, N., Mokhtar, S., Gani, A., Ullah Khan, S.: The rise of big data on cloud computing: review and open research issues. Inf. Syst. **47**, 98–115 (2015). https://doi.org/10.1016/j.is.2014.07.006, http://www.sciencedirect.com/science/article/pii/S0306437914001288, ISSN 0306–4379
28. Assuno, M.D., Calheiros, R.N., Bianchi, S., Netto, M.A.S., Buyya, R.: Big data computing and clouds: trends and future directions. J. Parallel Distrib. Comput. **79–80**, 3–15 (2015). Special Issue on Scalable Systems for Big Data Management and Analytics. https://doi.org/10.1016/j.jpdc.2014.08.003, http://www.sciencedirect.com/science/article/pii/S0743731514001452, ISSN 0743–7315
29. Dean, J., Ghemawat, S.: MapReduce: simplified data processing on large clusters. Commun. ACM **51**(1), 107–113 (2008). https://doi.org/10.1145/1327452.1327492, ISSN 0001–0782
30. Shvachko, K., Kuang, H., Radia, S., Chansler, R.: The Hadoop distributed file system. In: 2010 IEEE 26th Symposium on Mass Storage Systems and Technologies (MSST), May 2010, pp. 1–10 (2010). https://doi.org/10.1109/MSST.2010.5496972, ISSN 2160-195X
31. Shoro, A.G., Soomro, T.R.: Big data analysis: apache spark perspective. Global J. Comput. Sci. Technol. [S.l.] (2015). https://www.computerresearch.org/index.php/computer/article/view/1137, ISSN 0975–4172
32. Merkel, D.: Docker: lightweight Linux containers for consistent development and deployment. Linux J. **2014**(239), 2 (2014)
33. Dua, R., Raja, A.R., Kakadia, D.: Virtualization vs containerization to support PaaS. In: 2014 IEEE International Conference on Cloud Engineering, Boston, MA, pp. 610–614 (2014). https://doi.org/10.1109/IC2E.2014.41, http://ieeexplore.ieee.org/stamp/stamp.jsp?tp=&arnumber=6903537&isnumber=6903436
34. Docker Copy on Write Strategy. https://docs.docker.com/engine/userguide/storagedriver/imagesandcontainers/
35. Md Hasanul, F., Murshed, M., Calheiros, R.N., Buyya, R.: Network-aware virtual machine placement and migration in cloud data centers. In: Emerging Research in Cloud Distributed Computing Systems, pp. 42–91. IGI Global (2015). Web. 9 Jan 2018. https://doi.org/10.4018/978-1-4666-8213-9.ch002

A Capability and Compatibility Approach to Modelling of Information Reuse and Integration for Innovation

Valentina Plekhanova(✉)

Faculty of Computer Science, University of Sunderland, Sunderland, UK
Valentina.Plekahnova@sunderland.ac.uk

Abstract. This paper presents a new formal approach to the modelling of information reuse and integration for innovation. Not all information is useful for innovation, and many ideas do not become profitable. We believe that information resources should not only be available, but also should be capable and compatible with the required information/needs. Use of relevant tools for information management should improve the capacity for effective decision making for innovation. Use of data mining technologies for the extraction of potentially useful information may not always produce the required information. Hidden or previously unknown information may be found in datasets, but the required information for innovation may not be in the datasets. There is a need for the development of techniques to ensure that decision makers are provided with capable and compatible information. Profile Theory is used for the analysis and modelling of reuse and integration of available information.

1 Introduction

Innovation is an important problem in many fields such as business, technology, economics, and engineering. Nowadays innovation is considered as the sustainable basis for organisational growth [6]. There are diverse definitions of innovation [11, 12] that encompass several different perspectives on, and aspects of, innovation. The common aspect, which is addressed in these definitions, is that an idea, a change, a renewal or an improvement is only an innovation when it leads to effective, better, or more profitable outcomes or results. The key point in these definitions is that innovation must bring positive results in the areas where it is implemented.

There are varied degrees of innovation, e.g. radical, incremental, disruptive, etc. The required degree of innovation is defined by the need. Innovation can be considered to be the match between a need and ideas which already exist [17]. Thus, in business, innovation often results from the application of a scientific and/or technical idea in decreasing the gap between the needs or expectations of the customers. In the knowledge-based economy it is important to find a tool which can support innovation.

Innovation is based on the transformation of ideas and relevant information/knowledge into a benefit/need. A benefit may be new, or improved, products, processes or services. In a knowledge-based economy, information is used as a driver for any modern

L. Barolli et al. (Eds.): EIDWT 2018, LNDECT 17, pp. 383–393, 2018.
https://doi.org/10.1007/978-3-319-75928-9_34

business and organisational development. Therefore, use, reuse and integration of information become key tasks for innovation.

In this work we consider innovation from the viewpoint of information reuse and integration, which key parts of information and knowledge management.

Like any management process, information reuse and integration require specific tools to support the capacity of an organisation to innovate, improve, and grow. This paper addresses key aspects in innovation that define a rationale for the proposed formal framework for modelling information reuse and integration.

In this work, Profile Theory [22] is used as a tool for the analysis and modelling of reuse and integration of available information.

2 Key Issues in Innovation

The concept of innovation is discussed by many business disciplines [12, 15] and it has many different definitions that are associated with the relevant and dominant paradigms. Despite the varieties of discussions, theories and approaches, innovation management is still an immature science. The most critical aspect/question is what affects an organisation's ability to innovate. It is recognised that nowadays companies may gain advantage if they can innovate not just through cost reduction, and reengineering. Innovation is a key element in the process of aggressive top-line growth of businesses and organisations [6].

Innovation cannot occur without invention. Invention is a process of creation of new artifacts that are based on ideas and/or scientific research. An invention increases the volume of knowledge. Innovation is a process of implementation of invention to produce practical benefits. An artifact of innovation results from a combination of research, discovery, invention, experimentation, development, engineering, diffusion and management.

The critical task is to define or identify information that direct organisations to a profitable economic outcome. Not all information or knowledge is useful for innovation, and many ideas do not become profitable. In many cases this is not because of failure of the idea. The generation and implementation of new ideas are tasks of equal importance in innovation.

2.1 The Nature of Innovation

The key reason for innovation is growth. The specific nature of innovation is defined by different needs (or objectives, goals, aspirations, expectations) and/or different problems. New needs may involve new problems; and vise verse; new problems may involve new needs. Therefore, we believe that the nature of innovation is different for different needs and the "speed" of change of innovation nature is thus also different.

Several reports discuss the nature of innovations [4, 13]. "Innovation is the main lever for a more competitive national economy" [13]; and the notion of innovation is changing radically and rapidly:

"The nature of innovation is changing dramatically in the 21st century. Proprietary invention in search of purpose is out. Open, collaborative and multi-disciplinary approaches to innovation are taking center stage in the shaping of new ideas and creation of tangible value for business, individuals, and the world." [16].

Each period of time in a real-life situation defines specific aspirations and demands. Therefore, major contribution of this work is that it identifies milestones for current events and defines the most important aspects and relevant needs.

2.2 Innovation: Methodologies and Information Reuse and Integration

The following components reflect key aspects of innovation:

- *Wisdom* is a starting point in the innovation process; i.e. we need to identify needs (why?) and a rationale for information and knowledge management.
- *Information* is an essential resource for innovation which involves definitions, descriptions, directions (what/who/where/when?) that can be used for accomplishment of needs.
- *Knowledge* comprises/relates to "tools" (how?); e.g. methods, approaches, strategies, frameworks, for transformation of relevant information to useful results.
- *Idea* is a rational conception which defines possible courses of action and is relevant to something new; i.e. method, procedure, strategy, resources, market, new or improved products.
- *Various processes* (e.g. management including RDD - Research, Development, Diffusion) are involved in supporting the innovation process.
- *Performance measures* (e.g. cost, time, resources, better market, more sales turnover; profit) are used for the measurement of improvement(s) of system performance (e.g. business).

The initial task for innovation is the identification of problem followed by the development of a solution to the identified problem via research and the generation of creative ideas. These first and key steps involve the creation of new knowledge. There are several discussions of different ways for the creation of new knowledge. It is noticed [2, 9] that new knowledge can be created by combining old and new knowledge in an innovative way. Learning processes are considered as key contributors to the production of new and economically useful knowledge [9, 19]. That is, it is recognised that research and development, search and experimentation, learning, data mining and imitation are sources of new knowledge.

Many leading methodologies are used for innovation processes. Some well-known approaches are Brainstorming [21], Brainwriting [14], Delphi Method [3, 12], Heuristic Redefinition Process, Transformation of Ideal Solution Elements with Associations and Commonalities (TILMAG) [24], and Theory of Inventive Problem Solving (TRIZ) [1]. Innovation methodologies are mainly derived from the objectives and expectations of the main stakeholders.

If we closely analyse these methodologies we find that the key tasks are relevant to information and knowledge management and, in particular, to information reuse and integration. In general, these qualitative methods involve information reuse and integration tasks.

2.3 Information Management vs. Knowledge Management in Innovation

Knowledge creation, and the translation of this knowledge into products and services is a key determinant of economic success. It is recognised that knowledge becomes the most important economic recourse in the knowledge society [10, 18]. It leads to a new focus on the role of information, technology and learning in economic performance. In a knowledge-based economy it is important to find a tool which can support innovation. The term "knowledge-based economy" is used to confirm that knowledge management plays an important role in modern economies [10].

Traditionally information management is considered as a process which is about the collection, storage, retrieval, classification, security and distribution/diffusion of data.

Knowledge management is about the use of information which involves modelling of real-life systems, management and application of relevant information, critical evaluation of its application results and decision making. It leads to the creation of new information or knowledge. Knowledge creation is, in fact, a process of value addition to previous knowledge through innovation [8, 20].

It is important to note that innovation is not just about any new knowledge, i.e. not any new knowledge leads to innovation. An innovation generates new knowledge, which is unique since it leads to innovation.

Information management and knowledge management are interrelated and are of equal importance. Information management nurtures knowledge management and knowledge management is used for information management.

The key contributors to volume and quality of knowledge are individuals who are knowledge creators, developers, carriers, and users [5]. Information is an input and people translate/transfer this information into outputs. Knowledge cannot be explicitly managed because it resides in a person's mind. The same information can produce different outputs.

Thus, knowledge involves the link between information and its potential applications. People/stakeholders and the use of information technology define the quality of this link. Information/knowledge management technologies provide stakeholders with decision making support, i.e. allocate the relevant information to the right problem, at the right time for the right request. It increases and improves the capacity for effective decision making/actions for innovation. Note that irrelevant information makes decision making difficult, leads to confusion, affects performance, and reduces the opportunity for innovation.

Therefore, information can be identified as the critical resource for decision making for innovation; and management can be considered as an information-intensive activity. It is crucial for managers to be aware of what information they require, what the quality of this information is; and how to get best/optimal benefits from the use of it in order to survive in the modern knowledge-based economy. Information and knowledge management are key management processes in innovation, and information reuse and integration is an important part of these processes.

2.4 Data Mining vs. Information Reuse and Integration

According to a recent study [7], one of the leading topics in information reuse and integration is data mining and knowledge discovery, acquisition and management.

Data mining technologies are widely used for the extraction of potentially useful information from datasets. Data mining gives information that would not be otherwise available. Data Mining is concerned with the extraction of hidden information. Hidden or previously unknown information can be found in datasets, but the required information may not be in the dataset. If it does not exist (i.e. if it is not there) we cannot extract the required information.

The outcome of the applications of a data mining technique is information, which becomes available for decision making. However, this information may not satisfy the required need. Therefore, there is a need for reuse and integration of information in order to reduce the gap between required and available information. If hidden or unknown information is available the next step is that it should be used/reused, and integrated with other elements of available information to ensure that the required information is allocated to the needs/tasks. Therefore, the following important problems must be addressed in the process of information reuse and integration: how relevant the information is, and how well it satisfies a required need.

That is, information resources should not be only available; but should also be capable and compatible with the required information/needs for innovation. Therefore, there is a need for the development of techniques to ensure that decision makers are provided with capable and compatible information.

Innovation requires new knowledge which uses available information. Data mining technology can be used for the extraction of potentially *useful information* from datasets; but for innovation it may be not enough since we are interested in potentially *utile information.* We define *utile/innovative information* as (profitable) information, when is used, leads to benefits/innovation.

2.5 The Need for Formal Modelling of Information Reuse and Integration

As discussed, information reuse and integration are a part of information and knowledge management. It is important to note that innovation can be identified after it has occurred. Modelling is a tool that supports decision making for innovation. Information nurtures knowledge through models. A model with the relevant (i.e. capable and compatible) information is used for decision making, analysis, explanations, and developments. The quality of information is a crucial factor for model performance. Therefore, modelling of information reuse and integration is one of the most important tasks in decision making. The integration of descriptive, normative and learning models contributes to the quality of decision making for innovation. It is recognised that innovation is a fundamental determinant of organisational performance. Since innovation has to provide effective, better and more profitable results, normative and learning models for information reuse and integration become “must have” tools that can contribute to effective development of innovation.

3 Profile Theory: Brief Introduction

In this work, Profile Theory is used as a tool for the analysis and modelling of reuse and integration of available information where information capability and compatibility problems are addressed. Profile Theory [22, 23] is designed to address an important class of problems, which are not addressed by traditional theories (i.e. including complex systems theories), in which the central issues relate to capability and compatibility problems of complex systems and/or their elements; and determination of complex system structures. The importance of Profile Theory derives from the fact that in the real world such problems are the norm rather than the exception. Key research issues include: the analysis and modelling of complex systems where capabilities and compatibilities of their elements are critical factors.

We define the information resource *integration problem* as a problem of allocation of capable and compatible information resources to tasks in order to provide information resource utilisation at the desired performance and technology levels. Thus, the determination of information resource capabilities and compatibilities is a focus of an integration problem. In order to provide information resource integration, we determine an *allocation strategy*, which is used for the allocation of capable information resources to the tasks, and then an *integration strategy*, which is applied for the allocation of compatible information resources. Analysis and reuse of available information is considered as a key task in the information integration process.

3.1 Profile Definition

In Profile Theory an object is described by its internal characteristics. That is, an object is described by a set of factors, and in turn each factor may be defined by multiple characteristics. A set of such factors forms a profile. More factors are used for an object description, and more explicit identification and definition of object is provided. We represent a factor by qualitative and quantitative information. A quantitative description of the *i*th profile factor is defined by an indicator characteristic ε_i, property v_i, and weight w_i[1]. In particular,

- ε_i - is the indicator characteristic, that indicates and expresses, by factor presence in the object description, the existence of certain conditions. In particular,
    1. ε_i may be defined as a time characteristic of the *i*th factor $\varepsilon_i = \varepsilon_i(t)$:

    $$\varepsilon_i\text{:T} \rightarrow \text{E}_i \text{ or } \text{T} \times \text{E}_i = \{(t, \varepsilon_i),\ t \in \text{T and } \varepsilon_i \in \text{E}_i\}$$

    where T is a set of time characteristics; E_i is a set of possible indicator characteristics of the *i*th factor.

[1] It should, however, be pointed out that each profile factor may be described by an N-dimensional tuple [22].

Domain constraints may define bounds, i.e. $\varepsilon_i^b \leq \varepsilon_i \leq \varepsilon_i^u$, where $\varepsilon_i^b \geq 0$, $\varepsilon_i^u \geq 0$ represent bottom (lower) and top (upper) values of the *i*th factor time range, respectively.

The time characteristic can represent the duration (or length) of experience or factor utilisation.

2. ε_i may also represent a number of times of factor utilisation (e.g. a number of projects (or tasks) where a particular knowledge/skill was utilised)
3. ε_i may also represent a binary case. For instance, factor existence $\varepsilon_i = 1$ or non-existence $\varepsilon_i = 0$; Boolean variable: $\varepsilon_i = 1$ if factor is true or $\varepsilon_i = 0$ if factor is false.

- v_i - is the property of the *i*th factor (e.g. depth, range, complexity, capability, degree, grade of compatibility or level of a factor): $v_i \geq 0$. Since a property may change with time, v_i can be defined as a function of time $v_i = v_i(t)$:

$$v_i\text{: T} \rightarrow \text{V}_i \text{ or } \text{T} \times \text{V}_i = \left\{(t, v_i),\ t \in \text{T} \text{ and } v_i \in \text{V}_i\right\}$$

where V_i is a set of property characteristics of the *i*th factor.

Domain constraints may define bounds, i.e. $v_i^b \leq v_i \leq v_i^u$, where $v_i^b \geq 0$, $v_i^u \geq 0$ represent bottom (lower) and top (upper) values of the *i*th factor property range, respectively.

- w_i - is the weight of a factor which defines either the factor importance or the factor priority: $w_i \geq 0$. Factor weights can vary, and therefore, w_i can also be considered as a function of time $w_i = w_i(t)$:

$$w_i\text{:T} \rightarrow \text{W}_i \text{ or } \text{T} \times \text{W}_i = \left\{(t, w_i),\ t \in \text{T} \text{ and } w_i \in \text{W}_i\right\}$$

where W_i is a set of possible weights of the *i*th factor.

4 Information Modelling

The strategic management of information capabilities *via* analysis, comparison and management of the information resource capabilities and their compatibilities is an important component in innovation. Thus, tasks are to analyse, reuse and integrate available information profiles in such a way that provides *innovative information*, i.e. information which is used for innovation. Let us consider a set of information profiles as a topological space. We need to find among available information profiles, a profile or a set of profiles that cover(s) the required information profile for a given task.

Let us semantically define a set of information profiles as a set of N available information profiles: $E = \left\{b^{(j)}, j = \overline{1, N}\right\}$ and B as a subset of the available information profiles of E for a given task: $B = \left\{b^{(j)}\text{:}b^{(j)} \in E,\ j \in \{1, 2, \ldots, N\}\right\}$.

For information space modelling we define the information profile space as a topological space (E, B), where the family B is a *topology* in the set E. The subsets B are the collections of the available information profiles for a given task that are the open sets of E.

In order to select all combinations of the available information profiles, that satisfy the required one, we represent the information profile space as a metric space. The concept of a metric is used to define a notion of distance on the information profile space and carries with it a certain structure. A metric space approach to information profile space modelling provides both the modelling of combination of the individual information profiles and the determination of a number of the viable alternative combinations of information profiles.

4.1 Information Profile Space as a Metric Space

In order to define a metric on a set E of the available information profiles of the information space, we need to define the distance from the information profile $b^{(i)}$ to the information profile $b^{(j)}$ [2].

A "deviation" of the available information profile $b \in E$ from a required information profile $b^{(0)}$ can be measured by the covered power index [22]:

$$\rho(b) = \rho(b, b^{(0)}) = \frac{m}{n},\ 0 \leq \rho(b) \leq 1$$

where m is a covered power of b; n is the required information profile power.

The distance $d(b^{(i)}, b^{(j)}) = |\rho(b^{(i)}) - \rho(b^{(j)})|$ between two covered power indices can be considered as the distance between the available information profiles $b^{(i)}$ and $b^{(j)}$ (or as a compatibility-length metric) [22].

Any L_p - metric can also be used as a distance between profiles:

$$\|b - b^{(0)}\|_p = \left[\sum_{i=1}^{n} \left|V_i - V_i^{(0)}\right|^p\right]^{1/p},\ p \in \{1, 2, 3, \ldots\} \cup \{\infty\}$$

or any weighted L_p - metric:

$$\|b - b^{(0)}\|_p^{\lambda} = \left[\sum_{i=1}^{n} \left(\lambda_i \left|V_i - V_i^{(0)}\right|^p\right)\right]^{1/p},\ p \in \{1, 2, 3, \ldots\} \cup \{\infty\}$$

where $\lambda \in R^n$ is a non-negative vector of weights; and V_i is available factor capability with respect to required factor capability and is defined as [22]:

[2] Distance is not always distance in the colloquial sense, for example cost, elapsed time, reliability, compatibility, capability, etc. can also be interpreted as a distance.

$$V_i = V(b_i) = w_i \left(\frac{\varepsilon_i}{\varepsilon_i^{(0)}} \right) \left(\frac{v_i}{v_i^{(0)}} \right)^2$$

Thus, the information profile space is represented by a metric space (E, d) with the metric $d(b^{(i)}, b^{(j)})$. This allows the definition of the metric on the information profile space, which generates the topologies for E, and defines the feasible combinations of information profiles (individual performers) for a given task. The topology for E is generated by metric.

We say, that *feasible combinations* of the information profiles for a given task are generated by metric and represented by the topologies.

A given or required distance we call a *radius* $r = r(b)$.

Thus, we can define a number of the possible *criterions for the acceptance an available* information *profile for a task*. In particular, an available information profile can be accepted for utilisation for a given task, if the distance between an available information profile and a required information profile is less than a given radius r

$$d(b^{(i)}, b^{(0)}) < r,\ r > 0.$$

where r can be represented by any metric or characteristic.

We may define the *ball with radius r and centre* $b^{(0)}$ (required information profile), that is the set of the available information profiles for which the distance from a required information profile $b^{(0)}$ is less than a given radius r [22]:

$$B_r(b^{(0)}) = \{b^{(i)} | b^{(i)} \in E: d(b^{(i)}, b^{(0)}) < r,\ r > 0,\ i \in \{1, 2, \ldots, N\}\}$$

It was theoretically proved that the available information profiles are accepted for a task, if they are elements of the ball with radius r and centre $b^{(0)}$ [22].

For instance, the *covered power distance* between an available information profile and a required information profile is defined as $d(b, b^{(0)}) = |\rho(b) - 1|$. An available information profile b *covers* the required information profile $b^{(0)}$ if the covered power distance is less or equal to the given radius: $d(b, b^{(0)}) \leq r(b)$.

The required information profile may be associated with many alternative available information profiles that may have identical information profiles but are relevant to different performance profiles. We need to find a set of information profiles which gives us optimal or better results for the given problem. An *innovative information profile* is a profile which determines an optimal or better result(s) to the given task/problem. Also it can define an optimal strategy for information reuse and integration for innovation.

5 Conclusion

In this work we consider innovation from the viewpoint of information reuse and integration; which are a key part of information and knowledge management. Information is considered as the critical resource for innovation. That is, information should not be only available but also should be capable and compatible with the required information/

needs, and relevant tools for information management should be provided in order to increase and/or improve the capacity for effective decision making for innovation. The need for a formal modelling approach to information reuse and integration is critically discussed. Data mining technologies may not deliver the required useful information for innovation; since it may not be in the datasets.

Normative and learning models for information reuse and integration are defined as "must have" tools that can identify (capable and compatible) information which could lead to innovation. In this work, Profile Theory is used as a tool for modelling of reuse and integration of available information. A number of metrics for measurement of capability and compatibility of information are introduced. An information profile space modelling provides both the modelling of a combination of individual information profiles and the determination of a number of the viable alternative combinations of information profiles. The required information profile may be associated with many performance profiles and may be introduced by the multiple presentations of alternatives that define the decision making environment for innovation. We define an innovative information profile as a profile which determines optimal or better result(s) to a given problem.

Thus in the proposed approach to modelling information reuse and integration, determination of information resource capabilities and compatibilities is a focus of an integration problem. There are two possible ways for the use of information profile(s) for the task, i.e. use of only one information profile or a combination of information profiles. In order to provide information resource integration, we determine an allocation strategy, which is based on an allocation of capable information resources to the tasks. The proposed formal approach to the modelling of information reuse and integration can be considered as a tool to support the capacities of organisations to innovate, improve, and grow.

References

1. Altshuller, G.: The Innovation Algorithm. Technical Innovation Center, Inc., Worcester (1999)
2. Bijker, W.E.: Of Bicycles, Bakelites, and Bulbs: Toward a Theory of Sociotechnical Change. The MIT Press, Cambridge (1995)
3. Cuhls, K., Blind, K., Grupp, H.: Innovations for our future. Delphi '98: new foresight on science and technology. Technology, Innovation and Policy, Series of the Fraunhofer Institute for Systems and Innovation Research (ISI), vol. 13. Physica Heidelberg (2002)
4. Danish Enterprise and Construction Authority, New Nature of Innovation, Copenhagen (2009)
5. Davenport, T.H., Prusak, L.: Working Knowledge: How Organisations Manage What They Know. Harvard Business School Press, Boston (1998)
6. Davila, T., Epstein, M.J., Shelton, R.: Making Innovation Work: How to Manage It, Measure It, and Profit from It. Wharton School Publishing, Upper Saddle River (2006)
7. Day, M.Y., Ong, C.S., Hsu, W.L.: An analysis of research on information reuse and integration. In: Proceedings of IEEE International Conference on Information Reuse and Integration, IRI-2009, Las Vegas, USA, pp. 188–193 (2009)

8. Duffy, J.: The tools and technologies needed for knowledge management. Inf. Manag. J. **35**(1), 64–67 (2001)
9. Edquist, C., Johnson, B.: Institutions and organizations in systems of innovation. In: Edquist, C. (ed.) Systems of Innovation: Technologies, Institutions and Organizations, pp. 41–63. Pinter Publishers, London (1997)
10. Foray, D., Lundvall, B.A.: The knowledge-based economy: from the economics of knowledge to the learning economy. In: Employment and Growth in the Knowledge-Based Economy, pp. 11–32. OECD, Paris (1996)
11. Freeman, C.: The national system of innovation in historical perspective. Camb. J. Econ. **19**, 5–24 (1995)
12. Gardner, J.: Innovation and the Future Proof Bank: A Practical Guide to Doing Different Business-as-usual. Wiley, Chichester (2009)
13. General Electric: The GE Global Innovation Barometer 2011: An Overview on Messaging, Data and Amplification (2011)
14. Geschka, H.: Creativity techniques in Germany. J. Creativity Innov. Manag. **5**(2), 87–92 (1996)
15. Kim, W.C., Mauborgne, R.: Value innovation: the strategic logic of high growth. Harvard Bus. Rev. **75**(1), 103–112 (1997)
16. IBM: IBM's Global Innovation Outlook 2.0, Innovation that Matters (2005)
17. Ijuri, Y., Kuhn, R.L.: New Directions in Creative and Innovative Management: Bridging Theory and Practice. Ballinger Publishing, Cambridge (1988)
18. Johnston, R., Rolf, B.: Knowledge moves to centre stage. Sci. Commun. **20**(1), 99–105 (1998)
19. Lundvall, B.A. (ed.): National Systems of Innovation: Towards a Theory of Innovation and Interactive Learning. Pinter Publishers, London (1992)
20. Narayanan, V.K.: Managing Technology and Innovation for Competitive Advantage. Prentice Hall, Englewood Cliffs (2001)
21. Osborn, A.: Your Creative Mind. Motorola University Press, New York (1991)
22. Plekhanova, V.: Capability and compatibility measurement in software process improvement. In: Proceedings of the 2nd European Software Measurement Conference, pp. 179–188. Federation of European Software Metrics, Amsterdam (1999)
23. Plekhanova, V.: Applications of the profile theory to software engineering and knowledge engineering. In: Proceedings of the Twelfth International Conference on Software Engineering and Knowledge Engineering, pp. 133–141. Knowledge Systems Institute, Chicago (2000)
24. Silverstein, D., Philip Samuel, P., DeCarlo, N.: The Innovator's Toolkit: 50+ Techniques for Predictable and Sustainable Organic Growth. Wiley, Hoboken (2008)

Vehicle Insurance Payment System Based on IoT

Elma Zanaj[1(✉)], Kristi Verushi[1], Indrit Enesi[1], and Blerina Zanaj[2]

[1] Polytechnic University of Tirana, Tirana, Albania
{ezanaj,krist.verushi,ienesi}@fti.edu.al
[2] Agricultural University of Tirana, Tirana, Albania
bzanaj@ubt.edu.al

Abstract. This study describes an application based on Arduino applied for Smart Driver as part of IoT. The purpose of such system comes as result of an evaluation for the payment of the vehicle insurance based on the way you drive the car. The technologies involved in our system are: Arduino, different sensors, Proteus simulation environment for verifying the circuit functionality, and a smart phone. The mobile technology will be used for the transmission of wireless data, also to complete the entire cycle for the application based on Android. The discussion about different use cases about the system functioning holds an important issue in the paper, here in this work are brought as well the common errors that might happen during the application of such project and those are the conclusions of this work.

1 Introduction

The applications in the near future will pass information from real objects toward the intelligent virtual network, to keep us informed about everything and also to help us to control the systems. Most part of nowadays applications fall under one of the following classifications like: smart city, smart home, connected health, connected cars, [1]. All of the these applications are within the IoT (Internet of Things), a term that was used for the first time by Ashton in 1998, [2], it is a paradigm and as well an ongoing process of study in different fields and industries where it is applied. IoT is based on the development of: microelectronics and information technology. One of its topologies is shown as in Fig. 1 below.

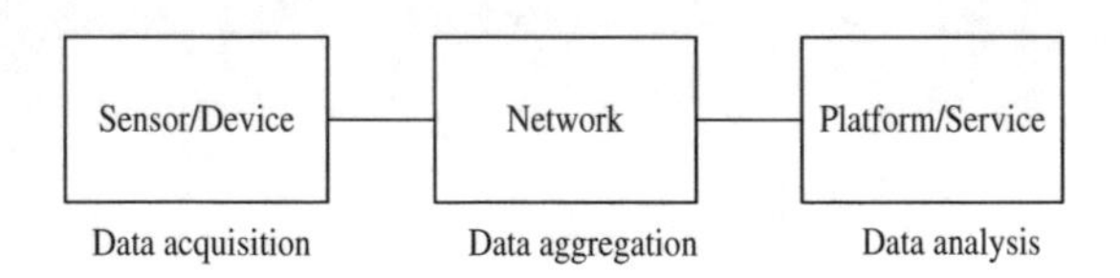

Fig. 1. IoT topology

The data acquisition in IoT system is done by small devices like sensors, that are able to sense from the environment and to communicate between them, Fig. 1, through network. The statistic provides that the worldwide market size of automotive sensors

L. Barolli et al. (Eds.): EIDWT 2018, LNDECT 17, pp. 394–402, 2018.
https://doi.org/10.1007/978-3-319-75928-9_35

from 2015 to 2021 is like shown in the Fig. 2 below. It is estimated that the global automotive sensors market size will increase from 24 billion U.S. dollars incomes in 2015 to some 43 billion U.S. dollars in 2021, Fig. 2. Sensors are a critical component in one of the emerging fields of Internet of Things (IoT). Driven by the use of IoT-connected devices, the global market for sensors is expected to grow considerably for the foreseeable future [3]. The sensors (data generators) are increasing their network connectivity, Fig. 3, by generating: different velocity, different veracity and different volumes of data, while their price is reduced as years go by.

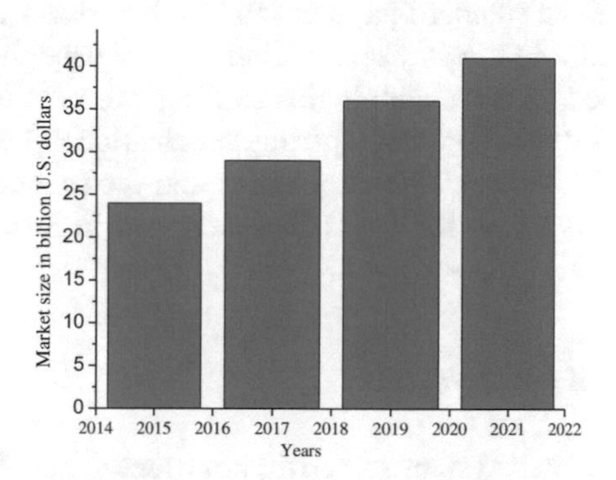

Fig. 2. Global automotive sensors market from 2015 to 2021 (in billion U.S. dollars)

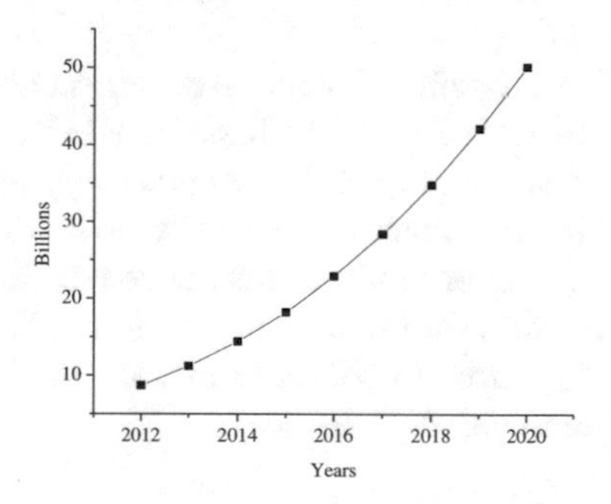

Fig. 3. Number of connected sensors [4]

So, the development of IoT is not just a necessity to fulfill all the requests generated from the existing services. But it can combine the opportunities to make the existing services more efficient. IoT is not composed only of transmitting devices or devices that forwards the packets to a gateway, now it realizes different functionalities needed for various areas of application around us. Although there are some aspects of such system that need to be taken in consideration like the security and the privacy.

IoT helps the components to be smart and to get the maximum of already existing systems like: transport, energy, transmission bandwidth especially in mobile communications, etc. Also the development of different components that compose the system, give more functionalities and opportunities to grow and improve the services offered to different companies. In this work the network includes different electronic devices that have a wireless communication and access into the network through the network card that they are supplied with. They communicate with each other through the system and they can interact with the humans through an interface. This research is focused on the architecture of one of the IoT application. This application is used by the insurance companies and it consists in developing a monitoring system that will observe the insured car's movement in order to set a rating point for the driver of the vehicles. The purpose of selecting the SmartDriver project for reflecting on an IoT system was the result of thinking of a vehicle insurance payment system based on the principle of meritocracy, rather than a fixed standard payment for all the drivers without distinguishing who is more careful and who is less. The remainder of the paper is organized as follows. In Sect. 2, are described all the technologies starting from Arduino hardware and the various sensors used. Section 3 presents the implementation in Proteus simulation environment for verification of circuit functionalities and the mobile technology; also the Android app will be discussed for different cases, and in Sect. 4 are brought some conclusions of this work.

2 Devices Used in the System

The main parts of considered system are: Arduino, Bluetooth, GPS (Global Positioning System), ultrasound sensor, vibration sensor, accelerometer and RGB led. In the following each of them will be described and their equivalent schematics in the Proteus environment will be shown.

Arduino Uno is a microcontroller board based on ATmega328. It has 14 digital inputs, 6 analog inputs, a 16 MHz crystal oscillator, one USB connection, a power jack, an ICSP header and a reset button, Fig. 4. It contains everything needed to support the microcontroller. So, we simply connect it to a PC with a USB cable or feed it with a battery. It offers a large number of facilities for the connection with the PC or another device. There are many projects for the communication between Arduinos in distance via a wireless module, [5]. Arduino software includes a serial monitor that allows sending data from/to Arduino board.

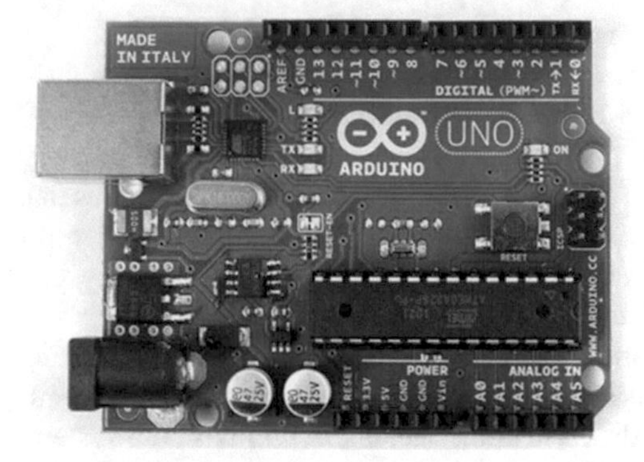

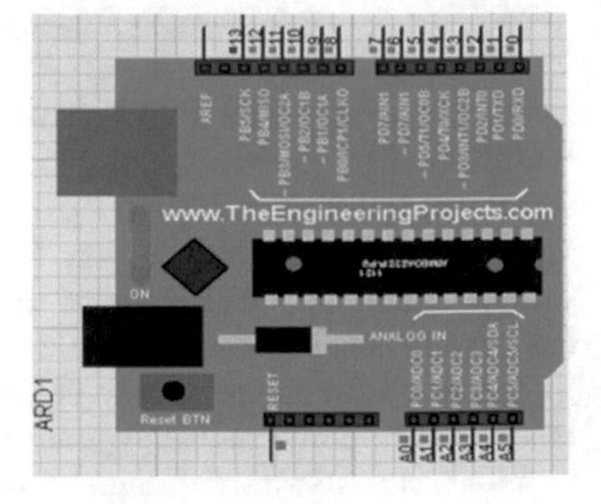

Fig. 4. Hardware Arduino Uno and its schematics design for simulation

In this project it will serve as a core element for the communication of the sensors with our software. Each signal collected by the sensors will be transmitted to the Smartphone through two main channels: Bluetooth and SMS by using GSM, [6].

Bluetooth is a standard technology for exchanging data in a short distance (using the low frequencies of UHF range and the wave in ISM band from 2.4 to 2.485 GHz) from fixed and mobile devices and in building Personal Area Networks (PANs). Bluetooth in our project will serve for the connection of the node with the Smartphone, Fig. 5. It constantly transmits geographic positions taken from Arduino via integrated GPS as well as other data obtained from distance sensors.

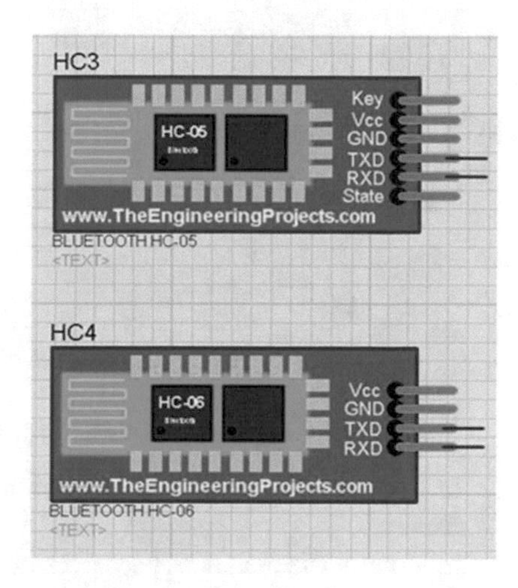

Fig. 5. Hardware Bluetooth and its schematics design for simulation

GPS (Global Positioning System) is an orbiting satellite network that sends accurate position details on the ground, Fig. 6. Signals are received from GPS receivers like the navigation devices, those are used to calculate the exact position, speed, and time for the vehicle location.

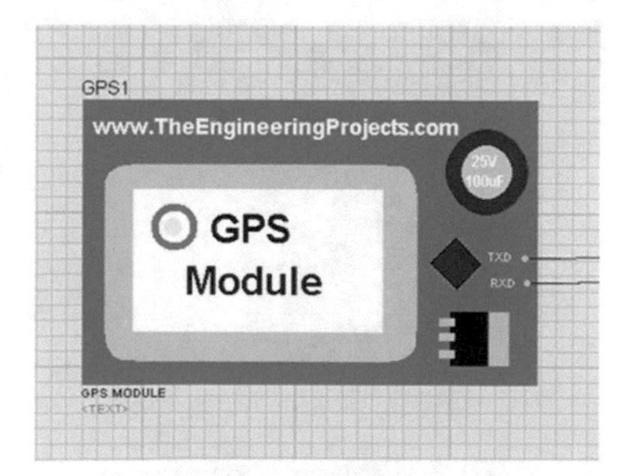

Fig. 6. Hardware GPS and its schematics design for simulation

GPS is one of the key elements of this project, for receiving the accurate geographical position of the vehicle in due time, that will help to calculate distances in average, as well an approximate data about the speed can be used. These data will serve to calculate the collected points for insurance payment for each user, Fig. 10.

An ultrasound sensor measures the distance from an object. The distance is measured by sending a sound wave at a certain frequency and waiting for the reflected wave. These sensors, Fig. 7, will be placed in different parts of the car, mainly in the most exposed parts in front and behind.

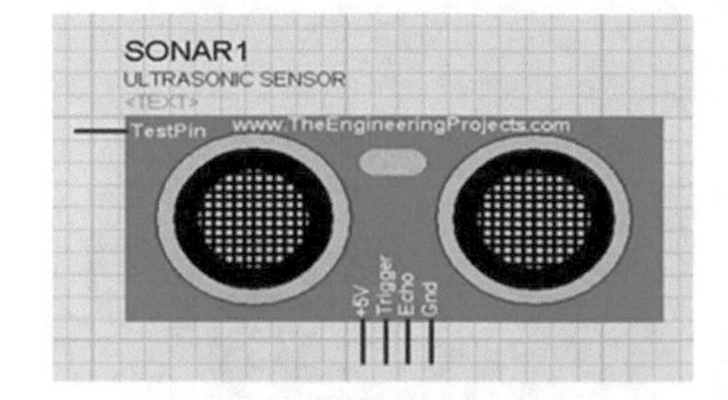

Fig. 7. Hardware ultrasound and its schematics design for simulation

This installation has two main purposes:

1. to find the distances from the surrounding objects for helping the user to park the car.
2. to monitor the distances kept from other vehicles, these data will serve as a reference for the evaluation of the driver's point.

The vibration sensor has the ability to detect vibrations in a particular area. This type of sensors will be used in our security systems, Fig. 8. The vibration sensor circuit will be placed in the front of the car next to the motor. It will serve as a confirmation if the car is switched on, that is a basic condition to start monitoring and at the same time to ask the user to connect to the hardware module.

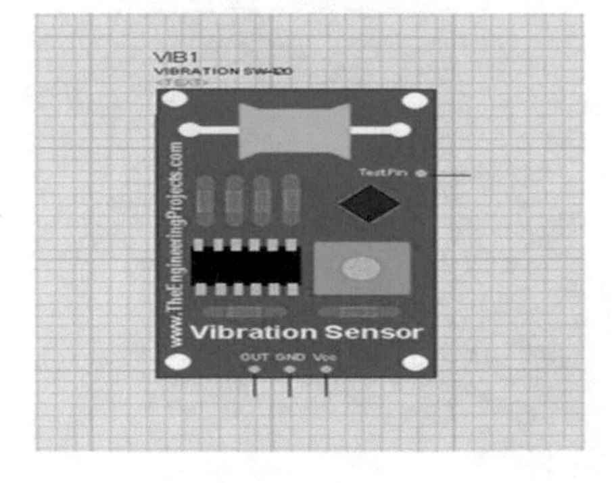

Fig. 8. Hardware vibration sensor and its schematics design for simulation

Accelerometer measures the acceleration of an object. Within a MEMS accelerator device there are microstructures that bend due to momentum and gravity. Any form of acceleration will make possible that these small structures to bend at a certain amount and to be electronically detected. The accelerometer is not an element for system security but only for monitoring the way the user drives the vehicle. So, in our project to maintain the lowest cost for this hardware, instead will be used the sensor integrated into a Smartphone Android. It will help to calculate the level of vehicle's acceleration at all the times the car is moving.

RGB led is a Light Emitting Diode (LED) that is used to show the condition of the equipment.

Smartphone. The main purpose for the design of mobile phones originally was for communication, but nowadays they are changing our way of living. There are many services offered by them not anymore only as means of communication. They have become a source of entertainment, information and services. Two of the most widespread operating systems for the smartphones are: Android and iOS. In this study is used Android Smartphone that will be connected with the device built in Arduino. So, it will receive a continuous information from the Arduino and it will notify on real-time the driver about the sensors data, but also it will send the data to a central server. This will make it possible to build an Android app that can be installed by any user of this system, [7].

3 Implementation of the System and Simulations

Our application has three main "Activity": DeviceListActivity, PaneliActivity, GPSActivity.

DeviceListActivity. It is the first activity running when the application starts, Fig. 9a. There is a list of all the devices that one can be connected via Bluetooth, one of them is also our equipment.

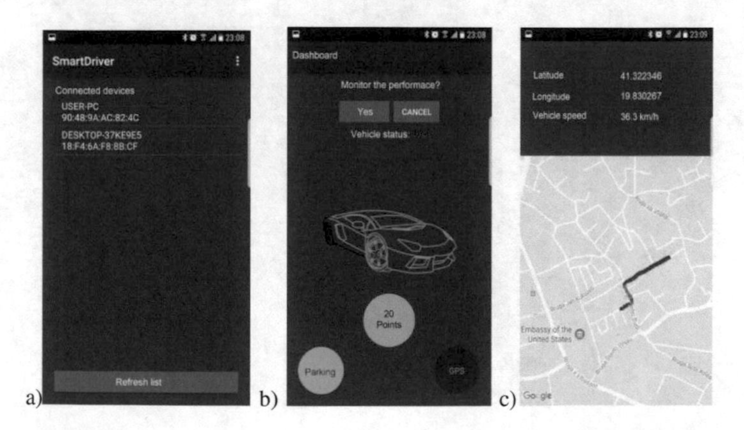

Fig. 9. (a) DeviceListActivity; (b) PanelActivity and (c) GPSActivity

PanelActivity. This panel shows the first communication with the device, Fig. 9b. It is compound of:

- Yes button: it is the reply of the question if the driver wants to permit the application to use the data for the evaluation;
- CANCEL button: it stops the system from monitoring the driver;
- Vehicle status: depending on the information received from the vibration sensor has two states ON/OFF;
- Points section (orange circle): the amount of points accumulated by the driver on the system.
- GPS section: sends the user to GPS Activity, which will be described in the following.
- Parking section: the driver can access it for receiving information about vehicle distances from the surrounding objects during a parking. Even if it not chosen, but a very close distance is detected by anyone of ultrasonic sensors, the application will send an alarm to the driver.

GPSActivity. GPSActivity, Fig. 9c, displays the driver's coordinates taken from the GPS and an approximate calculation for the vehicle's speed in different times, [8, 9]. There is also displayed the "route" on the map that the vehicle has followed. The main function used in this activity is: map display functionality by using Google play service, [10].

Our main purpose to design such a circuit, Fig. 10, by using Proteus, [11, 12], is to complete these three purposes:

- to monitor and evaluate the driver while he is using the vehicle;
- to protect and insure the vehicle;
- to assist the driver (for example: during the parking procedure).

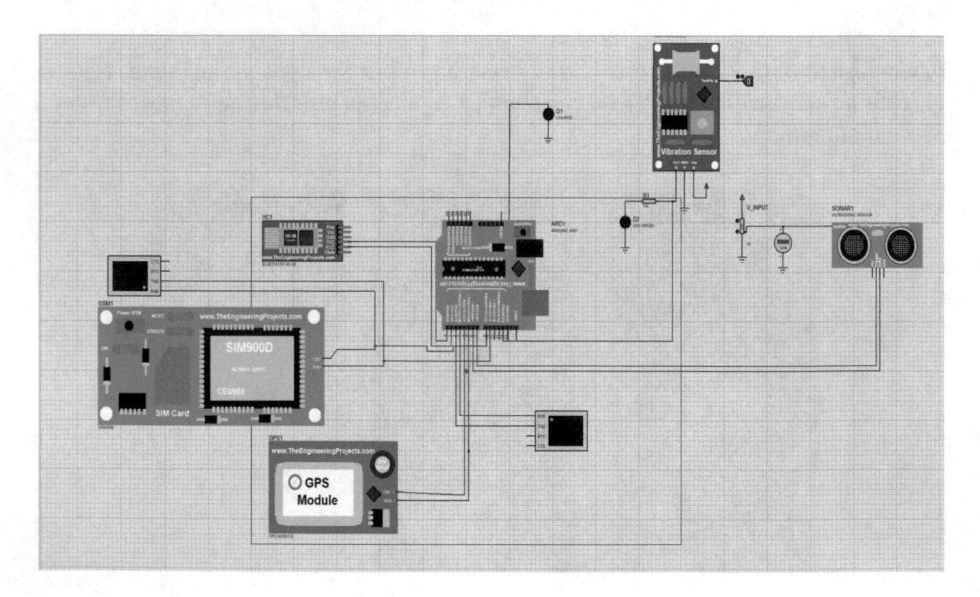

Fig. 10. System architecture for the connections in Proteus

CASE 1. *Monitoring and driver evaluation.* At the instance when the driver enters in the vehicle, it is required that he has to connect the Smartphone with the hardware through Bluetooth. When the car will be switched on, the module with the vibration sensor will be activated by giving a HIGH- positive voltage at its Vout pin. This voltage (pin 12) will activate Arduino for gathering information from GPS sensor in NMEA format. Through the TinyGPS library, Arduino passes the serial data taken from GPS (NMEA format), in a readable format such as latitude and longitude. After converting through Arduino TX serial port 1, the data are sent to Bluetooth module that will send them to the Smartphone. Our application receives this information and displays it to the driver, at the same time with the values obtained from the sensors integrated in Smartphone. Accelerometer sends data to the server that can be used further on to analyze the way the vehicle is used.

CASE 2. *Protect and insure the vehicle.* At the moment the driver enters in the car, it is required to connect the Smartphone to the hardware via Bluetooth. When the driver does not reply back and the car is switched ON. The module of vibration sensor will be activated by supplying into its pin of Vout a HIGH positive voltage. The conditions: the vehicle is switched on and the driver has neither responded positively nor negatively to the application, the device activates the *alarm state.* Alarm alert activation consists in sending a sms to the phone number of the owner of the vehicle with the vehicle location coordinates.

CASE 3. *Assist the driver.* At the moment the driver enters in the car, it is required to connect the Smartphone to the hardware via Bluetooth. After this moment, when the car will be switched ON, the module of the vibration sensor will be activated by supplying into its pin of Vout a HIGH positive voltage. This voltage (pin 12) will activate Arduino for gathering information. The user can choose from the application with the option Parking, and it turns to a new interface of the application. Arduino starts sending data from Ultrasonic sensors that gives the distance from the car's closest object and at the same time at some critical point alerts the driver to stop.

The design of the apps, demands to become familiar with most common in-app interactions that could lead to the error state. For example, it's impossible to properly synchronize the data if the device has a poor network connection. During the testing of our app for the system usage, we have found some bugs:

1. The user of the vehicle may have Bluetooth closed, so the sms will not be sent as there will be no reply to the request for monitoring.
2. In cases of robbery it may happen that in certain areas there will be no GSM coverage zone, so it will not be possible to send the location by *sms* to the owner in the specified interval.
3. One/some sensors can be damaged, it will bring to the system malfunction.
4. Data received from the GPS do not have the maximum precision.
5. Smartphone battery level become low, the driver will not be able to respond to the system.

4 Conclusions

During this study we explained some features of IoT as a system that is growing on, and described a real implementation for the drivers of some vehicle. There are chosen some technologies and programs to be used in the application like: Android operating system and Arduino, also some modules connected with it. It is shown how each module will contribute on the whole system and when each of them are useful for implementing together a driver's monitoring system but also assuring the security of the vehicle. We introduced the behavior of the system in different use cases. The project presented at the current level of development is a first step only for the moment, in the direction of IoT system and application of this paradigm. Some of the challenges for the next future of this project are: establishment of a server system based on Data Mining and Artificial Intelligence for analyzing the coordinates of each vehicle that will be part of this application as the most important feature of it for predicting the traffic in an area. The establishment of a Database based on road signs and limits to alert/suggest the driver when is in violation of any of them. This app gives the opportunity to add other modules or sensors at any moment.

References

1. Lueth, K.L.: The 10 most popular internet of things applications right now. IoT Analytics (2015)
2. Ashton, K.: That 'internet of things' thing in the real world, things matter more than ideas. RFID J. **22**(7), 97–114 (2009)
3. Global automotive sensors market from 2015 to 2021 (in billion U.S. dollars). https://www.statista.com/statistics/675275/automotive-sensors-market-size-globally/
4. National Cable & Telecommunications Association: Broadband by the numbers. https://www.ncta.com/broad-band-by-the-numbers. Accessed 22 Apr 2015
5. Evans, M., Noble, J., Hochenbaum, J.: Arduino in Action, 1st edn. Manning Publications, New York (2013)
6. Bell, C.: Beginning Sensor Networks with Arduino and Raspberry Pi. Apress Media, New York (2013)
7. JXTA Java™Standard Edition v2.5: Programmers GuideSeptember 10th (2007)
8. Nagi, R.S.: Cartographic visualization for mobile application. In: International Institute for Geo- information Science and Earth Observation, India (2004)
9. Kuhn, A., Erni, D., Loretan, P., Nierstrasz, O.: Software cartography: thematic software visualization with consistent layout. J. Softw. Maintenance Evol. Search Pract. **22**(3), 191–210 (2010)
10. GOOGLE MAP: Develop Your Own Location-Based Services Using Google Maps [Internet]. http://www.google.com/apis/maps/
11. Proteus 8. https://www.labcenter.com/
12. Arduino in Proteus. https://www.theengineeringprojects.com/2015/12/arduino-library-proteus-simulation.html

Software as a Service (SaaS) Service Selection Based on Measuring the Shortest Distance to the Consumer's Preferences

Mohammed Abdulaziz Ikram(✉) and Farookh Khadeer Hussain(✉)

Centre for Artificial Intelligence (CAI), School of Software, University of Technology, Sydney, Sydney, Australia
{mohommed.ikram, farookh.hussain}@uts.edu.au

Abstract. Software as a Service (SaaS) is a type of cloud service that runs and operates over the Platform as a Service (PaaS), which in turn works on the Infrastructure as a Service (IaaS). In the past few years, there has been an enormous growth in the number of SaaS services. It is estimated that the revenue of SaaS services will reach US$ 112.8 billion in 2019. This growth in the number of SaaS services makes the selection process difficult for a consumer who is looking to select the best service among the many services that have similar functionalities. In this article, we propose a Find SaaS framework to select a service based on measuring the shortest distance to the consumer's preferences. In order to explain how the Find SaaS framework works, a case study based on selecting a computer repair shop's SaaS application for the consumer has been presented.

1 Introduction

Software as a Service (SaaS) is a new paradigm of software applications that can be accessed through a web browser or application program interface (API) [1]. SaaS uses cloud infrastructure and Platform as a Service (PaaS) as a base of computing resources [2]. Data stored in SaaS is highly available to the SaaS consumers. Some SaaS services (such as Salesforce) provide the consumer with the ability to customise the service to fit with their requirements. SaaS services run and operate in the service provider's datacenter [3].

There are growing numbers of software applications that use cloud infrastructure to access them via the Internet. In addition to this, some of the big vendors, such as Oracle, SAP and IBM, have recently started to provide their services over the cloud infrastructure since the cloud application has many advantages to the consumer, such as avoiding the cost of building in-house IT. So, there are growing numbers of SaaS services. The last study conducted by the IDC (International Data Corporation) shows that the SaaS cloud market made US$ 48.8 billion in revenue in 2014 and it is expected to surpass $112.8 billion by 2019 [4].

Given the variety of SaaS services available to the consumer, selecting the best service can be confusing and tedious. The literature on SaaS service selection proposes a variety of approaches, which can summarized into two categories. The first category is called studying the quality of service (QoS), and it involves studying the main factors

L. Barolli et al. (Eds.): EIDWT 2018, LNDECT 17, pp. 403–415, 2018.
https://doi.org/10.1007/978-3-319-75928-9_36

that should be taken into account when selecting a service [5–9]. The second category is considering developing intelligent methods for selecting a SaaS service [10–14].

In this paper, we propose an intelligent method to rank and select SaaS services based on measuring the shortest distance to consumer preferences. This method is based on the TOPSIS technique that was developed by Hwang and Yoon in 1981 [15]. TOPSIS is a multi-criteria decision-making (MCDM) method that can find an optimal solution using two techniques: (1) determining the shortest distance from the positive-ideal solution; and (2) determining the longest distance from the negative-ideal solution. In our proposed method, we will rank and select a service based on measuring the shortest distance to the consumer-driven requirements, replacing the positive-ideal solution in TOPSIS with consumer-driven requirements. Therefore, the result of our proposed method will rank and sort the SaaS services, beginning with the service that has the shortest distance to the consumer's preferences. For evaluation purpose, a case study will be proposed based on selecting a computer repair shop's SaaS application from six services that have the same features.

The remainder of the paper is organized as follows. Section 2 discusses related works. The 'Find_SaaS' framework is presented in Sects. 3 and 4 introduces a case study to explain how the proposed framework works. Section 5 discusses the results of the case study and expected shortcomings and, finally, Sect. 6 concludes the paper with a summary.

2 Related Works

Given the dramatic growth in the number of services available to consumers, selecting a cloud SaaS service can be a difficult and tedious job. Selecting a SaaS service is based on functional and non-functional requirements. First, the consumer chooses the functional service that matches their requirements such as customer relationship management (CRM), enterprise resource planning (ERP) or hospital management software. Then, the service consumer will face the problem of selecting a service with different non-functional factors such as price, reputation and availability, which is known as the quality of service (QoS). In the literature, SaaS service selection is based on two approaches: (1) selecting a service based on studying the most important quality of service (QoS) that a consumer should consider when selecting a service [5–9]; and (2) selecting a service based on developing an intelligent method to select a service [10–14].

There are some studies that examine selection services based on studying the most important QoS that a service consumer should consider. Burkon [7] establishes a set of QoS attributes to manage SaaS cloud services. The study compares the SaaS characteristics with other types of software application. Therefore, the author concludes the main QoS that are used for managing a SaaS service are as follows: availability, performance, reliability, scalability, security, support, interoperability, modifiability, usability and testability. In addition to this, Repschlaeger et al. [16] propose a selection criteria for SaaS services to help consumers find an optimum service. The authors followed a 'Design Since' research approach and conducted a systematic literature review. They carried out an extensive market analysis of 651 SaaS providers and found that there are six criteria for selecting a SaaS provider: maintenance and service cycle, functional coverage, service category, user scaling, the probability of date and browser compatibility. Moreover, Al-Shammari and Al-Yasiri [5] provide valuable information on how to quantify the QoS in order to help a SaaS consumer and SaaS service provider obtain a service level agreement (SLA). However, most of the previous studies do not take into account the quantitative QoS that are available to the consumer when selecting the service. For example, the SaaS consumer may use some web monitoring tools in order to gain important factors that help to select an optimum service such as throughput, response time and availability.

Other works consider proposing an intelligent method to select a SaaS service and consider the important QoS that should be taken into account when making a SaaS service selection. The analytical hierarchy process (AHP) has been widely used to weight consumer requirements and to select a service. Godse and Mulik [11] propose an approach to weight SaaS criteria using the AHP format and then scored each service to make a selection. They consider the following criteria: functionality, architecture, usability, vendor reputation and cost. The authors present a case study to verify their approach based on sales force automation (SFA). In addition to this, Boussoualim and Aklouf [12] propose a method to weight the QoS of SaaS by using the AHP format in order to help SaaS consumers select the best service. They start by investigating the QoS that SaaS consumers consider when selecting a service. The authors considered the following criteria: reputation, cost, usability, structure, configurability and personalisation. Moreover, Limam and Boutaba [14] propose a framework that is reputation-aware when making a SaaS service selection. The approach proposed is based on aggregating the quality, cost and reputation parameters into a single metric in order to evaluate each service in order to make a selection. Also, Sundarraj [13] proposes a service selection framework of SaaS enterprise resource planning (SERP). The framework is based on two stages, which are studying the enterprise resource planning (ERP) characteristics and the SaaS characteristics. Finally, Badidi [10] proposes a framework for cloud service selection at the SaaS level. The proposed approach uses the concept of a cloud service broker (CSB) as a mediator service that helps to understand the user's requirements and then select the most appropriate service for the consumer. The framework is based on matching the user's requirements with the SaaS service providers and then ranking the services using an aggregate utility function. However, there is no work that considers service selection based on measuring the shortest distance to the consumer's preferences for ranking and selection a service.

This work proposes a SaaS service selection framework called Find_SaaS that is based on the consumer's preferences. Find_SaaS is composed of three main components. Firstly, the consumer's preferences process aims to collect all of the consumer's requirements regarding functional and non-functional service types. Secondly, the service registry repository aims to gain all of the QoS attributes in order to make a selection. In addition to this, a new QoS classification is proposed for making a decision. This classification is based on quantitative QoS and has been classified into three categories: monitoring tool QoS, service provider QoS and reputation QoS. Finally, the SaaS service selection component aims to select the best service based on the consumer's preferences. The selection process is based on the Technique for Order of Preference by Similarity to Ideal Solution (TOPSIS). However, in this work, we will make a service selection based on the shortest distance to the consumer-driven requirements rather than the positive-ideal solution that TOPSIS uses to make a decision. In the next section, we present and discuss the Find_SaaS framework.

3 Find SaaS Framework

This framework offers an intelligent way for SaaS consumers to select an optimum service among the huge number of similarly functional SaaS services satisfying their requirements. The main goal of the framework is to provide a ranking and select a service based on the shortest distance to the consumer's preferences regarding functional and non-functional parameters. The proposed framework is depicted in Fig. 1 and its main components are as follows:

(1) Consumer's preference processor
(2) Services registry repository
(3) SaaS selection processor.

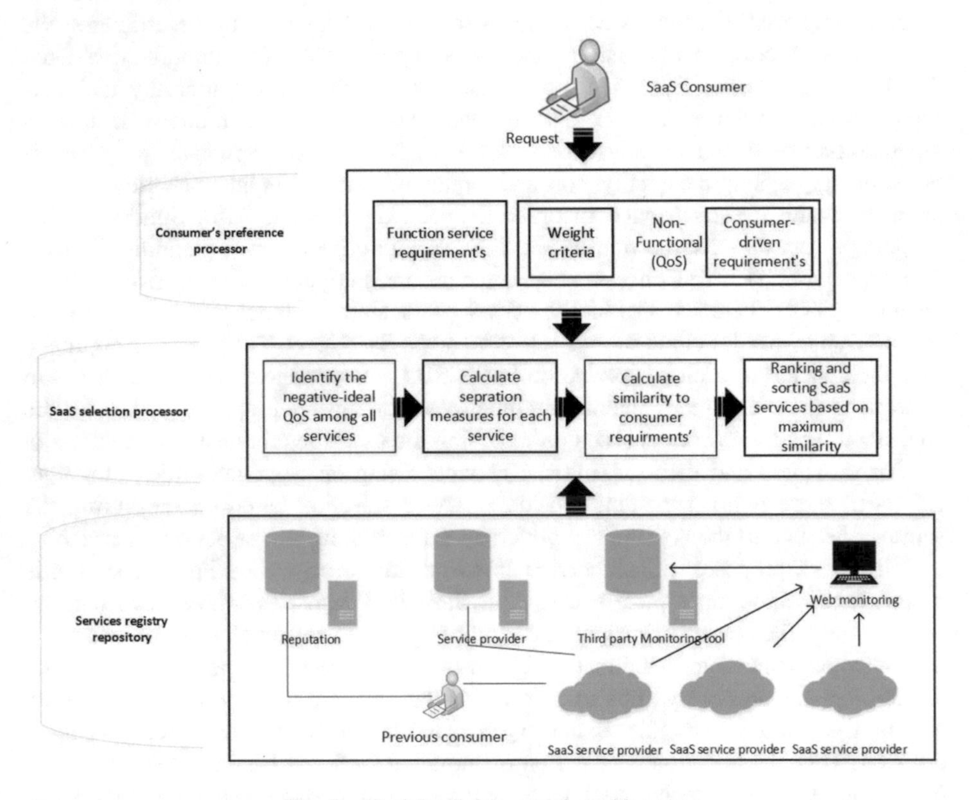

Fig. 1. Find SaaS framework architecture

3.1 Consumer's Preference Processor

The main aim of this component is to collect the consumer's preferences regarding the functional and non-functional components of a service. Functional services refer to the job of a service or what a service provider offers to the consumer. For example, Salesforce (www.salesforce.com) offers a CRM SaaS solution to their consumers. Non-functional services refers to the QoS that describes the characteristics of the service such as price and availability. In this proposed framework, the non-functional requirements have been divided into two parts:

I. ***Weighting criteria:*** This describes the priority of each QoS in comparison to other qualities. For example, if the price of service has a higher priority than other criteria such as availability and reputation, a consumer can weight their preferences of the price for 50% compared with other criteria such as reputation and availability.
II. ***Consumer-driven requirements:*** This refers the quantitative value of a quality that describes the consumer preferences for each criteria. For example, a consumer may need a SaaS service with a price of US$ 200 per month.

3.2 Services Registry Repository

This part of the Find_SaaS framework provides the pool of SaaS services that are available to the consumers. The selection of a SaaS service is based on the functional and non-functional components of the service. Therefore, each functional service, such as CRM category, has a number of SaaS services with similar functions and each SaaS service has a number of criteria (or QoS) that define the characteristics. However, we developed a new model of QoS that is available and easy to measure for the consumer. The quality of service is classified as outlined in the following steps.

I. ***Service provider:*** This refers to the non-functional quality attributes a service provider usually declares when they launch a service. It is composed of the following criteria: price and features. Price is the amount of money that a consumer will pay for using a service. The price of a SaaS service is a quantitative QoS and one of the most important criteria for selecting a SaaS service. A SaaS service provider usually launches a service to the consumer which is free for a month and then the consumer can decide whether or not to continue with that service. The price of the SaaS service is usually a monthly or annual payment. Also, the price of the service depends on the number of users and sometimes a service provider has an offer for consumers who need a service for more than one user. Features include a variety of qualitative QoS such as the location of the SaaS service, multi-language support and the date of establishment. Our proposed framework considers only the quantitative QoS.
II. ***Third party monitoring tool:*** This includes some qualities that can be collected by using a web monitoring tool available to the public. A number of web monitoring tools are available on the Internet such as up-down, statusOK and site24x7. All of these services are available by subscription and can be used to monitor some resources on the Internet. The majority of monitoring tools look at attributes such as availability, response time and throughput. This research

considers three attributes collected from third party monitoring tool: (1) availability (AV), which refers to the number of successful invocations/total invocations and used percentage as a unit of measure; (2) response time (RT), which relates to the time taken to send a request and receive a response (ms is used as the unit of measure); and (3) throughput (TH), which refers to the number of invocations for a given period.

III. ***Reputation:*** This refers to feedback from previous consumers. Since it can be difficult to obtain QoS user feedback due to privacy and security, the reputation rate refers to the average review rate a SaaS service obtained from other consumers.

3.3 SaaS Selection Processor

The main aim of this component is to rank and sort all of the SaaS services to the consumer after matching the functional requirement of the service. Therefore, this component helps a consumer to select an optimum service for them based on their requirements out of a variety of services that all have similar functional requirements. This research work aims to develop a ranking of SaaS services based on the TOPSIS technique. TOPSIS is an MCDM method that can find an optimal solution using two techniques: (1) determining the shortest distance from the positive-ideal solution; and (2) determining the longest distance from the negative-ideal solution. In addition to this, we will modify the TOPSIS method to add consumer-driven requirements, which are the minimum QoS that a consumer needs. We will also rank and sort services based on the similarity to the consumer's preference. We will enhance criteria decision-making by adding consumer requirements in each QoS; for example, if a consumer wants to select a SaaS service that has a maximum monthly cost of US$ 200 and their priority to meet to this price makes up 50% of their overall decision-making. Finally, the TOPSIS method will be used in order to find the best SaaS service that is most similar to the consumer's preferences.

Step 1: Calculate normalized ratings.

Equation (1) represents the vector normalization for the service attributes:

$$r_{ij} = \frac{x_{ij}}{\sqrt{\sum_{i=1}^{m} x_{ij}^2}} \quad i = 1, \ldots, m, \text{ and } j = 1, \ldots, n \tag{1}$$

Equation (2) represents the vector normalization for the price attribute:

$$r_{ij} = \frac{1/x_{ij}}{\sqrt{\sum_{i=1}^{m} 1/x_{ij}^2}} \quad i = 1, \ldots, m, \text{ and } j = 1, \ldots, n \tag{2}$$

Step 2: Calculate the weighted normalized ratings.

Equation (3) represents the weighted normalised ratings for all service attributes:

$$v_{ij} = w_j r_{ij} \quad i = 1, \ldots, m, \text{ and } j = 1, \ldots, n \tag{3}$$

Where w_j is the weight of the jth attributes.

Step 3: Identify the positive-ideal and negative-deal solutions.

In our approach, we will consider the positive-ideal solution as the consumer-driven requirements. Therefore, our method will rank and sort the services beginning from the service that has the most similar requirements to the consumer's preferences.

$$cr^* = \{\mathrm{cr}_1, \mathrm{cr}_2, \ldots\ldots\ldots, \mathrm{cr}_\mathrm{j}, \ldots\ldots\ldots\ldots, \mathrm{cr}_\mathrm{n}\} \tag{4}$$

Equation (4) represents the vector of consumer-driven requirements, where cr_j is the value for the jth attributes that define the consumer-driven requirements.

$$\begin{aligned} A^- &= \{v_1^-, v_2^-, \ldots\ldots\ldots, v_j^-, \ldots\ldots\ldots\ldots, v_n^-\} \\ &= \{(\min_i v_{ij} | j \in J_1), (\max_i v_{ij} | j \in J_2) | i = 1, \ldots\ldots, m\} \end{aligned} \tag{5}$$

Equation (5) represents the worst value for the jth attributes among all services.

Step 4: Calculate separation measures:

$$S_i^* = \sqrt{\sum\nolimits_{j=1}^{n} (v_{ij} - \mathrm{cons}_\mathrm{j})^2} \quad i = 1, \ldots, \mathrm{m}, \text{and } j = 1, \ldots., \mathrm{n} \tag{6}$$

Equation (6) represents the distance between each service from the consumer-driven requirement:

$$S_i^- = \sqrt{\sum\nolimits_{j=1}^{n} \left(v_{ij} - v_j^-\right)^2}, \quad i = 1, \ldots, \mathrm{m}, \text{and } j = 1, \ldots., \mathrm{n} \tag{7}$$

Equation (7) represents the distance between each service from the negative-ideal solution:

Step 5: Calculate similarities to consumer-driven requirements.

$$C_i^* = \frac{S_i^-}{S_i^* + S_i^-}, \quad i = 1, \ldots, \mathrm{m} \tag{8}$$

Equation (7) represents the measurement of similarity to the consumer-driven requirements. The result of this equation will be between 0 and 1. In the case that a similarity goes to 0, that means a service has a similarity to a negative-ideal solution, and, at the same time, if the similarity goes to nearly 1, that means the service has a maximum similarity to consumer-driven requirements.

Step 6: Rank the services that have the maximum $\underline{\mathrm{C}_\mathrm{i}^*}$.

Finally, the system will rank and sort the services based on the service that has a maximum value of similarity to the consumer's requirements.

4 Case Study

The purpose of the case study is to illustrate the process of using the Find_SaaS framework to select a SaaS service based on the consumer's preferences. Our case study includes six different SaaS services in the computer repair shop category that have been collected from the Capterra website (www.capterra.com). The Capterra website provides a variety of software applications to help individuals and businesses find software solutions. In addition to this, there are around 17 cloud SaaS services in the computer repair shop SaaS category. The reason that we chose those six services was that these services have the most selected to use comparing with other services and these services also have a number of reviews that will help us to use the reputation factor to select the right SaaS services for the consumer. Moreover, in order to gain the attributes from a web monitoring tool, we used site24x7 (www.site24x7.com) to collect the response time (RT), availability (AV) and throughput (TH). Site24x7 is a web monitoring tool that is able to monitor resources on the Internet. It provides a variety of services such as synthetic monitoring, FTP monitoring and SSL certificate monitoring. We monitored these six service from 21 September 2017 to 4 October 2017. Table 1 shows the six SaaS services in the computer repair shop category. It contains the services' names and the web service link as well as a number of criteria classified into three categories: service provider, reputation and monitoring tool.

A SaaS service provider offers their services with a price to use a service and the features. In terms of the price of the SaaS service, there are different classes of prices which are dependent on the number of users, the payment subscription (i.e. monthly or annually), features and support. We also consider the price of service per month and per user. In terms of features, all six of these services provide different features such as human resources (HR), CRM, billing services and more. Additionally, we consider factors such as the service location and the date of establishment that customers may take into account when selecting a service for further research work. The reputation of a service is based on the previous consumers' feedback reviews. As seen as in Table 1, the number of reviews is not fixed, which means that some services have more reviews in comparison to other services. Therefore, we proposed a previous-based reputation score which is measured via Eq. (9).

$$Reviewer - based\ reputation\ score = \frac{number\ of\ reviewers\ for\ the\ service}{Total\ number\ of\ reviewers} * review\ average \quad (9)$$

Finally, the three criteria that have been collected from a third party monitoring tool are used: availability (AV), response time (RT) and throughput (TH). However, the throughput factor does not appear for some services due to the short period the services were monitored for.

Table 2 shows an example of the input data for the Find_SaaS proposed framework to select a SaaS service among these six services. It contains a consumer's preferences of functional and non-functional (or QoS) of service. The consumer enters a computer repair shop SaaS service as a functional service. We will also consider five criteria for

Table 1. Six SaaS services in the computer repair shop category

Service number	Service name	SaaS website link	Number of reviewers	Review average	Reviewer-based reputation score	Service location	Date established	Service price	Availability (AV)	Response time (RT)	Throughput (TH)
1	Repair Shopr	http://www.repairshopr.com/	47	5	3.916666667	USA	2010	9.999	100	370	0
2	Repair Pilot	http://www.repairtrackingsoftware.com	5	3.5	0.291666667	UK	N/A	7.94271	99.91	1307	0
3	Open RMA	https://www.openrma.com/	3	4.5	0.225	Greece	2016	82.1788	100	2,709	0
4	RSRS	https://www.synolonsoft.com	3	4.5	0.225	Cyprus	2010	59.3047	100	7123	0
5	Busy Bench	https://busybench.com/	1	5	0.083333333	USA	2015	9.999	100	969	74
6	Cell Store	https://cellstore.co/	1	5	0.083333333	Canada	2012	29.99	100	660	121

selecting a SaaS service. There are two options to understand the consumer preferences of the non-functional criteria: firstly, consumer-driven requirements, which consider the value of QoS that a consumer needs; and secondly, weighting the QoS of criteria to rank and sort the services for a consumer.

Table 2. SaaS consumer requests

Consumer's preferences					
Functional	Computer repair shop SaaS services				
Non-functional (QoS)	Price	Reputation	AV	RT	TH
Consumer-driven requirements (CR)	$20	5 score	100%	20 ms	20 ms
Consumer weighting (w)	70%	20%	5%	2.5%	2.5%

Firstly, based on Eq. (1), we calculate a normalized rating vector for all services. Consumer-driven requirements are included since will be taking them into account as a positive-ideal solution.

Normalization	Price	Reputation	AV	RT	TH
Repair Shopr	0.500679549	0.615137405	0.378013056	0.047259154	0
Repair Pilot	0.630300591	0.045808105	0.377672844	0.166939767	0
Open RMA	0.060919541	0.035337681	0.378013056	0.346013642	0
RSRS	0.084416493	0.035337681	0.378013056	0.909802572	0
Busy Bench	0.500679549	0.01308803	0.378013056	0.123767892	0.5166243
Cell Store	0.166932138	0.01308803	0.378013056	0.084300112	0.8447505
CR	0.250314741	0.785281793	0.378013056	0.002554549	0.1396282

Secondly, based on Eq. (2), we calculate weighted normalized ratings for all services.

Weighting normalizing	Price	Reputation	AV	RT	TH
Repair Shopr	0.350475684	0.123027481	0.018900653	0.001181479	0
Repair Pilot	0.441210414	0.009161621	0.018883642	0.004173494	0
Open RMA	0.042643679	0.007067536	0.018900653	0.008650341	0
RSRS	0.059091545	0.007067536	0.018900653	0.022745064	0
Busy Bench	0.350475684	0.002617606	0.018900653	0.003094197	0.0129156
Cell Store	0.116852496	0.002617606	0.018900653	0.002107503	0.0211188
CR	0.175220318	0.157056359	0.018900653	6.38637E-05	0.0034907

Thirdly, we identify negative-ideal solutions for each criteria to rank and sort the services beginning from the SaaS service that has maximum similarity to consumer requirements and ending with the SaaS service that is the furthest away from consumer requirements.

QoS	Price	Reputation	AV	RT	TH
Negative-ideal	0.042643679	0.002617606	0.018883642	0.001181479	0

Fourth, we calculate the separation measures or distance via Eqs. (4) and (5) for each service to the consumer-driven requirements, which is the positive ideal solution, and to the negative-ideal solution.

S1+	0.178566071	S1−	0.330543616
S2+	0.304388993	S2−	0.398631683
S3+	0.200397417	S3−	0.008694025
S4+	0.191073705	S4−	0.027483135
S5+	0.233802873	S5−	0.308108772
S6+	0.166051365	S6−	0.077160926

Fifth, we calculate the similarity to the consumer-driven requirements via Eq. (8) in order to rank and sort all of the services that have the most matching similarity to the consumer's requirements.

C1	0.649258154
C2	0.567026969
C3	0.041580013
C4	0.125748225
C5	0.568559054
C6	0.317257510

Finally, all services are ranked from the service that has the highest similarity number to the service that has the lowest similarity number. Based our proposed framework, the Repair Shopr is the best SaaS solution for the consumer's requirements based on the request entered.

Final ranking of SaaS services based on the consumer preferences		Values of similarity to the consumer-driven requirements
C1	Repair Shopr	0.649258154
C5	Busy Bench	0.568559054
C2	Repair Pilot	0.567026969
C6	Cell Store	0.31725751
C4	RSRS	0.125748225
C3	Open RMA	0.041580013

5 Results and Discussion

The dataset of the case study is collected from the computer repair shop category on the Capttera website (www.capterra.com). There may be other SaaS services that have a similar functional type, since there is a lack of service registry on the cloud SaaS services which uses web service technology for service discovery. However, we recommend that the consumer could choose some services and then apply our method in order to find out which is the correct SaaS service for their requirements. This case study has also used site24x7 as a monitoring tool to collect some attributes to contribute to the selection process. Web monitoring tools such as site24x7 use multiple servers to ping the services that have been selected. However, all of these servers are located in the USA, including site24x7, which is why the response time for the services is less in the USA in comparison with other services. Therefore, we recommend that the consumer choose a monitoring tool that is in the same place that will deploy the service. The future direction of this research field may take into account other important attributes such as service location, supporting multiple languages and vendor history.

The proposed framework uses the quantitative QoS to make a selection and rank the services for the consumer. However, there are other important factors that contribute to the selection such as user interface and support. Since SaaS providers usually offer one free month to test and evaluate a service before making a decision, we should modify our framework to take into account additional factors based on the consumer's perspective of the selected services. Also, the consumer's request may contain some uncertainty in the request. For example, the consumer may search for a service with a low, medium, or high price. All these features will be considered in future research. Finally, when the consumer requests a service, the frame could consider a minimum and maximum value of attributes; for example, a price range of between US$ 20 to US $ 30. All these shortcomings will be considered in the future direction of the SaaS service selection.

6 Conclusion

Cloud SaaS service is a new type of software application that uses a cloud infrastructure to build their service. There are a growing number of cloud SaaS services, which means that selecting the best service can be a difficult and tedious task for consumers. This article presents the Find_SaaS framework, which can help a SaaS consumer select the best service for them based on their preferences. Our framework is based on three components: consumer's preferences processor, service registry processor and SaaS service selection processor. The selection process of this framework is based on measuring the shortest distance to the consumer-driven requirements. A case study has been presented in order to demonstrate how the framework works. This case study selects the best computer repair shop SaaS application from six services that have the same characteristics.

Further research will be conducted focusing on the consumer preferences regarding linguistic uncertainty requirements and maximum and minimum values of consumer-driven requirements. Also, we will evaluate and test our method by using a QWS dataset published by Al-Masri and Mahmoud [17] due to the inability to find a real SaaS dataset for testing and evaluating the new proposed method.

Acknowledgement. This research is supported by a scholarship from Umm Al-Qura University in Saudi Arabia.

References

1. Armbrust, M., et al.: A view of cloud computing. Commun. ACM **53**(4), 50–58 (2010)
2. Tsai, W., Bai, X., Huang, Y.: Software-as-a-service (SaaS): perspectives and challenges. Sci. Chin. Inf. Sci. **57**(5), 1–15 (2014)
3. Tyrväinen, P., Selin, J.: How to sell SaaS: a model for main factors of marketing and selling software-as-a-service. In: Regnell, B., van de Weerd, I., De Troyer, O. (eds.) Software Business: Second International Conference, ICSOB 2011, Brussels, Belgium, June 8–10, 2011, Proceedings, pp. 2–16. Springer, Heidelberg (2011)
4. Benjamin McGrath, R.P.M.: Worldwide SaaS and Cloud Software 2015–2019 Forecast and 2014 Vendor Shares (2015)
5. Al-Shammari, S., Al-Yasiri, A.: Defining a metric for measuring QoE of SaaS cloud computing. In: Proceedings of the 15th Annual PostGraduate Symposium on the Convergence of Telecommunications, Networking and Broadcasting (PGNET 2014) (2014)
6. Jagli, D., Purohit, S., Chandra, N.S.: Evaluating service customizability of SaaS on the cloud computing environment (2016)
7. Burkon, L.: Quality of service attributes for software as a service. J. Syst. Integr. **4**(3), 38 (2013)
8. Ghani, A.A.A., Sultan, A.B.M.: SaaS quality of service attributes. J. Appl. Sci. **14**(24), 3613–3619 (2014)
9. Monteiro, L., Vasconcelos, A.: Survey on important cloud service provider attributes using the SMI framework. Procedia Technol. **9**, 253–259 (2013)
10. Badidi, E.: A framework for software-as-a-service selection and provisioning. arXiv preprint arXiv:1306.1888 (2013)
11. Godse, M., Mulik, S.: An approach for selecting Software-as-a-Service (SaaS) product. In: 2009 IEEE International Conference on Cloud Computing (2009)
12. Boussoualim, N., Aklouf, Y.: Evaluation and selection of SaaS product based on user preferences. In: 2015 Third International Conference on Technological Advances in Electrical, Electronics and Computer Engineering (TAEECE) (2015)
13. Sundarraj, R.: A selection framework for SaaS-based enterprise resource planning applications (2011)
14. Limam, N., Boutaba, R.: Assessing software service quality and trustworthiness at selection time. IEEE Trans. Softw. Eng. **36**(4), 559–574 (2010)
15. Yoon, K.P., Hwang, C.-L.: Multiple Attribute Decision Making: An Introduction, vol. 104. Sage Publications, Thousand Oaks (1995)
16. Repschlaeger, J., et al.: Selection criteria for software as a service: an explorative analysis of provider requirements (2012)
17. Al-Masri, E., Mahmoud, Q.H.: QoS-based discovery and ranking of web services. In: 2007 Proceedings of 16th International Conference on Computer Communications and Networks, ICCCN 2007 (2007)

Efficient Content Sharing with File Splitting and Differences Between Versions in Hybrid Peer-to-Peer Networks

Toshinobu Hayashi[1] and Shinji Sugawara[2(✉)]

[1] Nagoya Institute of Technology, Nagoya 466-8555, Japan
toshinobu@sugawara-lab.org
[2] Chiba Institute of Technology, Narashino 275-0016, Japan
shinji.sugawara@it-chiba.ac.jp

Abstract. This paper proposes an efficient content sharing strategy using file splitting and difference between versions in hybrid Peer-to-Peer (P2P) networks. In this strategy, when a user requests a content item, he/she can get it from the network by retrieving the other version of the content item and the difference from the requested version, if the obtaining cost of the requested version is expensive. This way of content sharing can be expected to accomplish effective and flexible operation. Furthermore, efficient utilization of a peer's storage capacity is achieved by splitting each replica of a content item into several small blocks and storing them separately in the plural peers.

1 Introduction

In recent years, along with rapid development of computers and broadbandization of networks, content sharing using peer-to-peer (P2P) network has been actively researched [1]. In a P2P network, each terminal (peer) forming the network plays roles of both server and client against other peers, so that accumulation of shared information and concentration of accesses on a specific peer rarely occur, it is possible to distribute such loads on the network, and since multiple peers can provide the same service, it has excellent fault tolerance.

In content sharing using a P2P network, there is a possibility that due to a peer's failure or dropping off from the network, the content held by that peer can not be referred. As a countermeasure against this, a method of placing replicas of a content on plural peers is used. By this method, since one content can be referred from plural peers, even when a certain content holding peer can not be accessed, the possibility of referring to the same content from another peer can be enhanced. In addition, since plural peers hold a replica of the same content, it is possible to distribute accesses to the content to plural peers. Furthermore, because the possibility that a replica of the requested content exists near the content requesting peer becomes high, the load on the network can be suppressed [2,3].

L. Barolli et al. (Eds.): EIDWT 2018, LNDECT 17, pp. 416–428, 2018.
https://doi.org/10.1007/978-3-319-75928-9_37

In an environment where there exist plural replicas of a content on the network, when the content's update occurs at a peer, a situation occurs in which both old and new replicas exist at the same time. Thereafter if only updated content is used, it is necessary to quickly replace every replica with updated one so as to maintain consistency of the content [4]. On the other hand, if users need both old and new versions of the content, holding replicas of the two as they are requires a lot of storage and efficiency of the content sharing might be reduced. Therefore, in this case, it is conceivable to efficiently share the content in which there exist multiple versions by holding either version and the difference between the two, assuming the size of the difference is smaller than that of another version.

Various studies on content sharing [5–7] or version management [8,9] in P2P networks have been conducted so far. However, consideration on content sharing in environments where multiple versions of the same content exist has not been investigated sufficiently. Therefore in this paper, we propose a method to share content efficiently using the difference between versions of a content in hybrid P2P network under the environment where multiple versions of content replicas are shared. In addition, by dividing a content replica of each version into plural blocks and storing them in each peer, effective utilization of storage is also planned.

In the following, firstly we will briefly explain related researches in Sect. 2 and describe the environment assumed and the problem to be solved in this paper in Sect. 3. Then we will explain the proposed method in Sect. 4, evaluate the proposed method in Sect. 5, and lastly conclude this paper in Sect. 6.

2 Related Work

[5] written by an author is one of the researches on content sharing using P2P network, and the group including authors have proposed efficient replication deployment method so far. In this method, each peer holding a replica has the replica's reference range and there is a peer requesting the replica (hereinafter referred to as a requesting peer) outside the range, the requesting peer acquires the replica and uses it, and in addition, holds it in its storage and provides to the other peers. On the other hand, when the replica requesting peer is included in the reference range, it simply refers to or uses that replica and no replica is newly stored anywhere. In addition, this method repeatedly tries as necessary to control each replica to be located at an appropriate position on the network so that each replica always stays on the most advantageous peer for content sharing.

In Bit Torrent [10–13], each content is divided into smaller pieces of 256 kB and distributed in the network. When a content request occurs, the content requesting peer downloads pieces of a content replica from many peers and immediately uploads them to the other peers. This makes it possible for many peers to keep pieces of each content and to maintain peers' loads distribution, high speed downloading of content replicas, and equality of contribution to content sharing among peers.

However, these methods do not take into account situations where there exist multiple versions of a single content. In practice, such a situation can occur, and in that case it is inappropriate to keep all versions of the contents as they are from the viewpoint of effective use of storage.

3 Assumed Environment and Problem to Be Solved

In this paper, we assume an environment in which each peer requests a content held in the storage of another peer as its replica and can acquire it in a hybrid P2P network. In addition, we set some other assumptions as follows.

- Each peer provides its self-decided sized storage capacity to the content sharing system, and the system can freely use this for content sharing.
- There is no substantial difference between the original of each content and its replica.
- A replica of a content which is completed itself is called a full object, and a replica of a certain version can be reproduced from a full object replica of another version patched with the difference data between the two versions.
- A replica can be divided into small blocks and be stored in different peers separately.
- Each peer always may leave the network or join again according to the user's decision or the network environment.

The costs required for content sharing assumed in this paper is as follows.

(a) *Network Cost*
It is a load that occurs in the network when referring, deploying, and relocating contents, and in this paper it is assumed to be proportional to the product of the capacity of contents and the moving distance (number of hops) on the network.

(b) *Contents Loss*
It is the cost that occurs when specific content in the network disappears due to peer dropping out of the network or replicas' deletion and the content requested by the user can not be acquired (the user's purpose could not be achieved). In this paper, It is defined as the total number of times that the users could not obtain the requested contents' replicas, divided by the total number of times that the users requested contents.

In this paper, we will pursue a content sharing method that minimizes network cost and contents loss in the above environment.

4 Proposed Method

Let the set of peers ranging within a threshold $H_{\mathrm{th}}(\geq 1)$ hops from a peer be the reference range of the peer. Each peer becomes a peer having metadata of its possessing contents' replicas (hereinafter referred to as a replica management peer) in its initial state. A replica management peer manages full object replicas and differences related to the content to be managed within its own reference range. A newly joining peer shall be a replica management peer of the content held by the peer at the time of joining the network.

Each content replica is divided into plural blocks as shown in Fig. 1, and each block consists of a combination of a certain version's full object and some differences from the full object; the configuration of a block is different from others in general. The number of times of reference by users is counted for each version of each block, and the usefulness of each version is evaluated accordingly in each content. In the proposed method shown below, the usefulness of each version in each block of a content replica is defined in each peer as the reference frequency of each version in each block of the content in the past certain period $T(> 0)$.

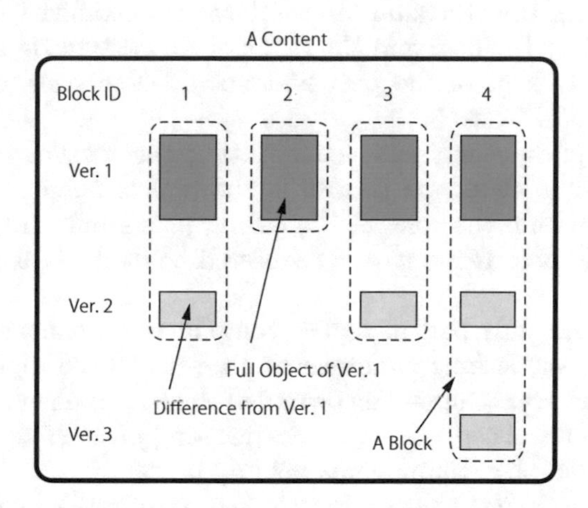

Fig. 1. An example of a content configuration

The proposed method consists of the following five procedures.

1. Procedure at the time of content request
2. Procedure at the time of content update
3. Procedure at peer leaving the network
4. Procedure for determining the destination of replica relocation
5. Procedure to change the version of replica to be kept as a full object

Procedures 1 through 3 are activated by a content request from a user, a content update, and a peer's leaving from the network, respectively. Procedure 4 is activated upon occurrence of a situation in which there is not enough capacity in the storage of the peer supposed to be placed a difference or a full object of a replica. Procedure 5 starts periodically every T.

When a peer joins in the network, if a replica management peer which manages the same content the joining peer holds exists within the reference range of the joining peer, the joining peer keeps possessing only the differences the replica management peer does not manage, and discards the other replica components. Thereafter, the replica management peer manages the difference.

Procedure at the time of content request

When a content request occurs at a peer, the following procedure is executed. However, the requested content is distinguished until the level of its version.

1. If the replica management peer of the requested content is not included within the reference range of the content requesting peer, go to 2. Otherwise go to 3.
2. If the storage capacity of the content requesting peer is available, a set of replica components (a full object and a difference) constituting the requested version is allocated to the requesting peer, the management authority of the content is given to the requesting peer, and the procedure is terminated.
   If there is no vacancy in the storage, the content requesting peer is given the management authority of the content, and the full objects and the differences are selected such that the sum of usefulness is maximized from the replicas originally held by the peer and the replica being assigned. The selected ones are continuously kept in the peer and not selected ones are treated to be relocated replicas. Go to 5.
3. If a set of replica components constituting the version requested by the requesting peer is under the control of the replica management peer of the same content within the content requesting peer's reference range, the content requesting peer refers to the requested content via a peer holding the replica.
   In the case where only part of replica components is managed by the replica management peer, lacking components are sent to the replica management peer so that the peer temporarily owns sufficient components to provide replicas of all versions. Then the content requesting peer refers to the requested content via the replica components holding peers.
4. The replica management peer relocates the set of replica components of the content to the neighboring peers according to the following sub-procedure. However, all the peers that exist within the reference range of the replica management peer except the peers that hold elements of the same content's components managed by another replica management peer are the candidates for relocation destinations.
   (i) It is assumed that one of the candidate peers is selected and a full set of replica components constituting the content is placed there.
   (ii) For each candidate peer, calculate (the number of hops up to the nearest replica of the content) $\times$ (the number of times that the content was referred to by the peer in the past), and find the sum of them as C.
   (iii) Change the peer that is assumed to be placed the content replica to another candidate peer and repeat the above processes of (i) and (ii) until getting the values of C for every case where a different candidate is assumed to be placed the replica.
   (iv) Specify the peer that minimizes C when being placed the content, and this peer is set as the relocation destination peer of the content.
   (v) If there is space in the storage capacity of the relocation destination peer, the full set of content replica components and its management authority

are transferred to the destination peer, and this procedure is terminated. If there is no vacancy, temporarily transfer the full set of content replica components and its management authority to the destination peer, and then select full objects and differences and keep them so that the sum of the usefulness in the storage is maximized. What is not selected is the content to be relocated. Go to 5.

5. Call another procedure 4 (to be described later: determining the destination of replica relocation). When the process returns here from the procedure, this procedure is terminated.

Procedure at the time of content update

When an update occurs in the content owned by a peer, the following procedure is executed.

1. A difference from the last to the latest version is transmitted to peers, that are not included in the reference range of the content updating peer and also are replica management peers of the updated content.
2. If there is space in the storage capacity of the peer that received the difference, the difference is placed in the peer, and this procedure is terminated.
   If there is no vacancy, select full objects and differences and keep them so that the sum of usefulness in the storage is maximized. What is not selected is the content to be relocated. Go to 3.
3. Call procedure 4 (determining the destination of replica relocation). When the process returns here from the procedure, this procedure is terminated.

Procedure at peer leaving the network

When a peer on the network leaves, the following procedure is executed.

1. In the case where a leaving peer holds a replica which is not managed by itself, If there is space in the storage of the management peer of the replica, the replica is supplemented from other peers on the network.
   If there is no vacancy, select full objects and differences from the replicas to be supplemented and ones possessed by the management peer and keep them so that the sum of the usefulness in the storage is maximized. What is not selected is the content to be relocated. Call procedure 4 (determining the destination of replica relocation). When the process returns here from the procedure, go to the next step.
2. In the case where a leaving peer holds a replica which is managed by itself, if the number of times that the replica is referred to by the peers within the leaving peer's reference range occupies a certain percentage C_{sup} or more of the total number of reference of the replica by all the peers within the same reference range, a neighboring peer of the leaving peer is newly set as a management peer of the replica.
   Otherwise, this procedure is terminated.

3. If there is space in the storage of the new management peer, the replica components that the leaving peer had are supplemented from the other peers on the network, and this procedure is terminated.
   If there is no vacancy, select full objects and differences from the replicas to be supplemented and ones possessed by the management peer and keep them so that the sum of the usefulness in the storage is maximized. What is not selected is the content to be relocated. Call procedure 4 (determining the destination of replica relocation). When the process returns here from the procedure, this procedure is terminated.

Procedure for determining the destination of replica relocation

The relocation destination candidates of the replica component to be relocated (i.e., full object or difference) are determined according to the following procedure. Here, all the peers included in the reference range of the management peer of the replica to be relocated correspond to the relocation destination candidates. However, in order to prevent different replica management peers from holding the same content, peers holding part of the same content managed by the other management peers are excluded from the relocation destination candidates. Note that C_{mov} is a complex variable that stores the content ID (i.e., unique identification number of the content, block, and version, as well as distinction between full object and difference) of each replica component to be relocated. The initial value of the variable is "NULL."

1. Among the candidate peers, if there are peers which have space in their storage, it is assumed that the relocating component is placed on one of the peers. If there is no do such peer, go to 4.
2. For all candidate peers, calculate (the number of hops up to the nearest replica component) $\times$ (the number of times that the component was referred to by the peer in the past), and find the sum of them as C.
3. Change the peer that is assumed to be placed the replica component to another candidate peer with space in its storage and repeat the above processes of 1 and 2 until getting the values of C for every case where a different candidate with space is assumed to be placed the replica component.
   After that, specify the peer that minimizes C when being placed the component, and the component is actually placed on this peer. Then, this procedure is terminated.
4. If $C_{\mathrm{mov}} = \mathrm{NULL}$, or C_{mov} and content ID of the relocating replica component do not match, go to 5. Otherwise, go to 7.
5. Among the candidate peers, select the peer with the smallest sum of the usefulness of its possessing components of replicas.
6. Input the content ID of each relocating replica component into its C_{mov} and select full objects and the differences from the components both originally held in the peer and relocating this time, and keep them in the peer so that the sum of the usefulness of the replica components becomes the maximum. What is not selected is the content to be relocated. Go back to 1.

7. Discard the relocating replica component and terminate this procedure. However, if the discarded component is a full object, all components constituting the content replica to which the discarded component belongs under the management of the same management peer are also discarded.

Procedure to change the version of replica to be kept as a full object

In each block including a full object managed by a replica management peer, the version to be held as a full object is reviewed in every $T(> 0)$ period according to the following procedure.

1. The number of times the version of the full object is referred to is compared with the numbers of times the other versions held as the differences are referred to. If the number of times the version of the full object is the largest, this procedure is terminated.
2. Select the version which is the most frequently referred to, reproduce the full object of the version based on the full object currently held and the difference to that version, and keep the newly reproduced full object in the peer.
3. Keep all other versions as differences with the new full object in the peer, delete the unnecessary differences, and terminate this procedure.

5 Evaluation

In this paper, we evaluate the effectiveness of the proposed method by computer simulations. The method that minimizes both network cost and content loss is excellent.

5.1 Methods to Be Compared

Two methods are used to be compared, i.e., owner replication method [14], and simplified BitTorrent which is added a mechanism to discard content replica components whose usefulness becomes minimum when the holding peer can not keep them because of lacking sufficient storage capacity (Hereinafter referred to as BT method).

5.2 Simulation Conditions

Table 1 shows simulation parameters.

First, a lattice-shaped network with an initially arranged 600 peers is created. As peers join and leave, the number of peers on the network changes around 600. 1200 kinds of content items are shared and in the initial state, a replica of each kind of contents are placed at one randomly selected peer in the network. Reference frequency of each replica is always given as the value calculated within recent T unit time, and T is set to 1500. In addition, the threshold H_{th} that defines the reference range of each peer in the proposed method is 2, considering

Table 1. Simulation parameters.

Parameters	Values
Network topology	Lattice
Simulation period	40000 [unit time]
Simulation runs	30 [times]
Number of initially deployed peers	600
Storage capacity of peers	Av. 5000 [MB] S.D. 1000 [MB]
Content variety	1200
Capacity of a content	Av. 500 [MB] S.D. 100 [MB]
Number of content's division	4 or 8
Capacity of a difference between two contents	Av. 1% - 4% of Content's Capacity
Threshold H_{th}	2
Content requests occurrence per unit time : λ_{req}	18
Updates occurrence per unit time : $\lambda_{\mathrm{update}}$	0.5 (Initial state) 0.05 (Stable state)
Peers' joining/leaving per unit time : λ_{mov}	0.5
Time period T	1500 [unit time]

the scale of the network. In the proposed method and the BT method, the number of divisions of one content is set to 4 or 8, and the threshold C_{sup} for complementing contents is set to 3%.

Contents requested by each peer is determined according to user's preference, which is defined by the following procedures 1 and 2.

Firstly, the breadth of each user's taste is determined (procedure 1). The breadth of each user's preference means the number of kinds of contents that match his/her preference, and in fact, it is defined as a value that follows a normal distribution with an average of 30 and a standard deviations of 5.

Secondly, the contents that match the preference of each user are selected by the number of preferences determined in procedure 1 (procedure 2). Specifically, using Zipf distribution, the probability that each content is selected is given, and the contents are assigned to each user so as not to exceed the breadth of the user's preference according to the probability. Conversely, it is assumed that each allocated content is required according to Zipf distribution.

For 500 unit time from the beginning of the simulation, in order to promote the spread of contents and ensure that the placement reaches a steady state at an early stage, peers are not leaving the network and not joining to, and content requests from randomly selected peers (i.e., users) occur for the number of times according to Poisson distribution with an arrival rate λ_{req} of 18 (3% of the total number of peers) per unit time. Content updates occur for the number of cases according to Poisson distribution with an arrival rate $\lambda_{\mathrm{update}}$ of 0.5 per unit time, for the contents selected in the same manner as the content requests by randomly extracted peers.

It is assumed that the simulation reaches the steady state when 500 unit time have elapsed from the start of the simulation, and the simulation is continued in the environment where peers' joining and leaving network occur, with the arrival rate of content request being maintained and that of content update setting to 0.05. Both peers' joining and leaving occur for the number of times according to Poisson distribution with arrival rate of 0.5 per unit time respectively, and a joining peer holds replicas of the number of contents according to the normal distribution with average 3 and standard deviation 1. The kinds of contents held at the time of joining is determined in the same manner as the content request.

For the sake of simplicity, joining peers are randomly selected from not yet joined peers and put into the positions to maintain the lattice structure in the network, and peers' leaving occur at randomly selected peers which keep joining the network.

Storage capacity of each peer is set to a value that follows the normal distribution with an average of 5 GB and a standard deviation of 1 GB. Size of a difference is set according to uniform distribution with a size of 1% to 50% of its full object in the block.

5.3 Simulation Results and Discussions

The simulation results are shown in Figs. 2 and 3 below. The vertical lines on the plotting points in the figures indicate the range of standard errors.

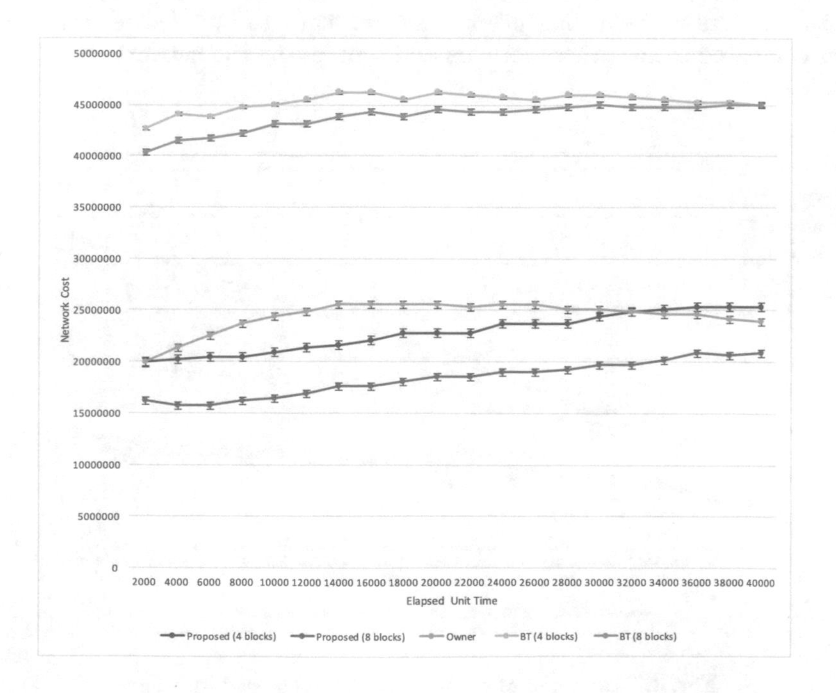

Fig. 2. Transition of network cost with elapsed unit time

As shown in Fig. 2, when dividing the contents into 8 blocks, proposed method always has the smallest network cost compared to other methods. When dividing the contents into four blocks, the cost is the smallest in the proposed method from the initial state to 30000 unit time, but during the last 6000 unit time, owner replication method achieves the smallest. Conversely, BT methods always take the largest cost. The reasons are as follows. In the proposed method, there exist not a few cases where it is sufficient for the users to request only differences between versions at the time of content requests, and this causes network cost small. In addition, since each content is divided into plural blocks, there are cases where it is possible to obtain all the necessary components of the content replica by merely requesting a part of the blocks. This is also a factor of reducing the network cost. Furthermore, as the number of content divisions increases, the storage capacity of each peer can be effectively utilized, the number of replicas that can be held throughout the network is large, so the possibility of acquiring the requested content from neighboring peers increases. This causes that proposed method with 8 blocks division minimizes the network cost in compared 5 methods. The effect of content division is the same for BT methods.

On the other hand, in owner replication method, when the number of content acquisition failures increases as time passes, the network cost of this method gradually decreases and finally becomes smaller than that of proposed method with 4 blocks division.

In the BT methods, since the content replica blocks are possibly requested not only to the requesting peers' neighbors but also to every peers possessing the blocks, the content providing peers are more likely to be selected from farther peers compared to the other methods and that is why the network cost tends to increase.

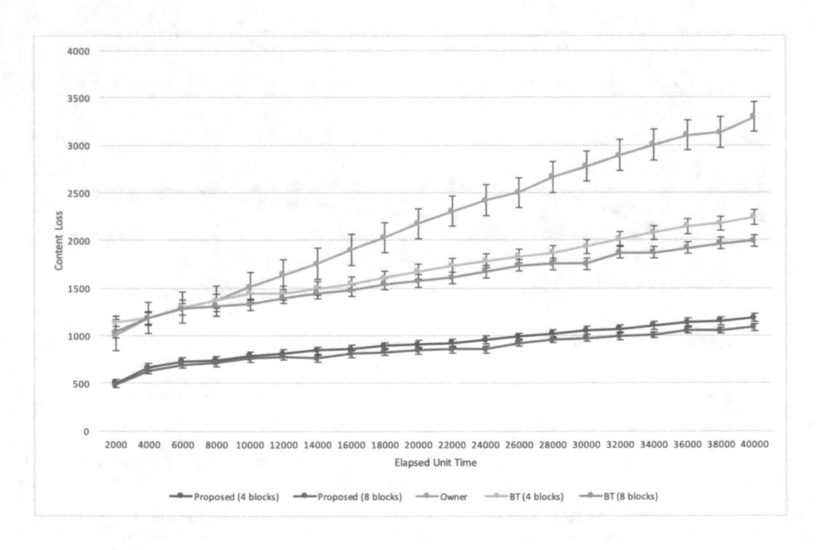

Fig. 3. Transition of content loss with elapsed unit time

As shown in Fig. 3, in the proposed method, the content loss values are always minimized compared with other methods. The reason for this is as follows. Since many versions of a content can be held as differences with small capacities when using the proposed method, necessary storage capacity to hold contents to be shared can be suppressed, and eventually, wide variety of versions of each content can be held on the network. Also, since each content is divided into small blocks, it is easy to keep them in small vacant areas of each storage.

Since the initial state, BT method keeps content loss lower than owner replication method. The reason for this is as follows. In the BT method, because each version of the full object exists for each block of the content, it is strongly influenced by the disappearance of the contents due to peers' leaving, and content loss becomes larger than the owner method for a while from the initial state. However, in the BT method, it is sufficient to hold only the blocks where the update occurred at the time of updating a content, whereas owner replication method must hold the full object of the updated content replica, and the storages of the peers are likely to be insufficient. Eventually, since BT method can be suppressed its content loss increase rate than that of owner replication method as the passage of time, content loss of BT method after 8000 unit time takes smaller values than that of owner replication method.

In the proposed method and BT method, the content loss decreases as the number of content divisions increases. This is because both methods can effectively utilize the storage of each peer by dividing the contents into smaller pieces, and it can be said that the division of contents is effective in reducing both network cost and content loss.

In all methods, content loss tends to increase as time passes. This is because in this simulation, all the versions that appeared even once on the network in the past are subjects to content requesting. Although the number of contents' versions increases monotonically for accumulating contents' updates, total storage capacity in the network is finite, and this result occurs.

6 Conclusion

In this paper, we propose a method that uses difference between contents and division of the content replicas in order to efficiently share contents with multiple versions in a hybrid P2P network, and evaluate its effectiveness by using computer simulations. As a result, we confirmed that our proposed method can sufficiently suppress network cost and content loss.

We plan to improve the proposed method further by evaluating using different network topology, investigating the influence of parameters such as storage capacity distribution and size of reference range on the effectiveness of the proposed method, and so on.

Acknowledgements. We thank Professor Yutaka Ishibashi at Nagoya Institute of Technology, Japan for precious advices and discussions. This work was partially supported by JSPS KAKENHI Grant Number JP17K00134.

References

1. Sunaga, H., Hoshiai, T., Kamei, S., Kimura, S.: Technical trends in P2P-based communications. IEICE Trans. Commun. **E87–B**(10), 2831–2846 (2004)
2. Kageyama, J., Shibusawa, S.: Replication that prevents low-demand files from disappearing in P2P network. DEIM Forum **C9–3** (2009)
3. Kawasaki, Y., Sato, T., Yoshida, N.: Popularity-based content replication in Peer-to-Peer network. In: Proceedings of Internet Conference 2005, pp. 106–113 (2005)
4. Yoichi, R., Sugawara, S.: Consistency preservation of replicas based on access frequency for content sharing in hybrid Peer-to-Peer networks. Int. J. Space-Based Situated Comput. (2017, in press)
5. Inoue, Y., Sugawara, S., Ishibashi, Y.: Efficient content replication strategy for data sharing considering storage capacity restriction in hybrid Peer-to-Peer networks. IEICE Trans. Commun. **E94–B**(2), 455–465 (2011)
6. Kato, T., Sugawara, S., Ishibashi, Y.: On efficient content sharing considering updating replicas in hybrid Peer to Peer networks. IEICE Technical Report, NS2010-80, pp. 61–64 (2010)
7. Kato, T., Sugawara, S., Ishibashi, Y.: Efficient content sharing taking account of updating replicas in hybrid Peer-to-Peer networks. In: Proceedings of IEEE 2011 International Communications Quality and Reliability (CQR) Workshop (2011)
8. Mukherjee, P., Leng, C., Terpstra, W.W., Schurr, A.: Peer-to-Peer based version control. In: Proceedings of ICPADS 2008, pp. 829–834 (2008)
9. Lin, X., Li, S., Shi, W., Teng, J.: A scalable version control layer in P2P file system. In: Proceedings of GCC 2005, pp. 996–1001 (2005)
10. Bit Torrent, http://www.bittorrent.com
11. Cohen, B.: Incentives build robustness in bit torrent. In: Proceedings of First Workshop on Economics of Peer-to-Peer Systems, pp. 251–260 (2003)
12. Shibuya, M., MIyazaki, N., Ogishi, T., Yamamoto, S., Suzawa, M.: Performance evaluation of BitTorrent based progressive download in broadband access networks. IEICE Technical Report, CQ2008-22, pp. 49–54 (2008)
13. Sugimoto, A., Tagashira, S., Fujita, S.: A method for effective matchmaking on BitTorrent-like parallel downloading system. IEICE Technical Report, IN2006-208, pp. 167–172 (2007)
14. Lv, Q., Cao, P., Cohen, E., Li, K., Sheker, S.: Search and replication in unstructured Peer-to-Peer network. In: Proceedings of International Conference on Supercomputing, pp. 84–95 (2002)

Improving Security with Cognitive Workflows

Giovanni Cammarata, Rao Mikkilineni, Giovanni Morana(✉), and Riccardo Nocita

C3DNA Inc., Cupertino, CA, USA
{gcammarata,rao,giovanni,rnocita}@c3dna.com

Abstract. Cloud Computing, SDN and virtual networking technologies have completely modified the relationship between the applications and the hardware resources that are used to execute them. They are no more tightly coupled to each other in a static context. However, elastic on-demand provisioning, auto-scaling and migration provided by cloud resources to address fluctuations in workload demands or available resource pools also bring with them new issues in managing security. In this paper, the authors propose a novel security system based on the concept of cognitive control overlay to proactively manage the security of service transactions. In particular, when the application components move, their configuration changes and the conventional intrusion detection systems (IDS) not aware of the mobility will fail. The cognitive overlay makes the IDS become aware of the mobility and take appropriate action. The solution addresses application security independent of server and network based security management systems.

1 Introduction

Clouds [1] and the "Everything as-a-service" approach, together with SDN [2] and virtual networking technologies [3], have completely transformed the way IT services are designed, managed and even consumed by users.

The components of each IT services are no more strictly coupled with each other or to specific hardware resources: they are designed to work in a distributed way, to be autonomous, stateless, scalable and portable.

This dynamicity has a significant impact on the management of the IT services.

On the one hand, a stateless and scalable design, associated with the flexibility offered by the API of Cloud providers, simplifies the creation and adoption of adaptive management strategies that can dynamically balance between performance, robustness, and cost.

On the other hand, operations such as migration or scaling in and out modify dynamically the structure of the services forcing the management system to continuative check of the configuration consistency for all the external software used for supporting the primary service (e.g., DNS, load balancer, proxies and so on). Among these last, the security system [4] is probably the most critical component to be affected.

Systems like this, in fact, base a significant portion of their activities on network monitoring, perform very well when they are used within a static context but suffer in all those operations that involve a dynamic change of IPs or endpoints' references.

L. Barolli et al. (Eds.): EIDWT 2018, LNDECT 17, pp. 429–436, 2018.
https://doi.org/10.1007/978-3-319-75928-9_38

Although new virtual network technologies are mitigating this kind of problems in resources managed by a single owner (i.e., homogeneous networking management), the distributed nature of the IT services forces the adoption of a solution able to work over networks managed with different policies.

In this paper, the authors propose a security system based on the concept of DIME Computing [5–8] and demonstrate how this system can adapt and evolve its behaviour to the changes in the structure of the monitored services.

In particular, it will be shown how the system manages the configuration of an open-source IDS [9] to adapt it to monitor the network activities of a 3-tier, distributed applications.

The rest of the paper is organized as follows.

A brief description of DIME computing will be given in Sect. 2. Section 3 will introduce the IDS. The details about the use case will be given in Sect. 4. The last section will provide the conclusions.

2 DIME, DNA and Self-managing Workflows

The DIME Network architecture mixes sensors and actuators along with knowledge-based agents to achieve a revolutionary goal: infusing cognition into computing.

It introduces three key functional constructs to enable process design, execution and management of distributed computing elements according to specific resilience policies:

1. Machines with an Oracle: Executing an algorithm, the DIME basic processor can choice to perform either the classical instructions cycle {read -> compute -> write} or its modified version {interact with a network agent -> read -> compute -> interact with a network agent -> write}
   Using DIME Technologies, the network agents can influence the further evolution of the computation, while the computation is still in progress.
   This is possible because: (i) DIME agent knows anything about algorithm: the context, constraints, communications, goal, and intent; (ii) the DIME basic processor has the visibility of available resources and the needs related to the execution of the tasks; (iii) the DIME agent also has the knowledge about a global (or partial) set of alternative paths available to facilitate the evolution of the computation, so to achieve its goal and realize its intent.
   All the information required by the DIME agent to manage the process evolution are formalized within a blueprint, analogous to a genetic specification in biology.
   According to this view, any computing node in the network with which the DIME unit interacts is an external agent
2. Blueprint: it captures the definition, configuration, and relationships of all application components that comprise a service or application workflow. The DIME agent uses the blueprint to configure, instantiate, and manage every aspect of the execution of the DIME basic processors (Fault, Control, Accounting, Performance, Security - FCAPS). Concurrent DIME basic processors both can coordinate and monitor their goal thanks to their own blueprints. In addition, blueprint allows the basic processors to implement various strategies to assure non-functional requirements (such as

availability, performance, security and cost management) even if the managed DIME basic processors are executing their intent.

3. DIME network management control overlay over the managed Turing oracle machines: DIME has two main communication channels: the data channel (which groups all read/write communication of the DIME basic processor) and the signaling channel (which groups the parallel communications among DIME basic processors). This allows the external DIME agents to influence the computation of any managed DIME basic processor in progress based on the context and constraints.
   The external DIME agents are DIMEs themselves. This implies that the changes in one computing element could influence the evolution of another computing element at run time. As a result, the network of DIME agents and their signal channel below can be programmed to execute a process, the intent of which can be specified in a blueprint. This feature sets complex scenarios within which each DIME basic processor can have its own oracle managing its intent, and groups of managed DIME basic processors can have their own domain managers implementing the domain's intent to execute a process. The specification, configuration, and management of the sub-network of DIME units is delegated to the management DIME agents. They are able to monitor and execute policies to optimize the resources at runtime (i.e. while delivering the intent).
   In the next section the authors describe an implementation of DNA to infuse cognition into Snort [9], an open-source IDS tool: this cognitive-aware version of Snort will able to guarantee the consistency of network security configuration also during operation such as migration or scaling.

3 IDS

One of the most used tools to build a security system is the Intrusion Detection System (IDS), a software able to carry out an accurate analysis to identify illegal traffic or malicious activities.

An IDS is useful in detecting suspect activities that occur on the network (N-IDS) or on the host (H-IDS), with its respective extension, the Intrusion Prevention System (IPS). This last, in particular, is also able to control, delete, reset them and initiate a series of related actions.

These tools adopt several approaches to discern the malicious traffic from the legitimate one, and they are classified based on their own detection approach.

Many IDSs have a detection mechanism based on a signature: they check the system and the network for a matching with a specific pattern that, usually, takes the form of a sequence of bytes or of a specific set of instructions sent over the network.

Another technique is purely based on anomaly detection: it is able to measure the degree of detachment from a normal data flowing condition called "baseline", earlier established and saved.

Other kinds of detection systems rely on specific protocol misuses.

The solutions based on the check of a signature it is only loosely coupled with the monitored traffic and needs only a signature's database constantly updated.

On the contrary, the approach based on anomaly detections requires accurate knowledge about the underlying network topology and its "standard traffic pattern" in order to ensure that no false positives are raised.

The solutions based on the check of a signature it is only loosely coupled with the monitored traffic and needs only a signature's database constantly updated.

Another important aspect to take care of for the IDS is its network visibility. The detection system needs to be placed in a position that maximizes its network visibility to inspect as much traffic as possible: if not correctly placed, the IDS could not cover a significant portion of the network and, as a consequence, it could not be able to guarantee an adequate protection.

4 Case Study: Dynamic Security Management

This section will describe how a DIME-aware IDS is able to adapt itself and, as a consequence, its security rules to the evolution of the dynamic structure of a 3-tier, distributed application.

The IDS used is Snort. Snort is an open-source software designed to act as a traffic sniffer, network intrusion and detection system. It has a rule-based intelligence and it can be extended to acting as Intrusion Prevention System (IPS), i.e., able to block and reset connections. Here Snort will be used as IDS, not as IPS.

The 3-tier application is a workload emulating a real-word e-commerce application composed of 3 basic components:

- A web page, i.e. a virtual host on Apache HTTP Server, acting as front-end or "presentation layer": it collects all the input coming from users and forward it to the business layer.
- A web application, i.e. a java program hosted by an Apache Tomcat, acting as back-end or "business layer": it elaborates all the requests coming from the users, reading and writing on database when needed.
- A database, hosted by MySQL DBMS, representing the "data layer": it stores all the data and represents the status of the application.

In order to improve scalability, the virtual host and the java application have to be stateless, interacting via asynchronous message-passing solution and their sessions have to be stored on client-side. All the three components, including the MySQL database, have to be addressable by logical names.

The considered use case will be focused on the most critical flow (see Fig. 1) of this application, i.e. the connections between the database and the controller. Snort, as shown in Fig. 2, is configured to monitoring that flow.

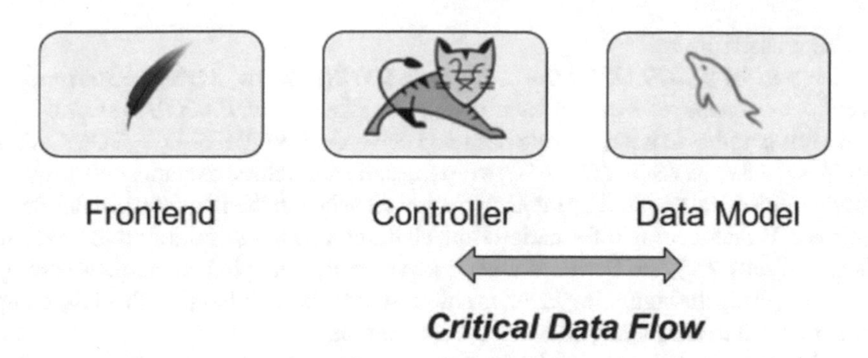

Fig. 1. The 3-tier application

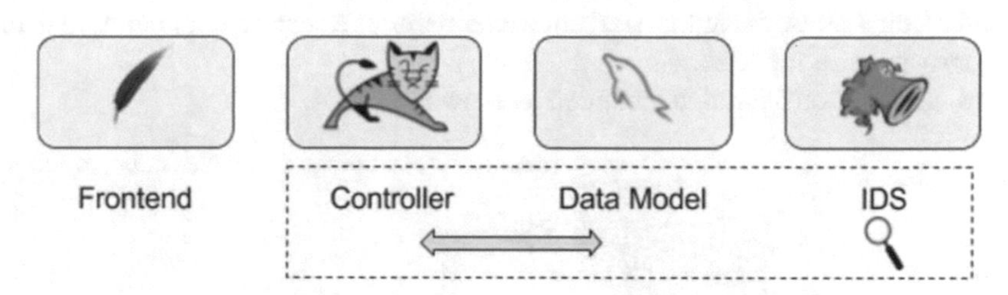

Fig. 2. The 3-tier application and IDS

All the components of the application and Snort are deployed in private network, each component is hosted on a specific machine and owns an IP address (Fig. 3).

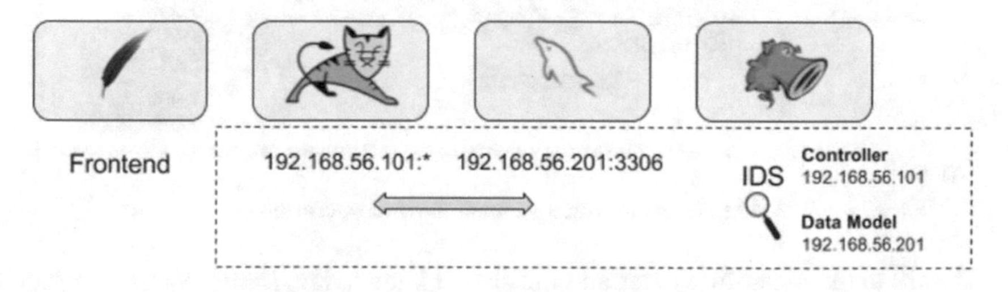

Fig. 3. Details on the network deployment

Snort can monitor MySQL at its default listening port (3306) and configured to be able to check all the connections requests at the database and accept only the ones marked as "controller", i.e., coming from the machines hosting the Tomcat component.

Snort is configured by means of rules files that contain the set of rules to be applied, according to an ad hoc syntax that takes into account the type of alarm, the source reference, the destination reference, the monitored flow and several additional parameters.

For this use case, to simplify the network dynamics, Snort is configured to raise an alarm each time it intercepts the creation of a new connection: according to the three-way handshake protocol, this happens filtering the syn packets on the network.

The related rule is:

alert tcp !$WORKFLOW_TOMCAT any -> $WORKFLOW_MYSQL 3306 (msg: "An unauthorized connection attempt was detected"; flags:S; sid:1000001; rev:1;)

As it is possible to note, the rule has two keywords: $WORKFLOW_TOMCAT and $WORKFLOW_MYSQL. These keywords, usually configured as static, will be dynamically modified and updated by the DIME system basing on the IP address of the Tomcat machines. In fact, getting information from all the components of the system (machines, OSes, and software), the DIME is always aware of the mapping <machine-ip-component> and, using the signalling layer, is able to share this knowledge with all the components involved in the management of the application.

Let's suppose that an unauthorized copy of the controller (i.e. Tomcat) tries to execute transactions to the MySQL database and, at the same time, let's suppose that, for load requirements, the authorized controller requires a scaling out operation in order to properly balance the load.

The scenario considered is amended as shown in Fig. 4.

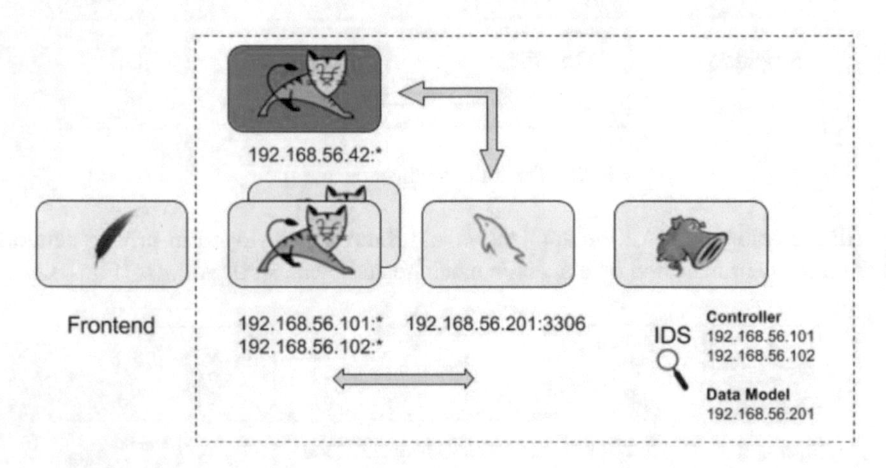

Fig. 4. Authorized vs unauthorized operations.

Thanks to the signalling layer and the shared knowledge, Snort is able to correctly map, in real-time, the operation executed over each controlled component. This type of knowledge enables Snort to detect the unauthorized copy of the controller and manage it correctly, raising an alert able to notify not only the IDS itself but all the components of the DIME network. This alert allows other involved applications to be able to react autonomously making decisions about the type of alert received.

The given example can also be extended in the case the database needs to modify its location dynamically.

The language of the Snort rules allows detailing numerous rules that make the system able to improve the dynamic traffic analysis, adding constraints about the packet flags, size, data content, time-to-live value, and so on. The described system enhances these rules by making them aware of location, since wherever these rules will be correctly translated, their content will be contextualized.

In this way, the IDS will have consciousness of the fact that a monitored application can be present, even if not active, locally rather than in a remote machine, for example focusing its attention on a specific interface rather than another one.

A fundamental consequence is the ability to replicate or migrate Snort itself together with the entire workflow (making it a cognitive-aware workflow, see Fig. 5), promptly allowing it to reproduce the same degree of security in different and independent networks thanks to the ability of the DIME system to provide location transparency at the level of IDS rules.

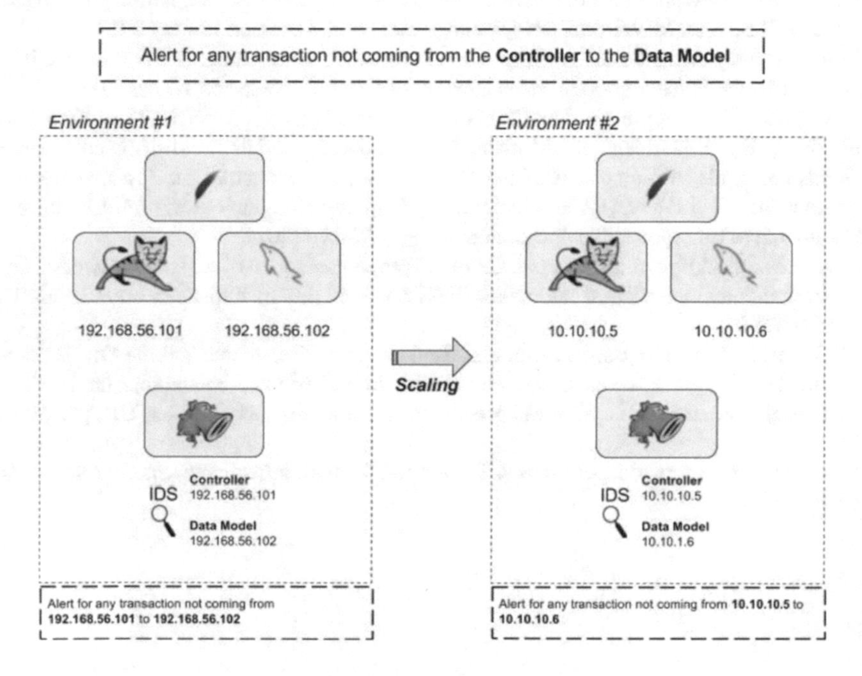

Fig. 5. Migration of IDS itself.

5 Conclusions

In this paper, the authors propose a security system based on the concept of DIME Computing and demonstrate how the proposed system is able to adapt its behaviour to the changes in the structure of the monitored service.

Taking as an example a 3-tier, distributed application, the authors described how it is possible to extend the capability of an open-source IDS integrating it within a DIME-aware environment.

As shown, the resulting security system is able to support operations such as migration or scaling, which modify the structure and - as a consequence - the network configuration of resources hosting the service, keeping consistent all the security rules associated with the IPs changes.

References

1. Buyya, R., Yeo, C.S., Venugopal, S., Broberg, J., Brandic, I.: Cloud computing and emerging IT platforms: vision, hype, and reality for delivering computing as the 5th Utility. FGCS **25**, 599–616 (2009). ISSN:0167-739X
2. Kreutz, D., Ramos, F.M.V., Veríssimo, P.E., Rothenberg, C.E., Azodolmolky, S., Uhlig, S.: Software-defined networking: a comprehensive survey. Proc. IEEE **103**, 14–76 (2015). https://doi.org/10.1109/JPROC.2014.2371999
3. ETSI Portal: Network Functions Virtualization (NFV) White Paper #3. https://portal.etsi.org/Portals/0/TBpages/NFV/Docs/NFV_White_Paper3.pdf. Accessed 15 May 2017
4. Cloud Security Alliance (CSA): Top Threats to Cloud Computing. (PDF version). https://cloudsecurityalliance.org/topthreats/csathreats.v1.0.pdf. Accessed on 10 Jan 2018
5. Mikkilineni, R.: Designing a New Class of Distributed Systems. Springer, New York (2011)
6. Eberbach, E., Mikkilineni, R., Morana, G.: Computing models for distributed autonomic clouds and grids in the context of the DIME network architecture. In: Proceedings of the International IEEE WETICE Conference Enabling Technologies: Infrastructure for Collaborative Enterprises, Toulouse, France, pp. 125–130 (2012)
7. Burgin, M., Mikkilineni, R., Morana, G.: Intelligent organization of semantic networks, DIME network architecture and grid automata. IJES **8**, 352–366 (2016). https://doi.org/10.1504/IJES.2016.077796
8. Mikkilineni, R., Comparini, A., Morana, G.: The turing O-machine and the DIME network architecture: injecting the architecture resiliency into distributed computing. In: The Turing Centenary Conference, EasyChair Proceedings in Computing, Manchester, UK, pp. 239–251 (2012)
9. Snort, Network Intrusion Detection & Prevention System. https://www.snort.org/. Accessed 10 Nov 2018

Threshold Model Based on Relative Influence Weight of User

Deyang Zhang[1], Xu An Wang[1(✉)], Xiaolong Li[1], and Chunfen Xu[2]

[1] Engineering University of CAPF, Xi'an, China
zhangdy0126@126.com, wangxazjd@163.com, 463387834@qq.com

[2] Jiaxing Vocational Technical College, Jiaxing, People's Republic of China

Abstract. Considering the existence of competition in the process of social network communication, and the change of sensitivity in the process of communication, this paper proposes a new relative influence weight function that combining with the existing linear threshold model, the sensitivity of information, and the threshold characteristic of the node, namely, URLT model. Which can measure the information communication ability. By simulating the spread of different networks, different sensitivity information and different node thresholds, comparing the final propagation situation, the experimental results show that the final influence range is consistent with the real spread situation. Therefore, the model has some reference value for the discovery and suppression of the law of information dissemination.

1 Introduction

With the rapid development of Internet, all kinds of network based social media have formed a completely new mode of information communication. Especially for some hot issues in today's society, the development of public opinion has reached an unprecedented trend. Whether domestic or international important events, people can get information through the Internet and related media in the first time. It is precisely because of the convenience of the network, the speed of receiving information and publishing information has been greatly improved. This kind of behavior that expresses the viewpoint through the network, tells the thought the behavior, without doubt has caused the very big pressure to the social stability, the public opinion guidance.

Social hot events generally includes two aspects: one is due to the formation of long-term social environment, cannot be resolved within a short time, has certain periodicity, such as the employment problem, open the "second child" problem, the global climate warming and so on, this kind of events in a period time has a high degree of attention, and then returned to normal level. Two, because of the domestic and international hot incidents, sudden events, and some of these events with intentional speculation ingredients, the purpose is to raise the attention of events.

So the research on hot events in the society is an urgent problem to be solved, the research on hot events already exist only at the theoretical stage, there will be specific in the dissemination of quantitative data and mathematical model. At present, the domestic research on the influence of hot spots and countermeasures have accounted

L. Barolli et al. (Eds.): EIDWT 2018, LNDECT 17, pp. 437–448, 2018.
https://doi.org/10.1007/978-3-319-75928-9_39

for the majority. This paper will focus on the issue of spread of public opinion hot events, have a comprehensive analysis through the existing communication model, this paper combines the advantages of the user's relative weight function and the linear threshold model, using the user's relative weight function $R(x, y)$ and information sensitivity P and threshold θ, this paper proposes a new model, *URLT* model.

The structure of this paper is as follows: in the first section, is an explanation about the existing research is and the work done in this paper is explained. The second section mainly defines the related concepts and introduces the background knowledge. In the third section, the model presented in this paper is introduced. The fourth section uses simulation software to do simulation experiments to analyze whether the model is the same as the actual propagation. The fifth section is a summary, pointing out the shortcomings and looking forward to the next step.

2 Related Concepts and Definitions

2.1 Mode of Transmission

Social networks are made up of thousands of individuals, everyone has a different family background, personality, and habits, so each person has a different way to solve the information. When receiving an interesting information, the individual will usually choose to accept and spread; when they are not good at or not interested in the information, will choose not to accept it.

Now, the way of social communication is more complex than any time in the history [2]. Thanks to the development of the Internet, people can get the information they want. However, it is also because of the convenience of the Internet, the forms of public opinion are becoming more and more complex, and the management of public opinion is becoming more and more difficult.

2.2 Influence

Social influence can only be manifested through the interaction between people:

(1) According to the behaviors between users, Rashotte [3] defines social influence as a change in people's thoughts, feelings, because of the behavior of users.
(2) According to the characteristics of statistics, some scholars define it as a part of the greater proportion of the distribution of influence intensity.

2.3 Sensitivity

The sensitivity is the trend of the individual affected by others, recent studies have found that sensitivity and influence both have an influence in spread of social networks, paper [4] have found the same user may not have high influence and high sensitivity at the same time. So, which factor plays a more important role in the process of communication and sensitivity, this is also a new problem.

2.4 Measurement of Influence

The traditional way of measurement influence is based on network topology, including measurement based on nodes and path. The social network is defined as $G = (V, E)$, and V is defined as the set of nodes, each node represents a social network of users. E is defined as the set of edges to represent the relationship between users (relatives, friends, enemies), $n = |V|$ defined as the number of nodes. The influence of node u on node v is b_{uv} (Table 1).

Table 1. Measure method of node influence

Type	Measurement method	Formula	Method description
Based on node degree	In degree	$Deg^{in}(v_i) = \sum_j a_{j,i}$ [5]	The extent to which users should be concerned
	Out degree	$Deg^{out}(v_i) = \sum_j a_{i,j}$	The extent to which the node should be concerned about the other nodes
	Degree centrality	$C^{DEG}(v_i) = \frac{Deg(v_i)}{n-1}$ [6]	Represents the average impact of a node on a neighbor node
Based on the shortest path	Tight centrality	$C^{CLO}(vi) = \frac{1}{\sum_{v_j \in V \backslash v_i} g'_{ij}}$ [7]	The impact of a node on other nodes reflects the social strength of the user
	Betweenness centrality	$C^{BET}(v_i) = \frac{\sum_{s<t} \lvert\{g^i_{st}\}\rvert / \lvert\{g_{st}\}\rvert}{n(n-1)/2}$ [8]	Location importance of nodes in a reactive social network

2.5 Model of Maximizing Influence

The propagation of social information in social networks is a process of spreading from the initial group that has been activated to the last one that can't be activated. The node has two states of activation and activation, and the active user activates the active user through a certain probability λ through activation, and the activation process is irreversible. That is to select different nodes, and at the same time in the network to spread, contrast the final impact area, select the largest range of impact, that is, to maximize the impact. The research on the influence maximization problem of greedy algorithm including [9], independent cascade model [9, 10] and the linear threshold model and [11, 12] based on the greedy algorithm and the linear threshold model of "accumulation", combined with the characteristics of a new kind of influence maximization algorithm HPG algorithm [13]. This paper reflects the propagation in the real world, when there are multiple sources of information, the recipient will choose the most affected are accepted, reflects the competition mechanism of information dissemination, combines the relative weight of users $R(x, y)$ sensitivity p and threshold θ by and information combination, this paper proposes a new *URLT* model.

2.5.1 Linear Threshold Model

The linear threshold model, referred to the LT model, is a model based on the specificity of the nodes. In Sect. 2.2, definitions of related parameters have been made, which are no longer explained here. The new parameter θ is defined as a threshold, indicating the critical weight value of a node being activated, which is determined by the node itself. Defined $A(v)$ as a collection of neighbor nodes that have already been activated, and all neighbor nodes have no more than 1 of the node weight, that is $\sum_{u \in A(v)} b_{uv} \leq 1$. Each node has its own specific threshold $0 < \theta \leq 1$. Only when $\sum_{u \in A(v)} b_{uv} \geq \theta$, the next moment of the node v becomes active, and the active state remains unchanged. It should be pointed out that in the LT model, when a node attempts to activate a node without success, the resulting influence is accumulated until the node v is activated or the propagation process is over.

2.5.2 Improved Weights in HPG Algorithm

In the existing linear threshold model, b_{uv} is used to express the influence of nodes u on nodes v, and is usually expressed in terms of formula (1). $d(v)$ represents the degree of the node v.

$$b_{uv} = \frac{1}{d(v)} \tag{1}$$

In the HPG algorithm, there is a mutual influence between the node u and the node v, and the influence of the node pair in the same network is different from that of the node pair. The following definitions are given.

In the past, the model ignored the role of the node itself, all the neighbors as equal status, ignoring the differences of users. Tian Jia Tang pointed out b_{uv} is the size of the node, reflects the feelings for an authoritative embodiment to its node, and not concern with other nodes, it only need to consider the nodes can be connected, which is the idea of adjacency graph. Only the node is composed of the neighbor nodes pointing to it. And improved estimation formula. $NG(v) = G'(V', E')$, $V' = \{v\} \cup N(v)$,

$$E' = \{(u, v) | x, y \in V', (x, y) \in E\},\ b_{uv} = \frac{Deg^{out}(u)}{\sum_{w \in N(v)} Deg^{out}(w)} \tag{2}$$

The representation of the influence between node u and node v is determined mainly by the adjacency structure. Define $N(v)$ as the set of the neighbor nodes of node v, the neighborhood refers to the neighborhood of the node.

3 Threshold Model URLT Based on Relative Influence Weight of User

3.1 Sensitivity of Information P

The process of information transmission can be divided into three stages, the heuristic stage of information, the stage of information rendering, and the stage of information dissemination [15]. at each stage, the sensitivity of information is different. In this paper, the sensitivity of the information is considered as the most sensitive in the process of communication, but the highest sensitivity of a message is short, so we can judge whether an information is sensitive through the sensitivity of information. Sensitivity is usually defined as sensitive information within the range of 0.5 to 1. As can be seen from Fig. 1, the propagation of a message has a certain period of survival, and when its sensitivity reaches its maximum, its sensitivity will gradually decline to 0.

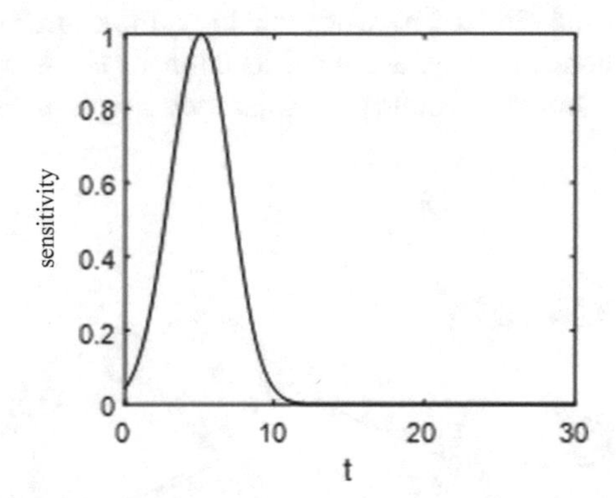

Fig. 1. The sensitivity of information varies with time

3.2 Node Specificity Threshold θ

The specific threshold node indicates that a user acceptance of a message, the value and the node's close contact with different types of people of different threshold, but in order to facilitate the experimental demonstration, while also taking into account the choice of most people, we will be specific threshold θ of each node is set to 0.5 to 1 of the interval in smaller threshold, the node receives the information may be greater, on the contrary, the less likely to receive information. In the simulation experiment in the fourth section, we compare the different thresholds of the nodes, and verify the authenticity of the theory.

3.3 Algorithm Definition

3.3.1 Algorithm Description

The existing model mainly considers the existing network topology in the social network, and takes little account of the user's own influence. The current work is basically each user equal treatment, given equal weight in the model, but in real life because each individual life experiences, education, the same level of experience, everyone's ideas but also by other people is affected by the neighbors, because this is put forward in this paper based on the influence function of [14] between users $R(x, y)$ to indicate the relative influence of node x to node y, since each node in a social network of the neighbors are different, so the true judgment between two nodes of the two greater influence, if the relative shadow which node ring more weight, so it has stronger influence. Combining the formula (2), this article defines the influence function as

$$R(u, v) = 2\frac{b_{uv}}{b_{uv} + b_{vu}} \tag{3}$$

What needs to be explained is that it is generally not equal, which is determined by the relative authority between the node and the node. The following example analysis.

3.3.2 Instance Analysis

As shown in Fig. 2 of the directed network topology, the degree of node 1 and node 7 is larger and can be considered as a critical location in the network. For node 1, its access node has =, so you can simplify its adjacency graph to Fig. 3

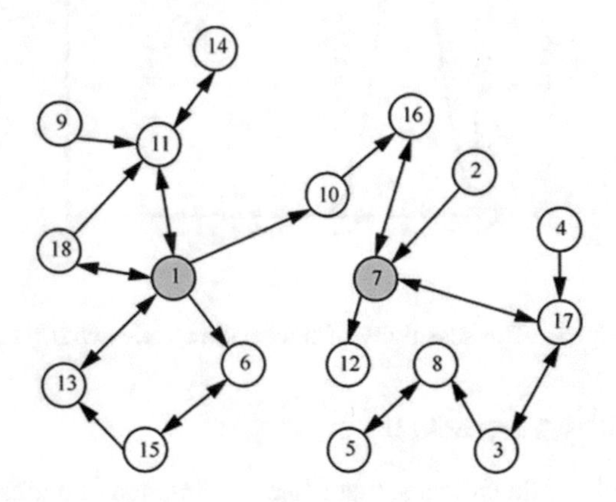

Fig. 2. Directed network topology

The output of node 1 Deg^{out} is shown as shown in Table 2.

So, we can get b_{ui1} namely $b_{13,1} = 0.111$, $b_{18,1} = 0.222$, $b_{11,1} = 0.222$, $b_{7,1} = 0.444$. Similarly, the adjacency graph of node 7 can be obtained, as shown in Fig. 4.

Can be calculated $b_{1,7} = 0.428$, $b_{10,7} = 0.143$, $b_{16,7} = 0.071$, $b_{2,7} = 0.071$, $b_{17,7} = 0.143$, $b_{8,7} = 0.143$. You can get it by type 3 $R(7, 1) = 1.018$, $R(1, 7) = 0.981$. Obviously, the role $R(u, v)$ is to adjust the weights of the connection node u and the node v according to the position of the two sides, which can better reflect the mutual influence weight size between the two nodes.

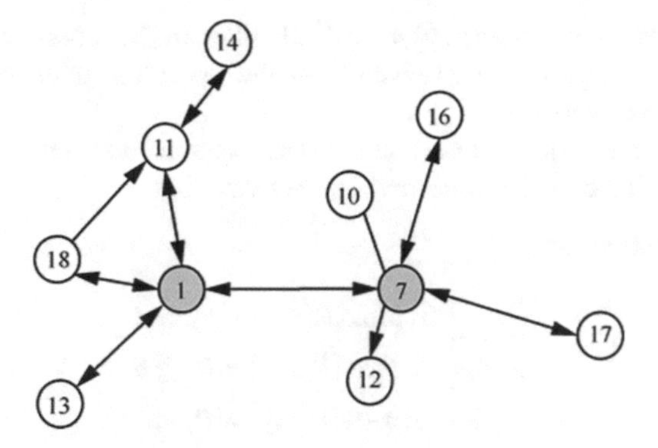

Fig. 3. Adjacency graph of node 1

Table 2. Output of node 1

u_i	13	18	11	7
$Deg^{out}(u_i)$	1	2	2	4

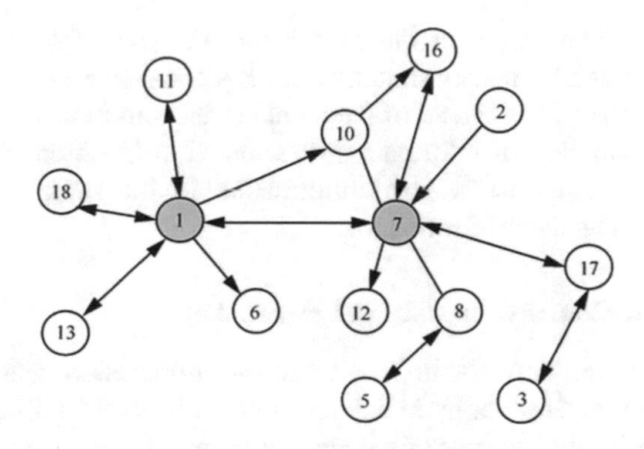

Fig. 4. Adjacency graph of node 7

3.4 URLT Model

According to the characteristics of information in network communication between information transmission and node not only influence but also with the information itself, so this paper proposes a new model combining user relative influence function and the linear threshold model based on. Based on the node characteristics in the network topology, we compare the user relative weight function $R(x, y)$ with the information sensitivity P, and obtain the actual propagation probability λ compared with the node threshold θ.

Specific provisions are as follows:

(1) The node is not activated if the maximum value $R(\mathrm{x}, \mathrm{y}) \bullet p$ of all other nodes connected to that node still does not exceed the threshold of the node.

(2) If the maximum value $R(\mathrm{x},\mathrm{y}) \bullet p$ of all other nodes connected to the node is greater than or equal to the threshold of the node, the node is considered to be successfully activated.
(3) Since θ is no more than 1 of all other nodes connected to the node, if it exceeds 1, then we specify that the node must be activated.

Can be represented as:

$$\lambda = \begin{cases} 1, Max(R(x,y) \bullet P) > 1 \\ 1, Max(R(x,y) \bullet P) \geq \theta \\ 0, Max(R(x,y) \bullet P) < \theta \end{cases} \quad (4)$$

The above rules can reflect the propagation in the real world, when there are multiple sources of information, the recipient will choose the most affected are accepted, reflects the competition mechanism of information dissemination.

4 Simulation Experiment Analysis

In this experimental environment are as follows: the use of MatlabR2015b in the Win10 system, through computer simulation in BA scale-free network and WS small world network, observed the spread of the model in the simulation of the network, and in contrast to the situation of information dissemination in different network models, comparing the different sensitivity of information in the same network under the condition of the final sphere of influence.

4.1 Propagation Contrast in Different Networks

Figure 5 is a scale-free network in BA, when the information sensitivity is 0.5, the threshold for the propagation of node 0.5, this figure reflects the initial node in different number of cases, with the increase of the number of initial nodes, the influence range of

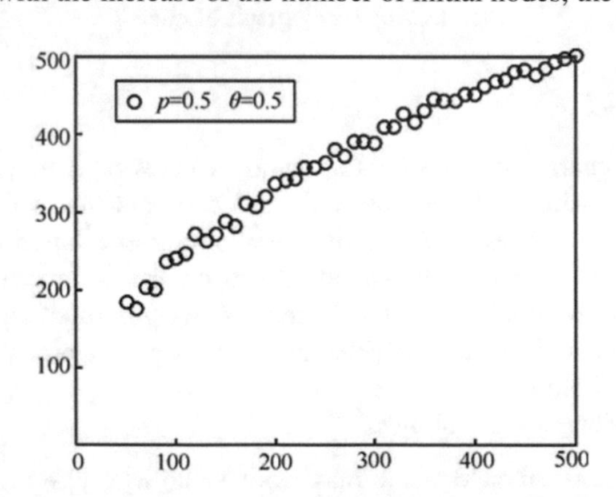

Fig. 5. Propagation diagram of URLT model in BA scale-free networks. The X axis is the initial number of nodes, and the Y axis is the number of final impact nodes

the final growth slows down trend. And from the local coordinates of the point of view, although the number of initial nodes increases, but its scope may also affect the final decrease, can reflect the regularity of information dissemination in the real world, that is not necessarily affect the initial node number more final range is bigger.

Figure 6 simulation is spread in the small world network, small world network between node and node of the strong coupling, neighbor nodes, each node has a fixed number, so the process of information dissemination is scale-free network speed. Compared with Fig. 5, the size of the network is 500 of the small world network, only 100 of the initial node can achieve the biggest impact, with the increase of the number of initial nodes of the affected areas also increases exponentially, viral spread with a small range of information presentation, also reflect in the crowded social network, sensitive information transmission speed is very quick.

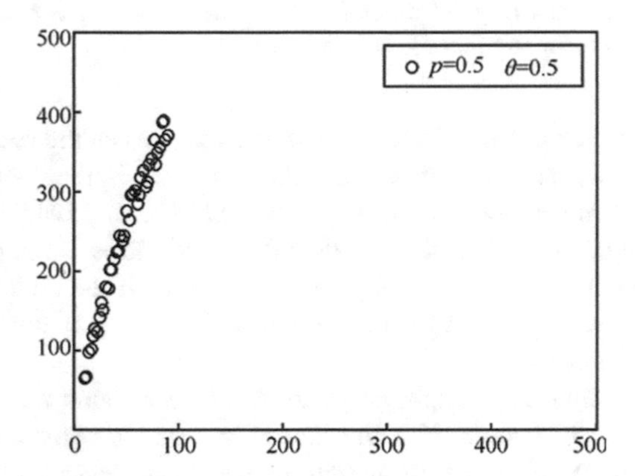

Fig. 6. Propagation of URLT model in WS networks. The X axis is the initial number of nodes, and the Y axis is the number of final impact nodes

4.2 Comparison of Different Sensitivity Information

Shown in Fig. 7, in the same network and the same node threshold, respectively with different sensitivity of information dissemination, can be seen in the initial number of nodes under the same conditions, the final impact range, high sensitivity and speed of information dissemination is significantly faster than the low sensitivity of the information. In combination with the propagation law of the real world, the information with high sensitivity is faster than the low sensitivity, but the trend of its propagation speed is gradually slow. On the other hand, ultimately affect the range of low sensitivity information showing the slope of the line is basically unchanged, the number of initial nodes increases, the ultimate impact range of growth rate remains unchanged, no growth speed and the final equilibrium.

4.3 Contrast in Different Thresholds

Similar to the previous experiments, we selected the information sensitivity is 0.5, the threshold for the two group of 0.5 and 1 of the spread curve, the speed of

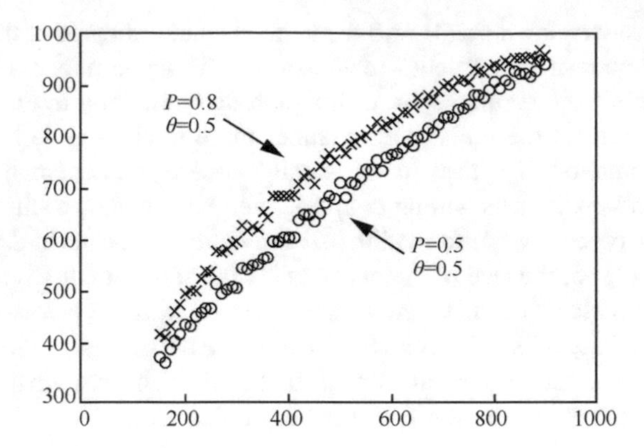

Fig. 7. Propagation contrast maps of different sensitivities. The X axis is the initial number of nodes, and the Y axis is the number of final impact nodes

communication here we use the slope of the curve changes as the speed of information dissemination, the experiment showed that the threshold is 0.5, which is when each node acceptance of information the time for half believe and half doubt the dissemination of information, speed is fast, and when the threshold is 1, the propagation speed increases with the initial node has no significant increase. If the node itself information judgment ability is strong, the information spread is slow, this conclusion consistent with the real social network.

Through several groups of experiments above, it was found that the URLT model proposed in this paper accords with the law of information transmission real social network, the sensitivity of node information threshold and other important parameters, but also on behalf of the attribute information and the node itself is a complex variable function, therefore, how these parameters from the theoretical arise to science, the next step is to pay attention to the problem (Fig. 8).

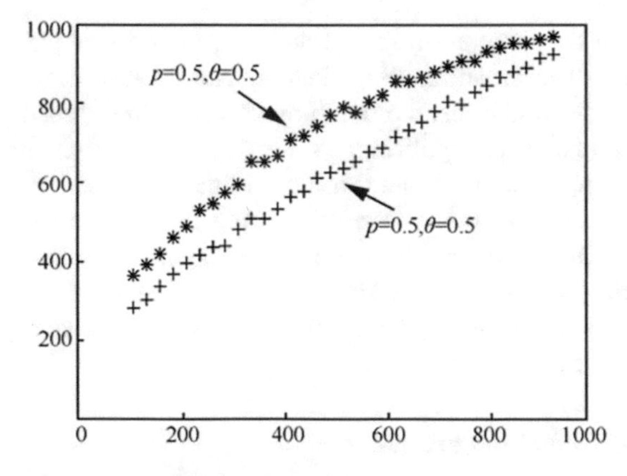

Fig. 8. Different threshold propagation contrast. The X axis is the initial number of nodes, and the Y axis is the number of final impact nodes

5 Conclusion

This paper takes social hot events as the research background, analyzes the similarities and differences between hot spot events and other information, defines relevant concepts, and explains the existing research. According to the propagation of hot events, on the basis of the existing research, this paper put forward the innovation point, not only to take into account the network topology, we should also take into account the interaction between nodes, is the relative impact of weight said. According to the law of information propagation, the sensitivity of information in different periods is defined within a certain range, so that the numerical calculation is convenient. Combined with the existing linear threshold model, the threshold model of relative influence weight is put forward. The simulation results show that the model can better reflect the law of information transmission. The research of this paper is only to analyze the static network, not to analyze the dynamic network, how to model and analyze it in the dynamic network is also the important content of the future research.

Acknowledgement. This work is supported by the National Cryptography Development Fund of China Under Grants No. MMJJ20170112, National Key Research and Development Program of China Under Grants No. 2017YFB0802000, National Nature Science Foundation of China (Grant Nos. 61772550, U1636114), the Natural Science Basic Research Plan in Shaanxi Province of china (Grant Nos. 2016JQ6037) and Guangxi Key Laboratory of Cryptography and Information Security (No. GCIS201610).

References

1. Liu, T., Zhong, Y., Chen, K.: Interdisciplinary study on popularity prediction of social classified hot online events in China. Telematics Inform. **34**(3), 755–764 (2017)
2. Jung, S.H., Kim, J.: A new way of extending network coverage: relay-assisted D2D communications in 3GPP. ICT Express **2**(3), 117–121 (2016)
3. Rasshotte, L.: Social Influence: The Blackwell Encyclopedia of Social Psychology, vol. IX. Blackwell Publishing, Malden (2007)
4. Aral, S., Walker, D.: Identifying influential and susceptible members of social networks. Sience **337**(6092), 337–341 (2012)
5. Xindong, W., et al.: Analysis of the influence of online social networks. J. Comput. Sci. **37**(4), 735–751 (2015)
6. Freeman, L.C.: Centrality in social networks conceptual clarification. Soc. Netw. **1**(3), 215–239 (1979)
7. Sabidussi, G.: The centrality index of a graph. Psychometrika **31**(4), 581–603 (1966)
8. Freeman, L.C.: A set of measures of centrality based on betweenss. Socialmetry **40**(1), 35–41 (1977)
9. Kempe, D., Kleinberg, J., Tardos, E.: Maximizing the spread of influence in a social network. In: Proceedings of the 9th ACM SIGKDD International Conference on Knowledge Discovery and Data Mining, Washington, USA, pp. 137–146 (2003)
10. Gruhl, D., Guha, R., Liben-Nowell, D., Tom-kins, A.: Information diffusion though blogspace. In: Proceeding of the 13th International Conference on World Wide Web, pp. 491–501 (2004)

11. Chen, W., Yuan, Y., Zhang, L.: Scalable in-fluence maximization in social networks under the liner threshold model. In: Proceedings of the 2010 IEEE International Conference on Data Mining, pp. 88–97 (2010)
12. Chen, W., Wang, C., Wang, Y.: Scalable influence maximization for prevalent viral marketing in large-scale social networks. In: Proceedings of the 16th ACM SIGKDD International Conference on Knowledge Discovery and Data Mining, pp. 1029–1038 (2010)
13. Jia-tang, T., Yi-tong, W., Xiao-jun, F.: A new hybrid algorithm for influence maximization in social networks. Chin. J. Comput. **34**(10), 1956–1964 (2011)
14. Long, W.J.: An online social network information propagation model based on relative weight of users. J. Phys. **64**(5), 050501 (2015)
15. Bicheng, L., Feng, D., et al.: Analysis of Network Public Opinion. Theory Technology and Application Strategy. National Defense Industry Press, Beijing (2015)

Design of a Low-Power Cold Chain Logistics Internet of Things System

Heshuai Shao(✉), Ronglin Hu, and Chengdong Ma

Huaiyin Institute of Technology, Huaian, China
Owen8848@126.com

Abstract. The temperature monitoring node based on MSP430F149 and CC1101 is designed, which has the low power consumption. The demo machine has been built and passed live test. The temperature monitoring node consists of MCU module, power module, CC1101 interface module and temperature acquisition module. The passive communication protocol is designed which can wake the CC1101 up on radio by polling. In order to collect temperature inside of the freezing truck at low-power consumption, the technology of function macro definition optimization and energy management based on context awareness is adopted. And the aim of monitoring temperature is realized by radio communication module sending temperature data to the sink node. The current is measured when the node run at different mode. When the node run at receiving mode, the measured current is 25 mA. When the node run at sending mode, the measured current is 9 mA. When the node run at sleeping mode, the measured current is 3 mA. The test results indicate that the temperature monitoring node runs stably, which lasts at least 90 days and achieves its objects.

1 Introduction

In the cold chain logistics Internet of things, we need to monitor and upload the ambient temperature in the refrigerated compartment in order to track whether the temperature data in the refrigerated compartment is in the specified range during the refrigerated transportation process. Therefore, it is necessary to design the temperature monitoring node in the refrigerated carriage to realize the real-time monitoring of the temperature in the refrigerated carriage. Because the transportation process of the cold chain logistics is long, the battery load of the temperature monitoring node is very limited, and the closed environment of the refrigerated compartment is not easy to replace the battery or recharge. Therefore, how to meet the low power consumption requirements of temperature monitoring nodes in low temperature environment has become an urgent problem for cold chain logistics enterprises.

With the rise of the Internet of things (IOT) industry, wireless sensor network (WSN) [1–4] technology has attracted more and more attention. Among them, due to the small size of the wireless sensor nodes, the normally portable battery is also very limited. And the area of its deployment is relatively complex, in some areas even human beings are unable to reach, recharging or replacing the battery is almost impossible [5]. Therefore, the low power design of wireless sensor networks (WSN) is one of the most important

L. Barolli et al. (Eds.): EIDWT 2018, LNDECT 17, pp. 449–461, 2018.
https://doi.org/10.1007/978-3-319-75928-9_40

issues to be considered [6]. In order to solve the above problems, from the perspective of embedded intelligent sensor system, we need to study the two aspects of hardware design and program design in order to study the low power design method of wireless sensor nodes [7].

For the microprocessor, the static power consumption of the CPU circuit is very low, which accounts for less than 1% of the total power consumption under the current process, which is not considered for the time being. The dynamic power calculation formula is as follows:

$$P_d = C_T V^2 f \tag{1}$$

In the form, the Pd is CPU's dynamic power consumption; CT is the load capacitor of CPU; V is the working voltage of CPU; F is the operating frequency [13] for CPU.

It is known from the upper level that the power consumption of CPU should be reduced and the power supply voltage and working frequency should be reduced as much as possible under the condition of normal operation. In addition, the current mainstream low power microprocessor generally supports a variety of working modes, and appropriate mode switching can also reduce power consumption.

In software, in order to solve the problem that wireless sensor nodes can not respond to requests of other nodes in real time after sleep, the time synchronization technology is introduced. Because of the traditional time synchronization protocol of NTP [14] and GPS [15] are not suitable for the use of unstable communications in wireless communication, a low cost and power consumption of the sensitive characteristic, many scholars have put forward various synchronization algorithms for wireless sensor networks, these algorithms can be divided into based on receiver receiver synchronization algorithm based on pairwise two-way synchronization, based on the synchronization algorithm and the sender receiver synchronization algorithm.

Compared with the three synchronization algorithms, a simple star topology network which consists of a converging node and four acquisition nodes in a refrigerated compartment is not applicable. Because we do not need to monitor temperature data at all times in the whole cold chain transportation process, we introduce the energy management strategy based on context aware to reduce the power consumption of temperature monitoring nodes. context aware [18] refers to the situation information that can be used to obtain its own environment, and can adjust the [19] according to the situational information. Context aware mainly includes scenario acquisition, context representation and context use. Context acquisition refers to obtaining context information from environment; context representation refers to storing and organizing context information in appropriate format; context use refers to using context information in appropriate ways. For the cold chain logistics transportation process, vehicle information and specific business processes can be stored as a context information in the database. The background system can choose appropriate scenarios based on specific context information to feedback to the temperature monitoring nodes to achieve low power consumption of nodes.

2 System Scheme Design

2.1 System General Description

This design is the temperature collection subsystem in the cold chain logistics traceability system, and the schematic diagram of the refrigerated vehicle monitoring system is shown in Fig. 1.

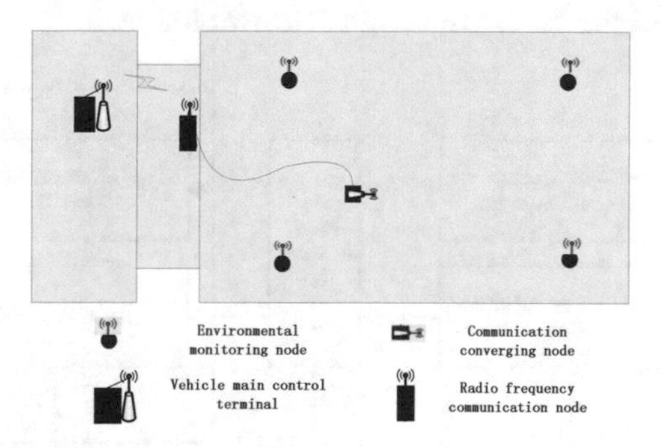

Fig. 1. Schematic diagram of a refrigerated vehicle monitoring system

The monitoring system for refrigerated vehicles consists of vehicle terminal, radio frequency communication node, converging node and temperature acquisition node. The temperature collection node is responsible for collecting the temperature in the refrigerated carriage and sending the temperature data to the gathering node in the mode of radio frequency communication. The aggregation node transmits the data of 4 temperature acquisition nodes in the refrigerated compartment to the radio frequency communication node in a wired way. The radio frequency communication node is sent to the vehicle terminal in the mode of radio frequency communication after receiving the data packet of the converging node.

2.2 System Requirement Analysis

The temperature monitoring node mainly completes the temperature collection and data transmission in the refrigerated carriage. In order to achieve the design requirements of easy installation and easy replacement of temperature monitoring nodes, the alkaline battery is selected to supply power, which eliminates the rewiring problem caused by vehicle power supply, and installs the tedious shortcomings. In addition, due to the continuous working time of rechargeable lithium battery under low temperature, it can not meet the requirements of cold chain logistics transportation, which is also an important reason for selecting alkaline batteries. Because the temperature collection nodes are powered by alkaline batteries, the design of low power consumption has become the focus of the design of temperature monitoring nodes in order to adapt to the long transportation time of the cold chain logistics and the inconvenient replacement of batteries in the transportation.

2.3 Overall Plan

2.3.1 Overview of the Scheme

The temperature monitoring node mainly includes two aspects of hardware design and program design.

2.3.2 Hardware Overall Design

The block diagram of the hardware design is shown in Fig. 2:

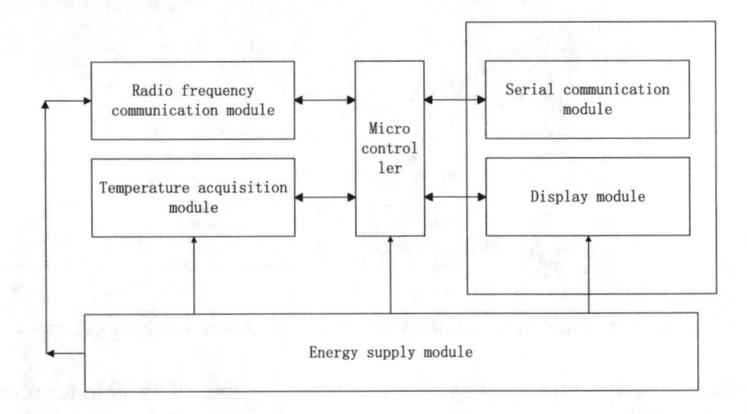

Fig. 2. Block diagram of hardware design

The hardware design includes the micro controller, the radio frequency communication module, the temperature acquisition module, the serial communication module, the display module and the energy supply module. The microcontroller controls temperature acquisition module to collect and transform the ambient temperature through the single bus protocol. After the microcontroller receives the temperature data, it encapsulates the specified packet format and sends it to the sink node through the radio frequency communication module. The energy supply module is responsible for providing the rated voltage required by other modules so that the nodes can work properly. In addition, the hardware design also includes serial communication module and display module, which is used for output node's current working state and data package content. It is mainly used for program debugging and testing. In real time, the two modules can be closed to reduce power consumption.

2.3.3 General Design of Program

Program development uses IAR Embedded Workbench as a development platform. The C language is a programming language. The program of temperature monitoring node is designed by using the layered and modularized programming idea. The general block diagram of the program is shown in Fig. 3:

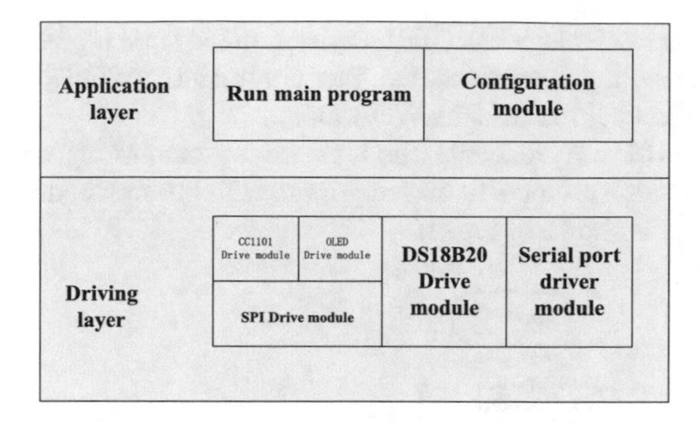

Fig. 3. General block diagram of program

The whole program is divided into the driving layer and the application layer. The driver layer mainly includes SPI driver module, CC1101 driver module, OLED driver module, DS18B20 driver module and serial port driver module, so that each module of temperature monitoring node can work normally and provide interface for application layer. The application layer mainly includes the configuration module and the main program module to realize the normal work of the temperature monitoring node according to the design requirements.

2.4 Low Power Design Scheme

2.4.1 Based Low Power Design Based on WOR

The full name of the WOR Wake On Radio (electromagnetic wave wake-up) is a low-power working mode supported by CC1101. The principle is to make CC1101 into sleep mode, use the timer periodic chip wake detection of CC1101 electromagnetic wave in the air, if the receiving set time to detect electromagnetic waves can receive data packets, or automatically re entering sleep mode with, as shown in Figs. 2, 3 and 4. Although WOR does not really use electromagnetic wave to wake up, it reduces the duration of the receiving state and reduces the power consumption of the system.

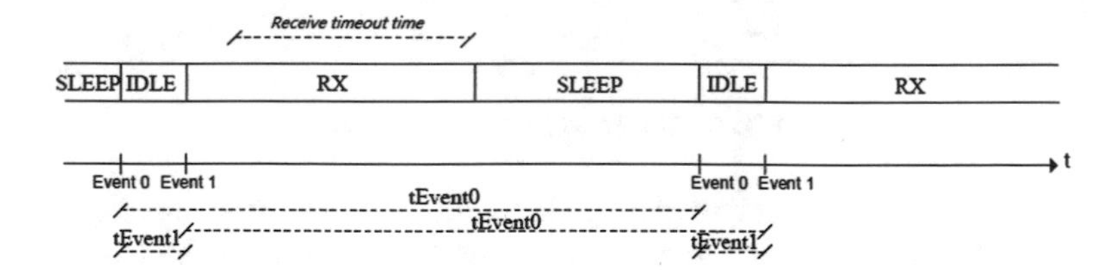

Fig. 4. WOR time diagram

In order to use the WOR function of CC1101, you need to configure the MCSM2, WOREVT1, WOREVT0, and WORCTRL registers. MCSM2 register set receive timeout duration; WOREVT1 register set Event1, duration from sleep state automatically enter the idle mode; the duration of the WOREVT0 register set Event0, two automatic wake-up intervals; WORCTRL register on the three register set weight. After the above register setting is received, the filter commands SWORRST and SWOR are sent in turn to make CC1101 into the WOR mode.

Based on CC1101's WOR mode, the temperature monitoring node uses a custom passive communication protocol to exchange data with the sink node. The flow chart of the collection node is shown in Fig. 5:

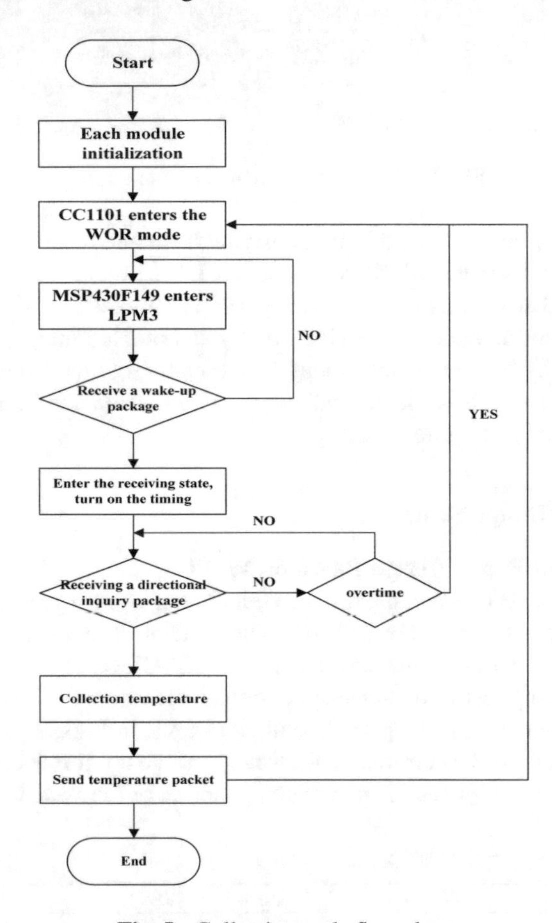

Fig. 5. Collection node flow chart

After the temperature acquisition node is powered on, first use the function provided by the driver layer to initialize each module, then make the CC1101 enter the WOR mode, and finally let MSP430F149 enter the LPM3 low power mode. When the acquisition node receives a wakeup packet sent by the sink node, MSP430F149 through the external interrupt wake-up from LPM3 low power mode, wait for the sink node to send packets and ask the directional timer timing opening, when directional reception to the sink node query package, temperature acquisition node obtains the temperature data through the CC1101 module to send to the sink node through the DS18B20, when the timeout timer detects a timeout after the temperature acquisition node to CC1101 into WOR mode, MSP430F149 LPM3 into a low power mode, waiting for the next wake (Fig. 6).

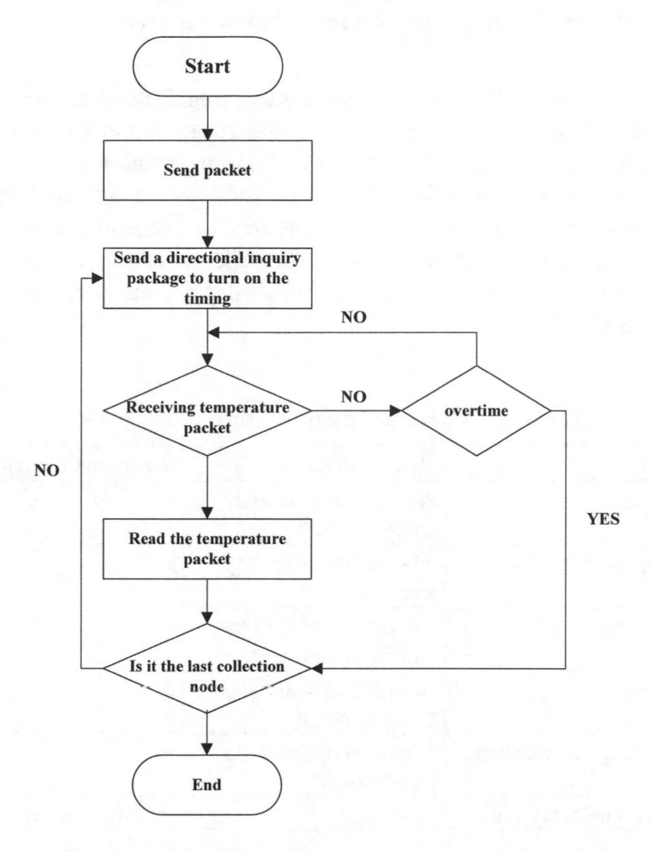

Fig. 6. The workflow diagram of the aggregation node

When the temperature of the data aggregation nodes need to obtain each acquisition node, sends the first wake-up packets wake up all the temperature acquisition nodes, whereas WOR work, wake up packets must continue to send Event0 set time to ensure that all temperature acquisition nodes are correct to wake up, and then send the directional inquiry package, and open the timeout timer, when the temperature data the received packet is read after the temperature the temperature node, continuing down a temperature acquisition node sends directional inquiry packets, when the timeout period has not yet received the temperature data packet is asked directly to a temperature acquisition node sends packet oriented.

2.4.2 Based Scenario Based Low Power Design

In the cold chain logistics and transport process, the temperature control node does not always need to monitor the temperature of the refrigerating compartment, refrigerated vehicles in different situations according to the model, the temperature control node can effectively reduce the power consumption by using different ways of working, prolonging the service life of the battery, reducing the number of battery replacement. The specific design of energy management based on context aware includes the following steps:

(1) All kinds of situational information gathered by vehicle monitoring terminal in cold chain logistics, including refrigerated vehicle location and speed information, is encapsulated into data packets and sent to the background server.
(2) The background system collects the scene information collected by the vehicle terminal, combines the current business process, and determines the user's current business scenario. According to the established scenario energy management model, it outputs the specific application strategy, namely energy management strategy, as shown in Table 1.

Table 1. Context-aware energy management strategy

Context	Node active state	Wake-up time interval
loading	Alternation of activity and dormancy	t_1
transportation-in	Alternation of activity and dormancy	t_2
Discharge cargo	Alternation of activity and dormancy	t_3
Parking rest	Alternation of activity and dormancy	t_4
Anchorage maintenance	Alternation of activity and dormancy	t_5
Empty cars return, idle	dormancy	Not Awakening

(3) When the vehicle monitoring terminal receives instructions from the system, it sends the instructions to the temperature monitoring node through the wireless communication mode through the relay node and the sink node. After receiving and parsing instructions, the temperature monitoring node set up operation parameters (node activity status, sampling interval, radio frequency communication time interval, etc.) using the scenario mode required by instruction to dynamically change the working mode of nodes.

2.4.3 Function Call Optimization

When a function is invoked, when the microcontroller enters the function, it will first press the value of the current register into the stack, and when the microcontroller exits from the function, it will re play the register value before pressing into the stack, so as to ensure the consistency of the register value before and after the function call. However, frequent function calls will generate too many stack and stack operations, prolonging the run time of the program and increasing the power consumption. Therefore, this design will replace some functions with simple functions by macrodefinition.

Macro definition will be converted to original code in program compile time, avoiding function calls, improving the operation efficiency of the program and reducing the power consumption.

3 Communication Protocol Establishment

The main program module realizes the initialization and related operations of each module by invoking the corresponding interfaces provided in the driver layer and configuring the corresponding configuration parameters in the configuration module, and completes the sending and receiving of the data packets and the realization of the communication protocol based on this. Three data package formats are defined, namely wakeup package, directional inquiry package and temperature data packet. For sender, data package format is shown in Figs. 7, 8 and 9:

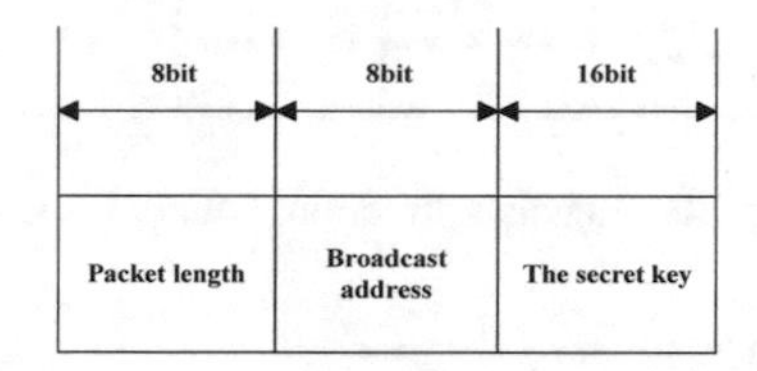

Fig. 7. Wake-up packet diagram

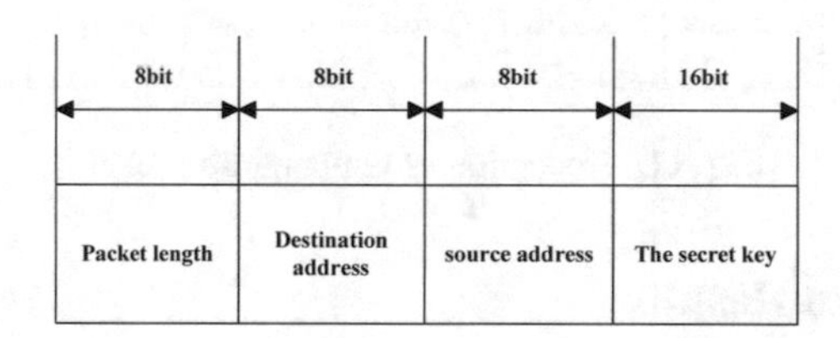

Fig. 8. Send a directional inquiry package

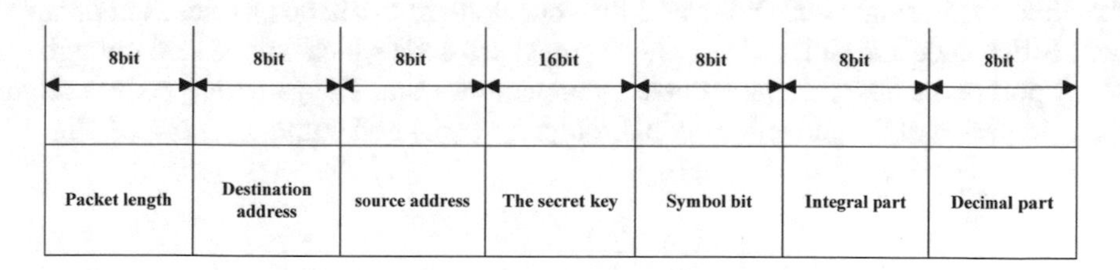

Fig. 9. Send temperature packet

Because the variable length packets are used in the CC1101 register configuration, the first byte of all packets must be data packet length, and the packet length is defined as the byte length outside the packet length. Because the address filter is opened in the CC1101 register configuration, the second byte of the packet is the destination address. For the wake-up package, all nodes need to be awakened, so the destination address is the broadcast address, 0x00. In order to prevent the temperature control node in different refrigerated vehicles mutual interference in radio communication, data packets into 16 bit key, each car refrigerator car has a unique secret key, the receiver receives the data packet after comparing the secret key, if the match is successful receiving data packets, packet or discarded. For a temperature packet, three bytes are used to represent the symbolic, integer and decimal parts of the temperature data. Among them, the symbol position is 0 is positive or negative.

For the receiver, two state bytes of RSSI and LQI are added on the basis of sending packets, respectively, to reflect the received signal intensity and link quality. As shown in Figs. 10 and 11:

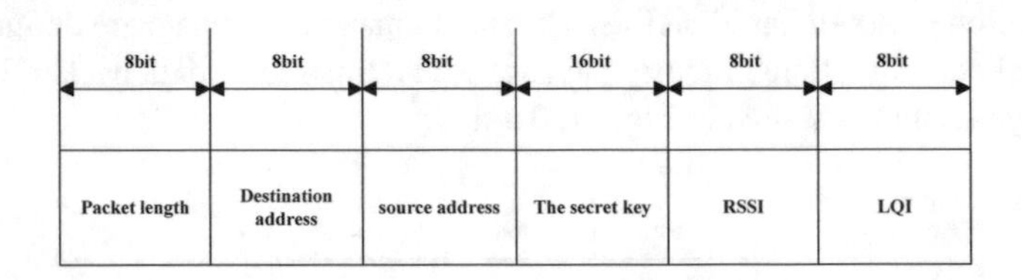

Fig. 10. Receive directional query package

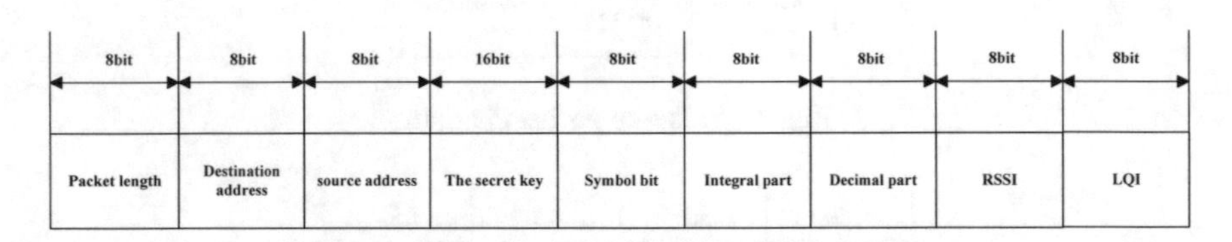

Fig. 11. Reception of temperature packet

4 Experimental Results

In order to test the temperature control node can carry out normal communication with the sink node using a sink node and the four node test data acquisition, temperature acquisition node and sink node to wake up get each node which will be sent through the serial port to the host computer display temperature data. The gathering node receives the data packet of the collection node through the serial port output as shown in Fig. 12:

```
d: 0x0A| s: 0x0D| key: 0x6F,0x6F| temp: 000,024,007| rssi: 063| lqi: 048
d: 0x0A| s: 0x0E| key: 0x6F,0x6F| temp: 000,024,010| rssi: 047| lqi: 048
Send wakeup
d: 0x0A| s: 0x0B| key: 0x6F,0x6F| temp: 000,024,007| rssi: 064| lqi: 050
d: 0x0A| s: 0x0C| key: 0x6F,0x6F| temp: 000,024,006| rssi: 071| lqi: 050
d: 0x0A| s: 0x0D| key: 0x6F,0x6F| temp: 000,024,008| rssi: 063| lqi: 049
d: 0x0A| s: 0x0E| key: 0x6F,0x6F| temp: 000,024,011| rssi: 047| lqi: 048
Send wakeup
d: 0x0A| s: 0x0B| key: 0x6F,0x6F| temp: 000,024,007| rssi: 064| lqi: 050
d: 0x0A| s: 0x0C| key: 0x6F,0x6F| temp: 000,024,006| rssi: 071| lqi: 050
d: 0x0A| s: 0x0D| key: 0x6F,0x6F| temp: 000,024,008| rssi: 062| lqi: 050
d: 0x0A| s: 0x0E| key: 0x6F,0x6F| temp: 000,024,010| rssi: 048| lqi: 049
Send wakeup
d: 0x0A| s: 0x0B| key: 0x6F,0x6F| temp: 000,024,007| rssi: 064| lqi: 047
d: 0x0A| s: 0x0C| key: 0x6F,0x6F| temp: 000,024,006| rssi: 071| lqi: 049
d: 0x0A| s: 0x0D| key: 0x6F,0k6F| temp: 000,024,008| rssi: 062| lqi: 049
d: 0x0A| s: 0x0E| key: 0x6F,0x6F| temp: 000,024,011| rssi: 047| lqi: 049
```

Fig. 12. Data packet reception diagram

The packet includes a destination address, source address, key, temperature, RSSI and LQI. The destination address identifying data packet sending source address; the source address identifies the data packet; the secret key for identification of different acquisition node in the refrigerating compartment; temperature temperature data acquisition node is obtained; RSSI said received signal strength; LQI link quality indication. The test results show that the temperature monitoring node can collect the temperature normally and upload it to the convergence node.

In order to test the power consumption of temperature monitoring nodes, this design is realized by measuring the power current of nodes. In the test, the digital multimeter is used to adjust the current file to connect with the temperature monitoring node circuit, so that the temperature monitoring nodes work separately in receiving mode, sending mode and sleeping mode. Each working mode current is shown in Table 2.

Table 2. Node power table

Working mode	Current value
Receiving mode	25 mA
Transmission mode	9 mA
Sleep pattern	3 mA

Receiving mode current is I1 (unit: mA), sleep mode working current of I2 (unit: mA), wake time interval T1 (unit: s), the waiting time for receiving T2 (unit: s), the temperature monitoring node average current (unit: mA) the expression:

$$\bar{I} = \frac{T_1 I_2}{T_1 + T_2} + \frac{T_2 I_1}{T_1 + T_2} \tag{2}$$

A battery of alkaline battery is Q (unit: mAh), and the number of batteries is n (unit: Section). Then the temperature monitoring node can work continuously after changing the battery once a day, D (unit: day):

$$D = nQ \div \bar{I} \div 24 \tag{3}$$

In accordance with the sink node every fifteen minutes to wake up a temperature monitoring node, temperature monitoring nodes are awakened after receiving the waiting time for 8 s, use 3 2300 mAh alkaline battery capacity can be calculated, the temperature control node of a battery replacement after days of continuous work for 90 days.

5 Conclusion

The design uses a low power micro controller MSP430F149 as the main control chip and the low power RF transceiver chip CC1101 as a wireless communication module. In the procedure of CC1101 WOR function is designed based on the communication protocol of a passive polling, the sink node wake-up temperature monitoring node, temperature monitoring nodes are awakened after waiting for sending and receiving directional acquisition node query package, the temperature control node receives the packet after the start of directional ask inside the refrigeration temperature data acquisition. The digital multimeter is connected with the temperature monitoring node circuit in series. The temperature monitoring node receiving mode, sending mode and current mode under sleep mode are tested. The test results show that the temperature monitoring node can work continuously for 90 days, which meets the requirements of low power consumption.

Acknowledgments. This work was supported by the Key research and development (industry and information) projects of Huaian, Jiangsu, China (Grant No.: HAG201604).

References

1. Tan, K.K., Huang, S.N., Zhang, Y., et al.: Distributed fault detection in industrial system based on sensor wireless network. Comput. Stan. Interfaces **31**(3), 573–578 (2009)
2. Zhang, X., Liu, Y., Liu, H., Li, Z.: An ultra-low power MAC protocol for wireless sensor networks. Chin. J. Sens. Actuators **40**(7), 2038–2048 (2007)
3. Tang, L.,Sun, Y.,Gurewitz, O., et al.: An energy-efficient predictive-wakeup MAC protocol for wireless sensor networks. In: IEEE INFOCOM, pp. 1305-1313 (2011)
4. Ganeriwal, S., Kumar, R., Srivastava, M.B.: Timing-Sync Protocol for Sensor Networks, pp. 138–149. ACM Press, New York (2003)
5. Allen, W.G., Johnson, J., Ruiz, M., et al.: Monitoring volcanic eruptions with a wireless sensor network. In: Proceeedings of the Second European Workshop on Wireless sensor Networks, pp. 108–120 (2005)
6. Noh, K., Serpedin, E., Qaraqe, K.A.: New approach for time synchronization in wireless sensor networks: pairwise broadcast synchronization. IEEE Trans. Wireless Commun. **7**(9), 3318–3322 (2008)
7. Marco, A., Casas, R., Ramos, J.L.S., et al.: Synchronization of multihop wireless sensor networks at the application layer. IEEE Trans. Wireless Commun. **18**(1), 82–88 (2011)
8. Elson, J., Girod, L., Estrin, D.: Fine-grained network time synchronization using reference broadcasts. In: Proceedings of the Fifth Symposium on Operating systems Design and Implementation (OSDI2002), pp. 147–163. ACM Press, New York (2002)

9. Hofmann-Wellenhof, B., Lichtenegger, H., Collins, J.: Global Positioning System: Theory and Practice. Springer, Berlin (1997)
10. Mills, D.L.: Internet time synchronization: the network time protocol. IEEE Trans. Commun. **39**, 1482–1493 (1991)
11. Lin, G., Stankovic, J.A.: Radio-triggered wake-up capability for wireless sensor networks. Real-Time Syst. **29**(2), 157–182 (2005)
12. Ma., W., Wu., D., Xu., D., et al.: Uniform identification system construction for agricultural IoT. China Stand. (1), 79–83 (2014). (in Chinese)
13. Liu, Y., Ma, R., Cao, W., et al.: Progress on the research of can bus in automatic navigation system of agricultural vehicles. J. Agric. Mechanization Res. **34**(8), 233–236 (2011). (in Chinese)
14. Gao, X., Ju, J., Jiang, M., et al.: Design on distributed agricultural greenhouse control system based on CAN bus. J. Chinese Agric. Mechanization **37**(4), 67–70 (2016). (in Chinese)
15. Ke, X., Zhang, W., Tang, K., et al.: Design of agricultural intelligent monitoring system based on GSM network and 485 bus. J. Chinese Agric. Mechanization **4**(5), 213–218 (2016). (in Chinese)
16. Zheng, N., Yang, X., Wu, S.: A survey of low-power wide-area network technology. Inf. Commun. Technol. **10**(1), 47–54 (2017). (in Chinese)
17. Liu, X., Zheng, H., Shi, N., et al.: Artificial intelligence in agricultural applications. Fujian J. Agric. Sci. **28**(6), 609–614 (2013). (in Chinese)
18. Zheng, J., Liu, P., Zhang, Z., et al.: Application of cabin intelligent compartments based on expert system in fishing vessel. Trans. Chinese Soc. Agric. Eng. **31**(6), 208–212 (2015). (in Chinese)
19. Wang, T., Zhang, X., Chen, W., et al.: RFID-based temperature monitoring system of frozen and chilled tilapia in cold chain logistics. Trans. Chinese Soc. Agric. Eng. **27**(9), 141–146 (2011). (in Chinese)

Publicly Verifiable 1-norm and 2-norm Operations over Outsourced Data Stream Under Single-Key Setting

Yudong Liu[1,2], Xu An Wang[1,2(✉)], Arun Kumar Sangaiah[3], and Heshuai Shao[4]

[1] Key Laboratory for Network and Information Security of Chinese Armed Police Force, Engineering University of Chinese Armed Police Force, Xi'an 710086, Shaanxi, China
wangxazjd@163.com

[2] Department of Electronic Technology, Engineering University of the Chinese Armed Police Force, Xi'an 710086, Shaanxi, China

[3] School of Computer Science and Engineering, VIT University, Vellore 632014, Tamil Nadu, India

[4] Huaiyin Institute of Technology, Huaiyin, People's Republic of China

Abstract. With the advent of the big data era, the amount of data computation is getting larger and larger, and the computational load of clients is also increasing day by day. The advent of clouds allows clients to outsource their data to the cloud for computing services. Outsourced computation has greatly reduced the computational burden of clients, but also brings the issue of trust. Because the cloud is not trustworthy, clients need to verify the correctness of the remote computation results. In this paper, we mainly study the common norm operations, and propose two publicly verifiable schemes for 1-norm and 2-norm operations respectively, any client can publicly verify these two common norm operations under single-key setting by using our schemes.

Keywords: Outsourced data stream · 1-norm and 2-norm · Public verification

1 Introduction

In the information age, many data sources usually generate data in the form of a stream. These data streams, which grow exponentially, put a heavy burden on the clients' storage and computation. Currently, outsourcing their generated data to the cloud server for storage and computation services is a better way for data sources, and clients simply need to get the computation results from the cloud server. However, cloud servers and clients are not trusted for data sources, outsourcing naturally causes the trust issues [1, 2], so clients and data sources need to verify the correctness of outsourced computation results, the cause of efficiency hinders the application of fully homomorphic encryption in this area [3]. The existing outsourced computation verification schemes are mostly designed for single-key setting. In these schemes, outsourced data generally comes from the same data source [4, 8], or data from multiple data sources but using the same key [3, 9], our proposed schemes belong to the former.

L. Barolli et al. (Eds.): EIDWT 2018, LNDECT 17, pp. 462–469, 2018.
https://doi.org/10.1007/978-3-319-75928-9_41

Norm is a basic concept in mathematics. It has important applications in classical control theory, machine learning, error estimation of matrices or vectors and so on [10]. In particular, norm operations frequently occur in the outsourced computation of data streams. Concretely, 1-norm is the sum of the absolute value of each data, 2-norm represents the distance in the usual sense [11]. These two norm operations are also common, but there are not any good schemes to verify the results of them in outsourced computation. In this paper, we propose two outsourced computation schemes for 1-norm and 2-norm. Finally, we conduct a security analysis of our proposed schemes in the random oracle model and the results prove that our schemes are secure.

Organization

The rest of this paper is organized as follows: Sect. 2 introduces the system model of our proposed schemes. In Sect. 3, we present our schemes' algorithm formulation. The security definition and concrete construction are shown in Sect. 4 and Sect. 5 respectively. In Sect. 6, we give a security analysis for our proposals. Then, an evaluation of storage and computation is given in Sect. 7. Finally, Sect. 8 summaries this paper.

2 System Model

The system model of our proposed schemes are shown in Fig. 1. In our system model, there are multiple data sources $M_1, M_2, \ldots, M_l$ and each data source has a unique pair of public and private keys, $\chi_{j,i,F}$ represents a new data generated by the data source M_j $(1 \leq j \leq l)$ at time i, the variant F in the hash functions h_1, h_2, h_3 or data stream χ denotes that data source M_j can outsource many different files (data streams) with the same index sequence $i = 1, 2, 3 \ldots$. To protect the outsourced data $\chi_{j,i,F}$, the data source M_j generates the corresponding publicly verifiable tag $\sigma_{j,i,F}$ and then outsources a tuple $\{i, \chi_{j,i,F}, \sigma_{j,i,F}\}$ to a third party server, clients request the server to compute the common norm operations of the machine M_j's outsourced data stream by sending a corresponding query. After receiving the query, the third party server makes the appropriate computation and gets the corresponding computation result *res*, at the same time, in order to facilitate clients to verify the correctness of the computation results, the server generates a proof π for every computation result *res*, and finally returns the computation result *res* and the corresponding proof π to clients.

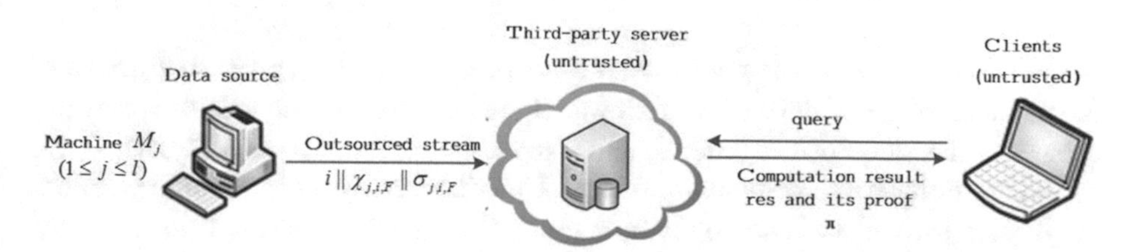

Fig. 1. System model

3 Algorithm Formulation

In this section, we provide the algorithm formulation of our schemes, our algorithm of publicly verifiable common norm operations schemes include five subsections as follows:

- *KeyGen*$(1^k) \rightarrow (pk_j, sk_j)$: Each machine runs a probabilistic algorithm by inputting a security parameter k, after that outputting a public key pk_j and a secret key sk_j.
- *TagGen*$(sk_j, i, \chi_{j,i,F}) \rightarrow \sigma_{j,i,F}$: The machine M_j runs a probabilistic algorithm by taking its secret key sk_j, the current discrete time i and data $\chi_{j,i,F}$ as input, and outputting a publicly verifiable tag $\sigma_{j,i,F}$.
- *Evaluate*(F_{NO}, χ_j): The outsourced data stream of machine M_j is denoted by $\chi_j = \{\chi_{j,1}, \chi_{j,2}, \cdots \chi_{j,n}\}$. The server runs this deterministic algorithm to compute 1-norm or 2-norm by inputting common norm operation function F_{NO} and the data stream χ_j, after that outputting a computation result *res*.
- *GenProof*$(F_{NO}, \sigma_i, \sigma_j, \chi_j) \rightarrow \pi$: The machine M_j generates the tag vector σ_j for the data source χ_j. After that the server runs this algorithm to generate a proof for the result *res* by taking the common norm operation function F_{NO}, the tag vector σ_j, as well as the data stream χ_j, and outputting a proof π.
- *CheckProof*$(F_{NO}, pk_j, res, \pi) \rightarrow 0, 1$: The client runs a deterministic algorithm to check the correctness of *res* by inputting the function F_{NO}, the result *res*, the proof π as well as the public key pk_j, and if outputting 1, the client accepts, otherwise the client rejects.

4 Security Definition

Our schemes are based on the *CDH* assumption: Knowing the values of $g, g_1 \in G$ and g^s $(s \in Z_q^*)$ and s is unknown, there not exists a polynomial-time algorithm which can compute g_1^s with negligible advantage [12].

5 Our Construction

In our schemes, we use the public parameters $\{e, G_1, G_2, q, g, g_1, g_2, g_3, h_1, h_2, h_3\}$ and these parameters are defined as follows: G_1 and G_2 are two multiplicative cyclic groups of the same prime order q, and e represents a bilinear map $G_1 \times G_1 \rightarrow G_2$ satisfying bilinearity, computability and Non-degeneracy [13]. Four generators $\{g, g_1, g_2, g_3\}$ are selected randomly from group G_1 by the data source. Three different collision-resistant hash functions are denoted by h_1:$\{0, 1\} \rightarrow Z_q^*$, h_2:$\{0, 1\} \rightarrow Z_q^*$ and h_3:$\{0, 1\} \rightarrow Z_q^*$ respectively.

5.1 The Concrete Scheme of 1-norm

(a) *KeyGen*(1^k): For $j = 1$ to l do: The data source chooses a random number $sk_j = s_j \in Z_q^*$ as the secret key, and computes $pk_j = g^{s_j}$, finally outputs (pk_j, sk_j).

(b) *TagGen*($sk_j, i, \chi_{j,i,F}$): In this phase, the data source computes the corresponding tag $\sigma_{j,i,F} = (g_1^{h_1(M_j,i,F)} g_2^{h_2(M_j,i,F)} g_3^{|\chi_{j,i,F}|} h_3(M_j, i, F))^{sk_j}$ for the data $\chi_{j,i,F}$, and outputs $\sigma_{j,i,F}$.

(c) *Evaluate*($F_{GS}, \chi_{j,F}$): The cloud server computes the result $res = \sum_{i\in\Delta} |\chi_{j,i,F}|$, and outputs *res*.

(d) *GenProof*($F_{on}, \sigma_{j,F}, \chi_{j,F}$):The cloud server computes the proof $\pi = \prod_{i\in\Delta} \sigma_{j,i,F}$ for the result *res*, and outputs π.

(e) *CheckProof*($F_{GS}, pk_j, \sigma_{j,F}, res, \pi$): Setting $S_\Delta = (S_1, S_2)$, $S_1 = \sum_{i\in\Delta} h_1(M_j, i, F)$ and $S_2 = \sum_{i\in\Delta} h_2(M_j, i, F)$. Because S_1 and S_2 has nothing to do with the data $\chi_{j,i,F}$, in order to reduce the verification cost, the client can compute them in advance, if the following *Eq*. (1) holds, then outputs 1(accept), else outputs 0(reject).

$$\begin{aligned}
&e(\pi, g)\\
&= e(\prod_{i\in\Delta} \sigma_{j,i,F}, g)\\
&= e((g_1^{\sum_{i\in\Delta} h_1(M_j,i,F)} g_2^{\sum_{i\in\Delta} h_2(M_j,i,F)} g_3^{\sum_{i\in\Delta} |\chi_{j,i,F}|} \prod_{i\in\Delta} h_3(M_j, i, F))^{sk_j}, pk_j)\\
&= e(g_1^{S_1} g_2^{S_2} g_3^{res}, pk_j) \prod_{i\in\Delta} e(h_3(M_j, i, F), pk_j)
\end{aligned} \tag{1}$$

5.2 The Concrete Scheme of 2-norm

(a) *KeyGen*(1^k): For $j = 1$ to l do: The data source chooses a random number $sk_j = s_j \in Z_q^*$ as the secret key, and computes $pk_j = g^{s_j}$, finally outputs (pk_j, sk_j).

(b) *TagGen*($sk_j, i, \chi_{j,i,F}$): In this phase, the data source computes the corresponding tag $\sigma_{j,i,F} = (g_1^{h_1(M_j,i,F)} g_2^{h_2(M_j,i,F)} g_3^{|\chi_{j,i,F}|^2} h_3(M_j, i, F))^{sk_j}$ for the data $\chi_{j,i,F}$, and then outputs $\sigma_{j,i,F}$.

(c) *Evaluate*($F_{GS}, \chi_{j,F}$): The cloud server computes the result $res = (\sum_{i\in\Delta} \left|\chi_{j,i,F}\right|^2)^{1/2}$, after that outputs *res*.

(d) *GenProof*($F_{on}, \sigma_{j,F}, \chi_{j,F}$):The cloud server computes the proof $\pi = \prod_{i\in\Delta} \sigma_{j,i,F}$ for the result *res*, and outputs π.

(e) *CheckProof*($F_{GS}, pk_j, \sigma_{j,F}, res, \pi$): Setting $S_\Delta = (S_1, S_2)$, like the scheme of 1-norm, the client can compute the values of $S_1 = \sum_{i\in\Delta} h_1(M_j, i, F)$ and $S_2 = \sum_{i\in\Delta} h_2(M_j, i, F)$ in advance, if the following *Eq*. (2) holds, then outputs 1(accept), else outputs 0(reject).

$$\begin{aligned} & e(\pi, g) \\ & = e(\prod\nolimits_{i \in \Delta} \sigma_{j,i,F}, g) \\ & = e((g_1^{\sum_{i\in\Delta} h_1(M_j,i,F)} g_2^{\sum_{i\in\Delta} h_2(M_j,i,F)} g_3^{\sum_{i\in\Delta} |\chi_{j,i,F}|^2} \prod\nolimits_{i \in \Delta} h_3(M_j, i, F))^{sk_j}, pk_j) \\ & = e(g_1^{S_1} g_2^{S_2} g_3^{res^2}, pk_j) \prod\nolimits_{i \in \Delta} e(h_3(M_j, i, F), pk_j) \end{aligned} \tag{2}$$

6 Security Analysis

In this section, we will prove the security of our proposed schemes in the random oracle model.

Theorem 6.1. Under the *CDH* assumption, our publicly verifiable computation schemes for 1-norm is secure against an adaptive chosen-message attack.

Proof: The security definition of our publicly-verifiable computation 1-norm scheme is similar to the *CDH* assumption, supposing that a result $res \neq \sum_{i\in\Delta} \left|\chi_{j,i,F}\right|$ is forged by the adversary A and passes the verification successfully. Next we will attempt to construct an adversary B that uses the information provided by A to solve the *CDH* problem.

According to our system model, the adversary A and B simulate our scheme of 1-norm as follows.

Setup: Machine M_j's public key $pk_j = g^{s_j}$, $g_1 = g \cdot g_3^{\alpha}$ and $g_2 = (g \cdot g_3)^{\beta}$ are set by the adversary B, α and β are two random numbers in Z_q^*. After that the adversary B gives the system parameters and the public key to A.

Query: The adversary B is queried by A for tags on the discrete time, data and file of its choice. Concretely, B gets a tuple $(M_j, \chi_{j,1,F}, 1)$ from A and then generates a tag $\sigma_{j,1,F}$, finally sends it back to A. The adversary A can continually make tag queries to B for the tags on $(M_j, \chi_{j,2,F}, 2))$, $(M_j, \chi_{j,3,F}, 3)$,…,$(M_j, \chi_{j,n,F}, n)$ of its choice. After receiving the tag queries, B makes the following response:

First of all, to record the tuples $(M_j, \chi_{j,i,F}, i, \gamma_{j,i,F}, \sigma_{j,i,F})$, B initializes an empty list L, and then:

- If $(\chi_{j,i,F}, i)$ has been queried before, B retrieves the tuple $(\chi_{j,i,F}, i, \gamma_{j,i,F}, \sigma_{j,i,F})$ from the list L and returns $\sigma_{j,i,F}$ to A.
- If i has not been queried, B selects a random number $\gamma_{j,i,F}$ from Z_q^* and sets $\sigma_{j,i,F} = pk_j^{\gamma_{j,i,F}} = g^{s_j \cdot \gamma_{j,i,F}}$. Then B adds $(\chi_{j,i,F}, i, \gamma_{j,i,F}, \sigma_{j,i,F})$ into the list L and returns $\sigma_{j,i,F}$ to A.
- Otherwise, i.e., i has been queried but $(\chi_{j,i,F}, i) \notin L$, B rejects this query.

Moreover, B returns $h_1(M_j, i, F) = \frac{\left|\chi_{j,i,F}\right| + \gamma_{j,i,F}}{1-\alpha}$ and $h_2(M_j, i, F) = \frac{\left|\chi_{j,i,F}\right| + \alpha\gamma_{j,i,F}}{\beta(\alpha-1)}$ to A for the hash queries. Under the public key $pk_j = g^{s_j}$, we can know the tag $\sigma_{j,i,F} = g^{s_j \cdot \gamma_{j,i,F}} \cdot h_3(M_j, i, F)^{s_j}$ on $(\chi_{j,i,F}, i)$ is valid because of the following *Eq*. (3):

$$\begin{aligned}
& e(g_1^{h_1(M_j,i,F)} g_2^{h_2(M_j,i,F)} g_3^{\left|\chi_{j,i,F}\right|} h_3(M_j, i, F), pk_j) \\
& = e(g^{h_1(M_j,i,F)} g_3^{\alpha h_1(M_j,i,F)} (g \cdot g_3)^{\beta h_2(M_j,i,F)} g_3^{\left|\chi_{j,i,F}\right|} h_3(M_j, i, F), pk_j) \\
& = e(g^{\gamma_{j,i,F}} h_3(M_j, i, F), g^{s_j}) \\
& = e(g^{\gamma_{j,i,F} \cdot s_j} h_3(M_j, i, F)^{s_j}, g)
\end{aligned} \tag{3}$$

Request: B sends a time set Δ to A and requests it to compute $\sum_{i\in\Delta} |\chi_{j,i,F}|$.

Forge: A returns a computation result *res* and the corresponding proof π to B. π is a valid proof because it can pass the algorithm *CheckProof*, but $res \neq \sum_{i\in\Delta} \left|\chi_{j,i,F}\right|$. Therefore, we know $\pi = (g_1^{S_1} g_2^{S_2} g_3^{res} \prod_{i\in\Delta} h_3(M_j, i, F))^{s_j}$, let $res' = \sum_{i\in\Delta} \left|\chi_{j,i,F}\right|$ be the correct result, we can obtain:

$$\begin{aligned}
\pi & = (g_1^{S_1} g_2^{S_2} g_3^{res} \prod\nolimits_{i\in\Delta} h_3(M_j, i, F))^{s_j} \\
& = (g_1^{\sum_{i\in\Delta} h_1(M_j,i,F)} g_2^{\sum_{i\in\Delta} h_2(M_j,i,F)} g_3^{res} \prod\nolimits_{i\in\Delta} h_3(M_j, i, F))^{s_j} \\
& = ((g \cdot g_3^{\alpha})^{\sum_{i\in\Delta} h_1(M_j,i,F)} (g \cdot g_3)^{\beta \sum_{i\in\Delta} h_2(M_j,i,F)} g_3^{res} \prod\nolimits_{i\in\Delta} h_3(M_j, i, F))^{s_j} \\
& = (g^{\sum_{i\in\Delta} h_1(M_j,i,F) + \beta \sum_{i\in\Delta} h_2(M_j,i,F)} \cdot g_3^{\alpha \sum_{i\in\Delta} h_1(M_j,i,F) + \beta \sum_{i\in\Delta} h_2(M_j,i,F) + res} \prod\nolimits_{i\in\Delta} h_3(M_j, i, F))^{s_j} \\
& = (g^{\sum_{i\in\Delta} \gamma_{j,i,F}} g_3^{res - res'} \prod\nolimits_{i\in\Delta} h_3(M_j, i, F))^{s_j} \\
& = (g^{\sum_{i\in\Delta} \gamma_{j,i,F}} \prod\nolimits_{i\in\Delta} h_3(M_j, i, F))^{s_j} \cdot g_3^{s_j (res - res')}
\end{aligned} \tag{4}$$

Due to $res' \neq res$, B can compute $g_3^{s_j} = \left(\frac{\pi}{(g^{\sum_{i\in\Delta} \gamma_{j,i,F}} \prod_{i\in\Delta} h_3(M_j, i, F))^{s_j}}\right)^{(res-res')^{-1}}$ from the above *Eq*. (4). In the experiment, all parameters and requirements satisfy our system model and scheme. Thus, under the *CDH* assumption, our 1-norm scheme is secure against an adaptive chosen-message attack in the random oracle.

Theorem 6.2. Under the *CDH* assumption, the public verifiable tags are unforgeable, no source can deny its tags outsourced to the server.

Proof: The proof is similar to the process of **Theorem** 6.1, supposing that B returns m_1, m_2 and m_3 to A for the hash queries $h_1(m_j, n+1, F)$, $h_2(m_j, n+1, F)$ and $h_3(m_j, n+1, F)$, and then the adversary A generates a valid tag $\sigma_{j,n+1,F} = (g_1^{m_1} g_2^{m_2} g_3^{\left|\chi_{j,n+1,F}\right|} .m_3)^{s_j}$ on data $\chi_{j,n+1,F}$ at time $n+1$. Given $\sigma_{j,n+1,F}$, the adversary

B can compute the value of $g_3^{s_j}$: $g_3^{s_j} = (\frac{\sigma_{j,i,F}}{g^{(m_1+\beta m_2)s_j} \cdot m_3^{s_j}})^{(\alpha m_1+\beta m_2+|\chi_{j,n+1,F}|)^{-1}}$, which contradicts the *CDH* assumption. Therefore, the public verifiable tags are undeniable [14].

The security analysis of our 2-norm scheme is similar to the process of 1-norm, here we will not repeat it.

7 Evaluation

In this section, we will make an assessment of our proposed schemes from two aspects: storage and computation.

7.1 Storage

Our system model consists of three main entities: data source, the third-party cloud server and clients. Data source only needs to store the system parameters, its own public and private keys, the data stream it generates and the corresponding tag of each data are all outsourced to the cloud server for storage. Clients simply store the data source's public key. Therefore, heavy storage burden is transferred to the cloud server, the data source and clients are lightweight.

7.2 Computation

For the data source and clients, the main computation of our proposed schemes focuses on the process of *TagGen* and *CheckProof* respectively. The concrete number of computation is shown in the above Table 1. Obviously, the time of generating a tag is very short. In the client side, S_1 and S_2 can be computed in advance because they are unrelated to the outsourced data, therefore, this way reduces verification cost to a certain extent. In our schemes, clients need to compute $(2 + \Delta)$ bilinear pairing operations, in view of the current computing capability, it is also acceptable.

Table 1. Computation cost

	Modular multiplication operation	Exponentiation operation	Bilinear pairing operation	Hash function
TagGen	6	4	0	3
CheckProof	$2 + \Delta$	3	$2 + \Delta$	3Δ

8 Conclusion

In this paper, we propose two novel publicly verifiable schemes for 1-norm and 2-norm which solve the verifiable problems of such operations in outsourced computation well, any keyless client can publicly verify the correctness of outsourced computation results. We conduct a security analysis for our schemes in Sect. 6 and the results show that our schemes are provably secure under the *CDH* assumption. The evaluation results in Sect. 7 also illustrate our schemes are practically feasible.

Acknowledgments. we are writing this few lines to express our sincere appreciation to all the reviewers and editors for their valuable comments and works. This paper is supported by National Natural Science Foundation of China (Grant Nos. 61772550, U1636114, 61572521), National Key Research and Development Program of China (Grant No. 2017YFB0802000), Natural Science Basic Research Plan in Shaanxi Province of China (Grant Nos. 2016JQ6037).

References

1. Sun, W., Liu, X., Lou, W., Hou, Y.T., Li, H.: Catch you if you lie to me: efficient verifiable conjunctive keyword search over large dynamic encrypted cloud data. In: 2015 IEEE Conference on Computer Communications (INFOCOM), pp. 2110–2118. IEEE (2015)
2. Liu, X., Zhang, Y., Wang, B., Yan, J.: Mona: secure multiowner data sharing for dynamic groups in the cloud. IEEE Trans. Parallel Distrib. Syst. **24**(6), 1182–1191 (2013)
3. Gordon, S.D., Katz, J., Liu, F.-H., Shi, E., Zhou, H.-S.: Multi-client verifiable computation with stronger security guarantees. In: Theory of Cryptography, pp. 144–168. Springer (2015)
4. Nath, S., Venkatesan, R.: Publicly verifiable grouped aggregation queries on outsourced data streams. In: International Conference on Data Engineering, pp. 517–528. IEEE (2013)
5. Gennaro, R., Wichs, D.: Fully homomorphic message authenticators. In: Advances in Cryptology-ASIACRYPT, pp. 301–320. Springer (2013)
6. Gennaro, R., Gentry, C., Parno, B.: Non-interactive verifiable computing: outsourcing computation to untrusted workers. In: Advances in Cryptology–CRYPTO, pp. 465–482. Springer (2010)
7. Thaler, J.R.: Practical verified computation with streaming interactive proofs. Ph.D. dissertation, Harvard University (2013)
8. Vu, V., Setty, S., Blumberg, A.J., Walfish, M.: A hybrid architecture for interactive verifiable computation. In: IEEE Symposium on Security and Privacy, pp. 223–237. IEEE (2013)
9. Papadopoulos, S., Cormode, G., Deligiannakis, A., Garofalakis, M.: Lightweight authentication of linear algebraic queries on data streams. In: International Conference on Management of Data, pp. 881–892. ACM (2013)
10. Chen, Z.: Introduction to Matrix. Beijing University of Aeronautics and Astronautics Press, Beijing (2012)
11. Xu, Z.: Matrix Theory Concise Tutorial. Science Press (2014)
12. Diffie, W., Hellman, M.E.: New directions in cryptography. IEEE Trans. Inf. Theor. **22**(6), 644–654 (1976)
13. Boneh, D., Franklin, M.: Identity-based encryption from the weil pairing. In: Advances in CryptologyłCRYPTO 2001, pp. 213–229. Springer (2001)
14. Liu, X., Sun, W., Quan, H., Lou, W., Zhang, Y., Li, H.: Publicly verifiable inner product evaluation over outsourced data streams under multiple keys. IEEE Trans. Serv. Comput. https://doi.org/10.1109/tsc.2016.2531665

Evaluation of Techniques for Improving Performance and Security in Relational Databases

Renalda Kushe(✉) and Kevin Karafili(✉)

Faculty of Information Technology, Polytechnic University of Tirana, Tirana, Albania
{rkushe,kevin.karafili}@fti.edu.al

Abstract. With the rapid development of technology, now it can not only be aimed a correct running of systems but also a high level of technical performance. In this paper, we will evaluate the database performance in terms of optimization and also the security it provides. We will present briefly the potential causes of an unsatisfactory performance of the database and also the solution for each determined issue. We will give in detail the optimization of TempDB, Cache and Index Fragmentation and how does those affect the overall DB performance. The optimization is intended to be achieved by simultaneous implementation of these optimization techniques. Also, we will determine some of the best security techniques in the DB, such as encryption or recognizing User Roles, and High Availability, which provides security in several database hierarchies. The combinations of these techniques provide overall database security.

1 Introduction

With the rapid development of the technology, now it can not only be aimed a correct running of systems, but a high level of technical performance. The database is the heart of the application because all the information on which the application operates is stored there. That is the reason why the database performance is so important to the overall application or system implementation. In this paper we are utilizing SQL Server for performance assessment research purposes. The goal of this paper is to determine the potential causes of a poor performance and delivering some of the best solutions that SQL Server database offers in improving it [1–3]. Previously research works evaluate the database performance in terms of three main factors that affect transaction throughput in a multiuser environment: multiprogramming level, degree of data sharing among simultaneously executing transactions, and transaction mix [4,5,7]. Other research works present the database performance by following best principles of hardware and software configuration and also various monitoring techniques under both operational and troubleshooting conditions [6]. At [7–10] there are given 4 main factors affecting the performance of database: Response time (the time interval from the instant a command is input to the system to the instant the corresponding reply begins to appear at the terminal); Throughput (the overall

L. Barolli et al. (Eds.): EIDWT 2018, LNDECT 17, pp. 470–478, 2018.
https://doi.org/10.1007/978-3-319-75928-9_42

capability of the computer to process data); Resources (hardware and software tools, which include memory, disk speed, cache controller etc.), Memory (the total space required by a query to complete its execution). The contribution of this paper is in evaluating the database performance in terms of optimization and also the security level it provides, during the database developing process. The optimization is intended to be achieved by simultaneously implementation of several optimization techniques such as optimization of TempDB, Cache, and the Index Fragmentation. Also in terms of security we will implement several security methods in databases such as encryption, recognizing User Roles and High Availability. In Sect. 2, are described database optimization techniques such as TempDB, Cache and Indexes Fragmentation. In Sect. 3 are given some security techniques integrated with the database and also some improved combined security techniques. Some conclusions and future work is presented in Sect. 4.

2 Database Optimization Techniques

Before we take steps to improve the database performance, first we need to evaluate its current state and to determine the CPU consumption related on two processes: user mode and kernel mode. There are many queries to get us info on the operation of SQL Server database. In Fig. 1, is presented the output of query on database operation.

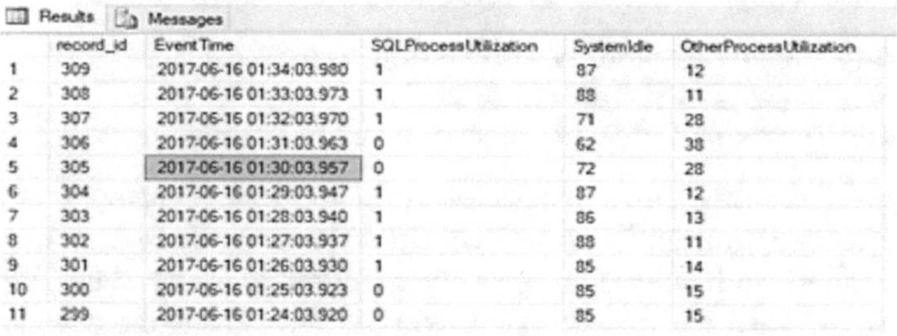

Results | Messages

	record_id	EventTime	SQLProcessUtilization	SystemIdle	OtherProcessUtilization
1	309	2017-06-16 01:34:03.980	1	87	12
2	308	2017-06-16 01:33:03.973	1	88	11
3	307	2017-06-16 01:32:03.970	1	71	28
4	306	2017-06-16 01:31:03.963	0	62	38
5	305	2017-06-16 01:30:03.957	0	72	28
6	304	2017-06-16 01:29:03.947	1	87	12
7	303	2017-06-16 01:28:03.940	1	86	13
8	302	2017-06-16 01:27:03.937	1	88	11
9	301	2017-06-16 01:26:03.930	1	85	14
10	300	2017-06-16 01:25:03.923	0	85	15
11	299	2017-06-16 01:24:03.920	0	85	15

Fig. 1. Info on database operation

The database application must be evaluated also in relation with "% Privileged Time" and "% User Time". In Fig. 2 is shown the user time assessment. If "% User Time" has the highest value, than most possible is that the problem is within the SQL Server, with one of the possible issues:

1. Executing a query constructed incorrectly;
2. The system tray is consuming a lot of CPUs;
3. Wrong compilation and wrong reconfiguration of queries.

Corresponding countermeasures for these issues:

1. Update Statistics as they streamline cache queue execution plans;

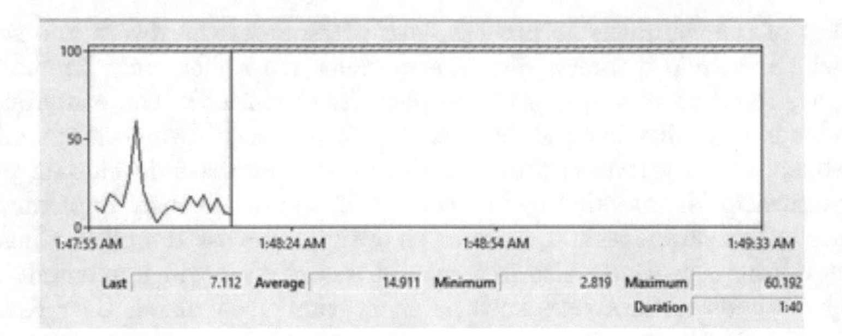

Fig. 2. User time assessment

2. Look if Optimizer suggests any index we miss from the XML plan;
3. It happens when the data has changed a lot in the tables, and these changes are not reflected in statistics. In this case, we can manually update these necessary statistics.

2.1 TempDB

One of the most implemented database optimization techniques has to do with TempDB, also known as scratch database. When the database is restarted, everything is deleted from TempDB because it is recreated every time the database is switched on. The information stored on TempDB is:

1. Temporary objects created by application users;
2. Temporary objects of the database itself;
3. Saving the versions of objects when changes are made.

The TempDB setting up is suggested to be done on a separate hard disk, as well as LogTransaction and also the core database itself. In this way jobs and loads are divided, so it makes an isolation of possible issues. The TempDB logs can be deployed with our database logs as they have the same data type. Keeping TempDB on a local disk increases its performance, especially with SSDs, and reduces I/O requirements on the shared HDD environment. Auto-growing happens when we at first don't consider the fact that our database could take large amount of data. When it does, and it is not configured for this scenario, it has problems in creating space in HDD, which leads to inaccessibility, fragmentation of files and as result, poor performance. As result, the database auto-growing must be taken into consideration since the beginning of database design and configuration. TempDB is affected from Auto-Growning since it has an initial size of 8 MB, so it is suggested to assign its initial size as the capacity of the HDD itself where it is located.

2.2 Cache

The SQL Server plan cache stores details on statements that are being executed over time. This way, when a query is executed multiple times, it does not have to pass all of the execution steps of a query, but executes its ready-made Cache plan. Meanwhile implementing optimization for Ad Hoc workloads, the entire cache execution plan can not be maintain, but only a small part of it, that can help in quick execution and enables to not fill cache with a lot of Ad Hoc type information. The SQL server limits the number of plans in each bucket and hash tables that they form to exclude the possibility of long reading times when the server is loaded. Cache uses buckets to save plans. The SQL Server plan cache can be a good tool to use in identifying current performance issues and in looking for new ways to improve the performance, in terms of plans that have missing indexes. In Fig. 3 is presented an examination inside plan cache to look up for any missing index.

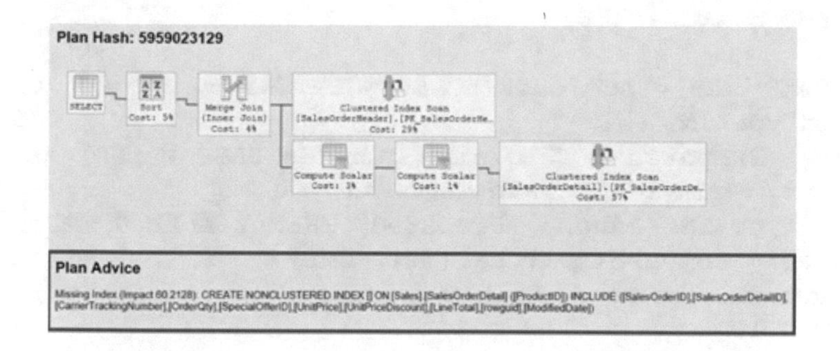

Fig. 3. An examination inside plan cache.

To find all current plans which have missing indexes, we execute the following query:

```
   FROM sys.dm_exec_cached_plans AS dec
CROSS APPLY sys.dm_exec_sql_text(dec.plan_handle) AS des
CROSS APPLY sys.dm_exec_query_plan(dec.plan_handle) AS deq
WHERE deq.query_plan.exist
(N'/ShowPlanXML/BatchSequence/Batch/Statements/
StmtSimple/QueryPlan/MissingIndexes/
MissingIndexGroup') <> 0
ORDER BY dec.usecounts DESC
```

After this we have to add manually missing indexes into appropriate tables. From cache examination we have noticed these issues:

1. There are many single-used plans;
2. There are many queries which need to be improved;
3. There are many queries that are logically the same for the execution mode.

Single-used plans are not an issue, but they risk filling up the cache (no place for other cache plans) and increasing the CPU consumption. To see if these plans are causing us problems, we need to compare how much space they have and how much space there is for the whole cache (depends on the size of the RAM allocated for the SQL Server). If it's over 50%, it is needed for optimization (or more than 2 GB single-used plans.) To prevent full cache filling with single-used plans, 'Optimize for Ad-Hoc Workloads' should be configured, which will allow us to not create a full plan during the execution of the script for the first time. If executed for the second time then a full cache plan is created. Then we can do a manual deletion of these plans or automatic delete by creating Jobs. Deleting SQL plans based on Ad Hoc/prepared single-used plans, is done by executing the SQL code below.

```
    DECLARE @MB decimal(19,3)
, @Count bigint
, @StrMB nvarchar(20)

    SELECT @MB = sum(cast((CASE WHEN usecounts = 1
AND objtype IN
(Adhoc , Prepared ) THEN size_in_bytes ELSE 0 END) as dec
, @Count = sum((CASE WHEN usecount = 1
AND objtype IN (Adhoc , Prepared) THEN 1 ELSE 0 END)
, @StrMB = convert(nvarchar(20), @MB)
FROM sys.dm_exec_cached_plans
IF @MB > 10
BEGIN
DBCC FREESYSTEMCACHE(SQL Plans)
RAISERROR (% MB was allocated to single-use plan cache.
Single-use plans have been cleared. , 10,1,@StrMB)
END
ELSE
BEGIN
RAISERROR(Only % MB is allocated to single
-use plan cache
no need to clean cache now. , 10, 1, @StrMB)
END
```

2.3 Index Fragmentation

An important task of table design in databases, is indexing. The column in a database table defined as index, help us to create the logic related to the data stored in the table, also it affects the performance when a simple query is implemented. The SQL Server retains data on the 8 kb page. When we throw data over a table, the SQL Server defines a page to get those data, unless that data is more than 8 kb, some pages will need to be sent to get row table information. Example: Let's pretend that first we have a 40-page table that is completely busy, and then, after some operation over the database, we have 50 pages which are only 80% fully occupied, due to multiple inserts and deletes of data on this table. This causes us a problem because now that we want to read from this table, we have to scan 50 pages instead of 40 pages, so we meet a performance decrease. The management plans are some solutions offered by the SQL Server itself, which can be quite useful. Index Management is one of them. We can rebuild indexes or reorganize them, but we cannot execute any logic over it. Some indexes may have minimal fragmentation, so their own minimal reorganization would be enough. By executing our customized scripts, we implement a logic on every action. So, based on the size of fragmentation, we apply reorganization, reconstruction, or no action. When the value of fragmentation is greater than 5% but less than 30%, the reorganization would be the appropriate action:

```
ALTER INDEX ALL ON TableName REORGANIZATION
```

If we want to perform reorganization only in a particular index of a table:

```
ALTER INDEX IndexName ON REORGANIZATION RefName
```

When the fragmentation value is greater than 30%, the rebuilding would be the appropriate action:

```
  ALTER INDEX ALL ON TableName
REBUILD WITH (ONLINE = ON)
```

If we want to perform reorganization only in a particular index of a table:

```
  ALTER INDEX IndexName
ON TableName REBUILD WITH (ONLINE = ON)
```

3 Database Security Aspects

When we evaluate the database performance, we should not forget one of its most important aspects: Database Security. Any work and optimization addressed so far can be discarded if we do not initially secure the database and the entire data store in it. SQL Server enables us to implement several database security techniques, each of which has its own features. These methods are based on High Availability, meaning that our system is always available even if one of the servers becomes not available. SQL Server Availability Types are:

1. AlwaysOn Failover Clustering Instances
2. AlwaysOn Availability Groups
3. Database Mirroring and Replication
4. Log Shipping

These security techniques are integrated in SQL Server, so the question is: which will we choose and what else can we improve. Some of techniques related to security improvement in SQL Server are:

1. Encryption;
2. Well-defined privileges on Users.

3.1 Availability Types

In this session, we will give shortly a description about availability types implemented in SQL Server database. AlwaysOn Failover Clustering Instance; this technique makes an automatic transition from one node to another when the server crashes or when we failover itself. We do not have data loss. Pending transactions are applied to the new instance. Delay in the failover process can take a few minutes, depending on the time it takes to launch the SQL server and to replicate halfway through transactions. AlwaysOn FCI requires shared type memory like: iSCSI or Fiber Channel SANs, which are accessible from all cluster nodes. AlwaysOn can be applied to the SQL Server physical systems or can be applied to systems running on the virtual machine. AlwaysOn Availability Groups; this is the most commonly used technique. We can create up to 8 replicas, two of which can be 'synchronous'. This enables high database protection when we have many users. We can simultaneously apply synchronized copies of 'high availability and automatic failover' as well as unsynchronized copy of 'disaster recovery'. Failover is automatic and takes only a few seconds of time. AlwaysOn AG transfers transactions from primary to secondary if the primary copy becomes unavailable. Database Mirroring and Replication; this technique is represented by an old technology and has its limitations. It is rarely used, so it will not be taken in consideration in this paper. Log Shipping; this technology can be used for Disaster Recovery. It is based on rebuilding on the last backup file and there may be loss of information so it is not recommended. We can combine AlwaysOn FCI and AlwaysOn AG. For example, we can use the first technology to ensure that the system will always be available and the second technology in case of system disaster (disaster recovery technology).

3.2 Database Encryption

The data encrypted in the database define the encryption techniques presented as below:

1. The transparent encryption of data;
2. Column level encryption;
3. Backup data encryption.

The transparent encryption of data does not provide security for the data exchanged between the database and the application, neither the data currently running. This technique encrypts the database files automatically on the page level. The database pages are encrypted before being written to disk and decrypted when read in memory. Column level encryption enables the encryption of defined columns in the database. This encryption technique goes through the following stages:

```
SELECT name KeyName,
  symmetric_key_id KeyID,
  key_length KeyLength,
  algorithm_desc KeyAlgorithm
FROM sys.symmetric_keys;
```

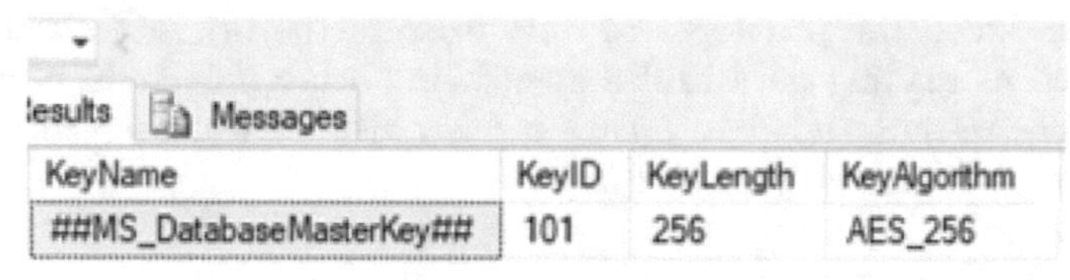

KeyName	KeyID	KeyLength	KeyAlgorithm
##MS_DatabaseMasterKey##	101	256	AES_256

Fig. 4. DMK key creation

1. Creating a Database Master Key (DMK); encryption begins with the creation of this key. DMK is a symmetric key used primarily to protect private keystrokes in database certificates and non-symmetric keys. DMK uses 256-bit AES encryption algorithm. DMK key represents the highest level in the encryption hierarchy. Also the database uses the SMK key (Service Master Key), and a password set by the user to save the DMK.
2. Creating an asymmetric encryption mechanism, in a form of an asymmetric key certification. Each uses a public key and a private key to provide asymmetric encryption, with the public key to provide data encryption, and the private one to do the opposite (the decryption).
3. The next step is to create a symmetric key that operates directly on the data itself in the hierarchy of encryption. Unlike asymmetric encryption, this encryption step uses only one key for both encryption and decryption. Here we define the symmetric key naming, and specify that the 256-bit AES algorithm will be used to encrypt the data.
4. At this phase, we have to implement the encryption hierarchy created in stages a, b and c (which stand for DMK creation, certificate, and symmetric key). The Fig. 4 gives the EncryptedID column, with encrypted data.

The backup encryption; is very important because the backup files store all the database data. The backup encryption is realized by AES 256 key algorithm. For the restore process the key or the certificate used for the encryption, must be available within the instance (Fig. 5).

	EmpID	NatID	EncryptedID
1	1	295847284	0x00B27C95596971469CF372A8EFFE6520010000000282B7B6...
2	2	245797967	0x00B27C95596971469CF372A8EFFE65200100000030A78D...
3	3	509647174	0x00B27C95596971469CF372A8EFFE652001000000003E9475C...
4	4	112457891	0x00B27C95596971469CF372A8EFFE6520010000002D605B...
5	5	695256908	0x00B27C95596971469CF372A8EFFE652001000000001667622...

Fig. 5. Encrypted data

3.3 Well Defined Privileges

A very important part about database security is the well-defined roles of different users who access the database. No user should be given more access than he needs, as it may compromise sensitive data. One very important point is to set privileges to the user that will be connected to our application. This user should be assigned the privileges to only execute the procedure and no other options, such as reading data tables or creating other objects in our database. It is therefore recommended to create a user with an executor role. This role should be created manually as below:

```
   CREATE ROLE db_executor
EXECUTE TO db_executor
```

4 Conclusions

The most common causes of a low database performance, which need optimization, are: The Cache and Indexes Fragmentation. "High Availability and Disaster Recovery" is the most appropriate technology for our system to be functional without disconnection throughout its work. This combined technology ensures system availability and also disaster recovery. The database security represents a sensitive issue, due to all the data stored on it. None of the security technologies integrated in databases fully provides us database security, but the combination of them with other security techniques gives us a pretty positive result.

References

1. LeBlanc, P., Jorgensen, A., Segarra, J., Nelson, A., Chinchilla, J.: Microsoft SQL Server Bible (2012)
2. Tripp, K.L.: https://www.sqlskills.com
3. https://logicalread.com
4. Boral, H., DeWitt, D.: A methodology for database system performance evaluation. In: ACM SIGMOD International Conference on Management of Data, pp. 176–185 (1984)
5. Subharthi, P., Jain, R.: Database systems performance evaluation techniques, a Project report, November 2008
6. Rees, S., Rech, T., Depper, O., Singh, N.K., Shen, G., B. Melnyk, R.: Best practices, tuning and monitoring database system performance. IBM publication, July 2013
7. Kaladi, A., Ponnusamy, P.: Performance evaluation of database management systems by the analysis of DBMS. Int. J. Modern Eng. Res. (IJMER) **2**(2), 067–072 (2012). ISSN 2249-6645
8. https://docs.oracle.com/cd/A95428_01/a86059/concpts.htm
9. Goodwell, M.: Application Performance Management (2014). https://www.dynatrace.com/blog/five-easy-steps-improvedatabase-performance/
10. Burleson, D.: High Throughput vs Response time (2015). http://www.dba-oracle.com/t_tuning_high_throughput_vs_fast_response_time.htm

E-learning Material Development Framework Supporting VR/AR Based on Linked Data for IoT Security Education

Chenguang Ma[1], Srishti Kulshrestha[4], Wei Shi[3], Yoshihiro Okada[1,3](✉), and Ranjan Bose[2]

[1] Graduate School of ISEE, Kyushu University, Fukuoka, Japan
okada@inf.kyusha-u.ac.jp
[2] Department of Electrical Engineering, Indian Institute of Technology Delhi, New Delhi, India
[3] Innovation Center for Educational Resources (ICER), Kyushu University, Fukuoka, Japan
[4] School of Information Technology, Indian Institute of Technology Delhi, New Delhi, India

Abstract. This paper treats one of the activities of a research project about IoT (Internet of Things) security education. In this project, the authors plan to provide SPOC (Small Private Online Course), MOOC (Massive Open Online Course) and education games (serious games) about IoT security. One of the key challenges is the development of educational materials that attract students to IoT security. As a first step, a database will be built that contains the IoT security information for educational materials. So, the authors have already been building a comprehensive database regarding the various kinds of IoT threats based on Linked Data. The other research agenda is to provide e-learning materials themselves using the database that attract students to IoT security. The use of VR (Virtual Reality) and AR (Augmented Reality) technologies enables e-learning materials to attract the students. In this paper, the authors propose e-learning material development framework supporting VA/AR based on Linked Data.

Keywords: E-learning · Educational materials · Linked Data · IoT security VR/AR

1 Introduction

This paper discusses with one of the activities of a collaborative research project titled "Security in the Internet of Things Space" between IIT Delhi and Kyushu University supported by DST (Department of Science & Technology, India) and JST (Japan Science & Technology Agency). Recently, Cybersecurity has become very popular as one of the research themes because there have been many serious cyberattacks in the world. Recent cyber attacks throughout the world have attracted a lot of attention from the researchers and government agencies. The growing connectivity and digitization have evolved the nature of cyber attacks and have drastically increased the attack surface. Similarly, IoT

L. Barolli et al. (Eds.): EIDWT 2018, LNDECT 17, pp. 479–491, 2018.
https://doi.org/10.1007/978-3-319-75928-9_43

(Internet of Things) security has become very important because today more-and-more things (devices or sensors) are connected to the Internet in order to collect various types of data, including human activity data. And by analyzing the data, these devices can be controlled remotely to provide a more safe and secure society. IoT space is sometimes referred as a smart society consisting of smart cities, smart grids, factory automation, intelligent transportation systems, smart buildings and smart homes. This leads to increase the demand of IoT security education. Consequently, it is regarded as a critical objective of the project to include IoT security education and awareness. Although IoT is related to various fields, this education-research focuses specifically on the 'Security of Smart Building/Smart Home' because these are closely related to our lives.

In this activity, we focus on IoT security education and consider developing such educational materials like SPOC, MOOC, educational games about IoT security. As a first step, we are building a database about the IoT security information. If a database is built based on Linked Data, it becomes easy to update in terms of its contents and to share them easily with other researchers/educators. Therefore, we have already proposed the use of Linked Data in order to build a database about the IoT security information [1]. The next step is to provide e-learning materials themselves about IoT security using the database. Such materials should be attractive for students. Recently, VR (Virtual Reality) and AR (Augmented Reality) have become popular again because of the technological advances for them, and the use of VR/AR enables e-learning materials to attract the students. We have already proposed an interactive educational contents development framework that supports VR/AR [2]. So, in this paper, we also propose the combinatorial use of this framework and the database of IoT security information realized as Linked Data, and introduce a prototype of the e-learning material system that support VR/AR for IoT security education.

The remainder of this paper is organized as follows: Sect. 2 treats related work. In Sect. 3, we explain our Linked Data and implementation of RDF stores for IoT security information, and introduce the overview of an authoring tool for educational materials and the current prototype system. Section 4 explains the interactive educational contents development framework that supports VR/AR and Sect. 5 introduces the prototype of the e-learning material using VR/AR. Finally, we conclude the paper and discusses our future work in Sect. 6.

2 Related Work

Berners-Lee coined the term Linked Data describing a set of best practices for publishing and connecting structured data on the Web [3, 4]. These practices are the foundation of the evolution of the Web of Documents to the Web of Data, a global data space connecting data from a multitude of different domains. In order to become part of a single global data space, Berners-Lee proposed a number of (technical) rules for publishing data on the Web that have become known as Linked Data Principles [3]:

1. Use URIs as names for things.
2. Use HTTP URIs so that people can look up those names.

3. When someone looks up a URI, provide useful information, using the standards (RDF, SPARQL).
4. Include links to other URIs, so that they can discover more things.

In summary, linked data is about data integration [5]. In the area of e-Learning, such data can be shared as learning objects (LO) or learning entities of any kind, respectively. Dicheva identifies three generations of Web-based educational systems [6]. The systems of the first generation provide a central entry-point for accessing learning materials and online course, e.g., LMS and educational portals. The systems of the second generation employ Web and AI technologies to support intelligently personalization and adaption. Such systems are called educational adaptive hypermedia systems. The third generation of Web-based educational systems is a class of ontology-aware software, using and enabling Semantic Web standards and technologies in order to grant scalability, reusability and interoperability of educational material that is distributed over the Web. Therefore, Linked Data technologies raise high expectations with respect to providing solutions in a field like e-Learning.

As for e-Learning material development systems support VR/AR, there are many development systems and tools for 3D contents. Some of them are commercial products like 3D Studio Max, Maya, and so on. Although these products can be used only for creating 3D CG images or 3D CG animation movies, usually, these cannot be used for creating interactive contents. As a development system for 3D interactive contents, there is *IntelligentBox*, a constructive visual software development system for 3D graphics applications [7]. This system seems very useful because there have been many applications actually developed using it so far. However, it cannot be used for creating web-based contents. Although there is the web-version of *IntelligentBox* [8], it cannot be used for creating story-based contents. With *Webble World* [9], it is possible to create web-based interactive contents through simple operations for authoring and of course, possible to render 3D graphics assets. However, it does not have same functionalities as that of *IntelligentBox*.

There are some electronic publication formats like EPUB, EduPub, iBooks and their authoring tools. Of course, these contents are used as e-Learning materials. However, basically, these are not available on the Web and do not support 3D graphics except iBooks. iBooks supports rendering functionality of a 3D scene and control functionality of its viewpoint. However, story-based contents cannot be created using it.

From the above situation, for creating web-based interactive 3D educational contents, we have to use any dedicated toolkit systems. The most popular one is Unity, one of the game engines, that supports creating web-contents. Practically, using Unity, we have developed 3D educational contents [10, 11] for the medical course students of our university because Unity is a very powerful tool that supports many functionalities. However, the use of Unity requires any programming knowledge and skills of the operations for it. Therefore, it is impossible to use Unity for standard end-users like teachers. As a result of the above, we proposed the framework [12, 13] and extended it for supporting VR/AR [2].

3 Linked Data of IoT Security Information

We integrated the knowledge and concepts of IoT security based on Linked data. As the same time, we built up the RDF store for storing designed RDF data based on Apache Jena. Thereby, the integration and retrieval of IoT security information which is designed based on RDF by SPARQL, a query language for RDF store, are implemented.

3.1 Design of RDF Store for IoT Security Information

There are three steps in designing the linked data for IoT security we are following. In the first step, we designed the schema of RDF store. An example of RDF store schema is shown in Fig. 1. As the figure shows, we designed a new prefix for IoT security, because there is no existed prefix for IoT security field according to the RDF schema document. Meanwhile, IoT security information contains the contents of Table 1 that are thought important as IoT threats information.

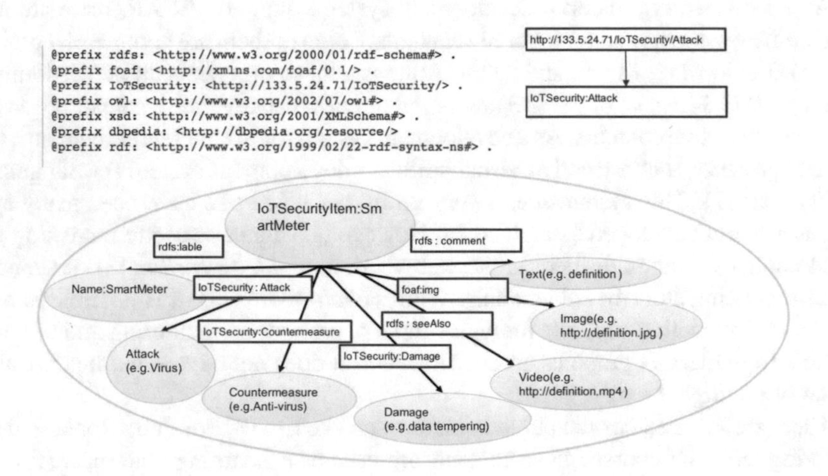

Fig. 1. An example of RDF store schema.

Table 1. IoT threats information

1.	What kinds of IoT (systems, devices, sensors) would be present in a Smart Building/Smart Home
2.	What kinds of attacks can be perpetrated on such IoT spaces
3.	What kinds of preventive measures can be taken to thwart such attacks
4.	What kinds of damages could be caused by such attacks and what would be the associated cost
5.	Textual explanation of IoT threats information
6.	Illustrated explanation of IoT threats information
7.	Video explanation of IoT threats information

In the second step, an excel file of IoT security is created according to the schema. In the third step, as Fig. 2 shows, the excel file of IoT security is converted to turtle format or RDF format by using OpenRefine, a free, open source, power tool for working with messy data. Figure 3 is the Turtle files of IoT security saved as RDF format.

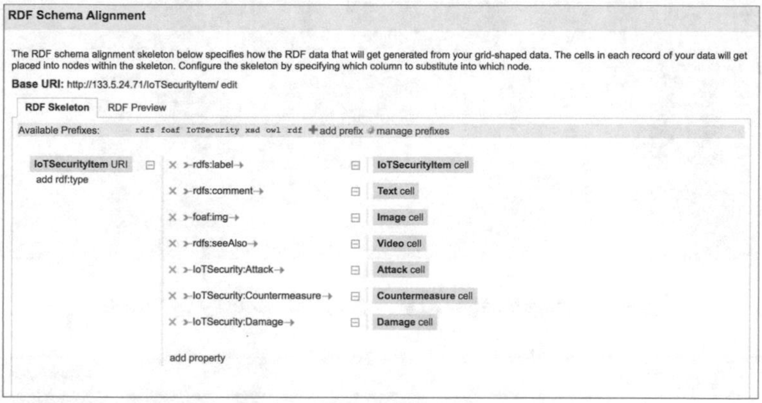

RDF Schema Alignment

The RDF schema alignment skeleton below specifies how the RDF data that will get generated from your grid-shaped data. The cells in each record of your data will get placed into nodes within the skeleton. Configure the skeleton by specifying which column to substitute into which node.

Base URI: http://133.5.24.71/IoTSecurityItem/ edit

RDF Skeleton | RDF Preview

Available Prefixes: rdfs foaf IoTSecurity xsd owl rdf + add prefix manage prefixes

IoTSecurityItem URI — rdfs:label → IoTSecurityItem cell
add rdf:type — rdfs:comment → Text cell
foaf:img → Image cell
rdfs:seeAlso → Video cell
IoTSecurity:Attack → Attack cell
IoTSecurity:Countermeasure → Countermeasure cell
IoTSecurity:Damage → Damage cell
add property

Fig. 2. Designation of RDF skeleton of IoT security.

```
<rdf:Description rdf:about="http://133.5.24.71/IoTSecurityItem/SmartMeter">
        <rdfs:label>SmartMeter</rdfs:label>
        <rdfs:comment>https://en.wikipedia.org/wiki/Smart_meter</rdfs:comment>
        <foaf:img>https://upload.wikimedia.org/wikipedia/commons/9/9a/Intelligenter_zaehler-_Smart_meter.jpg</foaf:img>
        <rdfs:seeAlso>https://www.youtube.com/watch?v=C4L01fejQJk</rdfs:seeAlso>
        <IoTSecurity:Attack>Unauthorized access(internet)</IoTSecurity:Attack>
        <IoTSecurity:Countermeasure>Vulnerability countermeasure</IoTSecurity:Countermeasure>
        <IoTSecurity:Damage>Measurement data is peeked.</IoTSecurity:Damage>
        <IoTSecurity:Countermeasure>User authentication</IoTSecurity:Countermeasure>
        <IoTSecurity:Damage>Measurement data is tampered with.</IoTSecurity:Damage>
        <IoTSecurity:Countermeasure>Firewall function</IoTSecurity:Countermeasure>
        <IoTSecurity:Damage>Measurement data can not be viewed.</IoTSecurity:Damage>
</rdf:Description>

<rdf:Description rdf:about="http://133.5.24.71/IoTSecurityItem/HERMS">
        <rdfs:label>HERMS</rdfs:label>
        <rdfs:comment>https://en.wikipedia.org/wiki/Energy_management_system</rdfs:comment>
        <foaf:img>https://research.ece.ncsu.edu/adac/wp-content/uploads/2015/10/total_prj.jpg</foaf:img>
        <rdfs:seeAlso>https://www.youtube.com/watch?v=Vi-16bO0Pn8</rdfs:seeAlso>
        <IoTSecurity:Attack>Virus infection</IoTSecurity:Attack>
        <IoTSecurity:Countermeasure>Vulnerability countermeasure</IoTSecurity:Countermeasure>
        <IoTSecurity:Damage>The data of device controlled by HEMS is peeked.</IoTSecurity:Damage>
        <IoTSecurity:Countermeasure>Anti- virus</IoTSecurity:Countermeasure>
        <IoTSecurity:Damage>The data of device controlled by HEMS is tampered with.</IoTSecurity:Damage>
        <IoTSecurity:Countermeasure>Software digital signature</IoTSecurity:Countermeasure>
        <IoTSecurity:Damage>The data of device controlled by HEMS can not be viewed.</IoTSecurity:Damage>
        <IoTSecurity:Countermeasure>Secure development</IoTSecurity:Countermeasure>
</rdf:Description>
```

Fig. 3. Exported RDF file of IoT security.

3.2 Implementation of RDF Store for IoT Security Information

There are three steps for the implementation of RDF store of IoT security. In the first step, RDF store is built up based on Apache Jena, which is capable to store the integrated data as shown in Fig. 4. In the second step, the exported Turtle File of IoT security is stored into the RDF store as shown in Fig. 5. In the third step, retrieval information of IoT security through virtuoso by using the query language is able to be achieved as shown in Fig. 6.

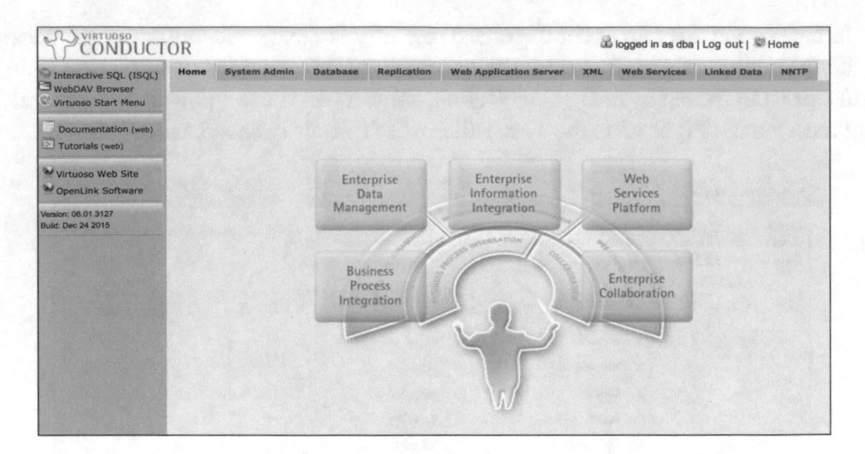

Fig. 4. Management of RDF data of IoT security by virtuoso.

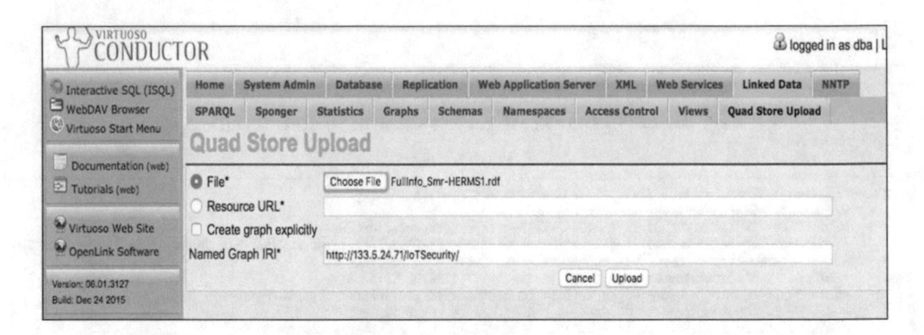

Fig. 5. Storage of RDF file.

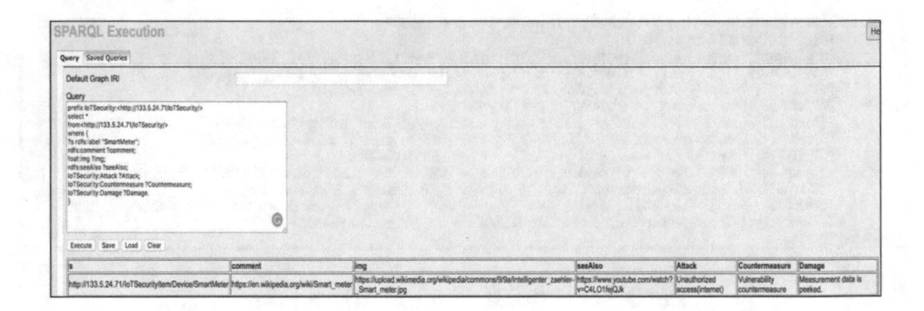

Fig. 6. Retrieval of RDF file.

3.3 Authoring Tool for Educational Materials of IoT Security

Figure 7 shows the overview of our authoring tool and Fig. 8 shows the screen image of our prototype system of the authoring tool [14]. By specifying a keyword 'Smart-Meter' (IoT device name) as the target of an educational material, the system retrieves text data, image data and video data related to the keyword from RDF stores. There are several candidate data for each of text, image and chart so the user can choose one of them by the operation on [back], [next] buttons. In this way, by using the authoring tool, educational material developers are able to edit the contents of the material regarding IoT security knowledge managed in RDF stores easily and efficiently.

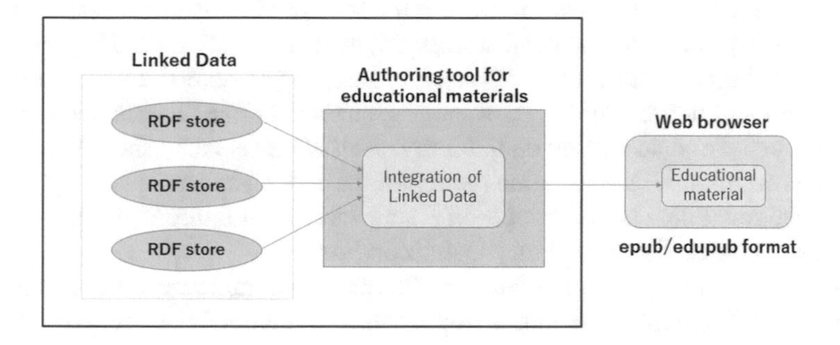

Fig. 7. Overview of authoring tool.

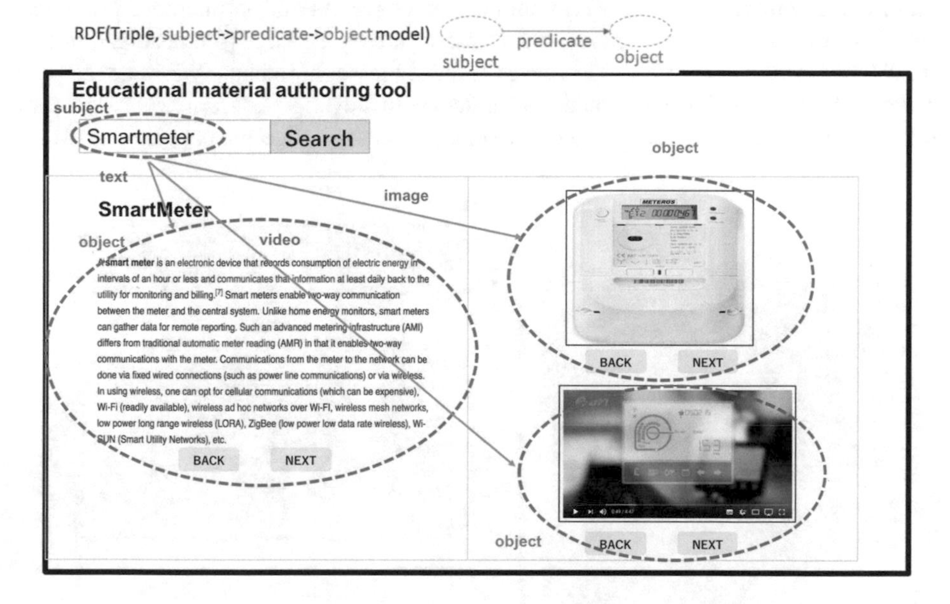

Fig. 8. Prototype system of authoring tool.

After achieving the authoring of IoT security educational materials, such educational materials are supposed to be shown as readable format like epub/edupub that facilitates the knowledge of IoT security to be learned on the Web.

4 E-learning Material Development Framework Supporting VR/AR

Our first targets of the framework were the contents of Japanese History study. In general, a history consists of several stories. So, the framework should support a story. Each story is realized with several 3D scenes consisting of several architecture objects like buildings and houses, and several moving characters like humans who have their own shape model and animation data. Firstly, such 3D assets data should be prepared. Next, contents creators have to define a story for the content as one JavaScript file called 'Story Definition File'. In our proposed framework, the requirements for creating a content are 3D assets data and 'Story Definition File'.

Figure 9 shows functional components of the proposed framework consisting of main components (Main.html) and sub components (AnimationCharacter.js). The main components include functions related to architecture objects and functions related to AnimationCharacter objects represented as moving characters. The sub components include the constructor of new AnimationCharacter class and its member functions. Besides main and sub components, our framework uses Three.js [threejs.org/], one of the WebGL based 3D Graphics Libraries as subsidiary components. When developing a story-based interactive 3D educational content with the proposed framework, a teacher has to prepare one story definition file and 3D assets for it. See the papers [12, 13] for more details.

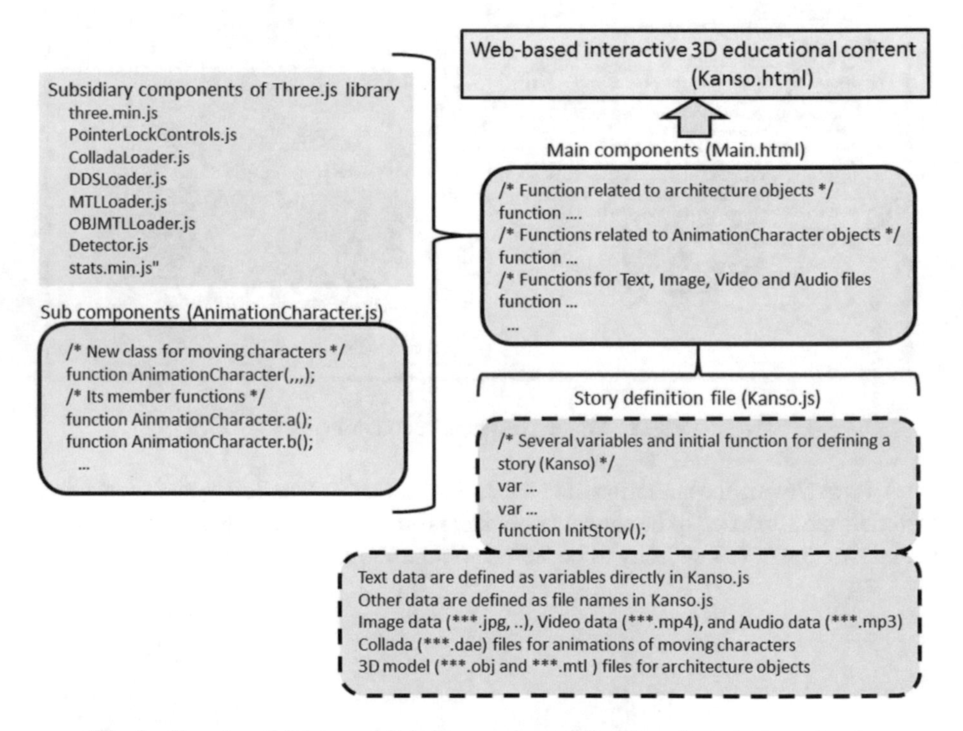

Fig. 9. Functional (Main and Sub) components of Framework, (ex) Kanso.html.

The followings are newly introduced functionalities to support VR/AR that are a stereo view, touch interfaces, device orientation/motion interfaces, geolocation interface and a camera interface for tablet devices and smart phones. See the paper [2] for more details.

(1) **Stereo view support**

The JavaScript program for the stereo view is very simple within Three.js because Three.js provides StereoEffect.js and you can read this file in your HTML file and can call effect.render(scene, camera) for the stereo view instead of calling renderer.render(scene, camera) for the standard view.

(2) **Touch interfaces**

The JavaScript program for touch interfaces is also simple because HTML5 supports 'touchstart', 'touchend', 'touchmove' events and you can access x-, y- positions of your touch fingers by event.touches[*].pageX and event.touches[*].pageY, here, * is the index of your fingers.

(3) **Device orientation/motion interfaces**

The device orientation/motion interfaces are simple but useful in a JavaScript program of HTML5 because there are 'deviceorientation' and 'devicemotion' events, and you can access the device orientation/motion of your smart phone by event.alpha, event.beta, event.gamma, event.acceleration.x, event.acceleration.y, event.acceleration.z.

(4) **Geolocation interface**

The JavaScript program for geolocation interface is also very simple within HTML5 because HTML5 supports navigator.geolocation variable and navigator.geolocation.getCurrentPosition() function, and you can access the device geolocation of your smart phone by position.coords.latitude and position.coords.longitude.

(5) **Camera interface**

For developing web-based AR applications, we have to manage video camera images in real-time on a web browser. There is WebRTC project that provides browsers and mobile applications with Real-Time Communications (RTC) capabilities via simple APIs. Using WebRTC, it is possible to manage video camera images on a web browser and to develop web-based AR applications.

5 E-learning Material Supporting VR/AR for IoT Security Education

Currently, we are developing e-learning material system for IoT security education that support VR/AR using the database of IoT threats information realized as Linked Data and the development framework explained in Sect. 4. In this section, we introduce its prototype.

Figure 10 shows a screenshot of the e-learning material as 3D CG for IoT security education executed on a PC browser. Since our current target is Smart Building/Smart Home, there are several IoT devices located inside the 3D model of a room as shown in the figure, e.g., a mobile phone, a room light, an air-conditioner, TV, PC, and so on, those are connected to the Internet. The purpose of this e-learning material is teaching the contents of Table 1 to students. Using this material, first of all, a student will read the explanation about IoT. Next, after understanding the meaning of IoT, he/she will start to look for any IoT devices in the room through the mouse operations interactively. If he/she find an IoT device and click the mouse on it, its detail information according to the contents of Table 1 will appear on the browser window as shown in the left part of the figure similarly to Fig. 8. In this way, the students will be able to learn IoT threats. For this, we newly added an object picking functionality to the framework. At present, unfortunately, the contents of Table 1 about various IoT devices should be entered manually into the e-learning material system. So, we are also developing authoring functionality to the system in order to make easier to enter the contents into it similarly to the authoring tool shown in Fig. 8.

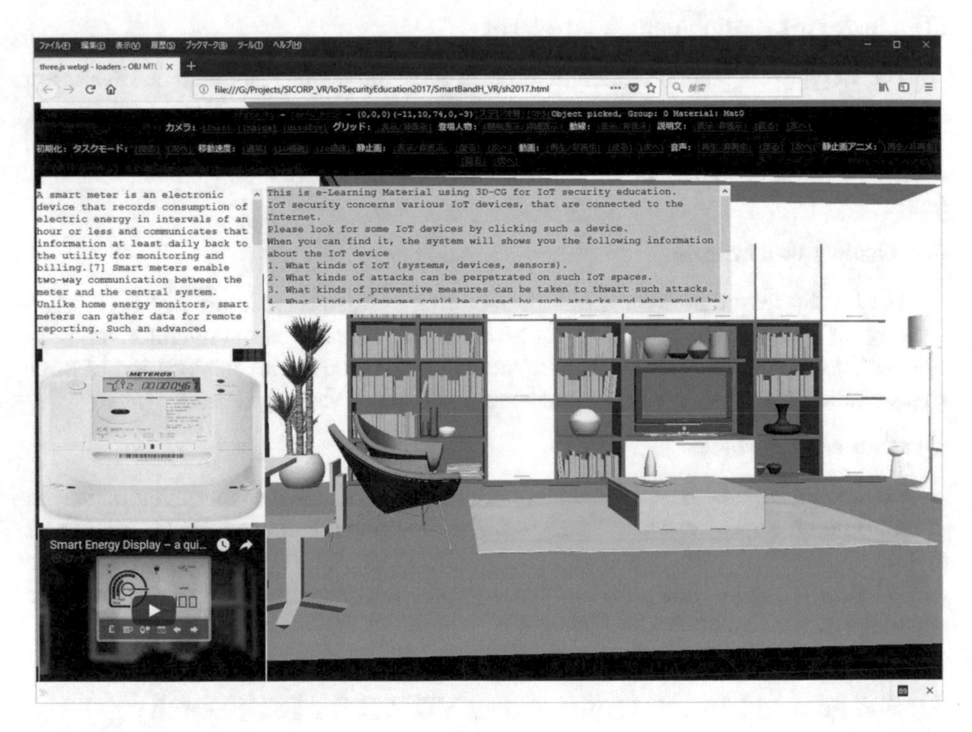

Fig. 10. A screenshot of the e-learning material on a browser for IoT security education.

As for VR aspect, the upper part of Fig. 11 shows a snap shot of the same e-learning material with a stereo view on a smartphone, the lower left part is a picture of VOX + 3DVR goggle and the lower right part is a picture of Poskey blue-tooth gamepad. Using this type of 3DVR goggle, web-applications on a smartphone that support device orientation interface and geolocation interface, and display the stereo images of 3D graphics, come to be Virtual Reality application. Although our framework supports touch interfaces for tablet devices, those interfaces cannot be available when using VOX + 3DVR goggle because a tablet device is put inside the goggle and the user cannot touch it. To compensate this inconvenience, our framework supports Poskey blue-tooth gamepad input. Instead of touch interfaces, the user can move in a virtual 3D space through the gamepad operations.

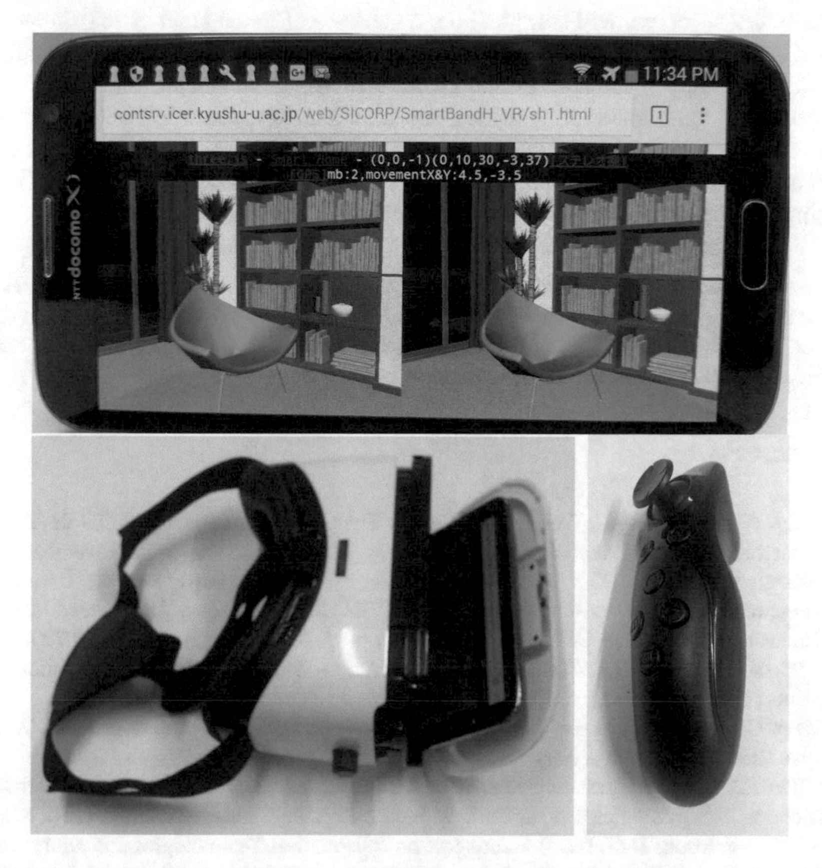

Fig. 11. Snap shots of the material for IoT security education with a stereo view on a smart phone (upper), VOX + 3DVR goggle (lower left) and Poskey blue-tooth gamepad (lower right).

As for AR aspect, our laboratory has EPSON Moveiro BT-300 smart glasses, a see-through type display glasses that supports 3D mode of side-by-side, touch interfaces and device orientation interface. Therefore, the walk-through in the e-learning material system of Fig. 10 similarly to the case of VOX + 3DVR goggle and Poskey blue-tooth gamepad is possible without any additional device. Furthermore, using geolocation interface, the same content becomes AR application because 3D-CG of IoT devices seen from the same viewpoint as the geolocation data can be automatically generated and those are displayed on the real room seen through the glasses.

6 Conclusions

In this paper, we have treated a novel framework for the development of educational materials related to IoT security. Especially, we proposed the combinatorial use of this framework and the database of IoT security information realized as Linked Data, and introduced a prototype of the e-learning material system that support VR/AR for IoT security education.

As future work, we will complete the database as Linked Data of IoT threat information as soon as possible and will finish the development of the e-learning material system that support VR/AR. Furthermore, we will evaluate the usefulness of the proposed e-learning material by asking several students to learn IoT security using it. In our collaboration project, we are supposed to develop a web-based viewer for educational materials provided as epub/edupub format files that has the ability to collect users' learning activity and history data for Learning Analytics.

Acknowledgements. This research was mainly supported by Strategic International Research Cooperative Program, Japan Science and Technology Agency (JST) regarding "Security in the Internet of Things Space", and partially supported by JSPS KAKENHI Grant Number JP16H02923, JP15K12170 and JP17H00773.

References

1. Ma, C., Kulshrestha, S., Wei, S., Okada, Y., Bose, R.: Educational material development framework based on linked data for IoT security. In: iCERI 2017, 16–18 November 2017
2. Okada, Y., Kaneko, K., Tanizawa, A.: Interactive educational contents development framework and its extension for web-based VR/AR applications. In: Proceedings of the GameOn 2017, Eurosis, 6–8 September 2017, pp. 75–79 (2017). ISBN 978-90-77381-99-1
3. Berners-Lee, T.: Design issues for the World Wide Web. Architectural and philosophical points (1998)
4. Bizer, C., Heath, T., Berners-Lee, T.: Linked data-the story so far. Int. J. Semant. Web Inf. Syst. (IJSWIS) **5**, 1–22 (2009)
5. Miller, P.: Sir Tim Berners-Lee: Semantic web is open for business, The Semantic Web (2008)
6. Dicheva, D.: Ontologies and semantic web for e-Learning. In: Adelsberger, H.H., Kinshuk, P.J.M., Sampson, D.G. (eds.) Handbook on Information Technologies for Education and Training, International Handbooks on Information Systems, pp. 47–65. Springer, Berlin, Heidelberg (2008)
7. Okada, Y., Tanaka, Y.: IntelligentBox: a constructive visual software development system for interactive 3D graphic applications. In: Proceedings of Computer Animation 1995, pp. 114–125. IEEE CS Press (1995)
8. Okada, Y.: Web Version of IntelligentBox (WebIB) and Its Integration with Webble World. Communications in Computer and Information Science, vol. 372, pp. 11–20. Springer (2013)
9. Webble World
10. Sugimura, R., et al.: Mobile game for learning bacteriology. In: Proceedings of IADIS 10th International Conference on Mobile Learning, pp. 285–289 (2014)

11. Sugimura, R., et al.: Serious games for education and their effectiveness for higher education medical students and for junior high school students. In: Proceedings of 4th International Conference on Advanced in Information System, E-education and Development (ICAISEED 2015), pp. 36–45 (2015)
12. Okada, Y., Nakazono, S., Kaneko, K.: Framework for development of web-based interactive 3D educational contents. In: 10th International Technology, Education and Development Conference, pp. 2656–2663 (2016)
13. Okada, Y., Kaneko, K., Tanizawa, A.: Interactive educational contents development framework based on linked open data technology. In: 9th annual International Conference of Education, Research and Innovation, pp. 5066–5075 (2016)
14. Takubo, H.: Educational material development system based on linked open data (in Japanese), Graduation thesis, Faculty of Science, Kyushu University (2017)

Automatic Test Case Generation Method for Large Scale Communication Node Software

Kazuhiro Kikuma[1(✉)], Takeshi Yamada[1], Kiyoshi Ueda[2], and Akira Fukuda[3]

[1] NTT Network Service Systems Laboratories, 9-11 Midori-cho 3-Chome, Musashino-Shi 180-8585, Japan
{kikuma.kazuhiro,yamada.t}@lab.ntt.co.jp

[2] Nihon University, 1 Nakagawara, Tokusada, Tamuramachi, Koriyama-Shi 963-8642, Japan
ueda@cs.ce.nihon-u.ac.jp

[3] Kyushu University, 744 Motooka, Nishi-ku, Fukuoka-Shi 819-0395, Japan
fukuda@f.ait.kyushu-u.ac.jp

Abstract. Emerging technologies driven by Network Function Virtualization (NFV) and Software Defined Network (SDN) should be implemented with high quality software. Lower service prices are necessary for carrier networks that provide conventional telephone services on an internet service network. However, the development and maintenance costs tend to remain high because the telephone services must be reliable and secure as they are valuable social infrastructure. Although the communication facilities that provided the backbone of communication infrastructure have become dramatically cheaper to run as hardware has been commoditized and virtualized, the software used in these facilities must guarantee sufficient quality and security. Thus the software development process requires various quality improvement measures, so the development cost remains high. To automate the software development process to solve this problem, we propose a method for automatically generating homogeneous test cases of the system testing phase that does not depend on the skills or know-how of engineers who interpret requirements specification documents written in natural language. We also implement a trial system for automatic test case generation to evaluate the effectiveness of the proposed method.

1 Introduction

As the mobile phone market has spread worldwide due to the expansion of global Internet of Things/machine-to-machine (IoT/M2M) services, the fixed broadband communication market has too [1,2]. Thus, many communications carriers must compete intensely to lower their service prices, because lower service prices are necessary for the carrier networks that provide internet service network including conventional telephone services.

L. Barolli et al. (Eds.): EIDWT 2018, LNDECT 17, pp. 492–503, 2018.
https://doi.org/10.1007/978-3-319-75928-9_44

Emerging technologies driven by Network Function Virtualization (NFV) and Software Defined Network (SDN) should be implemented with high quality software. Commoditizing or virtualizing hardware lowers the cost of communication infrastructure facilities. On the other hand, improving the quality of software used in these facilities is a challenge because the development period becomes longer and costs rise. Thus, formulation and the automation of software quality improvements in each phase of software development need to be studied.

Automation must be studied for all phases of the development. Specifically, engineers currently need to be highly skilled to analyze requirements and generate test cases in the requirement analysis phase because customers who are not engineers write requirements specification documents in natural language.

A Next Generation Network (NGN) [3] enabling fixed broadband communication provides an internet service and conventional telephone service (public switched telephone network (PSTN)) in a unified communications infrastructure. The telephone services must be reliable and secure as they are valuable social infrastructure. As a result, the NGN system requires its software to continually undergo many quality improvement processes to achieve sufficiently high reliability and quality.

To maintain high software quality continuously, a method is needed to detect software bugs unfailingly in the testing phase. To detect software bugs, we should understand specifications exactly and generate test cases comprehensively and precisely. When necessary test cases are not all completely chosen and tests are not performed comprehensively, we provide a customer with a system that contains software bugs and may break the continuity of the commercial service.

To achieve sufficiently high reliability in the development of a high-quality communications system, we develop the system by using a V-model (cf. Fig. 1).

Necessary test cases are generated on the basis of a document produced in the previous phase. Because the generation of appropriate test cases greatly depends on the skills and know-how of the engineers, it takes a long time to establish a test case generation standard and to review test cases to avoid useless test cases generated due to insufficient skills and know-how. Thus this operation increases the cost of the software development.

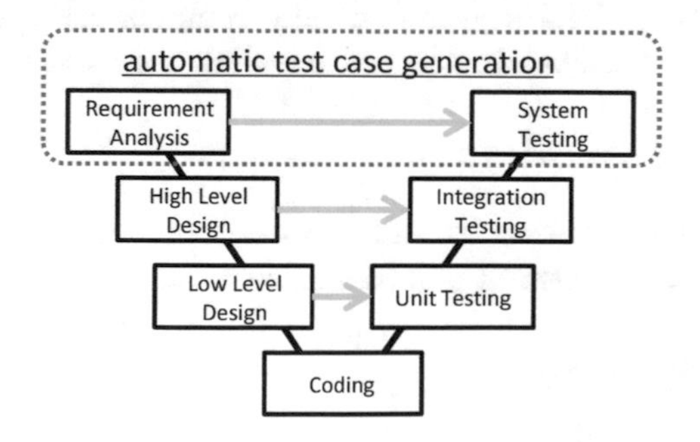

Fig. 1. V-model of software development lifecycle.

In this paper, we propose a method for automatically generating homogeneous test cases in the system testing phase by using machine learning. This does not depend on the skills or know-how of engineers who interpret the requirements specification written in natural language which is a part of the requirement analysis phase.

Previous studies formulated the description style of the requirements specification in a form with which a machine could deal and had customers use this style to describe the requirements specification. However, no method has been developed to generate test cases from natural language documents.

We also implement a trial system for automatic test case generation. We evaluate the effectiveness of the proposed method by seeing if the test cases generated automatically using the trial system are the same as those generated by a highly skilled engineer. If they are, that means we can cut the cost of employing highly skilled engineers to do this task.

2 Related Work

For unit testing, automatic testing and automatic test case generation have been well studied and many support tools were used commonly [6]. Almost all tools analyze the source code and then generate test cases. However, none of the tools can be practically used for system testing in communication node software development.

Researchers have started to use analysis design like Unified Modeling Language (UML) for test case generation [7]. UML use formal language to describe specification documents by using some diagrams. However, for communication node software development, stakeholders often use their natural language rather than UML to exchange their idea, business processes, business rules, and other specifications and to describe the specifications in documents [8,9].

Masuda et al. [8,9] proposed a specification documents analysis technique that can put each branch elements into a decision table [10] and transform the tables into test cases. For the analysis, they created an algorithm and applied it to specification documents.

Because specification documents written in natural language often lack many necessary of words, and have grammatically complex sentences, we introduce machine learning and artificial intelligence techniques for the analysis.

3 Proposed Method

We propose a method to generate test cases automatically for the system testing phase from an unstructured requirements specification document written in natural language.

The steps are as follows:

Part 1

(1) Attach tags manually to the document sentences corresponding to the test cases in the requirements specification document.

Part 2

(2) Perform machine learning with the tagged data.
(3) Apply the results of the machine learning to new requirements specifications, which are not tagged, and attach tags automatically.
(4) Generate test cases automatically from the new tagged data.

3.1 Tags for Structuring Requirements Specifications

The tag should be conscious of generation of test cases (Fig. 2). Test cases of the system are generally constructed by using information input into a system, the state of the system at that time, and information output from the system corresponding to the input. Therefore, for information of a minimum tag, we defined *<Input>*, *<Condition>*, and *<Output>*. Furthermore, considering detailed information of the input and output, we defined *<Input condition>* and *<Output condition>*. In addition, a communications network is comprised of multiple systems, so if the same input is used in multiple systems, they may produce different outputs. We defined the system information tag *<Agent>*, which is the target of the test. There are cases in which the details of process in the system are listed in the requirements specifications. The details are the points that we need to verify through the process, and we defined the check condition tag *<Check point>*.

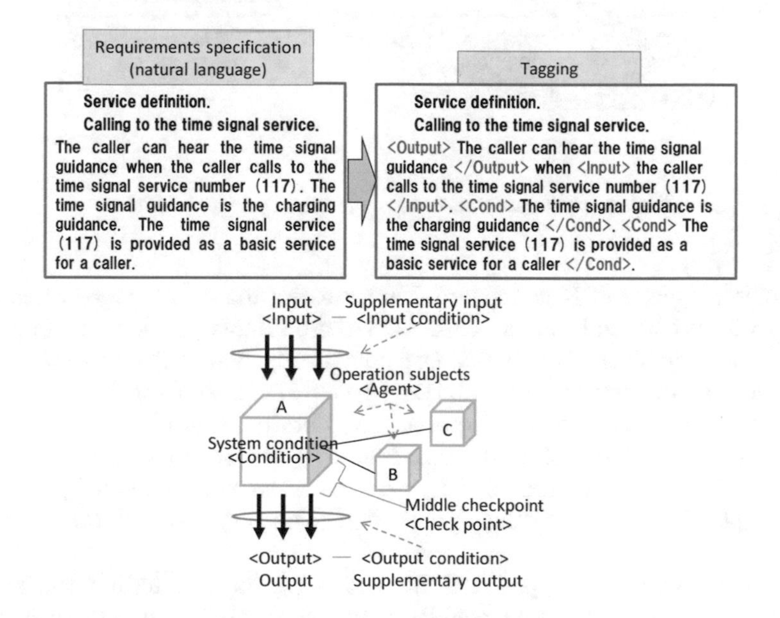

Fig. 2. Seven tags for structuring requirements specification.

3.2 Automatic Test Case Generation System

We implemented a trial system for generating test cases from requirements specifications. Part 2 in Fig. 3 shows the procedure to structure new requirements specifications by machine learning automatically.

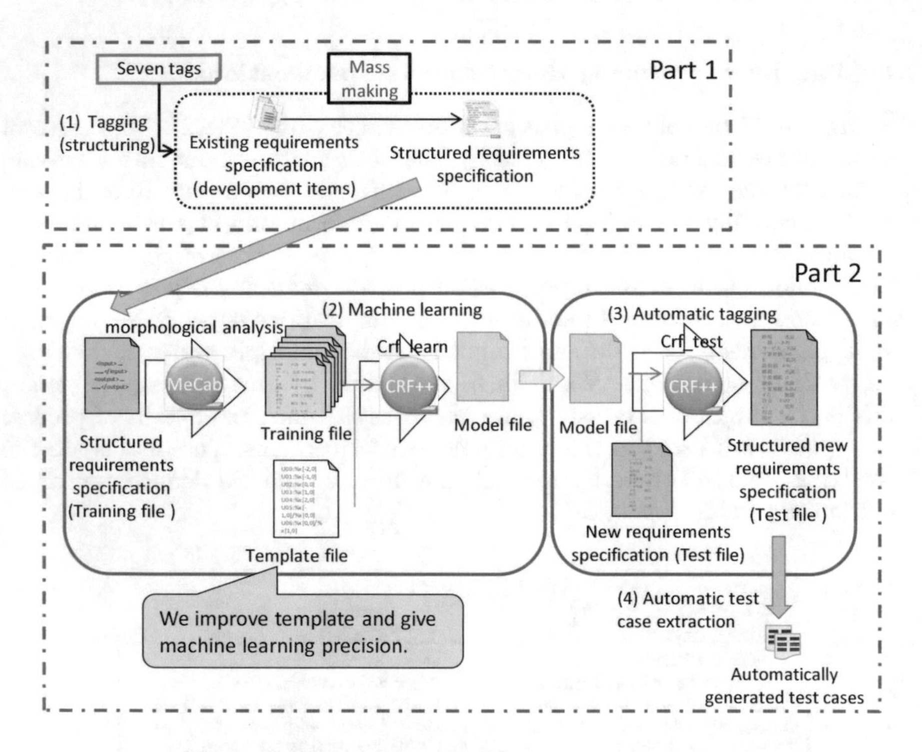

Fig. 3. Procedure of automatic test case generation.

Chinese, Japanese, Korean, and Thai are the main languages consistently written without space between words. For these languages, documents must be divided into words for (2) machine learning (in Fig. 5 above) by tagging using a morphological analysis tool MeCab [11]. MeCab is a morphological analyzer that represents surface form, part of speech, and detailed classification of each part and chunk (answer) tag of each word. MeCab gives word category information to word units extracted from the tagged structured requirements specification. For example, MeCab adds "IN(Input)", "OU(Output)", and "O(Other)" as no tag.

In the above way, we produce mass training data including the kinds of attribute the word holds and enables machine learning software to learn the data.

The training data and template are put in machine learning software CRF++ [12], which learns them in advance and outputs a model file as a learning result.

CRF++ is a simple, customizable, and open source implementation of Conditional Random Fields (CRFs) for segmenting/labeling sequential data. CRF++ will be applied to a variety of Neuro Linguistic Programming (NLP) tasks, such as named entity recognition, information extraction and text chunking.

We have to specify the feature templates in advance. These templates describes which features are used in training and testing. The templates specify the position of the observed token relative to that of the current focusing token in the training data.

The automatic tagging needs to be made more precise for it to be used in commercial system development. We can improve machine learning precision by arranging template files. Each line in the template file denotes a special macro %x[row, col] that is used to specify a token in the input data. The rows specify the relative position from the current focusing token and the cols specify the absolute position of the column. An example of a relative position of an observed token of a template is shown in Fig. 4.

The template in Fig. 4 appoints three observed tokens the relative position of "0, −1, −2" as seen from the current focusing token (Trigram) for a word. For a part of speech, the template appoints two observed tokens the relative position of "0, 1" as seen from the current focusing token (Bigram). For the detailed classification of each part, the template appoints one observed token the relative position of "−2" as seen from the current focusing token (Unigram).

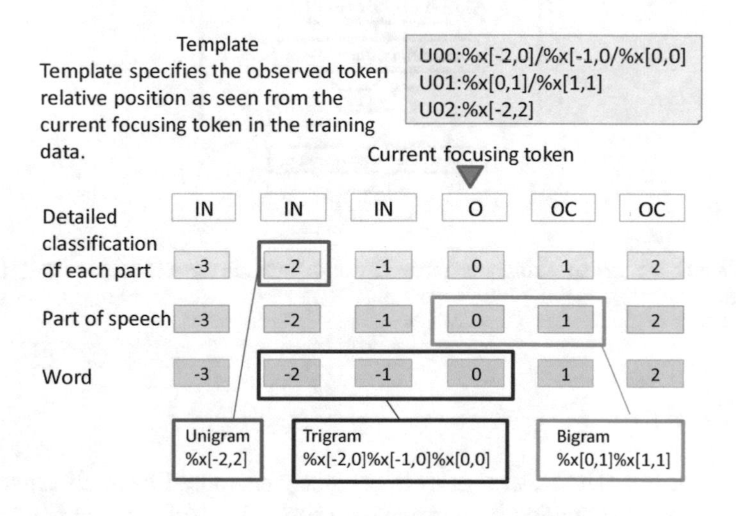

Fig. 4. Template for machine learning CRF++.

Then, we put the model file and a new requirements specification document preprocessed by morphological analysis into the machine learning software and they are automatically tagged in (3) automatic tagging.

3.3 Tagging Check and Test Case Extraction

We developed a method to extract test cases from many tagged documents for checking whether tagged results are correct as test cases. Using the Excel VBA (Visual Basic for Application), we made a function to extract a tag. However, correct results were sometimes omitted from the extracted test cases because there was a tagging error such as misspelling, so errors needed to be corrected. Therefore, the test cases extraction method has three phases: (1) extract tags, (2) detect tagging error, and (3) correct tagging error. Figure 5 shows the flow of this function.

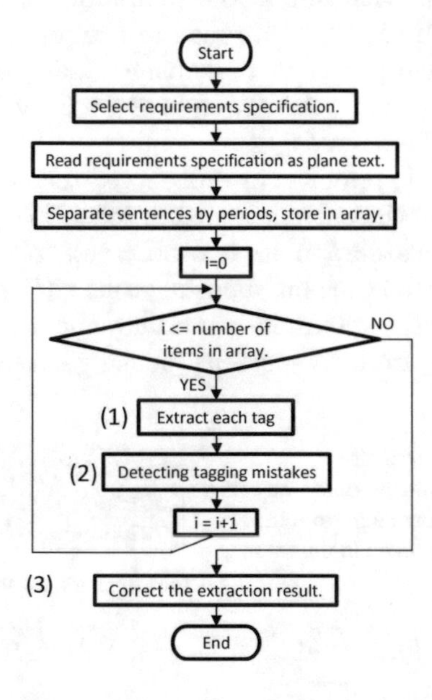

Fig. 5. Flowchart of checking structured requirements specification and test case extraction.

4 Evaluation

We judged whether the test cases were correctly extracted from all tagged documents by automatic tagging by comparing them with test cases extracted by an expert with over 20 years of experience of manual tagging (comparison 1). We also compared test cases generated by automatic tagging and the test cases of an actual development (comparison 2). We calculated the percent of correct test cases (the number of test cases judged to be correct/the number of all correct test cases) (Fig. 6).

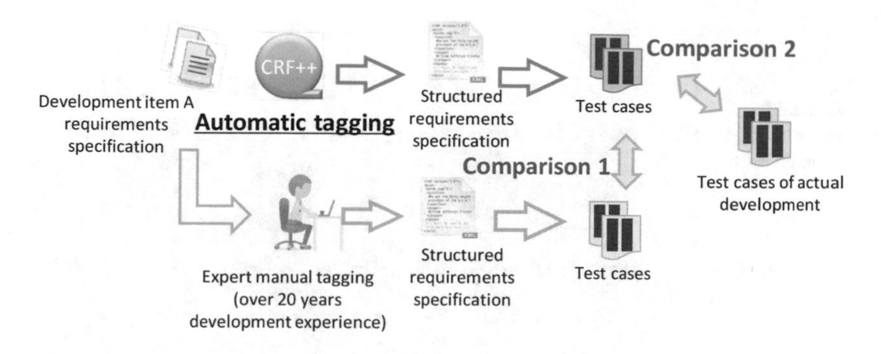

Fig. 6. Test case consistency comparison process.

4.1 Effect of Template Arrangement

We gradually expanded the observation range of training data by using the template for a part of speech and detailed classification of each part, and performed prior learning and automatic tagging. Examples of templates that we arranged are shown in Figs. 7 and 8.

Figure 9 shows a graph of comparison 1 results, the numbers of correct tags and correct no-tags (Other) when the observation range of template is gradually expanded. We found that the numbers of correct tags and correct no tags (Other) tended to rise with the expansion of the observation range of both parts of speech and detailed classification of each part in the template as shown in Fig. 9. We made a training dataset (35,307 words). The test dataset of a new requirements

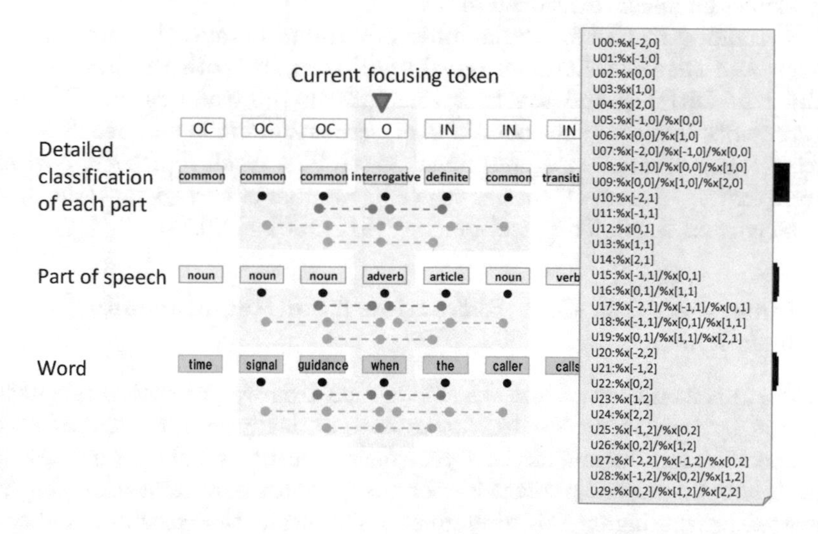

Fig. 7. Basic template containing in CRF++ as example (Template Number 3).

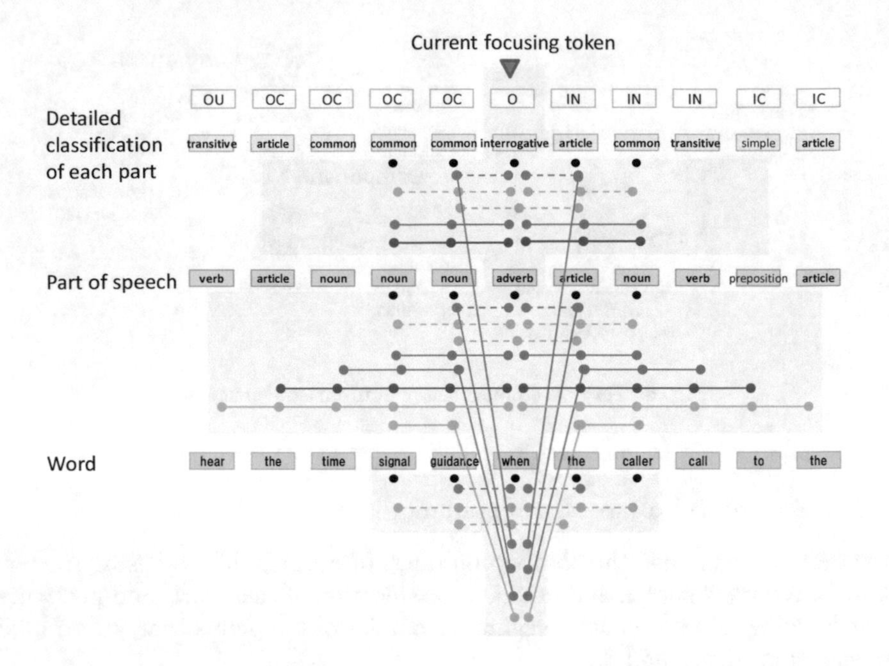

Fig. 8. Best template in examination (Template Number 11).

specification document was almost 3,000 words. We evaluated the effectiveness of proposed method by sceing if thc test cases generated automatically using the trial system with templates arranged were the same as those generated by the highly skilled engineer mentioned above.

Table 1 shows results for the number of automatic tags, the number of correct tags, and the precision, and recall of all tags, for both the basic template contained in CRF++ and the best template in the examination. The precision represents how many automatic tags were correct (the number of correctly extracted tags/the number of automatic tags). The recall represents how many tags were extracted from all correct tags (the number of correct tags/the number of the all correct tags). The precision was (915/1501=) 60.1.

4.2 Efficacy of Test Case Extraction from Requirements Specification

We evaluate the efficacy of test case extraction from requirements specification documents by comparing the test cases automatically extracted from requirements specification documents of a past development and the actual test cases of the development (comparison 2). The requirements specification document was tagged by the highly skilled engineer. The evaluation results are shown in Fig. 10.

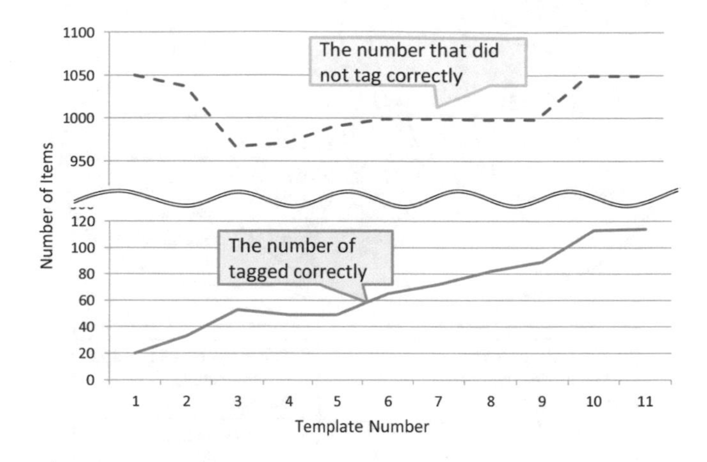

Fig. 9. Machine learning precision of each template.

Table 1. Machine learning precision and recall improvement of each tag by template.

Basic template containing in CRF++ as example (Template Number 3)

	Input	Output	Input condition	Output condition	Agent	Condition	No tag
The number of automatic tags	38	74	24	281	0	0	1084
The number of correct tags	7	42	23	37	0	0	806
Precision	18.4	56.8	95.8	13.2	0.0	0.0	74.4
Recall	8.0	43.3	21.3	26.4	0.0	0.0	76.4

Best template in examination (Template Number 11)

	Input	Output	Input condition	Output condition	Agent	Condition	No tag
The number of automatic tags	18	16	51	109	1	0	1306
The number of correct tags	6	13	42	52	1	0	1049
Precision	33.3	81.3	82.4	47.7	100.0	0.0	80.3
Recall	6.8	13.4	38.9	37.1	50.0	0.0	99.4

There were 65 test cases extracted from the tagged document, 22 of which were the same as actual test cases. These were normal process test cases of a new additional function. The other 43 test cases were the branches and leaves of the same test cases. The branches and leaves are not specified in the extracted test cases, but the extracted test cases are thought to almost all cover the actual test cases considering environment condition of the system testing.

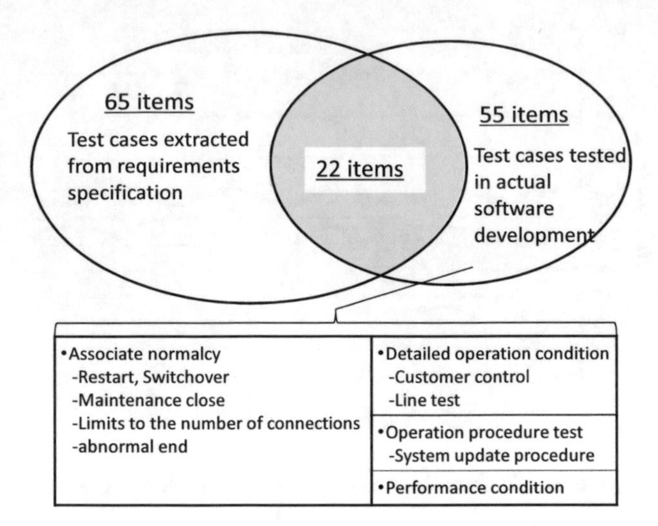

Fig. 10. Comparison between test cases extracted from requirements specification and those in actual software development.

There were 55 actual test cases, 33 of which were not extracted from the tagged document. These 33 test cases were subnormal process test cases, operative procedure confirmation test cases, and non-functional requirement test cases not generally listed in the requirements specification document. The highly skilled engineer used his know how to generate actual non-functional requirement test cases as test cases not listed in the requirements specification document but that experience says should be examined as tacit knowledge. They cannot all be covered only by tagging the requirements specification document.

It was revealed that the extraction of the test cases for system testing was enabled by tagging the requirements specification document. Approximately 40–50% of normal process test cases were correctly extracted. The system failed to extract test cases not listed in the requirements specification document such as subnormal process test cases, but we can cover these test cases by including non-functional requirements and rewrite requirements specification words as antonyms.

5 Conclusion

In this paper, we proposed a low-cost method for automatically generating homogeneous test cases that does not depend on the skills or know-how of an engineer. Our method generated test cases automatically for the system testing phase from an unstructured requirements specification document written in natural language. High skilled engineers attached tags manually to the document sentences corresponding to the test cases in the requirements specification document. We performed machine learning with the tagged data. Then we applied the results of the machine learning to new requirements specifications, which were not tagged, and attached tags automatically.

We implemented a trial system and evaluated the effectiveness of our proposed method. Experimental results revealed the feasibility of our automatic test cases generation system using machine learning. We also found the tag precision was improved by arranging a CRF++ analysis template.

We will produce more training data and apply them to machine learning. We will also raise precision to extract test cases from a tagged document automatically.

References

1. White Paper on Information and Communications in Japan. Economic Research Office, ICT Strategy Policy Division, Global ICT Strategy Bureau, Ministry of Internal Affairs and Communications, Japan (2017). http://www.soumu.go.jp/johotsusintokei/whitepaper/eng/WP2016/2016-index.html
2. Cisco Systems Inc.: Cisco Visual Networking Index: Forecast and Methodology, 2015–2020 (2016). http://www.cisco.com/c/en/us/solutions/collateral/service-provider/visual-networking-index-vni/complete-white-paper-c11-481360.pdf
3. ITU-T Recommendation Y.2012: Functional requirements and architecture of the NGN release 1. Telecommunication Standardization Sector of ITU (2007). https://www.itu.int/rec/T-REC-Y.2012-200609-S/en
4. Boehm, B.W.: Guidelines for verifying and validating software requirements and design specifications. In: Proceedings of EURO IFIP 79, London, pp. 711–719 (1979)
5. Bundesrepublok Deutschland: V-Model XT, Version 1.1.0 (2004). http://ftp.uni-kl.de/pub/v-modell-xt/Release-1.1-eng/Dokumentation/pdf/V-Modell-XT-eng-Teil1.pdf
6. Godefroid, P., Klarlund, N., Sen, K.: DART: directed automated random testing. In: Proceedings of the 2005 ACM SIGPLAN Conference on Programing Language Design and Implementation (PLADI2005), pp. 213–223 (2005)
7. Khandai, M., Acharya, A.A., Mohapatra, D.P.: A survey on test case generation from UML model. Int. J. Comput. Sci. Inf. Technol. (IJCSIT) **2**(3), 1164–1171 (2011)
8. Masuda, S., Hosokawa, N., Iwama, F., Matsuodani, T., Tsuda, K.: Semantic analysis technique of logics retrieval for software testing from specification documents. In: IEEE Software Testing, Verification and Validation Workshops (ICSTW), pp. 1–6 (2015)
9. Masuda, S., Matsuodani, T., Tsuda, K.: Detecting logical inconsistencies by clustering technique in natural language requirements. IEICE Trans. Inf. Syst. **E99-D**(9), 2010–2018 (2016)
10. ISO/IEC/IEEE JTC 1/SC7: Software and system engineering - software testing - part 4 Test techniques. ISO/IEC/IEEE JTC 1/SC 7, pp. 70–72 (2015)
11. MeCab: Yet Another Part-of-Speech and Morphological Analyzer (2013). https://github.com/jordwest/mecab-docs-en
12. CRF++: Yet Another CRF toolkit (2013). http://taku910.github.io/crfpp/#source

Big Data in Cloud Computing: A Review of Key Technologies and Open Issues

Elena Canaj[1(✉)] and Aleksandër Xhuvani[2]

[1] Faculty of Information Technology, Graduate School of ICT, Polytechnic University of Tirana, Tirana, Albania
elenacanaj@gmail.com

[2] Department of Computer Engineering, Faculty of Information Technology, Polytechnic University of Tirana, Tirana, Albania
axhuvani@fti.edu.al

Abstract. Academia, industry and government as well, are involved in big data projects. Many researches on big data applications and technologies are actively being conducted. This paper presents a literature review of recent researches on key technologies and open issues for big data management via cloud computing. Its goal is to identify and evaluate the main technology components and their impacts on cloud-based big data implementations. This is achieved by reviewing 40 publications published in the latest four years, 2014–2017. We classified the results based on the main technical aspects: frameworks, databases and data processing techniques, and programming languages. This paper also provides a reference source for researchers and developers, to determine the best emerging technologies for big data project implementation.

1 Introduction

In recent years, big data management has attracted a lot of attention. Big data, as a variety of data, structured, semi-structured and unstructured, requires high storage and high performance computing. The processing of big data is easier via cloud computing due to its distributed computational paradigm. The cloud computing architecture provides a good solution for large-scale data storage and processing, addressing two of the main requirements of big data. Although big data management as a service in cloud computing has solved many of the big data requirements, it also has raised many important issues, related to the data migration in cloud such as, data security and data privacy.

Many researches on the new methodologies and technologies for both cloud computing and big data are proposed and developed recently. In this paper we present a literature review of recent researches for big data management via cloud computing. This paper identifies and evaluates the key technologies and open research issues of big data management via cloud computing. Its contribution in that respect may be summarized as follows:

- A literature review of key technologies for big data deployment in cloud computing with respect to frameworks, databases and data processing techniques, and programming languages;

L. Barolli et al. (Eds.): EIDWT 2018, LNDECT 17, pp. 504–513, 2018.
https://doi.org/10.1007/978-3-319-75928-9_45

- An overview over the open research issues and challenges of big data management via cloud computing.

Within the context of this paper we provide a reference source for researchers and developers, to determine the best emerging technologies for big data project implementation via cloud computing. For a researcher who is exploring the big data deployment in cloud, is really critical to determine which tools to use during project implementation and the concerns that should be taken into considerations.

The paper consists in: Sect. 2, in which the research methodology is deployed; Sect. 3, in which results and analysis of the searches in a quantitative perspective are presented; Sect. 4, that gives a detailed description and evaluation of the reviewed papers; Sect. 5, that provides an overview of open issues and challenges for big data management in cloud computing; and the final section is the epilogue concluding our work.

2 Methodology

The objective of this paper is to identify and evaluate the main technology components and open issues for implementing big data management as a service in cloud by reviewing and structuring the existing literature. Over hundred publications were first extracted from searches made on three reference and citation-enhanced indexing databases, Google Scholar, Scopus, and Web of Science for the following keywords: big data, cloud computing, Hadoop, MapReduce, Spark, NoSQL, programming language. We paid a particular attention to publications of research results on digital libraries: ACM, IEEE Xplore, SpringerLink, and Elsevier. The time range for this search was limited from 2014 until November 2017. The challenges related to the cloud platforms and frameworks, techniques for big data storage, pre-processing and processing, databases, algorithms, and programming models were all within the scope of this review paper. We focused on reviewing researches of open source technologies but important development on commercial ones are analyzed as well. Papers that were purely focused on technical design were left out of this review. From the numerous research publications, at the end 40 researches and reports were selected and analyzed.

The selected publications are classified based on their main research focus; the categories are: Frameworks, Databases and Data Processing Techniques, and Programming Languages. We analyzed them from a quantitative and qualitative perspective.

3 Literature Review: Quantitative Results and Analysis

In this section we present the results of our work in a quantitative perspective. The selected publications are analyzed and evaluated based on their research contributions. They are noted by their type as Review, Survey, Technical Report, New Design Proposal, Comparative Study, and special attention is paid to real experiments, simulation/emulation and system implementations made by authors. Table 1 shows all the selected publications for this review.

Table 1. List of selected publications.

Ref.	Frameworks	Databases	Programming models	Type	Implement/ Experiments	Year
[1]	Main Topic		x	Survey		2017
[2]	Main Topic			Technical Report	x	2016
[3]	Main Topic		x	New Design	x	2014
[4]	Main Topic		x	New Design	x	2017
[5]	Main Topic		x	New Design	x	2015
[6]	Main Topic			New Design	x	2014
[7]	Main Topic			Review		2015
[8]	Main Topic		x	New Design	x	2016
[9]	Main Topic		x	New Design	x	2016
[10]	Main Topic		x	New Design	x	2015
[11]	Main Topic		x	Case Study	x	2015
[12]	Main Topic		x	New Design	x	2015
[13]	Main Topic			Comparative Study	x	2014
[14]	Main Topic			Survey		2015
[15]	Main Topic			New Design	x	2015
[16]	x	Main Topic	x	Review		2016
[17]		Main Topic		Review		2014
[18]		Main Topic		Comparative Study	x	2017
[19]	x	Main Topic	x	New Design	x	2016
[20]	x	Main Topic	x	Survey		2017
[21]		Main Topic		Survey		2015
[22]	x	Main Topic	x	Survey		2017
[23]		Main Topic		Survey		2017
[24]		Main Topic		New Design	x	2015
[25]		Main Topic		Review		2015
[26]	x	Main Topic		New Design	x	2015
[27]		Main Topic		Comparative Study		2016
[28]		Main Topic		New Design	x	2016
[29]	x	Main Topic	x	New Design	x	2015
[30]			Main Topic	Review		2017
[31]			Main Topic	Review		2017
[32]			Main Topic	Review		2014
[33]			Main Topic	Survey		2015
[34]	x		Main Topic	New Design	x	2017
[35]			Main Topic	Technical Report	x	2015
[36]	x		Main Topic	Comparative Study	x	2014
[37]			Main Topic	New Design	x	2017
[38]			Main Topic	Survey		2017
[39]			Main Topic	New Design	x	2014
[40]			Main Topic	New Design		2016

Based on the research contribution, we present the results on the total number of publications per category in Table 2. During that process we observed that framework component is mostly researched.

Table 2. Distribution of main technology aspects

Domain	Total no. of publications	Percentage Papers/Category (%)
1. Frameworks	15	37
2. Databases and data processing techniques	14	35
3. Programming languages	11	28
Total	40	100

In the Table 3, it can be noticed that the majority of the publications considered and analysed are original researches proposing new designs and improvements for big data management in cloud computing.

Table 3. Representation of the total number per publication type.

Publication type	Total no. of publications	Percentage Publication/Type (%)
Survey	8	20
Review	7	17
New Design	18	45
Technical Report	2	5
Comparative Study	4	10
Case Study	1	3
Total	40	100

4 Literature Review: Topics-Related Analysis

This section is an overview of each of the selected papers. The publications are mapped based on the main topic and their contributions on big data as a service implementation in cloud. Our work is focused more on reviewing researches of open source technologies, but important development on commercial ones are analyzed as well.

4.1 Frameworks

In this subsection, we will analyze and discuss recent researches of the most essential component of a big data system. To simplify the decision on choosing the best framework solution for big data implementation in cloud, we reviewed 15 publications that address development, improvements, experiments and open issues on various frameworks [1–15].

Firstly we analysed the publications found during our search for Hadoop framework. Hadoop is the backbone of several large scale applications in different domains for analyzing large scale data. In the first paper [1] Hadoop framework is used to analyze workload prediction of data from cloud computing. Hadoop was also used for analyzing the tweets on the large scale in paper [2]. Authors in [3] used Hadoop, and MapReduce parallel processing paradigm to propose a new solution - Keyword-Aware Service Recommendation method for analyzing data in service recommender systems. The results of these papers show that Hadoop framework is a good choice for batch processing that are not time-sensitive. Original researches proposing improvements and new designs for Hadoop are done. Authors in [4] propose an integrated Hadoop and MPI/OpenMP system for higher processing speed.

Open issues are raised in literature relating Hadoop security, due to missing encryption at the storage and network levels. According to this issue authors in [5] propose a new security model for G-Hadoop, which is based on public key cryptography and the SSL protocol.

In addition to Hadoop, there are several other frameworks studied and proposed in the following papers.

Authors in [6] investigated the performance of big data applications on Spark with different virtualization frameworks. Review paper [7] analyses the infrastructure of another open source framework, Apache Storm. In paper [8] authors discuss on the evolution of big data frameworks and propose a new design framework Scalation using the Scala programming language. Spark, Samza, Kafka and Scalation frameworks are evaluated through experiments. Authors in [9] propose a new framework design, High Performance Analytics Toolkit (HPAT). Their evaluation has demonstrated that HPAT is faster than Spark. Authors in [10] have implemented a cloud-based analytics service using Hadoop and Spark, and the results are being compared. In paper [11] authors investigated the Yahoo!S4 framework for processing of real time streams. Authors in [12] present a new cloud based framework for big data management in smart grids. In paper [13] a detailed comparative study of three different frameworks Apache Hadoop, Project Storm, Apache Drill is performed. Survey paper [14] also presents a theoretical comparison of the main frameworks for big data. Authors in [15] analyze DistributedWekaSpark, a distributed Spark framework for Weka workbench.

At the end of this subsection we outline our findings as following: Hadoop framework is suited for workload where time is not critical. It's a good choice for batch processing that are not time-sensitive. It is an open source and easier to implement than other solutions. For stream processing Storm and S4 are typical frameworks for real-time large scale streaming data. These frameworks have very low latency processing. Samza framework when is integrated with Kafka also is a good solution for stream processing. Flink frameworks support stream processing and also handle batch processing. It is heavily optimized, but it is still unstable. For interactive environment, Spark is very adequate. Apache drill is also best for interactive and ad-hoc analysis.

There are many options for processing data in cloud based big data system but the best fit for any implementation depends on the data to process and time requirements. Workbenches like the one analyzed in this section are available for testing the frameworks. For researchers and developers, it's possible to simulate some situations prior the final implementation of any project. The biggest concerns that should be taken into consideration while using frameworks in cloud computing are data security and data privacy.

4.2 Databases and Data Processing Techniques

In this subsection 14 recent publications on data storage and processing techniques are reviewed [16–29].

The first article reviewed on this topic [16] outlines the appropriate technologies for implementing big data project. This article includes technologies such as in-memory databases, NoSQL and NewSQL systems, and Hadoop based solutions.

A comparative study about the performance of NoSQL databases: BigTable, Cassandra, HBase, MongoDB, CouchDB, CrowdDB is done in paper [17]. Authors in [18] also compare the performance of HBase and MongoDB. Authors in [19] present a new solution that integrates Cassandra with MapReduce. Paper [20] is a review of the researches done for incorporation of data warehouse with MapReduce for handling of big data. Paper [21] is a survey of in-memory big data management systems. Authors in [22], also presents a survey on recent technologies developed on big data, for Data Processing Layer, Data Querying Layer, Data Access Layer and Management Layer. Paper [23] studies the solution of various types of unstructured data storage, analyzes all the problems existing in the storage system, and summarizes the key issues to achieve the unified storage of unstructured data.

Authors in [24] propose new solutions for implementation of big data warehouses under the column oriented NoSQL DBMS. In paper [25], a detailed classification for modern big data models is done. Paper [26] introduces the combination of NoSQL database HBase and enterprise platform Solr. Authors in [27] have compared SQL with NoSQL databases and the four NoSQL data models (document-oriented, key-value pairs, column-oriented or graphs). In paper [28], a new design allowing the automatic transformation of a multidimensional schema into a tabular schema is implemented in Hive. Last paper [29] evaluates the performance of Spark SQL.

At the end of this subsection we summarize our findings. Traditional relational database management systems are not suitable for big data. NoSQL database management systems are designed for use in high data volume applications in cloud environments. Several open source NoSQL databases exist, MongoDB, Cassandra, CouchDB, CrowdDB, Hypertable, HBASE, Couchbase, etc. NoSQL databases are highly scale able, flexible and good for big data storage and processing. The current issue with NoSQL databases is that they do not offer a declarative query language similar to SQL and there is no single, unified model of NoSQL databases. Integrated solutions are researched and proposed. Applications that combine relational and procedural queries run faster. We noticed that many researches are still going forward to optimize data storage and processing techniques. Heterogeneity of data is also a problem that is currently under study. We have review NoSQL data models that can process big data up to the petabyte range. Exabyte range data processing is still an open problem under-researched.

4.3 Programming Languages

We selected 11 publications from our search results that address latest researches on programming languages [30–40].

Firstly we have reviewed all existing literatures that summarize existing programming languages and paradigms available in cloud-based big data area. Review papers [30–32] discuss and compare various programming models, analyzing how they fit into the big data projects.

Authors in [33], presents a survey of programming models for big data implementations in grid and cloud. Nystrom in [34] describes a Scala framework that can be used for experimenting with supercompilation techniques. In this paper a supercompiler for JavaScript is implemented. Authors in [35] analyze the new programming language Julia which is appropriate for parallel computing. In paper [36] programming language R is integrated with Hadoop framework and they are used together for big data statistical programming. Authors in [37] propose a new design in order to manage language runtimes. Survey paper [38] analyzes the MapReduce-based algorithms for handling big RDF graphs. Paper [39] investigates the program transformations for Pig Latin. Authors in [40] propose a new approach JAVA2SDG, for stateful big data processing.

In the reviewed publications we found many programming languages like R, Python, Java, Scala, Hadoop languages (Pig Latin, Hive), Julia and new programming language proposals. We analyzed all of them in order to have a full understanding of the actual research process in the field of programming paradigm for cloud-based big data. Programming language R is adequate for executing large numbers of calculations. Python is a good choice for advanced analytics. Scala is also a good choice for large streaming since it is a hybrid programming language, which combine both Functional Programming with Object Oriented Paradigms. Java is used as a basic code for many frameworks but it has a very high learning curve. The new programming language Julia fits well for real-time streams applications.

We conclude that Functional Programming (FP) is considered to be the most adequate for big data implementations. Its difficult syntax has pushed researchers to try new hybrid solutions. During our search as it is shown in Sect. 3, we didn't find quite many recent researches on programming languages. As stated at IEEE Spectrum "The 2017 Top Programming Languages", it seems a period of consolidation in coding as programmers digest the tools created for cloud and big data applications [41].

5 Challenges and Open Research Issues

Although big data management in cloud computing is widely used, it still faces challenges and open issues. This section presents an overview of the open issues identified in the literature that affect big data deployment in cloud computing. We address the following open research issues for further improvement:

- Security and privacy: Data security and data privacy are critical issues when migrating big data to cloud environment. Many technical solutions using data

encryption are applied: Data encryption affects the performance of big data processing thus researchers are still working for further improvements to this issue.
- Data transmission: While transferring large-scale data to the cloud, the capacity of the network bandwidth is the main obstacle. Over the years many algorithmic proposals and improvements are being applied to minimize cloud upload time, however, this process still remains a major research challenge.
- Data volume: The exponential growth of data to the exabyte range raises lots of concerns for the big data storage and processing in the cloud environment. Due to limited network bandwidth, cloud computing is not suitable for exabyte data processing. Exascale computing is a major research challenge.
- Data storage and processing: The current issue with NoSQL databases is that they do not offer a declarative query language similar to SQL and there is no single, unified model of NoSQL databases. Integrated solutions that combine relational and procedural queries for better performance are still being researched.

6 Conclusion

In this paper we identified the key technologies and open research issues of big data management via cloud computing. We reviewed 40 publications in order to address the recent researches with respect to frameworks, databases, data processing techniques, and programming languages. We found that framework component is mostly researched. There are various framework solutions for big data in cloud but the best fit depends on the data to process and time requirements. Regarding databases many NoSQL integrated solutions are researched and proposed. Functional Programming (FP) is considered to be the most adequate programming paradigm for big data implementations.

We conclude that big data management via cloud computing has still open research issues related to the data transfer, data volume, data storage, data security and data privacy. All of these aspects makes cloud-based big data management a viable research field.

Within this paper we provide a reference guide for researchers and developers, to determine the best emerging technologies for implementing big data as a service in cloud computing.

References

1. Mallika, C., Selvamuthukumaran, S.: Hadoop framework: analyzes workload predicition of data from cloud computing. In: 2017 International Conference on IoT and Application (ICIOT), pp. 1–6. IEEE (2017)
2. Nodarakis, N., Sioutas, S., Tsakalidis, A., Tzima, G.: Using Hadoop for Large Scale Analysis on Twitter: A Technical Report. arXiv preprint arXiv:1602.01248 (2016)
3. Meng, S., Dou, W., Zhang, X., Chen, J.: KASR: a keyword-aware service recommendation method on MapReduce for big data applications. IEEE Trans. Parallel Distrib. Syst. **25**(12), 3221–3231 (2014)

4. Bhimani, J., Yang, Z., Leeser, M., Mi, N.: Accelerating big data applications using lightweight virtualization framework on enterprise cloud. In: High Performance Extreme Computing Conference (HPEC), pp. 1–7. IEEE (2017)
5. Ortiz, J.L.R., Oneto, L., Anguita, D.: Big data analytics in the cloud: spark on hadoop vs MPI/OpenMP on Beowulf. Procedia Comput. Sci. **53**, 121–130 (2015)
6. Zhaoa, J., Wang, L., Tao, J., Chen, J.: A security framework in G-Hadoop for big data computing across distributed Cloud data centres. J. Comput. Syst. Sci. **80**(5), 994–1007 (2014)
7. Huang, T., Lan, L., Fang, X., An, P., Min, J., Wang, F.: Promises and challenges of big data computing in health sciences. Big Data Res. **2**(1), 2–11 (2015)
8. Miller, J., Bowman, C., Harish, V., Quinn, S.: Open source big data analytics frameworks written in scala. In: 2016 IEEE International Congress on Big Data (BigData Congress), pp. 389–393 (2016)
9. Totoni, E., Anderson, T., Shpeisman, T.: HPAT: High Performance Analytics with Scripting Ease-of-Use. arXiv preprint arXiv:1611.04934 (2016)
10. Khan, Z., Anjum, A., Soomro, K., Tahir, M.A.: Towards cloud based big data analytics for smart future cities. J. Cloud Comput. **4**(1), 2 (2015)
11. Xhafa, F., Naranjo, V., Caballé, S.: Processing and analytics of big data streams with Yahoo! S4. In: 2015 IEEE 29th International Conference on Advanced Information Networking and Applications (AINA), pp. 263–270 (2015)
12. Baek, J., Vu, Q., Liu, J., Huang, X., Xiang, Y.: A secure cloud computing based framework for big data information management of smart grid. IEEE Trans. Cloud Comput. **3**(2), 233–244 (2015)
13. Chandarana, P., Vijayalakshmi, M.: Big data analytics frameworks. In: Proceedings of the International Conference on Circuits, pp. 430–434. IEEE (2014). ISBN: 978-1-4799-2494-3
14. Singh, D., Reddy, C.K.: A survey on platforms for big data analytics. J. Big Data **2**(1), 8 (2015)
15. Koliopoulos, A., Yiapanis, P., Tekiner, F., Nenadic, G., Keane, J.: A parallel distributed weka framework for big data mining using spark. In: 2015 IEEE International Congress Big Data (BigData Congress), pp. 9–16 (2015)
16. Zicari, R., Rosselli, M., Korfiatis, N.: Setting up a big data project: challenges, opportunities, technologies and optimization. In: Studies in Big Data, vol. 18, pp. 17–47. Springer (2016)
17. Sharma, S., Tim, U.S., Wong, J., Gadia, S.: A brief review on leading big data models. Data Sci. J. **13**, 138–157 (2014)
18. Matallah, H., Belalem, G.: Experimental comparative study of NoSQL databases: HBASE versus MongoDB by YCSB. Comput. Syst. Sci. Eng. **32**(4), 307–317 (2017)
19. Dede, E., Sendir, B., Kuzlu, P., Weachock, J., Govindaraju, M., Ramakrishan, L.: Processing Cassandra datasets with Hadoop-streaming based approaches. IEEE Trans. Serv. Comput. **9**(1), 46–58 (2016)
20. Ptiček, M., Vrdoljak, B.: MapReduce research on warehousing of big data. In: Mipro 2017 (2017)
21. Zhang, H., Chen, G., Ooi, B.C., Tan, K.L.: In-memory big data management and processing: a survey. IEEE Trans. Knowl. Data Eng. **27**(7), 1920–1948 (2015)
22. Oussous, A., Benjelloun, F.Z., Lahcen, A.A., Belfkih, S.: Big data technologies: a survey. J. King Saud Univ.-Comput. Inf. Sci. (2017)
23. Peng, S., Liu, R., Wang, F.: New Research on Key Technologies of Unstructured Data Cloud Storage. Francis Academic Press, UK (2017)

24. Dehdouh, K., Bentayeb, F., Boussaid, O., Kabachi, N.: Using the column oriented NoSQL model for implementing big data warehouses. In: Proceedings of the International Conference on Parallel and Distributed Processing Techniques and Applications (PDPTA), The Steering Committee of the World Congress in Computer Science, Computer Engineering and Applied Computing (WorldComp), p. 469 (2015)
25. Sharma, S.: An extended classification and comparison of NoSQL big data models. arXiv preprint arXiv:1509.08035 (2015)
26. Chang, B.R., Tsai, H.F., Chen, C.Y., Huang, C.F., Hsu, H.T.: Implementation of secondary index on cloud computing NoSQL database in big data environment. Sci. Program. 19 (2015)
27. Sitalakshmi Venkatraman, K.F., Kaspi, S., Venkatraman, R.: SQL versus NoSQL Movement with Big Data Analytics (2016)
28. Santos, M.Y., Costa, C.: Data warehousing in big data: from multidimensional to tabular data models. In: Proceedings of the Ninth International C* Conference on Computer Science and Software Engineering, pp. 51–60. ACM (2016)
29. Armbrust, M., Xin, R.S., Lian, C., Huai, Y., Liu, D., Bradley, J.K., Meng, X., Kaftan, T., Franklin, M.J., Ghodsi, A., Zaharia, M.: Spark SQL: relational data processing in spark. In: Proceedings of the 2015 ACM SIGMOD International Conference on Management of Data, pp. 1383–1394 (2015)
30. Siddiqui, T., Alkadri, M., Khan, N.A.: Review of programming languages and tools for big data analytics. Int. J. Adv. Res. Comput. Sci. **8**(5) (2017)
31. Wu, D., Sakr, S., Zhu, L.: Big data programming models. In: Handbook of Big Data Technologies, pp. 31–63. Springer (2017)
32. Dobre, C., Xhafa, F.: Parallel programming paradigms and frameworks in big data era. Int. J. Parallel Prog. **42**(5), 710–738 (2014)
33. Jackson, J.C., Vijayakumar, V., Quadir, M.A., Bharathi, C.: Survey on programming models and environments for cluster, cloud, and grid computing that defends big data. Procedia Comput. Sci. **50**, 517–523 (2015)
34. Nystrom, N.: A scala framework for supercompilation. In: Proceedings of the 8th ACM SIGPLAN International Symposium on Scala, pp. 18–28, October 2017
35. Edelman, A.: Julia: a fresh approach to parallel programming. In: 2015 IEEE International Conference on Parallel and Distributed Processing Symposium (IPDPS), p. 517 (2015)
36. Oancea, B., Dragoescu, R.M.: Integrating R and hadoop for big data analysis. arXiv preprint arXiv:1407.4908 (2014)
37. Maas, M., Asanović, K., Kubiatowicz, J.: Return of the runtimes: rethinking the language runtime system for the cloud 3.0 era. In: Proceedings of the 16th Workshop on Hot Topics in Operating Systems, pp. 138–143, May 2017
38. Cuzzocrea, A., Buyya, R., Passanisi, V., Pilato, G.: MapReduce-based algorithms for managing big RDF graphs: state-of-the-art analysis, paradigms, and future directions. In: Proceedings of the 17th IEEE/ACM International Symposium on Cluster, Cloud and Grid Computing, pp. 898–905 (2017)
39. James Stephen, J., Savvides, S., Seidel, R., Eugster, P.: Program analysis for secure big data processing. In: Proceedings of the 29th ACM/IEEE International Conference on Automated Software Engineering, pp. 277–288 (2014)
40. Fernandez, R.C., Garefalakis, P., Pietzuch, P.: Java2SDG: stateful big data processing for the masses. In: 2016 IEEE 32nd International Conference Data Engineering (ICDE), pp. 1390–1393 (2016)
41. The 2017 Top Programming Languages, IEEE Spectrum ranking. https://spectrum.ieee.org/computing/software/the-2017-top-programming-languages. Accessed 27 Oct 2017

A Novel Question Answering System for Albanian Language

Evis Trandafili(✉), Elinda Kajo Meçe, Kristjan Kica, and Hakik Paci

Department of Computer Engineering, Faculty of Information Technology, Polytechnic University of Tirana, Tirana, Albania
{etrandafili,ekajo,kristjan.kica,hpaci}@fti.edu.al

Abstract. The volume of unstructured data is constantly growing, drawing the attention of the research community toward Natural Language Processing tasks. Recent advances in Information Extraction have led to the implementation of different systems and tools for Question Answering. These approaches are mainly language dependent as they need information about the language structure and syntax to perform well. This paper proposes an approach to extracting answers of factoid questions for a given text in Albanian Language. As far as we know, this is the first attempt of a Question Answering system for Albanian language. Experiments show that this is an effective solution for single domain documents.

1 Introduction

The volume of available digital information is constantly growing and the way we live and work relies more and more on the universal information available. The freedom of publishing has led to replicated information usually mixed with other non-crucial data. Moreover, this large volume of information is mainly unstructured thus finding relevant information becomes more complicated.

The main objective of a question answering system is to provide the exact required information with much less human efforts. The users of digital information tend to query data in natural language. Traditional information retrieval systems respond to a query with a list of the most relevant documents where the user has to investigate for the required answer. This is time consuming. From a QA system point of view, information retrieval techniques are used to extract the exact information within the document which responds to the question. Most of question answering systems are implemented based on factoid questions. These are the type of questions whose answers are simple facts expressed with a short string which refers to a date, a place, a name, etc. [1]. For example, questions like *"Kur u shpall pavarësia e Shqipërisë?", (English: When was Albania proclaimed independent?), "Ku buron lumi Shkumbin?", (English: Where does the Shkumbin river originate?), "Kush e shkroi librin Meshari?", (English: Who wrote the book Meshari?),* correspond to some factoid questions in Albanian language.

A typical question answering system first determines the answer type by processing the query/question and then reformulates it in the appropriate query format of the

L. Barolli et al. (Eds.): EIDWT 2018, LNDECT 17, pp. 514–524, 2018.
https://doi.org/10.1007/978-3-319-75928-9_46

search engine. The outputs are the relevant ranked documents broken into passages. Finally, text answers are extracted and ranked. Our QA system is based on the above method. Language dependent rules are implemented to detect the type of factoid question and to highlight the answer type. We used Text-Based question answering paradigms to extract the answer type and formulate the query relying on the lexical and semantic matching of the key words present both in the question and in the document.

The organization of the paper is as follows: Sect. 2 presents the background and the related works on question answering systems; Sect. 3 presents some basic information about the Albanian language structure; Sect. 4 presents the architecture and some implementation details of our novel QA system; Sect. 5 discusses the evaluation metrics used to test the performance of our system; and we conclude in Sect. 6 with conclusions and future work.

2 Background and Related Works

Question answering is a subfield of Information Retrieval and Natural Language Processing which focuses on the extraction of passages within the document in response to the user's need for information expressed in natural language. Ongoing efforts are done to automate and improve this process. There are three approaches for implementing Question Answering systems; the simplest is the IR-based model which focuses on factoid questions; knowledge-based model relies on the semantic representation of the query; and the hybrid model uses text corpuses and structured knowledge databases [1]. Most of QA systems are based on factoid questions due to thc implementation simplicity, whereas knowledge-based approaches aim to answer questions about definitions or concepts and are much harder to implement.

The standard algorithm for QA based on factoid questions encompasses three main modules: the question processing module which extracts the answer type (identifying the entity of the answer) and the keywords for the IR system to retrieve passages in the document. Next, the passage retrieval module uses an answer type classification to filter out passages that contain the wrong answers and then rank the remaining passages using supervised machine learning algorithms. The final module is answer extraction where the extraction process is achieved using information about the expected answer type together with regular expression patterns or using N-gram tiling when QA is applied in web search [1].

Efforts in the implementation of QA systems have been made since 1961 with BASEBALL [2] that answered questions about baseball games and LUNAR [3] that provided answers about soil samples taken from Apollo exploration. Both these systems answered English written questions by transforming the question into a database query through pattern matching rules.

Ongoing attempts aim to improve the overall performance of QA systems. Watson is based on a parallel framework composed of two modules: Natural Language Processing module which deals with question analysis and Information Extraction module which retrieves the candidate answers [4].

Another approach on QA uses the star architecture by viewing the subtasks (question processing, passage retrieval and answer extraction) as nodes connected by a central node that generates the optimal strategy to find the answer depending on the type of question [5].

An interesting approach is a factoid-based QA system named Sybil, which works with spoken documents in English [6]. It uses natural language analyzers and linguistic information obtained with machine learning tools. Sybil was evaluated using the European Parliament Plenary Sessions English corpus and its performance is better than the state-of-the-art on this corpus.

A QA system which finds the answers from a single document based on the category space acquired from Wikipedia is proposed in [7]. A Natural Language Processing Toolkit is used for keyword extraction and the distance between categories is used to rank the answers.

There are also several attempts to connect different domains of artificial intelligence with question answering. The authors in [8] focused on the state of the art of Visual Question Answering, a field of study that combines Computer Vision with Natural Language Processing methods. They reviewed different approaches that aim to map questions and images in vector representation using a common feature space. They highlighted that a successful approach is the combination of convolutional neural networks that are trained on object recognition, with word embedding's, trained on large text corpora.

A new way of thinking regarding question answering is proposed in [9]. The proposed architecture automatically generates questions from sentences containing important facts or events and their respective answers. This QA system builds and maintains a database of pairs (question, answer) corresponding to the domain of documents. The main components used for the system are: sentence split, named-entity recognition, question generation, question filtering and question/answer indexing. The user is presented with a set of candidate query questions for his information needs.

Moreover, substantial works are made to handle questions in other than English languages. A QA monolingual system for searching French documents based on French questions is proposed by [10]. They used a named entity recognition technique and a syntactic analyzer to identify the candidate answers. Following, a matching strategy is applied to rank the answers. The bilingual module is able to answer questions written in Dutch, German, Italian, Portuguese, Spanish, English and Bulgarian. This is achieved by automatically translating the original question into French and then proceeding with the monolingual QA system.

Moreover, a bilingual question answering system is proposed in [11] and handles Bangla and English electronic documents. The authors claim that this system generates questions and answers efficiently. It has four main components: database initialization and processing, storage, question answering and query execution.

Several efforts are made to support questions made in Arabic language. A QA system for Arabic language is implemented in [12]. This system handles different types of questions and relies on a language dependent preprocessing step to increase its overall performance.

Furthermore, in [13] the authors introduce a QA system for Vietnamese language based on ontology. The system has two modules: the question analysis module and the answer retrieval module. The semantic structure of the question is captured through an intermediate representation and then it is matched with the target ontology.

3 Albanian Language Structure

Albanian language is considered a separate branch of the Indo-European language family and cannot conclusively be closely connected with any other Indo-European language. It is the official language of Albania, the co-official language of Kosovo, and the co-official language of many western municipalities of the Republic of Macedonia. Albanian is also spoken widely in some areas in Greece, southern Montenegro, southern Serbia, and in some towns in southern Italy and Sicily.

The Albanian language has two main dialects spoken in two major regions; the north dialect called Gheg and the southern dialect called Tosk. The official Albanian language is written in Roman alphabet and from 1909 till World War II was based on the south Gheg dialect. Since then it has been modeled on Tosk dialect [14]. For the purpose of this paper, the official Albanian language is used. It has 7 vowels and 29 consonants. The vowels are represented by single Latin letters (*a, e, ë, i, o, u, y*), and the consonants by single letters (*b, c, ç, d, f, g, h, j, k, l, m, n, p, q, r, s, t, v, x, z*), and combination of different letters (*dh, gj, ll, nj, rr, sh, th, xh, zh*).

Even though the most common word order in a sentence is Subject-Verb-Object (SVO), Albanian like German is relatively free [15] because the role of the word in a sentence is not determined by its occurring position, but by its inflectional ending and by its relative meaning. For example, the sentence *"Teuta read all the documents."* can have different possible role orders with only slight pragmatic differences. All the sentences listed below are grammatically correct.

SVO - *Teuta i lexoi të gjitha dokumentat.*
SOV - *Teuta, të gjitha dokumentat i lexoi.*
VOS - *I lexoi të gjitha dokumentat Teuta.*
VSO - *I lexoi Teuta të gjitha dokumentat.*
OVS - *Të gjitha dokumentat i lexoi Teuta.*
OSV - *Të gjitha dokumentat Teuta i lexoi.*

Albanian questions are linguistic expressions used to compose a demand for information. In this paper we focus only on factoid questions, which are questions that can be answered with simple facts expressed in short text answers [1]. One of the main attributes of the syntactic structure of factoid questions are the interrogative pronouns and adverbs that are also known as wh-words or function words. In Albanian language the most common wh-words are: *kush – who; cili – which; çfarë/çka/ç'/se – what; pse/përse – why; nga/ku - where*. There are other question words such as *si/qysh – how*; *sa – how much; i/e satë/i/a* asks for the order of something; *sejtë* asks about the composition of something; *kush* can be associated with a preposition and has the grammatical function of case; *cili* and *satë* have the grammatical function of case, number and gender [16].

4 Algorithmic Approach and System Design

In this section we introduce the design and the implementation details of the question answering system.

4.1 System Architecture

Our novel question answering system is composed by three main modules:

1. Document preprocessing and indexing module that encompasses language-dependent rules.
2. Question analysis module that extracts the answer type and generates the query taken as input by the next module.
3. Passage Retrieval and Answer Extraction module finds the candidate answers using the information catalogued in the preprocessing module, the query generated by the question analysis as well as additional information from an external database of language dependent data.

The above modules are implemented as a java desktop application (Fig. 1).

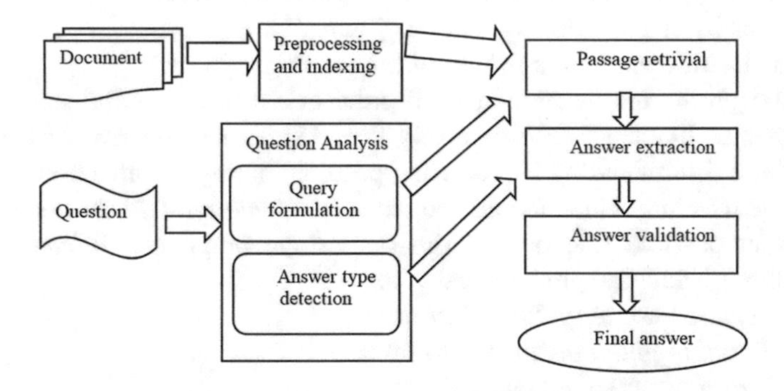

Fig. 1. Architecture of QA system for Albanian language

4.2 General Assumptions and Additional Resources

The proposed QA system is based on the assumption that the questions are factoid and somehow related to a single predefined document. Also we assume that the questions are correctly written in standard official Albanian language. Even though the system may answer correctly, it performs better when user types the letters *ë* and *ç*, instead of using *e* and *c*. Furthermore, our system aims to extract answers from a single predefined document.

4.3 Document Preprocessing

The text document is preprocessed beforehand. We extended Albanian abbreviations to their original word form. For example we converted *shek.* to *shekull,* (English : *century*), *dt.* to *datë,* (English : *date*), *etj.* to *e të tjerë,* (English : *etcetera*), *p.e.s.* to *para erës sonë,* (English : *Before the Common Era*) and *e.r.* to *era e re*, (English : *common era*). After that, named entities are extracted by hand and indexed for later faster searches. For the purpose of this paper, the named entities taken in consideration are: PERSON, LOCATION, DATE, TIME, REASON, NUMBER, MANNER, LANGUAGE and PUBLICATION. Furthermore, the document is preprocessed using tokenization, stop-words/common verbs removal and stemming algorithms [17]. As a result, the final text is cleaned from the inflectional endings and suffixes allowing a better word to word matching with the keywords extracted from the question. The previous mentioned algorithms are implemented in java programming language and are based on several rules comprising the Albanian language structure.

The document pre-processing algorithm:

```
Abbreviations' extension
NER
Tokenization
Stop-word and common verb removal
Stemming
    remove_inflectional_endings
    remove_sufixes
```

4.4 Question Analysis

This module includes answer type detection and query formulation. The answer type will be used to extract from the document the name entity that corresponds to the requested answer while the generated query will contribute in the extraction of passages that are most likely to contain the answer.

The most important attribute that we use to extract the answer type, is the wh-word. Since the standard question structure in Albanian is $<preposition> +$ $[question\,word] + [\ldots]$ we make use of the following wh-words: "*kush, ku, kur, pse, sa*", that gives us enough information to extract the answer types respectively, PERSON, LOCATION, TIME/DATE, REASON, NUMBER. If a more general wh-word is identified, such as *çfarë, ç', cili* we utilize the question headword which is the first noun phrase after the verb. To facilitate this task, we constructed a database of headwords that map to their respective answer type as shown in Table 1.

Table 1. Overview of the data collected to map headwords with their answer type.

Headword	Answer type
komandant, sulltan, djalë,vajzë (English: commander, sultan, boy, girl)	PERSON
vise, tokë, qytet, shtet, principatë (English: land, land, city, state, principality)	PLACE/LOCATION
datë, vit (English: date, year)	DATE/TIME/YEAR
mënyrë, metodë (English: way, method)	MANNER
…	…

The headwords give us additional information beyond the specific wh-words. For example: *"Në cilin vit u shkrua 'Historia dhe gjenealogjia e shtepisë së Muzakajve'?" (English: In which year was the 'Story and genealogy of the Muzaka family' written?)*, tells us that the user is asking for a YEAR, which is more specific than the wh-word "*kur*" which only tells us that a TIME entity is required. If a supplementary entity is not found, TIME entity is provided instead.

In case the wh-word is not located in the beginning but somewhere inside the question string, we also search for it inside the user question. If still the entity is not found, we search through the question for keywords associated with certain answer types. For example: *"Çfarë ka shkruar Marin Barleti për jetën e Skënderbeut?", (English: What has Marin Barleti written about Skanderbeg's life?),* using *shkruar* we can deduce that the user is asking about a PUBLICATION.

Furthermore, the question words are removed from the question string and the same preprocessing algorithms applied to the text document are also applied to the question string.

The answer type detection algorithm:

```
answer_type=null
Find the question_word
If general question_word
  Find headword
  If headword found and headword in database
    answer_type=get_answer_type_from_headword
  else
    Find words associated with answer types
    If associated_words
      answer_type=get_answer_type_from_associated_words
    else
      answer_type=default_from_general_question_word
else
  answer_type=answer_type_from_specific_question_word
```

4.5 Passage Retrieval and Answer Extraction

The tdf/if representation is generated and the cosine similarity between the query string and the text passages is calculated. Based on cosine similarity, we ranked the document passages by the probability of containing the correct answer. To account for the common user input mistakes with the letters *ë* and *ç*, we replace them with *e* and *c*.

At this stage, we have a list of previously ranked text passages. From the first 5 candidate answer passages, we extract the entities that comply with the answer type entity extracted previously. Since the search is being performed in a single document, the answer is unlikely expressed more than once. Instead of the lexical matching between the query and the passages we also expanded it with synonyms, homonyms and hyponyms [18] to account for different ways of asking the same question. These word extensions are maintained in a database constructed for the purpose of expanding the query words.

The candidate answers are ranked using the following criteria:

1. Cosine Similarity with idf between the passage containing the requested entity and the query string.
2. Number of query words in the candidate answer passage.

3. Average word distance between the query words that are present in the passage and the candidate entity.
4. Longest sequence of query words in the candidate passage.
5. Number of query words less than 4 words apart from the candidate in the passage.

We used the metrics discussed previously to evaluate the rank value. The candidate with the higher value assigned is most likely to be the right answer. The formula we used to calculate the final evaluation is as follows:

$$E = nr\_of\_words * cos\_similatity * 100 * (1 + longest\_sequence/5) * proximity. \quad (1)$$

where proximity is the inverse of the calculated value for average word distance.

The ranking algorithm applied to the candidate answers is illustrated in Tables 2 and 3 with two different example questions.

Table 2. Question 1- *Kur lindi Skënderbeu? (English: When was Skanderbeg born?)*

Candidate	në vitin 1405	në vitin 1468
Context	*Skënderbeu ose Skënderbej lindi në vitin 1405 dhe vdiq në vitin 1468* *(English: Skanderberg was born in 1405 and died in 1468)*	
Proximity	3.83	1.64
Number of words	2	2
Longest sequence	2	2
Cosine similarity	0.062	0.062
Evaluation	65.90	15.69

The evaluation is calculated based on the values assigned to proximity, number of words, longest sequence and cosine similarity. The answer is the candidate with the highest evaluation. In the example in Table 2, the evaluation is mainly influenced by the value of proximity, whereas in the example shown in Table 3 the ranking is more difficult.

5 Performance Evaluation

To evaluate our novel question answering system, we used a text document in Albanian language with 1300 words, regarding *Skanderbeg/Gjergj Kastrioti*, the Albanian national hero.

The first metric of evaluation used is the percentage of correct answers where we considered the candidates ranked first against the correct answers manually labeled. We use the Precision metric as in to the following formula:

Table 3. Question 2- *Sa motra kishte Skënderbeu? (English: How many sisters did Skanderbeg have?)*

Candidate	5	4	12 vjet
Context	*Gjergj Kastrioti ishte djali më i vogël i Gjon Kastriotit dhe i princeshës Vojsava, fëmija i fundit midis 4 djemve dhe 5 vajzave. (English: Gjergj Kastrioti was the youngest son of Gjon Kastriot and Princess Vojsava, the last child between four boys and five girls)*		*Shkrimet latinisht të Frangut të vitit 1480, 12 vjet pas vdekjes së Skënderbeut, mjerisht u përvetësuan nga të tjerë, dhe përkthimi dhe botimi i saj italisht u bë më vonë, pas vdekjes. (English: The Latin script of Frang in 1480, 12 years after Scanderbeg's death, was unfortunately taken by others, and its translation and publication to Italian was done after his death)*
Proximity	0.34	0.27	0.31
Number of words	0.70	0.70	1.00
Longest sequence	0.70	0.70	1.00
Cosine similarity	0.10	0.10	0.0064
Evaluation	2.78	2.18	0.24

$$\text{Precision} = \frac{\text{m}}{\text{n}} * 100\%. \quad (2)$$

In (2), n is the total number of questions, and m is the number of correctly answered questions.

We also applied the Mean Reciprocal rank (3) to the ranked list of candidates provided by the system. Each question is scored as the inverse of the first correct answer in the list of candidates. The following formula is used:

$$\text{MRR} = \frac{1}{\text{n}} * \sum\nolimits_{i=1}^{n} \frac{1}{j}. \quad (3)$$

In (3), n is the total number of questions and j is the rank of the correct answer in the ranked list previously generated by the QA system.

Out of 138 questions, we achieved a precision value of 69.5%, and a mean reciprocal rank scoring 73.5%.

6 Conclusions and Future Work

In this paper we introduced a novel question answering system which can answer to factoid questions based on a single text document written in Albanian language. As far as we know there are no previous works done in QA for Albanian language. We implemented a prototype of a QA system that is composed by three main modules:

Document preprocessing; Question analysis; Passage Retrieval and Answer Extraction. The majority of language dependent tasks are comprised in the preprocessing module even if some language dependent rules are also needed inside the question analysis module for identifying the answer type. Furthermore, we used the tf/idf representation of the document and calculated the cosine similarity to identify the similarity between the query string and the text passages.

We evaluated the QA system using a specific domain and by asking questions regarding that domain. For the evaluation process we used two different statistical metrics and achieved a precision value of 69.5% and a mean reciprocal rank value of 73.5%.

As a future work, our novel QA system can be further automated by using Machine Learning approaches for Named Entity Recognition in Albanian language. Some humble works are previously done in this context but it does not exist a formal tool that can be applied in different NLP tasks.

Moreover, the text preprocessing module should be enhanced with new grammatical and syntactic rules. An extended database of synonyms, homonyms and hypernyms should be comprised to support multiple domain questions. The performance of the QA system can also benefit from the incorporation of a Part of Speech tagger in the preprocessing module.

Furthermore, the question analysis module can be improved by using Machine Learning classification algorithms to best determine the answer type and to improve the system's precision.

As this QA system considers only factoid questions, additional efforts should be done to extend the range of questions with an alternative list of questions.

References

1. Jurafsky, D., Martin, J.H.: Speech and Language Processing, 2nd edn. Prentice-Hall, Inc., Upper Saddle River, New Jersey (2009). ISBN 0131873210
2. Green, B.F., Wolf, A.K., Chomsky, C., Laughery, K.: Baseball: an automatic question answerer. In: Proceedings of the Western Joint Computer Conference, vol. 19, pp. 219–224 (1961). Reprinted in Grosz et al. (1986)
3. Woods, W.: Progress in natural language understanding - an application to lunar geology. In: Proceedings of AFIPS Conference, vol. 42, pp. 441–450 (1973)
4. Ferrucci, D.A.: Introduction to "this is Watson". IBM J. Res. Develop. **56**(3/4), 1–15 (2012)
5. Nyberg, E., et al.: The JAVELIN question answering system at TREC2003: a multi-strategy approach with dynamic planning. In: The Proceedings of the 11th Text Retrieval Conference (2003)
6. Comas, P.R., Turmo, J., Màrquez, L.: Sibyl a factoid question-answering system for spoken documents. ACM Trans. Inf. Syst. **30**(3), Article 19, 40 (2012)
7. Wang, X., Xu, B., Zhuge, H.: Automatic question answering based on single document. In: 12th International Conference on Semantics Knowledge and Grids. IEEE (2016)
8. Wu, Q., et al.: Visual question answering: a survey of methods and datasets. Comput. Vis. Image Underst. **163**, 21–40 (2017). https://doi.org/10.1016/j.cviu.2017.05.001

9. Kim, M., Kim, H.: Design of question answering system with automated question generation. In: Fourth International Conference on Networked Computing and Advanced Information Management. IEEE (2008)
10. Perret, L.: A question answering system for French. In: Peters, C., Clough, P., Gonzalo, J., Jones, G.J.F., Kluck, M., Magnini, B. (eds.) Multilingual Information Access for Text, Speech and Images. CLEF 2004. Lecture Notes in Computer Science, vol 3491. Springer, Berlin, Heidelberg (2005)
11. Hoque, S., et al.: BQAS: a bilingual question answering system. In: Proceedings of International Conference on Electrical Information and Communication Technology. IEEE (2015). https://doi.org/10.1109/eict.2015.7392020
12. Kamal, A.I.: Enhanced Arabic question answering system. In: Sixth International Conference on Computational Intelligence and Communication Networks. IEEE (2014)
13. Nguyen, D.Q., et al.: A vietnamese question answering system. In: International Conference on Knowledge and Systems Engineering. IEEE (2009)
14. Hamp, E.P.: Albanian language, Encyclopedia Britannica (2016)
15. Kurani, A., Trifoni, A.: Syntactic similarites and differences between Albanian and English. Eur. Sci. J. **17** (2011)
16. Çabej, E.: Studime etimologjike në fushë të shqipes I, Tiranë (1982)
17. Sadiku, J., Biba, M.: Automatic stemming of Albanian through a rule-based approach. J. Int. Res. Publ. Lang. Individuals Soc. **6** (2012). ISSN-1313-2547
18. Dhrimo, A., Tupja, E., Ymeri, E.: Fjalor sinonimik i gjuhës shqipe: Mbi 30 000 zëra. Botimet Toena, Tiranë (2002)

A Thorough Experimental Evaluation of Algorithms for Opinion Mining in Albanian

Nelda Kote[1(✉)], Marenglen Biba[2], and Evis Trandafili[3]

[1] Department of Fundamentals of Computer Science, Faculty of Information Technology, Polytechnic University of Tirana, Tirana, Albania
nkote@fti.edu.al

[2] Department of Computer Science, Faculty of Information Technology, New York University of Tirana, Tirana, Albania
marenglenbiba@unyt.edu.al

[3] Department of Computer Engineering, Faculty of Information Technology, Polytechnic University of Tirana, Tirana, Albania
etrandafili@fti.edu.al

Abstract. Nowadays, analysis of opinions in online media such as newspapers, social media, forums, blogs, product review sites, has a key role in the human life. In this context, opinion mining is one of the fastest growing research areas in natural language processing that aims to extract and organize opinions from users. Machine Learning techniques represent a powerful instrument to analyze and understand correctly text data. In this paper we present a thorough experimental evaluation of machine learning algorithms used for opinion mining in Albanian language. The experimental results are interpreted with respect to various evaluation criteria for the different algorithms showing interesting features on the performance of each algorithm.

1 Introduction

An opinion is an unproven judgement or view about something from someone. It cannot be necessarily based on facts or knowledge but can be useful in different situations. Referring to [1] an opinion is a quintuple, *(ei, aij, sijkl, hk, tl)*, where *ei* is the name of an entity, *aij* is an aspect of *ei*, *sijkl* is the sentiment on aspect *aij* of entity *ei*, *hk* is the opinion holder, and *tl* is the time when the opinion is expressed by *hk*. The sentiment level will vary from the application and it will have different level as positive or negative; as positive, negative or neutral; as points for example 1–10 used in hotels, films rating, etc.

Nowadays there is a huge amount of people's review, opinions in online and social media and analyzing it will be important to make decision in the future. So, recently business and academia are focused in finding the best way to analyze this huge amount of online opinions using machine learning techniques.

Opinion mining is an ongoing field of text mining and natural language processing that aims to identity and extract subjectivity information in people's opinion. This is known also as sentiment analysis and involve the development of a system to collect,

L. Barolli et al. (Eds.): EIDWT 2018, LNDECT 17, pp. 525–536, 2018.
https://doi.org/10.1007/978-3-319-75928-9_47

analyze, summarize or categorize the opinions based on different criteria. Referring to the opinion definition in [1] the aim of opinion mining is to evaluate the quintuple of the opinion. Opinion mining can be performed in document level, in sentence level and in aspect-based level.

The aim of this paper is to evaluate through experiments the classification algorithms used to perform opinion mining in document level. For this we created a text corpus with opinions in Albanian language collected from well-known Albanian online newspaper. The corpus contains opinions categorized in five different subjects, and for each subject the opinions are categorized as positive and as negative opinions. First, we have cleaned the dataset by preprocessing the text data with a stop-word removal and a stemmer. Then the clean dataset is used to train and test the classification algorithms. We use Weka software to evaluate the performance of 53 classification algorithms.

The structure of the paper is as follows: Sect. 2 presents background and related work; Sect. 3 presents methodology for opinion categorization; Sect. 4 gives a short description of the classifiers; Sect. 5 present experiment results and the Sect. 6 conclude our work and give future work ideas.

2 Background and Related Work

In [1] the author gives an in-depth comprehensive study of all research fields of opinion mining. Since the last decade opinion mining is one of the fastest growing research areas in natural language processing. Companies and researchers are increasingly focused in opinion mining and to use the result from it in their daily work, as marked predicting, the feedback form a student for a given lesson, etc.

As opinion mining is a classification technique, the main machine learning techniques for text classification, supervised and unsupervised can be used on it. In supervised learning is used a leveled dataset to build the classification model, but in unsupervised learning the dataset is unleveled that make it not too appropriate to be used in opinion mining.

The work in [2] consist in performance evaluation of three supervised machine learning techniques, Maximum Entropy, Naïve Bayes and SVM for opinion mining in English Twitter dataset combined with the Semantic Orientation based WordNet, for extracting the synonyms of the content. From the results of the performed experiments we can conclude that Naïve Bayes algorithm used with unigram technique outperform the two other algorithms and the use of WordNet improve the accuracy.

To address the problem of unleveled dataset and use the unsupervised methods in opinion mining the work in [3] proposed to use a clustering algorithm. In [4] the author used an unsupervised machine learning technique as spectral clustering to cluster the English Twitter dataset as positive and negative. The proposed system has four main steps as following: cleaning and normalization of the data set; applying a test dataset to cluster the dataset using unsupervised machine learning algorithm; applying a k-means to normalize the generated matrix and finally applying a hierarchical clustering algorithm to merge the clusters in one. The experimental results indicate that unsupervised machine learning technique outperform the supervised machine learning.

Opinion mining would be performed in document level, in sentence level and in aspect-based level. At document level opinion, the whole document is indicated to be a positive or negative opinion. The documents can be from one domain or cross-domain; from one or more languages. At sentence level opinion, a sentence is indicated to be a positive or negative opinion [5]. The main task in aspect-based level opinion mining is the identification of the aspect, of the word related to this aspect and the orientation of this word [6]. The focus of our work is to evaluate the performance of classification algorithms in document level classified as positive or negative opinion.

One of the first study in opinion mining classification has been realized by [7] that performed an experimental evaluation of three machine learning algorithm, Naïve Bayes, SVM and Maximum Entropy in a movie review corpus classified as positive and negative opinions using unigrams and bigrams. They concluded that the algorithms do not performs as well as in topic-based categorizing and there are too many challenges to address.

To improve the performance of opinion classification to many researchers have performed experimental evaluation of machine learning algorithms including new features and testing different features combinations.

An interesting study was conducted in [8] to evaluate the performance of four machine learning algorithm, Naive Bayes, Stochastic Gradient Descent, SVM and Maximum Entropy, in reviews of IMDb dataset categorized as positive and negative opinions, using TF-IDT and Count Vectorizer technique. They tested all the algorithms for different n-gram models. At the end the results are compared with results of 20 previews works by different authors. The value on n in n-gram is in disproportional portion with the classification accuracy, so the result taken in unigram and bigram are better than in trigram, four-gram and five-gram. The used technique for converting the text document into matrix improve the accuracy.

In [9] the authors proposed a method that combine machine learning and semantic orientation based on the integration of knowledge resource to perform a polarity opinion classification based in majority voting system. They used three corpuses one with opinions in Arabic, the second the corresponding in English and the third corpus bilingual. They used SVM combined with different feature as using or no preprocessing data, and using different n-gram models. The three corpuses obtain the best score in terms of F1 with different combination of SVM with features, but the proposed technique has a better performance in bilingual corpus.

An impressive research work has been conducted in opinion mining in different languages as: English, German, Spanish, Turkish, Arabian, etc. [10–13]. The paper [12] is a review paper on opinion mining research conducted for categorizing opinions in Spanish language for the period 2012 to 2015. Comparing with research in English there is a sporadic research conducted in Spanish, but the results from this work is promising and encouraged to develop more research in this language. Another research work [13] present a multiple classifier system used for opinion mining in Turkish language. This system is based in a vote algorithm combined with three machine learning algorithms, Naive Bayes, SVM, and Bagging. The proposed solution showed better performance than the three classifiers used individually.

Insignificant research work has been done in Albanian language for opinion classification. The first paper in this fill is [14] that evaluate through experiments the application of machine learning algorithms in opinion classification.

All the research work discussed above and reviewed during the preparation of this paper, use and recommend a preprocessing phase to the data set before applying the machine learning algorithms for opining mining. There are few attempts to develop preprocessing tools for Albanian language. The authors in [15] provided a naïve-single-step stemming algorithm for Albanian language and a stop-words list. They have evaluated the algorithm through human experts. The experts indicate that there are more rules that are not included in the algorithm that could improve its capabilities. A rule-based stemmer for Albanian language implemented in java programming language is presented in [16]. This stemmer is an improved version of the compositor implemented in [17]. We used the stemmer introduced in [16] to preprocess our data set.

3 Methodology for Opinion Classification

The aim of this work is to evaluate through experiments the performance of 53 classification algorithms for opinion classification on document level in a corpus in Albanian language. We use Weka software to test and train the classification algorithms on our corpus. To perform this work, we followed the following steps.

3.1 Data Collection

We created a text corpus of 500 Albanian written opinions collected from different well-known Albanian newspapers. The corpus has five subjects related to nowadays discussion in Albania as follow: Tourism in Albania, Politics, the including of VAT tax in Small Business, waste import in Albania, the law of Higher Education. For each of the subjects we have collected fifty text document categorized as positive opinions and fifty text documents categorized as negative opinions.

3.2 Preprocessing

Referring to all the research work reviewed, the corpus is preprocessed before it is used to training and test the classification algorithms. Only in [9], one of the corpus is used without preprocessing the dataset. We used the rule-based stemmer implemented in [16] to perform the preprocessing step. This stemmer contains 134 rules based on the Albanian language morphology but to find the stem of the word it is not taken in consideration the linguistic meaning of the word. The stemmer is experimentally tested by the authors to classify text documents corpuses and the results demonstrate its effectiveness.

The preprocessing process include the following steps:

1. The first step is applying a stop-word that include:
   - All the data are set to lower case;
   - Remove the stop-word like: “dhe”, “ne” “sepse”, “kur”, “edhe” “ndonëse”, “mbase” etc. These are words that in Albanian language do not have any meaning.

The stop-word can be removed because they do not have any positive or negative connotation and do not affect the emotion of the opinion;
 - Special characters (including the punctuations) and numbers are removed.
2. The second step is applying the stemmer that finds the stems of all the words in the text document.

At the end of this process the text file is a file containing words that do not have any language structure.

3.3 Experiment Settings

As mentioned above, we create five corpuses contain opinion in Albanian language as. Each corpus contains fifty text documents of positive opinions and fifty text documents as negative opinions. The corpus firstly is preprocessed. Then the output text file from this phase is used as input for the algorithms. To evaluate the performance of the classification algorithm for opinion mining, we used WEKA software [18]. Weka software is implemented in Java programing language and contains a collection of machine learning algorithms and tools to perform data mining tasks.

To perform the experiments in Weka we created for each corpus a ARFF file. We load all text documents of one corpus in Weka using the *textDirectoryLoader* class. This class loads the text documents in a folder using the names of the sub-folders as class labels and store the content of it in a string attribute. Then we apply the *StringToWordVector* filter that convert all the string attributes into a word vector, that represent the occurrence of the word from the text contained in the strings.

We chose 53 classification algorithms implemented in Weka to evaluate the performance in each corpus. We use 10-folds Cross-validation for training a model and testing all the chosen algorithms.

4 The Classifiers

The classification of text is a widely study in data mining, text mining and natural language processing, that appoint categorical values to the text. Referring to [19] the problem of classification is defined as having a set of training records $D = \{X_1\ X_2, X_3, \ldots, X_n\}$, such that each record is labeled with a value from a set of k different discrete values indexed by $\{1\ldots k\}$. The trained data are used to build a classification model, that links the features of the record data to one of the labels. So, for each unknown test instance, the training model predict a label for it.

The classification methods are categorized in five categories: Probabilistic and Naïve Bayes Classifiers, Rule-based Classifiers, Proximity-based Classifiers, Linear Classifiers and Decision Tree Classifiers.

4.1 Probabilistic and Naïve Bayes Classifiers

Probabilistic classifiers use a mixture model and can predict for a given input a probability distribution over a set of classes, and not only the most likely class that the input should belong to.

Naïve Bayes Classifier is a straightforward and powerful algorithm, that uses two groups of models. These two models calculate the posterior probability of a class based on the distribution of the words in the text document. The models differ by each other in using or not the frequencies of the word and from the action taken to model the probability space.

The probabilistic and Naïve Bayes classifiers taken in consideration during this paper in Weka are: Bayesian Logistic Regression, Bayes Net, Complement Naïve Bayes, Naïve Bayes, Naïve Bayes Multinomial, Naïve Bayes Multinomial Text, Naïve Bayes Multinomial Updateable and Naïve Bayes Updateable.

4.2 Rule-Based Classifiers

Rule-based classifiers uses a set of rules to classify data sets. They use separate-and-conquer method, that is a repetitive process of generating a rule to cover a subset of the training data and then removing all the data covered by the rule form the training set. This process is repeated until there are no data left to be covered. These techniques are very used because it is easy to interpret and maintain them. One of the most used rule-based classifier is RIPPER implemented in Weka as JRip.

The rule-based classifiers taken in consideration during this paper in Weka are: Conjunctive Rule, Decision Table, JRip, NNge, OneR, PART, Ridor and ZeroR.

4.3 Proximity-Based Classifiers

To realize the classification, the proximity-based classifiers use distance-based measures. The documents are assigned into a class based on similarity of their measures such as the dot product or the cosine metric. A test data is classified by computing the similarity between this data and the training data, and assign this data to the class that contain the greater number of similar data.

The proximity classifiers taken in consideration during this paper in Weka are: IB1, IBK, Kstar and LWL.

4.4 Linear Classifiers

Linear classifiers identify the class belonging the data by making a classification decision based on the value of linear combination of its characteristics. The data characteristics are presented to the machine in a vector called feature vector.

The Support Vector Machines classifier, Regression-Based Classifiers and Neural Network Classifiers are included in this category.

The linear classifiers taken in consideration during this paper in Weka are: Logistic, RBF Classifier, RBF Network, SGD, SGD Text, Simple Logistic, SMO and Voted Perceptron.

4.5 Decision Tree Classifiers

A Decision Tree Classifiers use a hierarchical division of the main training dataset using a condition or a predicate on the attribute value. This division is designed to create partition of the dataset that are more skewed in terms of their dataset distribution. In text data these predicates are conditions on the absence or presence of one or more words in the document.

The decision tree classifiers taken in consideration during this paper in Weka are: Decision Stump, HoeffdingTree, J48, LMT, Random Forest, Random Tree, REPTree and SimpleCart.

We have taken in consideration and algorithms implemented in two Weka's class: Meta class and Misc class. The classifiers in Meta class are a combination of different classifiers: AdaBoost M1, Attribute Selected Classifier, Bagging, Classification Via Regression, Filtered Classifier, Interative Classifier Optimizer, Logit Boster, Multi Class Classifier, Multi Class Classifier Updateable, Random Committee, Randomizable Filtered Classifier, Random Sub Space, Real Ada Boost and Vote. The Misc Class have FLR, Hyper Pipes and Input Mapped Classifier classifiers.

5 Experiment Results

We performed all the experiments based on the specification in the Sect. 3. We evaluated the performance of classification algorithms on each corpus in term of percentage of correctly classified instances.

In Table 1 are shown the results of the experiments, highlighting in red the best percent of correctly classified instances and in blue the second-best percent of correctly classified instances for each corpus.

By analyzing the experimental results, we note that for different corpus we have different algorithms which have the best performance. So, the results are as follow:

1. For C1, there are two algorithms, Logistic and Multi Class Classifier, that have the best percent of correctly classified instances by 94%;
2. For C2 the best performing algorithm is Hyper Pipes with 92% correctly classified instances;
3. For C3 the best performing algorithm is RBF Classifier with 79% correctly classified instances;
4. For the two last corpuses RBF Network is the best performing algorithm but with different percent of correctly classified instances, 86% for C4 and 89 for C5.

There are five best performing algorithms with percent of correctly classified instances varying from 79% to 94%. An interesting fact is that exist a significant gap between the best and the worst percent of correctly classified instances for each

Table 1. Experimental results in terms of percent of correctly classified instances for each algorithm.

Algorithm	C1	C2	C3	C4	C5
BayesianLogisticRegression	89	85	77	74	78
BayesNet	87	59	68	49	71
ComplementNaiveBayes	87	89	77	79	85
NaiveBayes	86	83	71	74	78
NaiveBayesMultinomial	87	89	77	79	85
NaiveBayesMultinomialText	50	50	50	50	50
NaiveBayesMultinomialUpdateable	87	89	77	79	85
NaiveBayeslUpdateable	86	83	71	74	78
Logistic	94	84	66	82	79
RBFClassifier	85	77	79	67	75
RBFNetwork	49	88	66	86	89
SGD	88	84	75	75	81
SGDText	50	50	50	50	50
SimpleLogistic	84	74	66	63	71
SMO	88	82	74	75	78
VotedPerceptron	86	76	74	71	75
IB1	59	65	62	60	62
IBK	60	65	62	60	62
KStar	58	65	62	60	62
LWL	69	50	75	58	57
AdaBoostM1	80	71	70	52	76
AttributeSelectedClassifier	84	63	66	53	71
Bagging	83	62	75	62	66
ClassificationViaRegression	78	54	71	54	64
FilteredClassifier	85	56	68	49	60
InterativeClassifierOptimizer	79	51	75	50	71
LogitBoster	81	67	61	62	76
MultiClassClassifier	94	84	66	82	79
MultiClassClassifierUpdateable	88	84	75	75	81
RandomCommittee	77	71	69	67	69
RandomizableFilteredClassifier	55	60	58	55	49
RandomSubSpace	82	65	73	64	64
RealAdaBoost	78	70	67	58	68
Vote	50	50	50	50	50
FLR	89	91	66	85	81
HyperPipes	92	92	67	85	80
InputMappedClassifier	50	50	50	50	50
ConjunctiveRule	62	50	61	59	58
DecisionTable	69	54	71	57	61
JRip	76	53	70	64	70
NNge	85	82	69	64	65
OneR	69	49	75	56	65
PART	75	67	63	58	65
Ridor	75	65	57	52	64
ZeroR	50	50	50	50	50
DecisionStump	61	50	75	57	56
HoeffdingTree	86	81	74	73	82
J48	87	67	58	60	62
LMT	84	74	66	64	70
RandomForest	86	76	76	65	74
RandomTree	57	64	59	64	67
REPTree	74	51	71	54	54
SimpleCart	73	53	75	54	62

algorithm. The gap for each algorithm is: Logistic -> 28%; Multi Class Classifier -> 28%; Hyper Pipes -> 25; RBF Classifier -> 12%; RBF Network -> 40%. Logistic and Multi Class Classifier have the same performance. The RBF Classifier has his best percent of correctly classified instances under the C1 corpus and not under C3 where it is the best performing for this corpus.

To rank these five algorithms from the best performance to the lowest performance, for each corpus we rated each algorithm with a score of 5 to 1 based on the percent of correctly instances classified. We use this score to calculate the weighted average for each algorithm in terms of percent of correctly classified instances.

For example, for corpus C1 we order the algorithms based on their performance in term of percent of correctly classified instances and then each of them was rated with points. So, the best performing algorithms are Logistic and Multi Class Classifier with 94% of correctly classified instances, rated with 5 points. The next performing algorithm is Hyper Pipes with 92% of correctly classified instances, rated with 4 points. Then comes RBF Classifier with 85% of correctly classified instances, rated with 3 points. And the least performing algorithm is RBF Network with 49% of correctly classified instances, rated with 3 points. We applied the same marking scheme to each corpus. The results of this marking scheme and the calculation of the weighted average of percent of correctly classified instances for each algorithm are shown in Table 2.

Table 2. The rank of best performing algorithms in terms of weighted average of percent of correctly classified instances.

Algorithm	C1	C2	C3	C4	C5	Weighted average
Hyper pipes	92 * 4	92 * 5	67 * 4	85 * 4	80 * 4	83.62
Logistic	94 * 5	84 * 3	66 * 3	79 * 3	79 * 3	82.53
Multi class classifier	94 * 5	84 * 3	66 * 3	79 * 3	79 * 3	82.53
RBF network	49 * 2	88 * 4	66 * 3	89 * 5	89 * 5	80.16
RBF classifier	85 * 3	77 * 2	79 * 5	75 * 2	75 * 2	77.71

The algorithms are ranked from the algorithm that have the highest weighted average of percent of correctly classified instances to the lower one. Hyper Pipes is the best performing algorithm with 83.62% weighted average of percent of correctly classified instances, followed by Logistic and Multi Class Classifier with a difference of 1.09%. Then comes RBF Network with 80.16% weighted average of percent of correctly classified instances. And the least performing algorithm is RBF Classifier.

We performed another experimental evaluation to identify any statistical difference between the best performant algorithms using as comparison field percent_correct. Based on the result of Table 2, we choose Hyper Pipes as a base algorithm, and compare it with the other four. The cross-validation experiment was performed for each corpus using the Weka Experimenter tool with the five best performant algorithms, 10 cross-validation and 10 repetitions. In Table 3 are shown the results of this experiment.

Table 3. Cross-validation experiment results in term of percent_correct for each algorithm.

Algorithm	C1	C2	C3	C4	C5
Hyper Pipes	89.90	90.40	64.40	83.90	80.90
Logistic	92.20	84.60	66.70	81.10	83.70
Multi Class Classifier	92.20	84.60	66.70	81.10	83.70
RBF Network	90.50	88.80	63.90	83.80	89.70 v
RBF Classifier	85.80	83.50	76.60 v	69.70 *	75.80

For corpus C1 and C2, the experimental results indicate that there are no important differences in performance between the algorithms. The result of RBF Classifier for corpus C3 has a "*v*". This means that RBF Classifier has performed statistically significantly better than Hyper Pipes. The same sign is and in the result of RBF Network when running on corpus C5. Also, this means that RBF Network has performed statistically significantly better than Hyper Pipes. But the result of RBF Classifier for corpus C4 has a "*" meaning that it has performed statistically significantly worse than Hyper Pipes.

To conclude even with this experiment, we cannot define which algorithm perform statistically better.

For the five best performing algorithms we have performed different experiments changing the value of their parameters. The changes do not increase the performance, but sometimes the performance is even decreased.

We have performing experiments using unigram, bigram and trigram tokenizer instead of *WordToTokenizer* in *StringToWordVector* filter. When we used unigram the performance in terms of percent correctly classified instances is equal to the performance when *WordToTokenizer* is used. For some algorithm the value of percent correctly classified instances is insignificantly increased when we used bigram. And when is used trigram the value of percent correctly classified instances for all the algorithm is significantly decreased.

6 Conclusions and Future Work

In this paper we perform a thorough experimental evaluation of algorithms for opinion mining in Albanian language. At the beginning of our work we reviewed the state of art and the recent work done in this field.

We created a text corpus of 500 Albanian written opinions collected from different well-known Albanian newspapers. The corpus has five subject related to nowadays discussion in Albania. Each subject is used as a corpus to evaluate the performance of the algorithms. Each corpus has fifty text document categorized as positive opinions and fifty text documents categorized as negative opinions.

Firstly, we cleaned the dataset passing it in a preprocessing phase composed by a stop-word removal and a stemmer. Then these unstructured data are used as an input to train and test 53 classification algorithm in Weka.

The experimental results show that there are five different best performing algorithms in terms of percent of correctly classified instances. The best performing algorithm are:

Logistic and Multi Class Classifier for corpus C1; Hyper Pipes for corpus C2, RBF Classifier for corpus C3 and RBF Network for corpus C4 and C5.

We evaluated the performance using *WordToTokenizer*, unigram, bigram and trigram tokenizer. All the results showed in this paper are using *WordToTokenizer*. The performance when is used unigram is equal to the performance when *WordToTokenizer* is used. Some of the algorithms perform insignificantly better when is used bigram. And when is used trigram the performance is significantly decreased.

In the future, it would be of a great interest to evaluate the performance of the classification algorithms for Opinion Mining in a bigger corpus and in a cross-domain corpus in Albanian language. Also, it would be interesting to test the algorithms when the opinions are classified with more than two level, for example: as positive, neutral and negative; with points 1 to 10; etc.

References

1. Liu, B.: Sentiment Analysis and Opinion Mining. Morgan & Claypool Publishers, San Rafael (2012)
2. Gautam, G., Yadav, D.: Sentiment analysis of Twitter data using machine learning approaches and semantic analysis. In: Seventh International Conference on Contemporary Computing (IC3), 437–442 (2014)
3. Hastie, T., Tibshirani, R., Friedman, J.: Unsupervised learning. In: The Elements of Statistical Learning: Data Mining, Inference, and Prediction. Springer (2009)
4. Unnisa, M., Ameen, A., Raziuddin, S.: Opinion mining on Twitter data using unsupervised learning technique. Int. J. Comput. Appl. **148**(12), 12–19 (2016)
5. Shahana, H.P., Omman, B.: Evaluation of features on sentimental analysis. Procedia Comput. Sci. **4**, 1585–1592 (2015)
6. Chinsha, T.C., Shibily, J.: A syntactic approach for aspect based opinion mining. In: 9th International Conference on Semantic Computing (IEEE ICSC 2015) (2015). ISBN: 978-1-4799-7935-6
7. Pang, B., Lee, L., Vaithyanathan, S.: Thumbs up?: sentiment classification using machine learning techniques. In: Proceedings of the ACL 2002 Conference on Empirical Methods in Natural Language Processing, vol. 10, pp. 79–86 (2002)
8. Tripathy, A., Agrawal, A., Rath, K.S.: Classification of sentiment reviews using n-gram machine learning approach. Expert Syst. Appl. **57**, 117–126 (2016)
9. Perea-Ortega, M.J., Martín-Valdivia, T.M., Ureña-López, A.L., Martínez-Cámara, E.: Improving polarity classification of bilingual parallel corpora combining machine learning and semantic orientation approaches. J. Assoc. Inf. Sci. Technol. **64**(9), 1864–1877 (2013)
10. Lommatzsch, A., Butow, F., Ploch, D., Albayrak, S.: Towards the automatic sentiment analysis of German news and forum documents. In: Innovations for Community Services: 17th International Conference, I4CS 2017, Darmstadt, Germany, pp. 18–33 (2017)
11. Farra, N., Challita, E., Assi, A.R., Hajj H.: Sentence-level and document-level sentiment mining for arabic texts. In: IEEE International Conference on Data Mining Workshops (ICDMW) (2011)
12. Miranda, H.C., Guzmán, J.: A review of sentiment analysis in Spanish. Tecciencia **12**(22), 35–48 (2017)
13. Catal, C., Nangir, M.: A sentiment classification model based on multiple classifiers. Appl. Soft Comput. J. **50**, 135–141 (2017)

14. Biba, M., Mane, M.: Sentiment analysis through machine learning: an experimental evaluation for Albanian. In: Recent Advances in Intelligent Informatics, Part of the Advances in Intelligent Systems and Computing Book Series (AISC), vol. 235, pp. 195–203 (2014)
15. Karanikolas, N.N.: Bootstrapping the Albanian information retrieval. In: Fourth Balkan Conference in Informatics (BCI) (2009)
16. Sadiku, J., Biba, M.: Automatic stemming of Albanian through a rule-based approach. J. Int. Res. Publ. Lang. Individ. Soc. **6** (2012). ISSN-1313-2547
17. Biba, M., Gjati, E.: Boosting text classification through stemming of composite words. In: ISI, pp. 185–194 (2013)
18. Witten, H.I., Frank, E., Hall, A.M., Pal, J.C.: Data Mining, Practical Machine Learning Tools and Techniques, 4th edn. Elsevier Inc. (2017). ISBN: 9780128042915
19. Aggarwal, C.C., Zhai, C.: Mining Text Data. Springer, Boston (2012)

Performance Evaluation of Text Categorization Algorithms Using an Albanian Corpus

Evis Trandafili[1(✉)], Nelda Kote[2], and Marenglen Biba[3]

[1] Department of Computer Engineering, Faculty of Information Technology, Polytechnic University of Tirana, Tirana, Albania
etrandafili@fti.edu.al

[2] Department of Fundamentals of Computer Science, Faculty of Information Technology, Polytechnic University of Tirana, Tirana, Albania
nkote@fti.edu.al

[3] Department of Computer Science, Faculty of Information Technology, New York University of Tirana, Tirana, Albania
marenglenbiba@unyt.edu.al

Abstract. Text mining and natural language processing are gaining significant role in our daily life as information volumes increase steadily. Most of the digital information is unstructured in the form of raw text. While for several languages there is extensive research on mining and language processing, much less work has been performed for other languages. In this paper we aim to evaluate the performance of some of the most important text classification algorithms over a corpus composed of Albanian texts. After applying natural language preprocessing steps, we apply several algorithms such as Simple Logistics, Naïve Bayes, k-Nearest Neighbor, Decision Trees, Random Forest, Support Vector Machines and Neural Networks. The experiments show that Naïve Bayes and Support Vector Machines perform best in classifying Albanian corpuses. Furthermore, Simple Logistics algorithm also shows good results.

1 Introduction

The digital word is expanding not only with data that users create themselves, but even with data created about these users. A study conducted by IDC stated that from now until 2020, the digital universe will double every two years [1]. The increase of accessible textual data has caused a flood of information instead of providing knowledge. In this situation there is an urgent need to explore and upgrade Text Mining algorithms and design new methods to exploit this avalanche of text.

Furthermore, we have to consider that most of the digital information is composed by unstructured text data; as a result the process of knowledge discovery and analysis is becoming an issue. The aim of Data Mining techniques and methods is to extract patterns and/or analyze databases, structured, well organized data. However, text is unstructured as it is based on language syntax and structure and therefore much more difficult to handle.

L. Barolli et al. (Eds.): EIDWT 2018, LNDECT 17, pp. 537–547, 2018.
https://doi.org/10.1007/978-3-319-75928-9_48

Text Mining is similar to Data Mining, but works with unstructured or semi-structured data sets (such as full-text documents or HTML files). Starting with a collection of documents, a text-mining tool retrieves a particular document and preprocesses it by checking its format and character sets. It then goes through text analysis, sometimes repeating techniques until the targeted information is extracted [2].

For this paper, we created a text corpus in Albanian language by collecting information from different online portals and grouping the documents in twenty categories. This corpus can also be used for future experiments and other purposes in Text Mining. As a first step, the corpus is passed through a language dependent preprocessing task composed of stop-word removal and stemming providing 'cleaned' datasets. On the output dataset is performed the training and testing of the algorithms. Since text categorization improves the organization level of the corpus, we focused our work on the evaluation of the performance of text classification algorithms. The classification problems are used in different domains of data mining and information retrieval and are implemented in publicly available software systems. We will evaluate the performance of the following classification algorithms: Naïve Bayes, Logistic Regression, k-Nearest Neighbor, Decision Trees, Random Forest, Support Vector Machines and Neural Networks. As studied on [3], the preprocessing task provides significant improvement on classification accuracy depending on the domain and language. Under this point of view, we can assume that the results of our work may not be the same if the corpus used is not composed of Albanian text documents and a different language is used.

The organization of the paper is as follows: Sect. 2 presents the background and the related work on text preprocessing and classification; Sect. 3 presents some basic information about the Albanian language structure; Sect. 4 presents the structure of the corpus and the preprocessing steps applied to it; Sect. 5 presents and analyzes the experiments; Sect. 6 analyses the classification algorithms taken in consideration; and we conclude in Sect. 7 with conclusions and future work.

2 Background and Related Works

The leading Text Mining approaches are listed by [4] such as: Information Retrieval, Natural Language Processing, Information Extraction from text, Text Summarization, Unsupervised Learning Methods, Supervised Learning Methods, Probabilistic Methods for Text Mining, Text Streams and Social Media Mining, Opinion Mining, Sentiment Analysis and Biomedical Text Mining. The process of selecting the appropriate technique optimizes the efforts of extracting the most valuable information [5].

The main technologies for Text Classification are supervised, semi-supervised, and unsupervised approaches. Supervised learning and semi-supervised learning are broadly used for text classification, while unsupervised learning is mainly used for clustering. Some studies show that a hybrid method which combines supervised and unsupervised methods outperforms the supervised support vector machine (SVM) in terms of both F1 performance and classification accuracy [6].

The Albanian language has not been much explored from the perspective of Natural Language Processing and Computational Linguistics. There are some trivial works previously conducted by [7] who proposed a rule-based stemmer for Albanian language and [8] who enhanced the previous stemmer by supporting the composite words. The performance of the composite stemmer was tested by using text classification algorithms and showing that preprocessing the document with the stemmer of composite words significantly enhances the performance of the classifier.

Text classification algorithms are evaluated on text corpuses of different languages. For example, in [9] the authors tested some classification algorithms on Turkish written documents. Their experimental results estimated that the Random Forest classifier gives more accurate results than Naïve Bayes, Support Vector Machines, K-Nearest Neighbor and J48.

The authors in [10] evaluated the classification algorithms (decision trees, rule induction, naive Bayes, neural networks and support vector machines) for n-gram collocations in Croatian language and concluded that the best classifier for bigrams was SVM, while for trigrams the decision tree.

Another interesting work was conducted in [11] to compare the performance of Naïve Bayes and Support Vector Machines in literary domain. The algorithms were also combined with text pre-processing tools to study the impact on the classifiers' performance. NB and SVM achieved high accuracy in sentimental chapter classification, but the NB classifier outperformed the SVM classifier in erotic poem classification.

Furthermore, in [12] the authors investigated the preprocessing techniques that impact the performance of Support Vector Machines classification algorithm in English and European Portuguese languages. They treated the document representation as an optimization problem in terms of feature reduction, selection and term weighting.

Another text classification comparison is conducted by [13] to evaluate the performance of K-Nearest Neighbor and Naïve Bayes. The assessment is done on a corpus of XML documents and the optimal value of $k = 13$ that yield the best performance for K-NN was identified.

There are some other works which focus on Arabic languages. And interesting case study is conducted by [14] using an Arabic corpus and demonstrating that using an Artificial Neural Network model is effective in capturing the non-linear relationships between document vectors and document categories if used with feature reduction methods.

A novel approach is the exploitation of text classification methods for multilingual language classification with the use of Convolutional Neural Networks. The work carried out by [15] showed that the classifier does not require syntactic and semantic knowledge of the language and performs well even on new languages.

3 Albanian Language Structure

The Albanian language is considered the modern survivor of the Indo-European language family, mostly spoken in Albania, Kosovo and in other parts of the Balkans. Dacian and Illyrian have been considered its ancestors of ancient languages. There are two main dialects Gheg, spoken in the north, and Tosk, spoken in the south, which by now have been diverging to their most extreme and diverse forms. The official Albanian language is written in Roman alphabet and from 1909 till World War II was based on the south Gheg dialect. Since then it has been modeled on Tosk dialect [16]. For the purpose of this paper, the official Albanian language is used.

The official Albanian language has 7 vowels and 29 consonants. The vowels are represented by single Latin letters (*a, e, ë, i, o, u, y*), and the consonants by single letters (*b, c, ç, d, f, g, h, j, k, l, m, n, p, q, r, s, t, v, x, z*), and combination of different letters (*dh, gj, ll, nj, rr, sh, th, xh, zh*).

There are some words in Albanian which do not carry any meaning. These are the stop-words which for the purpose of this paper are identified and removed from the corpus. Some of the most useful stop-words in Albanian are: *"dhe"*, *"sepse"*, *"kur"*, *"edhe"*, *"në"*, *"prej"*, *"apo"*, *"ose"*, etc.

The main grammatical categories of Albanian are: nouns which show gender, number, case and are inflected with suffixes to show definite or indefinite meaning, e.g. *tavolinë – "table", tavolina – "the table"*. A large number of noun plurals have irregular stem formation; Adjectives follow the noun and are preceded by a particle that agrees with the noun, e.g. in *një njeri i fortë*, "a strong man," *burrë* "man" is modified by the adjective *fortë* "strong," preceded by *i*, which agrees with the noun "man"; verbs have a great variety of forms and are quite irregular in forming their stems. As a conclusion, the grammar and formal distinctions of Albanian are inherited by the Romance languages and of Modern Greek [16].

4 Data Collection and Preprocessing

It is difficult to work with text processing algorithms with Albanian texts because it doesn't exist a formal corpus where you can rely. So, as part of our work we had to create a text corpus of Albanian written documents. We collected text data regarding 20 domains as follows: Animals, Art, Astronomy, Biology, Charity, Chemistry, Culture, Curiosities, Economy, Environment, Fashion, Food, History, Literature, Medicine, Politics, Religion, Sport, Technology and Tourism. Each category has 40 documents made up by textual information chosen randomly on the web, respectively from the fields chosen previously.

Before running the experiments, the corpus of documents was passed through a preprocessing phase which consists of the tokenization, stop-word removal and the stemming algorithm [7]. The aforementioned algorithms are implemented in java programming language and the whole implementation is based on different rules comprising the Albanian language structure. After the preprocessing step the text files look like a bag of word and do not have language structure any more. This structure is then used as input for text classification algorithms. There are a variety of publicly available software systems which implement different machine learning algorithms and data mining tasks like WEKA [17] and MALLET [18]. For the purpose of this paper we chose the Weka software. For our text documents to be classified by Weka we converted the file format from .txt to .arff which is an ASCII text file that describes a list of instances sharing a set of attributes and stands for Attribute-Relation File Format [17]. For this purpose we used the textDirectoryLoader class implemented in Weka. Then the file was passed through the StringToWordVector filter which transforms all the string attributes into a vector that represents the word occurrence information from the text in strings.

5 Classifier Selection

Text classification is the process of assigning predefined categories to text documents. The classification problem is defined as follows: on a training set of documents, $D = \{d_1, d_2, \ldots d_n\}$, such that each document d_i is labeled with l_i from the set of categories $L = \{l_1, l_2, \ldots, l_k\}$. The goal is to find a classification function (classifier) f where $f(d) = l$ which assigns the correct category label to a new document d not previously used for training [4]. There are different methods for the classification task which are applied in domains such as quantitative or categorical data. Text data is modeled as quantitative data regarding frequencies and word attributes so most of the classification methods can be applied directly on text [19].

The classification techniques are divided in five main categories: Regression, Distance, Decision Trees, Rules and Neural Networks [20]. In order to handle text classification in breadth we selected five key classifiers, one for each classification technique, respectively: Logistic Regression, K-Nearest Neighbor, C4.5, Naïve Bayes and Back Propagation. Furthermore, Support Vector Machines (SMO) and Random Forest algorithms are also included in our experiments due to their popularity and the results obtained in similar works.

5.1 Logistic Regression

Logistic regression is a statistical machine learning algorithm that uses a logistic function, also called sigmoid function, to compute the probability for each class and then choosing the class with the maximum probability. It is considered a linear model because it assumes a linear additive relationship between the predictors and the log odds of a classifier. A key difference with the linear regression is that the output value being modeled is a binary value rather than a numeric value [21]. For the purpose of this paper the Simple Logistic algorithm in Weka is used.

5.2 Naïve Bayes

The Naive Bayes classifier is the most popular among generative classifiers and as the name suggests is based on Bayes rules. The algorithm computes the posterior probability of a class, based on the distribution of the words in the document by ignoring the actual position of the words in the document, and working with the "bag of words" assumption [19]. Despite the simplicity of this algorithm, it is fast and does not have big storage requirements.

5.3 K-Nearest Neighbor

The k-Nearest Neighbor classifier algorithm compares new items with all members in the training set based on the distance of the k most similar neighbors to predict the class of the new unlabeled document, X. The classes of the neighbors are weighted using the similarity of each neighbor to X. The similarity is measured by Euclidean distance or the cosine value between two document vectors. KNN does not rely on prior probabilities, since the main computation is the sorting of training documents in order to find the k nearest neighbors for the new document. It is computationally expensive to find the k nearest neighbor in high dimensions. KNN is implemented in Weka as IBk [17], (Instance Based Learner).

5.4 Decision Tree (C4.5)

C4.5 belongs to the category of statistical classifiers. It is an improvement of ID3 classification algorithm. This algorithm creates a decision tree by using the entropy to determine which attribute of a given instance will optimize the classification of the instances in the dataset and which values of these ranges will provide the best classifying results. Rules can be generated for each path in the tree. After building the tree from a training dataset, the algorithm receives new data and classifies it. In Weka this algorithm is implemented as J48 [17].

5.5 Neural Network with Back Propagation

Backpropagation is an algorithm based on supervised learning for training an Artificial Neural Network Classifier, ANN. Backpropagation is very efficient in recognizing complex patterns and performing nontrivial mapping functions. During the training phase, the connection weights of the neural network are given randomly initialized values. These training examples are then delegated to the ANN classifier which adjusts the connection weights using the back propagation algorithm. The procedure is repeated until the desired learning error is reached [22]. Multilayer Perceptron is the implementation of the Back Propagation algorithm in Weka software.

5.6 Support Vector Machines

Support vector Machines is a classifier based on statistical information theory and structural risk minimization. Sequential minimal optimization, SMO, is used for training a support vector classifier in weka. SMO breaks the quadratic programming problem into small quadratic programming problems and solves them analytically. The memory required by SMO is linear in the training set size; thereby the algorithm can handle large training sets. SMO is fastest for linear SVMs and sparse data sets. On real world sparse data sets, SMO can be more than 1000 times faster than the chunking algorithm [23].

5.7 Random Forest

Random forest algorithm creates the forest from a set of decision trees, each created by selecting random subsets of training data. The final class of the new object is assigned to the class with the highest value and is achieved as an outcome of all trees in the forest. Tree ensembles are a divide-and-conquer approach used to improve the performance. Random inputs and features yield good results in classification, run fast and are able to deal with missing data, but it is less beneficial in regression. It can't predict beyond the training range, resulting in an over fit in noisy data sets [24].

6 Experiments

In order to perform the classification experiments, a corpus of 20 different categories, each with 40 text Albanian documents is created. The classification can be affected by the type of categories and the similarity between them. For this purpose we created sub corpuses of different sizes and content of categories.

Furthermore, we also created corpuses of the same sizes and number of categories, differing from category names and text content inside the documents. We expect the same algorithm to slightly vary in performance based on the content of documents and type of class used. All the classification experiments are run on the corpuses listed in Table 1.

Table 1. Experimental corpus

Corpus code	Interpretation
C1	Corpus of 20 categories (Animals, Art, Astronomy, Biology, Charity, Chemistry, Culture, Curiosities, Economy, Environment, Fashion, Food, History, Literature, Medicine, Politics, Religion, Sport, Technology, Tourism) each with 20 documents.
C2.1	Corpus of 10 categories (Animals, Charity, Environment, Fashion, Food, Medicine, Politics, Religion, Technology, Tourism) each with 20 documents.
C2.2	Corpus of 10 categories (Art, astronomy, chemistry, economy, Food, literature, Politics, sport, Technology, Tourism) each with 20 documents.
C3	Corpus of 10 categories (astronomy, biology, chemistry, culture, curiosities, economy, history, literature, sport) each with 40 documents.
C4.1	Corpus of 5 categories (Environment, Fashion, Medicine, Technology, Tourism) each with 20 documents.
C4.2	Corpus of 5 categories (Animals, Charity, Fashion, Politics, sport) each with 20 documents.
C5.1	Corpus of 3 categories (chemistry, sport, Tourism) each with 20 documents.
C5.2	Corpus of 3 categories (Art, Food, literature) each with 20 documents.
C6.1	Corpus of 2 categories (Medicine, Tourism) each with 20 documents
C6.2	Corpus of 2 categories (Charity, Technology) each with 20 documents.

Each corpus is used for training a model and testing the chosen classification algorithms: Naïve Bayes, IBk, J48, SMO, Random Forest, Simple Logistic and Multilayer Perceptron. The Table 2 shown below gives the results of the experiments. We have highlighted in red the best percentage of correctly classified instances of each corpus.

Table 2. Experimental results with the classification accuracy for each algorithm.

Algorithm	C1	C2.1	C2.2	C3	C4.1	C4.2	C5.1	C5.2	C6.1	C6.2
Simple Logistic	64%	76%	76%	61%	81%	87%	96%	81%	100%	85%
Naïve Bayes	66%	82%	77%	57%	89%	93%	98%	81%	97%	87%
K-Nearest Neighbor (IBk)	18%	36%	21%	30%	52%	24%	41%	55%	85%	85%
Decision Tree (J48)	43%	52%	50%	55%	65%	67%	95%	72%	87%	80%
SVM (SMO)	65%	78%	78%	60%	91%	82%	93%	93%	97%	92%
Random Forest	58%	67%	69%	58%	77%	77%	95%	88%	92%	78%
ANN (Multilayer Perceptron)	5%	9%	9%	12%	19%	19%	48%	33%	82%	85%

To rate the algorithm from the best performant to the least performant, for each corpus we assessed each algorithm with a score from 6 to 0 based on the percent of correctly classified instances. For example, corpus C1 performs best with Naïve Bayes algorithm achieving 66% of correctly classified instances, so we rate this experiment with 6 points. The second best performant algorithm for C1 is SMO with 65% of correctly classified instances, so we rate SMO with 5 points. Next comes Simple Logistic with 64% scoring 4 points; Random Forest with 58% correctly classified instances scores 3 points; J48 scores 2 points with 43% correctly classified instances; IBk scores 1 point with 18% correctly classified instances and Multilayer Perceptron scores 0 points as the less performant of all. An equivalent scoring scheme is applied to every corpus listed in Table 1 and the result is shown in Table 3. The last column, Total, is calculated summing the scores in the rows and is considered as the score achieved by the algorithm.

Table 3. Algorithms evaluation scheme

Algorithm	C1	C2.1	C2.2	C3	C4.1	C4.2	C5.1	C5.2	C6.1	C6.2	Total
Simple Logistic	4	4	4	6	4	5	5	4	6	4	46
Naïve Bayes	6	6	5	3	5	6	6	4	5	5	**51**
K-Nearest Neighbor (IBk)	1	1	1	1	1	1	1	2	2	4	15
Decision Tree (J48)	2	2	2	2	2	2	4	3	3	3	25
SVM (SMO)	5	5	6	5	6	4	3	6	5	6	**51**
Random Forest	3	3	3	4	3	3	4	5	4	2	34
ANN (Multilayer Perceptron)	0	0	0	0	0	0	2	1	1	4	8

From the table shown below we can see that the best performing algorithms on our Albanian corpus are Naïve Bayes and Support Vector Machines, scoring both 51 points. Simple Logistics is the next best choice, and then comes Random Forest, J48, IBk and the worst performance is achieved by Multilayer Perceptron.

Based on the above results, we decided to evaluate the best performing algorithms to show if they have any statistical differences with one another. For this purpose we used the Experimenter in Weka for each dataset with Naïve Bayes, Support Vector Machines and Simple Logistics using the *Percent_correct* as comparison field. In this way we compared the percent of correctly classified instances of Naïve Bayes with Support Vector Machines and Simple Logistics. The summarized results are shown in the following Table 4.

Table 4. Results of the statistical evaluation

Corpus	*Naïve Bayes*	*SVM (SMO)*	*Simple Logistic*
C1	66.49	64.33	64.16
C2.1	81.86	78.60	76.67
C2.2	76.40	76.73	76.69
C3	57.24	59.43	61.20
C4.1	88.31	88.85	81.27
C4.2	92.07	83.22 *	86.18
C5.1	99.17	93.33	96.50
C5.2	81.83	92.33 v	87.83
C6.1	98.50	98.25	97.25
C6.2	90.40	92.15	82.60

The statistical comparison showed no significant differences except for corpuses C4.2 and C5.2. When the comparison is run on corpus C4.2, the result for SMO has a "*" next to it, meaning that SMO is statistically different with Naïve Bayes and the later performs better. From the other side, when the comparison is run on corpus C5.2, the result for SMO has a "v" next to it, meaning that SMO statistically outperforms Naïve Bayes. As a conclusion, we cannot determine statistically the best algorithm among Naïve Bayes, Support Vector Machines and Simple Logistics based on the percent of correctly classified instances.

7 Conclusions and Future Work

The main goal of this paper was the overall comparison of performance of text classification algorithms for Albanian language. For this purpose we reviewed the state of the art for text classification problems in different languages. Since there isn't any public corpus previously created for Albanian, we created a general corpus of 20 classes, each with 40 text documents and divided it in 10 different sets appropriate for our experiments. The corpus was preprocessed with stop words removal and JStem algorithm. We used Weka software to test Naïve Bayes, IBk, J48, SMO, Random Forest, Simple Logistic and Multilayer Perceptron algorithms. For each algorithm we tested the performance on

10 sets of documents. The best performing algorithms on our Albanian corpus are Naïve Bayes and Support Vector Machines both with the same score. From an overall projection of the results we can say that in general all the algorithms perform quiet well in classification problems when the number of classes is relatively small (2 or 3).

As a future work, it is of great interest to measure the effect of preprocessing phase on the performance of text classification. The stemming phase in Albanian language is a rule based algorithm that needs further improvements, and we believe that this will also improve the overall performance of the classification algorithms.

Furthermore, a public bigger corpus for Albanian needs to be created for further experiments in Natural Language Processing and Text Mining. The same algorithms ca also be compared using another bigger corpus.

References

1. Gantz, J., Reinsel, D.: The Digital Universe in 2020: Big Data, Bigger Digital Shadows, and Biggest Growth in the Far East. Technical Report 1. IDC, 5 Speen Street, Framingham, MA 01701 USA (2012)
2. Fan, W., Wallace, L., Rich, S., Zhang, Z.: Tapping the power of text mining. Commun. ACM **49**(9), 76–82 (2006)
3. Uysal, A.K., Gunal, S.: The impact of preprocessing on text classification. Inf. Process. Manage. **50**, 104–112 (2014)
4. Allahyari, M., et al.: A brief survey of text mining: classification, clustering and extraction techniques. In: Proceedings of KDD Bigdas, Halifax, Canada, 13 p., August 2017
5. Talib, R., et al.: Text mining: techniques, applications and issues. Int. J. Adv. Comput. Sci. Appl. **7**(11) (2016)
6. Zewen, X.U., et al.: Semi-Supervised Learning in Large Scale Text Categorization. Shanghai Jiao Tong University and Springer, Heidelberg (2017)
7. Sadiku, J., Biba, M.: Automatic stemming of Albanian through a rule-based approach. J. Int. Res. Publ. Lang. Individ. Soc. **6** (2012). ISSN 1313-2547
8. Biba, M., Gjati, E.: Boosting text classification through stemming of composite words. In: ISI 2013, pp. 185–194 (2013)
9. Kılıncx, D., et al.: TTC-3600: a new benchmark dataset for Turkish text categorization. J. Inf. Sci., 1–12 (2015)
10. Karan, K., Snajder, J., Basic, B.D.: Evaluation of classification algorithms and features for collocation extraction in Croatian. In: LREC 2012, Eighth International Conference on Language Resources and Evaluation (2012). ISBN 978-2-9517408-7-7
11. Yu, B.: An evaluation of text classification methods for literary study. Literary Linguist. Comput. **23**(3), 327–343 (2008)
12. Gonçalves, T., Quaresma, P.: Using IR techniques to improve automated text classification. In: Meziane, F., Métais, E. (eds.) Natural Language Processing and Information Systems, NLDB 2004. LNCS, vol. 3136. Springer, Heidelberg (2004)
13. Rasjida, Z.E., Setiawan, R.: Performance comparison and optimization of text document classification using k-NN and Naïve Bayes classification technique. In: 2nd International Conference on Computer Science and Computational Intelligence 2017, ICCSCI 2017, vol. 1314, Bali, Indonesia, October 2017

14. Al-Zaghoul, F., Al-Dhaheri, S.: Arabic text classification based on features reduction using artificial neural networks. In: UKSim 15th International Conference on Computer Modelling and Simulation. IEEE (2013)
15. Zaid Enweiji, M., Lehinevych, T., Glybovets, A.: Cross-language text classification with convolutional neural networks from scratch. Eureka: Phys. Eng., 24–33 (2017). https://doi.org/10.21303/2461-4262.2017.00304
16. Hamp, E.P.: Albanian Language, Encyclopedia Britannica (2016)
17. Witten, I.H., Frank, E.: Data Mining: Practical Machine Learning Tools and Techniques, 2nd edn. Morgan Kaufmann (2005)
18. McCallum, A.K.: Mallet: A Machine Learning for Language Toolkit (2002)
19. Aggarwal, C., Zhai, C.X.: Mining Text Data. Springer (2012)
20. Dunham, M.H.: Data Mining: Introductory And Advanced Topics. Pearson Education (2006)
21. Moreaux, M.: Text Classification with Generic Logistic-Regression Classifier (2015)
22. Ramasundaram, S., Victor, S.P.: Text categorization by backpropagation network. Int. J. Comput. Appl. (0975 – 8887) **8**(6), October 2010
23. Platt, J.: Fast training of support vector machines using sequential minimal optimization. In: Schoelkopf, B., Burges, C., Smola, A. (eds.) Advances in Kernel Methods - Support Vector Learning (1998)
24. Breiman, L.: Random forests. Mach. Learn. **45**(1), 5–32 (2001)

Battery Size Impact in Green Coverage of Datacenters Powered by Renewable Energy: A Latitude Comparison

Enida Sheme[1(✉)], Sébastien Lafond[2], Dorian Minarolli[1], Elinda Kajo Meçe[1], and Simon Holmbacka[2]

[1] Polytechnic University of Tirana, Tirana, Albania
{esheme,dminarolli,ekajo}@fti.edu.al
[2] Åbo Akademi University, Turku, Finland
{slafond,sholmbac}@abo.fi

Abstract. The use of renewable energy is a major trend to meet datacenters energy needs. However, its intermittent nature requires energy storage devices to store the over produced energy when not being used. Thus, the green coverage value, representing the fraction of total energy consumption covered by renewable energy, is increased. In this paper, we analyze the impact of using different battery sizes to optimize renewable energy usage. We have built a battery simulation tool able to provide the battery state, track the amount of stored and used energy by the battery as a function of the energy consumed by a datacenter and the energy produced by solar panels. We show the impact of battery size on the green coverage percentage, green energy loss, and brown energy taken from the traditional grid. A comparison of these metrics is made for three different geographical locations at 10°, 35°, and 60° latitude. We discuss the competitiveness of constructing datacenters in different geographical locations based on the results.

1 Introduction

Following an invariant growth in the required computational capacity, the total energy combustion of datacenters increases over time. In 2010 the energy consumption of data center was already estimated to represent between 1.1% to 1.5% of the worldwide electrical energy production [1]. A recent study [2] predicts the energy consumption of datacenters in US alone will reach 73 TWh in 2020, representing more than 5 times the production capacity of one new EPR nuclear reactor.

In order to decrease the CO_2 footprint of datacenters, the use of renewable energy is required. Nevertheless, the production of energy from renewable resources, such as sunlight and wind, is variable over time and dependent of weather conditions. One common way to address this problem is to use batteries, not only as a backup in case of energy outage or as a power peak shaving, but as an energy storage device. More and more studies [3–6] propose the usage of

L. Barolli et al. (Eds.): EIDWT 2018, LNDECT 17, pp. 548–559, 2018.
https://doi.org/10.1007/978-3-319-75928-9_49

batteries as a key source of energy for the energy system supply of a datacenter. In this case, the main question to be tackled is: how to choose the battery capacity to maximize the renewable energy usage thus minimize the green energy loss. The concept of green coverage mentioned in this paper represents the percentage of total energy consumption provided by renewable energy. Furthermore, we investigate how different geographical locations affect the required battery size to reach a desirable green coverage.

In November 2017, Tesla finished the construction of the world's largest energy storage installation with a capacity of 129 MWh [7]. This demonstrates the technology readiness of current large battery farms.

Figure 1 illustrates the energy sources system for our study.

We consider a datacenter consuming energy, which is provided by one of these three sources: renewable energy, battery energy and grid energy, according to this order of priority. The battery gets charged only by the renewable source and it charges only when the renewable energy is of greater amount than the datacenter energy needs. If the battery is full and there is still renewable energy produced and therefore not being used, this extra energy is considered overproduction. The battery discharges when datacenter energy consumption is higher than what is provided by renewable sources. In cases when both, renewable sources and the battery, are not enough to fulfill the datacenter energy requirements, additional energy is taken from the traditional grid (also referred to as brown energy).

The remainder of this paper is organized as follows. Section 2 presents recent work on battery usage as an energy source for datacenters. Section 3 describes solar energy analysis for three selected countries. Section 4 introduces a simulator

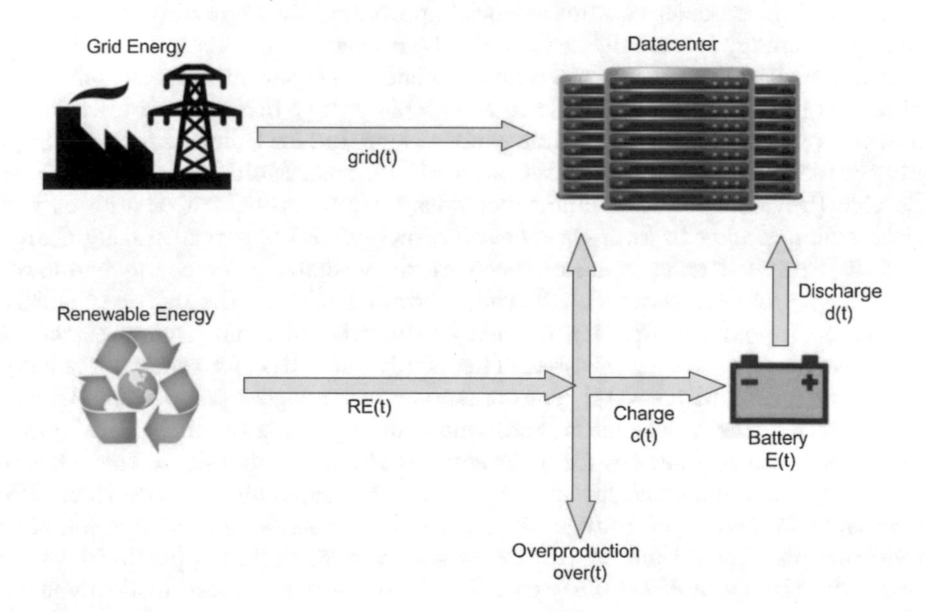

Fig. 1. Illustration of energy sources balance in our study.

tool we built to model and predict battery energy over time based on a battery as a finite state machine perspective. Section 5 describes the experimental setup and scenarios, as well as the results with the interpretations. Lastly, Sect. 6 concludes this paper and discusses future work we plan to advance and generalize our study.

2 Related Work

Energy storage devices (ESD) have largely been involved as a means of backup against power outages or to shave power peaks. However, several recent review papers focus on latest usages of batteries as a key supporting source of energy combined with renewable energy sources. The authors in [3] theoretically describe important characteristics of different ESD technologies, the trade-offs of placing them at different levels of the power hierarchy, and quantify trade-offs as a function of workload properties. [4] presents an overview of energy storage in renewable energy systems, describing advantages and drawbacks of different storage technologies. [5] categorizes existing research works according to their basic approaches used, like workload scheduling, virtual machine management, etc. Meanwhile [6] describes energy storage systems with detailed classification, features, advantages, environmental impacts, and implementation possibilities. The large number of references shows high interest on the topic of using batteries for several purposes. In our study instead, we present an approach to find the battery size impact on the green coverage for a datacenter and compare this impact for countries in typical northern, medium, and lower latitudes.

Innovative ways to use or study batteries are introduced at articles [8–13]. Authors at [8] consider two fundamental approaches for improving the usage of renewable energy in a small/medium-sized data center: opportunistic scheduling and usage of ESDs, which store renewable energy surplus at first and then provide energy to the datacenter when renewable energy becomes unavailable. [9] and [12] propose an efficient, online control algorithm for Datacenter Power Supply System called SmartDPSS and later MultiGreen. MultiGreen allows Cloud Service Providers to make online decisions on purchasing grid energy at two time scales, aiming to leverage renewable energy, and opportunistically charge and discharge batteries, in accordance with the available green energy and lower electricity prices. Authors at [10] study curtailment using the metric of energy return on investment (EROI), defined as the ratio of useful energy extracted from each unit of energy invested. They study the EROI for renewable energy farms when used with several types of storage technologies. [11] presents Green-Planning, a framework to hit balance among multiple energy sources, grid power and energy storage devices for a datacenter. Results demonstrate that Green-Planning can reduce the lifetime total cost and emission by more than 50% compared to traditional configurations, while still satisfying service availability requirement. The authors at [13] investigate cost reduction opportunities that arise with the use of ESDs. They consider the problem of opportunistically using these devices to reduce the time average electric utility bill in a datacenter by developing an online control algorithm. The innovation we bring in our paper is a simulation tool for battery energy modeling and prediction, and investigation of the battery size impact in green coverage percentage for different locations.

The issue of energy storage as a future trend for datacenter energy sourcing has been covered also in web articles, magazines, news, etc. The Wall Street Journal [14] claims that the giant battery is whats been missing in the renewable-energy revolution. In Times 2017 article [15], Logan Goldie-Scot, an energy-storage analyst at Bloomberg New Energy Finance cites Energy storage is being viewed by network operators as a potential tool in their toolbox, and that hasn't been the case up until now. While these articles express information on trends and opinions, we perform an experimental study aiming to help datacenter operators make wise decisions on choosing the battery capacity.

3 Solar Energy Analysis

In this section we describe the total amount of renewable energy provided over a year by each of the three selected countries: Finland, Crete, and Nigeria. In our study, only solar energy is considered as a type of renewable energy as it is dependent to geographical location e.g. latitude. Wind and other types of renewable energy are more irregular and have less possibility to be modeled based on latitude. Figure 2 graphically represents the total amount of solar energy produced monthly over a year by a 1 m^2 solar panel in Finland, Crete, and Nigeria. Nigeria shows an almost constant solar energy production. Since all days throughout the year last approximately 12 h, there is only a slight difference between winter and summer months. In Crete, solar intensity provides a large but varying energy production. The winter months provide less sunlight than the summer months, and have therefore a lower energy production than Nigeria. However, due to longer summer days, daily produced solar energy exceeds the energy productions of Nigeria despite of Nigeria's greater solar intensity.

The most varying results are measured in the Finnish location. The winter months produce almost no solar energy because of a very short time of sunlight during the day. On the other hand, during the summer the solar energy production can exceed both Greece and Nigeria because of the long duration of sunlight during the day. These data are also presented in Fig. 3 showing respectively highest energy value produced in 1 h and total amount produced over the whole year for Finland, Crete and Nigeria. We calculate the quantity of solar panels needed to achieve an arbitrary value of 50% green coverage over one year. We base our solar energy production data and assume a total energy consumption of 260 MWh for our datacenter as presented in [17]. Table 1 presents numerically the total amount of annual solar energy provided by a 1 m^2 solar panel and the required quantity of solar panels in m^2 for the three countries, aiming for half of the datacenter energy consumption to be provided by the solar energy source. On closer observance of the data, we notice a fair similarity between Nigeria and Crete, with only a 17% difference. Almost twice (45%) of the number of solar panels needed in Nigeria are needed in Finland to achieve same target of solar energy production.

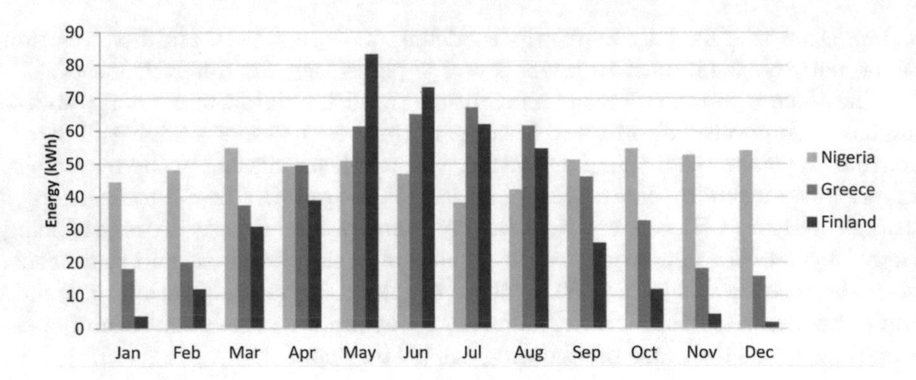

Fig. 2. Solar energy produced by $1\,\text{m}^2$ solar panel in three geographical locations monthly over a year [17]

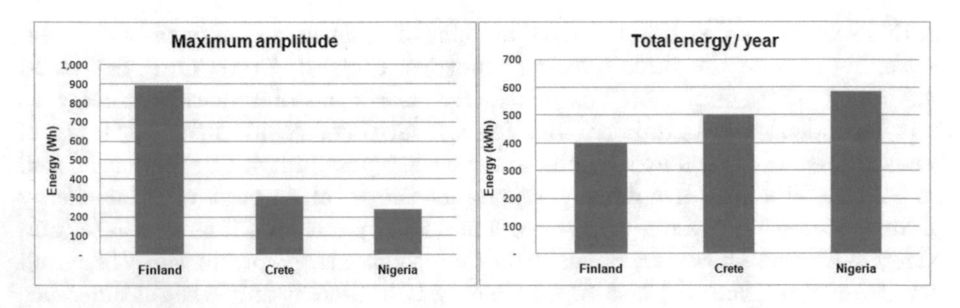

Fig. 3. Maximum amplitude in Wh per country during a year (left) and the total annual amount of solar energy produced per country (right).

Table 1. The total amount of annual solar energy provided by $1\,\text{m}^2$ solar panel and the number of solar panels in m^2 needed to cover 50% of energy consumption for Finland, Crete and Nigeria.

Country	Annual solar energy/$1\,\text{m}^2$ (kWh)	Required solar panels (m^2)
Finland	402	324
Crete	500	260
Nigeria	585	222

4 Battery Simulator Tool

In order to analyze and being able to predict the amount of available energy in the battery over specific moments in a chosen time period, we developed a simulation tool. The simulator is based on the concept of the battery as a finite state machine, whose trigger conditions are related to the available amount of solar energy and datacenter energy consumption at a specific time t.

4.1 Battery as a Finite State Machine

There are four possible states that the battery can be at any time t: Full, Discharging, Charging or Empty. Therefore, there are 16 possible state transition combinations. Practically, all of these combinations are feasible except for the Full - Empty and Empty - Full transitions which are generally limited by the charge/discharge rate of the battery during a certain period of time. Possible triggers from one state to the other depend on the amount of available solar energy in a moment of time t, referred to as *RE(t)* (Renewable Energy) and the datacenter energy needs in that moment t, referred to as a *consum(t)*. Other affecting factors are the value of the stored energy in the battery, *E(t)* and the maximum energy capacity of the battery named *Efull*.

4.2 Implementation

The pseudocode for developing the simulator is given below as a set of 8 steps. *BS(0)* refers to the initial Battery State assigned to Full, assuming the battery is fully charged when the simulation begins running. Each of the 16 'current - next' state combinations is assigned a combination number (named *combinationNr*), which calculates the energy (*E(t)*) the battery will have on every t. The green coverage is calculated according to Eq. 2 and printed out. The loop repeats 8760

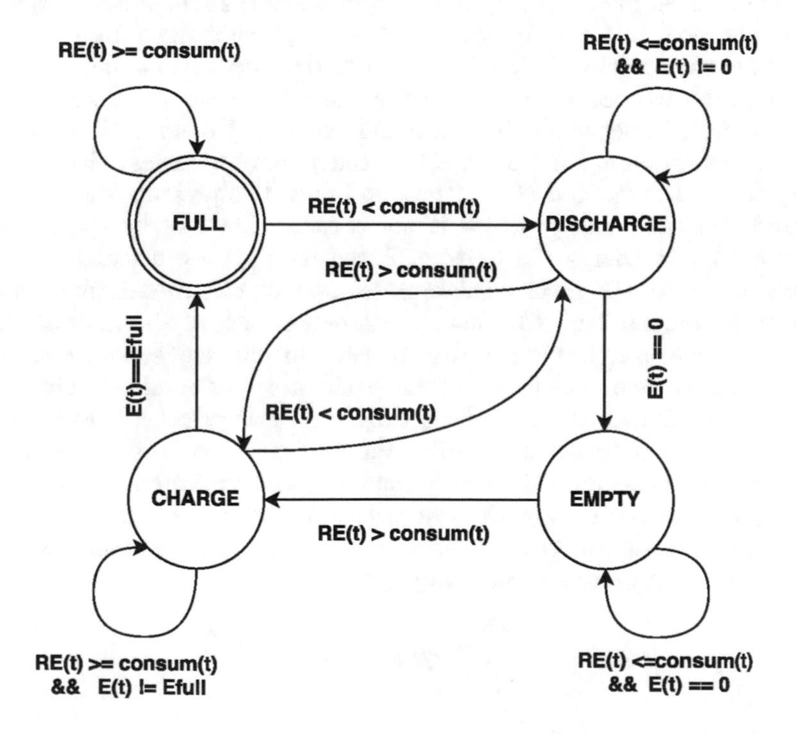

Fig. 4. Battery states over time represented as a finite state machine.

times for every hour of the year, resulting in outputs of total charging and discharging amount of the battery, overproduction, and grid energy for every hour.

Algorithm 1. Simulator tasks

$BS(0) \leftarrow FULL$, $t \leftarrow 1$
repeat
  Define $BS(t) = f[BS(t-1), RE(t), consum(t)]$
  Define $combinationNr = f[BS(t-1), BS(t)]$
  Calculate $E(t) = f[E(t-1), combinationNr]$
until $t \leq 8760$
$greenCoverage$ = annual $[RE(t) - over(t)]$/annual $[consum(t)]$
Print $charge(t), discharge(t), overproduction(t), grid(t), greenCoverage$

The Eq. (1) is checked by the simulator to be true for every moment of time t during the simulation time period.

$$RE(t) + d(t) + grid(t) = consum(t) + c(t) + over(t) \tag{1}$$

On the left side of Eq. (1) are listed the providing energy sources and on the right side is listed drawn energy. $RE(t)$ represents the renewable energy being produced at moment t, $d(t)$ is the amount of energy discharged from the battery in the time period $(t-1)$ to t, and $grid(t)$ represents the amount of energy taken from the traditional grid at that moment t, in case it is needed to fulfill the datacenter energy needs. The variable $consum(t)$ refers to the datacenter energy consumption at moment t, $c(t)$ is the amount of energy charged to the battery in the time period $(t-1)$ to t and $over(t)$ represents the part of the produced renewable energy which is not consumed neither by the datacenter nor from used to charge the battery. Based on Eq. 1 we calculate the green coverage metric of our specified datacenter over a year, as described in Eq. 2. The sum is composed of 8760 hourly values for each of the metrics: $RE(t)$, $over(t)$ and $consum(t)$. We can distinguish two different scenarios from this equation. First, theoretically we evaluate the green coverage simply as total renewable energy over the year divided by total energy consumption over the year. We have no information regarding the overproduction without running the simulations, so we assume it is zero. Second, experimental simulations show that the overproduction is greater than zero, meaning that the real green coverage is less than the theoretical calculated value. The battery size is the key element affecting the minimization of the overproduction value.

$$greenCoverage = \sum_{t=1}^{8760} [RE(t) - over(t)] / \sum_{t=1}^{8760} consum(t) \tag{2}$$

5 Experiments and Results

To perform the experiments we have run the simulation tool by changing the input of battery size from 0 to 10 MWh, separately for each of the countries. The first run under battery size equal to 0 shows how much from the produced solar energy is spent in vain as overproduction. Increasing the size of the battery means increased amount of stored solar energy. The yield of this is diminishment of the amount of overproduction and energy taken from the grid, translating to an overall higher renewable energy usage. The optimization concept based on the battery size is related to the fact that as battery size increases, the amount of energy taken from the grid and the overproduced energy is decreased. The decreasing rate depends on the battery capacity and of the energy production pattern, which differs in the selected geographical locations. The simulation time covers a period of 1 year (8760 h), taking as input the hourly solar energy records and the hourly energy consumption values over the year. The datacenter is composed of 100 servers and we assume in this paper a synthetic variable workload over one week repeated over the whole year. The annual energy consumption for this datacenter equals to 260 MWh. The used synthetic variable workload is obtained as presented in [17].

5.1 Experimental Scenarios

Experimental scenarios represent the number of simulations we have performed and the goal of running each of them. Our simulator tool asks for input the number of solar panels required to achieve a theoretical value of green coverage and the battery size supporting this energy system. For each of the 3 selected countries we have run the simulations requiring the number of m^2 solar panels according to Table 1 (222, 260 and 324). Also, the battery size is first chosen to be 0 and then increased to 1 kWh, 10 kWh, 100 kWh, 1 MWh and 10 MWh. We stopped increasing the battery size at 10 MWh as the calculated overproduction value over a year reaches zero and the theoretical annual green coverage equals the annual green coverage calculated from the simulator. After achieving this goal, increasing the battery size furthermore has no practical meaning and brings no change to the metrics we are studying.

5.2 Results and Interpretations

In this subsection we present the results and interpretations of our experiments through Figs. 5, 6 and 7. Through empirical simulations we found that to achieve no solar overproduction over a year, a 230 kWh battery capacity is needed for the datacenter located in Nigeria, a 292 kWh battery capacity for the Greece location, and a 9.3 MWh battery capacity if location is Finland. According to a typical Tesla Powerwall 2 battery chosen as a template with a capacity of 13.5 kWh [16], the overall battery size needed is equal to 17, 21, and 688 of such batteries for Nigeria, Crete, and Finland respectively. According to Eq. 2, the real green coverage is decreased from the theoretical one in proportion to

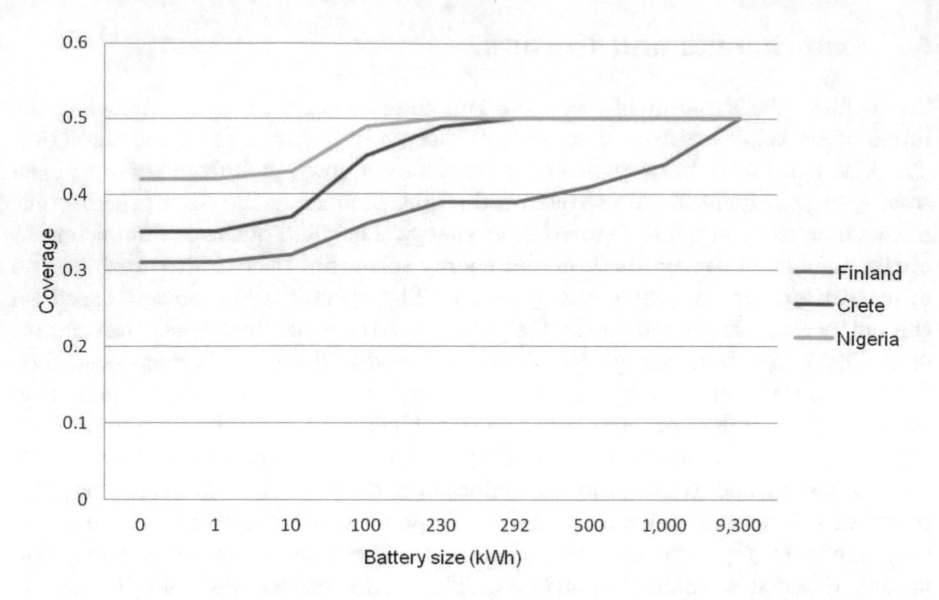

Fig. 5. The value of achieved average green coverage over a year with increasing battery size for Finland, Crete, and Nigeria, under the conditions of Table 1.

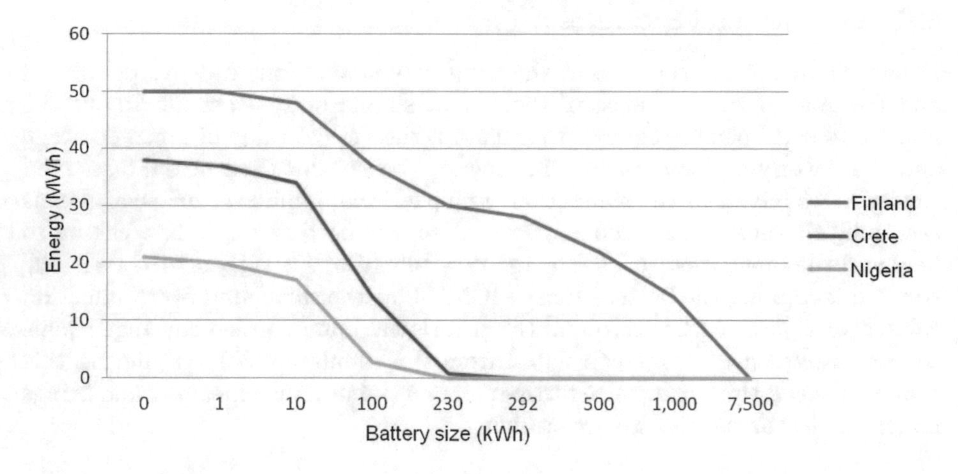

Fig. 6. The amount of overproduction in MWh for one year achieved by increasing battery size for Finland, Crete, and Nigeria, under the conditions of Table 1.

the overproduction amount. For a theoretical green coverage for the 3 countries, we chose to assign the same value of 50%. The first simulation with battery size 0 for each of the countries shows lower values of green coverage because of the overproduction energy, illustrated in Fig. 5. The achieved values are 42%, 35%, and 31%, for Nigeria, Crete, and Finland respectively. This means that out of a desired annual renewable energy usage of 50%, there is an 8%, 15%, and

19% drop from this desired usage because of the total overproduction energy over the whole year. The way to recover from this problem is increasing battery size. Figure 5 graphically shows higher values of green coverage rising towards 0.5 (50%) with increasing battery size. The rate by which this increase happens changes by country, referring to the annual values.

We can notice that Nigeria and Crete achieve 50% green coverage for much smaller battery size. Note also that the battery scale for the x-axis is not linear but rather intentionally distinguishes the values of 230 kWh, 292 kWh, and 9300 kWh to show the battery size needed for the overproduction to be equal to 0.

An increased value of green coverage is the result of a diminished amount of overproduction energy. Figure 6 shows the decreasing rate of overproduction by increasing the battery size for three countries. This happens because the battery stores more of the surplus solar energy, so there is less unused amount of it. Overproduction goes towards zero and reaches it when battery size is 230 kWh for Nigeria, 292 kWh for Crete, and 9300 kWh for Finland. We can notice that Finland is clearly distinguishable for very slow rate of impact by the battery size. This mainly happens because of its solar energy characteristics over the whole year, with high contrasts between winter and summer.

Higher green coverage leads to less amount of brown energy taken from the grid. This concept is shown in Fig. 7. Out of 260 MWh, only half of it (130 MWh) should theoretically be taken from the grid for a green coverage of 50%. The experiments show that for battery size equal to zero, the real amount of brown energy would be 150 MWh, 167 MWh, and 179 MWh for Nigeria, Crete, and

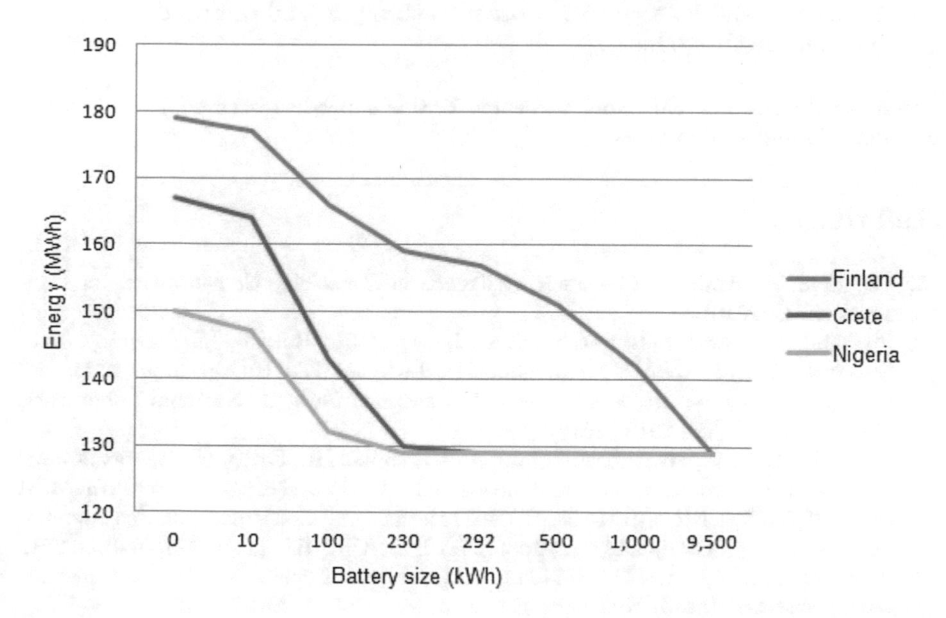

Fig. 7. The amount of energy in kWh taken from the traditional grid for a year with increasing battery size for Finland, Crete, and Nigeria, under the conditions of Table 1.

Finland. This amount reaches 130 MWh by increasing the battery size up to the key values of 230, 292, and 9300 kWh for each of the countries.

6 Conclusions and Future Work

The renewable energy sources are a promising means of energy supply for current datacenters. Nevertheless, they can be in excess or less than needed because of their variability, bringing the need of batteries as energy storage devices. In this paper, we investigate the impact of battery size in green coverage and green energy loss. We also study how different geographical locations affect the required battery size to reach a desirable green coverage.

We built a battery simulator to provide the amount of available battery energy every time t in a chosen time period. We chose a 260 MWh/year datacenter, 50% green coverage, and ran the simulator for three selected countries: Finland, Crete, and Nigeria. The results show that the solar panels needed in Crete and Finland are slightly more than Nigeria, 17% and 45% respectively. However, though Finland provides only 15% less annual amount of solar energy compared to Nigeria, it requires a battery size of 39 times greater to achieve wasted energy at level 0. While in Crete, battery capacity of only 27% greater than Nigeria is needed for this goal.

We plan to generalize the given solution for a required percentage of green coverage and a specific geographical location taken as an input from the user. We also intend to apply genetic algorithms for finding the best match of solar panels quantity and battery size in order to keep grid and overproduced energy at minimum possible values.

Acknowledgements. The work presented in this paper has been supported by the Erasmus Mundus programme.

References

1. Corcoran, P., Andrae, A.: Emerging Trends in Electricity Consumption for Consumer ICT (2013)
2. Shehabi, A., Smith, S.J., Sartor, D.A., Brown, R.E., Herrlin, M., Koomey, J.G., Masanet, E.R., Horner, N., Azevedo, I.L., Lint-ner, W.: United states data center energy usage report. Ernest Orlando Lawrence Berkeley National Laboratory, Technical report 06/2016 (2016)
3. Wang, D., Ren, C., Sivasubramaniam, A., Urgaonkar, B., Fathy, H.: Energy storage in datacenters: what, where, and how much? In: Proceedings of the 12th ACM SIGMETRICS/PERFORMANCE Joint International Conference on Measurement and Modeling of Computer Systems, SIGMETRICS 2012, pp. 187–198, June 2012
4. Amrouche, S., Rekioua, D., Rekioua, T.: Overview of energy storage in renewable energy systems. Int. J. Hydrogen Energ. **41**(45), 20914–20927 (2016)
5. SevketGuney, M.: YalcinTepe: classification and assessment of energy storage systems. Renew. Sustain. Energ. Rev. **75**, 1187–1197 (2017)

6. Kong, F., Liu, X.: A survey on green-energy-aware power management for datacenters. ACM Comput. Surv. (CSUR) **47**(2) (2015)
7. Ayre, J.: Tesla Completes Worlds Largest Li-ion Battery in South Australia (2017). https://cleantechnica.com/2017/11/23/tesla-completes-worlds-largest-li-ion-battery-129-mwh-energy-storage-facility-south-australia-notfree. Accessed 23 Nov 2017
8. Li, Y., Orgerie, A., Menaud, J.: Balancing the use of batteries and opportunistic scheduling policies for maximizing renewable energy consumption in a Cloud data center. In: 25th Euromicro International Conference on Parallel, Distributed and Network-Based Processing (2017)
9. Deng, W., Liu, F., Jin, H., Wu, C., Liu, X.: MultiGreen: cost-minimizing multi-source datacenter power supply with online control. In: Proceedings of the Fourth International Conference on Future Energy Systems (e-Energy 14), Berkeley, California, pp. 149–160, May 2013
10. Ghiassi-Farrokhfal, Y., Keshav, S., Rosenberg, C.: An EROI-based analysis of renewable energy farms with storage. In: Proceedings of the 5th International Conference on Future Energy Systems (e-Energy 14), Cambridge, United Kingdom, pp. 3–13, June 2014
11. Kong, F., Liu, X.: GreenPlanning: optimal energy source selection and capacity planning for green datacenters. In: 7th International Conference on Cyber-Physical Systems (ICCPS), Vienna, Austria, April 2016
12. Deng, W., Liu, F., Jin, H.: SmartDPSS: cost-minimizing multi-source power supply for datacenters with arbitrary demand. In: 33rd International Conference on Distributed Computing Systems (ICDCS), Philadelphia, PA, USA, July 2013
13. Urgaonkar, R., Urgaonkar, B., Neely, M., Sivasubramaniam, A.: Optimal power cost management using stored energy in data centers. In: Proceedings of the ACM SIGMETRICS Joint International Conference on Measurement and Modeling of Computer Systems, SIGMETRICS 2011, New York, USA, pp. 221–232 (2011)
14. This giant battery is whats been missing in the renewable-energy revolution. Wall Str. J. (2017)
15. Worland, J.: How Batteries Could Revolutionize Renewable Energy (2017). http://time.com/4756648/batteries-clean-energy-renewables. Accessed 1 May 2017
16. Wikipedia: Tesla Powerwall (2017). Last edited on 17 November 2017
17. Holmbacka, S., Sheme, E., Lafond, S., Frasheri, N.: Geographical competitiveness for powering datacenters with renewable energy. In: Carretero, J., Garcia-Blas, J., Margenov, S. (eds.) Third Nesus Action Workshop on Network for Sustainable Ultrascale Computing (NESUS) (2016)

A Comparison of Data Fragmentation Techniques in Cloud Servers

Salvatore Lentini[1], Enrico Grosso[2], and Giovanni L. Masala[1(✉)]

[1] Big Data Group, School of Computing, Electronics and Mathematics, University of Plymouth, Drake Circus, Plymouth PL4 8AA, UK
salvatore.lentini@postgrad.plymouth.ac.uk, giovanni.masala@plymouth.ac.uk

[2] Computer Vision Laboratory, Department of Political Science, Communication, Engineering and Information Technologies, University of Sassari, Viale Mancini, 5, 07100 Sassari, Italy
grosso@uniss.it

Abstract. Security and privacy are key issues in modern cloud computing. This paper focuses on data fragmentation techniques and it shows how these techniques can impact the global performance of a cloud service. Based on Amazon AWS, different fragmentation techniques are implemented and compared for several different types of file; the comparison is extended to AES encryption techniques in order to better evaluate the system performance and better understand potential effects on the overall security of the system. The presented results can be of great importance in the development of multi cloud and Fragmentation as a Service (FaaS) systems, targeting applications like large scale data mining and management.

1 Introduction

Cloud computing is rapidly growing and most of modern applications are now delivered as hosted services over the Internet. These services, which are broadly divided into three categories (IaaS Infrastructure-as-a-Service, PaaS Platform-as-a-Service and SaaS Software-as-a-Service) essentially rely on high-speed Internet and fully managed, scalable on demand, resources, made available by public or private providers [1]. Virtualization and distributed computing are the keystones of this scenario; they offer great benefits in terms of accessibility, flexibility and elasticity, but also introduce new and challenging problems; among these, security and privacy play a critical role because of the potential vulnerability of data stored and accessed through the Internet [2].

When considering Multi-cloud systems, that is the use of two or more cloud services to minimize the risk of downtime or data loss in the face of localized failures, the problem further complicates. In fact, while replication of data across different geographical regions certainly helps in order to ensure availability and continuous functioning of applications and services, it poses additional problems of data protection and data transaction management. This is especially true considering large databases, where the need of a distributed system is also related to the great amount of data to be stored. Data deduplication is a technology in cloud storage services which permits to save the storage

L. Barolli et al. (Eds.): EIDWT 2018, LNDECT 17, pp. 560–571, 2018.
https://doi.org/10.1007/978-3-319-75928-9_50

space by eliminating the duplicated data but provides some security issues analysed in [3]. Systems of extrusion detection of illegal files maliciously uploaded/downloaded, from the cloud, are considered in [4].

Data fragmentation techniques, including policies for fragments allocation and clustering network sites, are recently attracting the interest of researchers working in the cloud arena [5]. These technologies seem even promising in order to guarantee additional levels of security without leverage on complex standards of encryption [6] like AES encryption techniques [7]. However, authors in [8] propose more complex solutions, on powerful GPU hardware, combining fragmentation, encryption and dispersion.

This paper tries to investigate this aspect first defining a number of baseline methods for data fragmentation and implementing these methods on a virtual machine hosted by Amazon AWS [9]. Different techniques are thus compared for binary and text files and the overall performance of the system is evaluated by multiple trials of split and recovery. The comparison is finally extended to AES encryption techniques in order to better evaluate the system performance and better understand potential benefits and disadvantages of data protection realized exclusively by means of fragmentation.

2 Methods

Performance of data fragmentation techniques is evaluated through five *baseline* methods. In particular, we implemented one method related simple fragmentation, two methods based on permutation and two additional methods based on AES cryptography.

All the developed techniques are tested on the same dataset, including binary and images; multiple trials are used in order to assess statistical significance of results. In following illustrations and examples, we always consider sending the files to a single cloud provider; this is the worst possible case because the whole data are available for an attacker inside the cloud. The application to a Multi-cloud systems (multi-provider) is possible and there are not issues in the application of the methods in hybrid, private or public cloud, but we are considering the worst case, in terms of security, related to the public cloud.

2.1 Simple Fragmentation

Simple fragmentation is a basic and genuine version of the file fragmentation. The data are subdivided into N parts and each of these parts is stored on a separate file [10]. Individual file containing only one, or more parts of the fragmented file, is named *split file*. In this simple implementation of fragmentation of a file, each *split file* will contain a chunk (a fragment of the original file). The approach divides a file in N chunks sending the *split files* to the cloud. The fragmentation involves a complexity of reconstruction equal to N! since the attacker (e.g. a cloud provider insider) does not know the correct order of these files. Permutation of the chunks ("swapping" the order of the chunks) is thus required in order to reconstruct a meaningful file. Figure 1 better represents the fragmentation process related to this method. Note that in practice the fragmentation is implemented though the following steps:

- The user chooses the length of each single chunk *-len_chunk-*
- The number of N chunks with the same *len_chunk* are calculated.
- The file is divided in N chunks where each one is stored in a single *split file*

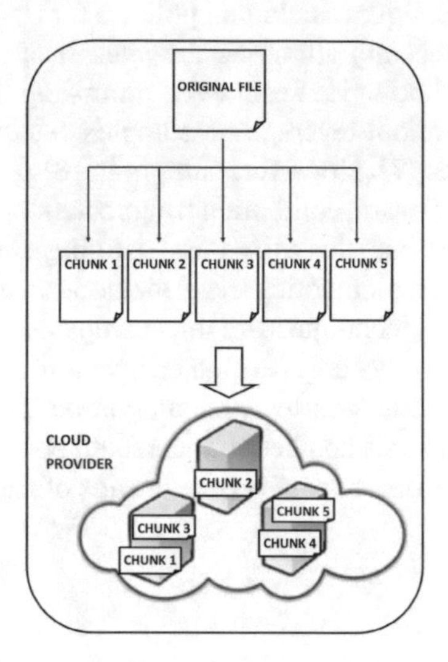

Fig. 1. Simple fragmentation method (fragmentation steps). The original file is divided on different chunks. Each chunk is stored in a single split file and neatly it is sent to the cloud.

In such easy implementation, the files are sent to the cloud in the same order of the fragmentation. Therefore, if the cloud attacker rescues the order of arrival of the individual files, he has the key to reconstruct the data in the recovery phase (Fig. 2). For these

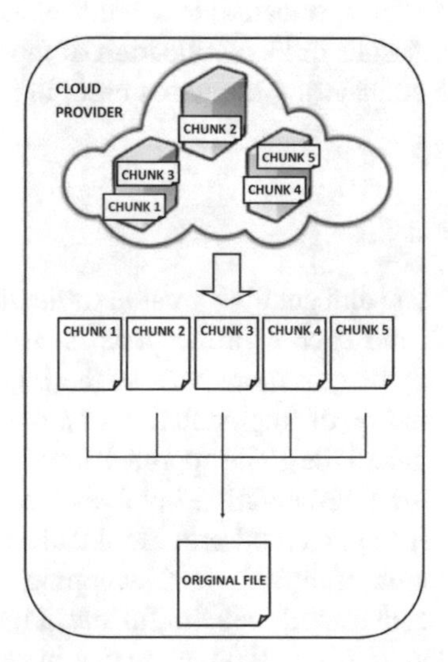

Fig. 2. Simple fragmentation method (recovery steps). The chunks are retrieved ordinately from the cloud to reconstruct the original file.

reasons, chunks are usually sent to a temporary buffer before starting the transmission, and from the buffer the *split files* are sent in a different order. This brings to the permutation technique described in the following section.

2.2 Permutation

In the light-weight permutation approach [6], the authors propose a method for mobile clients to store data on one or multiple clouds by using pseudo-random permutation based on chaos systems. The method, such as permutation, rather than using expensive operations, such as secret key or public-key encryptions,

The light-weight security method is based on simple operations, and provides a good balance between security and overall efficiency. In fact, several mobile devices have limited resources, such as limited power energy, low speed CPU and small capacity of RAM, and it is impossible to use expensive operations, like AES encryption, for each file when offload/download operations are required. However, authors show, in their work optimized for JPEG images, that the method achieve superior performance compared to over encryption methods, such as AES and encryption on JPEG encoders, while protecting the mobile user data privacy.

The method splits files into multiple files and uses a pseudo-random permutation to scramble chunks in each *split file*. The algorithm reads binary file rather than specific format (e.g. using a JPEG). There two phases to split files into multiple files and to recombine the files are as follows:

– *Disassembling* - the original file is splitted into multiple binary files, each containing multiple chunks of the original file. Chunks distribute through multiple files based on a *pattern*, chunks in each file randomly scramble by using the chaos system [11]. A *pattern* can be defined as a key by a user or it can be selected randomly as a predefined method. A user can define different *patterns* to provide different strategy for distribution.

  The output of this phase (split files) is stored in cloud.

– *Assembly* - all split files are recombined in order to reorganize the original file. In this phase the following steps will be performed:
  - read all scramble files from the cloud;
  - apply the chaos system random arrays (which are used for disassembling) to reorder the chunks in each split file;

As in the original approach, in our implementation of the method a user can configure his/her application to set:

- the number of split files,
- the size of chunks,
- the cloud user account(s) information to upload the files.

We implemented the method based *on pattern* in two different way, considering a predefined pattern fragmentation in Figs. 3 and 4, and a random pattern fragmentation in Figs. 5 and 6.

Predefined pattern fragmentation

The predefined pattern implementation, in Fig. 3, consists of an approach that leads to the creation of a split file whose chunks inside have an even or odd index. After dividing the file into N chunk each chunk will be written in a different split file according to the indexes in the pattern. The length of each single chunk *-len_chunk-* is calculated. Only two *split files* are created, inside which we find some chunks, presented in the order of parity or not.

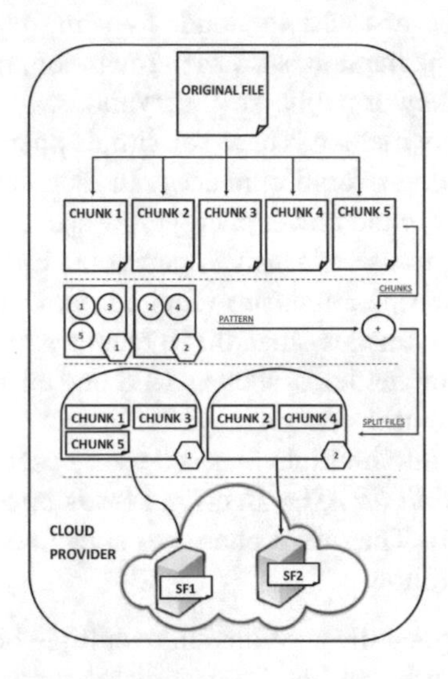

Fig. 3. Predefined pattern fragmentation method (fragmentation steps). The original file is divided on different chunks. Each chunk is stored in one of two predefined split files (even or odd). Finally, the split files are sent to the cloud.

In this method, the attacker will not be able to rebuild the file without knowing, the chunk length.

In the reconstruction phase, Fig. 4, the *split files* are opened in the order in which they are created, considering the relative chunk lengths. The chunks from each split file, are put in a *dictionary* data structure whose feature is to associate a *data* with a *key*. In this case, the *key* is the object contained in each pattern cell (being the pattern a list), which object is exactly the original chunk position in the initial file. Then, going to order the dictionary based on the keys, it is possible to retrieve the original file. The result of this reordering will then be saved within the file that will be the product of the recombination phase.

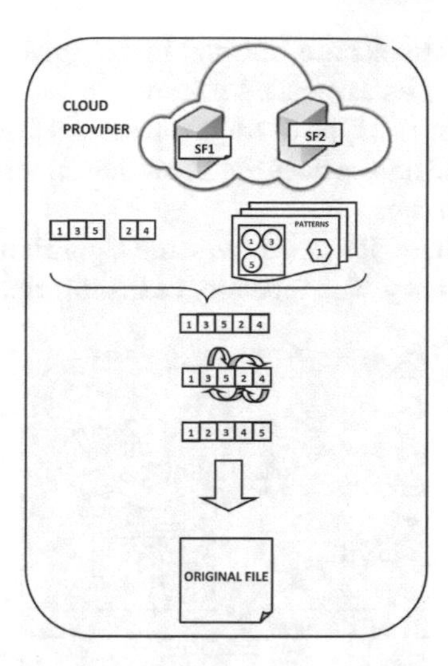

Fig. 4. Predefined pattern fragmentation method (recovery steps). The split files are retrieved from the cloud and the chunks are extracted on the base of the pattern reconstructing the original file.

Random pattern fragmentation

In the random pattern fragmentation, in Fig. 5, the pattern indexes are calculated with a random function i.e. a randomly permutation of a number N of elements selected by the user. The original file will be divided into N chunks. According to the random order of

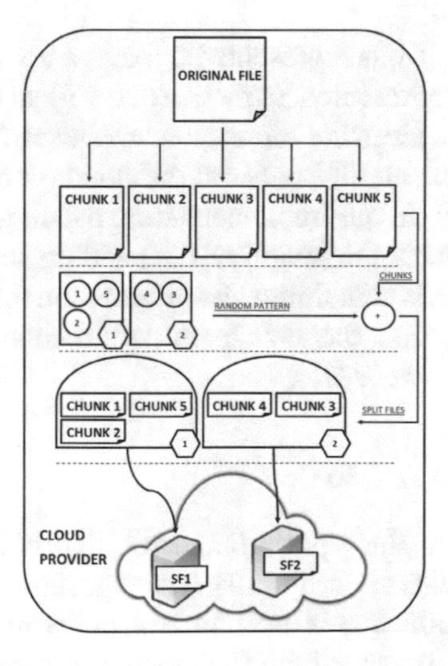

Fig. 5. Random pattern fragmentation method (fragmentation steps). The original file is divided on different chunks. Each chunk is stored in one of two split files through a random selection. Finally, the split files are sent to the cloud.

the patterns, the different chunks are inserted in the *split files*. Each *split file* has the length equal to the length of its associated pattern.

The interesting aspect of this type of fragmentation is that the attacker doesn't know the length of each chunk and he/she does not know the random order in which the chunks are distributed in each *split file*.

In the reconstruction phase, it is used the same algorithm of the simple pattern fragmentation, based on *dictionary*. The recovery of the information is shown in Fig. 6.

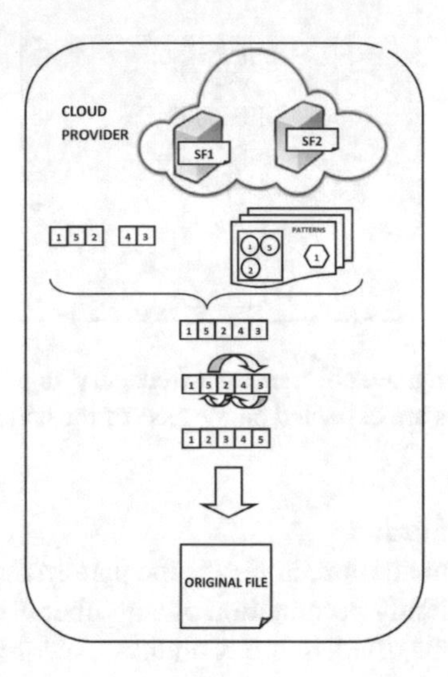

Fig. 6. Random pattern fragmentation method (recovery steps). The split files are retrieved from the cloud and the chunks are extracted on the base of the pattern, reconstructing the original file.

The original method in [6], is represented through a use case based on images. It is relevant how the header is represented. In fact, in the original method the header is stored alone in a separate file, with a smaller dimension with respect to the other split files, and it is also possible don't transmit the header to the cloud (or e.g. transmitting the header to a second cloud provider). In our implementation, padding bytes are added to the file containing the header to mask the length of the padding so that the attacker does not understand its length, which is containing less information than a normal *split file*. Using a random pattern fragmentation, the attacker doesn't recognize the *split file* containing the header among the others *split files*.

2.3 Simple Encryption AES 256

Nowadays, AES (Advanced Encryption Standard) algorithm is the most widely used encryption technique for data security [12]. The algorithm is defined as a symmetric encryption which means same key is used for both encryption and decryption process. Even though same key is used for both the process, it provides the required level of security by the way it encrypts the data. It supports block length of 128 bits and key size of 128,192 and 256 bits. In our experiments we use 256 bits.

We encode the whole original file with AES 256 before to send to the cloud. In the recovery phase we retrieve decoding the same file from the cloud, as represented in Fig. 7.

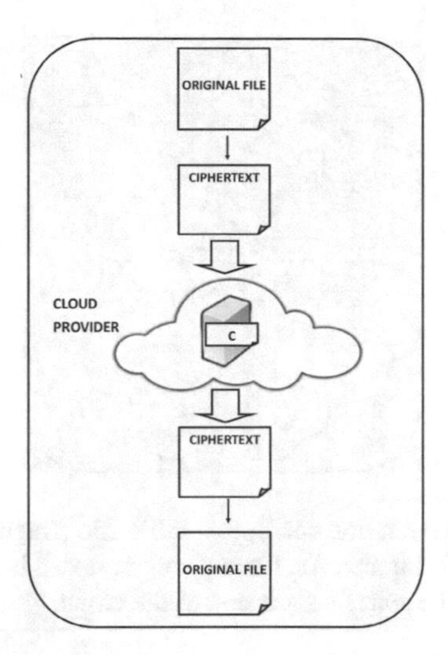

Fig. 7. Simple encryption AES 256 (Encoding and decoding steps). The whole original file is encrypted, and it sent to the cloud as a unique file. Vice versa in the decoding phase from the cloud only one single file produces the original file.

2.4 Random Pattern Fragmentation with Additional Encryption AES 256

The proposed implementation, in Fig. 8, involves the use of a *random pattern fragmentation* algorithm associated with the use of AES for data encryption. The basic idea is encrypting the original file through AES 256 CBC, dividing the cypher text in chunks. The chunks are arranged using a *pattern* randomly selected within the split files. Finally, each split file will then be sent to the cloud.

In the recovery phase, in Fig. 9, all split files in the cloud are read in sequence, until all the chunks are extracted. Finally using the pattern, the whole cypher file is recreated and after is decoded with the appropriate key obtaining the plain text. The proposed method has been designed to give a higher level of security than the previous ones, but having the burden of encryption (time consuming).

As general note, in all methods descripted in our work the chunks are always of the same dimension through the addition of few bytes of padding when it needs. All information about the fragmentation (e.g. lengths of chunks, number of chunks, rank, padding, etc.) are reported in a file in the user machine.

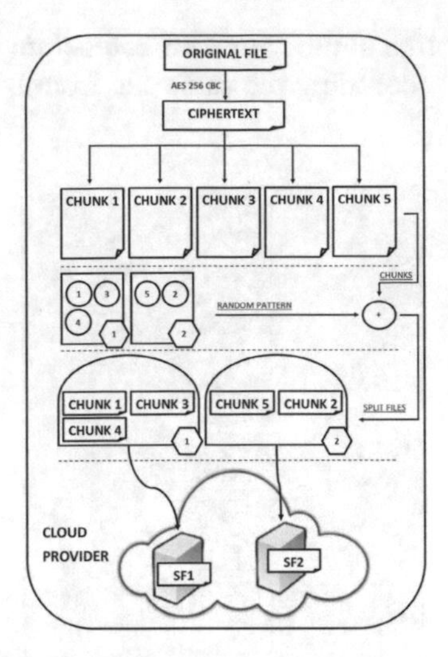

Fig. 8. Random Pattern fragmentation encryption AES 256 (fragmentation steps). The original file is encrypted and subdivided in chunks. Each chunk is stored in one of two split files through a random selection. Finally, the split files are sent to the cloud.

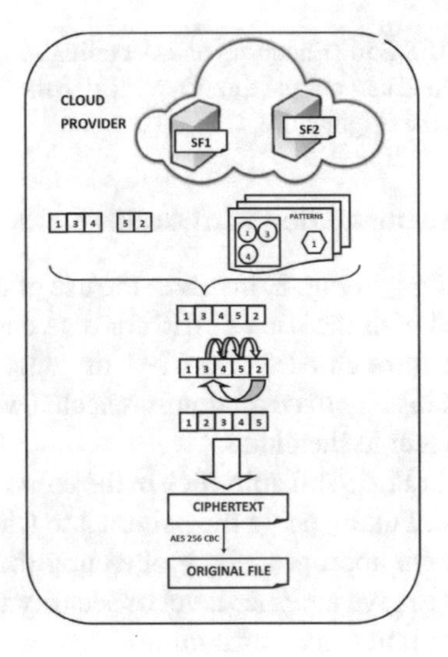

Fig. 9. Random pattern fragmentation encryption AES 256 (recovery steps). The split files are retrieved from the cloud and the chunks are extracted on the base of the pattern, reconstructing the original encrypted file. Finally, the original file is decoded.

3 Comparison

In this paragraph we first show a comparison of the general characteristics of the baseline methods and we thus provide a comparative analysis of the algorithm execution time. In Table 1 different kind of files are listed. We consider the files BPM, JPG, DOCX and

Table 1. Execution time (worst case expressed in seconds), varying the file format, the length of the chunks for the different methods used; it is also represented the mean time for the generic binary files with the standard deviation. Moreover, the BMP file is considered as example of binary file or image file, splitting the header alone.

	Format used	Simple	Predefined pattern	Random pattern	AES + random pattern	AES	Length chunks
	Files 100 KB	Seconds × 1 File	Seconds × 1 File	Seconds × 1 File	Seconds × 1 File	Seconds × 1 File	Bytes
Generic file	BMP	0.562	0.023	0.059	0.124	0.117	**1000**
		0.791	0.037	0.062	0.128	0.117	**750**
		1.052	0.047	0.091	0.149	0.117	**500**
		5.074	0.055	0.131	0.150	0.117	**10**
	DOCX	0.234	0.025	0.040	0.134	0.121	**1000**
		0.685	0.046	0.066	0.134	0.121	**750**
		0.979	0.057	0.080	0.168	0.121	**500**
		4.805	0.084	0.185	0.168	0.121	**10**
	PDF	0.466	0.045	0.045	0.142	0.125	**1000**
		0.547	0.048	0.045	0.148	0.125	**750**
		0.792	0.054	0.054	0.158	0.125	**500**
		5.039	0.068	0.093	0.161	0.125	**10**
	JPG	0.575	0.025	0.052	0.126	0.151	**1000**
		0.763	0.035	0.065	0.138	0.151	**750**
		0.983	0.037	0.083	0.139	0.151	**500**
		5.155	0.047	0.172	0.145	0.151	**10**
	MEAN	0.46 ± 0.16	0.03 ± 0.01	0.05 ± 0.01	0.13 ± 0.01	0.13 ± 0.02	**1000**
		0.70 ± 0.11	0.04 ± 0.01	0.06 ± 0.01	0.14 ± 0.01	0.13 ± 0.02	**750**
		0.951 ± 0.11	0.05 ± 0.01	0.08 ± 0.02	0.15 ± 0.01	0.13 ± 0.02	**500**
		5.02 ± 0.15	0.06 ± 0.02	0.15 ± 0.04	0.16 ± 0.01	0.13 ± 0.02	**10**
Splitting header	BMP	0.255	0.118	0.110	0.131	0.155	**1000**
		0.872	0.117	0.143	0.137	0.155	**750**
		0.974	0.131	0.182	0.148	0.155	**500**
		5.592	0.143	0.221	0.157	0.155	**10**

PDF as generic binary files. We show also the example of a BPM considered as image, splitting and coding the header alone. Furthermore, different lengths of the chunks are considered.

We consider, as cloud environment, the virtual machines available in Amazon AWS [9] for our experiments. However, with the aim to consider the performances of the algorithm independently by the network data rate, we show in Table 1, the processing time for the algorithm related to a computer desktop equipped with an Intel Core i7 - 6500U CPU 2.50 GHz, 8 GB of RAM on 64 Bit - Windows 10. The study takes into account the time needed for fragmentation and recovery, but not the time to transfer the data in cloud, which is a function of the networks speed.

As is reported in the Table 1, the simple pattern method is much slower of all the others approaches because it uses a split file for each chunk. The predefined and random pattern techniques perform better than AES, using instead the random pattern after AES, this proposed approach slightly increase the computation time. For all fragmentation methods if the length of the chunks is not optimized (big or very small) the calculation time increase. However, increasing the number N of the chunks the security increases. Considering the file simply as binary some differences arise in the computation time; for this reason, the mean and the standard deviation are reported. In addition, it is possible to note looking the two approaches applied to the BPM image that splitting the header of an image as separate chunk, with respect to considering the image as binary file, it increases the computing time. While the idea of maintaining the header file in local, without to transmit this information to the cloud could be interesting, this approach provides also practical issues related the management of the header in local, further increasing the processing time. The possibility to send instead the header to a second cloud provider, naturally increase the security aspect but it introduces additional complexity in the data management and additional time delay. Considering the image as binary file it seems to be the best option because considering a random distribution of the N chunks, the complexity to discover only the header is N but discovering the complete chunks rank is N! Moreover, the number N theoretically is not known for the attacker! In the Table 1 using e.g. a chunk length of 500 bytes, for a file of 100 KB correspond to N = 205.

The main properties of the different approach are summarized in Table 2.

Table 2. Comparison of the different methods

Descriptions	Simple	Predefined pattern	Random pattern	AES + random pattern	AES
Split files with only one chunk	v	x	X	x	v
Split files with multiple chunks	x	v	v	v	x
Encryption (key management)	x	x	x	v	v
Pattern subdivision	x	v	v	v	x
Random dispersion	x	x	v	v	x
Splitting header	v	v	v	v	v
Rank fast (slower 1… quicker 5)	1	**5**	4	3	2
The safest	AES + RANDOM PATTERN FRAGMENTATION				
The less safe	SIMPLE FRAGMENTATION				

Analyzing the Table 2, we note that the predefined and random pattern are good solutions to ensure the privacy of data considering devices with limited power with respect to the use of AES (e.g. several mobile models). However, the smartphones are increasing the computing performances and an AES solution could be acceptable. In this case the combination of AES and random pattern techniques can add a second layer of security, maintaining a good performance in possible future scenario (e.g. more powerful attack to decrypt AES).

4 Conclusion

In this paper we analyzed different fragmentation techniques, as a viable solution to ensure the data privacy in cloud server, also in case of untrusted providers or attackers inside the cloud. The approach based on random fragmentation is very fast compared to the standard encryption through AES. Looking forward the future, with new powerful machine and smartphones, the combination of AES with a random fragmentation, could be helpful to add further protection of data in cloud.

References

1. Velte, A., Velte, T., Elsenpeter, R.: Cloud Computing: A Practical Approach. McGraw-Hill, New York (2010)
2. Masala, G.L.C., Ruiu, P., Grosso, E.: Biometric authentication and data security in cloud computing. In: Computer and Network Security Essentials, pp. 337–355. Springer (2017)
3. Li, X., Shen, Y., Zhang, J.: The verifiable secure schemes for resisting attacks in cloud deduplication services. Inter. J. Grid Utility Comp. **7**(3), 184–189 (2016)
4. Hegarty, R., Haggerty, J.: Extrusion detection of illegal files in cloud-based systems. Int. J. Space-Based Situated Comput. (IJSSC) **5**(3) (2015)
5. Hababeh, I.: A novel cloud computing data fragmentation service design for distributed systems. In: 2011 International Conference on Parallel and Distributed Processing Techniques and Applications (2011)
6. Bahrami, M., Singhal, M.: A light-weight permutation based method for data privacy in mobile cloud computing. In: 3rd IEEE International Conference Mobile Cloud Computing, Services, and Engineering (MobileCloud), pp. 189–198 (2015)
7. Daemen, J., Rijmen, V.: The Design of Rijndael: AES - The Advanced Encryption Standard (2002)
8. Memmi, G., Kapusta, K., Lambein, P., Qiu, H.: Data Protection: Combining Fragmentation, Encryption, And Dispersion (ITEA2-CAP WP3 Final Report), November 2016
9. AWS Amazon. https://aws.amazon.com, updatc November 2017
10. Kapusta, K., Memmi, G.: Data protection by means of fragmentation in distributed storage systems. In: International Conference on Protocol Engineering (ICPE) and International Conference on New Technologies of Distributed Systems (NTDS) (2015). Télécom ParisTech, CNRS, LTCI, UMR 5141
11. Gharajedaghi, J.: Systems Thinking: Managing Chaos and Complexity: A Platform for Designing Business Architecture. Elsevier (2011)
12. Prabhu, M., Paramesha, K.: An approach for efficient utilization of public cloud storage and security data. In: IRJET, May 2017

SimpleCloud: A Simple Simulator for Modeling Resource Allocations in Cloud Computing

Dorian Minarolli(✉), Elinda Kajo Meçe, Enida Sheme, and Igli Tafa

Polytechnic University of Tirana, Tirana, Albania
{dminarolli,ekajo,esheme,itafa}@fti.edu.al

Abstract. Dynamic virtual machine resource allocation in cloud computing infrastructures is important in order to achieve optimal energy costs and minimal Service Level Agreement (SLA) violations. On the other hand, evaluating resource allocation approaches in real large-scale cloud infrastructures in repeatable and controlled manner, is difficult to achieve. For this reason, simulation and modeling tools are very important, especially in initial stages of development. Although existing state of art cloud computing simulation tools offer the possibility for virtual machine resource allocation evaluation, they are nonetheless complex to work with. In order to help speed up the process of evaluating new resource allocation approaches, we designed and implemented a new cloud computing simulator called SimpleCloud, that is simple to understand and efficient. Experimental evaluation through a resource allocation case study, show its efficiency and suitability for simulating large-scale cloud computing infrastructures.

1 Introduction

Cloud computing is becoming important as a new IT infrastructure management model. Clouds can be seen as large pool of easily usable and accessible virtualized resources that can be dynamically reconfigured and are typically exploited by a pay-per-use model [18]. One of the most important cloud computing service model is the Infrastructure as a Service (IaaS) model, in which raw computing resource such as CPU, memory and storage are offered in the form of Virtual Machines (VM).

An important issues in large-scale IaaS cloud infrastructures is dynamic management of VM resources according to workload variation. The main goal of VM resource management is to keep, Service Level Agreement (SLA) performance violation of consumer VMs to minimum while reducing energy consumption and resource costs of cloud infrastructures.

One of the most energy-efficient method for VM resource management in cloud infrastructures, is dynamic consolidation of VMs to few physical machines, creating the possibility to reduce energy by turning off idle machines. This is

L. Barolli et al. (Eds.): EIDWT 2018, LNDECT 17, pp. 572–583, 2018.
https://doi.org/10.1007/978-3-319-75928-9_51

done by moving VMs form under-loaded or overloaded physical machines to other machines through live migration technique [6].

There are a lot of resource management approaches proposed in literature that allocate resource to VMs in a dynamic way. Trying new resource management approaches in real large-scale cloud infrastructures is not feasible for different reasons. First, it is difficult to find large-scale cloud infrastructures to run experiments where the price can be a limiting factor. Second, even if there is an infrastructure available, running experiments in repeatable and controlled manner in a dynamic environment such as the cloud infrastructure, its a difficult undertaking. For these reasons simulation tools are becoming important for running large number of repeated experiments in large-scale cloud infrastructures, at least in the initial stages of development.

Although existing state of art simulation tools such as CloudSim [4] offer the possibility to evaluate new resource allocation approaches in large-scale clouds they are complex to work with. Some effort and time is required to understand their internal design and code before they can be extended to include new approaches. For this reason we developed a new cloud simulation tool called SimpleCloud, having as main feature the simplicity of design. Its simplicity can speed-up the process of implementing new resource allocation approaches. Its simplicity also contributes to its efficiency in simulating large-scale cloud infrastructures as shown by experimental evaluation through a resource allocation case study.

The paper is organized as follows. Section 2 discusses related work. Section 3 presents the cloud infrastructure that SimpleCloud simulates. Section 4 describes the SimpleCloud design and implementation. Section 5 presents the experimental evaluation. Section 6 concludes the paper and outlines areas of future work.

2 Related Work

One of the most popular cloud simulators available today is CloudSim [4]. It supports the simulation of a wide range of features such large data-centers, virtual machines, live migration, brokering and multiple cloud federation. These wide range of features are also its drawback by making it very complex to work with and implement new resource allocation algorithms.

There are different simulation tools that extend CloudSim in different ways to implement new features. CloudAnalyst [19] focus on simulation and evaluation of the performance of geographically distributed applications on different cloud data-centers. NetworkCloudSim [8] extends CloudSim to include support for modelling distributed applications communication patterns and network topologies. MR-CloudSim [10] supports modelling applications processing large amount of data using MapReduce paradigm. In contrast to the above works, SimpleCloud is not based on some external simulation engine but uses its own simple simulation core.

There are also simulation tools that are not based on CloudSim but use other simulation engines. GreenCloud [11], a packet level simulator, is based on the NS-2 [13] network simulation engine. This simulator is more concerned with detailed modelling of energy consumption of different data-center components such as servers, links, switches etc. and network communication simulation at packet level. SimIC [16] is a discrete event simulation tool based on SimJava simulation engine. Its main goal is to simulate an inter-cloud environment modelling components such as meta-brokers, local-brokers etc. needed for the interoperability of cloud services distributed across different data-centers. MDCsim [12] simulates a large-scale multi-tier data-center supporting three-tier web applications and communication link modelling. Its drawbacks are: it is commercial and not supporting virtualized IaaS infrastructure.

Some simulators use their own simulation engine. GroudSim [14] a discrete event based simulator, simulates both cloud and grid platforms. Its focus is on modelling scientific work-flow based applications. DCSim [17] similar to SimpleCloud is used for simulating a virtualized data-center offering IaaS cloud service model. Compared to DCSim, SimpleCloud has fewer components making it simpler.

3 Cloud Infrastructure

We simulate an IaaS cloud infrastructure composed of physical hosts each running several VMs as shown in Fig. 1. Each host runs an agent that is responsible for resource utilization monitoring and local resource allocation to all VMs. We focus on dynamic CPU resource allocation, since it is the most important resource that has main impact on application performance. To allocate CPU capacity to each VM we simulate the CPU CAP mechanism supported by modern virtualization technology [1]. The CAP, given as percentage of CPU capacity, is the maximum CPU capacity that each VM can consume. The CPU CAP implements what is called the non work-conserving mode of operation, supported by modern virtual machine CPU schedulers, which offers better performance isolation between VMs [9]. Although a machine can have several physical CPU cores we do not model the cores separately. We model the total CPU capacity as 100%. For example, if a CPU CAP of 50% is given to a VM on a physical machine with quad core, it means it is assigned 2 Virtual CPUs (VCPUs) that run on two physical CPU cores. This is a good approximation that models the behaviour of Xen Credit scheduler [5] that load balances the VCPUs (distributes CPU CAP capacity) between physical CPU cores.

The Agent allocates CPU CAP to each VM in discrete time intervals, where in each interval it decides the CAP value given for the next interval. The time interval value is chosen 5 s in our simulations. The CAP value is calculated based on average CPU utilization of the VM plus a 10% of CPU capacity. The Agent can resolve conflicts for CPU resource, by redistributing CPU CAPs between VMs when the total requested CPU CAPs of all VMs is greater than CPU capacity.

We model memory allocation by assigning a static RAM capacity to each VM at the start of simulation. Dynamic memory allocation can be implemented by including memory load utilization data form workload trace files and changing VM assigned RAM capacity at run time, similarly to CPU CAP.

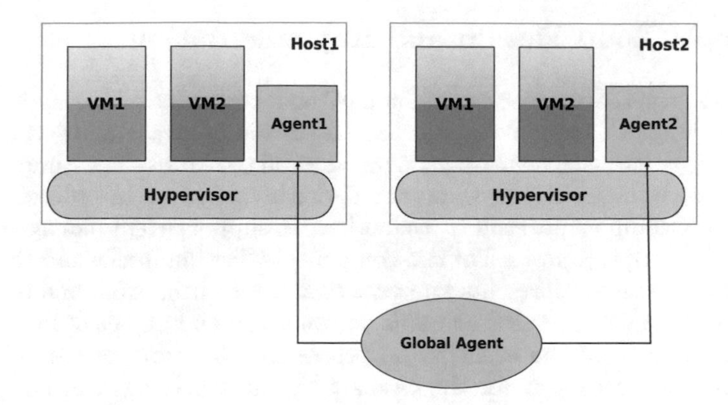

Fig. 1. Cloud infrastructure

For a case study we simulated a global resource allocation approach implemented by a global agent. The global agent is responsible for monitoring host CPU utilizations and making decisions for live migrating VMs form under-loaded or overloaded hosts to other hosts to minimize performance violations and reduce energy consumption. Similar to the host agents, the global agent makes allocation decisions in discrete time intervals but with larger interval values such as every 120 s. The global agent applies some form of Best Fit Decreasing (BFD) algorithm [3] to makes allocation decisions. To detect if a host is overloaded or under-loaded, upper and lower threshold CPU utilization values are used. A host is considered overloaded if the current and the past three host CPU utilization values are greater than the upper threshold. While a host is considered under-loaded if the current and the past three host CPU utilization values are less than the lower threshold.

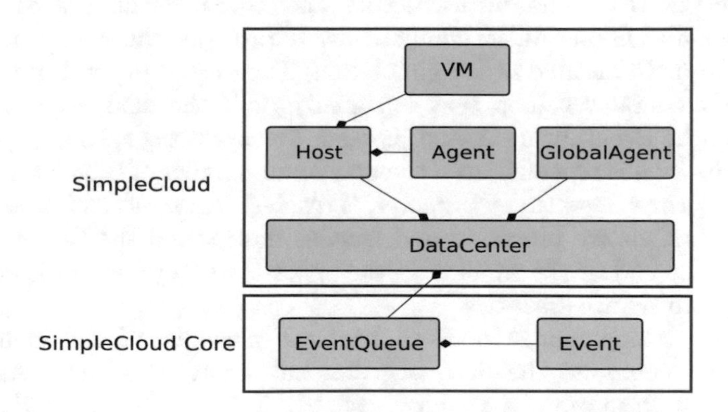

Fig. 2. SimpleCloud architecture

4 SimpleCloud Design and Implementation

In Fig. 2 is shown the architecture of SimpleCloud, composed of two layers: SimpleCloud and SimpleCloud core. SimpleCloud is written in Java language, about 1600 lines of code, where each component is represented by a Java class. SimpleCloud is an event-based simulation tool where simulation is carried out by processing events that different components send to each other. To support event management, the SimpleCloud core is composed of two components: EnventQueue and Event.

Event component represents an event that can be send from one component to another. Event component contains an event tag and object data. Event tag identifies the type of the event to be processed. Currently our simulator has three event tags necessary for the case study, but other tags can be added by user. $MIGRATING\_VM\_FINISHED$ tag identifies the event of processing VM live migration. $CPU\_UTILIZATION\_UPDATE$ tag identifies the event that is sent and processed every time interval (5 s) to update CPU utilization of all hosts and VMs, update host power consumption, keep track of VM SLA violations and to activate the host agent for making allocation decision. $GLOBAL\_AGENT\_ACTIVATED$ tag identifies the event that activates the global agent for making VM live migration decisions. Also this event is sent and processed periodically every time interval (120 s) of the global agent. Object data is used to contain information need to process the corresponding event. For example for the $MIGRATING\_VM\_FINISHED$ event, information needed to carry out the live migration is used such as source host, destination host and the VM to be migrated.

For each event that is sent, the time when it should be processed, given as the number of seconds after the moment of sending, is specified. Each sent event, is put in the EventQueue which sorts the events according to the time of processing.

The Datacenter is the key component of SimpleCloud. It represents the data center of the cloud infrastructure and contains the list of all hosts, EventQueue and GlobalAgent components. It provides the $run\_sim()$ method which is the main method of the simulation. The $run\_sim()$ method, shown in Algorithm 1, contains a loop that repeatedly pulls the next event to be processed form the EventQueue Q, and based on the event tag calls the appropriate method. The loop is repeated until the maximum number of time intervals, specified by $max\_time\_intervals$ parameter, is reached. An event can be sent by any component but all are processed and have as destination the Datacenter component. At the end of the simulation the $printResults()$ method is called that prints all performance statistics.

The Host component models a physical machine of the cloud infrastructure. It contains the list of VMs it runs, the host Agent and some other information such as, host id, RAM capacity, CPU utilization history and host idle power. Besides some helper methods, for example for adding VM or adding current host CPU utilization to the history, two important methods of Host component are $getHost\_Util()$ and $get\_power()$. $getHost\_Util()$ is used to get host CPU utilization

Algorithm 1. run_sim()

```
1  while getTimeInterval() < max_time_intervals do
2      Event e = Q.next_event()
3      switch e.get_tag() do
4          case CPU_UTILIZATION_UPDATE
5              process_update_utilization_event(e)
6              break
7          end
8          case MIGRATING_VM_FINISHED
9              process_migrating_vm(e)
10             break
11         end
12         case GLOBAL_AGENT_ACTIVATED
13             process_global_activation(e)
14             break
15         end
16     end
17 end
18 printResults()
```

at the current time interval, given as the sum of all VM CPU utilizations it runs. *get_power*() is used to get the average host power during current time interval. Since the CPU consumes most of the host power and as shown by previous research power consumption is approximately linear to CPU utilization [7] we use the power model [2] given below:

$$pw = ID\_POW + DYN\_POW \times (\frac{Host\_Util}{100}) \tag{1}$$

where, $Host\_Util$ is CPU utilization at current interval, ID_POW is power when host is idle, DYN_POW is dynamic power with 100% CPU utilization.

The VM component models the consumer VM running on cloud infrastructure. It contains information such as VM id, RAM allocated, CPU CAP allocated, accumulated VM SLA violation, and a history of past VM CPU utilizations. Three key methods of VM component are: *get_migration_time*(), *get_cpu_utilization*() and *set_sla_violation*(). *get_migration_time*() method returns the time it takes to live migrate the corresponding VM. Since the main factor that contributes to the migration time is the amount of RAM allocated to VM, we model migration time as RAM amount divided by network link bandwidth. *get_cpu_utilization*() returns the VM CPU utilization for the corresponding time interval. In this method, different utilization patterns of different applications running inside VMs, can be modelled. In our simulation the CPU utilization data are taken from workload trace files of thousands VMs running on PlanetLab [15] infrastructure. *set_sla_violation*() is called in each time interval, to estimate VM SLA violation metric for the current interval. We define the VM SLA violation metric as the following. There is a VM SLA violation if the difference between the CPU CAP and CPU utilization of a VM is less than 5% of the CPU capacity for 4 consecutive time intervals. The VM SLA violation metric is estimated as the CPU percentage by which CPU utilization exceeds the 5% threshold, for all 4 consecutive time intervals. The reasoning for the above metric is that we can have performance degradation if the VM CPU utilization is near CPU capacity and persist for several time intervals.

The Agent component contains information and methods related to CPU utilization monitoring and allocation decisions. Some of the key methods of Agent component are the following. $getUpperThr()$ estimates the upper threshold of the corresponding host. In general the upper threshold can be a fixed value such as 80% but in our simulation we defined a dynamic upper threshold that depends on the number of VMs and is related to VM SLA violations. More concretely it is given as $(100 - 5 * NR\_VMs)$ where 5 is the VM SLA violations 5% utilization threshold and NR_VMs is the number of VMs running on the host. The reasoning behind this formula is that the upper threshold represents utilization level beyond which VMs will experience SLA violations. $getLowerThr()$ returns the lower threshold, used for host underload detection, and is set to a fixed value of 10%. $VM\_decision()$ method is called in each time interval to allocate CPU CAP for all VMs running on the corresponding host as discussed in Sect. 3. The following methods are not used for local VM CPU allocation but needed by the GlobalAgent. $isOverloaded()$ and $isUnderloaded()$ return true if the host is overloaded or under-loaded respectively as explained in Sect. 3. $selectVM()$ is called whenever a VM live migration is needed to select a VM. In our simulation the VM with maximal average CPU utilization is selected in order to have few VM migrations but other selection policies can be implemented. $canAcceptVM()$ is called to estimated if a live migration destination host can accept a new VM. It returns true if average host CPU utilization plus average migrating VM CPU utilization is lower than the upper threshold and host RAM capacity can accommodate VM allocated RAM.

The GlobalAgent component contains all information and methods necessary to implement VM live migration decisions algorithm. The GlobalAgent together with host Agent are key components that can be changed by researchers to implement new resource allocation algorithms. The GlobalAgent implements, in the *activate*() method, the Best Fit Decreasing (BFD) algorithm [3] to make allocation decisions.

5 Experimental Evaluation

In this section we present the experimental results, evaluating SimpleCloud efficiency and suitability to model dynamic resource allocation in cloud computing environments. All experiments are run on a machine with 2 Intel(R) i5-3320M 2.6 GHz cores, 8 GB of RAM, running Ubuntu 14.04 operating system. The simulator is implemented in Eclipse platform version 3.7.2. We simulated a large data center where the number of hosts varied from 4000 to 40000. We started the simulation with 3 VMs per host. To have some diversity in memory resource

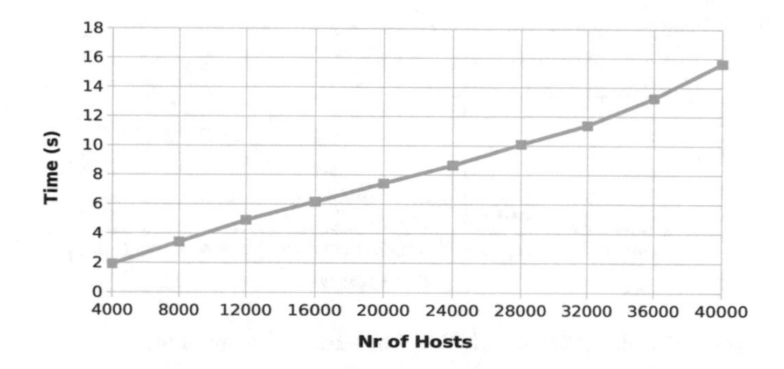

Fig. 3. SimpleCloud initialization time for different number of hosts

allocation we assigned 512 MB of RAM to half of the VMs and 1 GB to the other half. We run the simulations for 570 time intervals where the time interval is set to 5 s while the GlobalAgent time interval is set to 120 s. We run the experiments for 3 different load levels called *Load_0*, *Load_2* and *Load_4*. Each load level represents the CPU usage consumed by the VMs, where *Load_0* is the lowest load.

5.1 SimpleCloud Efficiency

In this section we present the evaluation of efficiency of SimpleCloud in large-scale cloud infrastructure, with respect to initialization time, simulation time and memory consumption. Initialization time is the time taken by the SimpleCloud to create the DataCenter, all the hosts and VMs and to load CPU usage data. Simulation time is the time taken by SimpleCloud to run the simulation excluding the initialization time. We should stress that simulation time depends not only on the simulator but also on the resource allocation algorithm applied. Memory consumption is the amount of RAM that SimpleCloud consumes during the simulation run excluding the memory consumed by the Eclipse platform itself. All simulations results are taken by running a global resource allocation approach that is a variant of Best Fit Decreasing (BFD) algorithm [3].

In Fig. 3 is shown SimpleCloud initialization time for different number of hosts. We can observe that the initialization time is low and not greater than 16 s for a large data center with 40000 hosts. We can also observe that the initialization time increases linearly with the number of hosts, showing the scalability of the simulator with respect to this metric.

In Fig. 4 is shown simulation time for different number of hosts. Although the simulation time does not increases linearly with the number of hosts it is still at acceptable levels (4.75 min) even for a large data center with 40000 hosts.

In Fig. 5 is shown SimpleCloud memory consumption for different number of hosts. It can be observed that the simulator consumes no more than 1.25 GB of RAM even for a large data center with 40000 hosts, which is acceptable for today typical computers with 8 GB of RAM. We can observe a slow linear increase in

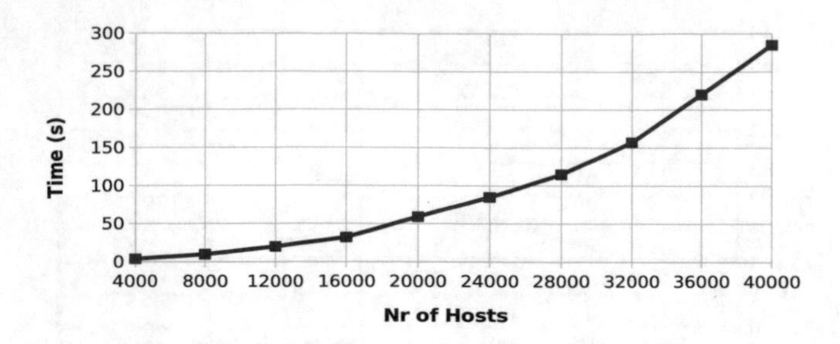

Fig. 4. SimpleCloud simulation time for different number of hosts

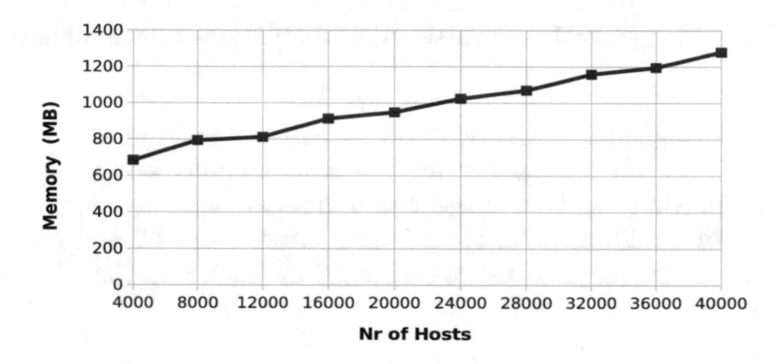

Fig. 5. SimpleCloud memory consumption for different number of hosts

memory consumption with the number of host, showing the scalability of the simulator with respect to this metric.

5.2 Dynamic Resource Allocation Case Study

In this section we run experiments to test the suitability of SimpleCloud for evaluating dynamic resource allocation approaches. In this case study we compared, a variant of Best Fit Decreasing (BFD) algorithm [3] for consolidation of VMs to a base case with no consolidation. The first approach is called "Resource Allocation" and the base case is called "No_Migration". We simulated a cloud infrastructure with 1000 hosts and 3 VMs per host. We evaluated SimpleCloud with respect to three performance metrics: (a) Cumulative SLA violation which is the sum of all VM SLA violations for the whole experimental time (b) Energy consumption of the data center for whole experimental time and (c) ESV metric that is the product of cumulative SLA violation with Energy consumption which represents the overall performance metric. We repeated the experiments 3 times and average values of performance metrics are shown in respective graphics.

In Fig. 6 is shown cumulative VM SLA violation for the two approaches and three load levels. Besides for low load *Load_0*, for other loads the "Resource Allocation" approach achieves lower SLA violations. This is because it migrates VM from overloaded hosts to other hosts reducing the performance violations.

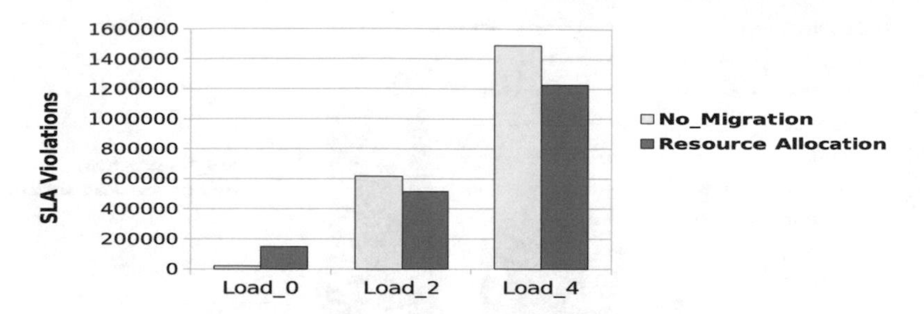

Fig. 6. Cumulative SLA violation over three load levels for 1000 hosts

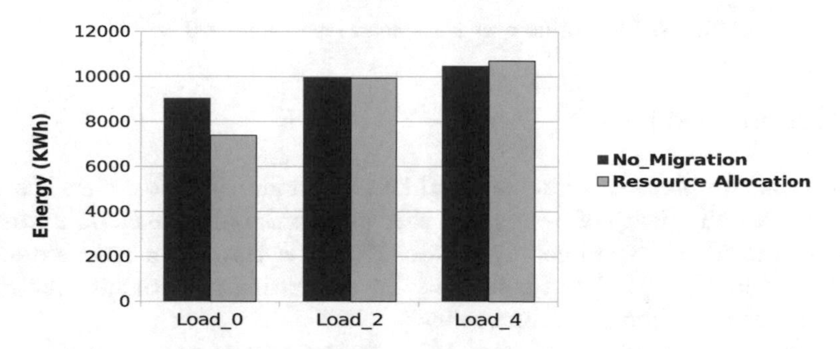

Fig. 7. Energy consumption over three load levels for 1000 hosts

For low load, since it consolidates VM to fewer hosts to save energy it results in higher performance violations.

In Fig. 7 is shown energy consumption of data center for the two approaches and three load levels. It can be observed that for low load *Load_0*, the "Resource Allocation" approach achieves energy savings by consolidating VMs to fewer hosts. For load *Load_2*, it achieves the same energy consumption while for high load *Load_4*, it consumes slightly more energy than the "No_Migration" approach.

In Fig. 8 is shown the ESV metric for the two approaches and three load levels. It can be observed that, besides low load, the "Resource Allocation" approach achieves better (lower) ESV values compared to "No_Migration" approach. For low load the "Resource Allocation" approach trade-offs a little SLA violations for energy savings thus resulting in higher ESV value.

This case study shows the suitability of SimpleCloud to evaluate and compare resource allocation approaches with respect to different performance metrics and understand better their behaviours.

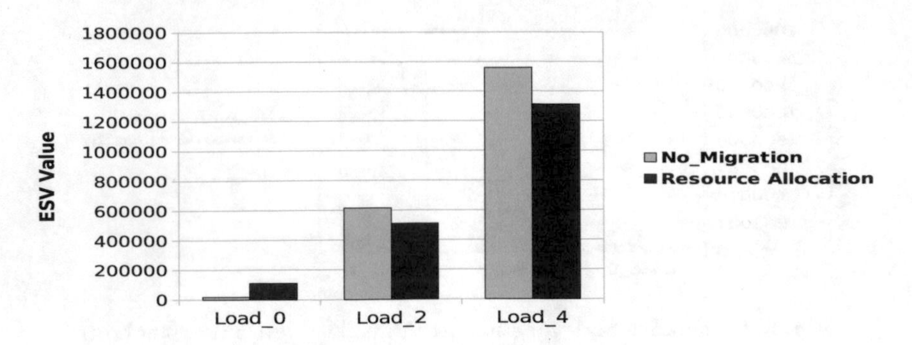

Fig. 8. ESV value over three load levels for 1000 hosts

6 Conclusions

In the paper we presented SimpleCloud an IaaS cloud simulator that has as its main goal the simplicity of design and efficiency for large-scale cloud infrastructures, as shown by experimental results. We show through a resource allocation case study, SimpleCloud suitability for evaluating performance metrics of dynamic resource allocation approaches.

In the future we will investigate which features to add to SimpleCloud and how to simulate allocation of other resources (e.g. storage) while keeping its design simple.

References

1. Barham, P., Dragovic, B., Fraser, K., Hand, S., Harris, T., Ho, A., Neugebauer, R., Pratt, I., Warfield, A.: Xen and the art of virtualization. In: Proceedings of 19th ACM Symposium on Operating Systems Principles, pp. 164–177. ACM Press, New York (2003)
2. Beloglazov, A., Abawajy, J., Buyya, R.: Energy-aware resource allocation heuristics for efficient management of data centers for cloud computing. Future Gener. Comput. Syst. **28**(5), 755–768 (2012)
3. Beloglazov, A., Buyya, R.: Optimal online deterministic algorithms and adaptive heuristics for energy and performance efficient dynamic consolidation of virtual machines in cloud data centers. Concurrency Comput. Pract. Experience **24**(13), 1397–1420 (2012)
4. Calheiros, R.N., Ranjan, R., Beloglazov, A., De Rose, C.A.F., Buyya, R.: CloudSim: a toolkit for modeling and simulation of cloud computing environments and evaluation of resource provisioning algorithms. Softw. Pract. Experience **41**(1), 23–50 (2011)
5. Cherkasova, L., Gupta, D., Vahdat, A.: Comparison of the three CPU schedulers in Xen. SIGMETRICS Perform. Eval. Rev. **35**(2), 42–51 (2007)
6. Clark, C., Fraser, K., Hand, S., Hansen, J.G., Jul, E., Limpach, C., Pratt, I., Warfield, A.: Live migration of virtual machines. In: Proceedings of 2nd Conference on Symposium on Networked Systems Design and Implementation, pp. 273–286. USENIX Assoc., Berkeley (2005)

7. Fan, X., Weber, W.D., Barroso, L.A.: Power provisioning for a warehouse-sized computer. SIGARCH Comput. Archit. News **35**(2), 13–23 (2007)
8. Garg, S.K., Buyya, R.: NetworkCloudSim: modelling parallel applications in cloud simulations. In: Proceedings of 4th IEEE International Conference on Utility and Cloud Computing, pp. 105–113. IEEE Press, Washington, DC (2011)
9. Gmach, D., Rolia, J., Cherkasova, L.: Satisfying service level objectives in a self-managing resource pool. In: Proceedings of 3rd IEEE International Conference on Self-Adaptive and Self-Organizing Systems, pp. 243–253. IEEE Press, Washington, DC (2009)
10. Jung, J., Kim, H.: MR-CloudSim: designing and implementing MapReduce computing model on CloudSim. In: Proceedings of 2012 International Conference on ICT Convergence, pp. 504–509. IEEE Press, Washington, DC (2012)
11. Kliazovich, D., Bouvry, P., Khan, S.U.: GreenCloud: a packet-level simulator of energy-aware cloud computing data centers. J. Supercomputing **62**(3), 1263–1283 (2012)
12. Lim, S.H., Sharma, B., Nam, G., Kim, E.K., Das, C.R.: MDCSim: a multi-tier data center simulation, platform. In: Proceedings of IEEE International Conference on Cluster Computing and Workshops, pp. 1–9. IEEE Press, Washington, DC (2009)
13. Ns2 Networking simulator (2016). http://www.isi.edu/nsnam/ns/
14. Ostermann, S., Plankensteiner, K., Prodan, R., Fahringer, T.: GroudSim: an event-based simulation framework for computational grids and clouds. In: Proceedings of Conference on Parallel Processing, pp. 305–313. Springer, Heidelberg (2011)
15. Park, K., Pai, V.S.: CoMon: a mostly-scalable monitoring system for planetlab. ACM SIGOPS Operating Syst. Rev. **40**(1), 65–74 (2006)
16. Sotiriadis, S., Bessis, N., Antonopoulos, N., Anjum, A.: SimIC: designing a new inter-cloud simulation platform for integrating large-scale resource management. In: Proceedings of IEEE 27th International Conference on Advanced Information Networking and Applications, pp. 90–97. IEEE Press, Washington, DC (2013)
17. Tighe, M., Keller, G., Bauer, M., Lutfiyya, H.: DCSim: a data centre simulation tool for evaluating dynamic virtualized resource management. In: Proc. 8th International Conference on Network and Service Management and 2012 Workshop on Systems Virtualization Management, pp. 385–392. IEEE Press, Washington, DC (2012)
18. Vaquero, L.M., Rodero-Merino, L., Caceres, J., Lindner, M.: A break in the clouds: towards a cloud definition. SIGCOMM Comput. Commun. Rev. **39**(1), 50–55 (2008)
19. Wickremasinghe, B., Calheiros, R.N., Buyya, R.: CloudAnalyst: a CloudSim-based visual modeller for analysing cloud computing environments and applications. In: Proceedings of 24th IEEE International Conference on Advanced Information Networking and Applications, pp. 446–452. IEEE Press, Washington, DC (2010)

Multimodal Attention Agents in Visual Conversation

Lorena Kodra(✉) and Elinda Kajo Meçe

Polytechnic University of Tirana, Tirana, Albania
lorena.kodra@gmail.com, ekajo@fti.edu.al

Abstract. Visual conversation has recently emerged as a research area in the visually-grounded language modeling domain. It requires an intelligent agent to maintain a natural language conversation with humans about visual content. Its main difference from traditional visual question answering is that the agent must infer the answer not only by grounding the question in the image, but also from the context of the conversation history. In this paper we propose a novel multimodal attention architecture that enables the conversation agent to focus on parts of the conversation history and specific image regions to infer the answer based on the conversation context. We evaluate our model on the VisDial dataset and demonstrate that it performs better than current state of the art.

1 Introduction

Artificial intelligence (AI) is becoming increasingly accurate and interactive. Rapid progress has been made in the intersection between computer vision and natural language, in particular in image captioning and answering questions about images [16, 25, 34] and videos [23, 24]. The next step, also the ultimate goal of AI and computer vision, is to develop systems that not only can "see" and "understand" the image, but also interact with humans through a meaningful conversation in natural language where future answers from the system are related to the context of the conversation. There are several uses for this kind of intelligent system, ranging from assisting visually impaired users to understand their surroundings, search and rescue missions where the human operator might not be able to see the whole scene [20], enabling users to interact in a natural language with intelligent assistants, helping analysts to process large quantities of surveillance data, etc.

Visual conversation agents are agents that understand and communicate with humans in a natural language about visual content. Their main difference from visual question answering (VQA) agents is that VQA represents only a single round of dialogue between human and machine. There is no conversation context and the machine does not memorize previous questions or answers. This prevents it from keeping track of a conversation and answering the next question based on conversation context. On the other hand, conversational models need to memorize conversation history and understand conversation context. They also need to overcome coreference ambiguities such as the use of pronouns "he", "she", "it", "they", etc., and correctly resolve and ground them in the image.

L. Barolli et al. (Eds.): EIDWT 2018, LNDECT 17, pp. 584–596, 2018.
https://doi.org/10.1007/978-3-319-75928-9_52

In this paper we propose a novel multimodal attention architecture for visual conversation that enables the agent to focus on specific parts of the conversation history and specific image regions to infer the answer based on the conversation context. Attention mechanisms have been successfully implemented in various areas such as image captioning [8, 27, 28, 30], neural machine translation [9, 11, 15], and visual question answering [6, 7, 14, 17–19]. They allow neural network models to use a question to selectively focus on specific parts of the input. Multimodal attention mechanisms allow models to use the question to simultaneously focus on specific part of the textual and visual input. The idea of using multimodal attention in visual conversation models has not been explored before.

The contributions of our work are as follows:

- We propose a novel multimodal attention architecture that enables the conversation agent to simultaneously focus on parts of the conversation history and image regions that are relevant to the current question and conversation context.
- We evaluate our model on the VisDial v0.9 dataset [2] and demonstrate that our model performs better than current state of the art.

The rest of the paper is organized as follows: In Sect. 2 we describe related work in this research area. Section 3 describes in detail our proposed model. In Sect. 4 we describe the experimental setup, evaluation results and analyze them quantitatively and qualitatively. Finally, in Sect. 5 we discuss about the conclusions.

2 Related Work

Visual Question Answering and Attention Mechanisms. Significant progress has been made on this task with state of the art models demonstrating competitive performance and improving human-machine interaction. The most promising solutions infer the answer under an encoder-decoder framework and use Convolutional Neural Networks (CNNs) to process images and extract image features, combined with Recurrent Neural Networks (RNNs) to process word sequences and generate the final answer [7, 18, 19, 22, 26, 32, 33]. Recently, attention mechanisms have been successfully implemented in VQA [6, 7, 14, 17–19]. Some models like [17, 18] perform image attention in multiple hops by querying the image multiple times to infer the answer progressively. Other models [7, 19] include attention into the standard RNN architecture.
Some authors [6, 14] combine visual and textual attention and create multimodal attention models which demonstrate state of the art performance on the VQA task.

Following this line of research, we propose a novel solution for the task of visual conversation that uses multimodal attention and enables the agent to leverage both attention modalities (visual and textual) to answer the question based on the conversation context. This kind of approach has not been explored before for this task.

Visual Conversation. This task has been very recently introduced [2] and explored in [1, 3–5]. The current state of the art for visual conversation is the model proposed in [2]. To tackle this task, the authors present various solutions under an encoder-decoder framework such as late fusion encoders, hierarchical recurrent encoders and memory networks with the latter being the most successful one. In order to improve the accuracy of the answer, they use attention over conversation history guided by question representation, and a joint representation of the question and image activations extracted from the penultimate layer of VGG-16 [31]. Different from them, we introduce attention over image, learnt end to end, guided by question representation. We extract convolutional feature maps from the fourth convolutional layer of VGG-16 [31]. This layer returns a 196 512-dimensional convolutional feature map for each image. This feature map will be used to calculate image attention where each of the 196 image regions contributes differently. Together with attention over conversation history, we build a multimodal attention model where, in order to answer the question, the agent uses both modalities simultaneously and learns to attend to specific parts of the conversation history, as well as specific part of the image that are relevant to the question and conversation context.

3 Multimodal Attention Model

The benefits of using attention mechanisms are twofold. First, they allow models to reduce the amount of processed information. Second, they increase answer accuracy as there is less input information that is irrelevant to the question. Our agent benefits from two separate attention modalities (textual and visual) that help it decide on the best answer. The agent consists of a model that uses the encoder-decoder framework described in [2]. Under this framework, an encoder embeds the input into a vector space and a decoder converts this embedded vector into an output prediction. Both encoder and decoder are implemented using LSTMs.

During testing, the model generates an image caption from the input image using [30]. In order to mimic real life situations where the human operator may be visually impaired, he has information only about the image caption and asks questions in order to gain an understanding about the image scene. During training the model takes as input the image I, the conversation history H, the question Q and 100 candidate answers $A_t = \left\{A_t^{(1)}, \ldots, A_t^{(100)}\right\}$ and is asked to return a probability distribution of A_t elements.

History H is composed of image caption C used as the first input token followed by previous ground truth question-answer pairs $H = \left(\underbrace{C}_{H_0}, \underbrace{(Q_1, A_1)}_{H_1}, \ldots, \underbrace{(Q_{t-1}, A_{t-1})}_{H_{t-1}}\right)$.

The model maintains previous question-answer rounds in its memory and learns to infer the answer based on its memory and the image.

Each question word is first transformed into its one-hot representation, a column vector the size of the vocabulary where there is a single one at the index of the token in the vocabulary. The words are then embedded into a real-valued word vector Q.

The same procedure is followed for generating the answer embeddings A and caption embedding C. We use 300 as the dimension of the word embedding space. For the image representation I we extract the output from the fourth convolutional layer of VGG-16 [31] which generates a 196 (14×14) 512-dimensional feature map of a 224×224 pixels input image. Figure 1 illustrates the dataflow in our model.

According to [2], the textual attention is calculated as follows:

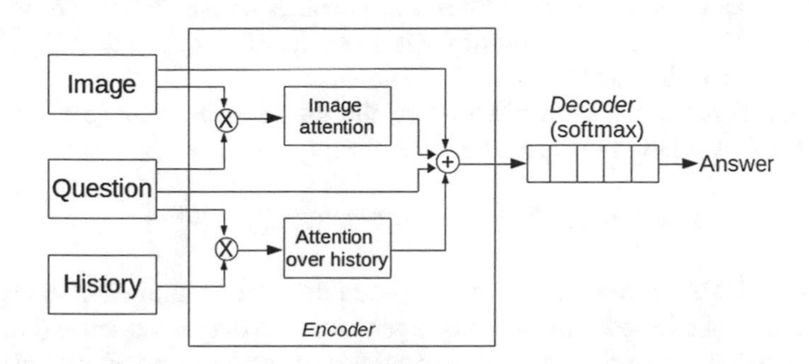

Fig. 1. Model dataflow. The encoder uses image, question and conversation history to generate the attention modalities which are added to its final output. The decoder uses a softmax classifier to generate the answer.

$$qh_t = Q_t H_t. \tag{1}$$

$$r_t^{txt} = softmax\left(W_r^{txt} qh_t\right). \tag{2}$$

$$a_t^{txt} = r_t^{txt} H_t. \tag{3}$$

$$l_t^{txt} = \tanh\left(W_a^{txt} a_t^{txt}\right). \tag{4}$$

The term r_t^{txt} represents the attention probabilities of each conversation round. Based on these attention probabilities the history attention vector is calculated as the product of the history representation and attention probabilities. The attention term a_t^{txt} decides the contribution of each conversation round at the t-th step.

We introduce a new attention modality (visual attention), thus transforming the existing model into a multimodal attention architecture.

The visual attention is calculated as follows:

$$qi_t = Q_t I. \tag{5}$$

$$r_t^{img} = softmax\left(W_r^{img} qi_t\right). \tag{6}$$

$$a_t^{img} = r_t^{img} I. \tag{7}$$

$$l_t^{img} = \tanh\left(W_a^{img} a_t^{img}\right). \tag{8}$$

The term r_t^{img} represents the attention probabilities of each image region. Based on these attention probabilities the image attention vector is calculated as the product of image features and attention probabilities. The attention term a_t^{img} is a 196-dimensional vector that decides the contribution of each image feature at the t-th step. The W coefficients are learnable parameters.

Both attention modalities are added to the question and image representation. Equation (9) shows how the input is encoded:

$$e_t = \tanh(W_l^{txt} l_t^{txt} \oplus W_l^{img} l_t^{img} \oplus W_e^{txt} Q_t \oplus W_e^{img} I). \tag{9}$$

The symbol "$\oplus$" represents element-wise addition. The output from the encoder in Eq. (9) is sent to the decoder LSTM which generates the output sequence (i.e. answer). We treat question answering as a classification task and use a softmax classifier to generate the answer. During training we also feed the ground truth encoded representation and maximize its log-likehood. During evaluation, the log-likehood scores are used to rank candidate answers.

4 Evaluation Results and Discussion

In this section we describe model implementation details, evaluation results and analyze them quantitatively and qualitatively.

4.1 Model Setup, Implementation Details, Hyper-parameters, Hardware

We use Torch [10] to develop our model. Before training, all questions are normalized to lower case and the question marks are removed. The model is initialized with Xavier initialization [13] except for the embeddings which used random uniform initialization, and is trained with Adam update rule [12] with a minimal learning rate of $5 * 10^{-5}$. We use cross-entropy loss to train the model with backpropagation. During testing we select the candidate answer with the largest log-likehood. We set batch size to be 16 and train for up to 40 epochs with early stopping if the validation accuracy has not improved in the last 5 epochs. The dimension of the LSTM network is set to 380 for all experiments. We set the embeddings size 300 as in [2] in order to make a fair comparison. We apply dropout with probability 0.5 on each layer and also gradient clipping to regularize the training process. We rescale the images to 224×224 pixels. We use the activations from the fourth convolutional layer of VGG-16 [31] for extracting image features used for calculating image attention.

We trained and tested our model on an Nvidia Jetson TX1 [29]. The Jetson TX1 module incorporates four, 64-bit ARM Cortex-A57 cores along with a 256 CUDA cores Maxwell architecture GPU (Nvidia Tegra X1 GPU) that delivers more than

1 TFLOPS of computing power. The module has 4 Gbytes of LPDDR4 memory with a 25.6 Gbyte/s bandwidth. It is not expandable. There is also 16 Gbytes of eMMC storage, plus additional SDIO and SATA storage interfaces.

Using the above settings, training lasted 4 days (90 h) and evaluation lasted 24 h.

4.2 Dataset and Evaluation Metrics

We evaluate the proposed model on the VisDial v0.9 dataset [2]. This dataset was used for fairness of comparison with the current state of the art model [2] and also because currently it is the only dataset available for the task of visual dialogue. The VisDial v0.9 dataset is constructed using the Microsoft COCO dataset [21] and contains 123'287 images, 1 dialog/image, 10 rounds of question-answers/dialog, and a total of 1.2 M dialog question-answers. It consists of two sets: training set (235 MB in size and 82'783 images) and validation set (108 MB in size and 40'504 images).

Instead of evaluating for a specific task as in a goal-oriented dialogue or evaluating the entire conversation as in a goal-free dialogue, we evaluate responses at each conversation round. We use retrieval metrics proposed by Das et al. [2] to evaluate the model. More specifically, we use mean rank of human response (the lower the better), recall @ k (existence of the human response in top-k ranked responses) and MRR (mean reciprocal rank) of the human response (the higher the better). All recall @ k metrics are expressed as percentages.

4.3 Results and Analysis

Table 1 shows the evaluation results and comparison with the state of the art model. For both models we use the setup and hyper-parameters as described in Sect. 4.1. We notice that our model performs better than state of the art on all retrieval metrics.

Table 1. Performance of models on the VisDial v0.9 dataset measured by mean reciprocal rank (MRR), recall @ k and mean rank. For MRR and recall @ k, higher values are better while for mean rank, lower values are better.

Model	MRR	R@1	R@5	R@10	Mean rank
Visdial [2]	0.4599	35.28	55.66	60.70	21.89
Ours	**0.5190**	**41.89**	**61.57**	**67.29**	**17.79**

MRR is given as the reciprocal rank of the first correct answer averaged over all questions. A MRR of 1 means that, on average, the correct answer is the first answer ranked by the model. From Table 1 we can see that our model performs 5.9% better than state of the art indicating that the position of the correct answer generated by our model is on average 5.9% higher and closer to the top ranked answer.

Recall @ k measures the presence of the human answer in the top k ranked answers generated by the model. We notice that our model performs better than state of the art by at least approximately 6% on this metric. According to Table 1, on 41.89% of the cases the top ranked answer generated by our model is the same as the human answer. This is equivalent to a 6.61% improvement compared to state of the art. As expected, with the increase of the number of answers taken into consideration, the presence of the human answer increases to 61.57% and 67.29% for top 5 and top 10 ranked answers generated respectively for each question. Compared to state of the art this counts for 5.91% and 6.59% improvement respectively for top 5 and top 10 ranked answers.

As regards the mean rank of the human answer we notice that our model performs 18.7% better than state of the art. This indicates that our model ranks 18.7% fewer incorrect answers higher in the list of generated answers.

In order to gain a better understanding on the behavior of our model we analyzed the answers generated by our agent during the conversations. Table 2 shows some example conversations based on the VisDial dataset.

Table 2. Conversation examples on the VisDial dataset.

Image + caption	Conversation
 1. A dog is sitting near a bike in the street	1. Is it a color photo? Yes
	2. What color is the dog? White
	3. What size is the dog? Small
	4. Are you able to tell what breed? No
	5. Does the puppy have a human with him? No
	6. Does he have a collar? No
	7. Does he have a leash? No
	8. Is the puppy facing the camera? No
	9. Are you able to see the color of his eyes? No
	10. Is the setting in a park? No
 2. A baseball player is swinging at a ball	1. What is the weather like? Sunny
	2. Do the fans look happy? Yes
	3. Are people wearing sunglasses? No
	4. What color are the baseball players' uniforms? Blue and white
	5. Can you see the whole ball field? No
	6. Can you see the score of the game? No
	7. Can you see the pitcher? No
	8. Are there kids there watching? Yes
	9. Do you see any cars? No
	10. Is it sunny out? Yes

(*continued*)

Table 2. (*continued*)

Image + caption	Conversation
3. Bikes parked near metallic bench	1. Are the bikes the same color? No
	2. Can you see any buildings? Yes
	3. What color are the bikes? Black
	4. Can you see any people? No
	5. Can you see the street? Yes
	6. Are there lots of buildings? Yes
	7. Can you see any vehicles? No
	8. Does it look nice out? Yes
	9. Can you see the sky? No
	10. Is it sunny? Yes
4. A bedroom with a bed and a desk	1. Is the photo in color? Yes
	2. Is it a professional photo? No
	3. Is it well lit? Yes
	4. Is it daytime? Yes
	5. What color is the chair? Brown
	6. What's the chair made of? Wood
	7. Does the chair have arms? Yes
	8. Is the desk large? No
	9. Does the desk look expensive? No
	10. Is there anything on the desk? Yes
5. A group of birds sitting on a tree branch	1. Is the photo in color? Yes
	2. Is there any people? No
	3. Are the birds the same size? Yes
	4. Are they the same color? Yes
	5. Are they in a cage? No
	6. Is this at a zoo or in the wild? Can't tell
	7. Is there any buildings? No
	8. Is it daytime? Yes
	9. Is there any poles around? No
	10. Is the sky visible? No

From the answers generated by our agent in these conversation examples we notice the following characteristics:

- Our agent has a good understanding of the image scene and object relations (e.g. image 2: questions 5, 6 and 7, image 3: questions 4, 5, 6, 7 and 8, image 5: question 6).
- Good understanding of conversation context and coreference. The agent is able to keep track of conversation context and correctly ground in the image object references done using pronouns. The agent resolves the reference and is able to focus

its attention on the correct object and give a correct answer (e.g. image 1: questions 4, 6, and 7, image 4: question 3, image 5: questions 4 and 5).

- Good ability to distinguish objects in the foreground and their characteristics. The agent is able to correctly focus its attention to differentiate foreground objects from the background and distinguish characteristics like color, size, etc.
- Good ability to understand image color setting. We notice that the agent answers correctly when asked about image color (e.g. image 1: question 1, image 4: question 1, image 5: question 1).

We can see from Table 2 that there are some errors present in the answers generated by our agent. We notice that the agent does not infer correctly on background objects (e.g. image 2: question 3, image 3: question 1 and 3, image 4: question 5 and 6). This indicates that attention performs better on foreground objects and is weakened for background objects or objects that appear to blend with the background.

Table 3. Error answer examples on the VisDial dataset.

Image + caption	Conversation
1. A group of bulls standing in a barn	1. Is this in color? Yes
	2. Do you see any people? There are no people
	3. How many bulls? 2
	4. Do any have horns? No
	5. What color are the bulls? Brown
	6. Are they indoors? Yes
	7. What color are the walls? White
	8. Do you see a fence? Yes
	9. Are the bulls standing? Yes they are
	10. Is the photo well lit? Yes it is
2. A man riding a skateboard on a ramp	1. How old is the skateboarder? Can't tell
	2. What is the skateboarder's gender? Male
	3. What is the skateboarder wearing? Shirt and pants
	4. Do you see any other people? No
	5. How many other people? 1
	6. Are the people watching the skateboarder? No
	7. Is the skateboarder in the air? Yes
	8. Is this at a skateboard park? No
	9. Is it a sunny day? Yes
	10. Does the skateboarder look like he's going to have a smooth landing? No

(*continued*)

Table 3. (*continued*)

Image + caption	Conversation
 3. A group of people riding skis down a snow covered slope	1. Is the photo in color? Yes
	2. How many people? 2
	3. Is the sun out? Yes
	4. How is the weather? Sunny
	5. What color is his board? Black
	6. What color is his coat? Black
	7. Is he wearing pants? Yes
	8. What color pants? Black
	9. Is there trees? Yes
	10. Are there leaves on the trees? Yes
 4. A group of children sitting at a table with a cake	1. How many kids? 3
	2. Is this birthday party? No
	3. Any adults? No
	4. Is this outdoors? No
	5. Is table round? Yes
	6. Is there tablecloth? No
	7. Are kids wearing hats? No
	8. Are smiling? Yes
	9. Does cake have candles? No
	10. Are they lit? No
 5. A train is coming down the tracks in a city	1. Is this a passenger train? Yes
	2. Is it moving? No
	3. Is it at a station? No
	4. Any people around? No
	5. Does the station have a roof? No
	6. What color is the platform? Green
	7. Any seats? No
	8. Is it daytime? Yes
	9. Is it sunny? Yes
	10. Any lights on the train? Yes

4.4 Qualitative Error Analysis

In order to gain a better insight on the errors that our agent makes and understand where we can improve it in the future, we analyzed the conversations where it gives wrong answers. Table 3 shows some examples.

From the answers generated by our agent in these conversation examples we notice the following characteristics:

- Difficulty in identifying objects in the background. The agent cannot differentiate correctly the objects in the background. The focusing attention is weakened for this kind of objects and the agent cannot infer correctly about them (e.g. image 1: question 3, image 2: question 4 and 6).
- Difficulty in counting objects in the background. The agent focuses and counts objects in the foreground correctly, but the counting ability is weakened as objects became blended with the background (e.g. image 1: question 3, image 3: question 2). This problem also comes from the difficulty of differentiating objects in the background.
- Counting ability is weakened for incomplete objects. The agent is able to focus, identify and count complete objects in the image, but has difficulty in distinguishing and inferring about objects that appear incomplete in the image (e.g. image 4: question 1 and 3).
- Difficulty in differentiating objects that appear blended with each-other. Attention does not work correctly for differentiating and inferring correctly about these objects in the image (e.g. image 1: question 3 and 4, image 5: question 5 and 6, image 4: question 9).

5 Conclusions

In this paper we proposed a novel multimodal attention model for visual conversation. Our conversational agent leverages both textual and visual attention simultaneously in order to keep track of conversation context and ground it in the image. We evaluated our agent on retrieval metrics and showed that it performs better than current state of the art. We analyzed the results qualitatively and noticed that our agent is able to use multimodal attention correctly to – (1) understand image scene and object relations, (2) understand conversation context, coreference and ground object references correctly in the image, (3) distinguish objects in the foreground and their characteristics like size, color, etc., (4) understand image color setting. We also noticed that our agent had difficulty in – (1) identifying objects in the background, (2) counting objects in the background, (3) counting incomplete objects in the image, (4) differentiating objects that appear blended with each-other in the image. These difficulties are indicative of the need to improve the attention mechanisms and solving them is subject to future work.

References

1. Das, A., Kottur, S., Moura, J. M.F., Lee, S., Batra, D.: Learning cooperative visual dialog agents with deep reinforcement learning. In: ICCV (2017)
2. Das, A., Kottur, S., Gupta, K., Singh, A., Yadav, D., Moura, J.M.F., Parikh, D., Batra, D.: Visual dialog. In: CVPR (2017)
3. de Vries, H., Strub, F., Chandar, S., Pietquin, O., Larochelle, H., Courville, A.: GuessWhat?! Visual object discovery through multi-modal dialogue. In: CVPR (2017)

4. Strub, F., de Vries, H., Mary, J., Piot, B., Courville, A., Pietquin, O.: End-to-end optimization of goal-driven and visually grounded dialogue systems. arXiv:1703.05423 (2017)
5. Chattopadhyay, P., Yadav, D., Prabhu, V., Chandrasekaran, A., Das, A., Lee, S., Batra, D., Parikh, D.: Evaluating visual conversational agents via cooperative human-AI games. In: CVPR (2017)
6. Hyeonseob, N., Jung-Woo, H., Jeonghee, K.: Dual Attention Networks for Multimodal Reasoning and Matching. arXiv:1611.00471 (2017)
7. Zhu, Y., Groth, O., Bernstein, M., Fei-Fei, L.: Visual7W: grounded question answering in images. In: CVPR (2016)
8. Hendricks, L.A., Venugopalan, S., Rohrbach, M., Mooney, R., Saenko, K., Darrell, T.: Deep compositional captioning: describing novel object categories without paired training data. In: CVPR (2016)
9. Delbrouck, J. B., Dupont, S.: Multimodal compact bilinear pooling for multimodal neural machine translation. In: ICLR (2017)
10. Collobert, R., Kavukcuoglu, K., Farabet, C.: Torch7: a matlab-like environment for machine learning. In: BigLearn, NIPS Workshop (2011)
11. Huang, P-Y., Liu, F., Shiang, Sz-R., Oh, J., Dyer, C.: Attention-based multimodal neural machine translation. In: Proceedings of the First Conference on Machine Translation (2016)
12. Kingma, D., Ba, J.: Adam: a method for stochastic optimization. arXiv preprint arXiv:1412. 6980 (2014)
13. Glorot, X., Bengio, Y.: Understanding the difficulty of training deep feedforward neural networks. In: AISTATS, pp. 249–256 (2010)
14. Lu, J., Yang, J., Batra, D., Parikh, D.: Hierarchical question-image co-attention for visual question answering. In: NIPS (2016)
15. Caglayan, O., Aransa, W., Wang, Y., Masana, M., García-Martínez, M., Bougares, F., Barrault, L., van de Weijer, J.: Does multimodality help human and machine for translation and image captioning? arXiv preprint arXiv:1605.09186 (2016)
16. Antol, S., Agrawal, A., Lu, J., Mitchell, M., Batra, D., Zitnick, C.L., Parikh, D.: VQA: Visual Question Answering. In: ICCV (2015)
17. Xu, H., Saenko, K.: Ask, attend and answer: exploring question-guided spatial attention for visual question answering. In: ECCV (2016)
18. Yang, Z., He, X., Gao, J., Deng, L., Smola, A.: Stacked attention networks for image question answering. In: CVPR (2016)
19. Xiong, C., Merity, S., Socher, R.: Dynamic memory networks for visual and textual question answering. In: ICML (2016)
20. Mei, H., Bansal, M., Walter, M. R.: Listen, attend, and walk: neural mapping of navigational instructions to action sequences. In: AAAI (2016)
21. Lin, T-Y., Maire, M., Belongie, S., Bourdev, L., Girshick, R., Hays, J., Perona, P., Ramanan, D., Zitnick, C. L., Dollar, P.: Microsoft COCO: Common Objects in Context. arXiv:1405. 0312 (2015)
22. Fukui, A., Huk Park, D., Yang, D., Rohrbach, A., Darrell, T., Rohrbach, M.: Multimodal compact bilinear pooling for visual question answering and visual grounding. In: EMNLP (2016)
23. Tapaswi, M., Zhu, Y., Stiefelhagen, R., Torralba, A., Urtasun, R., Fidler. S.: MovieQA: understanding stories in movies through question-answering. In: CVPR (2016)
24. Tu, K., Meng, M., Lee, M.W., Choe, T.E., Zhu, S.C.: Joint video and text parsing for understanding events and answering queries. IEEE Multimedia **21**(2), 42–70 (2014)
25. Zitnick, L., Agrawal, A., Antol, S., Mitchell, M., Batra, D., Parikh, D.: Measuring machine intelligence through visual question answering. AI Mag. (2016)

26. Ma, L., Lu, Z., Li, H.: Learning to answer questions from image using convolutional neural network. In: AAAI (2016)
27. Vinyals, O., Toshev, A., Bengio, S., Erhan, D.: Show and tell: a neural image caption generator. In: CVPR, pp. 3156–3164 (2015)
28. Xu, K., Ba, J., Kiros, R., Courville, A., Salakhutdinov, R., Zemel, R., Bengio, Y.: Show, attend and tell: neural image caption generation with visual attention. arXiv preprint arXiv: 1502.03044 (2015)
29. Berkeley Design Technology: A Test Drive of the NVIDIA Jetson TX1 Developer Kit for Deep Learning and Computer Vision Applications. https://www.bdti.com/MyBDTI/pubs: Nvidia_JetsonTX1_Kit.pdf. Accessed 06 Nov 2017
30. Karpathy, A., Fei-Fei, L.: Deep visual-semantic alignments for generating image descriptions. In: CVPR, pp. 3128–3137 (2015)
31. Simonyan, K., Zisserman, A.: Very deep convolutional networks for large-scale image recognition. In: ICLR (2014)
32. Noh, H., Hongsuck Seo, P., Han, B.: Image question answering using convolutional neural network with dynamic parameter prediction. In: CVPR (2016)
33. Li, R., Jia, J.: Visual question answering with Question Representation Update (QRU). In: NIPS (2016)
34. Ren, M., Kiros, R., Zemel, R.: Exploring models and data for image question answering. In: NIPS (2015)

Data Interpretation Using Mobile Uncalibrated Sensor for Urban Environmental Monitoring

Elson Agastra(✉), Bexhet Kamo, Ilir Shinko, and Renalda Kushe

Department of Electronic and Telecommunications, Polytechnic University of Tirana, Sheshi Nene Tereza 1, 1004 Tirana, Albania
{eagastra,bkamo,ishinko,rkushe}@fti.edu.al

Abstract. Intelligent Transportations Systems (ITS) are complex information and communication technologies platforms aimed for a better and cost effectiveness public transport organization. Our aim is to extend the basic of ITS platform for real time environmental data monitoring and processing. Mobile environmental sensing is becoming one of the best options to monitor our environment due to its high flexibility. In this context, the proposed system utilizes public transportation vehicle intelligence and sensing to monitor a set of environmental parameters over a large area by "filling in the gaps" where people go but environmental monitoring sensor infrastructure has not yet been installed. In this article, we show as well how to extract useful data from low cost and uncalibrated environmental sensor nodes. The presented architecture is focused on low cost and highly scalable architecture in term of type and number of sensors that can be installed and processed. This work presents some data from a testing deployed infrastructure in Albania and the measured data indicate the usefulness of the proposed architecture.

1 Introduction

In the modern and highly dynamic metropolitan cities, the authorities need to monitor a very wide phenomenon (environmental data, traffic congestions, road status, air quality and so on). Based on these data, a difficult decision making task can occur. Recently, air pollution monitoring has gained worldwide relevance due to the influence of air quality on our lives [1]. Traditionally, environmental monitoring is conducted by official authorities which usually have high quality but expensive monitoring stations which require also high cost on calibration instruments and maintenance. Besides of its highly cost, the actual environmental monitoring situation leads to often low spatial and temporal resolution data to meet the information demands from the public and organizations. In Tirana municipality area of 1110 km^2, and administrative area of 41 km^2, since 2012 are present only seven air quality monitoring stations operated by the National Institute for Public Health [2].

In this context, the proposed system utilizes public transportation vehicle intelligence and sensing to monitor a set of environmental parameters over a large area by "filling in the gaps" where people go but environmental monitoring sensor infrastructure has not yet been installed.

L. Barolli et al. (Eds.): EIDWT 2018, LNDECT 17, pp. 597–605, 2018.
https://doi.org/10.1007/978-3-319-75928-9_53

A key challenge is combining existing ITS technologies and state-of-the-art, sensor and positioning technologies, data fusion, traveler behavior, traffic modelling and emissions dispersion modelling techniques and vehicle/person-mounted sensors.

While some sensors are already commonly present in ITS platform (e.g., geolocation, gyroscopic, etc.) [3, 4], other types of compact, low-power sensors (e.g., air quality) are not yet commonly included even though they offer the ability to collect additional data of individual and social interest.

The present work shows the capability of using environmental sensors over public transport vehicles and cooperation of the ITS infrastructure for getting geo-located sensing data.

2 Overall System Architecture

The proposed ICT architecture employs bidirectional data communication distributed structure. Local collected information by environmental sensors is preprocessed by a local server that acts as gateway and upload them on-site via 3G network. The sensor node core is composed by Atmega 1281 microcontroller (8 bit AVR RISC architecture) equipped with 8 kB SRAM, 128 kB flash memory and 4 kB EEPPROM. Also the sensor node has the capabilities of over the air programming which permits a better and easier software development or system upgrade also when the system is running. The sensor nodes used are populated with carbon monoxide (CO), Liquefied Petroleum Gas (LPG) sensors as well as temperature and relative humidity sensors. Avoiding cabling and power issues, sensors nodes are powered by local battery and communicate wirelessly to the gateway. Local gateway is low cost and low power general-purpose computer with GNU/Linux installed with 3G modem network capabilities and global positioning system (GPS) device connected. This part of the architecture can be the main part of the basic of ITS infrastructure.

The client side web interface is based on RESTful web services which are REST architecture based web services. RESTful web services are lightweight, highly scalable and maintainable and are very commonly used to create APIs for web based applications. We use Slim, a PHP micro framework that helps to quickly write simple yet powerful web applications and APIs. In Fig. 1 is presented the flowing of information data from the ITS mobile architecture to the online database.

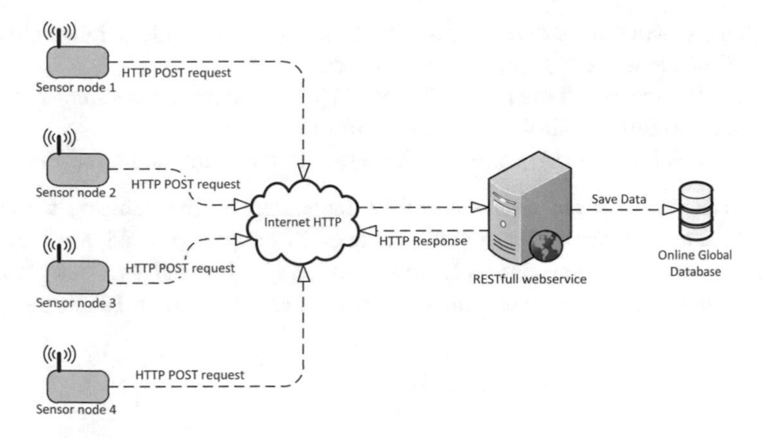

Fig. 1. Devices calling the web service to save sensors data on the online database.

To view and process data saved on the online database, a PHP web site was created. In this context, we can filter sensors data based on time, device and sensor. The filtered information is then shown in Google Maps using the Heat-Map layer (see Fig. 2). The geolocation information is derived from the GPS position of the ITS infrastructure.

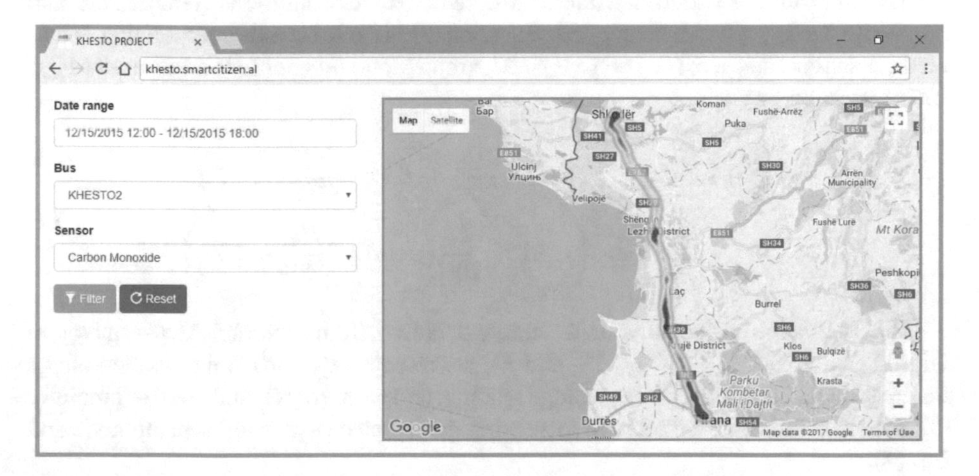

Fig. 2. Web application for post-processing and data visualization at georeferenced location: CO data presented.

3 Data Processing Module

Due to the typology of sensor used and the mobility of the measuring equipment, the obtain data need to be pre-elaborated before we can use them for environmental quality purposes. In this case we can estimate three major issues that need to be analyzed for performing accurate data analysis.

i. First, the sensor output data measurements are highly variable in range close to the real value due to the typology of sensor used.
ii. Second, the measured data are influenced by the instant temperature and relative humidity, so this variability should be removed.
iii. Third, the vehicle speed influences the sensor temperature and its reading.

To reduce the captured data variability depending on the sensors, we calculate directly at the sensing node, the average value of 25 samples ($n = 25$) with an interval of 10 ms between each consecutive sample as in Eq. (1). This corresponds of taking 25 samples in about 7 m distance computed by a moving vehicle with 100 km/h speed.

$$R_s = \frac{\sum_{i=1}^{n} R_i}{n} \tag{1}$$

In this equation, the R_i represents the instantaneous value captured from the sensor and R_s is the estimaded sensor reading. The R_s value is recorded to the web database. The Eq. (1) is general and R symbol will be substituted by the temperature, humidity, CO and LPG value or any symbol representing the sensors installed to the node. Using Eq. (1) improves the SNR (Signal to Noise Ratio) and removes uncorrelated noise affecting the readings.

To eliminate the variability due to different weather conditions (temperature and humidity), the Eq. (2) will be applied. Equation (2) is derived from interpolating manufacturer sensor datasheet (Winsen MQ-7B Carbon Monoxide Sensor and Winsen MQ-5 LPG Detection Sensor).

$$\begin{aligned} \tilde{V}_{co_s} &= V_{co_s} - 1.7 \cdot \log_{10}\left(-0.15 \cdot \frac{T_{\circ C}}{20^{\circ}C} - 0.12 \cdot \frac{H_{\%}}{55\%} + 1.27\right) \\ \tilde{V}_{LPG_s} &= V_{LPG_s} - 1.7 \cdot \log_{10}\left(-0.15 \cdot \frac{T_{\circ C}}{20^{\circ}C} - 0.12 \cdot \frac{H_{\%}}{55\%} + 1.27\right) \end{aligned} \tag{2}$$

Where $\tilde{V}$ is the corrected voltage sample data from 10 bit internal ADC (Analog to Digital Converter). V_{COs}, V_{LPGs}, $T_{\circ C}$ and $H_{\%}$ are respectively carbon monoxide voltage reading, liquefied petroleum gas voltage reading, temperature (°C) and relative humidity (%) stored at the database relative to sensing geo-located data using sensing node and Eq. (1).

The third variability of vehicle speed that needs to be analyzed, mostly influences the temperature and relative humidity of the sensing node. Sensors are placed at rooftop of the vehicle allowing during its trip a direct air flux to the sensors without direct impact of raining or aero-dynamic air flow modification do to vehicle speed. So, with the correct positioning of the sensors in the vehicle, the remaining factor to be corrected is only temperature and relative humidity as already analyzed by Eq. (2).

For referring the readings to the most common measuring unit for environmental gases, *PPM* (parts-per-million) the Eq. (3) is used. Similarly, as for deriving Eq. (2), the interpolating of manufacturer sensor datasheet is used in this case.

$$\begin{aligned} PPM_{CO_s} &= 3.027 \cdot \exp\left(1.0698 \cdot \tilde{V}_{CO_s}\right) \\ PPM_{LPG_s} &= 26.572 \cdot \exp\left(1.2894 \cdot \tilde{V}_{LPG_s}\right) \end{aligned} \tag{3}$$

In our test environment, we are not able to accurately reproduce the standard reference required by EU directive [5, 6] as our sensors are not calibrated by a reference authority. Also, due to the fabricated tolerance of the four sensor nodes realized for this project at our lab, the sensor outputs differ from each other of ±10% to the same environment at the same time. For this purpose, we will not present the data in *PPM* as by Eq. (3), but in relative reference (percentage %) to the average value of all readings for each vehicle trip as in Eq. (4) where *N* is the total sampled data during one vehicle trip.

$$\begin{aligned} Co_{\%} &= \frac{PPM_{CO_s}}{\frac{1}{N}\sum PPM_{CO_s}} \cdot 100\% \\ LPG_{\%} &= \frac{PPM_{LPG_s}}{\frac{1}{N}\sum PPM_{LPG_s}} \cdot 100\% \end{aligned} \tag{4}$$

Equations from (2) to (4) are applied in post-processing data analysis of the results stored at the online database of Eq. (1) values.

This strategy of presenting the results, cannot lead to misalignment data from different sensors and the official environmental sensor nodes. With this representation, we have removed the reference value, but not its trend and relative pollution concentration in different locations analyzed with the same sensor node. Anyway, the results are also very useful for analyzing and comparing different locations as will be shown in the next section.

4 Experimental Results

The proposed strategy of environmental monitoring is deployed in two different public bus lines. Two sensor nodes are installed in two different buses witch travels four time per day from Tirana city (TR) to Durres city (DR) in Albania and vice versa. Other two nodes are installed in respective two different buses operating two times per day from Tirana city to Shkoder city (SH), Albania and vice versa.

The data collected by the implemented sensors and installed in one of the bus lines TR-SH on December 15, 2015 are shown in Fig. 2. The data are displayed as overlay heat-map. The heat-map intensity is relative to data analysis with Eq. (4) as shown in the previous paragraph. This visual representation of the data is also used by authors in their previous work [7] for presenting measured RF signal quality for mobile telephony operators.

Analyzing the results presented in Fig. 2, some areas have stronger color intensity (red color) respect to areas with a lighter intensity. This means a higher CO concentration rather to other locations. This is only a first visual impact of the collected data. From the web interface developed for this project, we have the possibility to filter data by sensor, bus line and time.

A more useful data analysis is presented in Fig. 3. In this figure, data analyzed for the bus line TR-SH from December 15th, 2015 to January 15th, 2016 are presented. In Fig. 3a and b respectively are shown the CO and LPG concentration in percentage as by Eq. (4). The data are corrected by Eq. (2) using the information provided in Fig. 3c and d for temperature and relative humidity.

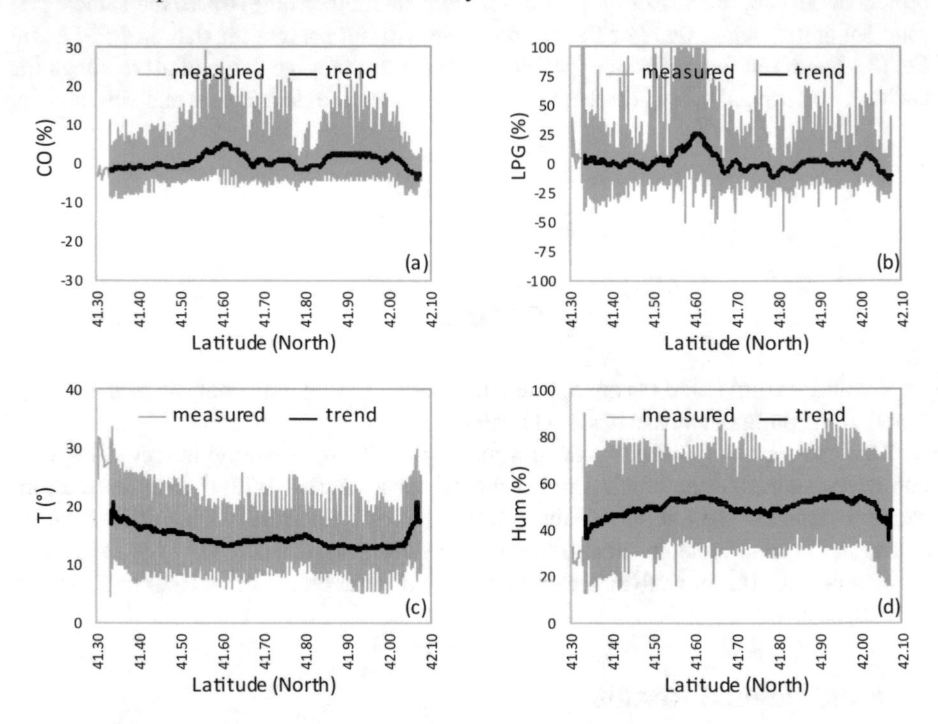

Fig. 3. Tirana (≈41.3 Latitude) – Shkodra (≈42.1 Latitude) measured environmental data from mobile sensor. Spatial variation Dec. 15–Jan. 16.

Analyzing the variability of the temperature and relative humidity, the results are in tendency with what we expect. In urban areas, we expect to have higher temperature rather than rural areas and our deployed sensor confirms this expectation for higher temperature in Tirana and Shkoder compared to the rural areas between these two cities (Fig. 3c).

For referring the elaborated data to their physical location, Fig. 4 presents a better output. In this case, as it was expected, we can refer the higher (%) of CO and LPG concentration to a physical location. In this case, Fig. 4, the area between Kruja and Laç have higher CO and LPG concentration.

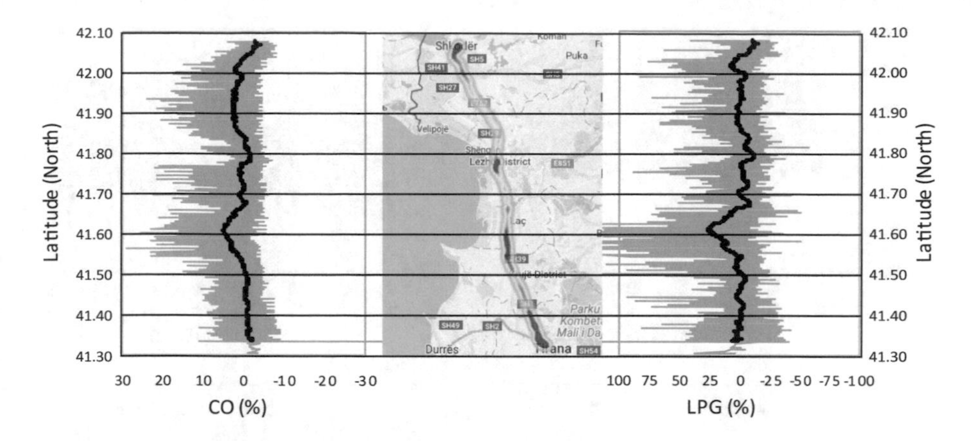

Fig. 4. CO and LPG statistical distribution over the analyzed path

Is not our purposes to find why in this areas there is higher CO and LPG gas concentration. This kind of analysis requires a deeper knowledge of the physical factors that contributes to CO and LPG air pollutants as traffic and road conditions, industrial area nearby, weather and wind flux ecc. Our purpose is to find an architecture able to constantly monitor a large area and also to help authorities for deeper knowledge of the air quality.

The same data analysis is also used for data collected from bus lines TR-DR and presented in Fig. 5. In this analysis, no hot-spots are present.

The data results presented in Figs. 3, 4 and 5 could not refer in a well-defined standard [5, 6], but we are able to evaluate areas that statistically have higher concentration of air pollution. This result can be useful for decision making official authority in analyzing the motivation of this higher concentration and takes actions to reduce this difference as suggested in [8–11]. Also, the proposed architecture allows decision making authority to evaluate in real time the effect of their environmental policies and their impact to the monitored area.

The proposed architecture permits different data analysis. We can make hourly, daily or monthly analysis for evaluating the gas concentration and correlate this data to other information's as road traffic concentration or industrial area influence or any other information that can be useful for environmental decision-making policies. This data representing interface, permits an immediate and also partial technical view of the data to the decision making for evaluating their strategy for reducing pollutions based in national and international air quality directives as in [5].

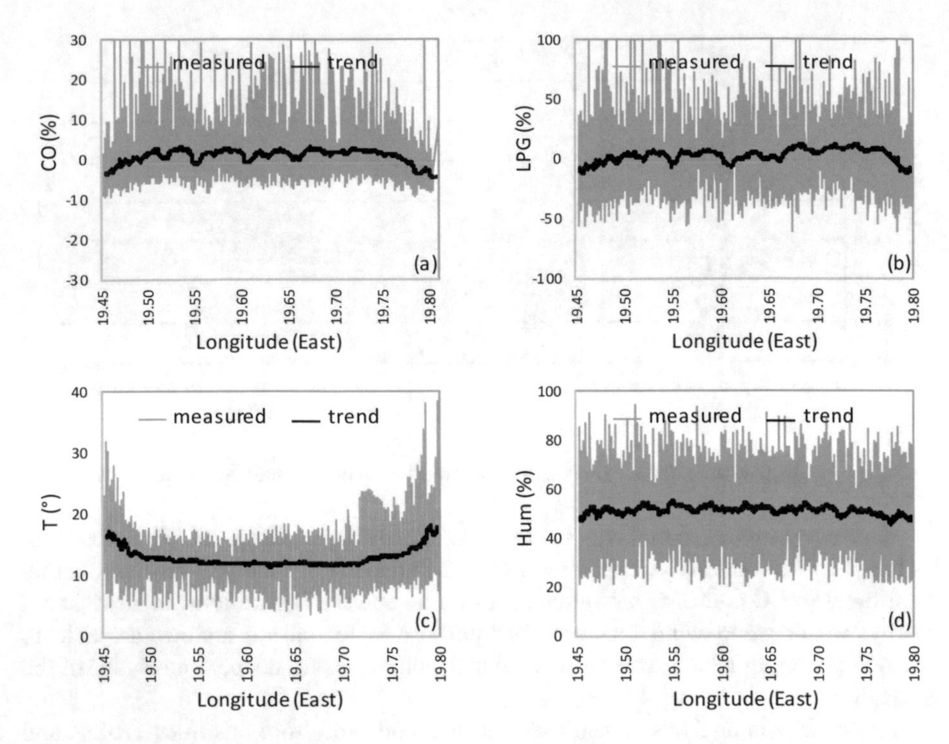

Fig. 5. Tirana (≈19.8 Longitude) – Durres (≈19.45 Longitude) measured environmental data from mobile sensor. Spatial variation Dec. 15–Jan. 16.

5 Conclusions and Future Work

The presented project demonstrated the practical application of a new system for sustainable traffic management that installs a network of multi-sensors placed at bus lines which may help to acquire early actions in order to manage the urban air pollution and ensure sustainable mobility in a city.

The sensor nodes implemented can be easily installed in an existing ITS infrastructure based to their low-cost sensor nodes, permitting a qualitative real time environmental monitoring.

It enables the local authorities to calculate the effects of various scenarios of traffic regulation and can compare the impact of the selected scenario on pollution at 'hotspots' with new data collected by the same measurement instruments. The pollution results enable traffic control measures to be fine tuned in real time.

References

1. Chen, T.-M., Gokhale, J., Shofer, S., Kuschner, W.G.: Outdoor air pollution: particulate matter health effects. Am. J. Med. Sci. **333**(4), 235–243 (2007)
2. ISHP: Annual air quality report, Tirana 2012 [original title: Raporti vjetor i cilesise se ajrit, Tirana 2012]. Institute of public health, Tirana, Albania (2012)
3. Xhafka, E., Teta, J., Agastra, E.: Mobile environmental sensing and sustainable public transportation using ICT tools. Acta Physica Polonica A **128**(2B), 122–124 (2015). ISSN 0587-4246
4. Agastra, E., et al.: Vehicle-to-Infrastructure communication using 3G/4G network for better public transport. In: 6th International Conference in Information Systems and Technology Innovations, Tirana, 05–06 June 2015
5. Ambient air quality and cleaner air for Europe. Directive 2008/50/EC of the European Parliament and of the Council of 21 May 2008
6. Quantitative Risk and Exposure Assessment for Carbon Monoxide – Amended. U.S. Environmental Protection Agency, EPA-452/R-10-009, July 2010
7. Agastra, E., Lala, A., Kamo, B., Kina, B.: Low cost QoS and RF power measuring tool for 2G/3G mobile network using Android OS platform. In: IEEE, 11th International Wireless Communications and Mobile Computing Conference (IWCMC 2015), Dubrovnik, Croatia, 24–28 August 2015, pp. 1324–1328 (2015). ISBN 978-1-4799-5343-1
8. Alvear, O., et al.: An architecture offering mobile pollution sensing with high spatial resolution. J. Sens. **2016**, Article ID 1458147 (2016)
9. Zoysa, K.D., Keppitiyagama, C., Seneviratne, G.P., Shihan, W.W.A.T.: A public transport system based sensor network for road surface condition monitoring. In: Proceedings of the 2007 Workshop on Networked Systems for Developing Regions, NSDR 2007, 27 August 2007, Kyoto, Japan (2007)
10. Kenta, I., Go, H., Yoshikazu, A., Yoshitaka, S.: A road condition monitoring system using various sensor data in vehicle-to-vehicle communication environment. Int. J. Space-Based Situated Comput. **6**(1), 21–30 (2016)
11. Riyadh, A., Sritrusta, S., Dadet, P., Nobuo, F.: Classification extension based on IoT-big data analytic for smart environment monitoring and analytic in real-time system. Int. J. Space-Based Situated Comput. **7**(2), 82–93 (2017)

Reducing Excess Traffic in Urban Areas with Microscopic Traffic Modeling in SUMO

Alban Rakipi(✉), Joana Jorgji(✉), and Desar Shahu(✉)

Polytechnic University of Tirana, Tirana, Albania
{arakipi,jjorgji,dshahu}@fti.edu.al

Abstract. Increased vehicular traffic in dense urban areas is an issue of major concern. With more cars hitting the road, the inherent results are higher pollution levels and car accidents, shrinking of parking areas and severe traffic congestion. Attempts must be made to encourage the use of fewer cars. Switching to public transport and walking suggests an improvement. In this paper, we present a method to estimate the rate of daily vehicle usage in the city of Tirana. With the use of SUMO (Simulator of Urban Mobility) micro-traffic simulator we simulate the city traffic and evaluate the average traveling distances. Given a distance/time threshold, destinations within less-than-threshold range are considered of ease of reach. The existence of a direct bus line and by-foot traveling time are considered when setting the threshold value. Tracking data are used to provide average daily driving information. Statistic results demonstrate the effectiveness of the proposed method.

1 Introduction

The inhabitants of urban areas can travel within the city for both work and leisure. Because of the recent growth in the surface area and population of Tirana, due to administrative changes, the city road planning has become a major issue. The contributing historical reasons mainly concerned to the unsupervised development of the city, give rise to complex problems like narrow roads, lack of parking areas, lack of cycle and bus lanes and the non-existence of other public transport means like metros, trains and trams. Traffic congestion is estimated having the most impact in the daily life of the citizens. This can be measured by the time spent on the road to travel relatively short paths [1]. Furthermore, traffic congestion leads to increased CO_2 emissions and in addition, resulting in economic costs of the increased fuel consumption, allied with the rising crude prices in the last decade.

To identify the critical problems of the city's road network, the urban traffic is generated by microscopic simulation using SUMO Simulator. We focus on the use of a simulator to interpret and simulate the city traffic by random movement of the vehicles in the road network of the city of Tirana. The use of a simulator provides a good method of road planning to serve as a testbed for future realistic traffic measurements using field collected data.

The remainder of this paper is organized as follows: Sect. 2 illustrates the literature review of the related work, Sect. 3 presents the subject being addressed in this paper, regarding to traffic simulations, Sect. 4 details the proposed method. Section 5

L. Barolli et al. (Eds.): EIDWT 2018, LNDECT 17, pp. 606–614, 2018.
https://doi.org/10.1007/978-3-319-75928-9_54

pinpoints and discusses the obtained results. Section 6 presents final remarks about the approach and discusses future work.

2 Related Works

While there are several methods to estimate and measure vehicular traffic in urban areas [11–15], no one of them has been yet deployed in Albania. The existing traffic data is almost non-existent and non-accurate. These methods require either high implementation cost of road sensors or extended research in the field. Under these circumstances, alternative ways have to be considered. The use of simulators makes it possible to access, analyze and predict the effects of various changes to the environment without altering the real world.

The simulation type that is the most adequate for this project is the microscopic traffic simulation that represents the process of creating a model of a certain network, providing a cheap and safe way to predict the effectiveness of the proposed modifications [9]. Using GNSS systems along with open-source simulators to estimate the state of the traffic is both time and cost efficient, compared to other approaches that require the implementation of a certain infrastructure. Although the recent map updates of the cities in Albania are resulting in more detailed and accurate data, many features of the sophisticated GPS Navigators and online digital maps such as public transport mode and traffic visualization are disabled in the case of Albania. The lack of traffic information limits the practical implementation of microscopic simulations in a realistic fashion, however it offers a testbed to be further extended with future researches.

As a result, our work differs from [11–15] in the fact that it is innovative in Albania and it combines various approaches such as the use of the GNSS system, traffic simulators and a method of transferring to other slower modes of transportation in order to reduce the recent dominance of cars in urban areas.

3 Urban Traffic Modelling

3.1 Simulation Environment

In this work, the microscopic road traffic simulation package "Simulation of Urban MObility" (SUMO) [2] is employed to model the urban traffic on arterial roads. SUMO is an open source tool and a microscopic road traffic simulation package that supports different types of transportation vehicles, each having its own route and moving individually through the network.

There are three main modules in the SUMO package:

- SUMO, reads the input information, processes the simulation, gathers results and produces output files. It uses a graphical interface called SUMO-GUI;
- NETCONVERT, imports digital road networks from different sources and converts them into the SUMO-format. It is responsible for creating traffic light phases;
- DFROUTER generates random routes and emits vehicles into networks.

SUMO road networks can be generated by importing a digital road map. The road network importer "N*etconvert*" allows to read common formats, such as Open Street Map [3].

3.2 Mode Transfer Procedure

The aim of this work is to determine the possible reduction extent of car usage in the urban area of Tirana. Using a simulation model that assigns «car tours» to alternative transport modes, it is intended to define the potential changes in the choice of mode regarding the time performance of the different transport modes. The following procedure is considered [1]:

- IF the travel-time budget is below the threshold that was set a priori THEN the transfer is possible and, the travel-time budget is changed accordingly;
- IF the travel-time budget exceeds the threshold, THEN the transfer fails for all trips of the car tour.

For the transferring procedure, two different scenarios are used: the usual traffic and heavy traffic conditions (with high traffic congestions). A map of the central urban area of Tirana is considered to estimate the average distance/time travel in the arterial roads starting from Scanderbeg's Square with destinations to the most frequented work and leisure places, using all the basic three means of transportation (driving, walking, public transport). With the help of Google Maps [4], the travelling information for the normal traffic conditions is provided, as given in Table 1.

Table 1. Average distance and time travels on arterial roads.

Destination	Distance (km)	Driving (min)	Bus (min)	Walking (min)
Mother Teresa Hospital, Rruga e Dibrës	3	11	–	28
Train Station	1.6	9	–	20
Karl Topia Square	1.4	6	–	17
21 Dhjetori	1.1	5	–	13
PUT	1.3	6	–	17
Ali Demi	1.8	8	–	23
Zoological Park	4.8	11	–	61
European University of Tirana	1.9	8	–	24
Qyteti Studenti	2.1	10	–	27
Average	**2.1**	**8.2**	**14.6**	**25.5**

Considering that no information could be provided from Google Maps in the case of using public transport, the average time in minutes is calculated using the following formula [5]:

$$Total\,travel\,time = \left(\frac{l}{v}\right) + \left(\frac{n_{stop}}{t_{stop}}\right) \cdot \left(\frac{1\,\text{h}}{3600\,\text{s}}\right) \tag{1}$$

where l is the average distance (2.1 km), v the average travelling speed (10 km/hr), n_{stop} the number of bus stops and t_{stop} the time a bus dwells in a station. The average time value obtained from Eq. (1) is 14.6 min (rounded up to the nearest minutes).

When calculating the total travel time, it is assumed an average distance between two bus stops of around 500 m and the average dwell time at a stop about 30 s [6] (there are around 4 stops in 2.1 km).

The transferring procedure from driving to walking and/or public transport takes into account the following assumptions:

- Public transport is available in all the arterial roads considered.
- Each road has a bus lane.
- The daily walking time of around 25.5 min to a one-way destination is acceptable for the average person and the transfer of the transport mode within this threshold is possible.

Using the information given in Table 1, the travel time threshold is set at 8.2 min of driving, under the usual traffic. For a 1-h long testing this is equal to 14% of the total travelling time.

For the second scenario, a threshold value of three times higher or around 25 min of driving is proposed, which makes 41.7% of the total amount of testing time. The setting of the threshold is based on people's daily experience of commute during the rush hours in the city of Tirana. Using the assumption that all roads include a bus lane, the time-travel budget of public transport remains unchanged even under heavy traffic conditions (Table 2).

Table 2. Average travelling time by bus.

Destination	Distance (km)	Driving (min)	Bus (min)	Walking (min)
Average	2.1	25	14.6	25.5

It is thus clear that, while the cost of walking or using public transport does not change, the cost of driving increases in terms of time spent on the road, fuel consumption, environmental pollution, parking areas and emotional stress. As a result, transfer to other slower, yet efficient modes of transportation, is reasonable.

4 Proposed Method

To run vehicular traffic tests in Tirana, it is firstly needed to obtain a map of the city, compatible with SUMO. The SUMO module Netconvert offers a way to import digital road networks from various sources. By accessing the OpenStreetMap website it is possible to search and easily export the city's map data into an XML file.

As shown previously in Fig. 1(a), the road network of downtown area in Tirana has been chosen as a simulation case study. The OpenStreetMap (OSM) XML file is firstly edited in Java OpenStreetMap Editor (JOSM) [7] in order to remove all the road edges which cannot be used by vehicles such as railway, roadways, and pedestrian.

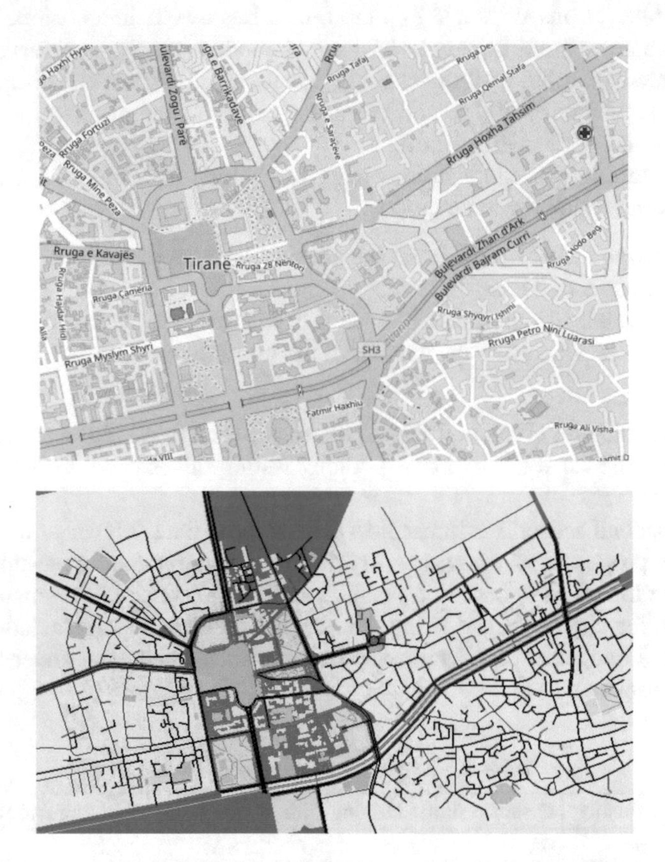

Fig. 1. Example conversion of an OpenStreetMap; (above) original OpenStreetMap view, (below) network imported into SUMO.

With the road network ready, the next step is to generate the correspondent traffic demand. Random routes of vehicles traveling on the road network are generated for a specific time interval using the DUAROUTER. Since no previous work of realistic traffic information was available the results of this paper are obtained from simulating random vehicular traffic generated while running the SUMO simulator. Having created the necessary output files, the number of cars to be reduced can be easily calculated, based on the previously set threshold for each of the above scenarios. While the results are only theoretical in this context, the proposed method gives a good approach to estimate at which extent car tours can be reduced daily in urban areas. To provide "ground truth" data, realistic traffic information has to be collected from prior field testing of the traffic in the urban area. This would serve as the input file for the SUMO simulator to generate traffic according to realistic field conditions.

The SUMO simulator also offers a way to measure other metrics such as fuel consumption or pollutant emission, based on the Handbook of Emission Factors for Road Transport (HBEFA) [8] database. It provides emission factors per traffic activity and measurements of CO_2 emissions and fuel consumption for various vehicle categories. This data is used to estimate the average reduction in the cost of travel when transferring to other modes of transportation other than driving.

5 Results and Discussion

The aim of this work is to study the vehicular traffic within the city, as shown in Fig. 2. The SUMO simulation is conducted for 1-h without incidents. As a result of the SUMO simulation, two useful data-sets were generated for further analysis. One is the location information of every vehicle called "vehroute output". It records at every timestamp the location of every vehicle in the simulated road network. Each record consists of a vehicle ID, time duration, depart and arrival time, lane ID and the vehicle's coordinates. The other one is the metrics information of fuel consumption and pollutant emissions called "emission output". It includes information such as vehicle ID, road edge ID, CO_2 emissions and fuel consumption of each vehicle.

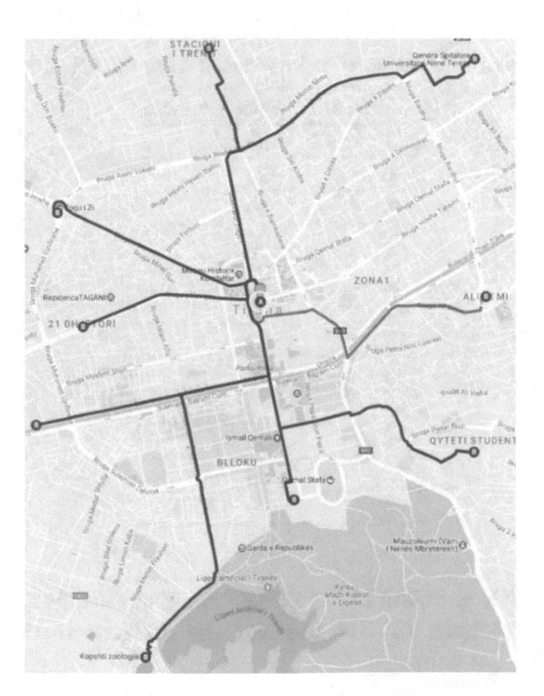

Fig. 2. The arterial roads of the central urban area of Tirana.

In SUMO, route definitions are as follows [9]:
Route Definition code in SUMO

```
<vehicle id = "0" type = " type1 " depart = "0" color =
"1 ,0 ,0">
<route edges = " id beginning id middle id end "/>
</ vehicle>
```

This means that each trip is defined as a collection of the ID of the edges that comprise that trip. During a 3600-second-interval, 100 vehicles have been generated with random trips.

Before running the simulation, the vehicle type "bus" was excluded from the network (only private cars are considered). The obtained results for both scenarios are described below.

5.1 Scenario 1: Normal/Average Traffic

Given the threshold of 14% of the total simulation time, 12 in 100 cars stayed in the network for less than the specified time duration. As a result, the cars travelling in the city of Tirana during the usual traffic are reduced to an extent of 12%. The results are given in Table 3.

Table 3. Vehicles to be reduced from the network.

Vehicle ID	Duration (sec)
1	100
0	200
34	200
58	100
6	300
2	300
17	300
15	500
92	300
85	100
76	200
95	200

The amount of CO_2 emission in the SUMO simulator is averaged to 0.24 mg/s per vehicle. Thus, by reducing 12% of the cars the decrease in the CO_2 emission is described by the Eq. (2) below:

$$CO_2\_emission = \sum_{t=0}^{t_{thresh}} 0.24 \cdot t \cdot n_t \quad (2)$$

Where t_{thresh} is the threshold value, t the average time each vehicle stays in the network and n_t the number of cars spending the same amount of time t in the road. Using Eq. (2) it is calculated a decrease of around 624 mg in the amount of CO_2 emission.

5.2 Scenario 2: Heavy Traffic

To simulate heavy traffic conditions the car delay is increased to its maximum value (1000 s). When the threshold value is set to 41.7% of the total amount of time, 47 out of 100 cars could be eliminated from the network or an impressive 47% reduce in the urban traffic. Similar to the previous scenario the amount of CO_2 emissions decreases by 8.688 g.

It must be highlighted that these values are only theoretical ones generated by the simulator. In practice, the average passenger vehicle emits about 411 g of CO2 per mile. Every gallon of gasoline creates about 8,887 g of CO2 when burned [10].

6 Conclusion

In this paper, a method of reducing the car traffic in urban areas was presented. It is described the representation of Tirana's arterial road network within the SUMO simulation environment. Initial data is imported from Open Street Maps and the issues raised from this process were analyzed. To test the simulation environment random traffic data generated by SUMO is employed. Although it lacks real traffic data gathered from the field tests in the city, the method potentially solves the problem, giving a theoretical approach to decreasing urban traffic by transferring to other potential modes of transportation such as walking or public transport when possible. Overall, the results indicate that SUMO was capable of simulating the Tirana urban traffic, thus providing useful results in the context.

Two different scenarios were considered, the usual traffic and heavy traffic conditions. It is found that, for the first scenario it was possible to eliminate 12% of the cars from the road network, resulting in a decrease of 0.624 g of CO_2 emissions. As for the second scenario, an impressive 47% of the cars could be reduced, resulting in 8.688 g less CO_2 emission. These results are obtained under the assumption that all arterial roads include a bus lane, thus the time-budget of using public transport does not change while the cost of driving increases significantly, in terms of time, pollution and other metrics such as fuel consumption and shrinking of parking areas.

Future work targets the simulation of Tirana's traffic in a realistic fashion, with field test data to generate "ground truth" traffic information, that employs the proposed urban traffic reduction method. Furthermore, the simulated network can be extended to cover a larger area of the city, apart from the arterial roads considered in this work. The system user can be alerted in case of low-efficiency use of their car, which if applicable, will result in a reduced number of cars daily and a cut in the overall inherent costs.

References

1. Massot, M.-H., Armoogum, J.: Speed and car traffic regulation in urban areas – the case of Paris. IATSS Res. **27**(2), 46–55 (2003)
2. Krajzewicz, D., Hertkorn, G., Wagner, P., Rössel, C.: SUMO (Simulation of Urban MObility) an open-source traffic simulation. In: 4th Middle East Symposium on Simulation and Modelling, pp. 1–4. SUMO. http://sumo.sourceforge.net/
3. OpenStreetMap. http://www.openstreetmap.org/
4. https://www.google.com/maps
5. Daganzo, C.F.: Public Transportation Systems: Basic Principles of System Design, Operations Planning and Real-Time Control. Institute of Transportation Studies University of California, Berkeley, October 2010
6. Walker, J.: Basics: The Spacing of Stops and Stations - Human Transit, 5 November 2010
7. https://josm.openstreetmap.de/
8. http://www.hbefa.net
9. Dias, J.C., Abreu, P.H., Silva, D.C., Fernades, G., Machado, P., Leitao, A.: Preparing data for urban traffic simulation using SUMO, May 2013
10. U.S. Environmental Protection Agency (EPA), Office of Transportation and Air Quality: Greenhouse Gas Emissions from a Typical Passenger Vehicle, May 2004
11. Leduc, G.: Road Traffic Data: Collection Methods and Applications. Working Papers on Energy, Transport and Climate Change, no. 1 JRC 47967 (2008)
12. Ferman, M., Blumenfeld, D., Dai, X.: An analytical evaluation of a real-time traffic information system using probe vehicles. J. Intell. Transp. Syst. **9**(1), 23–34 (2005)
13. Cheung, S.Y., Coleri, S., Dundar, B., Ganesh, S., Tan, C.-W., Varaiya, P.: Traffic measurement and vehicle classification with a single magnetic sensor California PATH Working Paper UCB-ITS-PWP-2004-7 California Partners for Advanced Transit and Highways, University of California, Berkeley. IEEE (2011)
14. Tao, S., Manolopoulos, V., Rodriguez, S., Rusu, A.: Real-Time Urban Traffic State estimation with A-GPS mobile phones as probes. J. Transp. Technol. **2**(01), 22–31 (2012)
15. Toral, S.L., Gregor, D., Vargas, M., Barrero, F., Cortes, F.: Distributed urban traffic applications based on CORBA event services. Int. J. Space Based Situat. Comput. **1**(1), 86–97 (2011). https://doi.org/10.1504/IJSSC.2011.039110

An Empirical Evaluation of Sequential Pattern Mining Algorithms

Marjana Prifti Skenduli[1(✉)], Corrado Loglisci[2], Michelangelo Ceci[2], Marenglen Biba[1], and Donato Malerba[2]

[1] University of New York Tirana, Tirana, Albania
{marjanaprifti,marenglenbiba}@unyt.edu.al
[2] Universita' degli Studi di Bari, Bari, Italy
{corrado.loglisci,michelangelo.ceci,donato.malerba}@uniba.it

Abstract. Sequence mining is one of the most investigated tasks in data mining and it has been studied under several perspectives. With the rise of Big Data technologies, the perspective of efficiency becomes prominent especially when mining massive sequences. In this paper, we perform a thorough experimental evaluation of several algorithms for sequential pattern mining and we provide an analysis of the results focusing on the different algorithmic choices and how these affect the performance of each algorithm. Experiments performed on real-world and synthetic datasets highlight relevant differences between existing algorithms and provide indications for Big Data scenarios.

1 Introduction

Sequences are elements arranged according to a total or partial ordering. They are very common in many real-world scenarios. For instance, click-streams and trajectories. Click-streams represent sequences produced by the web browsing activity of a user, the web pages visited by the user represent the elements of these sequence, while the order is established by the time-stamp when a web page is visited. Trajectories are sequences of geo-referenced positions produced by the movement, while the order is established by the motion of the moving objects [8]. The order is not necessarily related to time, but also to space. Likewise, in textual documents and biological studies, sequences are series of chars or series of nucleotides whose order is based on the position that they have with respect to the other elements. In many applications, the elements denote complex entities and often represent sets of single basic elements where no order relation holds. Market basket analysis is one of these applications, where a set of items corresponds to a purchase of a customer in a mall, while a sequence represents the series of purchases made in a week. Analyzing sequences can thus become profitable and advantageous for many real-life applications and one of the technologies adopted is represented by *Sequential pattern mining* (SPM), which comprises techniques for mining sequential data and discovering interesting sub-sequences in a set of sequences [4].

L. Barolli et al. (Eds.): EIDWT 2018, LNDECT 17, pp. 615–626, 2018.
https://doi.org/10.1007/978-3-319-75928-9_55

The blueprint for sequence mining algorithms proposed in the literature is enumerating all the interesting sub-sequences in the space of all the possible sub-sequences. Typically the notion of interestingness relies on the relative frequency, which provides statistical evidence to a sequential pattern and thus provides arguments about the regularity of a sub-sequence in a database [10]. Therefore, sub-sequences that have a frequency greater than a user-defined threshold are those selected as valid sequential patterns. To discover all the frequent sub-sequences, we should explore the whole search space by means of a generate-and-test strategy that builds all possible sub-sequences and tests the frequency against the minimum threshold.

This problem has been often tackled by approaches specifically designed for selected domains (e.g., [2,9,14,18]), while others have a more general characterization (e.g., [5,6,15]. However, regardless of the domain or specific approach, most of research studies mainly focus on three specific algorithmic features and on how these can be effectively and efficiently developed: *(i)* the method to explore the search space, *(ii)* the representation of the database of sequences, and *(iii)* the generation of the sequential patterns. Although, there are several theoretical studies and surveys [4,11,12], we ascertain the lack of empirical studies. An experimental viewpoint that highlights the characteristics of the three algorithmic features may be helpful when facing the sequence mining problem in the context of Big Data scenarios, where the necessity for methods able to analyze time-ordered and unbounded data produced at high rate becomes more and more pressing.

In this paper, we investigate how those three features have been developed in representative algorithms and propose a comparative evaluation on real and synthetic sequence datasets. Our contribution is not a theoretical discussion, but it should be intended as an empirical study that complements experiments presented in papers of specific, relevant sequential pattern mining algorithms.

This paper is organized as follows. In the next section, we introduce classical definitions and necessary notions to understand the problem of SPM and existing solutions to solve it. In Sect. 3, we discuss the most representative SPM algorithms by illustrating the core algorithmic decisions behind them. Then, in Sect. 4, we present the empirical evaluation upon real and synthetic datasets, further discussing on how the three features aforementioned work. Finally, the conclusions drawn in Sect. 5 mark the closure of this paper.

2 Background and Basics

SPM has originally been formalized in [15] and the subsequent research has inherited the same formulation, which revises one of the association rules mining problem.

Let $I : \{i_1, i_2, \ldots, i_m\}$ a set of elements with nominal values, termed *items*. An itemset X is a unordered set of items such that $X \subseteq I$, the cardinality of X, $|X|$ corresponds to the number of contained items. Without loss of generality, we assume that items of an itemset are sorted in lexicographic order. An itemset

X of cardinality $|X| = k$ is said to be of length k. For example, given the set of items $I : \{a, b, c, d, e, f\}$, the set $\{a, b, c\}$ has length 3 and consists of the items a, b and c.

A *sequence* is an ordered set of itemsets $s = \langle I_1, I_2, \ldots, I_n \rangle$, such that $I_h \subseteq I$ $(1 \leq h \leq n)$. A sequence s of cardinality $|s| = n$ is said to be of length n. An example of sequence is $s = \langle \{a, b\}, \{c\}, \{f, g\}, \{g\}, \{e\} \rangle$, where it is assumed an order relation which establishes the itemset $\{a, b\}$ to precede the itemset $\{c\}$. For instance, in the scenario of the click-stream, by assuming the order relation holding on the hours, the itemset $\{a, b\}$ indicates a set of two web-pages a and b visited in the same hour, while the itemset $\{c\}$ indicates a set consisting of the sole page c visited in a subsequent hour.

A *sequence database* is a list of sequences $SDB = \langle s_1, s_2, \ldots, s_p \rangle$, each assigned to an identifier. Sequential patterns are mined by searching the occurrences of sequences in a database SDB and computing their *support*. The *support* of a sequence s_a is the number of sequences of a database SDB that contain s_a. If this value exceeds a user-defined minimum threshold of support, denoted as $minSUP$, the sequence s_a is considered *frequent* and identified as valid sequential pattern. To check if a sequence $s_b : \langle B_1, B_2, \ldots, B_m \rangle$ contains a sequence $s_a : \langle A_1, A_2, \ldots, A_n \rangle$, we search for a series of integers $1 \leq i_1 < i_2 < \ldots < i_n \leq m$ such that $A_1 \subseteq B_{i1}$, $A_2 \subseteq B_{i2}$, ..., $A_n \subseteq B_{in}$.

The exploration of the search space, common to all the sequential pattern mining algorithms, consists on mining valid sequential patterns based on the shorter sequential patterns, starting with sequential patterns of length 1. This is typically done by means of two basic operations which extend the sequences by inserting either one itemset or one item. These two operations are termed *s-extension* and *i-extension*. A sequence $s_b : \langle I_1, I_2, \ldots, I_m, \{x\} \rangle$ is a s-extension of a sequence $s_a : \langle I_1, I_2, \ldots, I_m \rangle$ if s_a is a *prefix* of s_b and the item x appears in an itemset occurring after all the itemsets of s_a. A sequence $s_a : \langle A_1, A_2, \ldots, A_n \rangle$ is a prefix of a sequence $s_b : \langle B_1, B_2, \ldots, B_m \rangle$ if $n < m$, $A_1 = B_1, A_2 = B_2, \ldots, An - 1 = Bn - 1$ and A_n is equal to the first $|A_n|$ items of B_n according to the lexicographic order on the items. A sequence $s_c : \langle I_1, I_2, \ldots, I_m \cup \{x\} \rangle$ is a i-extension of a sequence $s_a : \langle I_1, I_2, \ldots, I_m \rangle$ if s_a is a prefix of s_c, the item x is appended to the last itemset of s_a and the item x is the last one in I_m according to the lexicographic order. Next we report two examples for illustration purposes. The sequences $\langle \{a\}, \{a\} \rangle$, $\langle \{a\}, \{b\} \rangle$ and $\langle \{a\}, \{c\} \rangle$ are s-extensions of the sequence $\langle \{a\} \rangle$. The sequences $\langle \{a, b\} \rangle$ and $\langle \{a, c\} \rangle$ are i-extensions of the sequence $\langle \{a\} \rangle$.

3 Approaches for Sequential Pattern Mining

Before illustrating the comparative analysis, we present and discuss the most representative solutions for implementing the three features that characterize the algorithms of SPM. The features are *(i)* method to explore the search space, *(ii)* representation of the database of sequences, and *(iii)* generation of the sequential patterns. Although, these are not independent aspects from each other when designing an algorithm, thereby we try to inspect them separately highlighting their properties. The detailed description of specific algorithms is not subject of the current paper, so the interested reader may refer to the relative papers.

Exploration of the search space. The space of all the possible sequences, where searching the sequential pattern, can be modeled as a lattice, which is a partially-ordered set, whose elements are the patterns defined on the set of the items (I) present in the database. The space is organized by levels, the patterns of a level are prefixes of the patterns of the next level and have the same length. The patterns of a level are thus obtained as s-extensions or i-extensions of the patterns of the previous level, so their length is increased by one. The exploration of the lattice can be done in two different ways, *breadth-first* search and *depth-first* search. The breadth-first search follows a level-wise strategy, in that it visits a level of the lattice at time and processes all the patterns of a level before considering the next level. Once a level has been visited, it will not be explored again. Procedurally, the search first considers all the patterns of length 1, then it visits the next level to process all the patterns of length 2 and, subsequently, the level with the patterns of length 3. It follows this strategy until it reaches the longest patterns. The most representative algorithm implementing the breadth-first search is GSP [15].

The depth-first search explores the lattice in depth and processes the patterns of a level without necessarily completing that level. When it reaches the leaves, that is, when there are no patterns, it backtracks to the first level (patterns of length 1) and re-starts visiting the remaining patterns of the levels which it had previously visited. Procedurally, the search first considers all the patterns of length 1, then it takes one and processes one pattern of length 2 originated by the pattern of length 1. Subsequently, the pattern of length 2 is used to process one pattern of length 3 of the next level. This procedure is recursively performed until no pattern can be visited. The most representative algorithms implementing the depth-first search are PrefixSpan [13], SPADE [17] and SPAM [1].

The exploration of the lattice of the sequential patterns is costly, especially when the number of items is very large, considering that we should generate a set of patterns with magnitude order equal to 2^i from a database with i elements. To solve this problem and make the exploration efficient, the algorithms (regardless of the space search technique) have implemented pruning techniques aiming at removing sub-spaces that could contain uninteresting patterns. The most used technique relies on the *anti-monotonicity* property of the support, according to which we can avoid to generate the sequence s_a if there exists a sequence s_b, which is contained in s_a, whose support does not exceed the minimum threshold $minSUP$. The sub-space containing sequences longer than s_a can therefore be pruned, since those sequences will not exceed the threshold (intuitively, if a sequence s_a is not frequent, then all the sequences which contain s_a will be not frequent).

Representation of the database. The representation format of the input sequences becomes relevant when counting the number of the occurrences of a pattern in the database. There are three main solutions, (i) horizontal databases, (ii) vertical databases, (iii) projected databases. In the horizontal format, we transform the original transaction database SDB in a list of sequences ordered by identifier. For each sequence, the itemsets are sorted by relation order (e.g., time). This way, in order to count the occurrences of a pattern, we should match the itemsets of the pattern against those of a sequence of the database SDB. In case of GSP implementation, this solution requires to access the database a number of times equal to the number of input sequences, which may significantly raise the time consumption, especially in massive databases.

In the vertical format, we transform the original transaction database in a set of lists (*IDLists*), each associated to one item. A list indicates the itemsets of the input sequences where the corresponding item occurs. These structures are built by accessing only once the database, exactly at the beginning of the process when mining the frequent items. It is not necessary repeating the access operation because the number of the occurrences of the patterns of length greater than 1 is determined by joining the $IDlists$ of the frequent items. This solution is particularly effective in the algorithms that perform depth-first search, for instance SPAM, but it losses efficiency when the IDlists are very large, as typically encountered in dense databases and databases with very long sequences. A popular optimization approach is to encode IDLists as bit vectors [1].

Projected databases are subsets of the database SDB and they provide the means to reduce the search space. They are built simultaneously with the mining process and contain the only input sequences in which a pattern, which has been previously mined, occurs. More precisely, once mined the patterns of length k, for each pattern s_a we scan the database to create a (reduced) database with the only sequences in which s_a is present, while counting the occurrences. Recursively, new databases are created with the sequences in which the patterns s_b, built from the pattern s_a ($s_a \subseteq s_b$), are present. This representation allows us to work on the sequences really appearing in the database, but it has the disadvantage of repeatedly scanning the databases previously created.

Generation of frequent sequential patterns. The generation of sequential patterns is a procedure that consists of the operative steps to build the lattice of the patterns, which we explore through space search methods. The existing works differ on the use of the *generate-and-test* strategy, which is defined in terms of two main steps, (i) generation of candidate patterns and (ii) selection of the only candidates that appear in the input database and that meet the threshold $minSUP$. A candidate pattern of length k is generated by joining two frequent patterns of length $k-1$ that have $k-2$ itemsets in common. The other two items are used to reach the length k. More precisely, we first find all the frequent patterns of length 1, then we generate those of length 2 by using those of length 1. This step goes on until no longer patterns can be generated. Therefore, the patterns of the level k of the lattice cannot be built if we have not completed the level $k-1$. This explains why the generate-and-test strategy is often coupled with breadth-first search, as in the GSP algorithm [15]. The generate-and-test strategy has two main limitations. First, it may generate candidates which do not appear in the database. In fact, they are derived from the patterns present in the lattice and not from the sequences contained in the database. This may clearly affect the efficiency of the mining process. Second, it is necessary to store all the patterns of a level prior to building the candidates of the next level. In turn, this may require huge memory.

Alternative solutions have been designed in order to (i) avoid generating candidates that do not appear in the database and (ii) work on a smaller search space. There is a category of algorithms that resorts to the depth-first search in order to generate a candidate from one frequent pattern, which is taken from those previously mined [1,17]. More precisely, once mined the frequent items (length 1), the candidates of length 2 are built by appending an itemset (by means of the operations s-extension and i-extension) to one frequent item, then recursively one more itemset is added to the frequent pattern previously mined until no further itemsets can be appended. The procedure re-starts with another frequent item, with which patterns of increasing length can be built by appending one itemset at time. This kind of algorithms holds the advantage of generating patterns of length k by keeping only one pattern of length $k-1$, contrarily to the generate-and-test strategy. Extensions to this solution have been addressed to make the candidate generation efficient. In [3], the authors upgrade the algorithms SPADE [17] and SPAM [1] in order to avoid infrequent candidates. More precisely, they propose a preliminary step in which sequential patterns of length 2 are discovered. These are used later to eliminate the candidates in which the items of length 2 patterns are not present.

Another category of algorithms combines depth-first search and projected databases. They extend the frequent patterns, mined in previously created data bases, by using the items present in the newly created databases as suffixes or prefixes for longer patterns [7,13]. This way, they avoid building uninteresting candidates and early prune (sub)spaces of the lattice, achieving a two-fold result (i) utilizing much less memory and (ii) keeping the mining process focused only on those subsets of the database which can give frequent patterns.

4 Empirical Evaluation

To empirically evaluate the differences between the three main features described above, in this section we present experiments conducted on some algorithms that adopt different solutions to implement the three features. To this end, we considered GSP, PrefixSpan, SPAM, Spade and CM-Spade. More detailed description of these algorithms can be found in [11,12]. In order to perform a fair analysis, we used the SPMF framework [16], which collects many sequential pattern mining algorithms implemented in Java using the same design pattern. The experiments aim at evaluating the time consumption (in milliseconds) and memory consumption (in MB).

We used real-word datasets and synthetically generated sequence datasets. As to the real-world datasets, the experiments were performed by manually tuning the minimum threshold $minSUP$. Three categories of real-world sequences

Table 1. Characteristics of the real-world datasets used for the experiments.

Dataset	No. of sequences	No. of distinct items	Avg no. of itemsets per sequence	Density
Sign	730	267	52	0,0037
Bible	36369	13905	21,64	7,19E−05
Kosarak	69999	21144	7,98	4,72E−05
Msnbc	989818	17	4,75	0,058
Fifa	20450	2990	36,24	0,00033
Pumsb	49046	2088	50,48	0,00048

were considered: textual data, click-streams and census data. In the textual data, sequences correspond to sentences, while items correspond to words. The structure of the discourse defines the order between the words. Two datasets of textual data were used. *Sign*, which contains transcriptions of sign language utterances. *Bible*, which contains the transcription of Bible. In the context of click-stream data, sequences correspond to sessions of browsing on the Web, while the items correspond to Web pages. The order is defined in terms of the time-stamp when clicking on the page link. Three datasets were used. *Kosarak*, which contains sequences of click-stream data from an hungarian news portal. *Msnbc*, which contains click-stream data collected from logs of www.msnbc.com and news-related portions of www.msn.com for the entire day of September 28, 1999. *Fifa*, which contains the data of the requests made to the 1998 World Cup Web site between April 30, 1998 and July 26, 1998. In the census data, sequences correspond to population and demographic statistics collected across time. We used the dataset *Pumsb* that contains federal census data collected for the IPUMS project from Los Angeles – Long Beach area for the years 1970, 1980, and 1990. The real-world datasets are available at the link http://www.philippe-fournier-viger.com/spmf/index.php?link=datasets.php. The experiments were performed on a machine equipped with Windows 10 operating system, i3 3.3 GHz processor and 8 GB main memory. Table 1 reports a summary of the characteristics of these datasets. The density variable is obtained as the average number of items present in the itemsets divided by the number of distinct items.

As to the synthetically generated sequences, we used the datasets available at the link http://www.di.uniba.it/~ceci/micFiles/systems/CloFAST/ (also used in [5]). The experiments were performed by manually regulating one characteristic of the data at a time while leaving the others fixed. In particular, three characteristics were considered: average number of itemsets per sequence, density and number of sequences (denoted as C, T, D afterwards). The experiments were executed on a machine equipped with CentOS Linux operating system, Intel Xeon 2.4 GHz processor and 64 GB main memory.

In Fig. 1, we report the time spent by the algorithms on the real-world datasets. As expected, the larger the value of *minSUP* the larger the time consumption (the scales are logarithmic). In particular, the running time of GSP crosses at least three magnitude orders for all the datasets, while the algorithms SPAM, PrefixSpan and CM-Spade cross two magnitude orders on Sign, Bible, Kosarak and three orders on Fifa and Pumsb. The performance related to running time on Fifa and Pumsb can be attributed to a particular conjunction of the properties of these datasets, they have simultaneously many items (no. of distinct items), many itemsets per sequence (no. of distinct items) and large density, compared to the other datasets. When these characteristics are not present simultaneously, the running time is smaller, for instance the plots of SPAM and PrefixSpan algorithms on Kosarak and Sign are basically linear. Indeed, Kosarak has small sets of itemsets per sequence, while Sign has small sets of distinct items. A specific consideration can be drawn for SPAM on Kosarak and Bible. Although their respective running time has not an exponential tendency, the magnitude orders are greater compared to the other algorithms, including GSP. This can be explained with the data representation implemented in SPAM, where a bit vector is associated with each item. Considering that Kosarak and Bible have the larger sets of distinct items, the running time of SPAM can be affected by the cost of the operations for building the bit vectors.

The memory consumption (Fig. 2) follows the tendency of the results on the running time, although it increases quite linearly. The algorithms that take more memory space are those compliant with the generate-and-test procedure, such as, GSP and SPAM, because they keep the frequent patterns previously mined. This is more evident in GSP, where the breadth-first search forces us to keep the patterns of the levels of the lattice already visited. In particular, GSP needs more memory (and more running time) on Sign, Fifa and Msnbc for small values of $minSUP$ because the lattice grows up significantly. On the contrary, the choice of PrefixSpan to keep smaller search spaces is successful in terms of running time and memory consumption. In the case of SPAM, it consumes memory on Kosarak and Bible because it needs a huge number of bit vectors. The output of CM-Spade deserves a specific consideration. In Bible and Kosarak, we have no result because CM-Spade stops due to insufficient memory. This is quite expected because the two datasets have the maximum number of distinct items (compared to the other datasets) and CM-Spade uses all the memory because it has to perform a preparatory operation on the itemsets of length 2 (Sect. 3). On the contrary, it performs faster on the datasets with lower number of items (Sign, Msnbc and Pumsb), although it uses more memory compared to other choices, for instance, PrefixSpan.

The experiments on the synthetic datasets have been set on the properties of the input data sequences, which we can control. Figures 3a and d illustrate the running time and memory usage obtained while increasing values of C ($minSUP = 0.4$). We note that when C (average number of itemsets per sequence) increases, the time consumption increases too and for all the algorithms the efficiency drops of four orders of magnitude. The explanation is that the value of C affects the length of the patterns and consequently the size of the lattice. Thus, when C increases, we need to perform more s-extension operations, which lead the width and depth of the lattice to grow. This is particularly

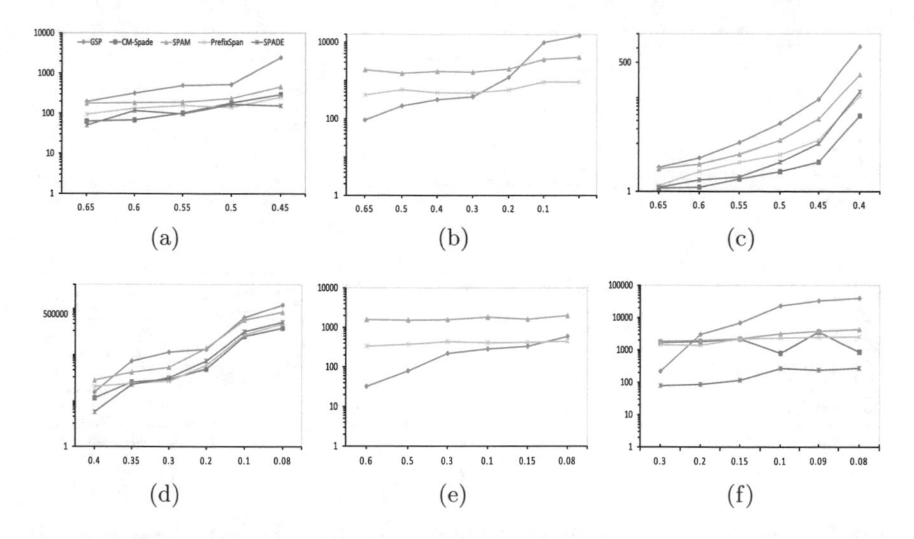

Fig. 1. Running time (in milliseconds) obtained on the real-world datasets (a) Sign, (b) Bible, (c) Pumsb, (d) Fifa, (e) Kosarak, (f) Msnbc.

evident on the algorithms that use the level-wise search, and candidate generation techniques, for instance, GSP, because they have to evaluate all the candidates of a level (which will be larger because there are more itemsets) before proceeding to the next level. A different behavior can be observed on the memory consumption, which grows linearly with C. This reveals a choice common to many algorithms, that is, limiting the use of the main memory at the cost of the running time. The exception is represented by CM-Spade, which asks for less time, but uses more memory for keeping the itemsets of length 2, especially for large values of C.

The performances obtained on T (density) follow those of C. With respect to running time, all the algorithms cross four orders of magnitude when T increases (Figs. 3b and e). The reason is that when the sequences are densest, the number of items present in the itemsets increases and the algorithms need more i-extension operations. This makes the lattice "more complex" and the pattern generation procedures costly. As to the memory, it is worth noting that the consumption of Prefix-Span is greater than GSP, especially for large densities. This is due to the fact that the size of the projected databases is not smaller than the input database for densest sequences.

The results obtained by varying D (Figs. 3c and f) allow us to evaluate the scalability properties of the algorithms, whose running time follows an exponential tendency. The algorithm GSP has at least one more magnitude order, compared to the others, and this is due to the algorithmic choice on the access and representation of the input sequences. In GSP, the computation of the occurrences is performed by scanning the whole database, hence the data representation relies on traditional transaction-based format, which requires costly I/O operations. Contrarily, PrefixSpan does not scan the whole database, but only partitions of the input sequences and this guarantees better efficiency. This gain

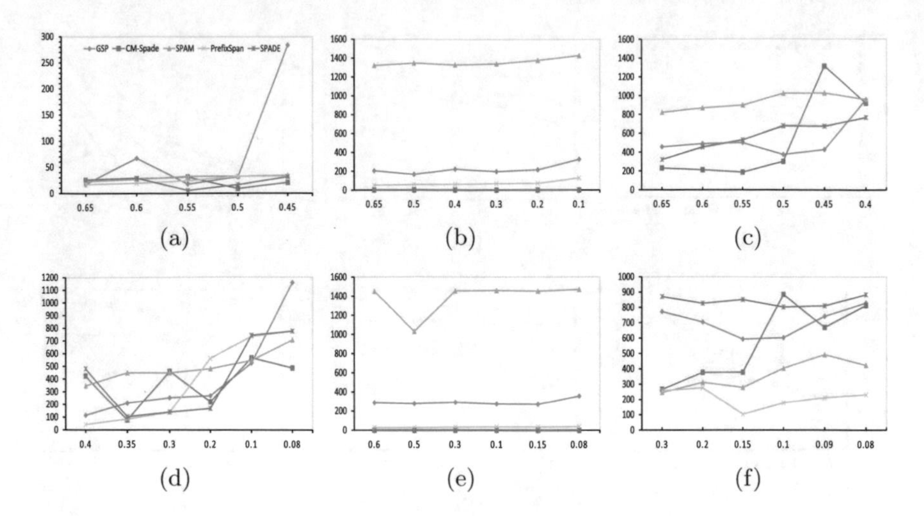

Fig. 2. Memory consumption (in MB) obtained on the real-world datasets (a) Sign, (b) Bible, (c) Pumsb, (d) Fifa, (e) Kosarak, (f) Msnbc.

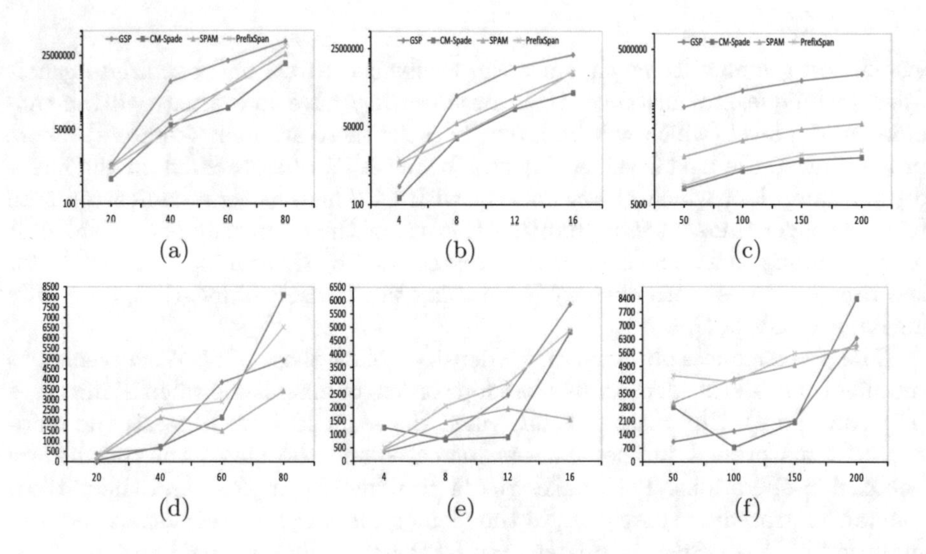

Fig. 3. Running time and memory consumption (in MB) obtained on the synthetic datasets (a), (d) Number of itemsets per sequences, (b), (e) Density and (c), (f) Number of sequences.

in running time implies a greater usage of the memory, as seen for the results of C. In the case of PrefixSpan, we need to create a large number of projected databases containing probably more sequences, while, in the case of SPAM and CM-Spade, we need very long IDlists. Finally, GSP asks for less because it uses memory essentially to store the lattice, whose size is related to the size of the input database.

5 Conclusions

In this work, we have conducted an experimental study to compare most representative algorithms designed for sequence mining. The main contribution lies in the empirical evaluation of the choices done in those algorithms for the *(i)* exploration of the lattice of the patterns, *(ii)* representation of the database of sequences, and *(iii)* generation of sequential patterns. This work can be intended as a preliminary investigation for the future design of a sequence mining approach on Big Data, considering solutions for data distribution and/or parallelization on different machines, including streaming approaches. Indeed, our evaluation, performed on real-world and synthetic datasets, has been set to draw indications on the performance in terms of time and memory consumption. We observed that the choices done to make the mining process efficient, for instance, in PrefixSpan and CM-Spade, imply greater memory consumption. Contrarily, solutions which are time consuming, for instance GSP, may perform relatively well in terms of main memory requirements. As future work, we plan to (i) investigate advanced data structures to adapt time-saving algorithms (for instance, PrefixSpan and CM-Spade) and (ii) consider distributed solutions for the generation of patterns in order to adapt memory-saving algorithms (for instance, GSP).

References

1. Ayres, J., Flannick, J., Gehrke, J., Yiu, T.: Sequential pattern mining using a bitmap representation. In: Proceedings of the Eighth ACM SIGKDD International Conference on Knowledge Discovery and Data Mining, KDD 2002, New York, NY, USA, pp. 429–435. ACM (2002)
2. Cheng, Y., Lin, Y., Chiang, K., Tseng, V.S.: Mining sequential risk patterns from large-scale clinical databases for early assessment of chronic diseases: a case study on chronic obstructive pulmonary disease. IEEE J. Biomed. Health Inform. **21**(2), 303–311 (2017)
3. Fournier-Viger, P., Gomariz, A., Campos, M., Thomas, R.: Fast vertical mining of sequential patterns using co-occurrence information. In: Advances in Knowledge Discovery and Data Mining - 18th Pacific-Asia Conference, PAKDD 2014, Proceedings, Part I, Tainan, Taiwan, 13–16 May 2014, pp. 40–52 (2014)
4. Fournier-Viger, P., Lin, J.C.-W., Kiran, R.U., Koh, Y.S.: A survey of sequential pattern mining. Data Sci. Pattern Recognit. **1**(1), 54–77 (2017)
5. Fumarola, F., Lanotte, P.F., Ceci, M., Malerba, D.: Clofast: closed sequential pattern mining using sparse and vertical id-lists. Knowl. Inf. Syst. **48**(2), 429–463 (2016)
6. Ge, J., Xia, Y., Wang, J., Nadungodage, C.H., Prabhakar, S.: Sequential pattern mining in databases with temporal uncertainty. Knowl. Inf. Syst. **51**(3), 821–850 (2017)
7. Han, J., Pei, J., Mortazavi-Asl, B., Chen, Q., Dayal, U., Hsu, M.: Freespan: frequent pattern-projected sequential pattern mining. In: Proceedings of the Sixth ACM SIGKDD International Conference on Knowledge Discovery and Data Mining, Boston, MA, USA, 20–23 August 2000, pp. 355–359 (2000)
8. Loglisci, C.: Using interactions and dynamics for mining groups of moving objects from trajectory data. Int. J. Geograph. Inf. Sci. 1–33 (2017)

9. Loglisci, C., Ceci, M., Impedovo, A., Malerba, D.: Mining spatio-temporal patterns of periodic changes in climate data. In: New Frontiers in Mining Complex Patterns - 5th International Workshop, NFMCP 2016, Held in Conjunction with ECML-PKDD 2016, Riva del Garda, Italy, 19 September 2016, Revised Selected Papers, pp. 198–212 (2016)
10. Loglisci, C., Ceci, M., Malerba, D.: Relational mining for discovering changes in evolving networks. Neurocomputing **150**, 265–288 (2015)
11. Mabroukeh, N.R., Ezeife, C.I.: A taxonomy of sequential pattern mining algorithms. ACM Comput. Surv. **43**(1), 3:1–3:41 (2010)
12. Mooney, C., Roddick, J.F.: Sequential pattern mining - approaches and algorithms. ACM Comput. Surv. **45**(2), 19:1–19:39 (2013)
13. Pei, J., Han, J., Mortazavi-Asl, B., Wang, J., Pinto, H., Chen, Q., Dayal, U., Hsu, M.: Mining sequential patterns by pattern-growth: the prefixspan approach. IEEE Trans. Knowl. Data Eng. **16**(11), 1424–1440 (2004)
14. Schweizer, D., Zehnder, M., Wache, H., Witschel, H.F., Zanatta, D., Rodriguez, M.: Using consumer behavior data to reduce energy consumption in smart homes: applying machine learning to save energy without lowering comfort of inhabitants. In: 14th IEEE International Conference on Machine Learning and Applications, ICMLA 2015, Miami, FL, USA, 9–11 December 2015, pp. 1123–1129 (2015)
15. Srikant, R., Agrawal, R.: Mining sequential patterns: generalizations and performance improvements. In: Apers, P.M.G., Bouzeghoub, M., Gardarin, G. (eds.) Advances in Database Technology - EDBT 1996, 5th International Conference on Extending Database Technology, Proceedings, Avignon, France, 25–29 March 1996, vol. 1057. Lecture Notes in Computer Science, pp. 3–17. Springer (1996)
16. Viger, P.F., Gomariz, A., Gueniche, T., Soltani, A., Wu, C., Tseng, V.S.: SPMF: a java open-source pattern mining library. J. Mach. Learn. Res. **15**(1), 3389–3393 (2014)
17. Zaki, M.J.: SPADE: an efficient algorithm for mining frequent sequences. Mach. Learn. **42**(1/2), 31–60 (2001)
18. Ziebarth, S., Chounta, I., Hoppe, H.U.: Resource access patterns in exam preparation activities. In: Design for Teaching and Learning in a Networked World - 10th European Conference on Technology Enhanced Learning, EC-TEL 2015, Proceedings, Toledo, Spain, 15–18 September 2015, pp. 497–502 (2015)

Towards Internet of Things and Cloud Computing for Management of Cars Network

Krzysztof Stepień and Aneta Poniszewska-Marańda(✉)

Institute of Information Technology, Lodz University of Technology, Łódź, Poland
aneta.poniszewska-maranda@p.lodz.pl

Abstract. XXI century is full of smart devices which amount is still substantially growing. Therefore there is a big need of context-aware platform that uses all of these devices and connect them together. However, clearing this way to accomplishing that problem requires a lot of research and development. Internet of Things (*IoT*) is the one of approaches which help in finding the solutions for present and future directions. Global idea of treating the cars like many other devices and try to interconnect each of them may cause in solving the serious problems that could lead to decrease the number of accidents and improve the traffic transportation. Addressing these tendencies the paper presents the developed solution which could provide a big network for all cars with the use of IoT and cloud computing concepts.

1 Introduction

The Internet is a powerful tool used in all kinds of the information systems. The network is available almost anywhere, at home, at work, also on mobile devices (phones, watches). People start to think to connect the Internet to almost all devices of everyday use, so they can communicate with each other by taking simple decisions for people and helping them in their life. Such idea is called the *Internet of Things (IoT)*.

A closer look suggest that problem of stating the definition of *IoT* is composed of two pillars: "Internet" and "Things". We can assume that "thing" is every device capable to connect to the Web and "Internet" is a common description of network. IoT instead of forming a huge network that connects people, forms a network that connects things. Of course that connection does not embrace only in computers, smartphones and tablets but also in any other devices that are capable to connect to the given network [1,3,17,18,22].

Internet of Things could not be developed without cloud solutions (CC). Cloud computing is general term which refers to the idea of sharing resources not over local servers, but over the global network accessible from every place in the world. Cloud computing is a collocation of two words: "cloud" as the metaphor of the Internet and "computing" as different services such as servers, storage and applications.

L. Barolli et al. (Eds.): EIDWT 2018, LNDECT 17, pp. 627–638, 2018.
https://doi.org/10.1007/978-3-319-75928-9_56

Most of the devices connected to the IoT are rather simple. The devices are not necessarily smart themselves but joined together with other connected devices become to be smarter than people might think. The most excited thing here is that network of these devices which work together, could create a coherent system that can act with its own type of intelligence. Internet of Things is a system of related computing devices, that could be either mechanical and digital machines. Recent years shows that any device that fulfils some requirements, can connect to the cloud [3,4,7,15,16,19].

Cars have been get progressively smarter constantly since even 30 years. New cars contain dozens of electronic units or smartphone-like devices, all working together to provide as comfortable mean of transport as it could be. All of these "computers" control in-car functions such as brakes, cruise control, heating/cooling and all entertainment systems [5,8].

Car field is one of those devices that might be not represented by people itself (primarily described as Internet of People). Instead of constant people access, the IoT enables them to access the data and communicate with one another. The future Internet starts to form the future branch called pipeline for non-human devices – *M2M* (*Machine-To-Machine*). Today many cars seamlessly connect to people smartphones, register real-time traffic jams, stream music and even offer emergency call (like Opel *OnStar* – 24-hour emergency call service). Even if systems are high complicated they still operate on one car as simple closed eco-system. But world has continuously changing and car manufacturers are on the brink of a revolution and the driving force is the suite of technologies known as the Internet of Things (IoT).

The problem presented in the paper concerns the solution that could communicate the independent systems offered separately by each car manufacturer and could provide a big network for all cars around the world. Developed solution embraces in server application which connects other smart cars and it allows to connects the cars to the cloud. Smart car is simply each model of car fulfilled the technical requirements.

The presented paper is structured as follows: Sect. 2 deals with the IoT and CC implementations in the cars field – it describes the present solutions of the newest cars and existing technologies served as IoT solutions while Sect. 3 presents the created approach and its application as the solution for cars network using IoT and CC concepts.

2 Internet of Things and Cloud Computing in the Cars Field

IoT in the automotive industry is very fresh term, where there is no one simple definition. Today there are three main branches. On the one hand, there is a connectivity between vehicles – from two up to more: *vehicle-to-vehicle* – cars are connecting with other cars. Second approach bases on connecting those vehicles to the infrastructure (*vehicle-to-infrastructure*) – cars are interacting with roads or external objects. Third idea connects external hardware with external devices (*vehicle-to-devices*) – it is a wireless communication to any device. Therefore IoT refers to the real-time data interaction between those elements (*vehicle-vehicle*, *vehicle-city*, *vehicle-road*) using mobile communication and enables them to interact by exchanging the information [2].

From the technical point of view more an more vehicles functions are implemented in the software. Newest cars are full of embedded computerized units, called *ECU – Electronic Control Unit. ECU* handles different operations, for example electronic mirrors, suspension mode, engine control. With the development of WiFi platform, a lot of sensors have ability to connect to the particular repair-unit in order to notify the broken part. The network inside vehicle is divided into smaller parts (sub-networks) using different bus systems. The most popular technologies are *CAN* (*Controlled Area Network*), *MOST* (*Media Oriented System Transport*) and *LIN* (*Local Interconnect Network*). They are chosen according to the given communication requirements, the specification of the component and security issues [8]. To get data from those sub-networks the *On-Board Diagnostic II* (*OBD-II*) is used. This port is mainly used to perform diagnostics, communicate with *ECU*s, send and receive information from many sensors (Fig. 1) [11].

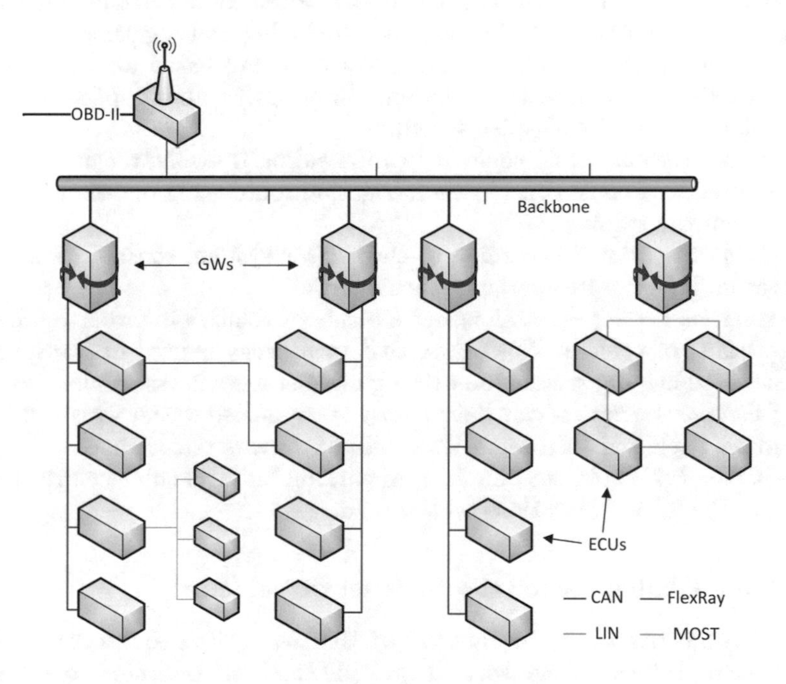

Fig. 1. Example of in-vehicle network [8]

2.1 Purpose of IoT Integration Inside Cars – Functional View and Benefits

Gathering all the information together forms a question: who already owns the road – the drivers themselves or some self-intelligent systems. Answer is somewhere between. The roads still need drivers but the drivers with increasing number of vehicles need some systems which could help them on the road.

There are some crucial points which even force the car producers to elaborate the smart innovations inside their cars [11]:

- improved safety – 93% of all accidents are caused by a human error,
- increased productivity – nearly 5.5 billion hours are lost in traffic,
- less fuel emission – cars could help drivers in driving more economical and ecological,
- enhanced freedom – it could help elderly or disabled people in order to drive car, in the few decades there will be three times more 80+ years old people.

According to the purpose of IoT integration inside the car, IoT tries to implement some systems which could really handle drivers' problems [8,9]:

- *Intersection control* – crucial point is how to schedule the traffic lights, taking into consideration traffic volume. Reducing waiting time is the most important thing here. Most of the algorithms are based on *V2V* (*Vehicle-To-Vehicle*) communication – detailed information about speed, position is collected to plan the effective schedule.
- *Route navigation* – IoT could help in avoiding drawbacks caused by GPS. Connected cars could construct a navigation route using real-time data gathered from the traffic.
- *Parking navigation* – according to the *onStar* system, consultant could help driver in finding a free parking place nearby.
- *Cooperative driving* – coordinating a queue of vehicles in order to create one big "train" of vehicles. Nowadays, that technology is used in transportation area where helping truckers in driving big distances is essential.
- *Self-diagnosis* – newest cars have plenty of connected sensor where each could diagnose itself and then sends that data into saved repair shop.
- *Co-pilot* – *IoT* could not only help in parking but also enhance driver awareness of the hazard situations on the road.

2.2 Main Challenges for the Internet of Vehicles

The main objective stated by *Internet of Vehicles, IoV* is to integrate the multiple vehicles, things and networks to provide the best connected community as it is possible [12]. Unfortunately due to the specification of IoT there are a lot of challenges which IoT should take into consideration [10]:

- *poor network connectivity and stability* – even though 4G LTE connectivity is easy to get, always there is a threat that in some places some link failures, network disconnections, message loss could occur,

- *high reliability requirements* – due to the complexity of the network, poor scalability of the tricky IoT network topology, it could be tricky to achieve high reliability in *IoV*,
- *high scalability requirements* – system should work for both big fleet of vehicles and a casual user,
- *security and privacy* – the most important factor here; data encryption, data channel access are the most crucial components of the IoT,
- *power and battery consumption* – car will be constantly connected to the Web, therefore car manufacturers should keep in mind how not do drain car's battery,
- *presence* – even if all the challenges could be solved, one more point is important: money – system should be accessible for all the customers around the world with appropriate car.

Internet of Things in the context of automotive branch is very deep field of industry which may succeed in the future. With all the advantages come also a lot of threats, how to construct a reliable system independent from hacker attacks.

3 Management of Cars Network with Use of Cloud Computing and Internet of Things

Many existing concepts and ideas were combined together to create the application which will be accessible for all the devices and all the cars capable to have OBD-II port and valid Internet connection. *ConnectYourCar* is not only the application, but also a new approach in creating the connection between the user and the car.

3.1 Server Application

The server application enables the administrator to govern all devices (cars) connected to the network. In the future, buying a new car will be already buying the data-center on the wheels. Most of them will receive, analyse and send gathered data automatically. However, there is still a big hope in *IoT* solutions that will govern all those processes and will be as easy as possible for the future administrator.

ConnectYourCar gathers missing and not well polished factors to combine one solution which will improve the drivers' life:

- *Map with the cars* – place where all connected cars are visible for the administrator, place where he can check whether something wrong had happened with the given car,
- *Car Status* – shows current data of the given car,
- *Vehicle* – list of vehicles connected and not connected to the system,
- *Alert* – logs about current and previous problems with the cars,
- *Events* – administrator can manage existing events and create new ones such as traffic congestion, objects on the road, accident.

Server application was created using *IBM Bluemix* cloud. This tool enables the user to create scalable ecosystem that is capable to use plenty of frameworks and services. Version control system was managed using *CloudFoundry* tool.

Application was written using *Node.js* environment – it bases on *JavaScript* and *HTML* files. It uses several modules which are helpful in creating the *IoT* solutions. First of them is *IBM Watson IoT* module which is responsible for providing the variety services for *IoT* platform. Second one is *Fleet4Vehicles* – a service responsible for variety solutions in areas such as in-vehicle services, advanced mobility, autonomous driving and vehicle data commerce. Finally, all the elements are tied up in *Cloudant NoSQL database* (all database requests could be accessed using JSON files and POST/GET requests) (Fig. 2).

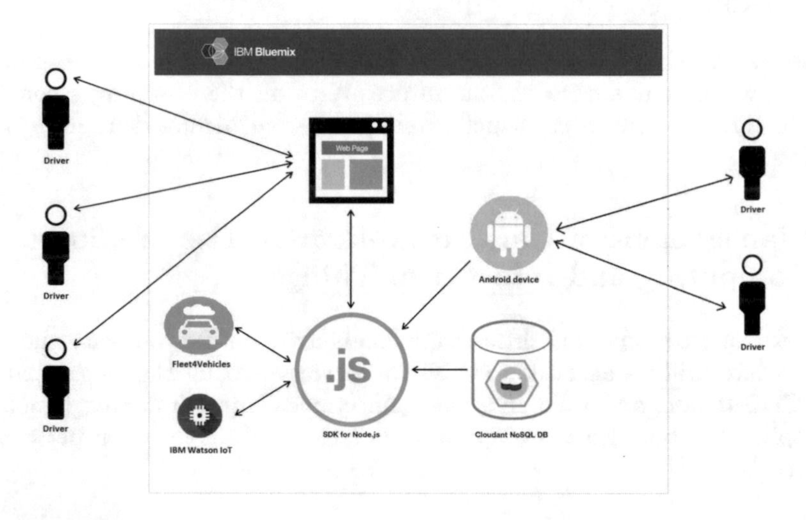

Fig. 2. IoT architecture diagram for the *ConnectYourCar* application server

Watson IoT is a cloud service and as such, as demand for whatever service increases, the capacity to meet that demand will be scaled. Devices connecting through a gateway could experience problems if the gateway was overloaded, but in case of presented solution (IBM Watson IoT) the administrator and the users should not experience such issues.

Differently from the other solutions available on the market (built-in modules placed by the car manufacturers) *ConnectYourCar* offers separate module for the cars in order to share the data with cloud server. Server application is accessible via Internet Browser (*Mozilla Firefox* or *Google Chrome*). User has only to pass the credentials to log in to the maintenance section.

All information received from each car are stored in the *Cloudant NoSQL* database. Data could be stored daily, weekly, monthly depending on user choice. This kind of database is designed strictly for web solutions and leverage a flexible *JSON* schema. All data is processed real-time.

Application consists of several components. Each section in the application is represented by the separate component. *Map* is the core element in the application, because it handles all actions that had happened on the road. It stores a current place of the car. *Alerts* keep all data about extraordinary situations on the road. *Car status* is virtualized car itself – it stores information about fuel, errors. *Events* tied up all situations related to the given car (Fig. 3).

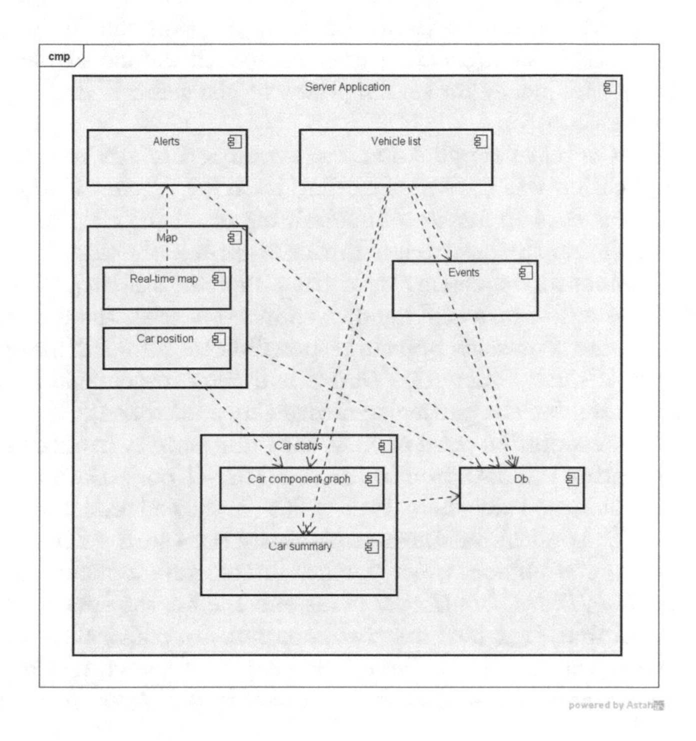

Fig. 3. Component of *ConnectYourCar* server application

From the administrator point of view, server application could be very helpful tool for managing all cars connected to the Web. *ConnectYourCar* presents basic *Internet of Vehicles* network where all cars are connected to the one cloud. Continuous need of integration all devices to the car demand on car producers to create their own clouds. Unfortunately, all the existing systems work separately, so only owner who has specific car model, can use this particular architecture. The created approach is the solution where not only new cars, but also used ones could be treated as the part of the *IoT* network.

3.2 Client Application

Created client application is not an application itself. It works mostly as a microcontroller hidden inside a car (or just like a casual smartphone that the user owns) which connects with the car via Bluetooth interface. Application is constantly sending data using WiFi or 3G/LTE network to the cloud about current position of the car and current car status (speed, fuel level, errors with any occurs). It avoids buying any specialized machinery, because the only need here is the Smartphone with Android 5.0 or newer.

GUI layer bases strictly on XML layout created in Android Studio, but it is not important here, because application is used mostly like a controller that sends the data, not application itself. Client application consists of three buttons: *Change network* (changes connected Bluetooth microcontroller), *Change frequency* (change frequency for sending data to the server), *End Session* (end session of sending data).

ConnectYourCar client application was developed in Android Studio IDE. The whole application was created according to *MVC* (*Model-View-Controller*) pattern. *MVC* was used to facilitate maintaining of all the classes.

The core of the application resists in the *MainActivity* class – there are all methods responsible for requesting data from the car, starting/stopping sending data to the cloud. *ObdBridge* handles connection with the car via *OBD-II* port. *ObdParameter* represents objects responsible for all particular errors and data taken from the car. Finally *IoTDevice* is a class responsible for managing authentication token for the particular device (key and token).

From the implementation point of view, the main activity in client application focuses on requesting the data from a car via *OBD-II* port. Data requested via OBD-II port is not stored anywhere. Part of data is showed on the screen live and sent to the cloud. Application stores temporarily data in the *List* Java object. That list receives the objects which further enable the application to request data via *OBD-II PID*'s. *OBD-II PID* is responsible for the specific data about car. It consists of four digit hex: first two numbers are responsible for receiving data mode (show data, request *Data Trouble Codes*), next two numbers are responsible for specific data which the user wants to get. Application constantly

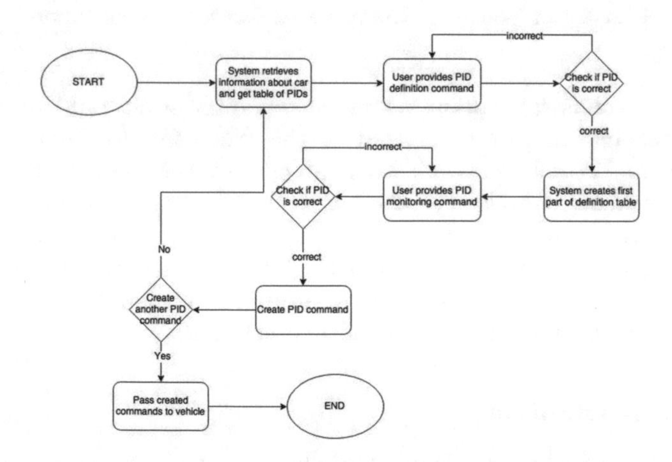

Fig. 4. Algorithm of requesting data via *OBD-II* by *PID*

sends requests to the car. Each request is represented by the separate 3 digits number (e.g. 504, 508). Then, if a connection could be established, the car sends *PID* code (Fig. 4). There is no data layer in the client application. All data is packed into list and constantly sent through the Web to the Server application.

Client application presents retrofit solution – to work, it needs smartphone located inside a car that handles all car activities by the *OBD-II* port. All data that is sent to the cloud could be used further in the business context, but also for the national purpose. All cars connected to the Web could help in creating self-intelligent systems in the cities, prevents thefts or accidents.

3.3 Main Features of *ConnectYourCar* solution

ConnectYourCar server application from the user point of view can run on every Internet browser with the Internet connection. Application consists of five sections. Navigation drawer located on the left provides all functionalities of the application. All events are displayed in the centre of the screen. Application enables the user to find and manage the cars using *OpenStreetMap*. User can use its location to open the nearest places next to him or choose a capital of the given voivodeship. The map is very accurate, so every house, shop, park is visible on the map. The information about all cars connected to the system for the chosen area and information how many of them have some troubled or are even in the critical state (e.g. car ran out of fuel, accident happened) is given in the left corner. In the right down corner, all voivodeships are listed with the same information about the cars (Fig. 5).

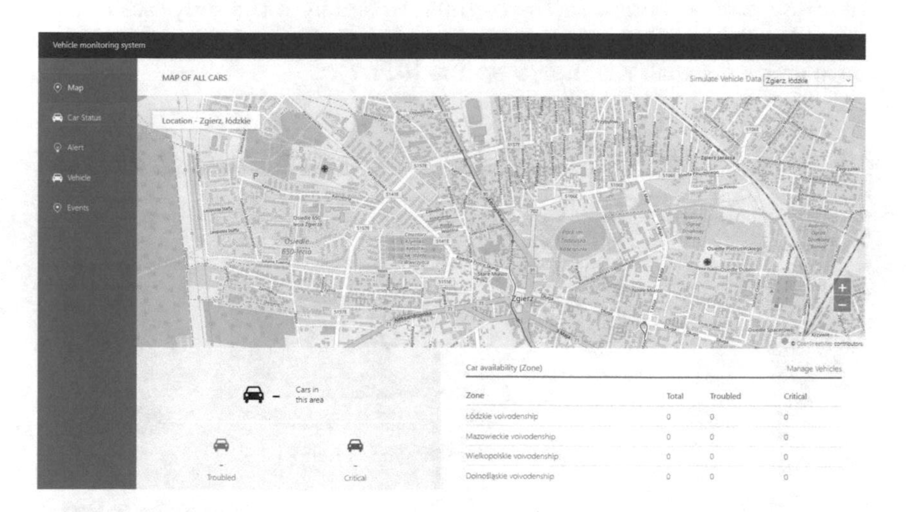

Fig. 5. Map with chosen area – user has the possibility to choose any city he wants and monitor number of problems in the given voivodenship

Administrator can manage the existing events and create the new ones such as traffic congestion, objects on the road, accidents. Each alert that had happened in the real time (car is run out of fuel or it has problem with the engine) is marked on the map (car icon changes its colour to orange or red) and it is listed on the events history list (Fig. 6).

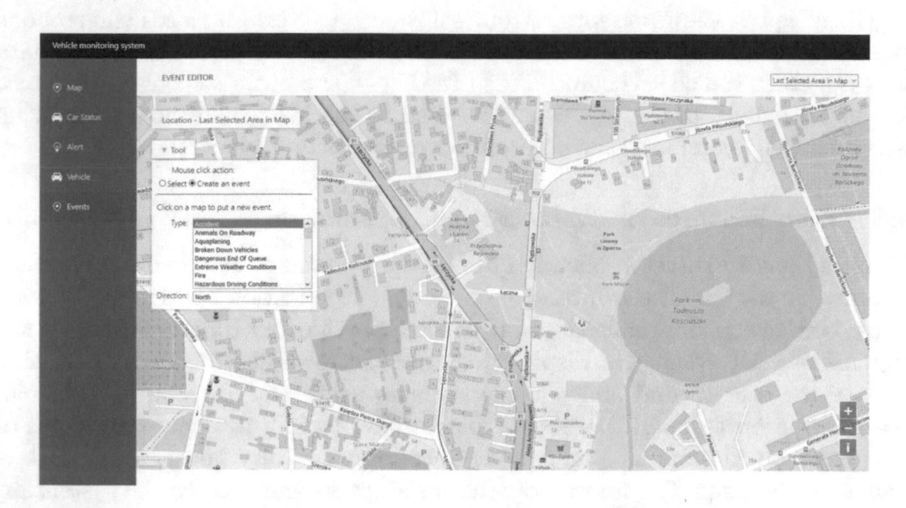

Fig. 6. Events creator and manager where the user can manage the events by simply dropping in the specific location

Each car sends specific data about fuel amount or speed. It also indicates whether any error occurred in the engine. To simulate the purposes a user can also run simulation where fake cars are going through the streets. They also simulate problems with the fuel or engine (Fig. 7).

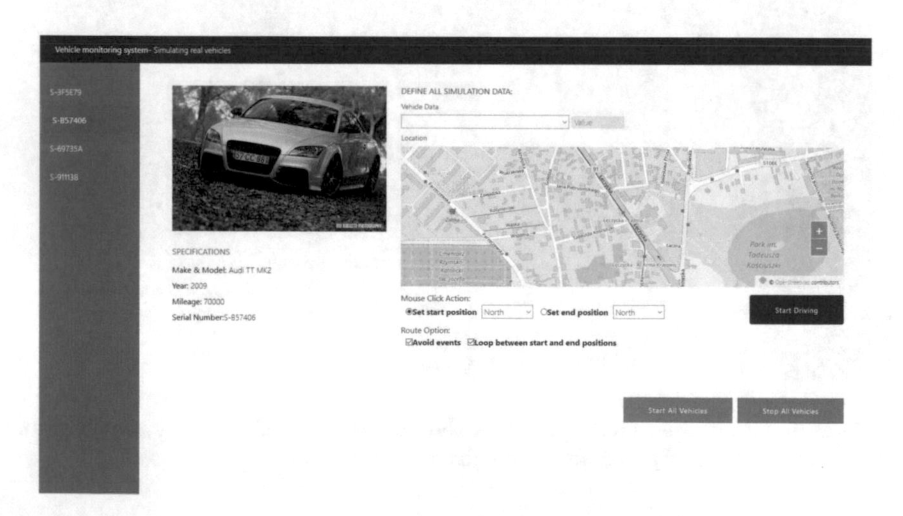

Fig. 7. To simulate purposes, user can simulate car behaviour and drop cars to the map with fake data

4 Conclusions

Internet of Things as a new separate field of computer science emerged into new paradigm which provides the new form of integration and communication of smart devices. That's why it is so much important to create a system which could integrate all of them into one working eco-system and hide their complexity from the user. Even though many challenges are still remaining (starting from communication needs up to middleware development) and need the further investigation.

Possibility of connecting the vehicles to IoT brings a lot of new functionalities and could make the transport faster and easier. The performance of the mobile phones, but also micro-controllers (like *Arduino*) spread like wildfire. Therefore the number of new mobile ecosystems will grow. The most crucial (and critical) point here is to ensure the safety, predictability and dependability [4]. From the technical point of view these systems will use the smart phones to gather onboard information, such as destination, speed, current engine status and interact with external city systems, such as traffic control systems, parking management.

From the cities point of view, smart transformation will bring more security and safety on the streets [7]. At the building level, smart ideas will lead to control the whole building and manage the accessory within one control unit which enables them (e.g. thermostats) to communicate with each other. Smart energy is slightly connected with transportation problems and could also solve the safety problems where human factor is the weakest link.

Presented solution might help to provide the widely open tool for the automotive field. It enables all the drivers to have a specific microcontroller connecting their cars to the Web and sharing their pieces of data [21]. Drivers are able to share the information about fuel level, car position, speed without any personal involvement. This data could be of hand while managing the traffic or preventing the car accidents.

ConnectYourCar represents a smart application that not only helps a particular driver, but also the society of drivers. In the future, such solutions might be enhanced by the intelligent module that can analyse the situation, (e.g. using *OpenCV* library and their own algorithms to detect cars, road) on the road and suggest or alarm the driver. *ConnectYourCar* is one of the number of steps that could be developed into universal platform capable to connect each car and satisfy each driver needs. Then we would state that the *IoT* revolution in the automotive field starts to become real.

References

1. Dastjerdi, A.V., Buyya, R.: Internet of Things, Principles and Paradigms. Elsevier, Morgan Kufmann (2016)
2. Dhanjani, N.: Abusing the Internet of Things. O'Reilly Media, Sebastopol (2015)
3. Ashton, K.: That 'Internet of things' thing. RFID J. **22**, 97–114 (2009)
4. Gubbia, J., Buyyab, R., Marusic, S.: Internet of things (IoT): a vision, architectural elements, and future directions. Future Gener. Comput. Syst. J. **29**(7), 1645–1660 (2013)

5. Stepien, K., Poniszewska-Maranda, A.: Management and control of smart car with the use of mobile applications. Inf. Syst. Manag. **6**(1), 70–81 (2017). WULS Press, ISSN: 2084-5537
6. Vermesan, O., Friess, P.: Internet of things: converging technologies for smart environments and integrated ecosystems. River Publishers Series in Communications (2013)
7. Miller, M.: The Internet of Things: How Smart TVs, Smart Cars, Smart Homes, and Smart Cities are Changing the World. Pearson Education, Inc. (2015)
8. Peng, H., Cakmakci, M., Ulsoy, A.G.: Automotive Control Systems. Cambridge University Press, New York (2012)
9. Ji, Z., Ganchev, I., O'Droma, M.: A cloud-based car parking middleware for IoT-based smart cities: design and implementation. Sens. J. **14**(12), 22372–22393 (2014)
10. Kleberger, P.: On Securing the Connected Car, Methods and Protocols for Secure Vehicle Diagnostics. Ph.D thesis, Chalmers University of Technology (2015)
11. Panga, G., Zamfir, S., Balanr, T.: IoT diagnostics for connected cars. Scientific Research and Education in the Air FORCE-AFASES, pp. 287–294 (2016)
12. Smith, C.: Car hackers handbook, A Guide for the Penetration Tester (2014)
13. McEwen, A., Cassimally, H.: Designing the Internet of Things. Wiley, Chichester (2014)
14. da Costa, F.: Rethinking the Internet of Things. A Scalable Approach to Connecting Everything. Apress open, California (2013)
15. Arsénio, A., Serra, H., Francisco, R., Nabais, F., Andrade, J., Serranol, E.: Internet of Intelligent Things: Bringing Artificial Intelligence into Things and Communication Networks. Springer, Heidelberg (2014)
16. Ruggieri, M., Nikookar, H., Vermesan, O., Friess, P.: Internet of Things: Converging Technologies for Smart Environments and Integrated Ecosystems. River Publishers (2013)
17. Patra, L., Rao, U.P.: Internet of things - architecture, applications, security and other major challenges. In: Proceedings of the 3rd International Conference on Computing for Sustainable Global Development (INDIACom), New Delhi, pp. 1201–1206 (2016)
18. Billure, R., Tayur, V.M., Mahesh, V.: Internet of things - a study on the security challenges. In: Billure IEEE International Advance Computing Conference (IACC), Banglore, pp. 247–252 (2015)
19. Ahamed, J., Rajan, A.V.: Internet of things (IoT): application systems and security vulnerabilities. In: Proceedings of the 5th International Conference on Electronic Devices, Systems and Applications (ICEDSA), pp. 1–5 (2016)
20. Poniszewska-Maranda, A., Kaczmarek, D.: Selected methods of artificial intelligence for Internet of Things conception. In: Annals of Computer Science and Information Systems, FedCSIS 2015, vol. 5, pp. 1343–1348 (2015)
21. Poniszewska-Maranda, A., Gebel, L.: Retrieval and processing of information with the use of multi-agent system. J. Appl. Comput. Sci. **24**(2), 17–37 (2016)
22. Kaczor, S., Kryvinska, N.: It is all about services - fundamentals, drivers, and business models. J. Serv. Sci. Res. **5**(2), 125–154 (2013)

An Edge Computer Based Driver Monitoring System for Assisting Safety Driving

Toshiyuki Haramaki(✉) and Hiroaki Nishino

Division of Computer Science and Intelligent Systems, Faculty of Science and Technology, Oita University, Oita, Japan
{haramaki,hn}@oita-u.ac.jp

Abstract. Driver Monitoring System (DMS) is a promising IoT application in Intelligent Transport Systems (ITS) research field. DMS assists car drivers by monitoring their driving activities, sensing incidents to cause possible dangers, and alerting the drivers to prevent accidents. We aim to realize a new DMS that is inexpensive and highly effective. This paper proposes a method for detecting any incidents based on machine learning. The proposed method firstly configures a detector by training in-car environment data and driver's vital signs gathered from multiple sensors. Then, the detector is embedded in a self-contained edge computer for monitoring a driver in a car. The device is always connected to the information communication network by radio waves. Those data obtained by monitoring are stored in the cloud server. The server learns and analyzes the stored data using processing such as machine learning. As a result, we acquire knowledge leading to safe driving. The edge computer uses these knowledge to process the sensor data in real time, observe the driver, sense the danger, and call attention. These mechanisms prevent occurrence of troubles such as traffic accidents. The paper describes the proposed system overview, implementation method, and initial evaluations.

1 Introduction

Recently, research on existing driving assistance for driving a car is actively studied to control a car. What kind of control is based on the result of recognition of a situation. For example, outside the car with a plurality of sensors, a three-dimensional map or the like, such as a collision prevention brake or a follow-up running device. These researches are aimed at realizing a series of automatic driving technologies such as measurement and recognition of the outside environment, reflection of the results, and reflection on driving operation. On the other hand, I focused a little on the environment and situation of drivers and passengers in the car and changed my perspective. And, by looking at them, it is possible to detect danger and call attention to include a traffic accident. Based on that, we propose a system that prevents occurrence of traffic problems beforehand.

For that reason, we are working on research to prevent accidents by encouraging drivers and passengers. The purpose of the research is to prevent accidents due to problems with drivers and passengers, to secure passengers and to secure and improve the comfort of the indoor environment. As a concrete realization method, IoT equipment

L. Barolli et al. (Eds.): EIDWT 2018, LNDECT 17, pp. 639–650, 2018.
https://doi.org/10.1007/978-3-319-75928-9_57

equipped with various sensors for in-car monitoring is installed. The purpose is to contribute to safe driving by securing the comfort of the in-car environment. The object to be monitored is the driver, and is the passenger including the in-car environment. Various data acquired by the sensor are accumulated in the device and are completed. The IoT device is always wirelessly connected to the network, and you can store and share the various data acquired by sensors in a dedicated server or cloud. Learn and analyze data accumulated through machine learning and other processes and acquire knowledge leading to safe driving. In addition, we present information to the driver via multiple output channels (image, sound, vibration, scent, etc.) to draw attention to the high-risk event derived from the analysis result. Furthermore, if a very dangerous situation is detected, we will also consider a mechanism to present information and emergency calls to surrounding cars, pedestrians, police, etc.

However, it is problematic only by monitoring the driver and car interior by the conventional IoT device. The biggest problem is that a computing computer with high computing power is necessary to judge the danger level for safe driving from input data from various sensors. In general, IoT devices are inexpensive, have low computing power, and acquisition of sensor data is the main task. Therefore, these complex processes run on the cloud server. However, with this method, there is a high possibility that the real-time nature of the output with respect to the input is impaired.

As a method to solve this problem, we propose a mechanism to input, process and output sensor data by Edge Computer. Figure 1 shows the mechanism of sensor data processing by Edge Computer. The server acquires knowledge from the received data using a method such as machine learning. In that server, a knowledge model is acquired using a learned model such as machine learning from the received data. We transfer the acquired knowledge model to Edge Computer in a processable format. As a result, the edge computer can process the sensing data at a high speed and present an appropriate processing result.

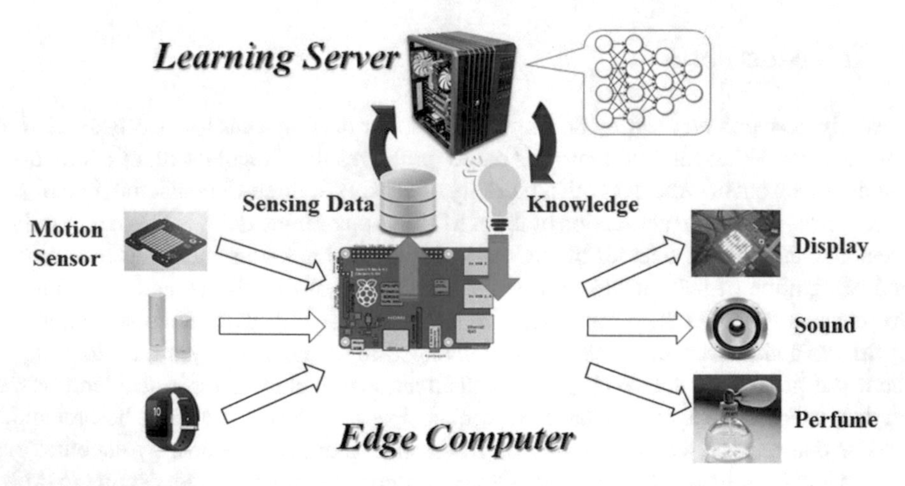

Fig. 1. Relationship between Edge Computer and Machine Learning Server

In this paper, we propose a driver monitoring system built by previous research, visualize information system. Machine learning has become a standard paradigm for detecting data analysis on various scales. In recent years, due to the innovation of machine learning, the data source targeted for machine learning has expanded to everything. Machine learning is useful not only for simple data processing but also for real world learning, understanding and knowledge extraction. Our aim in this research is to derive guidelines to support safe driving. As a matter of course, the guideline challenges to extract common understanding and knowledge which is an index for both human beings and machines even when automatic driving cars become popular, as a matter of necessity for human beings to drive a car.

2 Related Work

We have been studying methods of collecting data and providing information using sensors in response to various situations.

Figure 2 shows a network configuration visualizer using AR (Augmented Reality). To effectively manage the computer infrastructure, administrators need to understand the latest state of the system appropriately. When managing new servers and communication equipment, accurate grasp of situation is particularly important. In this paper, we propose a status visualization system based on AR which supports administrators in actual work environment. The proposed system graphically visualizes state information of a specific device such as a server or a network switch via a head mounted display (HMD) worn by a system administrator. It also visualizes logical and physical network structures connecting servers and switches. This system assists in efficiently executing computer device management tasks. In addition, it helps administrators to master practical knowledge and skills related to servers and networks [1–3].

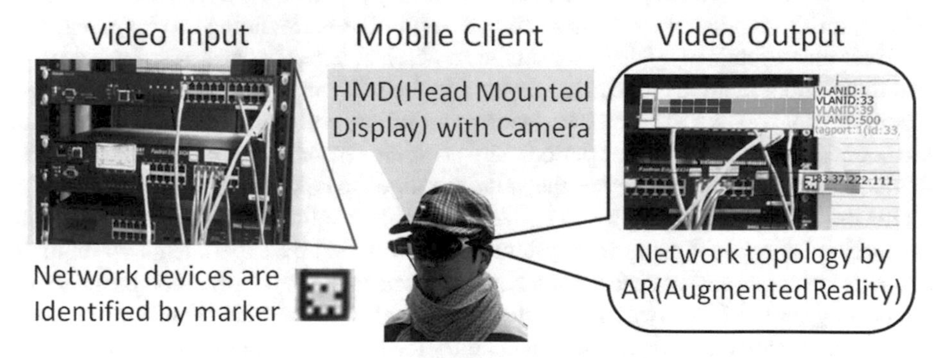

Fig. 2. Network Monitoring Agent for IT Infrastructure Visualizer.

Figure 3 shows a wireless network visualizer. Efficiently managing network infrastructures laid in corporate organizations and universities is a crucial task. Making wireless network segments stably running is particularly an expensive and time-consuming task. Providing a system for continuously monitoring the wireless network and simply visualizing up-to-date network conditions should be a big help for administrators and even for end users. In this paper, we propose a wireless network visualizer based on signal strength observation using IoT devices and lightweight communication protocol. We describe a concept and an implementation approach of the wireless network visualization system [4].

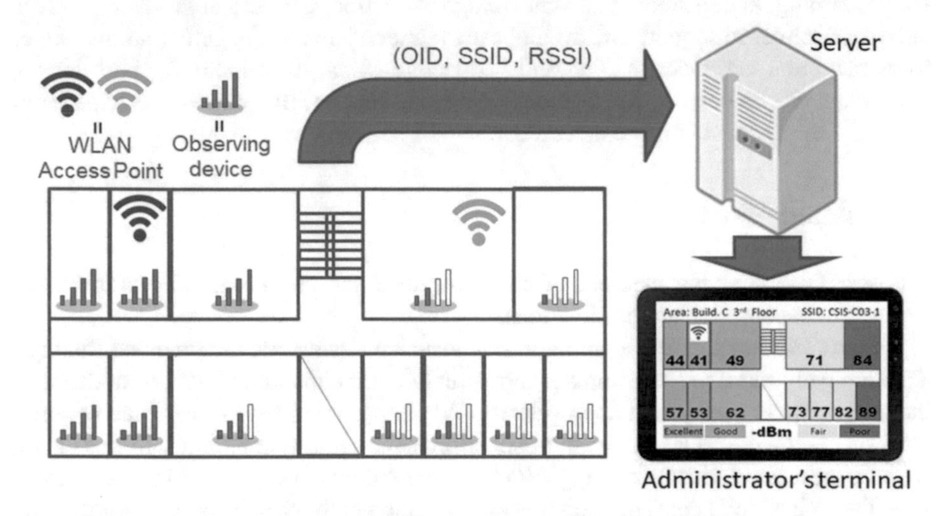

Fig. 3. Configuration of the Wireless Network Visualizer.

Figure 4 shows a heatstroke warning device. For the controller, use Raspberry Pi which is a single board computer of open source hardware and Netatmo Weather Station which is IoT environmental observation device of Netatmo Inc. This sensor has a plurality of sensors (a temperature sensor, a humidity sensor, an atmospheric pressure sensor, a CO_2 concentration sensor) for observing the surrounding environment, and obtains values obtained by these sensors via a wireless network can do. Actuators used, and their output method are realized by turning the LED on/off using GPIO, generating a beep sound, rotating the motor to generate vibration, and using the speaker to sound the sound It was. A system that periodically acquires sensor values observed by the IoT device via the network, processes the value to judge the risk of heat stroke, and manipulates the actuator according to the risk level to present information to construct. In the evaluation experiment using this system, it was revealed by experiments by multiple subjects that when subjects were not paying attention to LEDs, information presentation by turning on/off LEDs was not perfectly noticed. On the other hand, information presentation by warning or vibration can notify the presentation of information without paying any attention at all [5].

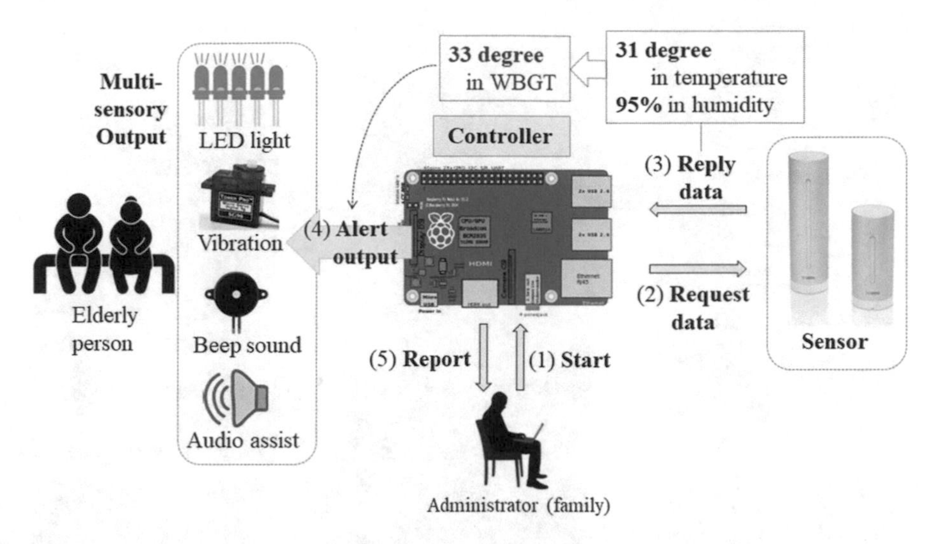

Fig. 4. Configuration of the Heatstroke Warning System.

3 System Design

In this section, we describe the system setting for supporting safe driving using Edge Computer. This design is based on the application of some of the research results of visualization and knowledge of sensor data we have done so far. Figure 5 shows the overall configuration of the system proposed in this paper. This system consists of edge computer and machine learning server.

3.1 Sensors of the Edge Computer

This system consists of a sensor device, a controller, a display, an in-car device main body having a wireless communication function and the like, a sensor which is attached to a driver, a seat, a steering wheel or the like to communicate wirelessly with the main body to collect information, a machine learning configure with server. In this research, we classify the sensors into three categories according to the relationship between the monitored object and the sensor, and cite sensor candidates to be used for each type below.

1. Direct contact sensor: A wrist band type heart rate monitor that acquires vital information, an activity meter, a perspiration meter
2. Indirect contact sensor: Pressure sensor and vibration (IMU) sensor (attached to the seat), Heart rate sensor and perspiration sensor (attached to handle)
3. Non-contact sensor: Infrared type radiation thermometer (non-contact driver's body temperature measurement), Sound sensor (inferring health condition from driver's voice), Visible light camera, Near infrared camera, Ultrasonic sensor (driver's behavior), Temperature Humidity sensor, CO_2 concentration sensor, sound sensor, light sensor, vibration (IMU) sensor, Visible light/near infrared camera, ultrasonic sensor (behaviors other than driver)

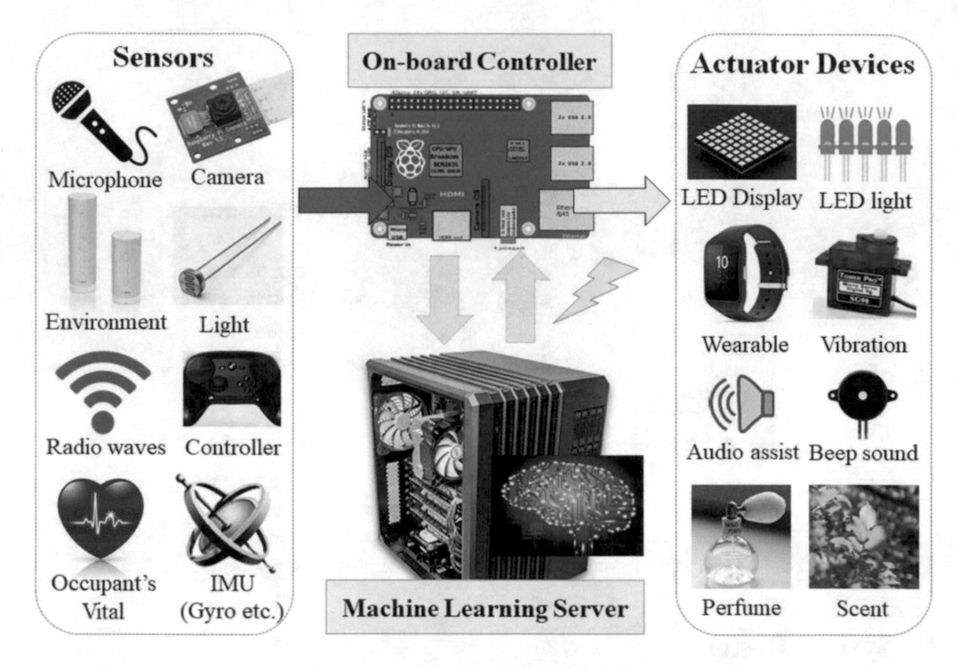

Fig. 5. Overall configuration of the proposed system.

In this system, three different types of sensors are used to monitor the inside of the car and the driver. Figure 6 shows the relationship between these sensors and the objects to be monitored. The direct contact sensor acquires the biological data of the driver. The proximity sensor acquires the motion data of the car. The non-contact sensor acquires the environmental data of the car. The controller processes these sensor data and presents safe driving support information to the driver.

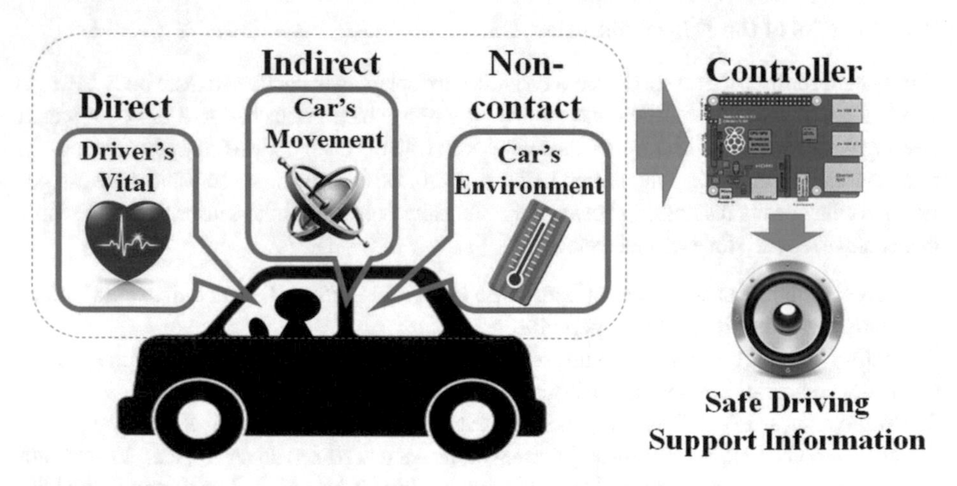

Fig. 6. The relationship between these sensors and the objects to be monitored

3.2 Processing Flow of the System

Figure 7 shows the processing flow diagram of the proposed system. The system runs in a learning mode and a monitoring mode. In the learning mode, the controller uploads the measurement data of the driver's vital sign, car's movement and car's environment to the leaning server via the wireless communication network. These data are stored in the learning server, and a safe driving support model is created using a method such as machine learning. In the monitoring mode, the controller processes the sensing data to determine the degree of risk assessment for the driver. And the controller sends information to the output device according to the result of risk assessment.

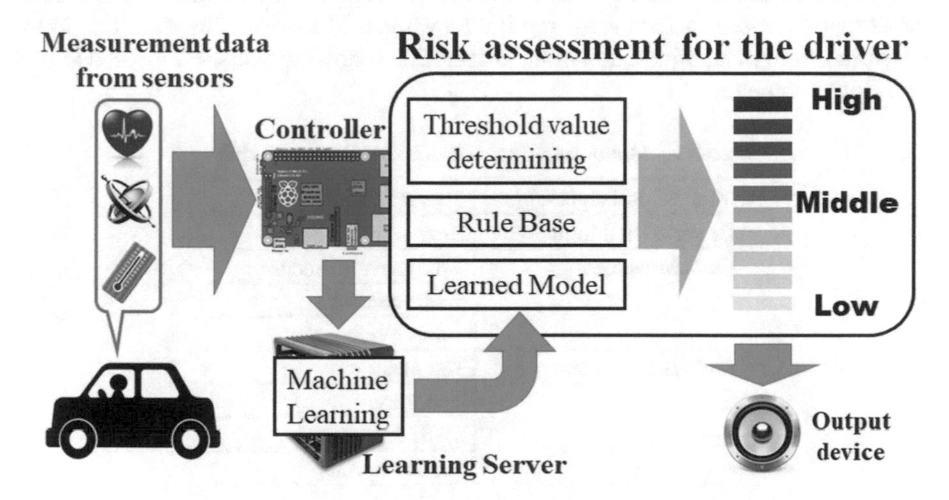

Fig. 7. The flow diagram of the proposed system

3.3 Methods of the Controller and the Learning Server

The controller used in this system calculates the risk by processing the sensor measurements received from the edge computer. There are three ways to calculate the risk calculation. The first method is a method using simple threshold judgment. This is used to determine whether air temperature, humidity, etc. exceed each normal range. The second method is based on rule base. In this method, the result is determined by processing a combination of sensor data using knowledge. The third method is to use a learned model of machine learning. Machine learning leads to new rules by analyzing the obtained data. By processing using a new rule, the learning model, the controller can make appropriate risk assessment from sensor data.

The function of the learning server is to learn the sensor data received from the controller and generate the learned model. There are several ways to generate a learning model, but in this research, teacher data is not prepared to generate a learning model. In order to handle time series information of sensor data, we plan to generate a learning model using RNN (Recurrent Neural Network).

4 System Implementation

In this section, we describe how to acquire in-car and driver data in the system designed in the previous section. For this time, we installed devices installed in the car in three devices for each function. The advantage of being distributed in three devices is that it can be installed in a location suitable for obtaining each sensor. In addition, these devices exchange information with each other by wireless communication.

Table 1 shows monitoring targets and their sensor modules in the system. When implementing this system, the Edge computer was divided into three parts. The first one is equipped with a crew and a sensor to measure the in-car environment. The second one comprises a sensor for measuring the movement of the car. The third one is for measuring biological information of the driver. The following Sects. 4.1 to 4.3 describe each Edge computer.

Table 1. Monitoring target objects and Sensor Modules

Monitoring Target Object	Sensor Module
Driver's Vital Sign	Heart rate
Car's Movement	3-axis accelerometer
	3-axis compass
	3-axis gyroscope
Car's Environment	Temperature
	Humidity
	Light dependent
	Sound intensity

4.1 Edge Computer for Driver's Vital Sign Sensing

In this research, to obtain driver's vital data, use a wearable device. This device can measure heart rate and activity by wearing it on the wrist like a wrist watch. There are methods for measuring the heart rate, respiration, and skin surface temperature of the subject using non-contact sensors. However, in this research, a wrist watch type sensor is used for obtaining biological information by a low cost and highly reliable method. In our system, we use the activity meter provided by Fitbit. This activity meter can measure heart rate and three-axis acceleration. By using this device, data on the heart rate and arm movement during driving is acquired. Figure 8 shows a graph of heart rate data acquired by a device attached to the wrist and a sensor of the device.

This device is always connected with smartphone and Bluetooth Low Energy. Therefore, it is possible to acquire and use sensing data in real time when necessary.

4.2 Edge Computer for Car's Movement Sensing

Figure 9 shows an edge computer for measuring a car's movement. It contains Raspberry Pi to install onboard car operation measurement devices. For Sensor, use Sense Hat which is an add-on type built-in sensor for Raspberry Pi and connect it to GPIO of

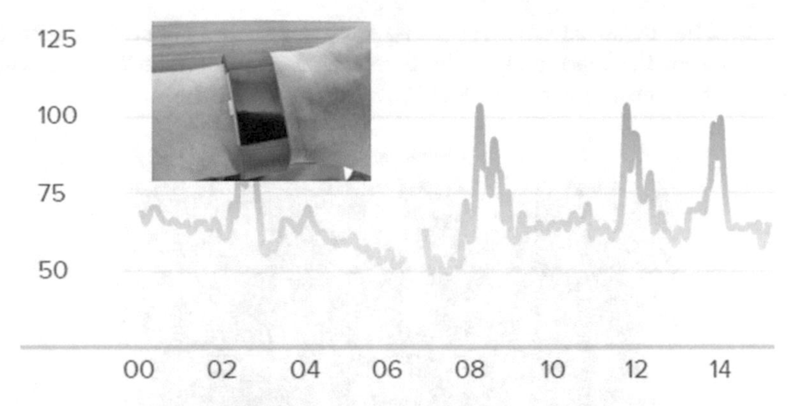

Fig. 8. The monitoring device of vital signs and a graph of heart rate data

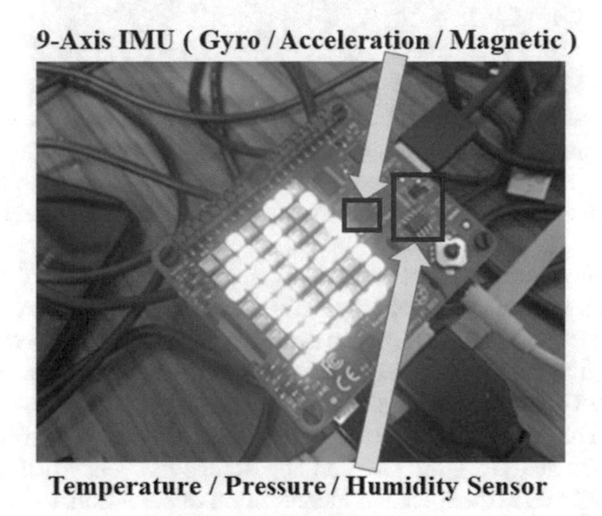

Fig. 9. An edge computer for measuring a car's movement

Raspberry Pi. Sense Hat is equipped with 9-axis multiple sensors (gyro sensor, acceleration sensor, magnetic sensor) that detect the movement of the main body, and sensors (temperature sensor, humidity sensor, atmospheric pressure sensor) that observe the environment around the main body. Based on the sensor value acquired by Sense Hat, the status of the device itself and the surrounding environment are detected, the result is displayed on the LED display, and information is presented to external devices via video output and audio output.

4.3 Edge Computer for Car's Environment Sensing

To measure the in-car environment, the controller uses the Raspberry Pi and the sensor uses the sensor of the Grove System. The Grove System is a sensor module provided by seeed studio that can be used by merely inserting it in a connector. In this time, we mainly record the in-car environment using temperature and humidity sensor, optical sensor, sound sensor, ultrasonic type distance sensor. Figure 10 shows a photograph of edge computer connected with sensors handled in this research.

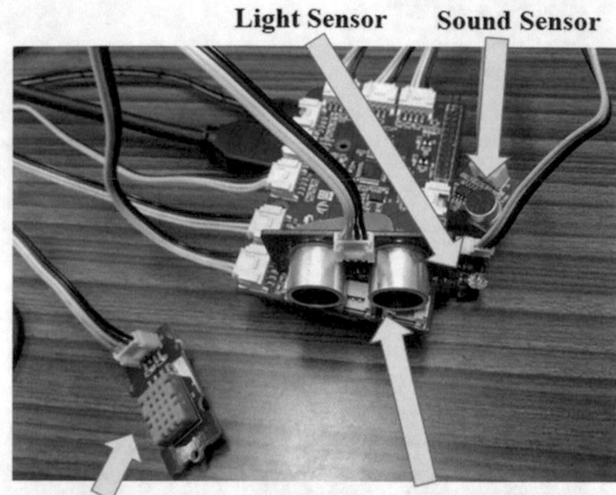

Fig. 10. An edge computer for measuring a car's environment

Temperature/humidity sensors are used to measure the comfort of the in-car environment. When the temperature and humidity inside the car exceeds the appropriate range, the ability required for driving such as concentration tends to decrease. Therefore, when they exceed the appropriate range, a warning is issued. Also, by analyzing the data and the movement of the car, it is possible to investigate whether the in-car environment affects driving. The sound sensor senses sounds emitted from the driver, the car itself, and the driver. By analyzing the size and appearance frequency of the sound on the server, it is thought that it will lead to identification of sounds that may interfere with safe driving. The ultrasonic distance sensor is installed facing the driver. By analyzing the distance data to the object, there is a possibility that the inside of the car or the driver discriminates the state different from normal.

5 Preliminary Experiments

We conducted experiments to acquire sensor information simultaneously using three different Edge Computers described in the previous chapter. As a result, five sensor data (Temperature, Humidity, Light, Sound, Distance) related to the in-car environment and nine sensor data (3-axis gyro, 3-axis acceleration, 3-axis magnetic), and we could simultaneously acquire driver's pulse data and uploaded it to the server.

In addition, we created a processing model that identifies the situation by machine learning using the acquired 15 types of sensor data. We confirmed that a mechanism to send the model from the server to the edge computer side and present the information by judging the situation from the sender data has been functioning correctly. We also confirmed that edge computers and friends can communicate via Bluetooth communication without going through the server on the clad side, so that the data of each edge computer can be exchanged mutually.

6 Conclusions

We proposed a system to support driver monitoring and safe driving support information by edge computer implemented by inexpensive IoT device as a safe driving support method for drivers in operation. In this proposal, we constantly monitored the interior of the car including the driver, in addition to researching the IoT equipment we have worked on so far, and worked to estimate the risks leading to traffic accidents from the relationship with the movement of the car.

In the system proposed in this paper, several preliminary experiments were conducted to verify the effectiveness of presentation of safe driving support information. We conducted preliminary experiments using prototype in-car equipment, quickly detected onboard environment and drivers that changed with multiple sensors and confirmed whether advice is possible. As a result, we were able to acquire sensor data from multiple edge computers.

It was confirmed that by creating a learning model with an edge computer, it was possible to judge the input value of the sensor, and to issue a warning that the state is different from the usual state. In the future, it is necessary to verify the operation of the public road driving by installing it in the car of the actual passenger car, and further verify the usefulness of the safe driving support function of this system. Also, among the functions proposed this time, by utilizing the function of accumulating the data acquired by the Edge and periodically sending it to the cloud server, it is possible to graphically display the acquired data and graphically predict the operation of the user can. In the future, to realize more efficient system management, we plan to realize a mechanism that predicts occurrence of management risks and accidents with high probability by processing by machine learning based on a large amount of observation data.

Acknowledgments. This work was supported by JSPS KAKENHI Grant Number 15K00277.

References

1. Haramaki, T., Nishino, H.: A network topology visualization system based on mobile AR technology. In: Proceedings of the 29th IEEE International Conference on AINA-2015, pp. 442–447 (2015)
2. Haramaki, T., Nishino, H.: A device identification method for AR-based network topology visualization. In: Proceedings of the 10th International Conference on BWCCA-2015, pp. 255–262 (2015)
3. Haramaki, T., Nishino, H.: A sensor fusion approach for network visualization. In: Proceedings of IEEE International Conference on 2016 ICCE-TW, pp. 222–223 (2016)

4. Haramaki, T., Shimizu, D., Nishino, H.: A wireless network visualizer based on signal strength observation. In: Proceedings of IEEE International Conference on 2017 ICCE-TW, pp. 23–24 (2017)
5. Yatsuda, A., Haramaki, T., Nishino, H.: An unsolicited heat stroke alert system for the elderly. In: Proceedings of IEEE International Conference on 2017 ICCE-TW, pp. 345–346 (2017)

A Mobile Wireless Network Visualizer for Assisting Administrators

Dai Shimizu, Toshiyuki Haramaki, and Hiroaki Nishino(✉)

Division of Computer Science and Intelligent Systems, Faculty of Science and Technology, Oita University, Oita, Japan
{v16e3009,haramaki,hn}@oita-u.ac.jp

Abstract. As a performance index of wireless network, a heat map function for visualizing the distribution of signal strengths received from APs (Access Points) is a useful tool. Existing technologies, however, have some problems such as insufficient automation for acquiring signal information to visualize and inflexibility for dealing with on-demand monitoring requests issued by administrators on site. We propose a practical method for stably monitoring signal conditions in a managed site and efficiently visualizing the observation results. We implement the proposed method based on a light-weight publisher-subscriber communication framework applicable for various IoT applications. It can handle the on-demand monitoring requests for immediately visualizing the latest condition while it is constantly monitoring whole area as a background process. In this paper, we describe background and purpose, implementation detail, and preliminary evaluations of the proposed system.

1 Introduction

Wireless networks are prevailing access gates to backbone networks managed by companies, universities, and ISPs. Since they can easily be used without wiring, users tend to place additional APs (Access Points) as needs arise. Some users may activate temporary devices such as personal WiFi routers in conjunction with stationary APs in offices and workplaces. These facts make wireless networks more complex and harder to manage than wired networks for network administrators. They consume enormous amounts of time for isolating a problem source when a communication failure occurs. They also pour efforts for finding an optimum spot when placing a new AP to improve the stability and performance of the network in a managed area. Additionally, it is a common situation where end users without technical skills and experiences need to manage wireless networks due to the lack of full-time administrators. Therefore, an effective tool for visualizing communication status of wireless networks in an easy-to-understand manner is desired to efficiently supporting administration tasks.

A direct indicator for measuring the performance and stability of wireless communication is to observe the strength of radio waves (signal strength) received from installed APs. There are some free software tools for measuring the signal strength in the office and visualizing it as a colored heat map. Figure 1 shows example heat maps monitored

L. Barolli et al. (Eds.): EIDWT 2018, LNDECT 17, pp. 651–662, 2018.
https://doi.org/10.1007/978-3-319-75928-9_58

in our laboratory environment by using a tool called WiFi-mireru [1]. It displays a colored image where detected signals are strong in blue and green colors, weak areas in yellow and red. The users can intuitively grasp the differences of the radio signal distribution in some time zones such as early morning or late night when a few users present, and daytime when many users are working.

When the signal condition in a certain area deteriorates, adjusting neighboring APs' position or their antenna direction may improve the situation [2]. In such case, the administrator may confirm the improved condition by comparing the heat maps before and after the adjustments. He/she must, however, travel around the area and manually measure the strength of all receivable radio waves at multiple locations to get a shot whenever he/she makes an adjustment. Enforcing such measurement action every time to get a heat map increases the administrator's workload. To solve this problem, an *active monitoring* function is required. It is a function activated by the administrator on demand, concurrently observing and gathering the latest signal strengths at multiple locations, and immediately visualizing the result as an updated heat map on the administrator's terminal. It allows the administrator to efficiently confirm the effect of the adjustment task without travelling the area.

Meanwhile, a *passive monitoring* function, which periodically monitors and visualizes the signal condition in the management area as a background process, is also an important function for practical network management. It provides clues to detect some anticipated serious conditions in advance such as degradations and fluctuation of observed signal strengths.

In this paper, we propose a method for implementing the above-mentioned signal monitoring functions, active and passive monitoring functions. The proposed system allows the administrator for easily and quickly checking the signal condition on his/her mobile terminal whenever necessary. Additionally, the system constantly collects the latest signal information from multiple monitoring points with low communication overheads, and provides a function to perform trend analysis over a certain period from the history of monitoring data.

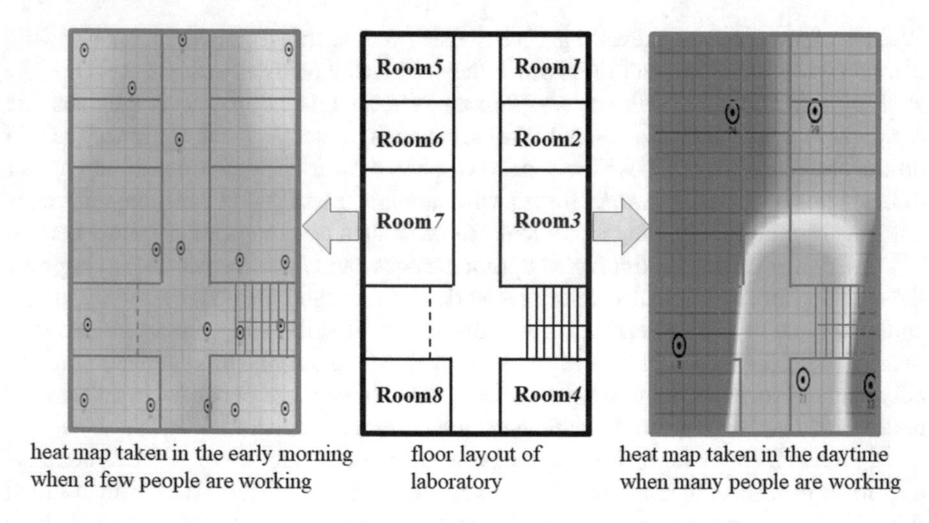

Fig. 1. Heat map examples taken in different time of day.

2 Related Work

There are some previous research activities for aiding administrators by automatically monitoring signal conditions. There is a publicly available software tool for manually creating a heat map as mentioned in Sect. 1 [1]. However, making a shot needs to take at least a few minutes for moving and manually measuring in some locations in our laboratory as shown in Fig. 1.

Kim et al. proposed a method for automatically surveying signal conditions in a specific site using a mobile robot [3]. They installed a monitoring device on a mobile robot, and implemented a control system for automatically running the robot on a planned route and performing measurement at some reference points. They set multiple operation modes with different number of measuring points such as a complete mode to fully perform the measurement in the night and a selective mode to partially measure in the daytime. Although this is an effective method to automate the passive monitoring function, it cannot deal with on-demand requests such as the active monitoring function.

WiFi based indoor localization techniques estimate the current position of a measurement point by using observed signal strengths. There is a method to improve the positioning accuracy by deploying a large number of sensors [4]. This is also an effective approach for automating the signal condition monitoring, so we devised and implemented an initial solution by deploying sensors specific for the monitoring task [5]. However, some problems became apparent such as additional costs for the dedicated sensor placements and unstable monitoring results by the wireless sensors. Although detecting the deterioration of wireless signal condition particularly is a vital issue to solve, we faced the problem that the solution could not guarantee for monitoring the wireless network by itself.

Using computers in operation with wired connections as monitoring nodes is a solution to this problem. Bahl et al. proposed a method for using desktop computers in operation as monitoring nodes [6]. Although the method can stably monitor the signal condition, it cannot flexibly dealing with the dynamic nature of wirelessly connected devices such as notebook PCs and mobile terminals.

While the proposed system in this paper is similar to the method of Bahl et al., we use portable notebook PCs in operation with both wired and wireless connections as monitoring nodes. Additionally, we devised and implemented an efficient system architecture based on a light-weight publisher-subscriber communication framework usable for various IoT applications consisting of many mobile nodes.

3 System Implementation

3.1 Publisher-Subscriber Communication Model

Because the proposed system must constantly be up and running for watching the received signal strengths from all APs installed in a management area, it needs to operate at low overhead and minimize the impact on user's computing power and network traffic. Additionally, the scale and topology of wireless network are frequently changed depending on organizational restructurings, the system should be scalable and flexible for effectively dealing with these dynamic changes.

We implement the system based on the publisher-subscriber model as shown in Fig. 2 to satisfy the above-mentioned requirements. As shown in the figure, an arbitrary number of publishers and subscribers exchange messages via a broker. The publishers are modules for sending messages and the subscribers are dedicated for receiving messages. They exchange each message in a specific class set individually in advance by both sides and the broker manages the message exchange between the publishers and subscribers. Because the publishers and subscribers are loosely coupled based on the asynchronous communication model via the broker, it enables the system to be more scalable and flexible for handling dynamic network configuration than popular synchronous protocols used in many IoT applications such as http and ftp.

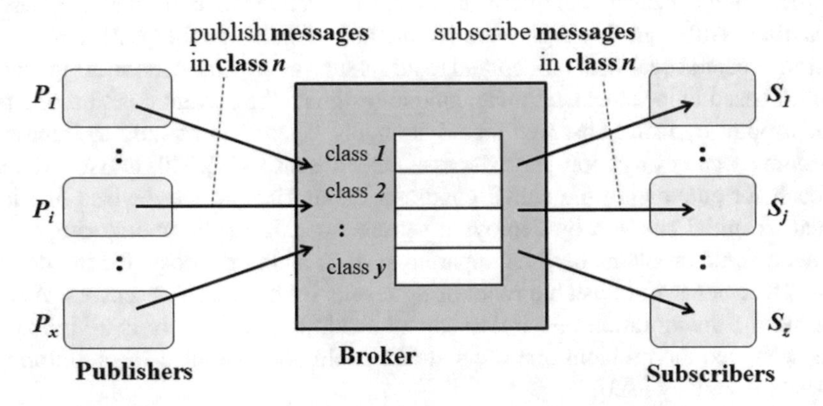

Fig. 2. A system architecture based on publisher-subscriber model.

3.2 System Organization

Figure 3 shows an example system organization. The system consists of three components such as a *monitor node* installed in each room for continuously monitoring signal strength information, a *server* to act as a broker for managing signal information exchange in the system, and an *administrator's mobile terminal* for visualizing the signal condition as a heat map in an management area.

Robustly detecting any spots where their signal conditions become unstable is particularly an important requirement for generating an effective heat map. The proposed system, therefore, uses running notebook PCs with both wired and wireless connections as monitor nodes. We adopt a light-weight network framework called MQTT (Message Queue Telemetry Transport) [7] which is a type of the publisher-subscriber protocol as explained in Sect. 3.1. It is an application level protocol running on TCP/IP in the protocol stack. In the server, a publicly available MQTT broker module called mosquitto [8] is running. The server also acts as a logger for recording the signal information collected from the monitor nodes and an analyzer for visualizing the transitional trends of signal states in the entire management area or for a specific AP.

In the passive monitoring mode, the administrator's mobile terminal acts as a subscriber for periodically receiving the signal information from all of the monitor nodes via the server and continuously updating a heat map image. In the active monitoring mode, the administrator's mobile terminal initiates an on-demand signal monitoring request as a publisher through the server. The implementation method of the two monitoring modes is elaborated in the next section. The administrator usually carries the mobile terminal for watching the signal status in the management area and finding an optimum adjustment condition by requesting the active monitoring function when he/she adjusts a specific AP to improve the signal condition. He/she can timely check the signal status in the management area by using the server's analyzing function on the mobile terminal or on the server's monitor.

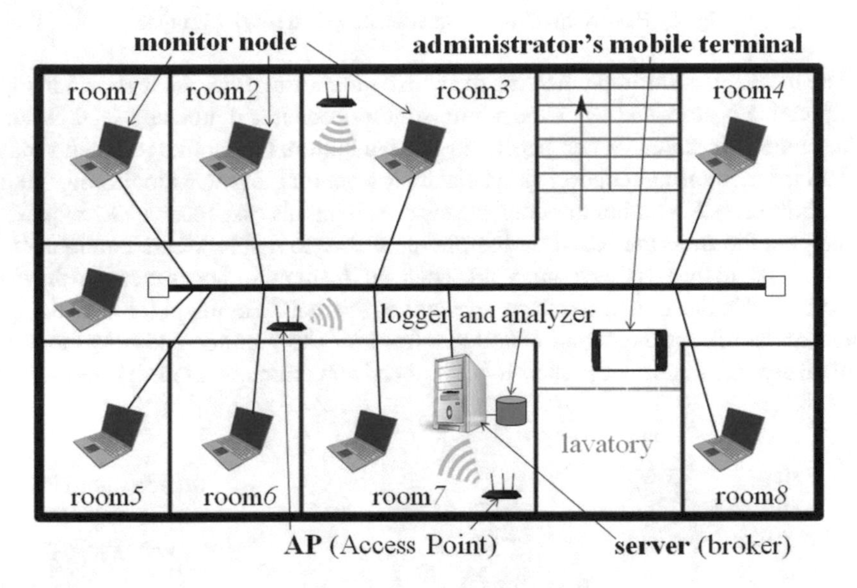

Fig. 3. Example system organization.

3.3 Scalable Communication Based on MQTT Protocol

Figure 4 shows the mechanism for implementing the passive monitoring function. The broker manages information exchange between publishers and subscribers via classified topics. Prior to starting information exchange, both publisher and subscriber need to agree on which topic to exchange signal information as a message. The system assigns each room to measure with a different topic name. Each monitor node periodically sends a signal condition status to the topic assigned to its monitoring room as a publisher. Then, the administrator's mobile terminal receives messages from all the topics as a subscriber and visualizes the results as a heat map. Therefore, the passive monitoring function uses the n-publishers to one-subscriber communication style as shown in Fig. 4.

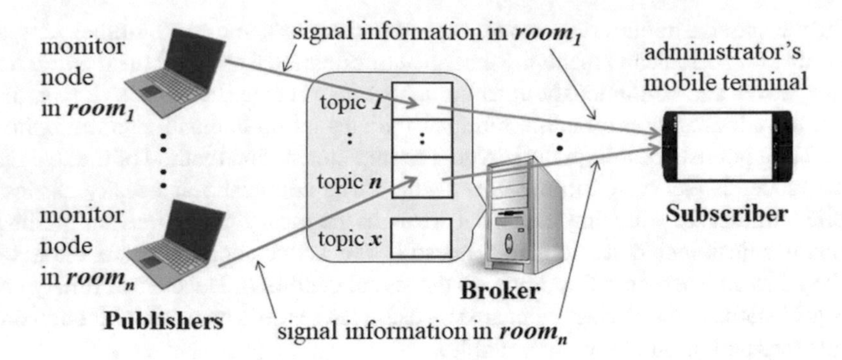

Fig. 4. Passive monitoring function based on MQTT protocol.

The mechanism for implementing the active monitoring function is shown in Fig. 5. This function is activated when the administrator needs it and presses a GUI button on his/her mobile terminal. When he/she pushes the button, the mobile terminal sends an on-demand monitoring request to a special topic x set for the active monitoring function as a publisher. All the monitor nodes as subscribers in this case receive this requests by exchanging the message based on the one-publisher to n-subscribers communication style opposite to the passive monitoring. Then, all the monitor nodes measure the signal conditions at that time, and sending the results to the mobile terminal for visualizing the heat map. Notifying the signal conditions from all the monitor nodes to the mobile terminal uses the same n-publishers to one-subscriber communication style as illustrated in Fig. 4.

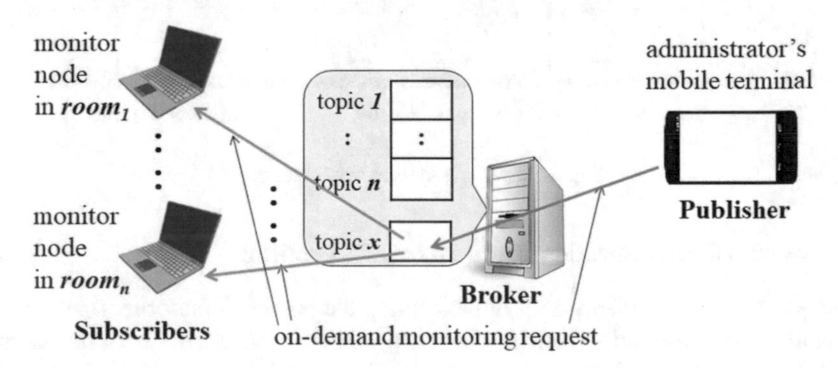

Fig. 5. Active monitoring function based on MQTT protocol.

3.4 Software Architecture and Processing Flow

Figure 6 shows all the software modules running on the monitor node, the server, and the administrator's mobile terminal. Both the monitor node and the mobile terminal are equipped with a pair of publisher and subscriber modules because they need to play both roles depending on the monitoring mode (active or passive) as explained in Sect. 3.3.

Additionally, the monitor node has a sensing module for measuring the signal condition and the mobile terminal has a graphics module for rendering the heat map based on the collected data, respectively. The monitor node uses a free software tool called WirelessNetView [9] as the sensing module. It is a small program running in the background, and gathering various information of detected wireless networks such as SSID (Service Set IDentifier), RSSI (Received Signal Strength Indicator), signal quality, MAC Address, channel frequency, and channel number. We coded the modules running on the monitor node by using Python and the modules on the mobile terminal by using Java, respectively. In addition to the mosquitto broker, the server is equipped with the logging module for accumulating and managing the signal information exchanged between the monitor nodes and the mobile terminal, and the analyzation module for visualizing the trends of signal conditions and predicting any possible failures based on the log data file.

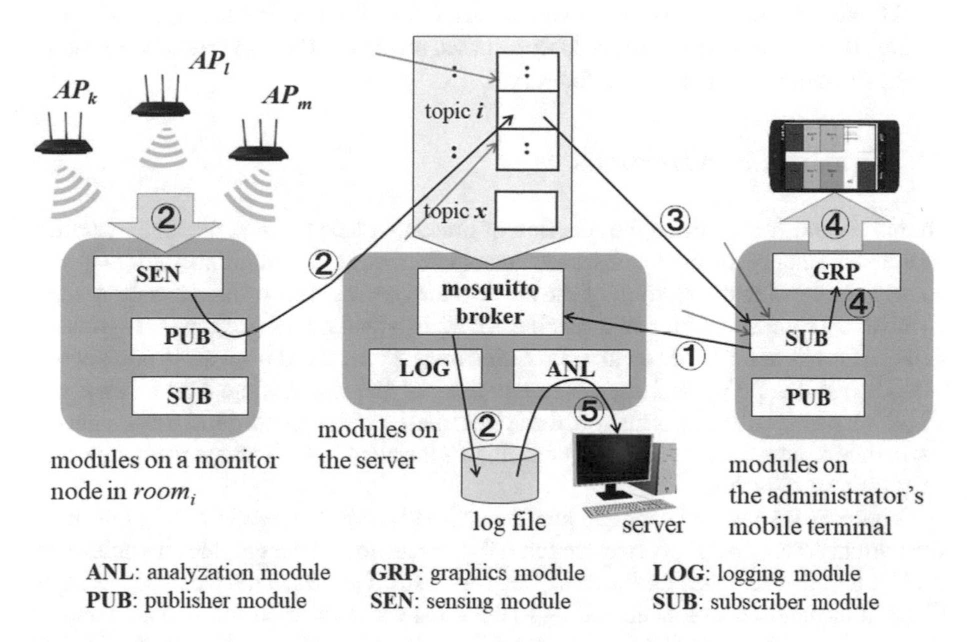

Fig. 6. System software architecture and processing flow in the case of passive monitoring.

Figure 6 also shows the processing flow in the case of the passive monitoring function. The following numbered items correspond to the numbers shown in the Fig.

1. Immediately after system startup, the subscriber module on the mobile terminal requests the broker for subscribing the signal information and waits for the information to be published. Because the system exchanges the information by setting separate topics for different rooms to monitor, the terminal subscribes all the topics to visualize. In the case of active monitoring function, the subscriber module on each monitor node requests to the broker for subscribing the on-demand monitoring request activated by the mobile terminal as explained in Sect. 3.3.

2. The publisher module on each monitor node periodically requests the sensing module to get the wireless activity information and sends it as a message to its corresponding topic. The logging module on the server concurrently saves the published message to the log file in a format usable for later analysis. In the case of active monitoring function, the publisher module on the mobile terminal sends the on-demand monitoring request to the topic x when the administrator activates the function.
3. The broker sends the message to the subscriber module on the mobile terminal waiting for it to be published via the designated topic. The terminal receives all messages to be published through the subscribing topics.
4. The subscriber module notifies the graphics module for receiving the latest signal information as messages through all subscribing topics. Then, the graphics module draws and displays a heat map to the monitor using the information.
5. The administrator can use the analyzation module supported on the server whenever he/she needs to perform trend analysis based on the accumulated signal information on the mobile terminal or on the server.

4 Visualization Functions

In this section, we describe the function of drawing a heat map on the administrator's mobile terminal based on the observed signal information. The administrator may wish to monitor the signal condition of the entire management area, or he/she only needs to monitor the signal condition of a specific AP when adjusting the position or the antenna orientation of the AP. The signal information message received from each monitor node includes SSID, RSSI, and measurement time of the monitoring APs as shown in Fig. 7(a). Based on this information, the system makes and presents the above-mentioned two kinds of heat maps, a heat map for all APs installed in the management area and a map for a specific AP.

Figure 7(b) shows an example heat map visualized for a specific AP. In this mode, the administrator firstly indicates which AP to visualize and the graphics module on the mobile terminal searches for the designated AP's RSSI values in the received messages. Then, it paints room colors according to the RSSI values. We set the signal strength in four classes as shown in Table 1 in accordance with a grouping rule adopted by a WiFi scanning software tool called inSSIDer [10]. We use cold colors (blue and green) for stronger signal classes and warm colors (yellow and red) for weak ones, resulting in the similar coloring as shown in Fig. 1. Figure 7(c) shows an example heat map visualized for all APs installed in the management area. In this mode, the graphics module searches for the maximum RSSI values in the received messages and paints all room colors according to the values. Because this mode simultaneously displays the SSID of the strongest signal AP when coloring each room, the administrator can easily confirm which AP to receive the strongest signal in each room. The administrator can select which visualization mode to use by GUI buttons on the mobile terminal.

```
SSID    : AP1
RSSI    : -65
Time    : 2017/10/23 14:06:23
SSID    : AP2
RSSI    : -45
Time    : 2017/10/23 14:06:24
SSID    : AP3
RSSI    : -57
Time    : 2017/10/23 14:06:24
```

(a) example signal information message

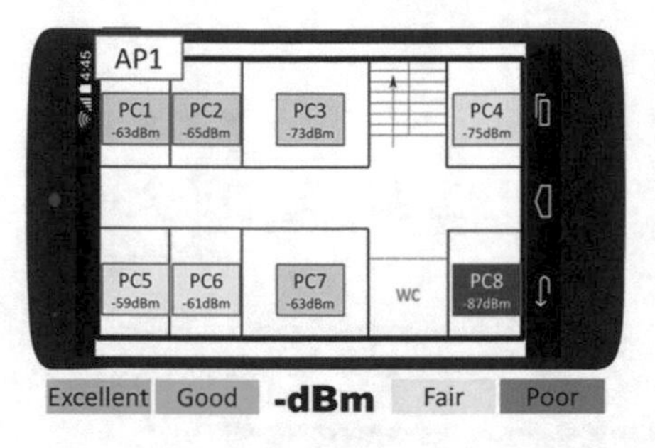

(b) heat map for a specific AP

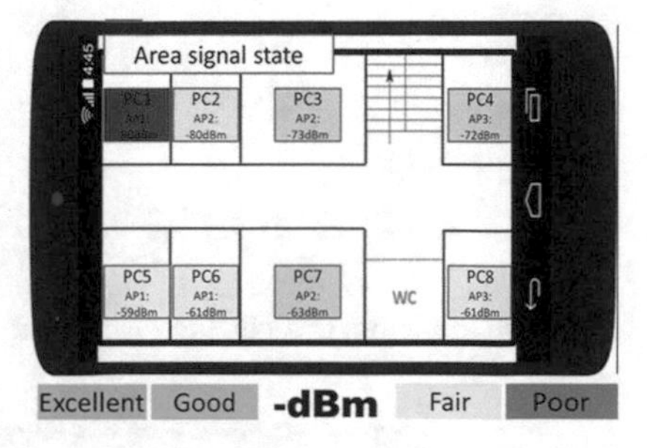

(c) heat map for all APs in the area

Fig. 7. Signal condition visualization function.

Table 1. Classification of signal strength for coloring.

Signal quality	Color in a map	RSSI value (dBm)
Excellent	Blue	−61 or higher
Good	Green	−74 through −62
Fair	Yellow	−86 through −75
Poor	Red	−87 or lower

5 Preliminary Evaluations

We conducted test operations for several weeks to verify the effectiveness and practicality of the system in the laboratory environment as shown in Fig. 3. We had been continuously monitored the signal conditions received from three APs installed in room 3, room 7, and hallway as illustrated in the figure. We deploy a running windows notebook PC with wired and wireless connections in each room as a monitor node, a desktop PC (HPE 560) with guest Linux OS as a server, and a Nexus 7 android tablet as an administrator's mobile terminal, respectively.

When checking signal conditions in a specific room during the daytime by using the passive monitoring function, the signal strength observed from an AP placed in the same room is stable and stronger than the signal from other AP in the adjacent room across the hallway. A graph in Fig. 8(a) shows the histogram of RSSI values observed in the former case, indicating that strong and stable signals in the range of −45 to −47 (dBm) are always detected. On the other hand, we can confirm that multiple weaker range signals are observed in the histogram of Fig. 8(b) showing the latter case. We set one second as an update frequency for the periodical monitoring and observed no explicit performance degradation on the monitor nodes.

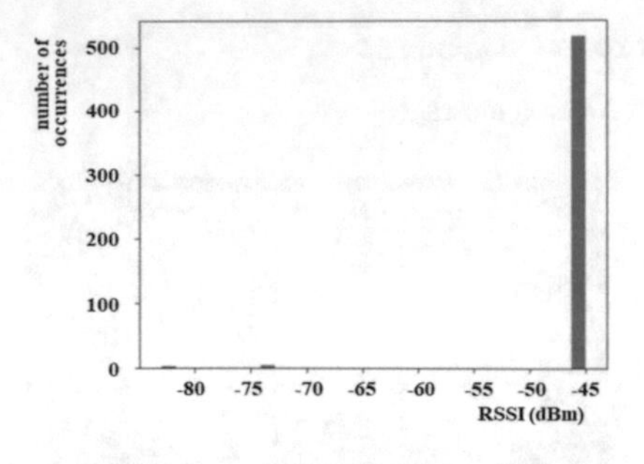

(a) RSSI histogram in room 3 received from an AP installed in the same room

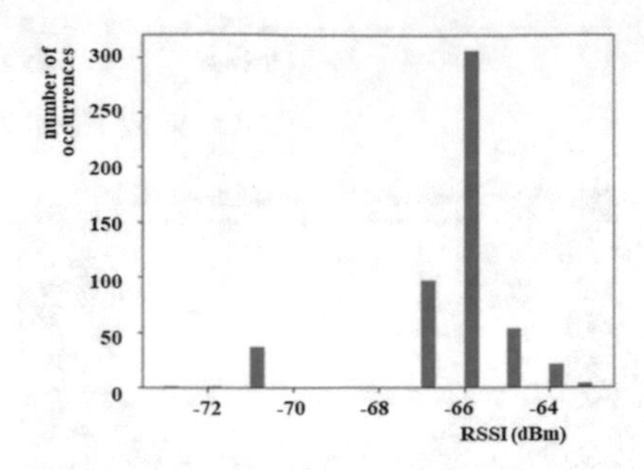

(b) RSSI histogram in room 3 received from an AP installed in the neighbor room

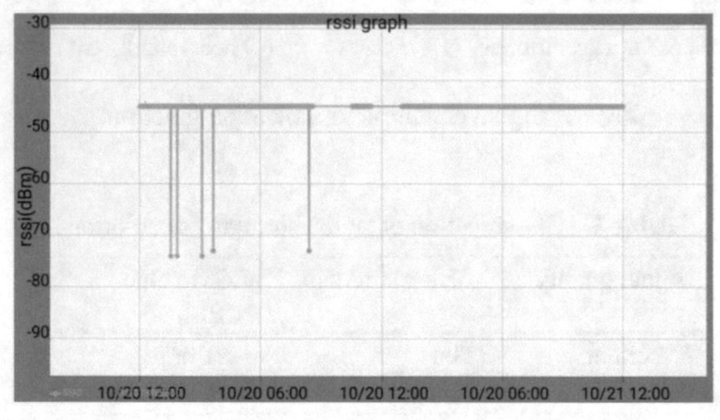

(c) Stable trend example of signal strength over 24 hours

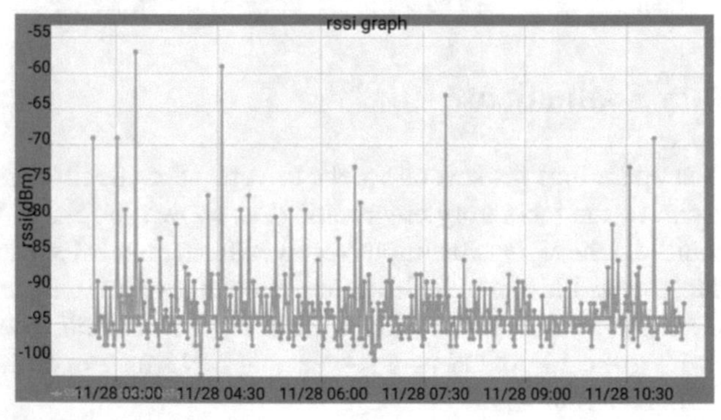

(d) Unstable trend example of signal strength in time zone where many users are working (2: 30 pm to 23: 30 pm)

Fig. 8. Observed signal conditions by passive monitoring function.

When we were observing the signal condition in the whole floor by using the heat map shown in Fig. 7(c), we could confirm whether the communication quality is good or bad by analyzing the trend of signal strength over a certain period of time. Figure 8(c) shows a graph of a room where a stable strong signal is constantly observed. The horizontal axis shows the monitoring time (24 h) and the vertical axis shows the observed RSSI values. A monitor node in this room have always been received a strong signal of roughly −45 dBm throughout the day and we confirmed that wireless connection in this room is quite stable. Meanwhile, Fig. 8(d) is an unstable trend example. In the graph, varied RSSI values around the range of −90 to −97 dBm are observed in time zone where many users are working (2:30 pm to 23:30 pm). It should be difficult for stably connecting to a wireless segment under these RSSI values. We confirmed that this is the room 8 and unstable connections were frequently occurring in this room. These graphs are produced by using the analyzation module supported on the server and the administrator can take a look at these graphs on a mobile terminal or on a server monitor.

After we found the signal condition in the room 8, we tried to improve the situation by adjusting the antenna direction of an AP installed in the hallway. We could monitor the latest signal condition in the whole floor by using the active monitoring function during the adjustment task. After the administrator activated the active monitoring function from the mobile terminal, the latest heat map was rendered on the terminal in a few seconds. As a result, we could improve the signal condition about 10 dBm stronger for the room 8 in a few minutes task while maintaining other rooms' signal conditions. Dealing with this kind of management task requiring immediacy is a difficult issue for existing management tools.

6 Conclusions

We proposed a method for constantly monitoring signal strength received from installed APs in a management area. We designed the system architecture based on a light-weight publisher-subscriber communication framework usable for various IoT applications and implemented a prototype system based on the architecture. The system can deal with on-demand whole area monitoring request frequently required by administrators while it is constantly monitoring the whole signal condition in a background process. Because the administrators can check the monitoring results as a heat map visualized on a mobile terminal at their working place, the system can significantly reduce their time and cost for performing maintenance tasks. They can also look at various trends in some graph formats created by stored monitoring data for analysis. We operated the system in our laboratory environment to evaluate its effectiveness and the results are promising.

In the near future, we would like to conduct more comprehensive experiments in a larger site for quantitatively testing the scalability and applicability of the system in a real network administration environment.

Acknowledgments. This work was supported by JSPS KAKENHI Grant Number 15K00277.

References

1. I/O Data Device, Inc. http://www.iodata.jp/news/2016/information/wifimireru.htm. Accessed Dec 2017
2. Funabiki, N., Soe Lwin, K., Kuribayashi, M., Lai, I.-W.: Throughput measurements for access-point installation optimization in IEEE 802.11n wireless networks. In: Proceedings of the IEEE ICCE-TW 2016, pp. 218–219 (2016)
3. Kim, K.-H., Min, A.W., Shin, K.G.: Sybot: an adaptive and mobile spectrum survey system for WiFi networks. In: Proceedings of the Sixteenth Annual International Conference on Mobile Computing and Networking, pp. 293–304 (2010)
4. Yin, J., Yang, Q., Ni, L.M.: Learning adaptive temporal radio maps for signal-strength-based location estimation. IEEE Trans. Mob. Comput. **7**(7), 869–883 (2008)
5. Shimizu, D., Haramaki, T., Kagawa, T., Nishino, H.: Development of a low cost wireless network condition visualization system for administrators. In: Proceedings of the JCEEE Kyushu - International Session, pp. 83–84 (2017)
6. Bahl, P., Padhye, J., Ravindranath, L., Singh, M., Wolman, A., Zill, B.: DAIR: a framework for managing enterprise wireless networks using desktop infrastructure. In: Proceedings of the Fourth Workshop on Hot Topics in Networking (HotNets-IV), 1–6 (2005)
7. MQTT. http://mqtt.org/. Accessed Dec 2017
8. Mosquitto. https://mosquitto.org/. Accessed Dec 2017
9. WirelessNetView http://www.nirsoft.net/utils/wireless_network_view.html. Accessed Dec 2017
10. MetaGeek: inSSIDer User Guide. http://www.4gon.co.uk/documents/metageek_inSSIDer_wifi_scanner_user_guide.pdf. Accessed Dec 2017

C++ Memory Detection Tool Based on Dynamic Instrumentation

Siran Fu[1,2(✉)], Baojiang Cui[1,2], Tao Guo[1], and Xuyan Song[1,2]

[1] School of Cyberspace Security, Beijing University of Posts and Telecommunications, Beijing, China
{fusiran,cuibj}@bupt.edu.cn, guotao@163.com
[2] National Engineering Laboratory for Mobile Network, Beijing, China

Abstract. C++ language has the characteristics of flexible programming, high execution efficiency, but there are also a large number of undefined behavior, which is easy to cause security risks. In this paper, focus on the memory-use error in C++ program, designed and implemented a memory check tools named MemDetect, based on dynamic instrumentation platform, which is platform-cross, efficiency and accuracy. MemDetect can detect memory leaks, cross-border access memory and memory does not match the release problems effectively, the validity and efficiency of MemDetect are proved by comparing with other detection tools.

1 Introduction

C++ language is an object-oriented high-level language. It has high abstraction ability, and its execution efficiency is also very high, so its application scope is quite extensive. As shown in Fig. 1, C++ has been the top three in the programming language rankings over the past ten years.

Compared with C++ and Java and other high-level languages, developers need to manage its own memory in programming [1], need to be involved in program memory operation is very clear, so if the developer is not clear of the memory operation, it's easy to cause memory management C++ application error. Compared with other program defects, memory problems are hard to detect, especially memory leaks. It is different from other memory errors, such as multiple release, illegal access, or the release of more exposed, memory leaks are in general better hidden, looking for memory leaks is tantamount to tens of thousands of lines of code to look for a needle in the ocean. Therefore, it is very practical to check memory errors with memory analysis tools, which not only improves the quality of software, but also shortens the development cycle of software.

According to the C++ program memory management errors, analysis in the basic research and the existing memory detection methods, based on dynamic binary instrumentation framework Pin, tracking C++ programs' memory operations by using the shadow memory technology, we design and implements a detection tool-MemDetect. This tool can detect memory leak, memory access violation, not match the release and release of multiple memory errors, not only supports C++ dynamic memory allocation,

L. Barolli et al. (Eds.): EIDWT 2018, LNDECT 17, pp. 663–674, 2018.
https://doi.org/10.1007/978-3-319-75928-9_59

but also the memory management functions to a user-defined monitor and management. Finally, an experiment is designed to verify the function and performance of MemDetect. At the same time, it is compared with some existing memory detection tools. The results show that MemDetect can achieve dynamic memory leak detection in C++ programs [2], and has good performance indicators. The main feature of this tool is that it does not need the source code of the tested program, and directly analyzes and detects binary programs dynamically. Therefore, it does not need to be recompiled to provide transparency for the tested code.

Programming Language	2017	2012	2007	2002	1997
Java	1	1	1	2	17
C	2	2	2	1	1
C++	3	3	3	3	2
C#	4	4	6	10	-
Python	5	7	7	16	27
PHP	6	5	4	6	-
JavaScript	7	9	8	7	20
Visual Basic .NET	8	24	-	-	-
Perl	9	8	5	4	3
Assembly language	10	-	-	-	-
Lisp	28	12	12	9	7
Ada	30	15	16	15	10

Fig. 1. Programming language rankings

2 Related Works

This section mainly introduces C++ memory management defects, dynamic binary piling platform Pin, and shadow memory technology.

2.1 Introduction of C++ Memory Management Defects

C++ language memory management is not like Java language which is automatic allocation and recovery by the system responsible for memory, but by the developers themselves to manage memory allocation and recovery, the security mechanism of the C++ language and flexibility, but there is memory management defects.

Memory management defects mainly include memory leakage, memory mismatch release, and memory duplication release [3].

Refers to a memory leak, and no release to have any memory after use, with long time running, large memory did not apply for a release operation causes the system memory resources available to reduce until exhausted, resulting in the program is running slow or even termination or collapse.

Memory mismatch release means that the library function/operator of memory application does not match the library function/operator that releases the block memory, or the size of the requested memory is different from the size of the released memory, resulting in memory errors.

Memory duplication is a memory error caused by the release of the memory that has been released again.

2.2 Introduction of Pin

Pin is a dynamic binary piling framework developed by Intel. It can run on Linux, Unix, Mac OS and Windows systems, supporting ARM and IA-32 instruction set. Pin is a closed source binary piling framework [4], but it has an active developer community (PinHeads), and provides a wealth of API for developers to develop detection and analysis tools-Pintool, and Pintool interacts with Pin in the form of plug-ins. The system architecture of Pin is shown in Fig. 2.

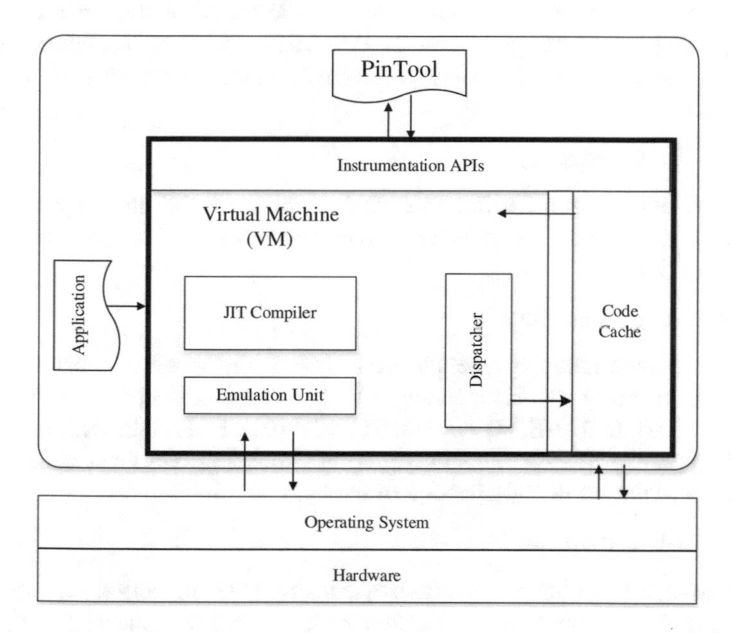

Fig. 2. System structure of Pin

2.3 Shadow Memory

Shadow memory technology is an important technology in the field of program dynamic detection, the technology is to open a special memory space to map the actual program memory usage, by inserting the detection and analysis of code in the target program, the actual operation of the actual operating sequence tracking memory in the process of the target program, and relevant information will be recorded into the shadow track in memory, program of defect detection program according to the shadow memory information detection [5]. At present, many program dynamic analysis tools use shadow memory technology, including Valgrind's MemCheck, memory boundary detection tool Annelid, data flow visualization tool Redux and so on.

The detection effect of shadow memory is very good, but it is not easy to achieve in practical engineering practice. It is difficult to realize the efficiency and accuracy of detection. Because the shadow memory is the mapping of memory space to the actual operation of the program, so considering the accuracy of detection, we first need to insert a lot of analysis code to track all the memory operation instructions of the target program and update the corresponding shadow memory in real time. It is also necessary to map the memory space of the target program at the bit level, which will have a great impact on the execution efficiency of the detection program. Therefore, taking into account the efficiency of the detection program and the acc-uracy of the detection, it is necessary to carry out a good design to achieve the balance of the two.

In this paper, the memory management defects of the C++ program are detected, so it is necessary to use the shadow memory to map the register and internal control space of the target program. In order to balance the efficiency and accuracy of detection program execution, we designed different shadow memory mapping meth-ods for C++ program memory layout.

(1) Single bit-byte mapping

Single bit-byte mapping refers to each byte of the actual memory space, in the shadow memory using a bit to map, the map is the main way of variable/memory block boundary identification.

(2) Double bits-byte mapping

Double bits-byte mapping refers to each byte of the actual memory space, in the shadow memory using two bit mapping, two bits can represent four states one byte: NOACCESS (00), DEFINED (11), UNDEFINED (01), PARTDEFINED (10). For the bytes corresponding to PARTDEFINDE, a corresponding two level table is also designed to record the actual initialization of the byte.

(3) Eight bits-byte mapping

Eight bits-byte mappings refer to each byte in the actual memory space, and use eight bits to map in the shadow memory. Because of the accuracy of the detection, for some key memory areas, such as registers, bit level information needs to be recorded.

For the mapping of register, it is a part of the central processor unit (CPU). Registers are high speed storage components of limited storage capacity, which can be used to store instructions, data and addresses. In general, the register has a small capacity. According to the different CPU types, the register capacity is generally between several to dozens of bytes. The sensitive information such as instructions, operands and addresses stored in registers is not large. Therefore, eight bit-byte mapping is used to save the information of registers in shadow memory.

For the BBS section, DATA section and TEXT section of these memory areas, they are a block of contiguous memory space, and in the program is loaded and address allocated at runtime and never change, so you can put them in the mapping process together. For BBS segment, the uninitialized global variables and static variables are stored, and the memory space of BBS segment is very small. Therefore, eight bit-byte mapping is used to store information in shadow memory. For the DATA segment, the global variables and static variables that are initialized are stored, which can be directly used by the program without any memory management errors. Therefore, a pair of pointers can be used to save the corresponding starting addresses. The TEXT segment is stored program instructions and procedures is mirrored in memory, only allows the program to read, not allowed to modify, so there is no memory management errors, so a pointer to the starting address and save the corresponding set flag is not writable can also use.

For stack area, it is a continuous memory that changes continuously during program running. It stores information about function parameters, local variables and return addresses. It is responsible for distributing and releasing stack memory. In general, the operating system limits the size of the stack, taking the Windows system as an example, and the stack size is 2 M. Due to the sensitive information stored in the stack such as function parameters and return address, if these information is destroyed, it will seriously affect the normal operation of the program, even lead to the premature termination or collapse of the program. Therefore, there are two ways to map the stack. The first is a eight bits-byte mapping, save memory stack information in the shadow memory to bit size, at the same time in order to prevent stack overflow, the introduction of single bit-byte mapping, followed by record adjacent variable boundary.

For a heap, heap area is generally not continuous, and heap area than other areas of memory stack area other than the most important is the heap memory by the developers themselves responsible for the application and release, if the developer on the heap memory operation is not clear or in the process of development is not careful enough, it is easy to happen memory error. In a lot of practical engineering practice, heap memory is a serious disaster area of memory error. Because the reactor area is not continuous and the heap area is also large, the mapping of the memory in the reactor area is relatively complex. We use two level tables similar to the operation system page table and map the heap area memory by using double bit-byte mapping. First, the memory is divided into 64 K memory basic block-basic_block, the first level table is a global pointer array, each element stores a pointer to a basic_block, which can cover all the address space. The two level table is also an array that stores the shadow memory information of the 64 K basic block memory that is mapped by two bit bytes.

3 Design and Implementation of MemDetect

3.1 Framework of MemDetect

Through inserting pile analysis to the C++ target program by using the binary piling platform Pin, tracking function call of the application and the memory operation sequence and updating the shadow memory [6] in real time, and corresponding defect detection rules are used to determine whether there is a defect in the target program, finally, the results are displayed to the user.

For defect detection tool MemDetect, it mainly includes piling analysis module, shadow memory management module, function call tracking module, defect detection module and result output module. According to the application and detection purpose, the piling analysis module first decides the level of pile insertion, mainly including function level piling and instruction level piling, and inserts the corresponding analysis code for different pile level and specific execution branch. The function call tracking module is mainly the method of recording the program execution sequence, and assists locating the location of the defect. The shadow memory management module is used to obtain and update the memory state of the program, providing data for defect detection. The defect detection module is based on the memory operation state and the execution state of the program provided by the shadow memory, and the defect detection method is applied to detect the actual defects of the target program. The result output module is a module that interacts with the user, and the final result is displayed in the form of visual or detection report to the user.

The defect detection module is the core module of this tool. The following will focus on the design and implementation of the defect detection module. The organization diagram of the MemDetect modules is shown as follows (Fig. 3):

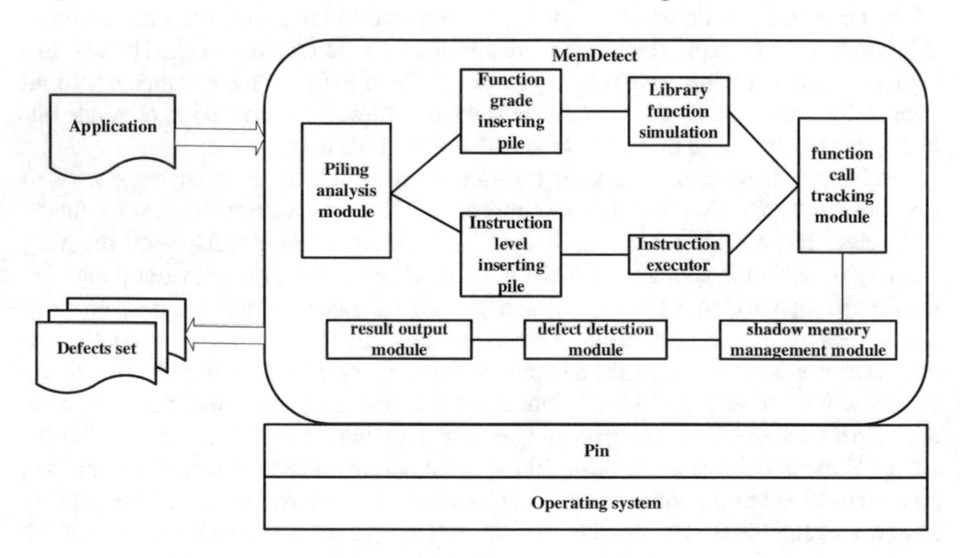

Fig. 3. Module diagram of MemDetect

3.2 Design of Defect Detection Module

The defect detection module detects the application defect based on the memory operation sequence information of the shadow memory and the execution state of the program, and the location of the defect is further located according to the information provided by the function call tracking module.

The difficulty and emphasis of defect detection module is to detect memory leak. This paper uses memory block reachability analysis algorithm to detect program memory leak. For a heap memory, we design a special shadow memory to map the memory operation state, taking into account the heap memory is not continuous, we use basic_block as the unit division, to achieve full coverage of the heap. What's more, we design a hash table to store the memory allocation block and the eight bits of the initialized byte high precision mapping in basic_block.

MemDetect will track memory blocks applied by methods such as malloc/new, so when the program exits, you can know which memory blocks are not released yet. When the program dynamically applies memory to the operation system through malloc/new or other methods, the operation system returns to the pointer to the corresponding memory after allocated memory. There are two types of pointers: one is start-pointer, which points the start of the heap block, the other is interior-pointer, which points the middle of the heap block. In general, most of the pointers are start-pointer, but system will return interior-pointer in some special cases. For example, a pointer to an array of characters in C++ std:string [7], some compilers add 3 words to the starting position of the std:string to store string length, capacity, and reference count. The compiler then returns the pointer after the three words to point to the array of characters. According to the accessibility of different memory blocks, we divide the memory blocks into three types: (1) definitely access, which has a start-pointer that points to the starting position of the block. (2) possibly access, which has a interior-pointer that points to the middle position of the block. (3) Deny access, which has no pointer that points to any position of the block.

For memory leak detection, we need to find the set R of pointers held by the program firstly. For of the loss of variable type information at the binary machine code level, it's impossible to judge whether the memory stored in memory is a pointer. But most compilers currently implement the rules of memory alignment, the position where all the pointers can appear under this rule is regular. So it is possible to traverse the memory space of the program and try to interpret the data that conforms to the rules of the memory alignment, initialization, and the size of the machine word as a pointer, if it points to the corresponding memory block, it means that this is a pointer variable. The position that the pointer can appear in the memory space is also fixed, namely the register, the stack space, the global variable area, and the reachable heap memory.

The key to the above detection method is to find the set R of pointers held by the program. When R is found, the accessibility of the memory block can be judged according to the different pointer types, and the results of the memory leak detection are given. Its detection algorithm is described as follows (Table 1):

Table 1. Memory leak detection algorithm

```
Input: Register-memory, Stack-memory, Global-memory, Heap-memory
Output: All-Pointer-List = R, May-Lost-List = M, Definite-Access-List = K,
Deny-Access-List = N
1:function: Memory-Leak-Check(Register-memory, Stack-memory, Global-
memory, Heap-memory)
2:    R = Φ, M = Φ, K = Φ, N = Φ;
3:    for P in getPointer(Register-memory U Stack-memory U Global-memory):
4:        R.add(P);
5:    end for
6:    for P in R:
7:        delete P from R;
8:        if P is start-pointer then
9:            K.add(P);
10:           tmp-heap-memory = getHeapMemroy(P, Heap-memory);
11:           for P1 in getPointer(tmp-heap-memory):
12:               R.add(P1);
13:           end for
14:       end if
15:       else
16:           M.add(P);
17:       end else
18:       remain-heap-memory = getRemainHeapMemory(M, K, heap-memory);
19:       for M1 in remain-heap-memory:
20:           N.add(M1);
21:       end for
22:end function
```

The method for detecting mismatch release and repeated release of memory is consistent. The program's memory application/release operating function can be wrapped and replaced by the dynamic insert of Pin. When a program performs a dynamic memory application, the program's call to the original application function is redirected to the replacement function address [8], completing the memory application and record the relevant information of the application which mainly includes the starting address, memory size, application method, etc. And the information is stored in the application information table of the shadow memory. When the program releases the heap memory, it also redirects the call of the original release function of the program to the replacement function address, firstly, detecting whether the memory block to be released is reachable (that is, its address is accessible) to detect repeated release. If reachability detects the matching of the mode of release and the application mode, the detection process is as follows (Fig. 4):

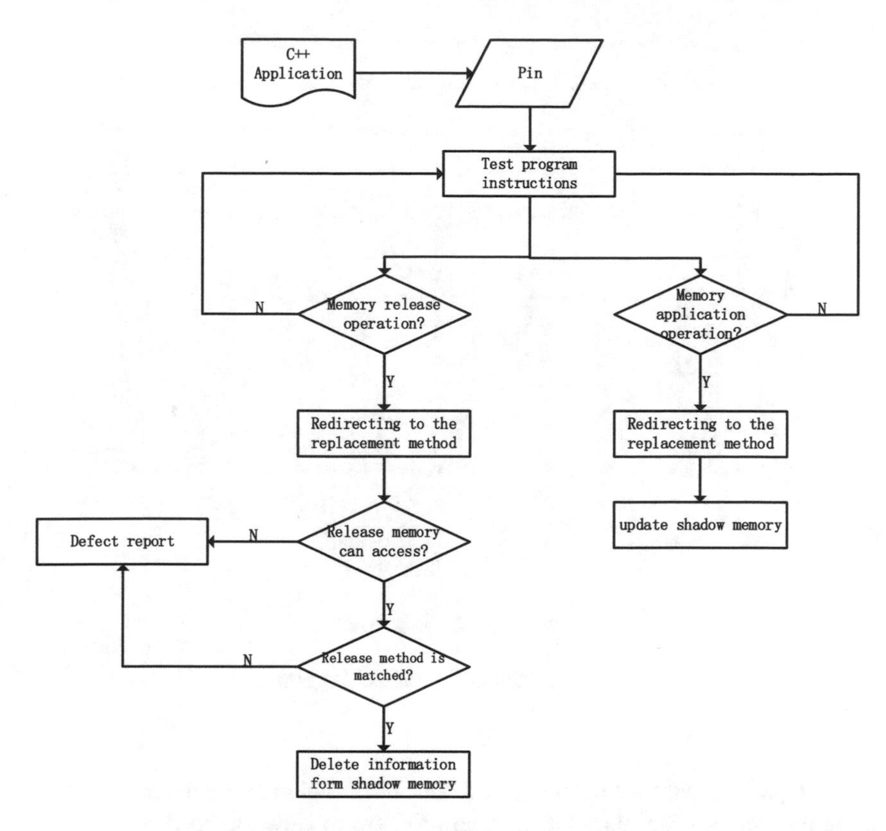

Fig. 4. Memory release error detection process

4 Experimental Analysis

This section verifies the actual detection effect of the C++ memory defect detection tool, MemDetect, by experimentation, the experiment is mainly divided into two aspects: One is to test the running time of a detection tool by testing some well-known open source programs, the other is to write a special test program to test the accuracy of the tool detection. At the same time, there is a horizontal contrast with the memory defect detection tool MemCheck to enhance the persuasiveness of the detection results. Test objects are 8 open source projects, including msgpack, envoyproxy, lilydict, lifeograph, libsnark, sparsehash, bzip2, Onion, etc. and 5 test procedures: DefectTest01 ~ DefectTest05.

The runtime test is mainly for the time consumed by tool detection, recording the running time of the original program, detecting the running time of MemDetect and running time of MemCheck. In order to minimize the random error, the experimental data are the average of the 15 times after the operation. The contrast diagram with MemCheck is shown as follows (Fig. 5):

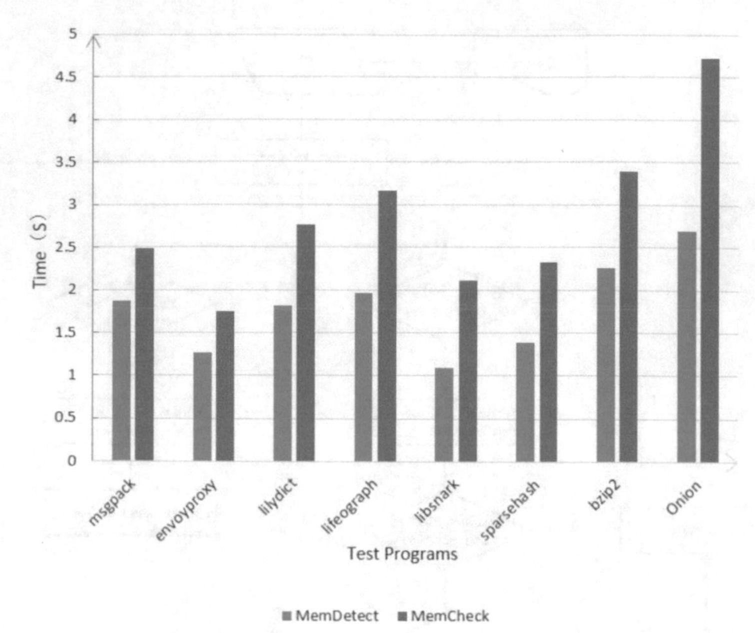

Fig. 5. Run time contrast diagram

By comparison with the open source MemCheck tool memory defect detection, the running time of this tool than the original program to slow 10–35 times, the running time of MemCheck than the original program to slow 15–45 times, which proved the tool compared superiority in terms of the running time of heavy instrumentation Valgrind implementation based on MemCheck framework.

The test accuracy test means that the defect testing tool is used to test the corresponding defect test program to test the accuracy of the test results. In this experiment, 5 specially written test procedures are covered by the memory leak, not match of release, repeated memory release memory defects, and the defects were combined in different test procedures, the use of MemDetect defect detection tool for defect detection of test procedures, test results are shown as follows (Fig. 6):

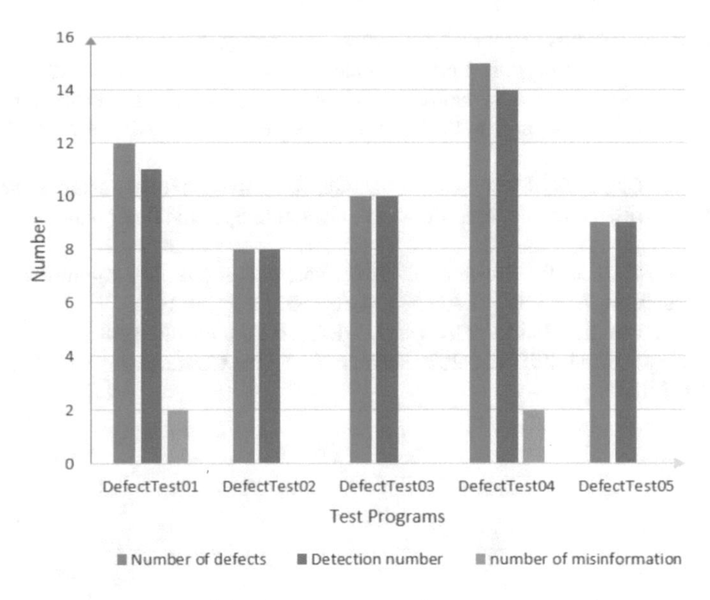

Fig. 6. Accuracy of detection

By testing the accuracy of tool detection results, we can find that the detection method proposed in this paper for C++ memory management defects is of high accuracy.

5 Conclusion

This paper studies the C++ memory management mechanism, based on dynamic binary Instrumentation Platform Pin and shadow memory to design and implement a defect detection tool MemDetect. This tool can effectively detect the C++ memory leak, repeated release, memory does not match the release, and through the experiments verify the accuracy of detection, and similar tools compared with the time running overhead of this tool is relatively small, so that the tool has strong practicability, can effectively help C++ developers to program memory management.

Acknowledgment. This work was supported by National Natural Science Foundation of China (No. U1536122).

References

1. Ball, T., Rajamani, S.K.: The SLAM project: debugging system software via static analysis, pp. 1–3 (2002)
2. Godefroid, P., Levin, M., Molnar, D.: Active property checking. In: Proceedings of the 8th ACM International Conference on Embedded Software, pp. 19–24 (2008)
3. Nethercote, N., Seward, J.: Valgrind: a framework for heavyweight dynamic binary instrumentation. In: Proceedings of PLDI 2007, San Diego, California, USA, June 2007

4. Turkboylari, M.: Implementation of a secure computing environment by using a secure bootloader, shadow memory, and protected memory: US, US 7313705 B2 (2007)
5. Offutt, A.J., Hayes, J.H.: A semantic model of program faults. In: Proceedings of ACM SIGSOFT International Symposium on Software Testing and Analysis. ACM Press, San Diego (2013)
6. Engler, B.D., Chelf, B., Chou, A., et al.: Checking system rules using system-specific, programmer-written compiler extensions. In: Operating Systems Design and Implementation (2010)
7. Zeng, Q., Wu, D., Liu, P.: Cruiser: concurrent heap buffer overflow monitoring using lock-free data structures. ACM SIGPLAN Not. **46**(6), 367–377 (2011)
8. Seward, J., Nethercote, N.: Using Valgrind to detect undefined value errors with bit-precision. In: Proceedings of the USENIX 2005 Annual Technical Conference, Anaheim, California, USA, April 2005

Flying Ad Hoc Network for Emergency Applications Connected to a Fog System

Dan Radu[1(✉)], Adrian Cretu[1], Benoît Parrein[2], Jiazi Yi[3], Camelia Avram[1], and Adina Aştilean[1]

[1] Technical University of Cluj-Napoca, Cluj-Napoca, Romania
{dan.radu,camelia.avram,adina.astilean}@aut.utcluj.ro,
adriancr.ro@gmail.com
[2] LS2N, Polytech'Nantes, Nantes, France
benoit.parrein@polytech.univ-nantes.fr
[3] Laboratoire d'Informatique, Ecole Polytechnique, Palaiseau, France
yi@lix.polytechnique.fr

Abstract. The main objective of this paper is to improve the efficiency of vegetation fire emergency interventions by using MP-OLSR routing protocol for data transmission in Flying Ad Hoc NETwork (FANET) applications. The presented conceptual system design could potentially increase the rescuing chances of people caught up in natural disaster environments, the final goal being to provide public safety services to interested parties. The proposed system architecture model relies on emerging technologies (Internet of Things & Fog, Smart Cities, Mobile Ad Hoc Networks) and actual concepts available in the scientific literature. The two main components of the system consist in a FANET, capable of collecting fire detection data from GPS and video enabled drones, and a Fog/Edge node that allows data collection and analysis, but also provides public safety services for interested parties. The sensing nodes forward data packets through multiple mobile hops until they reach the central management system. A proof of concept based on MP-OLSR routing protocol for efficient data transmission in FANET scenarios and possible public safety rescuing services is given.

1 Introduction

The main objective of this paper is to introduce MP-OLSR routing protocol, that already proved to be efficient in MANET and VANET scenarios Yi et al. (2011a), Radu et al. (2012), into FANET applications. Furthermore, as a proof of concept, this work presents a promising smart system architecture that can improve the saving chances of people caught in wildfires by providing real-time rescuing services and a temporary communication infrastructure. The proposed system could locate the wildfire and track the dynamics of its boundaries by deploying a FANET composed of GPS and video enabled drones to monitor the target areas. The video data collected from the FANET is sent to a central management system that processes the information, localizes the wildfire and

L. Barolli et al. (Eds.): EIDWT 2018, LNDECT 17, pp. 675–686, 2018.
https://doi.org/10.1007/978-3-319-75928-9_60

provides rescuing services to the people (fire fighters) trapped inside wildfires. The data transmission QoS (Quality of Service) for data transmission in the proposed FANET network scenario is provided to prove the efficiency of MP-OLSR, a multipath routing protocol based on OLSR, in this types of applications.

Wildfires are unplanned events that usually occur in natural areas (forests, prairies) but they could also reach urban areas (buildings, homes). Many such events occured the last years (e.g. Portugal and Spain 2017, Australia 2011). The forrest fire in the north of Portugal and Spain killed more than 60 people. During the Kimberley Ultra marathon held in Australia in 2011 multiple persons were trapped in a bush fire that started during a sports competition.

The rest of this paper is structured as folows. Section 2 presents the related works in the research field. Section 3 introduces the proposed system design. Section 4 shows and discusses the QoS performance evaluation results. Finally, Sect. 5 concludes the paper.

2 Related Works

Currently there is a well-known and increasing interest for providing Public Safety services in case of emergency/disaster situations. The US Geospatial Multi Agency Coordination[1] provides a web service that displays fire dynamics on a map by using data gathered from different sources (GPS, infrared imagery from satellites). A new method for detecting forest fires based on the color index was proposed in Cruz et al. (2016). Authors suggest the benefits of a video surveillance system installed on drones. Another system, composed of unmanned aerial vehicles, used for dynamic wildfire tracking is discussed in Pham et al. (2017).

This section presents the state of the art of the concepts and technologies used for the proposed system design, current trends, applications and open issues.

2.1 Internet of Things and Fog Computing

Internet of Things (IoT), Fog Computing, Smart Cities, Unmanned Aerial Vehicle Networks, Mobile Ad Hoc Networks, Image Processing Techniques and Algorithms, and Web Services are only some of the most promising actual, emerging technologies. These all share a great potential to be used together in a large variety of practical applications that could improve, sustain and support people's life.

There are many comprehensive surveys in the literature that analyse the challenges of IoT and provide insights over the enabling technologies, protocols and possible applications Al-Fuqaha et al. (2015). In the near future, traditional cloud computing based architectures will not be able to sustain the IoT exponential growth leading to latency, bandwidth and inconsistent network challenges. Fog computing could unlock the potential of such IoT systems.

[1] https://www.geomac.gov.

Fog computing refers to a computing infrastructure that allows data, computational and business logic resources and storage to be distributed between the data source and the cloud services in the most efficient way. The architecture could have a great impact in the emerging IoT context, in which billions of devices will transmit data to remote servers, because its main purpose is to extend cloud infrastructure by bringing the advantages of the cloud closer to the edge of the network where the data is collected and pre-processed. In other words, fog computing is a paradigm that aims to efficiently distribute computational and networking resources between the IoT devices and the cloud by:

- allowing resources and services to be located closer or anywhere in between the cloud and the IoT devices;
- supporting and delivering services to users, possibly in an offline mode when the network is partitioned by example;
- extending the connectivity between devices and the cloud across multiple protocol layers.

In the near future, traditional cloud computing based architectures will not be able to sustain the IoT exponential growth leading to latency, bandwidth and inconsistent network challenges. Fog computing could unlock the potential of such IoT systems.

Currently the use cases and the challenges of the edge computing paradigm are discussed in various scientific works Lin et al. (2017), Al-Fuqaha et al. (2015), Ang et al. (2017). Some of the well known application domains are: energy, logistics, transportation, healthcare, industrial automation, education and emergency services in case of natural or man made disasters. Some of the challenging Fog computing research topics are: crowd based network measurement and interference, client side network control and configuration, over the top content management, distributed data centers and local storage/computing, physical layer resource pooling among clients, Fog architecture for IoT, edge analytics sensing, stream mining and augmented reality, security and privacy.

There are numerous studies that connect video cameras to Fog & IoT applications. The authors of Shi et al. (2016) discuss a couple of practical usages for Fog computing: cloud offloading, video analytics, smart home and city, and collaborative edge. Also, some of the research concepts and opportunities are introduced: computing stream, naming schemes, data abstraction, service management, privacy and security, and optimization metrics. Authors of Shi and Dustdar (2016) present a practical use case in which video cameras are deployed in public areas or on vehicles and they could be used to identify a missing person's image. In this case, the data processing and identification could be done at the edge without the need of uploading all the video sources to the cloud. A method that distributes the computing workload between the edge nodes and the cloud was introduced Zhang et al. (2016). Authors try to optimize data transmission and ultimately increase the life of edge devices such as video cameras.

Fog computing could be the solution to some of the most challenging problems that arise in the Public Safety domain. Based on the most recent research studies and previous works concerning public safety Radu et al. (2012), Yi et al. (2011b) it can be stated that real time image & video analysis at the edge of a FANET network could be successfully implemented in the public safety domain, more specifically for fire detection and for rescuing emergency services provisioning.

One of the most important advantages of Fog computing is the distributed architecture that promises better Quality of Experience and Quality of Service in terms of response, network delays and fault tolerance. This aspect is crucial in many Public Safety applications where data processing should be done at the edge of the system and the response times have hard real-time constraints.

2.2 Flying Ad Hoc Networks

Unmanned Aerial Vehicles (UAV's, commonly known as drones) become more and more present in our daily lifes through their ease of deployment in areas of interest. The high mobility of the drones, with their enhanced hardware and software capabilities, makes them suitable for a large variety of applications including transportation, farming and disaster management services. FANET's are considered as a sub type of Mobile Ad Hoc Networks networks that have a greater degree of mobility and usually the distance between nodes is greater as stated in Bekmezci et al. (2013).

A practical FANET testbed, build on top of Raspberry Pi©, that uses two WiFi connections on each drone (one for ad hoc network forwarding and the other for broadcasted control instructions is described in Bekmezci et al. (2015). Another FANET implementation that consists of quadcopters for disaster assistance, search and rescue and aerial monitoring as well as the design challenges are presented in Yanmaz et al. (2018).

2.3 Routing Protocols

OLSR (Optimized Link State Routing) protocol proposed in Jacquet et al. (2001) is an optimization of link state protocol. This single path routing approach presents the advantage of having shortest path routes immediately available when needed (proactive routing). OLSR protocol has low latency and performs best in large and dense networks.

In Haerri et al. (2006) OLSR and AODV are tested against node density and data traffic rate. Results show that OLSR outperforms AODV in VANETs, providing smaller overhead, end-to-end delay and route lengths. Furthermore there are extensive studies in the literature regarding packets routing in FANET's. Authors of Oubbati et al. (2017) give a classification and taxonomy of existing protocols as well as a complete description of the routing mechanisms for each considered protocol. An example of a FANET specific routing protocol is an adaptation of OLSR protocol that uses GPS information and computes routes based on the direction and relative speed between the UAV's is proposed in Rosati et al. (2016).

In this paper authors use MP-OLSR (Multiple Paths OLSR) routing protocol based on OLSR proposed in Yi et al. (2011a), that allows packet forwarding in FANET and MANET networks through spatially separated multiple paths. MP-OLSR exploits simultaneously all the available and valuable multiple paths between a source and a destination to balance traffic load and to reduce congestion and packet loss. Also it provides a flexible degree of spatial separation between the multiple paths by penalizing edges of the previous paths in an original Dijkstra algorithm execution.

Based on the above considerations, a system architecture that can improve the saving chances of people caught in wildfires by providing real-time rescuing services and a temporary communication infrastructure is proposed.

3 System Design

One of the main objectives of this work is to design and develop a smart system architecture, based on FANET networks, which integrates with the numerous emergent applications offered by the Internet of Things, that is:

- extensible: the system architecture should allow any new modules to be easily plugged in;
- reliable: the system should support different levels of priority and quality of service for the modules that will be plugged in. For example, the public safety and emergency services that usually have real-time hard constraints should have a higher priority than other services that are not critical;
- scalable: the architecture should support the connection of additional new Fog components, features and high node density scenarios;
- resilient: the system will be able to provide and maintain an acceptable level of service whenever there are any faults to the normal operation.

The overview of the proposed model, in the context of Internet of Things & Fog Computing, is given in Fig. 1.

The system could locate the wildfire and track the dynamics of its boundaries by deploying a flying ad hoc network composed of GPS and video enabled drones to monitor the target areas. The fire identification data collected from the FANET is sent to a central management system that processes the data, localizes the wildfire and provides rescuing services to the people (fire fighters) trapped inside the wildfires. The proposed system intends to support and improve emergency intervention services by integrating, based on the real-time data collected from the Fog network, multiple practical services and modules such as:

- affected area surveillance;
- establishing the communication network between the disaster survivors and rescue teams;
- person in danger identification and broadcasting of urgent notifications;
- supporting the mobility of the first responders through escape directions;
- rescuing vehicle navigation.

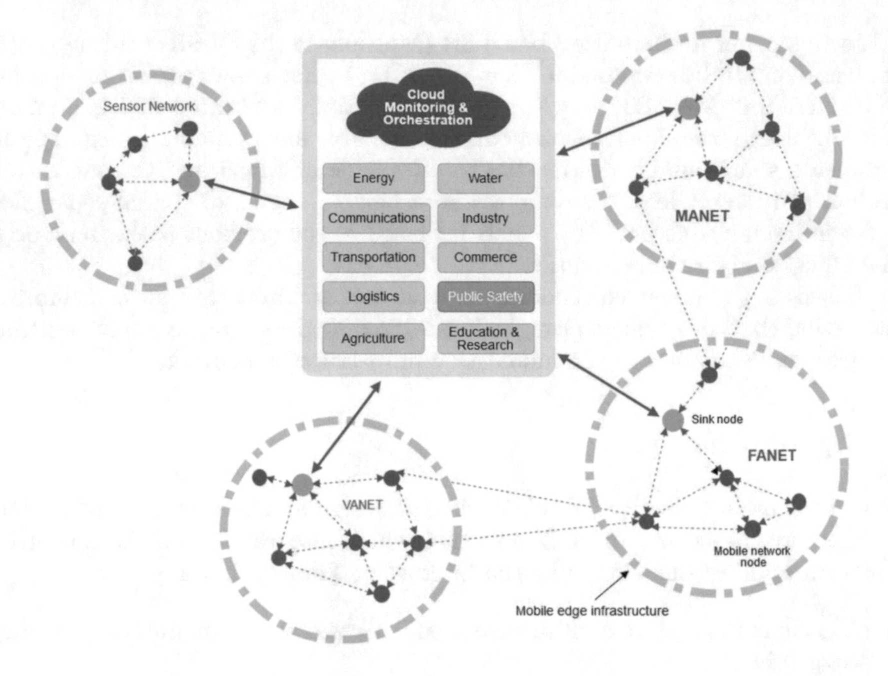

Fig. 1. System overview

Our FEA (FANET Emergency Application) network topology is presented in Fig. 2 and it is composed of three main components:

- A MANET of mobile users phones;
- FANET - video and GPS equipped drones that also provide sufficient computational power capabilities for fire pattern recognition;
- Fog infrastructure that supports FANET data collection at the sink node located at the edge of the network. This provides data storage, computational power and supports different communication technologies for the interconnection with other edge systems.

This last component can be done through an object store as proposed in Confais et al. (2017b) where a traditional Bittorrent P2P network can be used for storage purpose. Combined with a Scale-out NAS as in Confais et al. (2017a), the Fog system avoids costly metadata management (even in local accesses) and computing capacity thanks to an intensive I/O distributed file system. Moreover, the global Fog system allows to work on a disconnected mode in case of network partitionning with the backbone.

FEA uses a FANET network, to collect fire identification data from drones (GPS and video enabled), and a MANET network composed of users smartphones. Sensing nodes periodically transmit data to the central management system where the fire dynamics is determined for monitoring purposes. If a fire has been detected by a sensing drone, based on the dynamics of the fire, rescuing information will be computed and broadcasted back into the FANET and

MANET so that the people trapped in the fire to be able to receive the safety information on their smartphones in real-time. We make the following assumptions, that will be taken into account for the simulation scenario modelling, regarding the FEA message forwarding:

- when fire is detected by sensing drones they start to periodically forward data packets with the information regarding fire dynamics over multiple hops in the mesh network towards the sink node;
- the central management system processes the fire detection data received from FANET nodes and computes the fire dynamics using the GPS coordinates that are included in the received data messages;
- the central management system sends back into the mobile network (FANET and MANET) rescuing information that will be received by people in danger on their smartphones.

In FEA system FANET nodes are responsible for: fire identification based on video recording, forwarding the processed information (alongside with GPS coordinates) towards the collector node and rescuing information forwarding to the MANET nodes. The proposed network architecture could also serve as a temporary communication infrastructure between rescuing teams and people in danger.

One of the many advantages of FEA is the ease of deployment, all the technologies and components of the system are widely available, inexpensive and easy to provide. Also the Quality of Service in the FANET network, which is essential in emergency services where delays and packet delivery rations are very important, is enhanced by using MP-OLSR routing protocol that chooses the best multiple paths available between source and destination.

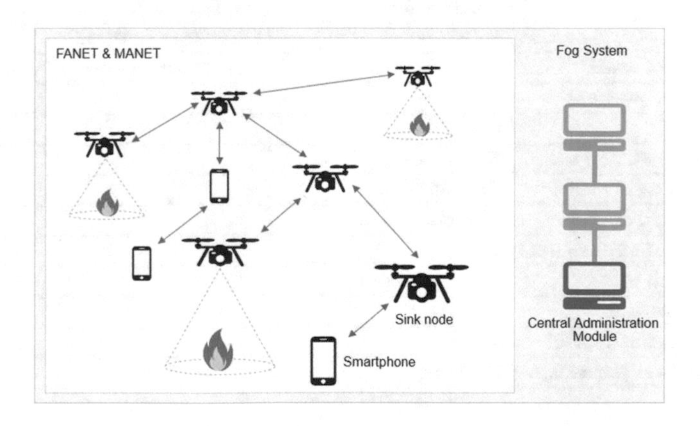

Fig. 2. Emergency system architecture

4 System Evaluation

The simulations are performed to evaluate MP-OLSR in the proposed FANET scenario. This section is organized as follows. The simulation environment configuration and scenario assumptions are given in Sect. 4.1 and then the Quality of Service performance are compared between OLSR and MP-OLSR in Sect. 4.2.

4.1 Simulation Scenario

For the simulations we designed a 81 nodes FANET & MANET hybrid topology placed in a 1480 square meters grid topology. The Random Waypoint Model mobility pattern was used for different maximal speeds suitable for the high mobility of drones: 1–15 m/s (3.6–54 km/h). We make the assumption that only a subset of nodes (possibly the ones that detect fire or the smart phones of people in danger) need to communicate with the Fog edge node through the mesh network so the data traffic is provided by 4 Constant Bit Rate (CBR) sources. Qualnet 5 was used as a discrete event network simulator. The detailed parameters for the Qualnet network scenario and routing protocols configuration parameters are listed in Table 1. The terrain altitude profile is shown in Fig. 3.

Table 1. Simulation parameters.

Simulation parameter	Value	Routing parameter	Value
Simulator	Qualnet 5	TC Interval	5 s
Routing protocols	OLSRv2 and MP-OLSR	HELLO Interval	2 s
Area	1480 × 1480 x 34.85 m^3	Refresh Timeout Interval	2 s
Number of nodes	81	Neighbor hold time	6 s
Initial nodes placement	Grid	Topology hold time	15 s
Mobility model	Random Waypoint	Duplicate hold time	30 s
Speeds	1–15 m/s	Link Layer Notification	Yes
Number of seeds	10	No. of path in MP-OLSR	3
Transport protocol	UDP		
IP	IPv4		
IP fragmentation unit	2048 bytes		
Physical layer model	PHY 802.11b		
Link layer data rate	11 Mbits/s		
Number of CBR sources	4		
Sim duration	100 s		
CBR start-end	15–95 s		
Transmission interval	0.05 s		
Application packet size	512 bytes		

4.2 Simulation Results

For each routing protocol a number of 80 simulations were executed (10 different seeds/speed ranges). To compare the performances of the protocols, the following metrics are used:

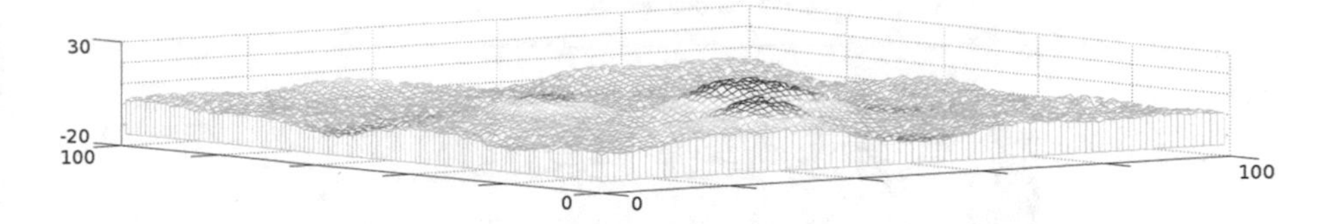

Fig. 3. Qualnet altitude profile pattern for 100 m^2

- *Packet delivery ratio (PDR)*: the ratio of the data packets successfully delivered at all destinations.
- *Average end-to-end delay*: averaged over all received data packets from sources to destinations as depicted in Schulzrinne et al. (1996).
- *Jitter*: average jitter is computed as the variation in the time between packets received at the destination caused by network congestions and topology changes.

Figures 4, 5 and 6 show the QoS performance of MP-OLSR and OLSR in terms of PDR, end-to-end delay and Jitter results with standard deviation for each point. From the obtained results it can be seen that PDR decreases slightly with the mobility as expected. For the proposed FANET scenario MP-OLSR delivers an average of 10% higher PDR than OLSR protocol. As expected, when the speed increases to values closer to the high mobility of FANET scenarios the links become more unstable so OLSR performance decreases while MP-OLSR provides a much better overall delivery ratio than OLSR (around 9% in average at higher speeds). MP-OLSR also performs much better than OLSR in terms of end-to-end delay and Jitter. The delay of OLSR is around 2 times higher at the highest speed while Jitter is 50% higher. This aspect is very important for the proposed emergency application where the response time must be provided as quickly as possible. Furthermore, the MP-OLSR standard deviation for all the results is smaller than for OLSR.

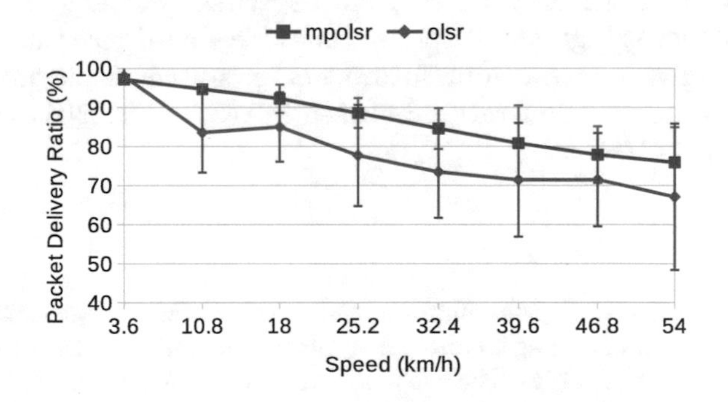

Fig. 4. Delivery ratio

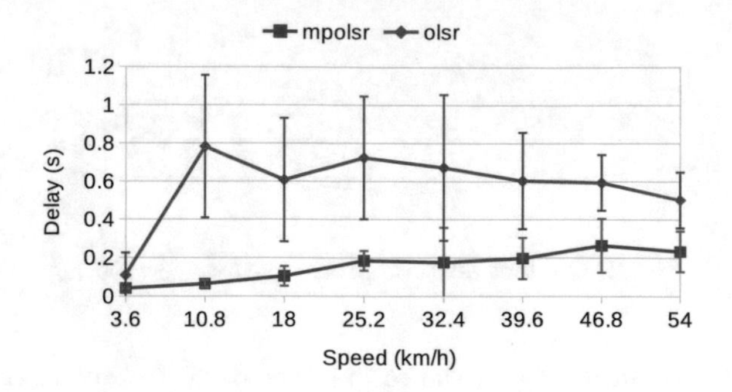

Fig. 5. End-to-end delay

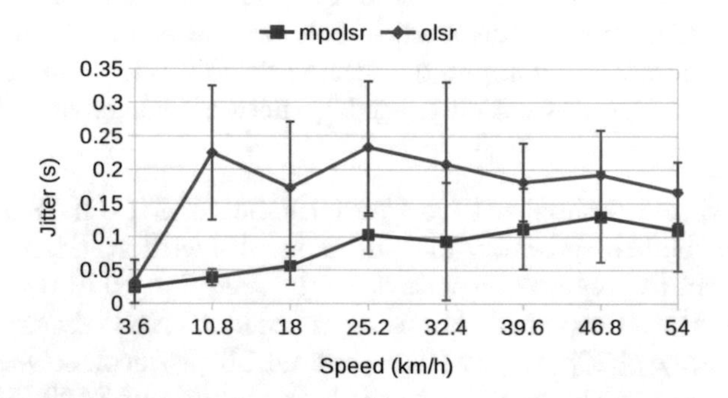

Fig. 6. Jitter

5 Conclusion and Future Work

We described FEA system as a possible emergency application for MP-OLSR routing protocol which uses a FANET network to collect fire dynamics data from drones and through a central management system it provides safety instructions back to the people in danger. The performance evaluation results show that MP-OLSR is suitable for FANET scenarios, most specifically emergency applications, where the mobility is high and response times have hard real-time constraints.

The folowing are some of our future works: system deployment on a real testbed, analysis of the cooperation between MANET & FANET, data analysis based on thermal cameras.

References

Al-Fuqaha, A., Guizani, M., Mohammadi, M., Aledhari, M., Ayyash, M.: Internet of things: a survey on enabling technologies, protocols, and applications. IEEE Commun. Surv. Tutorials. **17**(4), 2347–2376 (2015). https://doi.org/10.1109/COMST.2015.2444095

Ang, L.M., Seng, K.P., Zungeru, A.M., Ijemaru, G.K.: Big sensor data systems for smart cities. IEEE Int. Things J. **4**(5), 1259–1271 (2017). https://doi.org/10.1109/JIOT.2017.2695535

Bekmezci, I., Sahingoz, O.K., Temel, Ş.: Flying ad-hoc networks (fanets): a survey. Ad Hoc Netw. **11**(3), 1254–1270 (2013). https://doi.org/10.1016/j.adhoc.2012.12.004. http://www.sciencedirect.com/science/article/pii/S1570870512002193

Bekmezci, I., Sen, I., Erkalkan, E.: Flying ad hoc networks (fanet) test bed implementation. In: 2015 7th International Conference on Recent Advances in Space Technologies (RAST), pp. 665–668 (2015). https://doi.org/10.1109/RAST.2015.7208426

Confais, B., Lèbre, A., Parrein, B.: An object store service for a fog/edge computing infrastructure based on IPFS and scale-out NAS. In: 1st IEEE International Conference on Fog and Edge Computing - ICFEC 2017, Madrid, Spain (2017a). https://hal.archives-ouvertes.fr/hal-01483702

Confais, B., Lèbre, A., Parrein, B.: Performance Analysis of Object Store Systems in a Fog and Edge Computing Infrastructure. Transactions on Large-Scale Data- and Knowledge-Centered Systems. Springer, Heidelberg (2017b). https://doi.org/10.1007/978-3-662-55696-2_2. https://hal.archives-ouvertes.fr/hal-01587459

Cruz, H., Eckert, M., Meneses, J.M., Martínez, J.: Efficient forest fire detection index for application in unmanned aerial systems (UASS). Sensors **16**(6), 893 (2016)

Haerri, J., Filali, F., Bonnet, C.: Performance comparison of AODV and OLSR in vanets urban environments under realistic mobility patterns. In: 5th IFIP Mediterranean Ad-Hoc Networking Workshop, Med-Hoc-Net 2006, June, Lipari, ITALY, pp. 14–17 (2006)

Jacquet, P., Muhlethaler, P., Clausen, T., Laouiti, A., Qayyum, A., Viennot, L.: Optimized link state routing protocol for ad hoc networks. In: Proceedings of IEEE International Multi Topic Conference, 2001, IEEE INMIC 2001, Technology for the 21st Century, pp. 62–68 (2001)

Lin, J., Yu, W., Zhang, N., Yang, X., Zhang, H., Zhao, W.: A survey on internet of things: Architecture, enabling technologies, security and privacy, and applications. IEEE Int. Things J. **4**(5), 1125–1142 (2017). https://doi.org/10.1109/JIOT.2017.2683200

Oubbati, O.S., Lakas, A., Zhou, F., Güneş, M., Yagoubi, M.B.: A survey on position-based routing protocols for flying ad hoc networks (FANETS). Veh. Commun. **10**, 29–56 (2017). https://doi.org/10.1016/j.vehcom.2017.10.003. http://www.sciencedirect.com/science/article/pii/S2214209617300529

Pham, H.X., La, H.M., Feil-Seifer, D., Deans, M.: A distributed control framework for a team of unmanned aerial vehicles for dynamic wildfire tracking. CoRR abs/1704.02630 (2017). http://arxiv.org/abs/1704.02630

Radu, D., Avram, C., Aştilean, A., Parrein, B., Yi, J.: Acoustic noise pollution monitoring in an urban environment using a vanet network. In: Proceedings of 2012 IEEE International Conference on Automation, Quality and Testing, Robotics, pp. 244–248 (2012). https://doi.org/10.1109/AQTR.2012.6237711

Rosati, S., Kruželecki, K., Heitz, G., Floreano, D., Rimoldi, B.: Dynamic routing for flying ad hoc networks. IEEE Trans. Veh. Technol. **65**(3), 1690–1700 (2016). https://doi.org/10.1109/TVT.2015.2414819

Schulzrinne, H., Casner, S., Frederick, R., Jacobson, V.: RFC 1889: RTP: A transport protocol for real-time applications. Network Working Group Memorandum (1996). http://www.faqs.org/rfcs/rfc1889.html

Shi, W., Dustdar, S.: The promise of edge computing. Computer **49**(5), 78–81 (2016). http://doi.ieeecomputersociety.org/10.1109/MC.2016.145

Shi, W., Cao, J., Zhang, Q., Li, Y., Xu, L.: Edge computing: vision and challenges. IEEE Int. Things J. **3**(5), 637–646 (2016). http://dx.doi.org/10.1109/jiot.2016.2579198

Yanmaz, E., Yahyanejad, S., Rinner, B., Hellwagner, H., Bettstetter, C.: Drone networks: communications, coordination, and sensing. Ad Hoc Netw. **68**, 1–15 (2018). https://doi.org/10.1016/j.adhoc.2017.09.001. http://www.sciencedirect.com/science/article/pii/S1570870517301671. Advances in Wireless Communication and Networking for Cooperating Autonomous Systems

Yi, J., Adnane, A., David, S., Parrein, B.: Multipath optimized link state routing for mobile ad hoc networks. Ad Hoc Netw. **9**(1), 28–47 (2011a)

Yi, J., Parrein, B., Radu, D.: Multipath routing protocol for manet: application to h.264/svc video content delivery. In: 2011 The 14th International Symposium on Wireless Personal Multimedia Communications (WPMC), pp. 1–5 (2011b)

Zhang, Q., Yu, Z., Shi, W., Zhong, H.: Demo abstract: Evaps: Edge video analysis for public safety. In: 2016 IEEE/ACM Symposium on Edge Computing (SEC) 00(undefined), pp. 121–122 (2016). http://doi.ieeecomputersociety.org/10.1109/MC.2016.145

An Improved Informative Test Code Approach for Code Writing Problem in Java Programming Learning Assistant System

Nobuo Funabiki[1(✉)], Khin Khin Zaw[1], Ei Ei Mon[1], and Wen-Chung Kao[2]

[1] Okayama University, Okayama, Japan
funabiki@okayama-u.ac.jp
[2] National Taiwan Normal University, Taipei, Taiwan
jungkao68@gmail.com

Abstract. The *Java Programming Learning Assistant System (JPLAS)* has been studied to enhance Java programming educations by offering advanced self-learning environments. As one problem type in JPLAS, the *code writing problem* asks a student to write a source code to satisfy the specifications described in a *test code* that verifies the correctness of the code on *JUnit*. Previously, we proposed an *informative test code* approach to help a novice student to complete a complex source code using concepts in the object-oriented programming. It describes the necessary information to implement the code, such as names, access modifiers, and data types of classes, methods, and variables, in addition to behaviors. Unfortunately, it has drawbacks in handling input/output files for an assignment. In this paper, we propose an improved informative test code approach by adopting the *standard input/output* to solve them. For evaluations, we generated improved informative test codes for five graph algorithms and requested three students in our group to write the source codes, where all of them completed the source codes with high software metrics.

1 Introduction

To advance Java programming educations, we have developed the Web-based *Java Programming Learning Assistant System (JPLAS)* [1–7]. JPLAS inspires students by offering sophisticated learning environments via quick responses to their answers for self-studies. It supports teachers by reducing loads of evaluating codes. JPLAS is implemented as a Web application using *JSP/Java*. For the server, it adopts the operating system *Linux*, the Web application server *Tomcat*, and the database system *MySQL*. For the browser, currently, it assumes the use of *Firefox* with *HTML*, *CSS*, and *JavaScript*.

JPLAS has several types of problems to cover a variety of students at different learning levels. Among them, the *code writing problem* [2] asks a student to write a source code satisfying the specification of a given assignment, based on the *test-driven development (TDD) method* [8] using an open source framework *JUnit* [9].

L. Barolli et al. (Eds.): EIDWT 2018, LNDECT 17, pp. 687–698, 2018.
https://doi.org/10.1007/978-3-319-75928-9_61

JUnit automatically tests the codes on the server to verify their correctness using the *test code* when they are submitted by students. Thus, students can repeat the cycle of writing, testing, modifying, and resubmitting codes by themselves, until they can complete the correct codes for the assignments.

To register a new assignment for the code writing problem in JPLAS, a teacher has to prepare a *problem statement* describing the code specification, a *reference source code*, and a *test code* using a Web browser. It is noted that the reference source code is essential to verify the correctness of the problem statement and the test code. Then, a student should write a source code for the assignment while referring the statement and the test code, so that the source code can be tested by using the given test code on *JUnit*.

For a novice student, it is hard in general to write a source code that is composed of multiple classes/methods with several arguments. The detailed information can help him/her to implement the code. Therefore, we proposed an *informative test code* approach for the code writing problem with describing any important information to implement the code, such as names, access modifiers, and data types of classes, methods, and variables, in addition to behavior specifications [6]. For example, this test code for a graph theory algorithm should test the methods in the class for handling the graph data for a given graph and those in the class for finding the answer. By writing the source code passing the test, a student is expected to complete the qualitative code using the proper classes/methods/variables. However, to test a code that handles the file input/output, the corresponding input/output files must be prepared in addition to the test code, and their paths must be shared by the test code and the student's source code.

In this paper, we propose an improvement of the *informative test code* approach for the code writing problem in JPLAS by adopting the standard input/output as the data input/output functions in the source code [7]. Then, this test code can test a source code handling input/output files, without preparing these files for software test on *JUnit* besides the test code. This means that every necessary data/procedure to test a source code, including the input/output data, can be described in the test code only. It can reduce the load of a teacher using JPLAS in writing and managing many test codes for programming assignments. For evaluations, we generated the improved informative test codes for five graph theory algorithms, and asked three students studying Java programming in our group to write the source codes using them. They successfully completed the codes that do not only pass the test codes but also give high software metrics.

The rest of this paper is organized as follows: Sects. 2 and 3 review the TDD method and our previous work respectively. Section 4 presents the improved informative test code approach. Section 5 shows evaluations of the proposal. Section 6 concludes this paper with future works.

2 Test-Driven Development Method

In this section, we introduce the TDD method along with its features.

In the TDD method, the test code should be written before or while the source code is implemented, so that it can verify whether the current source code satisfies the required specifications during its development process. The basic cycle in the TDD method is as follows:

(1) to write the test code to test each required specification,
(2) to write the source code, and
(3) to repeat modifications of the source code until it passes each test using the test code.

In JPLAS, *JUnit* is used as an open-source Java framework to support the TDD method. *JUnit* can assist the unit test of the source code unit or a *class*. *JUnit* has been designed with the Java-user friendly style, where a test is performed by using a given method whose name starts from *assert*. Here, *assertThat* method is adopted to compare the output of the source code with its expected value.

A test code should be written using libraries in *JUnit*. Here, by using the following **source code 1** for *MyMath* class, we explain how to write a test code. *MyMath* class returns the summation of two integer arguments.

source code 1

```
1 public class Math {
2   public int plus(int a, int b) {
3     return( a + b );
4   }
5 }
```

Then, the following **test code 1** can test the *plus* method in the *MyMath* class.

test code 1

```
1 import static org.junit.Assert.*;
2 import org.junit.Test;
3 public class MathTest {
4   @Test
5   public void testPlus() {
6     Math ma = new Math();
7     int result = ma.plus(1, 4);
8     assertThat(5, is(result));
9   }
10 }
```

The names in the test code should be related to those in the source code so that their correspondence becomes clear:

- The class name is given by the *test class name + Test.*
- The method name is given by the *test + test method name.*

The test code imports *JUnit* packages containing test methods at lines 1 and 2, and declares *MathTest* at line 3. *@Test* at line 4 indicates that the succeeding method represents the test method. Then, it describes the test method.

The test code performs the following functions:

(1) to generate an instance for the *MyMath* class,
(2) to call the method in the instance in (1) using the given arguments,
(3) to compare the result with its expected value for the arguments in (2) using the *assertThat* method, where the first argument represents the expected value and the second one does the output data from the method in the source code under test.

3 Informative Test Code Approach

In this section, we review the *informative test code* approach in JPLAS [6].

3.1 Goal of Informative Test Code

The *informative test code* helps a student to complete the qualitative source code for a harder code writing problem that requires the use of multiple classes/methods, and/or the adoption of advanced concepts of the object-oriented programming such as *encapsulation*, *inheritance*, and *polymorphism*. It describes the necessary information to implement the code, such as the names, data types, and access modifics of the classes, methods, and member variables, and the exception handlings.

3.2 Assignment Generation with Informative Test Code

In general, an assignment for the code writing problem requires the test code file, the input data file, and the expected output data file that be prepared by a teacher, in addition to the statement in natural language. Then, a student is requested to write the source code that passes every test described in the test code on *JUnit*. This means that a student writes the source code by referring to the detailed specifications described in the test code. A teacher generates a new assignment for the code writing problem using the informative test code by the following procedure:

(1) to prepare the statement and the *input data file* for the assignment,
(2) to prepare the *reference source code* that does not only satisfy every specification of the assignment but has the high quality design,
(3) to obtain the *expected output data file* by running the reference source code, where this file is used for comparison with the output data file of the student source code to check the correctness, and
(4) to generate the *informative test code* describing the necessary information to implement the source code by a student.

3.3 Drawbacks of Previous Approach

The previous approach requires the input data file to the source code and the expected output data file from the code, in addition to the test code file, for each assignment. Then, the following drawbacks can be observed:

- A teacher must prepare and manage the test code that are consistent with the input data file, where some tests usually depend on the content of the input file.
- A teacher must describe the input/output file paths in the test code that are actually implemented in the JPLAS server, whereas they usually should not be disclosed to a user.
- A student must describe the input/output file paths in the source code that are actually implemented in the JPLAS server, whereas they usually should not be disclosed to a user.

To solve these drawbacks, the test code should contain the necessary information in the input/output files for the assignment. In this paper, this requirement is achieved by adopting the standard input/output.

4 Proposal of Improved Informative Test Code

In this section, we propose the improved informative test code by adopting the standard input and output to read and write the data, instead of using the input and output data files.

4.1 Handling of Standard Input/Output in Test Code

The proposed approach in this paper reads and writes the data through using the standard input and output. However, in general, when a test code runs on *JUnit*, it cannot read data from the standard input (keyboard) or write data to the standard output (console). To solve it, the test code in our approach adopts the classes defined in [10], which feeds the data described in the test code into the standard input to the source code, and intercepts the standard output data from the source code to the test code, as follows:

- To describe the standard input data to the source code, *Inputln* method in *StandardInputSnatcher* class is adopted in the test code. It is noted that *StandardInputSnatcher* class is extended from *InputStream* class.
- Any standard input data is described in the argument of *Inputln*.
- To receive the standard output data from the source code, *readLine* method in *StandardOutputSnatcher* class is adopted in the test code. It is noted that *StandardOutputSnatcher* class is extended from *PrintStream* class.
- To obtain the expected standard output data from the code for each input data, the reference source code is executed with this input data.
- Each pair of the standard input and output data is embedded into the test code.

4.2 Requirements in Source Code

The source code for an assignment needs to satisfy the following requirements:

- to read the input data from the input data file or the standard input (console), and
- to write the output data to the output data file or the standard output (display).

In the following source code for the *breadth first search algorithm (BFS)*, line 133 describes the *standard input* of the input data to the BFS algorithm, and line 143 describes the *standard output* of the output data from it. They can be replaced by the file input and output if necessary. This code has two classes, one constructor and six methods for class *SimpleGraph*, and four methods including the *main* method for class *BFS*. Thus, the implementation is hard for a novice student.

source code for BFS

```
1  package Standardinout;
2  import java.util.ArrayList;
3  import java.util.Scanner;
4
5  class SimpleGraph {
6      public boolean[][] edges;// adjacency matrix
7      public String [] labels;
8      public SimpleGraph(int n) {
9          edges = new boolean[n][n];
10         labels = new String[n];
11     }
12     public int size() {
13         return labels.length;
14     }
15     public void setLabel(int vertex, String label) {
16         labels[vertex] = label;
17     }
18     public Object getLabel(int vertex) {
19         return labels[vertex];
20     }
21     public void addEdge(int source, int target) {
22         edges[source][target] = true;
23     }
24     public boolean getEdge(int source, int target) {
25         if (edges[source][target]) {
26             return true;
27         }
28         return false;
29     }
30     public int [] neighbors(int vertex){
31         int count = 0;
32         for (int i = 0; i < edges[vertex].length; i ++){
33             if (edges[vertex][i]) count ++;
34         }
35         final int[] answer = new int[count];
36         count = 0;
```

```
37            for (int i = 0; i < edges[vertex].length; i ++) {
38                if (edges[vertex][i]) answer[count ++] = i;
39        }
40            return answer;
41        }
42    }
43
44    public class BFS {
45    // traverse nodes using BFS algorithm
46        public static String findBFS (SimpleGraph G, int s) {
47            final int [] dist = new int [G.size()];
48            final int [] pred = new int [G.size()];
49            final int [] Queue = new int[G.size()];
50            final boolean [] visited = new boolean [G.size()];
51            int source = 0;
52            int target = 1;
53            String bfsout = " ";
54            for (int i = 0; i < dist.length; i ++) {
55                dist[i] = Integer.MAX_VALUE;
56            }
57            dist[s] = 0;// initialize distance for source node by zero
58            Queue[0] = s;
59        // if head and tail are not equal, it's not empty
60            bfsout += "sele_node pre_node"+"\n";
61            while (source != target) {
62                final int u = Queue[source]; // get the head in Queue
63                bfsout += (String) G.getLabel(u)+"  "+G.getLabel(pred[u])+"\n";
64                visited[u] = true;
65            // each vertex adjacent to u
66                int []m = G.neighbors(u);
67                for (int j = 0; j < m.length; j ++) {
68                    if (dist[v] == Integer.MAX_VALUE) {
69                        dist[v] = dist[u]+1;
70                        pred[v] = u;
71                        Queue[target] = v;
72                        target ++;
73                    }
74                }
75                source ++;
76            }
77            return bfsout;
78        }
79    // check the data form the input string whether is edge or not?
80        public boolean isEdge(String[] inputStr) {
81            boolean isEdge = false;
82            for (int i = 0; i < inputStr.length; i ++) {
83                try {
84                    Integer.parseInt(inputStr[i]);
85                    isEdge = true;
86                } catch(Exception e) {
87                    isEdge = false;
88                    break;
89                }
90            }
91            return isEdge;
92        }
93
94        public SimpleGraph findGraph(String result) {
95            BFS bfs = new BFS();
96            ArrayList<String[]> NodeList = new ArrayList<String[]>();
97            ArrayList<String[]> EdgeList = new ArrayList<String[]>();
98        // check the data from the input string whether is edge or node
99            for (String w:result.split("\n", 0)) {
```

```
100             if (!(w.equals("node node_Name") || w.equals("source target"))
        ){
101                 String[] element = w.split("\\s+");
102                 if (bfs.isEdge(element)) {
103                     EdgeList.add(element);// edge
104                 }
105                 else {
106                     NodeList.add(element);// node
107                 }
108             }
109         }
110     // add the node and edge data to the graph using SimpleGraph class
111         SimpleGraph G = new SimpleGraph(NodeList.size());
112         for (int i = 0; i < NodeList.size(); i ++) {
113             String [] node = NodeList.get(i);
114             int index = Integer.parseInt(node[0]);
115             String Label = node[1];
116         // add the node to graph
117             G.setLabel(index, Label);
118         }
119         for (int i = 0; i < EdgeList.size(); i ++) {
120             String [] edges = EdgeList.get(i);
121             int source = Integer.parseInt(edges[0]);
122             int target = Integer.parseInt(edges[1]);
123         // add the edge to graph
124             G.addEdge(source, target);
125         }
126     //return the graph object
127         return G;
128     }
129
130     public static void main (String args[]) {
131         BFS bfs = new BFS();
132     // read the data from the standard input
133         Scanner scan = new Scanner(System.in);
134         String subPattern = "";
135         while(scan.hasNext()) {
136             subPattern += scan.nextLine();
137             subPattern += "\n";
138         }
139     // find the node and edge, and add them to the graph
140         SimpleGraph G = bfs.findGraph(subPattern);
141     // apply BFS algorithm and return the string for BFS output
142         String strbfs = bfs.findBFS(G, 0);
143         System.out.println(strbfs);
144     }
145 }
```

4.3 Requirements in Test Code

The test code needs to satisfy the following requirements:

- to feed the standard input data to the source code,
- to intercept the standard output data from the source code,
- to describe any input data to test the source code, and
- to describe the expected output data for each input data.

In the following test code for the source code for *BFS*, line 73 describes feeding the standard input data to the source code, and lines 74–80 describe intercepting the standard output data from the code.

test code for BFS

```
1  package Standardinout;
2  import static org.junit.Assert.*;
3  import static org.hamcrest.CoreMatchers.is;
4  import static org.junit.Assert.assertThat;
5  import java.io.InputStream;
6  import org.junit.Before;
7  import org.junit.Test;
8  import java.io.BufferedReader;
9  import java.io.ByteArrayOutputStream;
10 import java.io.IOException;
11 import java.io.PrintStream;
12 import java.io.StringReader;
13 import java.util.Arrays;
14 import org.junit.rules.TemporaryFolder;
15
16 public class BFSTest {
17     private StandardInputSnatcher in = new StandardInputSnatcher();
18     private StandardOutputSnatcher out = new StandardOutputSnatcher();
19
20     @Before
21     public void setUp() {
22         System.setIn(in);
23         System.setOut(out);
24     }
25
26     @Test
27     public void testSimpleGraph() {
28         SimpleGraph G = new SimpleGraph (5);
29         boolean a = G.labels instanceof String[];
30         boolean b = G.edges instanceof boolean[][];
31         assertEquals(true, a);
32         assertEquals(true, b);
33         assertEquals(5, G.labels.length);
34         assertEquals(5, G.edges.length);
35         assertEquals(5, G.edges[0].length);
36     }
37
38     @Test
39     public void testSetLabel() {
40         SimpleGraph G = new SimpleGraph(2);
41         G.setLabel(1, "a");
42         assertEquals("a", G.labels[1]);
43     }
44
45     @Test
46     public void testGetLabel(){
47         SimpleGraph G = new SimpleGraph(2);
48         G.setLabel(1, "b");
49         String label = (String)G.getLabel(1);
50         assertEquals("b", label);
51     }
52
53     @Test
54     public void testAddEdgeandGetWeight() {
55         SimpleGraph G = new SimpleGraph(3);
56         G.addEdge(1, 2);
57         assertEquals(true, G.getEdge(1, 2));
58     }
```

```
59
60      @Test
61      public void testNeighbours() {
62          SimpleGraph G = new SimpleGraph(3);
63          int [] expectedNode = {1, 2};
64          G.addEdge(0, 1);
65          G.addEdge(0, 2);
66          assertTrue(Arrays.equals(expectedNode, G.neighbors(0)));
67      }
68
69      @Test
70      public void testStandardInOut() throws Exception {
71          StringBuffer bf = new StringBuffer();
72          String actual,line,expected;
73          in.Inputln("node node_Name\n0 s\n1 r\n2 w\n3 t\n4 x\n5 v\
        n6 u\n7 y\n" + "source target\n0 1\n0 2\n1 5\n2 3\n2 4\n3
         6\n4 7\n");
74          BFS.main(new String[0]);
75          System.out.flush();
76          while ((line = out.readLine()) != null) {
77              if (bf.length() > 0) bf.append("\n");
78              bf.append(line);
79          }
80          actual = bf.toString();
81          expected = "sele_node pre_node\ns s\nr s\nw s\nv r\nt w\
        nx w\nu t\ny x\n";
82          assertThat(actual, is(expected));
83      }
84
85      public class StandardInputSnatcher extends InputStream {
86          private StringBuilder buffer = new StringBuilder();
87          private String crlf = System.getProperty("line.separator");
88          public void Inputln(String str) {
89              buffer.append(str).append(crlf);
90          }
91
92          @Override
93          public int read() throws IOException {
94              if (buffer.length() == 0) {
95                  return −1;
96              }
97              int result = buffer.charAt(0);
98              buffer.deleteCharAt(0);
99              return result;
100         }
101     }
102
103     public class StandardOutputSnatcher extends PrintStream {
104         private BufferedReader buffer = new BufferedReader(new StringReader(""));
105         public StandardOutputSnatcher() {
106             super(new ByteArrayOutputStream());
107         }
108
109         public String readLine() {
110             try {
111                 String line = "";
112                 if ((line = buffer.readLine()) != null) {
113                     return line;
114                 } else {
115                     buffer = new BufferedReader(new StringReader(out.toString()));
116                     ((ByteArrayOutputStream) out).reset();
117                     return buffer.readLine();
118                 }
119             } catch (IOException ex) {
120                 throw new RuntimeException(ex);
121             }
122         }
123     }
124 }
```

5 Evaluation

In this section, we evaluate the effectiveness of our proposal. We generated the improved informative test codes for *BFS*, *DFS (depth-first search)*, *Prim*, *Dijkstra*, and *Kruskal* algorithms in the graph theory, and asked three students in our group to write the source codes. Then, all of them completed the source codes that pass the tests.

To evaluate the quality of the source codes, their software metrics are measured by *Metrics plugin for Eclipse* [11]. Table 1 shows the results. The metrics of any source code are good or at least acceptable.

Table 1. Software metrics results.

Algorithm	BFS			DFS			Prim			Dijkstra			Kruskal		
Student	S1	S2	S3	S1	S2	S3	S1	S2	S3	S1	S2	S3	S1	S2	S3
NOC	6	3	2	8	4	2	13	6	2	9	5	2	12	5	3
NOM	26	11	10	31	12	9	49	21	15	37	18	17	44	13	12
VG	6	6	5	6	4	6	6	10	5	6	10	5	6	9	13
NBD	4	3	4	4	3	5	4	5	4	4	5	4	4	5	5
LCOM	0.722	0.5	0.5	0.722	0.5	0.5	0.8	0.5	0.5	0.722	0.33	0.5	0.8	0.75	0.5
TLC	239	132	129	281	128	125	437	278	141	362	214	153	410	173	195
MLC	146	86	98	170	78	93	278	192	100	226	142	106	257	106	146

S1 adopted other classes than the expected in the test code to represent the graph in the source codes. Then, NOC (number of classes), NOM (number of methods), TLC (total lines of code), and MLC (method lines of code) of his codes are larger than the others. Besides, *S1* inherited *BFS* from *DFS* and *Dijkstra* from *Prim*, because they are similar algorithms. At the same time, his codes use unnecessary statements to pass the given test codes, which makes LCOM (lack of cohesion in methods) larger. They indicate that our informative test code approach may lose the flexibility in the code implementation. The provision will be in future works.

S2 commented that the behaviors of *Inputln* method in *StandardInputSnatcher* class and *readLine method* in *StandardOutputSnatcher* class in the informative test code for the standard input/output are hard to understand. This student has the sufficient level of Java programming. Thus, it is necessary to provide supplementary explanations to let novice students easily understand them, such as a sample test code and source code that only contain them, which will also be in future works.

S3 uses a lot of conditional statements for *Kruskal*, which increases VG (Cyclomatic Complexity).

6 Conclusion

In this paper, we proposed the improved *informative test code* approach for the harder code writing problem in JPLAS of adopting the *standard input/output* to solve drawbacks in handling the file input/output. We evaluated the proposal through applications to three students in our group, where all of them completed the high-quality source codes passing the test codes. In future works, we will improve the flexibility and comprehensibility of the informative test code approach for novice students, generate informative test codes to other assignments, and apply them in Java programming courses. We will also investigate the extension of the proposal to integration testing.

References

1. Funabiki, N., Tana, Zaw, K.K., Ishihara, N., Kao, W.-C.: A graph-based blank element selection algorithm for fill-in-blank problems in Java programming learning assistant system. IAENG Int. J. Comput. Sci. **44**(2), 247–260 (2017)
2. Funabiki, N., Matsushima, Y., Nakanishi, T., Amano, N.: A Java programming learning assistant system using test-driven development method. Int. J. Comput. Sci. **40**(1), 38–46 (2013)
3. Zaw, K.K., Funabiki, N., Kao, W.-C.: A proposal of value trace problem for algorithm code reading in Java programming learning assistant system. Inf. Eng. Express. **1**(3), 9–18 (2015)
4. Ishihara, N., Funabiki, N., Kao, W.-C.: A proposal of statement fill-in-blank problem using program dependence graph in Java programming learning assistant system. Inf. Eng. Express. **1**(3), 19–28 (2015)
5. Ishihara, N., Funabiki, N., Kuribayashi, M., Kao, W.-C.: A software architecture for Java programming learning assistant system. Int. J. Comput. Soft. Eng. **2**(1) (2017)
6. Zaw, K.K., Funabiki, N.: A design-aware test code approach for code writing problem in Java programming learning assistant system. Int. J. Space-Base. Situated Comput. **7**(3), 145–154 (2017)
7. Funabiki, N., Kusaka, R., Ishihara, N., Kao, W.-C.: A proposal of test code generation tool for Java programming learning assistant system. In: Proceedings IEEE International Conference on Advanced Information Networking and Applications (AINA2017), pp. 51–56 (2017)
8. Beck, K.: Test-Driven Development: By Example. Pearson Education, Boston (2002)
9. JUnit. http://www.junit.org/
10. Diary of kencoba. http://d.hatena.ne.jp/kencoba/20120831/1346398388
11. Metric Plugin. http://metrics.sourceforge.net

Tourism Support System Using AR for Tourists' Migratory Behaviors

Haruna Sonobe[1(✉)], Hiroaki Nishino[2], Yoshihiro Okada[1,3,4], and Kousuke Kaneko[2,4]

[1] Graduate School of Information Science and Electrical Engineering, Kyushu University, Fukuoka, Japan
2IE17029N@s.kyushu-u.ac.jp
[2] Faculty of Science and Technology, Oita University, Oita, Japan
hn@oita-u.ac.jp
[3] Kyushu University Library, Innovation Center for Educational Resources, Kyushu University, Fukuoka, Japan
okada@inf.kyushu-u.ac.jp
[4] CybrerSecurity Center, Kyushu University, Fukuoka, Japan
kaneko.kosuke.437@m.kyushu-u.ac.jp

Abstract. This paper treats tourism support systems and the authors propose a tourism support system for tourists' migratory behaviors using AR (Augmented Reality). The target users of this system are mainly young tourists who frequently use one of the mobile devices like smartphones and tablets. The users of this system can receive tourism information about shops and restaurants in his/her sightseeing spots as push-typed data, sometimes called passive information, that are automatically sent to the target users by Bluetooth and Beacon. The sent information will be displayed using AR (Augmented Reality). The purpose of this system is to activate users' migratory behaviors during their tours. In this paper, the authors also show experimental results to evaluate the usefulness of the proposed system. From the results, it can be said that the propose system will be useful.

1 Introduction

In these years, tourism called community-based tourism is attracting a lot of attention for a variety of tourist needs. This is a new approach that local residents actively provide local information to tourists differently in the case of conventional tourism. In order to make this approach successful, it is important to increase tourist satisfaction.

It needs to enable tourists to get their required information as easily as possible. However, conventionally tourism information is mainly given from travel-guidebooks and certain Web sites, tourists cannot get any useful information unless they actively search. Therefore, tourists cannot get their required information, or it may take much time or need great effort even if they could get the information. This is one burden for travelling and regarded as one of the reasons that decrease tourists' satisfaction. Although a lot of information can be offered about sightseeing spots and historic places, the information about shops and restaurants are not enough. This current situation leads

L. Barolli et al. (Eds.): EIDWT 2018, LNDECT 17, pp. 699–710, 2018.
https://doi.org/10.1007/978-3-319-75928-9_62

us to the requirement of a system that flexibly responds to the regional features and the various needs of tourists.

In this paper, our research goal is to develop a tourism system that can easily collect regional information about shops and restaurants and temporal information like time-limited sales, and can actively provide such information to tourists through their smart-phones/tablets. Especially, the targets of this system are young generation tourists who get used to the operations on a smartphone/tablet. If tourists can passively get the information without their positively searching it, their costs to to do so will be reduced. We expect that this can activate tourists' migratory behaviors.

The remainder of this paper is organized as follows: the next section overviews related works. In Sect. 3, we explain details of the proposed system and prototyping as a mobile application. Section 4 shows experimental results to evaluate the usefulness of the prototype. Finally, we conclude the paper and discuss about future work in Sect. 5.

2 Related Work

Tourism systems that recommend sightseeing spots corresponding to user needs have been widely studied in the information providing systems [1]. Regarding information presentation to tourists, there are many researches those use AR [2, 3]. The effectiveness of information presentation by AR has been clarified from these researches. Such systems provide information on tourist attractions such as historical heritage sites. However, it is difficult to continue attracting tourists' attention. As tourists repeatedly visit, interest in tourist attractions is weakened [4]. There are meals and shopping that tourists do on their travels. They are the activities performed regardless of the number of visits. In this research, we focus on shopping and meals. We propose a system that provides shop information using AR because there are few studies focusing on shop information.

In this research, not only outdoor use but also indoor use of the system is taken into consideration from the viewpoint of shop information. There are many researches those use Bluetooth beacons as an indoor positioning system [5, 6]. Bluetooth beacon is a class of Bluetooth low energy device that becomes a trigger for any action when a device such as a tablet or a smartphone comes near the beacon. This technology is compatible with customer tracking to notify any information. Therefore, we decided to employ Bluetooth beacons for push-type notification to tourists and for information collection about their migratory behaviors.

3 Proposed System and Prototyping as Mobile Application

In this section, we explain the details of the proposed tourism support system. We developed one mobile application for Android as its prototype. Figure 1 shows the overview of the system.

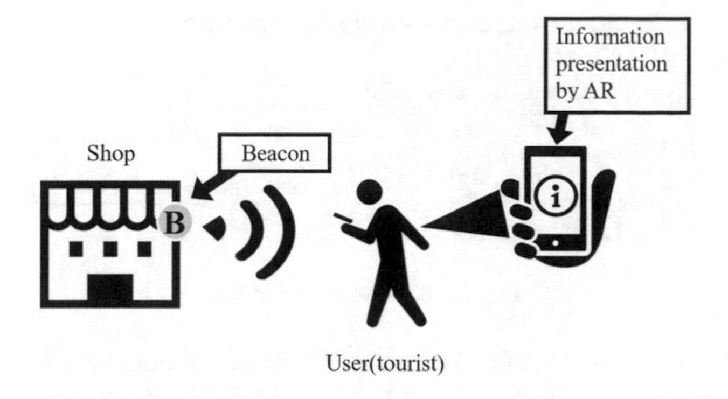

Fig. 1. The proposed system overviews.

First of all, the users walk around sightseeing spots with a smartphone on that this application was installed and launched. While their walking, by signals sent from Bluetooth beacons located at each shop, they are notified the shop information. Users start to look for the shop according to the received information and find out the shop's landmark (e.g. signboards, illustrations) works as a trigger for ARs. Users hold their smartphone camera over the landmark to make the camera recognize the landmark, and then, a certain object containing a name and an inside picture of the shop will appear on the smartphone screen. By tapping on the displayed object, more detail information on the shop will be showed. Users resume taking a walk after confirming the showed information. By repeating the above, users can passively get the information. In addition, users can get message that shops send in real time. Each notification can also be received in the background after the application has been activated. The details of each function are explained in the next subsections.

3.1 User Interface

The explanation about each function is as follows.

(1) **Bluetooth beacon Log.** The screen shifts to an image for displaying the log of the number of times that users entered into the area for each Bluetooth beacon terminal as shown in Fig. 2. This screen displays the identifier of the received Bluetooth beacon besides the number of times that users entered into the Bluetooth beacon area.

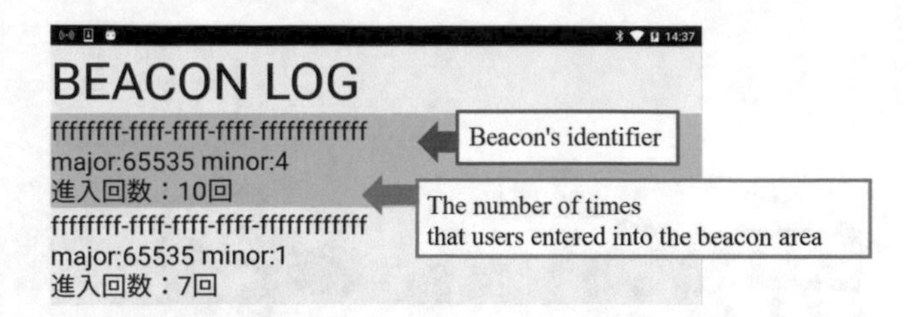

Fig. 2. Bluetooth beacon log.

(2) **Start AR Camera.** Start the AR camera on the user's terminal.

(3) **Message Log.** The screen shifts to an image for displaying messages received by the user's terminal as shown in Fig. 3. It displays received each message that includes its title, text, received date and store name. By selecting shops in the pull-down menu, it is possible to display only the preferable shops information.

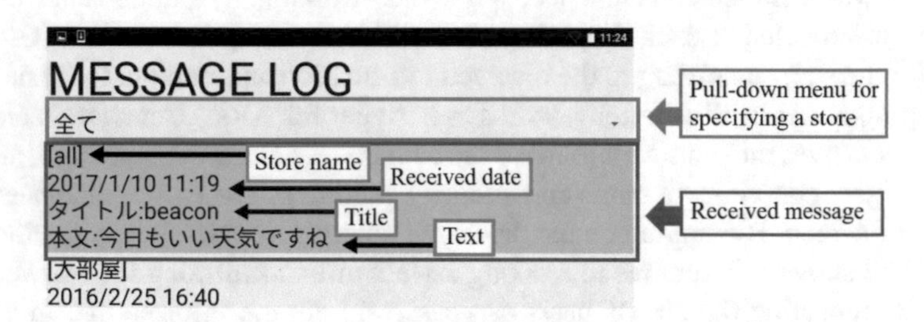

Fig. 3. Message log.

(4) **Action Log.** The screen shifts to an image for displaying all Bluetooth beacons received by the user's terminal in the order of the reception. This displays information of the received Bluetooth beacon's identifier and the signal received date and the shop name where each Bluetooth beacon is installed as shown in Fig. 4.

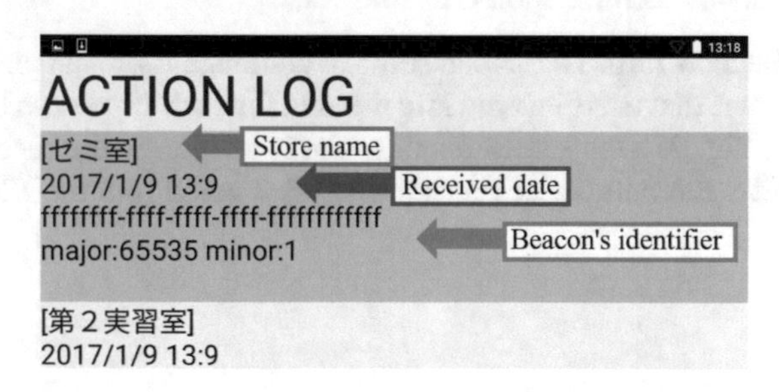

Fig. 4. Action log.

3.2 Bluetooth Beacon Notice

At the same time a user enters within 15 m from the Bluetooth beacon installed in the shop, a notification is sent to the user's smartphone. The content of the notification is the shop name and the simple message set in advance. If you are more than 15 m away from the Bluetooth beacon, the notification automatically disappears. A flowchart of the Bluetooth beacon notification is shown in Fig. 5.

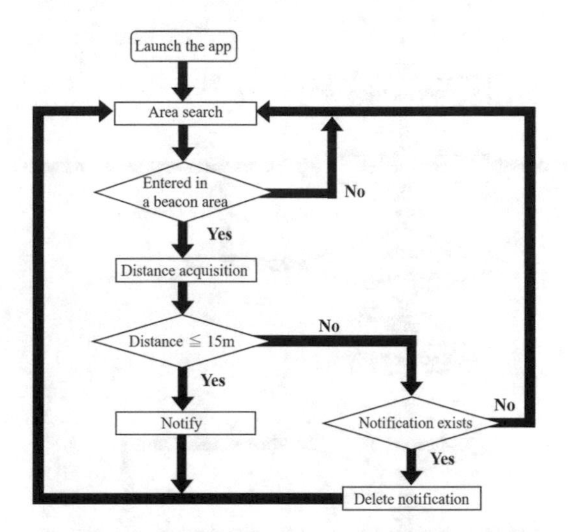

Fig. 5. Flowchart of the Bluetooth beacon notification.

3.3 AR View

Figure 6 shows how to use AR camera. The object including the name of the installed shop and the picture is displayed when the user's smartphone recognizes the landmark for AR display. The screen of the smartphone shifts to an image for detail information by tapping on the displayed object.

Fig. 6. How to use AR camera.

3.4 Message

The application receives messages using mBaaS (mobile Backend as a Service). The user who receives the message is managed by the topic. A flowchart of topic registration is shown in Fig. 7. The user can receive only the information of a nearby shop in real time with this function.

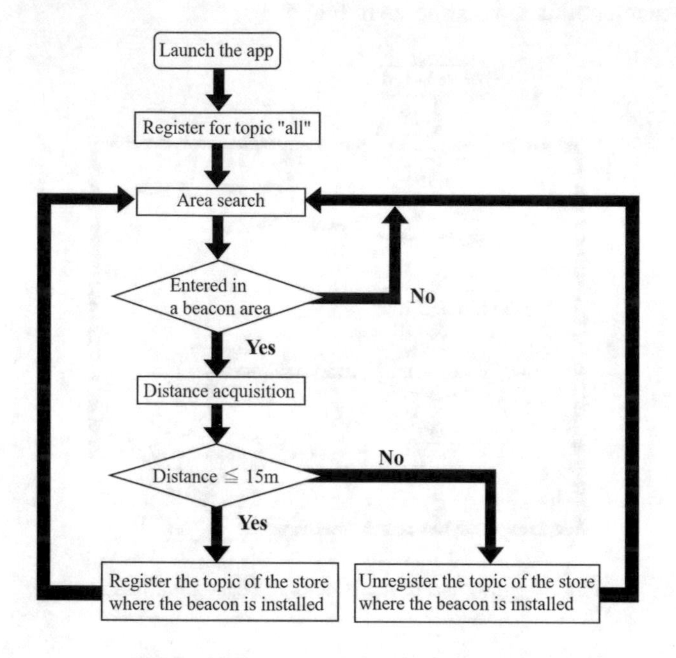

Fig. 7. Flowchart of the topic for message.

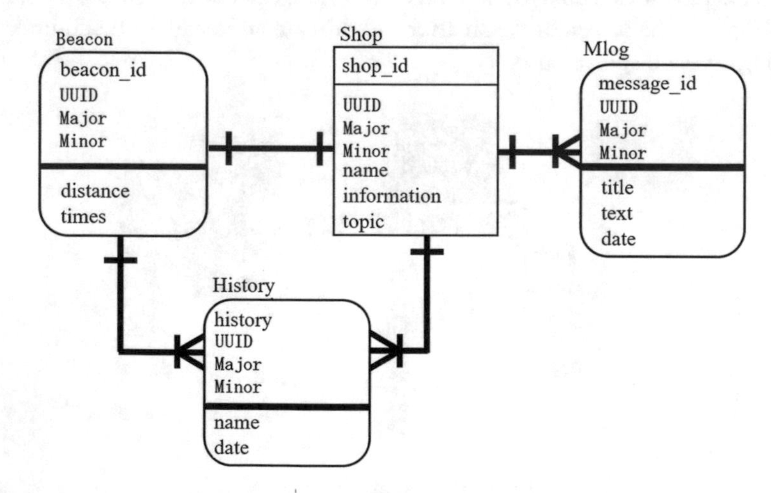

Fig. 8. ERD of shopDB.

3.5 Database

With SQLite, we created action history for each user's smartphone, a database for storing message log as shopDB. Figure 8 shows the ER diagram of shopDB that is currently implemented.

4 Experiments

We performed experiments to evaluate the proposed tourism support system.

4.1 Experiment Outline

In order to verify whether this system is effective in providing shop information, we asked eight participants to answer the evaluation questionnaire after using the system. The participants are young generation who get used to a smartphone. In the experiment, participants use the system by walking around inside the building after explaining how to operate the system. The participants used Android tablet that is installed the application developed in this research.

4.2 Experiment Environment

We conducted experiments on the 2nd and 3rd floors of the Oita University Computer System Building. Since landmarks do not exist near the room, we used the printed illustration (size: 15 cm by 20 cm) as landmarks for displaying AR. In this case, Landmarks and Bluetooth beacons were installed at the four places in the building as shown in Fig. 9. Each Bluetooth beacon is installed on the wall behind the target.

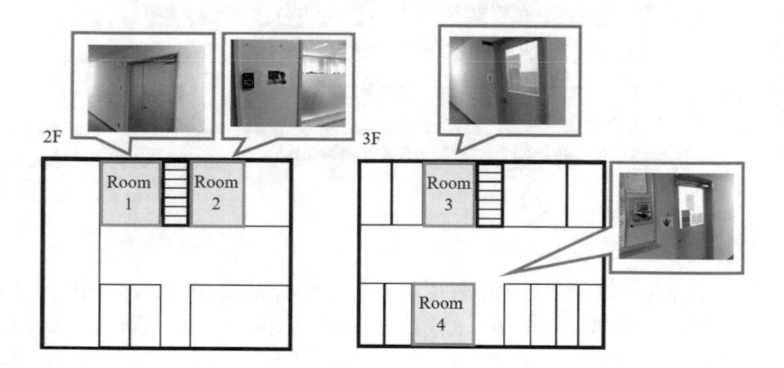

Fig. 9. Experiment environment.

4.3 Experiment Method

The procedure of the experiment is shown in Table 1.

Table 1. Procedure of experiment.

1.	Activate the system at the entrance of the building and go for a stroll in the building
2.	Receive Bluetooth beacon notifications
3.	After finding out the marker, activate the tablet's AR camera and make it recognize the AR landmark
4.	The object appears that consists of the room name and its indoor photo
5.	Tap the object, then more information on the room will appear
6.	Quit the tablet's AR camera
7.	Repeat steps 2–6
8.	Receive message when the two landmarks were found
9.	Check the message log after finding out 4 landmarks
10.	Answer a questionnaire

4.4 Evaluation Content

We conducted a questionnaire survey to evaluate the effectiveness of information notified by the system and usability. The questions are shown in Table 2. It consists of six questions. For question 1 to 5, participants answer each question by scoring 1 (worst) to 5 (best), and question 6 ask participants to freely write their opinions and comments.

Table 2. Questionnaire

Q1.	Do you think you can receive effective information from the system?
Q2.	Is the location notification by Bluetooth beacon useful?
Q3.	Is the information using AR impressive?
Q4.	Do you think that the receiving a message is effective?
Q5.	How is the operability?
Q6.	Opinions and comments on the system.

4.5 Results

Table 3 shows results about the answers from the participants for questions 1 to 5. The average score for each participant and the average score for each question are also included in the table. Figure 10 shows the results for questions 1 to 5 as charts. The average score among questions 1 to 5 exceeded 4 for any participant, and those mean this system can almost satisfy the participants. However, as for question 2, we found that the evaluation is lower than other questions. *Q2* is "*Is the location notification by Bluetooth beacon useful?*". As the evaluation of *Q2*, there were many opinions that too many notifications prevent users from understanding. On the other hand, there was a positive opinion that it was able to find the existence of the room by notice, to know what kind of a room is in the vicinity and helped exploration.

Table 3. Questionnaire results

	Q1	Q2	Q3	Q4	Q5	Average
Participant 1	4	5	5	5	4	4.6
Participant 2	5	4	5	4	4	4.4
Participant 3	5	4	5	5	4	4.6
Participant 4	5	3	5	5	5	4.6
Participant 5	5	3	5	4	5	4.4
Participant 6	5	4	5	5	5	4.8
Participant 7	5	3	5	5	5	4.6
Participant 8	3	5	5	5	5	4.6
Average	4.625	3.875	5	4.75	4.625	

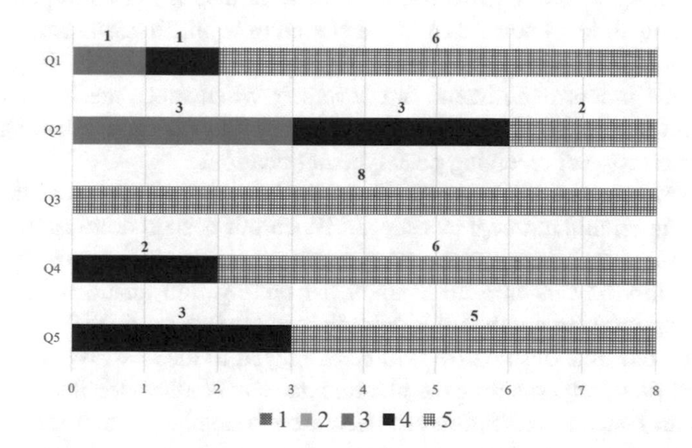

Fig. 10. Questionnaire results

The highest scored question is *Q3* that "*Is the information using AR impressive?*". The scores are all *5*. From this result, it was found that the information using AR of the system is very effective. There were many positive opinions that displaying an indoor photo was helpful to understand the room visually and interesting.

Q6 is "*Opinions and comments on the system.*". We obtained various opinions and comments. For instance, displaying not only pictures but also movie is helpful to make tourists more interested in sightseeing spots. The consumed time and effort can be reduced by a Bluetooth beacon notification even while the AR camera is being activated. It becomes easy to find a room if there is distance information in the Bluetooth beacon notification.

4.6 Discussion

This subsection discusses about the problems of the current system revealed from the evaluation results and the improvement points of it. In the followings, we discuss about (1) push type Bluetooth beacon notification and (2) push type information notification in our system.

(1) **Push type Bluetooth beacon notification:** In the system, push type notification was performed when the distance between the Bluetooth beacon and the user becomes less than a certain value. This does not become a problem when shops are located away from each other. However, when shops are located closely to each other, too many notifications come to be displayed, especially at sightseeing spots. This becomes the problem that received information becomes hard to understand. Therefore, it is considered necessary to improve the notification method.

The first proposal is to adjust the range to send push type notification by Bluetooth beacon. In this system, notification was pushed when the distance between the Bluetooth beacon and the user become less than 15 m. In the questionnaire, there are many opinions that the notification transmission range is too wide in this experimental environment. However, contrary to the negative opinion, there is also a positive opinion that it is convenient to be able to know there are what kind of rooms in advance. Depending on sightseeing spots, it seems that there are several patterns, such as when shops are located closely like this indoor experiment, or conversely when shops are located widely. In other words, it can deal with the diversity of tourist spots by implementing a function that adjusts the scope of receiving push type notifications.

The second proposal is to change Bluetooth beacon notification method. We can implement a function that integrates several Bluetooth beacon notifications in order to solve the problem that there are too many notifications. In the current system, simple shop information appears in a notification bar but we will abandon it. Then, in the notification bar, only the number of Bluetooth beacons that received the signal, in other words the number of shops nearby, will be displayed to the user. By tapping the integrated notification, implementing the function that can confirm detailed information on each Bluetooth beacon notification can decrease the number of notifications. Furthermore, it is considered that the user can easily find his/her preferable shop information by implementing a function for selecting the type of a shop to be displayed on the confirmation screen of the shop information.

(2) **Push Type Information Notification.** The push type notification using a Bluetooth beacon of the system displays only a shop name and a simple message. However, there is the problem that the information is insufficient to find the shop which sends it. Therefore, we have to enrich the information to be displayed by the push type notification. Specifically, we can propose to add the actual distance to the Bluetooth beacon to be displayed on the confirmation screen as an additional information. When a user finds out a shop that interests the user, and taps the shop name on the screen, the user can know the shop's appearance photos and the distance to the shop. As a result, it is thought that users will be able to find out their own preferable shops readily.

5 Conclusions

In this paper, we proposed and implemented the tourism support system for tourists' migratory behaviors. The system gives shop information on sightseeing spots to the

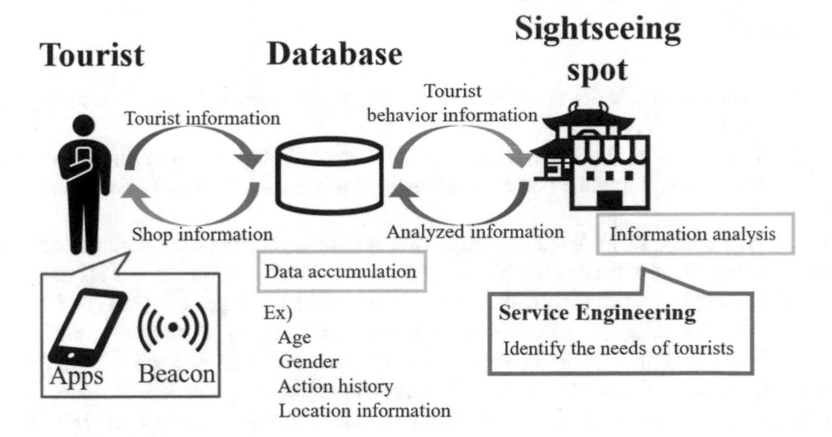

Fig. 11. Future system overview

tourists by push type notification with Bluetooth beacon, visual information presentation using AR, and received messages in real time. Next, we performed the experiment that uses the current system as a room guide in order to clarify its usefulness, problems and improvement points. The results showed that the system is effective as an information notification tool when searching for shops. It can be clarified that the information notification by AR is impressive to users. However, regarding the push type notification with the Bluetooth beacon revealed that there are problems such that it is difficult to understand due to too many notices and the obtained information is insufficient.

As future issues, it is necessary to implement the function of notifying after organizing the information rather than the function of just notifying the received Bluetooth beacon. It is thought possible to implement a system that makes it easier for users to passively receive shop information by this. In addition to this, in the current push type notification, it can be clarified from the experiment that the information content is insufficient and that it cannot reach the AR landmark with just that. Therefore, we propose to add information such as the appearance picture of the shop and the shop type and the actual distance to the shop so that the shop can be found at a glance. It is considered that the user can easily find the shop because of these pieces of information. In this study, we focused on tourists and implemented the system. However, in order to revitalize the tourism business, it seems necessary to create a system to support the provision of information on the sightseeing side which is the information providing side. Figure 11 shows a schematic of the system including the information provider. In the current system, the action log is collected for each terminal. It becomes possible for the sightseeing spot side to grasp how visitors using the system are conducting their migratory behaviors by accumulating that information in a database on a server. The service will develop further because the sightseeing spot side can identify the needs of tourists by analyzing the behavior histories from the concept of service engineering.

Acknowledgments. This research was partially supported by JSPS KAKENHI Grant Number JP16H02923 and JP15K12170.

References

1. Borràs, J., Moreno, A., Valls, A.: Intelligent tourism recommend der systems: a survey. Expert Syst. Appl. **41**(16), 7370–7389 (2014)
2. Shi, Z., Wang, H., Wei, W., Zheng, X., Zhao, M., Zhao, J., Wang, Y.: Novel individual location recommendation with mobile based on augmented reality. Int. J. Distrib. Sens. Netw. **12**(7), (2016)
3. Fukuda, H., Funaki, T., Kodama, M., Miyashita, N., Ohtsu, S.: Proposal of tourist information system using image processing-based augmented reality. IPSJ SIG Technical report Information Processing Society of Japan, vol. 2011-IS-115(13), pp. 1–8 (2011)
4. Okamura, K., Fukushige, M.: How to promote repeaters? Empirical analysis of tourist survey data in Kansai Region. Discussion Papers in Economics and Business, vol. 07(42) (2007)
5. Chiba, T., Uetake, T., Horikawa, M.: Construction of indoor guide system using behavior measurement by smart glass and BLE localization. In: Proceedings of 78th National Convention of IPSJs, vol. 2016(1), pp. 443–444 (2016)
6. Kudou, D., Horikawa, M., Furudate, T., Okamoto, A.: The proposal of indoor positioning system by area estimation using BLE beacon. In: Proceedings of 78th National Convention of IPSJs, vol. 2016(1), pp. 425–426 (2016)

Some Improvements in VCP for Data Traffic Reduction in WSN

Ezmerina Kotobelli[1(✉)], Mario Banushi[2], Igli Tafaj[1], and Alban Allkoçi[1]

[1] Information Technology Faculty, Polytechnic University of Tirana, Tirana, Albania
{ekotobelli,itafaj,aallkoci}@fti.edu.al
[2] Netconomi, Graz, Austria
mariobanushi@hotmail.com

Abstract. Many of today's applications use Wireless Sensor Networks (WSNs) to collect data from a particular phenomenon. Studying WSNs includes many aspects such as routing, security, evaluation of energy etc. Since data transmission is the operation that causes the biggest consumption of the node residual energy, our work is focused on the problem of data management on WSN in order to reduce data traffic. For this we have chosen to study the VCP (Virtual Cord Protocol) because it is an efficient routing scheme that also provides data management methods such as identifying, storing, and retrieving data. Our goal has been to improve VCP in order to reduce data traffic. During the analysis of the VCP we noticed some problems and proposed the respective solutions. The simulation results of the new algorithm have shown that routing is optimized and data traffic is reduced, facilitating data lookup process.

1 Introduction

Various applications related to environmental monitoring have as their main goal collecting data that characterize the environment under study as much as possible. Therefore, managing these data remains an essential task of these monitoring systems, especially when network size continues to grow [1]. WSNs are used very well in such applications because they fit in with requirements such as low-cost, simple infrastructure, scalability and efficient use of routing techniques [2, 3]. Standard routing protocols rely on techniques that have not aimed at efficient use of resources as sensory nets require [4]. The combine of Distributed Hash Table (DHT) to maintain and retrieve data from network with routing based to peer-to-peer standard led to the creation of basic routing techniques in virtual coordinates [5]. Most of these algorithms are complicated and do not guarantee the delivery of packages in the shortest way. Moreover, only a few of them are practically applied to the sensor nodes. Routing by '*greedy forwarding*' scheme that use the relative addresses of the nodes connected to a virtual cord best meets the conditions to be an efficient method for managing data on large WSN networks. VCP is a virtual-based routing protocol designed specifically for WSNs, which also provide data management modes such as: identification, storage and data lookup [6]. It uses the *greedy forwarding* scheme for routing [7].

L. Barolli et al. (Eds.): EIDWT 2018, LNDECT 17, pp. 711–722, 2018.
https://doi.org/10.1007/978-3-319-75928-9_63

Since the tiny sensor nodes are powered by limited battery resources, energy efficiency is one of the primary challenges to the successful application of WSNs [8]. Usually energy is consumed during three processes which are sensing, processing and communication process. Energy consumption during communication process prevails over the other two processes. Therefore, our goal was to improve VCP routing in order to reduce data traffic, which means less data communication, less energy consumption. Some of the modifications we have made are considering the RSSI parameter for the cord construction and the use of casual timers for sending *hello* messages to reduce congestion.

In the following section we will give a brief description of the VCP and related work to it, then in Sect. 3 we have analyzed the VCP and have suggested some modifications. In Sect. 4 we have presented the results from the simulation of the improved algorithm and the conclusions of this work are given in Sect. 5.

2 VCP Description

VCP is specially designed for WSN by providing basic routing between sensor nodes. Each node maintains a small amount of routing information that is independent of the number of nodes in the network. To provide a similar interface to the hash table, each node is required to support a single operation: by giving a key on the entering, a node should be able to route message to the node responsible to that key. As such, the design of the VCP mainly addresses the issues related to the support of this data-orientation routing operation in a completely distributed way (does not require any form of centralized control, coordination or configuration) that is scalable (nodes maintain routing tables that are independent of the number of network nodes), which is simple and easy to build and robust to node failures.

2.1 The Virtual Cord

The idea behind VCP is to combine data lookup with routing techniques in an efficient way. The VCP achieves this by placing all nodes in a virtual cord that is also used to connect the data to them. A hash function is used to relate data with values in a predefined range [S, E], which is fully covered by the participating nodes. Thus, each node holds a portion of the whole range. The cord begins with a given minimum value (for example: 0.0) and ends with a maximum value (for example: 1.0). Each intermediate node has a unique value and communicates at least with the preceding node in value and the successor node in the cord. VCP organizes all nodes in a structured cord and each node maintains a small amount of routing information. If a node generates a data, it stores it in a predefined node. Consequently, all queries related to this data will be forwarding deterministically to this node. The predefined node is determined by a simple hash function.

2.2 Routing

The routing mechanism relies on two concepts: First, the virtual cord can be used to find a route to each destination network. Second, neighborhood information that is available locally is used for efficient *greedy* routing towards the destination.

For the routing in VCP each node must first recognize the physical neighbors. Then, a *greedy* algorithm is used to send packets to the neighbor node that has physical position closer to the destination until no progress is possible and the value lies between the positions of the ancestors and descendants. In static networks such as those for environmental monitoring, this method provides the best guarantee of finding the way between any two network nodes.

2.3 Related Work to VCP

In this section, similar works to VCP are described, which have in common the treatment of data management problem in WSN and the relation to DHT.

Chord is an example to show the problems and gaps of DHT implementation in WSN [9]. It is an efficient distributed lookup system based on hash functions and implemented as an overlay layer that means a hop in Chord involves some hops on the real network. Wherefore, Chord requires a routing protocol that provides underlay services.

Virtual Ring Routing (VRR) is a routing protocol inspired by DHT overlay layer [10]. In addition to routing, it provides the traditional functionality of DHT. VRR uses a unique key to identify the nodes. This key is an integer independent of the position. VRR organizes nodes such as Chord, into a virtual ring in the growing order of identifiers. To route the packet to the destination, the node with the identifier closest to the destination is chosen and the message is forwarded to that node.

Compared to the Chord, VRR reduces the number of routing tables from two to one. But the problem with these protocols is that the adjacent nodes in the ring can be very far in the real network. As a result, forwarding messages to the nearest node may result in a very long route. Moreover, scalability is a problem because by increasing the network, the protocol should maintain a largest routing table.

Geographic Hash Tables (GHT) connect keys with geographic locations using hash function, so data is stored in the sensor node geographically closer to the key hash [11]. The stored data are replicated locally to ensure continuity when nodes fail. As common DHTs, GHTs are built as an overlay and rely on underlay routing. For underlay routing is used the GPSR, which uses the physical location of the nodes for routing [12]. Thus, it is assumed that all nodes in the network know their position. The GRWLI (Geographic Routing without Location Information) protocol does not use real coordinates. Rather, it forms a virtual n-dimensional coordinate system [13]. However, the formation of virtual coordinates in many dimensions brings an additional energy expenditure on communication.

In the hop ID routing scheme in virtual coordinates, each node holds a hop ID, which is a multidimensional coordinate based on the distance of some landmark nodes [14]. Each node adjusts its hop ID periodically and transmits the new hop ID using a *hello* message. But joining and removing nodes can affect a large number of nodes in the network.

2.4 Comparisons with VCP

Compared to the abovementioned algorithms, VCP borrows from them, but has its own specifics. From Chord, VCP borrows a simple one-dimensional identifier, but however VCP has only one routing table compared to two routing tables in overlay implementations. From VRR, VCP borrows the combination of routing services and DHT, although the details of the addressing and forwarding are completely different. Also unlike VRR, the VCP routing table carries only local information. From GRWLI, VCP borrows the notion of using virtual coordinates, but (unlike GRWLI) it uses a very simple algorithm for coordinate creation, from which *greedy forwarding* guarantees packets distribution.

3 Some Improvements to VCP

VCP simulation in a simulator like Castalia, noted the VCP's weaknesses in terms of a real-level radio communication that this simulator offers. VCP is a routing protocol, and as such belongs to the layer of network in the protocol stack. By sticking to this fact, the protocol has been implemented in routing area of the Castalia's stack. VCP uses some types of packets for signaling in the cord. Loss of these packets is not critical to the cord performance, because the nodes send *hello* packets periodically to keep the status of the latest update. So the loss of signaling packets generally leads to the delay in building the cord, but not in its sabotage.

However, we have noticed a specific case when the loss of a packet leads to the destruction of the cord. It is a very specific moment of cord creation, when the cord has only one initial node and the second node is contacting it. The initial node sends to the second node a special ACK packet called *firstack*. After sending this packet, the initial node sets up a locking flag called *SentF* that does not allow it to send this *firstack* packet anymore. If this packet loses for any physical reasons, the creation of the cord is eventually sabotaged, because when *SentF* flag is set up, there is no turning back.

3.1 The Need for MAC Layer Control Mechanisms

The VCP does not foresee any technique for controlling the delivery of packets to the destination. If the situation described in the paragraph above is evidenced, the network is unable to get out of the wrong state. A possible solution to the problem would be to enrich the VCP with a confirmation/resend mechanism for critical packets, or just for the *firstack* packet. This would lead to the complexity of the protocol. The most practical solution (which we have used concretely) is the delegation of control mechanisms to other layers of the network. In general, Link's layer is equipped with such techniques. In Castalia this layer is known as "MAC Layer".

In Sect. 4 we presented a simulation that proves the wrong behavior of the VCP in terms of losing the *firstack*.

3.2 Improvement of the Cord

In the VCP, the node of the cord broadcast *hello* packets periodically around themselves. If a node not joined to the cord catch such a packet, it starts a certain procedure to become part of the cord. Initially, the new node (the node who wants to join the cord) checks if it is in direct contact with the first node of the cord. Then the new node checks if it is in contact with the last node. The check is done using the neighborhood tables that each node holds locally. After not verifying any of the above cases the new node checks if it is in contact with two successive nodes on cord. In the end it checks if the new node communicates only with one node of the cord. This last case is not very desirable as it leads to the creation of virtual nodes, which reduces the quality of the cord.

Since the initial node is the first record in the tables of all the nodes that received its message, there is a tendency of the cord to "gather" nodes around this node. This tendency can create non-optimal links between nodes. The situation is explained in Fig. 1. where is the case when the new node seeking to join the cord has communicated with two nodes that have the extreme values on cord (Fig. 1a).

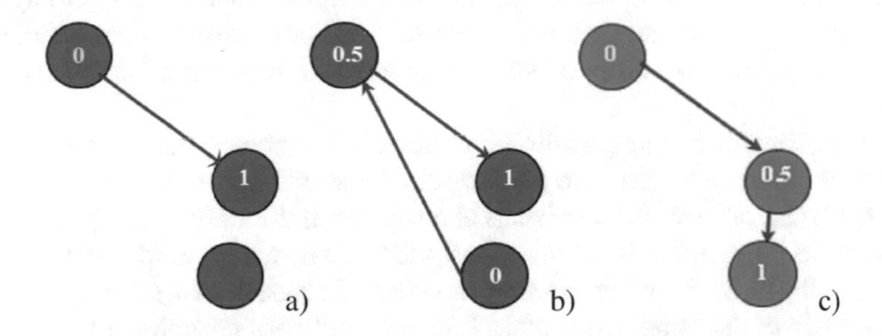

Fig. 1. (a) A node that wants to connect to the cord communicates with two extremity nodes. (b) The connected nodes according to VCP. (c) The improved virtual cord.

According to VCP, the incoming node[1] becomes adjacent to the initial node[2] (Figure 1b) although it is a physical neighboring with the second node that is closest to it. It is obvious that this situation is not optimal, since it would be more appropriate for the incoming node to become neighbors in cord with the second node, but the fact that the initial node is the first in the incoming node table gives priority to this connection. The consequences are reflected in the cost of data routing.

Our proposal for improving the situation is to add a new field to the records of the table that the nodes hold under the VCP. Concretely, we propose the addition of the Received Signal Strength Indication (RSSI) value. Each time the node table is refreshed with data (from *hello* packets); the node measures the signal strength of RSSI and stores

[1] Incoming node - means that the current node that seeks to join the virtual cord.
[2] Initial node is the extremity node in the cord that has the minimum value.

it in the created record. Directly after that, the node makes a descending order of records according to RSSI. Thus, priority is given to that node with which the incoming node has higher communication quality (Fig. 1c), against the chronological priority presented in the original VCP proposal. Simulations have shown that this change decreases the number of lost data packets during the routing.

3.3 Congestion Handling

Signaling packets in VCP are sent to certain periodic setup phase moments. The first signal in the network about the formation of the virtual cord is the *hello* packet, which "awakens" the nodes, provoking a series of response signals. If we assume a *hello* period of 1 s, the initial node of the network will send *hello* packets at moments 0, 1, 2, etc. The second network node will become a member of the cord at a very close time of 0 or 1, and will continue to send its *hello* packet, let say, at moments 0.01, 1.01, 2.01, etc. This rule is worth to all nodes of the cord. By the nature of VCP signaling, the whole network is characterized by very short intervals and periodic of rattling activity (burst packets). So around seconds 1, 2, 3 etc., the network faces with congestion conditions.

The first problem stemming from this situation is the loss of packets due to the condition of the radio module of the node. When the node radio is transmitting a packet, the status *TX* is declared. When the radio is in *TX* mode, node cannot get any packets from nearby nodes. Since all nodes are activated very close to each other in time to send signals, most packets are dropped because destination nodes themselves are sending packets.

Another problem that aggravates congestion is the appearance of the interference phenomenon. Since all nodes are sending almost simultaneously, the receiving node radios are bombarded by several signals at once. When the amount of signal strength at the node radio input exceeds a limit value, the node is unable to identify each signal. From the point of view of the routing module, this means that all packets reaching simultaneously to the same node have a high probability of dropping all.

Our proposal to resolve or at least to shrink congestion is the use of casual timers for breaking the strict periodicity of *hello* messages. The nodes do not send *hello* packets every 1 s, but time fluctuates for example at intervals 1 s ± 0.3 s, i.e. [0.7 s, 1.3 s]. The casual timer uses the pseudorandom function (a, b) that produces as output a random number inside [a, b] interval.

3.4 The Hash Function for Generating the Keys

The hash functions are varied but in our work is used a function selected by the linear family of hash functions called the hash multiplication function defined by Eq. 1:

$$h(k) = \lfloor n * ((f(k) * r)(mod 1)) \rfloor \quad 0 < r < 1 \tag{1}$$

Multiplication of a real number r between 0 and 1 with an integer key k gives another real "random" number. The modal hash partition limits the output to a range of values.

If that output is multiplied with the size of the hash table (n) is obtained a random index on it.

By this function implemented in the application layer, data are linked to a key whose value extends to the interval [0, 1]. For the packet routing, the node uses this key and the relative positions of the nodes that also are values in the interval [0, 1]. The key produced by the hash in application layer will serve for node as an indication of the destination, so it will route the packet to the neighbor that having the closest relative address to this destination key.

4 Simulations and Evaluations

The following simulations make the VCP assessment in the original form as well as shed light on the efficiency of the improvements made by us. For all simulations we have used a simple network with 9 nodes which are uniformly distributed in a quadratic area.

4.1 Case 1

Initially we did a network simulation by extinguishing control/retransmission mechanisms for lost packets. This is accomplished by setting in 0 the parameter value of the MAC module for the retransmission of the lost packets. The same simulation is repeated by reactivating the transmission of failed packet after setting the parameter to the value of 3 (up to three attempts).

The following tables show the status of the nodes at the end of the simulation for each case. They give the values of the relative positions that the nodes receive in the cord, and provide information on who are the predecessor and the successor to the node.

Apparently from the results of Table 1, the algorithm has failed, since the nodes do not form a linked list between them, and some nodes have remained untouched by the wave of *hello* packets (in these case their value are by default −1).

Table 1. Without MAC layer control mechanisms

Node	1	2	3	4	5	6	7	8	9
Value	0.5	0	−1	−1	0.55	0	−1	−1	1
Predecessor	0	−1	−1	−1	0.5	−1	−1	−1	0.5
Successor	1	0.55	−1	−1	0.5	0.55	−1	−1	−1

In the second simulation (Table 2), the nodes form a correctly aligned cord. Node 5 results to have a node with a value of 0.5 as a predecessor and successor simultaneously. This means that node 5 has as predecessor a virtual node, which is simulated by the node 4. The visual configuration of the cord in the second simulation is given in Fig. 2.

Table 2. With MAC layer control mechanisms

Node	1	2	3	4	5	6	7	8	9
Value	0.45	0.405	0.3645	0.5	0.525	0	0.55	0.595	1
Predecessor	0.405	0.3645	0	0.45	0.5	−1	0.5	0.55	0.595
Successor	0.5	0.45	0.405	0.55	0.5	0.3645	0.595	1	−1

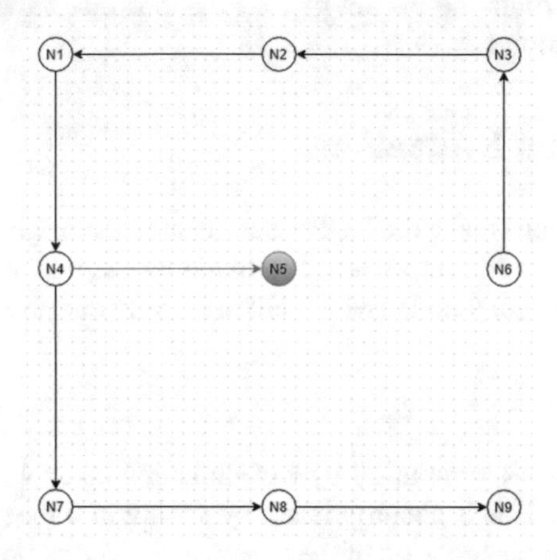

Fig. 2. Connecting nodes during simulation with MAC layer control mechanisms activated

4.2 Case 2

In the list of VCP parameters, there are two coefficient values that are used in the formula for determining the position of the node in the cord. These two values are intended to "soften" the values in the cord, in such a way that they are uniformly distributed in the interval [0, 1]. Uniform distribution provides equal sharing of load for information store to nodes. Values defined under VCP are: CORD_INTERVAL = 0.1 and CORD_VIRTUAL_INTERVAL = 0.9.

Considering the values of Table 2, we will notice that the intermediate nodes of the cord extend to the interval [0.36, 0.6]. In practice, this means that node 0, 0.36, 0.6 and node 1 will be responsible for most data, while other nodes will distribute a small load to even smaller parts among them. By changing the smoothing coefficients values of the interval, a fairer load distribution is guaranteed: CORD_INTERVAL = 0.2 and CORD_VIRTUAL_INTERVAL = 0.8. The values are as shown in Table 3.

Table 3. Uniform values of relative positions of the nodes in the cord

Node	1	2	3	4	5	6	7	8	9
Value	0.4	0.32	0.256	0.5	0.55	0	0.6	0.68	1
Predecessor	0.32	0.256	0	0.4	0.5	−1	0.5	0.6	0.68
Successor	0.5	0.4	0.32	0.6	0.5	0.256	0.68	1	−1

With the new values of the softening coefficients, the cord configuration remains the same as in Fig. 2, but the distribution of values is more uniform. The coverage on the intermediate nodes of the cord moves through the interval [0.256, 0.68].

4.3 Case 3

One of our proposals for VCP improvement was the ordering of records on routing tables according to RSSI of the signal received by the neighbor. We apply the proposed change over the network of Table 2. The simulation result gives us the configuration of the cord according to Table 4 and shown in Fig. 3.

Table 4. The relative positions of the nodes according to RSSI

Node	1	2	3	4	5	6	7	8	9
Value	0.45	0.405	0.3645	0.5	0	0.2775	0.55	0.595	1
Predecessor	0.405	0.3645	0.2775	0.45	−1	0	0.5	0.55	0.595
Successor	0.5	0.45	0.405	0.55	0.2775	0.3645	0.595	1	−1

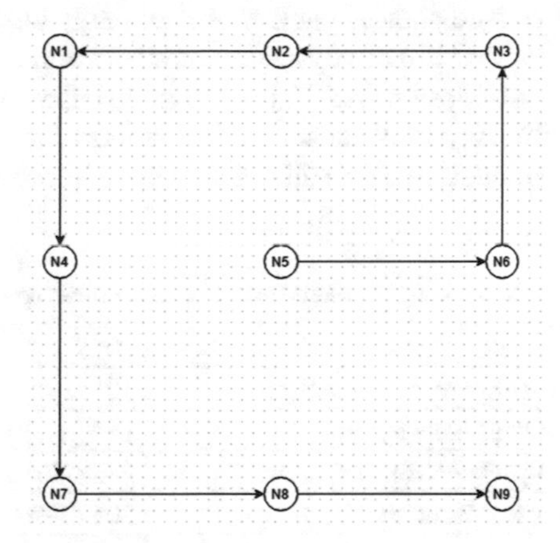

Fig. 3. The virtual cord created using the RSSI parameter

The change is noticeable, as node 5 is no longer connected via a virtual node, but is an "organic" part of the cord. This proves that the nodes are freer to be organized according to the physical proximity they have, without being under the "pressure" of the initial node of the network (N1). From the theoretical predictions made for the standard VCP version, network performance is proportional to the number of virtual nodes created. This can be proven by estimating the average number of rebroadcast for lost data packets. Before improvement, the average number of lost data packets rebroadcast was 2.21, while after improvement, it dropped to 1.68.

4.4 Case 4

We also have done another simulation where we inserted a random element (a casual timer) in the periodicity of sending *hello* messages. For this we have used the same network as in Table 2, maintaining the improvement introduced in Case 3. As we explained in Sect. 3 with the casual timer we understand the change of the *hello* message period from fixed value 1 s to a variable value in the interval [0.7 s, 1.3 s]. The average statistic for the packets obtained in the 9 nodes is as shown in Table 5.

Table 5. Using of casual timer for *hello* messages

Packets sent to the node	723
Successfully received packets	309 (42.7%)
"Missed" packets by distance	268 (37%)
Missed packets from *TX* status	61 (8.4%)
Packets lost from interference	85 (11.7%)
The average number of rebroadcasts	1.37

We think that "missed by distance" packets are not really missed. This is Castalia's way of identifying the lack of direct physical communication between the two nodes located outside of their transmission area. Failure to receive these packets does not cause re-transmission to the MAC layer. We note that the number of rebroadcasts has dropped from 1.68 to 1.37. The simulation data without the above improvement is presented in Table 6.

Table 6. Using a fixed timer for *hello* messages

Packets sent to the node	1053
Successfully received packets	201 (19%)
"Missed" packets by distance	259 (24.6%)
Missed packets from *TX* status	288 (27.4%)
Packets lost from interference	305 (29%)
The average number of rebroadcasts	1.68

It seems from the values that the effect of the loss by interference and by the wrong status of the radio module is significantly reduced, more than double. Reducing re-transmission reduces the total number of packets sent to the network. So the total traffic has reduced, the congestion has reduced, thus supporting our improvements.

5 Conclusions

Challenges affecting sensor network systems are diverse. This paper addressed the problem of routing that provides efficient data management. Virtual cord is one of the effective routing algorithms that support data management in terms of naming and storage, as well as lookup. Our implementation in a special simulator for WSN, such as Castalia, allowed us to study the behavior of the protocol on a real-world model of the wireless channel. During his analysis, we made some modifications in order to reduce data traffic.

The VCP has a tendency to create the cord around the initial node. This can lead to the creation of non-optimal routing. A solution is to consider the RSSI parameter as a ranking criterion in selecting the neighbors of the cord from each node. Also, VCP creates high traffic events in the network (congestion) periodically, whenever nodes send packets to detect neighboring nodes. Our proposal for introducing a casual timer on the periodicity of sending messaging decreases the probability of collisions that can cause interference, packet re-transmission and congestion.

We have analyzed the improved algorithm performance by estimating the average number of rebroadcasts for a given missing packet, which for the original version of the VCP module was 2.21, with the addition of RSSI as expected is decreased and even more with the use of casual timer for *hello* messages. From the packet analysis we see that the improved with casual timer has reached the target of reducing packet loss.

References

1. Oliveira, L.M.L., Rodrigues, J.J.P.C.: Wireless sensor networks: a survey on environmental monitoring. J. Commun. **6**(2), 143–151 (2011)
2. Yick, J., Mukherjee, B., Ghosal, D.: Wireless sensor network survey. Comput. Netw. **52**(12), 2292–2330 (2008)
3. Arampatzis, T., Lygeros, J., Manesis, S.: A survey of applications of wireless sensors and wireless sensor networks. In: The 13th Mediterranean Conference on Control and Automation Limassol, Cyprus, 27–29 June 2005
4. Bhattacharyya, D., Kim, T., Pal, S.: A comparative study of wireless sensor networks and their routing protocols. Sensors J., ISSN 1424-8220 (2010)
5. Tanh, V.V., Chan, H.N., Viet, B.P., Hu, T.N.: A survey of routing using DHT over wireless sensor networks. In: The 6th International Conference on Information Technology and Applications, ICITA (2009)
6. Awad, A., Sommer, C., German, R., Dressler, F.: Virtual Cord Protocol (VCP): a flexible DHT-like routing service for sensor networks. IEEE (2008). 978-1-4244-2575-4/08/$20.00c
7. Xing, G., Lu, C., Pless, R., Huang, Q.: On greedy geographic routing algorithms in sensing covered networks. In: MobiHoc (2004)
8. Wang, J., Ma, T., Cho, J., LeeAn, S.: Energy efficient and load balancing routing algorithm for wireless sensor networks. ComSIS **8**(4), Special issue, October 2011
9. Stoica, I., Morrisy, R., Liben-Nowelly, D., Kargery, D., Frans Kaashoeky, M., Dabeky, F., Balakrishnany, H.: Chord: A Scalable Peer-to-Peer Lookup Service for Internet Applications. University of California, Berkeley (2002)
10. Caesar, M., Castro, M., Nightingale, E.B., O'Shea, G., Rowstron, A.: Virtual ring routing: network routing inspired by DHTs. In: ACM SIGCOMM 2006, Pisa, Italy. ACM, September 2006
11. Ratnasamy, S., Karp, B., Shenker, S., Estrin, D., Govindan, R., Yin, L., Yu, F.: Data-centric storage in sensornets with GHT, a geographic hash table. ACM/Springer Mobile Netw. Appl. (MONET) Spec. Issue Wirel. Sens. Netw. **8**(4), 427–442 (2003)
12. Karp, B., Kung, H.T.: GPSR: greedy perimeter stateless routing for wireless networks. In: 6th ACM International Conference on Mobile Computing and Networking, ACM MobiCom 2000, Boston, MA, pp. 243–254 (2000)

13. Rao, A., Ratnasamy, S., Papadimitriou, C., Shenker, S., Stoica, I.: Geographic routing without location information. In: 9th ACM International Conference on Mobile Computing and Networking, ACM MobiCom 2003, San Diego, CA, September 2003
14. Zhao, Y., Chen, Y., Li, B., Zhang, Q.: Hop ID: a virtual coordinate-based routing for sparse mobile ad hoc networks. IEEE Trans. Mob. Comput. **6**(9), 1075–1089 (2007)

Indoor Self Localization Method for Connected Wheelchair Based on LED Optical Frequency Modulation

Kazuyuki Kojima(✉)

Saitama University, 255 Shimo-Okubo, Sakura-ku, Saitama 338-8570, Japan
kojima@mech.saitama-u.ac.jp

Abstract. This paper describes development of our self-localization method in indoor environment based on LED optical frequency modulation. Full color LEDs are used as markers for position estimation. The characteristic of this system is that red, green and blue led's optical patterns are frequency modulated independently and used them for including some kinds of information. By using the information of these optical patterns, the system can acquire all positions of the markers before the calibration. Then the time and labor for the calibration will be eliminated. In this paper, we conduct basic experiments which confirm the method to acquire the information which is provided from LED optical patterns in an actual environment.

1 Introduction

We have been developing a wheelchair which is used for improvement of nursing care and patients' quality of life (QoL) inside and outside of nursing and caring facilities (Fig. 1). In order to provide better support to both caregivers and patients, we require a system for positioning and tracking of patients in their daily life. Generally, a global positioning system known as GPS is used for determining a position, however, such technology basically has a problem which we can not use it in indoor environment. To solve this problem, various methodolgies have been proposed. Vicon [5] and OptiTrack [6] are two famous motion capturing system which can be used for determining patient location, however, they are quite costly at the implementation. LiDAR, SLAM or other vision based sensors [7,8] are also used for indoor localization, but they have a problem of calibration. Wireless signals such as Wi-Fi or Bluetooth are one of candidates to solve the cost problems, but for these techniques, WiFi or Bluetooth devices are needed.

In this paper, we use LED optical frequency modulation for determining a position of the wheelchair. Full color LEDs consist of three basic color LEDs (red, green and blue LEDs) are used as markers for position estimation. The characteristic of this system is that red, green and blue LED's optical patterns are frequency modulated independently and used them for including some kinds information. By using the information of these optical patterns, the system can acquire all positions of the markers before the calibration. Instead of using the color LED, infrared LED can be used for the same purpose. This is effective in

L. Barolli et al. (Eds.): EIDWT 2018, LNDECT 17, pp. 723–731, 2018.
https://doi.org/10.1007/978-3-319-75928-9_64

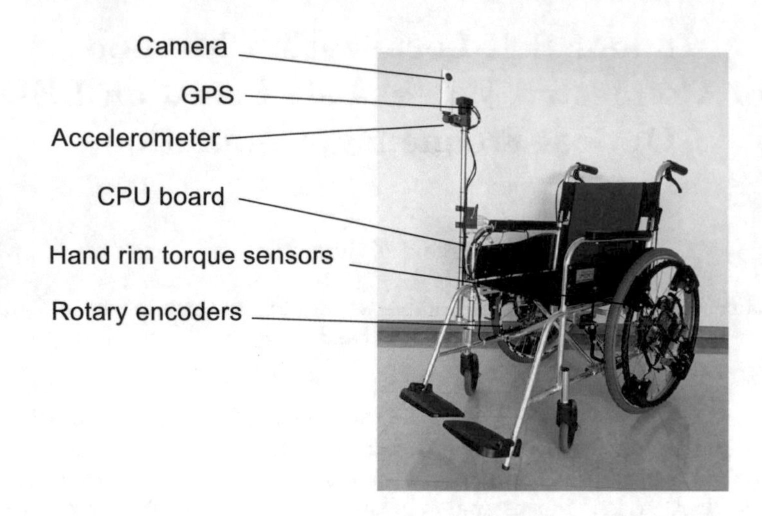

Fig. 1. Our developed manual wheelchair has some sensory devices including a camera, a GPS unit, an accelerometer, a CPU board, torque sensors, and rotary encoders. Using these devices, life log of patients are recorded for their QoL improvement [1–4].

the situation where visible light is objectionable or under the environment with noisy visible light.

We develop the positioning system using color LEDs and conduct experiments to confirm the basic functionality of our proposed system.

2 Experimental Apparatus

2.1 LED Luminescence Device

Figure 2 shows our developed LED marker which is composed of two color LEDs and a micro controller. The micro controller controls frequency of flashes of

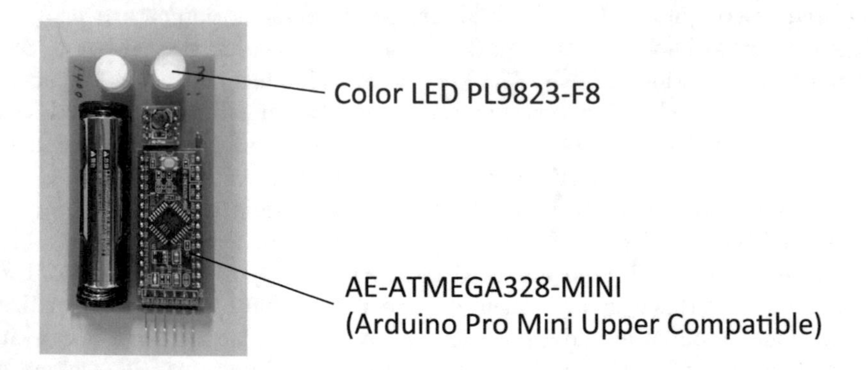

Fig. 2. LED Luminescence device is controlled by using CPU board. There are two LEDs on the board which can be independently controlled. In this paper, two LEDs are controlled to emit same color simultaneously.

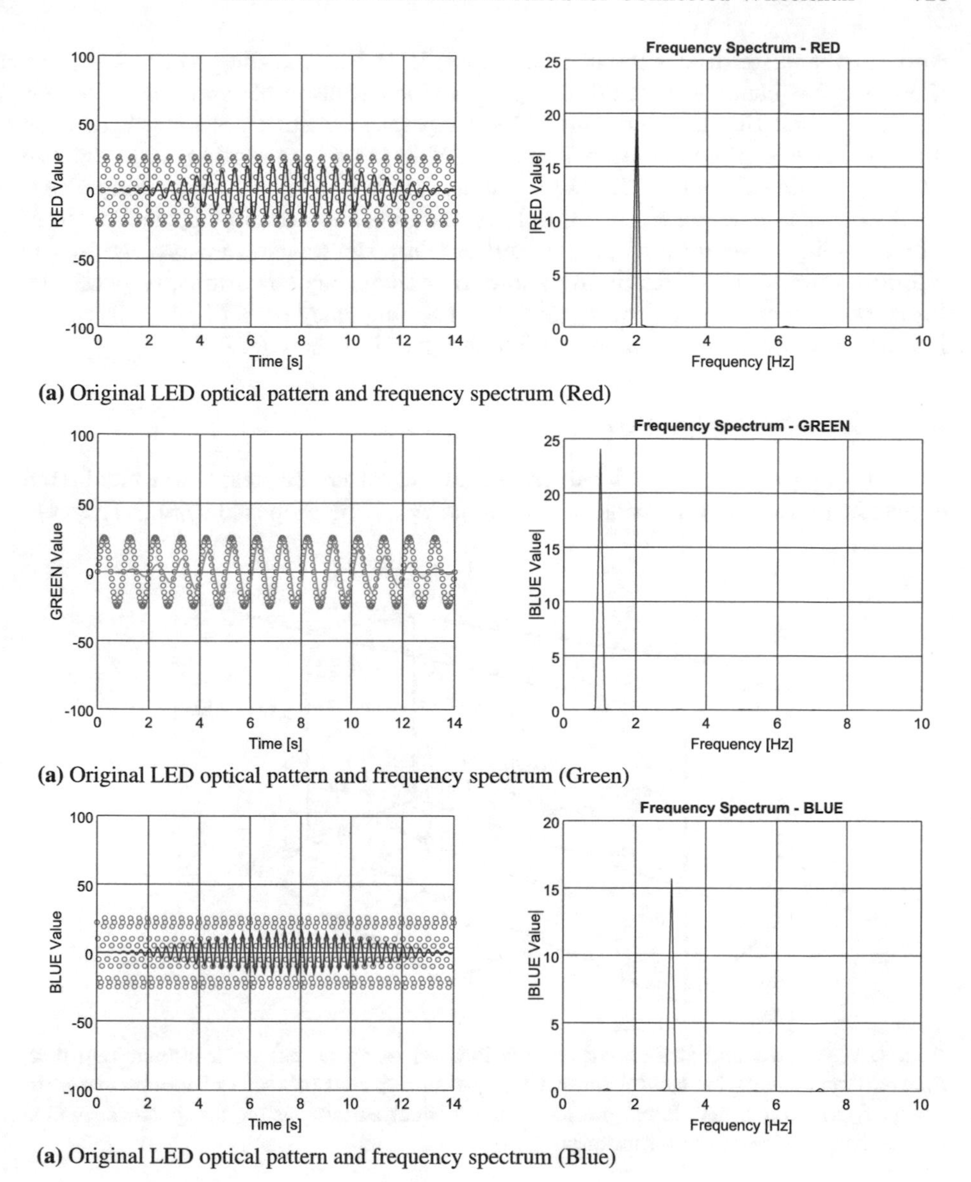

Fig. 3. Red, green and blue LED inside the color LED are controlled independently according to the position of the LED device. Through the process of FFT with the hanning window, each frequency can be calculated.

the LEDs according to the position of the LED luminescence device. This LED includes three color LED, red, green and blue LEDs. The values of these three color LEDs are controlled independently, and in our system, their frequency has some kinds of information. The amount of information is dependent on a specifications of a camera used as a photodetector. In this case, since we use a 30 fps camera, the cutoff frequency of this system is 15 Hz according to the sampling theorem. Frequency resolution is dependent on duration of sampling time. For instance, when the duration time is 1 s, frequency resolution is 1 Hz. When the duration is 10 s, the resolution is 0.1 Hz. This means, resolution and response time is a trade-off relation. In Fig. 3, the left three figures show an example case of LED optical pattern. In each graph, blue circles show original optical pattern. The solid line colored by red, green and blue are the waveform processed by the hanning window.The right figures show their frequency spectrums, respectively. From these figures, we can recognize that the frequency of red LED is 2 Hz, the frequency of green is 1 Hz, and the frequency of blue is 3 Hz.

2.2 Experimental Setup

We set three markers and a camera in an actual environment and conducted experiments for the validation of functionality of our proposed system (Fig. 4).

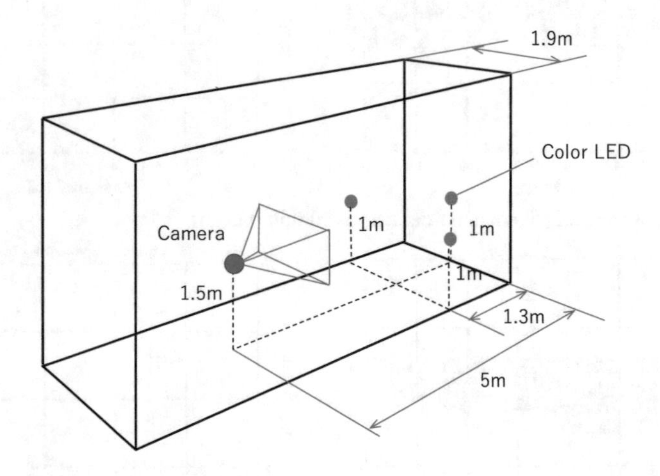

Fig. 4. A camera and three markers are located in an actual environment and conducted experiments for validation of this system. Optical patterns of markers are different from each other. Each marker can be identified by using the patterns which means frequencies of these markers.

3 Signal Processing

3.1 Image Processing

In order to extract the RGB values in each frame, captured images are processed by following steps:

Step 1 Difference image extraction
Step 2 Grayscale conversion

Step 3 Binarization
Step 4 Denoising
Step 5 Clustering
Step 6 Masking
Step 7 Tracking
Step 8 Averaging

Clusters in each image are tracked referring to positions and feature points in the image. Through the steps above mentioned, RGB values in each cluster

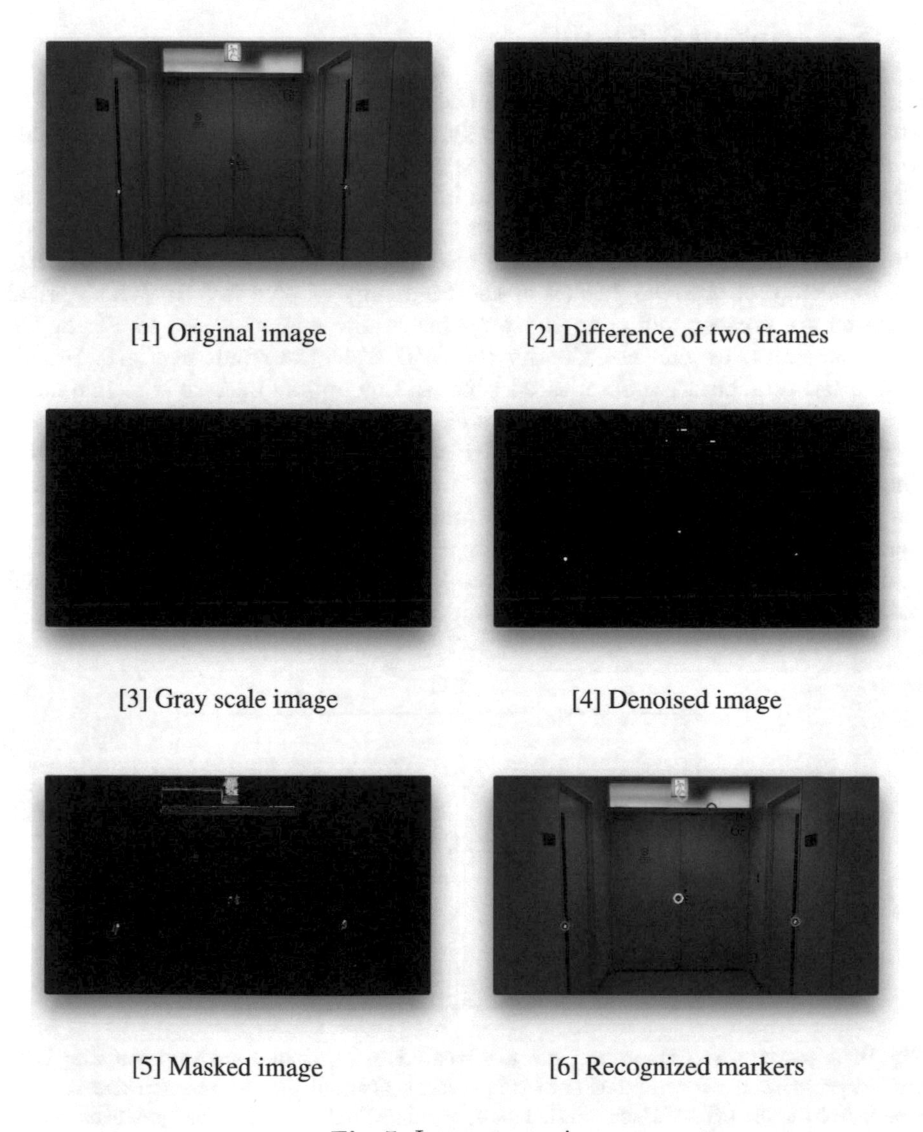

Fig. 5. Image processing

in each frame could be acquired (Fig. 5). However, markers are not identified at this time. After this image processing, signal processing is needed.

3.2 FFT

After image processing, some time series waveforms are acquired. Red, green and blue values are extracted from clusters which are generated by using the image processing. After that, these values are processed by FFT with hanning window as time series signals. Through this process, frequency of each cluster is calculated.

4 Experimental Result

Figure 7 shows an example case when the signal is identified as a marker. The left graphs show RGB values, respectively, just like as Fig. 3. In each graph, blue circles show calculated optical pattern based on the procedure above mentioned. The solid line colored by red, green and blue are the waveform processed by the hanning window. The right figures show their frequency spectrums, respectively. From these figures, we can recognize that the frequency of red LED is 2 Hz, the frequency of green is 1 Hz, and the frequency of blue is 3 Hz. This optical patterns are corresponding to the patterns of the example case in Fig. 3. On the other hand, we can not identify the LED from the result in Fig. 8. In this case, a few blue circles are appeared because this optical pattern is visible noise in the experimental environment. Therefore, characteristic frequencies are not appeared in the right graphs and then, this pattern is not recognized as a marker. Finally, we get three markers as shown in Fig. 6. Positions of the three markers are derived as (116, 258), (321, 210) and (497, 248) in the 640×360 image. Accordingly, considering of the 35 mm equivalent focal length of 24 mm, the position of the wheelchair is estimated as 4.56 m from the front wall. Since actual position is 5 m, this result is reasonable even though there are some error.

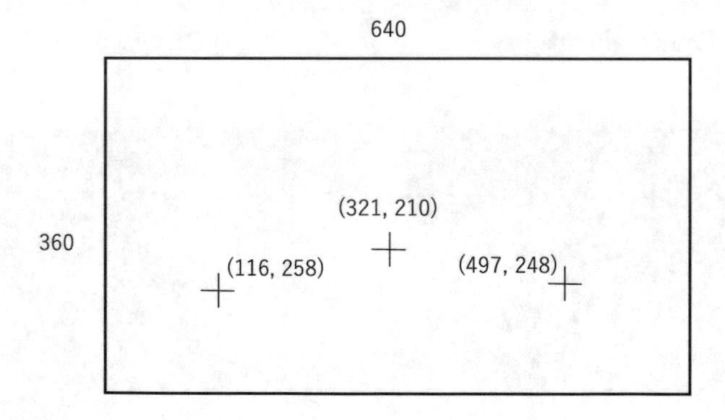

Fig. 6. A camera and three markers are located in an actual environment and conducted experiments for validation of this system. Optical patterns of markers are different from each other. Each marker can be identified by using the patterns which means frequencies of these markers.

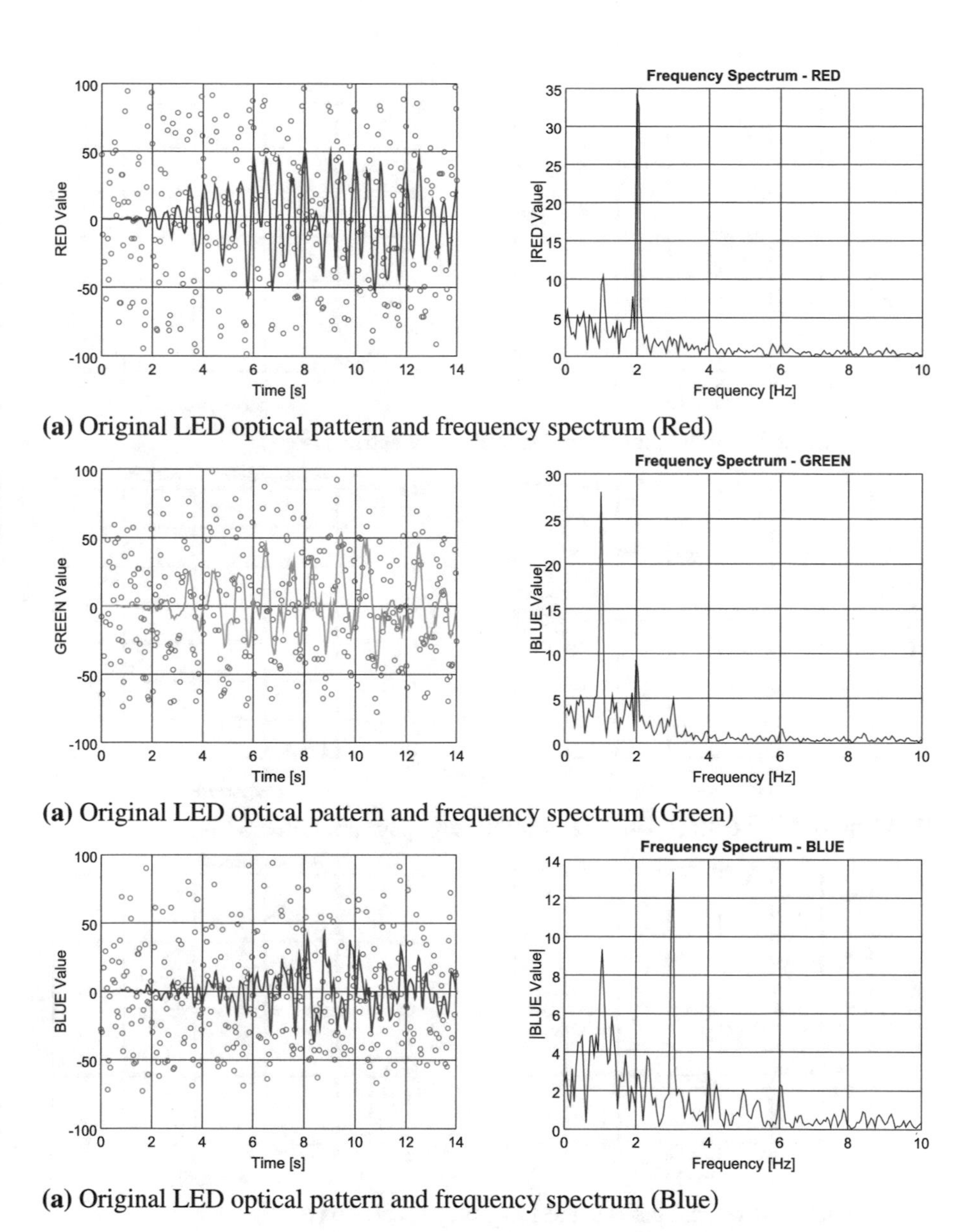

(a) Original LED optical pattern and frequency spectrum (Red)

(a) Original LED optical pattern and frequency spectrum (Green)

(a) Original LED optical pattern and frequency spectrum (Blue)

Fig. 7. Through the process of FFT with the hanning window, RGB frequencies can be calculated as 2 Hz, 1 Hz, and 3 Hz, respectively. Apparently, this result show that this optical pattern can be identified as the same LED in Fig. 3

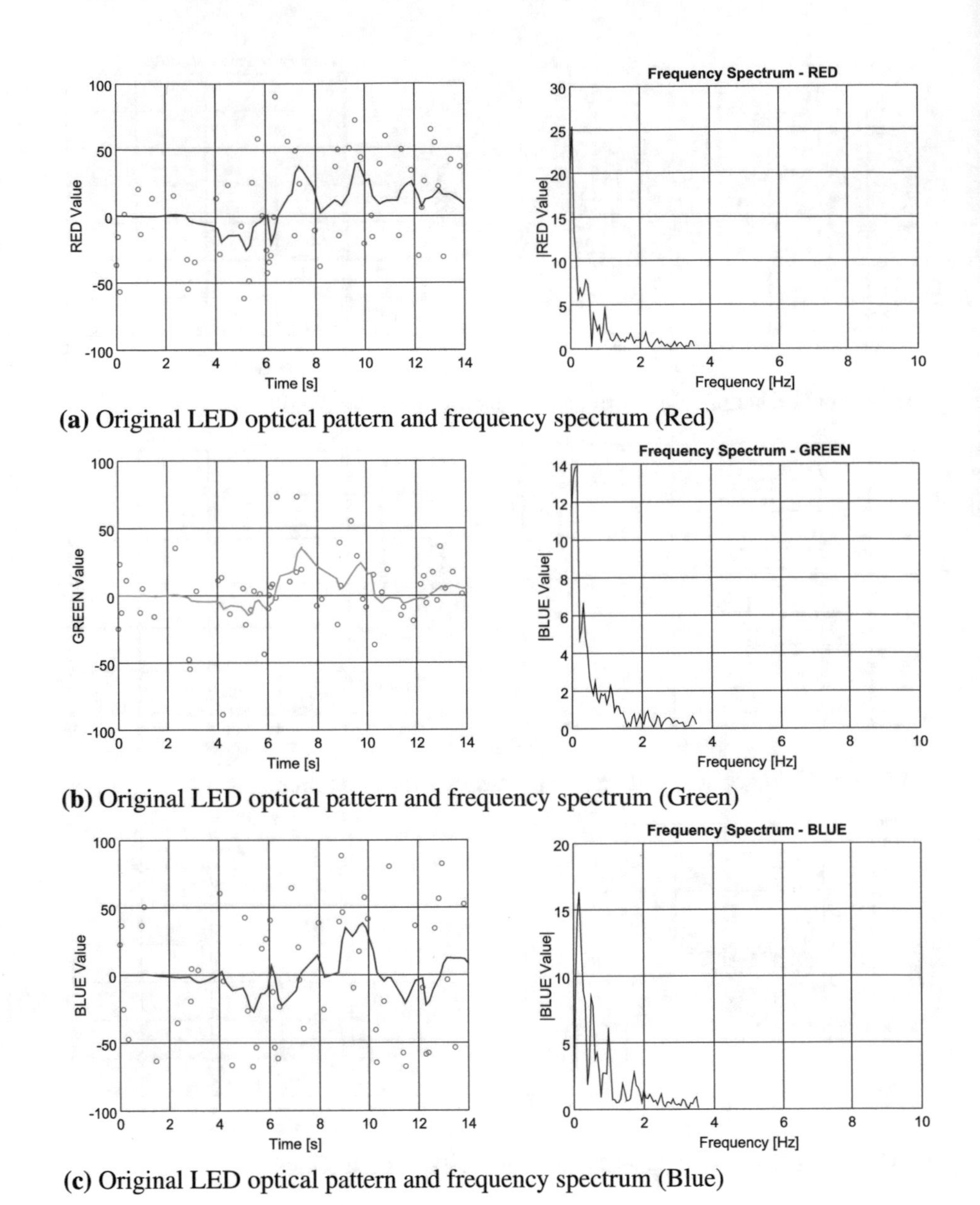

Fig. 8. Frequency spectrum is quite different from the one in Fig. 3. This means this optical pattern is not the LED in Fig. 3.

5 Conclusions

In this paper, a self-localization method in indoor environment based on LED optical frequency modulation was proposed. We used full color LEDs as markers and their red, green and blue optical patterns are frequency modulated. We acquired video images in our experimental environment and processed the image in order to define the positions of markers by using our proposed method. Finally, we could estimate the position of camera and found the result was reasonable. There are some errors in our experimental results. Improving the accuracy of this method is the next step of this development.

References

1. Kojima, K., Kaneko, J.: Fault tolerant calculation method of predicting road condition for network-connected wheelchair. In: Proceedings of 2017 IEEE International Conference on Consumer Electronics - Taiwan (ICCE-TW 2017), pp. 345–346 (2017)
2. Kojima, K., Taniue, H., Kaneko, J.: Mahalanobis distance-based road condition estimation method using network-connected manual wheelchair. In: Proceedings of 2016 IEEE International Conference on Consumer Electronics-Taiwan (ICCE-TW 2016), pp. 284–285 (2016)
3. Kojima, K., Taniue, H., Kaneko, J.: Development of road condition categorizing system for manual wheelchair using mahalanobis distance. In: ROMANSY21 - Robot Design, Dynamics and Control, Proceedings of the 21st CISM-IFToMM Symposium. Springer, pp. 377–384 (2016)
4. Sato, M., Kojima, K., Kaneko, J.: Development of pavement surface inspection system for wheel chair comfortability. In: Proceedings of 2014 IEEE 3rd Global Conference on Consumer Electronics (GCCE 2014), pp. 219–220 (2014)
5. Vicon Motion Capturing System Introduction. https://www.vicon.com/products Subordinate document. Cited 30 Nov 2017
6. OptiTrack product Introduction. https://www.optitrack.com/ Subordinate document. Cited 30 November 2017
7. Fontanelli, D., Ricciato, L., Soatto, S.: A fast RANSAC-based registration algorithm for accurate localization in unknown environments using LIDAR measurements. In: Proceedings of 2007 IEEE International Conference on Automation Science and Engineering, pp. 597–602. IEEE (2007)
8. Wan, K., Ma, L., Tan, X.: An improvement algorithm on RANSAC for image-based indoor localization. In: Proceedings of 2016 International Conference on Wireless Communications and Mobile Computing Conference (IWCMC), pp. 842–845. IEEE (2016)
9. He, S., Chan, S.-H.G.: Wi-Fi fingerprint-based indoor positioning: recent advances and comparisons. IEEE Commu. Surv. Tutorials **18**(1), 466490 (2016)
10. Matsuo, K.: Implementation and experimental evaluation of an omnidirectional wheelchair for sports and moving in rooms with narrow spaces. Int. J. Space-Based Situated Comput. **7**(1), 1–7 (2017)
11. Hayashi, M., Goshi, K., Sumida, Y., Matsunaga, K.: Development of a route finding system for manual wheelchair users based on actual measurement data. In: Proceedings of Ubiquitous Intelligence and Computing and 9th International Conference on Autonomic and Trusted Computing (UIC/ATC), pp. 17–23 (2012)

An Auction Framework for DaaS in Cloud Computing

Anjan Bandyopadhyay[1(✉)], Fatos Xhafa[2], and Sajal Mukhopadhyay[1]

[1] Department of Computer Science and Engineering, National Institute of Technology, NIT Durgapur, Durgapur, India
anjanmit@gmail.com, sajmure@gmail.com

[2] Department of Computer Science, Universitat Politècnica de Catalunya, Barcelona, Spain
fatos@cs.upc.edu

Abstract. Data as a Service (DaaS) is the next emerging technology in cloud computing research. Small clouds operating as a group may exploit the DaaS efficiently to perform substantial amount of work. In this paper an auction framework is studied when the small clouds are strategic in nature. We present the system model and formal definition of the problem. Several auction DaaS-based mechanisms are proposed and their correctness and computational complexity analysed. To the best of our knowledge, this is the first and realistic attempt to study the DaaS in strategic setting.

1 Introduction

Take a specialist's substantial infrastructure instead of creating your own individual set up, has been the key for cloud computing success. In the last decade there has been a significant research to deal with the allocation of resources in cloud computing [1,26]. Several companies (such as Amazon EC2, Microsoft Azure etc.) have come up with technologies to support the key viewpoint of cloud computing. Being the industry standard, cloud computing brings new challenges as well. One of such challenges is to acquire data on the fly for some business specific queries, leading to the Data-as-a-Service model.

Two broad solutions could be provided:

- Access the data already web-crawled by the big giants (Google, Microsoft, etc.) by their enormous infrastructure but at a high price.
- Some Small to Medium size Enterprises (SMEs) may join hand in hand and collect the data for future use and there by serving themselves independently several times.

In this paper this later viewpoint is addressed and an auction framework is proposed. Currently there are many existing small clouds (representing SMEs) and some of the clouds (henceforth we will use micro-clouds) may collaborate and form a bigger clouds. This collaboration will help collecting a substantial

L. Barolli et al. (Eds.): EIDWT 2018, LNDECT 17, pp. 732–741, 2018.
https://doi.org/10.1007/978-3-319-75928-9_65

amount of data. Whenever a query is made to any micro-cloud, then two cases may occur:

- Either the data is available with the bigger cloud where it belongs.
- The data may be available within some of the other bigger clouds.

If the data is available within the bigger cloud, there is an infrastructural cost it has to pay to access the data. If it is outside the bigger cloud, we run an auction to set the price.

The reminder of this paper is organized as follows: Sect. 2 presents the literature review of the previous work and then find out the research gap that motivates the need for the evaluation of cloud data-as-a-service business models. Section 3 proposes the system model and problem formulation. Section 4 we have discussed the proposed mechanism, then in Sect. 5 we have discussed the analysis of the proposed algorithm and last we present a summary of our work and highlight some future directions in Sect. 6.

2 Related Work

Data-as-a-Service(DaaS) is coming up as an alternative school of thoughts in cloud computing where data are obtainable as a service through network [1–3]. In [1] a DaaS architecture is presented for data discovery, storing and moving data, and processing of the data with the consideration of big data as a service. A pricing scheme for a query processing in DaaS platform is addressed in [3]. No auction based work is proposed, to the best of our knowledge, in DaaS environment. In this paper an auction framework is discussed when DaaS is in operation. However in literature several incentive schemes (mostly in monetary aspects) have been proposed in IaaS framework for stimulating the service providers to provide the best possible services to the users. Nielsen [4] proposes pricing technique for allocating computing resources. Sutherland [5] proposes a game theoretic auction model approach (based on auction theory) for allocating the processor time in a single computer. An auction mechanism proposed by Amazon is called spot marketing which is prevailing in the current cloud computing market. In [6] a strategy-proof mechanism for the allocation of multiple resources to a buyer in a large scale distributed system is proposed. In auction theory the setup with single buyer and multiple sellers falls into the class of reverse auction. In Combinatorial auctions (CA) which has been widely studied by the researchers [7–11] allow service providers to sell bundle of items rather than individual item and the users (buyers) to bid on any combination of items or services. Das and Grosu [19] proposed a combinatorial auction-based protocol for resource allocation in grids. They considered a model where different grid providers can provide different types of computing resources. The third party auctioneer collects this information about the resources and runs a combinatorial auction-based allocation mechanism where users participate by requesting bundles of resources. However, when multiple buyers and multiple sellers are present in the market a double auction mechanism is visible [13–15]. The double auction mechanism is extended into the online double auction environment in [15,16,26]. In [25] the cloud market ready for computing jobs with completion deadlines and designing an efficient online auction mechanism where cloud user gives their bids for future cloud resources to execute its job is studied. In their work each bid includes:

- A utility, the amount of payment the user is willing to pay for executing its job, and
- A deadline for finishing a preferred job. If the job is not finish in the stipulated time then a penalty should be given for violating the deadline.

In this paper an auction framework is discussed when DaaS is in operation. In the model so far in IaaS participating agents were individual in nature. However in our DaaS participating agents may be largely the individual groups. So, the earlier models may not be directly applied.

3 System Model and Problem Formulation

In this paper we have a set of clouds depicted as $C = \{C_1, \ldots, C_n\}$. Each of C_i consisting of some micro clouds. Depending on the number of micro clouds in each C_i two cases are possible for any pair of (C_i, C_j):

1. $\mid C_i \mid = \mid C_j \mid$
2. $\mid C_i \mid \neq \mid C_j \mid$

We can generalize the notation and can write $C_i = \{mC_1^i,\ mC_2^i, \ldots,\ mC_{k_i}^i\}$ where $k_i \in \{1, 2, \ldots, m\}$ and $i \in \{1, 2, \ldots, n\}$. This fact is depicted in Fig. 1.

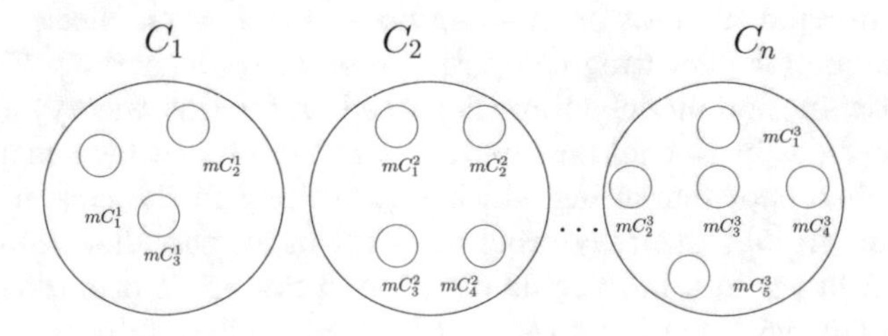

Fig. 1. Different micro-cloud forming the cluster

Each micro clouds $mC_j^i \in C_i$ joins C_i with some collection of data and that micro-cloud may be referred at any point of time. To promote the participation in this model it is assumed that when a mC_j^i joins a cloud C_i, it needs only its data and no registration fees. The data is maintained by the existing infrastructure of C_i. However when a query is made, in future, mC_j^i is charged a fixed amount if the data is available inside that C_i as the infrastructure cost. Otherwise, to get the data an auction is run with the other clouds C_js, where the data are available. So, the payment of mC_j^i can be formulated as $p_j^i = x_j^i\ \sigma_j^i + \bar{x}_j^i\ \bar{\sigma}_j^i$ where x_j^i is the indicator function defined as:

$$x_j^i = \begin{cases} 1, & \text{if data is inside } mC_j^i. \\ 0, & \text{otherwise.} \end{cases} \tag{1}$$

and

$$\bar{x}_j^i = \begin{cases} 1, & \text{if data is not inside } mC_j^i. \\ 0, & \text{otherwise.} \end{cases} \tag{2}$$

A micro cloud may earn some money also when an auction is run. If in the auction cloud C_i wins, then the money $mC_j^i \in C_i$ earns is: $\bar{p}_j^i = y_j^i(\frac{1}{2}a^i) + \bar{y}_j^i(\frac{1}{2}a^i w_j^i)$. Here, like the previous case, y_j^i and $\bar{y}_j^i$ are the indicator functions defined as:-

$$y_j^i = \begin{cases} 1, & \text{if } mC_j^i \text{ is the contributor in } C_i. \\ 0, & \text{otherwise.} \end{cases} \tag{3}$$

and

$$\bar{y}_j^i = \begin{cases} 1, & \text{if } mC_j^i \text{ is not the contributor in } C_i. \\ 0, & \text{otherwise.} \end{cases} \tag{4}$$

In this case a^i is taken as the money won by cloud C_i. If, mC_j^i is the contributor in C_i then it is taken that $\frac{1}{2}a^i$ will be given to mC_j^i and $\frac{1}{2}a^i$ will be divided based on the weight w_j^i corresponding to mC_j^i. It is not a bad idea to give the contributor, the half of the amount won, which will definitely boost the contributor and give him the luxury of providing quality services. It is assumed that each mC_j^i is associated with a weight. When a new mC_j^i joins the cloud C_i, then a small weight w_j^i is assigned to it and its score increases by 1 if its data is invoked by other cloud. There by the remaining amount is divided by a proportional share mechanism. So, the total payment made by a mC_j^i over all transactions can be defined as

$$\hat{p}_j^i = \sum_{i=1}^{\hat{k}_1} p_j^i - \sum_{i=1}^{\hat{k}_2} p_j^i \tag{5}$$

Where $\hat{k}_1$ is the number of transactions for which mC_j^i had to contribute money and $\hat{k}_2$ is the number of transactions where it earned money.

4 Proposed Mechanism

In this section we propose an auction based DaaS algorithm for the framework discussed in the earlier sections. The algorithm is termed as Auction Based DaaS (ABDaaS). The ABDaaS algorithm has five main components:

- `Main Routine`
- `Identify`
- `Run Auction`
- `Set Buyer Price`
- `Set Seller Price`

In the `Main Routine` (see Algorithm 1) the query is made for some cloud data center (micro-clouds).

Algorithm 1. Main Routine

1: **for** $t = 1, 2, \ldots$ **do**
2: Query is made for $mC_j^i \in C_i$
3: **if** data $\in C_i$ **then** payment of
/* Update the payment of mC_j^i */
4: $mC_j^i \cdot p \leftarrow mC_j^i \cdot p + \text{FC}$
/* FC is some fixed cost to maintain infrastructure */
5: **else**
/* Search for the other clouds for data */
6: $\underline{i} = i$
7: $S \leftarrow$ identify $(\text{C} \setminus C_i)$
8: S', v', mC_j, $i' \leftarrow$ Run_Auction(S)
/* $S' \rightarrow$ which cloud will provide the data
$v' \rightarrow$ Second price
$mC_j \rightarrow$ which micro-cloud inside S' contributed
$i' \rightarrow$ The index of S' */
9: Set_Buyer_Price $(mC_j^{\underline{i}}, v')$
10: Set_Seller_Price $(S', \bar{v}, mC, i')$
11: **end if**
12: **end for**
13: **end**

If the data is available within the bigger cloud where the micro-cloud belongs a fixed payment is made by the micro-cloud. In the else part auction is run. First `Identify` routine (see Algorithm 2) searches for the cloud agents who can provide the data and then an auction is run `Run Auction` (see Algorithm 5) based on the framework of Vickrey auction cast into reverse auction setting [27]. The `Set Buyer Price` and `Set Seller Price` set the payment of the buyer and the seller micro clouds.

Algorithm 2. Identify($\bar{S}$)

1: $S \leftarrow \phi$
2: **for** $i = 1$ *to* $|\bar{S}|$ **do**
3: if data available to $\bar{S}_i$
4: $S \leftarrow S \cup \bar{S}_i$
5: **end for**
6: return S

Algorithm 3. Run-Auction(S)

1: $v \leftarrow \phi$
/* Collection of Bids */
2: **for** $i = 1$ to $|S|$ **do**
3: $v_i \leftarrow bid(S_i) \in S$
4: $v \leftarrow v \cup v_i$
5: **end for**
/* Selecting Winner */
6: $i = \text{argmin}_{j:v_j \in v} v_j$
7: $i' = i$ ▷ *Remember the index of the cloud selected*
8: $S' \leftarrow S_i$
9: $S = S - S_i$
/* Selecting the second highest bidder for payment */
10: **for** $i = 1$ to $|S|$ **do**
11: $v_i \leftarrow bid(S_i \in S)$
12: $v \leftarrow v \cup v_i$
13: **end for**
/* Selecting the second highest bidder */
14: $i = \text{argmin}_{j:v_j \in v} v_j$
15: $v' = v_i$
16: $mC \leftarrow$ extract (S') ▷ *Which micro cloud served the data*
17: $return(S',\ v',\ mc,\ i')$

Algorithm 4. Set_Buyer_Price(mC_j^i, v')

1: Set_Buyer_Price $(mC_j^i,\ v')$
2: $mC_j^i \cdot p = mC_j^i \cdot p + v'$

Algorithm 5. Set_Seller_Price(S', $\bar{v}$, mC, i')

1: Set_Seller_Price(s', v', mC, i')
2: **for** $j = 1$ to $|S'|$ **do**
3: **if** $mC = mC_j$ **then**
4: $mC_j^{i'}.\bar{p} = mC_j^{i'}.\bar{p} + \frac{1}{2}v'$
/* $\bar{p}$ is the payment earned and p is the payment spend */
5: $mC_j^{i'} \in S' - mC$
6: $mC_j^{i'} \cdot \bar{p} + \frac{1}{2}v' \cdot \frac{1}{|S'|-1}$ (*equal share*)
7: or $mC_j^{i'} \cdot \bar{p} = mC_j^{i'} \cdot \bar{p} + \frac{v'}{2} \cdot \frac{mC_j^{i'} \cdot w_j^{i'}}{\sum_{mC_j^{i'} \in S' - mC} mC_j^{i'} \cdot w_j^{i'}}$ (*proportional share*)
8:
9: **end if**
10: **end for**

5 Analysis of the Proposed Method

5.1 Time Complexity

In each round time complexity can be measured as follows: The algorithm `Identify` will take $O(n)$ time. The `Run-Auction` will take $O(nlogn)$ time as the simplest implementation of $argmin()$ may be a sorting. For pricing schemes two subroutines are used: (1) `Set_Buyers_Price` and (2) `Set_Seller_price`. `Set_Buyers_Price` will take $O(1)$ time, whereas `Set_Seller_price` will take $O(m)$ time where $m < n$. So, the average time complexity is $O(nlogn)$.

5.2 Correctness of the Proposed Algorithm

For correctness of the algorithm we have to give emphasis on the main aspect of the proposed algorithm *i.e* pricing. It is to be shown that, when a cloud agent i is securing information, its payment being updated properly. Think all micro-clouds in a two dimensional array; where each row corresponds to a micro-cloud mC_i and the column corresponds to the two attributes $mC_j.p$ and $mC_j.p$ along with others.

When a query is made, $mC_i.p$ and $\bar{mC}_i.p$ either incremented by 0 or some quantity Δ (where, $\Delta = \frac{1}{2}v' \cdot \frac{1}{|S'|-1}$ or $\frac{v'}{2} \cdot \frac{mC_j^{i'} \cdot w_j^{i'}}{\sum_{mC_j^{i'} \in S' - mC} mC_j^{i'} \cdot w_j^{i'}}$) as can be observed from the `Main Routine` or from `Set_Buyers_Price` and `Set_Seller_price`.

- In the main routine:
  $mC_i.p = mC_i.p + FC$, /*an increment*/
  or $mC_i.p = mC_i.p + 0$, /*not the corresponding cloud.*/
- In the `Set_Buyers_Price`
  $mC_i.p = mC_i.p + v'$, /*an increment by the auction price*/
  or $mC_i.p = mC_i.p + 0$, /*if not the corresponding cloud*/
- In the `Set_Seller_price`
  $mC_i.\bar{p} = mC_i.\bar{p} + \frac{1}{2}v'$, /*an increment if the micro-cloud is the main contributor */
  or $mC_i.\bar{p} = mC_i.\bar{p} + \Delta$, /*if the micro-cloud is not the main contributor but belongs to C_i */
  or $mC_i.\bar{p} = mC_i.\bar{p} + 0$, /* if the micro-clouds does not belong to C_i. */

This argument shows that the payment of each micro-cloud is updated when the corresponding micro-cloud is involved in the transaction.

Lemma 1. *Fixed payment doesn't affect the auction.*

Proof. The payment of any micro-cloud mC_j^i at any round $t \in T$ is given by $p_j^i = x_j^i\ \sigma_j^i + \bar{x}_j^i\ \bar{\sigma}_j^i$. The if part of ABDaaS is responsible for $x_j^i.\sigma_j^i$ and else part contributes $\bar{x}_j^i.\bar{\sigma}_j^i$ and the indicator function depicts the fact that at any round either $x_j^i.\sigma_j^i$ will be resulted or $\bar{x}_j^i.\bar{\sigma}_j^i$ will be resulted. This confirms that fixed payment does not affect the auction. □

Theorem 1. *Auction proposed in ABDaaS is Truthful.*

Proof. By truthful it is meant that in the auction cloud agents cannot gain by manipulation. Here the utility for any cloud agent i is defined as:

$$u_i = \begin{cases} p_i - v_i, & \text{reverse auction settings.} \\ 0, & \text{otherwise.} \end{cases} \tag{6}$$

where p_i is the payment mode to the seller who wins and v_i is his original valuation.

Now, we observe that:

Case 1. If a seller wins and gives a valuation less, he still wins and utility $\hat{u}_i = u_i$. If he gives a valuation $\hat{v}_i > v_i$, then again two cases arise: he can win or loss. In the win case, his utility $\hat{u}_i = u_i$ and we can say no gain is there. In the losing case, $\hat{u}_i = 0 < u_i$ and hence no gain.

Case 2. If a seller loses by reporting her true valuation. With a similar argument of Case 1, it can be proved that no gain is achieved in this case also. □

6 Conclusion and Future Works

In this paper we have proposed an auction frame work for Data-as-a-Service (DaaS) in cloud computing. The DaaS model has emerged as an important model in cloud computing to provide data on demand to the users. This model is attractive to data consumers, because it enables the separation of data cost and of data usage from the cloud infrastructure cost. In our proposal, how a subset of smaller clouds collaborate and exchanges information (data) is oriented in an auction framework.

In our future work we will address the issue when a micro-cloud needs data from two or more cloud agents. This setting will lead us to the combinatorial auction. Additionally, we plan to evaluate the proposed auction framework under various multi-provider cloud settings.

Acknowledgements. We acknowledge all the departmental faculty members and research scholars for inspiring us. Supported by the PhD scholarship provided in the Visvesvaraya Scheme.

References

1. Terzo, O., Ruiu, P., Bucci, E., Xhafa, F.: Data as a Service (DaaS) for sharing and processing of large data collections in the cloud. In: 2013 Seventh International Conference on Complex, Intelligent, and Software Intensive Systems, pp. 475–480 (2013)
2. Magoules, F., Pan, J., Teng, F.: Cloud Computing, Data-Intensive Computing and Scheduling, Multidimensional Data Analysis in a Cloud Data Center, pp. 63–84. CRC Press (2012)

3. Oliveira, A.C., Fetzer, C., Martin, A., Quoc, D.L., Spohn, M.: Optimizing query prices for data-as-a-service. In: IEEE International Congress on Big Data, pp. 289–196 (2015)
4. Nielsen, N.R.: The allocation of computer resources— is pricing the answer? Commun. ACM **13**(8), 467–474 (1970)
5. Sutherland, I.E.: A futures market in computer time. Commun. ACM **11**(6), 449–451 (1968)
6. Meng, T.Y., Mihailescu, M.: A strategy-proof pricing scheme for multiple resource type allocations. In: International Conference on Parallel Processing, pp. 172–179, Vienna (2009)
7. Mashayekhy, L., Grosu, D.: Strategy-proof mechanisms for resource management in clouds. In: 14th IEEE/ACM International Symposium on Cluster, Cloud and Grid Computing, pp. 554–557 (2014)
8. Milgrom, P.R.: Putting auction theory to work. J. Polit. Econ. **108**(2), 245–272 (2000)
9. Nisan, N., Roughgarden, T., Tardos, E., Vazirani, V.V.: Algorithmic Game Theory. Cambridge University Press, Cambridge (2007)
10. Baranwal, G., Vidyarthi, D.P.: A fair multi-attribute combinatorial double auction model for resource allocation in cloud computing. J. Syst. Softw. **108**, 60–76 (2015)
11. Fujiwara, I., Aida, K., Ono, I.: Applying double sided combinatorial auctions to resource allocation in cloud computing. In: 10th Annual International Symposium on Applications and the Internet, pp. 7–14 (2010)
12. Ibrahim, S., He, B., Jin, H.: Towards pay-as-you-consume cloud computing. In: 2011 IEEE International Conference on Services Computing, pp. 370–377. Washington, DC (2011)
13. Lehmann, D., O'callaghan, L.I., Shoham, Y.: Truth revelation in approximately efficient combinatorial auctions. J. ACM **49**(5), 577–602 (2002)
14. Archer, A., Papadimitriou, C., Talwar, K., Tardos, E.: An approximate truthful mechanism for combinatorial auctions with single parameter agents. In: Proceedings of SODA, pp. 205–214, Philly, PA, USA (2003)
15. Mualem, A., Nisan, N.: Truthful approximation mechanisms for restricted combinatorial auctions. Games Econ. Behav. **64**(2), 612–631 (2008)
16. Bartal, Y., Gonen, R., Nisan, N.: Incentive compatible multi unit combinatorial auctions. In: Proceedings of TARK, New York, USA, pp. 72–87 (2003)
17. Bandyopadhyay, A., Mukhopadhyay, S., Ganguly, U.: Allocating resources in cloud computing when users have strict preferences. In: 2016 IEEE International Conference on Advances in Computing, Communications and Informatics (ICACCI), PP. 2324–2328, Jaipur, India (2016)
18. Bandyopadhyay, A., Mukhopadhyay, S., Ganguly, U.: On free of cost service distribution in cloud computing. In: 2017 IEEE International Conference on Advances in Computing, Communications and Informatics (ICACCI), Mangalore, India (2017)
19. Das, A., Grosu, D.: Combinatorial auction-based protocols for resource allocation in grids. In: Proceedings of 19th International Parallel and Distributed Processing Symposium, 6th Workshop on Parallel and Distributed Scientific and Engineering Computing (2005)
20. Roughgarden, T.: CS364A: Algorithmic game theory, Lecture 9: Beyond quasi-linearity, October 2013
21. Roth, A.E., Sotomayor, M.: Two sided matching. In: Aumann, R., Hart, S. (eds.) Handbook of Game Theory with Economic Applications, pp. 485–541. Elsevier, Haarlem (1992)

22. Gale, D., Shapley, L.S.: College admissions and the stability of marriage. Am. Math. Mon. **69**, 9–15 (1962)
23. Jiang, C., Chen, Y., Wang, Q., Liu, R.: Data-driven auction mechanism design in IaaS cloud computing. IEEE Trans. Serv. Comput. **PP**, 1 (2015)
24. Mazrekaj, A., Shabani, I., Sejdiu, B.: Pricing schemes in cloud computing: an overview. Int. J. Adv. Comput. Sci. Appl. **7**(2), 80–86 (2016)
25. Zhou, R., Li, Z., Wu, C., Huang, Z.: An efficient cloud market mechanism for computing jobs with soft deadlines. IEEE/ACM Trans. Netw. **25**(2), 793–805 (2017)
26. Zhang, H., Jiang, H., Li, B., Liu, F., Vasilakos, A.V., Liu, J.: A framework for truthful online auctions in cloud computing with heterogeneous user demands. IEEE Trans. Comput. **65**(3), 805–818 (2016)
27. Vickrey, W.: Counterspeculation, auctions and competitive sealed tenders. J. Financ. **16**(1), 8–37 (1961)

A Fingerprint Enhancement Algorithm in Spatial and Wavelet Domain

Indrit Enesi(✉), Algenti Lala, and Elma Zanaj

Polytechnic University of Tirana, Tirana, Albania
{ienesi,alala,ezanaj}@fti.edu.al

Abstract. Fingerprinting is one form of biometrics, which people's physical characteristics to identify them. Fingerprints are ideal for this purpose because they're inexpensive to be collected and analysed. They never change, even as people grow old. The performance of a fingerprint image-matching algorithm depends heavily on the quality of the input fingerprint images. The acquired fingerprint images from the scanner are often with low contrast, noisy and the ridges are blurred. The enhancement is an essential step required to improve the quality of the fingerprint image. In this paper, we propose an enhancement method in spatial and wavelet domain. The fingerprint image contrast is increased, the histogram is equalized and ridges are deblurred. The image is then filtered by Gabor filters and denoised in wavelet domain. Experimental results show that this method increases the number of true minutiae extracted.

1 Introduction

Fingerprint is the first biometric system adopted by law enforcement agencies, and now is also the most widely used system. Different types of fingerprint recognition devices can be found for network access and physical access entry configurations; it is the primary tool utilized for Automated Fingerprint Identification System (AFIS) databases; and it is also the biometric tool of choice for financial institutions. Most AFISs are based on minutiae matching. The major minutiae features used by AFISs, are endings and bifurcations, which represent terminations and intersections of fingerprint ridge line flows. Although the automatic fingerprint recognition and identification have wide and long practical application, there still exist a lot of challenging and established image processing and pattern recognition problems. Fingerprint image quality is of much importance to achieve high performance in AFIS. Enhancement of fingerprint images can be performed on either binary ridge images or direct grey images. Binarization before enhancement will generate more spurious minutiae structures and lose some valuable original fingerprint information; it also poses more difficulties for later enhancement procedure. Therefore, most enhancement algorithms are performed on grey images directly [1]. In order to ensure that the performance of an automatic fingerprint identification system will be robust with respect to the quality of input image, it is essential to incorporate a fingerprint enhancement algorithm [2] in the feature extraction module. The purpose is to reduce above distortion by using pre-processing algorithm

L. Barolli et al. (Eds.): EIDWT 2018, LNDECT 17, pp. 742–753, 2018.
https://doi.org/10.1007/978-3-319-75928-9_66

and to approach input raw image to ideal image, which have ideal properties. Ones receive that the ideal fingerprint may be described by distorted circular grating and the properties of ideal fingerprint image will be the same as properties of distorted circular grating. Purpose of pre-processing is to decrease False Acceptance Probability and increase Authentic Acceptance Probability. The improved Gabor filtering [3, 8] can protect the ridge structure and avoid the spurious ridges effectively. In addition, a Wavelet-Transformation based method [4] is proposed to reduce the noise in the image. A hybrid fingerprint enhancement algorithm is based on spatial domain by increasing the contrast and equalizing the histogram components [4], the Lucy-Richardson algorithm deblures the ridges. The Gabor filter can repair the damaged ridges by removing the noise from the ridges and keeping the ridge structure [5].

Wavelet transform is used to analyse the image at different scale, the multiresolution analysis provides the information about the signal at more than one resolution. Image de-noising by wavelet transform is used to get rid of the additive noise while keeping hold of as much as possible the important features. Wavelet thresholding procedure removes noise by thresholding only the wavelet coefficient of the details coefficients, by keeping the low-resolution coefficients unaltered [6, 7].

2 Noise in Fingerprint Images

A digitized fingerprint is often noisy. Irregularities in the digitization process and in the fingerprint itself are caused by several factors: Fingerprint imperfections such as ridge gaps, usually caused by skin folds, or injuries to the skin, such as cuts, burns and abrasions, affect the alignment ridges and may even result in destruction of some ridges. Prints could be noisy due to smudging, over inking, excessive pressure while taking fingerprints. There may be many false minutiae caused by variations in the amount of ink and pressure or by smearing during finger rolling. Minutiae details of the print itself, such as pore holes on the ridges, can cause false gaps or breaks in the ridge lines. Contiguous and smeared ridges may be introduced by improper recording. Distortion introduced by the digitization operation can affect the accuracy [9].

3 Image Enhancement

An important step in fingerprint recognition is to automatically and reliably extract minutiae from the input fingerprint images. The performance of a minutiae extraction algorithm relies on the quality of the input fingerprint images. In order to ensure that the performance of an AFI system will be robust with respect to the quality input fingerprint images, it is essential to implement a fingerprint enhancement algorithm in the minutiae extraction module [10]. We present an enhancement algorithm, which can adaptively improve the clarity of ridge and furrow structures of input fingerprint images on spatial and wavelet domain.

3.1 Image Adjustment

Contrast adjustment remaps image intensity values to the full display range of the data type. An image with good contrast has sharp differences between black and white [11]. The image with poor contrast has intensity values limited to the middle portion of the range. In the high contrast image, highlights look brighter and shadows look darker. To change the contrast or brightness of an image, the Adjust Contrast tool performs *contrast stretching*. In this process, pixel values below a specified value are mapped to black and pixel values above a specified value are mapped to white. The result is a linear mapping of a subset of pixel values to the entire range of display intensities (dynamic range). This produces an image of higher contrast by darkening pixels whose value is below a specified value and lightening pixels whose value is above a specified value. The Adjust Contrast tools adjust brightness by moving this window over the display range, without changing its size. In this way, the pixel values map to lighter or darker intensities.

3.2 Adaptive Histogram Equalization

Ordinary histogram equalization uses the same transformation derived from the image histogram to transform all pixels. This works well when the distribution of pixel values is similar throughout the image. However, when the image contains regions that are significantly lighter or darker than most of the image, the contrast in those regions will not be sufficiently enhanced.

Adaptive histogram equalization (AHE) improves on this by transforming each pixel with a transformation function derived from a neighbourhood region. In its simplest form, each pixel is transformed based on the histogram of a square surrounding the pixel. The derivation of the transformation functions from the histograms is exactly the same as for ordinary histogram equalization: The transformation function is proportional to the cumulative distribution function (CDF) of pixel values in the neighbourhood [12].

3.3 Deblurring the Image Using Lucy-Richardson Algorithm

An image recorded from an image acquisition device, is generally blurred by a point spread function. Non-point sources are effectively the sum of many individual point sources, and pixels in an observed image can be represented in terms of the point spread function and the latent image as:

$$d_i = \sum_j p_{ij} u_j \tag{1}$$

where p_{ij} is the point spread function (the fraction of light coming from true location j that is observed at position i, u_j is the pixel value at location j in the latent image, and di is the observed value at pixel location i. The statistics are performed under the assumption that u_j are Poisson distributed, which is appropriate for photon noise in the data.

The basic idea is to calculate the most likely uj given the observed di and known p_{ij}. This leads to an equation for uj which can be solved iteratively according to:

$$u_j^{(t+1)} = u_j^{(t)} \sum_i \frac{d_i}{c_i} p_{ij} \tag{2}$$

Where

$$c_i = \sum_j p_{ij} u_j^{(t)} \tag{3}$$

It has been shown empirically that if this iteration converges, it converges to the maximum likelihood solution for u_j [13].

3.3.1 Deconvolution with Lucy-Richardson Algorithm

The deconvlucy function is used to deblur the image using the accelerated, damped, Lucy-Richardson algorithm. The algorithm maximizes the likelihood that the resulting image, when convolved with the PSF, is an instance of the blurred image, in our case the Gaussian noise statistics is considered. Noise amplification is a common problem of maximum likelihood methods that attempt to fit data as closely as possible. After many iterations, the restored image can have a speckled appearance, especially for a smooth object observed at low signal-to-noise ratios. These speckles do not represent any real structure in the image, but are artifacts of fitting the noise in the image too closely. To control noise amplification, the deconvlucy function uses a damping parameter, DAMPAR. This parameter specifies the threshold level for the deviation of the resulting image from the original image, below which damping occurs. For pixels that deviate in the vicinity of their original values, iterations are suppressed.

H = fspecial('gaussian', HSIZE, SIGMA) returns a rotationally symmetric Gaussian lowpass filter of size HSIZE with standard deviation SIGMA (positive). HSIZE can be a vector specifying the number of rows and columns in H or a scalar, in which case H is a square matrix. In our case the value of HSIZE is 5, the value of SIGMA is 5. 10 iterations are applied.

3.3.2 Lucy-Richardson Algorithm to Handle the Camera Read-Out Noise

Noise in charge coupled device (CCD) detectors has two primary components:

- Photon counting noise with a Poisson distribution
- Read-out noise with a Gaussian distribution

The Lucy-Richardson iterations intrinsically account for the first type of noise. You must account for the second type of noise; otherwise, it can cause pixels with low levels of incident photons to have negative values. The deconvlucy function uses the READOUT input parameter to handle camera read-out noise. The value of this parameter is typically the sum of the read-out noise variance and the background noise (e.g., number of counts from the background radiation). The value of the READOUT parameter specifies an offset that ensures that all values are positive.

3.4 Gabor Filters

Gabor filtering is one of the most common fingerprint enhancement algorithm. The principle of the algorithm is based on the mathematical model of fingerprint, in the local area fingerprint can be thought of as a set of parallel with certain frequency linear, along the ridge orientation using the Gabor window function to filter the image, so that the ridge information was enhanced. Because it is along the ridge orientation to filter, which along the ridge line has a smooth function, so that some fault ridge can be repaired to its original state; at the same time the Gabor filtering has good frequency selectivity, it can effectively remove the noise on the ridge and keep the ridge structure [14].

The mathematical expression of Gabor function is shown as Eq. 4:

$$G(x,y) = \frac{1}{2\pi\sigma_x\sigma_y}\exp\left[-\frac{1}{2}\left(\frac{x^2}{\sigma_x^2}+\frac{y^2}{\sigma_x^2}\right)\right]\exp(-2\pi fix) \tag{4}$$

By Eq. 4 can be seen, in the x direction, Gabor filter is a band-pass filtering, y direction is a low-pass filtering. When filtering, filter only need rotating, in according with the fingerprint orientation, which can make the fingerprint information greatly enhancing, vertical to the ridge orientation information is weakened relatively.

According to Euler's formula, get the real part R(x, y) and imaginary part V (x, y) of G(x, y)

$$R(x,y) = \frac{1}{2\pi\sigma_x\sigma_y}\exp\left[-\frac{1}{2}\left(\frac{x^2}{\sigma_x^2}+\frac{y^2}{\sigma_x^2}\right)\right]\cos(2\pi fx) \tag{5}$$

$$V(x,y) = \frac{1}{2\pi\sigma_x\sigma_y}\exp\left[-\frac{1}{2}\left(\frac{x^2}{\sigma_x^2}+\frac{y^2}{\sigma_x^2}\right)\right]\sin(2\pi fx) \tag{6}$$

It is obviously to see that R(x, y) is an even function, in the region of the current point (x, y) as the centre of the symmetry, the information of (x, y) will be enhanced,

Therefore, the general form of even symmetric Gabor filtering is as follows:

$$R(x,y,f,\varphi) = \frac{1}{2\pi\sigma_x\sigma_y}\exp\left[-\frac{1}{2}\left(\frac{x_\varphi^2}{\sigma_x^2}+\frac{y_\varphi^2}{\sigma_x^2}\right)\right]\cos\left(2\pi fx_\varphi\right) \tag{7}$$

In Eq. (7) $\begin{bmatrix} x_\varphi \\ y_\varphi \end{bmatrix} = \begin{bmatrix} cos\varphi & sin\varphi \\ -sin\varphi & cos\varphi \end{bmatrix}\begin{bmatrix} x \\ y \end{bmatrix}$ where φ is the orientation of the Gabor filter, f is the frequency of a sinusoidal plane wave, [x ϕ, y ϕ] stands for axis [x, y] along the counter clockwise rotation ϕ degrees.

3.5 Wavelet Transform

The wavelet transform provides the time-frequency representation of signal. Wavelet transform is based on small waves known as wavelets of varying frequency and limited duration. Temporal information is not lost in transformation process as it is lost in Fourier transformation process. Wavelet bases are constructed from translation and scale of a mother wavelet as described by the Eq. 1 [15]:

$$\psi_{ab}(t) = \frac{1}{\sqrt{a}}\psi\left(\frac{t-a}{a}\right) \tag{8}$$

where a and b are the scaling and translation parameters, respectively. Wavelet transform of a function f(t) ∈ L2 is defined as (Eq. 2),

$$Wf(a,b) = \langle f, \psi_{(a,b)} \rangle = \int_{-\infty}^{+\infty} f(x)\frac{1}{\sqrt{a}}\psi\left(\frac{t-b}{a}\right)dt \tag{9}$$

The reconstruction relation can be expressed as (Eq. 10):

$$f = \frac{1}{C_\psi}\int_0^{\infty}\int_{-\infty}^{+\infty} Wf(a,b)\psi_{(a,b)}(t)db\frac{da}{a^2} \tag{10}$$

where, Cψ known as Admissibility constant is defined by Eq. 11,

$$C_\psi = \int_0^{\infty}\frac{|\psi(\omega)|^2}{\omega}d\omega < +\infty \tag{11}$$

Where ψ(ω) is the Fourier transform of ψ(t).

Wavelet transform is best suited for localized frequency analysis, because the wavelet basis function have short-time resolution for high frequencies and long-time resolution for low frequencies. It can be used to analyse the signal or image at different scale i.e. the multiresolution analysis which provides the information about the signal at more than one resolution [17] (Fig. 1).

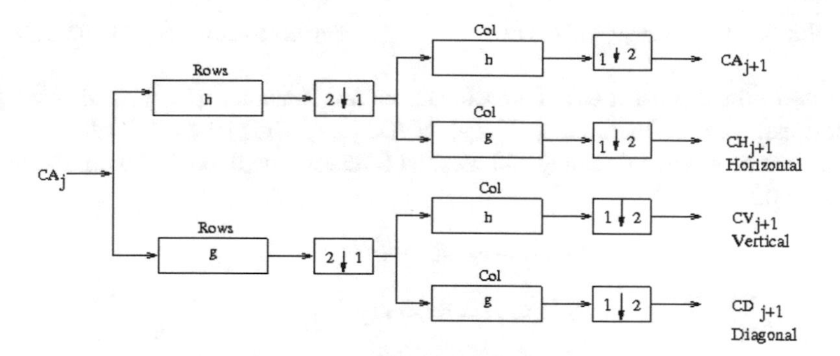

Fig. 1. 2D wavelet decomposition

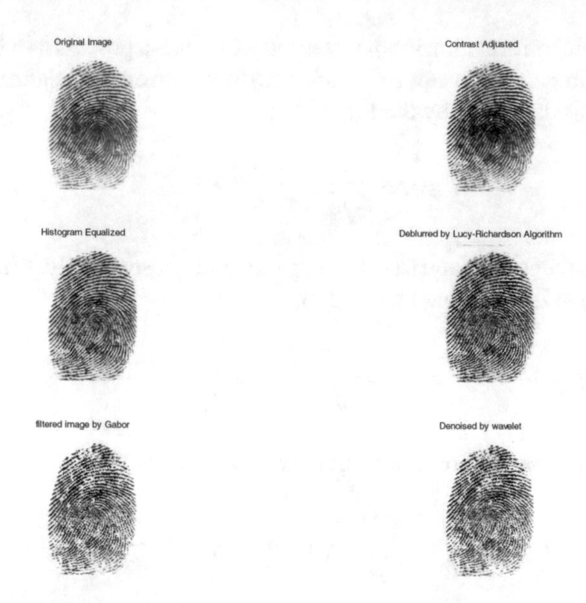

Fig. 2. Fingerprint image 107_3.tif from database FVC2004. Results after each step.

Fig. 3. Original image 107_3.tif

Fig. 4. Binary image of 107_3.tif

In two dimensions, a two-dimensional scaling function, $\varphi(x, y)$ and three two-dimensional wavelet functions, $\psi H(x, y)$, $\psi V(x, y)$, $\psi D(x, y)$ are required.

The two-dimensional scaling and wavelet functions are described by the following equations (Eqs. 12–15):

$$\Phi(x, y) = \Phi(x)\Phi(y) \tag{12}$$

$$\Psi H(x, y) = \Psi(x)\Phi(y) \tag{13}$$

$$\Psi V(x, y) = \Psi(y)\Phi(x) \tag{14}$$

$$\Psi D(x, y) = \Psi(x)\Phi(y) \tag{15}$$

These wavelets measures functional variations-intensity variations for images-along three different directions viz. horizontal, vertical and diagonal. The scaled and translated basis functions are defined as (Eqs. 9 and 10):

$$\emptyset_{j.m.n}(x,y) = 2^{\frac{j}{2}}\emptyset\left(2^{j}x - m, 2^{j}y - n\right) \tag{16}$$

$$\Psi^{i}_{j,m,n}(x,y) = 2^{\frac{j}{2}}\Psi^{i}\left(2^{i}x - m, 2^{j}y - n\right) \tag{17}$$

where index i = H, V, D identifies the directional wavelets in Eqs. 13–15. The discrete wavelet transform of image f(x, y) of size M × N is then expressed as follows

$$W_{\emptyset}\left(j_0, m, n\right) = \frac{1}{\sqrt{MN}}\sum_{x=0}^{M-1}\sum_{y=0}^{N-1} f(x,y)\emptyset_{j_0,m,n}(x,y) \tag{18}$$

$$W^{i}_{\Psi}(j.m.n) = \frac{1}{\sqrt{MN}}\sum_{x=0}^{M-1}\sum_{y=0}^{N-1} f(x,y)\Psi^{i}_{j,m,n}(x,y) \tag{19}$$

where i = H, V, D.

Wavelet analysis can be considered to be a time-scale method embedded with the characteristic of frequency. It gives its best performance when it is applied to the detection of short time phenomena, discontinuities, or abrupt changes in signal. The decomposition of the input signal into approximation and detail space is called multi-resolution approximation, which can be realized by using a pair of finite impulse response (FIR) filters h and g, which are lowpass and high-pass filters, respectively as shown in Fig. 4. The analytical description and details regarding the wavelets can be found in Refs. [15, 16].

3.6 Wavelet Thresholding

Image de-noising is used to get rid of the additive noise while keeping hold of as much as possible the important features. Wavelet thresholding is an effective method which is achieved via thresholding. Wavelet thresholding procedure removes noise by thresholding only the wavelet coefficient of the details coefficients, by keeping the low-resolution coefficients unaltered. There are two thresholding methods commonly used as: soft thresholding and hard thresholding [18].

The hard-thresholding TH can be defined as:

$$T_H = \left\{ \begin{array}{c} x\,for\;|x| \geq t \\ 0\;in\;all\;other\;regions \end{array} \right\}$$

Here t is the threshold value.

Soft thresholding is where the coefficients with greater than the threshold are shrunk towards zero after comparing them to a threshold value. It is defined as follows in all other regions.

$$T_S = \left\{ \begin{array}{c} sign(x)(|x| - t) \quad for|x| > t \\ 0 \; in \; all \; other \; regions \end{array} \right\}$$

In practice, it can be seen that the soft method is much better and yields more visually pleasant images. This is because the hard method is discontinuous and yields abrupt artifacts in the recovered images.

4 Fingerprint De-noising

A fingerprint image consists of non-ridge area, high quality ridge area, and low quality ridge area. It is well known that low quality ridge area in the fingerprint images would cause serious effects, which deteriorate the quality of the image. The Fingerprint image is infected with the Gaussian and Salt & Pepper noise. Many dots can be spotted in a Photograph of fingerprint taken with a digital camera or fingerprint reader under low lighting conditions or the machine hardware problem. Actually this type of noise is the uniform Gaussian noise. Facade of dots is due to the real signals getting corrupted by noise (superfluous signals). On loss of reception or recover any Fingerprint image from the storage device random black and white snow-like patterns can be seen on the Fingerprint images. The resulting subimage is extracted from the original fingerprint image with noise in the complex wavelet transform domain. Then, according to the characteristics of the sub-image data, thc dc-noised fingerprint is being used for further reference purposes.

5 Algorithm

Fingerprint images usually are with low and nonuniform contrast. Their histograms usually are not equalized, the ridges are blurred. Some fault ridge can be repaired to its original state using the bandpass Gabor filters, they can effectively remove the noise on the ridge and keep the ridge structure. Multiresolution analysis provides information about the image at more than one resolution. Wavelet thresholding procedure removes noise by thresholding only the wavelet coefficient of the details coefficients, by keeping the low-resolution coefficients unaltered.

The first step does an image contrast adjustment.
The second step does adaptive histogram equalization
The third step deblurs the fingerprint image using Lucy-Richardson Algorithm
The fourth step implements the Gabor filters bank.
The fifth step denoises the fingerprint image in wavelet domain.

6 Experimental Results

To estimate the efficiency of the proposed method, the fingerprint images from public database Fingerprint Verification Competition 2004 [19] are considered. The experimental fingerprint images are with low quality or degraded, where the reliable enhancement is necessity (Fig. 2).

For evaluating the experimental results, the visual inspection and the true extracted minutiae are used before and after the enhancement. The true minutiae number is calculated by the Matlab code "Fingerprint minutiae extraction" [20]. From the visual inspection of the resulted images is noticed that the proposed algorithm gives a good performance in recovering the ridges with low quality, it reduces the noise and in general the accuracy of ridges is reached (Figs. 3 and 5).

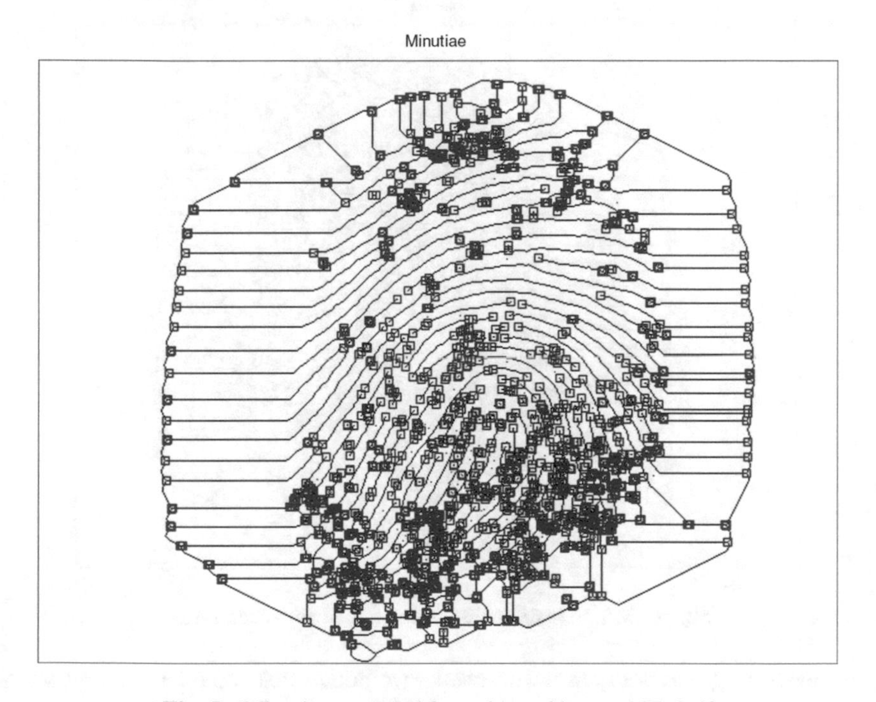

Fig. 5. Minutiae extracted from thinned image 107_3.tif

After image enhancement the results are as following (Figs. 6, 7 and 8).

Fig. 6. Enhanced image 107_3.tif

Fig. 7. Binary image of enhanced one

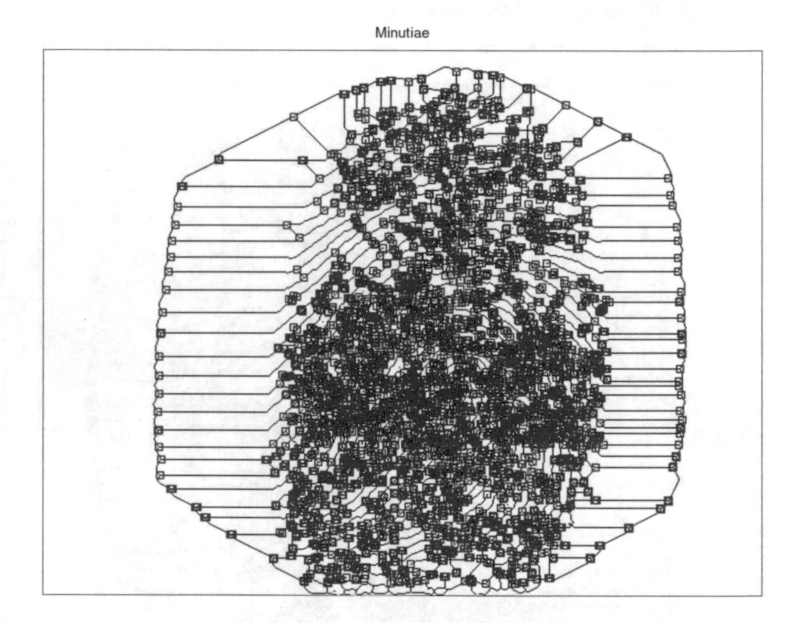

Fig. 8. Minutiae extracted from thinned enhanced image

By evaluating and comparing the results we notice that the enhancement method increases significantly the number of true minutiae extracted. The number of true minutiae is increased and the number of false minutiae is decreased. The ratio depends from the damage scale of the fingerprint image and the noises occurred in the image (Table 1).

Table 1. True minutiae extracted from fingerprint image database

FVC2004 database	Before enhancement	After enhancement
Image 103 FVC2004DB1	119	130
Image 105 FVC2004DB2	101	109
Image 102 FVC2004DB3	107	114

7 Conclusions

This paper proposes an enhancement algorithm in spatial and wavelet domain. Estimating the results we notice that the fingerprint image is enhanced, the noise is reduced, the ridges are sharper and the number of true minutiae is increased.

Enhancement in spatial domain followed by applying by Gabor filter and wavelet denoising, the performance of the fingerprint recognition system is improved.

References

1. Vidhya, T., Thivakaran, T.K.: Fingerprint image enhancement using wavelet over Gabor filters. Int. J. Comput. Technol. Appl. IJCTA **3**(3), 1049–1054 (2012)
2. Rusyn, B., et al.: Fingerprint image enhancement algorithm. In: Proceedings of IEEE Conference on CAD Systems in Microelectronics, Ukraine, pp. 193–194 (2000)
3. Kim, B.-G., et al.: New enhancement algorithm for fingerprint images. In: Proceedings of IEEE Conference on Pattern Recognition, Korea, pp. 879–882 (2002)
4. Hashad, F.G., et al.: A hybrid algorithm for fingerprint enhancement. In: Proceedings of IEEE Conference on Finger Enhancement, Menoufia University, Menouf, pp: 57–62 (2009)
5. Bo, F., Zhi, H., et al.: A novel fingerprint enhancement method based on Gabor filtering. In: Proceedings of IEEE Conference on Image and Signal Processing, China, pp. 66–69 (2009)
6. Hadhoud, M.M., et al.: An adaptive algorithm for fingerprints image enhancement using Gabor filters. In: Proceedings of IEEE Workshops on Fingerprints, Menoufia University, Menoufia, pp. 57–62 (2007)
7. Hong, L., et al.: Fingerprint enhancement. In: Proceedings of 1st IEEE Conference on WACV, Sarasota, FL, pp. 202–207 (1996)
8. Hong, L., et al.: Fingerprint image enhancement: algorithm and performance evaluation. IEEE Trans. PAMI **20**(8), 777–789 (1998)
9. Balaji, S., Venkatram, N.: Filtering of noise in fingerprint images. Int. J. Syst. Technol. **1**(1), 87–94 (2008). ISSN 0974-2107
10. Hong, L., Wan, Y., Jain, A.: Fingerprint image enhancement: algorithm and performance evaluation. www.math.tau.ac.il/~turkel/imagepapers/fingerprint.pdf
11. https://www.mathworks.com/help/images/contrast-adjustment.html
12. https://en.wikipedia.org/wiki/Adaptive_histogram_equalization
13. https://en.wikipedia.org/wiki/Richardson%E2%80%93Lucy_deconvolution
14. Ke, H., Wang, H., Kong, D.: An improved Gabor filtering for fingerprint image enhancement Technology. In: 2nd International Conference on Electronic and Mechanical Engineering and Information Technology (EMEIT 2012) (2012)
15. Mallat, S.G.: A Wavelet Tour of Signal Processing, 2nd edn. Academic Press, London (1999)
16. Gonzalez, R.C., Woods, R.E.: Digital Image Processing, 3rd edn. Prentice-Hall, New Delhi (2008)
17. Kakkar, V., Sharma, A., Mangalam, T.K., Kar, P.: Fingerprint image enhancement using wavelet transform and Gabor filtering. Acta Technica Napocensis Electron. Telecommun. **52**(4), 17–25 (2011)
18. Dass, A.K., Shial, R.K., Gouda, B.S.: Improvising MSN and PSNR for finger-print image noised by GAUSSIAN and SALT & PEPPER. Int. J. Multimedia Its Appl. (IJMA) **4**(4), 59–72 (2012)
19. http://bias.csr.unibo.it/fvc2006/
20. https://www.mathworks.com/matlabcentral/fileexchange/31926-fingerprint-minutiae-extraction

Android OS Stack Power Management Capabilities

Olimpjon Shurdi(✉), Luan Ruçi(✉), Vladi Kolici(✉), Algenti Lala(✉), and Bexhet Kamo(✉)

Polytechnic University of Tirana, Tirana, Albania
{oshurdi,vkolici,alala,bkamo}@fti.edu.al,
luan.rucci@gmail.com

Abstract. During the last decade, mobile communications and smartphone technology increasingly became part of people's daily routine. Nowadays, Android smartphones and iPhone are more and more pervasive and widely used even from people with less incomes. Lots of new applications are daily introduced or updated. Such high usage brings new challenges regarding devices' battery lifetime. For smartphone mobile devices and embedded system device, Power Management (PM) it's getting more and more importance because of very limited battery power. There are more and more sensors, I/O and OS SW updates introduced recently in these mobile devices that can be used to improve the effectiveness of PM. In this paper, we do not want to show only the Android smartphones standards regarding power management the system uses, but also to compare the design of Android PM with possibilities that OS middleware libraries/kernel stack offers to reduce it. So, our aim in this paper is to show how kernel level solution can be realized and how. They solutions seems to be CPU less intensive interactions than any other user space or API ones.

Keywords: Application · Smartphone · Power management · Android OS
Kernel

1 Introduction

Nowadays, wirelessly connected smartphone devices are present almost everywhere at any time. The rapidly increase of mobile Internet due to technological convergence and social adoption trends is driving substantial mobile device growth. Featuring a wide range of PDA functionality on a single mobile device, made smartphones have increased their significance role in people's professional and private lives. But such growing popularity of it is accompanied by the cost of higher energy consumption, which brings power supply to a crucial issue in the development of advanced functionalities. Currently, the limited battery lifetime on smartphones is still unable to live up to the demands of sustaining the full functionality on smartphones without frequent recharge. Therefore, the mobile communications industry like operators, vendors and programmers are focused on UE power management techniques or features to extend power supply for increasing the reliability of smartphones. Some significant contributors to the fast battery drain are the intensive use of Wi-Fi and 3G/LTE radio interfaces on smartphones [1, 2]. However, the power use of 3G/LTE and Wi-Fi on smartphones which is

L. Barolli et al. (Eds.): EIDWT 2018, LNDECT 17, pp. 754–764, 2018.
https://doi.org/10.1007/978-3-319-75928-9_67

measured and studied seems almost "inefficient". In this view, our goal is to utilize power efficiency of mobile phones operating on Android OS, by means of power efficiency management of packet transmission at network kernel level inside OS. The presented work is organized as the following. We first investigate and characterize the network data traffic on an Android platform [3], their problems and fast battery drain root causes. We talk about applications and data transmission periodicity together with their "negative" impact on power consumption. We have studied the basic concept and functionality of Android power management and its "inefficiency" for a whole power save mode. For this we present some power measurements analysis and highlight the most contributors in fast battery drain. Based on this we studied and analyzed algorithms for smart data transmission for power save purposes, but this is not considered here. What was important is to distinguish and show how to implement any applicable power save algorithm. For example managing the UE traffic of an Android smartphone through kernel by dynamically scheduling the packets on the current condition. For this we must go deeper inside the Android kernel networking stack and analyze commands and the benefits on implementing any suitable algorithm for our power save purpose.

a. API periodicity and its problems

We can classify the network data traffic into a few major kinds: non-periodic, periodic, and DNS query initiated periodic communication. For example, Gmail traffic is an example of a non-periodic communication application whose next transmission (on UL) is unpredictable. Periodic communication, on the other hand, is a communication that keeps updating periodically like sending/receiving the packets on background. Such communication type is a kind of "non-stop" connection between the application or UE and the remote server which "never" initiates a new communication with a DNS query packet if it is left running in background. As for DNS query initiated periodic communication, it is adopted by most popular applications like Weather, Facebook, and Flickr, in which the application sends a DNS query packet to expect in the returning a response packet to retrieve the answering IP to set up connection and receive data from the corresponding server. But furthermore, these applications are operating independently from the interference of the sync behavior of other applications, based on which the UE networking interface and protocols used are remaining on active status by causing an extra energy spent continuously. As a result, next step is to investigate the kernel of networking in Android OS to modify the existing system behavior regarding packet transmission by delaying or transmissions in burst so as resulting in better battery performance.

The result shows that these applications have a periodicity, and their periods can be tuned in the settings or OS can do it for all depending where any solution is applied. It will be great that Android system aligns these sync applications, but this is not OS decision since each API has its own design and functionality/protocols rules. It will be a good approach toward power saving if the applications are synchronizing from different time in the beginning, at the same time after several trials, or gradually becoming closer and finally identical. We find that the applications do not have same behavior and are not mutually affected as there are different protocols and predefined settings. The measuring difference between two applications for example was found 10 s apart from each other at the beginning, but the difference remains the same also at the fourth synchronization. This is affecting the way how the system or UE handles the 3G/4G network interface status, which gives a huge contribution to total power waste [4] (Fig. 1).

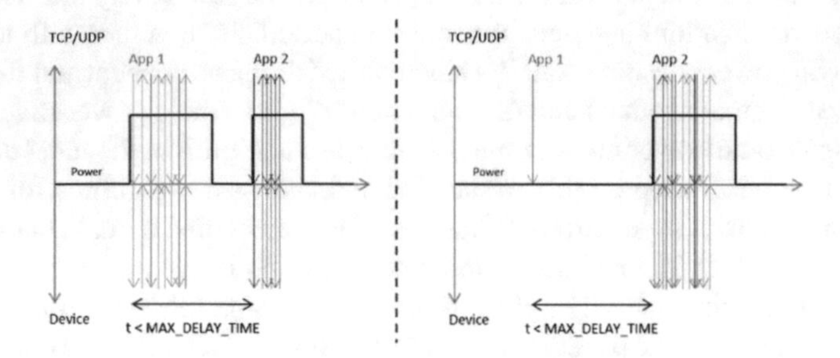

Fig. 1. An example of aligning two near applications

2 Android Power Management Concept

Android software stack/OS is composed of four main components: Operating System (OS), Middleware, Application Framework and Applications. First Android OS design is based on Linux kernel. Linux has its own power management that we have described in previous section. The following diagram (Fig. 2) shows the main components of the Android OS.

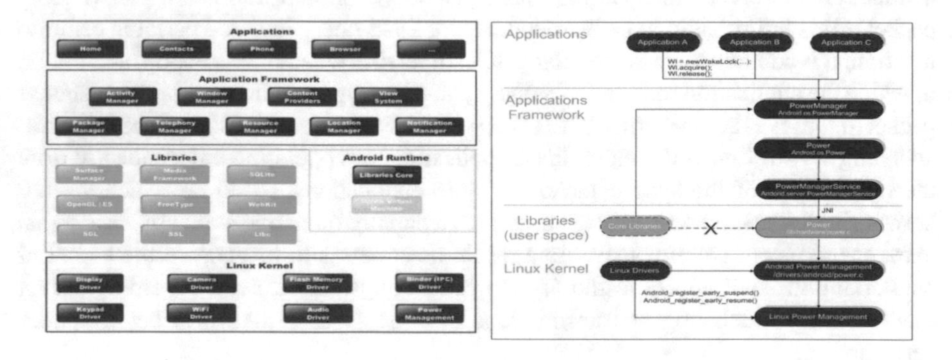

Fig. 2. Android architecture and power management

Android inherits many kernel components from Linux including power management component. Original power management of Linux is designed for personal computers, so there are some power saving status such as suspend and hibernation. However, these mechanisms of Linux PM do not satisfy and suitable for mobile devices or embedded systems. Mobile devices such as cell phones are not as same as PCs that have unlimited power supply. Because mobile devices have a hard constraint of limited battery power capacity, they need a special power management mechanism. Therefore, Android has an additional methodology for power saving. The implementation of Android power management was sitting on top of Linux Power Management. Nevertheless, Android has a more aggressive Power Management policy than Linux, in which app and services must request CPU resource with "wake locks" through the Android application framework and native Linux libraries in order to keep power on, Android will shut down the CPUs. Referring to Fig. 2, Android try not to modify the Linux kernel and implements an applications framework on top of the kernel called Android power management applications framework which is like a driver. It is written in Java language which connects to Android power driver through JNI.

However, what is JNI? JNI (Java Native Interface) is a framework that allows Java code running in a Java Virtual Machine (JVM) to call native C applications and libraries. Through JNI, the PM framework written by Java can call function from libraries written by C.

Android PM has a simple and aggressive mechanism called "Wake locks". The PM supports several types of "Wake locks". Applications and components need to get "Wake lock" to keep CPU on. If there is no active wake locks, CPU will turn off. Android supports different types of "Wake locks" (WAKEUP, RELEASE, WAKELOCK). Finally the main concept of Android PM is through wake locks and time out mechanism to switch state of system power, so that system power consumption will decrease. The Android PM Framework provides a software solution to accomplish power saving for mobile devices.

Still this is not enough as per the fact that Android can't control the users or applications behaviors (at least considering background activities), so an ongoing improvement on power save methods based on better smart networking capabilities is a must (Fig. 3).

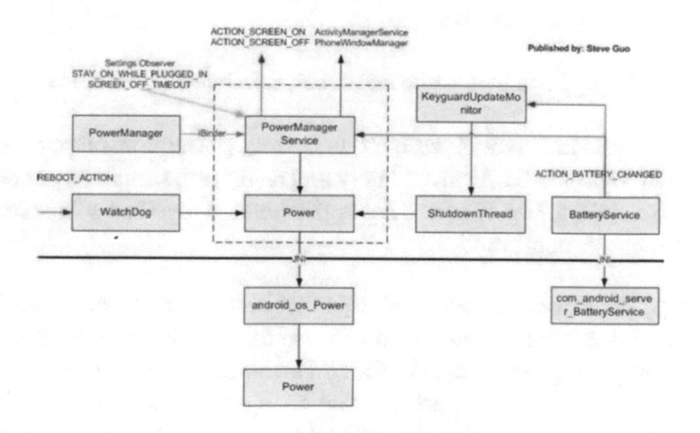

Fig. 3. Android power management architecture

There is a need handling a set of measurement to understand where and how the power is consumed. This will be used also to compare to several other studies on this area and as a guide toward power saving techniques.

3 Pre-measurements Setup and Results

The measurement environment is composed of a notebook, a group of API's, and an Android smartphone, in which the notebook serves as a sniffer to capture real traffic for further analysis. Instead of this some Android sniffer or packet capturing API's can be used [5]. Figure 4 shows physical measurements composed by a voltmeter, precise resistor (~0.2 Ω) placed between battery and smartphone. On this way, every activity on smartphone causing a voltage drop can be measured by a voltmeter on resistor. On the right is shown the 2nd method where some specific API's on a laptop or smartphone can measure or store power measurements data for post processing.

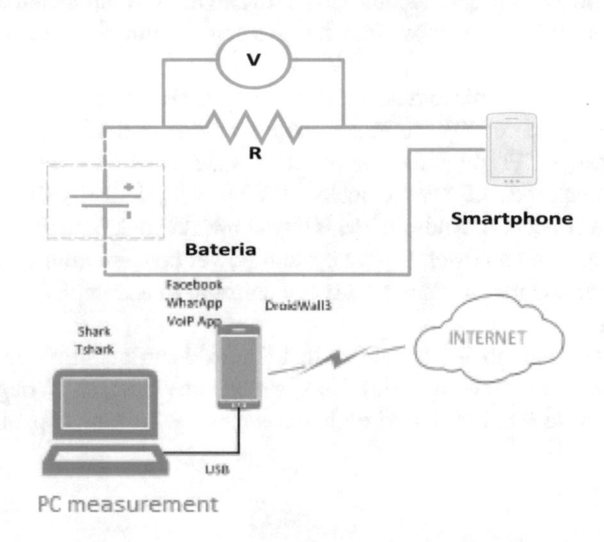

Fig. 4. Setup to measure real or simulated smartphone traffic

In one of our previous works we have measured power consumption on a Smart-phoneNokia and Samsung S2 Android lately and results are similar as per other authors fingers. For more refer to Fig. 5 where network interface together with screen has a big contribution on power consumption.

Measurement of power consumption: Many other authors on the area find that 2G and 3G/4G incur a high tail energy overhead because the UE power state remains high for a period after completing a data transfer [6, 7]. Furthermore, for background traffic there is found that so many network interface state changes from Idle to FACH or to DCH where different power consumption occurs, and can't be avoided by system except cases where manual system settings are set to prevent from such behavior.

Reduction of power consumption: There exist several proposals to reduce power consumption by using cellular radio's to forward notifications of incoming VoIP calls, waking up the WiFi radio to receive the call with tiny delay, switching multiple radio interfaces such as Wi-Fi and Net interface to achieve seamless transmission and increase battery lifetime, use a second lower-power radio as a wake-up channel for the main radio so that the main radio interface can be shut down when there's no transmission. By tuning parameters of power save mode in 802.11, it's also a way to reduce power consumption. Delaying packets that belongs to background traffic and then transmitting on burst, or transferring on channels which consume less power whenever possible are some power saving techniques not to be neglected. In case of streaming packets, they can be buffered and burst at a time so meaning of delaying transferring somehow. Several protocols solutions and are proposed, like for example TailEnder, Tailtheft, parallel TCP sessions, computation offloading, batching which tend to reduce the tail or extra overhead energy produced during data transmission like author had proposed at [8].

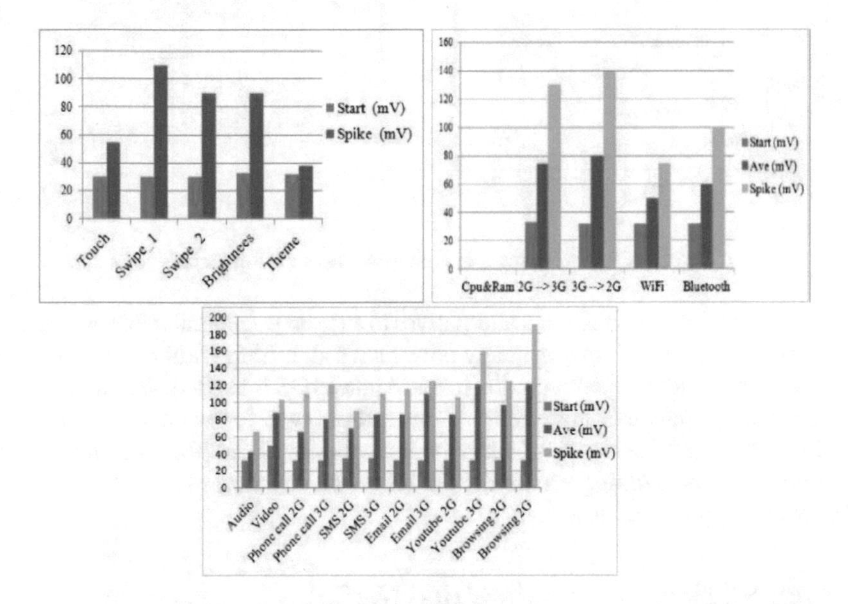

Fig. 5. Power consumptions measurements results on a smartphone

4 State of Art

Battery saver applications offers "power save" based on combinations of settings by measuring and preventing many activities that tend to reduce CPUs and other UE networking activities. But better performance can be offered if solution is provided as middleware or inside kernel of Android OS. We have selected Android as the dominant OS for nowadays smartphones and as open source software for further development. Power saving techniques can be split in two main parts: HW and SW. Our focus is for SW solutions and this not on API section (user level) but as middleware or inside kernel network stack. There exist several studies regarding where improvements from SW point of view can be made. As seen in measurement, network interface (state changes) is one of the main contributors and this in regards with small and so often data transmission which happening without user knowledge (for background applications). Reducing as much as possible network interface state changes with acceptable QoS is on our focus. For example, the transmission of network packets according to an energy-efficient algorithm requires the handling of network traffic at network level. Going deeper on data communications there exist three kind of data traffic channels with different power level and utilizing them as per power saving needs seems that FACH channels can be utilized for small, slow or background traffic. For this traffic must be saved and delayed for transmission to reduce network interfaces transitions.

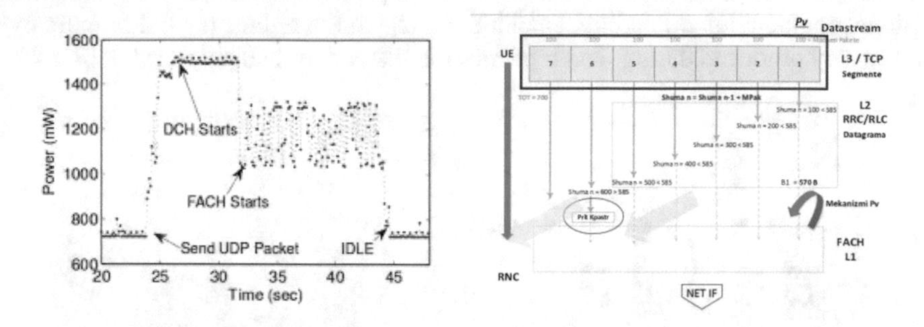

Fig. 6. Channels power and simple power saving technique

This section describes the main alternatives to support communications within Linux kernel and user space to implement any power save algorithm within kernel space. For this a better know-how implementation inside Android OS is needed. We have consulted introductions and valuable information's on Android web forums and official website regarding kernel and the results of our previously papers like at [9, 10]. Solution can be implemented by performing modifications inside several libraries or as one new file covering all required changes.

5 Android Networking Stack and Netlink

The Linux networking stack is based on the TCP/IP model, which basically defines 4 layers: Application, Transport, Internet and Data link. Figure 8 shows how this model is implemented in Linux. User space interacts with lower layers through system calls. These system calls allow the application layer to establish communication between kernel and user space by means of sockets. One of three main type of sockets in Linux is Netlink sockets [11]: it uses the AF NETLINK protocol family.

Netlink is one of the interfaces that Linux provides to user-space. Netlink is a socket family that supplies a messaging facility based on the BSD socket interface to send and retrieve kernel-space information from user-space. Netlink sockets, are flexible and extensible messaging system that provides communication between kernel and user-space. Netlink sockets have become one of the main interfaces that kernel subsystems provide to user-space applications in Linux. Netlink is portable, highly extensible and it supports event-based notifications. Netlink is a datagram-oriented messaging system that allows passing messages from kernel to user-space and vice-versa (Fig. 7).

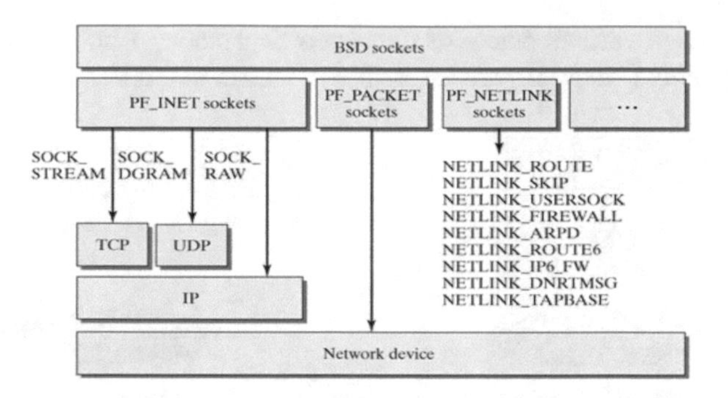

Fig. 7. Netlink sockets on Linux

JNI [12] is a programming framework that allows Android JAVA applications to call native programs and libraries written in other programming languages, such as C or C++. *Transport layer* is the highest layer within kernel space. It uses sockets to pass network data packets from kernel space to user space and vice versa. *Internet layer* is responsible for verifying and routing incoming and outgoing packets from lower to higher layers, and vice versa. The information of a packet is copied into a socket buffer structure (sk_buff) which is moved layer by layer. Finally a link layer protocol is used to send and receive traffic. Moreover it copies the information of the packets received by the network card into sk_buff and sends it to the Internet layer. The packets received from the Internet layer are sent through the network card and this for UL directions. The transmission of network packets according to an energy-efficient algorithm requires the handling of network traffic.

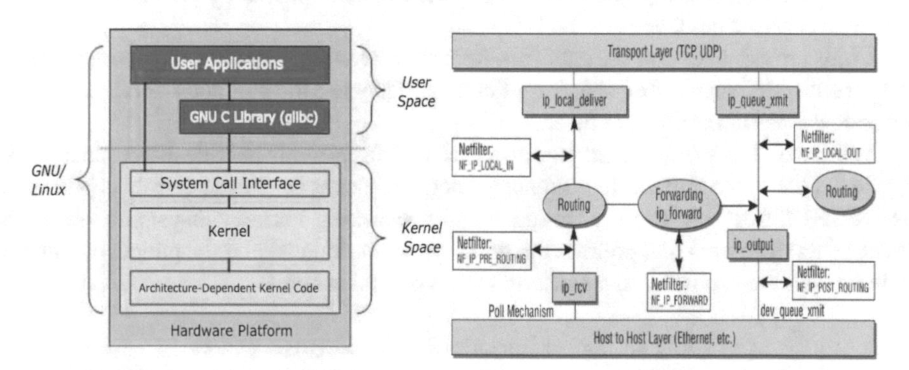

Fig. 8. The fundamental architecture of the GNU/Linux OS and networking stack

One of the main alternatives to detect and modify network traffic within Linux kernel: by Netfilter modules (.ko) for capturing and modifications of traffic and Netlink for communication with user space. Netfilter is a framework within the Linux kernel that provides network capabilities such as network address translation (NAT) and packet interception and modification [13]. Therefore, the flow of the network traffic can be controlled and modified according to our requirements. Netfilter consists of different predefined hooks in various places of the networking stack. A hook is a decision point in a specific protocol stack. IP version 4 (IPv4) defines five different hooks, see Fig. 9.

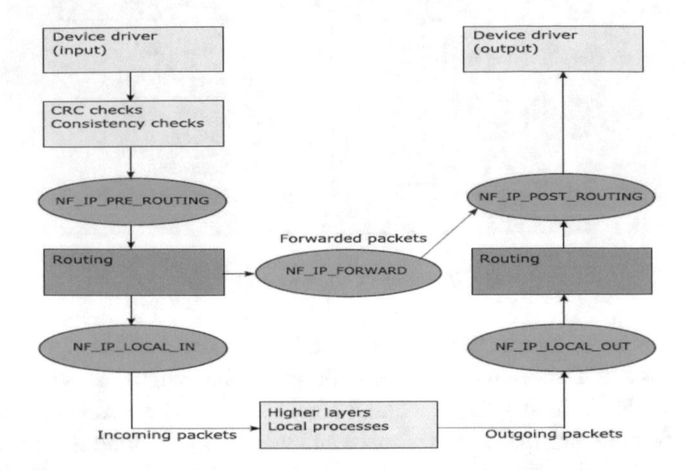

Fig. 9. Netfilter architecture

NF_IP_PRE_ROUTING: this is the first hook passed by an incoming packet.NF_IP_LOCAL_IN: this hook is passed by all incoming packet directed to the local host. NF_IP_FORWARD: this hook is passed by all incoming packet not directed to the local host. NF_IP_LOCAL_OUT: this hook is passed by all outgoing packet created on the local host.NF_IP_POST_ROUTING: this is the last hook passed by all outgoing packets before leaving the device.

Linux provides several virtual file system interfaces that can be used to communicate kernel subsystems and user-space applications. They are the character and block driver interfaces, /proc files and sys files.

We have also summarized these kernel interfaces and their design properties in Table 1. User space applications create a datagram socket in some specific domain, such as AF INET forIPv4. Then, user-space and networking kernel subsystems use fixed-layout data structures that contain the networking configuration to be added, updated or deleted; and they transfer these information's by means ioctls.

Table 1. Linux kernel interfaces by design properties

Type	Architectural portable	Event-based signaling	Easily extensible	Large data transfers
System call	No	No	No	Yes
/dev	No	No	No	Yes
/proc	Yes	No	No	No
Sysfs	Yes	Yes	No	No
Sockets_	No	No	No	Yes
Netlink	Yes	Yes	Yes	Yes

Netfilter provides the ability to handle network traffic at user space by combining a Netfilter LKM, Netlink sockets and libnetfilter queue user space API.A combination of them offers many options to modify traffic based on our conditions would bring benefits in terms of energy saving for UE. Modifications on existing Netfilter modules or new modules are the options to start the packet interception and modification as per our needs. Libnetfilter queue API allows user space to receive the queued packets.

6 Implementations on Android

As per 2nd part of Fig. 6, is presented a simple rule how to treat packets in transmitting or UL (since DL data can't be controlled from UE side) directions by analyzing them and treating as small or large and storing them temporarily on 2 buffers for delaying purposes. By delaying them we can reduce network interface switching and from this a small valuable % of power can be saved.

Therefore these stored packets can be modified and re injected into the stack as per our conditions. Benefits from such solution are also due to fact the CPU involvement is less compared to any other method as it required less CPU involvement. Implementing in Android as new LKM requires that new module as C or C++ to be compiled with right Android OS and proper ARM tool chain as described [14, 15]. The Android OS must have enabled the loading of new kernel modules otherwise it must be enabled manually. After the new module can be ported to Android via a group of commands and it can be monitored via Android SDK tool for OS logs. Main steps are as per (Fig. 10):

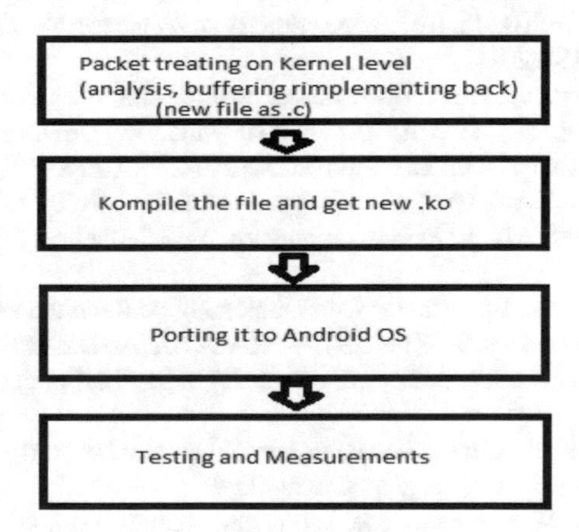

Fig. 10. Main steps implementing a power save kernel solution

7 Conclusion

It is worth to use energy to save more energy. Solution offered on kernel by handling data packets handling mechanism does not consume much and saves large amounts of energy. Therefore, the life of the battery of the phone can be prolonged. As applications, do not distinguish background traffic from the rest of traffic classes, also for kernel proposed solutions this would be a key point for future studies. Kernel level modifications or addons can better improve to any other methods the energy saving on UE. A close knowledge of networking stack in Linux kernel is a must for such cases.

References

1. Abdelmotalib, A., Wu, Z.: Power consumption in smartphones (hardware behaviourism). IJCSI Int. J. Comput. Sci. Issues **9**(3) 2012. ISSN (Online): 1694-0814
2. Balasubramanian, N., Balasubramanian, A., Venkataramani, A.: Energy consumption in mobile phones: a measurement study and implications for network applications. In: Proceedings of the 9th ACM SIGCOMM Conference on Internet Measurement Conference (2009)
3. Nokia Siemens Solutions Smart_Lab_WhitePaper on smartphones energy consumption. www.nsn.com/system/files/…/Smart_Lab_WhitePaper_27012011.low-res.pdf
4. Deng, S., Balakrishnan, H.: Traffic-aware techniques to reduce 3G/LTE wireless energy consumption. In: Proceeding CoNEXT 2012. pp. 181–192. ACM, New York (2012)
5. Wireshark tool and tips. http://www.wireshark.org/
6. Carroll, A., Heiser, G.: An analysis of power consumption in a Smartphone. In: Proceedings of the 2010 USENIX Conference on USENIX Annual Technical Conference (2010)
7. Huang, J., Xu, Q., Tiwana, B., Mao, Z.M., Zhang, M., Bahl, P.: Anatomizing application performance differences on Smartphones. In: Proceedings of the 8th International Conference on Mobile Systems, Applications, and Services (2010)
8. Könönen, V., Paakkonen, P.: Optimizing power consumption of always-on applications based on timer alignment. In: The 4th IFIP International Conference on New Technologies, Mobility and Security (NTMS) (2011)
9. Android OS for Samsung. https://opensource.samsung.com/reception/
10. Ruci, L., Karcanaj, L., Shurdi, O.: Nokia windows mobile's power consumption measurements and analysis. In: ICT Innovations 2014, Web Proceedings © ICT ACT, 2014, 6th ICT Innovations Conference, Skopje, pp. 1–12 (2014). ICT ACT http://ict-act.org. On line edition published on http://ictinnovations.org. 9–12 September 2014, Ohrid, Macedonia. ISSN 1857-7288
11. Ayuso, P., Gasca, R.M., Lefevre, L.: Communicating between the kernel and user-space in Linux using Netlink sockets. SOFTWARE—PRACTICE AND EXPERIENCE, Softw. Pract. Exper. (2010). 23 September 2002, v2.2. University of Seville, Spain and University of Lyon, France
12. Android Developers JNI. https://developer.android.com/index.html
13. Netfilter. https://www.netfilter.org/documentation/
14. ADB Android. https://developer.android.com/studio/command-line/adb.html
15. Android and JNI. https://developer.android.com/training/articles/perf-jni.html

Signal Routing by Cavities in Photonic Crystal Waveguide

Hiroshi Maeda[1(✉)], Xiang Zheng Meng[2], Keisuke Haari[2], and Naoki Higashinaka[3]

[1] Department of Information and Communication Engineering, Fukuoka Institute of Technology, 3-30-1 Wajiro-Higashi, Fukuoka 811-0295, Japan
hiroshi@fit.ac.jp
[2] Master's Course of Information Networking, Graduate School of Engineering, Fukuoka Institute of Technology, Fukuoka, Japan
[3] Department of Information and Communication Engineering, Fukuoka Institute of Technology, Fukuoka, Japan

Abstract. A design of all-optical signal routing circuit for the internet or telecommunication system are proposed by photonic crystal waveguide, which is composed of line of defects in periodic structure by dielectric pillars. Cavities are introduced in or by the waveguide for filtering and switching by making use of Fabry-Perot resonance for a typical carrier signal. Experimental results for the model waveguide in microwave frequency are shown to demonstrate the filtering and switching characteristics depending on length of the cavity. Proposed circuit design is applicable for high speed network switches, which is free from electronic-optical (E/O) or O/E conversion with time delay and is also free from signal labeling by substitutive use of the carrier frequency.

1 Introduction

High speed signal routing for faster data transmission in network technology has been intensively studied. Most of modern network switching system equips optical fiber cable, signal processor unit, and optical/electronic (O/E) or O/E signal converters. The key point is conversion time in O/E and E/O converters with regeneration of the signal form. Current telecommunication systems hiring hybrid of electric and optical signals must include the E/O or O/E signal conversion circuit in the switch. Those switches are necessary to maintain the signal waveform as well as the input waveform for lower bit error ratio (BER), though the signal conversion causes time delay especially for higher bit rate telecommunication. Generally, throughput of the internet and telecommunication systems are regulated by this time delay in the switch.

In optical fiber communication system over several thousand kilometers, Erbium doped fiber amplifier (EDFA) is implemented practically. As EDFAs give gain to the optical signal directly in optical frequency domain, the time delay due to the O/E and E/O conversion was drastically improved with decrease of

L. Barolli et al. (Eds.): EIDWT 2018, LNDECT 17, pp. 765–772, 2018.
https://doi.org/10.1007/978-3-319-75928-9_68

power consumption in the system. Together with dispersion compensation by fiber grating, the pulse distortion can be corrected to obtain higher bit rate.

For further high speed switching, we propose a circuit for switching/routing system without signal frequency conversion to achieve higher throughput. For this purpose, we introduce photonic crystal structure as the fundamental transmission line and the signal process circuits.

Photonic crystal or electromagnetic band gap structure is a key component for optical communication systems, because of the unique propagation characteristics based on photonic band gap theory [1–4]. In the system, cavity structure and its application will provide important role as a filtering device, especially in optical wavelength division multiplexing (WDM) system. For the other application, it is also important, for example, to laser oscillation in active device for all-optical signal processing system [3].

As scaling rule generally holds in photonic crystal structure with respect to normalized frequency $\omega P/2\pi c = P/\lambda$, where P is periodicity or lattice period of photonic crystal, c and λ is velocity of light and its wavelength in vacuum, respectively, it is possible to investigate its characteristics in microwave frequency in place of optical frequency. This means that we can make use of magnified models [5,6] in microwave frequency range with lattice period in the order of a few centimeters.

In this paper, we first summarize measurement result of single cavity, referring to our previous researches [7,8], which is situated in waveguide with some variation, for microwaves around 4 GHz. The waveguide is fabricated as a line defect in two-dimensional, square-lattice, pillar-type photonic crystal [5].

The resonance characteristics of single cavities situated in the middle of waveguide was measured as S_{11} and S_{21}. The cavity was put as defects of crystals in the photonic crystal waveguide with a pair of surrounding single rod. It is also possible to put a cavity aside the waveguide. Then characteristics of cavity resonance were measured in reference to Ref. [6]. Finally, we propose T-shaped branch waveguide with cavities to obtain carrier selective operation characteristics due to the resonance. Application of the waveguide to signal routing depending on the carrier frequency is explained.

2 Measurement of Photonic Crystal Waveguide

Experiment is done in accordance with Refs. [5–8]. Periodic square array of ceramic dielectric rod with $\varepsilon_r = 36.0$ is put between aluminum plates. The top view is illustrated in Fig. 1(a) with parameters. All of surrounding plates are grounded to assume that the structure is equivalently two dimensional along with vertical axis. Photonic crystal waveguide is depicted in Fig. 1(b). A line of rods were removed to be waveguide [8] in the figure. At the input and output port in Fig. 1(b), S-parameters S_{11} as reflection and S_{21} as transmission were measured by *Agilent E5071C* vector network analyzer. The transmission and reflection spectra are shown in Fig. 1(c) to show that the structure operates as a waveguide with large transmission and little reflection between 3.9 and 4.2 GHz.

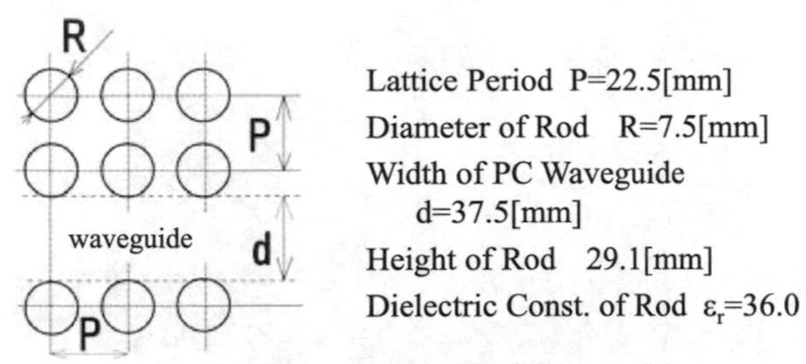

(a) Parameters of PC Waveguide

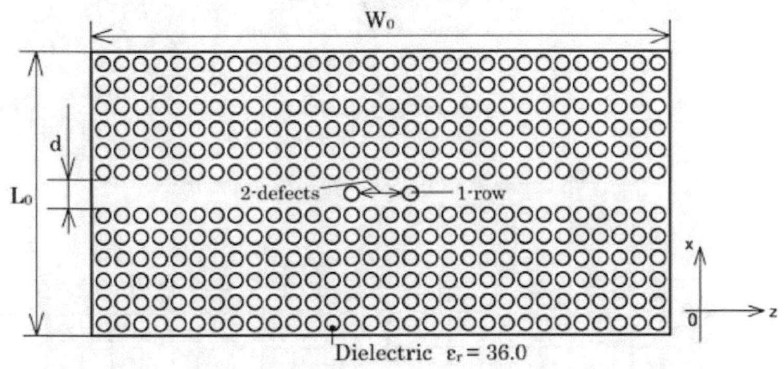

(b) Illustration of 2D Cavity Structure in Waveguide from Top View

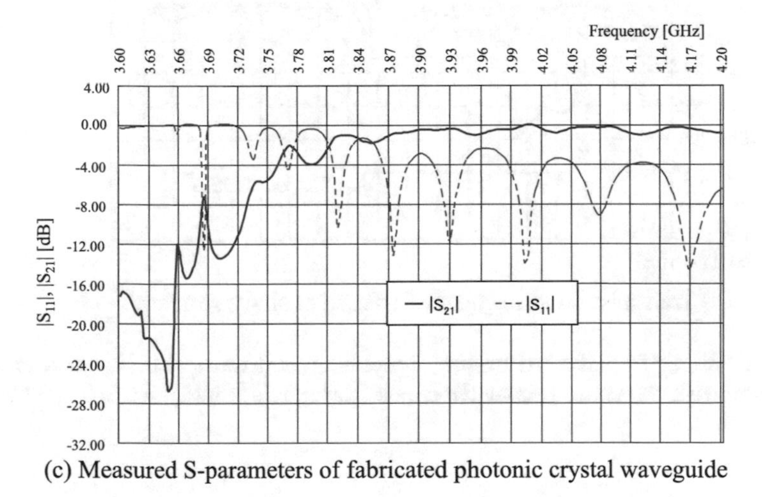

(c) Measured S-parameters of fabricated photonic crystal waveguide

Fig. 1. Illustration of top view of photonic crystal waveguide with line defect and the transmission spectra.

To put a cavity for resonance in the structure, we set a pair of rods [7] as depicted in Fig. 2(a). The transmission and reflection spectra is shown in Fig. 2(b) to show a sharp resonant peak in the transmission S_{21}. We found that the structure has a function of transmission filter.

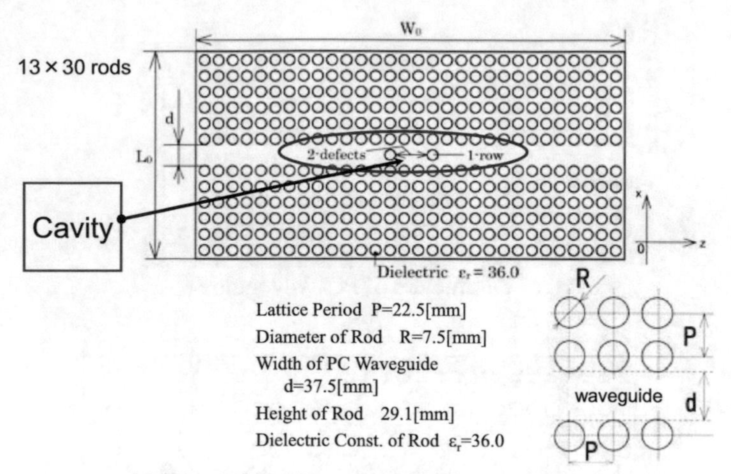

(a) Waveguide with a pair of rods as cavity.

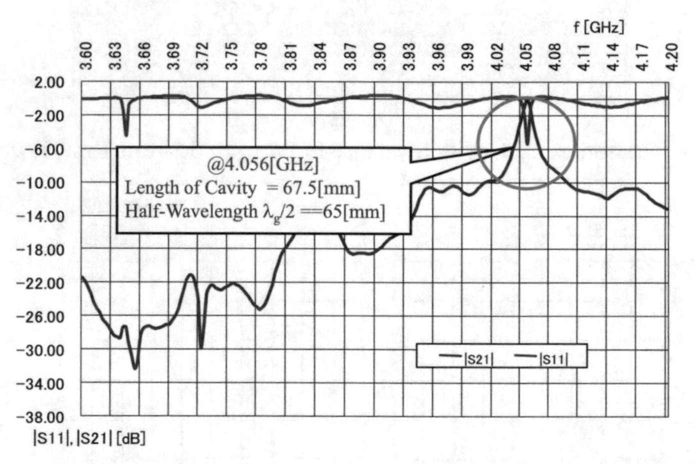

(b) Transmission and reflection spectra of above structure.

Fig. 2. Variety of cavities situated beside photonic crystal waveguide. Type A and B are asymmetric with respect to propagation axis (waveguide length), while Type C and D are symmetric.

Next, we removed a pair of rods beside waveguide [7] to make a defect as is in Fig. 3(a). Then, we measured S-parameters for single cavity structures as is depicted in Fig. 3(b). The structure shows a keen dip at 4.142 GHz in transmission with quite high reflection.

As a measure of sharpness of resonance, Q factor is defined by

$$Q = \frac{f_c}{f_{3dB}}, \tag{1}$$

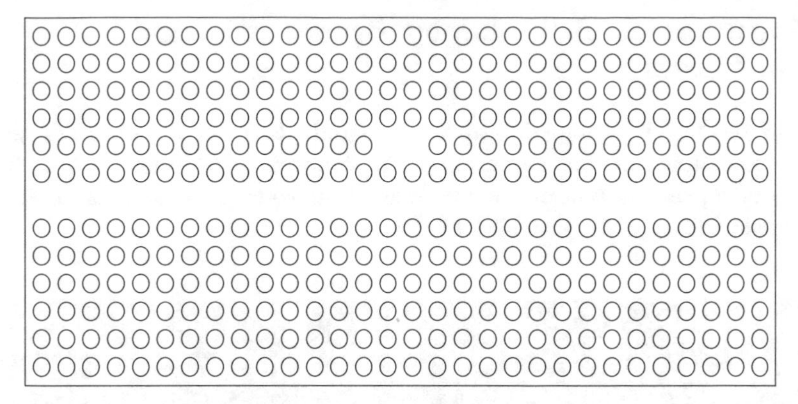

(a) Cavity as defect by waveguide.

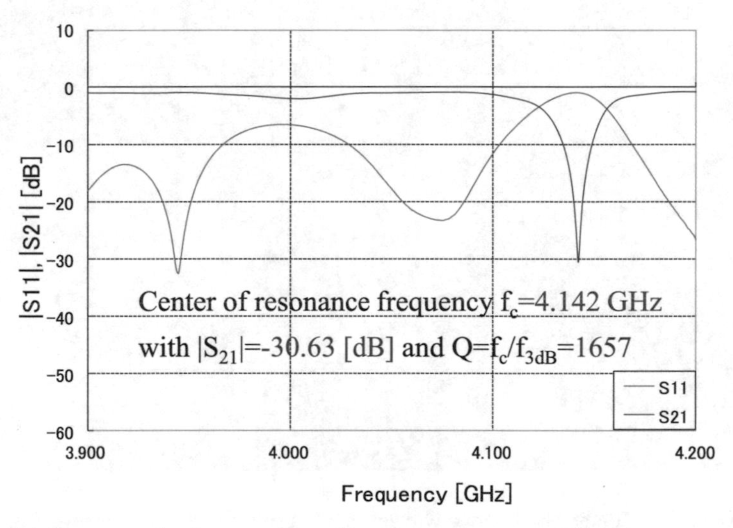

(b) Transmission and reflection spectra of above structure

Fig. 3. Photonic crystal waveguide with a cavity aside, and the transmission/reflection spectra.

where f_c is center frequency of resonance and f_{3dB} is the 3dB band width, respectively. For a structure in Fig. 3(a), $Q = 1657$ was obtained at $f = 4.142\,\text{GHz}$.

The cavity in Fig. 2 is named as *'Band-Transmission Filter(BTF)'*, while the structure in Fig. 3 as *'Band-Reflection Filter(BRF)'*.

3 Measurement of T-Shaped Branch Waveguide with Filters

Picture of proposed T-shaped branch waveguide with cavity filters is shown in Fig. 4. The structure has same circuit parameters and dimensions with the experiments in previous section. As is expected, sharper band stop characteristics was observed in S_{21} with $Q = 3420$ at $f = 4.104$ GHz.

Fig. 4. Photograph of T-shaped branch waveguide. From top-right to counterclockwise, port #1, #2, and #3 are situated. A BTF is seen in horizontal waveguide, while a BRF is put in adjacent of vertical waveguide.

The relation of open ports and the channel frequency are summarized in Fig. 5. It is found that a channel is open between port #1 and #2 at 4.09 GHz, while the other channel is open between port #2 and #3 at 3.77 GHz. This means that a pair of channel with different carrier signal can be separated by proposed T-shaped branch. frequency division multiplexing system, and the channel between port #2 and #3 are insulated. In data transmission like the internet, when packets modulated by carrier frequency of 4.09 GHz is given to port #1, it is automatically transported to port #2. Similarly, packets with 3.77 GHz is delivered from port #1 to port #3.

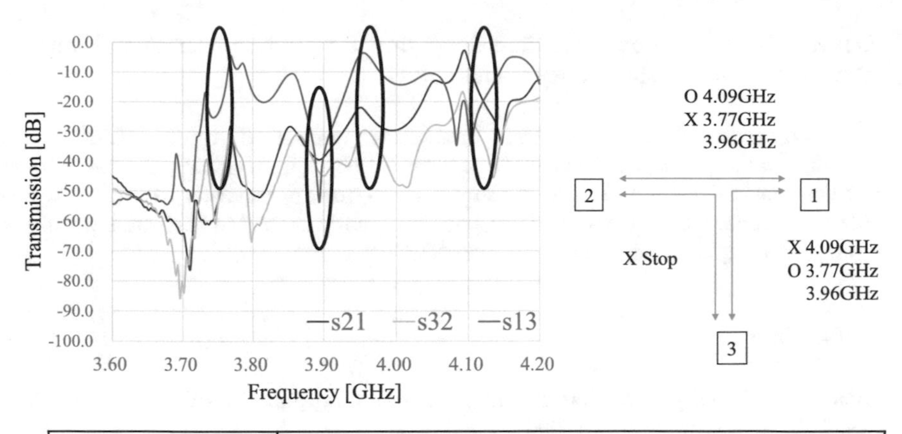

Channel Frequency		From		
		Port #1	Port #2	Port #3
To	Port #1	---	4.09GHz	3.77GHz 3.96GHz 4.16GHz
	Port #2	4.09GHz	---	---
	Port #3	3.77GHz 3.96GHz 4.16GHz	---	---

Fig. 5. Experimental result of transmission among three ports. Typical filtering characteristics are indicated by ellipsoids. Illustration at the right shows channels and the frequency. Open ports are indicated by circles with frequency, while stopped channels are shown by X.

4 Conclusion

A design of signal routing circuit without O/E or E/O conversions are proposed by making use of photonic crystal waveguide and cavities. Those cavities which performs as BTF or BRF are introduced in the waveguide and in adjacent of T-branch by Fabry-Perot resonance for a typical carrier signal. Experimental results for the proposed model in microwave frequency showed that the filtering and switching characteristics are obtained as expected. Proposed circuit design is applicable for high speed network switches, which is free from electronic-optical

(E/O) or O/E conversion with time delay and is also free from signal labeling by substitutive use of the carrier frequency.

Acknowledgment. This work was financially supported by KAKENHI No. 15K06043, Grant-in-Aid for Scientific Research (C) by Japan Society for the Promotion of Science(JSPS) in 2017. The authors express our appreciation to Mr. Y. Eguchi and Mr. N. Okuzono of Fukuoka Institute of Technology as part of their undergraduate research under supervising by H. Maeda in 2017-18.

References

1. Yasumoto, K. (ed.): Electromagnetic Theory and Applications for Photonic Crystals. CRC Press, New York (2006)
2. Inoue, K., Ohtaka, K. (eds.): Photonic Crystals - Physics, Fabrication and Applications. Springer, New York (2004)
3. Noda, S., Baba, T. (eds.): Roadmap on Photonic Crystals. Kluwer Academic Publishers, Dordrecht (2003)
4. Joannopoulos, J.D., Meade, R.D., Winn, J.N.: Photonic Crystals. Princeton University Press, New Jersey (1995)
5. Temelkuran, B., Ozbay, E.: Experimental demonstration of photonic crystal based waveguides. Appl. Phys. Lett. **74**(4), 486–488 (1999)
6. Beaky, M., et al.: Two-dimensional photonic crystal fabry-perot resonators with lossy dielectrics. IEEE Trans. Microw. Theor. Tech. **47**(11), 2085–2091 (1999)
7. Maeda, H., Hironaka, S.: Experimental study on improvement of filtering characteristics of cavities in photonic crystal waveguide. In: Proceedings of ISMOT 2009, pp. 229–232, December 2009
8. Maeda, H., Nemoto, R., Satoh, T.: Experimental study on cavities in 2D photonic crystal waveguide for 4 GHz band. In: Proceedings of ISMOT 2007, pp. 459–462, December 2007

Image Semantic Segmentation Algorithm Based on Self-learning Super-Pixel Feature Extraction

Juan Wang[1,2(✉)], Hao Shi[1,2], Min Liu[1,2], Wei Xiong[1,2], Kaiwen Cheng[1,2], and Yuhan Jiang[1,2]

[1] Hubei Key Laboratory for High-Efficiency Utilization of Solar Energy and Operation Control of Energy Storage System, Hubei University of Technology, Wuhan, People's Republic of China
happywj@hbut.edu.cn

[2] Hubei Collaborative Innovation Center for High-Efficiency Utilization of Solar Energy, Hubei University of Technology, Wuhan 430068, People's Republic of China

Abstract. Image semantic segmentation is a challenging task, influenced by high segmentation complexity, increased feature space sparseness and the semantic expression inaccurate. This paper proposes a stacked deconvolution neural network (SDN) based on adaptive super-pixel feature extraction to degrade computational cost and improve segmentation effectiveness. Firstly, the super-pixel segmentation is accomplished by simple linear iterative cluster (SLIC). Secondly, we add texture information as an optimization information to the evaluation function to guide the super-pixel segmentation and ensure the integrity of the super-pixel segmentation. Finally, we train a Stacked Deconvolution Neural Network (SDN) on the ISPRS Potsdam and the NZAM/ONERA Christchurch datasets and learn the sample data with weak annotation information to realize the accurate and fast super-pixel segmentation. Segmentation tests show that the proposed method can achieve the accurate segmentation of image semantics.

Keywords: Image semantic segmentation · Stacking neural networks
Super-pixel · Texture extraction

1 Introduction

In the face of the rapidly growing image data, it is a problem need to be solved that make computer have the ability to recognize and segment the image information. Image Semantic Segmentation is a key part of image processing and analysis which is a classic research branch in the field of computer vision.

Shotton et al. used semantic texel forest as a new visual feature and applied the constructed semantic texel model to image semantic segmentation. Girshick et al. proposed a target detection algorithm based on convolutional neural network, using the top-down detection method to locate and segment the objects in the image. Liu et al. Used a deformable template to determine the boundaries of objects in the image. Pawan and others combine the star structure of the object model to segment the foreground and background of image. From the research results, the academic community has been committed to solve the problem of image segmentation [1].

L. Barolli et al. (Eds.): EIDWT 2018, LNDECT 17, pp. 773–781, 2018.
https://doi.org/10.1007/978-3-319-75928-9_69

From the initial N-Cut and Grab-Cut to the CNN or FCN, which quote deep learning methods, Image Semantic Segmentation through a long process of development. However, due to the semantic richness and cognitive complexity of the image content, the existing research methods are still not perfect in technology.

Convolutional Neural Networks (CNN) break the performance barrier of many tasks in the field of scene classification, such as object segmentation, object classification and object recognition. CNN promote the research progress in the field of recognition, which not only improves the classification of the whole image, but also promotes the structured output of local tasks, including the generation of the bounding box in object recognition, test of part and key points [2]. However, the classification network with downsampling operations sacrifices the spatial resolution of feature maps to obtain the invariance to image transformations. The resolution reduction results in poor object delineation and small spurious regions in segmentation outputs [3].

The advent of deconvolutional neural networks is a milestone because it shows how training end-to-end with CNN to solve visual problems, it is the cornerstone of Image Semantic Segmentation using deep learning [4]. However, it still lacks the perception of different features, semantic representation is inaccurate, and can't achieve real-time processing speed at high resolution. A good image semantic segmentation method should have several advantages: First, segmentation of different semantic regions of a nature such as grayscale and textures are similar, the region is relatively flat. Secondly, it must has the low segmentation complexity and high segmentation efficiency. Lastly, the boundaries of different semantic regions are clear and regular [5]. On this basis, this paper proposes a Stacked Deconvolution Neural network (SDN) based on adaptive super pixel feature extraction, which connects SLIC and SDN in series through the evaluation function. Accurate and fast image semantic segmentation can be achieved.

2 Work Process

The work process of image semantic segmentation based on self-learning superpixel feature extraction is shown in Fig. 1.

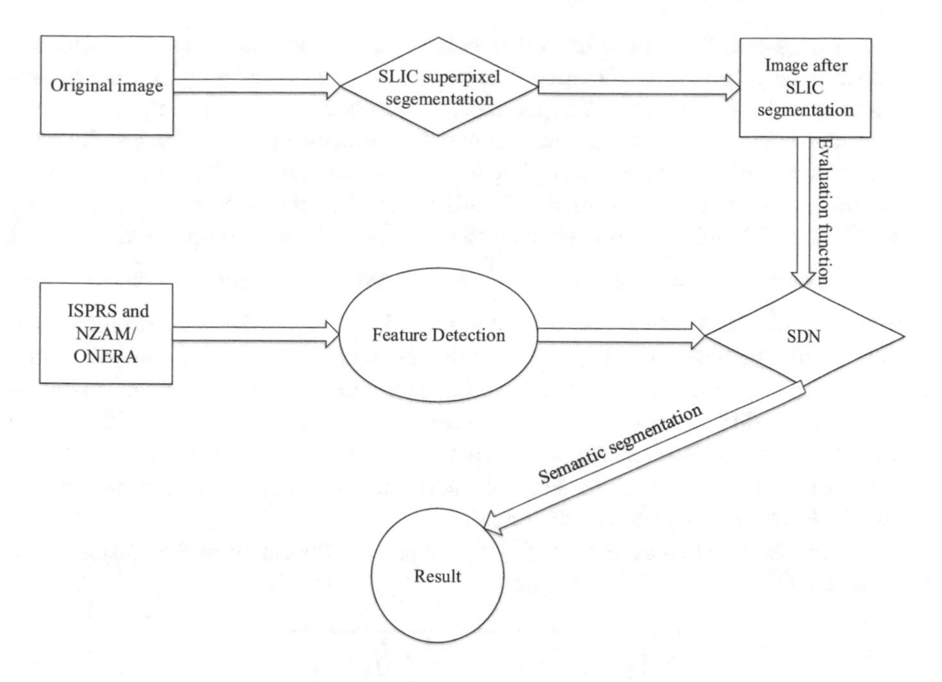

Fig. 1. The process of our method

3 SLIC Superpixel Segmentation of the Original Image

Image segmentation is the basic step in the application of image processing, the purpose is to extract the part of the image that people are interested in, providing the basis for subsequent processing and analysis. The most commonly used algorithm for image segmentation method with a threshold segmentation, edge division method, the area dividing method, a segmentation algorithm based on artificial neural networks, etc. [6]. However, most of these algorithms take too long, and can't get the desired segmentation results.

Rent and Malik proposed the concept of superpixels segmentation, the algorithm can take advantage of some features of pixels in the image segmentation, the image is divided into a large number of irregular small blocks (superpixels), and then do the subsequent processing. Superpixel segmentation provides a way of image preprocessing and greatly reduces the complexity of subsequent image processing tasks [7].

Due to the importance of superpixel segmentation, a large number of algorithms for generating superpixels have been proposed. These algorithms can solve some problems, but most of them are complicated and the generated superpixels have poor matching with the original image. On this basis, Achanta et al. proposed a simple linear iterative clustering algorithm (SLIC), which uses an improved K-means clustering algorithm to generate superpixels. After segmentation, the superpixel border has a strong dependency on the original boundary of the image, its processing speed and storage efficiency are superior to other superpixel segmentation algorithms [2].

From the segmentation results and calculation costs and other factors to consider, we use SLIC superpixel segmentation. It is easy to implement and only the only parameter that specifies the number of superpixels needed is very easy to apply in practice. Divide shape rules and seamlessly accommodate grayscale and color images. First, the cluster center is initialized, and according to the set number of pixels, the cluster center is uniformly distributed in the image. In this paper, the total number of pixels in the image is N = 4006002, which is planned to be split into K = 200 superpixels with the same size. Each superpixel has a size of $\frac{N}{K}$ and the distance between each cluster center is about $S = \sqrt{N/K}$. Compact coefficient m is used to measure the color similarity and spatial proximity. When m is large, the resulting superpixels are more compact, while when m is smaller, the superpixels have a higher degree of edge matching. Taking all factors into consideration, we set the coefficient to m = 20. In order to avoid the seed point falling on the gradient contour border and affecting the subsequent clustering effect, We re-select the cluster centers in the neighborhood of cluster centers $(n * n)$ and define the labels for the pixels in each field.

For each pixel, the color distance from each pixel to the center of the cluster is first calculated by Eq. (1).

$$d_c = \sqrt{(l_j - l_i)^2 + (a_j - a_i)^2 + (b_j - b_i)^2} \tag{1}$$

Secondly, calculate the spatial distance from the pixel to the center of the cluster using Eq. (2).

$$d_s = \sqrt{(x_j - x_i)^2 + (y_j - y_i)^2} \tag{2}$$

Lastly, the measurement distance D' is calculated by the Eq. (3).

$$D' = \sqrt{\left(\frac{d_c}{N_c}\right)^2 + \left(\frac{d_s}{N_s}\right)^2} \tag{3}$$

From the similarity between cluster centers, the labels of the most similar cluster centers are assigned to the pixels. The process is repeated until convergence, and we set the number of iterations to 10. After 10 iterations, we redistribute discontinuous and under-sized superpixels to neighboring superpixels, traverse all the pixels and give them the corresponding label, we get the split image.

As can be seen from the division result as a pre-treatment steps in this article, the image on the edge of goodness of fit, split speed, and performance division, in line with our expected results, as shown in Fig. 2.

Fig. 2. Original image and the image after SLIC segmentation

4 Training a Stacked Deconvolution Neural Network (SDN)

Convolutional Neural Network is a kind of neural network, which has become a research hotspot in the field of image segmentation. Its weight sharing network structure makes it more similar to biological neural network, reducing the complexity of the network model and reducing the number of weights [8]. Its multi-layer structure can automatically learn features, and can learn more than one level of features. Shallow convolutional layer perception area is small, you can learn some local features, deep convolution layer has a larger sensing area, can learn more abstract features. These abstractions are less sensitive to the size, position, and orientation of the object, helping to improve recognition performance. These abstract features are very helpful for classification, which can well determine what kind of objects are contained in an image. Convolutional networks drive the advancement of recognition, and for the whole image convolution not only improves classification but also advances local tasks with structured output [9]. However, because of missing some details of objects, the object can't be given a good outline, pointing out which pixel belongs to which object, so it is difficult to achieve accurate segmentation.

Compared with the Convolutional Neural Network (CNN), Fully Convolutional Neural networks (FCN) perform well in semantic segmentation, FCN transforms the fully connected layer in traditional CNN into convolutional layers [10]. It upsamples to get finer results, can accept input images of any size, without having to require that all training images and test images have the same size. Because it avoids the problem of duplicate storage and computational convolution due to the use of pixel blocks, and is more efficient. Nonetheless, the result of FCN upsampling is still vague and smooth, insensitive to the details in the image. At the same time, the relationship between pixels and pixels is not fully considered, and the spatial regularization steps used in the usual pixel-based segmentation methods are ignored, which lacks spatial consistency [11].

4.1 Overview

This paper proposed SDN, which is a deeper deconvolution but easier to optimize than most convolutions. The deconvolutional network is composed of deconvolution and unpooling layers, which identify pixel-wise class labels and predict segmentation masks. In SDN, there are several shallow deconvolution neural networks called SDN units, stacked to consolidate front and back information and ensure good recovery of local information. At the same time, inter-unit and intra-unit connections are designed to facilitate network training and enhance fusion of features that can improve information and gradient propagation throughout the network. In addition, layered monitoring is applied during the upsampling of each SDN unit, which guarantees the identification of the feature representation and facilitates network optimization (Fig. 3).

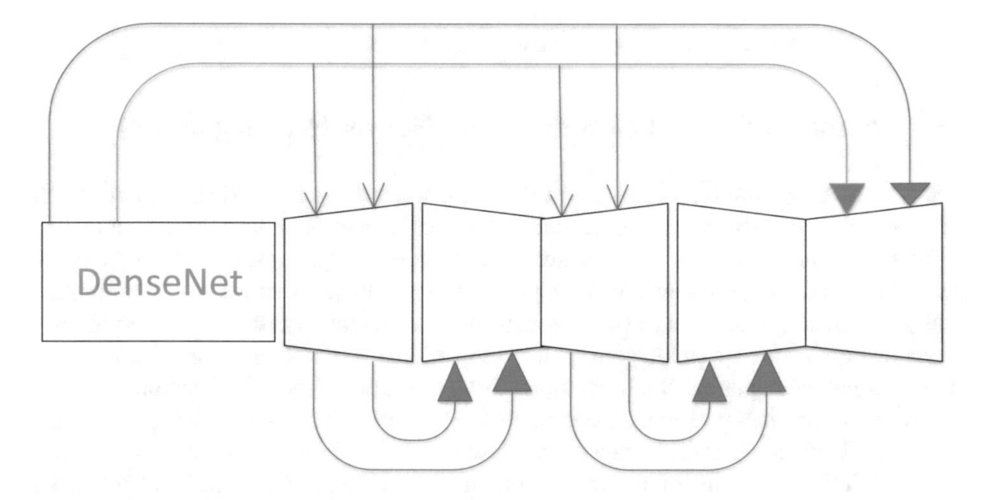

Fig. 3. The structure of stacked deconvolutional network

4.2 Deconvolutional Network

We design an efficient shallow deconvolution network called (SDN unit) and stacked multiple SDN unit in sequence, making the proposed SDN capture more contextual information and easily optimized, but as the number of stacked SDN unit increases, the difficulty of model training becomes a major issue. In order to solve this problem, first of all, we stratify oversight to be added to each SDN unit of the upsampled block. Specifically, the compression layer is mapped to pixel-labeled textures by the classification layer. With this structure, the web can be learned in more sophisticated ways, with no loss of ability to discriminate. Second, we import unit-to-unit connections to help optimize the network. Secondly, we import intra-unit and inter-unit connections to help the network optimization. The connections are shortcut paths from early layers to later layers, and they are beneficial to the flow of information and gradient propagation through out the network.

Taking into account the different intentions of information transfer, two types of connection between cells can be used. One is to promote network optimization by connecting decoders and encoder modules between any two adjacent SDN units. The other is to connect the multi-scale feature mapping of the encoder module of the first SDN unit to the corresponding decoder module of each SDN unit to maintain the detail of the low-resolution prediction.

For a given SDN unit, taking the output of the previous unit as input yields a low resolution feature map with a larger receptive field. We use downsampling twice to import $\frac{1}{16}$ spatial resolution of the image, as shown in Fig. 4.

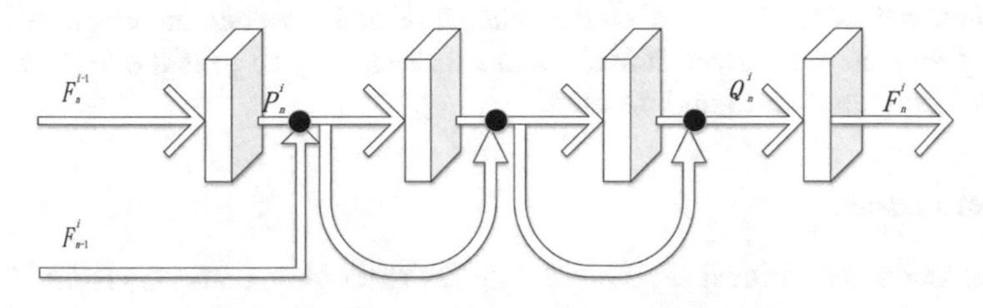

Fig. 4. The structure of downsampling block

The operations can be summarized by Eq. (4).

$$\begin{aligned} P_n^i &= \text{Max}\left(F_n^{i-1}\right), \\ Q_n^i &= \text{Trans}\left(\left[P_n^i, F_{n-1}^{i'}\right]\right), \\ F_n^i &= Comp\left(Q_n^i\right) \end{aligned} \tag{4}$$

where $\left[P_n^i, F_{n-1}^{i'}\right]$ stands for the connection of the feature maps P_n^i and $F_{n-1}^{i'}$. Max(.) means a max-pooling operation. Trans(.) means a transformation function of the densely connected structure. *Comp*(.) means a 3×3 convolutional operation.

In the decoder, we apply the up-sampling approach to up-sampling the feature map up to a much larger resolution. Use two magnifications to bring the resolution back to $\frac{1}{4}$ spatial resolution of the input image, as shown in Fig. 5.

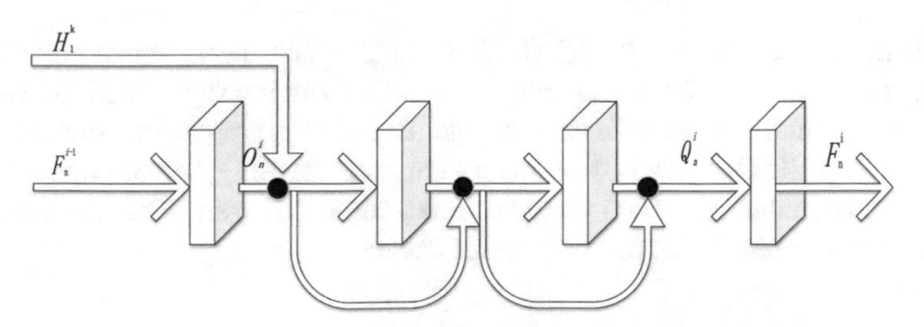

Fig. 5. The structure of upsampling block

The operations can be calculated by Eq. (5).

$$
\begin{aligned}
O_n^i &= Deconv\left(F_n^{i-1}\right), \\
Q_n^i &= Trans\left(\left[O_n^i, H_1^k\right]\right), \\
F_n^i &= Comp\left(Q_n^i\right).
\end{aligned} \tag{5}
$$

where *Deconv*(.) means the deconvolutional operation. Because of a quarter of the picture resolution, we can reduce GPU memory usage for a single SDN unit so that more units can be stacked.

Our deconvolutional network offers significant improvements over traditional deconvolutional networks. The simple encoding and decoder architecture and two simple upsampled and downsampled blocks make it easy to pass the SDN unit backwards and facilitate end-to-end training.

4.3 Work Details

We stack four convolutional layers at the lowest resolution, using the ReLU function, which computes a 3×3 convolutional operations with a probability of loss of 0.2 and the number of convolutional filters set to 36. The downsampling process, a 3×3 convolutional operations is performed, and the upsampling process also uses a 3×3 convolutional operations, and each SDN unit uses a ReLU operation. Finally, we train the Stacked Deconvolution Neural Network (SDN) on the ISPRS Potsdam and the NZAM/ONERA Christchurch datasets and learn the sample data with weak annotation information. We use the expanded bounding box to crop the window for training examples. The class labels for each crop area are provided based on only the centrally located objects, while all other pixels are marked as the background. In order to increase training data, we converted the input image to a 250×250 image, cropped the image to 224×224, and optionally flipped horizontally. Also, we provide sufficient training samples enough to train stack deconvolution neural networks from scratch, initial learning rates, momentum and weight decay of 0.01, 0.9 and 0.0005, respectively.

5 SDN for Image Semantic Segmentation

We evaluate our network on PASCAL VOC 2012 bench-mark, which contains 1456 test images and involves 20 object categories [12]. From the segmentation results, our method is one of the most advanced image semantic segmentation methods and is competitive. By SLIC superpixel segmentation preprocessing, semantic segmentation of the image becomes faster and more accurate, compared with FCN, the average loU is expected to increase 1.6%, reaching excellent accuracy.

6 Conclusion

In this paper, we propose a new method for image semantic segmentation. Firstly, the image is preprocessed by SLIC subpixel segmentation, and then the image is imported into the trained SDN network as an evaluation function. The stacking structure of SDN network and the connection between units promote the optimization of the network and make the network have better segmentation performance. We achieve fast and accurate segmentation of image semantics.

Acknowledgments. This research was supported by Program of International science and technology cooperation (2015DFA10940); Science and technology support program (R&D) project of Hubei Province (2015BAA115); PhD Research Startup Foundation of Hubei University of Technology (No. BSQD13037, No. BSQD14028); Open Foundation of Hubei Collaborative Innovation Center for High-Efficiency Utilization of Solar Energy (HBSKFZD2015005, HBSKFTD2016002).

References

1. Liu, Y., Liu, J., Li, Z., Tang, J., Lu, H.: Weakly-supervised dual clustering for image semantic segmentation. In: Proceedings of the IEEE Conference on Computer Vision and Pattern Recognition, pp. 2075–2082 (2013)
2. Achanta, R., Shaji, A., Smith, K., Lucchi, A., Fua, P., Süsstrunk, S.: SLIC superpixels compared to state-of-the-art superpixel methods. IEEE Trans. Pattern Anal. Mach. Intell. **34**(11), 2274–2282 (2012)
3. Fu, J., Liu, J., Wang, Y., Lu, H.: Stacked deconvolutional network for semantic segmentation. arXiv preprint arXiv:1708.04943 (2017)
4. Garcia-Garcia, A., Orts-Escolano, S., Oprea, S., Villena-Martinez, V., Garcia-Rodriguez, J.: A review on deep learning techniques applied to semantic segmentation. arXiv preprint arXiv: 1704.06857 (2017)
5. Shen, J., Du, Y., Wang, W., Li, X.: Lazy random walks for superpixel segmentation. IEEE Trans. Image Process. **23**(4), 1451–1462 (2014)
6. Liu, M.Y., Tuzel, O., Ramalingam, S., Chellappa, R.: Entropy rate superpixel segmentation. In: IEEE Conference on Computer Vision and Pattern Recognition (CVPR), pp. 2097–2104. IEEE (2011)
7. Ren, C.Y., Reid, I.: gSLIC: a real-time implementation of SLIC superpixel segmentation. Technical report, University of Oxford, Department of Engineering (2011)
8. Long, J., Shelhamer, E., Darrell, T.: Fully convolutional networks for semantic segmentation. In: Proceedings of the IEEE Conference on Computer Vision and Pattern Recognition, pp. 3431–3440 (2015)
9. Chen, L.C., Papandreou, G., Kokkinos, I., Murphy, K., Yuille, A.L.: Deeplab: semantic image segmentation with deep convolutional nets, atrous convolution, and fully connected CRFs. arXiv preprint arXiv:1606.00915 (2016)
10. Noh, H., Hong, S., Han, B.: Learning deconvolution network for semantic segmentation. In: Proceedings of the IEEE International Conference on Computer Vision, pp. 1520–1528 (2015)
11. Chen, L.C., Papandreou, G., Schroff, F., Adam, H.: Rethinking atrous convolution for semantic image segmentation. arXiv preprint arXiv:1706.05587 (2017)
12. Everingham, M., Van Gool, L., Williams, C.K., Winn, J., Zisserman, A.: The pascal visual object classes (VOC) challenge. Int. J. Comput. Vis. **88**(2), 303–338 (2010)

Research and Implementation of Indoor Positioning Algorithm for Personnel Based on Deep Learning

Hanhui Yue[1,2], Xiao Zheng[1,2], Juan Wang[1,2(✉)], Li Zhu[1,2], Chunyan Zeng[1,2], Cong Liu[1,2], and Meng Liu[1,2]

[1] Hubei Key Laboratory for High-Efficiency Utilization of Solar Energy and Operation Control of Energy Storage System, Hubei University of Technology, Wuhan 430068, People's Republic of China
happywj@hbut.edu.cn

[2] Hubei Collaborative Innovation Center for High-Efficiency Utilization of Solar Energy, Hubei University of Technology, Wuhan 430068, People's Republic of China

Abstract. A real-time indoor position algorithm based on deep learning theory for many complicated situations is proposed to satisfy the current demands for collection of position information efficiently. Firstly, the video images captured by the camera in real time are input into the network, ZCA (Zero-phase Component Analysis) whitening preprocessing is used to reduce the feature correlation and reduce the network training complexity. Secondly, deep network feature extractor is constructed based on convolution, pooling, multi-layer sparse auto-encoder. Then, the extracted features are classified by the Softmax regression model. Finally, the collected feature is accurately identified by the face recognition module. The algorithm is evaluated on the Indoor Multi-Camera data set, the experimental results are expected to improve the positioning accuracy greatly and implement indoor precise positioning.

Keywords: Deep learning · Deep convolution network · Softmax
Indoor precise positioning

1 Introduction

With the development of science and technology, people's demand for positioning service is increasing gradually. Especially in complex indoor environment, such as buildings, airport halls, mines and other environments, it is often necessary to determine the exact location of mobile workers indoors and to achieve the dispatching management of personnel, personal safety, emergency assistance [1].

The indoor positioning algorithm for personnel based on HOG + SVM has three stages: the body detection, the face detection and the face recognition [2]. However, the algorithm has the following problems: the performance of target anti-occlusion is poor [3]; the same person is marked as two targets which means the positioning is blocked; due to the influence of light and shadow around the target, the range of the target box given by the algorithm is much larger than the actual size of target and so on [4].

L. Barolli et al. (Eds.): EIDWT 2018, LNDECT 17, pp. 782–791, 2018.
https://doi.org/10.1007/978-3-319-75928-9_70

In this paper, a deep convolution network is used to construct an indoor real-time positioning algorithm which is suitable for a variety of complex conditions. Firstly, the ZCA (Zero-phase Component Analysis) whitening preprocessing for the video captured by the camera in real time is used to reduce the feature correlation and the network training complexity. Convolution and pooling are used to extract the features of the interested target, and the position of the object of interest is extracted. Then, the target of the human body is accurately boxed out according to the non-maximum suppression, the appearance features of the moving target are obtained by using the deep network feature extractor. The indoor location algorithm based on deep convolution network can effectively locate the target with the background occlusion, multi-target occlusion or the change of target appearance. This algorithm is quicker to adapt to the environment compared with the traditional method. The experimental scheme is carried out, the traditional method and algorithm based on convolution neural network are compared. Some useful conclusions are drawn.

2 Indoor Positioning Algorithm for Personnel Based on HOG + SVM

The flow chart of Indoor Positioning Algorithm for Personnel based on HOG + SVM is shown in Fig. 1 [5]. As a traditional indoor positioning method, human can be detected by this algorithm, but it is easily influenced by background interference. The test results show that there is erroneous detection, missed detection and so on. Indoor Positioning Algorithms for Personnel based on Deep CNN have great potential to solve these problems.

The videos are used as input for the first stage of human detection. A descriptor of Histogram of Oriented Gradient (HOG) is created in the next step, the feature descriptor is used to detect objects with the help of Support Vector Machine (SVM). HOG and SVM are combined to detect the human on the basis of the OpenCV library. A rectangle is drawn around the human body when humans are detected. In some cases, multiple rectangles for a single person is drawn due to false positives. Non-maximum suppression is used to select the rectangle based on the maximum overlap criterion to minimize the case of false positive.

Face detection is the second stage. The Haar feature with the face features is loaded into memory to detect the face. Then, the detected face is displayed in a separate window. Face detection is independent of body detection. But faces are also detected when no human is detected, it is the result that only part of the human body is visible instead of a complete body in some videos [6].

Face recognition is the last stage. There is a precondition for this stage, the machines must be trained to get some images of that person who need to be recognized. Haar Cascades and Local Binary Pattern Histograms (LBPH) is used for training.

The Haar feature is used to detect human faces in a video frame. Each human face detected is used to train machine as a dataset and the LBPH features corresponding to each human face are saved in a file. Face recognition is finished by importing the features into memory which is saved in a file. The module of face recognition attempts to recognize faces with the help of LBPH features. A label that identifies the face is

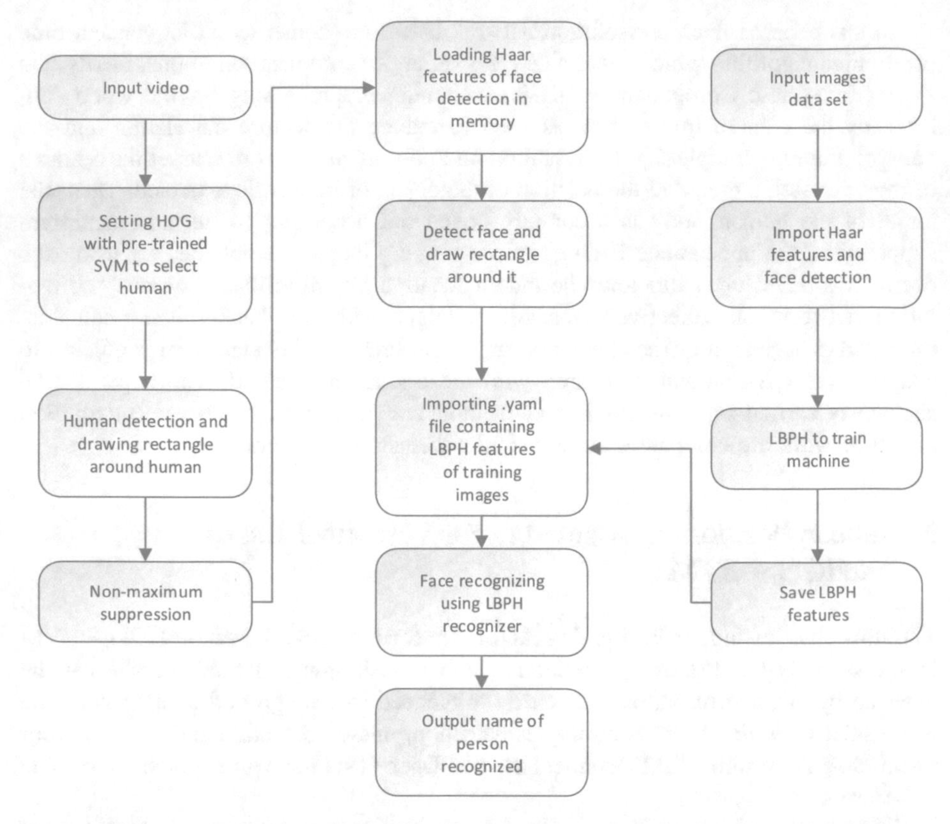

Fig. 1. The data flow frame of Indoor Positioning Algorithm for Personnel based on HOG + SVM

given while the face is identified. So the label is assigned to every person who is approved. The names or labels of personnel who are admitted are given as output along with confidence factors. The confidence factor is a zero-valued parameter when the image in the training set is same with that in the test set. The given image is similar to the human if the value of confidence factor is low enough. The threshold of confidence is set, the given image is not similar to the person who is marked if the threshold is exceed [7].

3 Indoor Positioning Algorithm for Personnel Based on Deep CNN

The main concern of personnel positioning with indoor monitoring is to accurately find the target personnel, thus the technology of human detection and identification has become the key for the effective positioning for personnel.

Although personnel positioning has obtained good results in some public places, such as the body's profile information is made full use in the algorithm based on HOG.

But the occlusion and other factors is still Large interference for the human profile information, thereby the detection performance of this method is reduced [8]. The discriminating features of human detection can automatically be learned and those with partial occlusion in a non-crowded scenario can be detected more than the artificially designed features by the technologies of deep learning, such as Convolution Neural Networks (CNNs). However, in the case of the crowded scene of the actual monitoring video, the performance of the personnel positioning methods based on CNNs is still an urgent need to be improved while serious personnel cover occurs (such as the 2/3 of the body are blocked) [9]. In view of this problem, this paper proposes a personnel positioning algorithm based on deep convolution networks, and gives its structure in Fig. 2. The frame is characterized by:

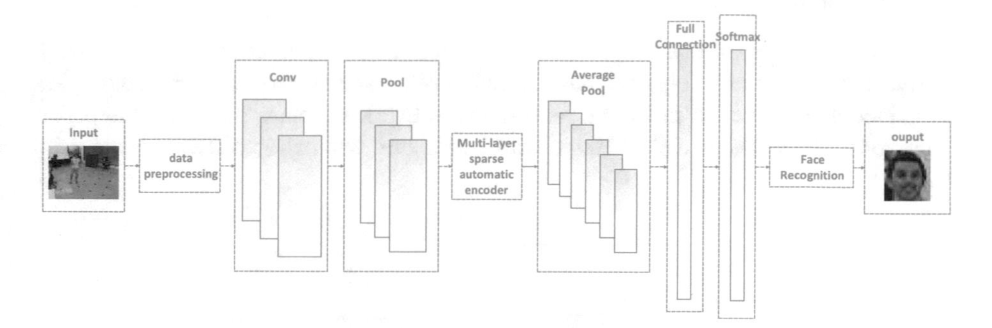

Fig. 2. The frame of Indoor Positioning Algorithm for Personnel proposed in this paper

(1) ZCA whitening and normalization is introduced, the original video image are pre-processed.
(2) The multi-layer sparse automatic encoder is used to construct the deep network feature extractor to extract the deep features of video images.
(3) The Softmax regression model is introduced to classify the features obtained by the deep network feature extractor.
(4) Face recognition module is used to identify facial features.

Specifically, the framework of the proposed algorithm can be divided into 4 parts: data preprocessing, neural network architecture for building deep network feature extractor, Softmax regression classification and face recognition.

3.1 Data Preprocessing

In this paper, based on a deep network structure built by a multi-layer sparse auto-encoder, we introduced ZCA whitening and normalization to preprocess the experimental data. Normalization uses standardization of scale, normalization of grayscale and histogram equalization to reduce the computational complexity of the whole network, while the useless information based on the original information is reduced [10]. First of all, ZCA whitening is shown in Eq. (1), the correlation between the various features is removed which is transformed by PCA; then Eq. (2) is used to output characteristics with unit variance; and the data is rotated which is back to get ZCA whitening results, As shown in Eq. (3), the input redundancy is reduced.

$$x_{rot,i} = U^T x_i \tag{1}$$

$$x_{PCAWhite,i} = \frac{x_{rot,i}}{\sqrt{\lambda_i}} \tag{2}$$

$$x_{ZCAWhite} = Ux_{PCAWhite} \tag{3}$$

3.2 Deep Convolution Neural Network with Multi-layer Sparse Automatic Encoder

Based on multi-layer sparse automatic encoder deep convolution neural network architecture is shown in Fig. 3. Convolution, pooling and sparse automatic coding are combined and improved to form a multi-layer network structure model in this paper. Therefore, this model is called deep network structure, the extracted features are deep features.

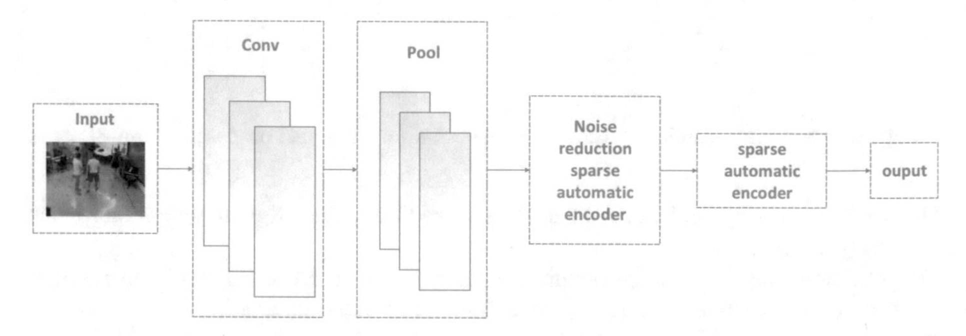

Fig. 3. The frame of deep feature extractor

Deep network is a multi-layer neural network structure, it has excellent feature learning ability. The data is described by learned features more fundamentally which helps to classify. As shown in Fig. 3, convolution layer, pooling layer, noise reduction sparse auto-coding layer, sparse auto-coding layer and output layer are included in the deep network.

Automatic encoder is one of the hot algorithms for deep learning. The idea is to capture the main components that can represent the input information, so it can reproduce the input as much as possible. Noise reduction sparse auto-encoder is the first layer of the automatic encoder in this paper, noise is added to the training data artificially. The network learn to remove the noise to obtain input which is not contaminated by noise, so it has a more generalization Ability to learn more robust representation of input signals [11]. The second Layer of auto-encoder uses a general sparse auto-encoder which noise does not need to be added. The specific implementation process is as follows, it includes pre-training and fine-tuning.

The steps of pre-training are as follows:

(1) Unlabeled data is given, the methods of unsupervised learning is used to learn characteristics.

As shown in Fig. 4(a), samples are input into the single-layer noise reduction sparse auto-encoder after preprocessing, convolution and pooling. The code h can be obtained after unsupervised learning of X. Then the input is rebuilt by the decoder, the reconstruction error is calculated between the input and the reconstruction results. The cost function $L(X;W)$ is constructed.

$$h = W^T X \tag{4}$$

$$L(X;W) = \|Wh - X\|^2 + \lambda \sum_j |h_j| \tag{5}$$

The parameters of the encoder and decoder are adjusted by using the L-BFGS algorithm, the cost function $L(X;W)$ is minimized. The parameters of first layer of network are saved at this time.

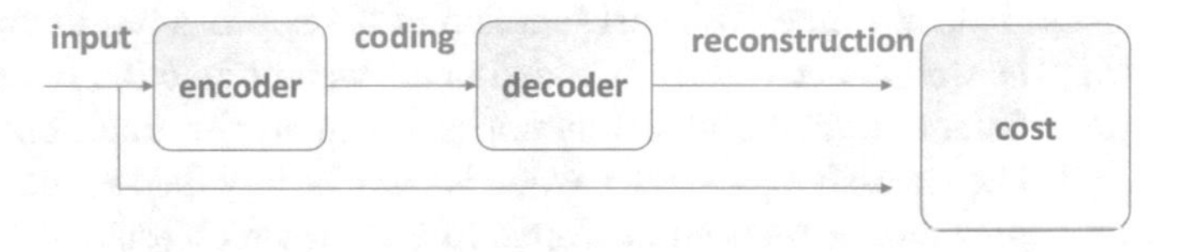

Fig. 4(a). The training frame of single-layer noise reduction sparse automatic encoder

(2) The next layer is trained for further training layer by layer.

The second layer of network is trained after that training of the first layer of network is completed. As shown in Fig. 4(b), the code 1 which is output by the first layer of network is used as the input signal of the second layer. The first step is repeated to obtain the code 2 of the second layer of the original information of input, the parameters of the second layer of network are saved.

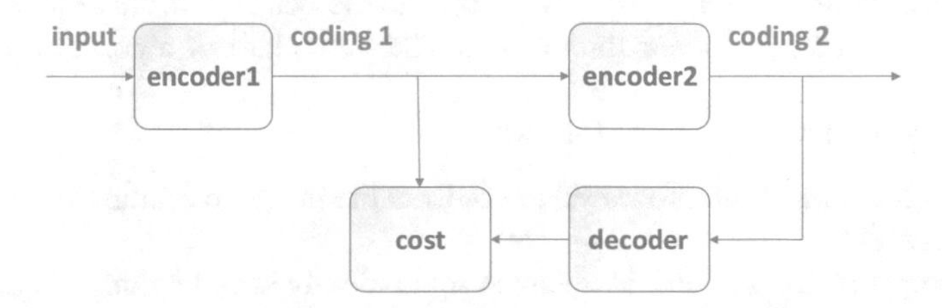

Fig. 4(b). The training frame of double layer noise reduction sparse automatic encoder

The training set in the selected standard library is used to fine-tune. As shown in Fig. 4(c), the code 2 is input into the Softmax regression model and trained by the method of supervised training of the standard multi-layer neural network. This paper uses a learning method of end-to-end to fine tune the entire system by the labeled sample. The network can be used to extract deep features after completing the above steps.

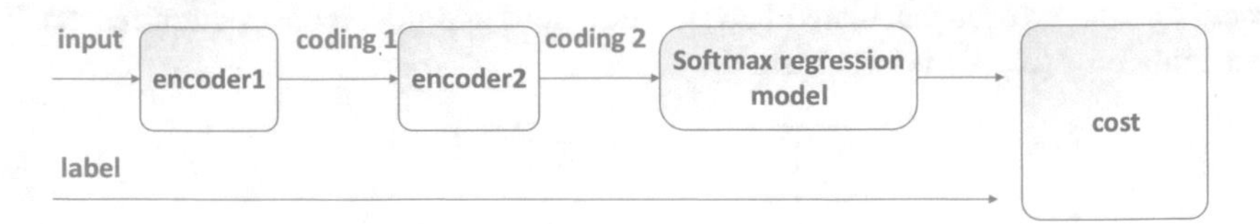

Fig. 4(c). The training frame of supervised fine-tuning

3.3 Softmax Regression Classification

The Softmax regression model is learned by supervision. It requires that the classification categories are strictly exclusive each other, that is, categories can not be occupied by one sample at a time. The cost function of the Softmax regression model is given in Eq. (6). The weight decay term is added in the weight function selected of this paper which is different from the cost function selected in the traditional Softmax regression model. The excessive parameter value is punished by the second addition in Eq. (6). The cost function becomes strict Convex function which guarantee to get the only solution by that addition, so as to avoid falling into the situation of partial optimal solution while the global optimal solution is obtained [12].

$$J(\theta) = -\frac{1}{m}[\sum_{i=1}^{m}\sum_{j=1}^{k} 1\left\{y^{(i)} = j\right\} \log \frac{e^{\theta_j^T x^{(i)}}}{\sum_{l=1}^{k} e^{\theta_l^T x^{(i)}}}] + \frac{\lambda}{2}\sum_{i=1}^{k}\sum_{j=0}^{n} \theta_{ij}^2 \tag{6}$$

3.4 Face Recognition

The face recognition module is used to recognize the detected facial features based on the face detection under indoor monitoring. The flow chart of algorithm is shown in Fig. 5.

The steps of algorithm are as follows:

(1) The target image is preprocessed to conform the image to training requirements of the network;
(2) The proper amount of data is obtained by randomly sampling the image in (1), the weights of CNN initialization filters are obtained by unsupervised pre-training with sparse self-encoder;

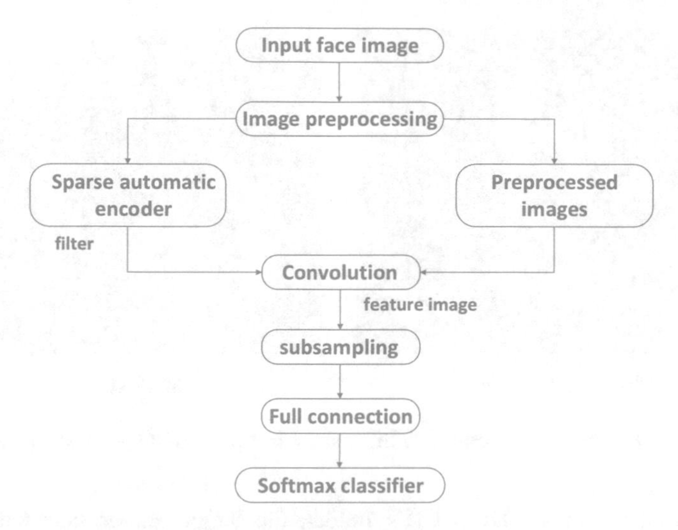

Fig. 5. The frame of face recognition algorithm

(3) A number of feature maps which is predetermined are obtained by convolution between the filter obtained in (2) and the image of training set in (1);
(4) The generalized image is obtained by maximizing the signature obtained in (3);
(5) The required feature map is obtained by quadratic convolution, subsampling for the feature map which is output in (4);
(6) All feature maps in (5) are transformed into a single column vector, which is used as input to the full connection layer to calculate the difference between the recognition result and the marker. The network parameters are adjusted and updated from the top to the bottom by the back propagation algorithm;
(7) The test image is classified by using the parameters of trained set of network and the weight parameters of fully connected network when the image test set is input, and the recognition result of the image is obtained through the Softmax classifier.

4 Experiment and Analysis

This experiment using the Ubuntu16.04 system, TensorFlow platform, using Indoor Multi-Camera Datasets of Institute of Computer Graphics and Vision (training database of human detection: 9137 images, testing database of human detection: 3916 images) to different model in the experimental verification the same data set.

(1) The experimental results of HOG + SVM algorithm and Deep CNN algorithm are shown in Fig. 6.
(2) In the results based on the HOG + SVM model, the targets can be detected more accurately without occlusion, however, the object was also detected as a human by mistake; the two human is detected as the same target erroneously in the presence of background occlusion or multi-target occlusion. In the estimated test

(a) HOG + SVM (No occlusion)

(b) HOG + SVM (occlusion)

Fig. 6. Part of the results on the test set based on HOG + SVM model

results based on the Deep CNN model, the target can be detected accurately without occlusion, targets can be accurately detected when the occlusion exists, detection speed can be faster than algorithm based on the HOG + SVM model, and the size of the positioning box is more accurate.

5 Conclusion

This paper presents a new deep convolution network structure based on multi-layer sparse auto-encoder indoor positioning for personnel algorithm network. Its main idea lies in: Proposed a new network structure which is tested under the Indoor Multi-Camera Datasets of Institute of Computer Graphics and Vision, is expected to have better accuracy of indoor positioning for personnel and higher positioning efficiency than the method of HOG + SVM. First of all, we introduce ZCA whitening and normalization algorithm to preprocess the experimental data. Then, this paper puts forward two kinds of structural design on the basis of the framework:

(1) ZCA whitening and normalization are used to pre-process raw video image data,
(2) The deep network structure is introduced and the multi-layer sparse automatic encoder is proposed.

Acknowledgments. This research was supported by Program of International science and technology cooperation (2015DFA10940); Science and technology support program (R & D) project of Hubei Province (2015BAA115); PhD Research Startup Foundation of Hubei University of Technology (No. BSQD13037, No. BSQD14028); Open Foundation of Hubei Collaborative Innovation Center for High-Efficiency Utilization of Solar Energy (HBSKFZD2015005, HBSKFTD2016002); Open Research Fund Project of High-Efficiency Utilization of Solar Energy and Energy Storage Operation Control Key Laboratory of Hubei Province (HBSEES201701).

References

1. Coleman, D.J., Rajabifard, A., Kolodziej, K.W.: Expanding the SDI environment: comparing current spatial data infrastructure with emerging indoor location-based services. Int. J. Digit. Earth **9**(6), 629–647 (2016)
2. Llorca, D.F., Arroyo, R., Sotelo, M.A.: Vehicle logo recognition in traffic images using HOG features and SVM. In: 2013 16th International IEEE Conference on Intelligent Transportation Systems-(ITSC), pp. 2229–2234 (2013)
3. Li, K., Liu, Y., Li, N.: Variable scale and anti-occlusion object tracking method with multiple feature integration. In: 2016 7th IEEE International Conference on Software Engineering and Service Science (ICSESS), pp. 426–433 (2016)
4. Nguyen, D.T., Li, W., Ogunbona, P.O.: Human detection from images and videos: a survey. Pattern Recogn. **51**, 148–175 (2016)
5. He, Z., Xu, M., Guo, A.: Multi-channel feature for pedestrian detection. In: Advanced Computational Methods in Life System Modeling and Simulation, pp. 472–480. Springer (2017)
6. Aguilar, W.G., Luna, M.A., Moya, J.F., Abad, V., Ruiz, H., Parra, H., Lopez, W.: Cascade classifiers and saliency maps based people detection. In: International Conference on Augmented Reality, Virtual Reality and Computer Graphics, pp. 501–510. Springer (2017)
7. Xu, C., Liu, Q., Ye, M.: Age invariant face recognition and retrieval by coupled auto-encoder networks. Neurocomputing **222**, 62–71 (2017)
8. Kim, Y., Moon, T.: Human detection and activity classification based on micro-Doppler signatures using deep convolutional neural networks. IEEE Geosci. Remote Sens. Lett. **13**(1), 8–12 (2016)
9. Redmon, J., Divvala, S., Girshick, R., Farhadi, A.: You only look once: unified, real-time object detection. In: Proceedings of the IEEE Conference on Computer Vision and Pattern Recognition, pp. 779–788 (2016)
10. Yang, D., Lai, J., Mei, L.: Deep representations based on sparse auto-encoder networks for face spoofing detection. In: Chinese Conference on Biometric Recognition, pp. 620–627. Springer (2016)
11. Jia, B., Feng, W., Zhu, M.: Obstacle detection in single images with deep neural networks. SIViP **10**(6), 1033–1040 (2016)
12. Zagoruyko, S., Lerer, A., Lin, T.Y., Pinheiro, P.O., Gross, S., Chintala, S., Dollár, P.: A multipath network for object detection. arXiv preprint arXiv:1604.02135 (2016)

Verifiable Outsourced Attribute Based Encryption Scheme with Fixed Key Size

Cong Li[1,2], Xu An Wang[1,2(✉)], Arun Kumar Sangaiah[3], Haibing Yang[1], and Chun Shan[4]

[1] Key Laboratory for Network and Information Security, Engineering University of Chinese Armed Police Force, Xi'an 710086, Shaanxi, China
wangxazjd@163.com

[2] Department of Electronic Technology, Engineering University of the Chinese Armed Police Force, Xi'an 710086, Shaanxi, China

[3] School of Computer Science and Engineering, VIT University, Vellore 632014, Tamil Nadu, India

[4] Guangdong Polytechnic Normal University, Guangdong People's Republic of China

Abstract. The limited storage capacity of small devices, such as mobile phone, has become a bottleneck for the development of many application, especially for security applications. The Ciphertext Policy Attributes Based-Encryption (CP-ABE) is a promising cryptographic scheme that allows encryption to choose an access structure that protects sensitive data. However, one of the problems with current CP-ABE scheme is the length of the key, whose size increases linearly with the number of attributes. In this paper, we propose a CP-ABE scheme for a constant size key. By modifying the modulus index in the key generation algorithm, the computational cost is reduced to a constant. Compared with other schemes for the literature, the private key is independent of the number of attributes.

1 Introduction

In 2005, Sahai and Waters [1] firstly proposed the concept of ABE from fuzzy identity based encryption (FIBE). The ABE schemes fall into two categories based on access control mechanisms: Key Policy Attributes Based Encryption (KP-ABE) and Ciphertext Policy Attributes Based Encryption (CP-ABE). In KP-ABE [2], the ciphertext is related to the attribute set, and the decryption key is related to the access policy. In CP-ABE [3] contrary to KP-ABE, the ciphertext is related to the access structure and the decryption key is related to the attribute set.

The ABE scheme effectively implements fine-grained sharing with public keys [4], but the practicality is still limited by the high complexity of encryption and decryption. The ABE algorithm requires an exponential operation of O(n), where n denotes the number of attributes in the access structure. In order to reduce the computational load of users, the concept of outsourcing is introduced into attribute based encryption. The core idea is to outsource a large number of complex operations to the cloud server. In 2011, Green et al. [5]. firstly proposed the outsourced decryption scheme of ABE,

L. Barolli et al. (Eds.): EIDWT 2018, LNDECT 17, pp. 792–800, 2018.
https://doi.org/10.1007/978-3-319-75928-9_71

which transforms its CP-ABE ciphertext into a constant-size ELGamal ciphertext by outsourcing transformation, which is proved as replayable chosen ciphertext attack (RCCA) security. In 2012, Zhou and Huang [6] proposed a privacy protection CP-ABE scheme. In 2012, Li et al. [7]. proposed an effective outsourced ABE scheme with great improvements in safety and performance. In 2013, Lai et al. [8] proposed a verifiable outsourced decryption scheme to improve the security of the Green outsourced decryption scheme, and for the first time proposed the verifiability of outsourced decryption. In 2014, Li et al. [9] ensured the correctness of decryption by adding multiple Key Generating Service Provider (KGSP). As long as one KGSP is honest, it can effectively verify the decryption result. In 2015, Qin et al. [10]. proposed to use symmetric encryption scheme to encrypt plaintext and asymmetric encryption scheme to encrypt symmetric key. At the same time, two hash functions are used to verify the output of outsourced decryption, which improves the applicability and safety of the system. In 2016, Wang et al. [11]. proposed a verifiable outsourced encryption scheme based on CP-ABE. The exponential operation in the encryption algorithm is reduced to a fixed constant, but the outsourced method increased ciphertext length. In 2017, Li et al. [12]. proposed a fuzzy encryption algorithm that verifies the outsourced attributes and implements cryptographic exponents into fixed constants and full verification. However, the outsourced key generation time increases from the increase in attributes.

Our Contributions
In order to solve the problem that the decryption key of CP-ABE scheme is too large. In this paper we divide the keys into outsourced decryption transformation key (DTK) and outsourced encryption transformation key (ETK). we proposed an efficient algorithm for outsourced key generation, in our new algorithm outsourced decryption transformation key only needs to conduct a modulus exponentiation operation, without increasing the size of the ABE ciphertext. Hiding the exponentiation of encryption operation in the ETK reduces the computational complexity of the encryption operation and index to a constant. Hiding indexing of encryption operations in ETK reduces the computational complexity of encryption operations to a constant. At the same time ensuring the authenticity of the plaintext and the correctness of outsourced results.

2 Paper Preparation

2.1 Bilinear Maps

Theorems 1. Let G and G_T be two groups of primes p. Suppose g is a generator of group G, the bilinear maps $e\ :\ G \times G \rightarrow G_T$. Bilinear mapping satisfies the following properties [13]:

(1) Bilinearity: $e(g_1^a,\ g_2^b) = e(g_1,\ g_2)^{ab}$, $g_1, g_2 \in G$ and $a, b \in Z_p$;
(2) Non-degeneracy: If $g_1, g_2 \neq 1_G$, then $e(g_1, g_2) \neq 1_{G_T}$;
(3) Computability: For $\forall : g_1, g_2 \in G$, there is a valid algorithm for calculating $e(g_1, g_2)$.

2.2 Linear Secret-Sharing Schemes(LSSS)

Definition 1. Let $\mathcal{P} = \{P_1, P_2, \ldots, P_n\}$ be a group of participants in a secret sharing scheme, the access structure, defined as $\mathcal{T}$ [14], is a subset of the authorized parties that can recover the shared secret.
A linear secret sharing scheme [15] includes an access structure, a monotone boolean function ϕ and a set of participants. If and only if $A \in \mathcal{T}$, then $\phi(A) = 1$. Definition **M** is a matrix of $\ell \times n$, the ith row of the matrix $\mathbf{M_i}$ is related to the participant P_i, the access structure $\mathcal{T}$ is defined as $\mathcal{T}(\mathbf{M}, \phi)$. Supposing that s is the secret value to be shared, selecting a vector $\boldsymbol{v} = (s, v_2, \ldots, v_n) \in Z_p^n$,and calculating $\{\lambda_i\}_{i \in I} = \boldsymbol{v}\mathbf{M}$, there exists $\{\omega_i \in Z_p\}_{i \in I}$, and $\sum_{i \in I} \omega_i \mathbf{M}_i = (1, 0, \ldots, 0)$, then the parties of A can recover the secret by $\sum_{i \in I} \omega_i \lambda_i = s$.

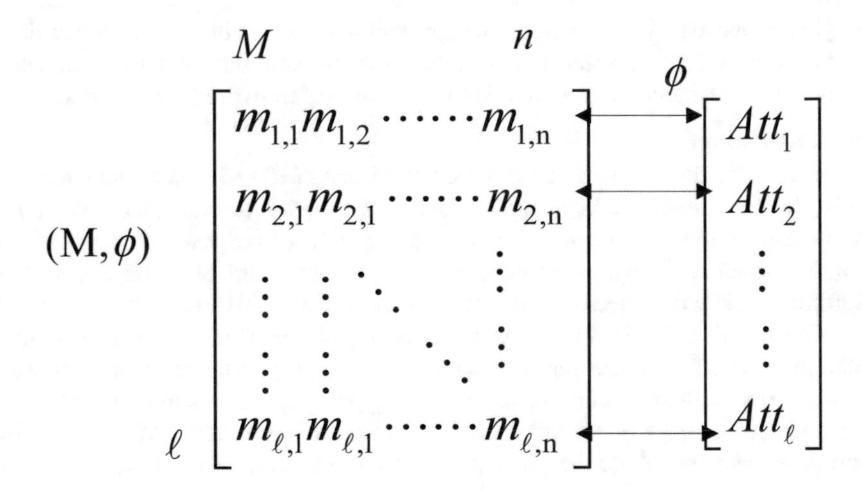

Fig. 1. LSSS access matrix

3 The Scheme Model and Security Model

3.1 Outsourced ABE Model

The proposed scheme enhances the efficiency of outsourced transformation key generation algorithm. The outsourced decryption of ABE is composed of the following six algorithms:

1. Setup: The authorization center inputs security parameters λ and attributes set U, and it outputs the public parameters MPK and the master private key MSK.
2. KeyGen: Authorization center inputs the master private key MSK and a group of attributes $S \subseteq U$, returns private key Key.

3. KeyGen$_{DTK}$: The decryption transformation key generation algorithm inputs the public key MPK, the key Key and the master private key MSK. It outputs decryption transformation key DTK.
4. KeyGen$_{ETK}$: It runs outsourced encryption key generation algorithm, the user inputs a set of attributes set S, returns the outsourced encryption transformation key ETK.
5. Encrypt: The encryption algorithm consists of three phases:
   (1) Partial encryption phase: The user inputs the public key MPK, the plaintext message m and the access structure $(\mathbf{M}, \phi)$, and outputs a partial ciphertext decryption C_1.
   (2) Outsourcing parameters generation phase: The user inputs the master public key MPK, the master private key MSK and the outsourced encryption transformation key ETK, and calculates the outsourcing parameter OP.
   (3) Outsourcing encryption phase: The cloud service provider inputs the outsourcing parameter OP and outputs the outsourced ciphertext C_2.

   Finally, it outputs ciphertext $C = (C_1,\ C_2)$.

6. Decrypt: The decryption algorithm is composed of two sub-algorithms •Transform and •Dec.
   •Transform: The cloud service provider inputs the decryption transformation key DTK and the ciphertext C. When the attributes set satisfies the access structure, it outputs the transformation ciphertext C_T.
   •Dec: The user enters the transformation ciphertext C_T and the decryption key SK if $f(MPK, \mathrm{C}_1, m) = 1$ then calculates the plaintext m, otherwise it outputs $\perp$.

3.2 Security Model

Definition: Replayable Chosen Ciphertext Attack (RCCA) security model, the RCCA model includes the adversary $\mathcal{A}$ and the challenger simulator $\mathcal{B}$, allowing the adversary access to challenge outsourcing parameters OP during the challenge phase.

Init: The adversary $\mathcal{A}$ selects a challenge access structure $(\mathbf{M}^*,\ \phi^*)$, $\mathbf{M}^*$ is the matrix of $\ell^* \times n^*$, and ϕ^* maps each row in matrix $\mathbf{M}^*$ to a attribute.

Setup: The challenger simulator $\mathcal{B}$ runs initialization algorithm, the public parameters MPK passed to the adversary $\mathcal{A}$.

*Phase*1: The simulator creates an empty table T and D. The adversary $\mathcal{A}$ repeats the following query:

-*Create*(S): The simulator inputs the set of properties S, runs a key generation algorithm, and generates (SK, DTK). Then lists (j, S, SK, DTK) in table T and sends DTK to the adversary.

-*Corrupt*(i): If the ith entity is in table T, the simulator adds the attribute set S to table D and returns the key SK to the adversary. Otherwise, outputs $\perp$.

-*Decrypt*(i, CT): If the ith entity is in table T, the simulator decrypts it and returns the corresponding plaintext to the challenger. Otherwise, it outputs $\perp$.

Challenge: The adversary $\mathcal{A}$ selects two equal-length plaintexts m_0 and m_1 for the challenger, while the adversary $\mathcal{A}$ selects one of the access structures $(\mathbf{M}^*, \phi^*)$ and $S \notin (\mathbf{M}^*, \phi^*)$. The challenger $\mathcal{B}$ randomly chooses a bit $b \in \{0, 1\}$, sends the challenge ciphertext CT^* to the adversary $\mathcal{A}$. The adversary can get outsourced parameters that challenge ciphertext OP^*.

*Phase*2: A similar inquiry as in phase 1 is performed when the following conditions are met.

(1) If $S \notin (\mathbf{M}^*, \phi^*)$, then terminate the inquiry.
(2) The ciphertext in the decryption query includes m_0 and m_1.
Guess : The simulator outputs guess b' of b, the adversary's advantage is defined as:$\mathrm{Adv}_{\mathcal{A}}^{\mathrm{RCCA}} = |\Pr[\mathrm{b}' - \mathrm{b}] - 1/2|$.

4 Verifiable Outsourcing Data Encryption Scheme with Fixed Key Size

In this paper, we proposed a new CP-ABE encryption scheme, which effectively shortens the size of outsourced decryption keys and outsourced a large number of modular exponentiations to the cloud server.

1. Setup: The initialization algorithm inputs a large number of attributes $U = \{1, 2, \cdots, |U|\}$ and security parameters $\lambda \in N$. Selecting a bilinear group G, with prime p and generator g. Selecting a bilinear mapping $e : G \times G \to G_T$,and a hash function $H : G^3 \to Z_p$. Randomly choosing α, β as a key, and randomly selecting the attribute-related $h_1, \ldots, h_{|U|}$ as the public key. The master private key is $MSK = (\alpha, \beta)$, the master public key is as follows:

$$MPK = \left(G, G_T, g, e(g, g)^\alpha, g^\beta, H, h_1, \ldots, h_{|U|}\right)$$

2. KeyGen: The key generation algorithm inputs the master private key MSK, the master public key MPK and a set of attributes $S \subseteq U$, and selects the random number $t \in Z_{p^*}$. Then it outputs the private key $Key : K = g^\alpha g^{\beta t}, K_0 = g^t, K_i = h_i^{-t}, i \in S$.
3. KeyGen$_{DTK}$: The decryption transformation key generation algorithm inputs the master private key MSK, the master public key MPK, the private key Key and a set of attributes $S \subseteq U$. And selects the random number $z \in Z_{p^*}$. Then it outputs the the decryption transformation key $DTK = \left(K' = K \cdot g^z,\ K'_0 = K_0,\ \{K'_i\} = \{K_i\}\right)$ and the private key $SK = z$.
4. KeyGen$_{ETK}$: For the encryption transformation key generation algorithm, the user inputs attribute sets S, for the user X, $x \in Z_P$ is selected at random, and outputs the user's outsourced encryption transformation key $ETK = \left(x, h_1^x, h_2^x, \ldots, h_{|S|}^x\right)$.

   Note that, We open DTK. Keep ETK confidential.

5. Encrypt: The encryption algorithm includes three sub-algorithms:
   (1) Partial encryption algorithm: The user inputs the plaintext message m and the master public key MPK. Meanwhile, Choosing an access structure $(\mathbf{M}, \phi)$, the function ϕ maps each row in the matrix $\mathbf{M}$ to the attributes. In Fig. 1, the matrix

$\mathbf{M}$ is the access matrix of $l \times n$. Then, the algorithm randomly chooses elements $s, d \in Z_p$, and a vector $\mathbf{v} = (s, v_2, \ldots, v_n) \in Z_p^n$. For $I \subset \{1, 2, \ldots, \ell\}$, then it calculates $\{\lambda_i\}_{i \in I} = (\mathbf{Mv})_i$ for ℓ as shares of s. Then it calculates the partial ciphertext:

$$C_1 = m \cdot e(g, g)^{\alpha s},\ C_2 = g^s,\ C_3 = g^{sr}$$

and the blinding factors $\left\{g^{\beta d}, h^x_{\phi(i)}\right\}$, $\phi(i) \in S$, where $r = H(C_1, C_2, m)$.

(2) Outsourcing parameters generation algorithm: the user inputs sr, x, g^β, λ_i, $h_{\phi(i)}$, then calculates $sr - x$, $\lambda_i - d$, $i = 1, \ldots, \ell$. And it outputs the outsourcing parameters $OP = (sr - x, \lambda_i - d, g^{\beta d} \cdot h^x_{\phi(i)})$ to the cloud server.

(3) Outsourcing encryption algorithm: The cloud service provider inputs the outsourcing parameter OP and outputs the outsourced ciphertext $\{C_i = (g^\beta)^{\lambda_i - d} \cdot g^{\beta d} \cdot h^x_{\phi(i)} \cdot h^{sr-x}_{\phi(i)} = g^{\beta\lambda_i} \cdot h^{sr}_{\phi(i)}\}$.

Finally, the ciphertext is outputed $C = (C_1,\ C_2,\ C_3,\ \{C_i\})$. $C_1 = m \cdot e(g, g)^{\alpha s}$, $C_2 = g^s$, $C_3 = g^{sr}$, $C_i = g^{\beta\lambda_i} \cdot h^{sr}_{\phi(i)}$, $i = 1, \ldots, \ell$.

6. Decrypt: The decryption algorithm is composed of two sub-algorithms •Transform and •Dec.

•Transform : The cloud service provider inputs the decryption transformation key DTK and ciphertext C. If the attribute set satisfies the access structure, let $I \subseteq \{1, 2, \ldots, \ell\}$, $I = (i : \phi(i) \in S)$, there is a constant $\{\omega_i\}_{i \in I}$ such that $\sum_{i \in I} \omega_i \lambda_i = s$.

Then it calculates the partial decryption of ciphertext:

$$U = \frac{e(K', C_2)}{\prod_{i \in I} \left(e(C_i, K'_0) e(C_3, K'_{\phi'(i)})\right)^{\omega_i}}$$

$$Q = e(C_2, g)$$

Finally, it outputs the partial decryption of ciphertext: $CT_{out} = (Q, U)$.

•Dec: The user inputs the partial decryption of ciphertext CT_{out} and the decryption key SK, and then calculates $m = C_1 \cdot Q^z / U$. If $C_3 = C_2^{H(C_1, C_2, m)}$ we accept the plaintext. Otherwise, it outputs $\perp$.

5 Analysis of Our Scheme

5.1 Security Analysis

Theorem 2. This scheme is built on the basis of scheme [12], and its security is also RCCA security.

Proof: The difference in the RCCA security game is that the outsourced parameter OP^* of challenges message in the challenge phase is also passed to the adversary.

The challenge ciphertext $C^* = (C_1^*, C_2^*, C_3^*, \{C_i^*\})$, the simulator randomly selects $r_1, r_2 \in Z_p$ and sets the outsourcing parameter as $OP^* = (r_1, r_2, g^{-r_1} C_i^* h^{-r_2})$. It is valid when the outsourced parameter satisfies $g^{OP_1^*} \cdot OP_{3,i}^* \cdot h^{OP_2^*} = C_i^*$.
Although the outsourced encryption algorithm is introduced, the outsourcing parameters $OP = (sr - x, \lambda_i - d, g^{\beta d} \cdot h_{\phi(i)}^x)$ do not reveal any information to the cloud server in this scheme. Next we will discuss the security of encryption transformation key x, $\{h_{\phi(i)}^x\}$, the security of encryption operations sr, $\{\lambda_i\}$ and the security of outsourced operations.

The security of the transformation encryption key: The adversary cannot factorize the blinding factor $\left\{g^{\beta d}, h_{\phi(i)}^x\right\}$ into $g^{\beta d}$ and $h_{\phi(i)}^x$ for random (d, x) to the adversary. Hence, $h_{\phi(i)}^x$ is secure.

The security of encryption operations: The secret encryption exponents λ_i and sr are blinded by d and x respectively. So the outsourced parameters $\lambda_i - d$ and $sr - x$ do not reveal any useful information, so it is security.

The security of outsourced operations: Computing transformation ciphertext

$$U = \frac{e(K', C_2)}{\prod_{i \in I}\left(e(C_i, K_0)e\left(C_3, K_{\phi'(i)}\right)\right)^{\omega_i}}$$

$$Q = e(C_2, g)$$

In order to obtain information from the outsourced key, the cloud server must decompose the discrete logarithm. So the outsourced operations are security.

5.2 Performance Analysis

The performance comparison between the proposed scheme and the existed outsourced ABE scheme [5, 10–12] is shown in Table 1. In the table, n refers to the number of attributes in the access structure of the scheme, L denotes the length of a group, and Out key.C denotes the number of computation mode exponents of the outsourced decryption key.

In addition, the outsourcing transformation key generation time of scheme [5, 12] increases from the increase in attributes. Compared with the scheme [5, 12], this paper has realized that the number of outsourced transformation key modulus index operation

Table 1. Performance analysis

Scheme	Ciphertext size	Out key.C	Full verifiable
Green [5]	(2 + |S|)L	2n	No
Qin [10]	(2 + 2|S|)L	1	Yes
Wang [11]	(3 + 3|S|)L	n + 2	No
Li [12]	(3 + |S|)L	2n + 1	Yes
Ours	(3 + |S|)L	1	Yes

is fixed. Through the above analysis, we know that this scheme is more suitable for electronic information security.

6 Conclusions

This paper proposed a verifiable outsourced ABE scheme for fixed key size. Firstly the authority center generates an outsourced encryption transformation key and the corresponding outsourced decryption transformation key. The user uses the key to reduce the encryption exponent from $O(n)$ to $O(1)$ in the key generation phase. At the same time, the number of modulo exponentiation of the decryption transformation key is reduced to a fixed constant and achieves a fixed size of the outsourced decryption key, which effectively shortens the storage space and the generation time of the outsourced decryption key.

Acknowledgments. The authors thank all the reviewers and editors for their valuable comments and works. This paper is supported by National Key Research and Development Program of China (Grant No. 2017YFB0802000), National Natural Science Foundation of China (Grant Nos. U1636114, 61772550, 61572521), Natural Science Basic Research Plan in Shaanxi Province of China (Grant Nos. 2016JQ6037).

References

1. Sahai, A., Waters, B.: Fuzzy identity-based encryption. In: Eurocrypt, vol. 3494, pp. 457–473 (2005)
2. Goyal, V., Pandey, O., Sahai, A., et al.: Attribute-based encryption for fine-grained access control of encrypted data. In: Proceedings of the 13th ACM Conference on Computer and Communications Security, pp. 89–98. ACM (2006)
3. Goyal, V., Jain, A., Pandey, O., et al.: Bounded ciphertext policy attribute based encryption. In: Automata, Languages and Programming, pp. 579–591 (2008)
4. Yang, K., Han, Q., Li, H., et al.: An efficient and fine-grained big data access control scheme with privacy-preserving policy. IEEE Int. Things J. **4**(2), 563–571 (2016)
5. Green, M., Hohenberger, S., Waters, B.: Outsourcing the decryption of ABE ciphertexts. In: USENIX Security Symposium, vol. 2011, no. 3 (2011)
6. Zhou, Z., Huang, D.: Efficient and secure data storage operations for mobile cloud computing. In: Network and Service Management (CNSM), 2012 8th International Conference and 2012 Workshop on Systems Virtualiztion Management (SVM), pp. 37–45. IEEE (2012)
7. Li, J., Jia, C., Li, J., et al.: Outsourcing encryption of attribute-based encryption with mapreduce. In: Information and Communications Security, pp. 191–201 (2012)
8. Lai, J., Deng, R.H., Yang, Y., et al.: Adaptable ciphertext-policy attribute-based encryption. In: International Conference on Pairing-Based Cryptography, pp. 199–214. Springer, Cham (2013)
9. Li, J., Huang, X., Li, J., et al.: Securely outsourcing attribute-based encryption with checkability. IEEE Trans. Parallel Distrib. Syst. **25**(8), 2201–2210 (2014)
10. Qin, B., Deng, R.H., Liu, S., et al.: Attribute-based encryption with efficient verifiable outsourced decryption. IEEE Trans. Inf. Forensics Secur. **10**(7), 1384–1393 (2015)

11. Wang, H., He, D., Shen, J., et al.: Verifiable outsourced ciphertext-policy attribute-based encryption in cloud computing. Soft. Comput. **21**, 1–11 (2016)
12. Li, J., Li, X., Wang, L., et al.: Fuzzy encryption in cloud computation: efficient verifiable outsourced attribute-based encryption. Soft. Comput. **22**, 1–8 (2017)
13. Waters, B.: Ciphertext-policy attribute-based encryption: an expressive, efficient, and provably secure realization. In: Public Key Cryptography, vol. 6571, pp. 53–70 (2011)
14. Beimel, A.: Secure schemes for secret sharing and key distribution. Technion-Israel Institute of technology, Faculty of computer science (1996)
15. Liu, M., Liu, S., Wand, Y., et al.: Optimizing the decryption efficiency in LSSS matrix-based attribute-based encryption without given policy. Acta Electronica Sin. **43**(6), 1065–1072 (2015)

Subjective Annoyance Caused by Low Frequency Noise

Ling Lu(✉), Hong-Wei Zhuang(✉), and Liang Xu(✉)

Engineering University of the People's Armed Police, Xi'an 710086, China
616783347@qq.com, 215548800@qq.com, suoduoma19861216@sina.com

Abstract. Noise exposure has adverse effects on the physiology and psychology of the human body. This paper selects three typical noise sources in life according to the loudness level of different frequency pure tone of subjective annoyance common relative size, feature extraction from the frequency noise spectrum, pure tone synthesis of several noise samples will have the characteristic frequency, of subjective annoyance research on actual noise and synthetic noise by using the paired comparison method, for improving the environmental noise in the future.

Keywords: Perception mining · Dormant behavior · Stealth attack

1 Introduction

In the 1960s, with the rapid development of expressways and airports, traffic noise pollution was strongly protested, and noise pollution was widely concerned. As a result, the United States and Europe officially listed noise as an important environmental pollution factor, Established the corresponding laws and regulations, followed by other countries. The physical and psychological effects of noise are a very important basis in the development of standards for products and environmental noise.

The research on the influence of noise on physiology is very rich, which can be roughly divided into the auditory system and the non-auditory system. The impact on the non-auditory system mainly includes the cardiovascular system, the endocrine system and the nervous system.

The impact of noise on the psychology is mainly manifested in the subjective annoyance of people to noise [1–3], including irritation, dissatisfaction, annoyance, displeasure, torture, anger, annoyance, distress, anger, discomfort, anxiety, sadness, Hate and other negative feelings. The research shows that the subjective annoyance of noise has a good correlation with its loudness, and the loudness has thus become an effective index for predicting the subjective annoyance of noise. However, with the deep research on the subjective worry of noise, people have found that in addition to the intensity loudness and the sound pressure level, the spectral characteristics and time-domain characteristics also have a greater impact on it. However, different emotions will make a certain change in the behavior of people, so only the subjective annoyance to measure the impact of noise on the psychology of people is not accurate, so this article on the emotional specific anger to study its noise The relationship between.

L. Barolli et al. (Eds.): EIDWT 2018, LNDECT 17, pp. 801–809, 2018.
https://doi.org/10.1007/978-3-319-75928-9_72

In the study of the subjective worry about noise, many scholars at home and abroad [4, 5] have done it through social investigation and subjective evaluation experiment. Subjective evaluation of noise in the experiment commonly used methods are semantic subdivision and paired comparison method [6], which paired comparison method has the advantages of simple operation, the evaluator easy to make judgments.

In the experiment of subjective evaluation of noise, the experimental results are difficult to be compared due to the differences in the time-frequency and frequency-domain characteristics of the actual sound sources used and the fact that the frequency components contained in the actual sound sources are complicated [7]. Pure tone, white noise and pink noise and other standard sound sources have a good grasp of the frequency components, so their comparability of subjective worry is better, and the relevant research conclusions are more consistent [8]. Therefore, it is of great significance to study the relationship of anger and emotion between pure noise and pure tone combination with its related characteristic parameters and explore the feasibility of using pure tone or pure tone combination to simulate the anger caused by actual noise.

2 Experiment

The printing area is 110 mm × 195 mm. The text should be justified to occupy the full line width, so that the right margin is not ragged, with words hyphenated as appropriate. Fill pages so that the length of the text is no less than 170 mm.

Use 10-point type for the name(s) of the author(s) and 9-point type for the address(es) and the abstract. For the main text, please use 10-point type and single-line spacing. We recommend the use of Computer Modern Roman or Times. Italic type may be used to emphasize words in running text. Bold type and underlining should be avoided.

2.1 Experimental Materials and Methods

The 18–22-year-old college students were chosen as subjects in the experiment. The number of the subjects in each evaluation group was 8, and the ratio of men to women was 1:1. Pre-audiometric tests showed that the audiophile listener was normal. The experimental site is a 5*6 analog office. The office adopts good sound insulation treatment. The office furniture is symmetrically placed on the desk and chair. The dodecahedral sound source is suspended in the center. The sound receiver seat has the same straight line distance as the sound source.

In equal loudness level pure tone subjective annoyance experiments, a total of 8 pure tone selected, the corresponding frequencies were 100–500 Hz within 1/3 octave center frequency.

Actual noise and synthetic noise subjective annoyance contrast experiment, select the three actual noise in life, respectively, for noise pure tone in distribution room, pump noise pure tone and diesel generator noise pure tone, channel acoustic analyzer acquisition and adjustment, to achieve 100–16000 Hz Band no obvious distortion, to ensure that does not affect the reliability of the experiment. Based on the spectrum analysis of the actual noise, the synthetic noise is extracted based on the experimental results of subjective annoyance of pure tones at different loudness levels, and the three characteristic frequencies F1, F2 and F3 with relatively large subjective annoyances are extracted and the actual noise Each characteristic frequency sound pressure level difference, will have a characteristic frequency, a combination of three pure tone noise samples. Any combination of above three pure tone can form 7 noise samples, as shown in the Table 1.

Table 1. Actual noise and composite noise sample

Number	A	E	F	B	C	G	D	H
Noise samples	F1	F2	F3	F1F2	F1F3	F2F3	F1F2F3	Actual noise

In the experiment, the equal loudness adjustment was performed on each acoustic sample so that the actual measured loudness levels of the ear of the receiver were 60phon, 65phon, 70phon and 75phon, respectively. During the process of equal loudness adjustment, the relative sound pressure level of each frequency component in the noise sample will not be changed.

2.2 Experimental Methods

In the experiment, a pair of actual noise and synthetic noise were used to make a comparison of subjective annoyances. This method is widely used in acoustic quality research. In the experiment, the acoustic samples to be evaluated are arranged in groups of 2 in each group. During the evaluation, the sound samples were randomly played by group, and the evaluators made a comparison of the sensory degree of two sound samples in each group. If you think the former is more troublesome than the latter, the evaluation value is 1 point; the same annoyance is 0 point; the latter is 1 point more annoying than the former [8, 9].

Related research shows that in the subjective annoyance evaluation experiment, the 5 s length acoustic sample is more suitable for the speaker to make an accurate judgment. In this study, there were 8 acoustic samples in each evaluation group, with 5 s for each acoustic sample, 3 s for acoustic sample switching and consonants judgment, and-less than 30 min for evaluation.

False judgment is unavoidable when the speaker evaluates his own feelings. Therefore, the correctness of the experimental data needs to be tested and the data set with a higher proportion of misjudgment should be excluded. In this study, the weighting consistency coefficient [9] is used as the criterion of data validity.

After eliminating the invalid data by false positive test, the evaluation matrix will be established to get the subjective annoyance ranking of each acoustic sample [8]. Quantitative subjective annoyance is calculated on the basis of this formula:

$$D = \left[\sum\nolimits_{1}^{m} (n - ri + 1)\right]/m \tag{1}$$

where D is the subjective annoyance of the acoustic sample, m is the number of valid speakers, n is the number of acoustic samples in the assessment group, and ri is the subjective annoyance order of the acoustic sample in the assessment group. For example, the ordering of the annoyances of the two samples to the acoustic sample A is the third and the fourth, respectively, whereas for the group of eight acoustic samples in the evaluation group, the subjective annoyance of A is 5.5.

2.3 Experimental Procedure

In the experiment, the subjects were asked to sit quietly in the laboratory and explain the experiment to the subjects through the recorded experiments. The experimental contents are as follows:

"Imagine you are sitting quietly in an office, during which time you will hear multiple pairs of paired sound samples, and after listening to a paired sound sample, indicate which of the sound samples made you feel More annoyed and scored according to the following rules: If you think the first voice is more annoying than the second voice, in the comparison result column filled with '1'; if you think the second voice is more annoying than the first voice, then in comparison Fill '−1' in the result column, and '0' in the comparison result field if you think the two voices are troublesome. Since the purpose of this experiment is to examine your personal attitudes toward various acoustic samples, the result is not wrong Please stay tuned while experimenting, and you will hear several sets of practice vocals before the experiment begins.

In the experiment, the pure tone at the same loudness level was firstly evaluated, and a total of 4 groups were obtained, resulting in subjective annoyance curves of equal loudness level pure tone. Then each of the actual noise and the corresponding synthetic noise evaluation, a total of 12 groups.

3 Results

3.1 Equal Loudness Subjective Annoying Pure Tone Experiment

It can be seen from Fig. 1 that subjective annoyances of 8 1/3 octave center frequencies in 100–500 Hz at different loudness levels (60 phon, 65 phon, 70 phon and 75 phon) are basically the same, with a 1 A subjective distress maximum, 160–200 Hz and 400 Hz near the emergence of a subjective distress minimum, more than 400 Hz pure tone subjective annoyance generally increases with increasing frequency. Zhu et al. [6] and other studies have shown that the case of equal loudness, 500–8000 Hz octave center frequency corresponding to the pure tone of the subjective annoyance increases with increasing frequency. This shows that the equal loudness level, the frequency of pure tone of the subjective annoyance more significant impact. This is in line with the conclusions of the relevant studies that the frequency characteristics outside of loudness are important factors that affect the subjective worry of noise [10, 11].

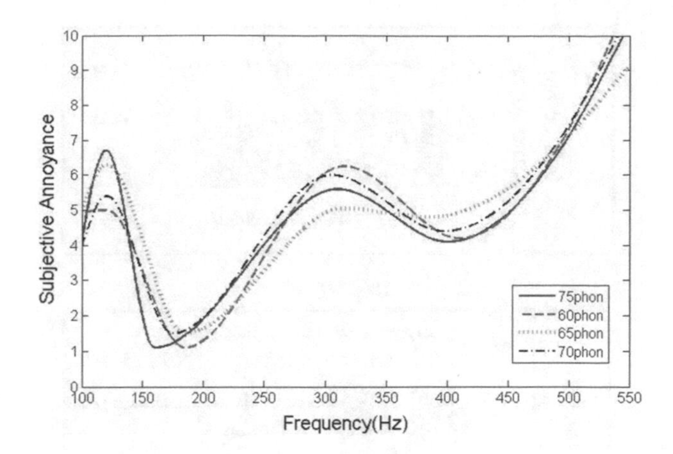

Fig. 1. Subjective annoyance of pure tone at equal loudness level.

3.2 Actual Noise and Synthetic Noise Subjective Worry Comparison Experiment

(1) The actual noise characteristic peak and acoustic energy distribution

Based on the spectral analysis of the actual noise, referring to the tendency of subjective annoyances of equal loudness levels shown in Fig. 2, three characteristic frequencies (F1, F2 and F3) with relatively large subjective annoyances are selected from the three kinds of actual noises. See Table 2 for details.

At the same time, Table 2 also gives the statistical results of energy distribution in different frequency bands of actual noise. It can be seen from the table that the sound energy of the low frequency band (20–250 Hz) in the power distribution substation accounts for 92.8% of the total acoustic energy; the sound energy in the low frequency band of the refrigerating unit only accounts for 46.9% of the total acoustic energy while the sound energy of 52% (250–1000 Hz). The noise of direct-fired turbines is similar to the noise of refrigerating units. 42.1% of the acoustic energy is distributed in the low frequency band, 54.4% of the acoustic energy is distributed in the middle frequency band, and the other is in the high frequency band (1000–20000). It can be seen that there are significant differences in the distribution of acoustic energy among the three types of actual noise in different frequency bands. Among them, the noise energy of the power distribution and distribution house is concentrated in the low frequency band. The noise energy of the direct combustion engine and the noise of the freezer are mainly distributed in the low and mid frequency bands, and the mid-band acoustic energy is higher than the low frequency band.

(2) The actual noise of a subjective frequency of pure tone annoyance

As shown in Table 1, there are C, D, G3 pure tones among the seven synthesized noises associated with each actual noise. From Table 3, we can see that subjective annoyances of the three characteristic frequencies pure tones of power distribution and distribution houses are classified as C > D > G(120 Hz > 301 Hz > 199 Hz); subjective annoyances of the noise characteristic frequencies of the freezer are D > G > C (797 Hz > 599 Hz > 199 Hz).

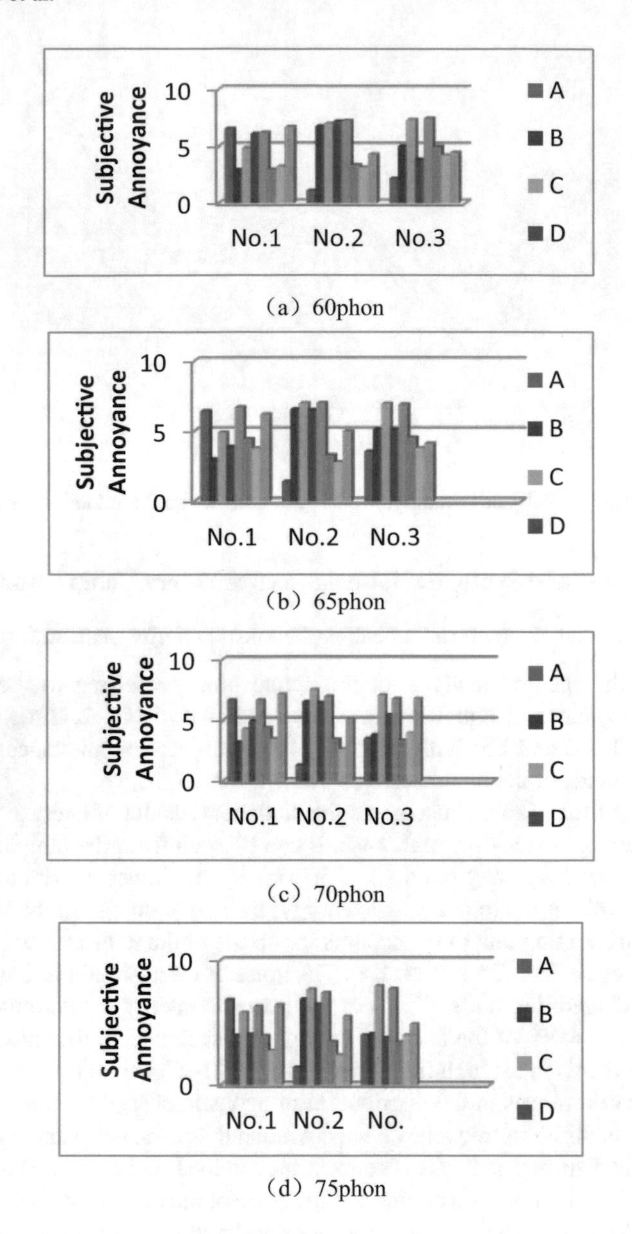

Fig. 2. Subjective annoyance ranking of actual noise and synthetic noise

The subjective annoyances of the three characteristic frequencies of direct engine noise were ranked as D > G > C(1131 Hz > 931 Hz > 124 Hz). The above results are in line with the trend of subjective annoyances of equal loudness pure tones (Fig. 1), which confirms the reliability of this experimental result to a certain extent. The relative magnitude of the subjective annoyances of pure tones above 500 Hz is also a powerful complement to the trend of subjective annoyances of equal loudness pure tones (Fig. 1). That is, the subjective

Table 2. Actual noise frequency characteristics table

Sound source	Characteristic peaks	Corresponding frequency (Hz)	Relative sound pressure level difference (dB)	Acoustic energy division (%)		
				20–250 Hz	250–1000 Hz	1000–20000 Hz
distribution room noise	F1	102	−10.4	92.8	6.4	0.7
	F2	199	0			
	F3	301	−7.1			
Pump noise	F1	199	0	46.9	52.0	1.1
	F2	599	−3.4			
	F3	797	−9.9			
Diesel generator noise	Pump noise pure tone (Hz)	Diesel generator noise pure tone (Hz)	−3.5	42.1	54.4	3.4
	F2	937	0			
	F3	1131	−3.3			

annoyances of equal loudness pure tones increase with increasing frequency in the range of 500–1131 Hz.

(3) Comparison of subjective worry of actual noise and synthetic noise

Table 3. Comparison of subjective annoyance from pure tones in a composite noise

Loudness level (phon)	Noise pure tone in distribution room (Hz)			Pump noise pure tone (Hz)			Diesel generator noise pure tone (Hz)		
	102	199	301	199	599	797	124	931	1131
60	6.55	2.98	4.88	1.10	6.77	7.00	2.20	5.03	7.34
65	6.43	3.00	4.89	1.41	6.57	6.98	3.55	5.13	6.94
70	6.70	3.28	4.38	1.36	6.57	7.55	3.56	3.88	7.10
75	6.86	4.00	5.83	1.45	6.91	7.68	4.03	4.12	7.96

As shown in Fig. 2, the subjective annoyances of the same kind of real noise and synthetic noise are in good order of rank at different loudness levels (55phon, 60phon, 65phon and 70phon), and the composite noise close to the actual noise subjective annoyance basically the same. Among them, the sound sample C (102 Hz pure tone) corresponding to the noise of the power distribution room is similar to the subjective annoyance of the actual noise. The sound sample A (containing 797 Hz, 599 Hz, 199 Hz pure tone) corresponding to the noise of the freezer is similar to the actual noise subjective worry, The acoustic sample A (containing 124 Hz, 931 Hz, 1131 Hz pure tone) corresponding to the machine noise is similar to the actual noise subjective worry.

From the above analysis, it can be seen that the low-frequency component of the power distribution substation dominates, that is, most of the acoustic energy is located in the low frequency band, and the sound energy distribution in the chiller and the direct-fired machine noise is relatively dispersed with the main acoustic energy located in the low frequency band and middle frequency band. However, the acoustic sample C, which is closest to the subjective annoyance of the actual distribution of noise in the power distribution house, contains only one pure tone in the low frequency band. The acoustic sample A, which is closest to the subjective annoyance of the actual noise of the refrigerating unit and the direct-A pure tone is located in the low band and two pure tones are in the middle band. It can be seen that the more concentrated the actual acoustic noise energy distribution, the less pure tones the synthesized noise closest to its subjective annoyance will be, and the more dispersed the actual acoustic noise energy distribution, the more pure tones the composite noise closest to the subjective annoyance. This shows that if the energy distribution of the actual noise in different frequency bands can be considered, pure tones with relatively large subjective annoyances are selected in several frequency bands with a larger proportion of acoustic energy respectively, and the synthesized noise of these pure tones can be better simulated Subjective distress of actual noise.

4 Conclusion

Subjective troubles trend pure tones of different frequencies within the loudness level 100–500 Hz pure tone level subjective loudness results show troubles, etc. is basically the same, a maximum value of each of a subjective troubles occur in the vicinity of 125 Hz and 300 Hz, and 400 Hz nearby 160–200 Hz Each of the subjective annoyances showed a minimum value. The subjective annoyance of pure tone above 400 Hz increased with increasing frequency, and pure tone frequency had a more significant impact on its subjective annoyances.

In the continuous spectrum of noise by pure tone combined analog subjective annoyance experiments with pure tones subjective annoyance actual noise characteristic frequency consistent relative results subjective annoyance size and the like loudness level pure tone, 500 Hz or more characteristic frequency pure tones subjective annoyance relative size is 100–500 Hz Equivalent loudness pure subjective annoyance test results, a powerful supplement, that 500–1131 Hz equal loudness subjective annoyance still increases with increasing frequency.

On the basis of the noise spectrum analysis, considering the noise energy distribution in different frequency bands, and the like with reference to pure tones of different frequencies relative sizes of the subjective loudness levels troubles were selected on subjective annoyance acoustic energy band occupied by a large proportion of relatively few Large characteristic frequency and relative difference of SPL of each characteristic frequency are preserved. The pure noise synthesized noise samples with these characteristic frequencies can better simulate the subjective annoyance of actual noise.

References

1. Persson-Waye, K., Bjorkman, M., Rylander, R.: Loudness, annoyance and the dBA in evaluating low frequency sounds. J. Low Freq. Noise Vib. Active Control **9**, 32–45 (1990)
2. Tempest, W.: Loudness and annoyance due to low frequency sound. Acustica **29**(4), 200–205 (1973)
3. Rylander, R.: Physiological aspects of noise-induced stress and annoyance. J. Sound Vib. **277**(3), 471–478 (2004)

4. Landstrom, U., Akerlund, E., Kjellberg, A., Tezars, M.: Exposure levels, tonal component and noise annoyance in working environment. Environ. Int. **3**(21), 265–275 (1995)
5. Moller, H.: Annoyance from audible infrasound. J. Acoust. Soc. Am. **78**, S32 (1985)
6. Huang, Y.F., Di, G.Q., Zhu, Y.T., et al.: Pair-wise comparison experiment on subjective annoyance rating of noise samples with different frequency spectrums but same a-weighted level. Appl. Acoust. **69**, 1205–1211 (2008)
7. Wiser, R., Bolinger, M.: 2010 Wind Technologies Market Report. U.S. Department of Energy, Oak Ridge (2011)
8. Otto, N., Amman, S., Eaton, C., et al.: Guidelines for jury evaluations of automotive sounds. Sound Vibr. **35**(4), 24–47 (2001)
9. Zhu, Y.T., Qi, G.Q.: The influence of low frequency environmental noise on the ability of thinking judgment. Environ. Sci. **4**(29), 1143–1147 (2008)
10. Skinner, C.J., Grimwood, C.J.: The UK noise climate 1990–2001: population exposure and attitudes to environmental noise. Appl. Acoust. **66**, 231–243 (2005)
11. Bridgeman, E., Popoff-Asotoff, P.: Local Government Noise Complaints Survey Report, Perth (2010)

The Relationship Between Personality Traits and Aggressive Behavior in People with Long Term Noise

Ling Lu(✉), Hong-Wei Zhuang(✉), and Liang Xu(✉)

Engineering University of the People's Armed Police, Xi'an 710086, China
616783347@qq.com, 215548800@qq.com, suoduoma19861216@sina.com

Abstract. At present, violent incidents occur frequently in the society and show a trend of getting younger and younger. Lead to explicit offensive aggressive behavior has many obvious incentives to study the relationship between the two in order to effectively prevent the occurrence of aggressive behavior. Noise as a common source of pollution in life, noise stimulation will affect people's cardiovascular, endocrine, etc., but also have a great impact on human emotions, easy to produce anxiety, irritability and other emotions. Noise stimulation easily lead to negative emotions, such as anger, frustration, etc., these emotions are generated with the typical occurrence of aggressive behavior. Therefore, the study of the relationship between aggressive behavior and noise provides a theoretical basis for the effective prevention of such events as violence.

Keywords: Noise stimulation · Personality · Aggressive behavior Canonical correlation

1 Introduction

The attack is caused by the combination of individual personality and situation, in which the individual personality is the decisive factor in the attack behavior, and the high attack behavior easily lead to criminal behavior. Mei Chuanqiang believes that the main psychological cause of criminal behavior is the personality traits of people [1]. Xiao Linlin and other studies have shown that the occurrence of crime, most of them outgoing, emotional instability, a spiritual personality traits [2]. Dai Chunlin and other studies have found that the individual's aggressive behavior has the characteristics of automation [3], imprisonment personality implicit aggressive personality [4]. Wang Junjie and other studies found that aggressive male juveniles more personality deviation, impulsive and poor stability-based [5]. Personality in the formation process will be affected by the environment, experience, parental personality, emotions and other aspects. As a kind of common environmental pollution, noise can induce people to have different emotions and influence the formation of personality.

Based on the previous studies, this paper studies the relationship between different personality traits and aggressive behavior under the long-term influence of noise. As the airport noise is more common, the impact on the surrounding residents and there are

L. Barolli et al. (Eds.): EIDWT 2018, LNDECT 17, pp. 810–817, 2018.
https://doi.org/10.1007/978-3-319-75928-9_73

many hazards, so this paper chose airport noise as a noise source. Using the Big Five Personality Scale, Impulsive Personality Scale and Aggressive Behavior Scale to investigate the long-term impact of airport noise, SPSS 16.0 was used to count the data by Pearson correlation, canonical correlation and multivariate linear regression analysis.

2 Affected by the Long-Term Impact of Airport Noise on the Population

Living environment will have an impact on people's personality. Therefore, when studying the relationship between noise, personality and aggressive behavior, we first analyze the personality traits of people under the influence of long-term noise.

2.1 Method

This experiment mainly uses the population affected by the noise near the airport for more than 10 years, and fill in the test scale to compare the personality characteristics of other groups. A total of 153 subjects were recruited in the vicinity of the airport of a city aged 35–40 years with random sample method. They were affected by airport noise for more than 10 years, 83 were males and 70 were females without any medical history. Taking the gender of the long-term population affected by airport noise as the independent variable and taking the Big Five Personality Factor and Impulsive Personality Factor as the dependent variables, the differences were compared.

Among them, there are 60 items of 5 factors including neuroticism, extroversion, openness, pleasantness and sense of responsibility, using Likert Five-grade Score. There are many versions of the impulsivity scale [7]. Among them, the Chinese version of the Barratt Impulsivity Scale, revised by Li Xiangyun, is a commonly used impulsiveness assessment tool. There are 30 items in total, ranging from 1 to 5 in terms of item score. The higher the impulse, the stronger.

2.2 By the Long-Term Impact of Airport Noise on the Personality Characteristics of the Results of the Analysis

Found that in the long-term impact of airport noise, men scored significantly lower than women in the factor of pleasantness, while scores in unplanned impulsivity, impulsivity and impulsive total score were significantly higher than those in women.

The long-term impact of airport noise on the impulsive personality dimensions of each dimension compared with Li Xiangyun and other surveyed community and impulsive personality of college students personality dimensions scores [3], can be seen from Table 1, the long-term impact of airport noise population groups The scores of unplanned impulsivity, cognitive impulses and impulsive scales were significantly higher than those of rural residents, urban residents and college students.

Table 1. Comparison of impulsive personality scale between criminals and rural residents, urban residents and College Students.

	Airport noise affects population (n = 153)	Rural residents (n = 603)	t values	Urban residents (n = 548)	t values	College students (n = 627)	t values
Impulse of action	37.9 ± 18.8	26.7 ± 17.1	0.93	32.7 ± 17.9	5.01	35.6 ± 15.7	0.82
Unplanned impulsivity	40.0 ± 17.9	29.6 ± 17.9	11.98	27.9 ± 15.9	13.93	33.7 ± 14.6	7.27
Cognitive impulsivity	44.5 ± 15.9	30.4 ± 16.3	18.25	34.5 ± 15.0	12.95	36.5 ± 11.8	10.36
The total score	41.13 ± 14.3	28.9 ± 13.1	18.40	31.7 ± 13.2	14.08	35.2 ± 11.5	9.30

3 Correlation Analysis Between Personality and Aggression in Long-Term Affected Airport Noise

In the measurement of aggressive behavior, the modified version of the Li Yuhe et al. [8] attack scale was adopted. As the scale has been improved according to Chinese customs and habits, it is more suitable for Chinese physique. The scale of 21 items in total, four factors, namely anger, hostility, body attacks, speech attacks. Yang Wenji et al. investigated this revised version among criminals and found that the half-confidence level was 0.85, the internal consistency coefficient Cronbach α was 0.875 and the test-retest reliability was 0.881 [9], so the scale can be used to measure the aggressive behavior.

Table 2. General correlation analysis between personality and aggressive behavior

	Anger	Hostility	Somatic aggression	Verbal aggression	Total attack score
Unplanned impulsivity	0.40	0.365	0.39	0.15	0.43
Cognitive impulsivity	0.33	0.26	0.22	0.02	0.28
Impulse of action	0.45	0.31	0.42	0.22	0.53
Impulsiveness Scale total score	0.51	0.40	0.48	0.19	0.52
Nervous	0.38	0.43	0.31	0.21	0.36
Extroversion	–0.10	–0.26	–0.19	–0.07	–0.23
Openness	0.06	0.06	0.06	0.03	0.08
Agreeableness	–0.34	–0.40	–0.46	–0.29	–0.51
Rigidity	–0.23	–0.23	–0.23	–0.11	–0.28

Pearson's simple correlation analysis was conducted on all dimensions of aggression and dimensions of Big Five personality and impulsive personality. The results showed that there was a significant positive correlation between unplanned impulsivity, impulsivity and impulsiveness scores and all dimensions of aggressive behavior. There was no significant correlation between cognitive impulses and verbal attacks, and other factors of aggression There was a significant positive correlation between neuroticism and aggressive dimensions. Extroversion, pleasantness and rigor were negatively correlated with total scores of anger, hostility, body attack and attack in Table 2.

In order to further investigate the overall relationship between personality traits and aggressive behaviors, take the Big Five personality and impulsive personality as independent variables, and attack behavior as the dependent variable to make a typical correlation analysis. The results showed that there was a significant Canonical Correlation Coefficient of Big Five Personality and Aggression, with a specific value of 0.62, reaching a moderate correlation. Big Five Personality and Aggression mainly through a pair of typical factors reflect the correlation between the two. Long-term impact of airport noise on the big five personality group variables can be explained by their own typical variable was 34.3%, can be explained by the relative variation of typical variables was 13%, while the aggressive behavior of this group of variables can be typical of their own. The variance explained by the variable was 64.8%, and the variation explained by the typical representative variable was 24.6%. The square of the typical correlation coefficient is 0.38, which shows that the variance of the two groups is 38%. The results are shown in Table 3.

Table 3. Typical correlation between Big Five personality and aggressive behavior

Canonical variables	1		
	Typical coefficient		Typical load
Big Five personality			
Nervous	−0.48		−0.46
Extroversion	0.13		0.27
Openness	−0.13		−0.04
Agreeableness	0.66		0.53
Rigidity	0.03		0.30
Variance representing proportion (%)		34.3	
Redundancy index		13.0	
Aggressive behavior			
Anger	2.64		−0.82
Hostility	1.75		−0.90
Somatic aggression	3.05		−0.80
Verbal aggression	1.47		−0.50
Variance representing proportion (%)		64.8	
Redundancy index		24.6	
Canonical correlation coefficient R		0.62	
R2		0.38	
F-measure		25.00	

For the Big Five, the typical loads and standard typical coefficients of neuroticism and pleasantness are high, and the typical loads and standard typical coefficients of anger, hostility, verbal attacks and body attacks are high for aggressive behavior. Therefore, it can be considered that the neuroticism and pleasantness play a greater role in the explanation of anger, hostility, speech attacks and physical attacks on the relationship between the Big Five personality and aggression.

As can be seen from Table 4, unplanned impulsivity, cognitive impulses and behavioral impulses and anger, hostility, body attacks and verbal attacks have a higher negative load on the first pair of canonical variables, and the correlation of the first pair of canonical variables The coefficient is 0.63. Cognitive impulsivity in impulsive personality has a higher negative load on the second pair of canonical variables, body attacks and verbal attacks in aggression have a higher positive load on the second pair of canonical variables, and second The correlation coefficient for a typical variable is 0.22.

Table 4. Typical correlations between impulsive personality and aggressive behavior

Canonical variables	1			2		
	Typical coefficient		Typical load	Typical load		Typical load
Impulsive personality						
Unplanned impulsivity	–0.52		–0.76	0.57		–0.27
Cognitive impulsivity	0.07		–0.60	–1.37		–0.81
Impulse of action	–0.72		–0.90	0.37		0.12
Variance representing proportion (%)		56.8			25.0	
Redundancy index		22.2			1.2	
Aggressive behavior						
Anger	–0.47		–0.95	–0.44		0.04
Hostility	–0.03		–0.70	–0.29		–0.05
Somatic aggression	–0.06		–0.83	0.94		0.43
Verbal aggression	0.33		–0.44	1.12		0.71
Variance representing proportion (%)		62.7			15.5	
Redundancy index		24.5			0.08	
Canonical correlation coefficient R		0.63			0.22	
R2		0.39			0.05	
F-measure		15.0			8.0	

Impulsive Personality The variables that can be explained by 2 pairs of their own typical variables are 81.8% and those that can be explained by their relative typical variables are 23.4%. While the aggressive behavior variables of this group can be explained by their own typical variables as 78.2%, and the relative variability explained by typical variables as 25.3%. The square of the Canonical Correlation Coefficient is 0.44, indicating that the shared variance of the two groups of variables is 44%. This shows that there is a moderate correlation between impulsive personality traits and aggressive behavior.

4 Regression Analysis on Personality Traits and Aggressive Behaviors of Persons Affected by Airport Noise for a Long Time

Taking the Big Five personality and impulsive personality as independent variables, the total score of aggressive behavior was taken as the dependent variable, and the stepwise regression analysis was carried out in Table 5.

Table 5. Regression analysis of personality and aggressive behavior

	Non standardized regression coefficient 系数		Standardized regression coefficient 归系数	t value	P value	R2	F value	P value
	B	Standardized						
Constant term	46.88	7.68		6.10	0.00	0.449	66.25	0
Impulse of action 动性	0.25	0.04	0.29	6.64	0.00			
Agreeableness	−10.32	1.54	−0.28	−6.70	0.00			
Unplanned impulsivity 性	0.22	0.05	0.27	4.85	0.00			
Nervous	6.29	1.48	0.18	4.26	0.00			
Cognitive impulsivity	−0.13	0.05	−0.14	−2.56	0.01			

5 Results

The long-term impact of airport noise on men in non-plan impulsivity, impulsivity and impulsivity scores was significantly higher than that of female criminals, revealing that male population is more impulsive than women. When the scores of impulsive personality influenced by the long-term impact of airport noise are compared with the scores of rural residents, urban residents and college students, it is found that the long-term impact of airport noise on non-planned impulsivity, impulsivity, cognitive impulses And impulsive personality scores were significantly higher than the other three groups of scores, indicating that the population as a whole than the average person more prominent impulsive personality traits. Impulsivity is a multidimensional concept. There is seldom studied the relationship between the impulse and attack of different components in the previous research, and the impulse is more investigated as a whole [10]. Adolescent studies of attempted suicide by Aldis et al. Have shown that adolescents who have attempted suicide have more pronounced impulsiveness and aggressiveness [11].

The results of this study suggest that there is a difference between offensive and impulsive aspects. Unplanned impulsivity, cognitive impulses, and impulsivity of actions were significantly and positively correlated with total scores of anger, hostility, body attacks and aggression. Cognitive impulsivity was not significantly correlated with verbal aggression. There was a significant positive correlation between neuroticism and attack dimensions and aggression scores, while the pleasantness and rigor were negatively correlated with all dimensions of aggression.

A typical correlation analysis was used to examine the overall relationship between personality traits and aggressive behaviors. The results show that the Big Five personality can explain 24.6% of the variation in aggressive behavior through a pair of typical variables, and aggressive behavior variation can account for 13% of the Big Five personality by a pair of typical variables. Because interpretation rates are standardized data that allows for direct comparison, personality traits have a greater impact on aggression. Impulsive personality can account for 25.3% of the variation in aggressive behavior by 2 pairs of typical variables, and aggressive behavior variation can explain 23.4% of impulsive personality by 2 pairs of typical variables. Therefore, impulsive personality and aggressive behavior are closely related and influence each other. From the results of canonical correlation analysis, it can be seen that there is significant correlation between the Big Five Personality Variables group and aggressive behavioral group, and the neuroticism and pleasantness are more helpful to interpret anger, hostility, verbal attacks and body attacks. Impulsive personality variable group and aggression behavior variable group overall typical correlation is significant, unplanned impulsivity, cognitive impulses and behavioral impulsivity of anger, hostility, body attacks and speech attacks explained a greater role.

Regression analysis found that impulsivity, pleasantness, unplanned impulsivity, neuroticism and cognitive impulses have some predictive power on the total aggressive behavior, with an explanation rate of 44.9%.

Through this study we can find that:

(1) Long-term by the impact of noise makes it easy to produce impulsive personality, of which mainly unplanned impulses and cognitive impulses more obvious.
(2) Impulsive personality, neurotic personality traits and the impact of long-term impact of airport noise attacks have a close relationship.

References

1. Mei, C.Q.: Psychological basis of core issues - the focal research of criminal psychology criminal responsibility. Mod. Law **25**(2), 72–77 (2003)
2. Xiao, L., Pan, Y., Yao, J.: Investigation of personality characteristics of adult male criminals. Chinese J. Clin. Psychol. **1**(9), 50–51 (2001)
3. Yang, Z., Dai, C., Wu, M.: Experimental study on implicit aggression. Psychol. Sci. **28**(1), 96–98 (2005)
4. Dai, C., Sun, X.: Study on, the implicit aggression of prisoners. Psychol. Sci. **30**(4), 955–957 (2007)
5. Wang, J., Sun, X., Lu, H., et al.: Analysis of personality characteristics of male delinquent adolescents with aggressive behavior. Chin. Minkang Med. **24**, 1325–1327 (2012)
6. Liang, L., Yao, J.: NEO-PI-R simplified version (NEO-FFI) application analysis in college students crowd Chinese. J. Clin. Psychol. **18**(4), 457–459 (2010)
7. Li, X., Fei, L., Xu, D., et al.: Reliability and validity of the Chinese version of the Barratt Impulsiveness Scale used in community and college students. Chin. J. Ment. Health **25**(8), 610–615 (2011)
8. The revised questionnaire of Li Yu He attack. Jinan University master thesis, Guangzhou (2005)

9. Yang Wenji attack questionnaire - revision of reliability and validity. Master's thesis of Jinan University, Guangzhou (2007)
10. Mathias, C.W., Stanford, M.S., Marsh, D.M., et al.: Characterizing aggressive behavior with the impulsive/premeditated aggression scale among adolescents with conduct disorder. Psychiatry Res. **151**, 231–242 (2007)
11. Aldis, L., Putnins, A.L.: Correlates and predictors of self-reported suicide attempts among incarcerated youths. Int. J. Offender Ther. Comp. Criminol. **49**(4), 143–157 (2005)
12. Ramsden, S.R., Hubbard, J.A.: Family expressiveness and parental emotion coaching: their role in children's emotion regulation and aggression. J. Abnorm. Child Psychol. **6** (2002)
13. Siqueland, L., Kendall, P.C., Steinberg, L.: Anxiety in children: perceived family environments and observed family interaction. J. Clin. Child Adolesc. Psychol. **2** (1996)
14. Sukhodolsky, D.G., Ruchkin, V.V.: Association of normative beliefs and anger with aggression and antisocial behavior in Russian male juvenile offenders and high school students. J. Abnorm. Child Psychol. **2** (2004)
15. Waschbusch, D.A., Pelham, W.E., Jennings, J.R., Greiner, A.R.: Reactive aggression in boys with disruptive behavior disorders: behavior, physiology, and affect. J. Abnorm. Child Psychol. **6** (2002)
16. Fredrickson, B.L., Mancuso, R.A., Branigan, C., Tugade, M.M.: The undoing effect of positive emotions. Motiv. Emot. **4** (2000)
17. Fredrickson, B.L., Maynard, K.E., Helms, M.J., Haney, T.L., Siegler, I.C., Barefoot, J.C.: Hostility predicts magnitude and duration of blood pressure response to anger. J. Behav. Med. **3** (2000)
18. Thompson, R.A.: Emotional regulation and emotional development. Educ. Psychol. Rev. **4** (1991)
19. Eisenberg, N., Fabes, R.A.: Empathy: conceptualization, measurement, and relation to prosocial behavior. Motiv. Emot. **2** (1990)
20. Hendricks, K., Liu, J.: Childbearing depression and childhood aggression: literature review. MCN Am. J. Matern. Child Nurs. **4** (2012)
21. Herts, K., McLaughlin, K., Hatzenbuehler, M.: Emotion dysregulation as a mechanism linking stress exposure to adolescent aggressive behavior. J. Abnorm. Child Psychol. **7** (2012)
22. Salvas, M.-C., Vitaro, F., Brendgen, M., Lacourse, É., Boivin, M., Tremblay, R.E.: Interplay between friends' aggression and friendship quality in the development of child aggression during the early school years. Soc. Dev. **4** (2011)
23. Yeo, L.S., Ang, R.P., Loh, S., Fu, K.J., Karre, J.K.: The role of affective and cognitive empathy in physical, verbal, and indirect aggression of a Singaporean sample of boys. J. Psychol. **4** (2011)
24. Roberton, T., Daffern, M., Bucks, R.S.: Emotion regulation and aggression. Aggression Violent Behav. **1** (2011)
25. McLaughlin, K.A., Hatzenbuehler, M.L., Mennin, D.S., Nolen-Hoeksema, S.: Emotion dysregulation and adolescent psychopathology: a prospective study. Behav. Res. Ther. **9** (2011)
26. Grusec, J.E.: Socialization processes in the family: social and emotional development. Annu. Rev. Psychol. (2011)
27. Conway, A.M.: Girls, aggression, and emotion regulation. Am. J. Orthopsychiatry. **2** (2010)
28. Salmivalli, C.: Bullying and the peer group: a review. Aggression Violent Behav. **2** (2009)

Efficient Expanded Mixed Finite Element Method for the Forchheimer Model

Yanping Li(✉) and Qingli Zhao

Shandong Jianzhu University, Jinan, China
wlyiran@126.com, shizilu@126.com

Abstract. In this article, expanded mixed finite element method is used to approximate the Forchheimer model. This method extends the traditional mixed element technique. Existence and uniqueness are proved. Optimal L^2-error analysis is obtained. Numerical simulations are given to validate the theoretical derivation.

1 Introudction

Darcy's flow is an important mathematical model. Darcy's law,

$$\frac{\tilde{\mu}}{\rho}\tilde{k}^{-1}u = -\nabla p + g,$$

draws the connection between velocity and pressure. This relationship is got by lots of experimental data under the hypothesis that the creeping velocity and permeability are small [1–3].

When the velocity is higher, a new relationship is established. The Forchheimer equation is as following

$$\begin{cases} \frac{\tilde{\mu}}{\rho}\tilde{k}^{-1}u + \frac{\beta}{\rho}|u|u = -\nabla p + g, & x \in \Omega, \\ \nabla \cdot u = f, & x \in \Omega, \\ p = -p_D, & x \in \partial\Omega. \end{cases} \tag{1}$$

In (1) p denote pressure. u denote velocity. $\Omega \subset R^2$ denote a bounded domain. $|\cdot|$ denotes the norm, $|\boldsymbol{u}|^2 = \boldsymbol{u} \cdot \boldsymbol{u}$. $\tilde{k}$ is the permeability.

Several scholars have proven that it has a unique weak solution [4,5]. Form (1) can be written as follows [6]

$$\begin{cases} -\nabla \cdot \left(\dfrac{2(\nabla p - g)}{\frac{\tilde{\mu}}{\rho\tilde{k}} + \sqrt{\left(\frac{\tilde{\mu}}{\rho\tilde{k}}\right)^2 + 4\frac{\beta}{\rho}\left|\nabla p - g\right|}} \right) = f, & x \in \Omega, \\ p = -p_D, & x \in \partial\Omega. \end{cases} \tag{2}$$

For elliptic problems, MFEM for linear case has received research [7–11]. The main numerical approximations for solving the Eq. (1) is MFEM [6,12,13] and

L. Barolli et al. (Eds.): EIDWT 2018, LNDECT 17, pp. 818–827, 2018.
https://doi.org/10.1007/978-3-319-75928-9_74

block-centered finite difference method [14]. The expanded mixed finite element method (EMFEM) was presented in [15,16]. The EMFEM for linear and quasi-linear problem was considered in [17,18]. In deriving existence, uniqueness and error estimates, traditional technique [19–26] does not work for model (1).

The rest of this article is listed as following: In Sect. 2nd, we consider weak formulation. In Sect. 3rd, existence and uniqueness are given. In Sect. 4th, L^2-error analysis are established. In Sect. 5th, several numerical simulations are carried out. Throughout this paper C denote a generic positive constant.

2 Weak Formulation

Suppose $g = 0$. Using (2), we have the model

$$\begin{cases} -\nabla\cdot\left(\dfrac{2\nabla p}{\frac{\tilde{\mu}}{\rho\tilde{k}} + \sqrt{\left(\frac{\tilde{\mu}}{\rho\tilde{k}}\right)^2 + 4\frac{\beta}{\rho}\left|\nabla p\right|}}\right) = f, \;\; x\in\Omega, \\ \qquad\qquad\qquad\qquad\qquad\qquad p = -p_D, \;\; x\in\partial\Omega. \end{cases} \tag{3}$$

The following conditions of coefficients is supposed to be hold

$$0 < \tilde{\mu}_{min} \le \tilde{\mu}(x) \le \tilde{\mu}_{max}, 0 < \rho_{min} \le \rho(x) \le \rho_{max}, 0 < \beta_{min} \le \beta(x) \le \beta_{max}.$$

Let $\langle , \rangle$ be the L^2-inner product, $\langle \xi, \eta\rangle = \int_{\partial\Omega}\xi\eta ds$. Let us introduce two variables

$$s = -\nabla p, \quad u = -\frac{2\nabla p}{\frac{\tilde{\mu}}{\rho\tilde{k}} + \sqrt{\left(\frac{\tilde{\mu}}{\rho\tilde{k}}\right)^2 + 4\frac{\beta}{\rho}\left|\nabla p\right|}} = -k(x, |\nabla p|)\nabla p = k(|s|)s,$$

here

$$k(x, |\nabla p|) = \frac{2}{\frac{\tilde{\mu}}{\rho\tilde{k}} + \sqrt{\left(\frac{\tilde{\mu}}{\rho\tilde{k}}\right)^2 + 4\frac{\beta}{\rho}\left|\nabla p\right|}}.$$

Then model (1) can be rewritten as

$$\begin{cases} \nabla\cdot u = f, \;\; x\in\Omega, \\ u - k(|s|)s = 0, \;\; x\in\Omega, \\ s + \nabla p = 0, \;\; x\in\Omega, \\ p = -p_D, \;\; x\in\partial\Omega. \end{cases} \tag{4}$$

So (3) is formulated as: seek $(p, s, u) \in W\times\Lambda\times V$, such that

$$\begin{cases} (\nabla\cdot u - f,\; \omega) = 0, \quad \forall\omega\in W, \\ (u - k(|s|)s,\; \mu) = 0, \quad \forall\mu\in\Lambda, \\ (s,\; v) - (p,\; \nabla\cdot v) = < p_D,\; v\cdot n >, \quad \forall v\in V \end{cases} \tag{5}$$

where
$W = L^2(\Omega), \Lambda = L^2(\Omega)^2, V = H(div, \Omega) = \left\{v\in L^2(\Omega)^2 \mid \nabla\cdot v\in L^2(\Omega)\right\}.$

Let $V_h \times \Lambda_h \times W_h$ denotes space. $V_h \times \Lambda_h \times W_h$ is an approximation to $V \times \Lambda \times W$. We get the EMFEM: find $(p_h, s_h, u_h) \in W_h \times \Lambda_h \times V_h$, such that

$$\begin{cases} (\nabla \cdot u_h - f,\ \omega_h) = 0,\ \forall \omega_h \in W_h, \\ (u_h - k(|s_h|)s_h,\ \mu_h) = 0,\ \forall \mu_h \in \Lambda_h, \\ (s_h,\ v) - (p_h,\ \nabla \cdot v_h) = < p_D,\ v_h \cdot n >,\ \forall v_h \in V_h \end{cases} \tag{6}$$

here

$$\begin{aligned} W_h(E) &= \{\omega \in W : \omega|_E \in W_h(E), \forall\ E \in T_h\}, \\ \Lambda_h(E) &= \{\mu \in \Lambda : \mu|_E \in V_h(E), \forall\ E \in T_h\}, \\ V_h(E) &= \{v \in V : v|_E \in V_h(E), \forall\ E \in T_h\}. \end{aligned}$$

The theoretical analysis will use $\pi_h : V \rightarrow V_h$ meets

$$(\nabla \cdot (v - \pi_h v),\ \omega_h) = 0,\quad \forall \omega_h \in W_h.\|v - \pi_h v\| \leq Ch^r \|v\|_r,\quad 1/2 \leq r \leq k+1,$$
$$\|\nabla \cdot (v - \pi_h v)\|_r \leq Ch^r \|\nabla \cdot v\|_r,\quad 0 \leq r \leq k+1.$$

The other two projections are P_h and R_h

$$(\omega - P_h \omega,\ \nabla \cdot v_h) = 0,\quad \forall \omega \in W, v_h \in V_h.(\mu - R_h \mu,\ \tau_h) = 0,\quad \forall \mu \in \Lambda, \tau_h \in \Lambda_h.$$

They satisfy

$$\|\omega - P_h \omega\| \leq Ch^r \|\omega\|_r, \|\mu - R_h \mu\| \leq Ch^r \|\mu\|_r,\quad 0 \leq r \leq k+1.$$

3 Existence and Uniqueness

Let us simplify assume that $p \in W^{1,\infty}(\Omega)$, note that $k \in C^1\left(\overline{\Omega} \times R^+\right)$ and there exists $k_0 > 0, k_1 > 0$ such that $(x, \tilde{s}) \in \overline{\Omega} \times R^+)$

$$k_0 \leq k(x, \tilde{s}) \leq k_1, \tag{7}$$

$$k_0 \leq k(x, \tilde{s}) + \frac{\partial}{\partial \tilde{s}} k(x, \tilde{s}) \leq k_1, \tag{8}$$

$$|\nabla_x k(x, \tilde{s})| \leq k_1. \tag{9}$$

Further

$$|k(x, \tilde{s})\tilde{s} - k(x, \tilde{t})\tilde{t}| \leq k_1 |\tilde{s} - \tilde{t}|, \tag{10}$$

$$k(x, \tilde{s})\tilde{s} - k(x, \tilde{t})\tilde{t} \geq k_0(\tilde{s} - \tilde{t}), \qquad for \quad \tilde{s} \geq \tilde{t}. \tag{11}$$

Now we define three operators.

$$A_1 : \Lambda \rightarrow \Lambda',\ B_1 : V \rightarrow V',\ B_2 : W \rightarrow W',$$

and give functions $F_1 \in \Lambda',\ G_1 \in V',\ H_1 \in W'$ as follows [27,28]

$$[A_1(s),\ \mu] = \int_\Omega (k(|s|)s) \cdot \mu dx,$$

$$[B_1(s),\ v] = -\int_\Omega s\cdot v dx, [B_2(u),\ \omega] = -\int_\Omega (\nabla\cdot u) w dx,$$

$$F_1 = 0,\quad [G_1,\ v] = -< p_D,\ v\cdot n >,\quad [H_1,\ \omega] = \int_\Omega -f\omega dx,$$

Then Eq. (4) will be redescribed as: seek $(p, s, u) \in W \times \Lambda \times V$ such that

$$\begin{cases} [A_1(s),\ \mu] + [B_1{}'(u),\ \mu] = [F_1,\ \mu], \forall \mu \in \Lambda, \\ [B_1(s),\ v] + [B_2{}'(p),\ v] = [G_1,\ v], \forall v \in V, \\ \qquad\qquad [B_2(u),\ \omega] = [H_1,\ \omega], \forall \omega \in W. \end{cases} \tag{12}$$

Lemma 1. *Suppose that*

(i) $A_1 : \Lambda \to \Lambda'$ *satisfies*

$$\|A_1(s_1) - A_1(s_2)\|_{\Lambda'} \le C_1\|s_1 - s_2\|_\Lambda,\quad \forall s_1, s_2 \in \Lambda,$$
$$[A_1(s_1) - A_1(s_2)\cdot(s_1 - s_2)] \ge C_0\|s_1 - s_2\|_\Lambda^2,\quad \forall s_1, s_2 \in \Lambda.$$

(ii) $B_1 : \Lambda \to \Lambda'$ *satisfies*

$$\sup_{v\in\mathbf{V}, v\neq 0} \frac{[B_1(s), v]}{\|s\|_\Lambda\|v\|_V} \ge \gamma_0. \tag{13}$$

(iii) $B_2 : W \to W'$ *satisfies*

$$\sup_{v\in\mathbf{V}, v\neq \mathbf{0}} \frac{[B_2(v), w]}{\|\omega\|_W\|v\|_\mathbf{V}} \ge \gamma. \tag{14}$$

There exists unique $(p, s, u) \in W \times \Lambda \times V$ *and*

$$\|p\| + \|s\| + \|u\|_\mathbf{V} \le C\left(\|f\| + \|g\|_{H^{\frac{1}{2}}(\partial\Omega)}\right). \tag{15}$$

Lemma 2. *There exists* $\gamma > 0$ *satisfy*

$$\inf_{\omega\in W}\sup_{v\in\mathbf{V}} \frac{(\omega, \nabla\cdot v)}{\|\omega\|_W\|v\|_\mathbf{V}} \ge \gamma. \tag{16}$$

Proof. *Let* p *meets*

$$\begin{cases} -\nabla\cdot(k(x, |\nabla p|)\nabla p) = \omega,\ \ x \in \Omega, \\ \qquad\qquad\qquad p = 0,\ \ x \in \partial\Omega. \end{cases} \tag{17}$$

There exists $\alpha > 0$ *satisfy*

$$\|p\| + \|\nabla p\| \le \alpha\|\omega\|.$$

So

$$\begin{aligned} \|v\|_\mathbf{V} = \|v\| + \|\nabla\cdot v\| &= \left(\int_\Omega(-k(|\nabla p|)\nabla p)^2 dx\right)^{\frac{1}{2}} + \|\nabla\cdot v\| \\ &\le k_1\|\nabla p\| + \|\omega\| \le k_1\alpha\|\omega\| + \|\omega\| \le \gamma\|\omega\|. \end{aligned} \tag{18}$$

Hence

$$\|v\|_{\mathbf{V}} \leq C\|\omega\|,$$
$$(\omega, \nabla \cdot v) = \|\omega\|^2.$$

Further we get

$$\frac{(\omega, \nabla \cdot v)}{\|v\|_{\mathbf{V}}\|w\|_W} \geq \frac{\|\omega\|^2}{\gamma\|\omega\|^2} = \gamma,$$

□

Lemma 3. *Suppose* $f \in W^{1,\infty}, g \in W^{2,\infty}$. *Then, there exists unique* $(p_h, s_h, u_h) \in W_h \times \Lambda_h \times V_h$ *of (5). Further*

$$\|p_h\| + \|s_h\| + \|u_h\|_{\mathbf{V}} \leq C\left(\|f\|_{1,\infty} + \|g\|_{2,\infty,(\partial\Omega)}\right). \tag{19}$$

Lemma 4. *There exists* $\gamma > 0$ *satisfy*

$$\inf_{\omega_h \in W_h} \sup_{v_h \in \mathbf{V}_h} \frac{(\omega_h, \nabla \cdot v_h)}{\|\omega_h\|_W \|v_h\|_{\mathbf{V}}} \geq \gamma. \tag{20}$$

4 Error Estimates

Subtracting (4) and (5), error equation is as follows

$$\begin{cases} (\nabla \cdot u - \nabla \cdot u_h,\ \omega_h) = 0,\ \forall \omega_h \in W_h, \\ (u - u_h,\ \mu_h) - (k(|s|)s - k(|s_h|)s_h,\ \mu_h) = 0,\ \forall \mu_h \in \Lambda_h, \\ (s - s_h,\ v) - (p - p_h,\ \nabla \cdot v_h) = 0,\ \forall v_h \in V_h. \end{cases} \tag{21}$$

Lemma 5. *If* $\tilde{s}, \tilde{t} \in R^2$, *then*

$$|k(|\tilde{s}|)\tilde{s} - k(|\tilde{t}|)\tilde{t}| \leq k_1|\tilde{s} - \tilde{t}|, \tag{22}$$

$$(k(|\tilde{s}|)\tilde{s} - k(|\tilde{t}|)\tilde{t}) \cdot (\tilde{s} - \tilde{t}) \geq k_0|\tilde{s} - \tilde{t}|^2. \tag{23}$$

Proof. *From (7) and (8), we get*

$$\begin{aligned} |k(|\tilde{s}|)\tilde{s} - k(|\tilde{t}|)\tilde{t}|^2 &= |k(|\tilde{s}|)|\tilde{s}| - k(|\tilde{t}|)|\tilde{t}||^2 + 2k(|\tilde{s}|)k(|\tilde{t}|)(|\tilde{s}||t| - \tilde{s} \cdot \tilde{t}) \\ &\leq {k_1}^2||\tilde{s}| - |\tilde{t}||^2 + 2{k_1}^2(|\tilde{s}||\tilde{t}| - \tilde{s} \cdot t) = {k_1}^2|\tilde{s} - \tilde{t}|^2, \end{aligned}$$

which proves (22). By (10) and (11), we know

$$\begin{aligned} &(k(|\tilde{s}|)\tilde{s} - k(|\tilde{t}|)\tilde{t}) \cdot (\tilde{s} - \tilde{t}) \\ &= k(|\tilde{s}|)|\tilde{s}|^2 - k(|\tilde{t}|)|\tilde{t}|^2 - (k(|\tilde{s}|) - k(|\tilde{t}|)\tilde{s} \cdot \tilde{t}) \\ &= (k(|\tilde{s}|)|\tilde{s}| - k(|\tilde{t}|))(|\tilde{s}| - |\tilde{t}|) + (k(|\tilde{s}|) + k(|\tilde{t}|))(|\tilde{s}||\tilde{t}| - \tilde{s} \cdot \tilde{t}) \\ &\geq k_0(|\tilde{s}| - |\tilde{t}|)^2 + 2k_0(|\tilde{s}||\tilde{t}| - \tilde{s} \cdot \tilde{t}) = k_0|\tilde{s} - \tilde{t}|^2. \end{aligned}$$

□

Lemma 6. *If $\tilde{s}, \tilde{t} \in R^2$, then*

$$\left(\int_\Omega \left|k(|\tilde{s}|)\tilde{s} - k(|\tilde{t}|)\tilde{t}\right|^2 dx\right)^{\frac{1}{2}} \leq k_1\|\tilde{s} - \tilde{t}\|, \tag{24}$$

$$\int_\Omega (k(|\tilde{s}|)\tilde{s} - k(|\tilde{t}|\tilde{t}) \cdot (\tilde{s} - \tilde{t})dx \geq k_0\|\tilde{s} - \tilde{t}\|^2. \tag{25}$$

Proof. *(24) and (25) are direct corollaries of (22) and (23).* □

Theorem 1. *There exists $C > 0$ satisfy*

$$\|p - p_h\| + \|s - s_h\| + \|u - u_h\| \leq Ch^r(\|p\|_r + \|s\|_r + \|u\|_r), \quad 1 \leq r \leq k+1,$$
$$\|\nabla \cdot (u - u_h)\| \leq Ch^r\|\nabla \cdot u\|_r, \quad 1 \leq r \leq k+1.$$

Proof. *By lemma 4.2, we have $(k(|s|)s - k(|s_h|)s_h, s - s_h) \geq k_0\|s - s_h\|^2$. Further*

$$(k(|s|)s - k(|s_h|)s_h, s - s_h)$$
$$= (k(|s|)s - k(|s_h|)s_h, s - R_hs) + (k(|s|)s - k(|s_h|)s_h, R_hs - s_h).$$

Using ε-inequality

$$(k(|s|)s - k(|s_h|)s_h, s - R_hs) \leq \varepsilon\|s - s_h\|^2 + C\|s - R_hs\|^2.$$

From (21) we see

$$(k(|s|)s - k(|s_h|)s_h, R_hs - s_h) = (u - u_h, R_hs - s_h) = (s - s_h, u - u_h) - (s - R_hs, u - u_h)$$
$$= (s - s_h, u - \pi_hu) + (s - s_h, \pi_hu - u_h) - (s - R_hs, u - u_h)$$
$$= (s - s_h, u - \pi_hu) - (p - p_h, \nabla \cdot (\pi_hu - u_h)) - (s - R_hs, u - u_h)$$
$$\leq \varepsilon\|s - s_h\|^2 + C(\|s - R_hs\|^2 + \|u - \pi_hu\|^2) \leq \varepsilon\|s - s_h\|^2 + Ch^{2r}.$$

So we get

$$\|s - s_h\| \leq Ch^r(\|s\|_r + \|u\|_r), \quad 1 \leq r \leq k+1.$$

Note that

$$\|\pi_hu - u_h\|^2 = (\pi_hu - u, \pi_hu - u_h) + (u - u_h, \pi_hu - u_h).$$

Using ε-inequality

$$(\pi_hu - u, \pi_hu - u_h) \leq \varepsilon\|\pi_hu - u_h\|^2 + Ch^{2r},$$
$$(u - u_h, \pi_hu - u_h) \leq C(s - s_h, \pi_hu - u_h) \leq \varepsilon\|\pi_hu - u_h\|^2 + Ch^{2r}.$$

So we have $\|\pi_hu - u_h\| \leq Ch^r(\|s\|_r + \|u\|_r)$, $\|u - u_h\| \leq Ch^r(\|s\|_r + \|u\|_r)$. Using condition (9),

$$\gamma_h\|P_hp - p_h\| \leq \sup_{v_h \in \mathbf{V}_h} \frac{(P_hp - p_h, \nabla \cdot v_h)}{\|v_h\|_\mathbf{V}}$$
$$= \sup_{v_h \in \mathbf{V}_h} \frac{(s - s_h, v_h)}{\|v_h\|_\mathbf{V}} \leq \|s - s_h\| \leq Ch^r(\|s\|_r + \|u\|_r)$$

Further

$$\|p - p_h\| \leq Ch^r(\|p\|_r + \|s\|_r + \|u\|_r).$$

So we get

$$\|\nabla \cdot (u - u_h)\| \leq Ch^r \|\nabla \cdot u\|_r, \quad 1 \leq r \leq k+1.$$

□

5 Numerical Experiments

We now give four experiments in 2D domain Ω.

Example 5.1. Let $\frac{\tilde{\mu}}{\rho \tilde{k}} = 1$, $\rho = 1$ and $\beta = 1$. The analytical solution is $p = sinxsiny$. We consider the following problem

$$\begin{cases} -\nabla \cdot \left(\dfrac{2\nabla p}{1 + \sqrt{1 + 4|\nabla p|}} \right) = f, & x \in \Omega, \\ \qquad\qquad\qquad\qquad p = 0, & x \in \partial\Omega. \end{cases} \tag{26}$$

here $\Omega = [0, \pi]^2$. The calculation data are listed in following Table 1.

Table 1. Error of Example 5.1

h	$\|\|p - p_h\|\|_{0,2}$	*rate*	$\|\|s - s_h\|\|_{0,2}$	*rate*	$\|\|u - u_h\|\|_{0,2}$	*rate*
$h = \pi/3$	1.3587	—	2.0602	—	1.3751	—
$h = \pi/6$	0.7202	0.92	1.1131	0.89	0.7463	0.88
$h = \pi/12$	0.3498	1.04	0.5513	1.01	0.3721	1.00
$h = \pi/24$	0.1667	1.07	0.2664	1.05	0.1806	1.04
$h = \pi/48$	0.0809	1.04	0.1302	1.03	0.0885	1.03

Example 5.2. Let $\frac{\tilde{\mu}}{\rho \tilde{k}} = 0.7$, $\rho = 1$ and $\beta = 1$. We consider the following equation

$$\begin{cases} -\nabla \cdot \left(\dfrac{2\nabla p}{0.7 + \sqrt{0.7^2 + 4|\nabla p|}} \right) = f, & x \in \Omega, \\ \qquad\qquad\qquad\qquad p = 0, & x \in \partial\Omega. \end{cases} \tag{27}$$

where $\Omega = [0, \pi]^2$. The solution is $p = \sin x \sin y$. The caculation data are listed in the following Table 2.

Table 2. Error of Example 5.2

h	$\|\|p-p_h\|\|_{0,2}$	*rate*	$\|\|s-s_h\|\|_{0,2}$	*rate*	$\|\|u-u_h\|\|_{0,2}$	*rate*
$h=\pi/3$	1.3566	—	2.0516	—	1.6046	—
$h=\pi/6$	0.7094	0.94	1.0966	0.90	0.8649	0.89
$h=\pi/12$	0.3486	1.02	0.5483	1.00	0.4368	0.99
$h=\pi/24$	0.1663	1.07	0.2650	1.05	0.2126	1.04
$h=\pi/48$	0.0807	1.04	0.1294	1.03	0.1041	1.03

Example 5.3. Let $\frac{\tilde{\mu}}{\rho\tilde{k}}=1$, $\rho=1$ and $\beta=1$. The solution is $p=xy-xy^2-y^2x-x^2y^2$. We consider the following problem

$$\begin{cases} -\nabla\cdot\left(\dfrac{2\nabla p}{1+\sqrt{1+4\,|\nabla p|}}\right)=f, & x\in\Omega,\\ \qquad\qquad\qquad\qquad\qquad p=0, & x\in\partial\Omega, \end{cases} \tag{28}$$

where $\Omega=[0,1]^2$. The numerical results are listed in the following Table 3.

Table 3. Error of Example 5.3

h	$\|\|p-p_h\|\|_{0,2}$	*rate*	$\|\|s-s_h\|\|_{0,2}$	*rate*	$\|\|u-u_h\|\|_{0,2}$	*rate*
$h=1/3$	0.014119	—	0.051047	—	0.044219	—
$h=1/6$	0.007350	0.94	0.025746	0.99	0.022509	0.97
$h=1/12$	0.003706	0.99	0.012783	1.01	0.011206	1.01
$h=1/24$	0.001856	1.00	0.006368	1.01	0.005587	1.00
$h=1/48$	0.000928	1.00	0.003179	1.00	0.002790	1.00

Example 5.4. Let $\frac{\tilde{\mu}}{\rho\tilde{k}}=0.7$, $\rho=1$ and $\beta=1$. We consider the following problem

$$\begin{cases} -\nabla\cdot\left(\dfrac{2\nabla p}{0.7+\sqrt{0.7^2+4\,|\nabla p|}}\right)=f, & x\in\Omega,\\ \qquad\qquad\qquad\qquad\qquad p=0, & x\in\partial\Omega, \end{cases} \tag{29}$$

where $\Omega=[0,1]^2$. The true solution is $p=xy-xy^2-y^2x-x^2y^2$. The calculation results are listed in the following Table 4.

Table 4. Error of Example 5.4

h	$\|\|p - p_h\|\|_{0,2}$	*rate*	$\|\|s - s_h\|\|_{0,2}$	*rate*	$\|\|u - u_h\|\|_{0,2}$	*rate*
$h = 1/3$	0.014209	—	0.051637	—	0.057489	—
$h = 1/6$	0.007402	0.94	0.026194	0.98	0.029641	0.96
$h = 1/12$	0.003726	0.99	0.012960	1.02	0.014743	1.01
$h = 1/24$	0.001865	1.00	0.006447	1.01	0.007344	1.01
$h = 1/48$	0.000933	1.00	0.003216	1.00	0.003666	1.00

6 Conclusion

Four examples are given to confirm the analysis results. Five different levels are computed. It can be seen that the error rates are in good agree with above theorems.

Acknowledgements. The work of first author is supported by the key technologies of intelligent manufacturing integration for large scale assembly line of construction machinery of the key research and development plan of Shandong province (2016ZDJS02A12), the research and application of key technologies for intelligent assembly data acquisition and processing of construction machinery based on industrial Internet of things of the major scientific and technological innovation projects in Shandong (2017CXGC0603).The work of second author was supported by the Science and Technology Project for the Universities of Shandong Province (Nos. J14LI53), the Doctoral Foundation of Shandong Jianzhu University (No. XNBS1441).

References

1. Aziz, K., Settari, A.: Petroleum Reservoir Simulation. Applied Science Publishers LTD, London (1979)
2. Neuman, S.P.: Theoretical derivation of Darcy's law. Acta Mech. **25**, 153–170 (1977)
3. Whitaker, S.: Flow in porous media I: a theoretical derivation of Darcy's law. Transp. Porous Media **1**, 3–25 (1986)
4. Fabrie, P.: Regularity of the solution of Darcy-Forchheimers equation. Nonlinear Anal. Theor. Methods Appl. **13**, 1025–1049 (1989)
5. Ruth, D., Ma, H.: On the derivation of the Forchheimer equation by means of the averaging theorem. Transp. Porous Media **7**, 255–264 (1992)
6. Pan, H., Rui, H.: Mixed element method for two dimensional Darcy-Forchheimer model. J. Sci. Comput. **52**, 563–587 (2012)
7. Brezzi, F.: On the existence, uniqueness, and approximation of saddle point problems arising from Lagrangian multipliers. RAIRO Anal. Numer. **2**, 129–151 (1974)
8. Roberts, J.E.: Global estimates for mixed methods for second order elliptic equations. Math. Comp. **44**, 39–52 (1985)
9. Duran, R.: Error analysis in $L^P, 1 < p < \infty$, for mixed finite element methods for linear and quasi-linear elliptic problems. RAIRO Model. Math. Anal. Numer. **22**, 371–387 (1988)

10. Falk, R.S., Osborn, J.: Error estimates for mixed methods. RAIRO Anal. Numer. **14**, 249–277 (1980)
11. Johnson, C., Thomee, V.: Error estimates for some mixed finite element methods for parabolic type problems. RAIRO Anal. Numer. **14**, 41–78 (1981)
12. Park, E.J.: Mixed finite element methods for generalized Forchheimer flow in porous media. Num. Methods Partial Differ. Equ. **21**, 213–228 (2005)
13. Girault, V., Wheeler, M.F.: Numerical discretization of a Darcy-Forchheimer model. Numer. Math. **110**, 161–198 (2008)
14. Rui, H., Pan, H.: A block-centered finite difference method for the Darcy-Forchheimer model. SIAM J. Numer. Anal. **50**, 2612–2631 (2012)
15. Arbogast, T., Wheeler, M.F., Yotov, I.: Mixed finite elements for elliptic problems with tensor coefficients as cell-centered finite difference. SIAM J. Numer. Anal. **34**, 828–852 (1997)
16. Arbogast, T., Dawson, C.T., Keenan, P.T., Wheeler, M.F., Yotov, I.: Enhanced cell-centered finite differences for elliptic equations on general geometry. SIAM J. Sci. Comput. **19**, 402–425 (1998)
17. Chen, Z.: Expanded mixed finite element methods for linear second-order elliptic problems. RAIRO Model. Math. Anal. Numer. **32**, 479–499 (1998)
18. Chen, Z.: Expanded mixed finite element methods for quasilinear second-order elliptic problems. RAIRO Model. Math. Anal. Numer. **32**, 500–520 (1998)
19. Milner, F.A.: Mixed finite element methods for quasilinear second-order elliptic problems. Math. Comp. **44**, 303–320 (1985)
20. Milner, F.A., Park, E.J.: A mixed finite element method for a strongly nonlinear second- order elliptic problem. Math. Comp. **64**, 973–988 (1995)
21. Park, E.J.: Mixed finite element methods for nonlinear second-order elliptic problems. SIAM J. Numer. Anal. **32**, 865–885 (1995)
22. Adams, R.A.: Sobolev Spaces. Academic Press, New York (1975)
23. Raviart, P.A., Thomas, J.M.: A mixed finite element method for 2-nd order elliptic problems. In: Mathematical Aspects of the Finite Element Method. Lecture Notes in Mathematics. Springer, Berlin (1977)
24. Brezzi, F., Douglas, J., Marini, L.D.: Two families of mixed finite elements for second order elliptic problems. Numer. Math. **47**(2), 217–235 (1985)
25. Brezzi, F., Douglas, J., Fortin, M., Marini, L.D.: Efficient rectangular mixed finite elements in two and three space variables. Math. Model. Numer. Anal. **21**, 581–604 (1987)
26. Ciarlet, P.G.: The finite element method for elliptic equations. North-Holland, Amsterdam (1978)
27. Gatica, G.N., Heure, N.: On the numerical analysis of nonlinear twofold saddle point problems. IMA J. Numer. Anal. **23**, 301–330 (2003)
28. Heise, B.: Analysis of a fully discrete finite element method for a nonlinear magnetic field problem. SIAM J. Numer. Anal. **31**, 745–759 (1994)

Research on Indoor Location Method Based on WLAN Signal Location Fingerprints

Tao Wang(✉), Tan Wang, Huanbing Gao, and Yanping Li

School of Information and Electrical Engineering,
Shandong Jianzhu University, Jinan 250101, Shandong, China
wlyiran@126.com

Abstract. Since the outdoor positioning technology has matured, people pay more attention to the development of indoor positioning technology in recent years. The use of existing WLAN signal for indoor positioning is a convenient way to realize. Aiming at the location fingerprint location algorithm based on WLAN signal, an in-depth study has been carried out in this paper. The K-means clustering algorithm and fuzzy logic are used to optimize the traditional algorithms in off-line database creation and on-line location phrase, which is expected to reduce the positioning error while improving the positioning efficiency. In the field simulation experiment, the actual effect of several similar algorithms is analyzed and compared, which proves that the research of this paper is effective for the optimization of indoor location algorithm.

Keywords: Indoor location · Fingerprint match · K-means clustering
Fuzzy logic

1 Introduction

In recent years, wireless network information technology and intelligent mobile terminal such as intelligent mobile phone, tablet computer has been in rapid development. Under these realistic conditions, positioning and navigation service has attracted more and more attention. In the outdoor environment, the mature, popular GPS system and the rapidly developing Beidou navigation system have basically met the needs of people. However, in the indoor environment, Location Based Service (LBS), has not been widely popular. Due to the frequency characteristics of the satellite signal, it has been greatly attenuated when it reaches the interior of the building, and it is difficult to obtain the necessary information. Besides, the indoor environment is complex and diverse, serious multipath effect and non-line-of-sight propagation interference are ubiquitous normality. At the same time, indoor positioning and navigation applications have higher requirements for precision, so they need targeted special positioning methods, which has also become a hot topic of relevant academic institutions at home and abroad [1, 2].

After years of extensive research, indoor positioning technology has formed lots of complex category. Current indoor positioning technology mainly includes infrared positioning, Bluetooth positioning, ultra-broadband positioning, RFID (radio frequency identification) positioning, ZigBee technology positioning, etc. [3]. Among those

L. Barolli et al. (Eds.): EIDWT 2018, LNDECT 17, pp. 828–840, 2018.
https://doi.org/10.1007/978-3-319-75928-9_75

technology, the localization method based on WLAN (Wireless Local Area Networks) has wide application prospect because of its use of the existing common signal network, so many terminals, easy installation, low cost and meeting certain precision requirements [4].

In the algorithm aspect of WLAN positioning technology, there are two main methods: geometric positioning method and scene analysis. The geometric positioning method needs to measure the distance between the target point to be measured and the reference point to determine the position of the target. The most commonly used methods to obtain the distance are Time of Arrival (TOA) method, Time Difference of Arrival (TDOA) method and the method of using the wireless signal propagation model (Shadowing model). The range accuracy of arrival time method depends on the measurement error of time, because the general radio magnetic signal propagation speed is very fast, and indoor space distance is limited, so subtle time deviation will seriously affect the accuracy of positioning. The method of arriving at the time difference improves the method of arrival time, narrowing down the requirement of strict synchronization of time to a range among base station only, so reduces the difficulty of realization. However, both of these methods require high precision of hardware devices and signal transmission quality. The method based on the wireless signal propagation model relies too much on the empirical model, except that the accuracy of the mathematical model is less than satisfactory and lacks universality for different environments. All above method can not overcome the influence of serious reflection, scattering and multipath phenomenon generally existing in the indoor environment, so the positioning performance is often less than ideal. The scene analysis method, also known as location fingerprinting, comes from the theory of pattern recognition. It can overcome the influence of irregular signal propagation to a certain extent so is more suitable for indoor location scenarios [5].

2 Location Fingerprint Localization Algorithm

The location fingerprint localization algorithm is based on the mutual mapping relationship between the collected physical signal features and the different locations of the region. For wireless electromagnetic signals reaching to each location within the space, path structure and signal propagation attenuation degree itself is different. This difference forms a "fingerprint" feature that distinguishes each position point from other points [6]. Therefore, collecting the radio signal characteristics at different positions in the space, and constructing the database before positioning, is equivalent to drawing an electronic map of the wireless signal intensity distribution [7]. Of course, the finer the map is, the higher the positioning accuracy should be. However, this will greatly increase the workload of previous data collection and database construction and make the positioning operation complicated and lengthy. Afterwards, in the real-time positioning stage, the numerical value of the signal characteristic actually received at the location is matched with the signal fingerprint in the database to find the similar point of the fingerprint and a specific operation on the coordinates of single or several similar points is carried out to realize the estimation of the target location [8]. Combined with

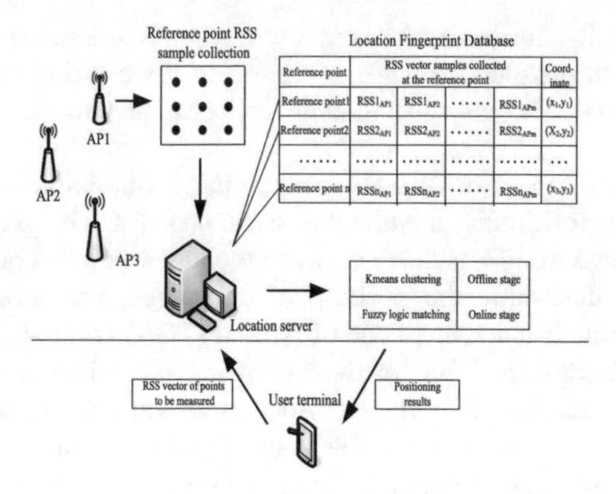

Fig. 1. Principle of positioning method based on position fingerprint

the algorithm used in this paper, the basic principle of localization based on the location fingerprints is as shown in Fig. 1.

Location fingerprinting methods generally include offline and online two processes [9]. The first offline stage, during which the collection and arrangement of the location fingerprint data of the spatial signal is in process, is the preparatory phase of system before positioning. First of all, a certain number of position reference points also called fingerprint nodes, are set in the positioning area according to a reasonable interval distribution, and the coordinates of these position points are measured. Then detect the feature value of the wireless signal at each position reference point and record the value of these Received Signal Strength (RSS) values from different sources. The random error of data can be reduced by means of multiple measurement and taking average. These RSS values from different AP (Wireless Access Point) make up the RSS vector of reference point, which can be expressed as:

$$\boldsymbol{RSSi} = \left(RXi_{A1}, RXi_{A2}, \cdots, RXi_{Aj}, \cdots, RXi_{An}\right) \tag{1}$$

Among this expression, "*i*" is serial number of reference point, "RXi_{Aj}" is the received signal strength value corresponding to the "*j*"th AP, "A*j*" represents the access point numbered "*j*".

Finally, the coordinate information of this point and RSS vector are combined into a record in the fingerprint database. Repeat this process at each fingerprint node to get the fingerprint database for that area.

The second stage is online positioning stage, which is the real-time positioning stage based on the signal characteristic vector detected at the location. The mobile terminal device carried by the user sends the signal characteristic parameters of the location to the positioning server through the positioning application software, and the server program searches and matches the data with the fingerprint information in the database by a certain matching algorithm, and selects the most similar one or several fingerprint data of the result of the comparison using their location information to estimate the position coordinates of the mobile terminal.

3 Filtering Processing of Sampled Signals

The initial job of setting up the fingerprint database in off-line stage is to collect the wireless signal data of the predetermined fingerprint point in the field. During sampling, we will find that the data continuously read at the same position is in a fluctuating state, and unexpected singular values are also common. In addition to the factors of the source itself, the various noise interference in the indoor environment and the change of the transmission route caused by the movement of personnel and goods are the important reasons of the data instability. The interference of these objective conditions is unavoidable, and the usual approach is collecting a set of data to fetch its average value. The algorithm of moving average filter with the same simplicity but better effect is used here.

The specific implementation method is: In a data sequence of an AP collected at the same location, the first N data are selected as a cyclic array, and N is always kept as a fixed value. The following data join the array in turn, while discarding the data at the first position in the array, the output remains the average of the existing data in the array, until the entire sequence is completed.

Mathematical operation formula is as follows:

$$FLout(n) = \frac{1}{N}\sum\nolimits_{i=n-N+1}^{n} s(i) = FLout(n-1) + \frac{s(n) - s(n-N)}{N} \quad (2)$$

In the above formula, $FLout(n)$ represents the output of the filter, $s(n)$ is the nth sample value in the sequence. Not the larger the value of N, the better the filtering effect, which depends on the total length of the data sequence.

4 Clustering of Databases

4.1 Clustering Algorithm

In the traditional method of location fingerprint localization, fingerprint vector matching can be performed online immediately to determine the position of the mobile terminal after the fingerprint database is created. However, the fingerprint database is often quite large, especially in the wider space such as the parking lot. In this case, comparing the signal intensity vector obtained from the point to be measured and the record in the fingerprint database one by one will consume a large amount of computing time and reduce the real-time performance of positioning. This article makes an improvement on this direct method of operation. In the offline phase, the original database is pre-processed by clustering method to reduce the workload of online positioning phase.

Clustering is a powerful tool for mining information of mass data. By clustering, we can systematically understand the status of data distribution, furthermore summarize the macroscopic information of large data, or concentrate some data analysis program within certain homogeneous range, improving processing efficiency and making deep information extraction. Specifically, clustering takes a certain variable (such as distance) that can express the similarity between the data as a standard and divides a data set containing a large amount of unidentified data into several categories. The goal is to make the similarities between individuals within the same class as large as possible, while the similarity between classes is as small as possible, to achieve the effect of "Birds of a feather flock together".

The clustering algorithm has three key points:

1. Select the criterion variable that measures the similarity between individuals.
2. Select the standard function evaluating clustering effect.
3. Update the cluster center in each category.

The fingerprint database gets a few clusters (Cluster) after the clustering operation. The signal data of the points to be measured only matches in the class, which will greatly reduce the times of comparison, reduce the amount of matching operations, and can avoid adverse impact of some points with long distance but smaller difference between the RSS vectors.

In order to give consideration to the efficiency of data preprocessing, this paper adopts the K-means algorithm in the traditional clustering method, which is a centroid-based heuristic classic clustering method with simple and efficient features and is suitable for rapid processing of large data sets.

The overall process of the algorithm is like this: Select randomly K individuals in the set as the original centers first, then calculate the values of the variable that reflect the degree of difference between the remaining individuals and the K centers, thereby classifying all the individuals into K initial clusters. After that, the new central individuals are selected within these clusters, and then assigned again according to these cores. Such iteration makes the criterion function for evaluating the clustering effect converge, and the criterion function of the K-means algorithm is usually the Sum of Squares Function of Error. Or after such a number of dynamic allocation of all individuals, the center of each cluster will no longer change, indicating that the subsequent clustering process becomes a complete repeat, the algorithm can be completed [10].

4.2 Application of K-Means Algorithm

Before the implementation of clustering algorithm, the premise is to determine the number "K" of clusters, depending on the specific circumstances of the positioning area. At the same time, the Euclidean distance is used to measure the similarity between the RSS vectors of the reference points in this paper. The concrete process Follows the steps below:

First, select K fingerprint records as the initial cluster center. The K-means clustering algorithm is sensitive to the initial cluster center, so the initial center point should be as evenly distributed as possible in the positioning area. K-means algorithm has the selection method to alleviate this sensitivity, but it needs a certain amount of calculation, which is not recommended here.

Secondly, calculate the Euclidean distances between the fingerprint vectors of other reference points and each cluster center one by one, and assign them to the nearest neighbor clusters according to the principle of maximum similarity.

Thirdly, after this round of distribution is completed, an update of the database clustering is obtained, then take the average of all the vectors in each cluster as the new clustering center. The updating formula of clustering center is as follows:

$$\boldsymbol{RSSq}(\boldsymbol{k}) = \frac{1}{|C_q|}\sum\nolimits_{i \in C_q} \boldsymbol{RSSi} \tag{3}$$

Repeat the process of the above two steps, iterate the loop, until all the cluster centers no longer change, then the iteration can be terminated, the clustering process is over. Thus each fingerprint has its own category, and each category is treated as a relatively independent subset of the fingerprint database.

After the clustering operation, the fingerprint data of different locations are recorded in the database in the form of the following information:

$$\left(RXi_{A1}, RXi_{A2}, \cdots, RXi_{Aj}, \cdots, RXi_{An}, x_i, y_i, \mathrm{C_q}\right)$$

Where "RXi_{Aj}" is the signal strength received from the "j"th AP at the "i"th reference point, "x_i" is the abscissa of the reference point, "y_i" is the ordinate of the reference point, "$\mathrm{C_q}$" indicates that the current fingerprint belongs to a cluster labeled "q".

In this way, during the online positioning phase, the Euclidean distances between the RSS vector obtained at the location point and each cluster center vector, are compared each other, and the cluster category of the locating point is determined according to the nearest neighbor principle. Then the fingerprint matching and the coordinate calculation process are only carried out in this cluster.

5 Algorithm for Determining Position Coordinates

Deterministic algorithm and Probabilistic algorithm are two kinds of methods to estimate position coordinate after adjacent points are obtained, the deterministic algorithm includes nearest neighbor(NN) method, K-Nearest neighbor(KNN) method, weighted K-Nearest neighbor(WKNN) method and Neural network method [11]. Probabilistic algorithm is mainly Bayesian probability algorithm.

If "d_i" is used to represent the Euclidean distance between the RSS vector of the point to be located and the RSS vector of the "i"th reference point, the NN algorithm directly considers the coordinates of the reference point with the least "d_i" value as the coordinates of the point to be located. The KNN algorithm selects K adjacent points with the least "d_i" value and calculates the average value of their position coordinates to replace the coordinates of the point to be located. On the basis of KNN algorithm, the WKNN algorithm transforms calculating average value to the weighted sum method, and the weighting coefficient is calculated generally by the following formula:

$$w_i = \frac{d_i^{-1} + \varepsilon}{\sum_{i=1}^{k} d_i^{-1} + \varepsilon} \tag{4}$$

In the formula, "ε" is set to a small positive value to avoid the situation that the denominator is zero [12].

In most cases, the KNN algorithm and the WKNN algorithm have a better positioning effect when the K value is 3 or 4, which is obviously higher than that of the NN algorithm [13].

Neural network is a kind of machine learning algorithm, which is a more effective method to realize nonlinear innuendo. The whole neural network generally consists of an input layer composed of a series of perceptrons, hidden layers containing one or more computing units and an output layer, which can achieve that input the necessary data into the network then directly obtain the coordinate information at the output of the network.

The Bayesian probability algorithm calculates the posterior probability value of the signal received at the point to be located relative to the corresponding position of each fingerprint reference point in the location region, as the weighted coefficient to estimates the coordinates of the locating point.

Previous research and practice show that WKNN algorithm has advantages compared to other algorithms in various positioning accuracy indicators.

In this paper, the WKNN algorithm in deterministic algorithm is adopted, but the weighted coefficient value is determined by the method of fuzzy logic instead of the general formula in order to obtain better positioning precision.

6 Fuzzy Logic Algorithm of Weighted Coefficients

The fuzzy logic algorithm in the intelligent algorithm, which simulates the reasoning thinking mode on the concept of uncertainty of mankind, has certain advantages over the precision algorithm in solving the complex strong nonlinear problem, so it is suitable for judging the matching similarity.

As an essential nonlinear model, the Sugeno model in fuzzy logic can be used to represent the characteristics of complex dynamic systems, and the results after inference are clear numbers which can be used directly as weights [14]. Therefore we use Sugeno reasoning system with double input and single output to calculate the weighted value in WKNN algorithm. The Fig. 2 shows the structure of the fuzzy inference system we used.

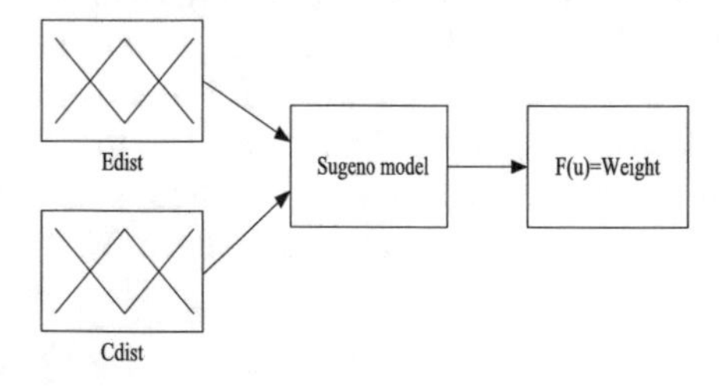

Fig. 2. Structure of fuzzy inference system

First, the input data is the RSS vector uploaded by the mobile terminal at its position, similar to the form of above formula, the vector is expressed as:

$$\boldsymbol{RSSx} = \left(RXx_{A1}, \cdots, RXx_{Aj}, \cdots, RXx_{An}\right) \tag{5}$$

Where, the order of each RSS component corresponding to each AP should be consistent.

As mentioned above, after the K-means clustering, the fingerprint database is divided into several clusters, and the above vectors has been compared with each cluster center to determine their own categories. So the Euclidean distance as the input of fuzzy logic is limited to between it and other RSS vectors in the cluster, denoted as "*Edist*". The formula is:

$$Edist_i = \sqrt{\sum\nolimits_{j=1}^{n} \left(RXx_{Aj} - RXi_{Aj}\right)^2} \tag{6}$$

Then take normalized scale transformation on this variable, the normalized formula is as follows:

$$normEdist_i = \frac{Edist_i - \min\left(Edist_{C_q}\right)}{\max\left(Edist_{C_q}\right) - \min\left(Edist_{C_q}\right)} \tag{7}$$

In the above formula, "$Edist_{C_q}$" represents the Euclidean distance between the RSS vectors of the point to be located and all reference point in the cluster. "$normEdist_i$" is the result of normalization.

In order to use fuzzy rules to make the judgment processing, it is necessary to fuzzify the input variables that have been make the domain transformation. The fuzzy sets and membership function of the fuzzy process are shown in Fig. 3.

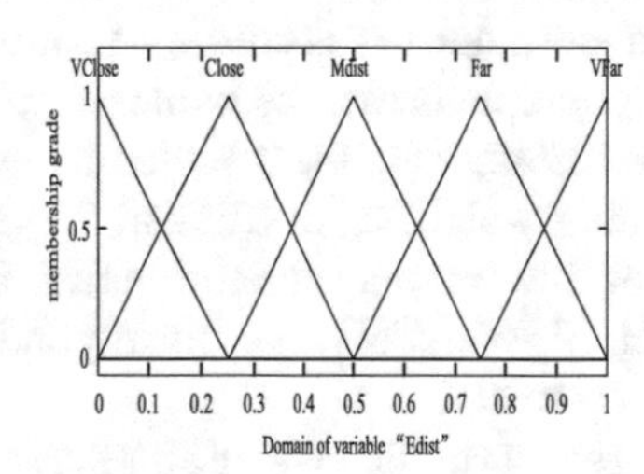

Fig. 3. Fuzzy sets and membership function of input variable "*Edist*"

The Euclidean distance between the RSS vector of the point to be located and the reference point and the spatial distance between the two points are not strictly monotonic correspondences, two points close relatively to each other may not be similar in Euclidean distances on signal. Therefore, it is not ideal to judge the similarity degree by Euclidean distance alone [15].

For this reason, the cosine distance between the two vectors is calculated to be a reference variable. The formula is as follows:

$$Cdist_i = \frac{RSSi \cdot RSSx}{||RSSi||^2 * ||RSSx||^2} = \frac{\sum_{j=1}^{n} \left(RXi_{Aj} * RXx_{Aj}\right)}{\sqrt{\sum_{j=1}^{n} RXi_{Aj}^2} * \sqrt{\sum_{j=1}^{n} RXx_{Aj}^2}} \tag{8}$$

Also normalized, the formula is:

$$\text{norm}Cdist_i = \frac{\left|Cdist_i - \max\left(Cdist_{C_q}\right)\right|}{\max\left(Cdist_{C_q}\right) - \min\left(Cdist_{C_q}\right)} \tag{9}$$

In the formula "$Cdist_{C_q}$" is the cosine distance between the RSS vector of the location point and all reference points in the cluster.

The fuzzy set settings and membership functions of these input variables are shown in Fig. 4.

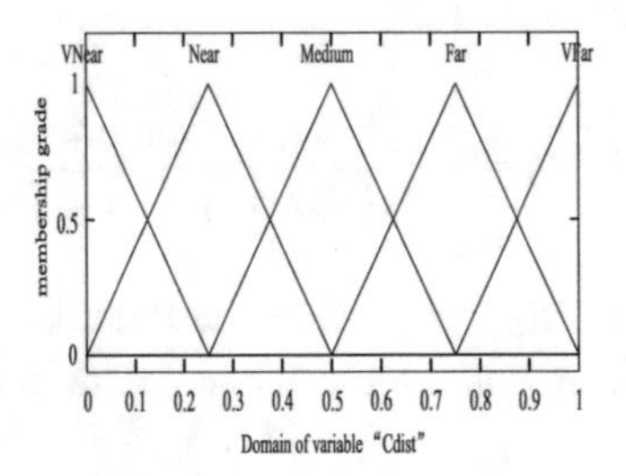

Fig. 4. Fuzzy sets and membership function of input variable "*Cdist*"

Fuzzy reasoning rules using the above two logical variables will not be described in detail, and should be adaptively adjusted according to the specific locating environment. However, the general principle is that the value of "*normEdist*" is the major one, refer to the value of "*normCdist*", and the smaller the value of "*normEdist*" and "*normCdist*", the corresponding weighting coefficient is greater.

The fuzzy value set of the output variable after fuzzy inference is set to (NonWei, LowWei, MedWei, HigWei, FulWei), corresponding to the output weights of (0, 0.25, 0.5, 0.75, 1) respectively.

After calculating the weighting coefficients of all reference points in the cluster, the K reference points with the largest coefficients are selected, and the position coordinates of the location points are obtained by weighted summation operation on this reference points. The calculation formula for position coordinates is as follows:

$$(\mathrm{x}, \mathrm{y}) = \sum_{i=1}^{k} \frac{w_i * (x_i, y_i)}{\sum_{i=1}^{k} w_i} \tag{10}$$

In the formula (x_i, y_i) is the coordinate of the reference point, "w_i" is the weighting coefficient of this reference point. Here, the coordinates (x, y) of the location point are finally obtained, and the positioning process ends.

The overall flowchart of the position fingerprint location algorithm described in this paper is shown in Fig. 5.

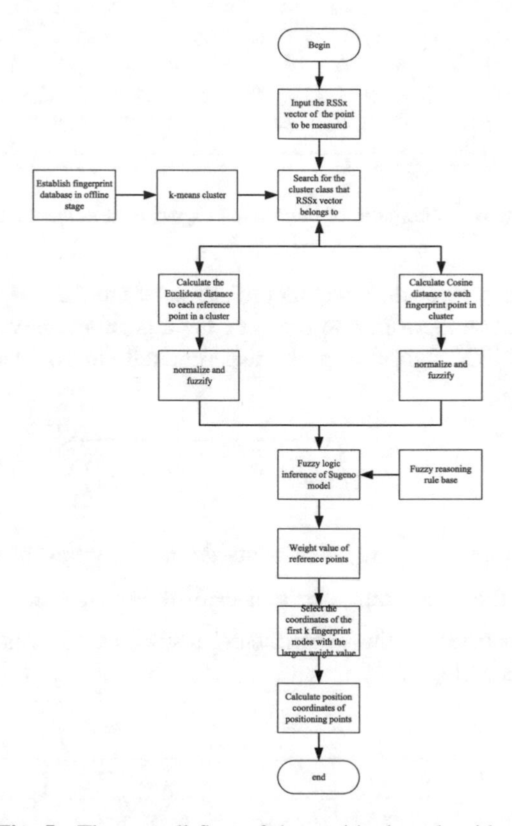

Fig. 5. The overall flow of the positioning algorithm

7 Simulation Experiment and Result Analysis

In order to verify the validity of the above algorithm, a positioning simulation experiment was conducted in the indoor working area on the first floor of the Research Institute in which isolating an irregular area of 16 m in length and 9.5 m in width. At the geometric vertex of the experimental area, 5 wireless routers are placed as AP, and 37 fingerprint reference points are calibrated in the region to sample the wireless signal for fingerprint database establishing, and 27 detection points with measured coordinates are used to verify the positioning accuracy. The plane sketch and the coordinate system design of the experimental environment are shown in Fig. 6.

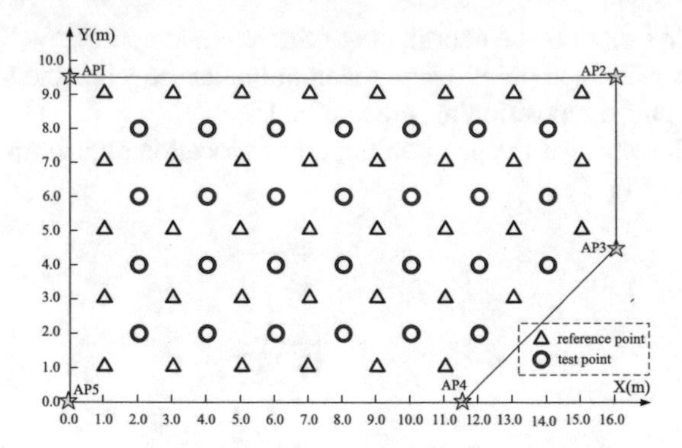

Fig. 6. Schematic of simulation experiment environment

In the same system, we adopt the simplification of the database and the method of simplifying the location algorithm to compare the algorithm above and the traditional NN algorithm and KNN algorithm in the experimental environment. Define the positioning error as:

$$err_p = \sqrt{\left(x_p - x_p'\right)^2 + \left(y_p - y_p'\right)^2} \tag{11}$$

In the above formula, (x_p, y_p) represents the actual coordinate of the test point, $\left(x_p', y_p'\right)$ represents the positioning coordinates of the test point.

The error comparison of the experimental results of 10 typical positions in the experimental area is as Fig. 7.

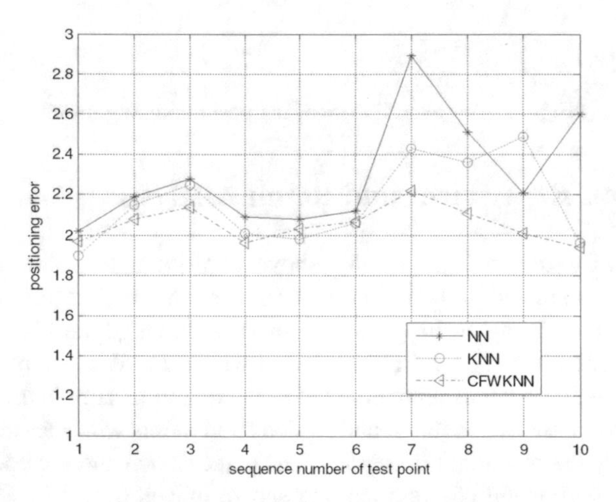

Fig. 7. Positioning error of fixed test point

It can be seen that the experimental results are basically consistent with the expected situation, and the coordinates obtained by algorithm in this paper are closer to the actual coordinates in most cases, which shows that fuzzy logic is effective for the localization algorithm optimization.

The cumulative positioning error index can more comprehensively reflect the positioning effect of various positioning algorithms in the whole experimental area. For this reason, the test points are randomly selected for 100 repeated experiments, and the statistic results show the percentages in different error distances, as shown in the Fig. 8.

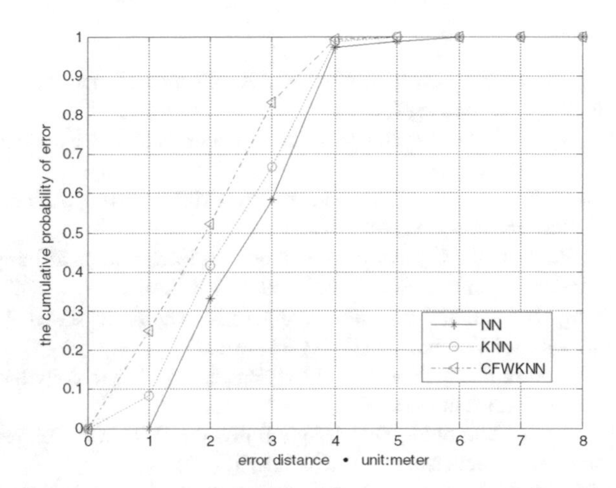

Fig. 8. Comparison of cumulative positioning error probability of different methods

The results of repeated experiments show that the experimental results of algorithm in this paper appear in the range of less than 3 m with higher probability. It is proved that the design of the algorithm is effective and feasible.

In addition, the average running time of the traditional KNN algorithm in experiment is 0.3271 s, and the average running time of the locating algorithm described in this paper is 0.2959 s. It proves that the clustering for reference data is also effective for shortening the positioning time.

8 Conclusion

In this paper, aiming at the fingerprint matching location algorithm based on WLAN signal, an improved method is proposed. In view of the large and complex location fingerprinting databases in large-scale space environment, a concise K-means clustering algorithm is used to classify fingerprint database in advance, which simplifies the calculation of fingerprint matching. In the online location phase, the weighted coefficients of the adjacent points are calculated by using the fuzzy logic Sugeno model based on the Euclidean distance and Cosine distance reflecting the similarity of vectors from different angles. We hope to improve the precision of positioning with more reasonable operation mode. The validity of the proposed algorithm is verified by the later simulation experiment.

Acknowledgments. The authors thank all the reviewers and editors for their valuable comments and works. This research is Supported by the Key research and development plan of Shandong Province (Major key technology) (No. 2016ZDJS02A12), the Major scientific and technological innovation project of Shandong Province (No. 2017CXGC0603).

References

1. Wan, Q., Guo, X., Chen, Z.: Theory, Method and Application of Indoor Positioning Method. Electronic industry press, Beijing (2012)
2. Wang, Y., Zhao, H.-D.: Overview and prospect of indoor location techniques. Meas. Control Technol. **35**(07), 1–3+8 (2016)
3. Wang, S.: Research on the application of hybrid wireless positioning technology. Informatization Res. **36**(3), 43–45, 48 2010
4. Shi, G., Wang, B., Wu, B.: Overview of indoor localization method based on WiFi and mobile smart terminal. Comput. Eng. **41**(09), 39–44+50 (2015)
5. Dong, Y., Zhang, H., Chen, J.: Location fingerprint algorithm based on Wi-Fi indoor positioning. Ind. Control Comput. **28**(1), 72–74 (2015)
6. Du, S.: Research and Implementation of Indoor Positioning Technology based on Location Fingerprint. Yunnan University (2013)
7. Lei, L.: Application on Positioning Technology Based on RFID in Warehouse Management. HuaZhong University of Science and Technology (2012)
8. Wu, D.: Tourist Attractions Mobile Phone Intelligent Navigation System Based on WIFI Scan. Jilin Agricultural University (2016)
9. Tang, N., Xiao, X., Chen, Z.: A method of multi-mode switching for SVC based on Sugeno Fuzzy Inference. Power Syst. Technol. **35**(08), 140–143 (2011)
10. Yang, M., Liu, K., Shao, D.: PCA clustering algorithm for indoor positioning in WLAN. Telecommun. Sci. **32**(07), 21–26 (2016)
11. Wang, Y., Ba, B., Cui, W., et al.: Indoor positioning algorithm based on Markov Monte Carlo. J Xidian Univ. (Sci. edn.) **43**(02), 145–149 (2016)
12. Miao, H., Wang, J., Li, C., et al.: A fuzzy logic-based indoor location approach. Control Instrum. Chem. Ind. **41**(04), 387–391+396 (2014)
13. Mao, Q., Zeng, B., Ye, L.-F.: Research on improved indoor mobile robot fuzzy position fingerprint localization. Comput. Sci. **42**(11), 170–173 (2015)
14. Cai, Z., Xia, X., Hu, B., et al.: Improvement of indoor signal strength fingerprint localization algorithm. Comput. Sci. **41**(11), 178–181 (2014)
15. Guo, W.: Research on Indoor Positioning Algorithm Based on Fuzzy Inference. University of Electronic Science and Technology of China (2015)

Homomorphic Authentication Based on Rank-Based Merkle Hash Tree

Ping Bai[1], Wei Zhang[1,2], Xu An Wang[1,2(✉)], Yudong Liu[1], HaiBin Yang[1], and Chun Shan[3]

[1] Electronic Technique Department, Engineering University of CAPF, Xi'an, China
wangxazjd@163.com
[2] Key Laboratory of Information Security, Engineering University of CAPF, Xi'an, China
[3] Guangdong Polytechnic Normal Universities, Guangzhou, People's Republic of China

Abstract. Under the settings of cloud storage, user's private data is distributed and sent to different servers for storage service, thus authentication systems are required to ensure data integrity. In this paper, combining the idea of Dario Catalanno's arithmetic circuit with Rank-based Merkle Hash Tree structure, a novel homomorphic authentication scheme is proposed. The main advantage of the proposed scheme is that the integrity of data transmission can be validated between different servers.

1 Introduction

With the continuous development of cloud computing technology [1], cloud storage gradually arouses people's attention. Compared with traditional storage methods, cloud storage has the following advantages: First, from a functional point of view, the cloud storage system is a collection of multiple online storage services in the network, while traditional storage systems are geared toward hardware. Second, from the performance point of view, the most important indicators that cloud storage services need to consider are data security, reliability and efficiency. Meanwhile, because of the complex network environment, high-quality cloud storage services put forward greater technical requirements. Third, from the efficiency point of view, it can provide huge storage resources without the need for local data storage and maintenance. These unique features attract many people to explore. There have been many research results in this area [2–6]. In the meantime, since different cloud servers have different functions, users may transmit data between different cloud servers to ensure the maximum benefits of users. However, the cloud server is an institution that cannot be completely trusted. User data may be maliciously tampered with or even lost due to various reasons, which brings great security risks to user data and poses new challenges to cloud storage. In order to ensure the safety of user data, a safe and efficient mechanism need be designed to ensure the interests of users.

Homomorphic MACs [7] were originally proposed by Gennaro and Wichs. The homomorphic message operation certification has the following three properties:

L. Barolli et al. (Eds.): EIDWT 2018, LNDECT 17, pp. 841–848, 2018.
https://doi.org/10.1007/978-3-319-75928-9_76

(1) The adversary can obtain the label of the news which it wants to know by inquiring, etc., but other labels of the message cannot be calculated by these labels. (2) Homomorphic message authentication should satisfy conciseness, that is, the cost of authenticating the tag is less than the cost of sending the original data message. (3) Homomorphic message authentication should be complex, that is, the previous verification tags can be reused in the next calculation. Compared with other verification methods [6], homomorphic MACs has unique advantages. All in all, in order to verify the integrity of data, homomorphic message authentication allows the user to use the private key to generate a label for the authentication message.

Combined with Rank-based Merkle Hash Tree [8], we use the idea of Encodings with Limited Malleability in the arithmetic circuit based homomorphic message authentication scheme proposed by Dario Catalanno [9] and construct a scheme of homomorphic authentication based on Rank-based Merkle Hash Tree during cloud data transportion. Compared with the literature [9], our scheme has some deficiencies in the degree of complexity. However, this scheme can verify the data integrity between different cloud servers, and enhances the security of user data storage. Homomorphic authentication allows users to authenticate data messages without knowing the private key, which is largely user-friendly. As homomorphism certification receives more and more attention, many research results have emerged [10–16].

In Sect. 2, we give the preliminary knowledge of this paper. We present the solution model and construction of our proposed scheme in Sect. 3. In Sect. 4, its security and performance analysis are described respectively. In the final section, we make a conclusion.

2 Preliminares

2.1 Bilinear Map

Let G and G_T be two multiplicative cyclic groups of prime order p. Let g be generator of G. Definition: $e : G \times G \to G_T$ be a bilinear map with the following properties:

(1) Bilinearity: For all $\mathrm{r}, \mathrm{s} \in G$ and $a, b \in Z_p$, then $e\left(r^a, s^b\right) = e(r, s)^{ab}$.
(2) Non-degeneracy: For arbitrary $\mathrm{r}, \mathrm{s} \in G$, $e(\mathrm{r}, \mathrm{s}) \neq 1$.
(3) Computable: For all $\mathrm{r}, \mathrm{s} \in G_T$, $e(\mathrm{r}, \mathrm{s})$ can be computed in polynomial time.

2.2 Labeled Program

The notion of labeled programs was introduced by Gennaro and Wichs. A labeled program **P** is defined by a tuple $\mathbf{P} = (f, \tau_1, \tau_2, \ldots, \tau_n)$ where $f : M^n \to M$ is a circuit, and the binary strings $\tau_1, \tau_2, \ldots \tau_n \in \{0, 1\}^*$ are the labels of the input nodes of f. Given some labeled programs $\mathrm{P}_1, \mathrm{P}_2, \ldots \mathrm{P}_t$ and a function $g : M^t \to M$, the composed program $\mathrm{P}^* = g(\mathrm{P}_1, \mathrm{P}_2, \ldots \mathrm{P}_t)$ is the circuit which evaluates a circuit g on the outputs of $\mathrm{P}_1, \mathrm{P}_2, \ldots \mathrm{P}_t$ respectively. While the scheme [6] defined labeled programs for Boolean circuits (i.e. $f : \{0, 1\}^n \to \{0, 1\}$), here we consider its extension to the case of arithmetic circuits $f : M^n \to M$.

2.3 Homomorphic Authenticator Scheme

The homomorphic message authenticator scheme (HomMAC) includes 4-tuple of algorithms as follows:

KeyGen(1^{λ}): Input the security parameter λ, the key generation algorithm outputs a secret key *sk* and a public evaluation key *ek*.

Auth(sk, τ, m): Given the secret key *sk*, an input-label τ and a message $m \in \mathrm{M}$, it outputs a tag σ.

Ver$(sk, m, \mathrm{P}, \sigma)$: Given the secret key *sk*, the message *m*, a program $\mathrm{P} = (f, \tau_1, \tau_2, \ldots \tau_n)$ and tag σ, the verification algorithm outputs 0 or 1.

Eval(ek, f, σ): Input the evaluation key *ek*, a circuit $f : M^n \rightarrow M$ and a vector of tags $\boldsymbol{\sigma} = (\sigma_1, \ldots, \sigma_n)$, the evaluation algorithm outputs a new tag σ.

2.4 Rank-Based Merkle Hash Tree

Merkle Hash Tree [8] is a common structure for data transfer authentication. RMHT is a promotion of this structure, the main purpose is to be able to validate the data deletion ability. The input of nodes in the RMHT structure is not only the value of the left and right children nodes, but also the number of leaf nodes in the node. As shown in Fig. 1, $h_i = H(m_i \| 1)(i = \{1, 2, \cdots, 8\})$ represents the node has a leaf node. $h_c = H(h_1 \| h_2 \| 2)$ shows h_c has two leaf nodes. $h_a = H(h_c \| h_d \| 4)$ shows h_d has four leaf nodes. In addition, the node's Auxiliary Authentiation Information (AAI) provides a great convenience for the calculation of the root node. Auxiliary verification information is mainly formed by the siblings of the upper level, for example, the node m_2 of AAI can be expressed as $\Omega_i = \{h_1, h_d, h_b\}$.

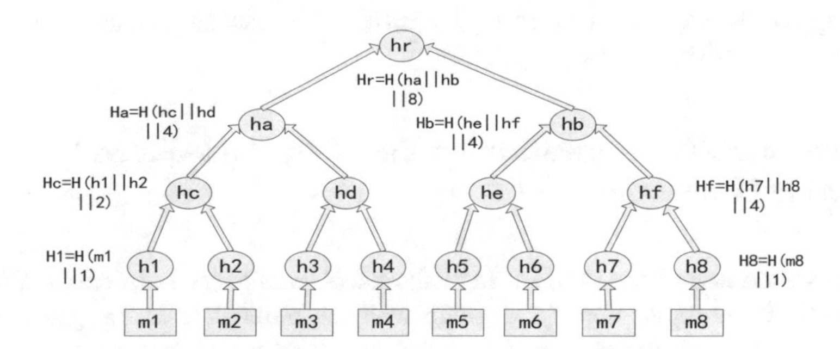

Fig. 1. The schematic of diagram of RMHT

2.5 Limited Extensibility Coding

The limited coding algorithm ξ consists of three sub-algorithms, as defined below:

Encryption generation algorithm (1^{λ}): Input the security parameters λ, then outputting the public key *pk* and private key *dk*. The encoding algorithm's message space is Z_p, where p is a prime with λ bits. Encoding space is T.

Encryption Algorithm (pk, m): Input the message $m \in Z_p$ into a random algorithm and output the value $t \in T$.

Test algorithm (dk, m, t): Input $m \in Z_p$ and $t \in T$ into a master algorithm to check if t equals the output of the above encryption algorithm. If the result is true, the output is 1, otherwise, the output is 0.

It can be seen from the above algorithm that the limited coding algorithm ξ is decisive, we do not need to use the private key *dk* for verification in the test algorithm, we only need to repeat the encoded message *m*, the test output is equal to the output of the encryption algorithm.

Definition 1. The coding algorithm has the property of additive homomorphism, for any $m, m' \in Z_p$, $t \in Enc(pk, m)$, $t' \in Enc(pk, m')$, we have $t \times t' \in Enc(pk, \mathrm{m} + m')$.

2.6 The Instance Model of Homomorphic Authentication Based on Rank-Based Merkle Hash Tree

The storage of outsourced data during the cloud environment mainly consists of three kinds of physical structures: Ordinary users, cloud servers, trusted third parties.

Ordinary users: The data storage capacity is weak relatively, and tends to take complicated data resources to the cloud server for storage or computation. However, they hope their data not be stolen or corrupted by the cloud server.

Cloud server: The cloud server has a strong ability of storage and computation, and can provide users with cloud storage and computing services. But there is an issue that the data on the cloud server may be maliciously attacked by hackers, so we must validate the stored data in order to ensure safety.

Third party (TPA): As the intermediate medium between users and cloud server, the third party will get the final result feedback to the user in advance, ensuring the security of transmission data.

3 Homomorphic Authentication Based on Rank-Based Merkle Hash Tree

Key generation algorithm $(1^{\lambda}, D)$**:** Input the safety parameters λ, the depth of poly(λ) is D. G, G_T is multiplication cycle group with the order p. g is the generator and $e : G \times G = G_T$ is a bilinear map. Run encryption algorithm (1^{λ}) to get the parameters pk, dk, p, T. Randomly select the parameters $u \in G$, $x \in Z_p$. $\alpha, \beta \in Z_p^*$ is a pseudo-random function $F_K : \{0, 1\}^* \rightarrow Z_P$ with seeds K, k is the symmetric key for encryption *AES*. Define the hash function $H : \{0, 1\}^* \rightarrow Z_P^*$ and it is used to construct the Merkle Hash Tree structure. Define public-private key authentication key generation algorithm $(spk, ssk) \leftarrow Sign()$ and compute $h_i = Enc(\mathrm{pk}, \mathrm{x}^i), \mathrm{i} = 0, 1, \ldots \mathrm{D} - 1$. Assuming that all operations are on Z_p, public key is $PK = \{G, G_T, g, H, spk, \mathrm{pk}\}$, private key is $SK = \{\alpha, \beta, \mathrm{ssk}, \mathrm{k}, \mathrm{K}, \mathrm{dk}, x\}$, evaluation key is $ek = (h_0, \cdots, h_{D-1})$.

Label generation algorithm (ek, f, σ)**:** Input evaluation key *ek*, arithmetic circuit $f: Z_p^n \rightarrow Z_p$ and verification tag $\sigma = (\sigma_1, \ldots, \sigma_n)$. Verification labels σ_i can be calculated $y(X) \leftarrow f(y^{(1)}(X), \ldots, y^{(n)}(X))$, where $y(X)$ has the coefficients $y_0, \cdots, y_d$. The message m to be verified is the corresponding message label $\lambda \in \{0,1\}^{\lambda}, \gamma_{\lambda} = F_K(\lambda)$ is calculated. At the same time, let $y_0 = m, y_1 = (\lambda - m)/x \bmod p$. When $d = 1$, $\sigma = (y_0, y_1)$. Otherwise, let $\Lambda = \prod_{i=0}^{d-} h_i^{y_{i+1}}$, then $\sigma = \Lambda$.

Storage algorithm $(\mathrm{M}, \sigma, \theta)$**:** The outsourced data M is divided into n' blocks, which are then further broken down into s sections. n' Sentinel blocks are randomly inserted in $n - n'$ blocks, which are stored in tables encrypted *PF* with symmetric keys k, that is $\mathrm{E} = AES_{enc}(k, PF)$. You can use the message block $m_{ij}, 1 \leq i \leq n, 1 \leq j \leq s$. Generate a polynomial validation label σ_i for each message block m_{ij}. Assuming that the message block $fname \in Z_p^*$ on the cloud server C, a message tag $\lambda = fname \| n \| \mathrm{E} \| Sign_{ssk}(fname \| n \| \mathrm{E})$ is generated, where $\lambda \in \{0,1\}^{\lambda}$. Using n block message to construct *RMHT*, all leaf nodes is $h_i = H(m_i \| 1)$, then obtain the RMHT root node R and compute the root node's label $\theta = Sign_{ssk}(\mathrm{H}(R))$, and the server C saves $(\lambda, \mathrm{M}, \sigma_1, \ldots \sigma_n, \theta)$.

Transmission algorithm (Q, λ^*)**:** Supposing that the user wants to transfer data from the cloud server C to the cloud server D. First, the user checks the integrity of the data through a third-party TPA invoking verification algorithm. If the verification result is correct, then the user uses the symmetric key k to decrypt the sentinel block table PF, and then generates a data transmission request $Q = fname \| \zeta \| sign_{ssk}(fname \| \zeta)$ and $\lambda^* = fname \| n^* \| \mathrm{E}^* \| sign_{ssk}(fname \| n^* \| \mathrm{E}^*)$ to cloud server C, where ζ is the marked message blocks, and n^* indicates the number of message blocks that need to be transmitted to the cloud server D, E^* indicates the location of the sentinel block in the message block to be transmitted. After confirming the validity of these requests, the cloud server C will send them to the cloud server D. The cloud server D will test and validate the messages sent by the cloud server C. If the test is correct, the cloud server will accept the data transmission. Otherwise, discard these messages.

Verification algorithm $(sk, m, \mathrm{P}, \sigma)$**:** Input $\mathrm{P} = (f, \lambda_1, \ldots, \lambda_n)$ and message $m \in Z_p$. Depending on the parameters d in the tag generation algorithm, the verification tag can be divided into two different forms: $\sigma = \Lambda$ and $\sigma = (y_0, y_1) \in Z_p^2$. First calculate $\mathrm{P} = f(\gamma_{\lambda_1}, \ldots, \gamma_{\lambda_n})$ where $\gamma_{\lambda_i} \leftarrow F_K(\lambda_i)$. Then according to the different forms of σ, we have the following checks:

(1) If $\varphi = (y_0, y_1) \in Z_p^2$, then output is 1 only and only if $\rho = y_0 + y_1 \cdot x \wedge y_0 = m$.
(2) If $\sigma = \Lambda$, then let $t = \Lambda^x$, return to the test algorithm $(dk, \rho - m, t)$ for verification.

4 Analysis of Our Proposal

4.1 Security Analysis

In this scheme, we combine the finite coding algorithm ε with the homomorphic message authentication organically, and construct the cloud data authentication scheme based on the homomorphic Rank-based Merkle Hash Tree (RMHT). Therefore, the security of our scheme only need verify whether separately two programs are safe or not, and you can show that the program is safe. We cite [13] to prove the security of the finite coding algorithm in this scheme.

We define the following difficult assumption: Computing $t = Enc(pk, 1/x)$ is a difficult problem, if knowing $t_i = Enc(pk, x^i), i = 0, \ldots, l$. Even if we consider in the prediction machine $O(dk, \eta)$, only when $\lambda \in Enc(pk, 0)$, then the output of the prediction machine is "YES", otherwise, the output is "NO". For any adversary A, we consider the following games $E - INV_A(\lambda, l)$:

(1) Run crypto-generating algorithm (1^{λ}) to generate a prime p length of at least λ bits, encoding space T, public key pk and private key dk;
(2) Choose a random parameter $x \in Z_p^*$;
(3) The public key pk and random parameters x are input into the encryption algorithm (pk, m) to generate h, that is $h_i \leftarrow Enc(\mathrm{pk}, x^i), i = 1, \ldots, l$;
(4) The generated $pk, p, T, h_0, \ldots h_l$ is input to an oracle, and generate a parameter $t \in T$, that is $t \leftarrow A^{O(dk,\cdot)}(pk, p, T, h_0, \ldots h_l)$;

If and only if $Adv_A^{\xi-INV}(\lambda, l) = \Pr[t = Enc(pk, 1/x)]$, the adversary A can win the game $E - INV_A(\lambda, l)$.

Definition 2. When input λ, the $Adv_A^{\varepsilon-INV}(\lambda, l)$ is negligible for any adversary, the encoding scheme is inversion-resistant.

The homomorphic message authentication security definition is proposed by Catalano and Fioro [13]. The specific game process is as follows:

Initialization: The challenger B runs the key generation algorithm (1^{λ}) to obtain the private key sk and the authentication key ek, and sends the authentication key ek to the adversary A. At the same time, initializes $T = \emptyset$.

Message Label Inquiry: The adversary can query the message label continuously, which can be divided into the following three situations: (1) If the adversary A sends repeated inquiries to the challenger B, the challenger B sends the same reply. (2) The adversary A sends an inquiry to the challenger B, that is, two different messages are labeled with the same label, and the challenger B ignores the query. (3) The adversary A sends an inquiry to the challenger B, then the challenger B computes the verification ticket generation algorithm to generate a new verification ticket σ while updating .

Verification Challenge: The adversary A sends an inquiry (m, P, σ) to the challenger B, who validates it using a verification algorithm $(sk, m, \mathrm{P}, \sigma)$ and outputs the result 1 or 0.

Forgery: If the adversary A finds that (m, P, σ) is counterfeit during the verification challenge, the adversary A immediately stops asking.

Definition 3. If there exits $\mathrm{PR}\left[HomUF - CMA_{\mathrm{C,HomMAC}}(\lambda) = 1\right] \leq \varepsilon(\lambda)$, where $\varepsilon(\lambda)$ is a negligible function, the homomorphic authentication scheme can be judged as safe.

Theorem 1. The finite coding algorithm ε is $(D-1) - inversion$ resistant to security, while the homomorphic authentication scheme game $HomUF - CMA_{\mathrm{C,HomMAC}}(\lambda)$ is also secure, you can determine the program is safe.

4.2 Performance Analysis

Catalanno et al. proposed a homomorphic message authentication scheme based on arithmetic circuit in [9]. This section we will compare our scheme with the schemes of CF13-2 [17] and Catalanno [9] in the aspects of label size, whether to support compound degree calculation and whether to support inter-server data erasure verification. Specific comparison results are shown in Table 1:

Table 1. Performance analysis

Scheme	Label size	Multiplictis	Validation query	Support for deletion verification
CF13-2 [17]	$O(1)$	×	√	NO
Dario Catalanno [9]	$O(k)$	√	√	NO
Our scheme	$O(n)$	×	√	YES

5 Conclusion

This paper constructs a Rank-based Merkle Hash Tree homomorphic authentication scheme by merging the data verification structure Merkle Hash Tree into the homomorphic authentication scheme. Through the Rank-based Merkle Hash Tree architecture, the solution can effectively verify the integrity of data transmission between the cloud servers, and improves the security of user data. Further work is to explore the application of the Homology verification scheme in the cloud transmission process and to construct a more efficient and practical cloud computing protocol with a homomorphism verification scheme.

Acknowledgement. This work is supported by the National Cryptography Development Fund of China Under Grants No. MMJJ20170112, National Key Research and Development Program of China Under Grants No. 2017YFB0802000, National Nature Science Foundation of China (Grant Nos. 61772550, U1636114), the Natural Science Basic Research Plan in Shaanxi Province of china (Grant Nos. 2016JQ6037) and Guangxi Key Laboratory of Cryptography and Information Security (No. GCIS201610).

References

1. Karim, M., Faouzi, A.: Factors explaning is managers attitudes toward cloud computing adoption. IGI Global **12**(1), 1–20 (2016)
2. Reddy, B., Paturi, R.: Research and application of cloud storage. In: Intelligent Workshop on intelligent System & Applications, pp. 1–5 (2010)
3. Shenchuan, W.: Research on storage technology based on cloud computing. Petrol. Ind. Comput. Appl. **2**, 53–55 (2011)
4. Hen, H., Wei, X., Peng, L., et al.: Admission control mechanism against Sybil attack for P2P cloud storage system. China Sci. Paper **02**, 150–158 (2015)
5. Clark, C., Fraser, K., Hansen, J.: Live migration of virtual machtices. In: Proceedings. of the 2nd Symposium on Networked System Design and Implementation, Berkeley, pp. 273–286 (2005)
6. Zhou, K., Wang, H.: Cloud storage technology and its application. ZTE Commun. **16**(4), 24–27 (2010)
7. Gennaro, R., Wichs, D.: Fully homomorphic message authenticators, vol. 8270, pp. 301–320. Springer, Berlin, Heidelberg (2013)
8. Dan, B., Lynn, B.: Short signatures from the weil pairing. J. Cryptol. **17**(4), 297–319 (2004)
9. Catalano, D., Fiore, D., Gennaro, R., Nizzardo, L.: Generalizing Homomorphic MACs for Arithmetic Circuits, vol. 8383(3), pp. 538–555. Springer, Berlin, Heidelberg (2014)
10. Liang, X., Ni, J.: Provable data transfer from provable data possession and deletion in cloud storage. Comput. Stand. Interfaces **54**(P1), 46–54 (2016)
11. Benabbas, S., Gennaro, R., Vahlis, Y.: Verifiable delegation of computation over. In: Large Datasets. Conference on Advances in Cryptology, vol. 52(3), pp. 111–131 (2012)
12. Wang, Y.X., Yang, Q., Chen, W., et al.: Application of lattice-based linearly homomorphic signatures in cloud storage dynamic verification. China Sci. Paper **20**, 2381–2386 (2016)
13. Gennaro, R., Gentry, C., Parno, B.: Non-interactive verifiable computing: outsourcing computation to untrusted workers, vol. 6223(3), pp. 465–482. Springer, Berlin, Heidelberg (2010)
14. Parno, B., Raykova, M.: How to delegate and verify in public: verifiable computation from attribute-based encryption. In: International Conference on Theory of Cryptography, vol. 7194, pp. 422–439 (2012)
15. Bowers, K.D., Juels, A.: Proof of retrievability: theory and implementation. In: Proceedings of ACM Workshop on Cloud Computing Security, New York, pp. 43–45. ACM Press (2009)
16. Varalakshmi, P., Deventhiran, H.: Integrity checking for cloud environment using encryption algorithm. In: Proceedings of International Conference on Roceut Treuds in Information Technology, pp. 228–232 (2012)
17. Catalno, D., Fiore, D.: Practical homomorphic MACs for arithmetic circuits. In: International Workshop on Public Key Cryptography, vol. 7881(3), pp. 336–352 (2013)

Behavioral Security in Cloud and Fog Computing

Marek R. Ogiela[1(✉)] and Lidia Ogiela[2]

[1] Faculty of Electrical Engineering, Automatics, Computer Science and Biomedical Engineering, AGH University of Science and Technology, 30 Mickiewicza Avenue, 30-059 Krakow, Poland
mogiela@agh.edu.pl

[2] Cryptography and Cognitive Informatics Research Group, AGH University of Science and Technology, 30 Mickiewicza Avenue, 30-059 Krakow, Poland
logiela@agh.edu.pl

Abstract. In this paper will be described several possible applications of cognitive and behavioral approaches in Fog and Cloud computing. In particular will be presented the ways of using selected behavioral characteristics for creation of security and cryptographic protocols, dedicated to secure distribution and management of strategic data in Fog and Cloud environment. Some possible applications of using movement and gesture features will be also presented.

1 Introduction

Cognitive informatics may have a strong impact on security areas, and thanks to the creation of several different classes of cognitive vision systems, allow to define a new area of advanced cryptography called cognitive cryptography [11, 12]. Cognitive systems allow to imitate the human behavior and thinking processes, and also perform resonance processes, which are based on knowledge based perception paradigm [4–6]. In this model mental processes are conducted, what allow to evaluate and semantically understand analyzed patterns. Application of cognitive systems may be also connected with using personal and behavioral features in security areas and cryptographic protocols. Such features enable to create strong encryption keys and define a novel security solutions like behavioral lock [12]. So cognitive cryptography as a new computational paradigm combines security procedures with selected personal information characteristic for users. Such connection of security approaches with cognitive processes may be very promising for creation of security solutions dedicated to Fog and Cloud computing.

In this paper will be presented several ideas of application of personal features and behavioral characteristics for security purposes, and cryptographic procedures. Presented examples will be based on extraction of unique personal features and theirs application in Cloud and Fog security [1, 2]. Such applications seems to be very important in creation of advanced security models and infrastructures, based on behavioral patterns, which can also describe human habits, motion features, and other human movement actions. Such new procedures should extend existing cryptographic methodologies, and create a new branch of cognitive cryptography [10, 11].

L. Barolli et al. (Eds.): EIDWT 2018, LNDECT 17, pp. 849–854, 2018.
https://doi.org/10.1007/978-3-319-75928-9_77

2 Fog and Cloud Computing Infrastructures

Cloud computing offers unlimited computing and storage infrastructures to facilitate data acquisition and processing. It is also connected with services and information management areas. One of the most important problem is the network traffic, generated by the great number of nods and computers towards the Cloud. It is very important for many applications oriented on fast data processing and secure information sharing.

To solve this limitation and extend the Cloud computation to more efficient processing, the data exploration and fusion tasks can be moved closer to the original data sources, to the layer existed between the nodes and Cloud units i.e. Fog or Edge infrastructure.

Fog computing is able to perform initial information or data processing, and also perform data analytics tasks. Both Cloud and Fog computing move the data processing applications away from centralized nodes, making the Cloud closer to the sources which generate signals or data. Edge and Fog layers participate in information processing tasks, especially connecting data coming from multiple sources, and determining if data should be analyzed immediately or on higher level in the Cloud.

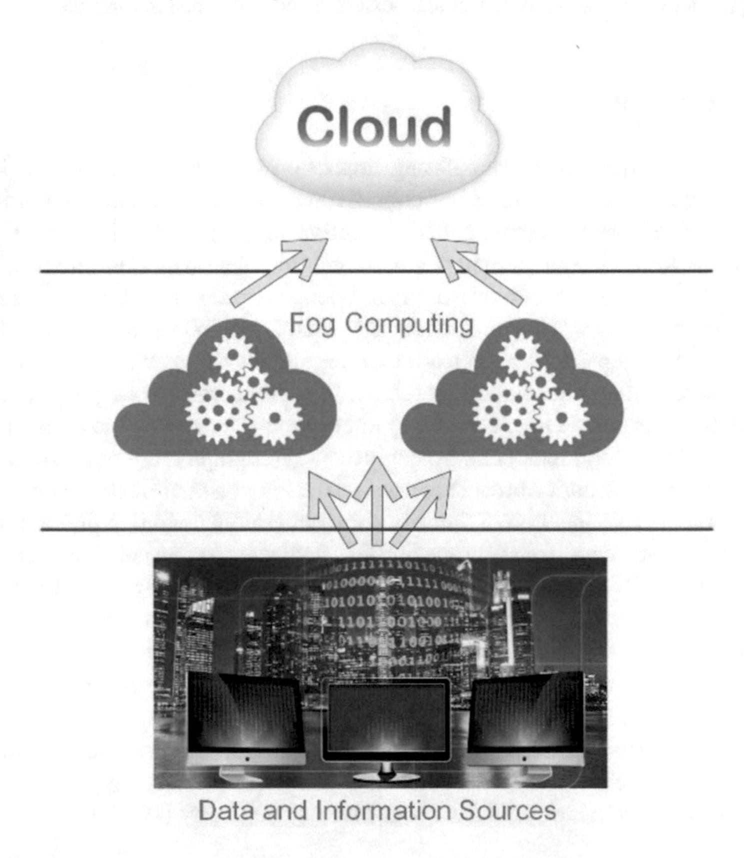

Fig. 1. Cloud and Fog computing relations.

Information and services available at Fog and Cloud units may be characterized at the lower levels, by a great number of sources generating and processing signals.

In Fig. 1 is presented a hierarchical architecture of Fog and Cloud computing structures. The lower level is connected with sensors and devices, which can be abstracted as nodes generating any information. The second level i.e. Fog are more distant from real objects, and in practice consist more powerful computing equipment, which can support big data analytics tasks, as well as storage of processed information. Fog nodes also enable to run Cloud application on their native architecture, and can be managed by Cloud providers.

Large-scale and complex structure of Fog and Cloud computing layers, which in fact may not belong to the one provider, requires that it is necessary to develop a secure and efficient protocols, which allow to manage transmitted and distributed data in efficient and secure manner.

3 Behavioral Patterns in Security Protocols

To collect personal features or behavioral patterns, which later can be applied in security protocols, we can use cognitive vision systems in connection with multimedia devices like Leap Motion, Kinect or Motion Capture sensors.

As examples of application of behavioral features in cryptographic procedures, we can consider different types of patterns like: finger and hand gestures, human motion patterns performed using other human body parts, and also complex motion patterns having the form of very specific exercises etc.

The most important areas of application of mentioned personal or behavioral patterns in security areas are connected with: secret division and steganography [13], human-oriented data management procedures [6, 7], and fuzzy vault and crypto biometric procedures [3, 8, 9].

The simplest solution in this area is application of finger or palm gesture for authentication procedures. Analysing hand movements we can extract very unique features, which next can be used in creation of behavioral lock. Analysis of such movements has many advantages among which we can noticed: fast data acquisition, possibility of recognition of different movement types like fix movement patterns, natural gesture, user specific etc. Analysis of finger and palm activities also can be performed in real-time. It is also possible to consider more complex gestures, which application may be better from security reasons. While analysing more complex gesture patterns it is necessary to extract some specific and distinctive parameters, characteristic only for particular user. To register enough complex patterns we can analyze movements performed using several fingers, or using the whole palms or hands.

In our research we implemented procedures, which involve cognitive systems for hand and finger analysis [12]. Such system firstly creates a learning set, which contain particular number of predefined motion patterns, also representing different gesture classes. The second stage allows to analyse a new motion pattern in real time, and extract specific for particular person feature vector. Classification of new patterns can be performed by theirs comparison with elements from learning set. Having evaluated personal features it is possible to determine the type gesture, and additionally it is possible to use feature vector as a specific behavioral key in different security applications. Among such applications we can consider personalized keys, multi secret steganography, fuzzy vault etc.

For security applications it is also possible to consider more complex movements performed while dancing, practicing sport or simply walking (Fig. 2). Analysis of such complex movements is much more difficult and usually is connected with application of professional motion analysis equipment, which allow to register and extract specific personal parameters. Analysis of complex movements, characteristic for different persons, may be dependent on personal habits, health conditions, age, and specific skills. For example persons practicing particular dances or ballet forms can perform flexible figures in characteristic personal manner. Such performance allows extracting personal features, which next can be used for security purposed. In the same way we can consider sport and gym activities and extract specific features from recording presenting very specific acrobatic or sport skills etc. (Fig. 2).

Fig. 2. Examples of advanced sport, dance and acrobatic techniques.

Analysis of sport, dance or acrobatic techniques for extraction of specific motion feature, can be done in the same way as in [12]. Having such specific motion parameters it is possible to use them for security purposes in the same manner as described previously for simple gestures.

4 Security Features of Behavioral Cryptographic Solutions

Application of behavioral features in security procedures, seems to be very important in future IT solutions, dedicated for Cloud and Fog computing. In behavioral cryptographic procedures we can use all the mentioned personal features, and also other biometrics or nonstandard personal parameters connected with morphometric values of human body parts, health records etc.

Presented solutions have also some important security features, which can be described as follows:

- Presented protocols are secure from cryptographic point of view, because they are based on the native procedures, which may be extended using personal features or behavioral patterns.

- Such techniques are efficient, and the complexity of behavioral algorithms are at the same level as the complexity of native procedures. Application of personal features requires to evaluate personal parameters, but this stage can be performed once, and extracted personal features may be stored in large personal record applicable over longer time.
- Behavioral characteristic can be evaluated for particular person, and should be stored in secure manner, as a specific encrypted record.
- Personal features should be evaluated in real time, when the particular gesture or movement are performed. This allow to prevent using counterfeit recordings, not representing live movements.

5 Conclusions

This paper has described a new possibilities of application behavioral features and personal motion parameters in security application dedicated for Cloud and Fog solutions. Behavioral patterns extracted from motion sequences represent personal features, which can be used in security protocols oriented for information management, transmission and storage. Besides the simplest hand motion patterns, in cryptographic application, it is also possible to consider more complex patterns.

Evaluation of behavioral features can be done with the application of cognitive vision systems, which allow extract unique features from different motion patterns or specific human body movements. Application of such specific and unique parameters proved that personal features and cognitive systems can be successfully applied in creation of advanced cryptographic protocols for key exchange, data management, and establishment of behavioral lock.

Acknowledgments. This work has been supported by the AGH University of Science and Technology research Grant No. 11.11.120.329.

References

1. Gomez-Barrero, M., Maiorana, E., Galbally, J., et al.: Multi-biometric template protection based on homomorphic encryption. Pattern Recogn. **67**, 149–163 (2017)
2. Hahn, C., Hur, J.: Efficient and privacy-preserving biometric identification in cloud. ICT Express **2**, 135–139 (2016)
3. Jin, Z., Teoh, A.B.J., Goi, B.-M., Tay, Y.-H.: Biometric cryptosystems: a new biometric key binding and its implementation for fingerprint minutiae-based representation. Pattern Recogn. **56**, 50–62 (2016)
4. Ogiela, L.: Cognitive Information Systems in Management Sciences. Elsevier, Academic Press, London (2017)
5. Ogiela, L., Ogiela, M.R.: Data mining and semantic inference in cognitive systems. In: Xhafa, F., Barolli, L., Palmieri, F., et al. (eds.) 2014 International Conference on Intelligent Networking and Collaborative Systems (IEEE INCoS 2014), Salerno, Italy, 10–12 September 2014, pp. 257–261 (2014)

6. Ogiela, L., Ogiela, M.R.: Management information systems. LNEE, vol. 331, pp. 449–456 (2015)
7. Ogiela, L., Ogiela, M.R.: Insider threats and cryptographic techniques in secure information management. IEEE Syst. J. **11**, 405–414 (2017)
8. Ogiela, M.R., Ogiela, U.: Linguistic approach to cryptographic data sharing. In: The 2nd International Conference on Future Generation Communication and Networking, FGCN 2008, Hainan Island, China, 13–15 December 2008, vol. 1, pp. 377–380 (2008)
9. Ogiela, M.R., Ogiela, U.: Grammar encoding in DNA-like secret sharing infrastructure. Lecture Notes in Computer Science, vol. 6059, pp. 175–182 (2010)
10. Ogiela, M.R., Ogiela, U.: Secure Information Management using Linguistic Threshold Approach. Advanced Information and Knowledge Processing. Springer, London (2014)
11. Ogiela, M.R., Ogiela, L.: On using cognitive models in cryptography. In: The IEEE 30th International Conference on Advanced Information Networking and Applications, IEEE AINA 2016, Crans-Montana, Switzerland, 23–25 March 2016, pp. 1055–1058 (2016). https://doi.org/10.1109/aina.2016.159
12. Ogiela, M.R., Ogiela, L.: Cognitive keys in personalized cryptography. In: The 31st IEEE International Conference on Advanced Information Networking and Applications, IEEE AINA 2017, Taipei, Taiwan, 27–29 March 2017, pp. 1050–1054. IEEE (2017). https://doi.org/10.1109/aina.2017.164
13. Shih, F.Y.: Digital Watermarking and Steganography: Fundamentals and Techniques. CRC Press, Boca Raton (2017)

Voice Quality Testing Using VoIP Applications over 4G Mobile Networks

Desar Shahu[1(✉)], Alban Rakipi[1], Joana Jorgji[1], Ismail Baxhaku[2], and Irena Galić[2]

[1] Polytechnic University of Tirana, Tirana, Albania
{dshahu,arakipi,jjorgji}@fti.edu.al
[2] J.J. Strossmayer University of Osijek, Osijek, Croatia
{ibaxhaku,irena}@etfos.hr

Abstract. Nowadays, many mobile applications in the telecommunication market provide Voice over IP (VoIP), video and data services. This paper presents the performance evaluation of a 4G mobile network in the city of Pristina. This was accomplished by testing two of the most popular VoIP applications, Skype and Viber. Stationary and dynamic scenarios have been considered to evaluate the voice quality and the Quality of Experience (QoE) of both applications under the same mobile network radio conditions, throughput capacity and coverage requirements. The Mean Opinion Score (MOS) method was used to evaluate the recorded speech files based on the feedback perceptions of different users. These results were found to be very useful for the continuously increasing demand on VoIP services and applications.

1 Introduction

Over the past decade subsequent expansions of the Global System for Mobile Communications (GSM) system were created. Data rates of 14 Mbps were achieved by using High Speed Packet Access (HSPA), increasing to a potential data rate of 42 Mbps for evolved HSPA+ standards if advanced Quadrature Amplitude Modulation (QAM) schemes are combined with Multiple Input Multiple Output (MIMO) techniques. Long Term Evolution (LTE) network was initiated as a project in 2004 and describes the standardization work by the Third Generation Partnership Project (3GPP) to define a new high-speed radio access method for mobile communication systems. The main radio access design parameters of this new system include Orthogonal Frequency Division Multiplexing (OFDM) waveforms to avoid the inter symbol interference that typically limits the performance of high-speed systems, and MIMO techniques to boost the data rates. Data rates are about to feature a range of 100 Mbps for high mobility scenarios up to 1 Gbps for low mobility scenarios using advanced LTE (LTE-A) networks [1]. In general, the LTE system has already become a world-wide standard. In western Balkans there are more than 10 million active users, introducing an economic growth and investments for the developing regions. Furthermore, it has created an adequate environment for mobile application services along with internet services such as VoIP, web browsing, and video streaming, with constraints on delays and bandwidth requirements [2]. An

L. Barolli et al. (Eds.): EIDWT 2018, LNDECT 17, pp. 855–862, 2018.
https://doi.org/10.1007/978-3-319-75928-9_78

efficient listen before talk (LBT) technique is analyzed in [3], to reduce interference and improve spectral efficiency of LTE network users.

To evaluate 4G mobile networks, it is generally necessary to investigate on different network radio conditions the Quality of Service (QoS) for voice, video and data traffic, as expected to be received by the users. VoIP applications present an interesting option as they combine voice and data communications and can be used to evaluate the performance of the mobile network in terms of QoS [4]. The mobile technology is evolving towards the 5G standard, which is expected to start its service in 2020. In LTE-A, the network delay budget for VoIP is about 100 ms [5] and the key challenge of 5G technology will be to reduce the delay less than 1 ms [6].

In this work, two of the most popular social mobile applications, such as Skype and Viber, have been considered and tested over the LTE mobile network operating within the city of Pristina. Skype is a telecommunication application introduced for the first time in 2003, mainly designed for personal computers (PC) as a tool for business. Skype was considered the best PC application on providing voice and video call services however it was limited due to mobility. To fit the market requirements, different Skype mobile device applications were developed under the trademark of Microsoft Corporation. Viber application was first lunched in 2010, initially for iPhone and later for Android platforms, as an alternative to Skype. It provides voice, SMS and video call services for mobile devices and recently available for PC, tablets and other electronic devices. Both applications are commonly used in the developing Balkan regions, where users are able to call their friends account without being charged or conveniently charged for landline or mobile number calls.

Stationary and dynamic experimental investigations over LTE mobile network have been considered using VoIP applications and MOS method was used to evaluate and compare the voice quality of both Skype and Viber applications. The MOS test has been used for decades in telecommunication networks to present the human's perceptions for the voice quality [7, 8]. The results of this experimental work were found to be interesting for the network operators in order to improve the LTE mobile network performance and for the mobile application users, always demanding for better voice quality with convenient costs.

2 Analysis of the Experimental System and Testing Scenarios

LTE technology has experienced some major changes of the network topology compared with the former 3G networks, as shown in Fig. 1. Node-B of 3G system was replaced by evolved Node-B (eNB), which is a combination of Node-B and radio network controller (RNC). The eNB communicates with the User Equipment (UE) and can serve one or several cells at one time. The serving gateway (SGW), as a part of the evolved packet core (EPC), is responsible for routing and forwarding packets between UE and packet data network (PDN) and charging. The mobility management entity (MME) manages UE access and mobility and establishes the bearer path for UE. Packet data network gateway (PGW) is a gateway to the PDN, and policy and charging rules function (PCRF) manages policy and charging rules [9].

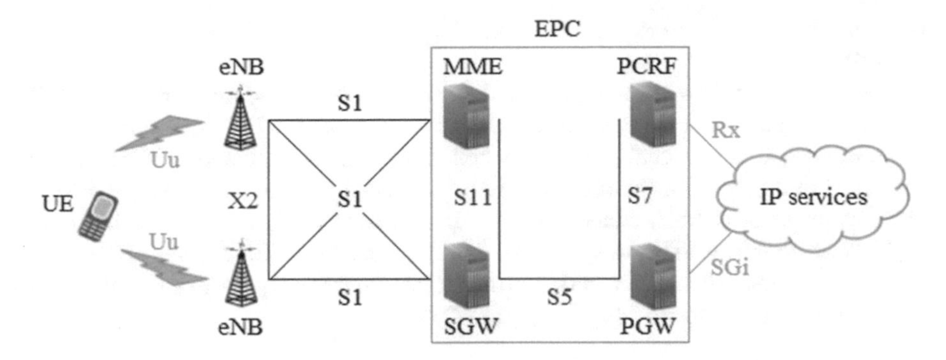

Fig. 1. LTE network topology.

Measurements of the required throughput for Skype and Viber applications were performed in the city of Pristina and the Test Mobile System (TEMS) discovery network equipment was used for data processing. The experiments are designed to conduct calls from a Samsung S5 as the calling side to the laptop's USB dongle 4G adapter in receiving mode, using Skype and Viber applications over a LTE network, as shown in Fig. 2.

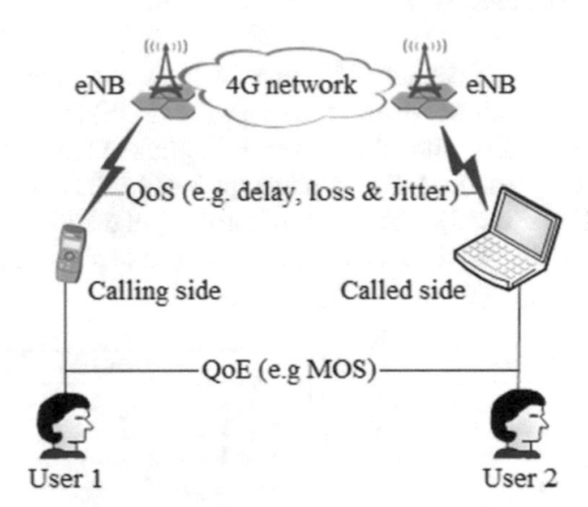

Fig. 2. QoS and QoE in LTE network.

Two different static test scenarios were considered during the measurement. The calling UE was first set at −105 dBm of signal strength, far from the antenna of the serving cell, and then it was set at −60 dBm of signal strength, near the antenna of the serving cell. Furthermore, high mobility calling tests were considered, with the calling UE moving with a constant speed of 30 km/h from a point having −120 dBm of signal strength to a point having −60 dBm of signal strength. The focus was to test the voice quality of the recorded calls under different network radio conditions by using the MOS method [7], which is based on a large number of users' perceptions. The better the QoS provided from the network operator, in terms of packet loss, jitter, delay and data rate, the better the QoE will be expected as a result, especially for real time applications such as VoIP.

Beside the required throughput for the VoIP applications, measurements of the Reference Signal Received Power (RSRP) parameter from the UE were reported to the core network to make decisions about cell selections or handovers. RSRP can be expressed in terms of the number of resource blocks (N), and the Received Signal Strength Indicator (RSSI), measured for a specific bandwidth as described in Eq. (1). RSRP typically levels in the range from −75 dBm, near the LTE's serving cell, to −120 dBm, far from the serving cell, and informs the system for the quality of signal in terms of Reference Signal Received Quality (RSRQ) [10]:

$$\mathrm{RSRP}\,(\mathrm{dBm}) = \mathrm{RSSI}\,(\mathrm{dBm}) + 10 * \log\left(12 * \mathrm{N}\right) \tag{1}$$

The signal quality is also indicated from the measured Signal to Interference and Noise Ratio (SINR), which can be expressed as the ratio of the measured power of the useful signal to the sum of the neighbor cell's interference and the background noise over the considered bandwidth.

3 Results and Discussion

Measurement results of the considered scenarios have been reported in this section. Figure 3 presents the probability distribution function (PDF) for the Physical Downlink Shared Channel (PDSCH) and the Physical Uplink Shared Channel (PUSCH) throughput, required by a static UE to conduct the VoIP calls in Skype, as in Fig. 3(a), (c) and Viber as in Fig. 3(b), (d). The results of throughput distribution in the case where the UE is distant from the serving antenna, at −105 dBm of signal strength, are reported

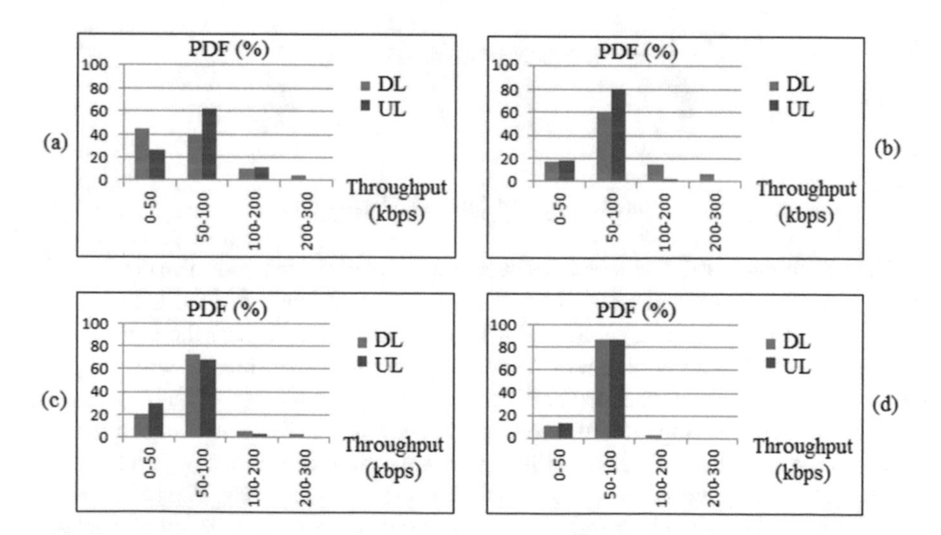

Fig. 3. Downlink/Uplink (DL/UL) throughput distribution for Skype and Viber VoIP applications.

in Fig. 3(a), (b) and the results for the case where the calling UE is at −60 dBm of signal strength are reported in Fig. 3(c), (d). In order to minimize the network traffic effect, measurements were conducted on the same time of the day, around the peak evening hours.

The modulation scheme used in both cases was the Phase Shift Keying (QPSK) and the required throughput for about 90% of the samples does not exceed the recommended value of 100 kbps. The better the radio conditions were the better data rates were achieved. The results show the minimum bandwidth that the VoIP applications need compared with the LTE cell capacity, especially when using Skype, which requires 30 kbps for about 40% of the samples.

The VoIP applications have also been tested under UE mobility conditions, as shown in Figs. 4 and 5, using Skype and Viber respectively. The measurements of the network parameters were conducted from a point of the cell having −120 dBm of signal strength, to a point of the cell having −60 dBm of signal strength.

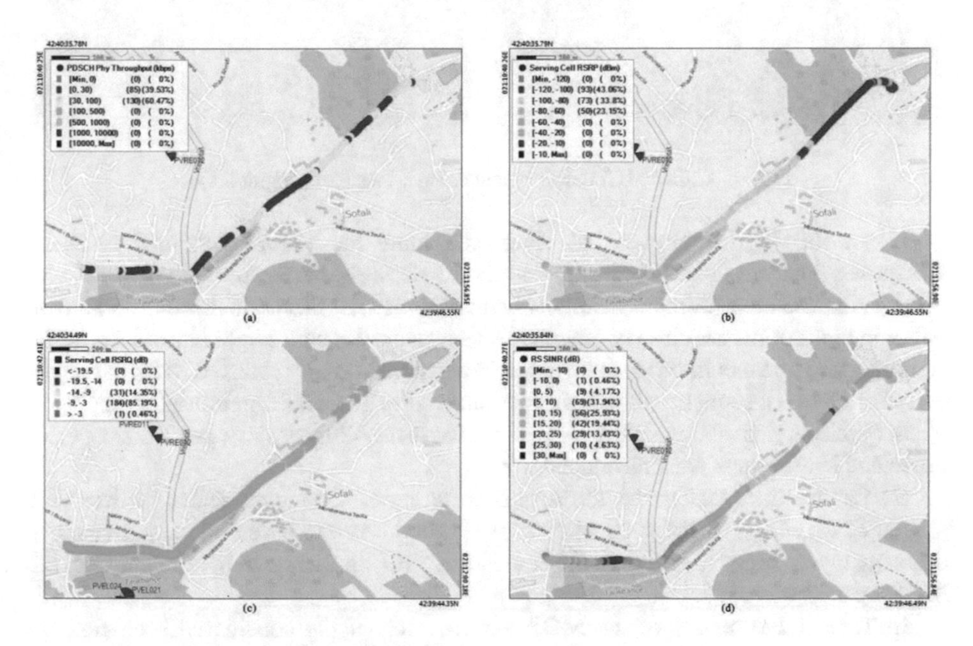

Fig. 4. Calling UE mobility test using Skype VoIP application.

Figure 4(a) shows that there was different throughput utilization during the considered scenario however it didn't exceed 100 kbps. The measured RSRP far from the serving cell, as reported in Fig. 4(b), commands the core network to make cell revelation or handovers. Based on the results of Fig. 4(c), the quality of signal has affected the quality of the recorded voice causing dropped calls and interference during the conversation on the reception side. It is obvious that the degradation of radio conditions directly affects the quality of VoIP calls in Skype. Figure 4(d) presents the SINR and shows that the cells are well optimized with acceptable noise and interference, except of a segment of the path where the SINR is between −10 dB and 0 dB.

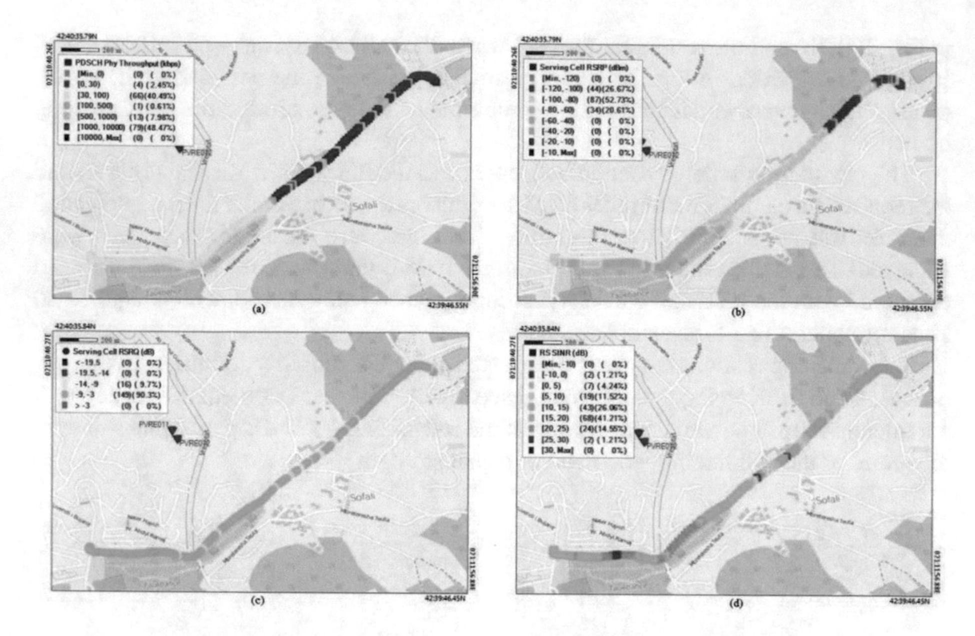

Fig. 5. Calling UE mobility test using Viber VoIP application.

Figure 5(a), (b), (c) relieves that Viber needs more than 1Mbps of throughput as the UE moves far from the serving cell, and the RSRP signs between −100 dBm to −80 dBm. From the measurement results we can observe that Skype is more rational than Viber in terms of the required bandwidth. This is mainly because Viber introduces more signaling and headers in order to keep the better signal quality referred to Skype application and as a consequence requires more bandwidth. The SINR measured for the VoIP Viber application, as shown in Fig. 5(d), is similar to the Skype tests, where an acceptable noise and interference was introduced.

So far, we have discussed the results in terms of QoS parameters, which have presented the ability of the network to provide good services based on the radio conditions quality. In order to compare the considered VoIP applications in terms of QoE the MOS method was used.

In Table 1 the results of the MOS test for each of the considered scenarios are presented. The voice quality of 15 min long recorded calls in Skype and Viber have been evaluated by 60 participants.

Table 1. MOS test results for the considered scenarios.

Scenario	Skype	Viber
Static (−105 dBm)	4.6	4.4
Static (−60 dBm)	3.4	4.6
UE in mobility	2.5	4.1

The MOS test results relive the better quality of signal for the VoIP Viber application. Usually a MOS result of 3.6 is mentioned as a limit for minimum acceptable quality of voice communication services. However, both platforms are excellent choices for the VoIP services.

4 Conclusions and Future Work

In this study the voice quality over a 4G LTE network in terms of QoS and QoE using Skype and Viber VoIP applications has been tested and analyzed. The results of this study were shown to be very useful for the users and operators of the considered 4G network in the city of Pristina. Based on the MOS method results the Viber voice quality seems to be better than Skype, especially for the mobility scenarios. However, in this case the throughput needed for the two applications was drastically different. The Viber VoIP application requires up to 10 Mbps of bandwidth, under the worst case of radio network conditions of -105 dBm of signal strength. That means that Skype is a more rational VoIP application in terms of bandwidth requirements than Viber, which didn't exceed more than 100kbps of throughput for all the considered scenarios. This difference between the considered VoIP applications is mainly because Viber introduces more signaling and headers in order to keep the better signal quality referred to Skype and as a consequence requires more bandwidth.

As a future work, we are looking to test the QoS and QoE of video call services and increase the number of participants in the MOS test, in order to improve the reliability of the voice and video quality perceptions. Comparing voice or video calls quality under different 4G network operators and considering 5G networks prospective would be of a great interest in the future works.

References

1. Zielinski, A., Zielinski, K.: Mobile telecommunication systems changed the electronic communications and ICT market. J. Telecommun. Inf. Technol. **2**, 5–13 (2013)
2. Chadchan, S., Akki, C.: A fair downlink scheduling algorithm for 3GPP LTE networks. J. Comput. Netw. Inf. Secur. **6**, 34–41 (2013)
3. Mushunuri, V., Panigrahi, B., Rath, H., Simha, A.: Efficient listen before technique for LTE-WiFi co-existence in unlicensed bands. Int. J. Space-Based Situated Comput. **7**(2), 108–118 (2017)
4. Wuttidittachotti, P., Daengesi, T.: Quality evaluation of mobile networks using VoIP applications: a case study with Skype and LINE based-on stationary tests in Bangkok. J. Comput. Netw. Inf. Secur. **12**, 28–41 (2015)
5. Mushtaq, M., Mellouk, A., Augustin, B., Fowler, S.: QoE power efficient multimedia delivery method for LTE-A. IEEE Syst. J., 1–12 (2015)
6. Mushtaq, M., Fowler, S., Augustin, B., Mellouk, A.: QoE in 5G cloud networks using multimedia services. In: IEEE Wireless Communication and Networking Conference (2016)
7. Ulseth, T., Stafsnes, F.: VoIP speech quality-Better than PSTN? Telektronikk, 119–132 (2006)

8. Fiedler, M., Shaikh, J., Collange, D.: Quality of experience from user and network perspectives. J. Ann. Telecommun. **65**, 47–57 (2010)
9. Salman, H., Ibrahim, L., Fayed, Z.: Overview of LTE-advanced mobile network plan layout. In: Fifth International Conference on Intelligent Systems, Modelling and Simulation (2014)
10. Atanasov, P., Kissovski, Z.: Measurement of the reference signal in 4G LTE network in Sofia. Telecom **24**, 29–36 (2016)

Service Management Protocols in Cloud Computing

Urszula Ogiela[1], Makoto Takizawa[2], and Lidia Ogiela[1(✉)]

[1] Cryptography and Cognitive Informatics Research Group, AGH University of Science and Technology, 30 Mickiewicza Ave., 30-059 Krakow, Poland
{ogiela,logiela}@agh.edu.pl

[2] Department of Advanced Sciences, Hosei University, 3-7-2, Kajino-cho, Koganei-shi, Tokyo 184-8584, Japan
makoto.takizawa@computer.org

Abstract. In this paper will be presented a description of service management protocols for Cloud infrastructure. Especially, presented solutions will be dedicated to Cloud and Fog Computing. The main idea will be described with application of semantic aspects, dedicated to service management application. Services management protocols in the cloud and in the fog can be realized with application of secure and strategic methods.

Keywords: Fog and cloud computing · Service management protocols
Data security

1 Introduction

The service management tasks are oriented to support planning and running a business processes [5]. The main idea of service management is the proper planning processes. These processes are dedicated to the fast company development. The processes of company development are oriented around the analysis of previous, current and possible future situation of the analysed company. The process of setting goals and appropriate actions to achieve them, should be defined. The goals setting processes concerns:

- determining the main planning goals,
- defining the intermediate planning goals.

In main planning procedures it's necessary to indicate the specified times and periods. In these periods we construct the main plans, dedicated to analysis the global situation and designed for company development. The main classification of planning processes is the following:

- strategic planning,
- long-term planning,
- medium-term planning,
- short-term planning,
- current planning.

L. Barolli et al. (Eds.): EIDWT 2018, LNDECT 17, pp. 863–869, 2018.
https://doi.org/10.1007/978-3-319-75928-9_79

By strategic planning we understand the planning processes lasted over five and more years. In this time the organization need define all important (strategic) goals. It's a future plan in which organization need determine the success procedures and all possible methods of realization them. What is going to be achieved in organization by has a success?

Next type of planning, is a long-term planning. It's also kind of future plan, but concerns a shorter time strategic planning. The long-term means from two to five year. In this planning company needs determine all projects aimed to achieve the most important task and goal. Determining the task which is needed for realization the most important goals, it's very important, because this process concentrate around the analysis of the current and future situation of organization.

Next type of planning, is the medium-term planning. The medium-term planning concerns from few months to maximum one year. In this process, is necessary to answer a question, how to realize a long-term plan? In this process, organization need to analyse all possible solutions and situations, and focus on the optimal choice, optimal ways.

The short-term planning up to three months. In this planning, organization need plans only short tasks and goals, due to the limited time of their implementation. In this period it's also necessary to understand the current situation of organization, and the indication of the most important tasks for quick realization.

The last type of planning is the current planning. This type concerns daily and weekly planning. Current planning processes determines the realization of current organization goals. By current planning, it's possible to indicate important things, which should be changed in the future (by creation strategic plan).

In this paper will be presented a new idea of service management protocol using the Cognitive Systems of Service Planning. This kinds of systems are dedicated to semantic description of service planning processes by evaluating semantic aspects of data interpretation.

2 Cognitive Service Management Protocols

The Cognitive Service Management Systems (CSrMS) dedicated to semantic description and service management in the cloud and fog computing processes were defined in [6, 7, 9, 10].

Cognitive service management systems are the systems working with application of semantic description evaluated in all services. Especially in planning, organizing, deciding and controlling processes. All defined service processes should include cognitive description aspects, semantic analysis and semantic meaning of analysed services. The main component (element) of proposed new classes of cognitive service management systems, is a stage of semantic description and analysis.

In CSrMS systems were describe and indicate the following subclasses [10]:

- **Cognitive Systems of Service Planning (CSSsP)** – systems for semantic description of service planning processes,
- Cognitive Systems of Service Organising (CSSsO) – systems for semantic understanding of service organizing processes,

- Cognitive Systems of Service Deciding (CSSsDe) – systems for semantic analysis of service management processes,
- Cognitive Systems of Service Motivating (CSSsM) – systems for motivating service analysis processes,
- Cognitive Systems of Service Controlling (CSSsC) – systems for analysing the meaning aspects of understanding basic of service controlling processes.

In this paper, Authors should concentrate to propose and describe the first class of service management systems – Cognitive Systems of Service Planning (CSSsP). Cognitive systems for service planning are the systems dedicated to semantic description [4] of service planning processes. In this class of systems is important to create a semantic aspects of service planning analysis.

The main idea of using semantic aspects in service planning processes, must consist semantic description and interpretation in planning processes. So, at all planning steps, we propose use semantic aspects of service analysis. Semantic description is dedicated to the proper analysis of the proposed solution.

The main stages in Cognitive Systems of Service Planning (CSSsP), are as follows:

- determining the main features of the plan:
  - **analysis of the situation on the basis of determining the significance of the analyzed condition – semantic description of analysed situation**,
  - indication of the planning period,
  - determining the minimum and maximum benefits,
- defining the purpose of the service planning processes:
  - **understanding the importance of the service planning processes**,
  - determining the long-term and short-term benefits,
- definition of the service planning primacy:
  - **determining the importance of individual services for the planning processes and for the final plan**,
  - determining the gradation of services, from the most important to the least,
- checking the completeness of the proposed plan of services:
  - **understanding the essence of the main goals of the service planning processes**,
  - including all stages of planning processes,
  - determination of the consequences of service planning processes,
  - indication of threats in case of omitting any stages of service planning,
- checking the effectiveness of the proposed solution:
  - **determining the usefulness of the propose solutions (plans) especially included analyses the service management processes**,
  - attempt to define and uses the proposed service planning solution.

The Cognitive Systems of Service Planning included semantic aspects in all possible planning steps. By used CSSsP systems, we proposed protocols dedicated to semantic description of service planning processes.

This protocol is the following:
Protocol 1

1. service semantic description and analysis of service offered,
2. defining the goals of the planning process,
3. identifying all problems and weak points in service offered set,
4. searching of optimal solutions,
5. defining an optimal service management processes,
6. analysis of favorable and unfavorable solutions in planning processes,
7. analysis of favorable and unfavorable solutions in service management processes,
8. choosing the optimal solution, expanding the set of services with new solutions (new services),
9. approved optimal solution,
10. secure important and strategic information, data and plans,
11. use of data sharing protocols for data security,
12. data distribution between all protocol participants according to the chosen protocol type,
13. realization and implementation of the approved plan of service management,
14. verification of the data security processes,
15. defining the level/levels of service management,
16. determining the relationship between service management levels,
17. distribution secret parts between the indicated levels of service management,
18. verification of the secret parts security in all levels of service management,
19. control of the implementation of the service management processes.

Protocols of cognitive service management for service planning, allow to analyse all factors relevant to the service management process. These analysis concern about interpretation meaning of all internal factors (directly referring to the services offered) and all external factors with direct and indirect meanings for service management processes.

Proposed cognitive service management protocol for semantic description of service planning processes, consists:

- planning for determining the development of services,
- programming for creating new system (IT) solutions,
- forecasting process for predicting the future state.

All planning processes consist aspects of semantic interpretation and analysis for thorough assessment of the current state.

3 Cognitive Service Management Protocols in the Cloud

Cognitive service management protocols can be realized in the fog and cloud. The most important is right selection of levels of service management processes – organization level, fog level, cloud level.

At the planning cognitive service management stage is necessary defining:

- all problems, which managers should solve – this stage refers to the semantic description of the current situation, analyzes of available services, assessing the possibility of enriching the set of proposed services,
- all effectiveness aspects – this level refers planning processes, assessment of organization development, security of all strategic plans, ideas or data,
- all possible strategic solutions for optimal functioning of the company – this level refers new solutions, propositions, development plans, repairs plans, etc.,
- ways to expand and diversify the service offers and improving the current situation of the company – all new directions, new offers (new services), control management processes, interpretation and analysis of all reasons of the current state, understanding the current situation, understanding of all reasons of the current state, secure the important/strategic/secret data.

Cognitive service management protocols for planning can be implemented in different levels in organization, from organization to the cloud level (Fig. 1).

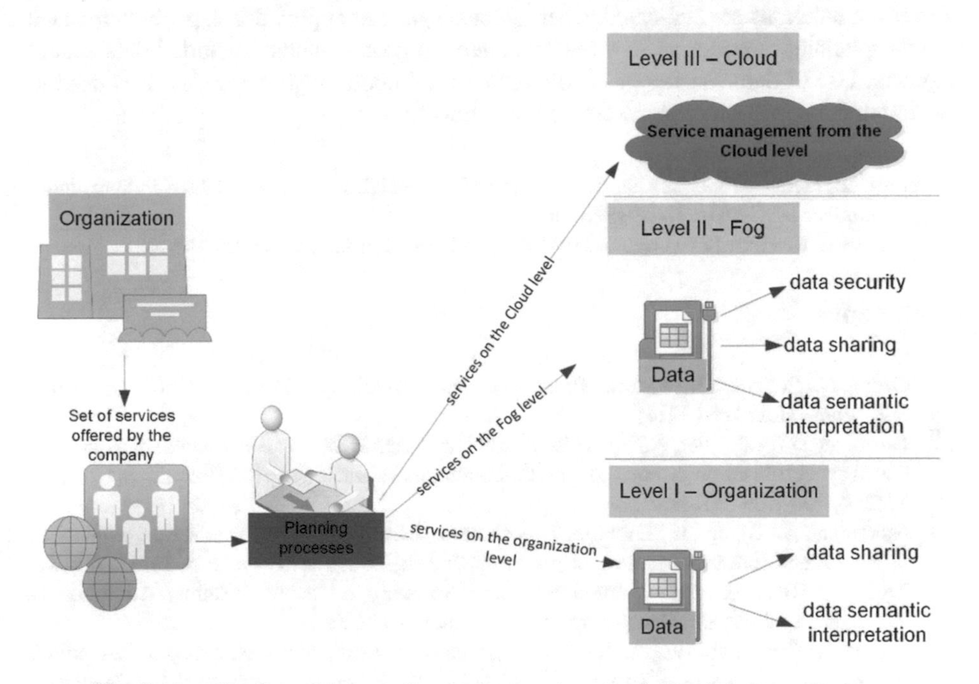

Fig. 1. Levels of cognitive service management protocols for planning processes.

Cognitive service management protocols for planning processes can be used for:

- semantic service management description,
- cognitive service management planning,
- service management processes in different levels in organizations and outside them (in the fog, in the cloud),

- secure important planning information by used sharing protocols,
- distribution all planning information, and all service data,
- improving service management processes by used protocols dedicated to realization of specific tasks.

Cognitive service management protocols dedicated to planning processes enrich the existing solutions of service management and service planning algorithms. Also, used the cryptographic algorithms for data security [1–3, 8, 11].

4 Conclusions

In this paper has been discussed new solutions in service management protocols. Presented algorithm was dedicated to the Cloud and Fog Computing, and for planning cognitive service management processes. The main idea of cognitive service management protocols for planning processes, concerns on application of semantic description and analysis of service sets, offers services and development of the services market. Proposed solutions are dedicated to service management application especially focused on the planning processes. Services management protocols also included data secure aspects. Use of data sharing protocols for secure important/strategic data it is used to secure data by unauthorized disclosure or acquisition.

Acknowledgments. This work has been supported by the National Science Centre, Poland, under project number DEC-2016/23/B/HS4/00616.

This work was partially supported by JSPS KAKENHI grant number 15H0295.

References

1. Gregg, M., Schneier, B.: Security Practitioner and Cryptography Handbook and Study Guide Set. Wiley, Hoboken (2014)
2. Duolikun, D., Aikebaier, A., Enokido, T., Takizawa, M.: Design and evaluation of a quorum-based synchronization protocol of multimedia replicas. Int. J. Ad Hoc Ubiquitous Comput. **17**(2/3), 100–109 (2014)
3. Nakamura, S., Ogiela, L., Enokido, T., Takizawa, M.: Flexible synchronization protocol to prevent illegal information flow in peer-to-peer publish/subscribe systems. In: Barolli, L., Terzo, O. (eds.) Complex, Intelligent, and Software Intensive Systems, Advances in Intelligent Systems and Computing, vol. 611, pp. 82–93 (2018)
4. Ogiela, L.: Cognitive computational intelligence in medical pattern semantic understanding. In: Guo, M.Z., Zhao, L., Wang, L.P. (eds.) ICNC 2008: Fourth International Conference on Natural Computation, vol. 6, Proceedings. Jian, Peoples Republic of China, pp. 245–247, 18–20 October 2008
5. Ogiela, L.: Towards cognitive economy. Soft. Comput. **18**(9), 1675–1683 (2014)
6. Ogiela, L., Ogiela, M.R.: Data mining and semantic inference in cognitive systems. In: Xhafa, F., Barolli, L., Palmieri, F., et al. (eds.) 2014 International Conference on Intelligent Networking and Collaborative Systems (IEEE INCoS 2014). Salerno, Italy, pp. 257–261, 10–12 September 2014

7. Ogiela, L., Ogiela, M.R.: Insider threats and cryptographic techniques in secure information management. IEEE Syst. J. **11**(2), 405–414 (2017)
8. Ogiela, M.R., Ogiela, U.: Grammar Encoding in DNA-Like Secret Sharing Infrastructure. Lecture Notes in Computer Science, vol. 6059, pp. 175–182 (2010)
9. Ogiela, M.R., Ogiela, U.: Secure Information Management Using Linguistic Threshold Approach. Springer-Verlag, London (2014)
10. Ogiela, U., Takizawa, M., Ogiela, L.: Classification of cognitive service management systems in cloud computing. In: Barolli, L., Xhafa, F., Conesa, J. (eds). Advances on Broad-Band Wireless Computing, Communication and Applications BWCCA 2017, Lecture Notes on Data Engineering and Communications Technologies, vol. 12, pp. 309–313, Springer International Publishing AG (2018) https://doi.org/10.1007/978-3-319-69811-3_28
11. Yan, S.Y.: Computational Number Theory and Modern Cryptography. Wiley, Hoboken (2013)

An Examination of CAPTCHA for Tolerance of Relay Attacks and Automated Attacks

Ryohei Tatsuda[1], Hisaaki Yamaba[1], Kentaro Aburada[1](✉), Tetsuro Katayama[1], Mirang Park[2], Norio Shiratori[3], and Naonobu Okazaki[1]

[1] University of Miyazaki, 1-1 Gakuen-Kibanadai-Nishi, Miyazaki 889-2192, Japan
aburada@cs.miyazaki-u.ac.jp
[2] Kanagawa Institute of Technology, 1030, Shimo-Ogino, Atsugi, Kangawa 243-0292, Japan
[3] Research and Development Initiative, Chuo University, 1-13-27 Kasuga, Bunkyo-ku, Tokyo 112-8551, Japan

Abstract. CAPTCHA is a type of challenge response test used to distinguish human users from malicious computer programs such as bots, and is used to protect email, blogs, and other web services from bot attacks. So far, research on enhance of CAPTCHA's resistance to bot attacks has been proceeded to counter advanced automated attacks method. However, an attack technique known as a relay attack has been devised to circumvent CAPTCHA. In this attack, since human solves CAPTCHA, the existing measures assuming bots have no effect on this attack. We designed a new CAPTCHA scheme for relay attacks tolerance and automated attacks tolerance. In this paper, we tested the robustness of the proposed method against several types of automated attacks. We constructed an experimental environment in which a relay attack can be simulated, and designed a series of experiments to evaluate the performance of the proposed method. As a result, we found that the proposed CAPTCHA scheme offers some of level of resistance to automated attacks and relay attacks.

1 Introduction

With the advancement of the Internet, web services are now widely accessible to everyone. However, exploitation of these services by automated programs called bots is a growing concern. For example, a bot will try to acquire e-mail addresses through accounts on the web service and use them to send unsolicited emails or spam. A reverse Turing test known as CAPTCHA is used to prevent such exploitation [1]. CAPTCHA is an easy test for humans to solve but hard for computers. By solving a CAPTCHA, the user confirms they are a human. On the other hand, if a user cannot solve a CAPTCHA, they are likely to be a bot abusing the web service. Typically, images of strings processed with distortion or noise are presented and the visitor must attempt to decipher the string into corresponding keystrokes. However, as the attack techniques used by bots have

L. Barolli et al. (Eds.): EIDWT 2018, LNDECT 17, pp. 870–879, 2018.
https://doi.org/10.1007/978-3-319-75928-9_80

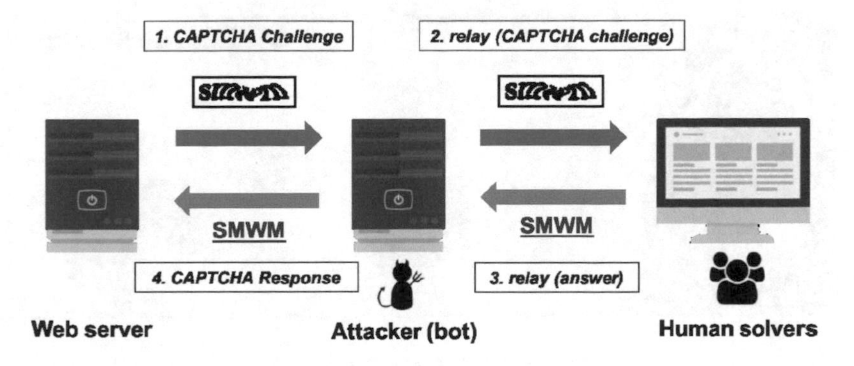

Fig. 1. An example of relay attack.

become more sophisticated, they are now capable of deciphering the CAPTCHA. Therefore, various CAPTCHA schemes have been investigated to strengthen resistance to bots. Research is being conducted to improve web service resistance against bots, but it is not enough for CAPTCHA to defend against bots alone. An attack method known as a relay attack may be used to maliciously break through CAPTCHA. The currently existing CAPTCHA is vulnerable to relay attack because humans, who are not susceptible to the same countermeasures as bots, are responsible for deciphering the CAPTCHA.

In this paper, we focus on the delay time that occurs on communication during a relay attack, and propose a CAPTCHA scheme that increases the difficulty of circumventing a CAPTCHA by using this attack method.

2 Relay Attack

Relay attacks are coordinated by sending CAPTCHA challenges to remote human-solvers who then decode the CAPTCHA and relay the answer back to break the CAPTCHA. The remote human-solver involved may be a hired hand or a regular Internet user. The flow of a typical relay attack is shown below (see Fig. 1).

1. An attacker accesses a website where a CAPTCHA is presented, and obtains the question image posed by the CAPTCHA.
2. The acquired image is automatically sent to a website where human-solvers organize to decipher CAPTCHA challenges.
3. As shown by the website above, human beings who exchange their services for a reward decipher the CAPTCHA and return the solution to the attacker.
4. The attacker then breaks the CAPTCHA using the obtained solution.

A series of processes occur involving the obtaining of the CAPTCHA challenge the transfer of the image to a human-solver, and the execution of the solution by an automatic program created by an attacker. Another method involves presenting the CAPTCHA to a regular Internet user on a fake website created by an attacker, and deceiving the user into solving the CAPTCHA challenge.

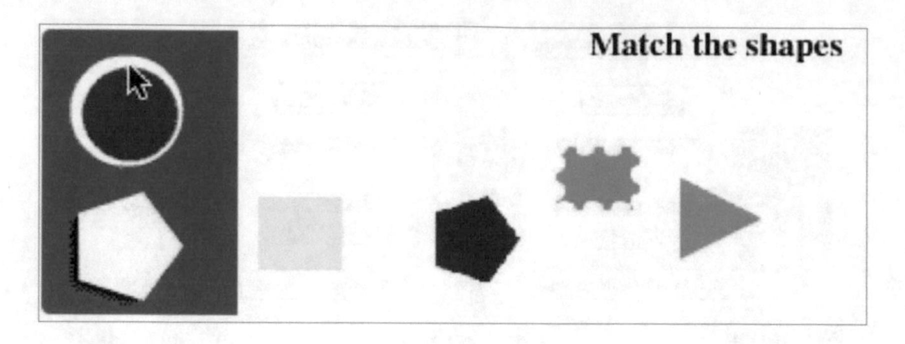

Fig. 2. DCG-CAPTCHA [2,3].

A Dynamic Cognitive Game (DCG)-CAPTCHA is an existing technique for defending against a relay attack [2,3]. Figure 2 shows an example of DCG-CAPTCHA. In the DCG-CAPTCH, a user selects an object suitable for instruction from among a plurality of objects, and performs a task of dragging and dropping the object to a designated position. For this kind of CAPTCHA, the user must complete a game-like cognitive task by interacting with a series of dynamic images (e.g., playing a simple object matching game). Because this CAPTCHA is of an interactive and dynamic nature, the DCG-CAPTCHA offers some level of resistance to relay attacks by causing network latency issues at the communication stages of a relay attack (e.g. transfer of the CAPTCHA challenge, decoding of the image by human-solvers).

However, the DCG-CAPTCHA has also been found to be vulnerable to bot attacks [2]. It is suggested that DCG-CAPTCHA may be broken by automated attacks based on image processing.

3 Design and Implementation of Proposed CAPTCHA

A CAPTCHA scheme which requires real-time responses, such as the DCG-CAPTCHA, has been suggested as a practical countermeasure against relay attacks [3]. However, it is necessary to investigate CAPTCHA systems which not only make it difficult to solve the CAPTCHA by a relay attack, but also are resistant to bots, against which the DCG-CAPTCHA is vulnerable.

In consideration of the above, we have proposed a new CAPTCHA scheme. Figure 3 shows the concept of the proposed method. In this proposed CAPTCHA scheme, the user identifies a moving object from an assortment of randomly appearing disturbing objects and tracks this moving object using the mouse cursor. The user can solve this CAPTCHA if he/she keeps a mouse cursor pointing on the moving object for over a given time length (e.g. 5 s) during a given trial span (10 s in the experiments). We refer to the time that a user keeps the mouse cursor pointing on the moving object as "tracking success time".

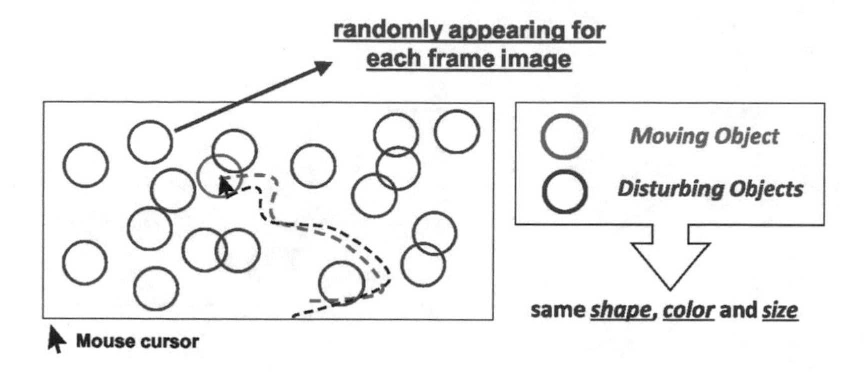

Fig. 3. Concept of proposed CAPTCHA.

4 Security of the Proposed CAPTCHA Scheme

If an attacker of our proposed method executes a relay attack, they would find it difficult to accurately track the moving object using solutions from a human-solver due to the delay time involved in communications during a relay attack. In addition, given that the moving object and the disturbing object of the proposed CAPTCHA scheme would be of the same shape, color and size, it would be difficult for the bot to detect the moving object from the frame image using a visual feature. For example, you can not track moving objects with a technique that uses patterns of tracked objects such as "template matching" (Since the visual feature does not differ between the disturbing object and the moving object, the disturbing object may be erroneously detected.).

In this section, we evaluate the resistance against automated attacks of the proposed CAPTCHA scheme.

4.1 Attack Using Difference Between Consecutive Frame Images

As an attack against the proposed CAPTCHA scheme, it is conceivable to extract several frames from a moving image and grasp the position of the moving object using the difference of the extracted frames. In the proposed CAPTCHA scheme has two features. First, the disturbing object changes its position randomly for each frame. Second, the moving object's change in position is small between consecutive frames. Therefore, it is possible to exclude disturbing objects by repeating the AND operation between consecutive frame images (binary images), and it is possible to extract the common area of moving objects. As a result, there is possibility that the position of the moving object may be specified (see Fig. 4).

As a countermeasure to this attack, we change the design of the object from the initial design. A new object is composed of 12 points (a circle is composed of 12 points), and the point to be displayed is switched for each frame (Example: display even numbered point, not display odd numbered point). With this

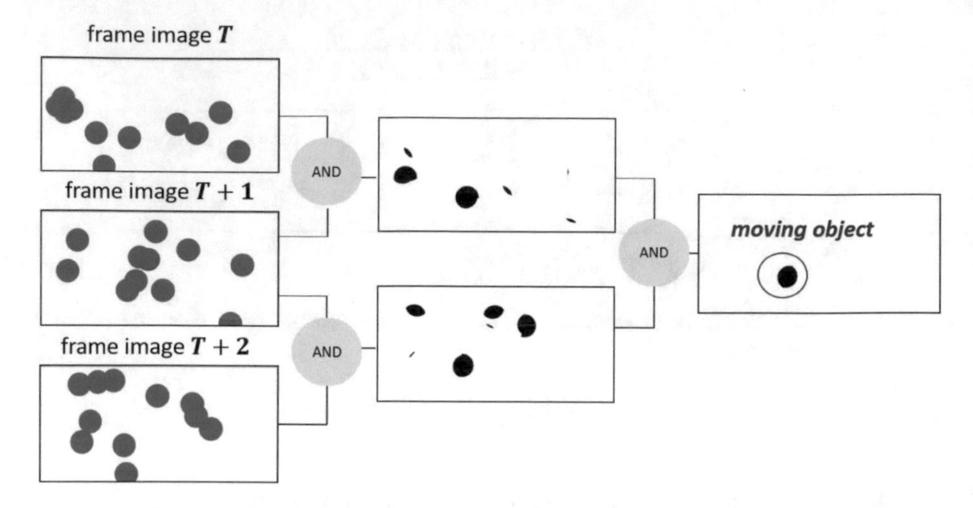

Fig. 4. Automated attack using difference of consecutive frame images (initial design).

countermeasure, it is possible to eliminate the occurrence of the common area of moving objects between consecutive frame images, and it is thought that it will be difficult to specify the position of the moving object even if the logical AND operation is performed (see Fig. 5).

4.2 Attack by Tracking with MeanShift Algorithm

In addition, we considered that there is an attack method by object tracking technique using the mean-shift algorithm as an another attack against the proposed CAPTCHA scheme. The mean-shift algorithm is an efficient approach to tracking objects whose appearance is defined by histograms [4].

The basic idea is to construct histograms of the certain features (such as gray scale intensity) of the object to be tracked and the candidate to be matched. Similarity between the two is measured using these histograms using certain metric, such as Bhattacharyya measure. In practical implementation of general tracking application, weights are assigned to histogram bins. These weights are back-projected to the candidate pixels to construct a back-projection image in which each pixel is assigned a value equal to the weight of the histogram bin to which it belongs. These values give the probability of the pixel belonging to the object being tracked. The shift in location of the object being tracked is found by finding the center of gravity (the location of mean value) of the back-projected weights in the tracking window. With movement of the object, the mean is shifted accordingly.

It seems like seemingly meaningless using the tracking technique based on color histogram for automated attack against the proposed CAPTCHA scheme (since moving object and disturbing objects are of the same color). However, because the moving object has a feature that changing position is small between consecutive

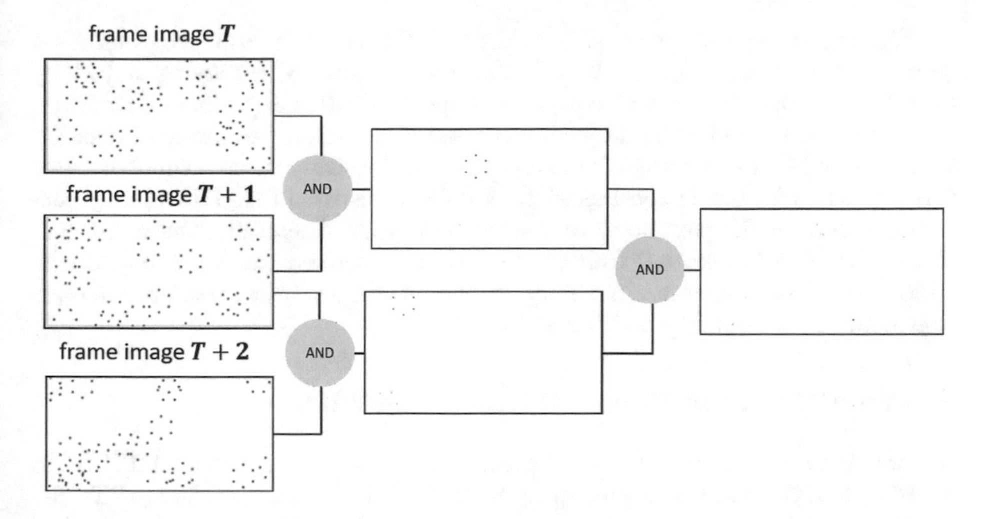

Fig. 5. Automated attack using difference of consecutive frame images (after changing the design of the object).

frame images, if a moving object enters the tracking window even once, it may track object (unless a moving object does not move so much as to move out of the tracking window in next frame image). In this time, we implemented a program of object tracking by mean-shift algorithm and verified that bot countermeasure by disturbing objects is how effective against this attack. Specifically, we attempt automatic tracking of a moving object using meanShift algorithm, and we measured "tracking success time". We executed 100 attacks against each for 5 patterns of CAPTCHA with 10, 20, 30, 40 and 50 disturbing objects.

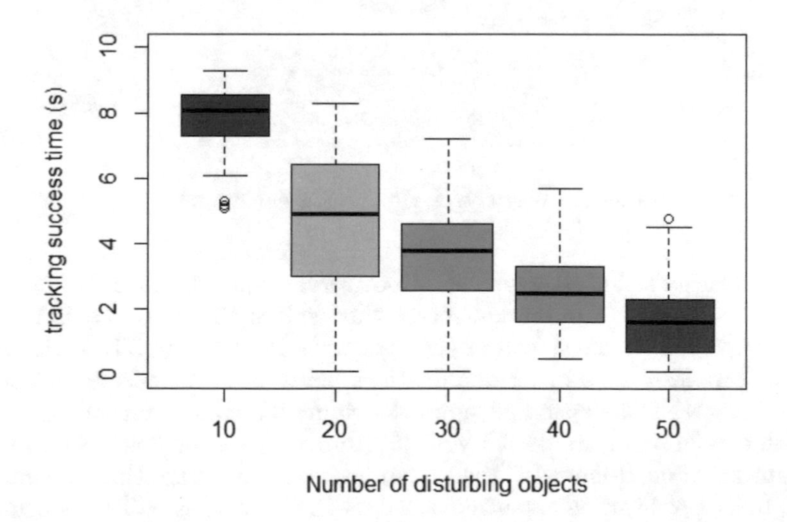

Fig. 6. "Tracking success time" by tracking method using meanShift algorithm.

Figure 6 shows the result of "tracking success time" by our program. As a result of the verification, we found that the tracking of a moving object by meanShift algorithm is effective against proposed CAPTCHA scheme. However, as the number of disturbing objects increased, the tracking accuracy of tracking a moving object by our program decreased and "tracking success time" became shorter. Therefore, it is considered that it becomes useful CAPTCHA by successfully setting the parameter of disturbing objects. Increasing the number of disturbing objects makes it difficult for humans to recognize a moving object, which leads to a decrease in usability. In the future, we must decide parameters that compatible usability and bot resistance.

5 Validation for Relay Attack Resistance

In this study, we simulated a relay attack against our proposed CAPTCHA to investigate if the additional time delay made it difficult to verify the CAPTCHA solution.

A virtual environment was used to simulate the relay attack using VirtualBox (see Fig. 7). In addition, we transferred the CAPTCHA challenge using VNC (Virtual Networking Computer) which has been used in previous studies for reproducing relay attacks [3].

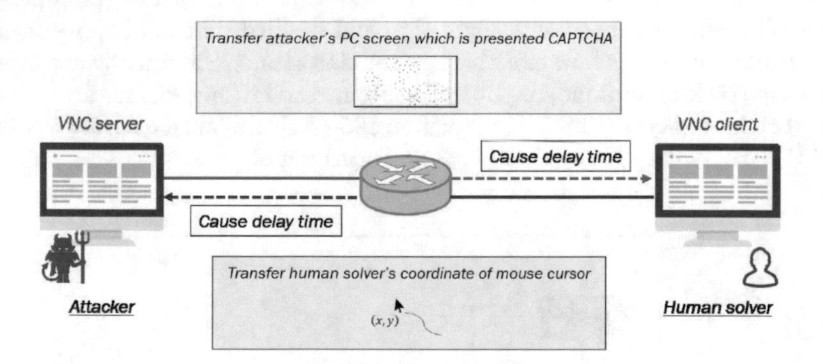

Fig. 7. Relay attack experiment environment.

The attacker's PC starts the VNC server and displays the proposed CAPTCHA scheme. The human-solver then solves the proposed CAPTCHA scheme displayed on the screen of the attacker's PC through VNC clients. The length of the delay time in communications between the attacker's PC and the human-solver's PC is dependent upon the communication environment.

With this in mind, we used VyOS [5] to investigate relay attacks in various communication environments. VyOS can generate any delay time of communication. In this verification experiment, we set the time delay (RTT) in communications between the attacker's PC and the human-solver's PC to 50 ms, 100 ms, 200 ms.

Ten students of University of Miyazaki participated in the experiment. They had previously solved the proposed CAPTCHA scheme five times by legitimate access and by relay attack, and their "the tracking success time" was measured. We simulated the relay attack using 4 different communication environments (No insertion of delay time, RTT = 50 ms, 100 ms, 200 ms).

6 Experimental Results

Figure 8 shows a histogram of the tracking success times of ten experiment participants who solved the proposed CAPTCHA scheme. Tracking a moving object by relay attack is more difficult by legitimate access as shown by the increase in delay time.

We determined the threshold at which it becomes difficult for a human-solver to track the moving object from "the tracking success time" data. We approximated the data of "the tracking success time" from legitimate access and relay

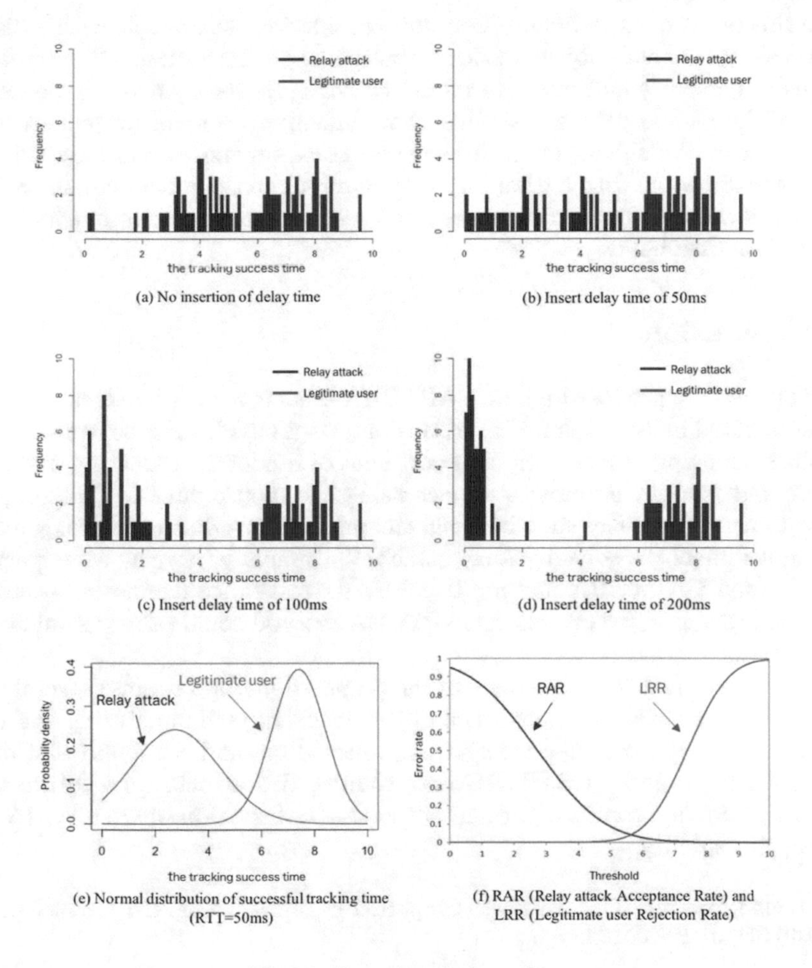

(a) No insertion of delay time

(b) Insert delay time of 50ms

(c) Insert delay time of 100ms

(d) Insert delay time of 200ms

(e) Normal distribution of successful tracking time (RTT=50ms)

(f) RAR (Relay attack Acceptance Rate) and LRR (Legitimate user Rejection Rate)

Fig. 8. Experimental results.

attack methods ($RTT = 50\,\text{ms}$) to obtain a normal distribution (show Fig. 8(e)). We calculated RAR (Relay attack Acceptance Rate) and LRR (Legitimate user Rejection rate) from the normal distribution (show Fig. 8(f)). We determined a threshold value of "the tracking success time" for preventing a relay attack based on RAR and LRR. The average success rate of reCAPTCHA which has been used in previous studies [6] is approximately 97%. For the proposed CAPTCHA scheme to have a comparable success rate for legitimate users, LRR should be 3%. When LRR is 3%, the threshold of successful tracking time is approximately 5.5 s.

Also, when setting the threshold value to 5.5 s, RAR is 5%. At this time, if a relay attack is performed in communication environment a 50 ms delay occurs and it is possible to prevent about 95% of relay attacks. According to investigative studies [7], human-solvers are paid a reward of $0.5 to $3.0 per every 1000 CAPTCHA challenges decoded. If RAR is 5%, the relay attack will not be economically feasible because even if 1000 CAPTCHAs are solved, only about 50 will be successful.

In this result of experiment, we aren't considering that people with a disability and eldery people's object tracking ability using mouse cursor. It is confirmed that elder persons's movement of mouse cursor is slower than young people [8]. Some of the people with a disability may difficult to control movement of the mouse cursor. We should check if they can track for more than 5.5 s. In addition, some of people with a disability may manipulate the mouse cursor with eye tracking devices [9]. In the future, we must assume various environments and set appropriate thresholds.

7 Conclusion

In this paper, we proposed a new CAPTCHA scheme for relay attacks tolerance and automated attacks tolerance. In this proposed CAPTCHA scheme, the user identifies a moving object from an assortment of randomly appearing disturbing objects and tracks this moving object using the mouse cursor. This proposed method focuses on delay time between communications during the relay attack, making it difficult to solve by relay attack. We simulated a relay attack against the proposed CAPTCHA scheme to verify its resistance against relay attack. The results suggest the proposed CAPTCHA method could prove useful against relay attacks.

We have verified the robustness of the proposed method against several types of automated attacks. Although the object tracking technique using the mean Shift algorithm is most effective against proposed method, we found that obfuscation of our proposed CAPTCHA can counter this attack. As a future work, we must decide appropriate parameters so that obfuscation of CAPTCHA does not impair usability.

Acknowledgments. This work was supported by JSPS KAKENHI Grant Numbers JP17H01736, JP17K00139.

References

1. von Ahn, L., Blum, M., Hopper, N.J., Langford, J.: CAPTCHA: telling humans and computers apart automatically. In: Advances in Cryptology, Eurocrypt 2003, Lecture Notes in Computer Science, vol. 2656, pp. 294–311 (2003)
2. Mohamed, M., Sachdeva, N., Georgescu, M., Gao, S., Zhang, C.: A three-way investigation of a game-CAPTCHA: automated attacks, relay attacks and usability. In: Proceedings of the 9th ACM Symposium on Information, Computer and Communications Security, pp. 195–206. ACM (2014)
3. Mohamed, M., Gao, S., Saxena, N., Zhang, C.: Dynamic cognitive game CAPTCHA usability and detection of streaming-based farming. In: The Workshop on Usable Security (USEC), Co-located with NDSS (2014)
4. Khan, I.R., Farbiz, F.: A back projection scheme for accurate mean shift based tracking. In: 2010 17th IEEE International Conference on Image Processing (ICIP), pp. 33–36. IEEE (2010)
5. Index of /software/vyos/iso/release/1.1.7 (2016). ftp.tsukuba.wide.ad.jp. http://ftp.tsukuba.wide.ad.jp/software/vyos/iso/release/1.1.7/. Accessed Oct 2016
6. Yan, J., EI Ahmad, A.S.: Usability of CAPTCHAs or usability issues in CAPTCHA design. In: Proceedings of the 4th Symposium on Usable Privacy and Security, pp. 44–52. ACM (2008)
7. Motoyama, M., Levchenko, K., Kanich, C., McCoy, D., Coelker, G.M., Savage, S.: Re: CAPTCHAs-Understanding CAPTCHA-Solving Services in an Economic Context. In: USENIX Security Symposium, Washington, pp. 1–18 (2010)
8. Bohan, M., Chaparro, A.: Age-related differences in performance using a mouse and trackball. Proc. Hum. Factors Ergon. Soc. Ann. Meet. **42**(2), 152–155 (1998)
9. Zende, S., Tambile, V., Thakur, A., Schendge, M., Rathi, S.: Mouse pointer movement using Gaze tracking system. Int. J. Comput. Appl. **140**(11), 1–4 (2016)

A Data Sharing Method Using WebRTC for Web-Based Virtual World

Masaki Kohana[1(✉)] and Shusuke Okamoto[2]

[1] Ibaraki University, Hitachi, Ibaraki, Japan
masaki.kohana.gopher@vc.ibaraki.ac.jp
[2] Seikei University, Musashino, Tokyo, Japan
okam@st.seikei.ac.jp

Abstract. This paper proposes an information sharing method for a web-based virtual world. Our system uses computing resources on Web browsers and does not use server resources. The Web browsers construct a ring-type peer-to-peer network using WebRTC and share the information using this network. The WebRTC allows the browsers the communication with the other browsers without any backend servers. Therefore, the browsers can collect the information among browsers. We measure the average transfer time and show the characteristic of the topology.

1 Introduction

This paper proposes a way to build a virtual world by using computing resources on Web browsers. A virtual world is an essential thing for video games and virtual reality applications. We focus on a type of multi-player online game. In this type of game, a user creates an avatar in the virtual world. The user controls the avatar and interacts with the other avatars and non-player characters (NPCs). A client software displays the virtual world and the other characters. Therefore, each client shares the character information.

A traditional type of online game forms server-client model. A game server stores all the game data and provides it to all client software. Each user controls the own avatar, and the client sends the information to the game server. In this type, as the number of users increases, the server load also increases.

To solve this problem, the most virtual world of online games consists of some blocks such as Fig. 1. In this figure, the virtual world consists of 9 blocks, and there are three game servers. Each block is managed by a server with the same color. An avatar exists in one block at a time. When the avatar reaches an edge of the block, the avatar moves to the neighbor block. The client software needs the information on the same block. The avatars disperse to the several blocks, which means that the users connect to the several servers. Therefore, the server load also disperses.

On the other hand, an open-world virtual world attracts attention. An open-world is not divided into blocks seemingly. The user can move the game world without the reloading map information, while the traditional world needs to

L. Barolli et al. (Eds.): EIDWT 2018, LNDECT 17, pp. 880–888, 2018.
https://doi.org/10.1007/978-3-319-75928-9_81

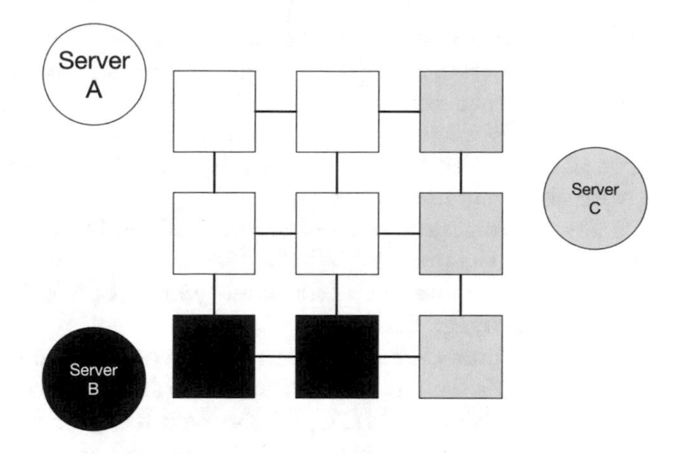

Fig. 1. Dividing world

reload the map when the user reaches the edge of the block. The user gets the higher degree of freedom of the movement. However, the open-world needs a lot of information and computing resources. Furthermore, load distribution is difficult for online games because the world is not divided.

One of the ways to get the computing resource, we used multiple servers in our previous work [6,7]. In this work, we divide the world into small blocks and allocate each ownership of the blocks to a server. The client gets the information that related to the display area on the client. The system moves the ownership to the other server according to the user requirement to reduce the frequency of the communication.

However, to get the computing resource, we need to increase the number of servers. To solve this problem, we focus on the resource on client computers. The number of client computers is the same as the number of users. If we can use the client computer as a computation node, we can get the computing resources according to the number of users. We also focus on a Web-based virtual world because a Web application can run the same code on the several platforms. Furthermore, new features are developed for Web browsers such as the Web Workers, the Web Sockets, and the WebRTC.

This paper surveys how to share the data among Web browsers and shows a preliminary experimental result. Section 2 introduces three key technologies of our system. Section 3 shows the literature surveys. Section 4 describes an overview of our system and Sect. 5 shows the experimental result. Finally, Sect. 6 concludes our paper.

2 Key Technologies

This section introduces three key features to build our system. These are Web browser features, and we can use them from JavaScript code.

Web Workers is a way to run a JavaScript code in background threads. W3C defines the specification of API [1]. A worker thread cannot handle a user interface. However, the thread runs independently of any user interface. In our system, we use the Web Workers to separate a rendering virtual world from a computation process. The Web browser can display the virtual world without any interfering with the computation process.

WebSocket is a communication protocol between a Web browser and a Web server. IETF standardizes the protocol [4]. The WebSocket provides a real-time data transfer. In our system, we use it for the signaling of WebRTC and to update static files such as image files.

WebRTC(Web Real-Time Communication) is a communication protocol among Web browsers. W3C defines the API specification [2], and IETF standardizes the protocol [3]. Using WebRTC, all the Web browsers can construct a peer-to-peer network. On this network, a browser can communicate with the other browser without any backend server. Using WebRTC, we can develop applications such as a video chat, file transfer, desktop sharing, and so on. In our system, all the Web browser share the character information on the WebRTC network, which can decrease the server load.

3 Literature Survey

This section introduces some studies related to our paper such as multi-server systems for online game and Web technologies.

In our previous work, we use Web Workers [8]. The Web browsers of our system undertake the tasks of the server. The browser separates the rendering task and the computing task using Web Workers. A worker thread performs the computing task, and the main thread has responsibility for the rendering.

The another our work uses WebRTC [9]. In this work, our system runs an MPI program. MPI is a parallel and distributed computing method. There are some computation nodes, and they exchange the data using a message passing. We transfer an MPI program written in C to JavaScript code. The multiple Web browsers run the code and exchange the data using WebRTC.

Ito et al. proposed a webcast system using Web browsers in class [5]. A teacher shares a screen on a computer, and students receive the screen and display on their computer. The computers of the teacher and the students are connected with the WebRTC and form a binary tree topology.

The webcast system uses the WebRTC. However, the original data is managed by the master node, and all the client nodes relay the data. On the other hand, in our study, all the client nodes need to broadcast the data because they update and need to share the data.

De Vleeshauwer et al. and Van Den Bossche et al. proposed dynamic microcell assignment [10]. This system divides a virtual world into microcells. The cells are allocated to a set of servers to address the high, dynamically varying player density.

The dynamic microcell system is an efficient way to build a virtual world using multiple servers. Our system uses only one server, and we use the computing resource on Web browsers.

4 System Overview

This section describes an overview of our system. The system includes only one Web server and multiple Web browsers as clients. The server keeps a topology. All the Web browsers construct a peer-to-peer network. When a browser participates in the network, the server selects a client, and the new client connects to the selected client.

Figure 2 is a screenshot of our online game. There are two avatars, the elephant and the frog, which means that two users shares this region of the game world. The elephant is located at the center of the screen. The user who see this screen controls the elephant, and the frog is controlled by the other user.

The Web browser that controls the elephant stores the information about the elephant to the Local Storage. The Local Storage is a type of Web Storage. The Web Storage has two types of the storage, the Local Storage and the Session Storage. The Local Storage stores the data permanently, while the Session Storage stores the data during the session. The information of the frog is stored on the other Web browser. To get the frog information, the Web browser shares the data using the WebRTC connection.

4.1 Signaling Process

In WebRTC, a signaling means that the server relays two Web browsers and they exchange the information that is needed for WebRTC connection.

Figure 3 shows the process of the signaling. There are three Web browsers and one Web server. In this figure, we assume that the client 0 and the client 1 are connected each other. When the client 2 participates the network, it sends a request to the server as step (a). The server knows the client 2 should connect to the client 0 and 1 to create a ring form. Therefore, the server replies the client number 0 and 1 to the client 2.

As the step (c), the client 2 sends the offer message with the client identifier. The server receives the message and transfers it to the client 0 and client 1 at the step(d). The offer message includes the connection information. The WebRTC connection needs the information. When the client 0 receives the offer message, it replies the answer message that also includes the connection information. The server receives the answer message and also transfers the message to the client 2. Then, the client 0 and the client 2 exchange the connection information, which means that the client 2 connects to the client 1 with the WebRTC protocol. The client 1 also performs the same process. Then, the client 2 connects to the client 1 with the WebRTC protocol.

At the signaling process, the clients that already participate the peer-to-peer network need the disconnection process as the Fig. 4. Since the browsers form the

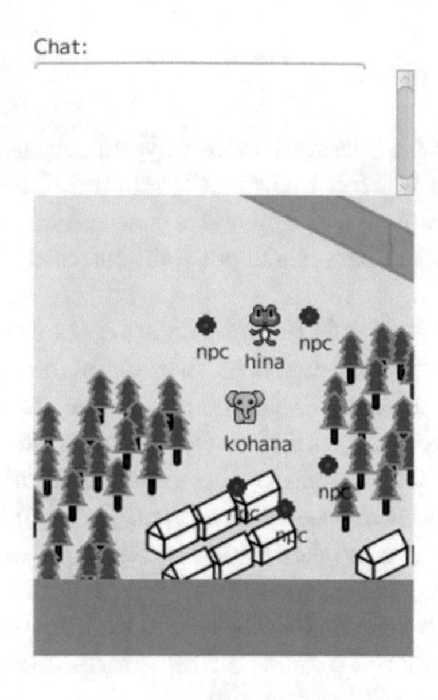

Fig. 2. Screenshot

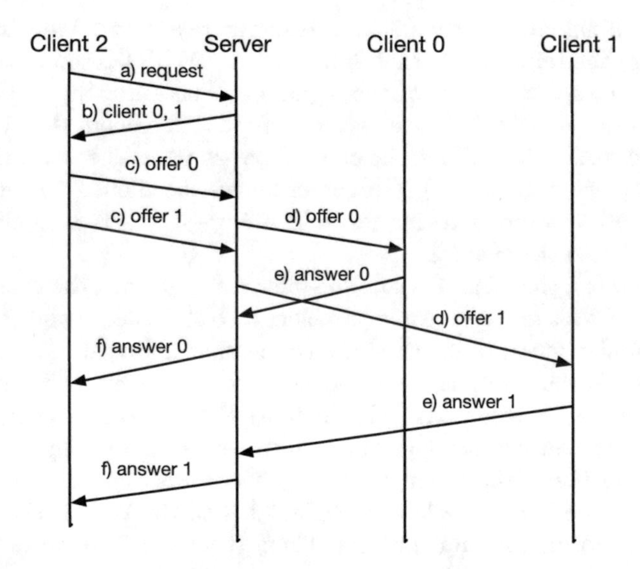

Fig. 3. Signaling process

ring topology, each browser has the left and the right connection. To construct a ring, the right connection of the client 0 connects to the left connection of the client 1. The left of the client 0 connects to the right of the connection 1. When the client 2 participates the network, the right of the client 2 should connect to the left of the client 0, and the left should connect to the right of the client 1. Therefore, the connection between the right of the client 1 and the left of the client 0 should be disconnected. After that, the client 2 makes the left and the right connection.

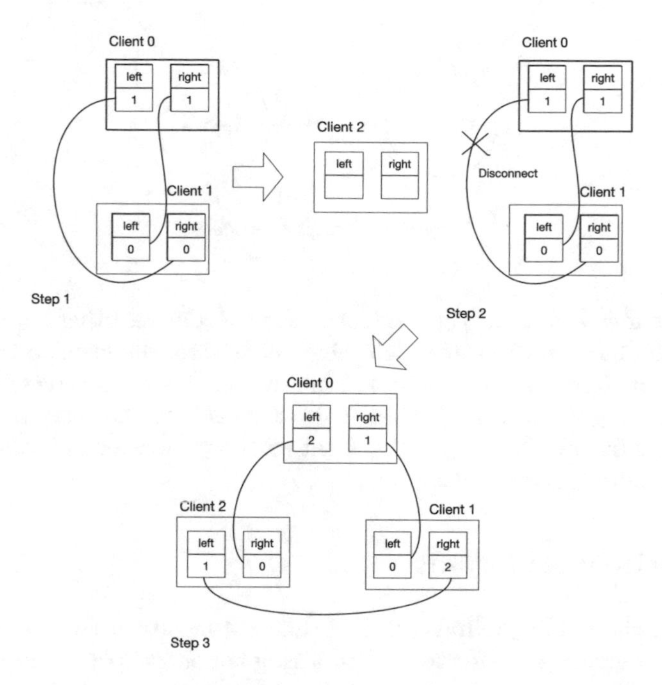

Fig. 4. Disconnection process

4.2 WebRTC Messaging Process

To share the information about avatars, each Web browser transfers the avatar information using the WebRTC connection. There are two timings for the sending the information. The one timing is when the browser updates the own avatar information by the user input. The another is when the browser receives the information from the other browsers.

Figure 5 shows an overview of the transfer information. In this figure, there are 6 Web browsers. When the user of the client 0 updates the avatar status, the client 0 is the start node. The start node sends the message to both the right and the left node. In this case, the client 1 and the client 5 receives the message from the client 0. Since the client 1 receives the message from the left side, it

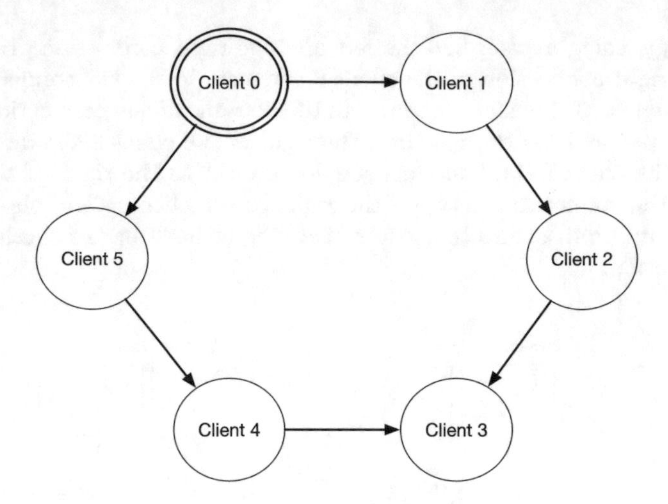

Fig. 5. Messaging process

transfers the message to the right side, the client 2. On the other hand, the client 5 receives the message from the right side. Therefore, the client 5 transfers the message to the left side, the client 4. Finally, the client 3 receives the message from both the right and the left side, which means that the client 3 receives the same message from both sides. If the client receives the same message, the client does not transfer the message.

5 Experimental Result

This section shows our preliminary experimental result. In this experiment, we measure the average transfer time with a ring topology. We use only one computer to avoid an effect by a network connection parameter such as bandwidth. This experiment shows a characteristic of a topology.

Table 1 shows our runtime environment. We open 10 tabs on the Web browser, and they form a ring topology. We set the client 0 as a start node. The client 0 sends a message to both sides of the node every 500 msec. The total number of messages is 1000. The client 0 sends a message with the current time. Each client receives the message and stores the difference between the time of the message and the time of the receiving. We take the average of the 1000 messages for each client.

Figure 6 shows the experimental result. The x-axis indicates the client number, while the y-axis indicates the average transfer time. The right side of the client 0 is the client 2, and the left side is the client 10.

The client 5 and 6 has the longest time. The client 6 is the farthest node from the client 0. Therefore, the number of the intermediate nodes is the largest. The client 5 has the same number of intermediate nodes as the client 7. However, the transfer time is longer than the client 7. Furthermore, both the client 2 and

Table 1. Runtime environment

CPU	Intel Core i5 1.3 GHz
RAM	8 GB 1867 MHz LPDDR 3
OS	macOS Sierra 10.12.6
Browser	Google Chrome 62.0.3202.94 64bit

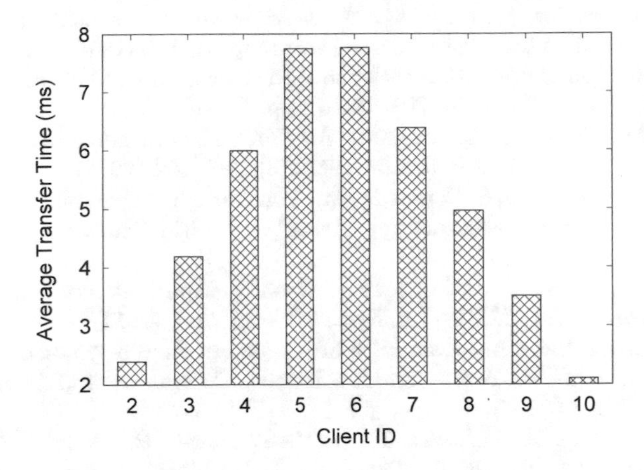

Fig. 6. Average transfer time

the client 10 are neighbor nodes to the client 0. However, the time for client 10 is less than the client 2. The reason for these things is our implementation. The start node, the client 0, sends a message to the left side node firstly. Then, it sends to the right side. Other clients have the same order. Therefore, the left side nodes get a message a little early.

6 Conclusion

This paper proposes a way to build a virtual world using Web browser resources. Our system constructs a peer-to-peer network with WebRTC protocol. The WebRTC provides a connection between browsers. Using WebRTC, the browsers can communicate without any backend server. We build a ring-form topology with the Web browsers and measure the transfer time from the start node to each node. The result shows that the message sending order affects the transfer time. As a result, the left side nodes get a message a little early.

In this system, all the clients share the same information. However, each client needs the different information from the other clients. Therefore, the client should get the necessary information. In our future work, we make a group and share the information with the group to share the necessary information.

References

1. Web workers. https://www.w3.org/TR/workers/. Accessed 13 Nov
2. Webrtc api. https://www.w3.org/TR/webrtc/. Accessed 13 Nov
3. Webrtc protocol. https://tools.ietf.org/html/draft-ietf-rtcweb-overview-19. Accessed 13 Nov
4. Websocket. https://tools.ietf.org/html/rfc6455. Accessed 13 Nov
5. Ito, D., Niibori, M., Kamada, M.: A real-time web-cast system for classes in the byod style. In: The 5th International Workshop on Web Services and Social Media (WSSM-2016) in Conjunction with the 19th International Conference on Network-Based Information Systems (NBiS2016), pp. 520–525 (2016)
6. Kohana, M., Okamoto, S., Ikegami, A.: Optimal data allocatin and fairness for online game. Int. J. Grid Util. Comput. **5**(3), 183–189 (2014)
7. Kohana, M., Okamoto, S., Kamada, M., Yonekura, T.: Dynamic reallocation rules on multi-server web-based morpg system. Int. J. Grid Util. Comput. **3**(2/3), 136–144 (2012)
8. Okamoto, S., Kohana, M.: Load distribution by using web workers for a real-time web application. Int. J. Web Inf. Syst. **4**(7), 381–395 (2011)
9. Okamoto, S., Kohana, M.: Running a MPI program on web browsers. In: 2017 IEEE Pacific Rim Conference on Communciations, Computers and Signal Processing (PACRIM 2017) (2017)
10. Van Den Bossche, B., Verdickt, T., De Vleeschauwer, B., Desmet, S., De Mulder, S., De Turck, F., Dhoedt, B., Demeester, P.: A platform for dynamic microcell redeployment in massively multiplayer online games. In: Proceedings of the 2006 International Workshop on Network and Operating Systems Support for Digital Audio and Video, NOSSDAV 2006, pp. 3:1–3:6. ACM, New York (2006). https://doi.org/10.1145/1378191.1378195

A Study on a User Identification Method Using Dynamic Time Warping to Realize an Authentication System by s-EMG

Tokiyoshi Kurogi[1], Hisaaki Yamaba[1(✉)], Kentaro Aburada[1], Tetsuro Katayama[1], Mirang Park[2], and Naonobu Okazaki[1]

[1] University of Miyazaki, 1-1 Gakuen Kibanadai-nishi, Miyazaki 889-2192, Japan
yamaba@cs.miyazaki-u.ac.jp

[2] Kanagawa Institute of Technology, 1030 Shimo-Ogino, Atsugi, Kangawa 243-0203, Japan

Abstract. At the present time, mobile devices such as tablet-type PCs and smart phones have widely penetrated into our daily lives. Therefore, an authentication method that prevents shoulder surfing is needed. We are investigating a new user authentication method for mobile devices that uses surface electromyogram (s-EMG) signals, not screen touching.

The s-EMG signals, which are detected over the skin surface, are generated by the electrical activity of muscle fibers during contraction. Muscle movement can be differentiated by analyzing the s-EMG. Taking advantage of the characteristics, we proposed a method that uses a list of gestures as a password in the previous study. In this paper, we introduced dynamic time warping (DTW) for improvement of the method of identifying gestures.

1 Introduction

This paper presents an introduction of dynamic time warping (DTW) to the user authentication method for mobile devices by using surface electromyogram (s-EMG) signals, not screen touching.

At the present time, mobile devices such as tablet type PCs and smartphones have widely penetrated into our daily lives. Consequently, an authentication method that prevents shoulder surfing, which is the direct observation of a user's personal information such as passwords, comes to be important.

In general, authentication operations on mobile devices are performed in many public places, so we have to ensure that no one can view our passwords. However, it is easy for people who stand near such a mobile device user to see login operations and obtain the users authentication information.

And also, it is not easy to hide mobile devices from attackers during login operations because users have to see the touch screen of their mobile device, which do not have keyboards, to input authentication information. On using a touchscreen, users of a mobile device input their authentication information through simple or multi-touch gestures. These gestures include, for example,

L. Barolli et al. (Eds.): EIDWT 2018, LNDECT 17, pp. 889–900, 2018.
https://doi.org/10.1007/978-3-319-75928-9_82

designating his/her passcode from displayed numbers, selecting registered pictures or icons from a set of pictures, or tracing a registered one-stroke sketch on the screen. The user has to see the touch screen during his/her login operation; strangers around them also can see the screen.

To prevent this kind of attack, biometrics authentication methods, which use metrics related to human characteristics, are expected. In this study, we investigated application of surface electromyogram (s-EMG) signals for user authentication.

S-EMG signals, which are detected over the skin surface, are generated by the electrical activity of muscle fibers during contraction. These s-EMGs have been used to control various devices, including artificial limbs and electrical wheelchairs. Muscle movement can be differentiated by analyzing the s-EMG [1]. Feature extraction is carried out through the analysis of the s-EMGs. For example, fast Fourier transform (FFT) can be adopted for the analysis. The extracted features are used to differentiate the muscle movement, including hand gestures.

In the previous researches [2–5], we investigate the prospect of realizing an authentication method using s-EMGs through a series of experiments. First, several gestures of the wrist were introduced, and the s-EMG signals generated for each of the motion patterns were measured [2]. We compared the s-EMG signal patterns generated by each subject with the patterns generated by other subjects. As a result, it was found that the patterns of each individual subject are similar but they differ from those of other subjects. Thus, s-EMGs can confirm ones identification for authenticating passwords on touchscreen devices. Next, a method that uses a list of gestures as a password was proposed [3,4]. And also, experiments that were carried out to investigate the performance of the method extracting feature values from s-EMG signals (using the Fourier transform) adopted in [5]. The results showed that the Fourier transform has certain ability to extract feature values from s-EMG signals, but further accuracy was desired.

In this paper, a dynamic time warping (DTW) was introduced to identify gestures from s-EMG signals. The false rejection rate and the false acceptance rate of this method were evaluated from the results of the experiments. We also discussed the security evaluation of the proposed method.

2 Characteristics of Authentication Method for Mobile Devices

It is considered that user authentication of mobile devices has two characteristics [2].

One is that an authentication operation is performed when a user wants to start using their mobile devices. The authentication often takes place around strangers. Therefore, the strangers around the user can possibly see the user's unlock actions. Some of these strangers may scheme to steal information for authentication such as passwords.

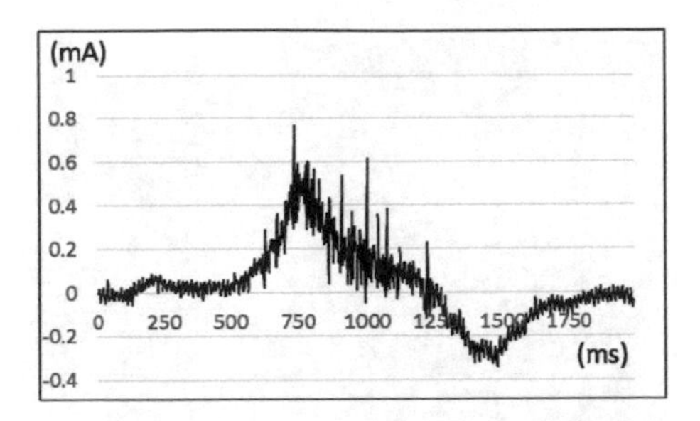

Fig. 1. A sample of an s-EMG signal

The other characteristic is that user authentication of mobile devices is almost always performed on a touchscreen. Since many of current mobile devices do not have hardware keyboards, it is not easy to input long character based passwords into such mobile devices. When users want to unlock mobile touchscreen devices, they input passwords or personal identification numbers (PINs) by tapping numbers or characters displayed on the touchscreen. Naturally, users have to look at their touchscreens while unlocking their devices, strangers around them also can easily see the unlock actions. Besides, the user moves only one finger in many cases. So, it becomes very easy for thieves to steal passwords or PINs.

To prevent shoulder-surfing attacks, many studies have been conducted. The secret tap method [6] introduces a shift value to avoid revealing pass-icons. The user may tap other icons in the shift position on the touchscreen, as indicated by a shift value, to unlock the device. By keeping the shift value secret, people around the user cannot know the pass-icons, although they can still watch the tapping operation. The rhythm authentication method [7] relieves the user from looking at the touchscreen when unlocking the device. In this method, the user taps the rhythm of his or her favorite music on the touchscreen. The pattern of tapping is used as the password. In this situation, the users can unlock their devices while keeping them in their pockets or bags, and the people around them cannot see the tap operations that contain the authentication information.

3 Surface Electromyogram Signals

The s-EMG signals (Fig. 1) are generated by the electrical activity of muscle fibers during contraction and are detected over the skin surface (Fig. 2) [2]. Muscle movement can be differentiated by analyzing the s-EMG.

However, since measured s-EMG signals vary by subject, the extracted features do not show enough performance to correctly differentiate the muscle movement in multiple subjects. Therefore, researchers have explored other methods to improve the performance of feature extraction. Since some methods demonstrate good performance for some subjects but other methods show better performance for other subjects, a feature that can be used to distinguish gestures for everyone is desired. For example, a method that uses the maximum value and the minimum value of raw s-EMG signals was proposed [8].

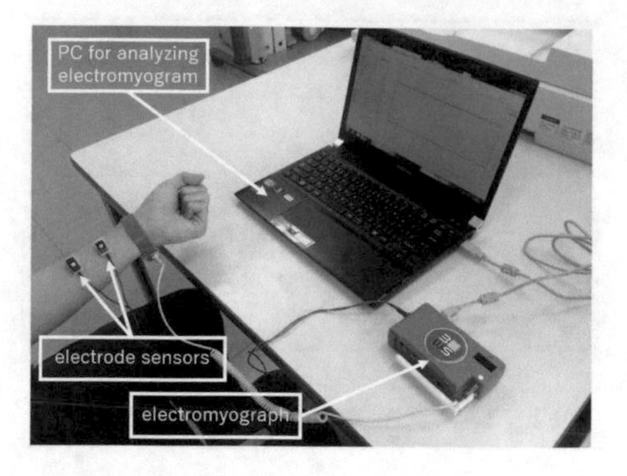

Fig. 2. Measuring an s-EMG signal

4 Propsed Method

4.1 User Authentication System Using S-EMG

In the previous research, the method of user authentication by using s-EMGs, which do not require looking at a touchscreen, was proposed [3,4]. Figure 3 shows the sketch of the authentication system using s-EMG signals. First, wearable devices such as a smart watch measures s-EMG signals and send the signals to mobile devices such as smartphones using Bluetooth. Next, the feature values of the measured raw signals are extracted. Then, the mobile device estimates gestures made by a user of a mobile device from the extracted features. Small devices is now available to measure s-EMG signals and communicate with other devices using wireless Bluetooth. It is expected that the authentication system using s-EMG will be realized using such kind of devices.

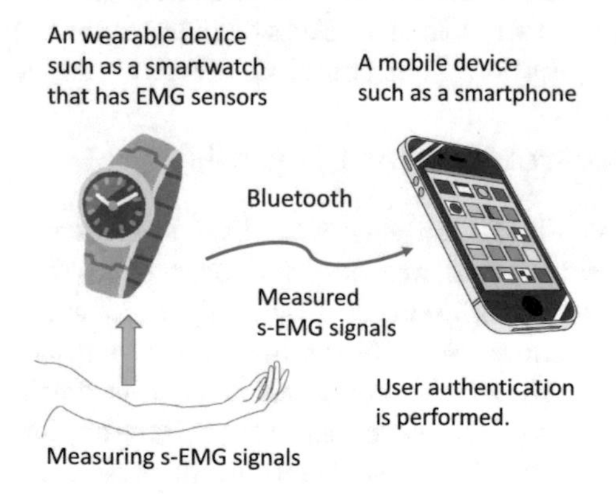

Fig. 3. A sketch of user authentication using s-EMG

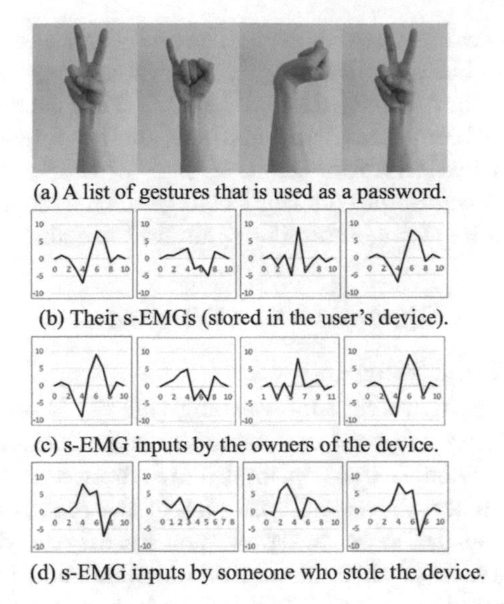

(a) A list of gestures that is used as a password.

(b) Their s-EMGs (stored in the user's device).

(c) s-EMG inputs by the owners of the device.

(d) s-EMG inputs by someone who stole the device.

Fig. 4. The user authentication method using s-EMG signals

4.2 User Authentication Method Using S-EMG

In this study, combinations of the gestures are converted into a code for authentication. These combinations are inputted into the mobile device and used as a password for user authentication.

1. At first, pass-gesture registration is carried out. A user selects a list of gestures that is used as a pass-gesture. (Fig. 4 (a))
2. The user measures s-EMG of each gesture, extracts their feature values, and register the values into his mobile device. (Fig. 4 (b))
3. When the user tries to unlock the mobile device, the user makes his pass-gesture and measures the s-EMG.
4. The measured signals are sent to his mobile device.
5. The device analyzes the signals and extracts the feature values.
6. The values are compared with the registered values.
7. If they match, the user authentication will succeed. (Fig. 4 (c))
8. On the other hand, an illegal user authentication will fail because a list of signals given by someone who stole the device (Fig. 4 (d)) will not be similar with the registered one.

Adopting s-EMG signals for authentication of mobile devices has three advantages. First, the user does not have to look at his/her device. Since the user can make a gesture that is used as a password on a device inside a pocket or in a bag, it is expected that the authentication information can be concealed. No one can see what gesture is made. Next, it is expected that if another person reproduces a sequence of gestures that a user has made, the authentication will not be successful, because the extracted features from the s-EMG signals are usually not the same between two people. And then, a user can change the list of gestures in our method using an s-EMG signal. This is the advantages of our method against other biometrics based methods such as fingerprints, an iris, and so on. When authentication information, a fingerprint or an iris, come out, the user can't use them because he/she can't change his/her fingerprint or iris. But the user can arrange his/her gesture list again and use the new gesture list.

4.3 Introduction of Dynamic Time Warping

4.3.1 Dynamic Time Warping

In time series analysis, dynamic time warping (DTW) is one of the algorithms for measuring similarity between two temporal sequences, which may vary in speed. This method calculates an optimal match between two given sequences (e.g. time series) with certain restrictions. One of the remarkable characteristics of this method is that DTW is "warped" non-linearly in the time dimension to determine a measure of their similarity independent of certain non-linear variations in the time dimension. DTW is a technique for efficiently achieving this warping.

In addition to data mining [9–11], DTW has been used in gesture recognition [12], robotics [13], speech processing [14], manufacturing [15,16].

In this research, we investigated whether gestures can be identified using DTW. All values in a sequence of a measured signal are used as the feature values in the user authentication method described in Sect. 4.1. A similarity between the registered s-EMG signal pattern and an input signal pattern using DTW. If they match (a value of the similarity exceeds some threshold value), the user authentication will succeed.

4.3.2 Selection of Speceimen Waveforms

Our method chooses a specimen waveform for each gesture of each users. Specimen waveforms will be registered in the authentication system and will be used to judge that an input waveform is correct one made by the correct user.

A specimen waveform of each gesture of each user is selected from n waveform data (n is the number of measurement). The similarities between each waveform and other $n - 1$ waveforms are calculated using DTW and their averages are obtained. The waveform that has the lowest average value is selected as the specimen waveform.

4.3.3 Determination of a Threshold Value

In order to realize the proposed method here, we have to introduce a procedure to determine an appropriate threshold value for identifying gestures. In this research, we adopted the procedure to select a threshold that can accept all waveform of the same gesture made by the same user. to reduce false acceptance rates.

First, similarity between the specimen waveform and a number of data of the same gesture made by the same user were calculated using DTW. Then, the largest value among DTW distances is adopted as the threshold value of the gesture; however, values that seem to exceed permissible range because of failure in measurement are removed in advance. In the experiments described in Sect. 5, decision to select failure data is left to the experimenter.

5 Experiments

5.1 Objectives

First, we carried out experiments to evaluate the performance of the gesture identification method using dynamic time warping. Concretely, false rejection rates (FRR) and false acceptance rates (FAR) of each gesture of each subject were investigated. FRR is the rate that the number of waveform inputs of the same gesture of the same subject that are regarded as wrong. FAR is the rate that the number of waveform inputs of the same gesture of other subjects that are regarded as correct. We also investigated the resistance against an accidental success. This is the rate that the number of waveform inputs of all gestures of other subjects that are regarded as correct.

Next, we simulated a scene of user authentication. In this experiment, four gestures were selected from the measured gestures and they were used as a four-digit password. We also evaluated the FRR and FAR of these passwords.

In order to compose the passwords of a subject, we calculated the averages of DTW distances between each of the specimen waveform and other waveforms for all gestures. Then, we selected four gestures and lined them up in descending order of the average values.

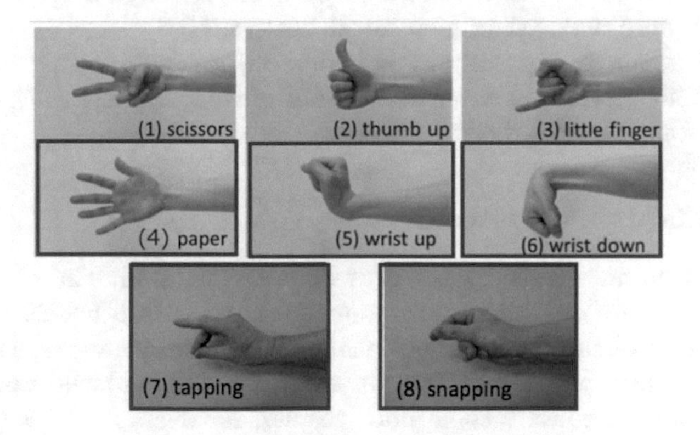

Fig. 5. Gestures used in the experiments

5.2 Conditions

5.2.1 Measurement Data and Instruments

It is assumed that an attacker does not know which his gesture generates s-EMG signals that is similar with a specific gesture of a victim user.

Eleven students of University of Miyazaki participated as experimental subjects (Subject A-H.) And the eight hand gestures shown in Fig. 5 were introduced. But the five red bordered gestures were selected in the experiments because their s-EMG signals were clearer than other gestures.

The set of DL-3100 and DL-141 (S&ME Inc.) that was an electromyograph used in the previous researches also used to measure the s-EMG of each movement pattern in this study.

The measured data were stored and analyzed on a PC. The subjects repeated each gesture ten times and their s-EMG signals were recorded. This measurement was carried out 3 times and 30 signals were obtained for each subject and for each gesture. However, we eliminated waveforms with high noise and distortion that seemed not to be suitable for the experiments. We adopted 1643 data.

5.2.2 Selection of Specimen Waveform

We have to choose the standard waveforms for each gesture of each users. The similarity average was calculated using DTW from 30 data of each gesture. The waveform with the lowest average value was adopted as the specimen waveform.

5.2.3 Determination of Threshold

In order to select threshold values for identifying each of the gestures similarity between the specimen waveform and 29 other data were calculated using DTW. The highest value among the 29 sequences of data was selected as the threshold value (Values that seemed to exceed permissible range because of failure in measurement were removed as mentioned in Sect. 4.3.3).

In addition, the average value was compared for each gesture of each users. We selected four gestures in descending order of their average values. These four gestures are four-digit passwords.

5.3 Idendification of Gestures

First, false acceptance rates by gestures were investigated. The specimen waveform of each gesture of each subject was compared with other 29 waveforms of the same gesture of the same subject using DTW. The average of false rejection rates of 55 specimen waveforms (five gestures of eleven subjects) was very small, 0.94%. However, there were some poor results, for example 7.1% (gesture 6 of subject K). It is considered that such failures were caused by unsteadiness of waveforms.

Next, false rejection rates by gestures were investigated. The specimen waveform of each gesture of each subject was compared with other 29×10 waveforms of the same gesture of other subjects using DTW. Table 1 shows the best gesture

Table 1. Best FAR gesture of each subjects

Subject	A	B	C	D	E	F	G	H	I	J	K
A		(5) 0%	(5) 0%	(8) 0%	(7) 0%	(7) 0%	(5) 0%	(5) 0%	(8) 0%	(6) 0%	(5) 0%
B	(5) 0%		(5) 0%	(4) 0%	(4) 0%	(4) 0%	(5) 0%	(7) 0%	(5) 0%	(4) 0%	(5) 3%
C	(4) 0%	(5) 0%		(4) 0%	(4) 0%	(4) 0%	(4) 0%	(5) 0%	(5) 0%	(4) 0%	(4) 0%
D	(5) 0%	(5) 0%	(4) 0%		(4) 0%	(4) 0%	(5) 0%	(4) 0%	(4) 0%	(4) 0%	(4) 0%
E	(8) 0%	(5) 0%	(5) 0%	(6) 0%		(5) 0%	(7) 0%	(5) 3%	(5) 6%	(6) 0%	(5) 0%
F	(7) 0%	(5) 0%	(7) 0%	(6) 0%	(4) 0%		(5) 0%	(5) 0%	(7) 0%	(7) 0%	(7) 0%
G	(6) 0%	(5) 0%	(6) 0%	(4) 0%	(4) 0%	(4) 0%		(6) 0%	(5) 0%	(4) 0%	(4) 0%
H	(4) 0%	(5) 0%	(5) 0%	(4) 0%	(4) 0%	(4) 0%	(4) 0%		(5) 0%	(4) 0%	(5) 0%
I	(4) 0%	(5) 0%	(7) 3%	(4) 0%	(4) 0%	(4) 0%	(4) 0%	(5) 63%		(4) 0%	(6) 0%
J	(4) 0%	(4) 0%	(4) 0%	(4) 0%	(4) 0%	(4) 0%	(4) 0%	(4) 0%	(4) 0%		(4) 0%
K	(5) 0%	(7) 0%	(4) 0%	(4) 0%	(5) 0%	(7) 0%	(5) 0%	(4) 0%	(7) 0%	(5) 0%	

Table 2. False rejection rates and false acceptance rates

Subject	A	B	C	D	E	F	G	H	I	J	K
FR rate(%)	0.000	13.546	0.000	0.000	13.206	0.000	3.448	0.000	0.000	3.846	10.344
FA rate(%)	0.000	0.082	0.000	0.000	0.236	0.000	0.000	0.000	3.304	0.000	0.062

Table 3. Resistance against an accidental success

Subject	A	B	C	D	E	F	G	H	I	J	K
Accidental success(%)	0.449	0.606	0.000	0.0001	0.708	0.000	0.215	0.002	0.217	0.000	0.207

and the FRR value of a combination of a subject of a specimen and a subject of a input data. Many of the FRR value are 0%, and most of subjects has such gesture. This means that a password that includes such gesture will be safe against spoofing attacks. However, there are undesirable combinations, for example subject H and I (the best FRR value is 63%). The password of subject I will be vulnerable against subject H because the password does not include confident gesture.

Finally, resistance against accidental success was evaluated. Waveforms of all gesture of other subject were compared with each of the specimen waveform using DTW. The average rate of successful matching of all specimen waveform was 0.22%.

5.4 Authentication of Passwords

In the experiments of this subsection, user authentication succeeds when all four gestures in a password match with those in an input passwords.

First, false rejection rates and false acceptance rates were investigated. In this experiment, we arranged 29 measured waveforms (except the specimen) of the gestures that were included in a password and prepared 29^4 password data for input. 29^4 inputs arranged from waveforms by the same subjects were used to evaluate FRR; $29^4 \times 10$ inputs by other subjects were used to evaluate FAR.

Table 3 shows the false rejection rates and false acceptance rates of the eleven subjects. Since the FRR of some gestures included in the passwords were not zero, the FRR of the passwords were not so good for some subjects.

On the other hand, almost all of the FAR were quite small except subject I because such the passwords included at least one gesture that has zero FAR. As for subject I, had false acceptance rates of 3.304%, with the result that one subject was 33.77% and one subject was 2.57% and the remaining 8 subjects were 0%. It means that all the gestures set only for that subject were similar.

Next, security evaluation was carried out for resistance against an accidental success. $29^4 \times_5 P_4 \times 10$ inputs were arranged from waveforms by other subjects for this experiment.

Table 4. The false rejection rates and false acceptance rates of subject I

Subject	I
FR rate(%)	6.896
FA rate(%)	0

The results are shown in Table 3. The obtained values are larger than the security level of four-digit PIN (0.0001.) This means that introduction of other gestures, longer pass-gesture, or improvement of accuracy of gesture identification are needed.

5.5 Discussion

Experimental results showed good results were obtained for many subjects, but inaccurate authentication results were obtained for some subjects. As for the performance of the proposed method, selection of an appropriate threshold is expected to improve the false acceptance rate very effectively.

In order to improve the performance of this method, we attempted to change the threshold value of one gesture so as to reduce its FAR. Because, in our method, the FAR of a password can be reduced by reducing FAR of only one gesture included in the password. Table 4 shows the results of false rejection rates and false acceptance rates of subject I. The threshold of one gesture was changed by the arbitrary decision of the experimenter.

It is expected that the user authentication using s-EMG can be realized by the approach used in this study. And also, it is need to explore other gesture candidate that are more suitable to distinguish s-EMG signals appropriate length of pass-gestures, select an appropriate threshold.

6 Conclusion

We investigated a new user authentication method that can prevent shoulder-surfing attacks in mobile devices. To realize such an authentication method, we adopted an DTW to identify gestures by s-EMG signals. A series of experiments was carried out to investigate the performance of DTW. False rejection rates, false acceptance rates, and resistance against an accidental access were also investigated. The results showed that this approach involves a problem to be resolved, but the method using DTW is promising. We are planning to select an appropriate threshold, introduce other gesture candidate, explore appropriate length of pass-gesture and so on.

Acknowledgments. The authors would like to thank H. Tamura for his helpful supports in measuring s-EMG signals. This work was supported by JSPS KAKENHI Grant Numbers JP17H01736, JP17K00186.

References

1. Tamura, H., Okumura, D., Tanno, K.: A study on motion recognition without FFT from surface-EMG (In Japanese). IEICE Part D J90-D(9), pp. 2652–2655 (2007)
2. Yamaba, H., Nagatomo, S., Aburada, K., et al.: An authentication method for mobile devices that is independent of tap-operation on a touchscreen. J. Robot. Netw. Artif. Life **1**, 60–63 (2015)
3. Yamaba, H., Kurogi, T., Kubota, S., et al.: An attempt to use a gesture control armband for a user authentication system using surface electromyograms. In: Proceedings of 19th International Symposium on Artificial Life and Robotics, pp. 342–245 (2016)
4. Yamaba, H., Kurogi, T., Kubota, S., et al.: Evaluation of feature values of surface electromyograms for user authentication on mobile devices. Artif. Life Robot. **22**, 108–112 (2017)
5. Yamaba, H., Kurogi, T., Aburada, A., et al.: On applying support vector machines to a user authentication method using surface electromyogram signals. Artif. Life Robot. **22**, 1–7 (2017). https://doi.org/10.1007/s10015-017-0404-z
6. Kita, Y., Okazaki, N., Nishimura, H., et al.: Implementation and evaluation of shoulder-surfing attack resistant users (In Japanese). IEICE Part D J97-D(12), pp. 1770–1784 (2014)
7. Kita, Y., Kamizato, K., Park, M., et al.: A study of rhythm authentication and its accuracy using the self-organizing maps (In Japanese). Proc. DICOMO **2014**, 1011–1018 (2014)
8. Tamura, H., Goto, T., Okumura, D., et al.: A study on the s-EMG pattern recognition using neural network. IJICIC **5**(12), 4877–4884 (2009)
9. Keogh, E., Pazzani, M.: Scaling up dynamic time warping for datamining applications. In: 6th ACM SIGKDD International Conference on Knowledge Discovery and Data Mining (2000)
10. Yi, B., Jagadish, H., Faloutsos, C.: Efficient retrieval of similar time sequences under time warping. In: International Conference of Data Engineering (1998)
11. Berndt, D., Clifford, J.: Using dynamic time warping to find patterns in time series. In: AAAI-94 Workshop on Knowledge Discovery in Databases (KDD-94) (1994)

12. Gavrila, D.M., Davis, L.S.: Towards 3-D model-based tracking and recognition of human movement: a multi-view approach. In: International Workshop on Automatic Face- and Gesture-Recognition. IEEE Computer Society (1995)
13. Schmill, M., Oates, T., Cohen, P.: Learned models for continuous planning. In: Seventh International Workshop on Artificial Intelligence and Statistics (1999)
14. Rabiner, L., Juang, B.: Fundamentals of Speech Recognition. Prentice Hall, Englewood Cliffs (1993)
15. Gollmer, K., Posten, C.: Detection of distorted pattern using dynamic time warping algorithm and application for supervision of bioprocesses. In: Morris, A.J., Martin, E.B. (eds.) On-Line Fault Detection and Supervision in the Chemical Process Industries (1995)
16. Caiani, E.G., Porta, A., et al.: Warped-average template technique to track on a cycle-by-cycle basis the cardiac filling phases on left ventricular volume. In: IEEE Computers in Cardiology, vol. 25 (1998). Cat. No.98CH36292

The Analysis of MATE Attack in SDN Based on STRIDE Model

Abeer E. W. Eldewahi[1(✉)], Alzubair Hassan[2], Khalid Elbadawi[1], and Bazara I. A. Barry[1]

[1] Department of Computer Science, Faculty of Mathematical Science, University of Khartoum, Khartoum, Sudan
abeereldewahi@gmail.com, khalid.badawi@gmail.com, bazara.barry@outlook.com

[2] Center for Cyber Security, School of Computer Science and Engineering, University of Electronic Science and Technology of China, Chengdu 611731, China
alzubairuofk@gmail.com

Abstract. Software defined network (SDN) is an emerging technology that decouples the control plane from data plane in its network architecture. This architecture exposes new threats that do not appear in the traditional IP network. And probably One of the main serious attacks is man-at-the-end (MATE) attack. This paper addresses the existing preventing methods based on selected criteria and analyzes the Spoofing, Tampering, Repudiation, Information Disclosure, Denial of Service and Elevation of Privilege (STRIDE) Model to determine the related attacks to MATE and which one can be the entry point to MATE. Furthermore, a new method to solve the MATE problem is proposed.

1 Introduction

No doubt network devices have a key role that allows information to travel around the world. The traditional IP network is complicated and hard to manage. It needs to configure each network device separately by using low level or vendor specific commands [13]. Also, current network is vertically integrated. The control plane and data plane are bundled inside the networking devices, which make it harder to improve in term of software. On the other hand, Software defined network (SDN) is an emerging networking paradigm that gives greater chance of improvement. It separates the control plane and data plane that made the control of network centralize in controller [1]. Figure 1 shows a simple view of SDN architecture.

The control logic is moved to an external entity called SDN controller or Network Operating System (NOS) that facilitates the programming of forwarding devices. SDN has no build in security mechanisms in its network architecture, this new architecture of SDN exposes new threats that do not appear in the previous paradigm. These threats appear according to the nature of SDN that centralizes the logic in the controllers and the ability to control the network by software.

L. Barolli et al. (Eds.): EIDWT 2018, LNDECT 17, pp. 901–910, 2018.
https://doi.org/10.1007/978-3-319-75928-9_83

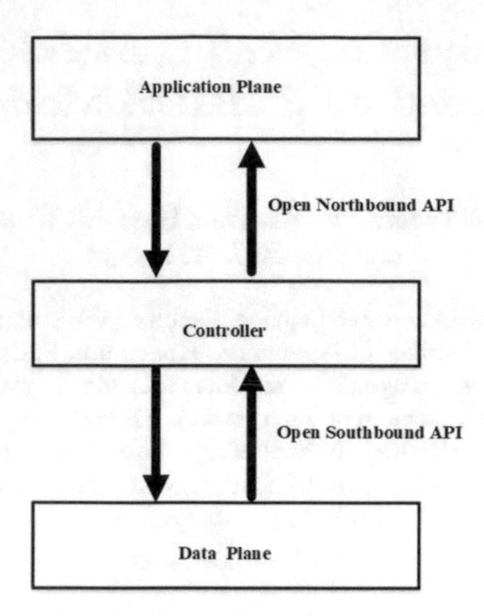

Fig. 1. Simplified view of SDN architecture [1]

1.1 SDN Related Threats

Kreutz et al. [12] divide the threats into seven dimensions three of them are special for SDN and the other related to both traditional IP network and SDN. First, forged or faked traffic flows; it may trigger by malicious or non malicious (faulty) devices that cause DoS attack against OpenFlow switches and controller. Second, attacks on vulnerabilities in switches, one switch can slow down or drop the packets. Third, attacks on control plane communications, which can be used to launch DoS and data theft, these communications are secured by TLS/SSL which has known vulnerabilities [9] also TLS/SSL is not enough to establish and assure trust between controllers and switches. Fourth, vulnerabilities in controllers, which is the most dangerous attack on SDN, a malicious or faulty controller that can compromise all the network. Fifth, lack of mechanisms to the controller is controllers and applications cannot establish trusted relationships. Sixth, attacks on and vulnerabilities in administrative stations, this attack is used to access the controller and compromise the entire network. It is known in the traditional network but has less effect. Seventh, lack of trusted resources for forensics and remediation. The system needs reliable information from all components to ensure the correct detection and fast recovery.

1.2 Vulnerabilities and Threats in SDN Controller

Some researchers focus on the vulnerabilities and threats that affect the network controller. Akhunzada et al. [2] further elaborate in the vulnerabilities and attacks of network controller which stated here: Packet-in controller manipulation attacks: packet-in is a packet that does not match any flow rule in the dataplane, so the switches decide to send it to the controller which has no security mechanisms to deal with it. The packet-in can cause many attacks on the controller like DoS. Configuration conflicts: there are various SDN controllers, and OpenFlow switches which lead to configuration conflict and also the controller cannot remember the past event so the state full applications cannot work correctly. Controller capability of proper auditing and authenticating diverse applications: a real challenge is to facilitate the auditing and authentication of the resources consumed by the applications that run on the controller. Controller scalability challenges: the controller becomes a bottleneck which causes a serious problem like a single point of failure.

One of the serious attacks on SDN controller is the man at the end attack (MATE). The man at the end (MATE) is very liberal attack model that the software protection researcher must cope with: it is assumed that the attacker has all power to access the software and hardware to modify, examine and utilize his/her capabilities [2]. MATE happens when an adversary gain access to the device to eavesdrops the hardware or software. Figure 2 depicts the attack in MATE scenario [3], where the digital asset is protected by asset sentries, in this scenario the digital asset and asset sentries are under the control of an attacker which makes this attack different.

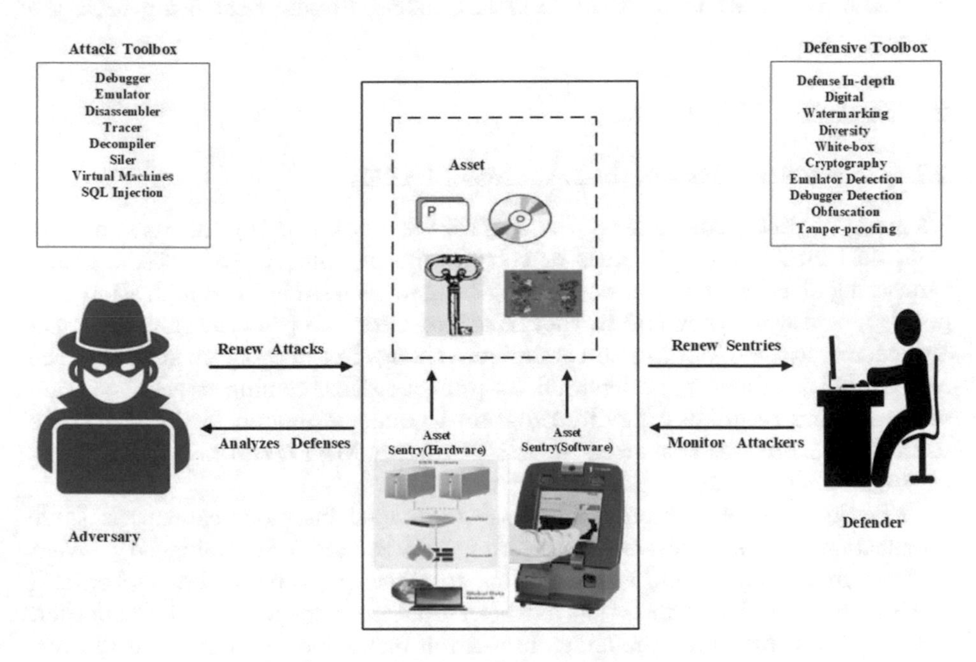

Fig. 2. MATE attack [2]

MATE can take many forms. First, tampering attack the opponent can violate the integrity of software under his control by modifying it in a way that the vendor doesn't intend [10]. Second, in malicious reverse engineering the attacker violates the confidentiality of the vendor by extracting information not revealed by the developer [17]. Finally, in a cloning attack, the attacker distributes an illegal copy of the software by violating the copy right low. MATE attack the end points of transmission rather than the data in transmission. An increasing number of applications are vulnerable to MATE, so we need strong detection model to deliver the nontrivial level of security against MATE [2]. MATE can be skilled and trusted MATE who are educated human have specialized technical expertise. They have a good understanding of different part on the system and know a sophisticated tool for different analysis. They have full access to a machine. Or sly and motivated RMATE who are highly intelligent but may have inadequate information of the different parts of the system. They constantly try to take benefit of exploiting an existing weakness in the system. In this paper, we present an analysis of the existing preventing methods against MATE based on certain criteria. It also analyzes the STRIDE model to determine the related attacks to MATE and which one is the entry point to MATE. Also we propose new methodology to solve the MATE attack.

The paper is organized as follows: Sect. 2 displays the preventing techniques against MATE and STRIDE model. Section 3 presents the proposed methodology. Section 4 shows a discussion on the existing gab in the preventing techniques against MATE and the analysis of STRIDE model. Finally, Sect. 5 concludes the paper.

2 Related Work

2.1 Preventing Techniques Against MATE

Methods of protection against the MATE are known as digital asset protection, anti tampering techniques or software protection [7]. To protect against tampering attacks, the tamper-proofing techniques search for modification in a program and when they find it, they take a reaction. To protect against reverse-engineering attacks, obfuscation techniques change in the program such that the MATE find it difficult to understand. To protect against cloning attacks, software watermarking techniques modify program to make it unique to the legitimate user. The techniques that used to protect against MATE can be software only or software with tamper-resistant hardware [17].

The Software protection techniques are divided into four categories: Code obfuscation to resist reverse engineering attack, tamper-proofing to prevent against tampering attack, watermarking to allow programs to be tracked and birth marking to detect code that has been lifted from one program into another. Cohen [7] first discussed the code obfuscation technique to create multiple versions of a program that makes the mission of malware challenging to analyze the program. The code obfuscation techniques include splitting code into smaller pieces, randomizing code placement, etc. [16]. Obfuscation techniques are based on adds additional complexity to the source code through different kinds of code transformations both regarding programs control flow and/or data structures [8]. Some code cannot be obfuscated in most cases breaking obfuscation is depended on time and attackers skills [5]. Some code cannot be obfuscated in most cases breaking obfuscation is depended on time and attackers skills [11].

Tamper-proofing techniques have different techniques. Introspection techniques check the integrity of the executable code and compare it with the expected value [6] environment-checking techniques verify the program is running on operating system and hardware that wont violate code integrity [10]. The Tamper-proofing techniques can combine with obfuscation to increase program complexity.

Moreover, each real copy includes watermarking techniques with a fingerprint for tracing purpose by the identified owners. To encode the fingerprint the watermarking uses simple program structures where nonfunctional code can be inserted, and a part of the program can be reordered. Watermarking can be combined with tamper-proofing techniques to make the fingerprint difficult to removing and also with obfuscation to make it harder to locate [10]. Another solution to the MATE attack is fleet [14]. Fleet solve the problem of degradation of network performance by the administrator who misconfigures the controller even if the controller is functioning correctly with no problem in the flow rule that installs in it.

2.2 Spoofing, Tampering, Repudiation, Information Disclosure, Denial of Service and Elevation of Privilege (STRIDE) Model

The STRIDE model attacks are widely applied to threat modeling of computer, software and network systems. Almost all attacks fall in these categories [4,15]. In [4] they applied the STRIDE model in SDN to find the vulnerabilities in SDN controller. This section gives a brief analysis of STRIDE model to show how the MATE attacker controls over the systems as follow:

1. Spoofing is a process of forging the network information to hide the identity of the attacker and traffic origin. The primary goal of spoofing is to cover the attacker tracks by misleading the server using e.g. a fake IP address. the SDN spoofing attack include Address Resolution Protocol (ARP) spoofing which links the attacker MAC address to a legitimate IP address and IP spoofing. Spoofing can be used to achieve man in the middle attack, not MATE.
2. Tampering is an intentional and unauthorized modification or destruction of network information such as flow rules or flow tables, for example, the attacker can inject flow rule to deny a legitimate host. In SDN the controllers communicate with different information through channels that must be secured to avoid the tampering. Man in the middle can tamper with information in the transition and can be MATE if the attacker tamper software under his/her control.
3. Repudiation is the ability to deny the participation in the communication or part of it. For example, the attacker can log into the system that does not have a log or tracing program running, so there is no evidence to decide who does what. Non repudiation is to ensure that this repudiation does not occur. The sender needs to verify that the packet is sent to the actual receiver and

the receiver need to verify that the packet is sent from the actual sender. In SDN the administrator must create an account for every user instead to have one account for all users, so the logs give a decision about who does what. Repudiation is not a way to do MATE it maybe is a result of a man in the middle if the attacker masquerades for each party to be another party.
4. Information Disclosure is spying the information by attackers rather than his/her direct intention to destroy the system. For example, when the web server is a crash there is an error message used by an administrator to discover the problem, but this error may give the attacker a chance to attack the server. In SDN if the attacker reaches the switches he/she can tamper with flow rule cause the traffic to go to the wrong destination. Information Disclosure is MATE if the attacker can gain information allows him/her to log into the system as an administrator and reach the controller. Man in the Middle attack is an information disclosure attack that targets the information in the transit.
5. Denial of service is flooding the computers with traffics that cannot handle by attackers. It is about to drop the performance of the network or stop the server. In SDN DoS can be destructive because the attacker can push flows between the controller and the switches to interrupt the normal network activities. DoS is not a man in the middle or MATE.
6. Elevation of Privilege is a process to obtain privileged access to a computer system that is not supposed to have or try to have administrator privilege. In SDN we can elevate the Privilege in two ways: first, turn the encryption off, so the user name and password are sent in clear, and any attacker use packet sniffer can catch them. Second, when the programmer made the browser interface, he/she didn't turn the auto complete off. If the attacker has access to the computer he/she can easily take the file of user name and password and extract it. Elevation of privilege is exactly the way for MATE when the attacker gets administrator privilege he/she can access the controller and do whatever he/she want.

3 Proposed Method

To solve the problem of MATE, we propose the following methodology (see also Fig. 3)

1. Decide how MATE enters the system by analyzing STRIDE model.
2. Determine exact behavior for MATE by monitor the controller.
3. Build detection model based on the exact behavior. The Application programming interface (API) call graph is proposed to detect MATE by reducing the stored amount of data, improving memory usage and employing dataset for MATE attack and other attacks and then applying the new model to the dataset to distinguish the MATE behavior.
4. Compare the detection methods with similar existing methods if it is better stop the MATE attacker, otherwise enhance it.

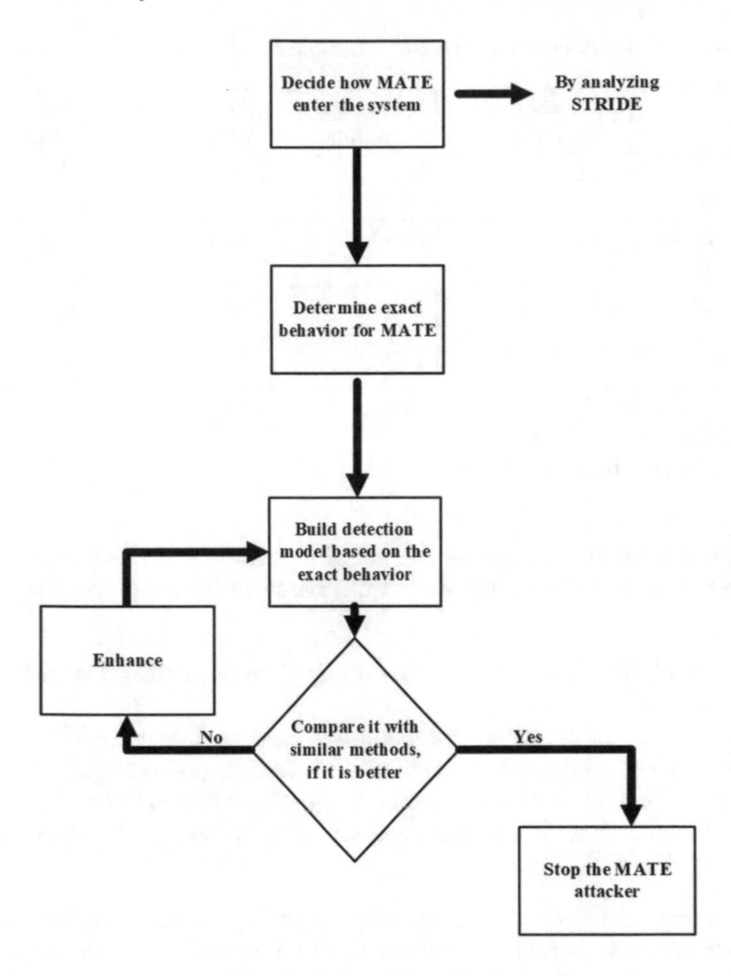

Fig. 3. Proposed method

4 Discussion

All techniques above were used to defeat MATE attack in different ways: The methods that used to defeat reverse engineering attack do not include or emphasize the MATE behavior. They rather solve the problems related to the confidentiality service. The methods used to resist against tampering attacks which violate the integrity of the software do not emphasize the behavior of the MATE it searches to find modification in the code to make a reaction. The Watermarking techniques that were used to defeat the distribution of an illegal copy of software also do not emphasize the MATE behavior, Finally, Fleet used to solve the problem of degradation of network performance. There is no yet a perfect/certain technique depends on the MATE behavior or characterizes it and no detection level imposed. The tamper proving techniques used to detecting

Table 1. Comparison between the techniques used to defeat MATE attack

	Structure of the code	Detection technique	Prevention technique	Behavior of MATE
Obfuscation techniques	*Yes*	*No*	*Yes*	*No*
Tamper-proofing techniques	*Yes*	*Yes*	*Yes*	*No*
Watermarking techniques	*Yes*	*No*	*Yes*	*No*
Fleet	*No*	*No*	*Yes*	*No*
Proposed method	*No*	*Yes*	*Yes*	*Yes*

Yes means satisfy the criteria
No means do not meet the criteria

the modified code and gathering the suitable response. Table 1 shows a simple comparison between the above techniques based on specific evaluation criteria such as:

1. Structure of the Code: when there is any change in the code to handle the problem
2. Detection Technique: when the method is used to detect the MATE attack.
3. Prevention Technique: when the method is used to prevent the MATE attack.
4. Behavior of MATE: when the method characterizes and determines behavior of MATE attacker. We emphasized the gab no technique depends on the behavior of MATE.

Table 2 can summarize the possibility to apply the STRIDE model in SDN and the relation with MATE and man in the middle (MiM). We find the elevation of privilege is the entry point for MATE attack.

Table 2. Analysis of STRIDE model

	Ability to apply in SDN	MiM	MATE
Spoofing	*Yes*	*Yes*	*No*
Tampering	*Yes*	*Yes*	*Yes*
Repudiation	*Yes*	*Yes*	*No*
Information disclosure	*Yes*	*Yes*	*Yes*
Denial of service	*Yes*	*Yes*	*No*
Elevation of privilege	*Yes*	*No*	*Yes*

Yes means satisfy the criteria
No means do not meet the criteria

5 Conclusion

We have tested the existing ways to prevent from MATE attack and emphasized that no method yet depends on the behavior of MATE. We also analyzed the STRIDE model and found that tampering and information disclosure can be MATE attack if the attacker tampers with software under his/her control or obtain information allowing him/her to enter the system as administrator. Elevation of privilege enables the attacker to elevate his/her privilege to be an administrator and then can control the software and hardware.

References

1. Akhunzada, A., Ahmed, E., Gani, A., Khan, M.K., Imran, M., Guizani, S.: Securing software defined networks: taxonomy, requirements, and open issues. IEEE Commun. Mag. **53**(4), 36–44 (2015)
2. Akhunzada, A., Gani, A., Anuar, N.B., Abdelaziz, A., Khan, M.K., Hayat, A., Khan, S.U.: Secure and dependable software defined networks. J. Netw. Comput. Appl. **61**, 199–221 (2016)
3. Akhunzada, A., Sookhak, M., Anuar, N.B., Gani, A., Ahmed, E., Shiraz, M., Furnell, S., Hayat, A., Khan, M.K.: Man-at-the-end attacks: analysis, taxonomy, human aspects, motivation and future directions. J. Netw. Comput. Appl. **48**, 44–57 (2015)
4. Alsmadi, I., Xu, D.: Security of software defined networks: a survey. Comput. Secur. **53**, 79–108 (2015)
5. Barak, B.: On the (im)possibility of software obfuscation. In: Proceedings Advances in Cryptology (CRYPTO 2001), pp. 1–18 (2001)
6. Chang, H., Atallah, M.J.: Protecting software code by guards. In: DRM 2001, vol. 2320, pp. 160–175. Springer, Heidelberg (2001)
7. Cohen, F.B.: Operating system protection through program evolution. Comput. Secur. **12**(6), 565–584 (1993)
8. Collberg, C.S., Thomborson, C.: Watermarking, tamper-proofing, and obfuscation-tools for software protection. IEEE Trans. Software Eng. **28**(8), 735–746 (2002)
9. Eldewahi, A.E., Sharfi, T.M., Mansor, A.A., Mohamed, N.A., Alwahbani, S.M.: SSL/TLS attacks: analysis and evaluation. In: 2015 International Conference on Computing, Control, Networking, Electronics and Embedded Systems Engineering (ICCNEEE), pp. 203–208. IEEE (2015)
10. Falcarin, P., Collberg, C., Atallah, M., Jakubowski, M.: Guest editors' introduction: software protection. IEEE Softw. **28**(2), 24–27 (2011)
11. Falcarin, P., Di Carlo, S., Cabutto, A., Garazzino, N., Barberis, D.: Exploiting code mobility for dynamic binary obfuscation. In: 2011 World Congress on Internet Security (WorldCIS), pp. 114–120. IEEE (2011)
12. Kreutz, D., Ramos, F., Verissimo, P.: Towards secure and dependable software-defined networks. In: Proceedings of the Second ACM SIGCOMM Workshop on Hot Topics in Software Defined Networking, pp. 55–60. ACM (2013)
13. Kreutz, D., Ramos, F.M., Verissimo, P.E., Rothenberg, C.E., Azodolmolky, S., Uhlig, S.: Software-defined networking: a comprehensive survey. Proc. IEEE **103**(1), 14–76 (2015)

14. Matsumoto, S., Hitz, S., Perrig, A.: Fleet: defending SDNs from malicious administrators. In: Proceedings of the Third Workshop on Hot Topics in Software Defined Networking, pp. 103–108. ACM (2014)
15. Unger, W., et al.: Evaluating the security of software defined network controllers (2015)
16. Wang, C., Knight, J.: A security architecture for survivability mechanisms. University of Virginia (2001)
17. Wang, C., Suo, S.: The practical defending of malicious reverse engineering (2015)

Performance Evaluation of Lévy Walk on Message Dissemination in Unit Disk Graphs

Kenya Shinki[1] and Naohiro Hayashibara[2(✉)]

[1] Graduate School of Frontier Informatics, Kyoto Sangyo University, Kyoto, Japan
i1658092@cc.kyoto-su.ac.jp

[2] Faculty of Computer Science and Engineering, Kyoto Sangyo University, Kyoto, Japan
naohaya@cc.kyoto-su.ac.jp

Abstract. Random walks play an important role in computer science, spreading a wide range of topics in theory and practice, including networking, distributed systems, and optimization. Lévy walk is a family of random walks whose the distance of a walk is chosen from the power law distribution. There are lots of works of Lévy walk in the context of target detection in swarm robotics, analyzing human walk patterns, and modeling the behavior of animal foraging in recent years. According to these results, it is known as an efficient method to search in a two-dimensional plane. However, all these works assume a continuous plane, so far. In this paper, we show the comparison of Lévy walk with various scaling parameters on the message dissemination problem. Our simulation results indicate that the smaller scaling parameter of Lévy walk diffuses a message efficiently compared to the larger one.

1 Introduction

It is well-known that random walks become an important building block for designing and analyzing protocols in mobile ad hoc networks [1–3], searching and collecting information [4,5] in P2P networks, and solving global optimization problems [6,7]. In particular, Lévy walk (also called Lévy flight) has recently attracted attention due to the optimal search in animal foraging [8,9] and the statistical similarity of human mobility [10]. It is a mathematical fractal which is characterized by long segments followed by shorter hops in random directions. The pattern has been proposed by Paul Lévy in 1937 [11], but the similar pattern has also been evolved and sophisticated as a naturally selected strategy that gives animals an edge in the search for sparse targets to survive [9].

Although the most of the works on Lévy walk have been done on a continuous plane, we assume unit disk graphs as the underlying environment. Unit Disk Graphs are widely used for routing, topology control, analysis of virus spreading

L. Barolli et al. (Eds.): EIDWT 2018, LNDECT 17, pp. 911–921, 2018.
https://doi.org/10.1007/978-3-319-75928-9_84

in ad hoc networks (e.g., [3,12,13]). Roughly speaking, the difference between a graph and a continuous plane is the freedom of movement. It means that the movement is restricted by the link structure of the graph. We suppose an agent as a mobile entity. Agents can only move to a neighbor node of the current node on the graph although agents move anywhere on the continuous plane. It means that agents are restricted their movement by the link structure of a graph. There are several research results on random walks on graphs [14,15] but there is no result on Lévy walk on graphs.

Mizumoto et al. [16] measured the encounter probability of Lévy walk. They assume agents are divided into two groups and find another one in the different group (e.g., sexually dimorphic movements). They found that the dimorphic movement pattern is efficient than the monomorphic movement pattern on mating in the infinite and borderless space. It means that the parameters of Lévy walk should be given independently for each group of agents to improve the encounter probability.

In this paper, we evaluate the message dissemination capability of mobile agents with the Lévy walk movement pattern on unit disk graphs which are the intersection graphs of equal sized circles in the Euclidean plane. Only one agent has a message at the beginning, and then agents exchange the message by encountering one another at the same node in a graph. Intuitively, the encounter probability correlates to the message dissemination rate. However, there is no result on the same configuration as this work. We show the performance results on message dissemination in varied parameters, which determine the behavior of Lévy walk, on unit disk graphs with various node degrees.

2 Related Work

We now introduce several research works on Lévy walk in a continuous plane and on random walks in a graph.

Birand et al. proposed the Truncated Levy Walk (TLW) model based on real human traces [17]. The model gives heavy-tailed characteristics of human motion. Authors analyzed the properties of the graph evolution under the TLW mobility model.

Valler et al. analyzed the impact of mobility models including Lévy walk on epidemic spreading in MANET [3]. They adopted the scaling parameter $\lambda = 2.0$ in the Lévy walk mobility model. From the simulation result, they found that the impact of velocity of mobile nodes does not affect the spread of virus infection.

Thejaswini et al. proposed the sampling algorithm for mobile phone sensing based on Lévy walk mobility model [18]. Authors showed that proposed algorithm gives significantly better performance compared to the existing method in terms of energy consumption and spatial coverage.

Fujihara et al. proposed a variant of Lévy walk which is called Homesick Lévy Walk (HLW) [19]. In this mobility model, agents return to the starting point with a homesick probability after arriving at the destination determined by the power-law step length. As their result, the frequency of agent encounter obeys the power-law distribution.

There are several papers on random walks on finite graphs. Ikeda et al. showed the impact of local degree information of nodes in a graph [14]. They proposed a random walk, called the β random walk, which uses the degree of neighboring nodes and moves to a node with a small degree in high probability compared to the ones with a high degree. The hitting time becomes $O(n^2)$ and the cover time becomes $O(n^2 log\ n)$ in the β random walk despite the hitting and cover times are both $O(n^3)$ in the simple random walk without local degree information.

Nonaka et al. presented the hitting time and the cover time in the Metropolis walk which obeys the transition probability produced by the Metropolis-Hastings algorithm [15]. It is a typical random walk used in Markov chain Monte Carlo methods. They showed that the hitting time is $O(n^2)$ and the cover time $O(n^2 log\ n)$.

Although there are several papers on random walks on graphs and Lévy walk on plane, there is no work on Lévy walk on graphs, so far.

3 System Model and Problem Description

In this paper, we assume a unit disk graph $UDG = (V, E)$ with a constant radius r. Each node $v \in V$ is located in the Euclidean plane and an undirected edge (link) $\{v_i, v_j\} \in E$ between two nodes v_i and v_j exists if and only if the distance of the link is less than $2r$. We call r the communication radius of each node in the rest of the paper. Note that r is the Euclidean distance in the plane. We assume that any pair of nodes in the graph has a path (i.e., connected graph).

We also assume computational entities that are called *agents*. Each agent can move between nodes in the graph. They can communicate with other agents only in the same node. Practically speaking, they have the short-range communication capability such as Bluetooth, ad hoc mode of IEEE 802.11, Near Field Communication (NFC), infrared transmission, etc.

Also, we assume that each node knows own position (e.g., obtained by GPS) which is accessible from agents, and each agent has a compass to get the direction of a walk. Each node has a set of neighbor nodes and those information (i.e., positions of neighbors) is also accessible. Moreover, every agent has no prior knowledge of the environment.

3.1 Problem Description

First, we assume to have agents initially located at random positions on a unit disk graph. Each of them can communicate with other agents when they are on the same node. It means that agents have some short-range communication capability such as Bluetooth, ad hoc mode of IEEE802.11, NFC, and infrared communication.

The goal of the problem is to spread an identical message to all agents. In the beginning, only one agent has the message and then it is gradually diffused by encountering one another.

There are many simulation results on the encounter of mobile entities using Lévy walk such as [3,16–19]. However they are on continuous fields, and hardly any results on graphs are available, so far.

4 Lévy Walk on Unit Disk Graphs

We explain the algorithm proposed in [20] for Lévy walk on unit disk graphs.

Lévy walk is a variation of random walks where each node selects a direction uniformly from within $[0, 2\pi)$ and a step length of a walk is determined the Lévy probability distribution described as follows.

$$p(d) \propto d^{-\lambda} \tag{1}$$

d is a step length and λ is the scaling parameter to draw the different shape of the probability distribution. Now we show the behavior of Lévy walk with $\lambda \in \{1, 2, 1.5, 2.0\}$ in a 2D plane (see Figs. 1, 2 and 3).

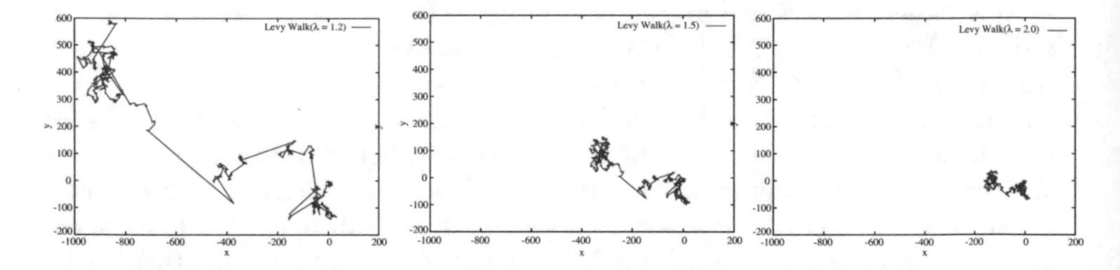

Fig. 1. Lévy walk with $\lambda = 1.2$ on 2D plane.

Fig. 2. Lévy walk with $\lambda = 1.5$ on 2D plane.

Fig. 3. Lévy walk with $\lambda = 2.0$ on 2D plane.

According to the figures, the maximum step length is getting longer as λ decreases. In general, Lévy walk becomes similar to the random walk in terms of its behavior when λ is greater than 3.0 [21].

We describe the algorithm for Lévy walk on unit disk graphs.

The main difference between continuous planes and graphs is freedom of mobility. Agents can only move to the neighbor node of the current node in graphs. It means that the movement of agents in a graph is more restricted than the one in a continuous plane.

4.1 Lévy Walk Algorithm

We explain the algorithm for Lévy walk on unit disk graphs proposed in [20].

In general, an agent selects a destination node v from a set of neighbors $N(u)$ of the current node u randomly and moves to v in random walks on graphs. In contrast, an agent determines the orientation o from $[0, 2\pi)$ at random and the step length d by the power law distribution (see Eq. 1) at the beginning of Lévy walk. Then, it selects v according to the information and moves to it d times. Note that d is the distance (hops) in a graph.

We now define *a walk* as the sequence of the movement in a direction and *a step* as the movement with the step length $d = 1$. Thus, a walk consists of a sequence of steps in Lévy walk though a walk equals to a step in random walks.

The algorithm described in Algorithm 1 is for a walk of Lévy walk. It means that the algorithm is repeated until the termination condition is satisfied.

Algorithm 1. A walk of Lévy walk on unit disk graphs

```
1: Initialize:
       c ← u                                  ▷ the walk starts from u
       o ← 0                                  ▷ orientation of the walk
       PN(c) ← ∅                              ▷ possible neighbors to move
2: d is determined by the power-law distribution
3: o is randomly chosen from [0, 2π)
4: while d > 0 do
5:     PN(c) ← {x|abs(θ_ox) < δ, x ∈ N(c)}
6:     if PN(c) ≠ ∅ then
7:         d ← d − 1
8:         move to v ∈ PN(c) where v has the minimum abs(θ_ov)
9:         c ← v
10:    else
11:        break                              ▷ no possible node to move
12:    end if
13: end while
```

In every walk, each agent determines the step length d by the power-law distribution described in Eq. 1, and selects the orientation o of a walk randomly from $[0, 2\pi)$.

Each agent can obtain a set of neighbors $N(c)$ and a set of possible neighbors $PN(c) \subseteq N(c)$, to which agents can move, from the current node c. In other words, a node $x \in PN(c)$ has the link with c that the angle θ_{ox} between o and the link is smaller than δ which is a given parameter, called a *permissible error*.

Each agent selects the next node $v \in PN(c)$ which has the minimum θ_{ov} and move to v. Thus, each agent moves towards the determined o with δ, d times in a walk.

5 Performance Evaluation on Message Dissemination

We measure time for message dissemination by agents using Lévy walk and discuss the behavior of them. For this purpose, we start from describing the message dissemination problem. Then, we explain the environment and parameters used in the simulation.

The goal of the problem is to spread an identical message to all agents. In the beginning, only one agent has the message and then it is gradually diffused by encountering one another.

We measure the simulation time (i.e., discrete time) for message dissemination on the discrete event simulator implemented in C++. Each agent can take a step in one time unit in our simulations.

5.1 Environment

In our simulation, we located 1,000 nodes at random in the 1,000 $\times$ 1,000 Euclidean plane. Each node has the constant sized circle r as its communication radius. We automatically generated unit disk graphs such that two nodes have a undirected link if these circles have an intersection. We have conducted simulation runs 1,000 *times* for each configuration.

5.2 Parameters

5.2.1 Number of Agents n

It is set $n \in \{5, 10, 20, 30, 40, 50\}$ for $r = 35, 50$. Initially, only one agent holds a message and then agents exchange the message with one another in the same node.

5.2.2 Scaling Parameter λ

Agents are divided into two groups. The scaling parameter is given for each group independently. We set pairs of the scaling parameters $\{(1.2, 1.2), (1.2, 3.0), (3.0, 3.0)\}$ to determine the step length according to the power-law distribution described in the Eq. 1. We started from $\lambda = 1.2$ because Lévy walk is efficient with this parameter in resource exploration and target detection [22,23]. We then carried out the simulations with $\lambda = 3.0$, it has a similar behavior of random walks.

5.2.3 On the Probability of Failure on Walks

In fact, the movement of each agent is restricted by the permissible error δ not to move on contrary side to o in a graph. We set the permissible error $\delta = 90$ in our simulations.

5.2.4 Communication Radius r

We set $r \in [35, 50]$ as a parameter in our simulations. r is a compatible parameter with the average degree of nodes in a unit disk graphs. The average degree of nodes $\overline{deg}(UDG)$ in a unit disk graph UDG is represented by the following equation.

$$\overline{deg}(UDG) = \pi r^2 \times \frac{N}{A}$$

Since $\overline{deg}(UDG)$ is proportional to r, we can control $\overline{deg}(UDG)$ with r. In our simulations, $\overline{deg}(UDG) = 14.6$ when $r = 35$ and $\overline{deg}(UDG) = 28.9$ when $r = 50$ in unit disk graphs. The diameter of a graph is 24 with $r = 35$ and 16 with $r = 50$.

5.3 Simulation Results

We explain our simulation results regarding the message dissemination. We measure the average message dissemination rate, which indicates how many agents share the same message.

5.3.1 Results with $r = 35$

We show the simulation results of the message dissemination problem using Lévy walk on unit disk graphs with the communication radius $r = 35$. Figures 4, 5, 6, 7, 8 and 9 show the average message dissemination rate by simulation steps with various number of agents. The horizontal line is a logarithmic scale.

In the case of $n \in \{5, 10\}$ shown in Figs. 4 and 5, the message dissemination rate increases slowly, but the difference between three configurations is negligible at earlier time (i.e., up to 80 steps). Then they gradually increase regarding the dissemination rate. The Lévy walk with the scaling parameters $(1.2, 1.2)$ is most efficient than the one with $(3.0, 3.0)$.

According to the increase of n, the message dissemination rate quickly increase and it finally reaches to 1.0 with a smaller number of steps. Therefore, the difference regarding the message dissemination rate is largest at the intermediate time (e.g., 320 steps on $n = 15$, 160 steps on $20 \leq n \leq 40$ and 80 steps on $n \leq 45$).

The Lévy walk with the scaling parameters $(1.2, 1.2)$ is most efficient regarding the message dissemination than the one with $(1.2, 3.0)$ and $(3.0, 3.0)$. It indicates that the straight-line movement of Lévy walk contributes to the increase of the message dissemination rate.

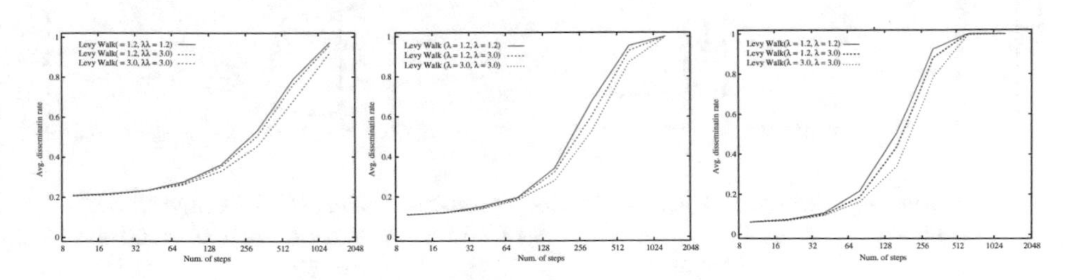

Fig. 4. Message dissemination with $n = 5$ and $r = 35$

Fig. 5. Message dissemination with $n = 10$ and $r = 35$

Fig. 6. Message dissemination with $n = 20$ and $r = 35$

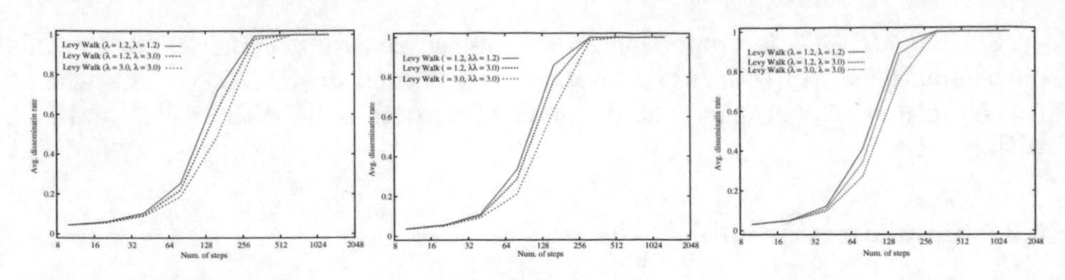

Fig. 7. Message dissemination with $n = 30$ and $r = 35$

Fig. 8. Message dissemination with $n = 40$ and $r = 35$

Fig. 9. Message dissemination with $n = 50$ and $r = 35$

5.3.2 Results with $r = 50$

We show the simulation results with $r = 50$. The restriction of movement of agents with $r = 50$ is weaker than the one with $r = 35$. Figures 10, 11, 12, 13, 14 and 15 show the average message dissemination rate by simulation steps with various number of agents.

They have the same tendency on the message dissemination as the one with $r = 35$. The difference is that the message dissemination rate increases faster than the one with $r = 35$. For instance, the dissemination rate with $n = 5$ and $r = 50$ almost equals to the one with $n = 10$ and $r = 35$.

In fact, agents abandon its movement during a walk with a designated distance d because there is no neighboring node in the specified orientation. This sort of cessation of walks occurs frequently with $r = 35$ compared to the one with $r = 50$ because of the restriction of movement. It means that the flexibility of movement of agents makes an impact on diffusing messages.

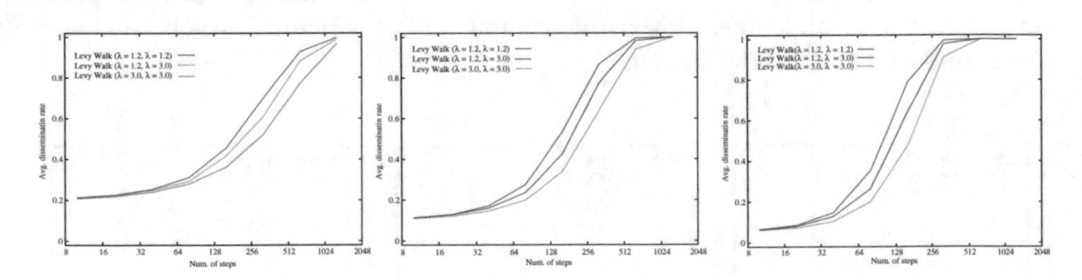

Fig. 10. Message dissemination with $n = 5$ and $r = 50$

Fig. 11. Message dissemination with $n = 10$ and $r = 50$

Fig. 12. Message dissemination with $n = 20$ and $r = 50$

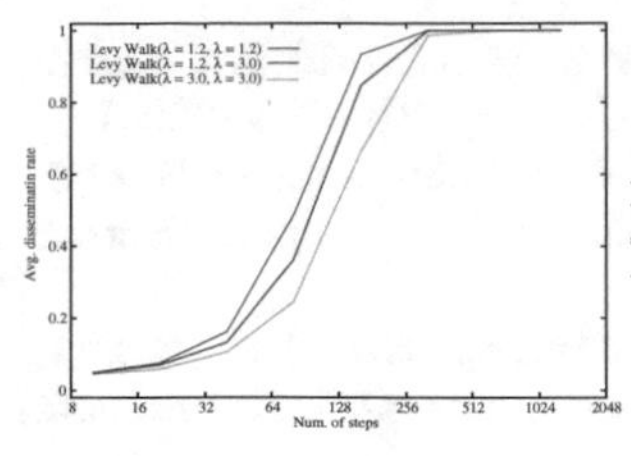

Fig. 13. Message dissemination with $n = 30$ and $r = 50$

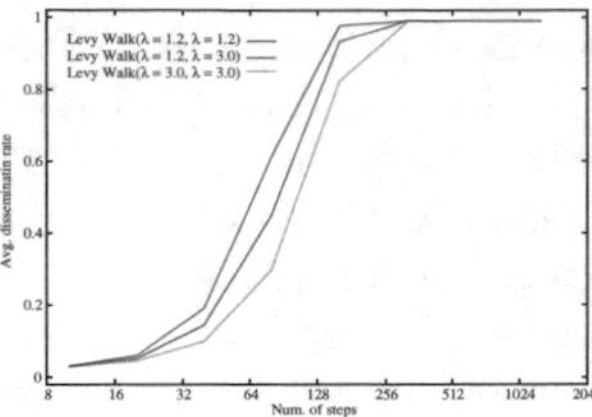

Fig. 14. Message dissemination with $n = 40$ and $r = 50$

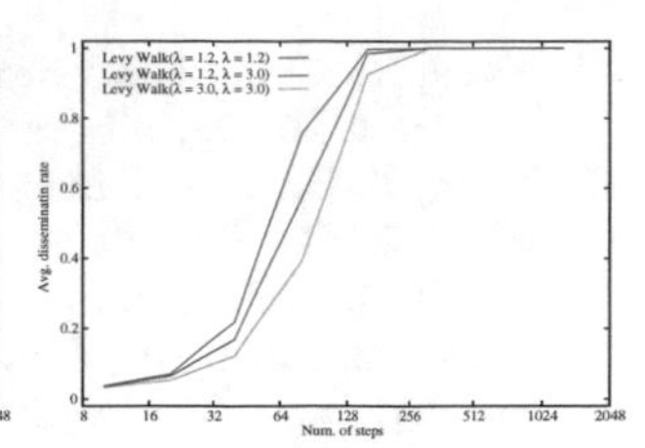

Fig. 15. Message dissemination with $n = 50$ and $r = 50$

6 Conclusion

In this paper, we showed the performance analysis on the message dissemination by Lévy walk on unit disk graphs obtained by our simulations. We first divided agents into two groups and configured the scaling parameter of Lévy walk for each group. So, agents can move in different patterns when the two groups have different parameters. We observed that how the pattens of given parameters influence the message dissemination on unit disk graphs.

The simulation results showed that a pair of the smaller parameters (i.e., (1.2, 1.2)) is always efficient than others. We also found that the node degree has a significant impact to the message dissemination rate on unit disk graph. According to increase the node degree (improve the flexibility of movement), a straightforward movement (i.e., parameters (1.2, 1.2)) is more efficient than the random walk movement pattern (i.e., parameters (3.0, 3.0)).

References

1. Dolev, S., Schiller, E., Welch, J.L.: Random walk for self-stabilizing group communication in ad hoc networks. IEEE Trans. Mob. Comput. **5**(7), 893–905 (2006). https://doi.org/10.1109/TMC.2006.104
2. Draief, M., Ganesh, A.: A random walk model for infection on graphs: spread of epidemics & rumours with mobile agents. Discrete Event Dynamic Syst. **21**(1), 41–61 (2011). https://doi.org/10.1007/s10626-010-0092-5
3. Valler, N.C., Prakash, B.A., Tong, H., Faloutsos, M., Faloutsos, C.: Epidemic spread in mobile ad hoc networks: determining the tipping point. In: Proceedings of the 10th International IFIP TC 6 Conference on Networking - Volume Part I, NETWORKING 2011, pp. 266–280. Springer, Heidelberg (2011). http://dl.acm.org/citation.cfm?id=2008780.2008807
4. Baldoni, R., Beraldi, R., Quema, V., Querzoni, L., Tucci-Piergiovanni, S.: TERA: topic-based event routing for peer-to-peer architectures. In: Proceedings of the 2007 International Conference on Distributed Event-based Systems, pp. 2–13 (2007)
5. Bisnik, N., Abouzeid, A.A.: Optimizing random walk search algorithms in P2P networks. Comput. Netw. **51**(6), 1499–1514 (2007). https://doi.org/10.1016/j.comnet.2006.08.004

6. Yang, X.-S.: Cuckoo search via Lévy flights. In: Proceedings of World Congress on Nature & Biologically Inspired Computing (NaBIC 2009), pp. 210–214 (2009)
7. Yang, X.-S.: Firefly algorithm, Lévy flights and global optimization. In: Bramer, M., Ellis, R., Petridis, M. (eds.) Research and Development in Intelligent Systems XXVI, pp. 209–218. Springer, London (2010)
8. Viswanathan, G.M., Afanasyev, V., Buldyrev, S.V., Murphy, E.J., Prince, P.A., Stanley, H.E.: Lévy flight search patterns of wandering albatrosses. Nature **381**, 413–415 (1996)
9. Edwards, A.M., Phillips, R.A., Watkins, N.W., Freeman, M.P., Murphy, E.J., Afanasyev, V., Buldyrev, S.V., da Luz, M.G.E., Raposo, E.P., Stanley, H.E., Viswanathan, G.M.: Revisiting Lévy flight search patterns of wandering albatrosses, bumblebees and deer. Nature **449**, 1044–1048 (2007)
10. Rhee, I., Shin, M., Hong, S., Lee, K., Kim, S.J., Chong, S.: On the Levy-walk nature of human mobility. IEEE/ACM Trans. Netw. **19**(3), 630–643 (2011). https://doi.org/10.1109/TNET.2011.2120618
11. Lévy, P.: Théorie de L'addition des Variables Aléatoires. Gauthier-Villars, Paris (1937)
12. Alzoubi, K.M., Wan, P.-J., Frieder, O.: Message-optimal connected dominating sets in mobile ad hoc networks. In: Proceedings of the 3rd ACM International Symposium on Mobile Ad Hoc Networking & Computing, MobiHoc 2002, pp. 157–164. ACM, New York (2002). http://doi.acm.org/10.1145/513800.513820
13. Kuhn, F., Wattenhofer, R.: Constant-time distributed dominating set approximation. In: Proceedings of the Twenty-Second Annual Symposium on Principles of Distributed Computing, PODC 2003, pp. 25–32. ACM, New York (2003). http://doi.acm.org/10.1145/872035.872040
14. Ikeda, S., Kubo, I., Yamashita, M.: The hitting and cover times of random walks on finite graphs using local degree information. Theoret. Comput. Sci. **410**(1), 94–100 (2009)
15. Nonaka, Y., Ono, H., Sadakane, K., Yamashita, M.: The hitting and cover times of metropolis walks. Theoret. Comput. Sci. **411**(16–18), 1889–1894 (2010)
16. Mizumoto, N., Abe, M.S., Dobata, S.: Optimizing mating encounters by sexually dimorphic movements. J. R. Soc. Interface **14**(130), 20170086 (2017)
17. Birand, B., Zafer, M., Zussman, G., Lee, K.-W.: Dynamic graph properties of mobile networks under Levy walk mobility. In: Proceedings of the 2011 IEEE Eighth International Conference on Mobile Ad-Hoc and Sensor Systems, MASS 2011, pp. 292–301. IEEE Computer Society, Washington, DC (2011). https://doi.org/10.1109/MASS.2011.36
18. Thejaswini, M., Rajalakshmi, P., Desai, U.B.: Novel sampling algorithm for human mobility-based mobile phone sensing. IEEE Internet Things J. **2**(3), 210–220 (2015)
19. Fujihara, A., Miwa, H.: Homesick Lévy walk and optimal forwarding criterion of utility-based routing under sequential encounters. In: Proceedings of the Internet of Things and Inter-Cooperative Computational Technologies for Collective Intelligence 2013, pp. 207–231 (2013)
20. Shinki, K., Nishida, M., Hayashibara, N.: Message dissemination using Lévy flight on unit disk graphs. In: IEEE 31st International Conference on Advanced Information Networking and Applications (AINA 2017), Taipei, Taiwan (2017)
21. Buldyrev, S.V., Goldberger, A.L., Havlin, S., Peng, C.-K., Simons, M., Stanley, H.E.: Generalized Lévy-walk model for DNA nucleotide sequences. Phys. Rev. E **47**(6), 4514–4523 (1993)

22. Koyama, H., Namatame, A.: Comparison of efficiency of random walk based search and Levy flight search. Inf. Process. Soc. Jpn. Tech. Rep. **20**, 19–24 (2008). (in Japanese)
23. Katada, Y., Nishiguchi, A., Moriwaki, K., Watanabe, R.: Swarm robotic network using Levy flight in target detection problem. In: Proceedings of the First International Symposium on Swarm Behavior and Bio-Inspired Robotics (SWARM 2015), Kyoto, Japan, pp. 310–315 (2015)

Human vs. Automatic Annotation Regarding the Task of Relevance Detection in Social Networks

Nuno Guimarães(✉), Filipe Miranda, and Álvaro Figueira

CRACS-INESC TEC and University of Porto,
Rua do Campo Alegre 1021/1055, Porto, Portugal
nuno.r.guimaraes@inesctec.pt, miranda.filipe@fe.up.pt, arf@dcc.fc.up.pt

Abstract. The burst of social networks and the possibility of being continuously connected has provided a fast way for information diffusion. More specifically, real-time posting allowed news and events to be reported quicker through social networks than traditional news media. However, the massive data that is daily available makes newsworthy information a needle in a haystack. Therefore, our goal is to build models that can detect journalistic relevance automatically in social networks. In order to do it, it is essential to establish a ground truth with a large number of entries that can provide a suitable basis for the learning algorithms due to the difficulty inherent to the ambiguity and wide scope associated with the concept of relevance. In this paper, we propose and compare two different methodologies to annotate posts regarding their relevance: automatic and human annotation. Preliminary results show that supervised models trained with the automatic annotation methodology tend to perform better than using human annotation in a test dataset labeled by experts.

1 Introduction

Social Networks have grown in recent years and placed themselves as the number one medium for constant connectivity amongst users worldwide. Due to the permanent connectivity (facilitated through mobile devices) and quick propagation of information, many relevant events are reported first through social networks posts (from users witnessing it in real-time) and, only after, through journalistic media sources [8,13]. It is important to track and detect these posts at early stages and throughout the enormous quantities of personal or irrelevant data that is published. A possible approach studied in previous works [5,14] was the use of machine learning algorithms to detect possible relevant information in social media.

However, to build models capable of detecting relevant posts, a solid *ground truth* is required. Nevertheless, the ambiguity of the task for human annotators (what is relevant for some may not be relevant for others) makes this task particularly hard. In fact, to have a solid inter-domain relevance dataset, it would

L. Barolli et al. (Eds.): EIDWT 2018, LNDECT 17, pp. 922–933, 2018.
https://doi.org/10.1007/978-3-319-75928-9_85

require a significant number of workers to evaluate each post due to the ambiguity of the task and a large number of entries in the dataset, for a good learning (and consequently prediction) of the models. The combination of these two requirements makes the use of paid human-annotators costly and the use of volunteers unfeasible.

In this work, we compare two different experiments to assess the best methodology to create a dataset for learning relevance detection in social network posts. Our goal is to develop models to automatically label series of social media posts. In the manual approach, we will use one Crowdflower worker per post and develop quality metrics to evaluate each annotation and worker, abdicating the inter-worker agreement to be able to label a large dataset in a less costly manner. In the automatic approach, we will rely on the similarity between news sources and social network posts to label what is relevant or not. We will then create models using the datasets extracted from both methodologies. Finally, we will test their performance using a dataset annotated by experts.

2 Related Work

Related work on detecting relevance on social networks has targeted Twitter, more specifically event detection. The work in [13] presents a system that monitors, detects, and predicts the trajectory of events on Twitter. The authors used a spatiotemporal model and tested the system on the detection of earthquakes in Japan with 96% of accuracy. Another work studied the activity on the social network in the days after Chile's earthquake in 2010 [8]. The authors concluded that rumors and news propagation differ and can be detected. In fact, earthquake detection through Twitter is a well-studied area [4,9,12].

Several authors also tried to use Twitter to detect flu epidemics. For example, the authors in [3] used SVM classifiers and a Bag-of-Words approach to detect flu epidemics using Twitter with a 0.89 Pearson correlation with the defined gold-standard. This and other studies like the ones in [1,6,7] provide solid evidence that Twitter can help to predict in early stages the flu epidemics.

Both these topics can be considered journalistically relevant. However, they are very specific. Our goal is to assess journalistic relevance in a wider scope. A more related work to what we intend to achieve is described in [15], where a system to detect news in tweets is presented. The authors used 2000 hand-picked accounts that belong to news media combined with clustering methods to distinguish newsworthy information from irrelevant one. However, this work is limited regarding its source since it only uses Twitter.

3 Data Acquisition Methodology

We present the two methodologies proposed to extract and label social network posts regarding their newsworthiness relevance.

3.1 Posts Retrieval

In this experiment, three types of posts were extracted: Tweets, Facebook posts and Facebook comments. We selected keywords related to relevant topics (which appeared on the recent news) and that would eventually include newsworthy and relevant information. The keywords selected were "terrorism", "refugees", "elections", "paralympic", "champions league", "emmys" and "wall street". The extraction methodology differed due to the limitations imposed by each of the social networks' APIs. While Twitter Search API allows the extraction of tweets regarding a keyword, Facebook API does not. Therefore, on Twitter, these keywords were passed down directly to the Search API. However, on Facebook, we first analyzed what were the top most popular pages in different categories from the United States, according to the LikeAlyzer tool. Then, we extracted posts from those pages, excluding the ones that did not contain any of the keywords mentioned previously.

The crawling procedures ran between September 7th to September 14th, 2016. From the data retrieved, a random sample of 4994 Facebook posts and comments, and 4994 tweets were selected, ultimately constituting our dataset for analysis.

3.2 Crowdflower HIT

We designed a Human Intelligence Task with several questions to test the coherency, focus, and commitment of each worker (in addition to the goal of classifying a post as relevant or not). The HIT included questions regarding the relevance of the post, user's awareness of news, sentiment of the post and the knowledge of the user regarding the content of the post. The totality of the questions included in the HIT are presented in Table 1. Using this methodology, we could expand the number of posts evaluated (in comparison to previous experiments [5,14]) without increasing significantly the cost of fees.

Questions 1.1 and 4.1 were used to evaluate the commitment of the user with the task since in most cases due to the low payment for each HIT, workers tend to focus on completing tasks the faster they can to efficiently monetize their time. With open questions, we can evaluate if the worker was careful enough on the answers provided. Questions 1 and 3 were used to evaluate the user itself and to give a degree of confidence in the answers provided. In addition, we also established a 30 s minimum time to complete the task. This way, annotators were warned that filling the HIT quicker would not award any advantage on monetization.

Only the highest level workers from the US and United Kingdom could participate in the experiment due to the difficulty of the task, the language constraint, and geographical scope of the posts extracted. Finally, each worker was only able to fill 25 HITs to ensure the diversity of the contributors.

After the Crowdflower task was over, we proceeded to analyze the labeled dataset and the answers provided by the users.

Table 1. Questions of Crowdflower HIT

#	Question	Type of answer
1	Have you had knowledge of this content through another source?	Yes/No
1.1	Choose (from the provided text) three distinct words that best summarize it:	3 text input fields
2	Can you find, in the provided text, information that can be considered relevant?	Yes/No
3	Do you consider yourself to be a person who is aware of the news?	Likert Scale
4	The sentiment expressed in this text is	Likert Scale
4.1	Choose (from the provided text) the word that best supports your previous answer:	Text input field

3.3 Crowdflower User Filtering

The designed HIT made for Crowdflower allowed us to define a set of quality metrics regarding each worker. We combined not only the answers given in this HIT but also the history of workers' answers from previous jobs. Therefore, we used data regarding previous experiences in Crowdflower [5,14] where workers have also participated. The metrics used to filter unreliable annotators are described below.

3.3.1 User Agreement Rate Deviation

The intuition behind this measure is that if a user is mostly disagreeing with the general opinion is because (a) he/she is randomly choosing the answers, or (b) what he/she classifies as relevant is not relevant for the general audience. Therefore, and since our goal is to detect what is journalistically relevant (i.e. to a general audience), this type of users potentially affect negatively our learning algorithms.

In order to understand which users are almost constantly disagreeing with the general opinion, we introduce a metric we call "user agreement rate deviation". This is calculated by dividing the frequency of agreement between a user's opinion and the majority's opinion by the number of posts evaluated by the user.

Using data from previous experiments, we compared the posts' evaluations of a user with the "agreements" for those posts (if the majority of the evaluations consider a post "relevant", then the agreement is "relevant"). Therefore, we can obtain the number of occurrences in which the user's opinion is the same as the agreement (frequency of agreement). The number of posts evaluated by a user can be obtained by simply counting the evaluations of that user in our dataset. For example, a user evaluated 10 posts in which 8 of them are in accordance with the agreement. In this case, the ratio will be 0.8.

Finally, these agreement deviation rates were used as an argument for a quality measure function. This quality measure function takes into account both the ratio of the user and the number of evaluations he/she completed. The formula used is a true Bayesian estimate and it's given by

$$\mathrm{QM} = (v \div (v + m)) \times R + (m \div (v + m)) \times C$$

where: R is the frequency of agreement ratio, v the number of evaluations made by the user, m the minimum number of posts to consider the user evaluation to have a superior weight than the average evaluations, and C the ratio's average. Using this metric, users with fewer evaluations (and therefore whose ratio is less perceived as credible) are closer to the average of all users' ratio.

We defined a threshold value of 0.1 and removed all posts from users which were below that value. This was selected after an analysis of the variation of the ratio (and consequently, on the number of posts to be excluded). A plot comparing the variation of the threshold with the number of posts remaining in the dataset is shown in Fig. 1.

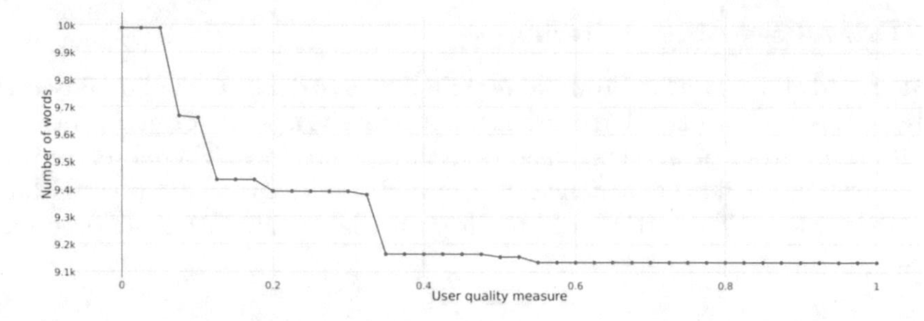

Fig. 1. Variation of threshold and the effect it has on the dataset size using the first quality measure

3.3.2 User's Correct Words Percentage

In this metric, we used the answer provided in question 1.1 to assess the credibility and focus of each user.

We assigned a score ranging from 0 to 3 for each user based on the words that were given as answers in 1.1 and that belonged to the post. Although this method does not allow to verify if it is the correct justification, it provides a direct way to double-check if the user did not write random characters on the text field to finish the task faster.

We used "Levenshtein Distance", a string similarity metric, to prevent users that committed typos were automatically excluded from the sample. In addition, we also tolerated users that wrote two words in the same text field. Users writing 1 or 2 words from the text in the input field score 1 point. Otherwise, if the user writes more than 2 words in the same field or leaves it empty, it gets 0 points. Next, we used the average points of each user as a metric and defined a threshold value of 2.5. This value was determined experimentally based on a careful analysis of the ratio of the users and the effect it has on the dataset size. Figure 2 shows the variation on the size of the dataset regarding the threshold values.

3.3.3 User's Consistency on the Field Regarding "News-Awareness"

To understand how randomly users were answering these questions, we listed all "news-awareness" answers for each one and calculated the mean and standard

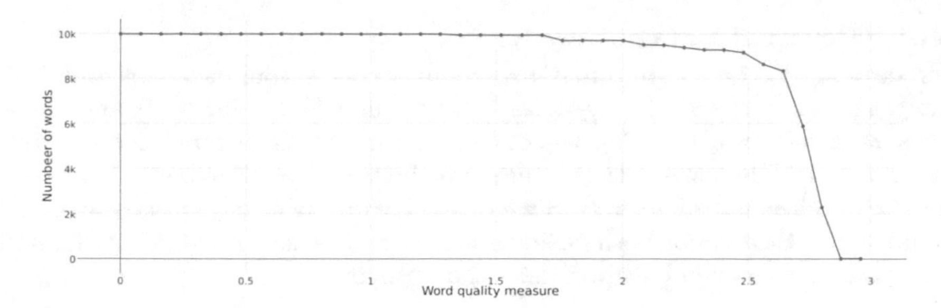

Fig. 2. Variation on the size of the dataset regarding the threshold for the second quality measure

deviation. A higher standard deviation value shows that a user was less consistent in the answers. To define the strictness of our filter we had to choose the maximum value allowed for the standard deviation. The plot presented in Fig. 3 shows the strictness in terms of number of valid posts (i.e. more strictness means less valid posts) as a function of the maximum standard deviation allowed. Based on it, we defined a 0.2 threshold. This way, users are given the benefit of the doubt if, by mistake, their answers are not in accordance with previous ones.

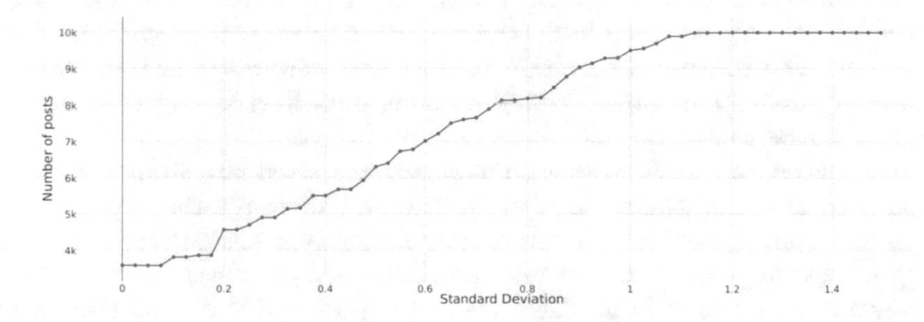

Fig. 3. Standard deviation threshold and the impact it has on the size of the dataset

Using the quality criteria presented in this section, we obtained a filtered dataset with 8852 entries where 4079 are labeled as not relevant and 4473 as relevant.

3.4 Automatic Assessment

The second labeling approach is an automatic assessment. This classification assumes that if the information of a post is present in the news then the post is newsworthy and consequently, journalistically relevant. The methodology can be divided into three steps.

3.4.1 Collected Data

The data labeled by this automatic assessment comes from Facebook and Twitter: posts shared by media agencies (Twitter and Facebook API) and shared posts that reference trending topics at the time in the United States (using Twitter API). The news articles, that provide the knowledge base for this system, were collected using the RSS feeds provided by renown journalistic entities in the United States. For both posts and news, all the data provided by the APIs is extracted, such as publication data and authors.

3.4.2 Post's and Article's Features

To compare posts with news, information is extracted from both texts. Entities were extracted using the Stanford CoreNLP toolkit [16], allowing for the identification of people, locations, organizations and dates present in the text. Another type of information extraction applied to the news' text was keyword detection, using the library Newspaper [10]. With the extraction of keywords, the rationale was that it would be able to detect relevant words to the context that might not be entities (verbs and adjectives as an example).

3.4.3 Newsworthy Assessment

To label a post as newsworthy or not, the system compares the information of that post with the information present in the database of news. The matching of posts with news only occurs if both were published in the same day. The system was implemented this way to address the fact that newsworthiness depends on the date of publication (posts related to news published a long time ago are usually no longer relevant).

In conclusion, to match posts with news, we computed how similar the information is by assessing the number of entities and keywords that both objects have in common. If there is a match of information (entities and keywords) the post is labeled as journalistically relevant. Otherwise, the system did not find any relevant information in its database to the post and this is labeled as not relevant.

The final dataset was composed of 7583 posts from Facebook and 3831 tweets. There were 8024 entries labeled as not relevant and 3390 as relevant, which is far more imbalanced than the previous one.

4 Feature Extraction

To successfully build a classifier, we must extract features from each post that can give us information on its relevance.

We focused on textual features. The assumption is that since we want to detect relevant posts in early stages, social network features that are a temporal-variant such as the number of "shares", "comments" and "likes" should not be considered. In addition, we do not wish to feed our classifiers with Facebook-specific or Twitter-specific features. Table 2 presents the classes of extracted features.

Table 2. Features extracted

Feature type	Description	Method
Text	Number of words, length in characters, presence of exclamation marks, presence of interrogation marks, presence of a combination of both exclamation and interrogation marks, presence of pronouns in the 1st, 2nd and 3rd person, presence of verbs in the past tense	OpenNLP [2] to identify the verbs and pronouns
Domain	Identification in 7 different domains: Health, Sports, Entertainment, Technology, Business, Politics, and General	The categorization is done using, dictionaries of domains and identifying the frequency of words from each domain that is present in the post
Sentiment	Confidence in positive sentiment, confidence in neutral sentiment and, confidence in negative sentiment	The classification is performed by 12 different sentiment systems [11]. The confidence is computed by dividing the number of systems that assign a certain classification by the total number of systems
Entity	Number of persons, organizations, locations and dates	Identified using OpenNLP [2]
Entity characteristics	Sum of newsworthiness and controversy values for each type of entity found (persons, organizations, and locations)	For newsworthiness values, using the Guardian API we identified the number of news where each entity was mentioned in the week that preceded the post date. In addition, using the Wikipedia API, we count the number of reverts [17] (i.e. the times where a modification in content was reverted to the previous state) in each entities page. Then, for each type of entity, we sum all the controversy values
Word2Vec	100 features using a Word2Vec	Since the model gives the proximity of an input word to each one of the 100 word vectors, we computed the average of the words in the tweet for each one of the 100 word vectors. Therefore, each tweet has 100 features corresponding to the average proximity value to each one of word vectors defined by the model

5 Results

Each dataset, extracted and labeled with the different methodologies, was used to train a small set of machine learning algorithms with the features previously mentioned. The models were trained using a 10-fold cross-validation. To evaluate them, we built a new and smaller dataset using experts.

This validation dataset was composed of social network messages collected from both Facebook and Twitter and from news and non-news related accounts. The collection was done for 7 days, aggregating posts created by journals and news channels (news accounts) and posts published using trending topics at the time on Twitter (non-news accounts). This way, we could make sure that the final dataset would most likely be balanced. A subset of posts was selected from the collected data: for each day, 40 messages coming from news-related accounts and 10 from trending topics were selected, resulting in a final dataset of $(40+10)\,\mathrm{x}\,7=1050$ posts. Each expert was given the indication that each post had to be evaluated considering the newsworthy relevance of the content, disregarding personal opinions. In addition, each expert was briefed with the goal of the experiment and was familiar with the concept of newsworthiness relevance. The validation dataset was composed of 577 entries labeled as not relevant and 473 as relevant.

To evaluate the performance of the models, weighted F1-measure was used. The weighted F1-measure has in consideration the disproportion of classes and it is given by the following equation:

$$F1 = F1_r * w_r + F1_{nr} * w_{nr}$$

where r and nr refers to the relevant and not relevant classes (respectively) and w_r and $w_n r$ is the percentage of entries in that class in the dataset.

The results regarding the models using the two different methodologies are presented in Table 3.

Table 3. Comparison of the models regarding their learning data in terms of weighted F1-measure

Learning algorithms	Methodologies	
	Automatic annotation	Human annotation
SVM	**0.50**	0.28
Naive Bayes	**0.64**	0.59
Decision trees	**0.46**	0.28
Random forest	**0.39**	0.30
Gradient boosted trees	**0.57**	0.55
Auto MLP	**0.52**	0.38

It is clear that in all models, the automatic approach surpasses the human annotation approach (even with the quality criteria applied to the Crowflower workers). The model performing the better in both cases is the Naive Bayes achieving 64% and 59% in the weighted F1-measure metric in the automatic and human annotation, respectively.

Although Naive Bayes also performs the better in the human annotation methodology, the results are considerably lower. We presume that this is due to the fact that the classification of journalistic relevance is a hard and ambiguous task for human annotators since it depends mainly on the areas of interest of each individual. Therefore, if a Crowdflower worker does not care and/or is not interested in a specific area, he/she may label the content as not relevant/newsworthy whereas another individual may rate it as relevant, leading to disagreement and consequently to a weaker ground truth. In addition, this kind of disagreement surpasses our criteria since we are filtering trusted users based on their personal consistency.

Regarding the automatic approach, although the data used for training was imbalanced and the fact that the validation dataset was extracted during a different time period, the results are significantly better. However, they are still far from perfect and some suggestions for improving them are presented in Future Work.

6 Conclusion and Future Work

In this work, we developed, tested and compared two methodologies to build large size datasets for the task of detecting relevance in social media posts. The necessity for such methodologies arise from the fact that journalistic relevance is an ambiguous task and to classify a large dataset (with the ultimate goal of train learning algorithms) using Crowdsourcing platforms would require a high number of evaluations per post which translates into high cost.

The first methodology is human annotation without multiple evaluations per post but instead using a set of filters to guarantee quality workers. This way we could guarantee near 9000 posts evaluated without a significantly high cost. The second methodology is automatic annotation using similarity between news content and social posts to assess newsworthiness relevance.

A dataset was created using each methodology and a set of features was extracted with the purpose of building supervised models. Using each dataset as the learning basis for a small set of learning algorithms, we tested the models using a small validation dataset annotated by experts. The results achieved provide evidence that the automatic methodology is better for creating datasets to build journalistic relevance detection models. Our assumption is that Crowdsourcing workers cannot assess what is journalistic relevant without relying on personal opinion. Therefore, without a large number of evaluations per post (which can demand a high cost), the model learns with the user personal beliefs which lead to an ambiguity definition of relevance and consequently to a low performance in the testing dataset.

Although the results from the automatic methodology are significantly better, they are far from excellent. Therefore, our future work proposal is two fold:

on one hand, we propose to increase the number of news sources used to build up the database of news of this automatic system. We intend to continuously crawl posts from both Facebook and Twitter and store them in a non-relational database (such as MongoDB) to train incremental machine learning models, thus extending our knowledge on what is relevant and what is not. On the other, we propose to add new features other than entities and keywords to compare news and posts and increase the reliability of the automatic approach. We believe that with these improvements and the already extracted features from the post, the model can see an increase in its performance in classifying posts as newsworthy or not.

Another current problem with social networks is the large amount of disinformation or "fake news" that are shared through the network [18]. Therefore, in future work, we intend to tackle the problem using automatic fact-checking (through a knowledge-based system) to avoid content that is relevant, but false.

Acknowledgements. This work is partially funded by the ERDF through the COMPETE 2020 Programme within project POCI-01-0145-FEDER-006961, and by National Funds through the FCT as part of project UID/EEA/50014/2013.

References

1. Achrekar, H., Gandhe, A., Lazarus, R., Yu, S.-H., Liu, B.: Predicting flu trends using Twitter data. In: 2011 IEEE Conference on Computer Communications Workshops (INFOCOM WKSHPS), pp. 702–707, April 2011
2. Apache: Opennlp (2010)
3. Aramaki, E., Maskawa, S., Morita, M.: Twitter catches the flu: detecting influenza epidemics using Twitter. In: Proceedings of the Conference on Empirical Methods in Natural Language Processing, EMNLP 2011, pp. 1568–1576, Stroudsburg, PA, USA. Association for Computational Linguistics (2011)
4. Doan, S., Vo, B.-K.H., Collier, N.: An analysis of Twitter messages in the 2011 Tohoku earthquake, pp. 58–66. Springer, Heidelberg (2012)
5. Figueira, A., Guimaraes, N.: Detecting journalistic relevance on social media, a two-case study using automatic surrogate features. In: Proceedings of the IEEE/ACM International Conference on Advances in Social Networks Analysis and Mining (2017)
6. Lampos, V., De Bie, T., Cristianini, N.: Flu detector - tracking epidemics on Twitter, pp. 599–602. Springer, Heidelberg (2010)
7. Lee, K., Agrawal, A., Choudhary, A.: Real-time disease surveillance using Twitter data: demonstration on flu and cancer. In: Proceedings of the 19th ACM SIGKDD International Conference on Knowledge Discovery and Data Mining, KDD 2013, pp. 1474–1477. ACM, New York (2013)
8. Mendoza, M., Poblete, B., Castillo, C.: Twitter under crisis: can we trust what we RT? In: Proceedings of the First Workshop on Social Media Analytics, SOMA 2010, pp. 71–79. ACM, New York (2010)
9. Muralidharan, S., Rasmussen, L., Patterson, D., Shin, J.-H.: Hope for Haiti: an analysis of facebook and Twitter usage during the earthquake relief efforts. Publ. Relat. Rev. **37**(2), 175–177 (2011)

10. Newspaper, Python Library, January 2017. https://pypi.python.org/pypi/newspaper,
11. Ribeiro, F.N., Araújo, M., Gonçalves, P., Gonçalves, M.A., Benevenuto, F.: Sentibench - a benchmark comparison of state-of-the-practice sentiment analysis methods. EPJ Data Sci. **5**(1), 23 (2016)
12. Robinson, B., Power, R., Cameron, M.: A sensitive Twitter earthquake detector. In: Proceedings of the 22nd International Conference on World Wide Web, WWW 2013 Companion, pp. 999–1002. ACM, New York (2013)
13. Sakaki, T., Okazaki, M., Matsuo, Y.: Earthquake shakes Twitter users: real-time event detection by social sensors. In: Proceedings of the 19th International Conference on World Wide Web, WWW 2010, pp. 851–860. ACM, New York (2010)
14. Sandim, M., Fortuna, P., Figueira, A., Oliveira, L.: Journalistic relevance classification in social network messages: an exploratory approach, pp. 631–642. Springer, Cham (2017)
15. Sankaranarayanan, J., Samet, H., Teitler, B.E., Lieberman, M.D., Sperling, J.: Twitterstand: news in tweets. In: Proceedings of the 17th ACM SIGSPATIAL International Conference on Advances in Geographic Information Systems, GIS 2009, pp. 42–51. ACM, New York (2009)
16. Stanford Named Entity Recognizer: The Stanford Natural Language Processing Group, January 2017. http://nlp.stanford.edu:8080/ner/
17. Yasseri, T., Spoerri, A., Graham, M., Kertész, J.: The most controversial topics in Wikipedia: a multilingual and geographical analysis. CoRR, abs/1305.5566 (2013)
18. Figueira, Á., Oliveira, L.: The current state of fake news: challenges and opportunities. Procedia Comput. Sci. **121**, 817–825 (2017). CENTERIS 2017 - International Conference on ENTERprise Information Systems/ProjMAN 2017 - International Conference on Project MANagement/HCist 2017 - International Conference on Health and Social Care Information Systems and Technologies, CENTERIS/ProjMAN/HCist (2017)

Performance Evaluation of Support Vector Machine and Convolutional Neural Network Algorithms in Real-Time Vehicle Type Classification

Ali Şentaş, İsabek Tashiev, Fatmanur Küçükkayvaz, Seda Kul, Süleyman Eken(✉), Ahmet Sayar, and Yaşar Becerikli

Computer Engineering Department, Kocaeli University, 41380 Izmit, Turkey
alisentas96@gmail.com, isabek.tashiev@gmail.com, fatmanur.kucukayvaz@gmail.com, {seda.kul,suleyman.eken,ahmet.sayar,ybecerikli}@kocaeli.edu.tr

Abstract. Intelligent traffic management systems needs to obtain information about traffic with different sensors to control the traffic flow properly. Traffic surveillance videos are very actively used for this purpose. In this paper, we firstly create a vehicle dataset from an uncalibrated camera. Then, we test Tiny-YOLO real-time object detection and classification system and SVM classifier on our dataset and well-known public BIT-Vehicle dataset in terms of recall, precision, and intersection over union performance metrics. Experimental results show that two methods can be used to classify real time streaming traffic video data.

Keywords: Vehicle detection and classification · Video processing Tiny-YOLO · Intelligent traffic management systems

1 Introduction

Vehicle type recognition is an important research topic during the last decades. It has a wide range of applications such as automatic vehicle identification [1], road capacity [2], traffic density measurement [3], speed detection [4–6], and traffic violation detection [7]. It is also important to identify vehicle categories for crime prevention and transportation investors.

Vehicle related data was obtained by the sensors which are previously positioned so that the path can be seen from above. With the advances in vision systems and technologies, image processing and pattern recognition are used to differentiate vehicle types. Afterwards, traditional methods are beginning to be inadequate for powerful computation ability. Also, the most important reasons not to be able to put real time video processing applications into practice are scalability and performance problems. Processing of videos from traffic surveillance cameras is an example of such applications, in which video is processed for early warning or extracting information through some real time analysis. Video data get very large in size, therefore using central processing techniques to process data and get the required information in right time and correctly is not an easy task. It even gets worse in the case of real-time scenarios.

L. Barolli et al. (Eds.): EIDWT 2018, LNDECT 17, pp. 934–943, 2018.
https://doi.org/10.1007/978-3-319-75928-9_86

Moreover, in such systems, they need to be processed and analyzed in a reasonable time period. For the decision making, the right data need to be dispatched to the right places in the right time.

Advancement in real-time image processing over the Internet and the technology of deep learning attract attention in many application areas such as intelligent traffic/vehicle management systems, transferring traffic surveillance video data. The work presented in this paper is summarized as a real-time vehicle detection and classification over the video streams provided by traffic surveillance cameras.

We apply the state-of-the-art, real-time object detection system Tiny-YOLO (You Only Look Once) [8], which have proven a good competitor to Fast R-CNNs and SSDs both in terms of detections and speed. In this paper, we will apply Tiny-YOLO and SVM classifier on own dataset (TPSdataset) and well-known public dataset BIT-Vehicle. The contributions of this paper are as following:

- Training and applying the state-of-the-art, real-time object detection system Tiny-YOLO for vehicle type detection and classification.
- Filtering real time streaming video data according to vehicle types.
- End users (which are searching for a vehicle) have capability of searching for a vehicle defined in type property.
- Building TPSdataset.

The remainder of this article is organized as follows. "Related work" section presents the related work. The main architecture for vehicle detection and classification is given in "Proposed framework" section. The performance results and their analyses are given in "Experimental setup, results, and analysis" section. The last section concludes the article.

2 Related Works

Vehicle detection and classification is a challenging problem [9]. Mostly because there are lots of different vehicle types and dimensions. Also detecting them with low processing power in an uncontrolled, varying environments make the problem complicated. Solving this problem in real-time applications makes it even more difficult. There are two kinds of approaches to this problem. Mostly researches would use a CNN or an algorithm such as HAAR or Histogram of Oriented Gradients (HOG) to extract features and then train a classifier such as a Support Vector Machine (SVM) or AdaBoost using them.

HAAR features introduced by Viola *et al.* [10] and used by Han *et al.* [11]. Extracting features from images can also be done with HOG algorithm introduced by Dalal *et al.* [12] which is one of the algorithms we use in our paper. Scale Invariant Feature Transform (SIFT) is introduced by Lowe *et al.* [13] can be used to extract features and gives good results but it's inefficient for real-time applications and for this reason Bay *et al.* [14] introduced Speeded Up Robust Features (SURF) which gives similar results but increases performance. There is also Local Binary Patterns (LBP) approach introduced by Ojala *et al.* [15] and it can also be combined with HOG and increase detection performance in some datasets [16]. Wu and Zhang [17] used standard Principal Component Analysis (PCA) for feature extraction. After extracting the features, SVM [18] or AdaBoost [19] classifiers are trained and used with different scales on an image which is a vehicle (positive).

The methods described above are for object detection and classification in general. There are three main stages for traffic surveillance systems, detection, classification and tracking. For vehicle detection most methods [20, 21] assume camera is static. Background subtraction is a popular method for detection, Lu *et al.* [22] proposed a moving vehicle detection which uses fuzzy background subtraction algorithm and achieves high detection rates. One of the drawbacks of background subtraction method is that it is not applicable for static images. In our previous work [23] binary image features (width, height, major/minor axis length, and etc.) are used to classify vehicles. Three classification algorithms (ANN, SVM, and Adaboost) are tested. Their accuracy values are 87.5, 81.6, and 85.4, respectively. In another previous study [24], we introduce a middleware system based on pub/sub messaging protocol and a dispatcher to preprocess the streams in real time. Classified vehicle images are send to related subscribers. Experimental show that middleware may be utilized in different areas such as infrastructure planning, traffic management. In this paper, we don't use background subtraction because it decreases performance and doesn't improve vehicle detection. We prefer the sliding window technique over background subtraction because detection vehicle areas after background subtraction requires a lot more work whereas sliding window is easy to implement, scalable and also faster.

With the development of image processing, pattern recognition and deep learning, vehicle type classification technology based on deep learning has raised increasing concern.

Dong *et al.* [25] proposed a vehicle type classification method that uses a half-supervised convolutional neural network with front view images of the vehicles. Unlike traditional feature extraction methods the neural network gives the possibility of what class may belong to the vehicle image given to it. The developed system works well on complex images. Dong *et al.* Created the BIT-Vehicle data set consisting of 9850 high-resolution images with only the front views of the vehicles. Their system is working with 96.1% accuracy in daytime conditions and 89.4% accuracy in nighttime conditions.

Wang *et al.* [26] have developed a vehicle type classification system by using Faster R-CNN which is a deep learning method. The system aims to classify cars and trucks. It has over 90% accuracy. They test the system on an NVIDIA Jetson TK1 board with 192 CUDA cores. Their system takes around 0.354 s to detect an image which is shows that it can run in real-time classification systems.

Gao *et al.* [27] proposed a vehicle brand recognition system based on CNN. They first detected vehicles by using frame difference method then the resultant binary image is used to detect the frontal view of cars. The frontal view of vehicles used in CNN for train and test. Their system achieved 88% accuracy.

Lee *et al.* [28] proposed an ensemble of global networks and mixture of K local expert networks for vehicle classification. They used AlexNet, GoogLeNet and ResNet deep convolutional neural network structures. The MIO-TCD dataset is used for train and test. Although the dataset is challenging due to lighting, image resolution etc. their work achieved 97.92% accuracy.

Huo *et al.* [29] proposed a model which is using the multi-view classification of vehicles captured in real surveillance. Their aim was not just to classify vehicles by using the frontal view images but to use all aspects (rear, front, and side). Region-based Convolutional Neural Network (RCNN) framework used for classification which has four vehicle categories (car, truck, bus, and van). Experiments shows that their system achieved 83% accuracy.

Kim *et al.* [30] proposed a new vehicle type classification model by using multi-view surveillance camera. They combined four concepts by using Bagging and CNN to increase the performance of the classification systems. Their system achieved 97.84% accuracy over 103,833 images.

3 Vehicle Detection and Classification

3.1 Support Vector Machines

SVM consist of a set of learning methods used for classification and regression. SVM is used with HOG. The main purpose of the HOG method is to define the image as a group of local histograms. These groups are the histograms of the magnitudes of the gradients in the orientations of the gradients in a local region of images. The features of the vehicle images have been removed with HOG. The images in the data set we use for the HOG property extraction are square. Dimensions should be multiples of 8. The image is divided into 8×8 cells and the HOG vehicle detector uses a sliding detection window which is moved around the image. At each position of the detector window, a HOG descriptor is computed for the detection window.

Here, SVM is used for detection and classification of vehicles. HOG descriptors are obtained from positive (vehicle) and negative (non-vehicle) images and vehicle detection is done with linear SVM classifier using HOGs. The second usage of SVM is about classification process. It processes for more than one class (5 classes).

3.2 Convolutional Neural Networks

Convolutional Neural Networks (CNN) are the key players of object classification and detection tasks in nowadays. One of the main reasons why Convolutional Neural Networks weren't used in real world applications was that their required more powerful computational resources. With the significant improvements of GPU boosting technologies.

In the last few years, there were developed a lot of variations of convolutional neural networks like R-CNN and its modifications Fast R-CNN and Faster R-CNN. Each of them improved the previous one on especially important criteria like speed and accuracy of the classification. One of the advantages of convolutional networks is that their can do object classification and detection simultaneously. The main criteria of object classification and detection are that their should be fast, accurate and able to recognize a variety of objects.

CNN consists of these main layers like convolutional layer, pooling layer and fully connected layer. Depends on architecture CNNs use these layers in different variations. In our experiments we used Tiny-YOLO, a state-of-the-art and real-time object classification and detection architecture. It has a simpler model architecture and it needs small GPU resources appropriately. It uses model which named Darknet-19 and consists 9 convolutional layers, 6 max-pooling layers, one average pooling layer and the last one is softmax layer.

4 Experimental Analysis

4.1 Building TPSdataset

We created our own TPSdataset with 3 video files we shot in Kocaeli city in Turkey. Four thousand nine hundred and forty-four frames were obtained from the traffic videos. The frames are then labeled with the "LabelImg" graphical image annotation tool [31]. Using this tool, we get the vehicles in a rectangular area. We label the field belonging to the vehicle class. We record the action we made while passing the other frame. Annotations are saved as XML files in PASCAL VOC format, the format used by ImageNet. The XML file contains the coordinates, width, height, and class label of the vehicles in that image (see Fig. 1 for one frame and its XML file). This way we create our own dataset.

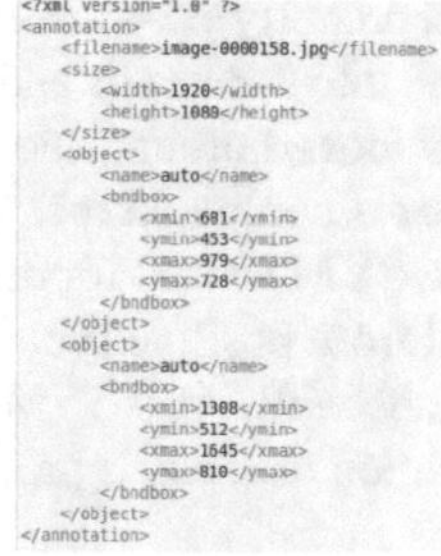


```
<?xml version="1.0" ?>
<annotation>
    <filename>image-0000158.jpg</filename>
    <size>
        <width>1920</width>
        <height>1080</height>
    </size>
    <object>
        <name>auto</name>
        <bndbox>
            <xmin>681</xmin>
            <ymin>453</ymin>
            <xmax>979</xmax>
            <ymax>728</ymax>
        </bndbox>
    </object>
    <object>
        <name>auto</name>
        <bndbox>
            <xmin>1308</xmin>
            <ymin>512</ymin>
            <xmax>1645</xmax>
            <ymax>810</ymax>
        </bndbox>
    </object>
</annotation>
```


Fig. 1. An example from TPSdataset and its XML file

Our dataset contains five classes: bus, minivan, minibus, truck, auto. Auto class is a broad class containing sedans, SUVs and hatchback vehicles. Truck contains trucks, long vehicles and smaller trucks which are called minivans in the BIT-Vehicle dataset. Minibuses are like minivans but bigger and have more space between roof and front window. Also in our dataset all samples are taken with day light. Table 1 shows the distribution of vehicle types in TPSdataset.

Table 1. Distribution of vehicle types

Type	#
Auto	719
Bus	147
Minibus	72
Minivan	187
Truck	266

4.2 Experimental Setup and Performance Metrics

In our experiments, we have used an Intel Core i5-4210U 1.7 GHz processor and Arch Linux OS. We test the system both on TPSdataset and another data set named BIT-Vehicle. Classification result is evaluated using recall and precision parameters as the performance measures.

4.3 Tests on BIT-Vehicle Dataset

BIT-Vehicle Dataset [25] is a dataset containing 9850 vehicle images, whose sizes are of 1600×1200 and 1920×1080 and captured from two cameras at the different time and places. Data set are divided into six categories: bus, microbus, minivan, SUV, sedan, truck. This dataset doesn't perform well with the HOG algorithm because HOG requires samples to be aligned similarly, however cars alignment vary very differently from one sample to another.

We use a linear SVM with C set to 0.3. Our image size for HOG is 128×128 pixels, block size, block stride and cell size are 16×16, 8×8, 8×8 pixels respectively. Also dataset is partition into 75% training and 25% test data. 3000 negative (non-vehicle) images are used. Table 2 shows performance results of SVM classifier on BIT-Vehicle dataset.

Table 2. Vehicle classification results on BIT-Vehicle dataset

Vehicle class	Sample count	Precision	Recall
Bus	555 (415/138)	98.5507	98.5507
Microbus	878 (659/219)	91.7431	91.3242
Minivan	474 (354/118)	87.7358	78.8136
SUV	1381 (1032/343)	89.7898	87.172
Sedan	5796 (4310/1436)	97.5762	98.1198
Truck	821 (615/205)	90	96.5854
Average	9905 (7385/2459)	92.5659	91.7609

It's noticeable that precision and recall of minivan is lower than other classes. The reason for this irregularity is that our trained model confuses minivans with trucks. 22 out of all 118 minivan test sample are misclassified as trucks. When you see all the vehicles from the same square window minivan and trucks does look alike. Also, minivan samples are less than others in BIT-Vehicle dataset.

YOLO can be used for classification and detection objects using bounding boxes. Intersection over Union (IOU) metric is an evaluation metric used to measure the accuracy of an our model on test dataset. In order to apply IOU for evaluation of our model we need: the ground-truth hand labeled bounding boxes which specify where in the our object is and the predicted bounding boxes from our model. Table 3 shows performance results of Tiny-YOLO performance on BIT-Vehicle dataset.

Table 3. Tiny-YOLO performance on BIT-vehicle dataset

Vehicle class	Sample count	Precision	Recall	IOU
Bus	558 (446/112)	100	100	90.03
Microbus	883 (706/177)	96.67	98.31	86.42
Minivan	476 (381/95)	98.95	98.95	85.51
SUV	1392 (1114/278)	97.19	99.64	89.59
Sedan	5921 (4737/1184)	97.20	99.75	90.13
Truck	823 (658/165)	87.41	100	100
Average	10053 (8042/2011)	97.90	99.60	89.29

4.4 Tests on TPSdataset

We use the same parameters for training SVM classifier to get the best results. Table 4 shows performance results of SVM classifier on TPSdataset.

Table 4. Vehicle classification results on TPSdataset

Vehicle class	Sample count	Precision	Recall
Bus	147 (111/36)	100	100
Minivan	187 (141/46)	91.8367	97.8261
Minibus	72 (54/18)	94.4444	94.4444
Truck	266 (200/66)	100	98.4848
Auto	719 (540/179)	99.435	98.324
Average	1391 (1046/345)	97.1432	97.8159

4 out of all 46 test samples for minivan are misclassified, 3 classified as auto and 1 as truck hence precision is low for minivan but this problem would be fixed by expanding the dataset which we aim to do so in the future. We test model with our test videos and it performed better at detecting and classifying cars than the model trained with BIT-Vehicle dataset. It's small compared to BIT-Vehicle but it's more suited to HOG approach because vehicles are aligned to each other properly and we aim to expand it and add night images.

As the results show on the Table 5, Tiny-YOLO demonstrates poor performance on TPSdataset compared to model which was trained on BIT-Vehicle Dataset. The result of this behavior is the lack of data.

Table 5. Tiny-YOLO performance on TPSdataset

Vehicle class	Sample count	Precision	Recall	IOU
Bus	568 (483/85)	40.00	75.29	58.96
Minivan	722 (639/83)	48.98	86.75	67.59
Minibus	297 (237/60)	36.27	61.67	49.63
Truck	540 (451/89)	44.74	76.40	61.52
Auto	3069 (2588/481)	52.53	92.93	76.05
Average	5196 (4398/798)	62.83	86.22	69.74

Another accuracy metric for evaluate models is Precision-Recall Curve (PR Curve). To find out trade off between precision and recall we calculate precision and recall values for threshold from 0.0 to 1.0 with step 0.1 and find suitable threshold (Fig. 2).

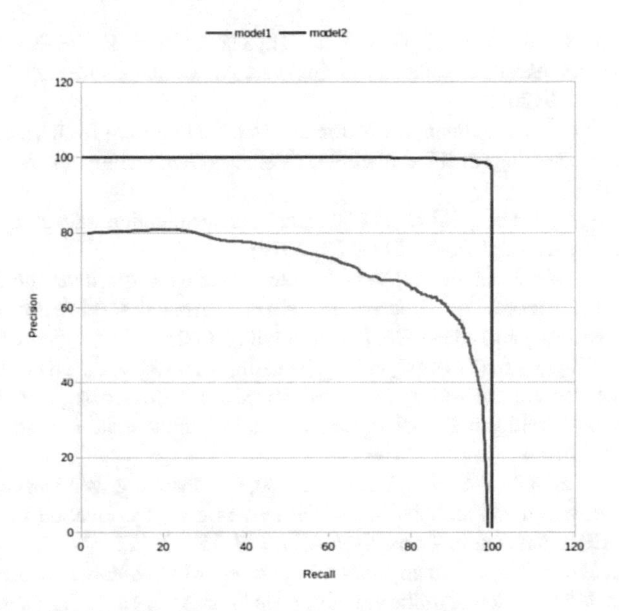

Fig. 2. Precision-recall curve for two models

5 Conclusion

Image processing and deep learning based video surveillance system in traffic management systems enables many studies such as license plate recognition, finding the number of vehicles, traffic density detection, vehicle speed calculation, detection of lane violations and vehicle classification. In this study, we firstly create own video surveillance dataset. Then, SVM classifier and Tiny-YOLO are tested on TPSdataset and well-known public dataset BIT-Vehicle in terms of recall and precision metrics. Also, we give IOU metric for Tiny-YOLO. Experimental results show that Tiny-YOLO outperforms SVM on BIT-Vehicle dataset. But, SVM is well than Tiny-YOLO on TPSdataset. Because, acquired dataset is imbalanced. This generally causes overfitting, which may affect the final results.

In the future, instead of classical approaches of centralized computation, a distributed scalable network of collaborating computation nodes is going to be developed to process streaming real time video data coming from traffic surveillance cameras. In this way, hierarchical topic-based publish-subscribe messaging middleware is going to be realized. Also, real time stream processing infrastructures such as Apache Kafka, Flink, and Storm will be considered.

Acknowledgments. This work has been supported by the TUBITAK under grant 116E202.

References

1. Ying-Nong, C., Chin-Chuan, H., Gang-Feng, H., Kuo-Chin, F.: Facial/license plate detection using a two-level cascade classifier and a single convolutional feature map. Int. J. Adv. Robot. Syst. **12**(12), 1–16 (2015)
2. Bas, E., Tekalp, A.M., Salman, S.: Automatic vehicle counting from video for traffic flow analysis. In: Proceedings of IEEE Intelligent Vehicles Symposium, pp. 392–397. IEEE Pres, Istanbul (2007)
3. Al-Sobkya, A.A., Mousa, R.M.: Traffic density determination and its applications using smartphone. Alex. Eng. J. **55**(1), 513–523 (2016)
4. Pelegri, J., Alberola, J., Llario, V.: Vehicle detection and car speed monitoring system using GMR magnetic sensors. In: Proceedings of IEEE Annual Conference in the Industrial Electronics Society, pp. 1693–1695. IEEE, Sevilla (2002)
5. Ryusuke, K., Shigeyuki, K., Tomoyuki, N., Takashi, A., Makoto, A., Hisao, O.: A punctilious detection method for measuring vehicles' speeds. In: Proceedings of the International Symposium on Intelligent Signal Processing and Communication System, pp. 967–970. IEEE, Tottori (2006)
6. Harry, H.C., Benjamin, D.S., Joe, P., Bin, L., Bo, C., Zhaoqing, W.: Development and field test of a laser-based nonintrusive detection system for identification of vehicles on the highway. IEEE Trans. Intell. Trans. Syst. **6**(2), 147–155 (2005)
7. Marikhu, R., Moonrinta, J., Ekpanyapong, M., Dailey, M.N., Siddhichai, S.: Police eyes: real world automated detection of traffic violations. In: Proceedings of ECTI-CON, pp. 1–6. IEEE, Krabi (2013)
8. Redmon, J., Farhadi, A.: YOLO9000: better, faster, stronger. In: Proceedings of Conference on Computer Vision and Pattern Recognition. IEEE, Honolulu (2017)
9. Kul, S., Eken, S., Sayar, A.: A concise review on vehicle detection and classification. In: Proceedings of the Third International Workshop on Data Analytics and Emerging Services, pp. 1–4. IEEE, Antalya (2017)
10. Viola P., Jones M.: Rapid object detection using a boosted cascade of simple features. In: Proceedings of Computer Vision and Pattern Recognition, pp. 511–518. IEEE, Kauai (2001)
11. Han, S., Han, Y., Hahn, H.: Vehicle detection method using Haar-like feature on real time system. World Acad. Sci. Eng. Technol. **59**(35), 455–459 (2009)
12. Dalal, N., Triggs, B.: Histograms of oriented gradients for human detection. In: Proceedings of IEEE Computer Society Conference on Computer Vision and Pattern Recognition, pp. 886–893. IEEE, San Diego (2005)

13. Lowe, D.G.: Distinctive image features from scale-invariant keypoints. Int. J. Comput. Vis. **60**(2), 91–110 (2004)
14. Bay, H., Tuytelaars, T., Van Gool, L.: Surf: speeded up robust features. In: Proceedings of Computer Vision–ECCV vol. 110, pp. 404–417 (2006)
15. Ojala, T., Pietikäinen, M., Harwood, D.: Performance evaluation of texture measures with classification based on Kullback discrimination of distributions. In: Proceedings of the 12th IAPR International Conference on Pattern Recognition, pp. 58–62. IEEE, Jerusalem (1994)
16. Wang, X., Han, T.X., Yan, S.: An HOG-LBP human detector with partial occlusion handling. In: Proceedings of IEEE 12th International Conference on Computer Vision, pp. 32–39. IEEE, Kyoto (2009)
17. Wu, J., Zhang, X.: A PCA classifier and its application in vehicle detection. In: Proceedings of International Joint Conference on Neural Networks, pp. 600–604. IEEE, Washington (2001)
18. Burges, C.J.C.: A tutorial on support vector machines for pattern recognition. Data Min. Knowl. Discov. **2**(2), 121–167 (1998)
19. Freund, Y., Schapire, R., Abe, N.: A short introduction to boosting. J. Jpn. Soc. Artif. Intell. **14**, 771–780 (1999)
20. Beymer, D., McLauchlan, P., Coifman, B., Malik, J.: A real-time computer vision system for measuring traffic parameters. In: Proceedings of IEEE Computer Society Conference on Computer Vision and Pattern Recognition, pp. 495–501. IEEE, San Juan, June 1997
21. Gupte, S., Masoud, O., Martin, R.F., Papanikolopoulos, N.P.: Detection and classification of vehicles. IEEE Trans. Intell. Transp. Syst. **3**(1), 37–47 (2002)
22. Lu, X., Izumi, T., Takahashi, T., Wang, L.: Moving vehicle detection based on fuzzy background subtraction. In Proceedings of IEEE International Conference on Fuzzy Systems (FUZZ-IEEE), pp. 529–532. IEEE, Beijing (2014)
23. Kul, S., Eken, S., Sayar, A.: Measuring the efficiencies of vehicle classification algorithms on traffic surveillance video. In: Proceedings of International Conference on Artificial Intelligence and Data Processing, pp. 1–6. IEEE, Malatya (2016)
24. Kul, S., Eken, S., Sayar, A.: Distributed and collaborative real-time vehicle detection and classification over the video streams. Int. J. Adv. Robot. Syst. **2017**, 1–12 (2017)
25. Dong, Z., Wu, Y., Pei, M., Jia, Y.: Vehicle type classification using a semisupervised convolutional neural network. IEEE Trans. Intell. Transp. Syst. **16**(4), 2247–2256 (2015)
26. Wang, X., Zhang, W., Wu, X., Xiao, L., Qian, Y., Fang, Z.: Real-time vehicle type classification with deep convolutional neural networks. J. Real-Time Image Process. **2**, 1–10 (2017)
27. Gao, Y., Lee, H.: Vehicle make recognition based on convolutional neural network. In: Proceedings of ICISS, pp. 1–4. IEEE, Seoul (2015)
28. Lee, J.T., Chung, Y.: Deep learning-based vehicle classification using an ensemble of local expert and global networks. In: IEEE Conference on Computer Vision and Pattern Recognition Workshops (CVPRW), pp. 920–925. IEEE, Honolulu (2017)
29. Huo, Z., Xia, Y., Zhang, B.: Vehicle type classification and attribute prediction using multi-task RCNN. In: Proceedings of 9th International Congress on Image and Signal Processing, BioMedical Engineering and Informatics (CISP-BMEI), pp. 564–569, Datong (2016)
30. Kim, P.K., Lim, K.T.: Vehicle type classification using bagging and convolutional neural network on multi view surveillance image. In: Proceedings of IEEE Conference on Computer Vision and Pattern Recognition Workshops (CVPRW), pp. 914–919, Honolulu (2017)
31. Tzutalin/LabelImg. https://github.com/tzutalin/labelImg

Cloud Orchestration with ORCS and OpenStack

Flora Amato[1](✉), Francesco Moscato[2], and Fatos Xhafa[3]

[1] DIETI University of Naples "Federico II", Naples, Italy
flora.amato@unina.it
[2] DiSciPol, University of Campania "Luigi Vanvitelli", Caserta, Italy
francesco.moscato@unicampania.it
[3] Department of Computer Science, Universitat Politècnica de Catalunya, Barcelona, Spain
fatos@cs.upc.edu

Abstract. During the past recent years there is an increasing interests in Cloud Services Orchestration. Efficient and even optimal allocation of Cloud resources is one of the main problems on which the scientific and development community has focused their effort. Some proposals for standards and middleware are now available for Cloud users and designers. However, the need for advancing on *composition* techniques is still requiring major efforts due to the new features, namely, composition of services at any layer of Cloud architecture, not only orchestration of resources. To that end, there have been proposed some Cloud patterns in order to describe composition of services. In a real setting, the composition is really complex and challenging, leading to *Orchestration of Cloud Service*, whose aim is to deal with both pattern-based composition and resource orchestration. In this paper, we show how the framework Orchestrator for Complex Services (OrCS) enables the use of pattern-based composition and resource orchestration. We also discuss its integration with the OpenStack Orchestrator (Heat).

1 Introduction

Cloud Orchestration is an important problem in the research agenda on Cloud Systems. Its aims are going far beyond the classical *Resource Orchestration* problem that addressed resources optimization in complex, federated or multi-cloud environment. Now orchestration deals with composition of services at any level of Cloud architecture, i.e. at resource layer too. Obviously, it is being benefited by former research findings and development activities on Web Services Orchestration, intended as workflows of activities. The Cloud orchestration however, has new features and requirements on automation. This explain somehow the lack of clear definitions, frameworks or languages to satisfactorily solve the *multi-level* Cloud Orchestration.

In [1,2] the authors report some scientific issues on orchestration of Cloud services as a *cross cutting* feature of the architectural levels of Cloud systems

L. Barolli et al. (Eds.): EIDWT 2018, LNDECT 17, pp. 944–955, 2018.
https://doi.org/10.1007/978-3-319-75928-9_87

(i.e. IAAS, PAAS and SAAS levels). Indeed, orchestration of Cloud services [3] requires both *intra-layer* (i.e. among services in the same level of Cloud architecture) and *inter-layers* composition (for example when a service in SaaS layer uses a PaaS or an IaaS service). The use of Cloud patterns, comprising design, architectural and programming patterns, are envisaged as a powerful means to cope with the composition of Cloud resources and services. The introduction of design patterns in Cloud computing [4] requires a concept of orchestration that is more complex than the one defined in [5]. As we will show in this work many design patterns with different purposes (from *QoS* improvement to special behaviours implementation) may be described as complex constructs of workflow languages. As a matter of fact, most of Cloud design patterns suggested by Big Vendors [6, 7] are described as workflows.

The main research and development efforts in composition during last years focused on the choice of the services and resources to use in a composite Cloud service mainly to improve the Quality of Service [8]. On the other hand, many works in the literature deal with optimization problems [9, 10]. Yet, there is still lacking a formal definition of the Orchestration problem, although there are some initiatives on providing a Cloud Orchestration Engine with an orchestration language such as in [11], where COPE (Cloud Orchestration Policy Engine) is presented. Peer-to-peer and collaborative approaches to composition have been also proposed [12, 13], where authors show the usefulness of the platform not only for efficient and reliable distributed computing [14, 15] but also for collaborative activities and ubiquitous computing [16]. From a platform development perspective, the work on orchestration made by OASIS in the Topology and Orchestration Specification for Cloud Application (TOSCA) is an important achievement [17]. It should be noted that several works have been reported in literature about Cloud pattern exploitation [18], but in general they contain only descriptions of different design patterns. Motivated by the need to advance on composition in terms of both Cloud computing patterns and workflow patterns, in this work we investigate how such patterns are closely related at different layers of abstraction [19, 20].

We observe that in order to achieve an automatic Cloud service composition, resources should carry out a semantic description of their functionality, as well as a semantic description of their parameters [21, 22]. Actually, several semantics-based approaches for *simple* web services composition exist (see [23] for a survey). Some of them [24, 25] exploit BPEL4WS orchestration language and OWL-based ontologies for services description. In addition, we need to enact Orchestration at all Cloud layers. This means that a Cloud orchestrator at service layer should be able to interact with different resource orchestrator in order to exploit their ability to manage and optimize resources during the execution of composite services. In this context, it is clear that a framework able to manage composition by orchestration is appealing, which explicitly takes into account of Cloud design patterns, resources and services orchestration.

In this paper, we introduce the main architecture of the Orchestrator for Complex Services (*OrCS*, hereafter) and show how it can be used in OpenStack

Heat Orchestrator[1]–a framework for automatic composition of Cloud services, which is driven by Cloud patterns definition. The framework enables both design-by-patterns of composite Cloud services, it implements run-time manager and monitors for services, and allows for automatic verification of composition soundness [26,27]; QoS requirements can be verified as well. The methodology of OrCS is based on formal definition of an orchestration language, on the use of ontologies to describe Cloud services interactions and on a rule-based reasoner that enacts composition. Furthermore, Model Driven Engineering techniques have been embedded into OrCS in order to fulfil validation and verification activities [14,28]. Finally, OrCS allows for building composite services in an easy way by simply declaring the composition type (i.e. the pattern) to achieve. Component services are automatically selected when declaring the semantics of services to implement and eventually the QoS to assure.

The rest of the paper is organised as follows. In Sect. 2 is presented the OrCS architecture. We present the orchestration in OrCS and OpenStack Heat Wrapper in Sect. 3. A case study to exemplify our approach is given in Sect. 4 and we conclude the paper in Sect. 5.

2 The OrCS Architecture

This section briefly introduces the architecture of the proposed OrCS framework, which aims to define, design, analyze and manage multi-layer orchestration.

Figure 1 depicts the main component of the OrCS framework.

A **User Interface** provides the tools to design Cloud services and eventually to define the main structure of resources needed to run the composite services. We work upon the existence of several, potentially heterogeneous, Cloud Providers (**Cloud1**, $\cdots$, **CloudN**). Each provider is able to instantiate common Cloud resources like computing nodes, virtual storages or virtual network (for inter and intra-Clouds). In addition, virtual storages can of course maintain different sets of data.

The architecture of the **Orchestrator** consists of:

- **Execution Scheduler:** the scheduler reads the description of a composite service and executes the proper services when needed, eventually scheduling data migration from a virtual storage to another.
- **Data Dispatcher:** It executes physical data migration and maintains information about data produced during the execution of the composite service.
- **Broker:** it enacts common service brokering actions. If the resource is not yet acquired on a provider, it provides for acquisition and management. It is also responsible for the provisioning of the resources and their configuration.
- **Deployer:** this component deploys needed services at SaaS (from a pool of available services) on proper resource in the Cloud.

[1] https://docs.openstack.org/heat/latest/.

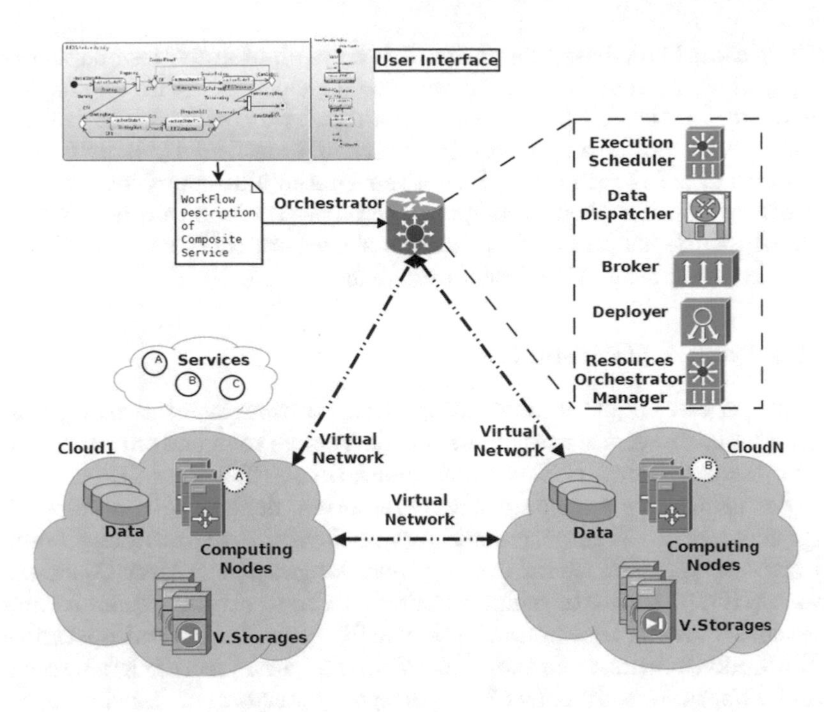

Fig. 1. The OrCS architecture

- **Resources Orchestrator Manager:** interface with existent orchestrators [2]. Currently this module supports COPE [11] and OpenStack HEAT Orchestrators.

The workflow-based language used for description of the whole service is called Operational Flow Language (OFL). OFL is rich enough to describe several patterns, as well as simple to be defined by means of clear operational semantics. Compositional rules enable patterns description. More details can be found in [29–31].

3 Orchestration in the OrCS and OpenStack Heat Wrapper

Our workflow language has simple constructs that can be used to define complex composition patterns. Patterns are described like source code skeletons for implementing composite services.

We describe orchestration by means of Cloud Patterns[2] [7,32,33]. Patterns are in turn defined by using a workflow-based language called Operational Flow Language (OFL) [19].

[2] https://cloudpatterns.org, http://en.clouddesignpattern.org/index.php/Main_Page.

OFL is a workflow-based language. It is a graph of activities and their relationships, where each activity represents one logical step within a process.

Therefore in the OFL language, workflow processes consist of a network of activity nodes and edges (transitions) identified by a pair of nodes (`FromActivity, ToActivity`). Activities represents atomic Cloud services or resource invocations, as well as composite activities (i.e. sub workflow processes).

Actually, OFL has many elements, we show here the part of the language that enables integration with OpenStack Heat.

3.1 Heat and HOT Template

HEAT [34] is the engine implementing Orchestration facilities in OpenStack. Mainly, HEAT is a service that allows: (1) developers to define and store requirements to execute Cloud (OpenStack) applications; (2) execute Cloud applications by creating (in a transparent way) resources, deploying (if necessary) and manage components at resources and services levels. Heat provides a template-based orchestrator; this means that a proper language, the Heat Orchestration Template (HOT), allows to define resources a Cloud service needs to run, and relationships between resources. A set of API allows for several operations on OpenStack components, from the simple setup and execution of virtual resources, to complex operations like copy of images, synchronization of distributed, redundant file systems, etc.

HEAT and HOT provide good means to manage resources for *one* (simple or complex) Cloud service, but there is few or none at all support for composition of services at SaaS level. In fact, during composition, it is possible that different services address the same resources (for example, a Cinder Storage in OpenStack). In order to share data and that the same services, contemporary, communicate each other by using REST interfaces. This is only possible in multi-level orchestration that addresses both composite services workflow and resource management.

In order to allow the OrCS using Heat to manage resources during orchestration, it has to provide: (1) in OFL, all the elements to be compliant to HOT templates; (2) the OrCS orchestrator has to be able to use Heat API to manage resources, depending on OFL declarations.

3.2 OFL and HOT

A description of the main elements of OFL can be found in [19]. Here we briefly discuss only the elements used to compile an HOT template for OpenStack.

Basic building blocks of HOT templates are: **parameters**, **resources**, and **outputs**. Parameters describe all *data* used in the Heat orchestration. They include the name and the type of images to execute on server, the type of instance (flavour) to use on server instances, the identifier of the private networks used to collect resources and other data like user names and passwords used for authentication, etc. Resources describe the configuration and the type of (virtual) servers, as well as other elements used during the orchestration process (like random strings used for authentication, etc.). Outputs section, declares the template outputs, usually provided for management or to collect results. Of course, each building block has many other sub-elements. We discuss in the following elements supported by OFL. In particular, sub-elements in resources are used to link images and other resources to servers.

Table 1 resumes the matching of OFL elements with HOT template blocks.

Table 1. Matching of OFL elements to HOT blocks

Hot block	Sub element	Description	OFL element
Parameter	Image	An OS image to load on a server	Relevant data
	Flavour	The type of instance to use	Relevant data
	Key	Key pairs for compute instances	Relevant data
	Private network	The Id of the private network connecting resources	Relevant data
Resource	Type	The type of resource in OpenStack (Nova, Cinder etc.)	Resource
	Properties	Several parameters used for resources, they vary depending on resource type	Relevant data
	Config	Element configuration script	Process data and route activity
	Deployment	Deployment information and link among config and resources	Relevant data, resource, deploy activity and dependencies

Hence, the HOT definition in the following,

It should be mentioned that this is not enough in OFL to reproduce all the features of an HOT template. The remaining information (i.e. association of activities, services with resources, deployment and configuration, etc.) must be included in the Orchestration process in proper activities (see [19] for further description of OFL). We show the remaining characteristics of OFL in the next section (Table 2).

Table 2. HOT template and OFL

Hot Template	OFL
parameters:	<Participants>
key_name:	<Resource Id="001" Name ="myinstance" type="compute" address="">
type: string	<param name="resourcTtype" type="string" value ="openstack" />
label: Key Name	<param name="resourceOS" type="string" value ="NovaServer" /> <input name="resourceImage" type="idref" value="d001" />
imageId:	</Resource>
type: string	...
label: Image ID	<RelevantData>
flavor:	<Data Id="d001" name="imageId value = "" type="string" interaction="input"/>
type: string	...
label: Instance Type	</RelevantData>
	...
resources:	
my_instance:	
type: OS::Nova::Server	
properties:	
key_name: {get_param: key_name}	
image: {get_param: image_id}	
flavor: {get_param: flavor}	
...	

4 An Application Example

We present here an application example where some Internet of Things (IoT) devices collect data from sensors, and asynchronously invoke a Store service on a set of resources implementing an *Hadoop as a Service* (*HaaS Store*) storage and analysis system. Another service (called *Analysis*), asynchronously with the other services, execute analytics on collected data. Data storage needed after a filtering process, implemented by a proper service (*Stream Filter*). Obviously, the HaaS service has to be deployed first (*Deploy HaaS*). Events that activate Filtering and Analysis are called here $Event : Sensor\_Input$ and $Event : Do\_Analysis$, respectively. Obviously, while the filtering and analysis services can be hosted on two different servers (**C1** and **C3** Compute nodes), HaaS and Analysis servers must share the same storage node **S1**. HaaS service is then installed on the compute node **S1**.

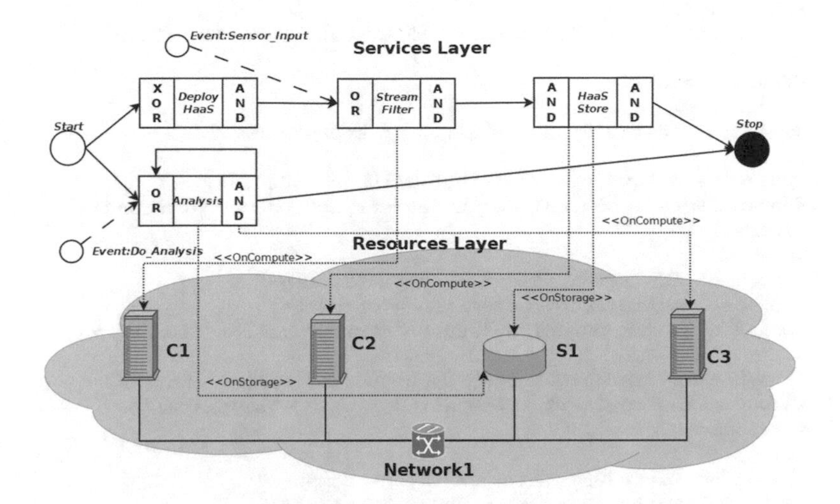

Fig. 2. An example

Figure 2 depicts the whole OFL process, where the top of the figure shows the services layer, and the bottom shows the resource layer. Proper transitions connect activities (in the boxes), defining precedences in the workflow of services. Dotted, stereotyped lines report connections between services and their resources.

Notice that this orchestrated process cannot be described by using HOT and Heat alone, since OpenStack Orchestration language is not able to define workflow services (e.g., it cannot describe precedences among filtering and storing services).

Data routing and some minor details are not reported for brevity.

The OFL description of this process has three main components: the first one describes activities, the second one describes resources, and the third one defines the relationships between activities, services and resources. The second and the third parts enables the creation of an HOT template for the process to implement on OpenStack.

The whole OFL process is too long to be reported here, but we report here for a flavour, some sketches of the OFL in order to show how the OrCS can generate and manage Heat Templates. Table 3 shows part of the OFL process.

The OrCS uses information stored in this process definition, in order to generate, at resource level, the HOT template in Table 4.

Table 3. OFL process

```
<Participants>
<Participant Id="001" Name ="MyProvider" type="provider" ...
/>
</Participants>
<Resources>
<Resource Id="r002" Name="virtNet" type="network" >
...
<param name="gw" type="string" value="10.224.84.1"/>
<param name="resourceType" type="string" value="neutron"/>
</Resource>
...
<Resource Id="r003" Name="S1" type="storage"
address="public:auto" networkRef="virtNet">
<param name="resourceType" type="string" value="cinder"/>
...
<input name="resourceImage" type="idref" value="image002"/>
<input name="keyPair" type="string" value="keypair01"/>
</Resource>
...
<Resource Id="r006" Name="C2" type="compute"
address="public:auto" networkRef="virtNet">
<param name="resourceType" type="string" value="nova"/>
...
<input name="resourceImage" type="idref" value="image001"/>
<input name="keyPair" type="string" value="keypair01"/>
<input name="OsUserName" type="idref" value="user001"/>
...
</Resource>
</Resources>
<RelevantData>
<Data Id="image0001" name="OsOnCompute" interaction="input"
value="novaTomcat" type="flavor"/>
<Data Id="user001" name="OpenStackUserName"
interaction="input" type="string"/>
<Data Id="C3PrivIP01" name="C3PrivIP" interaction="output"
type="string"/>
...
</RelevantData>
...
<Activities>
<Activity Id="A002" name="HaaSStore">
...
</Activity>
<Activity Id="A002" name="Analysis">
...
</Activity>
<ResourceLinks>
<OnCompute ActivityID="A003" ResourceID="r004"/>
<OnStorage ActivityID="A003" ResourceID="r003"/>
<OnStorage ActivityID="A002" ResourceID="r003"/>
<OnParticipant ResourceID="r002" ParticipantID="001"/>
...
</ResourceLinks>
```

Table 4. HOT template from OFL process

```
heat_template_version: 2017-12-23
parameters:
keypair_name:
type: string
label: SSH Keypair
flavor_name:
...
OS_user001:
type: string
OS_pass001:
type: string
hidden: true
OS_TENANT_NAME:
type: string
resources:
network:
type: OS::Neutron::Net
subnet:
type: OS::Neutron::Subnet
properties:
network_id: { get_resource: network }
ip_version: 4
cidr: 10.224.84.0/24
...
server:
type: OS::Nova::Server
properties:
key_name: { get_param: keypair_name }
image: 96c7fbd0-08c1-4afd-87bb-329e69c54f34
networks:
- network: { get_resource: network }
user_data_format: RAW
user_data:
str_replace:
template:
...
cat > /tmp/keystone.rc <<"EOF"
OS_TENANT_NAME=OS_TENANT_NAME
OS_user001=OS_user001
OS_pass001=OS_pass001
EOF
params:
OS_TENANT_NAME: { get_param: OS_TENANT_NAME }
OS_user001: { get_param: OS_user001 }
OS_pass001: { get_param: OS_pass001 }
...
outputs:
...
```

5 Conclusions and Future Works

This work describes the OrCS overall architecture: a framework for Orchestrate Cloud Composite Services at different Layers of Cloud Architecture. OrCS provides a Workflow language (OFL) that can be easily generated from Pattern-based description of composite services. In addition, its inner representation (the OFG graph) allows for analysis of composition. At the moment OrCS supports Openstack and HEAT, but future works include the management of more Cloud middleware in order to allow OrCS to be integrated in many multi-cloud and federated environments.

References

1. Ranjan, R., Buyya, R., Nepal, S., Georgakopulos, D.: A note on resource orchestration for cloud computing. Pract. Exp. Concurr. Comput. **27**, 2370–2372 (2014)
2. Ranjan, R., Benatallah, B., Dustdar, S., Papazoglou, M.P.: Cloud resource orchestration programming: overview, issues, and directions. IEEE Internet Comput. **19**(5), 46–56 (2015)
3. Verma, A., Kaushal, S.: Deadline constraint heuristic-based genetic algorithm for workflow scheduling in cloud. Int. J. Grid Util. Comput. **5**(2), 96–106 (2014)
4. Ye, X., Khoussainov, B.: Fine-grained access control for cloud computing. Int. J. Grid Util. Comput. **4**(2–3), 160–168 (2013)
5. VV.AA. Us government cloud computing technology roadmap release 1.0 (draft). In Special Publication 500–293, vol. 2, pp. 1–85. NIST (2011)
6. Wilder, B.: Cloud Architecture Patterns: Using Microsoft Azure. O'Reilly Media Inc., Newton (2012)
7. Fehling, C., Leymann, F., Rütschlin, J., Schumm, D.: Pattern-based development and management of cloud applications. Futur. Internet **4**(1), 110–141 (2012)
8. Jula, A., Sundararajan, E., Othman, Z.: Cloud computing service composition: a systematic literature review. Expert Syst. Appl. **41**(8), 3809–3824 (2014)
9. Gutierrez-Garcia, J.O., Sim, K.M.: Agent-based cloud service composition. Appl. Intell. **38**(3), 436–464 (2013)
10. Feng, G., Buyya, R.: Maximum revenue-oriented resource allocation in cloud. Int. J. Grid Util. Comput. **7**(1), 12–21 (2016)
11. Liu, C., Loo, B.T., Mao, Y.: Declarative automated cloud resource orchestration. In: Proceedings of the 2nd ACM Symposium on Cloud Computing, p. 26. ACM (2011)
12. Barolli, L., Xhafa, F.: JXTA-overlay: a P2P platform for distributed, collaborative, and ubiquitous computing. IEEE Trans. Ind. Electron. **58**(6), 2163–2172 (2011)
13. Xhafa, F., Fernandez, R., Daradoumis, T., Barolli, L., Caballé, S.: Improvement of JXTA protocols for supporting reliable distributed applications in P2P systems. In: Network-Based Information Systems, pp. 345–354. Springer, Heidelberg (2007)
14. Spaho, E., Mino, G., Barolli, L., Xhafa, F.: Goodput and PDR analysis of AODV, OLSR and DYMO protocols for vehicular networks using CAVENET. Int. J. Grid Util. Comput. **2**(2), 130–138 (2011)
15. French, T., Bessis, N., Xhafa, F., Maple, C.: Towards a corporate governance trust agent scoring model for collaborative virtual organisations. Int. J. Grid Util. Comput. **2**(2), 98–108 (2011)

16. Barolli, L., Xhafa, F., Durresi, A., De Marco, G.: M3PS: a JXTA-based multi-platform P2P system and its web application tools. Int. J. Web Inf. Syst. **2**(3/4), 187–196 (2007)
17. Tricomi, G., Panarello, A., Merlino, G., Longo, F., Bruneo, D., Puliafito, A.: Orchestrated multi-cloud application deployment in OpenStack with TOSCA. In: 2017 IEEE International Conference on Smart Computing (SMARTCOMP), pp. 1–6. IEEE (2017)
18. Fehling, C., Leymann, F., Retter, R., Schupeck, W., Arbitter, P.: Cloud computing patterns (2014)
19. Amato, F., Moscato, F.: Exploiting cloud and workflow patterns for the analysis of composite cloud services. Futur. Gener. Comput. Syst. **67**, 255–265 (2017)
20. Amato, F., Moscato, F.: Pattern-based orchestration and automatic verification of composite cloud services. Comput. Electr. Eng. **56**, 842–853 (2016)
21. Amato, F., Moscato, F.: A model driven approach to data privacy verification in e-health systems. Trans. Data Priv. **8**(3), 273–296 (2015)
22. Amato, F., Barbareschi, M., Casola, V., Mazzeo, A., Romano, S.: Towards automatic generation of hardware classifiers. In: International Conference on Algorithms and Architectures for Parallel Processing, pp. 125–132. Springer, Cham (2013)
23. Dustdar, S., Schreiner, W.: A survey on web services composition. Int. J. Web Grid Serv. **1**(1), 1–30 (2005)
24. Di Lorenzo, G., Mazzocca, N., Moscato, F., Vittorini, V.: Towards semantics driven generation of executable web services compositions. Int. J. Softw. JSW **2**(5), 1–15 (2007)
25. Di Lorenzo, G., Moscato, F., Mazzocca, N., Vittorini, V.: Automatic analysis of control flow in web services composition processes. In: PDP, pp. 299–306 (2007)
26. Wang, X., Huang, D., Akturk, I., Balman, M., Allen, G., Kosar, T.: Semantic enabled metadata management in PetaShare. Int. J. Grid Util. Comput. **1**(4), 275–286 (2009)
27. Xue, T., Ying, S., Wu, Q., Jia, X., Hu, X., Zhai, X., Zhang, T.: Verifying integrity of exception handling in service-oriented software. Int. J. Grid Util. Comput. **8**(1), 7–21 (2017)
28. Sawamura, S., Barolli, A., Aikebaier, A., Takizawa, M., Enokido, T.: Design and evaluation of algorithms for obtaining objective trustworthiness on acquaintances in P2P overlay networks. Int. J. Grid Util. Comput. **2**(3), 196–203 (2011)
29. Moscato, F., Aversa, R., Amato, A.: Describing cloud use case in MetaMORP(h)OSY. In: IEEE Proceedings of CISIS 2012 Conference, pp. 793–798 (2012)
30. Moscato, F., Amato, F., Amato, A., Aversa, R.: Model-driven engineering of cloud components in MetaMORP(h)OSY. Int. J. Grid Util. Comput. **5**(2), 107–122 (2014)
31. Amato, F., Moscato, F., Xhafa, F.: Multi-level orchestration of cloud services in OrCS. In: International Conference on P2P, Parallel, Grid, Cloud and Internet Computing, pp. 357–366. Springer, Cham (2017)
32. Falkenthal, M., Barzen, J., Breitenbücher, W., Fehling, C., Leymann, F.: From pattern languages to solution implementations. In: The Sixth International Conferences on Pervasive Patterns and Applications, PATTERNS 2014, pp. 12–21 (2014)
33. Microsoft Dev Net. Cloud design patterns: prescriptive architecture guidance for cloud applications (2014)
34. Kumar, R., Gupta, N., Charu, S., Jain, K., Jangir, S.K.: Open source solution for cloud computing platform using OpenStack. Int. J. Comput. Sci. Mob. Comput. **3**(5), 89–98 (2014)

Improving Results of Forensics Analysis by Semantic-Based Suggestion System

Flora Amato[1(✉)], Leonard Barolli[2], Giovanni Cozzolino[1], Antonino Mazzeo[1], and Francesco Moscato[3]

[1] DIETI, Universitá degli Studi di Napoli "Federico II", via Claudio 21, Naples, Italy
{flora.amato,giovanni.cozzolino,mazzeo}@unina.it
[2] Fukuoka Institute of Technology (FIT), Wajiro-higashi, Higashi-ku, Fukuoka, Japan
barolli@fit.ac.jp
[3] Universitá degli Studi della Campania, viale Ellittico 31, Caserta, Italy
francesco.moscato@unicampania.it

Abstract. Nowadays, more than ever, digital forensics activities are involved in any criminal, civil or military investigation and they are primary to support cyber-security. Detectives use a many techniques and proprietary forensic software to analyze (copies of) digital devices, in order to discover hidden, deleted, encrypted, and damaged files or folders. Any evidence found is carefully analysed and documented in "finding reports" that are used during lawsuits. Forensics aim at discovering and analysing patterns of fraudulent activities. In this work, we propose a methodology that supports detectives in correlating evidences found by different forensic tools and we apply it to a framework able to semantically annotate data generated by forensics tools. Annotations enable more effective access to relevant information and enhanced retrieval and reasoning.

1 Introduction

Cyber-security is becoming a hot topic in social, political, industrial and research fields. Information Technology (IT) systems are involved in almost all daily activities, related to business and industrial purposes, safety systems, education or entertainment, etc. [19]. The growing use of digital technologies increases the chances that a device should be the means or the witness of a crime. Thus we can use data generated, stored or transmitted by a digital device as evidences in a trial.

Birth and evolution of Computer Forensics is strictly related to the progress of information and communication technology. The development of new technologies promoted many changes in methodologies adopted to notice, collect, manage e analyse information during investigations (e.g. digital footprints, in a lawsuit [26]). The intangibility and volatility of these information, with the growing of volumes of data and the heterogeneity of digital media [20], complicate the work of Digital Forensic experts, because the manipulation and processing of data is related to a logical-inferential reasoning process, typical of scientific

L. Barolli et al. (Eds.): EIDWT 2018, LNDECT 17, pp. 956–967, 2018.
https://doi.org/10.1007/978-3-319-75928-9_88

research and forensic analysis. Moreover, often, the big amount of information has to be correlated with similar non-digital evidences, usually in a really short time.

According to ISO/IEC 27037, a digital evidence is every kind of information that can be stored or transmitted in a digital form and that can be considered an evidence. The goal of Digital Forensics is not only collection, acquisition and documentation of data stored on digital devices, but, above all, it is the interpretation of evidences. Notice that correlation of information is a crucial phase in forensics analyses, because examiner must take into account of all kinds of information (considering the broadest meaning of term), such as context of investigation, acquired clues, investigative hypothesis, and many other elements, not necessarily in digital form. Correlation of these information is the only mean to allow for the contextualization of digital evidences, promoting them as clues.

According to previous considerations, experts have to face many challenges during the different phases of forensic investigations:

- the admissibility of acquired data must be preserved: this assumes that examiner will not alter data avoiding to alter its probative value.
- the complexity of data, in terms of volume and heterogeneity, requires sophisticated data reduction computation;
- the lack of standard format for file produced by different practices and tools usually restricted to particular subset of activities, leads to consistency and correlation problems;
- the absence of tools that assist in integration of data gathered from heterogeneous sources, such as hard disks, memory, etc., in order to find traces and to reconstruct the events.
- the unification of timeline events from multiple sources or devices.

All these correlation problems reflect on the presentation phase too: since results of investigation have to be presented to a court's jury avoiding misunderstanding or ambiguities, the examiner have to agree on the set of terms and definitions with multiple parties involved in the analysis process, due to their different levels of expertise.

In this work we present a methodology to enrich the analysis and correlation process of a forensics investigation through the adoption of Semantic Web technologies. The addition of semantic assertion to data generated during various analysis phases, should improve the presentation results, enabling more accurate correlation of traces and more powerful searches.

The paper is structured as follows: Sect. 2 presents theoretical background in the area of Digital Investigations discussing basic principles of digital evidence and presenting some challenges in the field; Sect. 3 discusses the motivation for adopting a semantic-based approach, providing a high-level view of how these two disciplines can be combined with advantages; Sect. 4 presents the proposed methodology; in Sect. 5 we describe the system architecture we propose to implement the methodology; in Sect. 6 we present some results obtained in a simulated case of a digital investigation examination; in Sect. 7 there is a short presentation of related work that has also merged partially or fully these two areas; finally, in Sect. 8 we present some conclusions remarks.

2 Digital Investigations and Digital Evidences

Digital evidence is the main element of any forensics process. Literature offers many definitions of digital evidence, such as:

- any data stored or transmitted using a computer that support or refute a theory of how an offence occurred or that address critical elements of the offence such as intent or alibi [6]
- information of probative value stored or transmitted in digital form. [22]
- any digital data that contain reliable information that supports or refutes a hypothesis about an incident [5]

The intrinsic weakness of digital evidences make them easy prone to alterations or modifications, even from examiners that, if not experienced, may compromise and contaminate the *scena criminis* status [17]. Thus, its rigorous and authenticated management is of extremely importance.

In [25] author has identified three basic properties of digital evidence, namely fidelity, volatility and latency.

1. *Fidelity* refers to Chain of Custody forensic principle, and involves the adoption of techniques that grant the integrity of evidences through the process;
2. *Volatility* is related to the nature of the support were evidences are stored (disk, memory, registers, etc.): this property affects considerably the acquisition and analysis of evidences;
3. *Latency* involves the presence of additional information to contextualize and interpret a digital encoding. The focus of current work is on this property.

A Digital Forensic Investigation has to follow a set of principles and procedures, regulated by Budapest Convention on Cybercrime, to correctly manage the life-cycle of digital evidence, from acquisition to analysis and presentation [13,14].

This regulation can be considered a first set of *best practices* consisting the fundamental principles of Digital Forensics. It provide for adequate technical measures to adopt in order to guarantee a correct conservation and integrity of acquired data.

Main phases of a Digital Forensic Investigation can be summarised in the following:

- Identification: it consists in the choice of the devices or the systems to be examined;
- Preservation: it requires the preservation and the isolation of the examined evidence from external agents or, in general, the outside world;
- Acquisition: it is the most critical phase, since it must guarantee that the evidence doesn't experience modifications or alterations;
- Analysis and Correlation: despite of acquisition phase, characterized by the presence of standard and wide adopted procedures, the analysis phase strictly depends on the scenario and on requirements for searches;
- Documentation: it represents the conclusion of all the investigation process, defining and formalizing the examiners work, describing all the operation conducted.

3 Semantic Approach Advantages

Acquisition and analysis are critical phases in digital investigations: the heterogeneity of support that can store digital evidences, in addition to the technological evolution and the wide range of investigation scenarios, doesn't allow the identification of a unique procedure for digital evidences acquisition. Forensic tools play a major role in this phase, since they adapt acquisition processes to different digital devices (e.g. hard disks, USB flash drives, mobile devices, IDS Firewall logs, memory, etc.) [4]. The specialized and evidence-oriented design of forensic tools produces acquired data in different format and representations, that get more difficult the analysis process and so calls for advanced interoperability techniques for evidences correlation.

The development of automatic processes that assist detectives in data acquisition and analysis phases simplifies their work, especially when dealing with large amount of data. Currently, the most of forensic tools work on plain text data which does not allow advanced analysis processes. A semantic approach, eventually based on ontologies, use a unique and reliable representation of domain concepts, enabling data structuring on one hand, and standardised representation of data on the other hand.

Unlike structured data formats (i.e. relational databases), ontologies analysis tools easily make inference of new information, checking knowledge consistency, etc. This because an ontology explicitly represents relationships between entities.

The methodology proposed in this work improves, through semantic technologies, all the main phases of a digital investigation, with respect of evidence discovery, integrity and correlation. It can provide a framework able to describe, in a more expressive and formalized way, the representation of a given scenario. Figure 1 shows relationships between adopted semantic technologies and phases of digital investigation.

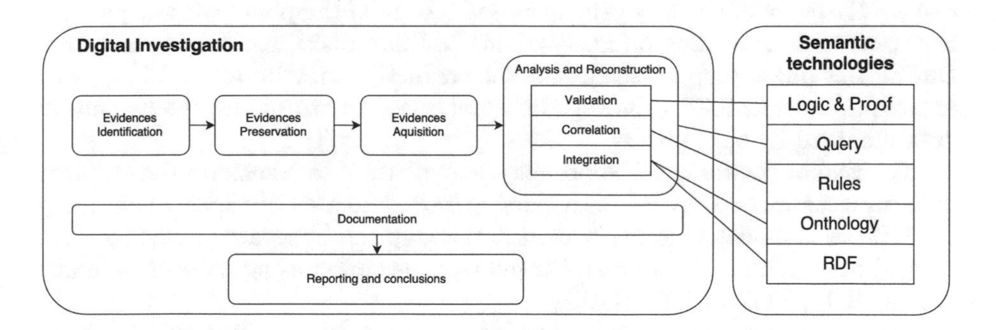

Fig. 1. Digital investigation phases and related semantic technologies.

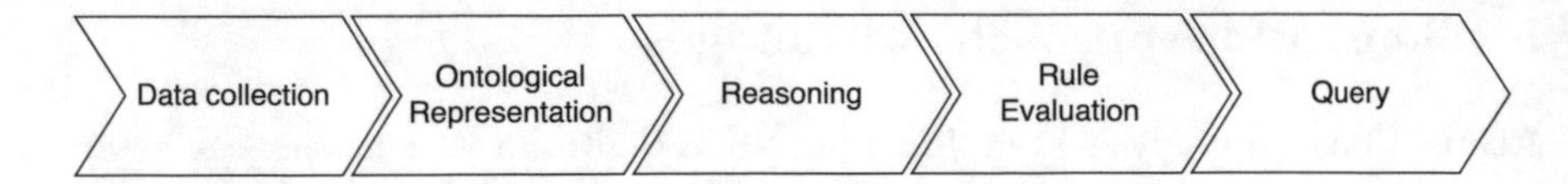

Fig. 2. Phases of the proposed methodology.

Main potential advantages of such an approach regard:

- Information Integration: the RDF data model simplify integration of data coming from multiple sources, due to its schema independence and standardized representation of knowledge in the subject-predicate-object form.
- Inference: the RDF/OWL combination can infer class membership and typing information from ontological definitions; a reasoner can them in order to infer dynamically class membership of the instances.
- Extensibility and Flexibility: RDF/OWL provide compatibility with data model of forensic tools input and output. OWL provides flexibility by defining custom ontologies according to the scope and integrating multiple existing ontologies through ontology mapping processes.
- Search: queries can be enhanced taking advantage of the reasoning engine and the semantic mark-up used along traditional keywords during document indexing.

4 Methodology

This section describes the proposed methodology for digital evidence integration and correlation. The methodology workflow is presented in Fig. 2.

The first "Data Collection" phase involves all the acquisition operations and aims at generating inputs for next phases. During this phase the examiner must respect the chain of custody principles and has to be compliant with acquisition best practices, depending on analysed media (hard disks, mobile devices, etc.). During this phase preprocessing and data reduction may be applied too, using techniques such as KFF (Known File Filter), in order to reduce the amount of data managed by next phases.

The goal of "Ontological Representation" phase is to transform the acquired data, by software or even hardware parsing tools, into a set of triplet constituting the RDF data model. Ontologies used in this step can be created ad-hoc or even fetched from shared repositories. Outputs can be stored using different formats, such as RDF/XML or RDF/OWL.

Ontological representation of data is processed during "Reasoning" phase, through an OWL-based reasoner that infers additional axioms on the basis of instances' relations. The reasoner can infer different types of new axioms, enriching the asserted instances with information regarding the definition of their class, properties or subclasses. Moreover, thanks to subclass hierarchy, property relations or property restrictions, reasoner can dynamically classify and correlate instances with higher precision compared to asserted data.

SWRL Rules can assert additional axioms that cannot be inferred through OWL. SWRL Rules are evaluated, during "Rule Evaluation" phase, by a rule engine in order to insert newly inferred axioms into the ontology. This operation is realized with the support of external Rule Engines that translate inferred axioms into RDF data model to permit ontology integration.

By using SPARQL language it is possible to query endpoint hosting for the sets of triplets constituting asserted and inferred axioms. This implements the "Query" phase.

5 System Architecture

In this section is presented an implementation of a system architecture of the proposed methodology, with a brief discussion of the tools and techniques used. Thanks to its flexibility, methodology can be implemented in different ways, so system architecture can be updated among the evolution of semantic technologies.

Our overall system consists of an ontology and five modules: Evidences Manager, Semantic Parser, Inference Engine, SWRL Rule Engine and a Query and Visualization module. An overview of the architecture is presented in Fig. 3. Its main components are further discussed below.

Evidence Manager

The evidence manager loads binary content of digital evidences, identifying the type of given source and verifying its integrity through hash values. This module provides tools to extract knowledge from a forensic image, like user files, browser history, Windows registry, etc. The extraction process uses forensics tools like Hachoir (for binary file manipulation) and Plaso (for timeline creation). The knowledge extracted consists of a set of attributes, including temporal information (date, time and timezone), a description of the information source (source and source type), a description of the event. Many of these fields are structured and ready to be used as instances attributes; instead, some of them, like description of the event, are non-structured fields and they require further processing based on regular expression or NLP techniques. The output of this module is a file containing all the footprints retrieved from the disk image.

Semantic Parser

Semantic parser module generates an OWL representation of knowledge extracted from digital evidence in the previous step. This module instantiate the ontology, combined from public domain ones, if available, or custom ones. Ad-hoc ontologies can be created to integrate all the referenced domain ontologies or to define new classes, additional restrictions, new object properties, etc. Ontology integration also promotes reuse of entities defined in other ontologies and increases system flexibility, since domain ontologies may not be easily modifiable.

Ontology population is made by creating an instance for each footprint item of acquired data, and by linking them each other according to formal object properties defined in the ontology schema.

Inference Engine

Inference engine performs automated reasoning, according to the OWL specifications, coming from domain ontologies or investigators knowledge. Reasoners can specify the granularity of inferences to be made, such as hierarchy relationship (an individual that is a member of a subclass is also a member of the parent class), or the generation of inverse object properties. Such kind of inferences can improve the performance of query execution. The goal of this step is to enrich the knowledge base with new inferred facts, increasing the examiner's knowledge on evidence. For example it is possible to infer a file type from its extension or to deduce the author of an action from user information in its active account.

SWRL Rule Engine

SWRL rules play a major role in the automated integration and correlation parts of the methodology. SWRL Rule Engine adopts SWRL rules in order to correlate different individuals or to establish relationships among individuals belonging to different ontologies but representing similar concepts. Saving SWRL rules in a separate text file promotes the decoupling and enables rules reuse. This module can be called before the reasoning engine make it able to process the axioms that can only be inferred by the SWRL rules, or you may execute the reasoning step twice, one before the rule engine and the other after it.

SPARQL Queries

The final module is responsible for accepting SPARQL queries from the user and evaluating them against a SPARQL query engine. The set of RDF triples that have been either asserted during the semantic parsing of the source data or inferred by the reasoning or the rule engine are loaded in-memory and SPARQL queries can be evaluated against it. Once more, the queries can be saved in separate files promoting reuse and decoupling.

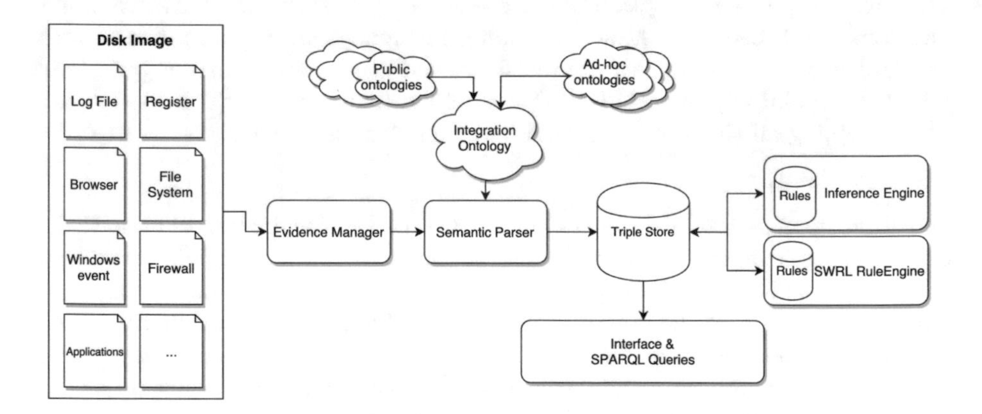

Fig. 3. System architecture for proposed methodology.

6 Experimentation and Results

This section aims at demonstrating the capabilities of our approach in the context of a digital investigation examination. For this purpose, we perform an analysis of effectiveness and efficiency in a simulated cases. We evaluate recall and precision of our system on query results, asking for a set of given evidences considered as ground truth.

The configuration of the machine used to run the experiment and hosting the triple store (Stardog) has a 3.20 GHz Intel Core i5-6400 processor and 8 GB RAM.

For our experimental campaign, we generated three disk image from a virtual machine running Windows 7. On these machines we performed a set of user actions to simulate a malicious behaviour caused by a malware.

After successfully completing the data collection phase, the ontological representation phase and the enhancement (reasoning and rule evaluation) phase, the investigator starts the analysis by searching all significant correlations between events. To search for all traces of potentially suspicious executables is possible to execute a SPARQL query to retrieve all executables. Among the results of the query can be identified entries with a suspicious name and then can be figured out who interacts with this malware and in which circumstances.

Table 1 provides information on the volumes with the number of entries extracted by the Evidence Manager, the number of triples generated during the enhancement phase and finally the number of significant correlations found. Table 2 gives the execution times (in seconds) for the different phases of the process.

The analysis phase is a time-consuming step that requires pre-filtering of non-critical information to reduce the amount of data to analyse. Moreover we

Table 1. Volumes of processed data through the reconstruction and analysis process

Criterion	Dataset No1	Dataset No2	Dataset No3
Extracted entries	6798	4263	15983
Generated triples	8786	10475	32498
Deduced triples	18	21	31
Correlations	1231	535	14982

Table 2. Volumes of processed data through the reconstruction and analysis process

Steps	Dataset No1	Dataset No2	Dataset No3
Collection	1.07	1.43	1.67
Representation	24.37	18.3	167.45
Reasoning	0.29	0.247	0.314
Analysis	776.3	241.3	12314.7

can see that the number of triples does not increase linearly with the number of the acquired evidences. It depends on the type of each evidence which can carry more or less information and therefore require a larger or smaller number of triples to be modelled in the ontology.

7 Related Work

Many approaches adopting Semantic Web technologies have been proposed in literature. However, although correlation has been described as one of the most important goal of the analysis phase, current support for correlation analysis is poor among forensic tools.

DFXML (Digital Forensics XML) [11] is an XML-based approach, providing a standardized XML vocabulary for representing meta-data of evidences, through various forensics-related XML tags such as `<source>`, `<imagefile>`, `<acquisition_date>`, `<volume>`, etc. Another XML-based approach is described in [18], where authors proposed the DEX (Digital Evidence Exchange) format. DEX can also adopt an XML-based representation of the output of various forensics tool, but with an additional capability of tracing the specific instructions given and the sequential order of the tools used. Although these projects promote a standardized format and tool interoperability, they are missing to convey the semantic content of what they represent.

RDF-based approaches have found limited usage in the area of digital investigations, despite their ability to express arbitrary meta-data. RDF makes an important improvement due to the support of creating attributes with standard or custom types instead of only string types. The most famous adoption of RDF is in the AFF4 forensic format project [9,10]. The Advanced Forensic Format (AFF) is a file format able to contain and store digital evidence. AFF includes both a copy of the original acquired data as well as arbitrary meta-data, such as system-related information or user-specified ones [21]. To improve the chain of custody, in [12] author introduces into AFF4 domain-specific concepts, such as examiner information, evidence access information [7,16], etc.

An ontology-based approach is FORE project. FORE (Forensics for Rich Events) architecture [24] is composed of a forensic ontology, an event log parser and a custom-defined rule language, FR3. The forensic ontology is based on two main concepts that represent tangible objects and their state changes over time. Rules expressed in FR3 language are evaluated against the knowledge base in order to add new causality links between instances. By taking advantage of the OWL capabilities of referencing external ontologies, the authors are able to express correlation rules that combined concepts and events from disparate domains [23]. In [15] authors use a blank ontology for encoding results retrieved from forensic tools emphasising relevant types of data and their relationships. Using an ontology query language (SQWRL) they extract additional information of probative value. In [8] authors introduce an approach based on a three-layered ontology, called ORD2I, to represent any digital events. ORD2I is associated with a set of operators to analyse the resulting timeline and to ensure the reproducibility of the investigation. Recent approaches uses modeling [2,3] and verification [1] systems in order to test the soundness of the results and data processing performances.

8 Conclusions and Future Work

In this work we propose the adoption of semantic technologies to support digital forensics investigators. The main goal is to take advantage of heterogeneous data integration, to support for automation and to improve analytical capabilities in the form of an expressive and flexible querying layer.

In this paper, we propose a reusable methodology based on semantic representation, integration and correlation of digital evidence and an architecture that implements it. The approach is based on ontologies defining domain of digital incidents and on a set of tools for extracting information from disk image, populating the ontology, inferring and analysing new facts. The use of an ontology allows for the representation of knowledge with a unified model and for simplifying the building of analysis processes.

Despite the aforementioned advantages, the approach has still some limitations that will be studied in future works. Performances can be improved, especially for what the execution time of the analysis phase concerns. Regarding the extraction phase, additional sources have to be integrated to reach a deeper analysis of an incident.

In future work, we plan to improve extraction phase by adding more information sources related to additional devices, such as Android and iOS. For the analysis, we plan to integrate new tools for event correlation based on pattern matching algorithm to detect illegal actions by identifying specific event sequences. Concerning the interface and query layer, will be proposed some enhancement regarding a graph visualization tool that easily show instances correlations.

In conclusion, we have seen that through the proposed approach a digital forensics examiner is able to easily extract the knowledge related to a digital incident. This knowledge can be enriched using a semantic representation that alleviate heterogeneity problem of forensic tools outputs. The reasoning process allows to deduce new knowledge from the existing one, improving analytical skills by allowing easy and fast ways to integrate and correlate forensic evidences. The use of a SPARQL interface allow to understand the interactions between facts.

References

1. Amato, F., Moscato, F.: Pattern-based orchestration and automatic verification of composite cloud services. Comput. Electr. Eng. **56**, 842–853 (2016)
2. Amato, F., Moscato, F.: Exploiting cloud and workflow patterns for the analysis of composite cloud services. Future Gener. Comput. Syst. **67**, 255–265 (2017)
3. Amato, F., Moscato, F., Moscato, V., Colace, F.: Improving security in cloud by formal modeling of IaaS resources. Future Gener. Comput. Syst. (2017)

4. Cahyani, N.D.W., Martini, B., Choo, K.-K.R., Al-Azhar, A.K.B.P.: Forensic data acquisition from cloud-of-things devices: windows smartphones as a case study. Concurrency Comput. Pract. Experience **29**(14), e3855 (2016)
5. Carrier, B., Spafford, E.H.: An event-based digital forensic investigation framework. In: Digital Forensic Research Workshop, pp. 11–13 (2004)
6. Casey, E.: Digital Evidence and Computer Crime: Forensic Science, Computers, and the Internet, 3rd edn. Academic Press, London (2011)
7. Castiglione, A., Cattaneo, G., De Maio, G., De Santis, A., Costabile, G., Epifani, M.: The forensic analysis of a false digital alibi. In: 2012 Sixth International Conference on Innovative Mobile and Internet Services in Ubiquitous Computing (IMIS), pp. 114–121. IEEE (2012)
8. Chabot, Y., Bertaux, A., Nicolle, C., Kechadi, T.: An ontology-based approach for the reconstruction and analysis of digital incidents timelines. Digital Invest. **15**, 83–100 (2015)
9. Cohen, M., Schatz, B.: Hash based disk imaging using AFF4. Digital Invest. 7(Suppl.), S121–S128 (2010). The Proceedings of the Tenth Annual DFRWS Conference
10. Garfinkel, S.L.: AFF: a new format for storing hard drive images. Commun. ACM **49**(2), 85–87 (2006)
11. Garfinkel, S.L.: Automating disk forensic processing with Sleuthkit, XML and Python. In: Fourth International IEEE Workshop on Systematic Approaches to Digital Forensic Engineering, SADFE 2009, pp. 73–84. IEEE (2009)
12. Giova, G.: Improving chain of custody in forensic investigation of electronic digital systems. Int. J. Comput. Sci. Netw. Secur. **11**(1), 1–9 (2011)
13. Horng, S.-J., Rosiyadi, D., Fan, P., Wang, X., Khan, M.K.: An adaptive watermarking scheme for e-government document images. Multimedia Tools Appl. **72**(3), 3085–3103 (2014)
14. Horng, S.-J., Rosiyadi, D., Li, T., Takao, T., Guo, M., Khan, M.K.: A blind image copyright protection scheme for e-government. J. Vis. Commun. Image Represent. **24**(7), 1099–1105 (2013)
15. Kahvedžić, D., Kechadi, T.: Semantic modelling of digital forensic evidence. In: International Conference on Digital Forensics and Cyber Crime, pp. 149–156. Springer (2010)
16. Kang, A., Lee, J.D., Kang, W.M., Barolli, L., Park, J.H.: Security considerations for smart phone smishing attacks. In: Advances in Computer Science and Its Applications, pp. 467–473. Springer (2014)
17. Khan, B., Alghathbar, K.S., Nabi, S.I., Khan, M.K.: Effectiveness of information security awareness methods based on psychological theories. Afr. J. Bus. Manage. **5**(26), 10862 (2011)
18. Levine, B.N., Liberatore, M.: DEX: digital evidence provenance supporting reproducibility and comparison. Digital Invest. **6**, S48–S56 (2009)
19. Liu, W., Srivastava, S., Lu, L., O'Neill, M., Swartzlander, E.E.: Are QCA cryptographic circuits resistant to power analysis attack? IEEE Trans. Nanotechnol. **11**(6), 1239–1251 (2012)
20. Liu, W., Srivastava, S., ONeill, M., Swartzlander Jr., E.E.: Security issues in QCA circuit design-power analysis attacks. In: Field-Coupled Nanocomputing, pp. 194–222. Springer (2014)
21. Miguel, J., Caballé, S., Xhafa, F., Prieto, J., Barolli, L.: Towards a normalized trustworthiness approach to enhance security in on-line assessment. In: 2014 Eighth International Conference on Complex, Intelligent and Software Intensive Systems (CISIS), pp. 147–154. IEEE (2014)

22. Scientific Working Group on Digital Evidence (SWGDE) International Organization on Digital Evidence (IOCE). Digital evidence: Standards and principles
23. Schatz, B., Mohay, G., Clark, A.: Rich event representation for computer forensics. In: Proceedings of the Fifth Asia-Pacific Industrial Engineering and Management Systems Conference (APIEMS 2004), vol. 2, pp. 1–16. Citeseer (2004)
24. Schatz, B., Mohay, G.M., Clark, A.: Generalising event forensics across multiple domains. In: School of Computer Networks Information and Forensics Conference. Edith Cowan University (2004)
25. Schatz, B.L.: Digital evidence: representation and assurance. Ph.D. thesis, Queensland University of Technology (2007)
26. Seo, H., Liu, Z., Choi, J., Kim, H.: Multi-precision squaring for public-key cryptography on embedded microprocessors. LNCS (including subseries Lecture Notes in Artificial Intelligence and Lecture Notes in Bioinformatics), vol. 8250, pp. 227–243 (2013)

Modeling of Radio Base Stations with the Numerical FDTD Method, for the Electromagnetic Field Evaluation

Algenti Lala(✉), Bexhet Kamo, Joana Jorgji, and Elson Agastra

Polytechnic University of Tirana, Tirana, Albania
{alala, bkamo, jjorgji, eagastra}@fti.edu.al

Abstract. Settling down an efficient and reliable procedure for the evaluation of the EMF exposure, from the Base Station Antennas, is important for mobile communications. In this work a calculation method of the exposure under radiofrequency, due to the presence of some antennas of the cellular Base Stations is introduced. The model of wave diffusion in free space, under ideal conditions gives ground for a convenient calculation of the exposure, even in cases of a considerable distance from the antenna, whose covering area is considerably larger, thus resulting in overestimation of the exposure. The calculation of the electrical intensity of the radiation is possible when the technical specifications of the given antenna (provided by the manufacturer) are known and by defining the position of the given point in relation to the antenna.

1 Scope

Fast spreading of the radio communication systems, in particular mobile technology related, has introduced a major concern with regard to evaluating EMF exposure from Radio Base Station Antennas. The scope of this work is to present a calculating model to quickly evaluate both the electrical and magnetic fields of the radio base stations. This method exploits the traditional EM modeling of the “far-field” radiation patterns for RF antennas. The radiated field is estimated at most anywhere around the antenna using a large number of evaluating field component samples. The method used for simulations is of particular interest as it allows to verify the results obtained from practical measurements and to interpolate these results outside the allowed space where measurements cannot be performed.

2 Introduction

There are currently four mobile operators in Albania: Telekom AL, Vodafone AL, Eagle Mobile and PLUS Communications. To evaluate the exposure and compare it with the safety limits two of the following options are possible:

- Using the calculating capabilities to simulate the wave propagation in the space of interest and calculate the E and H field values or the power density S.
- Performing measurements with the appropriate device and a reliable methodology.

L. Barolli et al. (Eds.): EIDWT 2018, LNDECT 17, pp. 968–977, 2018.
https://doi.org/10.1007/978-3-319-75928-9_89

Modeling of the radio base station antennas has been studied in free space and in the urban environment for the near field [1, 12–14]. In the far field region, which is defined as "the antenna's field region where the wave distribution is essentially independent of the distance from the antenna" [2], calculating the electromagnetic field is relatively simple as the required information is given in the theoretical antenna pattern and the radiated power. The "rigorous" numerical technique, the finite-difference time-domain (FDTD) method, also implemented in MATLAB, is widely used for the study of the field in vicinity and distant from the antenna. The problem generally associated with the application of this method is the antenna geometry recognition of the radio base stations [12].

For the electromagnetic field evaluation in a random point of the area close to the base station a numerical model is used, based on the FDTD and emphasize is given to a simple yet suitable calculation method for the radio frequency estimation. Evaluations of the EM field anywhere around the antenna from a considerable number of samples is then compared with the measured values performed in the same points in space using the NARDA SRM 3000 m. These values are gathered in planar surfaces of the far field region.

Interpretation of the results is given in the following framework:

- How much the results obtained from the evaluation model are approximated with the measurements performed.
- How and where to apply and the accuracy of the proposed model.

3 Simulation of Antenna Characteristics for the Proposed Calculation Model

Because of the lengthy shape of the base station cellular antenna a cylindrical geometry is generally more convenient to minimize the dimensions and timing of the assessed object. The NF-FF (near field to far field) transformation used in this work is grey color represented in Fig. 1. Both tangential components of the electrical field are assembled in a cylindrical surface. The spectrum of the cylindrical wave is estimated by the Fast Fourier Transform (FFT) and the far field model is simply its planar representation. From the radiation of the planar wave spectrum, it is possible to evaluate the electrical and magnetic fields in planar surfaces generally anywhere around the antenna. To calculate the electrical intensity of the radiated field from the antenna the far field equation is applied. What is further proposed consist in the following:

1. Using the theoretical method of choice, the intensity values of the radiated field are determined with respect to the distance, in the X and Y axis and the dependence is graphically represented.
2. Using the NARDA SRM 3000 selective radiation meter, the intensity values of the electrical field are measured along the distance and graphically represented in the same coordinate axis of the chart as in point 1 above.

The proposed method allows to make quick assessments of both near and far fields by simplifying the complex calculations. All four of the public mobile operators

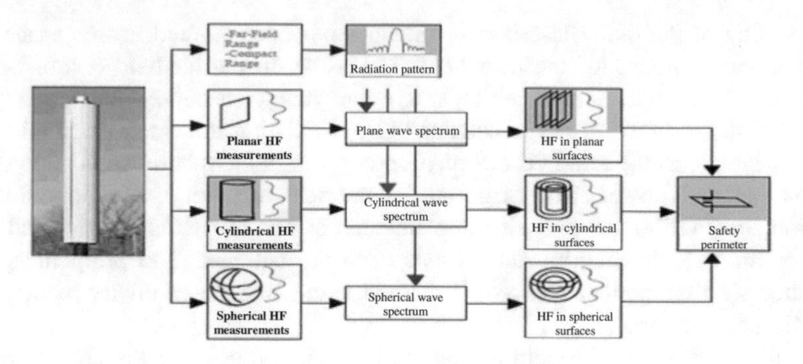

Fig. 1. Near field to far field transformation.

provide voice and data services commonly using the same intelligent antennas in the 2100 MHz and the 900 and 1800 MHz frequency bands, to support 3G and 2G services, respectively.

The Kathrein antennas [3] of model types 741344, 80010291, 80010292, 80010492, 80010670, 80010671, 80010672 are referred. Figure 2 depicts the scenario with three mobile operators in the area (with three antennas each) and a midpoint in space with cartesian coordinates of (X_{P1}, Y_{P1}, Z_{P1}) (X_{P2}, Y_{P2}, Z_{P2}) and (X_{P3}, Y_{P3}, Z_{P3}). Each operator uses the same antenna in 900/1800 and 2100 MHz frequency bands. The electromagnetic field is evaluated at the origin of the coordinate system given by point P (0.0.0). Equation (1) represents the conversion from cartesian to spherical coordinates and vice versa. Defining the cartesian coordinates of the midpoint is important as it affects the error rate.

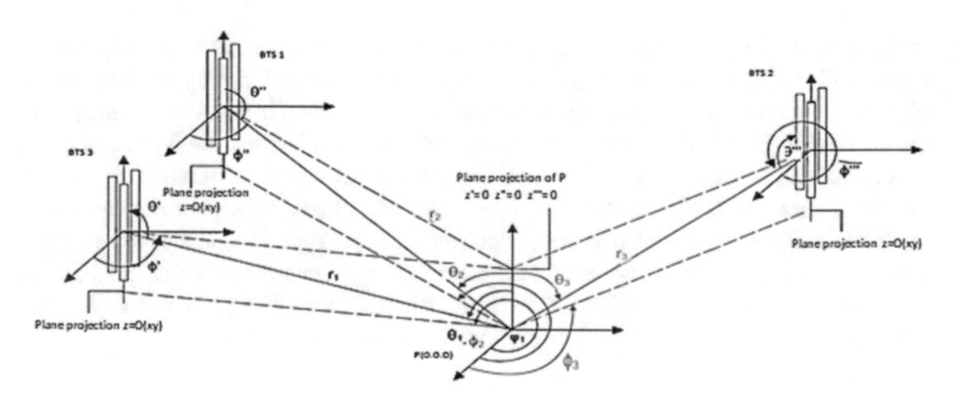

Fig. 2. Defining of the angles Θ_i and Φ_i.

Figure 2 presents the calculation method for n – antennas and the total electromagnetic field is obtained by superposition of the electrical intensities of the calculated electromagnetic fields of each individual antenna. The approximation of far field may result in overestimation of the measured electromagnetic field.

Table 1. Spherical coordinates of point P (0.0.0) from mobile operators 1, 2 and 3.

	Spherical coordinates of P from the 1st antenna	Spherical coordinates of P from the 2nd antenna	Spherical coordinates of P from the 3d antenna
Distance r	$r' = r_1 = \sqrt{x_{A_1}^2 + y_{A_1}^2 + z_{A_1}^2}$	$r'' = r_2 = \sqrt{x_{A_2}^2 + y_{A_2}^2 + z_{A_2}^2}$	$r''' = r_3 = \sqrt{x_{A_3}^2 + y_{A_3}^2 + z_{A_3}^2}$
Angle θ_i	$\theta' = -arctg \frac{\sqrt{x_{A_1}^2 + y_{A_1}^2}}{z_{A_1}}$	$\theta'' = -arctg \frac{\sqrt{x_{A_2}^2 + y_{A_2}^2}}{z_{A_2}}$	$\theta''' = -arctg \frac{\sqrt{x_{A_3}^2 + y_{A_3}^2}}{z_{A_3}}$
Angle Φ_i	$\Phi' = arctg \frac{y_{A_1}}{z_{A_1}}$	$\Phi'' = arctg \frac{y_{A_2}}{z_{A_2}}$	$\Phi''' = arctg \frac{y_{A_3}}{z_{A_3}}$
Angle relations	$\Phi + \Phi' = 2\pi$ $\theta + \theta' = \pi$	$\Phi + \Phi'' = 2\pi$ $\theta + \theta'' = \pi$	$\Phi + \Phi''' = 2\pi$ $\theta + \theta''' = \pi$

In Table 1 are given the spherical coordinates of point P as seen from the antennas of the mobile operator with reference to the midpoint, as shown in Fig. 2.

$$\begin{bmatrix} A_r \\ A_\theta \\ A_\Phi \end{bmatrix} = \begin{bmatrix} sin\theta cos\Phi & sin\theta sin\Phi & cos\theta \\ cos\theta cos\Phi & cos\theta sin\Phi & -sin\theta \\ -sin\Phi & cos\Phi & 0 \end{bmatrix} \begin{bmatrix} A_x \\ A_y \\ A_z \end{bmatrix} \tag{1}$$

Equation (2) is applicable in the far field for the calculation of the electromagnetic field [4]. The statistical study considered "The worst case is field vectors $E_{X1}, E_{X2}, \ldots E_{Xn}$ in phase as for EY and EZ".

$$E(d, \theta, \Phi) = \sum_{i=1}^{N} \frac{\sqrt{30 P_{in_i} G_e(\theta_i, \Phi_i)}}{d_i} \tag{2}$$

The total electromagnetic field is obtained by superimposing the electrical intensities of the calculated electromagnetic fields of each individual antenna. The approximation of far field may result in overestimation of the measured electromagnetic field. The calculation parameters are: P_n power of nth antenna, θ_n and Φ_n the angles giving the direction from the nth antenna to the reference point of calculation, $G_n(\theta_n, \Phi_n)$ the nth antenna gain in this direction (specified by the manufacturer) (calculations are made in each emitting space, in both horizontal and vertical planes, for 900/1800/2100 MHz frequency bands), r_n the distance between the nth and the point of calculation, E_n the electromagnetic field produced by the nth antenna at the point of calculation in the ith frequency, E_{Rn} the total electromagnetic field of all n antennas at the calculation point and N the number of samples required for the FDTD method. Equation (3) is valid for the electromagnetic field composed of individual field components of n antennas at the point of calculation and the given i^{th} frequency, indexed as $E_{Rn_{i^{th} frequency}}$.

$$Rn_{i^{th} frequency} = \sqrt{\sum_{1}^{n} \sum_{1}^{N} (E_x)^2 + \sum_{1}^{n} \sum_{1}^{N} (E_y)^2 + \sum_{1}^{n} \sum_{1}^{N} (E_z)^2} \tag{3}$$

Equation (4) is valid for the EM field of the total antennas at all three frequency bands 900/1800/2100 MHz, indexed as $E_{Rn_{TotalFrequency}}$ [5].

$$E_{Rn_{TotalFrequency}} = \sqrt{\sum_{1}^{n}\left(E_{Rn_{Frequency1}}\right)^{2} + \sum_{1}^{n}\left(E_{Rn_{Frequency2}}\right)^{2} + \sum_{1}^{n}\left(E_{Rn_{Frequency3}}\right)^{2}} \tag{4}$$

To apply Eq. (2) parameters θ and Φ are estimated from Eq. (1) and antenna gain $G(\theta, \Phi)$ in this direction is given by the antenna manufacturer in a two-columns structured (degree, gain) ASCII file, in 360° of radiation in both the horizontal and vertical planes. As previously mentioned, simulation of full waves requires more computer resources and time. To improve the computational efficiency a simple antenna geometry is proposed, using an optimization algorithm to achieve the required antenna specifications. Thus, applying the finite-difference time-domain (FDTD) method to estimate the field around the antenna, becomes easier and less time consuming. The antenna model used in this work consist in the form of several rows of cells. The cubic shaped cells are of size $\Delta x (= \Delta y = \Delta z) = 15\,\text{mm}$.

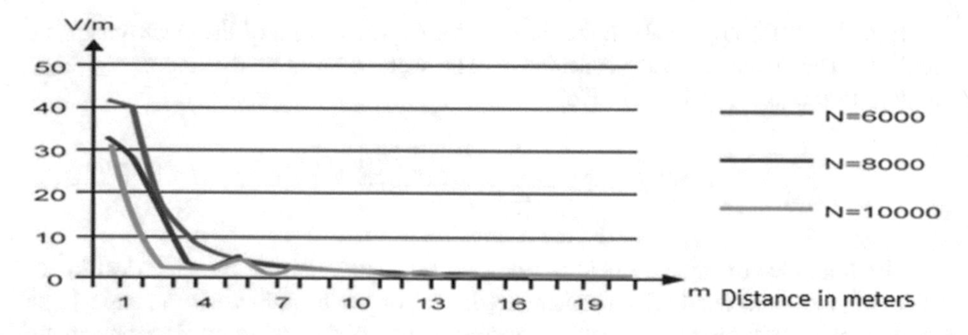

Fig. 3. Model of the considered antenna (Δx = Δy = Δz) = 15 mm.

The unit cells are used for designing the general antenna model referred to as the unit antenna. The selection criteria of N sample cells of choice for the FDTD method are given in Fig. 3. The number of samples considered is $N = 6000$. This estimate is recommended in [6] for the calculation of the near and far radiating field. Another application of the proposed methodology is the determination of the volume (space) of exclusion. The volume of exclusion is determined as the spatial volume where the electrical field intensity exceeds the recommended exposure guidelines. The algorithm used for calculations can be described as follows.

The volume of the safety zone is overestimated in a simple way to define a cuboid region S, which includes the safety volume B, of all the antenna system, as given in Fig. 4. A rule has been set to identify the points inside S with a calculated electromagnetic field equal to the Iso-Curve input levels in the program. Interpolation of the electromagnetic field is performed using a quadratic formula. The identified points are projected on the coordinate plane and the safety distances are estimated. The area is

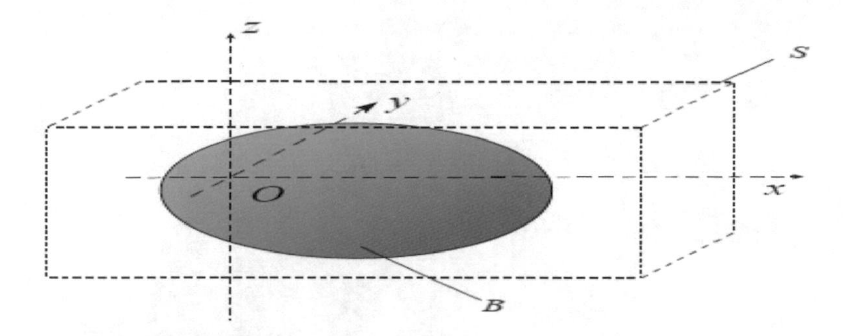

Fig. 4. Safety volume and respective distances.

then projected on the Oxy, Oxz and Oyz planes using the "worst case" criteria and considering that the point the projected area is the furthest from the origin of the coordinate system.

4 Evaluation of the Proposed Model with the Numerical FDTD Method

The Iso-Curve level of the electromagnetic field for each antenna of different operators is provided in the technical specifications of the antenna manufacturers. It is given: the antenna label (not required for calculations yet important for the graphical representation of the results), antenna gain (dBi), the mechanical down tilt (+/–) and the electrical down tilt of the antenna which must be included in the antenna radiation pattern diagram, the radiation pattern in both the horizontal and vertical planes, provided by the antenna manufacturer in a two-columns structured (degree, gain) ASCII file, the antenna's center in X coordinate, the antenna's center in Y coordinate, the antenna's center in Z coordinate, the file name containing the horizontal plane of the radiation pattern, the file name containing the vertical plane of the horizontal pattern and the input power of the antenna (Watt). At the end of the file a rule is set that requires accurate calculations of the electromagnetic field in some points identified with their respective cartesian coordinates. This rule generates the calculated electromagnetic field values printed in the output window.

Program conclusions are represented by the following data for each iso-curve level of the input file. The safety regions or iso-curves (the space where values $E \geq 41$ V/m) in the x-y plane (graphical window) represent the total contribution of the antennas and the individual safety zone for each antenna [7] (Fig. 5).

The safety regions or iso-curves (the space where values $E \geq 41$ V/m) in the x-z plane (graphical window) represent the total contribution of the antennas and the individual safety zone for each antenna.

The safety regions or iso-curves (the space where values $E \geq 41$ V/m) in the y-z plane (graphical window) represent the total contribution of the antennas and the individual safety zone for each antenna [8].

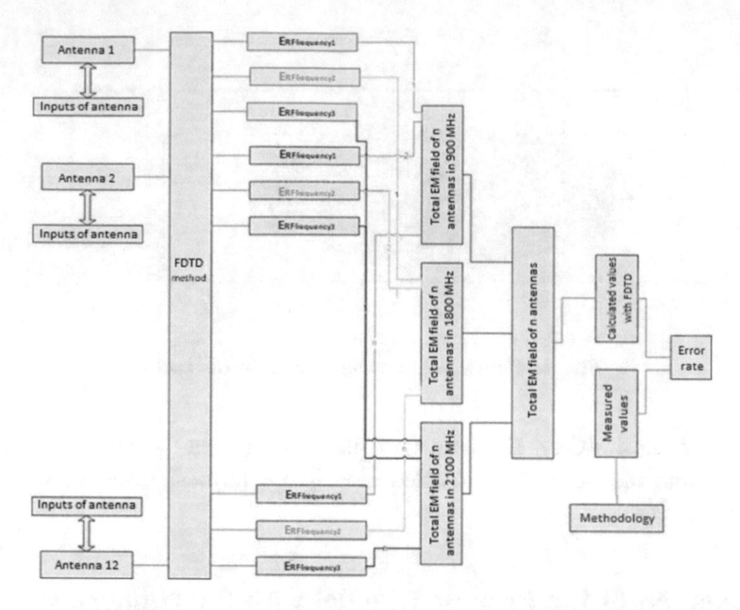

Fig. 5. The proposed model based on FDTD technique.

5 Analysis of the Experimental Study and Discussion of the Results

To prove the accuracy of the calculation method for the electromagnetic field evaluation at a random point in space in the presence of two or more GSM base stations the estimated results are compared with the measurements performed with the help of NARDA SRM 3000 m. The antennas of all the cellular base stations considered operate simultaneously at 900 MHz, 1800 MHz and 2100 MHz. Estimation of the radiated electrical field intensity with respect to distance is followed by the determination of the safety zones, whose value is equal to the radiation limits set by ICNIRP for public exposure. The case being considered has three different operators each using an antenna model of the Kathrein type [9]. The first operator uses K80010672 GSM antenna with X, Y, Z coordinates of the midpoint (5 m. 5 m. 5 m), respectively. The second operator uses K80010671 GSM antenna with X, Y, Z coordinates of the midpoint (–3 m. –3 m. –3 m) and the third operator uses K80010670 GSM antenna with X, Y, Z coordinates of the midpoint (15 m. 5 m. 5 m), respectively.

Gain and emitting areas in the horizontal and vertical planes for each antenna considered are provided by the antenna manufacturer in a two-columns structured (degree, gain) ASCII file. The antenna midpoint coordinates are defined with relation to the point where the electrical intensity is being estimated and taken at the origin of the coordinate system given by P (0.0.0).

The chart in Fig. 6 shows the comparison between the calculated value of the electrical field intensity (V/m) with the measured value, in the X axis, where the sampling step along X is every 0.5 m in a range from 0 to 16 m, while the Y and Z

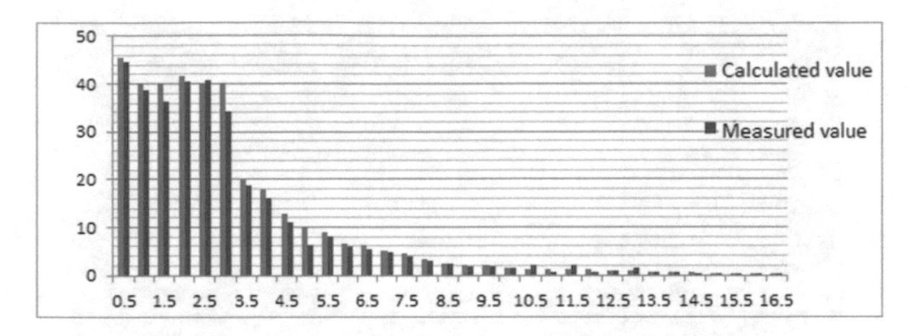

Fig. 6. Comparison between the calculated value and the measured value along X.

coordinates remain unchanged. The difference between the calculated and the measured values along X is very small, in particular, with the increase in distance being insignificant.

In general, the considered model tends to increase the calculated values of the field systematically. In this case it is possible that the calculations are estimated in areas where there are no obstacles and the free space model is valid and, in addition, the three main physical properties of electromagnetic waves that is reflection, absorption and interference are not considered.

The chart in Fig. 7 shows the comparison between the calculated value of the electrical field intensity (V/m) with the measured value, in the Y axis, where the sampling step along Y is every 0.25 m in a range from –3.25 m to 3.25 m, while the X and Z coordinates remain unchanged.

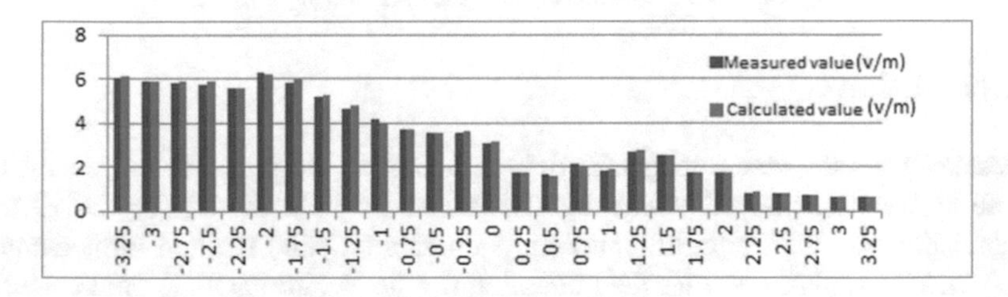

Fig. 7. Comparison between the calculated value and the measured value along Y.

Again, the theoretical model proposed above (based on FDTD technique) for the electromagnetic field evaluation converges to the measured field values along Y. Using the calculation algorithm based on FDTD as previously mentioned, the safety regions (part of space where $E \geq 41$ V/m) are calculated for the Oxy, Oxz and Oyz planes.

Figure 8 presents the distances along the Oxy, Oxz and Oyz planes where the intensity of the electrical field is equal to the radiation limits set by ICNIRP for public exposure, respectively being 41.25 V/m in 900 MHz, 58.33 V/m in 1800 MHz and 61 V/m in 2100 MHz, for each operator.

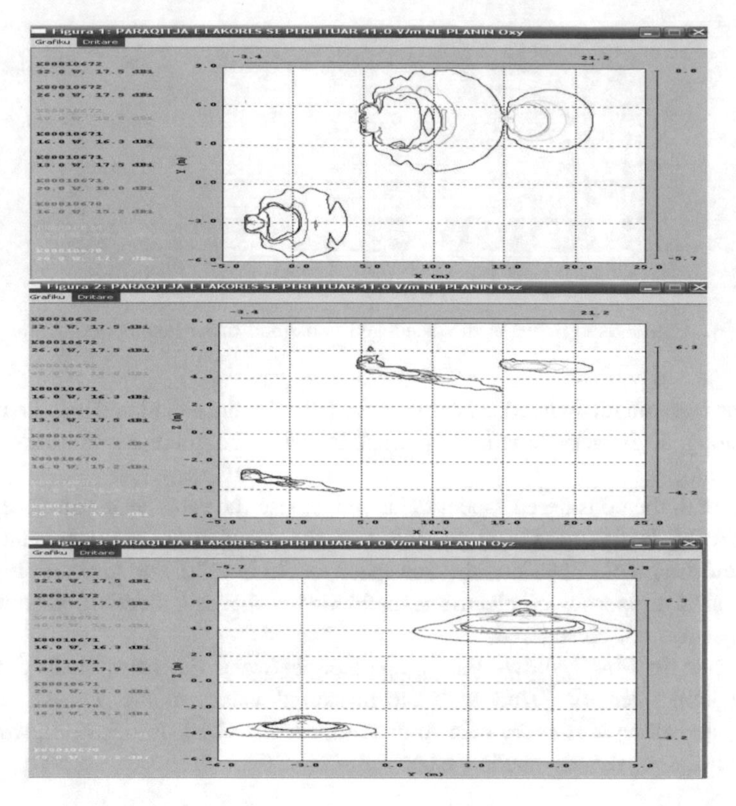

Fig. 8. Safety iso-curves in Oxy, Oxz and Oyz planes.

6 Conclusions

1. Evaluation of the electromagnetic field radiation in the presence of several GSM base station is achieved using the theoretical far field model for the EM field calculation and the finite-difference time-domain (FDTD) method implemented in MATLAB. Applying the far field equation for the electromagnetic field calculation requires the recognition of several parameters provided by the antenna manufacturer.
2. The comparative analysis of the calculated values with those measured proves that the proposed algorithm accurately estimates the field values near the GSM base stations (with up to 12 BTS antennas) in the urban environment, with an error rate of less than 6%.
3. For the considered cellular base stations, which operate simultaneously in 900/1800/2100 MHz, the safety regions or iso-curves (spaces where the value of $E \geq 41$ V/m) are calculated in the Oxy, Oxz and Oyz planes.
4. In a certain environment and in the presence of several GSM base stations operating at 900/1800/2100 MHz, it is possible to calculate the intensity of the electrical and magnetic fields and the power density in various distances from the antenna in a short time and with a reliable procedure.

References

1. Lala, A., Kamo, B., Cela, S., Shurdi, O.: Modelimi i antenave GSM të stacionit bazë për vlerësimin e fushës të ekspozuar në afërsi të antenës. Buletini i Shkencave Teknike, Maj 2011
2. Karwowski, A., Wójcik, D.: Numerical modelling calculations for evaluating exposure to radio-frequency emissions from base station antennas. In: Proceedings of the Microwaves, Radar and Wireless Communications, MIKON-2002, vol. 3, pp. 809–812 (2002)
3. KATHREIN-Werke KG: Technical Information and New Products: 790–2500 MHz Base Station Antennas for Mobile Communications, Catalogue Issue 02/06
4. Karwowski, A., Wójcik, D.: Evaluating exposure to radio-frequency emission from base station antennas. In: The 13th IEEE International Symposium on Personal, Indoor and Mobile Radio Communications - PIMRC 2008, vol. 4, pp. 1877–1881 (2008)
5. Wójcik, D., Karwowski, A.: Simply method for evaluation of the near field of the GSM base-station antennas. In: 17th International WROCLAW Symposium and Exhibition on Electromagnetic Compatibility, Wrocław, Poland, pp. 127–130 (2004)
6. Bernardi, P., Cavagnaro, M.: A UTD/FDTD model to evaluate human exposure to base-station antennas in realistic urban environments. In: Proceedings of 2003 IEEE MTT-S International Microwave Symposium, Philadelphia, Pennsylvania, USA, June 2008
7. Balanis, C.A.: Antenna Theory analysis and Design, pp. 212–216. Wiley (2005)
8. Barbiroli, M., Carciofi, C.: Analysis of field strength levels near base station antennas. In: Proceedings of VTC'99 - IEEE International Conference on Vehicular Technology, Houston, Texas, USA, May 2009
9. Barbiroli, M., Carciofi, C., Degli-Esposti, V.: Evaluation of exposure levels generated by cellular systems: methodology and results. IEEE Trans. Veh. Technol. **51**(6), 1322–1329 (2009)
10. Cela, S., Lala, A., Kamo, B., Biberaj, A., Mitrushi, R.M.: Estimation of simultaneous exposure to electromagnetic radiation of 2G and 3G base stations in Albania. J. Commun. Comput. **9**, 1142–1146 (2012). USA JCC-E20120401-1, ISSN: 1548-7709
11. Cela, S., Lala, A., Mitrushi, R.M.: Matja e fushës elektromagnetike në një mjedis urban përreth stacioneve celulare në qytetin e Tiranës. Buletini i Shkencave Teknike, Maj 2011
12. Adane, Y., Gati, A., Wong, M.-F., Dale, C., Wiart, J., Hanna, V.F.: Optimal modeling of real radio base station antennas for human exposure assessment using spherical-mode decomposition. IEEE Antennas Wirel. Propag. Lett. **1**, 215–218 (2002)
13. Richter, J., Al-Nuaimi, M., Ivrissimtzis, L.: Base station antenna design optimisation based on UTD ray-tracing models utilising radio site topographical data. In: 2001 Eleventh International Conference on Antennas and Propagation, vol. 1, pp. 257–259 (2001). (IEE Conference Publication No. 480)
14. Anton, R., Jonsson, H., Moshfegh, B.: Detailed CFD modelling of EMC screens for radio base stations: a parametric study. IEEE Trans. Compon. Packag. Technol. **32**(1), 145–155 (2009)

Evaluation of Mouse Operation Authentication Method Having Shoulder Surfing Resistance

Makoto Nagatomo[1], Yoshihiro Kita[2], Kentaro Aburada[3], Naonobu Okazaki[3], and Mirang Park[4](✉)

[1] Security Research Center, Kanagawa Institute of Technology, Atsugi, Japan
je.suis.mako@cco.kanagawa-it.ac.jp

[2] Department of School of Computer Science, Tokyo University of Technology, Hachioji, Japan
kitayshr@stf.teu.ac.jp

[3] Department of Computer Science and System Engineering, University of Miyazaki, Miyazaki, Japan
{aburada,oka}@cs.miyazaki-u.ac.jp

[4] Department of Information Network and Communication, Kanagawa Institute of Technology, Atsugi, Japan
mirang@nw.kanagawa-it.ac.jp

Abstract. Currently, typing character strings on a keyboard is used for personal authentication for PC login and unlocking. Although some graphical and biometric-based methods have been developed, most of them have weak authentication strength or weak shoulder surfing resistance. In this paper, we propose a personal authentication method that the user specifies positions on a $N \times N$ matrix using mouse clicks. The user can hide the mouse during authentication easily, so the method has shoulder surfing resistance and can be used in public places. We developed the proposed method (pattern method), and two variations (number and color, and combination method), and performed user testing and shoulder surfing experiment to validate the proposed method. The proposed method was tested by 20 subjects in pattern method, 23 subjects in the number and color, and the combination method. The authentication success rate is 63.1%, 56.3% and 87.5% on each method. The specific rate by shoulder surfing is 1.4% in pattern method.

1 Introduction

Currently, password entry using keyboard is the most common method for personal authentication on a PC to unlock it or sign into a service such as webmail in public places (e.g. an office or café). This method has some problems such that remembering characters increase user's memory burden and typing on a keyboard increase the possibility of password leakage by shoulder surfing attack or recording attack [1]. Therefore, various authentication methods that use images, positions or icons as passwords have been researched [2–6]. The passwords of these methods are easier to remember than keyboard one. For example, SECUREMATRIX [4] use positions on a matrix and Gaze

L. Barolli et al. (Eds.): EIDWT 2018, LNDECT 17, pp. 978–989, 2018.
https://doi.org/10.1007/978-3-319-75928-9_90

Gesture-Based Authentication System [5] use simple icons such as a square and a triangle.

Recently, biometric authentication, which employ human features such as fingerprints, handwriting and face recognition, is used. Among biometrics, authentication which use user's mouse movements are researched [7–9]. For instance, the mouse trajectory of clicks and drag and drops is used as features to authenticate the user in [7]. This method has the merit of no memory burden, but do not have shoulder surfing resistance because the mouse trajectory is observed by an attacker on the desk. Thus, it is difficult to use these method in public places.

In this paper, we propose a personal authentication which can be used in public places with mouse operations. We assumed that a user tries to unlock his/her PC in an office, or a café using mouse operations. The passwords in this method are positions on an $N \times N$ matrix, and the user move current positions on it by right click, left click, rotation upward, rotation downward and wheel click to specify positions. When unlocking his/her PC, the user can hide the mouse under a desk and perform the operations without seeing it handily. In addition, this method does not need special devices such as camera. We implemented the proposed method (called pattern method) and two variations (called number and color, and pattern method). We performed two experiments on the usability and shoulder surfing resistance of the proposed method. The number of subjects is 20 in pattern method, 23 subjects in the number and color, and the combination method. The authentication success rate was 63.1%, 56.3% and 87.5% on each method. The specific rate by shoulder surfing was 1.4% in pattern method.

2 Related Work

There are several researches of user authentication using icons or having shoulder surfing resistance. In order to compare with existing method, we list the related work on a PC or a smartphone below.

SECUREMATRIX [4] provides the password of positions denoted by numbers on three or four matrices. The user enters the numbers on matrices to specify positions by a keyboard. The password positions as a shape can be easy to remember (e.g. the shape of diagonal on a matrix), however, an attacker can identify passwords by watching the keyboard and the screen.

The method which utilizes gazing and following icons (Gaze Gesture-Based Authentication System) exist [5]. The user performs authentication by following password icons on the screen with their eyes. The dummy icons also move on the screen when the user performs authentication. It is difficult to identify the eye movements for people except for the user. Hence, this method has shoulder surfing resistance. However, the user must set a camera in front of monitor correctly, so this method cannot be used in public places.

In [6], a positional password, i.e. the sequence of coordinate pairs on the trajectory the user drew on the matrix by a mouse is proposed. The system provides three defense techniques to strengthen shoulder surfing resistance (decoy strokes, disappearing strokes, and line snaking), but does not have complete protection because the trajectory is detected by seeing the screen.

There are several researches which use features of mouse dynamics [7, 8]. In [7], the key feature are fine-grained angle-based metrics (direction, angle of curvature, curvature distance of mouse movements). This system used support vector machines (SVMs) for accurate and fast classification. In [8], the system uses SVMs, K-Nearest Neighbor and Naive Bayes for twelve features including number of changes in horizontal position and vertical position based on the angle of the trajectory. There is no memory burden in these system, but an attacker can imitate the mouse movement. Moreover, the user cannot hide the mouse handily to strengthen shoulder surfing resistance.

Some method with shoulder surfing resistance on smartphone [11, 12] is researched. The Secret Tap with Double Shift (STDS) method [11] provides graphical icons as password on a smartphone. In authentication phase, 16 icons including at least one password icon is displayed on a 4×4 matrix. The user can specify an icon by tapping the different icon on the screen by shift rules. Therefore, this method has shoulder surfing resistance, but password icons can be revealed by recording attack twice.

Circle Chameleon Cursor (CCC) [12] is an authentication system which use a smartphone's vibration. The authentication information are 4 numbers (as PIN), the user turns the dial until the registered number comes to selected position randomly when performing authentication. The smartphone vibrates when the indicator which continue to turn on the dial is on the position. This method has shoulder surfing resistance because it does not show the vibrated position, but it takes more time of authentication than PIN method because the user has to turn the dial.

As described above, there is no method which is easy to strengthen shoulder surfing resistance for a desktop PC or laptop in public places. Thus, we propose the method that can be used in public places with shoulder surfing resistance and that utilizes a mouse which most desktop PC user have.

3 Mouse Authentication Method

We introduce the authentication method with only mouse operations. The input interface is a common mouse with functions as follows (see Fig. 1(a)):

- Right and left click
- Upward and downward wheel rotation
- Wheel click
- Sensor for detection of mouse movements

The sensor is used in only a variation (combination method) of the proposed method.

The output interface is an $N \times N$ matrix (see Fig. 1(b)). The user performs authentication using these interfaces. The positions on the matrix and its order are password in this method.

Use case: The user unlocks a PC or signs into a service on a PC in public places while hiding the mouse. For example, the user signs into a webmail service on their own laptop in a café while hiding the mouse under the desk.

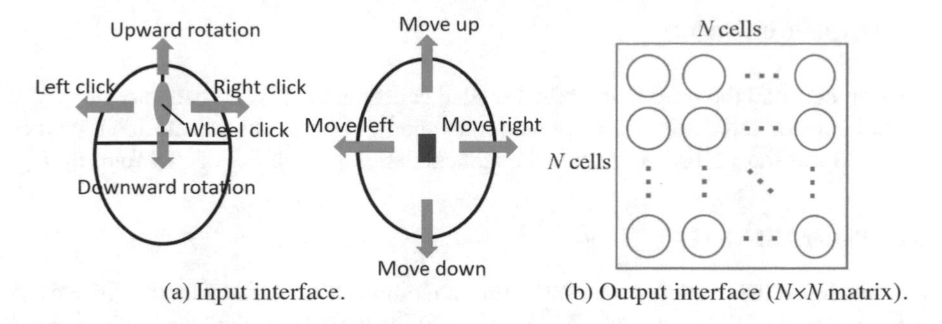

(a) Input interface. (b) Output interface ($N{\times}N$ matrix).

Fig. 1. Input and output interfaces used in the proposed method.

Strength: When the user registers m positions on an $N \times N$ matrix, the probability that an attacker succeeds in a random attack on this method (accidental authentication probability) is $1/N^{2m}$.

Registration phase: Registration is performed as follows.

1. The method displays an $N \times N$ matrix with an initial position randomly selected.
2. The user selects positions using mouse operations. Current positions are marked by a red circle. The current position moves left by left click, right by right click, up by upward wheel rotation, and down by downward rotation. The selected position is registered by wheel click.
3. The user registers m positions by repeating (1) and (2). The registered order of the cell is password.

Authentication phase: Authentication is performed as follows.

1. Authentication system displays an $N \times N$ matrix with a randomly selected initial position.
2. The user specifies the first registered position with mouse operations. The movements of the current position by mouse operations are the same as them in registration phase. Only the initial position is marked by red circle, and the current position is not display in order to strengthen shoulder surfing resistance.
3. The user specifies m registered positions in their registered order by repeating (1) and (2).

Benefits: The mouse operations and its effects are intuitive. Some people, for example older people, cannot type the keyboard without seeing it, but can use the mouse without seeing it. The user can perform authentication while hiding the mouse in order to strengthen shoulder surfing resistance. If our proposed method is applied to keyboard or touch screen system, it is bothersome to hide these in authentication phase. In addition, the proposed method is a challenge and response method. Therefore, users can use this method in public places safely.

4 Implementation

We implemented the proposed method (called pattern method) and two variations of it (called number and color, and combination method). We aimed for an accidental probability of less than 1/10,000 (accidental authentication probability of PIN four digits).

4.1 Pattern Method

We developed the proposed methods (pattern method) using a 5 × 5 matrix ($N = 5$). A position count of $m \geq 3$ meets $1/5^{2m} \leq 1/10{,}000$, hence, we set the length of password three positions or more. we decided that the user can register a position only after moving at least three times because an attacker can hear mouse sounds to detect the password when an initial position is near the registered positions.

Figure 2 shows the screen in registration and authentication phase. The red circle is the current position in registration screen, and the initial position in authentication screen. We set the back, registration/authentication and inverse button under the screen matrix. The user can delete one specified position, register or perform authentication by pushing back or registration/authentication button. By pushing inverse button, mouse operations are inversed (e.g. exchange right click with left click). The reason why we added the button is that the user may operate the mouse at the back of a desk when hiding the mouse. Moreover, we placed the explanation of mouse operations and its effects under the buttons so that the user can remind them. Here we give an example of this method as follows.

Registration Phase (see Fig. 3)

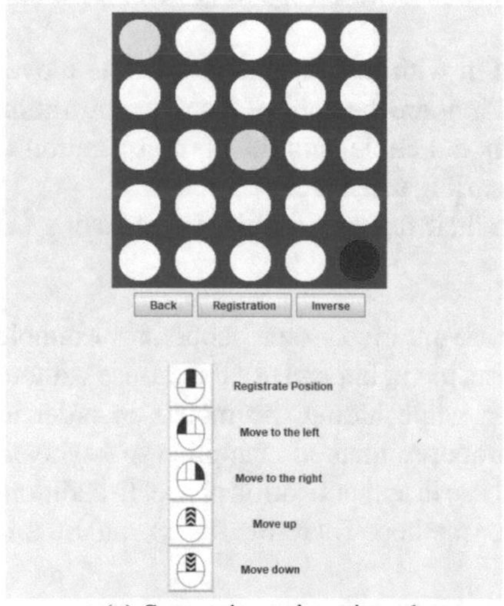

(a) Screen in registration phase.

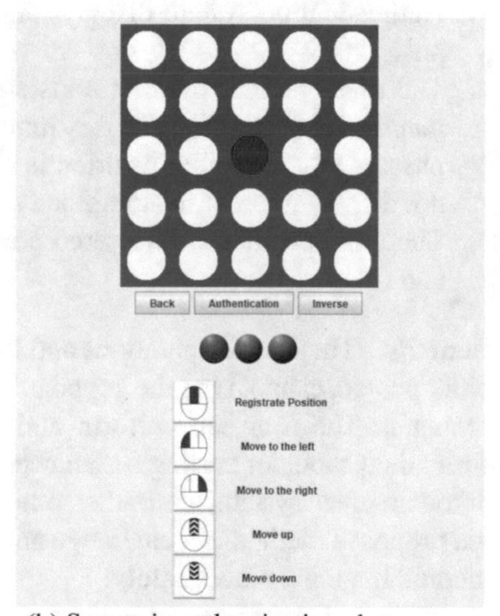

(b) Screen in authentication phase.

Fig. 2. Screen of pattern method in registration and authentication phase.

We assume that the user wants to register cells (1, 1), (3, 3) and (5, 5) ((x, y) represent x-th column and y-th row on the matrix).

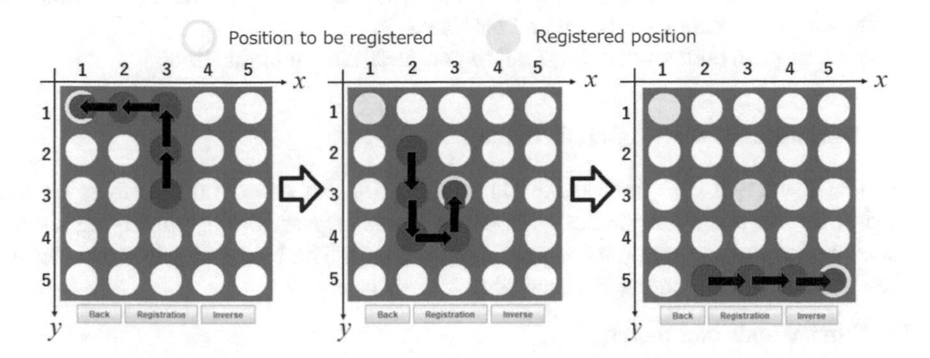

Fig. 3. Example of the registration phase.

1. First, displays a 5 × 5 matrix with the first initial position (3, 3) selected at random. The user register (1, 1) by rotating the wheel upward twice, clicking left twice and clicking the wheel.
2. Next, displays a matrix whose initial position is (2, 2). The user register (3, 3) by rotating the wheel downward twice, clicking right once, rotating the wheel upward once and clicking the wheel.
3. Finally, displays a matrix whose initial position is (2, 5). The user register (5, 5) by clicking right three times and clicking the wheel.
4. The user push registration button to register ((1, 1), (3, 3), (5, 5)).

Authentication Phase (see Fig. 4)

1. First, displays the matrix having the initial position (1, 2). The user specifies (1, 1) by clicking right once, rotating the wheel upward once, clicking left once and clicking the wheel.

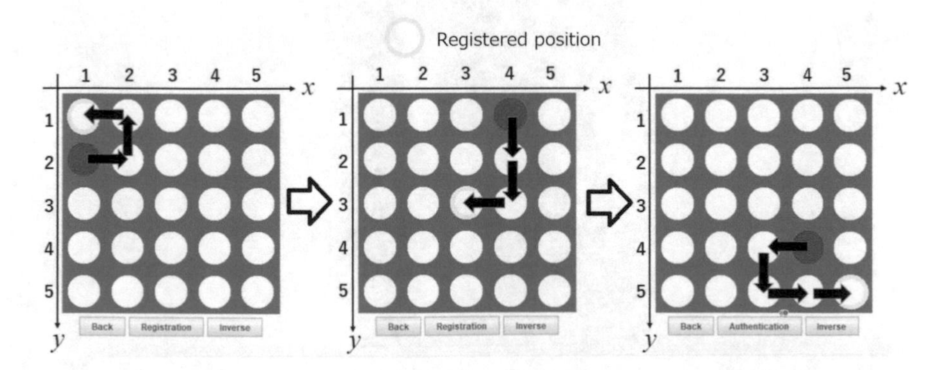

Fig. 4. Example of the authentication phase.

2. Next, displays the matrix with (4, 1) as the initial position. The user specifies (3, 3) by rotating the wheel downward twice, clicking left once and the wheel.
3. Finally, the matrix with (4, 4) is displayed. (3, 3) is specified by clicking left, rotating the wheel, clicking right twice and the wheel.
4. The user push authentication button to complete the authentication.

4.2 Variations of the Proposed Method

We developed two variations using (1) combinations of colors and numbers on the matrix (number and color method) and (2) positions as password by specifying it with mouse clicks and movements (combination method). The user can select a variation when performing authentication according to their taste.

(1) Number and color method

The user registers combinations of numbers and color as authentication information. In authentication phase, the matrix has cells that have combinations of numbers and color selected randomly. For example, in Fig. 5, the initial coordinate (2, 2) is marked with black circle, and each cell has a combination of a number and color. When the user wants to specifies the (4, yellow), the user clicks right, rotates upward, clicks left and clicks the wheel. The mouse operations and its effects are the same as that of pattern method. We use the number set $A = \{1, 2, 3, 4, 5, 6, 7, 8, 9\}$ and the color set $B = \{\text{red, blue, yellow, green}\}$. We decided the size of the matrix is 6×6 ($N = 6$) because the number of elements of $A \times B$ is 36. A combination count of $m \geq 3$ meets $1/6^{2m} \leq 1/10{,}000$, so we set the length of password three combinations or more.

(2) Combination method

The authentication information are positions (as pattern method). The user specifies the position using the combinations of mouse clicks and mouse movements. Figure 6(a) shows images of the combinations. In authentication phase, this method displays a 4×4 matrix whose each cell has an image corresponding to the combination of a mouse click

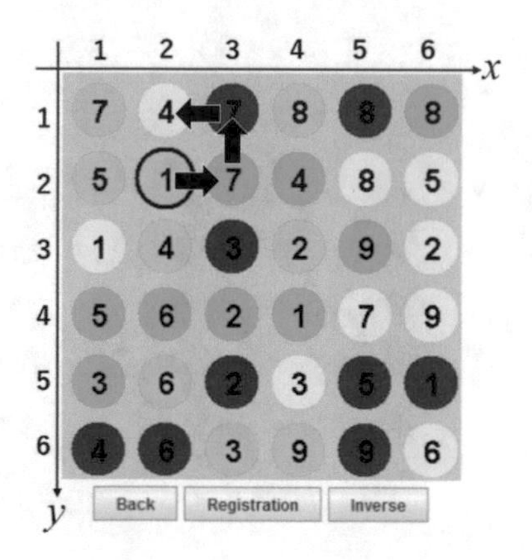

Fig. 5. Example of the number and color method.

and a mouse movement. We use the mouse click set A = {right click, left click, wheel click, left and right click, upward wheel rotation, downward wheel rotation} and the mouse movement set B = {right movement, left movement, upward movement, downward movement, no movement}. We do not use the set C = {right and left click, upward wheel rotation, downward wheel rotation} $\times$ (B – {no movement}) because it is difficult to operate the elements of the set C. Therefore, we use the set $M = (A \times B) - C$ as the set of the operations ($|M| = 18$), and use the 4×4 matrix ($N = 4$) in this variation. The combination is selected from the set M randomly in authentication. For example, in Fig. 6(b), the user can specify the coordinate (2, 3) by moving the mouse to the right while clicking the mouse wheel. A position count of $m \geq 4$ meets $1/4^{2m} \leq 1/10{,}000$, so we set the length of password four positions or more.

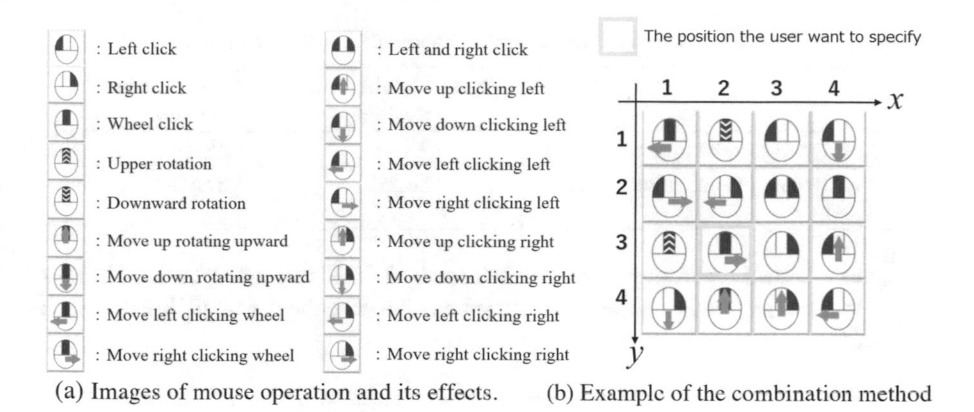

(a) Images of mouse operation and its effects. (b) Examplc of the combination method

Fig. 6. Example of the combination method.

5 Evaluation

We conducted two types of experiments on the usability and shoulder surfing resistance of the implemented methods. The subjects used a desktop PC and a common mouse with clicks and wheel rotations (Logicool wireless mouse M186). This mouse emits the sound of clicks and rotations clearly. We did not use the two variations in the shoulder surfing experiment reported here, but we plan to conduct experiments of these in the future.

5.1 Usability Test

We conducted the usability test as follows:

1. We explained how to use the method to the subjects and background of this method.
2. The subject completed a tutorial to become familiar with the mouse operations. The user can move current positions without displaying the position movements after this tutorial.
3. The subject registered password. The lengths of passwords are 3 (pattern), 3 (number and color) and 4 (combination) each.

4. The subject performed authentication on the desk. The subject performed three authentications successfully in pattern method (5 times authentication in number and color, and combination method).
5. The subject answered a questionnaire on comprehension, ease of use, easy of familiarization, safety for shoulder surfing, and user needs. These items were rated from 1 to 5 (very bad to very good).

The subjects were 20 Kanagawa Institute of Technology students in pattern method (23 students in number and color, and combination method). Table 1 shows the average tutorial time and authentication success rate. It takes time less than one minute, so the operations of our proposed methods are not difficult for users. In addition, the success rate in combination method is highest in our methods.

Table 1. Average tutorial time and authentication success rate.

	Pattern method	Number and color method	Combination method
Average of tutorial time	40.7 s	32.3 s	38.4 s
Average of authentication success rate	63.1%	56.3%	87.5%

Figure 7 shows the average time in each method. The average authentication time in combination method is smallest in our methods because the user can directly specify the positions.

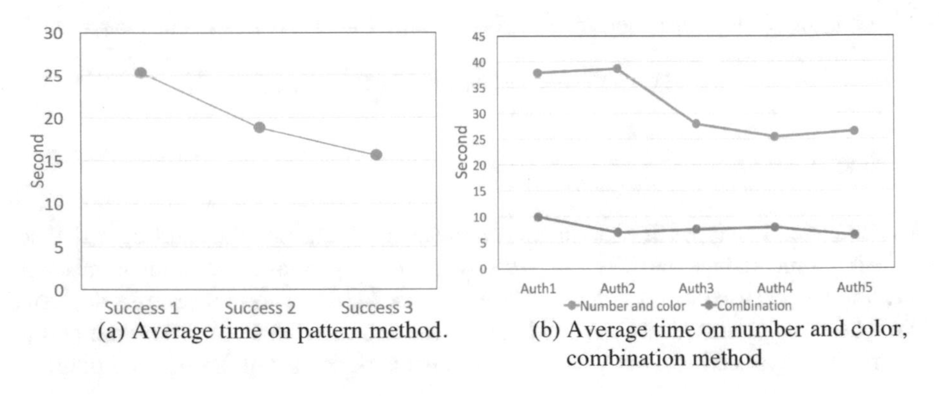

(a) Average time on pattern method. (b) Average time on number and color, combination method

Fig. 7. The average authentication time.

Figure 8 shows the result of the responses to the usability testing questionnaire. All average of items in all our proposed methods over 3, so they have good usability. As for understanding and safety, the scores are almost the same, but easily, familiarization and needs for combination method are higher than the other methods. Therefore, the combination method has higher usability than the other methods. Some subjects missed the current position in pattern, and number and color method.

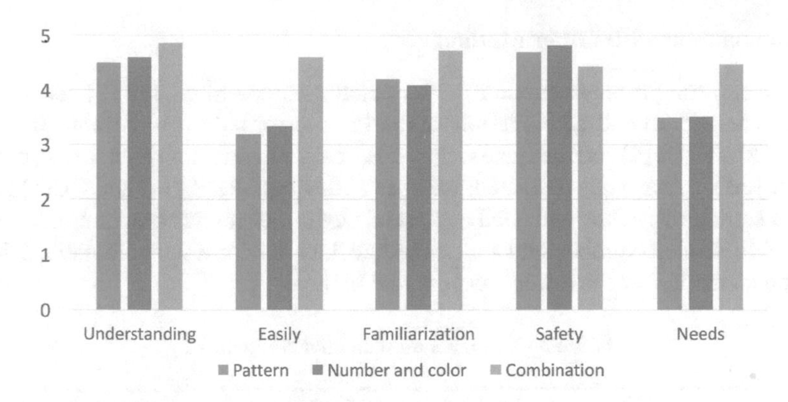

Fig. 8. The average score of the questionnaire.

5.2 Shoulder Surfing Resistance Test

We conducted the shoulder surfing experiment on only pattern method. Shoulder surfers were positioned 1 m behind the user so that the surfer could see the monitor and hear the mouse sounds. In addition, we eliminated ambient noise around the user in order that the shoulder surfers can hear the mouse sounds clearly. The procedure of this experiment was as follows:

1. We formed teams consisting of five or six people.
2. We choose a user from each team, and the user registered three positions in the pattern method. Shoulder surfers were the other people in the team, and not allowed to see the monitor during the registration.
3. The user performed ten authentication successes while hiding the mouse under the desk so that shoulder surfers cannot see the mouse. Shoulder surfers observed the authentication and tried to detect the registered positions; they were allowed to take notes.
4. We repeated steps 2–4 until everyone on the team was a user once.

The subjects were 16 Kanagawa Institute of Technology students. We made three teams with 5, 5, and 6 people each. Table 2 shows the result of the experiment. All three digits of only one subject is revealed by only one person in the same team. For the other 15 subjects, one or two positions were detected by one or two persons. Under real conditions in public places, ambient sound would interfere with the shoulder surfing and hence the results indicate that pattern method has shoulder surfing resistance. We will plan to perform experiment on shoulder surfing resistance on the other variations in the future.

Table 2. Specific rate by shoulder surfing for each digit.

	One digit	Two digits	Three digits
Specific rate	22.8%	8.6%	1.4%

5.3 Comparison with Other Methods

We compared the proposed method with related methods [4–8, 11, 12], as shown in Table 3. The pattern and combination method has fewer password combinations than SECUREMATRIX [4] and the Gaze Gesture-Based Authentication System [5], but have shoulder surfing resistance and needs no special device. STDS [11] and CCC [12] has shoulder surfing resistance and need no special device but fewer combinations than our proposed methods. Number and color method has the largest combinations in our proposed methods, but the lowest success rate in them.

Table 3. Comparison with related methods.

	SECURE MATRIX [4]	Gaze Gesture-Based Authentication System [5]	Method with mouse movement pattern [6]	Method with mouse dynamics [7, 8]
Combination (when password digits are 3)	32^3 = 32,768	36^3 = 46,656	-	-
Shoulder surfing resistance	-	x	-	-
Secret information	positions	icons	positions and its pairs	features of mouse dynamics
Special devices	none	camera	none	none

	STDS [11]	CCC [12]	Pattern method	Number and color method	Combination method
Combination (when password digits are 3)	16^3 = 4,096	10^3 = 1,000	25^3 = 15,625	36^3 = 46,656	16^3 = 4,096
Shoulder surfing resistance	x	x	x	x	x
Secret information	icons and shifts	numbers	positions	number and color	positions
Special devices	none	none	none	none	none

The method in [6] also use positions on a matrix as password, but its shoulder surfing resistance is weak because the mouse movements can be detected by seeing mouse trajectory. On the other hand, the method in [7, 8] has no memory burden, but the mouse movement can be imitated by an attacker.

6 Conclusion

In this paper, we proposed method with mouse operations and implemented it with two variations. We conducted the experiments on the usability and shoulder surfing resistance. The results show that the proposed method has good usability and shoulder surfing resistance. The averages of success rate of the proposed methods are 63.1%, 56.3% and 87.5% each, and the questionnaire result shows good user's needs. The variation using combinations of mouse clicks and movements (combination method) has great usability. However, it is difficult to operate the mouse in combination method under the desk. The specific rate by shoulder surfing ten times is 1.4% in pattern method. The problem in this method is that the mouse sounds reveal users' passwords, so we will develop the method having shoulder surfing resistance even when an attacker can hear the mouse sounds clearly. Another future work is improvement of the proposed method. The user does not always have the mouse, so we should develop the method which the user can also use with trackpad on the laptop, otherwise we need apply our method to MacOS mouse system. In addition, we performed experiments for only university students, hence, we should perform usability test for older or impaired people, in the future.

Acknowledgments. This work is supported by JSPS KAKENHI Grant Numbers JP17H01736, JP17K00139.

References

1. Balzarotti, D., Cova, M., Vigna, G.: Clearshot: eavesdropping on keyboard input from video. In: Proceedings of the 2008 IEEE Symposium on Security and Privacy, pp. 170–183 (2008)
2. Castelluccia, C., Durmuth, M., Golla, M., Deniz, F.: Towards implicit visual memory-based authentication. In: The Network and Distributed System Security Symposium (NDSS) (2017)
3. Biddle, R., Chiasson, S., van Oorschot, P.C.: Graphical passwords: learning from the first twelve years. ACM Comput. Surv. (CSUR) 1–25 (2012)
4. CSE: SECUREMATRIX. http://cse-america.com/index.htm. Accessed 15 Nov 2017
5. Rajanna, V., Polsley, S., Taele, P., Hammond, T.: A gaze gesture-based user authentication system to counter shoulder-surfing attacks. In: Proceedings of the 2017 CHI Conference Extended Abstracts on Human Factors in Computing Systems, pp. 1978–1986 (2017)
6. Zakaria, N.H., Griffiths, D., Brostoff, S., Yan, J.: Shoulder surfing defence for recall-based graphical passwords. In: Proceedings of the Seventh Symposium on Usable Privacy and Security (2011)
7. Karim, M., Heickal, H., Hasanuzzaman, M.: User authentication from mouse movement data using multiple classifiers. In: Proceedings of the 9th International Conference on Machine Learning and Computing, pp. 122–127 (2017)
8. Zheng, N., Paloski, A., Wang, H.: An efficient user verification system via mouse movements. In: Proceedings of the 18th ACM Conference on Computer and Communications Security, pp. 139–150 (2011)
9. Lei, H., Palla, S., Govindaraju, V.: Mouse based signature verification for secure internet transactions. In: Neural Networks and Machine Learning in Image Processing IX (2005)
10. Everitt, R.A.J., McOwan, P.W.: Java-based internet biometric authentication system. IEEE Trans. Pattern Anal. Mach. Intell. **25**(9), 1166–1172 (2003)
11. Kita, Y., Okazaki, N., Nishimura, H., Torii, H., Okamoto, T., Park, M.: Implementation and evaluation of shoulder-surfing attack resistant users. IEICE Trans. Inf. Syst. (Jpn. Edn.) **J97-D**(12), 1770–1784 (2014). (in Japanese)
12. Ishizuka, M., Takada, T.: CCC: shoulder surfing resistant authentication system by using vibration. In: IPSJ Interaction 2014, pp. 501–503 (2014). (in Japanese)

Phishing Detection Research Based on PSO-BP Neural Network

Wenwu Chen[1], Xu An Wang[1,2](✉), Wei Zhang[1,2], and Chunfen Xu[3]

[1] Key Laboratory for Network and Information Security of Chinese Armed Police Force, Engineering University of Chinese Armed Police Force, Xi'an, Shaanxi, China
wangxazjd@163.com

[2] Department of Electronic Technology, Engineering University of Chinese Armed Police Force, Xi'an, Shaanxi, China

[3] Jiaxing Vocational Technical College, Jiaxing, Zhejiang, People's Republic of China

Abstract. In order to effectively detect phishing attacks, this paper proposes a method of combining Particle Swarm Optimization with BP neural network to build a new phishing website detection system. PSO optimizes neural network parameters to improve the convergence performance of neural network detection model. Experimental results show that this algorithm can improve the prediction speed and the accuracy of detecting phishing websites by 3.7% compared with the conventional BP neural network algorithm.

1 Introduction

Phishing attacks are growing threats to cyber security in worldwide. According to the Phishing Activity Trends Report (the first half year of 2017) [1] released by the Anti-Phishing Working Group (APWG), from the fourth quarter of 2016 to the second quarter of 2017 with an increase of 65%, targeting a month more than 420 brands. This is the most frequent attack found since phishing was started in 2004 to track and report.

In order to obtain the user name, password, ID number, bank card number and other private information, the attackers attract unknown victims to click the fake websites and deceptive E-mails [2]. These criminals are usually profitable using phishing, so their goal usually is online banking, online payment platform, and mobile commerce applications. Researchers firstly developed the blacklist technology to combat phishing attacks [3]. Although URL blacklists have been somewhat effective, the attacker can bypass the blacklist system by slightly modifying the characters in the URL string, and the time of blacklist suspicious sites is relatively delayed and cannot effectively identify new phishing websites.

To make up for the shortcomings of blacklist technology, researchers have tried heuristic detection methods, such as CANTINA [4] and CANTINA+ [5], and the visual similarity test [6]. Recently, the use of machine learning algorithms to identify phishing links becomes the mainstream of current research [7–9]. BP neural network is a multi-layer feedforward network trained by error inverse propagation algorithm [10]. Its main characteristic is the signal is transmitted forward, the error propagates

L. Barolli et al. (Eds.): EIDWT 2018, LNDECT 17, pp. 990–998, 2018.
https://doi.org/10.1007/978-3-319-75928-9_91

backwards, the weights and thresholds are constantly adjusted by the back propagation of prediction error, so that the sum of squares of errors in the network is minimized. BP neural network has strong self-learning and self-adaptive abilities and with high fault tolerance. Its shortcomings mainly include slow convergence rate, easy fall into local optimum, sensitive to initial weights and thresholds, weaken network generalization ability and so on. Particle Swarm Optimization algorithm is a swarm intelligence optimization algorithm originated by Kennedy and Eberhart in 1995 [11], which is based on the simulation of bird flocks and fish prey. The algorithm has memory and it can improve the convergence speed of local area. Particle swarm optimization algorithm has been applied to traffic flow prediction, fault diagnosis, face recognition and many other aspects.

2 URL Feature Extraction and Analysis

2.1 Uniform Resource Locator Standard Format

The Uniform Resource Locator is a standard resource address on the Internet, and the entrance to a website. Uniform Resource Locator confusing is very common to phishing, to lure users to click on the URL to visit their phishing website is an important part of phishing. To increase the likelihood of users visiting phishing sites, phishing attackers often use deceptive URLs that are visually similar to the fake ones. The format of a standard URL is as follows:

Protocol://hostname[:port]/path/[;parameters][?query] #fragment

The common way to confuse URLs is to construct a phishing URL by partially modifying and replacing the host name part and the path part based on the target URL in order to confuse the user.

For example, the attackers using "www.amaz0n.com" as a fake Amazon website (the real URL is "www.amazon.com"), or using the "www.interface-transport.com/www.paypal.com/" as a fake PayPal website (the real URL is "www.paypal.com") and so on.

2.2 Extract the Features of the Uniform Resource Locator

The purpose of the attacker's phishing URL is to convince the user that this is a legitimate website. In this way, the cybercriminals can get the user's personal and leaked financial information [12]. In order to achieve this goal, attackers use some common methods to camouflage phishing links. Through the research on the common means of attacker, we have identified a set of features that can be used to detect if the URL is a phishing link:

Domain names exist in the Alexa ranking: Alexa ranking is a list of domain names ordered by the Internet. Most phishing sites are hacked into the legitimate sites or new domains. If the phishing attack is made on a hijacked website, then it is unlikely that the domain name will be a part of the TLD because the top-ranked domain names tend to have better security. If the phishing website is located in a newly registered domain name, the domain name will not appear in the Alexa rankings.

Subdomain length: The length of the URL subdomain. Phishing sites attempt to use their domain as their subdomain to mimic the URL of a legitimate website. Legitimate websites tend to have a short subdomain name.

URL length: Phishing URLs tend to be longer than legitimate URLs. Long URLs increase the likelihood of confusing users by hiding the suspicious part of the URL, which may redirect user-submitted information or redirect uploaded web pages to suspicious domain names.

Prefixes and suffixes in URLs: Phishers trick users by remodeling URLs that look like legitimate URLs.

Length ratio: Calculate the ratio between the length of the URL and the length of the path; phishing sites often have a higher proportion of legitimate URLs.

The "@" and "-" counts: The numbers of "@" and "-" in the URL. In the URL, the symbol "@" causes the browser to ignore inputs of previous and later redirects the users to the typed links.

Punctuation counts: The number of "! # $% &" in the URL. Phishing URLs usually have more punctuation.

Other TLDs: The number of TLDs displayed in the URL path. Phishing web links emulate legitimate URLs by using domain names and TLDs in the path.

IP address: The host name - part of the URL uses an IP address instead of a domain name.

Port Number: If a port number exists in the URL, verify that the port is included in a list of known HTTP ports, such as 21, 70, 80, 443, 1080 and 8080. If the port number is not in the list, mark it as a possible phishing URL.

URL Entropy: Calculate URL Entropy. The higher the entropy of the URL, the more complicated it is. Because phishing URLs tend to have random text, so we can try to find them by their entropy.

3 Algorithm Model

3.1 Particle Swarm Optimization Algorithm

The Particle Swarm Optimization (PSO) algorithm was proposed by Kennedy and Eberhart in 1995 on the predation behavior of birds. Imagine that there's a scene where a group of birds randomly search for food, in this area there is only one piece of food, all the birds do not know where the food is, and then what is the best strategy for finding food? The easiest and most effective way is to search for the area around the birds which are currently the closest to the food. PSO algorithm is inspired by this model to solve global optimization problems. In the PSO algorithm, a possible solution to each optimization problem is a bird in the search space, called a "particle". The fitness values of all particles are determined by an optimized function, and each particle have a velocity which determines the direction the birds fly and the displacement at each step. The particles then follow the current optimal particle search in the solution space. The PSO algorithm needs to initialize the particle set randomly (random solutions) and then find the optimal solution by iteration, in which particles update themselves by keeping track of the two "extrema". The first one is the best solution discovered by the particles themselves, which is called the extreme of the individual. The other is the best solution found for the entire population, which is called the global maximum.

Suppose that in an N-dimensional target search space, m particles form a community, and the i-th particle is represented as an N-dimensional vector:

$$\overrightarrow{x_i} = (x_{i1}, x_{i2}, \ldots, x_{iN}), i = 1, 2, \ldots, m$$

The i-th particle in the N-dimensional search space is $\overrightarrow{x_i}$. Each particle is a potential solution, $\overrightarrow{x_i}$ can be an objective function to calculate its fitness value, according to the size of the fitness value of measuring the pros and cons of $\overrightarrow{x_i}$. The i-th particle's velocity is also an N-dimensional vector, denoted $\overrightarrow{v_i} = (v_{i1}, v_{i2}, \ldots, v_{iN})$, and the velocity determines the displacement of particles in the search space unit iteration. The optimal position searched for the i-th particle swarm so far is: $\overrightarrow{p_i} = (p_{i1}, p_{i2}, \ldots, p_{iN})$. The optimal location searched for the entire particle swarm so far is:

$$\overrightarrow{p_g} = (p_{g1}, p_{g2}, \ldots, p_{gN}).$$

Particles update their speed and position according to the following formula:

$$v_{in}(t+1) = \omega \cdot v_{in}(t) + c_1 r_1 (p_{in} - x_{in}(t)) + c_2 r_2 (p_{gn} - x_{in}(t)) \quad (1)$$

$$x_{in}(t+1) = x_{in}(t) + v_{in}(t+1) \quad (2)$$

Among them, $i = 1, 2, \ldots, m; n = 1, 2, \ldots, N$, learning factors c1, c2 are learning factors, inertia weights, r1 and r2 are random numbers between [0, 1]. The basic principle of PSO algorithm is the constant set by the user. Iteration termination conditions are generally selected as the maximum number of iterations or PSO search to date to meet the adaptation threshold.

3.2 PSO-BP Neural Network

This paper uses the principle of optimization of PSO algorithm which is introduced into BP neural network, which makes the algorithm have stronger global optimal search abilities. In this paper, URL features are taken as input variables of neural network, the result of judgment is taken as output variable, the optimal initial parameters of neural network are searched by the movement and update of particles to overcome the disadvantage of BP neural network that it is easy to fall into the local optimum and the convergence speed is slow.

PSO-BP algorithm flow is as follows:

Step 1: Initialize BP neural network and PSO

Including setting the number of neurons in the input layer, hidden layer, output layer of the network and the learning rate α, and the input and output of the training samples, the size of the particle N and the position vector and velocity vector of each particle, the individual extremum and the global optimum of each particle, iteration error precision ε, constant coefficients c1 and c2, the maximum inertia weight $\omega_{\max}$, the minimum inertia weight $\omega_{\min}$, the maximum speed $v_{\max}$ and the maximum number of iterations.

Step 2: Iteration Update

(1) Update the velocity of each particle by using Eq. 1, and judge whether the updated velocity is larger than the maximum velocity $v_{\max}$. If it is larger than the maximum velocity, the updated velocity is taken as the maximum velocity $v_{\max}$; otherwise, it remains unchanged.
(2) Use Eq. 2 to update the position of each particle.
(3) Use Eq. 3 to calculate the fitness value of each particle:

$$f_i = \frac{1}{n_t}\sum_{q=1}^{n}(O_{iq} - T_{iq})^2 \tag{3}$$

Among which, n_t is the number of training samples; O_{iq}, T_{iq} are the network weights and the actual output and expected output of the network under the i-th particle position and the training sample q, respectively.

(4) Calculate the global minimum fitness $f_g = \min\{f_1, f_2, ..., f_N\}$ of the particle swarm optimization; if the current iteration number reaches the maximum number of iterations or $f_g < \varepsilon$ (the training error of the network reaches the precision requirement), the iteration stops and goes to step 3; Otherwise, the individual extreme $\overrightarrow{P}_i$ and global extreme $\overrightarrow{P_g}$ positions of each particle are calculated, go to Step 1 of the iterative update continue to update the speed and position of the particles.

Step 3: Output the network weight and threshold determined by the position of the global extremum, and the algorithm ends (Fig. 1).

3.3 Performance Evaluation

The purpose of phishing websites detection is to detect phishing instances from the test data set that contains phishing websites and legal websites, which is essentially a binary classification essence. In binary classification, a total of four kinds of classification, used to measure the accuracy of classification confusion matrix (Fig. 2).

Each URL falls into one of the four possible categories: true positive (TP, correctly classified phishing URL), true negative (TN, correctly classified as non-Phishing URL), false positives (FP, non-phishing URLs are incorrectly classified as phishing) and false negatives (FN, phishing URLs are incorrectly classified as non-phishing). Standard measures, such as accuracy, precision, recall, false negative rate, were determined using the following equation:

$$Accuracy = \frac{TN + TP}{TN + FP + TP + FN} \tag{4}$$

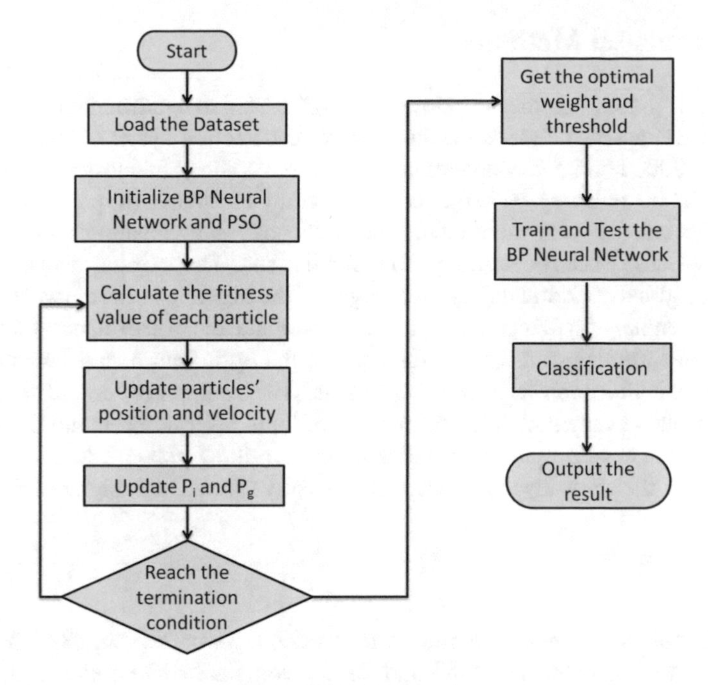

Fig. 1. PSO-BP flowchart

		Predicted Class	
		Positive	**Negative**
Actual Class	**Positive**	**TP**	**FN**
	Negative	**FP**	**TN**

Fig. 2. Confusion matrix

$$Precision = \frac{TP}{FP + TP} \tag{5}$$

$$Recall = \frac{TP}{FN + TP} \tag{6}$$

$$FNR = \frac{FN}{TP + FN} \tag{7}$$

4 Experimental Methods

This paper is based on the Windows platform Matlab environment to achieve a PSO-based BP neural network phishing website detection system. The dataset used consisted of 2000 legitimate websites collected from Yahoo Directory (http://dir.yahoo.com/) and 2,000 phishing websites collected from Phishtank (http://www.phishtank.com/). Collected data sets carry label values, “legal” and “phishing”. In this data set randomly selected 70% for training, 30% for the test. The training dataset is used to train the neural network and adjust the weight of the neurons in the network, while the test dataset remains unchanged and used to evaluate the performance of the neural network. After training, run the test data set on the optimized neural network.

Predict phishing websites using optimized BP neural network. The above ten features are taken as input, that is, the number of input layer nodes in the BP network is 10 and the number of output layer nodes is one. Training network to choose a strong adaptability of the three-layer BP network, incentive function is sigmoid function:

$$f(x) = \frac{1}{1 + e^{-x}} \tag{8}$$

The input layer uses linear neurons, with hidden layer nodes set to 8. After repeated experiments, the parameters of PSO and BP networks are set as follows: the learning rate of BP neural network is 0.7 and the momentum factor is 0.9; The PSO-based training algorithm has a population size of n = 30, an initial inertia weight of 0.9 and decreases linearly to 0.5 with the number of iterations, c1 = c2 = 2.

In order to better illustrate the accuracy of the algorithm in this paper, an ordinary BP neural network is used to test the experimental data set. By experimenting with the selected data set, the results show that PSO and BP neural network are better than normal BP neural network, and their prediction accuracy is higher than that of ordinary BP network (Table 1).

Table 1. Evaluations of PSO-BP neural network and BP neural network

Method	Accuracy	Precision	Recall	FNR
BP	0.9522	0.9518	0.9603	0.0397
PSO-BP	0.9895	0.9854	0.9881	0.0119

5 Conclusion and Discussion

In the phishing site testing process, many factors affect the test results, with a certain degree of non-linearity, if you simply use the BP algorithm, you need a longer convergence time, and it’s easy to fall into the local minimum. PSO algorithm has the characteristics of global search optimal solution, this paper uses particle swarm optimization algorithm to optimize BP neural network, and then uses BP neural network to train, so that the convergence speed of the site to speed up to avoid falling into a local minimum, and enhance the adaptive Process search capability. Experiments show that using BP algorithm to optimize BP neural network can significantly improve the detection accuracy of phishing websites.

Acknowledgments. The authors thank all the reviewers and editors for their valuable comments and works. This paper is supported by National Key Research and Development Program of China (Grant No. 2017YFB0802000), National Natural Science Foundation of China (Grant Nos. U1636114, 61772550, 61572521), Natural Science Basic Research Plan in Shaanxi Province of China (Grant No. 2016JQ6037).

References

1. Phishing Activity Trends Report 1st Half. Methodology (2017)
2. PhishTank: Free community site for anti-phishing service. http://www.phishtank.com/
3. Sinha, S., Bailey, M., Jahanian, F.: Shades of grey: on the effectiveness of reputation-based "blacklists". In: International Conference on Malicious and Unwanted Software, pp. 57–64. IEEE (2008)
4. Zhang, Y., Hong, J.I., Cranor, L.F.: Cantina: a content-based approach to detecting phishing web sites. In: International Conference on World Wide Web, WWW 2007, Banff, Alberta, Canada, DBLP, pp. 639–648, May 2007
5. Xiang, G., Hong, J., Rose, C.P., et al.: CANTINA+: a feature-rich machine learning framework for detecting phishing web sites. ACM Trans. Inf. Syst. Secur. **14**(2), 21 (2011)
6. Wenyin, L., Huang, G., Xiaoyue, L., et al.: Detection of phishing webpages based on visual similarity. In: Special Interest Tracks and Posters of the 14th International Conference on World Wide Web, pp. 1060–1061 (2005)
7. Ma, J., Saul, L.K., Savage, S., et al.: Beyond blacklists: learning to detect malicious web sites from suspicious URLs. In: ACM SIGKDD International Conference on Knowledge Discovery and Data Mining, Paris, France, 28 June–1 July 2009, DBLP, pp. 1245–1254 (2009)
8. Choi, H., Zhu, B.B., Lee, H.: Detecting malicious web links and identifying their attack types. In: Usenix Conference on Web Application Development, p. 11 (2011)
9. Ma, J., Saul, L.K., Savage, S., et al.: Identifying suspicious URLs: an application of large-scale online learning. In: International Conference on Machine Learning, pp. 681–688. ACM (2009)
10. Sadeghi, B.H.M.: A BP-neural network predictor model for plastic injection molding process. J. Mater. Process. Technol. **103**(3), 411–416 (2000)
11. Kennedy, J., Eberhart, R.: Particle Swarm Optimization. Springer, US (2011)
12. Ma, J., Saul, L.K., Savage, S., et al.: Learning to detect malicious URLs. ACM Trans. Intell. Syst. Technol. **2**(3), 1–24 (2011)
13. Sahoo, D., Liu, C., Hoi, S.C.H.: Malicious URL detection using machine learning: a survey (2017)
14. Lakshmi, V.S., Vijaya, M.S.: Efficient prediction of phishing websites using supervised learning algorithms. Procedia Eng. **30**(9), 798–805 (2012)
15. Marchal, S., Francois, J., State, R., et al.: PhishScore: hacking phishers' minds. IEEE (2014)
16. Chang, J.H., Lee, K.H.: Voice phishing detection technique based on minimum classification error method incorporating codec parameters. IET Sig. Process. **4**(5), 502–509 (2010)
17. Chang, J., Venkatasubramanian, K.K., West, A.G., et al.: Analyzing and defending against web-based malware. ACM Comput. Surv. **45**(4), 49 (2013)

18. Rao, R.S., Ali, S.T.: PhishShield: a desktop application to detect phishing webpages through heuristic approach. Procedia Comput. Sci. **54**, 147–156 (2015)
19. Shekokar, N.M., Shah, C., Mahajan, M., et al.: An ideal approach for detection and prevention of phishing attacks. Procedia Comput. Sci. **49**, 82–91 (2015)
20. Jain, A.K., Gupta, B.B.: A novel approach to protect against phishing attacks at client side using auto-updated white-list. EURASIP J. Inf. Secur. **2016**(1), 9 (2016)
21. Kim, D., Achan, C., Baek, J., et al.: Implementation of framework to identify potential phishing websites, p. 268. IEEE (2013)
22. Garera, S., Provos, N., Chew, M., et al.: A framework for detection and measurement of phishing attacks. In: ACM Workshop on Recurring Malcode, pp. 1–8. ACM (2007)
23. Olivo, C.K., Santin, A.O., Oliveira, L.S.: Obtaining the threat model for e-mail phishing. Appl. Soft Comput. J. **13**(12), 4841–4848 (2013)
24. Herzberg, A., Jbara, A.: Security and identification indicators for browsers against spoofing and phishing attacks. ACM Trans. Internet Technol. **8**(4), 1–36 (2008)
25. Pan, Y., Ding, X.: Anomaly based web phishing page detection. In: 2006 Computer Security Applications Conference, ACSAC 2006, pp. 381–392. IEEE (2006)
26. Fu, A.Y., Liu, W., Deng, X.: Detecting phishing web pages with visual similarity assessment based on earth mover's distance (EMD). IEEE Trans. Dependable Secure Comput. **3**(4), 301–311 (2006)

Evaluation of Index Poisoning Method in Large Scale Winny Network

Kentaro Aburada[1(✉)], Yoshihiro Kita[2], Hisaaki Yamaba[1], Tetsuro Katayama[1], Mirang Park[3], and Naonobu Okazaki[1]

[1] University of Miyazaki, 1-1 Gakuen-Kibanadai-Nishi, Miyazaki 889-2192, Japan
aburada@cs.miyazaki-u.ac.jp

[2] Tokyo University of Technology, 1404-1 Katakura, Hachioji, Tokyo 192-0982, Japan

[3] Kanagawa Institute of Technology, 1030 Shimo-Ogino, Atsugi, Kanagawa 243-0292, Japan

Abstract. In recent years, P2P file-sharing networks are used all over the world. This has led to social problems such as the illegal distribution of copyrighted material and the leakage of personal information through computer viruses because P2P does not have a control function for file distribution. To address these issues, a control method called index poisoning has been studied. However, problems such as pollution in the network index and a need to increment control traffic have been reported when index poisoning is applied to a P2P file-sharing network. We propose a method that implements dynamic clustering to limit the range of index poisoning as a solution for these problems that also maintains the effectiveness of the control function. In this study, we implement index poisoning on the large-scale Winny network and evaluate the system performance.

1 Introduction

In recent years, reliable broadband lines have spread as part of development of communication technology, making possible the distribution of a wide variety of content. Peer-to-peer (P2P) file-sharing software is one means of performing such distribution in the broadband era, as it can send and receive files efficiently by utilizing high-speed lines. Many research have focused on P2P, Grid and Cloud [1–4]. In Japan, pure P2P-type networks such as Winny [5] and Share have been widely used. Currently, however, there are some problems due to the leakage of personal information by computer viruses and the distribution of copyrighted software. Since the network itself does not have a mechanism to act as a content distribution control system, it cannot stem further distribution of such content. At the same time, the amount of Internet traffic has increased year by year, and Japanese domestic Internet traffic has reached 9.6 Tbps as of 2017. P2P file-sharing software is said to be one factor in this increase. In view of the above, it is necessary to consider a content distribution control system to prevent traffic from increasing excessively. As a solution, a technique called

L. Barolli et al. (Eds.): EIDWT 2018, LNDECT 17, pp. 999–1006, 2018.
https://doi.org/10.1007/978-3-319-75928-9_92

poisoning has been studied [6]. Poisoning extends the time to download illegal files. Thereby, it is a mechanism for causing users to give up on the download. In addition, it has a feature referred to as "measures for restoration" after such a file is found. Therefore, it is very useful for dealing with not only the distribution of copyrighted software but also the leakage of personal information. There are three types of poisonings: content poisoning, index poisoning, and parameter poisoning. Among them, index poisoning is relatively small the amount of traffic [7]. We previously proposed a method that implements dynamic clustering to limit the range of index poisoning in order to solve the abovementioned problems while maintaining an effective control function effective, and we implemented this method on the Winny network [8]. In this paper, we evaluate the system performance on the large-scale Winny Network.

2 Overview of the Winny Network

Winny is a pure P2P-type file-sharing network. In pure P2P, the network is constructed by interconnections between nodes. It has excellent fault tolerance and scalability because pure P2P does not have a central server for managing the network as in the case of hybrid P2P. Each node of the Winny network has a cache, the file information of the other nodes and files. The file information is referred as the key. Various metadata such as file ID and location information of a specific file are recorded in the key. A node exchanges keys with adjacent nodes in order to search for files. The key has a set lifetime, after which the key will gradually be erased from the network. New copies of the key are continuously being distributed while the owner of the file remains on the network. A key is erased after its owner is disconnected from the network. In other words, if other users cannot get a file from the owner, then the file key is also automatically deleted. Moreover, Winny has no mechanism that acts as a content distribution control system, which is also true of other pure P2P networks, and it is

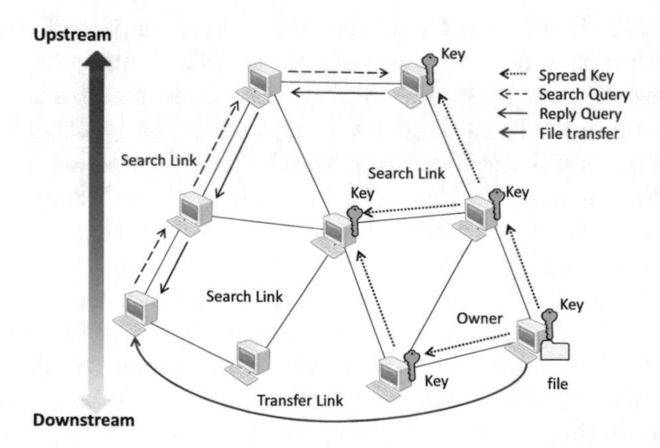

Fig. 1. Outline of Winny network.

very difficult to completely remove files once they have been distributed on the network. Figure 1 shows an outline of the Winny network. Winny has search links and transfer links. A search link is used to search for files and transfer the keys, whereas a transfer link is used for transferring the file caches. The nodes which have high-speed lines are designated as upstream nodes, whereas nodes which have slow lines are designated as downstream nodes. Upstream nodes have increased opportunity to exchange keys or file caches.

2.1 How to Distribute Files

The owner generates file caches and a key when distributing files. These are constructed from the owner's file which will be published. In order to get the owner's file, the request node must obtain the owner's key through a search link. The request node creates the search query with keywords. The search query is transferred to the 6-hop. The intermediate nodes which have the owner's key and have received the search query then transfer the key to the request node through the same path. The request node gets the file caches of the files of interest by using the location information from owner's key. There is an exception to this procedure. In Winny, some keys are rewritten to the address of a previous node with a certain probability. Therefore, this previous node who is not the owner may be requested to transfer file caches by the request node. This previous other node acts as a relay between the request node and the owner by using a transfer link. Moreover, the other node maintains the file caches after relaying it. By the above-described mechanism, the keys and file caches are distributed over all of the network, and the keys are rewritten to the addresses of these various nodes. It thus becomes very difficult over time to determine the original file owner in this network.

2.2 Clustering Mechanism

Winny has a clustering mechanism whereby a new node selects the nearest node by matching their tastes. For this purpose, up to three cluster words are set for each node, and the more cluster words of the nodes that coincide with the cluster, the higher of a priority is placed. Thus, it is possible for request nodes and nodes with the desired key to connect to each other effectively with fewer hops.

Table 1. Example of a dummy key.

Key	Contents
File name	The same name as the original
File size	The same size as the original
File ID	Random
File location	Fake IP address

3 Index Poisoning

In index poisoning, a control node modifies the keys for the purpose of falsification, and the request node becomes unable to obtain the files of interest. In Winny, index poisoning can be realized by setting a dummy key like shown in Table 1. The dummy key is set with the same name and size as the original key, but the file id and file location are fake. The request node cannot get the files of interest because the one accesses a node which does not exist. Although index poisoning can extend the time needed to download the target files, if target files increase, the index of the entire network is contaminated [9].

With our proposed method, we aim to solve the following two problems using index poisoning.

- As the number of control nodes increases, the index of the entire network becomes contaminated.
- A large amount of dummy keys are distributed over the Winny network.

First, we introduce a clustering function. As a rule, an index poisoning sets multiple nodes as control nodes. As the number of control nodes increases, control traffic and network pollution increase. In our proposed method, we introduce a clustering function in order to limit the target cluster (Fig. 2).

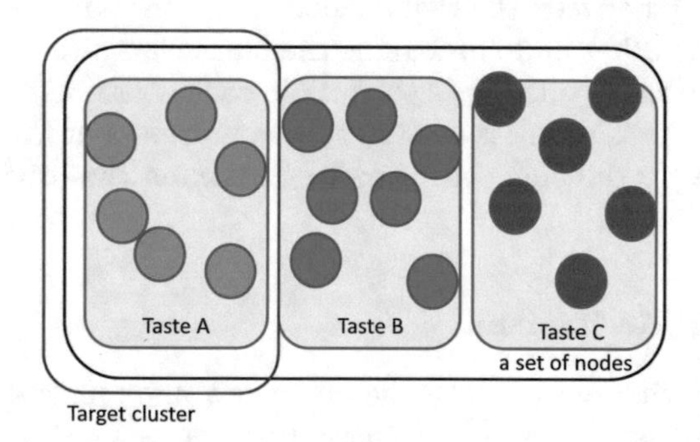

Fig. 2. Clustering function.

4 Proposed Method

4.1 Overview of the System

Our system rewrites memory and injects code into the official Winny client (Winny v2.0b7.1) to realize the index poisoning. We implement the system as a Hook DLL (Fig. 3).

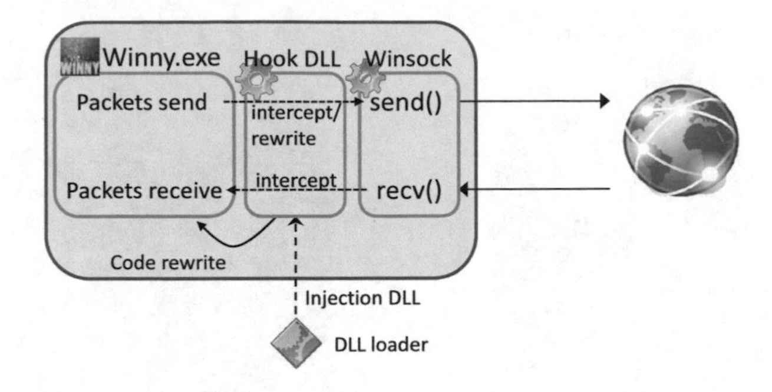

Fig. 3. Summary of Hook DLL.

4.2 Process of Index Poisoning

The process of index poisoning is explained below.

1. The control node sets the keywords most relevant to the target file.
2. The control node continues to collect the information of the nodes, and the node with the most relevant keywords is added to its cluster. Then the control node collects the keys from the adjacent nodes.
3. The control node sends the dummy keys of the target files.

5 Evaluation

We evaluate the proposed method in terms of the following two points.

- Average time required to download the target files.
- Index pollution rate.

In the Winny network, the number of nodes increases from upstream to downstream.

5.1 Simulation Environment

We create a virtual machine using VirtualBox. The key lifetime is set to 1500 s (default), 3600 s, and 5400 s. Figure 4 shows an evaluation environment with 23 nodes. The measure node downloads the target file. The control node performs index poisoning against the target file. The file owner shares the target file and 20 non-target files. The total number of files that exist on the network is 1000. This evaluation environment has 3 clusters. Each cluster has 2, 2, and 1 nodes from upstream to downstream. In addition, 5 general nodes exist which do not join any cluster.

Figure 5 shows an evaluation environment with 47 nodes. The file owner shares the target file and 40 non-target files. The total number of files that exist on the network is 2000. This evaluation environment has 3 clusters. Each cluster has 2, 4, and 7 nodes from upstream to downstream.

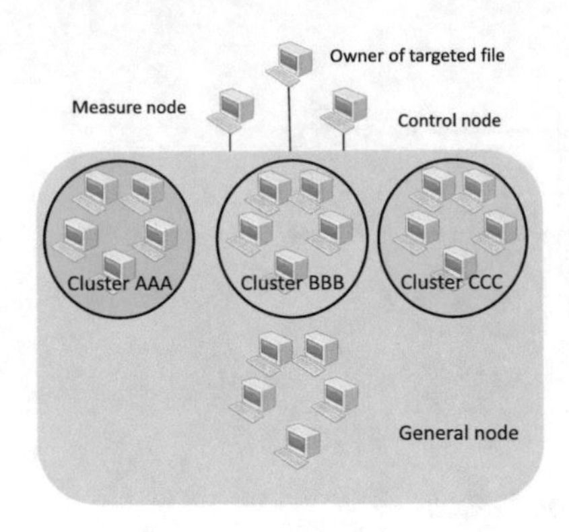

Fig. 4. Evaluation environment with 23 nodes.

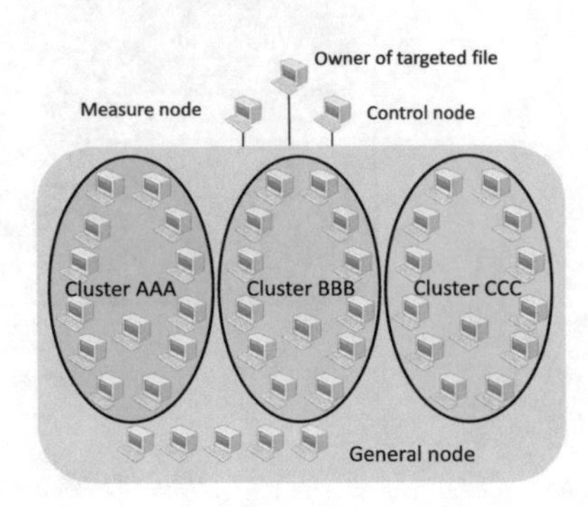

Fig. 5. Evaluation environment with 47 nodes.

5.2 Simulation Results

Figures 6 and 7 show the average time required to download the target files for 23 and 47 nodes, respectively. In the case of 23 nodes, it can be seen that having a key lifetime of 3600 s requires more time than either setting the key lifetime to 1500 or 5700 s. That setting the key lifetime to 5700 s gives the shortest download time was unexpected. In contrast, in the case of 47 nodes, setting the key lifetime to 5700 s gives the longest download time. In general, a key having a longer lifetime is more useful for poisoning.

For the case of 23 nodes, Figs. 8, 10 and 12 show the index pollution rate for key lifetimes of 1500 s, 3600 s, and 5700 s, respectively. In Figs. 8 and 12, the index pollution rate remains low, in the range from 0.6% to 1.5%. In contrast, Fig. 10 stays around 2.0% with ideal data.

In the case of 47 nodes, Figs. 9, 11 and 13 show the index pollution rate for key lifetimes of 1500 s, 3600 s, and 5700 s, respectively. As shown, a higher index pollution rate is maintained than in the case of 23 nodes. However, a number of large peaks appear, indicating that the index pollution rate is not stable.

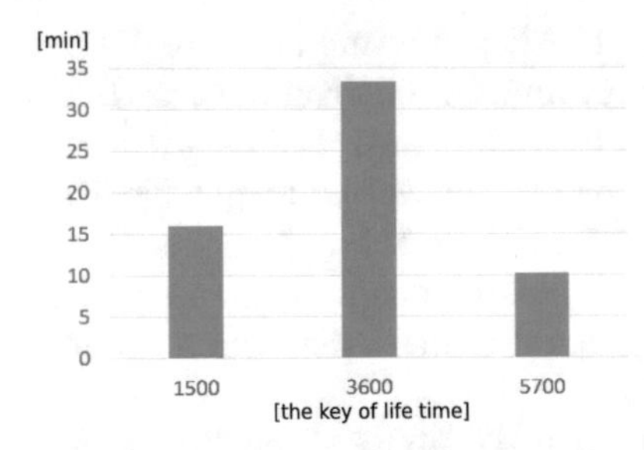

Fig. 6. Average time required to download the target files with 23 nodes.

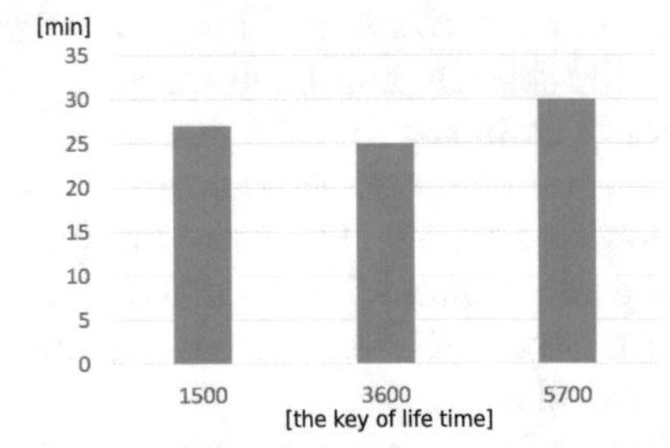

Fig. 7. Average time required to download the target files with 47 nodes.

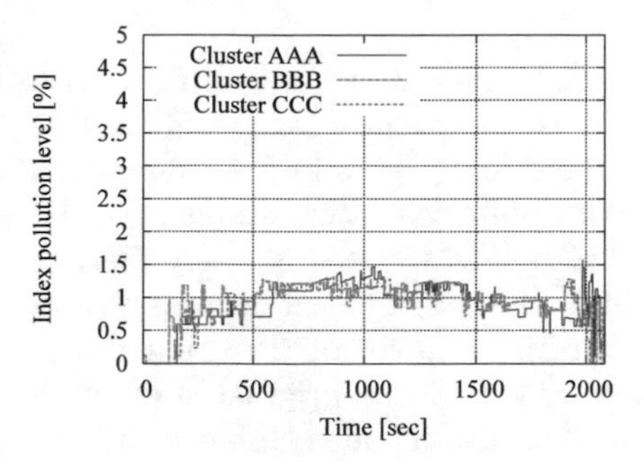

Fig. 8. Index pollution rate with 23 nodes and a key lifetime of 1500 s.

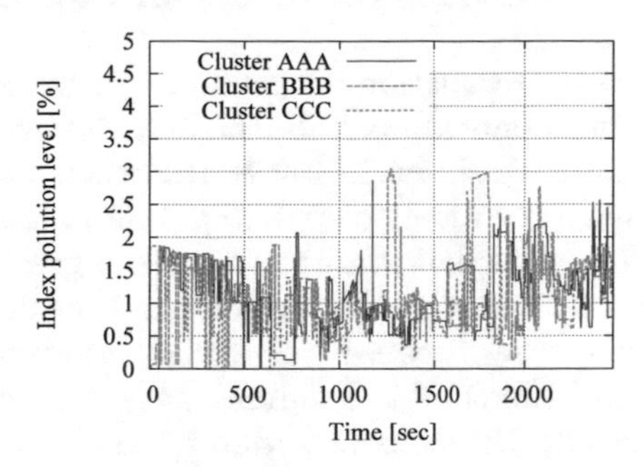

Fig. 9. Index pollution rate with 47 nodes and a key lifetime of 1500 s.

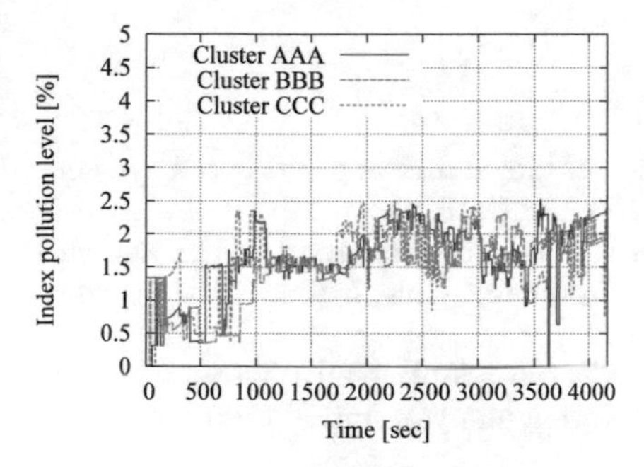

Fig. 10. Index pollution rate with 23 nodes and a key lifetime of 3600 s.

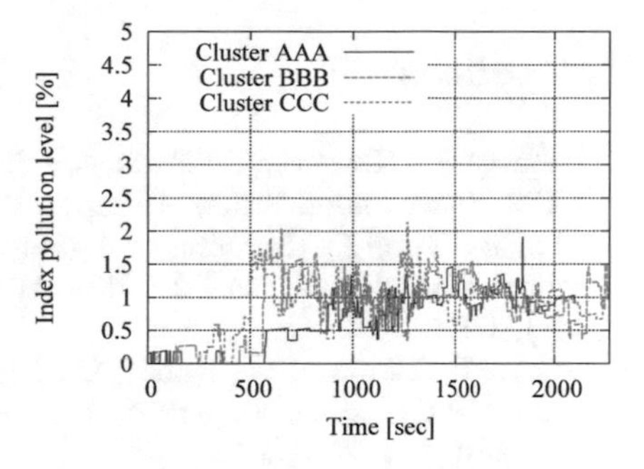

Fig. 11. Index pollution rate with 47 nodes and a key lifetime of 3600 s.

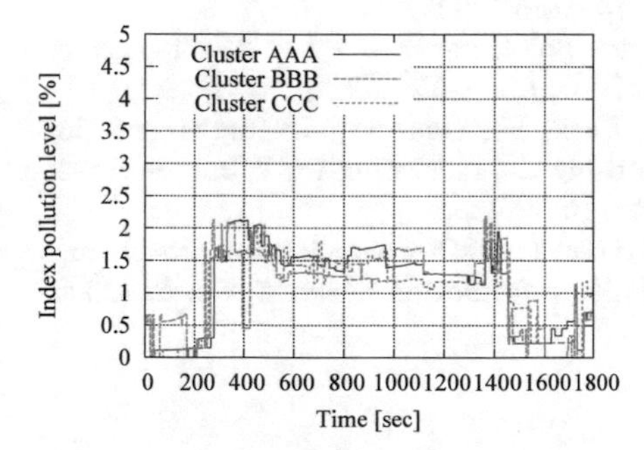

Fig. 12. Index pollution rate with 23 nodes and a key lifetime 5700 s.

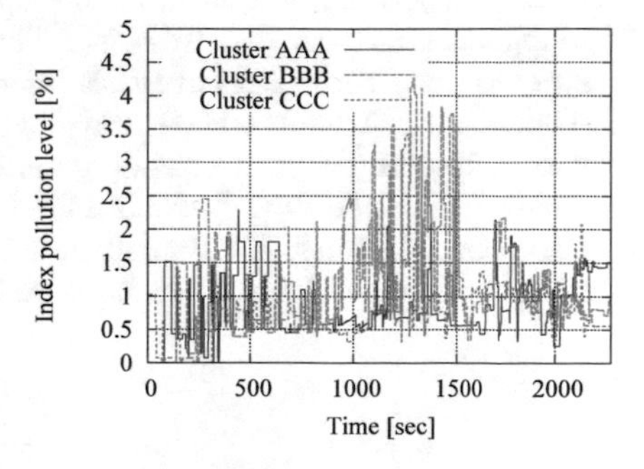

Fig. 13. Index pollution rate with 47 nodes and a key lifetime of 5700 s.

6 Discussion and Conclusions

In the evaluation of system performance for the large-scale Winny network with the proposed method, the index pollution rate is consistently low, from 0.6% to 1.5%, as shown in Fig. 8. It is thought that the dummy keys in the simulations were not widely distributed. In Fig. 12, the index pollution rate is stable from 400 to 1400 s; however, afterwards it greatly decreases, which is why a key lifetime of 5700 s results in the shortest download time. The index pollution rate greatly decreases because the owner periodically distributes copies of the original key, and therefore the dummy keys may be overwritten by the original keys. As a result, the index pollution rate for the case of 47 nodes remains stable with ideal data, regardless of the key lifetime. In future work, we will introduce multi-control nodes and evaluate the proposed method on a larger-scale network.

Acknowledgements. This work was supported by JSPS KAKENHI Grant Numbers JP17H01736 and JP17K00139.

References

1. Hegarty, R., Haggerty, J.: Extrusion detection of illegal files in cloud-based systems. Int. J. Space Based Situat. Comput. **5**(3), 150–158 (2015)
2. Cha, B., Kim, J.: Handling and analysis of fake multimedia contents threats with collective intelligence in P2P file sharing environments. Int. J. Grid Util. Comput. **4**(1), 1–9 (2013)
3. Gueye, B., Flauzac, O., Rabat, C., Niang, I.: A self-adaptive structuring for large-scale P2P Grid environment: design and simulation analysis. Int. J. Grid Util. Comput. **8**(3), 254–267 (2017)
4. Javanmardi, S., Shojafar, M., Shariatmadari, S., Ahrabi, S.S.: FR trust: a fuzzy reputation-based model for trust management in semantic P2P grids. Int. J. Grid Util. Comput. **6**(1), 57–66 (2015)
5. Winny. https://en.wikipedia.org/wiki/Winny
6. Liang, J., Kumar, R., Xi, Y., Ross, K.W.: Pollution in P2P file sharing systems. In: Proceedings of IEEE Infocom 2005, pp. 1174–1185 (2005)
7. Liang, J., Naoumov, N., Ross, K.: The index poisoning attack in P2P file sharing systems. In: Proceedings of IEEE Infocom 2006, pp. 1–12 (2006)
8. Aburada, K., Yamaba, H., Katayama, T., Park, M., Okazaki, N.: Implementation and evaluation of index poisoning method using the clustering for Winny network. Trans. IPS Japan **56**(12), 2395–2405 (2015). (in Japanese)
9. Naoumov, N., Ross, K.: Exploiting P2P system for DDos attacks. In: Proceedings of the 1st International Conference on Scalable Information Systems, vol. 47 (2006)

A Hybrid Technique for Residential Load Scheduling in Smart Grids Demand Side Management

Muhammad Hassan Rahim, Adia Khalid, Ayesha Zafar, Fozia Feroze, Sahar Rahim, and Nadeem Javaid(✉)

COMSATS Institute of Information Technology, Islamabad 44000, Pakistan
nadeemjavaidqau@gmail.com
http://www.njavaid.com

Abstract. Demand side management (DSM) and demand response (DR) are the key functions in smart grids (SGs). DR provides an opportunity to a consumer in making decisions and shifting load from on-peak hours to off-peak hours. The number of incentive base pricing tariffs are established by a utility for the consumers to reduce electricity consumption and manage consumers load in order to minimize the peak to average ratio (PAR). Throughout the world, these different pricing approaches are in use. Time of use tariff (ToU) is considered in this paper, to comparatively evaluate the performance of the heuristic algorithms; bacterial foraging algorithm (BFA), and harmony search algorithm (HSA). A hybridization of BFA and HSA (HBH) is also proposed to evaluate the performance parameters; such as electricity consumption cost and PAR. Furthermore, consumer satisfaction level in terms of waiting time is also evaluated in this research work. Simulation results validate that proposed scheme effectively accomplish desired objectives while considering the user comfort.

Keywords: Bacterial foraging algorithm (BFA)
Harmony search algorithm (HSA) · Smart grids (SGs)
Demand response (DR) · Peak to average ratio (PAR)

1 Introduction

Numerous challenges are faced by the electric power industry. The reliability of existing power grid is affected by the increase in power demand, a limited amount of natural resources, and aging infrastructure. Peak generators running on fossil fuels and natural gas are used by utilities to satisfy additional energy demand, ultimately causing environmental issues as these generators are a great source of emitting plenty of harmful gasses. Therefore, a need of more reliable, sustainable, and an efficient power grid system has emerged. In order to make power grids more reliable, sustainable, and robust. An intelligent and revolutionary smart

L. Barolli et al. (Eds.): EIDWT 2018, LNDECT 17, pp. 1007–1017, 2018.
https://doi.org/10.1007/978-3-319-75928-9_93

grid (SG) infrastructure needs is established. Advanced two-way communication capabilities such as advanced metering infrastructure (AMI), smart meter (SM) are incorporated in SG, to improve efficiency, safety, and control. Various methods; such as distributed energy, smart pricing, and demand response (DR) are devoted to facilitates this continuously evolving infrastructure. It is observed that more than 65% of the reduction in the electricity consumption is achieved by residential sector and small commercial building [1]. Home energy management (HEM) system plays a vital role to enhance the efficiency of SG.

Demand side management (DSM) and DR are the two major components of SG. DSM strategies are adopted by many utility companies. In order to motivate consumers to efficiently use electricity, monetary incentives are also provided to the consumers. So, consumers voluntary use electricity in an optimal way and avoid electricity wastage. This provides a balance between supply and demand.

DR programs are highly amenable and offers numerous benefits to the consumers [2]. It persuades consumers to modify their electricity usage pattern by shifting their high load from on-peak hours to off-peak hours in response to varying energy price. This facilitates in reducing the aggregated electricity consumption cost and PAR by efficiently managing power consumption pattern from which both consumers and utility get benefits.

Numerous energy pricing tariffs are established and are in used around the globe; such as real-time pricing (RTP), critical peak pricing scheme (CPP), day ahead pricing (DAP), time of use (ToU), inclined block rate (IBR), etc. Energy consumption minimization, minimization of green house gas emissions, efficient load management to reduce PAR and user comfort are some of the major problems in the residential sector of SGs.

Electricity consumption cost reduction, optimal management of gross load, making the grid more sustainable, and PAR reduction are some of the common objectives of SG. Different heuristic algorithms and DSM strategies are proposed to achieve these objectives in the past. Researchers use linear programming (LP), mixed integer linear programming (MILP), mixed integer nonlinear programming (MINLP), etc. In order to minimize electricity cost and optimally schedule household appliances to manage the load.

In this paper, we proposed a HEM system on the basis of two heuristic algorithms; bacterial foraging algorithm (BFA), harmony search algorithm (HSA). A hybrid of bacterial foraging and harmony search (HBH) is also proposed. These three heuristic approaches are used for comparative evaluation on the basis of performance parameters; electricity consumption cost reduction and PAR. User comfort in term of waiting time is also calculated. ToU energy pricing tariffs are used to calculate electricity consumption cost and PAR. For simulations, control parameters are kept same for all three algorithms. In the coming section, related work is discussed.

2 Related Work

In recent years, extensive research is going on in SG domain. To cope up with the challenges of this domain, such as, electricity consumption management, minimization of electricity consumption cost and PAR, and maximization of user comfort. Many heuristic techniques are in used to make SG more reliable, stable, and efficient. These techniques manages electricity load effectively and minimize electricity consumption cost. In this regard, some of the papers are discussed in this section.

A novel energy management system (EMS) for DR is proposed in [1] for residential and small commercial buildings. They formulate a fully automated EMS's rescheduling problem as a reinforcement learning (RL) problem, as this formulation does not require explicit modeling of the consumers dissatisfaction on job rescheduling. This enables the EMS to self-initiate jobs and allow the user to initiate more flexible requests. A mixed integer nonlinear optimization model is proposed on the base of ToU electricity tariff in [2]. Residential demand response is analyzed by the scheduling of home appliances, to minimize electricity cost. Consumers achieves more than 25% cost saving just by shifting their energy consumption in response to changing price. These relative incentives attract consumers to participate in DR programs. Electricity consumption cost minimization and efficient management of load consumption in on-peak hours is achieved.

Sherazi and Jadid proposed a home energy management with distributed energy resources (DERs) along with thermal and Electrical appliance scheduling (HEMDAS) [3]. They use dynamic pricing scheme to schedule the controllable appliances in off-peak hours and avoid on-peak hours in order to minimize electricity consumption cost. The energy management problem is modeled as MINLP using dynamic pricing scheme. 24 h time horizon is divided into 48 slots, 30 min each. Thermal resources are used during peak hours; consequently, minimizing the electricity consumption cost during on-peak hours and off-peak hours up to 22.2% and 11.7% respectively. Authors evaluate the performance of home energy management controller in [4], based on three heuristic algorithms; genetic algorithm (GA), binary particle swarm optimization (BPSO) and ant colony optimization (ACO). They also propose a generic architecture for DSM, to integrate residential area domain with smart area domain. Problem formulation is done by using multiple knapsack problem. ToU and inclined block rate (IBR) tariffs are used for energy pricing. Results validates successful achievement of objectives; consumption cost minimization and user comfort maximization. Trade-off between consumption cost and consumer comfort exists.

Authors in [5], developed a master controller for residential DR based on heuristic techniques. The objectives of optimization is to minimize cost along with user comfort maximization. They also compared the results of heuristic algorithms with linear programming. The proposed master controller is able to efficiently manage the load to minimize cost an user discomfort while minimizing PAR. In [6], proposed model based on ToU pricing signal to manage and control appliances for multiple consumers in a home. The proposed algorithm manages and schedule appliances based on multiple consumers preference. Two scenarios are implemented based on multiple consumers and their priorities to evaluate the performance parameters. Results show reduction in electricity consumption cost and energy consumption pattern. In [7], authors proposed a model for the scheduling problem focusing consumers satisfaction levels. Mixed integer programming (MIP) is used for defining a MOO problem. The Pareto front is considering on problem instances based on real-world household usage data and real-world electricity prices, in order to analyze the relation between two objectives. Authors divide households into five groups; family with children (FWC) and without children (FNC), multiple pensioners (MP), single pensioner (SP) and single non-pensioned (SNP). Energy consumption of each household and each individual appliance is monitored for every 10 min. Trade-off between the production cost and consumer comfort is observed. In order to minimize the consumption cost, consumers have to compromise their satisfaction level.

Proposed scheme is discussed in detail in the upcoming section.

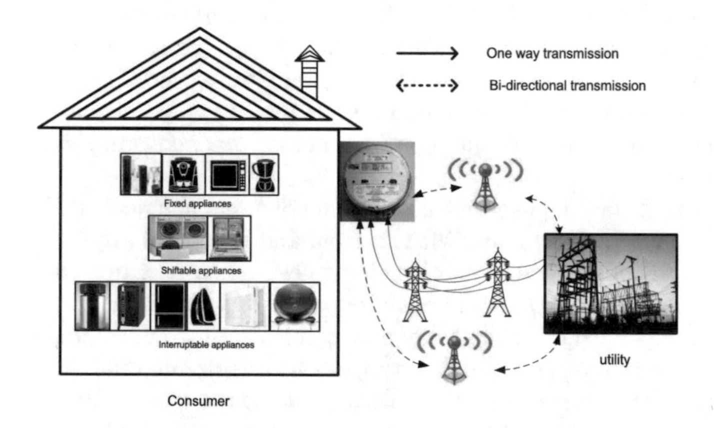

Fig. 1. System model

3 Proposed Scheme

There are many meta-heuristic approaches available to solve optimization and scheduling problems. In this research, two algorithms are used to evaluate the performance of proposed scheme; BFA and HSA respectively. A hybrid of BFA and HSA (HBH) is also proposed for comparative performance analysis of the system based on performance parameters. DSM manages energy consumption and controls demand side activities for end consumers. It persuade consumers to shift high load from on-peak hours and promote consumers to use energy in off-peak hours. A number of incentive based pricing tariff are established by utilities; such as, real time pricing (RTP), critical peak pricing (CPP), day-ahead pricing (DAP), incline block rate pricing (IBR), time of use ToU, etc. In this research, ToU price rates are considered to evaluate the performance of proposed system.

In this research, objectives are to minimize electricity consumption cost, reduce PAR, and to maximize user comfort. Major focus of this scheme is to efficiently manage energy consumption throughout a day, in order to minimize electricity consumption cost of end consumer and reduce PAR. Consumer satisfaction level in terms of waiting time is also considered. Figure 1, represent the system model design. All three heuristic algorithms implemented in this scheme are discussed in coming subsections.

3.1 BFA

Passino in 2000, introduced BFA for distributed optimization problems [8]. This is a bio-inspired algorithm based on foraging behavior of E.coli bacteria. In nature, animals having better foraging strategies tend to survive more than the animals having weak foraging strategies. In BFA, healthy bacterium splits and produce their clones and replace the weak bacterium to keep population size constant [9]. The swarm of bacterium stochastically move towards the optimal solution. This algorithm follow some basic steps which are discussed below; (1) Chemotaxis: This phase describes the movement of an E.coli cell. The E.coli bacterium can move in two alternate ways either it can swim in the same direction for a period of time or it may tumble. Bacterium follow these two ways for their entire lifespan. Life span of bacteria depend upon the number of chemotactic steps. Let's assume, (j, k, l) represents $i\text{-}th$ bacterium at $j\text{-}th$ chemotactic, $k\text{-}th$ reproductive and $l\text{-}th$ elimination-dispersal step. $C(i)$ is the step size taken in the random direction specified by the tumble. The movement of bacterium may represented by:

$$\theta_i[j, k, l] = \theta_i[j-1, k, l] + C(i)\frac{\Delta(i)}{\sqrt{(\Delta^t(i)\Delta(i))}}$$

where, 'Δ' represents a vector in random direction and its elements lie in $[-1, 1]$. (2) Reproduction: This phase deals with the elimination of the weak bacteria based on the fitness value and simultaneously split the healthy ones into two and replace them in the same place to keep the swarm size constant, so that they contribute in the next generation. (3) Elimination dispersal: In this phase, new random samples with low probability are inserted to compensate the discarded cells. At the end of aforementioned steps, best bacterium is selected for appliance scheduling based on the fitness, given as:

$$\jmath_i[j, k, l] = \jmath_i[j, k, l] + \jmath_{cc}(\theta_i[j, k, l], PoP[j, k, l])$$

where, $\jmath_{cc}$ is computed as:

$$\jmath_{cc} = \Sigma_{d=1}^{d-1}(100 \times (\theta(i, d+1) - (\theta(i, d))^2)^2 + (\theta(i, d) - 1)^2)$$

We adopt BFA foraging strategy to find optimal scheduling pattern for consumer in SG. This algorithm has exceptional attributes such as less computational burden, global convergence, require less computational time, and also suitable for handling multiple objective functions. Basic idea and working of HSA is given in upcoming section.

3.2 HSA

An overview of the basic HSA is presented in this section. This algorithm is proposed by Wang et al. [10]. It is inspired by the natural musical process which searches for an ideal state of harmony. The HSA uses a stochastic random search. It does not require initial values for decision variables. General working of the algorithm is as follows: Step 1. Define the decision variables and objective function. Input the system parameters. The optimization problem can be defined as:

$$Minimize\ f(x)$$

subject to:

$$x_{iL} \leq x_i \leq x_{iU}$$

where x_{iL} and x_{iU} are the lower and upper bounds for decision variables. HSA has following parameters: (i) harmony memory size (HMS) or the number of solution vectors in harmony memory (HM), (ii) HM considering rate (HMCR), (iii) distance bandwidth (bw), (iv) pitch adjusting rate (PA), and (v) number of improvisations (k) or stopping criterion. Where, k is same as the total number of alterations. Step 2. Initialize the HM. The HM is a memory where all solution vectors (sets of decision variables) are stored. Initially, HM is randomly generated based on the following equation:

$$x_i^j = x_{iL} + rand() \times (x_{iU} - x_{iL})j = 1, 2 ...\ HMS$$

where, $rand()$ is a random number function which give uniform distribution of [0, 1]. Step 3. Improvise a new harmony from the HM. Generating a new harmony x^{new} is called improvisation. Improvisation is based on 3 rules: memory consideration, pitch adjustment, and random selection. First of all, a uniform random number r is generated in the range [0, 1]. If r is less than the HMCR, the decision variable x_i^{new} is generated by the memory consideration; or else, x_i^{new} is produced by a random selection. Then, each decision variable x_i^{new} will undergo a pitch adjustment with a probability of pitch adjustment rate (PA), if it is produced by the memory consideration. The PA is given as follows:

$$x_i^{new} = x_i^{new} \pm r \times (bw)$$

Step 4. Update HM: After generating a new harmony vector x_i^{new}, the HM will be updated. If the fitness of the new harmony vector $x_i^{new} = (x_i^{new} + x_2^{new} ... x_n^{new})$ is better than that of the worst harmony, the worst harmony in the HM will be replaced with x^{new} and become a new member of the HM. Step 5. Repeat Steps 34 until the stopping criterion is met.

3.3 HBH

In this subsection, HBH is discussed. In HBH, all initial steps of BFA are implemented as explained in earlier section, except for elimination and dispersal step.

HSA steps of improvising new harmony are integrated with BFA steps in place of random elimination and dispersal step to get a hybrid, in order to achieve optimal solutions. Results for aforementioned algorithms are discussed in simulation and discussion section in detail.

4 Simulations and Discussions

In this section, simulation results are evaluated and discussed. In order to validate our simulation results, extensive simulations are conducted in Matlab. Heuristic algorithms BFA, HSA, and HBH are used for optimization of objective function on the basis of performance parameters, i.e., electricity consumption cost, energy consumption pattern, PAR, and user comfort. ToU price rates are used to analyze the behaviour of consumers and electricity consumption pattern. 24 h time horizon is divided into three segments off-peak hours, mid-peak hours, and on-peak hours based on consumer demand. Single home with eleven appliances is considered for the simulations. These appliances are further categorized into three classes while keeping in mind the consumers need. Appliance classes are; (i) fixed appliances, (ii) shiftable burst appliances, and (iii) interruptable appliances. Fixed appliances include light, fan, oven, and blender. Washing machine, cloth dryer, and dish washer are in shiftable burst appliances class. Interruptable appliances include AC, refrigerator, iron, and vacuum cleaner. Table 1, represents the appliances classes, their length of operational time and power rates.

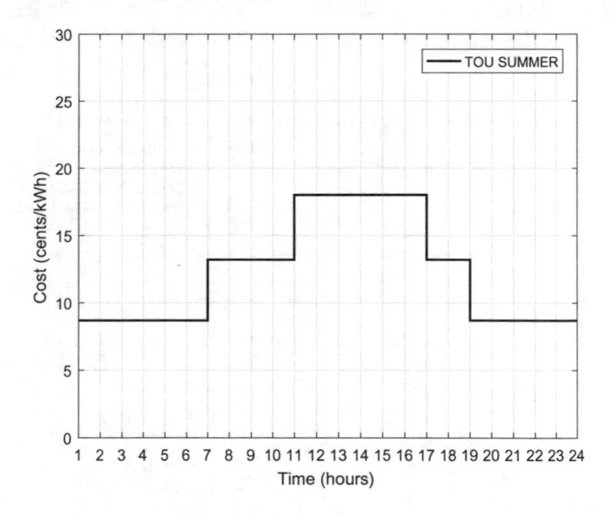

Fig. 2. ToU tariff model

Figure 2, represents TOU price rates that are used for the evaluation of earlier mentioned heuristic algorithms based on performance parameters. 11 am to 5 pm

Table 1. Appliances parameters used in summer

Classes	Appliances	LOT (h)	Power rate (kwh)
Fixed appliances	Light	12	0.1
	Fan	16	0.1
	Oven	9	3
	Blender	4	1.2
Shiftable appliances	Washing machine	5	0.5
	Cloth dryer	4	4
	Dish washer	4	1.5
Interruptible appliances	AC	12	1.1
	Refrigerator	12	1.2
	Iron	6	1.1
	Vacuum cleaner	9	0.5

are on-peak hours, mid-peak hours are from 7 am to 11 am and 5 pm to 7 pm, and the remaining hours are off-peak hours. Figure 3 represent the hourly electricity consumption cost for all three algorithm. These algorithms perform efficiently in minimizing the hourly electricity consumption cost after scheduling as shown in figure. Minimization in consumption cost is achieved by limiting the appliances on request in on-peak hours. Electricity consumption patter for each hour using BFA, HSA, and HBH is shown in Fig. 4. PAR is calculated after scheduling for each algorithm. Graphical representation of PAR is presented in Fig. 5. Difference in electricity bill per day that consumer has to pay is clearly observed from Fig. 6.

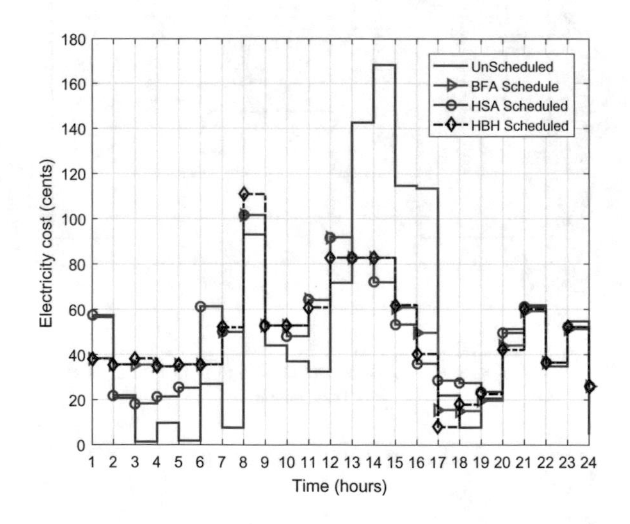

Fig. 3. Electricity cost per hour (cents)

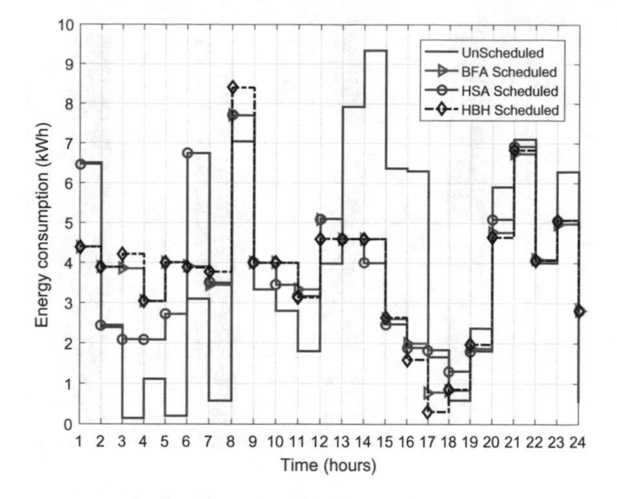

Fig. 4. Energy consumption per hour (kwh)

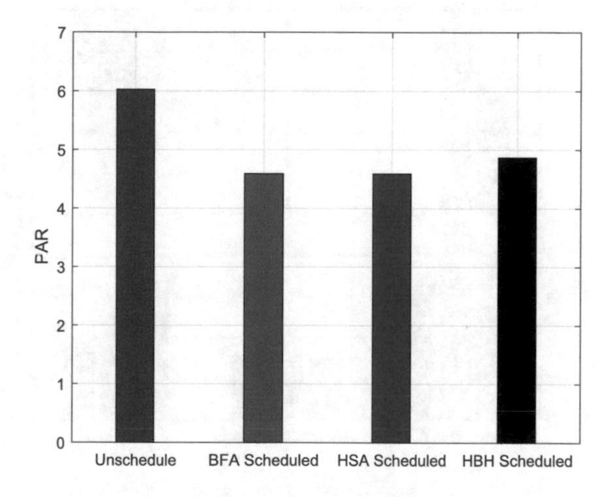

Fig. 5. PAR

In this research, consumer satisfaction level in term of waiting time is also considered. Figure 7 represent the consumer satisfaction level for shiftable, and interruptable appliances. Average waiting time is calculated as shown in figure.

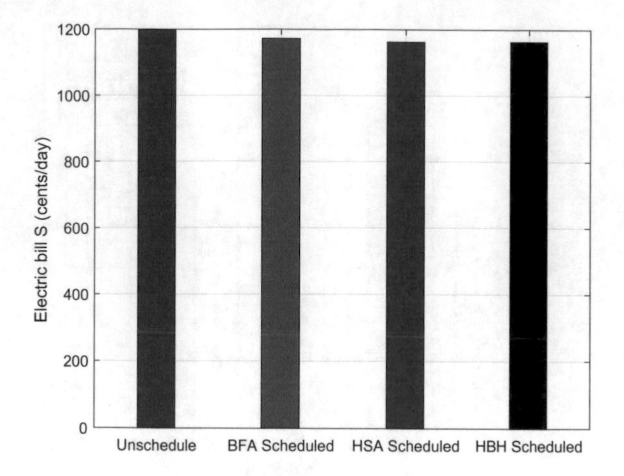

Fig. 6. Electricity cost per day (cents)

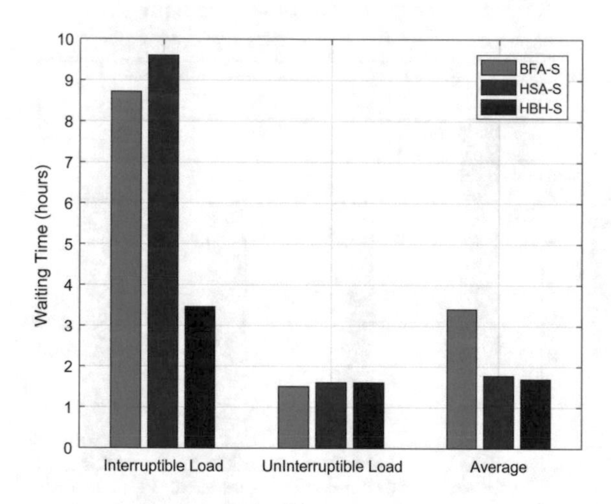

Fig. 7. User comfort

5 Conclusion and Future Work

In this paper, ToU based appliance scheduling scheme for residential DSM is proposed. Three heuristic algorithms; BFA, HSA, and HBH are used to evaluate the performance parameters; consumption cost, PAR and consumer satisfaction level in term of waiting time. Control parameters and classification of appliances are kept same to comparatively analyze the performance of all participating algorithms. Simulation results validates that all three algorithms achieve reduction in electricity consumption cost as compared to unscheduled scenario. Our proposed scheme efficiently achieved desirable trade-off to manage load in an optimal way to reduce electricity consumption cost, PAR and consumer satisfaction level.

Moreover, there is a trade-off between electricity consumption cost and consumer satisfaction level. In future, we will try to incorporate coordination between consumer and schedular in order to manage HEM system more efficiently to make it even more reliable, sustainable and robust.

References

1. Wen, Z., ONeill, D., Maei, H.: Optimal demand response using device-based reinforcement learning. IEEE Trans. Smart Grid **6**(5), 2312–2324 (2015)
2. Setlhaolo, D., Xia, X., Zhang, J.: Optimal scheduling of household appliances for demand response. Electr. Power Syst. Res. **116**, 24–28 (2014)
3. Shirazi, E., Jadid, S.: Optimal residential appliance scheduling under dynamic pricing scheme via HEMDAS. Energy Build. **93**, 40–49 (2015)
4. Rahim, S., Javaid, N., Ahmad, A., Khan, S.A., Khan, Z.A., Alrajeh, N., Qasim, U.: Exploiting heuristic algorithms to efficiently utilize energy management controllers with renewable energy sources. Energy Build. **129**, 452–470 (2016)
5. Manzoor, A., Ahmed, F., Judge, M.A., Ahmed, A., Tahir, M.A.U.H., Khan, Z.A., Qasim, U., Javaid, N.: User comfort oriented residential power scheduling in smart homes. In: International Conference on Innovative Mobile and Internet Services in Ubiquitous Computing, pp. 171–180. Springer, Cham (2017)
6. Abushnaf, J., Rassau, A., Grnisiewicz, W.: Impact on electricity use of introducing time-of-use pricing to a multi-user home energy management system. Int. Trans. Electr. Energy Syst. **26**(5), 993–1005 (2015)
7. Jovanovic, R., Bousselham, A., Bayram, I.S.: Residential demand response scheduling with consideration of consumer preferences. Appl. Sci. **6**(1), 16 (2016)
8. Passino, K.M.: Biomimicry of bacterial foraging for distributed optimization and control. IEEE Control Syst. **22**(3), 52–67 (2002)
9. Das, S., Biswas, A., Dasgupta, S., Abraham, A.: Bacterial foraging optimization algorithm: theoretical foundations, analysis, and applications. In: Foundations of Computational Intelligence, vol. 3, pp. 23–55. Springer, Heidelberg (2009)
10. Wang, X., Gao, X.Z., Zenger, K.: An Introduction to Harmony Search Optimization Method. Springer (2015)

Modeling Research on Time-Varying Impulse Resistance of Grounding Grid

Wang Tao(✉), Hu Xianzhe, Li Yangping, and Wang Qi

School of Information and Electrical Engineering, Shandong Jianzhu University, Jinan 250101, Shandong, China
wlyiran@126.com

Abstract. The behavior of grounding systems excited by high-current show great differs from that at low-frequency and low-current. Simulation of high-current draining to earth require accurate modeling of tower grounding system. The current model cannot meet this requirement. This paper proposed an accurate time-varying nonlinearly model of grounding systems established by ATP-EMTP, aiming at simulating the impulse characteristic of grounding systems. The results indicated that the strong agreement between the model and experimental values. At the end of this paper come to the calculation that the active length of the grounding bodies ray is 40 m in the grounding grid. And increasing the ray length infinitely is not conducive to reducing the impulse grounding resistance. The results above can be used for optimizing the grounding grid and selecting the length of grounding bodies ray.

1 Introduction

With the development of power system, the grounding technology of transmission tower is facing more and more challenges [1]. Reducing the grounding resistance of the tower is one of the effective measures to improve the stable operation of transmission lines. Currently, the primary way (improving the grounding structure [2] or adding the grounding module [3]) to reduce grounding resistance is directed to power frequency grounding resistance. In fact, when lightning stoke tower, the grounding body shows the inductive effect and spark effect, which makes it exhibit a time-varying non-linear characteristic [5, 6]. The traditional theory and method for reducing resistance cannot satisfy the actual requirements. Therefore, in order to optimize the structure of the grounding structure and improve the performance when high-current drain to earth, it is very meaningful to carry out the modeling research on grounding systems.

For the characteristics of grounding electrodes considering spark effect, domestic and foreign experts have made great research. Many scholars adopt matrix analysis and electromagnetic circuit theory to optimize the traditional transmission line model in time domain, and give the accurate model considering the spark effect. Simulation results show that: The impulse grounding model considering the spark effect has great agreement with the experiment values [7–11]. Related scholars use the method of software simulation to study the influence of sparking effects on the impulse grounding resistance under various factors (the shape of the grounding body and the radiation angle) [12–14], and give the optimal scheme of the grounding structure and the

L. Barolli et al. (Eds.): EIDWT 2018, LNDECT 17, pp. 1018–1026, 2018.
https://doi.org/10.1007/978-3-319-75928-9_94

radiation angle. Many scholars have carried out relevant researches on the influencing factors of impact characteristics (e.g. the crest value of the lighting current, soil type, soil resistivity), and draw corresponding conclusions [15–19]. In the literature [20], the equivalent circuit of grounding body is studied, the time domain analysis model of tower grounding device is established, and the algorithm flow of calculating grounding body impact characteristics is given. In the literature [21], the equivalent circuit model of single horizontal grounding body is built, and the influence of grounding body length, lightning current amplitude and soil resistivity on impulse grounding resistance is qualitatively studied. Literature [22] focuses on the study of various grounding forms and impact characteristics of horizontal grounding body under the condition of sand and clay, and puts forward some suggestions for optimizing the structure of grounding body. Literature [23] compares the impact characteristics of grounding bodies under 3 transient modeling methods, and gives recommendations for the selection of grounding body in low soil resistivity area.

To sum up, although the current research has analyzed various influencing factors on the impulse resistance of the tower, providing guidance for practical application. However, for the laying of horizontal grounding body, its length is usually based on the trade standard, the effective length and utilization rate of the grounding body are not considered.

In this paper the simulation model of grounding body is established, using TACS in electromagnetic transient analysis software ATP-EMTP, and given some advices for guiding the design of grounding body; the effective length of ground body ray in high resistivity soil is analyzed and the recommended value is given.

2 Preliminaries

Under the influence of impulse lightning current, the ray of the grounding body can be equivalent to the distribution parameter equivalent circuit (see Fig. 1). The equivalent circuit are characterized by their series resistance, series inductance, resistance to earth and capacitance to earth [23].

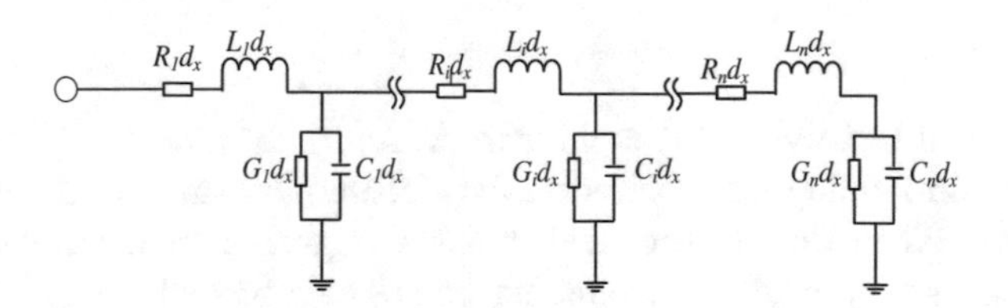

Fig. 1. Distribution parameters equivalent circuit of grounding body

In the picture above, the grounding body is divided into a finite number of small segments in theory(according to maximum wave length of excitation current),the more segments, the more accurate. In this paper the grounding body is divided into N segments (see Fig. 1). In each segment, Ri and Li represent the series resistance and series inductance. Gi and Ci represent the conductance and capacitance to earth. The values of the Ri, Li, Gi, Ci are respectively defined in the following Eq. [16]:

$$R_i = \frac{\rho_0 l_i}{\pi r^2} \tag{1}$$

$$L_i = \frac{\mu_0 l_i}{2\pi}\left(ln\frac{2l}{r} - 1\right) \tag{2}$$

$$G_i = \frac{2\pi l_i}{\rho\left(ln\frac{2l}{2hr} - 0.61\right)} \tag{3}$$

$$C_i = \varepsilon\rho G_i \tag{4}$$

where:

ρ_0 the resistivity of grounding body;
l_i the respectively length of grounding body segment;
r the radius of grounding body;
$\mu_0 =$ $4\pi \times 10^{-7}$H/m-the permeability of vacuum;
ρ soil resistivity;
$\varepsilon =$ $9 \times 8.86 \times 10^{-12}$F/m-the soil dielectric constant.

3 Grounding System Models

It is assumed a uniform ionized zone surrounding the grounding body. The soil breakdown phenomena are uniform and continuous (without considering the variation from point to point of water content, of grain size etc.). The simulation of soil breakdown must be a fundamental part of each model which has the characterization of grounding systems in the transient conditions as the main goal. When high impulse current is draining to earth, the electric field strength in the soil around the electrode at time t is described by the following function:

$$E(t) = \rho J(t) \tag{5}$$

where ρ is the soil resistivity, J(t) is the current density at time t.

In the soil surrounding the electrodes, the electric field strength increases with the current density. When the electric field strength is greater than the soil breakdown strength(E_c), the soil around the grounding body has a breakdown phenomenon and produce spark effect. Local transversal discharge starts from the electrodes' surface and stop at the points in which the electric field drops below the critical breakdown strength. At this time, the soil surrounding the grounding electrodes will produce four areas: (1) the spark discharge zone (2) ionized zone (3) deionization zone (4) without ionized zone [1] (see Fig. 2).

In the areas of spark discharge zone and ionized zone, the soil has a breakdown phenomenon, soil resistivity is 0 [15]. For simplicity, the resistivity of this zone can be

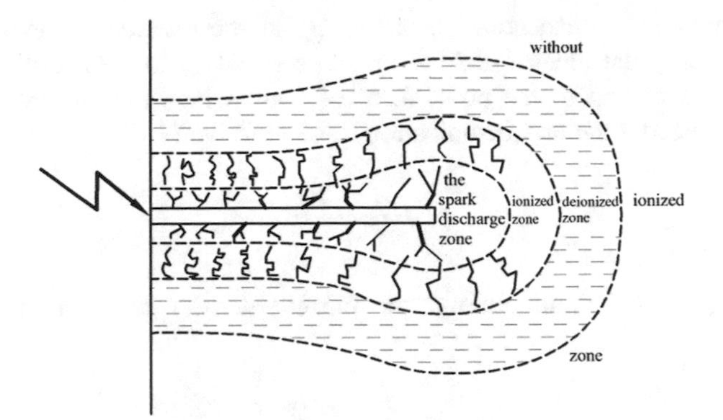

Fig. 2. The structure diagram of soil discharge

assumed equal to that fixed for the electrode because the influence of this value on the evolution of the phenomenon is negligible. This area can be simulated by means of a cylindrical sheath surrounding the conductors, regarding the ionized zone as the grounding body. The size of this area (apparent diameter) is variable during the impulse current drain to earth. At the current admitting point of the grounding body (where the probability of high current density is present), the radius of ionized zone is large. At the end of the electrodes (where the probability of low current density is present), the radius of ionized zone is small (see Fig. 3).

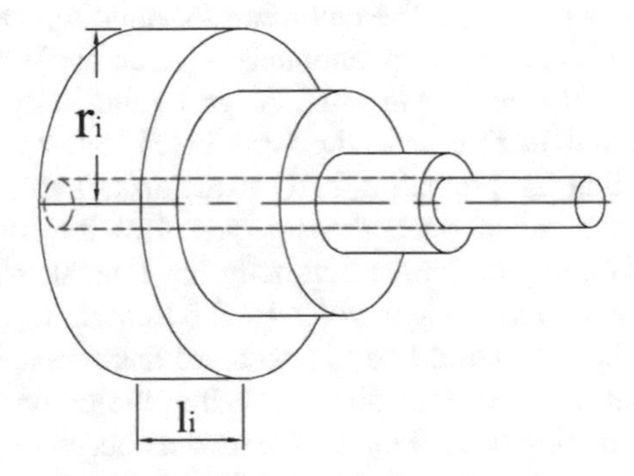

Fig. 3. Physical models of grounding bodies ray considering spark effect

On the bases of this assumption and using traditional approach, the apparent variation of the conductor's radius around each elementary tract of ground electrode can be simulated by the soil breakdown strength.

In the figure, r_i is the equivalent radius of the grounding body respectively for the segments considering the soil breakdown phenomenon, and l_i is the length of the grounding body under the equivalent radius. As mentioned above, the length of each segments must be less than the maximum wave length of excitation current. in this paper, the length of each segment is taken 5 m [3].

For the model establish above (see Fig. 3), at the interface of ionized zone and deionization zone, the electric field strength is equal to the soil critical breakdown strength(E_c). According to the point form of Ohm's law, the current density at the interface of ionized zone and deionization zone are given by:

$$J(t) = \frac{E_c}{\rho} \tag{6}$$

According to the circuit theory, the current density can also describe by the function:

$$J(t) = \frac{I_i(t)}{2\pi l_i r_i(t)} \tag{7}$$

Combining the two equations, the apparent radius of the electrode, due to ionization phenomena, can be computed according to the function gives:

$$r_i(t) = \frac{\rho I_i(t)}{2\pi l_i E_c} \tag{8}$$

where:

$r_i(t)$ the equivalent radius of each segment;
$I_i(t)$ the current drain to earth of each segment;

Based on the formula (1) to (8), The algorithm for building the model of grounding body considering the soil breakdown phenomena is given by: When the lighting stoke the tower, collecting the current flow through the grounding body, and then calculating the current density drain to earth utilizes the formula (7), then calculating the electrical field strength surrounding the grounding body uses formula (5), if the electrical field strength surrounding the grounding body is bigger than the soil critical breakdown strength, the soil surrounding the grounding body has a breakdown phenomenon, the variable diameter of the ionized zone is given by the formula (8). If the electrical field strength surrounding the grounding body is no more than the soil critical breakdown strength, the equivalent radius of the soil surrounding the grounding body is equal to the diameter of the grounding body. Finally, the model considering the soil breakdown phenomena established by means of ATP-EMTP shows in the following picture (Fig. 4).

The compared graph between the model data and the measured data are shown in Fig. 5.

4 Example Simulations

In the design process of power system towers, how to reduce grounding resistance in high soil resistivity area is a widely concerned research field. At present, the method of reducing the grounding resistance mostly by means of increasing the length of

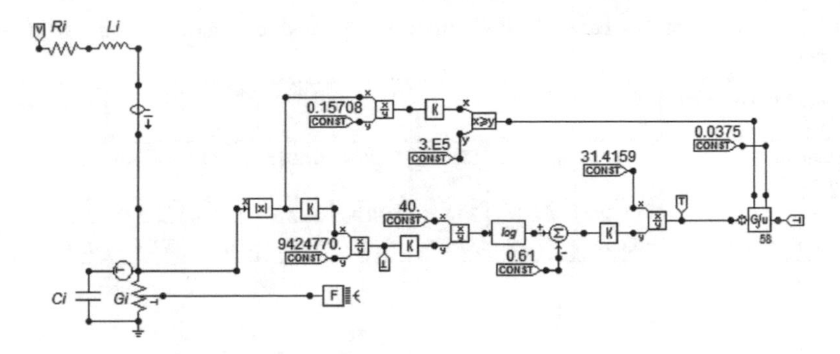

Fig. 4. Simulation model of sectional grounding bodies

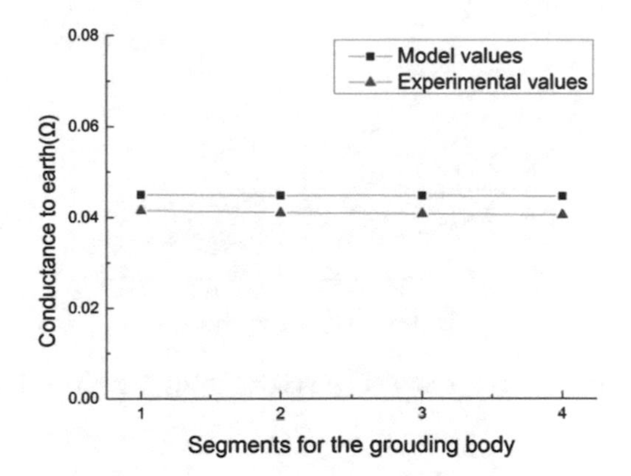

Fig. 5. Compared with the experimental values

grounding body. Because of the inductance effect, the irrational increasing of the grounding body length cannot effectively reduce the impulse grounding resistance. Therefore, in order to reduce the grounding resistance effectively, it is necessary to master the effective length of grounding body under different amplitude of lightning current, and obtain the effective quantitative criteria. Therefore, under the circumstance that the soil critical breakdown field strength is 300 kV/m and the soil resistivity is 1000Ω•m, this paper uses the simulation model of grounding body studying the impulse characteristics when the grounding body's ray length are respectively 10, 15, 20, 25, 30, 35, 40, 45 m, and getting the effective length of the grounding body. Table 1 is the comparative data between the power frequency grounding resistance and the impulse grounding resistance under different length grounding body. The reduction ratio is defined the difference between the resistance under impulse current and the resistance of power frequency. The characteristic curve between grounding body length and the impact grounding resistance is drawn in Fig. 6.

Table 1 shows that with the increase of the length of the grounding body, the reduction ratio decreases, and the minimum value is 35.76%. It indicates that the value of the power frequency grounding resistance gradually closes to the value of the

Table 1. The contrast data between power frequency ground resistance and impulse grounding resistance

The length of grounding body/m	10	15	20	25	30	35	40	45	50
Resistance in power-frequency/Ω	148.069	107.317	85.067	70.894	61.013	53.699	48.049	43.543	34.184
Impulse resistance/Ω	61.387	44.872	35.836	30.104	26.216	23.778	22.677	22.204	21.961
Reduction ratio/%	58.54	58.18	57.87	57.53	57.03	55.72	52.8	49.01	35.76

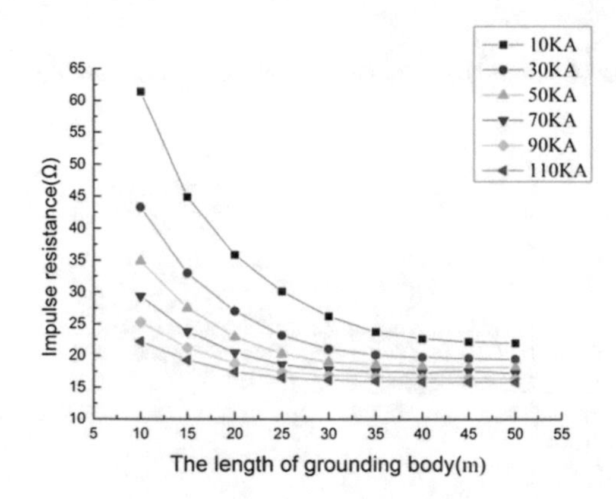

Fig. 6. The characteristic curve between grounding body length and impulse grounding resistance

grounding resistance. The longer grounding body is, the better impulse characteristics of power frequency grounding resistance reflecting the grounding body. As shown in Fig. 6, the magnitude of lightning current is inversely proportional to impulse grounding resistance. The lightning current amplitude decreases with the decrease of the grounding resistance of the grounding body, and the trend of decreasing gradually and finally becomes stable. It can be drawn that the lightning current with smaller amplitude shows larger impulse grounding resistance than the thunder current with larger amplitude. When the length of the grounding body is less than 40 m, the impulse grounding resistance decreases significantly with the increase of the grounding body length. When the grounding body length is greater than 40 m, the impulse grounding resistance hardly decreases with the increase of the grounding body length. This indicates that the effective length of the grounding body is 40 m.

5 Conclusions

In this paper a detail description of a computer model based on the circuital approach has been given. From the analysis of the results it is possible to conclude that the proposed model is an efficient and powerful tool for the transient analysis of grounding systems. In fact, it allows the investigation of grounding systems also when they are excited by high impulse current. At the end of the paper, the application of this model provides valuable reference and the effective length of grounding body was given.

Acknowledgments. The authors thank all the reviewers and editors for their valuable comments and works. This paper is supported by the Major Research Project of Shandong Provence (No. 2016ZDJS02A12), the Major Scientific and Innovation Project of Shandong Province (No. 2017CXGC0603), the Science and Technology Development of Shandong Province in 2014 (No. 2014GGX103011).

References

1. Geri, A.: Behavior of grounding systems excited by high impulse currents: the model and its validation. IEEE Trans. Power Deliv. **14**(3), 1008–1017 (1999)
2. Yuan, T., et al.: Simulation analysis of the impact characteristic of grounding electrode based on the current shielding effect. Trans. China Electrotech. Soc. **30**(1), 177–185 (2015)
3. Zhou, W.M., Guang-Ning, W.U., Cao, X.B., et al.: Study on effect of grounding module on grounding resistance. Insul. Surge Arresters **2**, 008 (2008)
4. Changzheng, D., et al.: Influence analysis of inductance effect and spark effect on lightning impulse characteristics of grounding conductors and its ambient soil. Gaodianya Jishu/High Volt. Eng. **41**(1), 56–62 (2015)
5. Zhang, B., et al.: Experimental analysis on impulse characteristics of grounding devices under high lightning current. High Volt. Eng. **37**(3), 548–553 (2011)
6. Sekioka, S., Sonoda, T., Ametani, A.: Experimental study of current-dependent grounding resistance of rod electrode. IEEE Trans. Power Deliv. **20**(2), 1569–1576 (2005)
7. Changzheng, X., Cixuan, C., Xishan, W.: Computation of impulse grounding resistance of extended grounding electrode. High Volt. Eng. **27**, 59–63 (2001)
8. Yang, L., et al.: Modeling of grounding electrode for lightning transient response analysis. In: Proceedings of the CSEE, vol. 31, no. 13, pp. 142–146 (2011)
9. He, Z.-Q., Wen, X.S., Wang, J.W.: Lightning impulse response characteristic computation and analysis of grounding grid. High Volt. Eng. **33**(3), 75–78 (2007)
10. Hua, X.U., et al.: Calculation of tower impulse grounding resistance. High Volt. Eng. **32**(3), 93–95 (2006)
11. Jingli, L., Jiang, J., Li, L.: Simulation and experiment study on resistance-reducing mechanism of grounding device with spicules. Power Syst. Technol. **1**, 036 (2013)
12. El Mghairbi, A., et al.: Technique to increase the effective length of practical earth electrodes: simulation and field test results. Electr. Power Syst. Res. **94**, 99–105 (2013)
13. Lei, C., et al.: Simulation studies on impulse characteristics of grounding device based on PSCAD. Insul. Surge Arresters (2014)
14. Mousa, A.M.: The soil ionization gradient associated with discharge of high currents into concentrated electrodes. IEEE Trans. Power Deliv. **9**(3), 1669–1677 (1994)
15. Mazzetti, C., Veca, G.M.: Impulse behavior of ground electrodes. IEEE Trans. Power Appar. Syst. PAS **102**(9), 3148–3156 (1983)
16. Ling, Z., et al.: Impulse grounding computation on extended electrode of transmission line tower based on spark discharge equivalent radius. High Volt. Appar. **39**, 22–24 (2003)
17. Xu, W., Liu, X., Huang, W.C.: Computation model of impulse grounding resistance of grounding devices for transmission towers based on spark discharge equivalent radius. Electr. Power Constr. **31**, 22–25 (2010)

18. Yulang, T., Wenhao, H., Qilin, Z., et al.: Influence study of soil nonlinear breakdown effect on dispersing characteristics of vertical grounding electrodes. Power Syst. Technol. **41**(5), 1689–1696 (2017)
19. Geri, A., et al.: Non-linear behaviour of ground electrodes under lightning surge currents: computer modelling and comparison with experimental results. IEEE Trans. Magn. **28**(2), 1442–1445 (2002)
20. He, J.-L., Kong, W.Z., Zhang, B.: Calculating method of impulse characteristics of tower grounding devices considering soil ionization. High Volt. Eng. **36**(9), 2107–2111 (2010)
21. Shengchao, J., et al.: The simulation study on time-varying grounding resistance of extended grounding electrode based on ATP-EMTP simulation. Insul. Surge Arresters **1**, 108–114 (2014)
22. Yang, S., et al.: Influence factor of impulse characteristics of box and ray grounding device. High Volt. Eng. **42**, 1548–1555 (2016)
23. Deng, C., et al.: Impulse characteristic analysis of grounding devices. Gaodianya Jishu/High Volt. Eng. **38**(9), 2447–2454 (2012)
24. Zanji, W.: Fitting of impulse voltage of arbitrary waveform. Transformer (6), 14–16 (1988)
25. Jinliang, H., Rong, Z.: Grounding technology of power system. The Science Publishing Company (2007)
26. Naor, M., Yung, M.: Public-key cryptosystems provably secure against chosen ciphertext attacks. In: 22nd ACM STOC. ACM Press, May 1990
27. DL/T621-1997 Guide on grounding design of AC electrical device (1997)

Research on Intelligent Parking Area Division and Path Planning

Yanping Li(✉), Boying Shi, Tao Wang, Qi Wang, and Linyan Wu

Shandong Jianzhu University, Ji'nan 250100, China
wlyiran@126.com

Abstract. Aiming at the problems of regional congestion and scramble for parking spaces in the current large parking lot, in this paper, a new parking area dynamic information guidance system is designed based on the new parking area division and parking guidance strategy; and the improved Dijkstra algorithm is applied in path planning. The model and algorithm have achieved good guidance effect in the installation and application of an underground parking lot. And this model improves the user's satisfaction; alleviates the congestion phenomenon of parking lot inside area; improves the operation efficiency and intelligent management level of the parking lot.

1 Introduction

In recent years, with the rapid growth of vehicle ownership in our country [1], the construction of the parking lot is also built on the larger scale, and a lot of parking lots are over a thousand parking spaces large parking lot [2]. In the parking lot, the characteristics of buildings with dense roads, numerous branches, similar structures and inadequate natural lighting have determined that the driver's view is limited; enter the independent search for parking spaces not only time-consuming and laborious, but also easy to lose direction, and it is easy to cause internal congestion, and traffic safety problems; therefore, the necessary intelligent parking management system and parking guidance system must be equipped to ensure the normal operation of the parking lot.

Based on the research and analysis of the existing parking guidance system and regional division method, this paper proposes a new parking guidance strategy and regional division method, and expounds the design of dynamic parking guidance system.

2 The Status of Parking Guidance System

At present, there are two kinds of parking spaces at home and abroad:

(1) Grading guidance of parking guide screen
In order to transfer parking spaces and direction information to users to complete the parking guide, set various levels of parking guiding display screen in the parking lot entrance and road fork, and the specific parking area, driving path and parking spaces are mostly selected by users themselves [3].

L. Barolli et al. (Eds.): EIDWT 2018, LNDECT 17, pp. 1027–1036, 2018.
https://doi.org/10.1007/978-3-319-75928-9_95

(2) Accurate parking guidance based on Positioning Technology
With the help of RFID, WIFI, Bluetooth and video recognition technology to locate the vehicle on the spot, and under the assistance of acoustic optic equipment to complete the vehicle's precise guidance and management.
However, RFID and other wireless models are prone to jam at the entrance of parking lot in the rush hour: the entrance vehicles queue to block the external road, and because there are so many parking spaces, the phenomenon of users losing cards often occurs [5].
Video positioning technology uses in the high-resolution cameras installed at each road intersection and parking space to accurately locate the vehicles inside the parking lot in real time [6]. For large and super large parking lots, the cost is too high and the latter is not easy to maintain. The low light environment of underground parking is not conducive to video capture.

3 Overall Design of Intelligent Parking System

The intelligent parking information guidance system is composed of six subsystems: parking area detection system, regional monitoring system, lighting guidance system, ZigBee communication system, license plate recognition system and host computer management system. The system structure diagram is shown in Fig. 1:

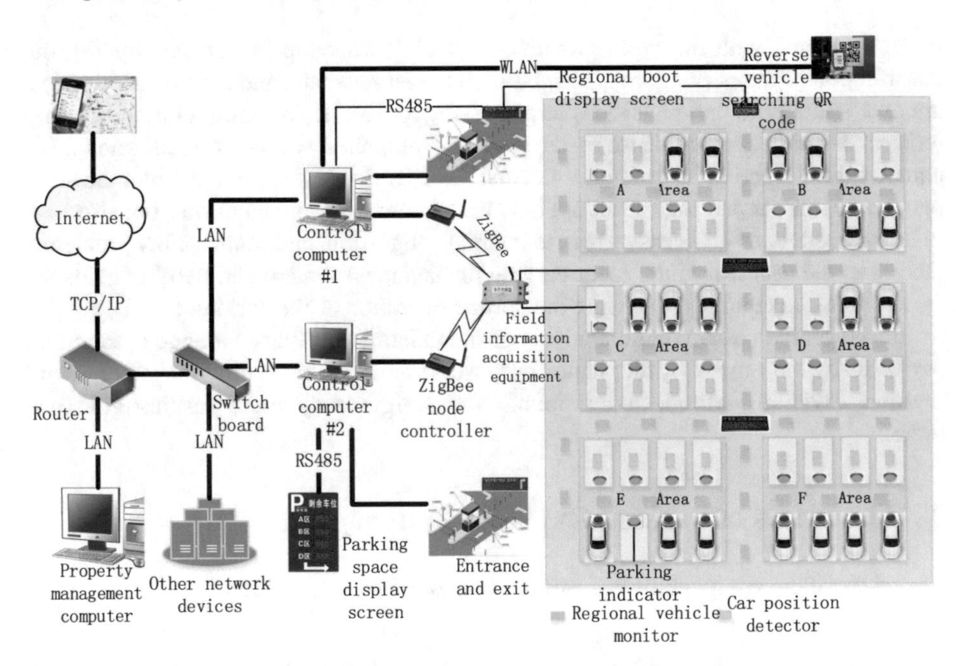

Fig. 1. Schematic diagram of system structure

4 Parking Lot Modeling Method

Parking environment modeling is the basis of parking area zoning and path planning. This system takes the region as the guidance terminal point, so the parking lot road network model is established based on the weighted undirected graph.

The model of road network is the corresponding network model based on a certain structure relation, which is to abstract the actual traffic road network into nodes and sides; Generally, it can be abstracted simply as Figure $G = (N, R)$, N represents the set of all nodes in a graph, R represents the set of all edges (arcs) in a graph. In this paper, a large underground parking lot in Shandong, Zibo is taken as the research object to study the system design, the parking lot plan is shown in Fig. 2.

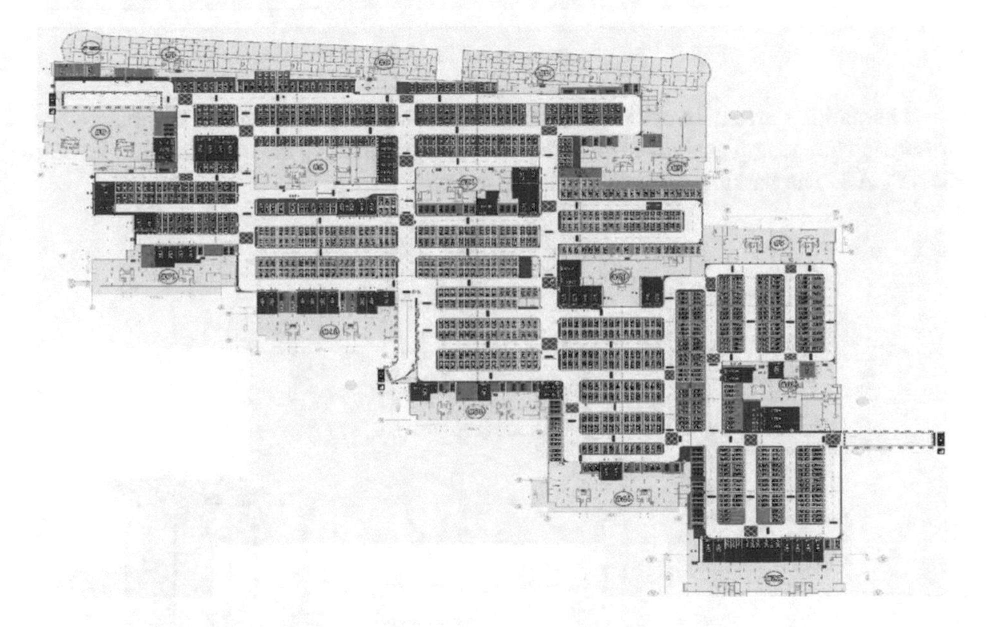

Fig. 2. Schematic diagram of a parking lot in Zibo

When establishing a road network model for parking lot, a rectangular coordinate system is established by scaling the parking area plan, the plane graph is abstracted to the path as the edge, and the entrance, intersection and road of the parking lot are nodes, building road network model by using MATLAB drawing tool.

Because the model node is relatively large, in order to save space and facilitate programming, the adjacency matrix is used to store the weight information between adjacent nodes, and the adjacency matrix [7] can be defined as follows:

According to the above formula, the adjacency matrix of the parking lot is shown in Fig. 3.

$$\mathrm{n}_{ij} = \begin{cases} \omega_{ij} & i, j \text{ adjacent}, \ \omega_{ij} \text{ is weigh} \\ 0 \text{ or } \infty & i, j \text{ non-adjacent} \end{cases} \tag{1}$$

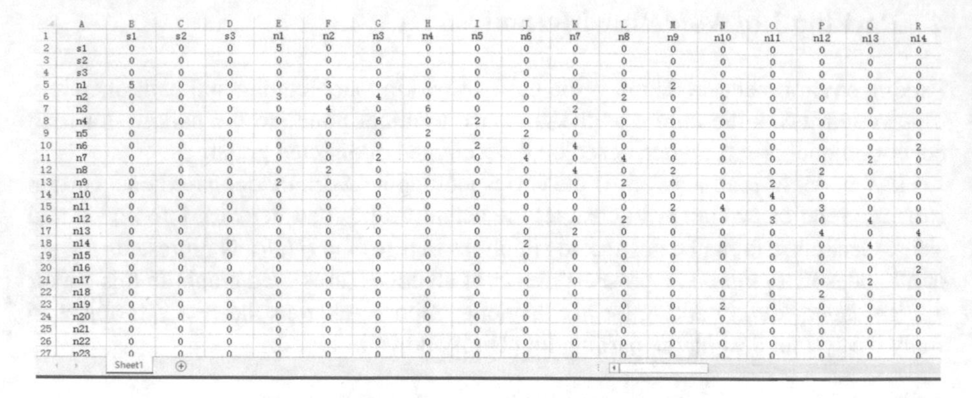

	A	B	C	D	E	F	G	H	I	J	K	L	M	N	O	P	Q	R
1		s1	s2	s3	n1	n2	n3	n4	n5	n6	n7	n8	n9	n10	n11	n12	n13	n14
2	s1	0	0	0	5	0	0	0	0	0	0	0	0	0	0	0	0	0
3	s2	0	0	0	0	0	0	0	0	0	0	0	0	0	0	0	0	0
4	s3	0	0	0	0	0	0	0	0	0	0	0	0	0	0	0	0	0
5	n1	5	0	0	0	3	0	0	0	0	0	0	2	0	0	0	0	0
6	n2	0	0	0	3	0	4	0	0	0	0	2	0	0	0	0	0	0
7	n3	0	0	0	0	4	0	6	0	0	2	0	0	0	0	0	0	0
8	n4	0	0	0	0	0	6	0	2	0	0	0	0	0	0	0	0	0
9	n5	0	0	0	0	0	0	2	0	2	0	0	0	0	0	0	0	0
10	n6	0	0	0	0	0	0	0	2	0	4	0	0	0	0	0	0	2
11	n7	0	0	0	0	0	2	0	0	4	0	4	0	0	0	0	2	0
12	n8	0	0	0	0	2	0	0	0	0	4	0	2	0	0	2	0	0
13	n9	0	0	0	2	0	0	0	0	0	0	2	0	0	2	0	0	0
14	n10	0	0	0	0	0	0	0	0	0	0	0	0	0	4	0	0	0
15	n11	0	0	0	0	0	0	0	0	0	0	0	2	4	0	3	0	0
16	n12	0	0	0	0	0	0	0	0	0	0	2	0	0	3	0	4	0
17	n13	0	0	0	0	0	0	0	0	0	2	0	0	0	0	4	0	4
18	n14	0	0	0	0	0	0	0	0	2	0	0	0	0	0	0	4	0
19	n15	0	0	0	0	0	0	0	0	0	0	0	0	0	0	0	0	0
20	n16	0	0	0	0	0	0	0	0	0	0	0	0	0	0	0	0	2
21	n17	0	0	0	0	0	0	0	0	0	0	0	0	0	0	0	2	0
22	n18	0	0	0	0	0	0	0	0	0	0	0	0	0	0	2	0	0
23	n19	0	0	0	0	0	0	0	0	0	0	0	0	2	0	0	0	0
24	n20	0	0	0	0	0	0	0	0	0	0	0	0	0	0	0	0	0
25	n21	0	0	0	0	0	0	0	0	0	0	0	0	0	0	0	0	0
26	n22	0	0	0	0	0	0	0	0	0	0	0	0	0	0	0	0	0
27	n23	0	0	0	0	0	0	0	0	0	0	0	0	0	0	0	0	0

Sheet1

Fig. 3. Adjacency matrix of parking lot network

The parking area network model of weighted undirected graph can be obtained by inputting the length weight and node coordinate value in the network model into MATLAB. The parking network model is shown in Fig. 4.

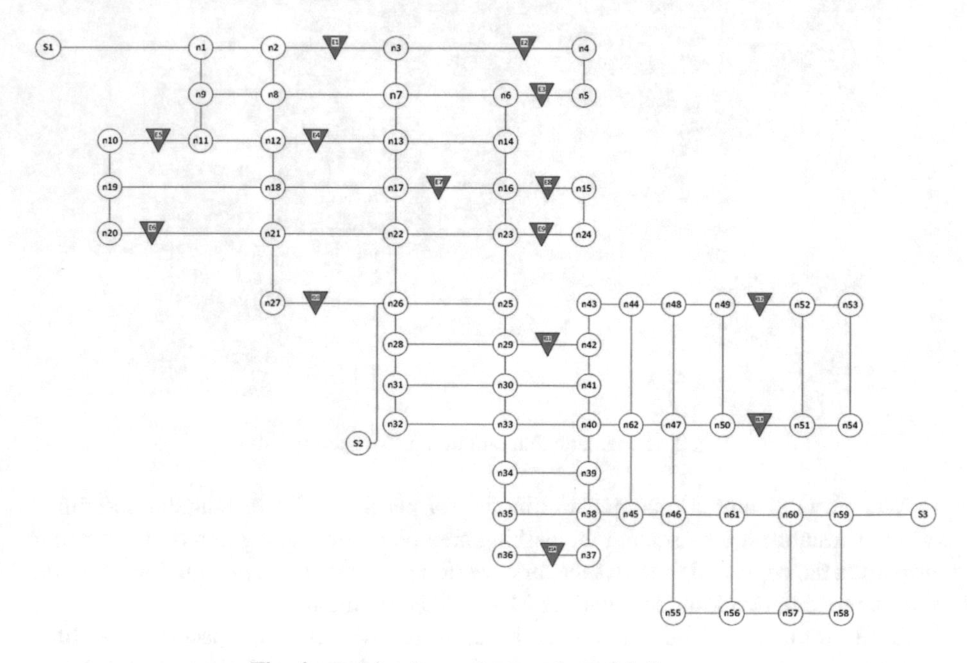

Fig. 4. Parking lot road network model diagram

$S1$, $S2$, $S3$ in Fig. 4 is the parking lot entrance, Figure ▽ is the elevator position identification of the ground building in the parking lot, this identifier is used when the region is divided.

5 Parking Area Division

In this paper, a new parking area zoning strategy [8] is proposed, which divides the area according to the influence size of different paths and the degree of regional dependence. The zoning procedure is as follows:

(1) Taking the import and export of the parking lot as the starting point, the elevator port, the import and export of parking spaces at other locations as the end point, establish the shortest path;
(2) The lane with high degree of path overlap boundary are initially classified as regional boundaries, then the internal boundary of parking lot and wall obstacle are used as constraints to determine the regional boundaries, the effect of zoning is shown in Fig. 5.

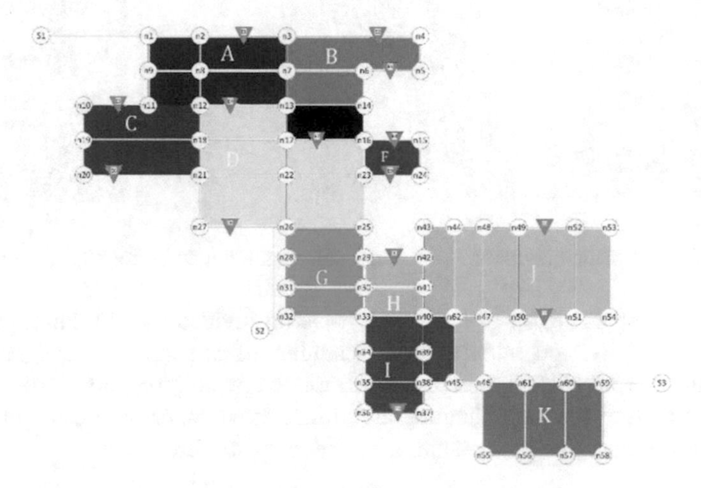

Fig. 5. Path dependent domain partition diagram

(3) After the shortest path is established, the path overlap degree of some lanes is much higher than that of the other lane, using this part of the lanes as the regional boundary can improve the parking efficiency and entry and exit efficiency in the area. According to the degree of path dependence, the parking area is divided into 11 areas, and Fig. 5 uses different color blocks to distinguish them. Among them, the black area is parking area hydropower equipment management area, no parking space; each parking area is a regular quadrilateral with its vertex marked regions, such as A, B regions, marked as follows:

$$A = (n_1, n_3, n_{13}, n_{11}), B = (n_3, n_4, n_5, n_7) \cup (n_7, n_6, n_{14}, n_{13}) \tag{2}$$

Since the regional division of the paper is involved in determining the driving direction of the vehicle in and out of a certain area, it is necessary to clarify the specific regional ownership of the vehicles on both sides of the lane. When there is no wall and obstacle natural segmentation in two adjacent areas, the parking area close to the shortest path and back to back is chosen as the regional boundary, as shown in Fig. 6.

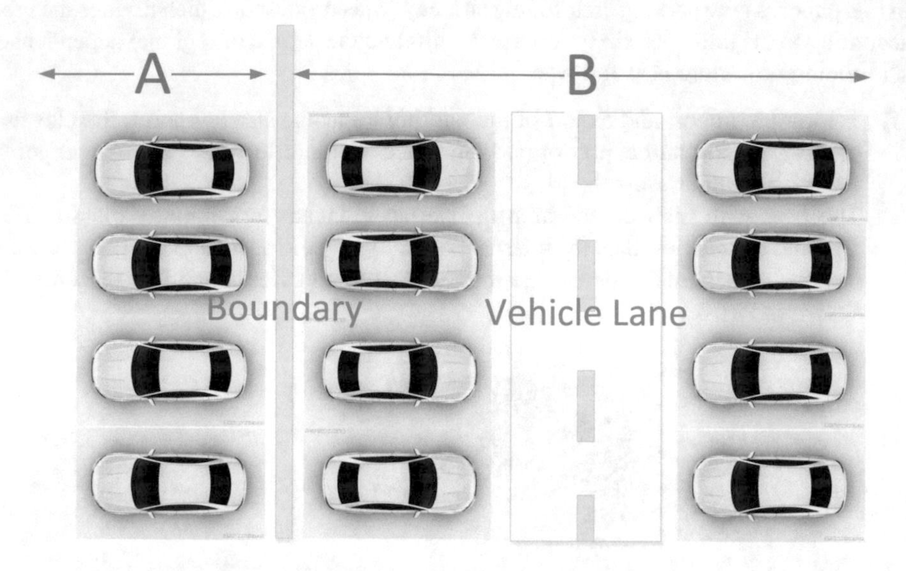

Fig. 6. Schematic diagram of boundary determination method for adjacent regions

The oversize parking lot is equivalent to being divided into 11 small parking lots after the regional division, and the average number of cars in each small parking lot is only 95. And the path calculation can be completed by only 63 nodes; moreover, after the improvement of the path planning algorithm, the number of nodes will be further reduced, the efficiency of path planning has been greatly improved.

6 Research on Path Planning Algorithm

There are many kinds of path planning algorithms [9, 10], this system uses regional guidance, and the requirements of the calculation quantity of which are not too high, therefore, the Dijkstra algorithm with low efficiency and convenient calculation is adopted and improved [11, 12].

6.1 Improvement of Dijkstra Algorithm

As a classical greedy algorithm, the efficiency of Dijkstra algorithm is low.

In order to improve the search efficiency and optimize this algorithm, the fan constraint algorithm is improved from the search direction and range.

The Dijkstra sector constraint algorithm is as follows:

(1) Definite axis
Take the S_1 entrance to the F area as an example, that is to find the shortest path of the node n_1 (From the entrance of S_1, you must go through n_1) to the F area. According to Euclidean distance formula:

$$d(n_i, n_j) = \sqrt{(n_i x_1 - n_j x_2)^2 + (n_i y_1 - n_j y_2)^2} \tag{3}$$

In the formula, $n_i x_1$ and $n_i y_1$ represent the Cartesian coordinates of the node n_i at the entrance n_4 as the origin, (n_4, s_1) as the axis, and (n_4, n_{37}) as the Y axis respectively.

According to Euclidean distance formula, calculate the length of straight line from each node in the F area to the starting node n_1. Take the straight-line segment (n_1, n_{24}) where the maximum value D_{max} is located as the axis of the fan-shaped restricted area.

$$D_{max} = \max\{|n_1, n_i|\} \qquad i \in (15, 16, 23, 24) \tag{4}$$

(2) Determine the search direction angle
In the triangle formed by the node n_1 and the arbitrary two nodes in F region, according to trigonometric function formula, the angle β with node n_1 as vertex in each triangle is calculated,

$$\beta = \arccos\frac{a^2 + b^2 - c^2}{2ab} \tag{5}$$

The maximum angle value of the β value is taken to be the search angle α of the sector restricted region. Thus, the search region from node n_1 to F region is determined as a sector restricted search region with (n_1, n_{24}) as the axis and 2α as the search direction angle. The calculation range of the improved algorithm is shown in Fig. 7.

As shown in Fig. 7, the search order in the improved algorithm is still from A to B, C. However, the search range is limited to the fan-shaped region with 2α as an angle, the search area and number of search nodes before and after the algorithm is improved are shown in Table 1.

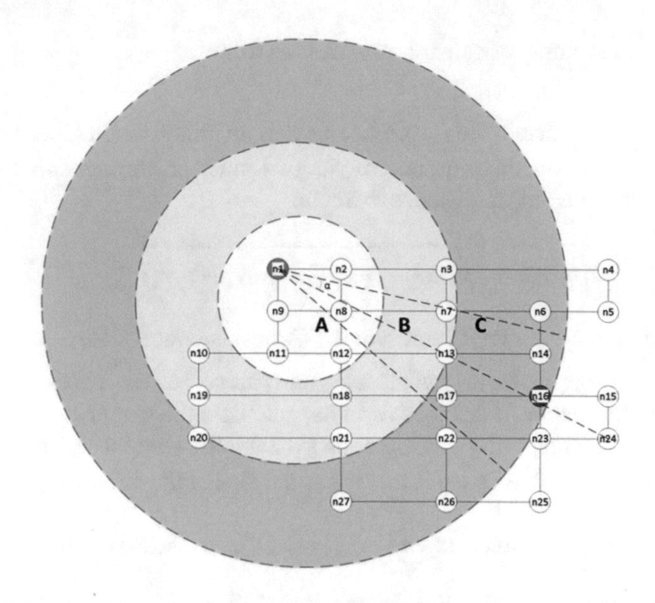

Fig. 7. Schematic diagram of operation range of Dijkstra sector constraint algorithm

Table 1. Comparison of search range

Algorithm	Search area	Node number
Traditional Dijkstra algorithm	πR^2	19
Dijkstra fan limited algorithm	αR^2	6

Among them, R is the search radius. Compared from the upper table, the search area and the number of nodes are reduced greatly after the fan restriction.

In order to further verify the effectiveness of the improved algorithm, the traditional Dijkstra algorithm and its fan limited algorithm are simulated, the number of nodes of the weighted graph is increased in turn while keeping the starting node and the target point unchanged, and the calculation time of the system under different number of nodes are shown in Table 2.

Table 2. Comparison table of computing time of algorithm

Node number	50		100		150	
Algorithm	Traditional algorithm	Fan limited algorithm	Traditional algorithm	Fan limited algorithm	Traditional algorithm	Fan limited algorithm
Computing time (ms)	89	51	169	78	377	156

The comparison of the time of simulation results show that the fan limited algorithm significantly improves the search efficiency compared with the traditional algorithm. Combine the results of Tables 1 and 2, the Dijkstra fan limited algorithm can greatly improve the speed of path planning and improve the response time of the system. Combined with parking lot network model, the number of nodes in the parking lot is less, you can choose to conduct real-time path planning; you can also choose to calculate the path between each node in advance, and save the corresponding path in the database when the system is installed.

6.2 Application of Dijkstra Fan Limited Algorithm in Parking Area Path Planning

The host computer management system selects the best parking area according to the distance between the current entrance and each region, the number of vehicles in the area and the traffic flow.

(1) The node set N of parking area network model G is divided into 4 subsets: path marker node set *Path*, optional tag node set *Select*, no marked node set *Unlabeled* and useless node set *Invalid*. When initializing, $\text{Path} = \{n_1\}, Select = \emptyset, Unlabeled = \emptyset$, then, the Euclidean distance formula and trigonometric function formula are used to determine whether all the nodes are in the sector region, if it is, put in set *Select*, if not, put in set *Unlabeled*.
(2) The node judgment method is as follows: Nodes n_1, n_{24} and arbitrary nodes n_i constitute a triangle, the angle between the edge (n_1, n_{24}) and the edge $(n_{1,} n_i)$ is γ; if $\gamma \leq \alpha$, the node n_i is the node within the sector restricted region, if $\gamma \leq \alpha$, the node is not a node within the restricted region.
(3) Find all the nodes in the set *Unlabeled* that are connected to thc node n_1 and put them into the set *Select*, at the same time, set the distance from this node to node n_1 as the weight of this node to node n_i, and remove it from the set *Unlabeled*. If it is not connected to the node n_1, nor in the sector restricted region, it is placed into the set *Invalid*.
(4) Find node n_i, which is the shortest distance from node n_1, from the optional tag node set *Select* (the sum of the weight from node n_1 to node n_i is minimal), then put the node n_i in the path marker node set *Path*, at the same time, delete the node from *Select*.
(5) One by one to search the node n_j that is connected with the node n_i, but not in the set *Path*, if n_j is in *Unlabeled*, then put it in *Select*, at the same time, delete it from the set *Unlabeled*;if n_j is in the set *Select*, then the sum of the weights in the set *Select* are modified.
(6) Determine whether the nodes in the F region appear in the set *Path*; if it is, then stop the computation, otherwise repeat (3), (4). Until the F region node appears in the set *Path*, the optimal path is obtained.

7 Conclusion

In this paper, the zoning strategy of intelligent parking area dynamic guidance system and the improvement of path planning algorithm are described, the hardware design of the system has not been discussed for space reasons. The improved Dijkstra fan-shaped constraint algorithm is analyzed by the comparison of the number of search nodes and verified by the algorithm simulation, and the number of search nodes is greatly reduced and the system path planning efficiency is improved; the dynamic guidance system provides a feasible solution to solve the regional congestion problem in large parking lot, improve the intelligent management level of parking lot.

Acknowledgments. This research is Supported by the key technologies of intelligent manufacturing integration for large scale assembly line of construction machinery of the key research and development plan of Shandong province (2016ZDJS02A12), the research and application of key technologies for intelligent assembly data acquisition and processing of construction machinery based on industrial Internet of things of the major scientific and technological innovation projects in Shandong (2017CXGC0603).

References

1. Traffic Administration Bureau of the Ministry of public security. In 2016, the nation's motor vehicles and drivers maintained rapid growth [EB/OL], 10 January 2017. http://www.mps.gov.cn/n2255040/n4908728/c5595634/content.html
2. Yanling, C.: Research on Guidance System of Underground Parking Lot Based on Ultrasonic Sensor, pp. 2–3. Beijing Jiaotong University, Beijing (2010)
3. Xun, M.: Research on Intelligent Control System of Underground Garage Path Planning and Lighting Navigation, pp. 16–18. Shandong Jianzhu University, Jinan (2014)
4. Longji, Z.: Research on Indoor Wireless Location Technology. Beijing Jiaotong University, Beijing (2013)
5. Ziming, W.: Research and Design of Intelligent Parking Management System Based on RFID. Beijing University of Posts and Telecommunications, Beijing (2010)
6. Zhaogong, S.: Intelligent Garage Management System Based on Video Recognition Technology. Shandong Jianzhu University, Jinan (2016)
7. Zhiqi, M., Hongwen, Y., Weidong, H.: A new topological sorting algorithm based on adjacency matrix. Comput. Appl. **27**(9), 2307–2309 (2007)
8. Honggang, W., Fengliang, W.: Algorithm and program implementation of ventilation network area division. Ind. Autom. **10**, 39–41 (2009)
9. Qing, L., Dingli, S., Shuangjiang, Z.: Two improved optimal path planning algorithms. J. Eng. Sci. **27**(3), 367–370 (2005)
10. Mengyin, F., Jie, L., Zhihong, D.: Distance shortest path planning algorithm for restricted search region. J. Beijing Inst. Technol. **24**(10), 881–884 (2004)
11. Fuhao, Z., Jiping, L., Qingyuan, L.: A shortest path optimization algorithm based on Dijkstra algorithm. Remote Sens. Inf. **2**, 38–41 (2004)
12. Yang, Y., Jianya, G.: An efficient implementation of Dijkstra shortest path algorithm. J. Wuhan Univ. (Information Science Edition) **24**(3), 209–212 (1999)
13. Yue, Y.: An efficient implementation of shortest path algorithm based on Dijkstra algorithm. J. Wuhan Tech. Univ. Surv. Mapp. **24**, 208–212 (1999)

Moving Applications to an On-demand, Software-as-a-Service Model in the Albanian Government Cloud

Enkeleda Kuka[1(✉)], Alba Haveriku[2], and Aleksander Xhuvani[2]

[1] ISDA Program, Minister of Innovation and Public Administration, Tirana, Albania
ekuka@icc-al.org
[2] Faculty of Information Technology, Polytechnic University of Tirana, Tirana, Albania
albahaveriku@gmail.com, axhuvani@yahoo.com

Abstract. The distinction between the three service models in cloud computing IaaS, PaaS and SaaS, especially between the last two, has diminished and will continue to do so with new cloud technology innovation happening every day. As the Albanian government has made progress on building the government Cloud, more challenges are faced, when it comes to the needs for more clouds services and resources by public institutions. The purpose of the paper is to estimate the impact of IaaS in the Government Cloud of Albania, and evaluate the advantages of moving services to the SaaS model in the overall IT applications of public organizations. We try to analyze the impact of moving to SaaS model in accordance with one of the most important usability attributes which is security, trying to set a stable equilibrium between the use and management of the resources available in the Cloud platform.

1 Introduction

Cloud computing is a relatively new and innovator concept for a lot of enterprises, governments and citizens. It provides a new model to the overall infrastructure and computing services. The cloud offers easy to use, scalable and customizable services. It eliminates the need to install and manage middleware and applications installed on user's computers, providing infrastructure as a platform and service, and making it easier to maintain software and hardware. The Fig. 1 presents cloud computing architecture according to NIST, which identifies the major actors, their activities and functions in cloud computing [1].

It is changing every day the way of working of big organizations and fortunately the governments and public sector are also considering adopting this new technology. In this consideration ENISA has given the definition of the Government cloud such as: '*A Gov Cloud is a deployment model to build and deliver services to state agencies (internal delivery of services), to citizens and to enterprises (external delivery of services to society)*' [2]. A large number of applications and different government agency's systems

L. Barolli et al. (Eds.): EIDWT 2018, LNDECT 17, pp. 1037–1048, 2018.
https://doi.org/10.1007/978-3-319-75928-9_96

are connected through a common governmental infrastructure. By using cloud, new possibilities and solution can be offered to the private and the public sector.

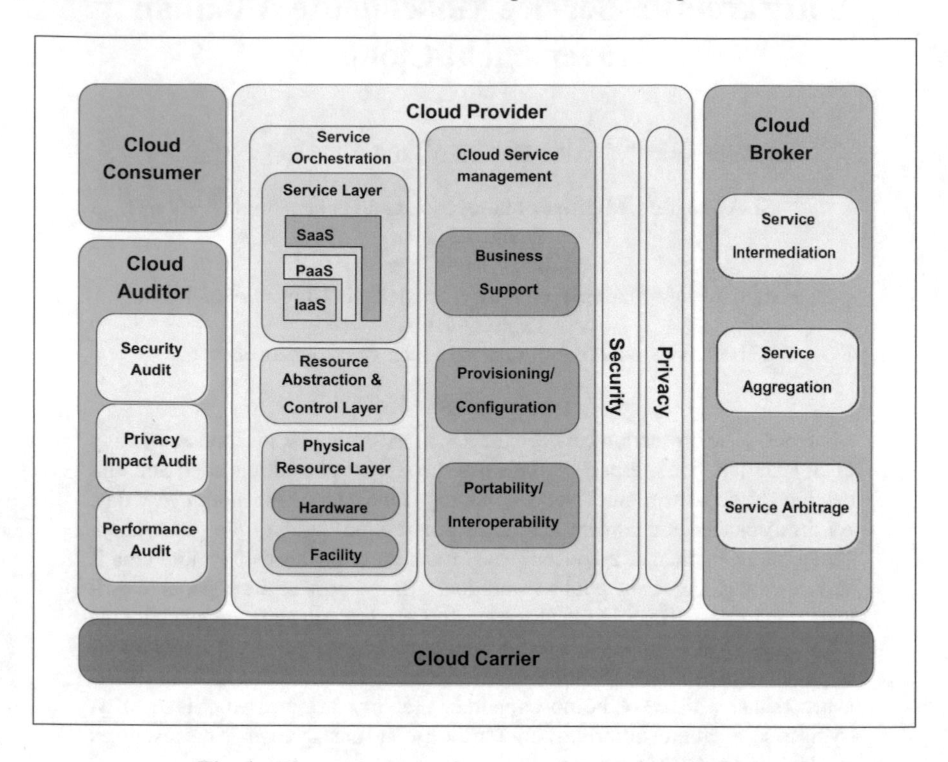

Fig. 1. The conceptual reference model cloud computing

The model of services adopted by different European countries in the Gov cloud, meets most of the request of the public administration for advanced technological solutions. It offers scalability, elasticity, high performance, replication and security, even though not all risks can be evaluated. The Table 1 below presents some comparison data related to Gov cloud implementation in different European countries.

The most frequent planned and developed Cloud Computing deployment models amongst the evaluated countries are the private and the community Cloud. On the other hand, when comparing Cloud computing service models, 75% of the evaluated countries rely on the combination of the three service models: Infrastructure as a Service (IaaS), Platform as a Service (PaaS), and Software as a Service (SaaS).

The Government of Albania has decided to address this concept by building its own governmental datacenter, so that to optimize their datacenters by building a private cloud, Infrastructure as a Service (IaaS). In this paper we aim to analyze the service model of the Albanian Gov Cloud and evaluate the advantages of moving to the SaaS model.

Table 1. Government Cloud implementation across European countries [2]

Gov cloud framework	Cloud computing national strategy	Deployment model	Service model	Status of deployment
Estonia	Yes	Public/private	IaaS/PaaS/SaaS	In planning phase
Spain	No	Public/private/ community/hybrid	IaaS/PaaS/SaaS	Deployed
UK	Yes	Private/community	IaaS/PaaS/SaaS	Deployed
Austria	Yes	Public/private/ community	IaaS/PaaS/SaaS	Planned
Denmark	No	Public/private/ community	SaaS	Planned
France	Yes	Community	IaaS	Development
Germany	Yes	Private/community	IaaS/PaaS/SaaS	Development
Ireland	Yes	Public/private/ community	IaaS/PaaS/SaaS	Planned

2 Government Cloud Framework in Albania

Cloud Computing is changing so fast that the new term X-aaS service category introduced will gradually take the place of many types of computational and storage resources used today [3]. The first layer is the IaaS solutions, which delivers IT infrastructure based on virtual or physical resources as a commodity to customers. PaaS offers the necessary environment for the whole application life cycle and can be considered as a set of programming languages, software and deployment tools. Software as a Service solutions are at the top end of the Cloud Computing stack and they provide end users with an integrated service comprising hardware, development platforms, and applications. The government of Albania has recently started to build its own Government Cloud and the Table 2 below presents the cloud framework:

Table 2. Gov Cloud framework

Gov cloud framework	
Cloud computing national strategy	No
What's in cloud	Public administration services
Deployment model	Hybrid private
Service model	IaaS/PaaS

2.1 What's in Cloud

Government of Albania has currently 13 ministries and more than 10000 employees that are using IT infrastructure on daily basis. These ministries and other government agencies require high quality services but rarely have capacity and knowledge to build and manage their own infrastructure or to provide every service needed in their environment. Centralizing IT services has provided these institutions flexibility to choose their own level of autonomy, to decide how they should deploy their own services without the burden of managing physical infrastructure. In this aspect, the government of Albania has recently started to build its own Government Cloud.

2.2 Deployment Model

Instead of choosing between public and private clouds, GoA chooses to combine the two methods to better serve public sector organizations. This way, the government cloud infrastructure can be used to extend the limits of internal resources and take advantage of the new offerings in the public service without giving up central control of services and security. As presented in the Fig. 2, a hybrid infrastructure allows GoA to establish a community cloud by making services available to other government institutions, rather than transferring that role to outside vendors. A good example is the Department of Public Administration which has created the HR module, and will be able to share it with other interested public institutions.

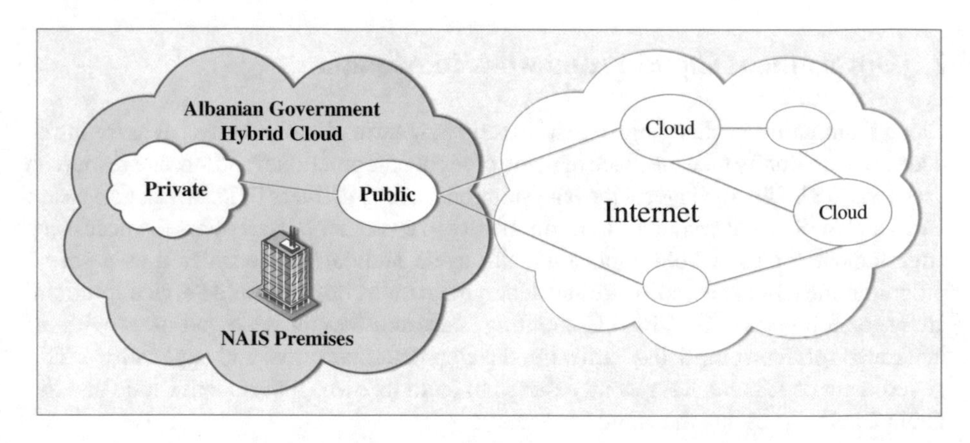

Fig. 2. Albanian government Cloud deployment model

The main operational advantages of the deployment of the hybrid private model are: *Workloads* - Applications and virtual machines can port between clouds in response to changing business conditions; *Performance* - Compute, memory, and storage resources may be dedicated to the government's specific needs, on shared or physically separated hardware; *Management* - A single dashboard can monitor the infrastructure, applications, operations, and processes.

While one of the big constrains in this deployment model become the Security, combination of private/community is lowest risk, and combination of public is greatest risk. By using a hybrid strategy that integrates with existing ICT technologies for management, GoA has avoided the 'secluded cloud' syndrome, which is, a cloud implementation that is cut off from the other part of the critical systems.

2.3 Service Model and Overall Deployment

The Gov Cloud delivers infrastructure which are made available as subscription-based services in a use-as-you-go model to government agencies. The National Agency for Information Society (NAIS), which is the main government agency responsible for building and running the government infrastructure, hosts the government datacenter. The IaaS infrastructure design is composed of Hyper-V Cloud which relies on three important components:

- Existing Datacenter environment and basic infrastructure components - This is the essential environment needed for laying out other Hyper-V cloud components which are: Active Directory Domain Services, DNS, DHCP and related Management tools.
- Fabric Management – Management components that are specifically providing Hyper-V Private cloud management functionalities;
- Resource Pool – Hosts servers, Storage and SAN, and Network Equipment. It uses Hyper–V technology for virtualization in the datacenter.

As seen in the Fig. 3, the recent configuration of Hyper-V is built on Windows Server 2012, which has a host of new features and improvements over Windows Server 2008 R2 [4]. This way, many virtual machines can be executed, and the hypervisor controls the hardware and allocates the sources of the virtual machine's operating system. Hyper-V maximizes the benefits of the chosen deployment model in our Gov Cloud by increasing availability and offering a better management. The Fig. 4 presents in a macro level the diagram of the government network currently up and running.

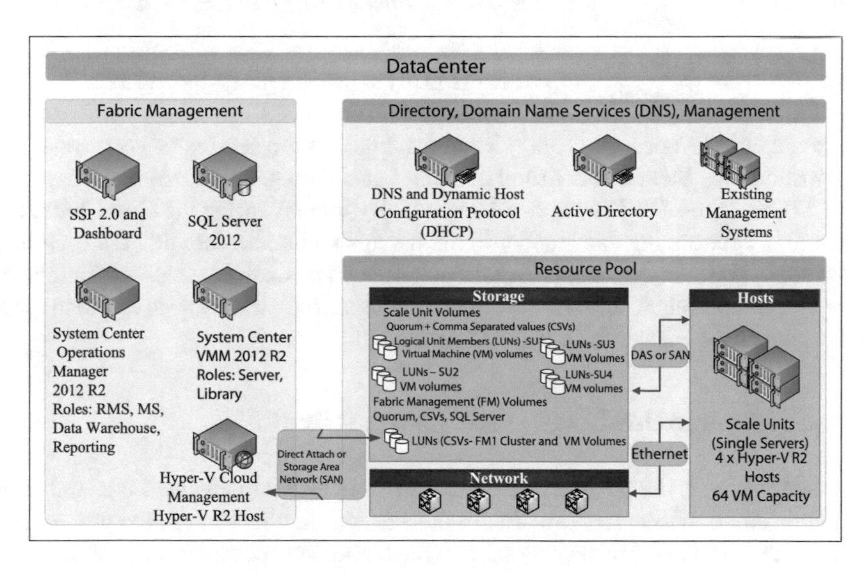

Fig. 3. IaaS infrastructure of Government Cloud

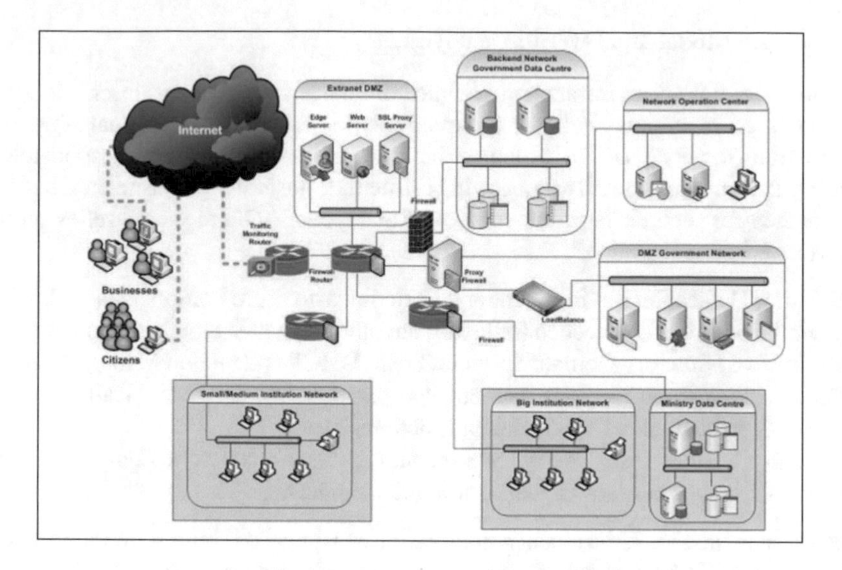

Fig. 4. Government network scheme Source: NAIS public documents

The following formulas are used in sizing Hyper-V hosts and to calculate the Total CPU and RAM requirements:

$$\text{Total CPU} = \sum\nolimits_{Institution\ 1}^{Institution\ n} \#\, Cores * CPU\ Speed\ (MHz) * CPU\ Util\ (\%). \tag{1}$$

$$\text{Total RAM} = \sum\nolimits_{Institution\ 1}^{Institution\ n} (Total\ RAM\ (MB) * RAM\ Util\ (\%) + 32\,MB. \tag{2}$$

Currently NAIS hosts over 300 virtual machines in the service of government and public institutions. Meanwhile virtual machines are heterogeneous and reach capacities up to 2 TB RAM, 64 GB Disk Space 16 vCPU. By implementing the Gov Cloud, GoA succeeded in centralizing and standardizing the application servers that are used in the government. The shared Cloud Infrastructure, which enables standardization, and sharing of computing resources and applications at the whole-of-government level, thereby generating cost savings to the Government.

3 SaaS Implementation in Gov Cloud Albania

As explained above, the government cloud infrastructure implemented is mainly IaaS, in a 'hybrid cloud' deployment model. In this section we will analyze the possibility of shifting to the third layer of service Model, SaaS and present the benefits of it.

Software as a Service (SaaS) provides to the public institutions of central government, the possibility to use the provider's (NAIS) applications running on a cloud infrastructure. This way the applications are accessible from various client devices through a thin client interface such as a web browser (e.g., web-based government email). Public institutions do not manage or control the underlying cloud infrastructure including network, servers, operating systems, storage, or even individual application capabilities, with the possible exception of limited user-specific application configuration settings.

Within this SaaS case, all the data should reside with the service provider, in our case NAIS, which can use public institutions data in any way it sees fit. The benefits of SaaS outweigh these concerns of data ownership and security. In general, for example, many of us use a free email service that controls and stores all our data. Although we might not know the provider's privacy policies, many of us will continue to use the service and feel relatively comfortable doing so. In the case of GoA, the official mail service which would be provided by NAIS controls and stores data for all users from the active directory and implements group policies in a centralized way. However, users are aware of the mail service features and policies because the official mail is used according to a regulation issued by the government. It is important to ensure that SaaS solutions comply with generally accepted definitions of cloud computing; in this architecture precisely, those principles are:

- Software will be managed from a central location: *gov datacenter*;
- Software will be delivered in a "one to many" model: *Software installed in the Cloud can be used by any institution on pay-per-use principle;*
- Users won't require handling software upgrades and patches: *NAIS IT team is going to do that in a centralized way.*

3.1 Moving Services Towards the SaaS Model

Referring to most of the literatures, SaaS is a method of delivering technology which is rapidly growing today. The most popular application categories for SaaS deployment are email applications, customer relationship management (CRM) applications, project management applications, financial and accounting applications, sales applications, expense management applications, sourcing and e-procurement applications, office management such as web conferencing applications, training applications [5].

As we mentioned above, the Government of Albania is currently using the official e-mail application service, which is managed in a central location, the Gov datacenter. However, with the success of this implementation, more application categories may be considered for SaaS deployment. We suggest that in the near future the following applications could be provided via SaaS solutions (Table 3).

Even cloud computing is designed to provide services with unlimited scalability, NAIS in cooperation with public institutions needs to take into consideration the cost-benefit analyses of each application mentioned above and services before moving to SaaS [6].

Table 3. Public administration application categories for SaaS deployment

Application categories for SaaS deployment	Is it Applied in GoA
Messaging (e-mail)	Yes
Customer Relationship Management (CRM)	No
Project management	No
HR applications	No
Supply chain management	No
Logistic management applications	No
Sourcing and e-procurement	No
E-learning Portal & Training Management System (TMS)	No
Integrated Administration and Control Systems (IACS)	No
Land Parcel Identification Systems (LPIS)	No
Land Management Information Systems (LMIS)	No
Cloud for farmers and citizens	No
Digital self service	No
Case management for public sector	No
Digital transaction systems in retail, transport, tourism, and payments	No
Public health and housing	No
Social care	No
Waste management and recycling	No
Planning and building control	No
Regulatory services	No
Public safety (police, fire and ambulance services, local authorities, health, utilities and transport providers.)	No
Cyber security	No
Network monitoring and managed services	No
Financial services	No

3.2 Benefits in SaaS Model

Software as a Service has become one of the top product on the IT market. Referring to the analysts from Forrester Research: "There is an estimation of the market value growth at 56.2% annually and forecast that it will continue to rise at a speed of 18.9% per year by 2020. Public cloud platforms, business services, and applications (SaaS) will grow at a 22% CAGR between 2015 and 2020, reaching $236B. Cloud platform revenues, whose 2020 total of $64B will be 45% higher than it was projected two years ago" [7].

It wasn't unexpected that SaaS would expand so fast. The model presented in Fig. 5 presents a diversity of advantages that attract a lot of new entrepreneurs. Different organizations are likely to choose SaaS because it is more beneficial and a more secure solution. Some of the most important benefits when moving services to the SaaS model are: **Cost -** There are no initial costs when purchasing a SaaS application because there are no license fees and clients also don't need to immediately purchase a whole product. Meanwhile, customers who need software for a limited period of time pay only for the service used during this period; **Customization -** Products which are delivered in the

SaaS model can be customized depending on client's needs. The price is determined individually in accordance to the customized software. SaaS providers also give access to their APIs which enables to integrate with existing systems. **Accessibility -** With SaaS we are able to access applications from every device with Internet connection at any time. This increases mobility and independence while working with our software. **No need for updates -** When we buy and install licensed software, we need to update it regularly, while in the case of SaaS, providers manage availability and all the updates, so that we don't need to add hardware or software with time.

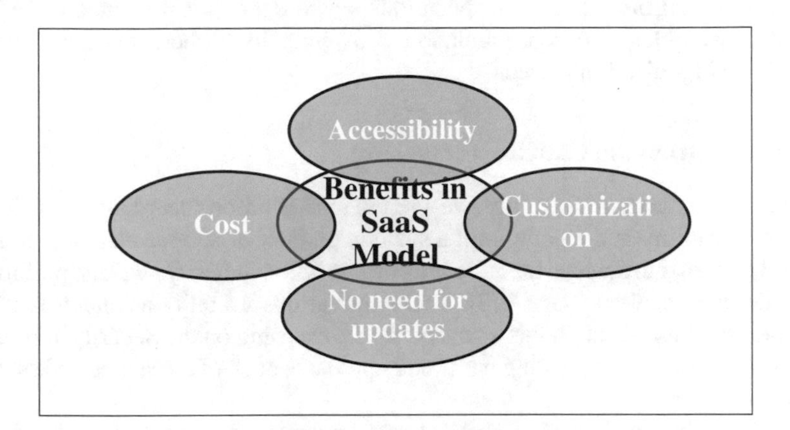

Fig. 5. Benefits in SaaS model

3.3 What Should We Consider When Moving the SaaS Cloud Model?

Cloud Services. Firstly, we should find what level of Infrastructure as a Service (IaaS) is necessary to support our SaaS app. For some public institutions, which don't demand high security, basic public cloud services might be sufficient. Other institutions such as Ministry of Defense, Ministry of Foreign Affairs, Ministry of Interiors Affairs, whose work involves storing sensitive data, more secure private cloud services might be needed. Flexibility is an important consideration because most SaaS apps are not written with cloud infrastructure architecture in mind. This means that NAIS (SaaS Provider) should host their services depending on the best type of IaaS option for the needs of public Institutions. Meanwhile, if NAIS needs help to monitor and manage their IaaS operations it should better consider managed cloud services.

Governance. Government Institutions need to adapt the governance issues and laws when considering adopting the SaaS model. If the processing of confidential data is necessary, then the location and data controls of NAIS should be independently accredited. Prior to moving to the SaaS model every institution should seek legal advice so that to be sure that the SaaS implementation would fulfill all legal requirements. Each government institution should consider some worst-case scenarios and ensure that the contract adequately covers them.

Training. Proper training is required to ensure the public institution employees can use SaaS adequately. It is important to make a contract which obligates NAIS to offer support continually and ensure that an appropriate training is provided to cover major updates. Such services should be rolled in the monthly charge providing scope for businesses to redeploy IT departments to more core ICT business projects. Training obligations should be separately set out and defined appropriately. While SaaS is very valuable in the Gov Cloud, there are certain situations where we believe it is not the best option. Examples where SaaS may not be appropriate are: 1. Applications where there is required a fast processing of real time data; 2. Applications where data isn't permitted to be hosted externally by any legislation or regulation; 3. Applications where an existing solution meets all the organization's needs.

3.4 Security from the Usability Perspective

In our paper, for the term 'Usability' we will use the definition according to ISO 9241-11 which is: "Usability is the point until a system, product or service can be used from a specified customer to achieve the assigned purpose with efficiency and high performance in a predestined context usage." [8]. There are various usability attributes that are of interest to cloud users, but in this analysis, we are focusing on the Security attribute [9]. Beyond usual concerns regarding the Cloud approach, each of these models has its own security concerns.

With IaaS, the developer has a better control over security, because applications run on virtual machines separated from other virtual machines running on the same physical machine. The only important thing in this model is to make sure that there are no security concerns in the virtual machine manager.

With PaaS, the provider might give some control to the people who build their applications on top of its platform. For example, developers might create their own authentication systems or data encryption. However, providers will still provide security below the application level and offer a strong assurance that the data is going to be inaccessible between applications.

With SaaS, the major concerns are security, loss of data control, data protection and compliance with government regulations [10]. Either maliciously or accidentally, Cloud provider's employees can leak the confidential data of a company. In order to save confidentiality, cloud customers might use encryption which is effective in securing data before it is stored at the provider, but it cannot be applied in services where data is to be computed. Security problems in the SaaS model usually fall into three areas: 1. Is the government institution data safe from malicious users (hackers)? 2. Is the data partitioned from other institutions (SaaS customers)? 3. Can the institutions get the data back if NAIS has a problem or in case they want to change the provider? NAIS should ensure the latest security protection systems towards the government institutions data, including state of the industry encryption, single sign-on (SSO), authentication, monitoring, alert and reporting systems. It should ensure that it's cloud service facilities are properly protected. It is also very important to have trained staff to administer all security systems, and formal policies so that to govern how these systems operate;

4 Conclusions

In this paper, we analyzed cloud implementation in GoA, the deployment and service model which have been chosen, and the features and general architecture of Gov Cloud in Albania. By implementing the Gov Cloud, GoA succeeded in centralizing and standardizing the application servers that are used in the government. As a result of the efficient use of the hardware, the overall IT costs might have been reduced, and the performance of services offered to government institutions has increased.

Then, we presented the Government of Albania benefits move from IaaS to the SaaS model. We have argued that SaaS provides many benefits to government institutions; no initial costs, maintenance and updates, greater accessibility, high availability, and pay-per use pricing. It is very important that the NAIS as a Gov Cloud provider ensures the latest security protection systems towards the confidential customer data. GoA should seek legal advice before moving services to the SaaS model because of legal requirements and Albanian laws regarding data confidentiality. It is also necessary to provide a proper training so that to ensure employees are able to use SaaS model in an adequate way. Finally, we mentioned a very important usability attribute of interest to cloud users which is security. With SaaS, users must rely heavily on their cloud providers for security and the provider must do all the work to keep users from seeing each other's data without permission. This way, the customers finds it very difficult to know if their data is being protected in the right manner. The SaaS model financial and functional advantages should become the main reasons for GoA to move from IaaS to SaaS.

References

1. Bohn, R.B., Messina, J., Liu, F., Tong, J., Mao, J.: NIST cloud computing reference architecture. In: Proceedings of the 2011 IEEE World Congress on Services. IEEE Computer Society, Washington (2011)
2. Security Framework for Governmental Clouds. © European Union Agency for Network and Information Security (ENISA), pp. 1–5 (2015)
3. Aymerich, F., Fenu, G., Surcis, S.: A real time financial system based on grid and cloud computing. In: Proceedings of ACM Symposium on Applied Computing, Honolulu, Hawaii, USA, pp. 1219–1220 (2009)
4. Windows Server 2008 R2 Customer Solution Case Study. Infosoft Systems. http://www.infosoftsystems.al/en/news-events/news/article/claud-computing-infosoft-systems-conclude-successfully-the-project-with-albanian-government/. Accessed 10 Dec 2017
5. Tan, C., Liu, K., Sun, L.: A design of evaluation method for SaaS in cloud computing. J. Ind. Eng. Manage. **6**, 50–72 (2013)
6. Implementing the Cloud Security Principles. National Cyber Security Centre (a part of GCHQ). https://www.ncsc.gov.uk/guidance/implementing-cloud-security-principles. Accessed 29 Nov 2017
7. Columbus, L.: Roundup of Cloud Computing Forecast. Forbes.com, https://www.forbes.com/sites/louiscolumbus/2017/04/29/roundup-of-cloud-computing-forecasts-2017/#282a79fd31e8. Accessed 10 Dec 2017
8. Guidance on Usability. ISO 9241-11 (1998)

9. Stanton, B., Theofanos, M., Joshi, K.P.: Framework for cloud usability. In: Tryfonas, T., Askoxylakis, I. (eds.) HAS 2015. LNCS, vol. 9190, pp. 664–671. Springer, Cham (2015). https://doi.org/10.1007/978-3-319-20376-8_59
10. Barillaud, F., Calio, C., Jacobson, J.A.: IBM Cloud Technologies: How the all fit together. IBM developerWorks. https://www.ibm.com/developerworks/cloud/library/cl-cloud-technology-basics/index.html. Accessed 15 Dec 2017

Performance Evaluation of an IoT-Based E-learning Testbed Considering Meditation Parameter

Masafumi Yamada[1(✉)], Kevin Bylykbashi[2], Yi Liu[1], Keita Matsuo[3], Leonard Barolli[3], and Vladi Kolici[2]

[1] Graduate School of Engineering, Fukuoka Institute of Technology (FIT), 3-30-1 Wajiro-Higashi, Higashi-ku, Fukuoka 811-0295, Japan
masafumi00835563@gmail.com, ryuui1010@gmail.com

[2] Faculty of Information Technology, Polytechnic University of Tirana, Mother Theresa Square, No. 4, Tirana, Albania
kevini-95@hotmail.com, vkolici@fti.edu.al

[3] Department of Information and Communication Engineering, Fukuoka Institute of Technology (FIT), 3-30-1 Wajiro-Higashi, Higashi-ku, Fukuoka 811-0295, Japan
{kt-matsuo,barolli}@fit.ac.jp

Abstract. Due to the opportunities provided by the Internet, people are taking advantage of e-learning courses and enormous research efforts have been dedicated to the development of e-learning systems. So far, many e-learning systems are proposed and used practically. However, in these systems the e-learning completion rate is low. One of the reasons is the low study desire and motivation. In this work, we present an IoT-Based e-learning testbed. We carried out some experiments considering meditation parameter with a student of our laboratory. We used Mind Wave Mobile (MWM) to get the data and considered four situations: Playing Game, Watching Movie, Listening Music and Reading Book. The evaluation results show that our testbed can judge the student situation by meditation parameter.

Keywords: Internet of Things · Testbed · Clustering algorithm
Meditation parameter

1 Introduction

The Internet is growing every day and the performance of computers is significantly increased [1,2]. Also, with appearance of new technologies such as ad-hoc networks, sensor networks, body networks, home networking, new network devices and application are appearing. Therefore, it is very important to monitor and control the network devices via communication channels and exploit their capabilities for the everyday real life activities. However, in large scale networks such as Internet, it is very difficult to control the network devices.

L. Barolli et al. (Eds.): EIDWT 2018, LNDECT 17, pp. 1049–1060, 2018.
https://doi.org/10.1007/978-3-319-75928-9_97

So for many e-learning systems are proposed and used practically. In [3], the authors presents a work-in-progress intending to enhance the learning experience for distance university students enrolled at the Open University of Catalonia (UOC). The UOC virtual campus has an integrated e-learning environment that allows students to pursue their studies completely online with the exception of taking final exams. By integrating the technologies of the IoT, they want to expand the learning environment and add a new learning place to the one existing on the computer. The authors hope to combine both the virtual and the physical environments in order to provide a better learning experience to their students. The authors consider two applications types: one related to the learning process and learning materials, the other related to creating a university community as well as fighting dropout and loneliness.

In [4], the authors present the context-aware and culture-oriented aspects of an adaptability approach called Adapt-SUR. Adapt-SUR is an international joint project between Argentina and Brazil. The approach is designed to be integrated into two distinct E-learning environments (ELEs): the AdaptWeb (Adaptive Web based learning Environment) system [5] and the eTeacher+SAVER (Software de Asistencia Virtual para Educacion Remota) environment [6]. This study describes the main features of the context-aware and culture-oriented aspects of a student profile and shows how to organize this contextual information in a multidimensional space where each dimension is represented by a different ontology, which may be handled separately or jointly. Finally the authors use some examples to discuss and illustrate how to use cultural information to provide context-based e-learning personalization.

In this work, we present an IoT-based e-learning testbed. We carried out some experiments considering meditation parameter with a student of our laboratory. We used MWM to get the data and considered four situations: Playing Game, Watching Movie, Listening Music and Reading Book.

The paper is organized as follows. In Sect. 2, we explain the overview of IoT and ULE. In Sect. 3, we present an overview of mean-shift clustering algorithm. In Sect. 4, we present the testbed. In Sect. 5, we discuss the experiments and simulation results. Finally, conclusions and future work are given in Sect. 6.

2 Overview of IoT and ULE

2.1 Internet of Things (IoT)

The Internet of Things (IoT) is a recent communication paradigm that envisions a near future, in which the objects of everyday life will be equipped with microcontrollers, transceivers for digital communication, and suitable protocol stacks that will make them able to communicate with one another and with the users, becoming an integral part of the Internet [7,8]. The IoT concept aims at making the Internet even more immersive and pervasive. Furthermore, by enabling easy access and interaction with a wide variety of devices such as, for instance, home appliances, surveillance cameras, monitoring sensors, actuators, displays, vehicles, and so on, the IoT will foster the development of a number of applications that make use of the potentially enormous amount and variety of data generated by such objects to provide new services to citizens, companies, and public administrations. This paradigm indeed finds application in many different domains, such as home automation, industrial automation, medical aids, mobile healthcare, elderly assistance, intelligent energy management and smart grids, automotive, traffic management, and many others [9].

2.2 Ubiquitous Learning Environment (ULE)

Ubiquitous learning is a seamless learning whenever it is in information space or in physics space, through ubiquitous computing information space and physics space are converged. In ULE Learning, learning demands and learning resources are everywhere; study, life and work are connected each other. When learners meet any practice problem ubiquitous computing help them to resolve it at anytime, anywhere. In the future, school, library, classroom, meeting room, museum, and the circulation fields send their information and knowledge to the learner through all kinds of technology, every learner immerse into information ecology surroundings that the real world and digital world intermingle. The learners can easily perception and obtaining learning objects detailed information and content through situational perception of mobile devices. Using dialogue, living community, cooperation studies, social process of internalization, participate in joint activity to realize social learning. An effective ubiquitous learning depends on founding of learning environment.

2.3 Role of IoT in ULE

According to learning environment classification, ubiquitous learning environment belong to a kind of learning environment that are deeper, and the highest flexibility. While the basic elements of constructing the learning environment mainly include three parts: ubiquitous communication network, learning terminal device, learning resources. The traditional single point centralized resource storage mode is unable to meet with the ubiquitous learning requirements whether the resource storage or the promptness of obtaining resources. IoT make not only real world are connected, but also the real world (physical narrow room) and virtual worlds (digital information space) are all interconnected, and it support effectively M2M interaction. IoT make every things of learning environment digital, intelligence and networking, make learning seamless integration, learner study what they need at any time, at anyplace, and adjust corresponding learning content, and make learning environment intelligence. For example, monitor and control light brightness by sensor; learn outdoor things by RFID, and so on.

3 Mean-Shift Clustering Algorithm

Mean-shift represents a general non-parametric mode finding/clustering procedure [10]. In contrast to the classic k-means clustering approach, there are no embedded assumptions on the shape of the distribution nor the number of modes/clusters. Mean-shift was first proposed by Fukunaga and Hostetler, later adapted by Cheng for the purpose of image analysis and more recently extended by Comaniciu, Meer and Ramesh to low level vision problems, including, segmentation adaptive smoothing and tracking.

The main idea behind mean-shift is to treat the points in the d-dimensional feature space as an empirical probability density function where dense regions in the feature space correspond to the local optima or modes of the underlying distribution. For each data point in the feature space, one performs a gradient ascent procedure on the local estimated density until convergence. The stationary points of this procedure represent the modes of the distribution. Furthermore, the data points associated (at least approximately) with the same stationary point are considered members of the same cluster.

Here is briefly described the variable bandwidth mean-shift procedure [11, 12]. Given n data points x_i on a d-dimensional space R^d and the associated bandwidths $h_i = h(x_i), i = 1, \ldots, n$, the sample point density estimator obtained with profile $k(x)$ is given by:

$$f(x) = \frac{1}{n} \sum_{i=1}^{n} \frac{1}{h_i^d} k\Big(\| \frac{x - x_i}{h_i} \|^2\Big). \tag{1}$$

It is utilized multivariate normal profile:

$$k(x) = e^{-\frac{1}{2}x}, x \geq 0. \tag{2}$$

Taking the gradient of Eq. (1), the stationary points of the density function satisfy:

$$\frac{2}{n} \sum_{i=1}^{n} \frac{1}{h_i^{d+2}} (x_i - x) g\Big(\| \frac{x - x_i}{h_i} \|^2\Big) = 0, \tag{3}$$

where $g(x) = -k'(x)$. The solution can be found iteratively via the fixed point algorithm

$$\overline{x} = \frac{\sum_{i=1}^{n} \frac{x_i}{h_i^{d+2}} g\Big(\| \frac{x - x_i}{h_i} \|^2\Big)}{\sum_{i=1}^{n} \frac{1}{h_i^{d+2}} g\Big(\| \frac{x - x_i}{h_i} \|^2\Big)}, \tag{4}$$

which is called mean-shift procedure. Comaniciu and Meer [13] show that the convergence to a local mode of the distribution is guaranteed when the mean-shift iterations are started at a data point. The mean-shift procedure for a given point x_i is shown in Algorithm 1.

Algorithm 1. The process of mean-shift algorithm.

1: Compute the mean-shift vector $m(x_i^t)$.
2: Translate density estimation window: $x_i^{t+1} = x_i^t + m(x_i^t)$.
3: Iterate steps 1. and 2. until convergence, i.e., $\nabla f(x_i) = 0$.

4 Testbed Description

In Fig. 1 is shown the structure of IoT-based e-learning testbed. Our testbed is composed of a Raspberry Pi B+ [14,15], MWM and SmartBox. The Raspberry Pi is a credit card-sized single-board computer developed by the Raspberry Pi Foundation. In the implemented system, we use MWM to get human EEG data. These data are sent to Raspberry Pi computer and are processed by mean-shift clustering algorithm. Then, the results are send to the SmartBox, which can change the mood of humans using its functions.

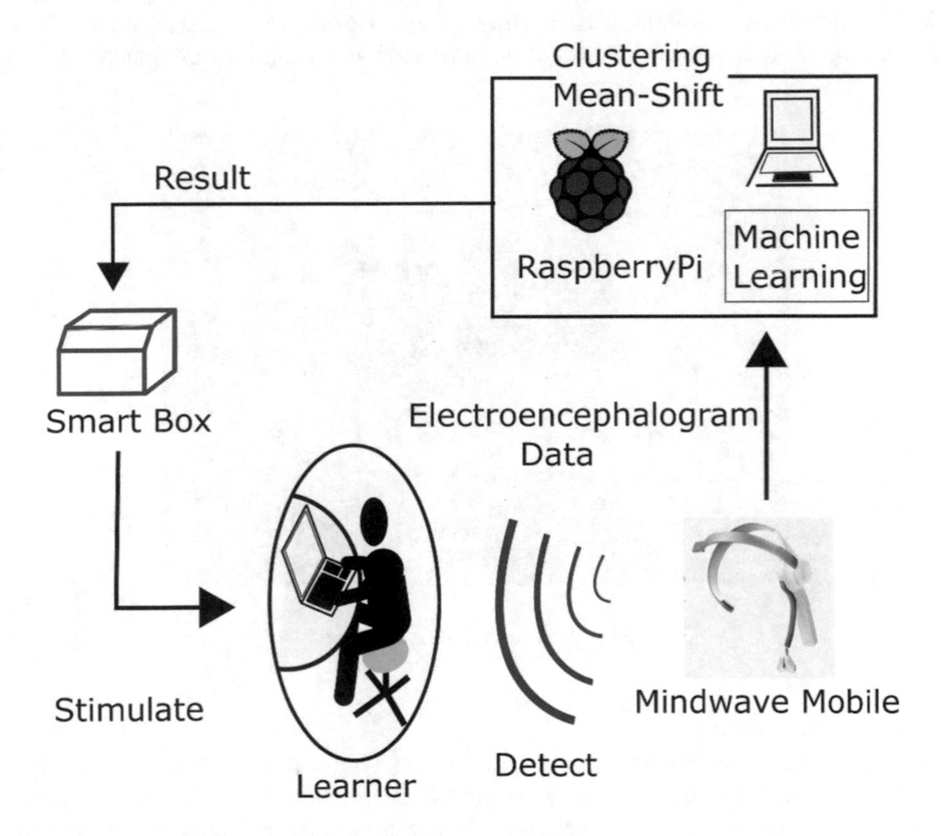

Fig. 1. Structure of IoT-based E-learning testbed.

4.1 MWM

A snapshot of the MWM is shown in Fig. 2. MWM is a device capable of acquiring the human EEG data [16]. The device measures the raw signal, power spectrum (Delta, Theta, Alpha, Beta and Gamma waves), Attention level, Mediation level and blink detection. The raw EEG data are received at a rate of 512 Hz. Other measured values are made every second. Therefore, raw EEG data are the main source of information on EEG signals using MWM. By MWM can be determined how effectively the user is engaging Attention (similar to concentration) by decoding the electrical signals and applying algorithms to provide readings on a scale of 0 to 100. These values are described in Table 1.

4.2 EEG

4.2.1 Delta Wave

Delta waves in humans are between 0 and 4 Hz. In general, low frequency oscillations dominate the human EEG during early developmental stages [17]. Low frequency oscillations are presumably generated when humans are sleeping.

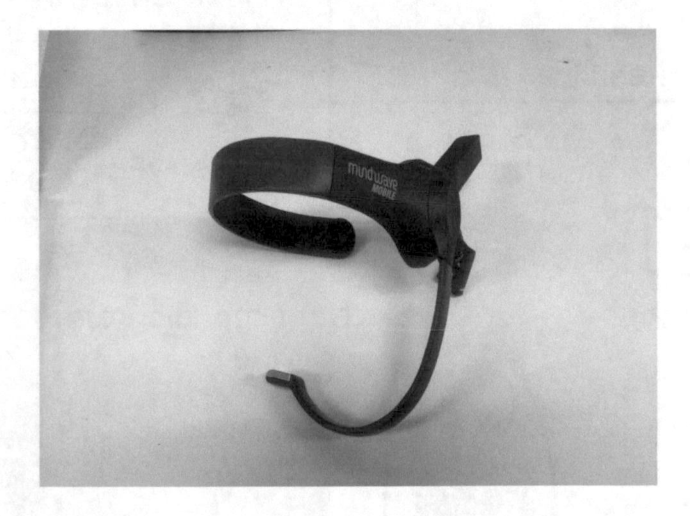

Fig. 2. A snapshot of MWM.

4.2.2 Theta Wave

Human theta wave lies between 4 and 7.5 Hz. The crucial finding is that with increasing task demands theta wave synchronizes. If EEG power in a resting condition is compared with a test condition, gamma power decreases and desynchronizes, while theta power increases and synchronizes.

4.2.3 Alpha Wave

Alpha wave is the dominant frequency in the human scalp EEG of adults [18]. When a healthy adult relaxes, with eyes closed, rhythmic electric activity of around 10 Hz can be recorded over the posterior scalp regions. The fact that alpha wave clearly is an oscillatory component of the human EEG has led to a recent "renaissance" in the interest of EEG alpha activity. Usually, alpha frequency is defined in terms of peak or gravity frequency within the traditional alpha frequency range of about 7.5–12.5 Hz.

4.2.4 Beta Wave

Human beta wave between 12 and 30 Hz. EEG is sensitive to a continuum of states ranging from stress state, alertness to resting state, hypnosis, and sleep [19]. During normal state of wakefulness with open eyes beta waves are dominant. In relaxation or drowsiness alpha activity rises and if sleep appears power of lower frequency bands increase. Sleep is generally divided into two broad types: Non-Rapid Eye Movement sleep (NREM) and REM sleep. NREM and REM occur in alternating cycles, when slower dominant frequencies responsiveness to stimuli decreases.

4.2.5 Gamma Wave

Human gamma wave lies between 30 and 80 Hz. A dramatic increase of activity in the gamma band in association with meditation-visible in the raw EEG-was reported in a study using trained practitioners of meditation [20].

Table 1. Descriptions of eSense meter values.

Values	Description
0–20	Strongly lowered levels
20–40	Reduced levels
40–60	Neutral/baseline levels
60–80	Slightly elevated/higher than normal levels
80–100	Elevated/heightened levels

4.3 SmartBox Description

We implemented a SmartBox device [1,2]. The size of the SmartBox is $35 \times 7 \times 12$ [cm]. The SmartBox is equipped with different sensors (for sensing learner situation) and devices (used for stimulating learner's motivation). The SmartBox has the following sensors and functions.

Fig. 3. A snapshot of SmartBox.

- Body Sensor: for detecting the learner's body movement.
- Chair Vibrator Control: for vibrating the learner's chair.
- Light Control: for adjusting the room light for study.
- Smell Control: for controlling the room smell.
- Sound Control: to emit relaxing sounds.
- Remote Control Socket: for controlling AC 100 [V] socket (on-off control).

A snapshot of the SmartBox is shown in Fig. 3.

5 Experiment and Simulation Results

We carried out some experiments considering meditation parameter. We collected data by Fast Fourier Transform (FFT) [21]. These data are collected by using the scikit-learn, which is a general purpose machine learning library for the Python [22].

We used MWM to get the data and considered four situations: Playing Game, Watching Movie, Listening Music and Reading Book. The experimental data are shown in Figs. 4, 5, 6 and 7. In Fig. 4 are shown the meditation values in the case of playing a game. At the beginning of the game, the meditation values are higher, so the student is concentrated on the game. However, after 600 s, the student concentration is decreased.

In Fig. 5 are shown the meditation values in the case of watching a movie. In general, the values of meditation are higher during all experimental time. This shows that the student is concentrated while he is watching the movie.

In Fig. 6, we show experimental results when the student is listening music. As can be seen from the figure, the meditation value are increased with increasing of the experimental time.

In Fig. 7, we show experimental data when the student is reading a book. At the beginning of the experiment, the meditation values are very high and after that are decreasing. This shows that the student concentration is higher when he starts to read the book and decreases with the increase of the experimental time. This is because the student become tired when he reads for a long time the book.

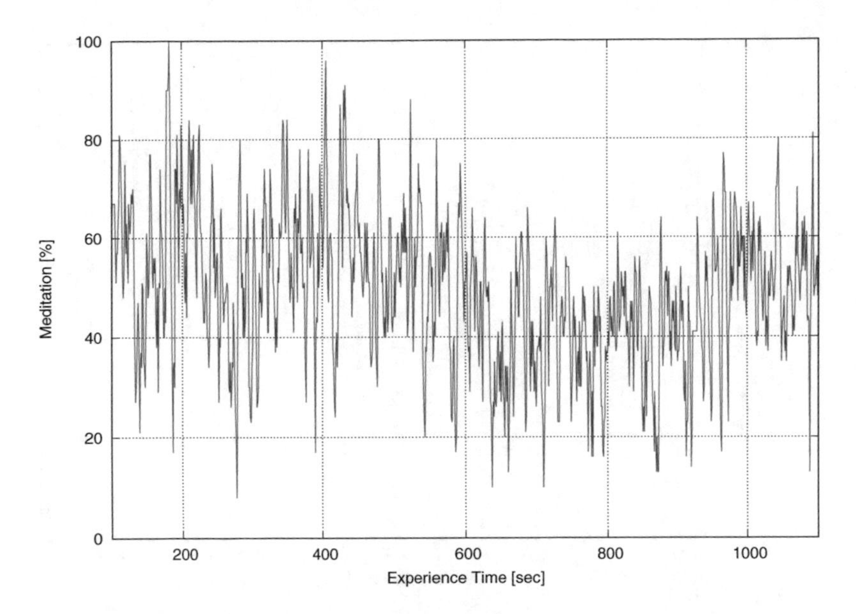

Fig. 4. Game

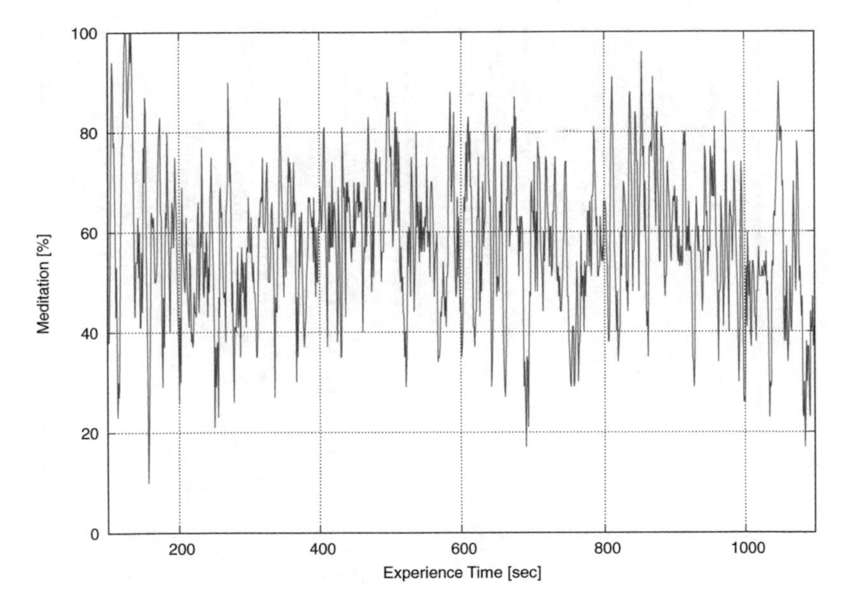

Fig. 5. Movie

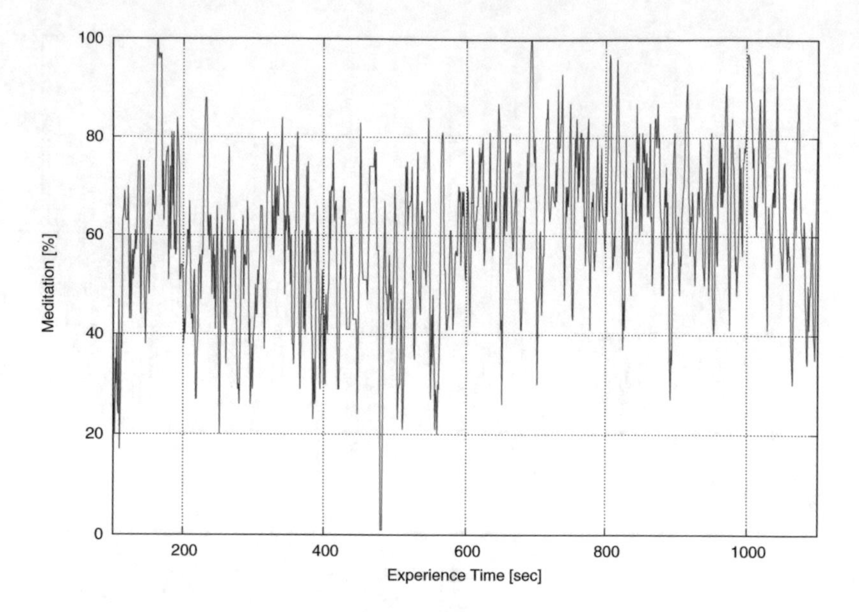

Fig. 6. Music

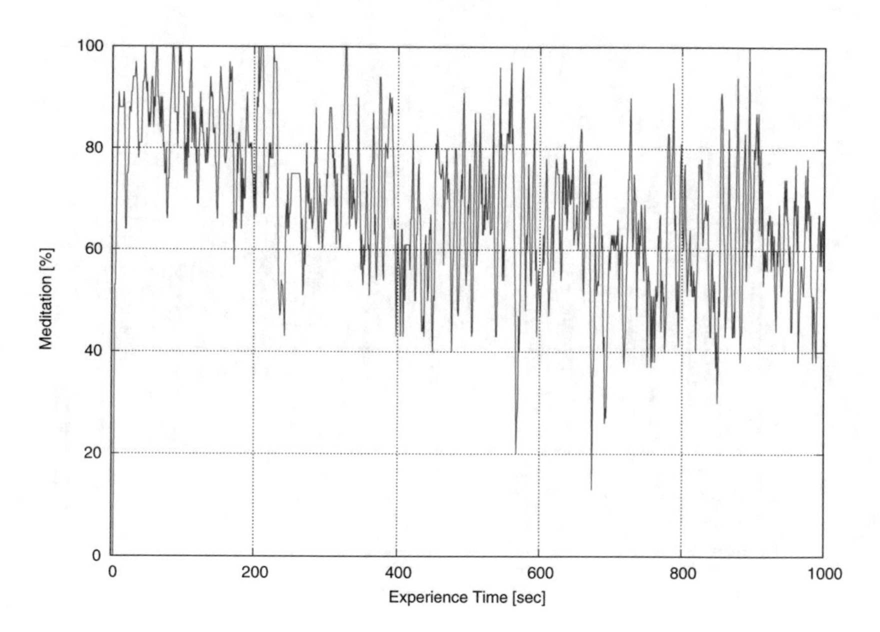

Fig. 7. Reading

6 Conclusions and Future Work

In this paper, we presented an IoT-based e-learning testbed. We evaluated its performance by carrying out some experiments with a student of our laboratory considering meditation parameter.

We used MWM to get the data and considered four situations: Playing Game, Watching Movie, Listening Music and Reading Book. The evaluation results show that our testbed can judge the student situation by considering meditation parameter.

In the future, we will carry out many experiments using the implemented testbed.

References

1. Matsuo, K., Barolli, L., Xhafa, F., Kolici, V., Koyama, A., Durresi, A., Miho, R.: Implementation of an e-learning system using P2P, web and sensor technologies. In: Proceedings of IEEE Advanced Information Networking and Applications (AINA-2009), pp. 800–807 (2009)
2. Matsuo, K., Barolli, L., Arnedo-Moreno, J., Xhafa, F., Koyama, A., Durresi, A.: Experimental results and evaluation of SmartBox stimulation device in a P2P e-learning system. In: Proceedings of Network-Based Information Systems (NBiS-2009), pp. 37–44 (2009)
3. Domingo, M.G., Forner, J.A.M.: Expanding the learning environment: combining physicality and virtuality - the Internet of Things for eLearning. In: Proceedings of 10th IEEE International Conference on Advanced Learning Technologies (ICALT-2010), pp. 730–731 (2010)
4. Gasparini, I., Eyharabide, V., Schiaffino, S., Pimenta, M.S., Amandi, A., de Oliveira, J.P.M.: Improving user profiling for a richer personalization: modeling context in e-learning. In: Intelligent and Adaptive Learning Systems: Technology Enhanced Support for Learners and Teachers, pp. 182–197 (2012). Chapter 12
5. de Freitas, V., Marcal, V.P., Gasparini, I., Amaral, M.A., Proenca Jr., M.L., Brunetto, M.A.C., Pimenta, M.S., Ribeiro, C.H.F.P., de Lima, J.V., de Oliveira, J.P.M.: AdaptWeb: an adaptive web-based courseware. In: Proceedings of International Conference on Information and Communication Technologies in Education (ICTE-2002), pp. 131–134 (2002)
6. Schiaffino, S., Garcia, P., Amandi, A.: eTeacher: providing personalized assistance to e-learning students. Comput. Educ. **51**(4), 1744–1754 (2008)
7. Zanella, A., Bui, N., Castellani, A., Vangelista, L.: Internet of Things for smart cities. IEEE Internet Things J. **1**(1), 22–32 (2014)
8. Atzori, L., Iera, A., Morabito, G.: The Internet of Things: a survey. Comput. Netw. **54**(15), 2787–2805 (2010)
9. Bellavista, P., Cardone, G., Corradi, A., Foschini, L.: Convergence of MANET and WSN in IoT urban scenarios. IEEE Sensors J. **13**(10), 3558–3567 (2013)
10. Derpanis, K.G.: Mean shift clustering. http://www.cse.yorku.ca/~kosta/CompVis-Notes/mean-shift.pdf. Accessed 14 Sept 2016
11. Comaniciu, D.: Variable bandwidth density-based fusion. In: Proceedings of IEEE Computer Vision and Pattern Recognition (CVPR-2003), vol. 1, pp. 59–66 (2003)

12. Tuzel, O., Porikli, F., Meer, P.: Kernel methods for weakly supervised mean shift clustering. In: Proceedings of 12th IEEE International Conference on Computer Vision, pp. 48–55 (2009)
13. Comaniciu, D., Meer, P.: Mean shift: a robust approach toward feature space analysis. IEEE Trans. Pattern Anal. Mach. Intell. **24**(5), 603–619 (2002)
14. Raspberry Pi Foundation. http://www.raspberrypi.org/
15. Oda, T., Barolli, A., Sakamoto, S., Barolli, L., Ikeda, M., Uchida, K.: Implementation and experimental results of a WMN testbed in indoor environment considering LoS scenario. In: Proceedings of 29th IEEE International Conference on Advanced Information Networking and Applications (AINA-2015), pp. 37–42 (2015)
16. NeuroSky to Release MindWave Mobile. http://mindwavemobile.neurosky.com
17. Knyazev, G., et al.: EEG delta oscillations as a correlate of basic homeostatic and motivational processes. Neurosci. Biobehav. Rev. **36**(1), 677–695 (2012). https://doi.org/10.1016/j.neubiorev.2011.10.002
18. Klimesch, W., et al.: EEG alpha and theta oscillations reflect cognitive and memory performance: a review and analysis. Brain Res. Rev. **29**(2–3), 169–195 (1999)
19. Teplan, M., et al.: Fundamentals of EGG measurement. Measur. Sci. Rev. **2**(2), 1–11 (2002)
20. Vialatte, F.B., Bakardjian, H., Prasad, R., Cichocki, A.: EEG paroxysmal gamma waves during Bhramari Pranayama: a yoga breathing technique. Conscious. Cognit. **18**(4), 977–988 (2009). https://doi.org/10.1016/j.concog.2008.01.004
21. Akin, M.: Comparison of Wavelet Transform and FFT methods in the analysis of EEG signals. J. Med. Syst. **26**(3), 241–247 (2002)
22. Pedregosa, F., et al.: Scikit-learn: machine learning in Python. J. Mach. Learn. Res. **12**(10), 2825–2830 (2011)

Towards Integrating Conversational Agents and Learning Analytics in MOOCs

Stavros Demetriadis[1], Anastasios Karakostas[2], Thrasyvoulos Tsiatsos[1], Santi Caballé[3(✉)], Yannis Dimitriadis[4], Armin Weinberger[5], Pantelis M. Papadopoulos[6], George Palaigeorgiou[7], Costas Tsimpanis[8], and Matthew Hodges[9]

[1] Aristotle University of Thessaloniki, Thessaloniki, Greece
[2] Centre for Research and Technology Hellas, Thermi, Greece
[3] Universitat Oberta de Catalunya, Barcelona, Spain
scaballe@uoc.edu
[4] Universidad de Valladolid, Valladolid, Spain
[5] University of Saarland, Saarbrücken, Germany
[6] Aarhus University, Aarhus, Denmark
[7] Learnworlds, London, UK
[8] Greek Universities Network, Athens, Greece
[9] Telefónica, Madrid, Spain

Abstract. Higher Education Massive Open Online Courses (MOOCs) introduce a way of transcending formal higher education by realizing technology-enhanced formats of learning and instruction and by granting access to an audience way beyond students enrolled in any one Higher Education Institution. However, although MOOCs have been reported as an efficient and important educational tool, there is a number of issues and problems related to their educational impact. More specifically, there is an important number of drop outs during a course, little participation, and lack of students' motivation and engagement overall. This may be due to one-size-fits-all instructional approaches and very limited commitment to student-student and teacher-student collaboration. This paper introduces the development agenda of a newly started European project called "colMOOC" that aims to enhance the MOOCs experience by integrating collaborative settings based on Conversational Agents and screening methods based on Learning Analytics, to support both students and teachers during a MOOC course. Conversational pedagogical agents guide and support student dialogue using natural language both in individual and collaborative settings. Integrating this type of conversational agents into MOOCs to trigger peer interaction in discussion groups can considerably increase the engagement and the commitment of online students and, consequently, reduce MOOCs dropout rate. Moreover, Learning Analytics techniques can support teachers' orchestration and students' learning during MOOCs by evaluating students' interaction and participation. The research reported in this paper is currently undertaken within the research project colMOOC funded by the European Commission.

L. Barolli et al. (Eds.): EIDWT 2018, LNDECT 17, pp. 1061–1072, 2018.
https://doi.org/10.1007/978-3-319-75928-9_98

1 Introduction

Understanding and designing with digital technologies has become a relevant competency across disciplines. Computational thinking and programming skills become increasingly important - also for non-programmers. European universities trail behind in offering these transversal competencies for students beyond the domain of computer science. Likewise, the current European Workforce often needs further training for digital literacy. For example, in-service teachers have a need to learn about instructional technologies that were not present during their study times. Last but not least, becoming digital literate may be a crucial springboard for unemployed groups with little access to higher education, e.g. single parents [1].

To help face these challenges, Higher Education Massive Open Online Courses (MOOCs) [2] arose as a way of transcending formal higher education by realizing technology-enhanced formats of learning and instruction and by granting access to an audience way beyond students enrolled in any one Higher Education Institution (HEI). However, the potential for European HEIs to scale up and reach an international audience of diverse backgrounds has not been realized yet. MOOCs have been reported as an efficient and important educational tool, yet there is a number of issues and problems related to their educational impact. More specifically, there is an important number of drop outs during a course, little participation, and lack of students' motivation and engagement overall. This may be due to one-size-fits-all instructional approaches and very limited commitment to student-student and teacher-student collaboration.

This paper introduces a new European project called colMOOC that aims to enhance the MOOCs experience by integrating (i) Collaborative settings based on Conversational Agents (CA) both in synchronous and asynchronous collaboration conditions; (ii) Screening methods based on Learning Analytics (LA) to support both students and teachers during a MOOC course. The colMOOC project aims also to reinforce European leadership by forming recommendations and policy guidelines. Conversational pedagogical agents guide and support student dialogue using natural language both in individual and collaborative settings [3, 4]. CAs have been produced to meet a wide variety of applications and studies exploring the usage of such agents have led to positive results. Integrating this type of CAs into MOOCs is expected to trigger productive peer interaction in discussion groups and, therefore, to considerably increase the engagement and the commitment of online students, reducing consequently, the overall MOOCs dropout rate. Moreover, this project proposes to use LA techniques as a method to support teachers' orchestration and students' learning during MOOCs by evaluating students' interaction and participation.

A relevant operative goal of the colMOOC project is to develop specific MOOCs for students of the humanities as well as employees and the unemployed to develop digital competencies that have not been covered in basic, mono-disciplinary studies. MOOCs are considered one of the key tools to address the problem of digital illiteracy and equal access to education. MOOCs have shown to not only provide expert and up-to-date knowledge to anyone with internet access, but have also repositioned Universities as beacons of knowledge with impact on society at large [5]. Therefore, the main motivation of the project is to develop MOOCs that would capture the interest of different target groups and develop transversal digital competencies in specific fields of burning, immediate needs for digital literacy. However, developing MOOCs for lifelong learning that capture the interest of university students, employed and unemployed adults requires innovative, engaging and motivational learning methods. To this end, collaborative learning arrangements and formative feedback can not only improve students' performance, but also minimize chances of dropout and foster engagement. There is an intense need to provide such collaboration settings to enhance active participation and social skills [6]. To achieve this, the project will employ collaborative activities supported by CA.

Considering the above, the main priority tackled in the colMOOC project identifies that there is a need to make MOOCs more collaborative in order to support teachers and engage and motivate students. Moreover, it is critical to ensure that assessment in MOOCs will monitor individual achievement and demand constant engagement [7]. Each learner performs specific individual and collaborative activities during a MOOC course, while his/her performance can be more easily measured and assessed as opposed to a traditional learning setting where individual contributions to a work project are usually unknown or difficult to assess. Furthermore with the usage of learning analytics more meaningful assessment may be achieved, since educators can evaluate the work's progress during each step and adapt the following steps accordingly. This also enables learners to reflect on each specific step and ensure they have obtained the necessary knowledge to progress effectively. Therefore, the project's outcome will allow educators and trainers to apply learning processes where learners participated actively and are engaged throughout the course.

The remainder of the paper is structured as follows: Sect. 2 reviews the key topics forming the context of the project and in particular the drawbacks and challenges of MOOCs. Section 3 shows the project methodology with the aims and work plan while Sect. 4 shows the expected results of the project. Finally, Sect. 5 summarizes the main ideas of the paper and outlines the following steps of the colMOOC project.

2 Background

Supporting MOOCs implies to identify the drawbacks, problems and several user requirements that they have not yet addressed. Since top-ranked academic community joined the MOOC hype, academic sectors have hold controversial discussions on the many MOOC challenges that must be faced before moving on [5, 8, 9]: (i) high learners' dropout rate, with only 5% to 15% of participants finishing the courses on average; (ii) limited teaching involvement during delivery stages; (iii) lack of adaptability to a great variety of specific needs.

The colMOOC project aims to provide innovative ways for promoting learners' interaction by enabling the teacher to configure a CA software component which attempts to trigger learners' discussion through appropriate interventions in dialogue-based collaborative activities. The development of the agent will consider all current research evidence on the value of both 'Academic Productive Talk' [17] and the 'Transactive dialogue' [18] frameworks as forms of productive peer dialogue. The main functional features of the agent component include: (a) a user friendly interface based on the concept map metaphor, for the instructor to easily model the knowledge domain necessary for the agent intervention and also configure the agent behaviour during students' discourse; (b) an appropriately designed MOOC interface for students' synchronous and asynchronous online discussions, where the agent also appears and intervenes in the discussion aiming to trigger productive students' social interaction, domain focused cognitive activity and deeper learning.

Assessment in MOOCs is about supporting student learning and achievement [7]. MOOCs bring a new dimension to assessing such large number of learners. The automated assessments that have evolved in recent years are specifically targeted to assess and evaluate the large enrolments since it is not possible to manually grade and provide feedback to all the learners. Learners can be assessed on time-on-task; learner-course component interaction; and a certification of the specific skills and knowledge gained from a MOOC [10]. While not the primary aim, these assessment techniques will provide an added incentive for the learners to persist and complete the MOOC.

Ultimately, the satisfaction gained from completing the course can be potential indicator of good learning experiences. Conversely, enhancing the learning experience can contribute to improving the MOOC completion rates [5]. As a result, assessment techniques that permit customization of content catered to the individual learner can track learner behavior and predict learning outcomes. Such a technique will further assist in developing and refining assessment procedures for improving learning outcomes. The colMOOC project will enhance MOOCs capabilities to this direction by exploiting LA based on previous projects and experiences [11, 12].

3 Project Methodology

In this section we present the project methodology, including a broad description of the project aims and the work plan, including the evaluation activities.

3.1 Project Aims and Objectives

The colMOOC project aims to deliver a highly innovative and beyond the current state-of-the-art MOOC model and implementation with the integration of services based on CA and LA. Especially for the CAs, it is emphasized that the latest studies systematically indicate that agent interventions during peer interaction (i.e. online discussions) trigger task relevant cognitive activity that leads to improved learning outcomes at many levels (domain-specific and domain-independent) [13–15]. The agent interventions are modelled according to what is known as "Academically Productive Talk" (APT) framework, which essentially refers to modelling the experienced teacher's "moves" (interventions) during students' dialogue to make students elaborate in the domain [16]. The learning experience that this type of agent provides to students is similar to having one more partner in their group trying to respond to this partner's prompts.

The strategic goal of the project is to make a significant contribution towards European universities in their development towards knowledge service providers to the economy and society at large and to target the now transversal competencies of programming, computational and design thinking as well as concrete application of new technologies for teaching. While the EU universities jointly offer a high level of education, they trail behind in offering novel and transversal competencies as well as in developing joint and novel approaches to MOOC education.

Ultimately, the project aims to provide novel methods for teaching and learning to address the problem of attrition in online learning. These goals will be achieved by pursuing the following objectives:

- Develop new learning and teaching methods for MOOCs, building on novel technologies in collaborative learning, such as CAs, that are capable of boosting learner interaction and facilitate learners' self-regulation and -assessment.
- To promote innovative solutions to current and future challenges and for sustainable impact on Europe's education and training systems by granting open access to the CAs built within the project.
- To demonstrate and validate the built capacity for innovative teaching and learning methods and mainstream them to the existing education and training systems, by the design, execution and assessment of several pilots that orchestrate individual and collaborative learning activities.
- Spread the best practice pilots built in the project to all participating universities and other economic and societal stakeholders (educational authorities and civil society organizations for further training for the unemployed, businesses for further training of the workforce) as well as granting access to the MOOCs on European and national providers for free, openly accessible MOOCs, e.g., platform.europeanmoocs.eu.

The application of the results in multiple countries and large number of users will collect and analyze substantive evidence on our approaches. The usage of a well-established pedagogical model and the gathering, processing and analysis of the generated data with novel LA tools will provide informative evidence that can set the foundations for future concrete methodologies. These methodologies will foster participatory learning in a structured environment where learners will be able to develop transversal as well as domain-specific skills, becoming equipped for their future employability.

Finally, the colMOOC project will enforce market stakeholders in educational technology sector with new tools and methods.

3.2 Starting Point

The experience and results achieved from two relevant research projects [12, 13] will input colMOOC regarding the development of the main components of CA and LA. These projects are briefly described here:

1. The research project "Promoting academically productive talk with conversational agent interventions" [13] set as a key objective to explore the impact of questioning interventions implemented by a conversational agent in the context of computer-supported collaborative activities in higher education. The type of interventions is modelled according to 'Academically Productive Talk', a model emphasizing the

orchestration of teacher-students talks and highlighting a set of useful discussion practices that can lead to reasoned participation by all students, increasing the probability of productive peer interactions to occur [17]. To achieve the desired objective this specific project:

- Designed and developed a teacher-configured CA component able to be integrated in chat tools.
- Implemented a number of research activities exploring various facets of the impact of the CA component on students' learning in a series of relevant studies [3, 11, 13, 15].

The overall outcome has been highly positive indicating that this type of APT-based automated form of conversational support can enhance students' explicit reasoning on domain concepts and improve individual and group learning in the context of a collaborative learning activity in higher education settings [13, 15]. However, all available research so far has been conducted in controlled experimental settings. The colMOOC projects aims to expand exploration of this approach 'in the wild', that is in actual MOOC settings.

2. ICT-FLAG (Enhancing ICT education through Formative assessment, Learning Analytics and Gamification) [12]. The main goal of this project was to design and build a set of eLearning tools and services to provide support to the learning process in university degrees in the field of ICT. The benefits had a repercussion on the students (improvement of the educational experience, greater participation and performance, lower drop-out rate) and on the lecturers, managers and academic coordinators (resources for monitoring a course, making decisions and predictions). To achieve these benefits, the contributions of this project focused on three axes:
   - Tools for Formative Assessment, which provided immediate feedback by means of automatic assessment.
   - Learning Analytics that monitored the activity and the progress of the student about the use of the mentioned tools and allowed for analyzing the learning results, identifying the critical points and defining actions of improvement. These analytics also incorporated other sources of academic and historical information to facilitate the course tracking and decision making processes to the teaching team.
   - Gamification, as an incentive scheme, motivated students to perform new activities and increased their engagement without sacrificing the academic rigor.

A relevant aspect considered by the eLearning tools developed in this project was the modularity and independence from technologies and particular learning systems, with the aim to facilitate its application to different courses and contexts. To this end, the functionalities of the ICT-FLAG platform [12] are offered as a set of services, using appropriate standards, with the ultimate aim that these services become feasible as part of both self-taught education (life-long learning) and traditional formal education as well as massive courses of on-line learning (MOOCs).

The colMOOC project will leverage the technological independence-based LA services of the ICT-FLAG platform to provide MOOC students with a broad set of general analytics on their collaborative learning progress.

3.3 Work Plan

In order to achieve the above aims and objectives, the colMOOC project methodology will be divided into three major phases:

- Analysis & Design: During this phase all the possible user requirements are to be identified with regards to the CA and LA components. Implementation: This phase regards the integration and practical application of the designed CA and LA components in real world settings, in particular MOOC platforms, and large number of end-users to test their validity and ability to enhance the quality of education and training.
- Evidence-based Policy Formation: This phase includes the accumulation, study and analysis of the evaluation results after the trials and the corresponding design of policy recommendations. These recommendations will be concrete and in accordance to the policy making lifecycle.

The relevant milestones as the corresponding main success indicators along with the major project results are:

- colMOOC educational approach and component design.
- Components models and tools and first version of integration mechanism released.
- Software modules configured, courses design and materials developed.
- Trials and evaluation completed. Validation activities in educational and training settings are performed and evaluated.
- Policy recommendations drafted.

In terms of work methodology, the project will follow a four-step loop approach where education and training policies are used as input in order to produce the project's objectives and then provide new policy actions as outputs based on validated evidence. This way, the project's potential will be maximized in successfully transferring the project results into European policy development. These steps include:

- Study of good practice projects under various projects that have dealt with and successfully raised awareness and translated results into policy actions.
- In-depth study or existing models for policy development and the education and training policy agendas. This will include study of the nature of policy and the particular nature of education policy within its wider social, political and economic contexts.
- Design of innovative approaches and test of their validity on large-scale multilingual and multidisciplinary settings.
- In-depth study of wider policy recommendations in a national and/or European level for the identification of best practices.

Finally, the evaluation plan of the project aims to develop several pilot MOOCs of different pedagogical disciplines:

- *Programming for Non-Programmers.* Programming skills are not only a major asset for getting hired for developing software, but also allow for better understanding and co-designing with a development team. Programming skills are required to use, build, and maintain digital infrastructures for managing and compiling data.

- *Computational and Design Thinking*. This pedagogical strategy fosters systematic and creative approaches to problem solving. This involves analysis and formalization of complex problems, understanding and developing feasible, technical solutions, as well as divergent and convergent ways of thinking.
- *Educational Technologies in the Classroom*. New technologies enhance learning through dynamic, interactive, multi-media formats in simulations, web-videos, and games. Advanced technologies enable new forms to aggregate and visualize data for feedback to students and teachers.

The above project activities lead to the preparation of input state-of-the-art papers (dissemination activities) that provide recommendations for the formulation of new policy actions in the university-level and training-level.

Considering all the above, Fig. 1 below shows the main activities of the colMOOC project.

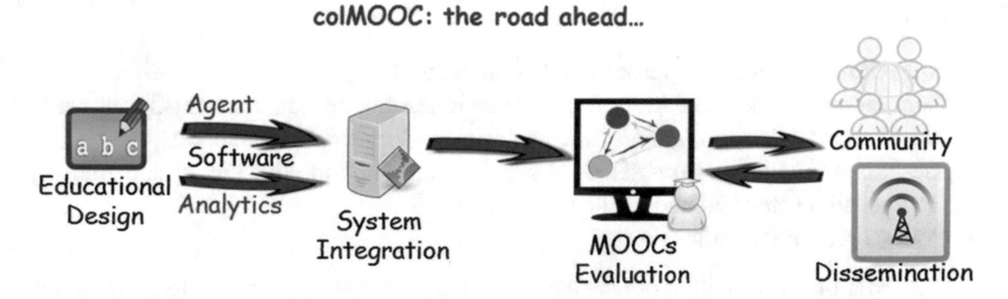

Fig. 1. colMOOC project main activities.

4 Project Outcomes

In this section, the expected benefits of the project will be first shown and then how these benefits will reach each of the identified target groups will be explained.

4.1 Expected Benefits

The outcomes of the colMOOC project are expected to benefit the higher education institutions in different ways:

- Enhanced MOOC courses: the project will provide prototype MOOC courses and services that will increase students' online engagement in conditions of productive dialogue and thus deeper learning. This, in turn, is expected to increase students' motivation and interest to follow the course. Moreover, university teachers will be able to use new and innovative pedagogical approaches such as collaborative learning supported by conversational agents to produce similar MOOCs for a diverse, larger audience.
- Minimize the dropout rate by integrating learning analytics based students' assessment and the cost of accreditation.

- As a marketing tool, increase the student enrolment by offering short MOOCs of different disciplines to potential students of formal academic programs, where they can see, try and understand the university pedagogical model and technological advances before the formal enrolment.
- New research opportunities, in terms of developing advanced CA based on dialogue modeling and interventions as well as integrating previous and current efforts of LA and CA into MOOCs.

In addition, enterprises are also going to be benefitted from the project outcomes as they will enforce their market strategy with new tools and methods. These results cannot be achieved without cooperation at national, regional or local level for the following reasons:

- The pilots will be implemented in four different countries (Greece, Spain, Germany, Denmark).
- In order to provide a more holistic approach that will be available to the market, the proposed solution should be able to be applied in different educational domains and approaches.
- The consortium includes experts in all fields (experts in pedagogy, MOOCs, conversational agents, learning analytics and educational scenarios).

4.2 Target Groups

The project targets three groups: the academic community, the training community, and the technology providers. As these groups transcend the project consortium, the project targets people through three networks. The first network is comprised by the project partners (e.g., researchers, educators, administrative staff, management of partner organizations), the second by the immediate networks of each of the project partners (e.g., municipalities, school/university networks, enterprises, associations), and the third by all other actors interested in MOOCs that could reach the project through the online presence and the outreaching activities planned in the project.

Next, the benefits from the project results for each of these three groups are explained:

Academic Community

- University students. Enrolled students require increased flexibility and support, something that cannot be always available in a university system that deals with rising number of students and limited funds.
- University teachers. As MOOC elements and online pedagogies are integrated into university curricula, the use of techniques, such as learning analytics, can provide a holistic view of the learning activity. This will be valuable for the teachers, in their effort to improve their courses.
- Academic institutions. The integration of conversational agents and learning analytics into MOOCs will allow the universities to offer more flexible and engaging ways of teaching. This, in turn, will allow the universities to reach wider demographic (based on age, occupation, etc.) and address the need for lifelong learning.

- Researchers. Cutting-edge original research will be explored during the project. To the best of our knowledge there are only two research groups currently worldwide developing this kind of agents with APT based dialogue modeling and interventions, one of these groups participating in the project [13].

Training Community

- Learners. The project will provide an enhanced learning experience for the online learner, offering at the same time different ways to formal education, lifelong learning, and training.
- Trainers. Trainers are in need of innovative, engaging and motivational learning methods in order to attract the attention of the employees of any sector during training sessions. In addition, trainers should be able to easily design courses and material and monitor their trainees' progress during sessions. This way, they will be able to adapt the process accordingly, change the real world scenarios used as problems that need solving and foster the employees' skills development.
- SMEs. European SMEs continuously need to train their employees to improve the competitive advantage of the enterprise. SMEs can be benefited by adopting a colMOOC training approach thus being able to provide personalized training to their employees and being able to monitor their progress.

Technology Providers

- Tool developers. Learning tool developers (e.g., MOOC plug-in developers) need to provide high quality personalized software products to their customers.
- Content providers. Learning content providers (e.g. MOOC providers) aim at making their content easily accessible to a great number of users.

Eventually, the project's target outcomes are in accordance to the main goals of the European policy agendas as this will enable and facilitate the dialogue between the consortium and the policy makers as well as their transfer to actual policy development.

5 Conclusions

This paper presents the scientific approaches of a newly started research project called "Integrating Conversational Agents and Learning Analytics in MOOCs (colMOOC)" funded by the European Commission. The main objective of the colMOOC project is to provide innovative ways for promoting learners' interaction by enabling the teacher to configure an agent software component, which attempts to trigger learners' discussion through appropriate interventions in dialogue-based collaborative activities.

The eLearning context as well as the main priority tackled in the project is the need to make MOOCs more collaborative in order to support teachers and engage and motivate students. Moreover, it is critical to ensure that assessment in MOOCs will monitor individual achievement and demand constant engagement. To this end, the project will leverage the power of learning analytics to achieve more meaningful assessment, thus allowing educators and trainers to apply learning processes where learners participated actively and are engaged throughout the course.

Acknowledgements. This research was funded by the European Commission through the project "colMOOC: Integrating Conversational Agents and Learning Analytics in MOOCs" (588438-EPP-1-2017-1-EL-EPPKA2-KA).

References

1. Lankshear, C., Knobel, M. (eds.): Digital Literacies: Concepts, Policies and Practices. Peter Lang, New York (2008). ISBN 9781433101687
2. Siemens, G.: Massive open online courses: innovation in education. In: Open Educational Resources: Innovation, Research and Practice, p. 5 (2013)
3. Tegos, S., Demetriadis, S.N.: Leveraging conversational agents and concept maps to scaffold students' productive talk. In: Proceedings of 6th International Conference on Intelligent Networking and Collaborative Systems (INCoS 2014), Salerno, Italy, pp. 176–183 (2014)
4. Bassi, R., Daradoumis, T., Xhafa, F., Caballé, S., Sula, A.: Software agents in large scale open e-learning: a critical component for the future of massive online courses (MOOCs). In: Proceedings of the Sixth IEEE International Conference on Intelligent Networking and Collaborative Systems, pp. 184–188. IEEE Computer Society (2014)
5. Daradoumis, T., Bassi, R., Xhafa, F., Caballé, S.: A review on massive e-learning (MOOC) design, delivery and assessment. In: Proceedings of the Eighth International Conference on P2P, Parallel, Grid, Cloud and Internet Computing, pp. 208–213. IEEE Computer Society (2013)
6. Barak, M., Watted, A., Haick, H.: Motivation to learn in massive open online courses: examining aspects of language and social engagement. Comput. Educ. **94**, 49–60 (2016)
7. Capuano, N., Caballé, S.: Towards adaptive peer assessment for MOOCs. In: Proceedings of the 10th International Conference on P2P, Parallel, Grid, Cloud and Internet Computing, pp. 64–69. IEEE Computer Society (2015)
8. Schuwer, R., Jaurena, I.G., Aydin, C.H., Costello, E., Dalsgaard, C., Brown, M., Teixeira, A.: Opportunities and threats of the MOOC movement for higher education: the European perspective. Int. Rev. Res. Open Distrib. Learn. **16**(6), 20–38 (2015)
9. Miguel, J., Caballé, S., Prieto, J.: Providing information security to MOOC: towards effective student authentication. In: Proceedings of the Fifth IEEE International Conference on Intelligent Networking and Collaborative Systems, pp. 289–292. IEEE Computer Society (2013)
10. Capuano, N., Caballé, S., Miguel, J.: Improving peer grading reliability with graph mining techniques. Int. J. Emerg. Technol. Learn. **11**(7), 24–33 (2016)
11. Tegos, S., Demetriadis, S.: Conversational agents improve peer learning through building on prior knowledge. Educ. Technol. Soc. **20**(1), 99–111 (2017)
12. Gañán, D., Caballé, S., Clarisó, R., Conesa, J., Bañeres, D.: ICT-FLAG: a web-based e-assessment platform featuring learning analytics and gamification. J. Web Inform. Syst. **13**(1), 25–54 (2017)
13. Tegos, S., Demetriadis, S., Karakostas, A.: Promoting academically productive talk with conversational agent interventions in collaborative learning settings. Comput. Educ. **87**, 309–325 (2015)
14. Karakostas, A., Demetriadis, S.: Adaptive vs. fixed domain support in the context of scripted collaborative learning. Educ. Technol. Soc. **17**(1), 206–217 (2014)

15. Tegos, S., Demetriadis, S., Tsiatsos, T.: A configurable conversational agent to trigger students' productive dialogue: a pilot study in the CALL domain. Int. J. Artif. Intell. Educ. **24**(1), 62–91 (2013). https://doi.org/10.1007/s40593-013-0007-3
16. Kumar, R., Rose, C.P.: Architecture for building conversational agents that support collaborative learning. IEEE Trans. Learn. Technol. **4**(1), 21–34 (2011)
17. Michaels, S., O'Connor, M.C., Hall, M.W., Resnick, L.B.: Accountable talk sourcebook: for classroom that works. University of Pittsburgh, Institute for Learning (2010). https://www.ortingschools.org/cms/lib/WA01919463/Centricity/domain/326/purpose/research/accountable%20sourcebook.pdf. Accessed 11 Jan 2018
18. Noroozi, O., Weinberger, A., Biemans, H.J.A., Mulder, M., Chizari, M.: Facilitating argumentative knowledge construction through a transactive discussion script in CSCL. Comput. Educ. **61**(2), 59–76 (2013)

Author Index

L. Barolli et al. (Eds.): EIDWT 2018, LNDECT 17, pp. 1073–1076, 2018.
https://doi.org/10.1007/978-3-319-75928-9

Z

Printed in the United States
By Bookmasters